Microwave Engineering

Sushrut Das

Assistant Professor
Indian School of Mines, Dhanbad

OXFORD

UNIVERSITY PRESS

OXFORD
UNIVERSITY PRESS

Oxford University Press is a department of the University of Oxford.
It furthers the University's objective of excellence in research, scholarship,
and education by publishing worldwide. Oxford is a registered trade mark of
Oxford University Press in the UK and in certain other countries.

Published in India by
Oxford University Press
22 Workspace, 2nd Floor, 1/22 Asaf Ali Road, New Delhi 110002, India

© Oxford University Press 2014

The moral rights of the author/s have been asserted.

First published in 2014

ISBN-13: 978-0-19-809474-6
ISBN-10: 0-19-809474-4

Typeset in Times New RomanPSMT
by Welkyn Software Solutions Pvt Ltd
Printed at Manipal Technologies Limited, Manipal

Third-party website addresses mentioned in this book are provided
by Oxford University Press in good faith and for information only.
Oxford University Press disclaims any responsibility for the material contained therein.

Dedicated to my twin daughters,
Spandita Das (Bub) and Sparshita Das (Sana)

FEATURES OF THE BOOK

Illustrations

Well-labelled illustrations with a three-dimensional feel will help students visualize concepts better.

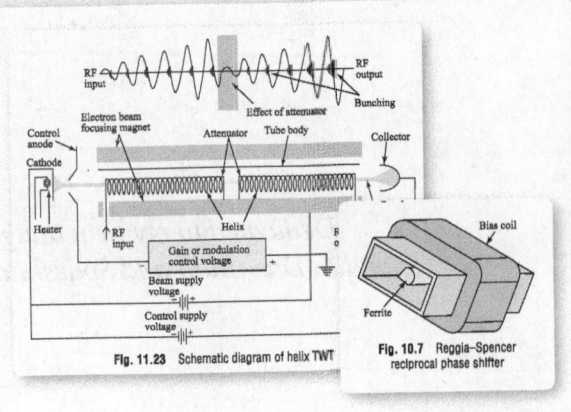

Fig. 11.23 Schematic diagram of helix TWT

Fig. 10.7 Reggia–Spencer reciprocal phase shifter

MATLAB Programs

MATLAB programs are interspersed in the book in relevant sections.

```
MATLAB Program
% Program to plot Z0 of strip line as a function of W/h for different er
% It assumes a zero strip thickness
% It neglects dispersion
% wbh=W/h;
clear all
clf
er=input('Enter relative permittivity of the dielectric medium:- ');
n=0;
for wbh=0.1:0.1:10
    n=n+1;
    wh(n)=wbh;
    if wbh>0.35
        webh=wbh;
    else
```

List of Formulae

A list of key formulae has been provided at the end of each relevant chapter for quick recapitulation.

Important Formulae

- The characteristic impedance of a parallel-plate transmission line for th
  $Z_0 = \eta b/a$.
- The phase constant of a parallel-plate transmission line for the TE ar
  $$\beta = \sqrt{k^2 - \left(\frac{n\pi}{b}\right)^2}.$$
- The cut-off frequency of a parallel-plate transmission line for the TE ar
  $$f_c = \frac{n}{2b\sqrt{\mu\varepsilon}}.$$
- The guided wavelength of a parallel-plate transmission line for the TE

Interspersed Examples and Problems

A large number of solved examples and practice problems are interspersed within the text for students to have a better understanding of the concepts discussed in the book. (Solutions to select problems are provided on the companion website.)

Exercises

A plethora of chapter-end exercises, such as objective-type questions, review questions, and problems (with hints) are provided, which will be a very useful resource for both faculty and students.

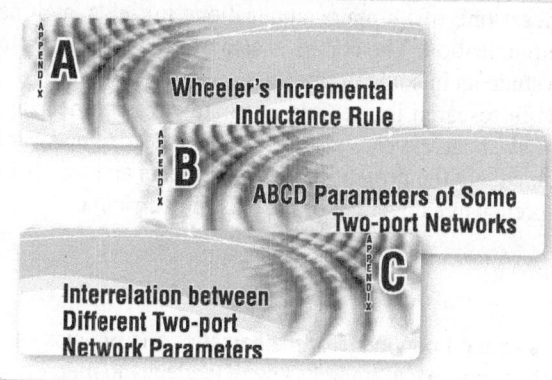

APPENDIX A
Wheeler's Incremental Inductance Rule

APPENDIX B
ABCD Parameters of Some Two-port Networks

APPENDIX C
Interrelation between Different Two-port Network Parameters

Appendices

Appendices at the end of the book provide information on Wheeler's incremental inductance rule and the different network parameters of two-port networks.

Preface

Imagination is more important than knowledge. Knowledge is limited. Imagination encircles the world.

— Albert Einstein

Since microwaves are a part of the electromagnetic spectrum, they obey the basic laws of electromagnetic radiation, which is dealt with in physics as well as in communication engineering. Microwave engineering also incorporates the theories of semiconductor devices and vacuum tubes—an essential part of electronic devices and circuits.

Being in the GHz range, microwaves possess some distinct features, such as high bandwidth, lower component size, and superior electrical characteristics. These properties have been put to use in mobile communication, satellite communication, and radar. Mobile communications generally use the lower end of the microwave range (0.9 GHz–2 GHz), whereas satellite communications generally operate in the 1–24 GHz band. Radars (once synonymous with microwaves) utilize that part of the frequency spectrum starting from a few MHz up to a few hundred GHz.

The microwave oven, which is now commonly used in most households, is a day-to-day application of microwaves. Other significant fields include medicine, astronomical research, industry, nuclear science, material technology, remote sensing, and warfare. The wide-ranging usefulness of microwaves has revolutionized the way we live and communicate, and this is the reason for their emergence as a separate subject in communication engineering for both undergraduate and postgraduate courses.

In practice, the concept and applications of microwaves are so vast that it is not possible to cover these in a single book. This is precisely why, the study of microwaves is offered by universities and institutes under different names, such as RF and microwave engineering, microwave and radar, microwave and antenna, and microwave and satellite communication, among others. The syllabi of these subjects include the basic theories of microwaves and at least one of its major applications in fields such as radars, antennas, or satellite communication. The course is also offered as a separate branch of study for postgraduate students in some universities.

With the rapid advancements in research in the field of microwave engineering, newer concepts and technologies have been developed, which readers need to be aware of. Bearing this in mind, the book covers the present-day developments and trends in this discipline, along with a comprehensive coverage of the existing curricula.

About the Book

The study of microwaves requires sound knowledge of mathematics and physics as it involves several derivations and theorems. Most students have a tendency to memorize as many of these as possible. However, this is not desirable, especially for a conceptual subject such as this. It is very important to have a step-by-step, clear understanding of the concepts. *Microwave Engineering* is a textbook that optimizes physical concepts and mathematical derivations to describe the different topics of RF and microwaves in a lucid manner. This book has been developed keeping in mind the different courses on microwave engineering. Hence it includes separate chapters on radar, antenna, and microwave communication, in addition to the basic theories of microwaves. The text also includes a detailed discussion on topics such as matching networks, planar transmission lines, filters, vacuum devices, microwave semiconductor devices, microwave amplifiers, microwave measurements, and hazards.

Key Features

- Thorough coverage of microwave antennas, radar, measurements, hazards, and the latest developments in this subject
- Figures with a three-dimensional feel for better understanding of the concepts
- Exhaustive pedagogy that includes solved examples, MATLAB programs, practice problems, objective-type questions, review questions, and exercise problems in every chapter
- List of important formulae at the end of relevant chapters to help in revision
- Hints for exercise problems to assist students in solving difficult problems

Coverage and Structure

The book is divided into 20 chapters and three appendices.

Chapter 1 describes the entire electromagnetic spectrum, the position of microwaves in the spectrum, its features, advantages, disadvantages, and applications. *Chapter 2* discusses the different aspects of transmission lines in detail. The chapter also deals with the Smith chart and its characteristics. *Chapter 3* talks about the different transmission line matching networks. Transmission line connectors and adaptors have been explained as well. *Chapter 4* consists of a general analysis of TE and TM modes, surface waves on dielectric slabs, and the different planar transmission lines. *Chapter 5* throws light on the different modes and their distribution in rectangular and circular waveguides, besides illustrating the different finlines.

Chapter 6 deals with the study of high-frequency resonators, Q-factor, transmission line resonators, and their classification and characteristics. *Chapter 7* analyses microwave network representations such as the [Z], [Y], [S], and ABCD parameters. *Chapter 8* explores the types of microwave power dividers and couplers in detail. *Chapter 9* is devoted to the study of periodic structures and the numerous microwave filters. *Chapter 10* lays emphasis on non-reciprocal devices such as isolators, phase shifters, circulators, gyrators, quarter-wave and half-wave plates, and attenuators.

Chapter 11 aids in the understanding of microwave linear beam (O-type) tubes. *Chapter 12* covers various microwave crossed field (M-type) tubes. *Chapter 13* presents a discussion on the different microwave solid state devices. *Chapter 14* is dedicated to the understanding of solid state transistors and MASERs. *Chapter 15* provides an extensive coverage of detectors and mixers, two-port transistor amplifiers, and MMIC.

Chapter 16 brings to the fore the effect of atmosphere on propagation, the different propagation modes, and microwave communication systems. *Chapter 17* gives a clear picture of radar systems, radar tracking, electronic countermeasures (ECM) and electronic counter countermeasures (ECCM), and the applications of radar. *Chapter 18* elucidates the radiation mechanism, antenna parameters, and the different types of microwave antennas. *Chapter 19* discusses the equipment used in microwave measurement and the measurement procedures of different microwave parameters. *Chapter 20* expounds on the hazards of electromagnetic radiation on fuel, ordnance, and personnel, radiation hazard limit and regulations, and protection from such harmful radiation.

Appendices A–C provides useful information on Wheeler's incremental inductance rule, ABCD parameters of some two-port networks, and the interrelations between two-port network parameters.

Online Resources

Adding value to the pedagogy-rich textbook is a plethora of online resources, available on the companion website http://oupinheonline.com/book/das-Microwave-Engineering/9780198094746, to help faculty and students.

For Faculty
- Solutions manual
- PowerPoint slides

For Students
- Additional appendices, which include a section on Microwave CAD describing microwave CAD software, such as CST Microwave Studio, Ansoft HFSS, FEKO, Mician Microwave Wizard, IE3D, and WiPLD.
- Key points from all chapters for quick recapitulation
- Additional MATLAB programs
- Additional objective-type questions
- Solutions to select practice problems from the textbook

Acknowledgements

At the outset I thank my teachers, Prof. Ajay Chakrabarty, Prof. Subrata Sanyal, Prof. Binay Kumar Sarkar, Department of Electronics and Electrical Communication Engineering, Indian Institute of Technology, Kharagpur; Prof. B.N. Biswas, Emeritus Professor, Chairman, Education Division, SKF group of institutions, Mankundu; Prof. B.C. Sarkar, Prof. S. Roy, Dr. Anida Bose, Dr. Debidas Mondal, Burdwan University, Burdwan; Dr. Partha Pratim Lahiri, Jadavpur University, Jadavpur; and Prof. P.N. Gupta, Banaras Hindu University, Varanasi, for their guidance and motivation in helping me understand this subject.

I am indebted to Prof. D.C. Panigrahi, Director, and all my colleagues at the Department of Electronics Engineering, Indian School of Mines, Dhanbad, for their support, encouragement, and important suggestions. Special thanks to all my students and research scholars who were always there to help me at every stage of this project.

In addition, I thank the authors of the several books—too many to mention in this space—I referred to, to better understand the subject, during the preparation of this book. All the reviewers deserve special appreciation for their valuable comments and constructive suggestions, which helped me in improving the manuscript.

I would also like to thank the editorial team at Oxford University Press, India, for their guidance and support.

I express my heartfelt gratitude, regards, and affection to my parents Mr Samir Kumar Das and Mrs Rajkumari Das, my father-in-law Mr Ardhangshu Bhusan Chakraborty, mother-in-law Mrs Swapna Chakraborty, my sister Mrs Malancha Sinha, and brother-in-law Dr Sangram Sinha, for their love and encouragement, which kept my passion for research alive.

I am obliged to my wife Mrs Ashmi Chakraborty Das, a faculty member in the Department of Applied Electronics and Instrumentation Engineering at Asansol Engineering College, for her dedication and constant support, which kept me going during the most difficult times in the development of the book. She was ever ready to shoulder responsibilities that were expected of me, so that I could work undisturbed.

I offer my special thanks and love to my twin daughters Spandita Das (Bub) and Sparshita Das (Sana) who always cooperated with me, in spite of not giving them enough attention, even at home. I therefore, dedicate this book to them. I also thank my nephews Samriddha Sinha (Kittu) and Samadrita Sinha (Tito).

Readers are welcome to share their valuable feedback and suggestions with me at sushrut_das@yahoo.com.

Sushrut Das

Brief Contents

Detailed Contents

Introduction

1.1 Electromagnetic Spectrum

In the 17th century, the term *spectrum* was first introduced into optics to explain the range of colours observed when white light is passed through a prism. It was soon applied to other waves such as sound waves and electromagnetic waves, and is now applied to any signal that can be decomposed into frequency components. Basically, the term *electromagnetic spectrum* refers to the range of all possible frequencies of electromagnetic radiations; it extends from low frequencies used for radio communication to high-frequency gamma radiation. Alternatively, it covers wavelengths from thousands of kilometres down to a fraction of the size of an atom.

Electromagnetic waves are generally classified into radio waves [very low frequency (VLF)–very high frequency (VHF)], microwaves and millimetre waves [ultra-high frequency (UHF) to extremely high frequency (EHF)], infrared [far infrared (FIR) to near infrared (NIR)], visible light, ultra-violet rays [near ultra-violet (NUV) and extreme ultra-violet (EUV)], X-rays [soft X-ray (SX) and hard X-ray (HX)], and gamma rays (γ-ray). Boundaries between these bands are not strict, and there may be some overlaps between the neighbouring bands. Figure 1.1 shows the locations of different bands in the electromagnetic spectrum. These classification schemes are summarized in Table 1.1 with some typical applications.

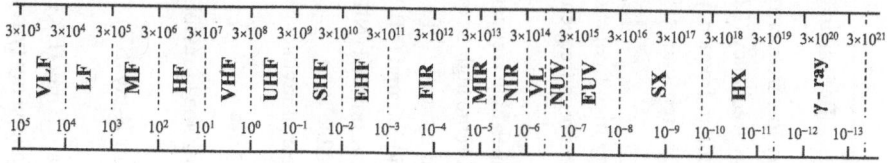

Fig. 1.1 Electromagnetic spectrum

The term *microwave* usually refers to that part of the EM spectrum that corresponds to wavelengths ranging from 1 m to 1 cm or frequencies from 300 MHz to 30 GHz. Above this frequency range, the wavelengths are of the millimetre order and are called *millimetre waves* (30–300 GHz). The term *micro* in the microwave stands for 'extremely small in scale', and it includes both the microwave and the millimetre wave spectrum, that is, UHF, SHF, and EHF bands.

Table 1.1 Electromagnetic frequency band designation

Frequency Band	Designations	Typical Applications
3 kHz–30 kHz	Very low frequency (VLF)	Navigation and sonar
30 kHz–300 kHz	Low frequency (LF)	Radio beacons and navigation
300 kHz–3 MHz	Medium frequency (MF)	AM broadcasting and coast guard communication
3 MHz–30 MHz	High frequency (HF)	Telephone, telegraph, fax, ship to coast and ship to aircraft communication, shortwave international broadcasting, amateur radio, and citizen's band
30 MHz–300 MHz	Very high frequency (VHF)	Air traffic control, police, television, FM, taxicab mobile radio, and navigational aids
300 MHz–3 GHz	Ultra high frequency (UHF)	Satellite communication, surveillance radar, mobile communication, television, and navigational aids
3 GHz–30 GHz	Super-high frequency (SHF)	Airborne radar, microwave communication, mobile communication, and satellite communication
30 GHz–300 GHz	Extreme high frequency (EHF)	Radar
300 GHz–6 THz	Far infrared (FIR)	Terahertz time domain spectroscopy and terahertz imaging
6 THz–100 THz	Mid infrared (MIR)	Guided missile and thermal imaging
100 THz–400 THz	Near infrared (NIR)	Fibre-optic telecommunication, night vision, and long-distance telecommunication
400 THz–750 THz	Visible light	Optical communication
750 THz–1 PHz	Near ultraviolet (NUV)	Optical sensors, UV-ID, label tracking, barcode, forensic analysis, drug detection, protein analysis, DNA sequencing, drug discovery, medical imaging of cells, solid-state lighting, curing of polymers and printer inks, light therapy in medicine, and bug zappers
1 PHz–30 PHz	Extreme ultraviolet (EUV)	Extreme ultra-violet lithography, optical sensors, disinfection, decontamination of surface and water, UV-ID, label tracking, barcode, protein analysis, DNA sequencing, and drug discovery
30 PHz–3 EHz	Soft X-ray (SX)	X-ray microscopic analysis, X-ray crystallography, medical imaging of bones, airport security, border control, and astronomy
3 EHz–30 EHz	Hard X-ray (HX)	Absorption spectroscopy, scanning microprobe, and radiotherapy
>30 EHz	Gamma ray (γ-ray)	Container security initiative, irradiation, gamma-knife surgery, and nuclear medicine

During World War II, the design and development of microwave radars prompted defence engineers to further subdivide the 1–12 GHz frequency spectrum into four frequency sub-bands and express them with the code letters L, S, C, and X for secrecy. Later, during the development of radar engineering, higher frequencies were used and the entire microwave range was subdivided into 16 sub-bands by the *Radio Society of Great Britain* (*RSGB*), as mentioned in Table 1.2.

Afterwards, different users used different frequency boundaries for letter designation of microwave bands. A few of them are listed in Table 1.3. However, the radar band classification provided by the RSGB is still in common use today, and until and unless specified we will follow the letter designation provided in Table 1.2.

Table 1.2 Letter designation of microwave bands as per the RSGB

Frequency Bands	Frequency Range (GHz)
L	1–2
S	2–4
C	4–8
X	8–12
Ku	12–18
K	18–26.5
Ka	26.5–40
Q	33–50
U	40–60
V	50–75
E	60–90
W	75–110
F	90–140
D	110–170
G	140–220
H	170–260

Table 1.3 Other letter designations of microwave bands

Band	US Navy Frequency Range (GHz)	International Telecommunication Union (ITU) Frequency Range (GHz)
L	0.390–1.55	1.215–1.400
S	1.55–3.90	2.300–2.500
		2.700–3.700
C	3.90–6.20	5.250–5.925
X	6.20–10.90	8.500–10.680
Ku	15.25–17.25	13.40–14.00
		15.70–17.70
K	10.90–36.00	24.05–24.25
		24.65–24.75
Ka	33.00–36.00	33.40–36.00
Q	36.00–46.00	
V	46.00–56.00	59.00–64.00
W	56.00–100.00	76.00–81.00
		92.00–100.00

1.2 Characteristic Features of Microwaves

In the UHF band, up to around a frequency of 1 GHz, most circuits are designed and constructed using lumped parameter circuit components. Above this frequency, the time an electrical signal requires to propagate from one point to another in a circuit becomes comparable with the time period of the signal. The lengths of the lumped parameter circuit components also become comparable with the wavelength of the signal. This results in a rapid variation in the amplitude and phase of the signal with distance along the circuit. The phase difference caused by the interconnection between different components and different parts of a single component is also not negligible at these high frequencies. As a result, at high frequencies the Kirchhoff's current law, Kirchhoff's voltage law, and voltage–current concepts are not applicable to describe the electrical phenomenon taking place in the circuit; thus, it becomes necessary to analyse the circuit in terms of associated electric and magnetic fields.

Although there is no distinct frequency boundary beyond which lumped circuit elements cannot be used, it is a common practice to consider 1 GHz as the limit. Beyond this frequency, the lumped circuit elements are replaced by distributed circuit elements. The distributed circuit elements are generally made of small transmission line/waveguide sections and are defined over an infinitesimal length so that the attributes of the circuit are distributed continuously throughout the length of the circuit. In this model, each circuit element is infinitesimally small and the wires connecting different elements are not perfect conductors. At high frequencies, the distributed circuit model is more accurate than the lumped element circuit model and also more complex in nature. The existence of non-uniform currents in the branches and non-uniform voltages at the nodes further complicates the analysis of the circuit. The use of infinitesimals in the distributed circuit model requires the application of calculus rather than linear algebra that is used in lumped element models. With the progress in modern technology at present, it is possible to construct very small printed circuit inductors that can retain their lumped parameter characteristics at frequencies as high as 10 GHz or even higher; however, the common use of distributed circuit elements in microwave region has not changed.

Similar to the designing of passive circuits, the high-frequency nature of microwaves has also introduced complexity and challenges in designing microwave-active components. At microwave frequencies, the transit time of the carriers through ordinary low-frequency triode and transistors becomes comparable with the time period of the wave that restricts its operation at these frequencies. Hence, a number of new principles of operation, namely, velocity modulation, interaction of space charge waves with EM fields, quantum mechanical tunnelling, avalanche breakdown, and transferred electron techniques, have been employed to generate microwave signals. The signal power level that active microwave components handle is usually very low and often very sensitive to loads.

At microwave frequencies, routine electronic measurements (such as voltage and current measurements) cannot be performed with a multimeter or any other low-frequency circuits because of the types of cables and connectors being used. At microwave frequencies, the parasitic impedance of such cables and connectors becomes quite large and frequently crosses the component values. Thus, special cables and connectors are required for microwave measurement. In addition, the multimeter impedance and capacitance also affect the measurement. The most common method to perform microwave measurement is to measure the field amplitudes, phase difference, and power carried by the waves. Another very commonly used method is based on the study of the standing wave pattern formed by the superimposition of two travelling waves propagating in opposite directions.

1.3 Advantages of Microwaves

In spite of its complex nature and difficult analysis procedure, the microwave band has several advantages, prompting the use of this frequency range; some of the advantages are listed as follows:

High bandwidth   Microwaves have a very high bandwidth compared to VHF or other lower-frequency bands, and thus allows more information to be transmitted through it in both analog and digital formats within a short time. Due to the high bandwidth, multiplexing of many users into a single wideband signal is possible, which allows a single transmitter and receiver to be used. This results in sharing of costs among many users and a much lower cost per user.

Improved gain/directive properties   The radiated power and the gain/directivity of an antenna are inversely proportional to λ^2. Therefore, it is very easy to achieve a high-gain antenna at microwave frequencies. The high gain of an antenna, in turn, results in the reduction of its beam width. Therefore, at microwave frequencies, antennas have a higher gain/directivity and narrow beam width compared to the lower-frequency band.

Reduction in antenna size   The size of an antenna is proportional to the wavelength. This means that the antenna size is reduced at microwave frequencies, allowing the formation of an antenna array.

Low-power requirement   Due to the availability of high-gain antennas, the power requirements of the microwave transmitters become very small compared to that at an MF/HF band.

Fading effect and reliability   In radio communication in the MF/HF band, the propagation of a signal takes place by reflections from the F and E layers of the ionosphere, whose characteristics vary widely with time and weather condition at any location. Therefore, the signal received varies widely with time and weather. This is known as the fading effect. At microwave frequencies, propagation of a signal takes place by line-of-sight propagation through neutral atmosphere, and hence microwave communication is subjected to less fading.

Transparency property of microwaves   The transparency of the ionosphere and atmosphere for microwave frequency allows a stable propagation of microwaves from the earth station to a satellite, and vice versa. This also facilitates the study of microwave radiation from sun and stars in radio astronomical research of space.

1.4 Disadvantages of Microwaves

The following are some of the disadvantages of the microwave band:

Line of sight propagation Microwave radio communication systems are based on line-of-sight communication. This requires a number of repeater stations between the source and the final destination, which increases the cost of the total communication system. Due to the line-of-sight propagation, microwave communication is also disrupted by obstacles such as mountains, multi-storied buildings, and aircraft. Microwave communication towers are also expensive to build.

Susceptibility to electromagnetic interference Microwave signals are easily affected by electromagnetic signals from different radio wave sources such as electric motor, electric power transmission line, wind turbines, television/radio stations, and mobile towers.

Sensitivity to bad weather Microwave signals are also affected by heavy moisture, snow, vapour rain, and fog.

Costly equipment Microwave communication equipment such as signal generator, network analyser, spectrum analyser, and power meters are costlier than their low-frequency counterparts.

1.5 Applications of Microwaves

Microwaves have a very broad range of applications, starting from short-distance to long-distance communications, from detecting a target at a short distance to identifying stars light years away, from identifying a small object to mapping the surface of a planet, from the study of a linear particle accelerator to that of plasma, from domestic cooking to industrial heating, and many more. It is therefore very difficult to discuss each of them in detail in a single section. A few important applications of microwave technology are briefly described in the following subsections.

Radio detection and ranging During World War II and even few years after that, the microwave technology became almost synonymous with radar (an acronym for radio detection and ranging). This is because at that time microwave technology was highly explored to design and develop high-resolution radars capable of detecting enemy planes and ships, as the situation demanded. Even today, radars, in their many varied forms, such as missile tracking radars, fire-control radars, weather detecting radars, missile-guidance radars, and airport traffic control radars, still use microwave frequencies. This use arises predominantly from the need to have compact-size antennas for easy installation of the system, narrow beam width for accurate determination of the position of the target, and high-power transmission efficiency with high gain/directivity for detection of the target from a long distance. In addition, stable propagation and low absorption of microwave signals in the atmosphere are also major reasons for choosing the microwave band in radar applications.

Terrestrial microwave link   Terrestrial microwave links have been used for many years. The TD-2 system was put in service in 1948. It operated in the 3.7–4.2 GHz band and had 480 voice circuits, each occupying a 3.1 kHz bandwidth. In 1974, the TN-1 system came into operation in the 10.7–11.70 GHz band; it had a capacity of 1800 voice circuits or one video channel with a 4.5 MHz bandwidth.

Transmission of multiple television channels over one link   Due to their high bandwidth, microwaves are used as carrier frequencies for efficient transmission of multiple television programmes over one link.

Satellite communication   The transparency of microwave signals in the iono-sphere and atmosphere has resulted in an interesting means of microwave communication through the use of a satellite as a microwave relay station. This mode of microwave communication is known as satellite communication. In addition, microwaves are widely used in wireless telecommunications such as direct broadcast satellite (DBS) television, cellular video (CV) systems, and global positioning system (GPS).

Radio astronomy   The transparency of a microwave signal in the ionosphere and atmosphere has also prompted its use in the study of radio astronomy. Sensitive microwave receivers are used in radio astronomy to detect electromagnetic radiations from the sun and the distant radio stars. The various mechanisms, responsible for plasma radiation, can also be predicted by measuring the noise radiated from plasmas with such receivers. Other applications of microwaves in *radio astronomy* are remote sensing and atmospheric temperature profile mapping.

Linear particle accelerator   It has been found that periodically shunt loaded waveguides support the propagation of slow waves. These waves can interact with charged particle beams very efficiently and can transfer its energy to the particle beam if both have the same velocity. This principle is used in linear particle accelerators to produce high-energy beams of charged particles for use in atomic and nuclear research. The opposite phenomenon, that is, transfer of energy from electron beam to electromagnetic wave, results in an amplification of the wave. This latter principle is used in the operation of travelling wave tubes.

Studies on basic properties of materials   Molecular, atomic, and nuclear systems exhibit various resonance phenomena under the influence of an applied microwave field. This has resulted in the use of the microwave field as a powerful experimental probe to study the basic properties of these materials. In the last few decades, such research has led to the design and development of some useful devices such as some non-reciprocal devices employing ferrites, solid-state microwave amplifiers, and oscillators.

Microwave oven   Another interesting and very familiar application of microwaves is found in the microwave oven, where a microwave signal is used for cooking. Many molecules, such as those of water and fat, behave like electrical dipoles. In the presence of an electromagnetic field of microwave frequency, these molecules absorb energy from the field and try to rotate themselves to align with the alternating electric field. This molecular movement represents heat, which is then dispersed as the rotating molecules hit other molecules and put them into motion.

Industry For industrial heating applications, such as drying grain, manufacturing wood and paper products, and material curing, the frequencies of 915 MHz and 2.45 GHz have been assigned. Other industries where microwave frequencies find their applications are food processing industry, rubber industry, plastic industry, chemical industry, forest product industry, etc.

Medical science Microwave radiation has found applications in medical hyperthermia or localized heating of tumours. Transmission of microwaves through human body can be used for monitoring of heartbeat and lung water detection.

Exercises

Objective-type Questions

1.1 A microwave frequency ranges from
   (a) 3 to 30 GHz
   (b) 30 to 300 GHz
   (c) 3 to 300 GHz
   (d) 300 MHz to 300 GHz.

1.2 As per the RSGB, an X-band refers to
   (a) 18–26.5 GHz
   (b) 4–8 GHz
   (c) 26.5–40 GHz
   (d) 8–12 GHz

1.3 Above 1 GHz, which of these statements holds true?
   (a) Propagation time is comparable with the time period.
   (b) Wavelength is comparable with the circuit length.
   (c) Distributed circuit elements are used.
   (d) All of these

1.4 A microwave frequency is
   (a) transparent to ionosphere
   (b) suffers less fading from environment
   (c) characterized by more stable propagation than low-frequency RF wave
   (d) all of these

1.5 Transparency property of microwaves through ionosphere helps in
   (a) astronomical research
   (b) satellite communication
   (c) all of these
   (d) none of these

Review Questions

1.1 Explain the different characteristics of microwaves.
1.2 Write the advantages and disadvantages of microwaves.
1.3 Write some major applications of microwave engineering.
1.4 Differentiate microwave engineering from other low-frequency RF engineering.

Transmission Line Theory

2

2.1 Introduction

In electronics and electrical communication engineering, a transmission line is defined as a conductor assembly of some predefined forms that is designed to carry radio frequency (RF). Some of the most common forms of transmission lines are coaxial cable, microstrip line, strip line, twisted pair, star quad, twin lead, etc. These are shown in Fig. 2.1. In a microwave system, transmission lines are used to carry electrical signals from one point to another in a circuit or to connect a transmitter/receiver with an antenna.

Ordinary electrical cables, used in home electrical appliances, are sufficient to carry signals of frequencies of only up to a few hundred Hz because at higher frequencies energy tends to radiate off the cables, causing power loss. The signal also tends to reflect from the discontinuities present in the line in the form of

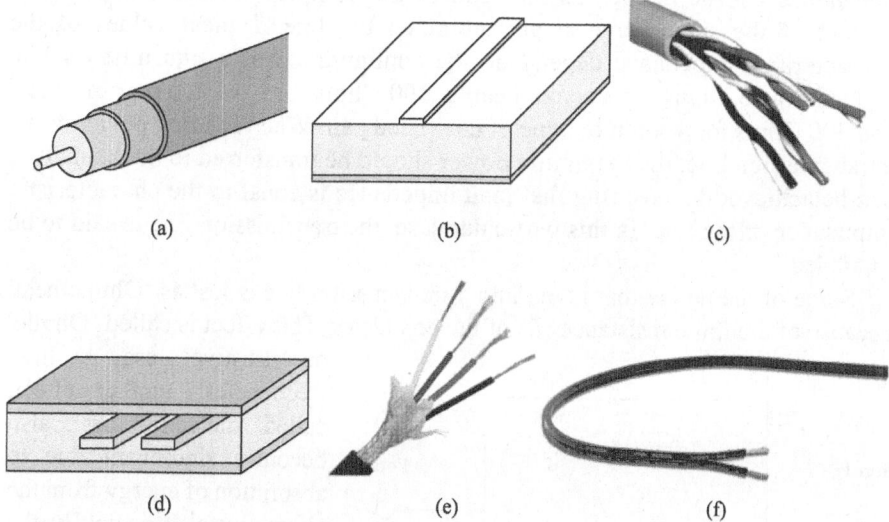

Fig. 2.1 Different forms of transmission lines (a) Coaxial cable (b) Microstrip line (c) Twisted pair (d) Strip line (e) Star quad (f) Twin lead

connectors, variations in wire cross-sectional areas, bends, etc. In contrast, a transmission line uses precise line geometry to carry electromagnetic waves with minimum reflections and power loss. The mathematical formulation of a transmission line resulted from the work of James Clerk Maxwell, Lord Kelvin, and Oliver Heaviside. The current in a submarine cable was first formulated using the diffusion model developed by Lord Kelvin in 1855. Later in 1885, the modern form of telegrapher's equations, consisting of linear differential equations that describe the voltage and current on an electrical transmission line in terms of distance and time, was developed by Heaviside during the analysis of the propagation of voltage waves in cables.

At high frequencies, the time period of the voltage signal becomes comparable or even less than the time it takes to travel down the line. In such cases, the length of the line becomes important because now the changes in voltage at the source (at a particular instant) are not reflected simultaneously at all points of the line. Under this circumstance, the wire is treated as a transmission line. In other words, if the signal wavelength is comparable to or less than the length of the line, then the line must be treated as a transmission line. A common rule of thumb is that the cable or line should be treated as a transmission line if the length is greater than $\lambda/10$, where λ is the wavelength. At this length, the interference of any reflections on the line and phase delay cannot be neglected because they can lead to unpredictable behaviour in systems.

For the purpose of analysis, an electrical transmission line can be modelled as a two-port network, as shown in Fig. 2.2. To simplify the analysis, the network is generally assumed to be linear. If the characteristics of a transmission line are uniform along its length, then its behaviour is largely described by a single parameter called the 'characteristic impedance', Z_0. The *characteristic impedance* can be defined as the ratio of the complex voltage and complex current of the same wave at any point on the line. Typical values of the characteristic impedance depend on the configuration of a transmission line and are 50–75 Ohms for a coaxial cable, 100 Ohms for a twisted pair of wires, and 300 Ohms for a common type of untwisted pair. When sending power down a transmission line, the maximum power should be transferred to the load. This can be achieved by ensuring that load impedance is equal to the characteristic impedance of the line. In this particular case, the transmission line is said to be 'matched'.

Some of the power that is fed into a transmission line is lost as 'Ohmic heat' because of the finite resistance (R) of the conductor. This effect is called 'Ohmic' or 'resistive' loss. At high frequencies, another effect, called 'dielectric loss', also becomes significant due to absorption of energy from the alternating electric field by the dielectric of the transmission line. The finite conductance

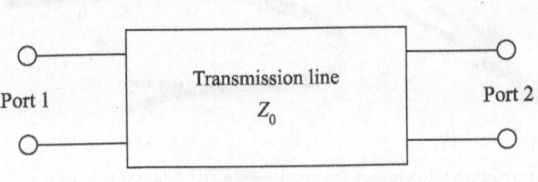

Fig. 2.2 Two-port network representation of transmission line

(G) of the dielectric is responsible for this. After absorption, the transmission line converts it into heat, causing dielectric heating in the line. For a transmission line to be lossless, both R and G must be equal to zero. In the transmission line theory, the finite resistance and conductance of the line are expressed in terms of their per-unit-length values and hence are called the per-unit-length resistance and per-unit-length conductance, respectively.

Apart from these, two other important parameters, per-unit-length inductance (L) and per-unit-length capacitance (C), also appear in the equivalent circuit of a transmission line. The per-unit-length inductance appears due to the finite inductance of conductors, whereas charges in the forward and return conductors, separated by the dielectric, develop some finite capacitance in the line.

2.2 Propagation of Voltages and Currents in Transmission Lines

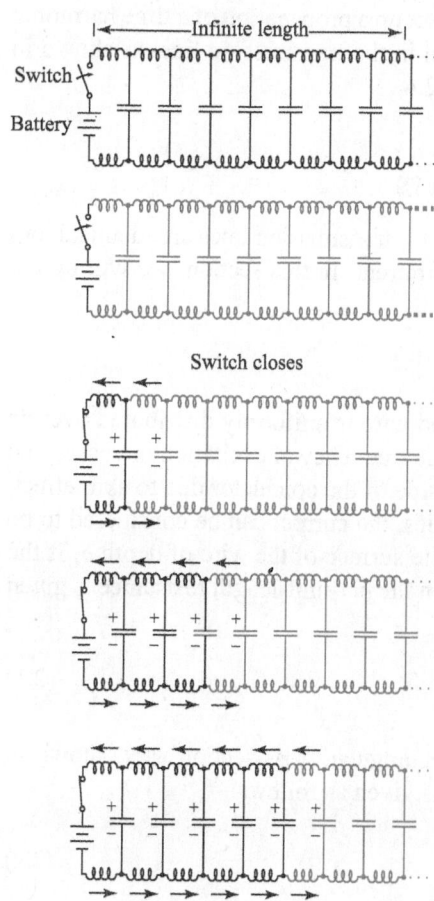

Fig. 2.3 Propagation of waves along a transmission line

Let us consider that an infinite lossless transmission line, connected to a voltage generator, is switched on at time $t = 0$. The line being lossless is characterized only by the per-unit-length shunt capacitance C and per-unit-length inductance L. As soon as a voltage is applied between the two conductors, the per-unit-length capacitance will react against the sudden voltage increase by charging up and drawing current from the generator. The amount of current drawn from the source will be limited by the presence of a series inductor. During the charging procedure, energy will be stored in the electric field, that is, in the capacitor, and in the magnetic field, that is, in the inductor. The storage of energy in the electric field opposes the change in voltage, whereas the storage of energy in the magnetic field opposes the change in current. This will set up a flow of current and voltage waves in the transmission line, as shown in Fig. 2.3. Now, as the time progresses, electrons will flow down the transmission line, and more and more per-unit-length capacitors will be charged up, resulting in the propagation of the current and voltage waves in the

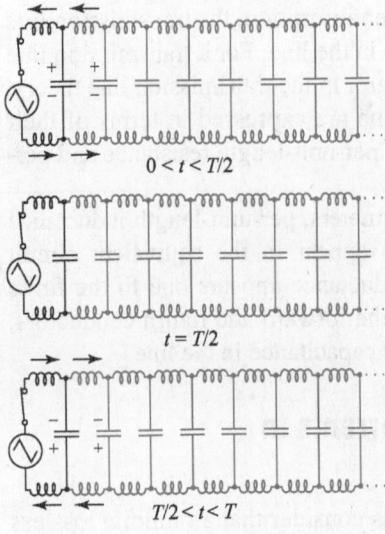

$0 < t < T/2$

$t = T/2$

$T/2 < t < T$

Fig. 2.4 Charging and discharging of distributive capacitor over a complete cycle

line, as shown in Fig. 2.3. Since the wire is infinitely long, all the distributed capacitors will be charged and the pair of wires will draw currents from the generator as long as the switch is closed.

If the generator produces a time harmonic signal with a time period T, then for the period from 0 to $T/4$ the capacitor will be charged, whereas in the duration from $T/4$ to $T/2$ it will be discharged. In the third quarter of the cycle, that is, from $T/2$ to $3T/4$, the capacitor will be charged again, but in the opposite direction. In the final quarter, it will be discharged again. The complete charging and discharging phenomena of the per-unit-length capacitors thus set up a propagation of a time harmonic signal in the transmission line, as shown in Fig. 2.4.

2.3 Transmission Line Parameters

Characteristic equations for all two-conductor transmission lines are identical, but their per-unit-length parameters may be different. In this section, we will derive the expressions for these parameters.

2.3.1 Per-unit-length Resistance (R)

At DC, the current flowing through a round wire is uniformly distributed over its cross-section, as shown in Fig. 2.5(a). As the frequency of excitation increases, the current tends to crowd closer to the outer side of the conductor due to skin effect, as shown in Fig. 2.5(b). At these frequencies, the current can be considered to be distributed uniformly over an annulus at the surface of the wire of depth δ. If the wire has a radius r_w and conductivity σ then the per-unit-length resistance is given by the following equation:[1]

$$R = \frac{1}{2\pi r_w \sigma \delta} \tag{2.1}$$

For a printed circuit board land with a rectangular cross-section $w \times t$, shown in Fig. 2.5(c), the per-unit-length resistance is given as follows:

$$R = \frac{1}{2\sigma\delta\,(w+t)} \tag{2.2}$$

---

[1]For detailed derivations see *Introduction to Electromagnetic Fields* by C.R. Paul, K.W. Whites, and S.A. Nasar, Tata McGraw-Hill, Third edition, New Delhi, pp. 494–498.

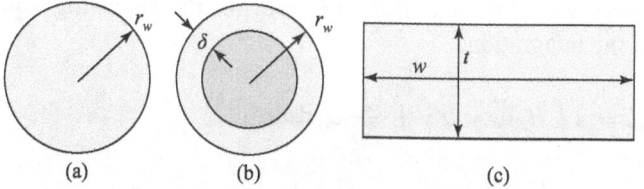

Fig. 2.5 Current distribution (a) Wire at low frequency (b) Wire at high frequency (c) Printed circuit board land

2.3.2 Per-unit-length Inductance (L)

Let us consider the two-wire transmission line shown in Fig. 2.6. The transverse magnetic field $\left(\overrightarrow{H_t}\right)$, generated due to the enclosed current in the conductors (I), contributes to the per-unit-length inductance of the transmission line. The portion of the magnetic flux interior to the wire contributes to per-unit-length internal inductance L_{int}, whereas the portion of the magnetic flux external to the wire contributes to per-unit-length external inductance L_{ext}. Therefore, the total per-unit-length inductance is as follows:

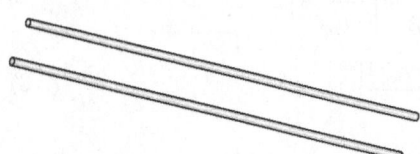

Fig. 2.6 Two-wire transmission line

$$L = L_{int} + L_{ext} \tag{2.3}$$

At high frequencies, the per-unit-length internal inductance of a round wire can be written as follows:[2]

$$L_{int} = \frac{1}{4\pi r_w}\sqrt{\frac{\mu_0}{\pi f \sigma}} \tag{2.4}$$

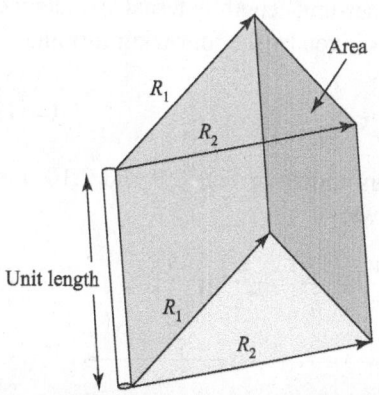

Fig. 2.7 Illustration of calculation of the flux of current through a surface

To calculate the per-unit-length external inductance, let us consider Fig. 2.7. The total magnetic flux ψ_{ext} external to the wire, penetrating a surface S of unit length along the wire direction, and lying between radii R_1 and R_2 ($R_2 \geq R_1$) is given by the following equation:

$$\psi_{ext} = \mu \int \overrightarrow{H_t}.\overrightarrow{ds} \tag{2.5}$$

For a round wire, the tangential magnetic field at a distance r is given by the following relation:

$$\overrightarrow{H_t} = \frac{I}{2\pi r}\hat{a}_\phi \tag{2.6}$$

[2]For detailed derivations see *Introduction to Electromagnetic Fields* by C.R. Paul, K.W. Whites, and S.A. Nasar, Tata McGraw-Hill, Third edition, New Delhi, pp. 494–498.

Substituting Eq. (2.6) into Eq. (2.5) we get the following equation, after carrying out the integration:

$$\psi_{ext} = \mu \int \overrightarrow{H_t}.\overrightarrow{ds} = \mu \int\limits_{0}^{1} \int\limits_{R_1}^{R_2} \frac{I}{2\pi r} \hat{a}_\phi . dz dr \hat{a}_\phi$$

or

$$\psi_{ext} = \frac{\mu I}{2\pi} \int\limits_{0}^{1} dz \int\limits_{R_1}^{R_2} \frac{dr}{r} = \frac{\mu I}{2\pi} [z]_{0}^{1} [\ln(r)]_{R_1}^{R_2} = \frac{\mu I}{2\pi} \ln\left(\frac{R_2}{R_1}\right) \qquad (2.7)$$

If the wires of the transmission line has radii r_{w1} and r_{w2}, respectively, and they are separated by a distance s such that $s \gg r_{w1}, r_{w2}$, then the proximity effect can be neglected and the total flux between the wires can be obtained from Fig. 2.8 as follows:

$$\psi_{ext,\ two\text{-}wire\ line} = \frac{\mu I}{2\pi} \ln\left(\frac{s-r_{w2}}{r_{w1}}\right) + \frac{\mu I}{2\pi} \ln\left(\frac{s-r_{w1}}{r_{w2}}\right)$$

or

$$\psi_{ext,\ two\text{-}wire\ line} = \frac{\mu I}{2\pi} \ln\left\{\frac{(s-r_{w1})(s-r_{w2})}{r_{w1} r_{w2}}\right\} \qquad (2.8)$$

Since $s \gg r_{w1}, r_{w2}$, Eq. (2.8) modifies to the following format:

$$\psi_{ext,\ two\text{-}wire\ line} = \frac{\mu I}{2\pi} \ln\left(\frac{s^2}{r_{w1} r_{w2}}\right) \qquad (2.9)$$

Therefore, the per-unit-length external inductance can be written as follows:

$$L_{ext,\ two\text{-}wire\ line} = \frac{\psi_{ext}}{I} = \frac{\mu}{2\pi} \ln\left(\frac{s^2}{r_{w1} r_{w2}}\right) \qquad (2.10)$$

The result presented in Eq. (2.10) is an approximate one and is valid for widely separated wires. The exact expression of the per-unit-length external inductance of a transmission line having a wire radius r_w is given by the following formula:

$$L_{ext,\ two\text{-}wire\ line} = \frac{\mu}{\pi} \cosh^{-1}\left(\frac{s}{2r_w}\right) \qquad (2.11)$$

For wires above a perfect ground plane, as shown in Figs 2.9 and 2.10, the per-unit-length external inductance is as follows:

$$L_{ext,\ wire\ above\ ground\ plane} = \frac{\mu}{2\pi} \ln\left(\frac{h}{r_w}\right) + \frac{\mu}{2\pi} \ln\left(\frac{2h}{h}\right) = \frac{\mu}{2\pi} \ln\left(\frac{2h}{r_w}\right) \qquad (2.12)$$

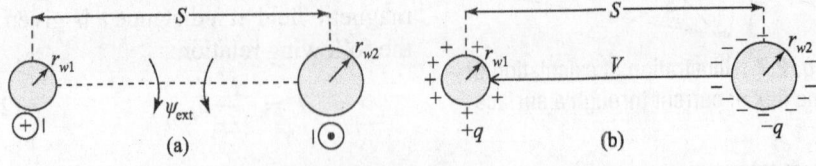

Fig. 2.8 Determination of per-unit-length parameters of a two-wire transmission line (a) Inductance (b) Capacitance

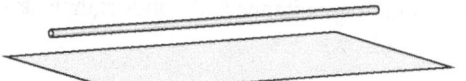

Fig. 2.9   Wire above ground plane

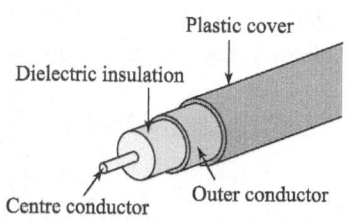

Fig. 2.11   Coaxial cable

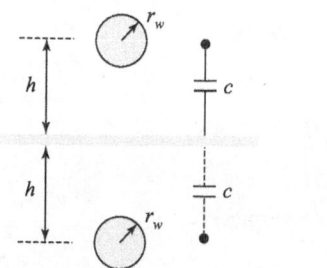

Fig. 2.10   Determination of the per-unit-length capacitance of a wire above ground plane

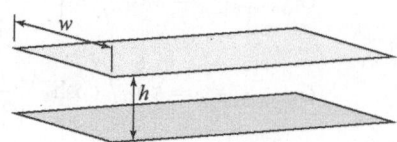

Fig. 2.12   Parallel-plate transmission line

The expression is again an approximate one. The accurate expression is as follows:

$$L_{\text{ext, wire above ground plane}} = \frac{\mu}{2\pi} \cosh^{-1}\left(\frac{h}{r_w}\right) \tag{2.13}$$

Finally, for a coaxial cable, shown in Fig. 2.11, the per-unit-length external inductance is given by the following equation:

$$L_{\text{ext, coaxial cable}} = \frac{\mu}{2\pi} \ln\left(\frac{r_s}{r_w}\right) \tag{2.14}$$

where r_w and r_s are the radii of the central and the outer conductors, respectively.

For parallel-plate line, shown in Fig. 2.12, the expression for the per-unit-length external inductance is as follows:

$$L_{\text{ext, parallel plate}} = \mu h/w \tag{2.15}$$

2.3.3 Per-unit-length Capacitance (C) and Conductance (G)

The transverse electric field $\overrightarrow{E_T}$ between the two conductor surfaces contributes towards the per-unit-length capacitance (C) and the per-unit-length conductance (G) in the line. The capacitance represents the displacement current, $\overrightarrow{J_d} = j\omega\varepsilon\overrightarrow{E_T}$, whereas the conductance represents the conduction current, $\overrightarrow{J_c} = \sigma\overrightarrow{E_T}$, flowing between the two wires through the surrounding lossy medium. For the transverse electro-magnetic (TEM) mode propagation through a homogeneous medium having conductance σ, permittivity ε, permeability μ, and per-unit-length external inductance L_{ext}, it can be shown that

$$L_{\text{ext}} G = \mu\sigma \tag{2.16}$$

$$L_{\text{ext}} C = \mu\varepsilon \tag{2.17}$$

Using Eqs (2.10)–(2.15), the per-unit-length capacitance and conductance can be obtained from the preceding equations as follows:

$$C_{\text{two-wire line}} = 2\pi\varepsilon \Big/ \ln\left(\frac{s^2}{r_{w1}r_{w2}}\right) \quad \text{(approximate)} \tag{2.18}$$

$$C_{\text{two-wire line}} = \pi\varepsilon \Big/ \cosh^{-1}\left(\frac{s}{2r_w}\right) \quad \text{(accurate)} \tag{2.19}$$

$$G_{\text{two-wire line}} = 2\pi\sigma_d \Big/ \ln\left(\frac{s^2}{r_{w1}r_{w2}}\right) \quad \text{(approximate)} \tag{2.20}$$

$$G_{\text{two-wire line}} = \pi\sigma_d \Big/ \cosh^{-1}\left(\frac{s}{2r_w}\right) \quad \text{(accurate)} \tag{2.21}$$

$$C_{\text{wire above ground plane}} = 2\pi\varepsilon \Big/ \ln\left(\frac{2h}{r_w}\right) \quad \text{(approximate)} \tag{2.22}$$

$$C_{\text{wire above ground plane}} = 2\pi\varepsilon \Big/ \cosh^{-1}\left(\frac{h}{r_w}\right) \quad \text{(accurate)} \tag{2.23}$$

$$G_{\text{wire above ground plane}} = 2\pi\sigma_d \Big/ \ln\left(\frac{2h}{r_w}\right) \quad \text{(approximate)} \tag{2.24}$$

$$G_{\text{wire above ground plane}} = 2\pi\sigma_d \Big/ \cosh^{-1}\left(\frac{h}{r_w}\right) \quad \text{(accurate)} \tag{2.25}$$

$$C_{\text{coaxial cable}} = 2\pi\varepsilon \Big/ \ln\left(\frac{r_s}{r_w}\right) \tag{2.26}$$

$$G_{\text{coaxial cable}} = 2\pi\sigma_d \Big/ \ln\left(\frac{r_s}{r_w}\right) \tag{2.27}$$

$$C_{\text{parallel plate}} = w\varepsilon/h \tag{2.28}$$

$$G_{\text{parallel plate}} = w\sigma_d/h \tag{2.29}$$

where σ_d is the conductivity of the dielectric material.

An alternative way to derive the per-unit-length capacitance is as follows. The voltage V, between two points at radial distances R_1 and R_2 ($R_2 > R_1$) is given by the following equation:

$$V = -\int \vec{E_T}.\vec{dl} = -\int_{R_2}^{R_1} \frac{q}{2\pi\varepsilon r} dr = \frac{q}{2\pi\varepsilon} \ln\left(\frac{R_2}{R_1}\right) \tag{2.30}$$

Therefore, the voltage between the wires of radii r_{w1} and r_{w2}, separated by a distance s ($s \gg r_{w1}, r_{w2}$), can be expressed as follows:

$$V = \frac{q}{2\pi\varepsilon} \ln\left(\frac{s - r_{w2}}{r_{w1}}\right) + \frac{q}{2\pi\varepsilon} \ln\left(\frac{s - r_{w1}}{r_{w2}}\right) = \frac{q}{2\pi\varepsilon} \ln\left\{\frac{(s - r_{w1})(s - r_{w2})}{r_{w1}r_{w2}}\right\} \tag{2.31}$$

or $\qquad C = \dfrac{q}{V} = \dfrac{2\pi\varepsilon}{\ln\left\{\dfrac{(s-r_{w1})(s-r_{w2})}{r_{w1}r_{w2}}\right\}}$ $\qquad\qquad$ (2.32)

Since $s \gg r_{w1}, r_{w2}$, Eq. (2.32) modifies to the following form:

$$C = 2\pi\varepsilon \bigg/ \ln\left(\dfrac{s^2}{r_{w1}r_{w2}}\right) \qquad\qquad (2.33)$$

which is the same as Eq. (2.18).

EXAMPLE 2.1 A coaxial transmission line has an inner conductor diameter of 0.81 mm and outer conductor diameter of 2.95 mm. The relative dielectric constant of the dielectric used is 2.4. If the frequency of excitation is 1 GHz then calculate the per-unit-length parameters. Assume that the conductivity of the conductor is 5.8×10^7 S/m and the conductivity of the dielectric is 10^{-15} S/m.

Solution Given: $r_s = 2.95 \times 10^{-3}/2$ mm $= 1.475 \times 10^{-3}$ mm, $r_w = 0.81 \times 10^{-3}/2$ mm $= 0.405 \times 10^{-3}$ mm, $f = 1$ GHz $= 10^9$ Hz, $\varepsilon_r = 2.4$, $\sigma_{Cu} = 5.8 \times 10^7$ S/m, and $\sigma_d = 10^{-15}$ S/m

The skin depth of the wire is as follows:

$$\delta = \dfrac{1}{\sqrt{\pi f \mu_0 \sigma_{Cu}}} = \dfrac{1}{\sqrt{\pi \times 10^9 \times 4\pi \times 10^{-7} \times 5.8 \times 10^7}} = 2.09 \times 10^{-6} \text{ m}$$

Therefore, the per-unit-length resistance of the wire is as follows:

$$R = \dfrac{1}{2\pi r_w \sigma \delta} = \dfrac{1}{2\pi \times 0.405 \times 10^{-3} \times 5.8 \times 10^7 \times 2.09 \times 10^{-6}} = 3.242 \ \Omega/\text{m}$$

The per-unit-length inductance is given by the following equation:

$$L_{\text{ext, coaxial cable}} = \dfrac{\mu}{2\pi}\ln\left(\dfrac{r_s}{r_w}\right) = \dfrac{4\pi \times 10^{-7}}{2\pi}\ln\left(\dfrac{1.475 \times 10^{-3}}{0.405 \times 10^{-3}}\right) = 2.585 \times 10^{-7} \text{ H/m}$$

The per-unit-length capacitance is as follows:

$$C_{\text{coaxial cable}} = \dfrac{2\pi\varepsilon}{\ln\left(\dfrac{r_s}{r_w}\right)} = \dfrac{2\pi \times 2.4 \times 8.8542 \times 10^{-12}}{\ln\left(\dfrac{1.475 \times 10^{-3}}{0.405 \times 10^{-3}}\right)} = 1.033 \times 10^{-10} \text{ F/m}$$

The per-unit-length conductance is given by the following expression:

$$G_{\text{coaxial cable}} = \dfrac{2\pi\sigma_d}{\ln\left(\dfrac{r_s}{r_w}\right)} = \dfrac{2\pi \times 10^{-15}}{\ln\left(\dfrac{1.475 \times 10^{-3}}{0.405 \times 10^{-3}}\right)} = 4.861 \times 10^{-15} \ \mho/\text{m}$$

Practice Problems

2.1 A two-wire transmission line has been constructed with copper wires as conductors and Teflon as the dielectric. If the copper wires have a radius of 0.91 mm and are separated by a distance of 2.225 mm, calculate the per-unit-length parameters. Assume that the signal propagating through the line has a frequency of 1 GHz. The conductivity of Teflon and copper are 4.27×10^{-23} S/m and 5.8×10^7 S/m, respectively. The dielectric constant of Teflon is 2.1. [$1.443 \, \Omega/$ m, 3.576×10^{-7} H/m or 2.621×10^{-7} H/m, 6.534×10^{-11} F/m or 8.914×10^{-11} F/m, 1.5×10^{-22} Ʊ/m or 2.047×10^{-22} Ʊ/m]

2.2 A copper wire of radius 1.31 mm is kept at a height 3.472 mm above an infinite copper plate. If a signal source of 3 GHz is connected between the wire and the plate, then calculate the per-unit-length parameter of the equivalent transmission line. The conductivity of copper is 5.8×10^7 S/m and that of air is 5.7×10^{-15} S/m. [$1.002 \, \Omega/$ m, 3.336×10^{-7} H/m or 3.260×10^{-7} H/m, 3.336×10^{-11} F/m or 3.413×10^{-11} F/m, 2.147×10^{-14} Ʊ/m or 2.197×10^{-14} Ʊ/m]

2.4 Transmission Line Equations

A transmission line can be analysed by solving either the Maxwell's field equations or the Kirchhoff's current and voltage equations in the equivalent distributed circuit model. The solution of Maxwell's equation involves three space variables in addition to the time variable, whereas the distributed circuit model involves only one space variable in addition to the time variable. Owing to this, analysis of a transmission line by the distributed circuit theory is much easier than the field theory and will be discussed here.

Based on the uniform distributed circuit theory, the schematic circuit of a two-conductor transmission line with distributed element parameters R, L, G, and C is shown in Fig. 2.13.

Since a transmission line operates at a frequency below 1 GHz and the respective circuit elements are distributed circuit parameters, the common Kirchoff's voltage and current laws and normal voltage–current relations can be used. Using Kirchoff's voltage law, we get the following relation:

$$v(z,t) = i(z,t)R\Delta z + \frac{\partial i(z,t)}{\partial t} L\Delta z + v(z + \Delta z, t) \tag{2.34}$$

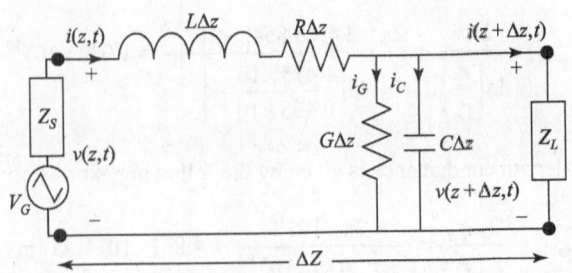

Fig. 2.13 Lumped circuit equivalent model of transmission line

Based on the mathematical definition of derivatives, we can write the following equation:

$$v(z+\Delta z,t)=v(z,t)+\frac{\partial v(z,t)}{\partial z}\Delta z \tag{2.35}$$

Substitution of Eq. (2.35) into Eq. (2.34) results in the following equation:

$$v(z,t)=i(z,t)R\Delta z+\frac{\partial i(z,t)}{\partial t}L\Delta z+v(z,t)+\frac{\partial v(z,t)}{\partial z}\Delta z$$

or

$$-\frac{\partial v(z,t)}{\partial z}\Delta z=i(z,t)R\Delta z+\frac{\partial i(z,t)}{\partial t}L\Delta z$$

or

$$-\frac{\partial v(z,t)}{\partial z}=Ri(z,t)+L\frac{\partial i(z,t)}{\partial t} \tag{2.36}$$

Now using Kirchoff's current law in the circuit, we get the following expression:

$$i(z,t)=i_G(z,t)+i_C(z,t)+i(z+\Delta z,t)$$

or

$$i(z,t)=v(z+\Delta z,t)G\Delta z+\frac{\partial v(z+\Delta z,t)}{\partial t}C\Delta z+i(z+\Delta z,t) \tag{2.37}$$

Again from the mathematical definition of derivatives, we can write the following equation:

$$i(z+\Delta z,t)=i(z,t)+\frac{\partial i(z,t)}{\partial z}\Delta z \tag{2.38}$$

Substituting Eqs (2.35) and (2.38) in Eq. (2.37), we get the following relation:

$$i(z,t)=\left\{v(z,t)+\frac{\partial v(z,t)}{\partial z}\Delta z\right\}G\Delta z+\frac{\partial}{\partial t}\left\{v(z,t)+\frac{\partial v(z,t)}{\partial z}\Delta z\right\}C\Delta z$$

$$+i(z,t)+\frac{\partial i(z,t)}{\partial z}\Delta z$$

or

$$-\frac{\partial i(z,t)}{\partial z}\Delta z=v(z,t)G\Delta z+\frac{\partial v(z,t)}{\partial t}C\Delta z+\frac{\partial v(z,t)}{\partial t}G(\Delta z)^2$$

$$+\frac{\partial^2 v(z,t)}{\partial t\partial z}C(\Delta z)^2 \tag{2.39}$$

Since Δz is very small, the terms incorporating $(\Delta z)^2$ can be neglected. Therefore, Eq. (2.39) reduces to the following form:

$$-\frac{\partial i(z,t)}{\partial z}=Gv(z,t)+C\frac{\partial v(z,t)}{\partial t} \tag{2.40}$$

Equations (2.36) and (2.40) are coupled differential equations that must be decoupled and then solved to find the branch currents and nodal voltages.

Differentiating Eqs (2.36) and (2.40) with respect to z and t, respectively, we get the following equations:

$$-\frac{\partial^2 v(z,t)}{\partial z^2} = R\frac{\partial i(z,t)}{\partial z} + L\frac{\partial^2 i(z,t)}{\partial t \partial z} \tag{2.41}$$

$$-\frac{\partial^2 i(z,t)}{\partial t \partial z} = G\frac{\partial v(z,t)}{\partial t} + C\frac{\partial^2 v(z,t)}{\partial t^2} \tag{2.42}$$

Substituting the expression of $\dfrac{\partial^2 i(z,t)}{\partial t \partial z}$ from Eq. (2.42) into Eq. (2.41), we get the following equation:

$$\frac{\partial^2 v(z,t)}{\partial z^2} = -R\frac{\partial i(z,t)}{\partial z} + LG\frac{\partial v(z,t)}{\partial t} + LC\frac{\partial^2 v(z,t)}{\partial t^2} \tag{2.43}$$

Again substituting the expression of $\dfrac{\partial i(z,t)}{\partial z}$ from Eq. (2.40) into Eq. (2.43), the following expression is obtained:

$$\frac{\partial^2 v(z,t)}{\partial z^2} = RGv(z,t) + (RC+LG)\frac{\partial v(z,t)}{\partial t} + LC\frac{\partial^2 v(z,t)}{\partial t^2} \tag{2.44}$$

In a similar fashion, differentiating Eqs (2.36) and (2.40) with respect to t and z, respectively; combining the result; and finally using Eq. (2.36), we can write the following equation:

$$\frac{\partial^2 i(z,t)}{\partial z^2} = RGi(z,t) + (RC+LG)\frac{\partial i(z,t)}{\partial t} + LC\frac{\partial^2 i(z,t)}{\partial t^2} \tag{2.45}$$

Equations (2.44) and (2.45) are now uncoupled and can be solved individually.

The instantaneous branch current and nodal voltage of the transmission line equivalent circuit can be expressed as follows:

$$v(z,t) = \mathrm{Re}\left\{V(z)e^{j\omega t}\right\} \tag{2.46}$$

and $\quad i(z,t) = \mathrm{Re}\left\{I(z)e^{j\omega t}\right\} \tag{2.47}$

where 'Re' stands for 'real part of'. The factors $V(z)$ and $I(z)$ are complex quantities of the sinusoidal functions of position z on the transmission line and are known as *phasors*.

In the frequency domain $\dfrac{\partial}{\partial t} \rightarrow j\omega$, and therefore, Eqs (2.36), (2.40), (2.44), and (2.45) modify to the following forms:

$$\frac{dV(z)}{dz} = -ZI(z) \tag{2.48}$$

$$\frac{dI(z)}{dz} = -YV(z) \tag{2.49}$$

$$\frac{d^2 V(z)}{dz^2} = \frac{d}{dz}\left[\frac{dV(z)}{dz}\right] = \frac{d}{dz}\{-ZI(z)\} = -Z\frac{dI(z)}{dz}$$

or $\quad \dfrac{d^2 V(z)}{dz^2} = -Z\{-YV(z)\} = ZYV(z) = \gamma^2 V(z) \tag{2.50}$

$$\frac{d^2I(z)}{dz^2} = \frac{d}{dz}\left[\frac{dI(z)}{dz}\right] = \frac{d}{dz}\{-YV(z)\} = -Y\frac{dV(z)}{dz}$$

or
$$\frac{d^2I(z)}{dz^2} = -Y\{-ZI(z)\} = ZYI(z) = \gamma^2 I(z) \tag{2.51}$$

where
$$Z = R + j\omega L \tag{2.52}$$
$$Y = G + j\omega C \tag{2.53}$$
$$\gamma = \alpha + j\beta = \sqrt{ZY} = \sqrt{(R + j\omega L)(G + j\omega C)} \tag{2.54}$$

Here, γ is a complex constant of the circuit and is known as the *propagation constant*. The real part (α) of γ, called the *attenuation constant*, is a function of R and G, whereas the imaginary part (β), called the *phase constant*, is a function of L and C. For a lossless line, $R = 0$ and $G = 0$ and thus $\alpha = 0$. Therefore, for a lossless line Eqs (2.48)–(2.51) reduce to the following forms:

$$\frac{dV(z)}{dz} = -j\omega LI(z) \tag{2.55}$$

$$\frac{dI(z)}{dz} = -j\omega CV(z) \tag{2.56}$$

$$\frac{d^2V(z)}{dz^2} = -\omega^2 LCV(z) \tag{2.57}$$

$$\frac{d^2I(z)}{dz^2} = -\omega^2 LCI(z) \tag{2.58}$$

Note Equations (2.48) and (2.49) are also known as the telegrapher's equation.

2.5 Solutions of Transmission Line Equations

One possible solution of Eq. (2.50) is as follows:

$$V(z) = V_+ e^{-\gamma z} + V_- e^{+\gamma z} = V_+ e^{-\alpha z} e^{-j\beta z} + V_- e^{\alpha z} e^{j\beta z} \tag{2.59}$$

The first term on the right-hand side (RHS) of Eq. (2.59) represents a positive z-directed wave, because for negative values of z it will take the form $V_+ e^{\alpha|z|}$ and hence will increase exponentially with −z, which is not practical. Similarly, we can say that the second term in Eq. (2.59) represents a negative z-directed wave. V_+ and V_-, in Eq. (2.59) are in general complex quantities and depend on the generator voltage, line characteristic impedance, source impedance, load impedance, line length, and propagation constant. V_+ and V_- are, however, independent of z. Substituting Eq. (2.59) into Eq. (2.48) we get the following expression:

$$I(z) = \frac{\gamma}{Z}\left(V_+ e^{-\gamma z} - V_- e^{+\gamma z}\right) \tag{2.60}$$

Substituting $\gamma = \sqrt{ZY}$ from Eq. (2.54) into Eq. (2.60), we get the following equation:

$$I(z) = \sqrt{\frac{Y}{Z}}\left(V_+ e^{-\gamma z} - V_- e^{+\gamma z}\right) = Y_0\left(V_+ e^{-\gamma z} - V_- e^{+\gamma z}\right) \tag{2.61}$$

where Y_0 can be expressed as follows:

$$Y_0 = \sqrt{\frac{Y}{Z}} = \sqrt{\frac{G+j\omega C}{R+j\omega L}} = \left\{\frac{\sqrt{G^2 +(\omega C)^2}\,\angle\tan^{-1}(\omega C/G)}{\sqrt{R^2 +(\omega L)^2}\,\angle\tan^{-1}(\omega L/R)}\right\}^{\frac{1}{2}} \tag{2.62}$$

and is called as the characteristic admittance of the line.

The characteristic impedance of a transmission line, Z_0, is the inverse of the characteristic admittance of the line and is equal to the load impedance, which when terminates the transmission line results in zero reflection from the termination. Since there is no reflection from the load, the forward traveling wave never comes back to the source and the transmission line behaves as a line that is infinitely extended. Therefore, virtually an infinitely extended transmission line can be designed by terminating a finite-length transmission line with its characteristic impedance. Mathematically, the characteristic impedance is expressed as follows:

$$Z_0 = \sqrt{\frac{Z}{Y}} = \sqrt{\frac{R+j\omega L}{G+j\omega C}} = \left\{\frac{\sqrt{R^2 +(\omega L)^2}\,\angle\tan^{-1}(\omega L/R)}{\sqrt{G^2 +(\omega C)^2}\,\angle\tan^{-1}(\omega C/G)}\right\}^{\frac{1}{2}} \tag{2.63}$$

The characteristic impedance, by definition, is a complex quantity. However, at microwave frequencies $\omega L \gg R$ and $\omega C \gg G$, and hence real parts of the numerator and denominator can be neglected in comparison to the imaginary part, thereby reducing Eq. (2.63) to the following form:

$$Z_0 = \sqrt{L/C} \tag{2.64}$$

Equation (2.64) is also valid for a lossless transmission line. Until and unless mentioned, we will assume that the characteristic impedance is purely real and the line is lossless.

From Eqs (2.10) and (2.18), the characteristic impedance of a two-wire line is given by the following relation:

$$Z_{0,\text{ two-wire line}} = \sqrt{\frac{L_{\text{ext, two-wire line}}}{C_{\text{two-wire line}}}} = \sqrt{\frac{\mu\ln\left(\dfrac{s^2}{r_{w1}r_{w2}}\right)}{2\pi} \cdot \dfrac{\ln\left(\dfrac{s^2}{r_{w1}r_{w2}}\right)}{2\pi\varepsilon}}$$

or $$Z_{0,\text{ two-wire line}} = \frac{1}{2\pi}\ln\left(\frac{s^2}{r_{w1}r_{w2}}\right)\sqrt{\frac{\mu}{\varepsilon}}$$

or $$Z_{0,\text{two-wire line}} = \frac{59.96}{\sqrt{\varepsilon_r}}\ln\left(\frac{s^2}{r_{w1}r_{w2}}\right) \tag{2.65}$$

Similarly, from Eqs (2.12) and (2.22), the characteristic impedance of a wire above a ground plane is given by the following equation:

$$Z_{0,\text{wire above ground plane}} = \sqrt{\frac{L_{\text{ext, wire above ground plane}}}{C_{\text{wire above ground plane}}}} = \sqrt{\frac{\mu \ln(2h/r_w)}{2\pi} \frac{\ln(2h/r_w)}{2\pi\varepsilon}}$$

or $\quad Z_{0,\text{wire above ground plane}} = \dfrac{1}{2\pi} \ln(2h/r_w) \sqrt{\dfrac{\mu}{\varepsilon}}$

or $\quad Z_{0,\text{wire above ground plane}} = \dfrac{59.96}{\sqrt{\varepsilon_r}} \ln(2h/r_w)$ $\qquad\qquad$ (2.66)

The characteristic impedance for a lossless coaxial transmission line can be obtained using Eqs (2.14) and (2.26) as follows:

$$Z_{0,\text{coaxial cable}} = \sqrt{\frac{L_{\text{ext,coaxial cable}}}{C_{\text{coaxial cable}}}} = \sqrt{\frac{\mu \ln(r_s/r_w)}{2\pi} \frac{\ln(r_s/r_w)}{2\pi\varepsilon}} = \frac{1}{2\pi} \ln(r_s/r_w)\sqrt{\frac{\mu}{\varepsilon}}$$

or $\quad Z_{0,\text{coaxial cable}} = \dfrac{59.96}{\sqrt{\varepsilon_r}} \ln(r_s/r_w)$ $\qquad\qquad$ (2.67)

Finally, the characteristic impedance for a parallel-plate transmission line can be obtained from Eqs (2.15) and (2.28) as follows:

$$Z_{0,\text{parallel plate}} = \sqrt{\frac{L_{\text{ext,parallel plate}}}{C_{\text{parallel plate}}}} = \sqrt{\frac{\mu h}{w} \frac{h}{\varepsilon w}} = \frac{h}{w}\sqrt{\frac{\mu}{\varepsilon}}$$

or $\quad Z_{0,\text{parallel plate}} = \dfrac{376.73h}{w\sqrt{\varepsilon_r}}$ $\qquad\qquad$ (2.68)

Equation (2.54) can be written as follows:

$$\gamma = \sqrt{(j\omega)^2 LC \left\{ \left(1+\frac{R}{j\omega L}\right)\left(1+\frac{G}{j\omega C}\right) \right\}^{\frac{1}{2}}}$$ $\qquad\qquad$ (2.69)

Expanding Eq. (2.69) in binomial series and neglecting the higher-order terms, we obtain the following equation:

$$\gamma = j\omega\sqrt{LC}\left\{ \left(1+\frac{1}{2}\frac{R}{j\omega L}\right)\left(1+\frac{1}{2}\frac{G}{j\omega C}\right) \right\}$$

or $\quad \gamma = j\omega\sqrt{LC}\left\{ 1+\frac{1}{2}\left(\frac{R}{j\omega L}+\frac{G}{j\omega C}\right) - \frac{1}{4}\frac{RG}{\omega^2 LC} \right\}$ $\qquad$ (2.70)

Since the third term within the { } of Eq. (2.70) is proportional to ω^{-2} and is much small compared to the other two terms, it can be neglected. Thus, Eq. (2.70) reduces to the following form:

$$\gamma = \alpha + j\beta = j\omega\sqrt{LC}\left\{ 1+\frac{1}{2}\left(\frac{R}{j\omega L}+\frac{G}{j\omega C}\right) \right\}$$

or $\qquad \gamma = \alpha + j\beta = \dfrac{1}{2}\left(R\sqrt{\dfrac{C}{L}} + G\sqrt{\dfrac{L}{C}}\right) + j\omega\sqrt{LC}$ (2.71)

Equating the real and imaginary parts of Eq. (2.71), we get the following simpler expressions for attenuation and phase constants:

$$\alpha = \frac{1}{2}\left(R\sqrt{\frac{C}{L}} + G\sqrt{\frac{L}{C}}\right) \qquad (2.72)$$

$$\beta = \omega\sqrt{LC} \qquad (2.73)$$

For the lossless case, $R = 0$ and $G = 0$, and we get the following relations:

$$\alpha = 0 \qquad (2.74)$$

and $\qquad \gamma = j\beta = j\omega\sqrt{LC}$ (2.75)

From Eqs (2.10) and (2.18), the propagation constant of a two-wire line can be given as follows:

$$\gamma_{\text{two-wire line}} = j\omega\sqrt{L_{\text{ext, two-wire line}}\, C_{\text{two-wire line}}}$$

or $\qquad \gamma_{\text{two-wire line}} = j\omega\sqrt{\dfrac{\mu}{2\pi}\ln\left(\dfrac{s^2}{r_{w1}r_{w2}}\right)\dfrac{2\pi\varepsilon}{\ln\left(\dfrac{s^2}{r_{w1}r_{w2}}\right)}} = j\omega\sqrt{\mu\varepsilon}$ (2.76)

Similarly, from Eqs (2.12) and (2.22), the propagation constant of a wire above a ground plane is given by the following equation:

$$\gamma_{\text{wire above ground plane}} = j\omega\sqrt{L_{\text{ext, wire above ground plane}}\, C_{\text{wire above ground plane}}}$$

or $\qquad \gamma_{\text{wire above ground plane}} = j\omega\sqrt{\dfrac{\mu}{2\pi}\ln(2h/r_w)\dfrac{2\pi\varepsilon}{\ln(2h/r_w)}} = j\omega\sqrt{\mu\varepsilon}$ (2.77)

The propagation constant for a lossless coaxial transmission line can be obtained using Eqs (2.14) and (2.26) as follows:

$$\gamma_{\text{coaxial cable}} = j\omega\sqrt{L_{\text{ext,coaxial cable}}\, C_{\text{coaxial cable}}}$$

or $\qquad \gamma_{\text{coaxial cable}} = j\omega\sqrt{\dfrac{\mu}{2\pi}\ln(r_s/r_w)\dfrac{2\pi\varepsilon}{\ln(r_s/r_w)}} = j\omega\sqrt{\mu\varepsilon}$ (2.78)

Finally, the propagation constant for a parallel-plate transmission line can be obtained from Eqs (2.15) and (2.28) as follows:

$$\gamma_{\text{parallel plate}} = j\omega\sqrt{L_{\text{ext,parallel plate}}\, C_{\text{parallel plate}}} = j\omega\sqrt{\frac{\mu h}{w}\frac{w}{h\varepsilon}} = j\omega\sqrt{\mu\varepsilon} \quad (2.79)$$

Based on Eq. (2.73), expressions for phase velocity and group velocity can be given in the following forms:

$$v_p = \omega/\beta = 1/\sqrt{LC} \qquad (2.80)$$

$$v_g = d\omega/d\beta = 1/\sqrt{LC} \tag{2.81}$$

The *phase velocity* of a wave is defined as the velocity at which the constant phase point of the wave propagates in space, whereas the *group velocity* of a wave is defined as the velocity with which the overall shape of the wave's amplitudes, also known as the modulation or envelope of the wave, propagates through space.

The product LC is independent of the size and separation of conductors and depends only on the permeability μ and permittivity ε of the insulating medium. If the lossless transmission line has an air dielectric and contains no ferromagnetic material, then $LC = \mu_0\varepsilon_0$ and

$$v_p = v_g = 1/\sqrt{\mu_0\varepsilon_0} = 3\times10^8 \text{ m/S} \tag{2.82}$$

If the relative permittivity and/or permeability of the insulating medium is greater than 1, then the phase and group velocities of the wave are less than the velocity of light in free space and is given by the following formula:

$$v_p = v_g = \frac{1}{\sqrt{\mu\varepsilon}} = \frac{1}{\sqrt{\mu_r\mu_0\varepsilon_r\varepsilon_0}} = \frac{1}{\sqrt{\mu_r\varepsilon_r}}\frac{1}{\sqrt{\mu_0\varepsilon_0}} = \frac{c}{\sqrt{\mu_r\varepsilon_r}} = v_r c \tag{2.83}$$

where $\quad v_r = 1/\sqrt{\mu_r\varepsilon_r}$ $\hspace{4cm}$ (2.84)

is called a relative velocity factor.

EXAMPLE 2.2 For the example of a coaxial transmission line given in Example 2.1, calculate the following:

(a) Characteristic impedance
(b) Propagation constant

Now assuming that the conductor and the dielectric are perfect, calculate these transmission line parameters using the transmission line dimensions given in Example 2.1 directly. In addition, calculate the phase velocity, group velocity, and relative velocity factor.

Solution Given: $R = 3.242\ \Omega/\text{m}$, $L = 2.585\times10^{-7}$ H/m, $C = 1.033\times10^{-10}$ F/m, $G = 4.861\times10^{-15}$ $\mho$/m, $f = 10^9$ Hz, $\varepsilon_r = 2.4$, $r_s = 1.475\times10^{-3}$ m, and $r_w = 0.405\times10^{-3}$ m From these data, we get the following relations:

$$R + j\omega L = 3.242 + j\times2\pi\times10^9\times2.585\times10^{-7} = 3.242 + j1624.203$$

$$= 1624.206\angle89.886°\ \Omega/\text{m}$$

and $\quad G + j\omega C = 4.861\times10^{-15} + j\times2\pi\times10^9\times1.033\times10^{-10}$

$$= 4.861\times10^{-15} + j0.649 = 0.649\angle90°\ \mho/\text{m}$$

Therefore, the characteristic impedance is given by the following equation:

$$Z_0 = \left[\frac{\sqrt{R^2+(\omega L)^2}\angle\tan^{-1}(\omega L/R)}{\sqrt{G^2+(\omega C)^2}\angle\tan^{-1}(\omega C/G)}\right]^{\frac{1}{2}} = \left[\frac{1624.206\angle89.886°}{0.649\angle90°}\right]^{\frac{1}{2}}$$

or $\quad Z_0 = [2502.629\angle-0.114°]^{\frac{1}{2}}$

or $\qquad Z_0 = 50.026\angle - 0.057 = 50.026 - j0.049 \ \Omega$

Similarly, the propagation constant is given by the following equation:

$$\gamma = \sqrt{(R + j\omega L)(G + j\omega C)} = \sqrt{(1624.206\angle 89.886°)(0.649\angle 90°)}$$

or $\qquad \gamma = \sqrt{1054.11\angle 179.886°}$

or $\qquad \gamma = 32.467\angle 89.943° = 0.032 + j32.467 \ \text{rad/m}$

Now, if the conductor is perfect then $R = 0$ and $G = 0$, and we can calculate the characteristic impedance as follows:

$$Z_0 = \frac{59.96}{\sqrt{\varepsilon_r}} \ln\left(\frac{r_s}{r_w}\right) = \frac{59.96}{\sqrt{2.4}} \ln\left(\frac{1.475\times 10^{-3}}{0.405\times 10^{-3}}\right) = 50.026 \ \Omega$$

Similarly, the propagation constant is calculated as follows:

$$\gamma = j\omega\sqrt{\mu\varepsilon} = j\times 2\pi\times 10^9 \sqrt{4\pi\times 10^{-7}\times 2.4\times 8.8542\times 10^{-12}} = j32.469 \ \text{rad/m}$$

The phase and group velocity is given by

$$v_p = v_g = \frac{1}{\sqrt{\mu\varepsilon}} = \frac{1}{\sqrt{4\pi\times 10^{-7}\times 2.4\times 8.8542\times 10^{-12}}} = 1.935\times 10^8 \ \text{m/s}$$

and finally the relative velocity factor is as follows:

$$v_r = \frac{1}{\sqrt{\mu_r\varepsilon_r}} = \frac{1}{\sqrt{2.4}} = 0.645$$

Practice Problems

2.3 For the two-wire transmission line described in Practice Problem 2.1, calculate the following parameters:

(a) Characteristic impedance
(b) Propagation constant

(c) Phase and group velocities
(d) Relative velocity factor

Recalculate the characteristic impedance and propagation constant from the transmission line dimensions given in Practice Problem 2.1 after assuming the conductor and the dielectric as perfect. $\qquad$ [73.938 − j0.025 Ω, 73.986 Ω, 0.01 + j30.389 rad/m, j30.372 rad/m, 2.069 × 10⁸ m/s, 0.69]

2.4 For the wire above a ground plane system, described in Practice Problem 2.2, calculate the following parameters:

(a) Characteristic impedance
(b) Propagation constant

(c) Phase and group velocities
(d) Relative velocity factor

Recalculate the characteristic impedance and propagation constant from the transmission line dimensions given in Practice Problem 2.2, after assuming that the conductor and the dielectric are perfect. $\qquad$ [99.906 − j0.024 Ω, 100.004 Ω, 0.005 + j20.98 rad/m, j20.958 rad/m, 2.998 × 10⁸ m/s, 1.0]

2.6 Reflection and Transmission Coefficients

Let us consider that a transmission line of characteristic impedance Z_0 and length l is terminated in a load impedance Z_L, as shown in Fig. 2.14. The voltage and current at any point z on the transmission line are given by Eqs (2.59) and (2.61), respectively. Using these two equations, the voltage and current at the load can be calculated as follows:

$$V_L = V_+ e^{-\gamma l} + V_- e^{+\gamma l} \tag{2.85}$$

$$I_L = \frac{1}{Z_0} \left(V_+ e^{-\gamma l} - V_- e^{+\gamma l} \right) \tag{2.86}$$

The ratio of Eqs (2.85) to (2.86) is equal to the load impedance and is as follows:

$$Z_L = \frac{V_L}{I_L} = Z_0 \left(1 + \frac{V_- e^{+\gamma l}}{V_+ e^{-\gamma l}} \right) \Big/ \left(1 - \frac{V_- e^{+\gamma l}}{V_+ e^{-\gamma l}} \right) \tag{2.87}$$

The term $V_- e^{+\gamma l} / V_+ e^{-\gamma l}$ represents the ratio of the reflected wave (negative z-directed) to the incident voltage (positive z-directed) at the load and is called the voltage reflection coefficient, or simply the reflection coefficient until otherwise mentioned, at the load (Γ_L). Therefore,

$$Z_L = Z_0 \left(\frac{1 + \Gamma_L}{1 - \Gamma_L} \right) \tag{2.88}$$

where $\quad \Gamma_L = \dfrac{V_- e^{+\gamma l}}{V_+ e^{-\gamma l}} \tag{2.89}$

Manipulating Eq. (2.88), we obtain the following relation:

$$\Gamma_L = \frac{Z_L - Z_0}{Z_L + Z_0} \tag{2.90}$$

Since the numerator of Eq. (2.90) is less than its denominator, Γ_L is always less than unity for passive loads. The values of Γ_L for different load conditions are given in Table 2.1.

Table 2.1 shows that the values of reflection coefficients range from -1 and $+1$. A negative reflection coefficient corresponds to an $180°$ phase change of the wave on reflection.

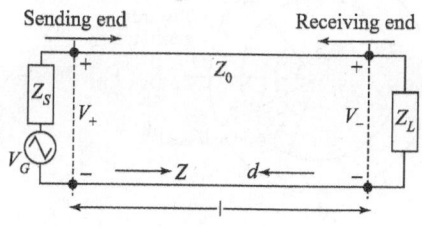

Fig. 2.14 Two-wire transmission line

Table 2.1 Values of Γ_L for different load conditions

Load Condition	Load Value	Γ_L
Short circuit	0	-1
Matched load	Z_0	0
Open circuit	∞	$+1$

If the load impedance and/or the characteristic impedance are complex quantities, as is usually the case, Eq. (2.90) indicates that the reflection coefficient at a load is also a complex quantity and hence can be expressed as follows:

$$\Gamma_L = |\Gamma_L| e^{j\theta_L} \tag{2.91}$$

where $|\Gamma_L|$ is the magnitude of reflection coefficient at the load, and θ_L is the phase difference between the incident and reflected voltages at the load and is also known as the phase angle of reflection coefficient at the load. As explained before, $|\Gamma_L| \leq 1$.

The voltage reflection coefficient at any point z on the transmission line can be expressed in analogy to Eq. (2.89) as follows:

$$\Gamma_z = \frac{V_- e^{+\gamma z}}{V_+ e^{-\gamma z}} \tag{2.92}$$

where $z = l - d$, as shown in Fig. 2.14. $\hspace{3cm}$ (2.93)

Substituting Eq. (2.93) into Eq. (2.92), we get the following relation:

$$\Gamma_z = \frac{V_- e^{\gamma(l-d)}}{V_+ e^{-\gamma(l-d)}} = \frac{V_- e^{\gamma l}}{V_+ e^{-\gamma l}} e^{-2\gamma d} \tag{2.94}$$

Substituting Eq. (2.89) into Eq. (2.94), we get the following equation:

$$\Gamma_z = \Gamma_L e^{-2\gamma d} \tag{2.95}$$

Further substituting Eq. (2.91) into Eq. (2.95), the following equation is obtained:

$$\Gamma_z = |\Gamma_L| e^{j\theta_L} e^{-2\alpha d} e^{-2j\beta d} = |\Gamma_L| e^{-2\alpha d} e^{j(\theta_L - 2\beta d)} \tag{2.96}$$

For a lossy line, both the magnitude and the phase of the reflection coefficient change in an inward-spiral way as shown in Fig. 2.15(a). For a lossless line, $\alpha = 0$; hence, the magnitude of the reflection coefficient remains constant and only the phase of reflection coefficient changes circularly towards the generator with an angle of $\angle -2\beta d$ as shown in Fig. 2.15(b). For $d = \lambda/2$, the total phase change of the reflection coefficient is 2π. Therefore, in a transmission line the phase pattern of the reflection coefficient repeats itself in each half-wavelength section.

To find an expression for the transmission coefficient, let us consider Fig. 2.16.

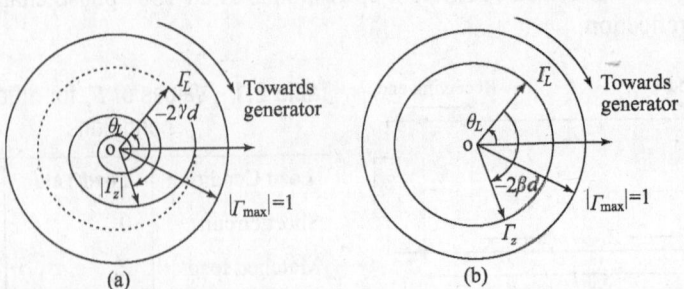

(a) $\hspace{7cm}$ (b)

Fig. 2.15 Polar plot of complex reflection coefficient (a) Lossy line (b) Lossless line

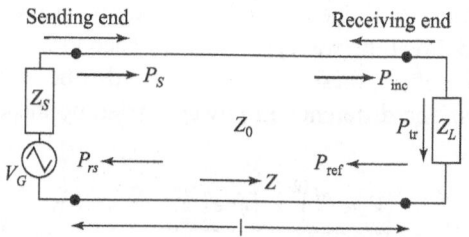

Fig. 2.16 Power transmission on a transmission line

The expression of the voltage travelling wave at the receiving end has already been given in Eq. (2.85). Since the load voltage is nothing but the transmitted voltage, we can write the following equation:

$$V_L = V_+e^{-\gamma l} + V_-e^{+\gamma l} = V_{tr}e^{-\gamma l} \text{ or } 1 + \frac{V_-e^{+\gamma l}}{V_+e^{-\gamma l}} = \frac{V_{tr}e^{-\gamma l}}{V_+e^{-\gamma l}} \tag{2.97}$$

The second term on the left-hand side (LHS) of Eq. (2.97) is the reflection coefficient, whereas the term on its RHS is the transmission coefficient (T_L) at the load. Therefore, Eq. (2.97) can be written as follows:

$$T_L = 1 + \Gamma_L \tag{2.98}$$

Substitution of the expression of Γ_L from Eq. (2.90) into Eq. (2.98) leads to the following expression:

$$T_L = 1 + \frac{Z_L - Z_0}{Z_L + Z_0} = \frac{2Z_L}{Z_L + Z_0} \tag{2.99}$$

Similar to the load impedance, if the source impedance is also not equal to the characteristic impedance of the line, then the reflected wave from the load will suffer further reflection by the source impedance upon arrival on the source terminal. Expressions for voltage reflection coefficient (Γ_S) and voltage transmission coefficient (T_S) at the source, in analogy with the voltage reflection coefficient and voltage transmission coefficient at the load, can be expressed as follows:

$$\Gamma_S = \frac{Z_S - Z_0}{Z_S + Z_0} \tag{2.100}$$

$$T_S = \frac{2Z_S}{Z_S + Z_0} \tag{2.101}$$

One can also find the same expressions for current reflection coefficients at the load or source or at any point z on a transmission line using the same procedure, starting with the solution of Eq. (2.61), which is as follows:

$$I(z) = I_+e^{-\gamma z} - I_-e^{+\gamma z} \tag{2.102}$$

The negative sign before the second term of the RHS of Eq. (2.102) indicates that the directions of the incident and reflected currents are opposite to each other.

It has already been explained that the first term on the RHS of Eq. (2.59) represents a forward travelling wave, whereas the second term represents a reverse travelling wave. Therefore, these two terms can also be interpreted as incident and reflected waves. The difference in power carried by these two waves can be derived as follows:

$$P' = P_{inc} - P_{ref} = \frac{\left(V_+ e^{-\alpha l}\right)^2}{2Z_0} - \frac{\left(V_- e^{\alpha l}\right)^2}{2Z_0} \tag{2.103}$$

Without any loss of generality, we can assume the transmitted wave as of the form $V_{tr} e^{-\gamma l}$. Therefore, the power carried to the load by the transmitted wave can be expressed as follows:

$$P_{tr} = \frac{\left(V_{tr} e^{-\alpha l}\right)^2}{2Z_L} \tag{2.104}$$

Since at the load point the incident power must be equal to the sum of the reflected power and the transmitted power, P' must be equal to P_{tr}. Therefore, equating Eq. (2.103) to Eq. (2.104), we get the following expression:

$$\frac{\left(V_+ e^{-\alpha l}\right)^2}{2Z_0} - \frac{\left(V_- e^{\alpha l}\right)^2}{2Z_0} = \frac{\left(V_{tr} e^{-\alpha l}\right)^2}{2Z_L} \text{ or } 1 - \left(\frac{V_- e^{\alpha l}}{V_+ e^{-\alpha l}}\right)^2 = \frac{Z_0}{Z_L}\left(\frac{V_{tr} e^{-\alpha l}}{V_+ e^{-\alpha l}}\right)^2$$

or $\quad 1 - \Gamma_L^2 = \dfrac{Z_0}{Z_L} T_L^2 \text{ or } Z_L\left(1 - \Gamma_L^2\right) = Z_0 T_L^2 \tag{2.105}$

Equation (2.105) is valid for any complex load and characteristic impedances.

EXAMPLE 2.3 Assume that the coaxial cable described in Example 2.1 is terminated in a dipole having an input impedance of $78.27 + j60.93 \ \Omega$. Calculate the following parameters:

(a) Reflection coefficient at the load
(b) Transmission coefficient at the load
(c) Reflection coefficient at a distance of 91.44 mm from the load

In addition, verify Eq. (2.105).

Solution Given: $Z_L = 78.27 + j60.93 \ \Omega$, $Z_0 = 50.026 - j0.049 \ \Omega$, $d = 91.44 \times 10^{-3}$ m, and $\gamma = 0.032 + j32.467$ rad/m

The reflection coefficient at the load is given by the following equation:

$$\Gamma_L = \frac{Z_L - Z_0}{Z_L + Z_0} = \frac{(78.27 + j60.93) - (50.026 - j0.049)}{(78.27 + j60.93) + (50.026 - j0.049)} = \frac{28.244 + j60.979}{128.3 + j60.881}$$

or $\quad \Gamma_L = \dfrac{67.202\angle 65.148°}{142.012\angle 25.385°} = 0.473\angle 39.763° = 0.364 + j0.303$

Similarly, the transmission coefficient at the load is as follows:

$$T_L = \frac{2Z_L}{Z_L + Z_0} = \frac{2 \times (78.27 + j60.93)}{(78.27 + j60.93) + (50.026 - j0.049)} = \frac{156.54 + j121.86}{128.3 + j60.881}$$

or $\qquad T_L = \dfrac{198.38\angle 37.899°}{142.012\angle 25.385°} = 1.397\angle 12.514° = 1.364 + j0.303$

The reflection coefficient at a distance of 91.44×10^{-3} m from the load is as follows:

$$\Gamma_z = \Gamma_L e^{-2\gamma d} = (0.364 + j0.303)e^{-\left[2\times(0.032 + j32.467)\times 91.44\times 10^{-3}\right]}$$

or $\qquad \Gamma_z = 0.238 + j0.406 = 0.471\angle 59.621°$

Now $\qquad Z_L\left(1 - \Gamma_L^2\right) = (78.27 + j60.93)\left\{1 - (0.364 + j0.303)^2\right\} = 88.5 + j41.213$

Whereas $Z_0 T_L^2 = (50.026 - j0.049)(1.364 + j0.303)^2 = 88.5 + j41.213$

Therefore, Eq. (2.105) is verified.

Practice Problems

2.5 Assume that the two-wire transmission line described in Practice Problem 2.3, is terminated by a load impedance of $43.72 + j87.63\ \Omega$. Calculate the following parameters:

(a) Reflection coefficient at the load
(b) Transmission coefficient at the load
(c) Reflection coefficient at a distance of 83.73 mm from the load

In addition, verify Eq. (2.105). $\qquad$ [0.632∠72.351°, 1.335∠26.812°, 0.631∠140.787°]

2.6 Assume that the transmission line system described in Practice Problem 2.4, is terminated by a load impedance of $165.21 + j6.97\ \Omega$. Calculate the following parameters:

(a) Reflection coefficient at the load
(b) Transmission coefficient at the load
(c) Reflection coefficient at a distance of 23.35 mm from the load

In addition, verify Eq. (2.105). $\qquad$ [0.248∠4.612°, 1.247∠0.915°, 0.248∠−51.55°]

2.7 Standing Waves

In terms of reflection coefficients, the general solutions of lossless transmission line voltage and current, given in Eqs (2.59) and (2.102), are given as follows:

$$V(z) = V_+ e^{-j\beta z} + V_- e^{+j\beta z} = V_+ e^{-j\beta(l-d)} + V_- e^{+j\beta(l-d)}$$

or $\qquad V(z) = V_+ e^{-j\beta l}\left(e^{j\beta d} + \Gamma_L e^{-j\beta d}\right)$ $\qquad\qquad$ (2.106)

$$I(z) = I_+ e^{-j\beta z} - I_- e^{+j\beta z} = I_+ e^{-j\beta(l-d)} - I_- e^{+j\beta(l-d)}$$

or $\qquad I(z) = I_+ e^{-j\beta l}\left(e^{j\beta d} - \Gamma_L e^{-j\beta d}\right)$ $\qquad\qquad$ (2.107)

Further considering the time harmonic variation of the voltage and current, these equations can be written as follows:

$$V(z,t) = \text{Re}\left\{V(z)e^{j\omega t}\right\} = \text{Re}\left\{|V_+|\left(e^{j(\omega t+\theta+\beta d)} + |\Gamma_L|e^{j(\omega t+\theta_L+\theta-\beta d)}\right)\right\}$$

or $\qquad V(z,t) = |V_+|\left[\cos(\omega t+\theta+\beta d) + |\Gamma_L|\cos(\omega t+\theta_L+\theta-\beta d)\right]$ (2.108)

and $\qquad I(z,t) = |I_+|\left[\cos(\omega t+\theta+\beta d) - |\Gamma_L|\cos(\omega t+\theta_L+\theta-\beta d)\right]$ (2.109)

where we have substituted $\Gamma_L = |\Gamma_L|e^{j\theta_L}$ from Eq. (2.91) and

$$V_+e^{-j\beta l} = |V_+|e^{j\theta} \tag{2.110}$$
$$I_+e^{-j\beta l} = |I_+|e^{j\theta} \tag{2.111}$$

Now, let us consider the following cases:

Case 1—Open-circuited transmission line Based on Table 2.1, we can write that for an open-circuited transmission line

$$|\Gamma_L| = 1 \tag{2.112}$$
and $\qquad \theta_L = 2n\pi \quad \text{where} \quad n = 0, \pm 1, \pm 2, \pm 3, \pm 4, \ldots$ (2.113)

Substituting the value of $|\Gamma_L|$ and θ_L from Eqs (2.112) and (2.113) into Eqs (2.108) and (2.109), we get the following expression:

$$V(z,t) = |V_+|\left[\cos(\omega t+\theta+\beta d) + \cos(\omega t+\theta-\beta d)\right]$$

or $\qquad V(z,t) = 2|V_+|\cos(\beta d)\cos(\omega t+\theta)$ (2.114)

and $\qquad I(z,t) = |I_+|\left[\cos(\omega t+\theta+\beta d) - \cos(\omega t+\theta-\beta d)\right]$

or $\qquad I(z,t) = -2|I_+|\sin(\beta d)\sin(\omega t+\theta)$ (2.115)

Equations (2.114) and (2.115) represent pure standing waves, shown in Fig. 2.17(a).

The voltage nodal point occurs at $\beta d = n\pi$, where $n = 0, 1, 2, 3, 4, \ldots$

or $\qquad d = \dfrac{n\pi}{\beta} = \dfrac{n\lambda}{2}$ (2.116)

whereas voltage antinodal point occurs at $\beta d = \dfrac{(2n+1)\pi}{2}$ where $n = 0, 1, 2, 3, 4, \ldots$

or $\qquad d = \dfrac{(2n+1)\pi}{2\beta} = \dfrac{(2n+1)\lambda}{4}$ (2.117)

Similarly, the condition for current nodal and antinodal points is as follows:

$$\beta d = \dfrac{(2n+1)\pi}{2}, \qquad \text{where } n = 0, 1, 2, 3, 4, \ldots$$

or $\qquad d = \dfrac{(2n+1)\pi}{2\beta} = \dfrac{(2n+1)\lambda}{4}$ (2.118)

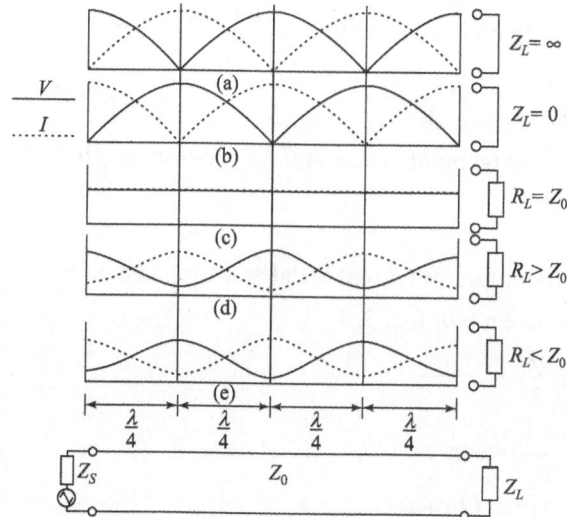

Fig. 2.17 Plot of SWR for different load conditions
(a) Open circuit (b) Short circuit (c) Matched load
(d) Load impedance > characteristic impedance
(e) Load impedance < characteristic impedance

and $\beta d = n\pi,$ where $n = 0, 1, 2, 3, 4, \dots$

or $d = \dfrac{n\pi}{\beta} = \dfrac{n\lambda}{2}$ (2.119)

Equations (2.116)–(2.119) show that the voltage and/or current nodal and antinodal points are spaced at a distance of $\lambda/2$ and also that the voltage maxima correspond to the current minima and vice versa.

Case 2—Short-circuited transmission line   Based on Table 2.1, we can write that for a short-circuited transmission line

$$|\Gamma_L| = 1$$ (2.120)

and $\theta_L = (2n+1)\pi$ where $n = 0, \pm 1, \pm 2, \pm 3, \pm 4, \dots$ (2.121)

Substituting the value of $|\Gamma_L|$ and θ_L from Eqs (2.120) and (2.121) into Eqs (2.108) and (2.109), we get the following expressions:

$$V(z,t) = |V_+|\big[\cos(\omega t + \theta + \beta d) - \cos(\omega t + \theta - \beta d)\big]$$

or $V(z,t) = -2|V_+|\sin(\beta d)\sin(\omega t + \theta)$ (2.122)

and $I(z,t) = |I_+|\big[\cos(\omega t + \theta + \beta d) + \cos(\omega t + \theta - \beta d)\big]$

or $I(z,t) = 2|I_+|\cos(\beta d)\cos(\omega t + \theta)$ (2.123)

Equations (2.122) and (2.123) represent pure standing waves, as shown in Fig. 2.17(b).

The voltage nodal point occurs at $\beta d = \dfrac{(2n+1)\pi}{2}$ where $n = 0, 1, 2, 3, 4, \ldots$

or $\qquad d = \dfrac{(2n+1)\pi}{2\beta} = \dfrac{(2n+1)\lambda}{4}$ $\hfill$ (2.124)

whereas voltage antinodal point occurs at $\beta d = n\pi$ where $n = 0, 1, 2, 3, 4, \ldots$

or $\qquad d = \dfrac{n\pi}{\beta} = \dfrac{n\lambda}{2}$ $\hfill$ (2.125)

Similarly, the condition for current nodal and antinodal points is

$\qquad \beta d = n\pi$ where $n = 0, 1, 2, 3, 4, \ldots$

or $\qquad d = \dfrac{n\pi}{\beta} = \dfrac{n\lambda}{2}$ $\hfill$ (2.126)

and $\qquad \beta d = \dfrac{(2n+1)\pi}{2}$ where $n = 0, 1, 2, 3, 4, \ldots$

or $\qquad d = \dfrac{(2n+1)\pi}{2\beta} = \dfrac{(2n+1)\lambda}{4}$ $\hfill$ (2.127)

Equations (2.124)–(2.127) show that the voltage and/or current nodal and antinodal points are spaced at a distance of $\lambda/2$ and also that the voltage maxima correspond to current minima and vice versa.

Case 3—Pure resistive load termination   Let us suppose that the line is terminated by a pure resistance R_L. Therefore,

$$\Gamma_L = \frac{R_L - Z_0}{R_L + Z_0} = \frac{r_L - 1}{r_L + 1} = -\frac{1 - r_L}{1 + r_L} \tag{2.128}$$

where Z_0 is the characteristic impedance and

$$r_L = R_L/Z_0 \tag{2.129}$$

Substituting the expression of Γ_L from Eq. (2.128) to Eq. (2.106), we obtain the following relation:

$$V(z) = V_+ e^{-j\beta l}\left[e^{j\beta d} - \left(\frac{1 - r_L}{1 + r_L}\right)e^{-j\beta d}\right]$$

or $\qquad V(z) = \dfrac{|V_+|}{(1 + r_L)}\left[(1 + r_L)e^{j(\beta d + \theta)} - (1 - r_L)e^{-j(\beta d - \theta)}\right]$ $\hfill$ (2.130)

where we have used the relation $V_+ e^{-j\beta l} = |V_+|e^{j\theta}$ from Eq. (2.110).

Similarly, substituting the expression of Γ_L from Eq. (2.128) into Eq. (2.107), we get the following expression:

$$I(z) = I_+ e^{-j\beta l}\left[e^{j\beta d} + \left(\frac{1 - r_L}{1 + r_L}\right)e^{-j\beta d}\right]$$

or $\qquad I(z) = \dfrac{|I_+|}{(1 + r_L)}\left[(1 + r_L)e^{j(\beta d + \theta)} + (1 - r_L)e^{-j(\beta d - \theta)}\right]$ $\hfill$ (2.131)

where we have used the relation $I_+ e^{-j\beta l} = |I_+|e^{j\theta}$ from Eq. (2.111).

Assuming a time harmonic variation, the transmission line voltage is given by $V(z,t) = V(z)e^{j\omega t}$ and therefore using Eq. (2.130), we obtain the following equation:

$$V(z,t) = \frac{|V_+|}{(1+r_L)} \text{Re} \left[(1+r_L)e^{j(\omega t+\theta+\beta d)} - (1-r_L)e^{j(\omega t+\theta-\beta d)} \right]$$

or

$$V(z,t) = \frac{|V_+|}{(1+r_L)} \left[(1+r_L)\cos(\omega t+\theta+\beta d) - (1-r_L)\cos(\omega t+\theta-\beta d) \right]$$

or

$$V(z,t) = \frac{|V_+|}{(1+r_L)} \left[(1+r_L)\cos(\omega t+\theta+\beta d) - (1-r_L)\cos(\omega t+\theta-\beta d) \right.$$
$$\left. + r_L\cos(\omega t+\theta+\beta d) - r_L\cos(\omega t+\theta+\beta d) \right]$$

or

$$V(z,t) = \frac{|V_+|}{(1+r_L)} \left[2r_L\cos(\omega t+\theta+\beta d) + (1-r_L)\cos(\omega t+\theta+\beta d) \right.$$
$$\left. - (1-r_L)\cos(\omega t+\theta-\beta d) \right]$$

or

$$V(z,t) = \frac{2|V_+|}{(1+r_L)} \left[r_L\cos(\omega t+\theta+\beta d) - (1-r_L)\sin(\beta d)\sin(\omega t+\theta) \right] \quad (2.132)$$

Similarly, the transmission line current is given by the following relation:

$$I(z,t) = \frac{2|I_+|}{(1+r_L)} \left[r_L\cos(\omega t+\theta+\beta d) + (1-r_L)\cos(\beta d)\cos(\omega t+\theta) \right] \quad (2.133)$$

In general for resistive loads, except when it is equal to characteristic impedance of the line, the incident voltage is neither completely absorbed nor completely reflected by the load, resulting in the existence of both a travelling wave and a standing wave on the transmission line. The first term within the bracket on the RHS of Eqs (2.132) and (2.133) represents the travelling wave, whereas the second term represents the standing wave.

For matched termination $R_L = Z_0$

or $r_L = 1$ (2.134)

Substituting the value of r_L from Eq. (2.134) into Eqs (2.132) and (2.133) we get the following equations:

$$V(z) = |V_+|\cos(\omega t+\theta+\beta d) \quad (2.135)$$

and $$I(z) = |I_+|\cos(\omega t+\theta+\beta d) \quad (2.136)$$

which are pure travelling waves and hence maximum power is transmitted to the load. For the open-circuit case, $r_L \to \infty$ and Eqs (2.132) and (2.133) reduce to Eqs (2.114) and (2.115), respectively, whereas for short-circuit case $r_L = 0$ and Eqs (2.132) and (2.133) reduce to Eqs (2.122) and (2.123), respectively.

Equation (2.106) can be written as follows:

$$V(z) = V_+ e^{-j\beta l} \left[e^{j\beta d} + |\Gamma_L|e^{-j(\beta d-\theta_L)} \right] \quad (2.137)$$

The voltage minimum occurs when the two terms within the bracket on the RHS of Eq. (2.137) adds out of phase, that is, the incident and reflected wave have a phase difference of $(2m+1)\pi$, where m is an integer. For the first minimum $m = 0$, and hence the incident and reflected waves have a phase difference of π. If the load is resistive and $R_L < Z_0$ then Eq. (2.90) shows that Γ_L is negative and real, that is, the incident and reflected voltages are out of phase. Therefore, if $R_L < Z_0$ then the first voltage minimum occurs at the position of the load.

Similarly, the voltage maximum occurs when the two terms within the bracket on the RHS of Eq. (2.137) adds in phase, that is, the incident and reflected waves have a phase difference of $2m\pi$, where again m is an integer. For the first maximum $m = 0$, and hence the incident and reflected waves have a phase difference of $0°$. If the load is resistive and $R_L > Z_0$ then Eq. (2.90) shows that Γ_L is positive and real, that is, the incident and reflected voltages are in phase. Therefore, if $R_L > Z_0$ then the first voltage maximum occurs at the position of the load.

The standing wave ratio (SWR) plots for $R_L = Z_0$, $R_L < Z_0$, and $R_L > Z_0$ are shown in Figs 2.17(c), (d), and (e), respectively.

Case 4—Complex load termination For complex loads, Eqs (2.106) and (2.107) can be written as follows:

$$V(z) = V_+ e^{-j\beta l}\left[e^{j\beta d} + |\Gamma_L| e^{j\theta_L} e^{-j\beta d}\right]$$

or
$$V(z) = V_+ e^{-j\beta(l-d)}\left[1 + |\Gamma_L| e^{j(\theta_L - 2\beta d)}\right] \tag{2.138}$$

and
$$I(z) = I_+ e^{-j\beta(l-d)}\left[1 - |\Gamma_L| e^{j(\theta_L - 2\beta d)}\right] \tag{2.139}$$

Taking the magnitudes of Eqs (2.138) and (2.139), we get the following expressions:

$$|V(z)| = |V_+|\left[1 + 2|\Gamma_L|\cos(2\beta d - \theta_L) + |\Gamma_L|^2\right]^{\frac{1}{2}} \tag{2.140}$$

and
$$|I(z)| = |I_+|\left[1 - 2|\Gamma_L|\cos(2\beta d - \theta_L) + |\Gamma_L|^2\right]^{\frac{1}{2}} \tag{2.141}$$

Equations (2.140) and (2.141) show that $|V(z)|$ and $|I(z)|$ will be maximum or minimum depending upon the value of $\cos(2\beta d - \theta_L)$. If $\cos(2\beta d - \theta_L) = 1$ then Eqs (2.140) and (2.141) reduce to the following forms:

$$|V(z)|_{max} = |V_+|(1 + |\Gamma_L|) \tag{2.142}$$

and
$$|I(z)|_{min} = |I_+|(1 - |\Gamma_L|) \tag{2.143}$$

Similarly, if $\cos(2\beta d - \theta_L) = -1$ then Eqs (2.140) and (2.141) reduce to the following forms:

$$|V(z)|_{min} = |V_+|(1 - |\Gamma_L|) \tag{2.144}$$

and
$$|I(z)|_{max} = |I_+|(1 + |\Gamma_L|) \tag{2.145}$$

The ratio of Eq. (2.142) to Eq. (2.144) is called the voltage standing wave ratio (VSWR), whereas the ratio of Eq. (2.145) to Eq. (2.143) is called the current standing wave ratio (ISWR). In general, VSWR and ISWR are identical and is simply called SWR, which is expressed as follows:

$$\rho = \frac{1+|\Gamma_L|}{1-|\Gamma_L|} \tag{2.146}$$

The value of ρ for different load values is given in Table 2.2. The plot of ρ versus $|\Gamma_L|$ is shown in Fig. 2.18.

The values of SWR range between 1 and ∞. When $\rho = 1$, the line is called a flat line. It is important to note that the SWR cannot be defined for a lossy line, as its pattern changes markedly from one point of the line to another. For a lossless line, the SWR is constant throughout the line.

Manipulating Eq. (2.146) we get the following relation:

$$|\Gamma_L| = \frac{\rho-1}{\rho+1} \tag{2.147}$$

The general solution of a transmission line equation consists of two waves travelling in the opposite directions with unequal amplitudes and can be written, based on Eq. (2.59), as follows:

$$V(z) = V_+ e^{-\gamma z} + V_- e^{+\gamma z} = V_+ e^{-\alpha z} e^{-j\beta z} + V_- e^{\alpha z} e^{j\beta z}$$

or

$$V(z) = V_+ e^{-\alpha z} \left[\cos(\beta z) - j\sin(\beta z)\right] + V_- e^{\alpha z} \left[\cos(\beta z) + j\sin(\beta z)\right]$$

or

$$V(z) = \left(V_+ e^{-\alpha z} + V_- e^{\alpha z}\right)\cos(\beta z) - j\left(V_+ e^{-\alpha z} - V_- e^{\alpha z}\right)\sin(\beta z) \tag{2.148}$$

This voltage equation can be written as follows:

$$V(z) = V_0 e^{-j\phi} \tag{2.149}$$

where $V_0 = \left[\left(V_+ e^{-\alpha z} + V_- e^{+\alpha z}\right)^2 \cos^2(\beta z) + \left(V_+ e^{-\alpha z} - V_- e^{+\alpha z}\right)^2 \sin^2(\beta z)\right]^{\frac{1}{2}}$ (2.150)

and

$$\phi = \tan^{-1}\left[\frac{V_+ e^{-\alpha z} + V_- e^{\alpha z}}{V_+ e^{-\alpha z} - V_- e^{\alpha z}} \tan(\beta z)\right] \tag{2.151}$$

V_0 is called the standing wave pattern of the voltage wave or the amplitude of the standing wave, and ϕ is called the phase pattern of the standing wave.

The maximum and minimum voltage amplitudes are thus can be expressed as follows:

$$V_{0,max} = |V_+| e^{-\alpha z} \left(1 + |\Gamma_z|\right) \tag{2.152}$$

and

$$V_{0,min} = |V_+| e^{-\alpha z} \left(1 - |\Gamma_z|\right) \tag{2.153}$$

Table 2.2 Value of Γ_L and ρ for different load conditions

Load Condition	Load Value	Γ_L	ρ
Short circuit	0	-1	∞
Matched load	Z_0	0	1
Open circuit	∞	$+1$	∞

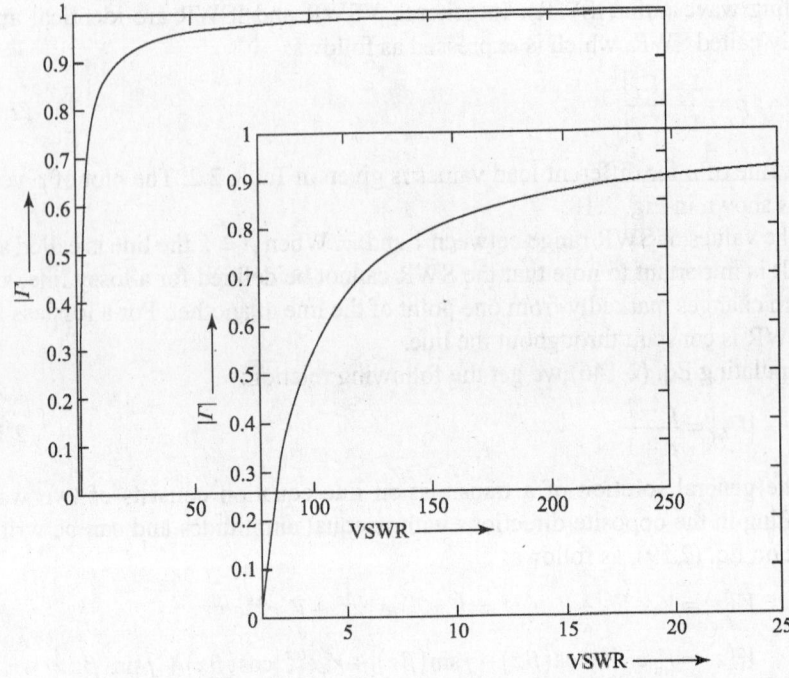

Fig. 2.18 Plot of SWR vs magnitude of reflection coefficient

Similarly, the following are the expressions for the maximum and minimum current amplitudes:

$$I_{0,\max} = |I_+| e^{-\alpha z} \left(1 + |\Gamma_z|\right) \tag{2.154}$$

and

$$I_{0,\min} = |I_+| e^{-\alpha z} \left(1 - |\Gamma_z|\right) \tag{2.155}$$

Since the maximum impedance corresponds to the minimum current and maximum voltage, the maximum normalized impedance can be given as follows:

$$|z_{\max}| = \frac{|Z_{\max}|}{Z_0} = \frac{1}{Z_0} \frac{V_{0,\max}}{I_{0,\min}} \tag{2.156}$$

Substituting the expressions for $V_{0,\max}$ and $I_{0,\min}$ from Eqs (2.152) and (2.155) into Eq. (2.156), we obtain the following equation:

$$|z_{\max}| = \frac{1}{Z_0} \frac{|V_+| e^{-\alpha z} \left(1 + |\Gamma_z|\right)}{|I_+| e^{-\alpha z} \left(1 - |\Gamma_z|\right)} = \frac{1 + |\Gamma_z|}{1 - |\Gamma_z|} = \rho \tag{2.157}$$

Similarly, the following expression represents the minimum normalized impedance:

$$|z_{\min}| = \frac{|Z_{\min}|}{Z_0} = \frac{1}{Z_0} \frac{V_{0,\min}}{I_{0,\max}} = \frac{1}{\rho} \tag{2.158}$$

If we assume that the line is lossless, as generally is the case, there will be no power loss in the line. The power delivered at the load is given by the following expression:

$$P_L = P_i - P_r \text{ or } \frac{P_L}{P_i} = 1 - \frac{P_r}{P_i} = 1 - \frac{\left|V_- e^{j\beta l}\right|^2}{\left|V_+ e^{-j\beta l}\right|^2} = 1 - \left|\Gamma_L\right|^2 \tag{2.159}$$

Substituting the value of $\left|\Gamma_L\right|$ from Eq. (2.147) into Eq. (2.159), we get the following equation:

$$\frac{P_L}{P_i} = 1 - \left(\frac{\rho - 1}{\rho + 1}\right)^2 = \frac{(\rho + 1)^2 - (\rho - 1)^2}{(\rho + 1)^2} = \frac{4\rho}{(\rho + 1)^2} \tag{2.160}$$

EXAMPLE 2.4 For the transmission line system described in Example 2.3, calculate the following parameters:

(a) VSWR
(b) Range of normalized impedance
(c) Percentage of power delivered at the load

Solution Given: $\Gamma_L = 0.473\angle 39.763°$
Therefore,

$$\rho = \frac{1 + \left|\Gamma_L\right|}{1 - \left|\Gamma_L\right|} = \frac{1 + 0.473}{1 - 0.473} = \frac{1.473}{0.527} = 2.795$$

$$\left|z_{max}\right| = \rho = 2.795 \text{ and } \left|z_{min}\right| = \frac{1}{\rho} = \frac{1}{2.802} = 0.358$$

Now, percentage of the incident power delivered at load will be as follows:

$$\frac{P_L}{P_i} \times 100 = \frac{4\rho \times 100}{(\rho + 1)^2} = \frac{4 \times 2.795 \times 100}{(2.795 + 1)^2} = 77.628\%$$

Practice Problems

2.7 For the transmission line system described in Practice Problem 2.5, calculate the following parameters:

(a) VSWR
(b) Range of normalized impedance
(c) Percentage of power delivered at the load   [4.435, 0.225–4.435, 60.056%]

2.8 For the transmission line system described in Practice Problem 2.6, calculate the following parameters:

(a) VSWR
(b) Range of normalized impedance
(c) Percentage of power delivered at the load   [1.66, 0.602–1.66, 93.844%]

2.8 Input Impedance of Transmission Lines

The voltage and current at any point z on a transmission line are given by Eqs (2.59) and (2.61), respectively. Using these two equations, the voltage and current at the source end ($z = 0$) are given by the following relations:

$$V_S = V_+ + V_-$$ (2.161)

$$I_S = \frac{1}{Z_0}(V_+ - V_-)$$ (2.162)

The voltage and current at the load end ($z = l$) are given by Eqs (2.85) and (2.86), respectively.

Applying Kirchoff's voltage law (KVL) at the source end of Fig. 2.19, we get the following relation:

$$V_S = V_G - I_S Z_S$$ (2.163)

Equating Eqs (2.161) and (2.163), the following relation can be obtained:

$$V_G - I_S Z_S = V_+ + V_-$$ (2.164)

Substitution of the expression of I_S from Eq. (2.162) into Eq. (2.164) leads to the following equation:

$$V_G - \frac{Z_S}{Z_0}(V_+ - V_-) = V_+ + V_- \text{ or } Z_0 V_G - Z_S V_+ + Z_S V_- = Z_0 V_+ + Z_0 V_-$$

or $\qquad (Z_S + Z_0)V_+ - (Z_S - Z_0)V_- = Z_0 V_G$ or $V_+ - \left(\dfrac{Z_S - Z_0}{Z_S + Z_0}\right)V_- = \dfrac{Z_0}{Z_S + Z_0}V_G$

or $\qquad V_+ - V_- \Gamma_S = \dfrac{Z_0}{Z_S + Z_0}V_G$ (2.165)

where Γ_S is the reflection coefficient at the source and is expressed as Eq. (2.100).

Applying KVL at the load end, the following equation is obtained:

$$V_L = I_L Z_L$$ (2.166)

Equating Eqs (2.166) and equation (2.85), we get the following relation:

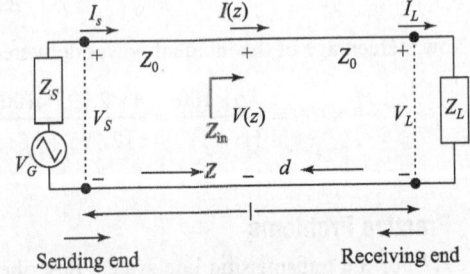

Fig. 2.19 Transmission line

$$I_L Z_L = V_+ e^{-\gamma l} + V_- e^{+\gamma l}$$ (2.167)

Substituting the expression for I_L from Eq. (2.86) into Eq. (2.167) yields the following equation:

$$\frac{Z_L}{Z_0}(V_+ e^{-\gamma l} - V_- e^{+\gamma l}) = V_+ e^{-\gamma l} + V_- e^{+\gamma l}$$

or $\qquad Z_L V_+ e^{-\gamma l} - Z_L V_- e^{+\gamma l} = Z_0 V_+ e^{-\gamma l} + Z_0 V_- e^{+\gamma l}$

or $\quad (Z_L - Z_0)V_+ e^{-\gamma l} - (Z_L + Z_0)V_- e^{+\gamma l} = 0$

or $\quad \left(\dfrac{Z_L - Z_0}{Z_L + Z_0}\right)V_+ e^{-\gamma l} - V_- e^{+\gamma l} = 0$

or $\quad V_+ e^{-\gamma l}\Gamma_L - V_- e^{+\gamma l} = 0$ $\qquad\qquad$ (2.168)

Multiplying Eq. (2.165) by $e^{+\gamma l}$ and Eq. (2.168) by Γ_S, and then subtracting the latter from the former, the following expression is obtained:

$$V_+ e^{+\gamma l} - V_+ e^{-\gamma l}\Gamma_S\Gamma_L - V_-\Gamma_S e^{+\gamma l} + V_- e^{+\gamma l}\Gamma_S = \frac{Z_0}{Z_S + Z_0}V_G e^{+\gamma l}$$

or $\quad V_+ e^{+\gamma l}\left(1 - \Gamma_S\Gamma_L e^{-2\gamma l}\right) = \dfrac{Z_0}{Z_S + Z_0}V_G e^{+\gamma l}$

or $\quad V_+ = \dfrac{Z_0}{(Z_S + Z_0)}\dfrac{V_G}{\left(1 - \Gamma_S\Gamma_L e^{-2\gamma l}\right)}$ $\qquad\qquad$ (2.169)

Multiplying Eq. (2.165) by $\Gamma_L e^{-\gamma L}$ and then subtracting Eq. (2.168) from it and rearranging the result, we obtain the following expression:

$$V_+\Gamma_L e^{-\gamma L} - V_-\Gamma_S\Gamma_L e^{-\gamma L} - V_+ e^{-\gamma l}\Gamma_L + V_- e^{+\gamma l} = \frac{Z_0}{Z_S + Z_0}V_G\Gamma_L e^{-\gamma L}$$

or $\quad V_- e^{+\gamma l}\left(1 - \Gamma_S\Gamma_L e^{-2\gamma L}\right) = \dfrac{Z_0}{Z_S + Z_0}V_G\Gamma_L e^{-\gamma L}$

or $\quad V_- = V_G\dfrac{Z_0}{(Z_S + Z_0)}\dfrac{\Gamma_L e^{-2\gamma l}}{\left(1 - \Gamma_S\Gamma_L e^{-2\gamma l}\right)}$ $\qquad\qquad$ (2.170)

Substituting the expression of V_+ and V_- from Eqs (2.169) and (2.170) into Eq. (2.59) and manipulating it, we get the following equation:

$$V(z) = V_G\frac{Z_0}{(Z_S + Z_0)}\frac{e^{-\gamma l}}{\left(1 - \Gamma_S\Gamma_L e^{-2\gamma l}\right)}\left[e^{\gamma(l-z)} + \Gamma_L e^{-\gamma(l-z)}\right]$$

or $\quad V(z) = V_G\dfrac{Z_0}{(Z_S + Z_0)}\dfrac{e^{-\gamma l}}{\left(1 - \Gamma_S\Gamma_L e^{-2\gamma l}\right)}\left[e^{\gamma d} + \Gamma_L e^{-\gamma d}\right]$ $\qquad$ (2.171)

Similarly, substituting the expression of V_+ and V_- from Eqs (2.169) and (2.170) into Eq. (2.61) and manipulating it, we get the following equation:

$$I(z) = \frac{V_G}{(Z_S + Z_0)}\frac{e^{-\gamma l}}{\left(1 - \Gamma_S\Gamma_L e^{-2\gamma l}\right)}\left[e^{\gamma d} - \Gamma_L e^{-\gamma d}\right]$$ $\qquad$ (2.172)

The ratio of Eq. (2.171) to Eq. (2.172) gives the expression for input impedance at a point z on a transmission line, which is as follows:

$$Z_{in}(z) = \frac{V(z)}{I(z)} = Z_0\frac{e^{\gamma d} + \Gamma_L e^{-\gamma d}}{e^{\gamma d} - \Gamma_L e^{-\gamma d}}$$ $\qquad$ (2.173)

where d is measured from the load. It may be noted that Eq. (2.173) can also be derived from the ratio of Eq. (2.59) to Eq. (2.61). The derivation is left for the readers as an exercise. Such a derivation, however, does not give any information about V_+ and V_-.

Substituting $e^{\pm \gamma z} = \cosh(\gamma z) \pm \sinh(\gamma z)$ in Eq. (2.173) and also substituting the expression of Γ_L from Eq. (2.90), we get the following expression:

$$Z_{in}(z) = Z_0 \frac{\left\{\cosh(\gamma d) + \sinh(\gamma d)\right\} + \left[\dfrac{Z_L - Z_0}{Z_L + Z_0}\right]\left\{\cosh(\gamma d) - \sinh(\gamma d)\right\}}{\left\{\cosh(\gamma d) + \sinh(\gamma d)\right\} - \left[\dfrac{Z_L - Z_0}{Z_L + Z_0}\right]\left\{\cosh(\gamma d) - \sinh(\gamma d)\right\}}$$

$$\text{or } Z_{in}(z) = Z_0 \frac{(Z_L + Z_0)\left\{\cosh(\gamma d) + \sinh(\gamma d)\right\} + (Z_L - Z_0)\left\{\cosh(\gamma d) - \sinh(\gamma d)\right\}}{(Z_L + Z_0)\left\{\cosh(\gamma d) + \sinh(\gamma d)\right\} - (Z_L - Z_0)\left\{\cosh(\gamma d) - \sinh(\gamma d)\right\}}$$

$$\text{or } Z_{in}(z) = Z_0 \frac{2Z_L \cosh(\gamma d) + 2Z_0 \sinh(\gamma d)}{2Z_L \sinh(\gamma d) + 2Z_0 \cosh(\gamma d)} = Z_0 \frac{Z_L \cosh(\gamma d) + Z_0 \sinh(\gamma d)}{Z_L \cosh(\gamma d) + Z_0 \cosh(\gamma d)}$$

$$\text{or} \qquad Z_{in}(z) = Z_0 \frac{Z_L + Z_0 \tanh(\gamma d)}{Z_0 + Z_L \tanh(\gamma d)} \tag{2.174}$$

For a lossless line $\alpha = 0$, $\tanh(\gamma d) = \tanh(j\beta d) = j\tan(j\beta d)$ and Eq. (2.174) reduces to the following form:

$$Z_{in}(z) = Z_0 \frac{Z_L + jZ_0 \tan(\beta d)}{Z_0 + jZ_L \tan(\beta d)} \tag{2.175}$$

Since $\tan(\beta d)$ is a periodic function of βd of period $\beta d = \pi$ or $d = \lambda/2$, the input impedance $Z_{in}(z)$ is also periodic and is repeated at every interval of $\lambda/2$.

For a generalized derivation, however, we will consider the lossy case. For a lossy line, Eq. (2.173) can be modified to the following form:

$$Z_{in}(z) = Z_0 \frac{1 + \Gamma_L e^{-2\gamma d}}{1 - \Gamma_L e^{-2\gamma d}} \tag{2.176}$$

Substituting Eq. (2.95) in Eq. (2.176), we get the following equation:

$$Z_{in}(z) = Z_0 \frac{1 + \Gamma_z}{1 - \Gamma_z} \tag{2.177}$$

Further substituting $\Gamma_z = |\Gamma_z|e^{j\theta_z}$ in Eq. (2.177), we get the following expression:

$$Z_{in}(z) = Z_0 \frac{1 + |\Gamma_z|e^{j\theta_z}}{1 - |\Gamma_z|e^{j\theta_z}} \tag{2.178}$$

$$\text{or} \qquad R_{in}(z) + jX_{in}(z) = |Z_{in}(z)|e^{j\psi_z} = Z_0 \frac{1 + |\Gamma_z|\{\cos(\theta_z) + j\sin(\theta_z)\}}{1 - |\Gamma_z|\{\cos(\theta_z) + j\sin(\theta_z)\}} \tag{2.179}$$

The RHS of Eq. (2.179) is of the form $(a + jb)/(c + jd)$, which can easily be manipulated to the form $(u + jv)$ after multiplying both the numerator and the denominator by $(c-jd)$, which can further be expressed in terms of magnitude and phase as $\sqrt{u^2+v^2}\ \tan^{-1}(v/u)$. Next, equating similar terms from both the LHS and the RHS of Eq. (2.179), we get the following expressions:

$$R_{in}(z) = Z_0 \frac{1-|\Gamma_z|^2}{1-2|\Gamma_z|\cos(\theta_z)+|\Gamma_z|^2} \tag{2.180}$$

$$X_{in}(z) = Z_0 \frac{2|\Gamma_z|\sin(\theta_z)}{1-2|\Gamma_z|\cos(\theta_z)+|\Gamma_z|^2} \tag{2.181}$$

$$|Z_{in}(z)| = Z_0 \sqrt{\frac{1+2|\Gamma_z|\cos(\theta_z)+|\Gamma_z|^2}{1-2|\Gamma_z|\cos(\theta_z)+|\Gamma_z|^2}} \tag{2.182}$$

and
$$\psi_z = \tan^{-1}\left(\frac{X_{in}(z)}{R_{in}(z)}\right) = \tan^{-1}\left[\frac{2|\Gamma_z|\sin(\theta_z)}{1-|\Gamma_z|^2}\right] \tag{2.183}$$

Equations (2.180)–(2.183) represent a family of circles that can be used to solve different transmission line problems graphically. Such a plot of a family of circles is known as a 'Smith chart' and will be discussed in detail in Section 2.14.

Based on Eq. (2.147), we can write the following relation:

$$|\Gamma_z| = |\Gamma_L| = \frac{\rho-1}{\rho+1} \tag{2.184}$$

Substituting Eq. (2.184) into Eq. (2.178), the following equation can be obtained:

$$Z_{in}(z) = Z_0 \frac{1+|\Gamma_z|e^{j\theta_z}}{1-|\Gamma_z|e^{j\theta_z}} = Z_0 \frac{1+\left(\dfrac{\rho-1}{\rho+1}\right)e^{j\theta_z}}{1-\left(\dfrac{\rho-1}{\rho+1}\right)e^{j\theta_z}} = Z_0 \frac{(\rho+1)+(\rho-1)e^{j\theta_z}}{(\rho+1)-(\rho-1)e^{j\theta_z}} \tag{2.185}$$

Case 1—Open-circuited transmission line   Let us consider that a lossless transmission line of length l is open circuited. For this case, $Z_L \to \infty$ and Eq. (2.175) reduces to the following form:

$$Z_{in}^{oc}(z) = -jZ_0 \cot(\beta d) \tag{2.186}$$

From Eq. (2.186), we get that for $2n\lambda/4 < d < (2n+1)\lambda/4$, $Z_{in}^{oc}(z) < 0$ and the line behaves as a capacitive load, whereas for $(2n+1)\lambda/4 < d < (n+1)\lambda/2$, $Z_{in}^{oc}(z) > 0$ and the line behaves as an inductive load, where $n = 0, 1, 2, 3, 4, \dots$. If L_{eq}^{oc} and C_{eq}^{oc} are the equivalent inductance and capacitance, then the following relations are obtained:

$$j\omega L_{eq}^{oc} = -jZ_0 \cot(\beta d) \text{ or } L_{eq}^{oc} = -\frac{Z_0}{\omega}\cot(\beta d) \tag{2.187}$$

and $\quad \dfrac{1}{j\omega C_{eq}^{oc}} = -jZ_0 \cot(\beta d)$ or $C_{eq}^{oc} = \dfrac{1}{\omega Z_0 \cot(\beta d)}$ $\qquad$ (2.188)

Since $\cot(\beta d)$ ranges from 0 to ∞, Eqs (2.187) and (2.188) reveal that inductance or capacitance of any value can be implemented using an open-circuited transmission line of proper length. This has facilitated the design of microwave filters using open-circuited transmission lines from their low-frequency LC prototype circuits. The details of such design have been provided in Chapter 9.

Note For a series resonant circuit $Z_{in} = j\omega L + \dfrac{1}{j\omega C} = j\omega L\left(1 - \dfrac{1}{\omega^2 LC}\right) = j\omega L\left(1 - \dfrac{\omega_0^2}{\omega}\right)$, which shows that $Z_{in} \rightarrow 0$ as $\omega \rightarrow \omega_0$, where $\omega_0 = 1/\sqrt{LC}$.

When the length of an open-circuited transmission line is an even multiple of $\lambda/4$, at $\beta d = n\pi$ or $d = n\lambda/2$, $Z_{in}^{oc}(z) \rightarrow \infty$ and the resultant circuit behaves as a parallel resonant circuit. For $\beta d = (2n+1)\pi/2$, or $d = (2n+1)\lambda/4$, $Z_{in}^{oc}(z) = 0$ and the resultant circuit behaves as a series resonant circuit. A plot of $Z_{in}^{oc}(z)$ as a function of βd is shown in Fig. 2.20(a).

Case 2—Short-circuited transmission line For a lossless short-circuited transmission line of length l $Z_L = 0$ and Eq. (2.175) reduces to the following form:

$$Z_{in}^{sc}(z) = jZ_0 \tan(\beta d) \qquad (2.189)$$

As we can see from Eq. (2.189), for $2n\lambda/4 < d < (2n+1)\lambda/4$, $Z_{in}^{sc}(z) > 0$ and a short-circuited transmission line behaves as an inductive load, whereas for $(2n+1)\lambda/4 < d < (n+1)\lambda/2$, $Z_{in}^{sc}(z) < 0$ and the line behaves as a capacitive load where $n = 0, 1, 2, 3, 4, \ldots$. If we assume that L_{eq}^{sc} and C_{eq}^{sc} are, respectively, the equivalent inductance and capacitance of a short-circuited transmission line, then the following relations hold:

$$j\omega L_{eq}^{sc} = jZ_0 \tan(\beta d) \text{ or } L_{eq}^{sc} = \dfrac{Z_0}{\omega} \tan(\beta d) \qquad (2.190)$$

and $\quad \dfrac{1}{j\omega C_{eq}^{sc}} = jZ_0 \tan(\beta d)$ or $C_{eq}^{sc} = -\dfrac{1}{\omega Z_0 \tan(\beta d)}$ $\qquad$ (2.191)

Therefore, like an open-circuited transmission line, a short-circuited transmission line can also be used to implement any arbitrary inductance or capacitance. In practice, a short-circuited transmission line is often preferred over an open-circuited line because the latter has a tendency to radiate power from its open end.

It may be noted that series or parallel resonator circuits can also be implemented using a short-circuited transmission line having a length of an even or odd multiple of $\lambda/4$. When the length of the transmission line is an even multiple of $\lambda/4$, $d = n\lambda/2$, or $\beta d = n\pi$ and $Z_{in}^{sc}(z) = 0$, and the resultant circuit behaves as a series resonant circuit. For $\beta d = (2n+1)\pi/2$ or $d = (2n+1)\lambda/4$, $Z_{in}^{sc}(z) \rightarrow \infty$ and

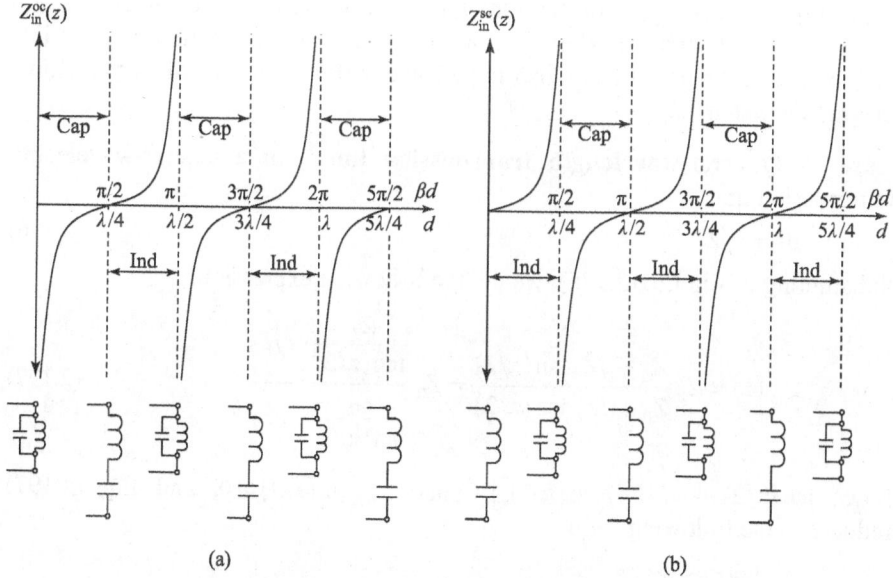

Fig. 2.20 Input impedance (a) Open-circuited transmission line
(b) Short-circuited transmission line

the resultant circuit behaves as a parallel resonant circuit. A plot of $Z_{in}^{sc}(z)$ as a function of βd is shown in Fig. 2.20(b).

Taking a ratio of the magnitude of Eq. (2.189) to Eq. (2.186), we get the following expression:

$$\left|Z_{in}^{sc}(z)\right| / \left|Z_{in}^{oc}(z)\right| = \tan^2(\beta d) \text{ or } \beta = \frac{1}{d}\tan^{-1}\sqrt{\left|Z_{in}^{sc}(z)\right| / \left|Z_{in}^{oc}(z)\right|} \qquad (2.192)$$

and multiplying Eqs (2.189) and (2.186), we get the following equation:

$$Z_{in}^{sc}(z)Z_{in}^{oc}(z) = Z_0^2 \text{ or } Z_0 = \sqrt{Z_{in}^{sc}(z)Z_{in}^{oc}(z)} \qquad (2.193)$$

Equations (2.192) and (2.193) are commonly used to determine the phase constant and characteristic impedance of a transmission line.

Note A short-circuited transmission line can be used in place of an open-circuited transmission line and vice versa, by changing its length by an amount of $\lambda/4$.

Case 3—One-eighth-wavelength transmission line For a one-eighth-wavelength transmission line terminated in a load impedance of Z_L,

$$\beta l = \pi/4 \qquad (2.194)$$

Substituting this into Eq. (2.175), we get the following equation:

$$Z_{in}(\lambda/8) = Z_0 \frac{Z_L + jZ_0 \tan(\pi/4)}{Z_0 + jZ_L \tan(\pi/4)} = Z_0 \frac{Z_L + jZ_0}{Z_0 + jZ_L} \qquad (2.195)$$

For real Z_0 and Z_L, both the numerator and the denominator of Eq. (2.195) have an equal magnitude. Therefore, a one-eighth-wavelength transmission line may be used to transform any load impedance to the magnitude of Z_0 to obtain a magnitude match.

Case 4—Quarter-wavelength transmission line For a quarter-wavelength transmission line,

$$\beta l = \pi/2 \tag{2.196}$$

Substituting this in Eq. (2.175), we get the following expression:

$$Z_{\text{in}}(z) = Z_0 \frac{Z_L + jZ_0 \tan(\pi/2)}{Z_0 + jZ_L \tan(\pi/2)} = Z_0 \frac{\dfrac{Z_L}{\tan(\pi/2)} + jZ_0}{\dfrac{Z_0}{\tan(\pi/2)} + jZ_L} \tag{2.197}$$

Since $\tan(\pi/2) \to \infty$, $Z_L/\tan(\pi/2) \cong 0$ and $Z_0/\tan(\pi/2) \cong 0$, and Eq. (2.197) reduces to the following form:

$$Z_{\text{in}}(\lambda/4) \cong Z_0^2/Z_L$$

or $\quad Z_0 = \sqrt{Z_{\text{in}}(\lambda/4)Z_L} \tag{2.198}$

Therefore, a quarter-wavelength transmission line with the characteristic impedance Z_0, given by Eq. (2.198), behaves as an impedance transformer and can be used to match a resistive load, Z_L, with the resistive input impedance $Z_{\text{in}}(z)$. That is why, quarter-wavelength transmission lines are often called quarter-wave transformers. This will be discussed in detail in the next chapter.

Quarter-wave transformers are extensively used to match an antenna with a transmission line. However, in practice, the range of antenna resistance and characteristic impedance of the lossless feed line that can be matched satisfactorily is limited approximately to a ratio of 10:1. Quarter-wave transformers can also be used when the load is not purely real. In such cases, a quarter-wave transformer should be connected between the points corresponding to voltage maxima or minima, or alternatively to current minima or maxima, where the input impedance of the transmission line is purely real.

A short-circuited quarter-wave transmission line has a very high input impedance and can be used as an insulator to support an open-wire transmission line or the centre conductor of a coaxial cable. These lines are also called 'copper insulators'. A basic copper line is shown in Fig. 2.21.

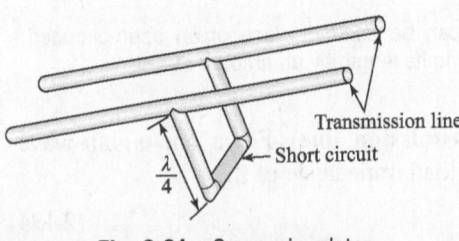

Transmission line
← Short circuit
$\frac{\lambda}{4}$

Fig. 2.21 Copper insulator

Case 5—Half-wavelength transmission line For a half-wavelength transmission line,

$$\beta l = \pi \tag{2.199}$$

Substituting this into Eq. (2.175), we get the following expression:

$$Z_{in}(\lambda/2) = Z_0 \frac{Z_L + jZ_0 \tan(\pi)}{Z_0 + jZ_L \tan(\pi)} = Z_L \qquad (2.200)$$

A half-wavelength transmission line behaves as a one-to-one transformer and is used in a system where extra space is required between two successive sections.

EXAMPLE 2.5 For the coaxial transmission line described in Example 2.3, calculate the input impedance at a distance of 91.44 mm from the load, provided the line is terminated by the following conditions:

(a) The same load impedance as provided in Example 2.3
(b) An open circuit
(c) A short circuit

If the line is assumed to be lossless then calculate the equivalent inductance/capacitance for cases (b) and (c). What will be the input impedance of the lossless transmission line if it is (i) one-eighth-wavelength, (ii) quarter-wavelength, and (iii) half-wavelength long and is terminated by a load impedance of 99.19 Ω?

Solution Given: $Z_L = 78.27 + j60.93$ Ω, $Z_0 = 50.026 - j0.049$ Ω, $d = 91.44 \times 10^{-3}$ m, $f = 1$ GHz $= 10^9$ Hz, and $\gamma = 32.467\angle 89.943° = 0.032 + j32.467$ rad/m

(a) The input impedance will be as follows:

$$Z_{in}(z) = Z_0 \frac{Z_L + Z_0 \tanh(\gamma d)}{Z_0 + Z_L \tanh(\gamma d)} \text{ or }$$

$$Z_{in}(z) = (50.026 - j0.049) \frac{(78.27 + j60.93) + (50.026 - j0.049) \tanh(0.003 + j2.969)}{(50.026 - j0.049) + (78.27 + j60.93) \tanh(0.003 + j2.969)}$$

or
$$Z_{in}(z) = (50.026 - j0.049) \frac{(78.27 + j60.93) + (0.146 - j8.721)}{(50.026 - j0.049) + (10.864 - j13.456)}$$

or
$$Z_{in}(z) = (50.026 - j0.049) \frac{(78.416 + j52.209)}{(60.89 - j13.505)} = 52.390 + j54.451 \ \Omega$$

(b) When the line is terminated open-circuited then the following equation is obtained:

$$Z_{in}(z) = Z_0 \coth(\gamma d) = (50.026 - j0.049) \times \coth(0.003 + j2.969) = 5.368 + j286.87 \ \Omega$$

(c) When the line is short-circuited, the following relation holds:

$$Z_{in}(z) = Z_0 \tanh(\gamma d) = (50.026 - j0.049) \times \tan(0.003 + j2.969) = 0.146 - j8.721 \ \Omega$$

The wavelength of the signal is as follows: $\lambda = 3 \times 10^8 / 10^9 = 0.3$ m
The length of the line in terms of wavelengths is as follows:

$$d = \frac{91.44 \times 10^{-3}}{0.3} \lambda = 0.305\lambda$$

Since the line is in the range $\lambda/4 < d < \lambda/2$ it will behave as an inductor for open-circuit termination and as a capacitor for short-circuit termination.

For the lossless case, the characteristic impedance of the line is $Z_0 = 50.026 \ \Omega$.

Therefore, the equivalent inductance for the open-circuited line is as follows:

$$L_{eq}^{oc} = -\frac{Z_0}{\omega}\cot(\beta d) = -\frac{50.026}{2\pi \times 10^9} \times \cot\left(32.467 \times 91.44 \times 10^{-3}\right) = 45.614 \ \text{nH}$$

The equivalent capacitance for the short-circuit case is as follows:

$$C_{eq}^{sc} = -\frac{1}{\omega Z_0 \tan(\beta d)} = -\frac{1}{2\pi \times 10^9 \times 50.026 \times \tan\left(32.467 \times 91.44 \times 10^{-3}\right)}$$

$$= 18.226 \ \text{pF}$$

(i) The input impedance for one-eighth wavelength is as follows:

$$Z_{in}(\lambda/8) = Z_0 \frac{Z_L + jZ_0}{Z_0 + jZ_L} = 50.026 \times \frac{99.19 + j50.026}{50.026 + j99.19} = 40.228 - j29.737$$

(ii) The input impedance for the quarter-wavelength line is as follows:

$$Z_{in}(\lambda/4) = \frac{(50.026)^2}{99.19} = 25.230 \ \Omega$$

(iii) The input impedance for the half-wavelength line is as follows:

$$Z_{in}(\lambda/2) = Z_L = 99.19 \ \Omega$$

Practice Problems

2.9 For the parallel two-wire transmission line described in Practice Problem 2.5, calculate the input impedance at a distance of 83.73 mm from the load provided the line is terminated by the following conditions:

(a) The same load impedance as provided in Practice Problem 2.5
(b) An open circuit
(c) A short circuit

If the line is assumed to be lossless, then calculate the equivalent inductance/capacitance for cases (b) and (c). What will be the input impedance of the lossless transmission line if it is (i) one-eighth-wavelength, (ii) quarter-wavelength, and (iii) half-wavelength long and is terminated by a load impedance of 97.93 Ω?

[18.7355 + j24.8323 Ω, 0.2239 + j108.75 Ω, 0.0695 − j50.2687 Ω, 17.307 nH, 3.165 pF, 71.111 − j20.248 Ω, 55.824 Ω, 97.93 Ω]

2.10 For the wire above a ground plane transmission line system described in Practice Problem 2.6, calculate the input impedance at a distance of 23.35 mm from the load provided the line is terminated by the following conditions:

(a) The same load impedance as provided in Practice Problem 2.6
(b) An open circuit
(c) A short circuit

If the line is assumed to be lossless, then calculate the equivalent inductance/capacitance for cases (b) and (c). What will be input impedance of the lossless transmission line if it is (i) one-eighth-wavelength, (ii) quarter-wavelength, and (iii) half-wavelength long and is terminated by a load impedance of $165.357\ \Omega$?

[$124.47 - j51.337\ \Omega$, $0.4062 - j187.35\ \Omega$, $0.1411 + j53.2757\ \Omega$, 0.8495 pF, 0.8407 nH,
$88.4394 - j46.4724\ \Omega$, $60.3616\ \Omega$, $165.357\ \Omega$]

2.9 Attenuation and Distortion in Transmission Lines

Technically, the term *attenuation* means a gradual loss of intensity of any kind of wave as it propagates through a medium. When a voltage wave propagates through a transmission line of finite conductivity or non-zero resistivity, its amplitude decreases, resulting in attenuation in transmission lines. Attenuation of the voltage wave is best described by the attenuation constant α that can be calculated from the expression of the propagation constant.

Squaring both sides of Eq. (2.54), we get the following expression:

$$\gamma^2 = (R + j\omega L)(G + j\omega C)$$

or
$$(\alpha^2 - \beta^2) + j2\alpha\beta = (RG - \omega^2 LC) + j\omega(LG + RC) \tag{2.201}$$

Equating the real and imaginary parts on both sides of Eq. (2.201), we get the following relations:

$$\alpha^2 - \beta^2 = RG - \omega^2 LC \tag{2.202}$$

and
$$2\alpha\beta = \omega(LG + RC) \text{ or } \beta = \frac{\omega}{2\alpha}(LG + RC) \tag{2.203}$$

Substitution of the expression of β from Eq. (2.203) into Eq. (2.202) leads to the following equation:

$$\alpha^2 - \frac{\omega^2}{4\alpha^2}\left(L^2 G^2 + R^2 C^2 + 2LGRC\right) = RG - \omega^2 LC$$

or
$$\alpha^4 - \alpha^2\left(RG - \omega^2 LC\right) - \frac{\omega^2}{4}\left(L^2 G^2 + R^2 C^2 + 2LGRC\right) = 0$$

This equation represents a quadratic equation of α^2, the roots of which can be expressed as follows:

$$\alpha^2 = \frac{1}{2}\left[\left(RG - \omega^2 LC\right) \pm \sqrt{\left(RG - \omega^2 LC\right)^2 + \omega^2\left(L^2 G^2 + R^2 C^2 + 2LGRC\right)}\right]$$

or
$$\alpha^2 = \frac{1}{2}\left[\left(RG - \omega^2 LC\right) \pm \sqrt{R^2 G^2 + \omega^4 L^2 C^2 + \omega^2 L^2 G^2 + \omega^2 R^2 C^2}\right]$$

or
$$\alpha^2 = \frac{1}{2}\left[\left(RG - \omega^2 LC\right) \pm \sqrt{\left(R^2 + \omega^2 L^2\right)\left(G^2 + \omega^2 C^2\right)}\right]$$

or
$$\alpha = \sqrt{\frac{1}{2}\left[\left(RG - \omega^2 LC\right) \pm \sqrt{\left(R^2 + \omega^2 L^2\right)\left(G^2 + \omega^2 C^2\right)}\right]} \tag{2.204}$$

In a similar way, we can show that

$$\beta = \sqrt{\frac{1}{2}\left[-\left(RG-\omega^2 LC\right) \pm \sqrt{\left(R^2+\omega^2 L^2\right)\left(G^2+\omega^2 C^2\right)}\right]} \tag{2.205}$$

Since α and β cannot be imaginary, we get the following expressions:

$$\alpha = \sqrt{\frac{1}{2}\left[\left(RG-\omega^2 LC\right)+\sqrt{\left(R^2+\omega^2 L^2\right)\left(G^2+\omega^2 C^2\right)}\right]} \tag{2.206}$$

and $$\beta = \sqrt{\frac{1}{2}\left[-\left(RG-\omega^2 LC\right)+\sqrt{\left(R^2+\omega^2 L^2\right)\left(G^2+\omega^2 C^2\right)}\right]} \tag{2.207}$$

Equations (2.206) and (2.207) reveal that the attenuation constant, α, and the phase constant, β, are functions of frequency. Therefore, when a complex voltage signal, such as pulses, that consists of a large number of frequency components, is transmitted through a transmission line, its different frequency components will suffer different attenuation and phase change during the passage. This is known as distortion. The *distortion* that occurs due to a variation of α with frequency is known as *frequency distortion*, whereas the distortion that occurs due to a variation of β with frequency is known as delay or *phase distortion*.

Frequency distortion can be reduced in transmission by the use of *equalizers* at the line terminals for high-quality radio broadcast programmes over transmission lines. These equalizers are networks whose frequency and phase characteristics can be adjusted to the inverse of those in the lines, resulting in an overall uniform frequency response over the desired frequency band. Phase distortion is of relatively minor importance to the voice signal transmission because of the characteristics of the ear. However, it can be very serious for video signal transmission through the line.

Frequency and phase distortion can also be reduced by choosing properly the transmission line parameters, that is, R, L, G, and C. To remove frequency distortion, α must be independent of ω, that is,

$$\frac{d\alpha}{d\omega}=0 \quad \text{or,} \; R^2 C^2 + L^2 G^2 + 2\omega^2 L^2 C^2 = 2LC\sqrt{R^2 G^2 + \omega^2(R^2 C^2 + L^2 G^2)+\omega^4 L^2 C^2}$$

or, $$R^4 C^4 + L^4 G^4 + 4\omega^4 L^4 C^4 + 2R^2 L^2 C^2 G^2 + 4\omega^2 R^2 L^2 C^4 + 4\omega^2 L^4 C^2 G^2$$
$$= 4R^2 L^2 C^2 G^2 + 4\omega^2 R^2 L^2 C^4 + 4\omega^2 L^4 C^2 G^2 + 4\omega^4 L^4 C^4$$

or, $$R^4 C^4 + L^4 G^4 - 2R^2 L^2 C^2 G^2 = (R^2 C^2 - L^2 G^2)^2 = 0$$

or, $$RC = LG \tag{2.208}$$

Similarly, to remove phase distortion, the following relation must hold:

$$\sqrt{\left(R^2+\omega^2 L^2\right)\left(G^2+\omega^2 C^2\right)}-\left(RG-\omega^2 LC\right)=0 \tag{2.209}$$

which again leads to the condition specified in Eq. (2.208).

In addition to distortion, attenuation loss also plays an important role to characterize the transmission line. Theoretically, if we can make per unit length resistance and conductance equal to zero, then attenuation will also be zero. However, it is never practically possible, as neither any conductor has infinite conductivity nor any dielectric has infinite resistivity. Therefore, in reality, no transmission line is lossless. However, it can be shown that a careful choice of the per-unit-length inductance/capacitance can minimize the attenuation loss of the line with the existing per-unit-length resistance and conductance.

If R, G, and C are constant then the condition for minimum attenuation is as follows:

$$\frac{d\alpha}{dL} = 0 \qquad (2.210)$$

Substituting the value of α from Eq. (2.206), we get

$$\frac{d}{dL}\left[\sqrt{\frac{1}{2}\left\{\left(RG - \omega^2 LC\right) + \sqrt{\left(R^2 + \omega^2 L^2\right)\left(G^2 + \omega^2 C^2\right)}\right\}}\right] = 0$$

or

$$\frac{1}{2}\frac{\dfrac{1}{2}\left[-\omega^2 C + \dfrac{\omega^2 L\left(G^2 + \omega^2 C^2\right)}{\sqrt{\left(R^2 + \omega^2 L^2\right)\left(G^2 + \omega^2 C^2\right)}}\right]}{\sqrt{\dfrac{1}{2}\left\{\left(RG - \omega^2 LC\right) + \sqrt{\left(R^2 + \omega^2 L^2\right)\left(G^2 + \omega^2 C^2\right)}\right\}}} = 0$$

or

$$\frac{1}{2}\omega^2\left[-C + L\sqrt{\frac{\left(G^2 + \omega^2 C^2\right)}{\left(R^2 + \omega^2 L^2\right)}}\right] = 0$$

or

$$L\sqrt{\left(G^2 + \omega^2 C^2\right)} = C\sqrt{\left(R^2 + \omega^2 L^2\right)} \text{ or } L^2 G^2 = C^2 R^2$$

or

$$L = CR/G \qquad (2.211)$$

Substitution of the expression of L from Eq. (2.211) into Eq. (2.206) yields the following equation:

$$\alpha_{min,L} = \sqrt{\frac{1}{2}\left[\left(RG - \omega^2 C^2\frac{R}{G}\right) + \sqrt{\left(R^2 + \omega^2 C^2\frac{R^2}{G^2}\right)\left(G^2 + \omega^2 C^2\right)}\right]}$$

or

$$\alpha_{min,L} = \sqrt{\frac{1}{2}\left[RG\left(1 - \frac{\omega^2 C^2}{G^2}\right) + \sqrt{R^2 G^2\left(1 + \frac{\omega^2 C^2}{G^2}\right)\left(1 + \frac{\omega^2 C^2}{G^2}\right)}\right]}$$

or

$$\alpha_{min,L} = \sqrt{\frac{1}{2}\left[RG\left(1 - \frac{\omega^2 C^2}{G^2}\right) + RG\left(1 + \frac{\omega^2 C^2}{G^2}\right)\right]}$$

Since $\omega^2 C^2/G^2 \approx 0$, this equation reduces to

$$\alpha_{min,L} = \sqrt{\frac{1}{2}RG[1+1]} = \sqrt{RG} \qquad (2.212)$$

Similarly, if R, G, and L are constant, then the condition for minimum attenuation is as follows:

$$\frac{d\alpha}{dC} = 0 \qquad (2.213)$$

Substituting the value of α from Eq. (2.206), we get the following expression:

$$\frac{d}{dC}\left[\sqrt{\frac{1}{2}\left\{\left(RG-\omega^2 LC\right)+\sqrt{\left(R^2+\omega^2 L^2\right)\left(G^2+\omega^2 C^2\right)}\right\}}\right]=0$$

or $\quad C = LG/R$ (2.214)

Similarly, substituting the expression of C from Eq. (2.214) into Eq. (2.206) we get the following relation:

$$\alpha_{min,C}=\sqrt{\frac{1}{2}\left[\left(RG-\omega^2 L\frac{LG}{R}\right)+\sqrt{\left(R^2+\omega^2 L^2\right)\left(G^2+\omega^2\left(\frac{LG}{R}\right)^2\right)}\right]}$$

or $\quad \alpha_{min,C} = \sqrt{RG}$ (2.215)

Now if we want to minimize attenuation by properly choosing R, then we can write the following equation:

$$\frac{d\alpha}{dR}=0$$ (2.216)

Substituting the value of α from Eq. (2.206), we get the following equation:

$$\frac{d}{dR}\left[\sqrt{\frac{1}{2}\left\{\left(RG-\omega^2 LC\right)+\sqrt{\left(R^2+\omega^2 L^2\right)\left(G^2+\omega^2 C^2\right)}\right\}}\right]=0$$

or $\quad G+R\sqrt{\frac{\left(G^2+\omega^2 C^2\right)}{\left(R^2+\omega^2 L^2\right)}}=0$ (2.217)

Similarly, if we want to minimize attenuation by properly choosing G, then we can write the following equation:

$$\frac{d\alpha}{dG}=0$$ (2.218)

Substituting the value of α from Eq. (2.206), we get the following expression:

$$\frac{d}{dG}\left[\sqrt{\frac{1}{2}\left\{\left(RG-\omega^2 LC\right)+\sqrt{\left(R^2+\omega^2 L^2\right)\left(G^2+\omega^2 C^2\right)}\right\}}\right]=0$$

or $\quad R+G\sqrt{\frac{\left(R^2+\omega^2 L^2\right)}{\left(G^2+\omega^2 C^2\right)}}=0$ (2.219)

The steps to derive Eqs (2.214), (2.215), (2.217), and (2.219) are left for the readers as an exercise. As R and G cannot be negative, Eqs (2.217) and (2.219) can never be equal to zero, unless both R and G are zero. In such a case,

$$\alpha_{min,R} = \alpha_{min,G} = 0$$ (2.220)

However, since neither R nor G can practically be zero, the minimum value of attainable attenuation is as follows:

$$\alpha_{min} = \sqrt{RG}$$ (2.221)

EXAMPLE 2.6 For the transmission line system described in Example 2.2, calculate the following parameters using Eqs (2.206) and (2.207):
(a) Attenuation constant   (b) Phase constant

Calculate (i) the value of L for which the attenuation will be minimum if R, G, and C are kept constant and (ii) the value of C for which the attenuation will be minimum if R, G, and L are kept constant. What will be minimum attainable attenuation with these modifications?

Solution Given: $R = 3.242\ \Omega/m$, $L = 2.585 \times 10^{-7}\ H/m$, $C = 1.033 \times 10^{-10}\ F/m$, $G = 4.861 \times 10^{-15}\ \mho/m$, and $f = 10^9\ Hz$

(a) The attenuation constant will be as follows:

$$\alpha = \sqrt{\frac{1}{2}\left[\left(RG - \omega^2 LC\right) + \sqrt{\left(R^2 + \omega^2 L^2\right)\left(G^2 + \omega^2 C^2\right)}\right]} = 0.141\ \text{Np/m}$$

(b) The phase constant will be as follows:

$$\beta = \sqrt{\frac{1}{2}\left[-\left(RG - \omega^2 LC\right) + \sqrt{\left(R^2 + \omega^2 L^2\right)\left(G^2 + \omega^2 C^2\right)}\right]} = 32.468\ \text{rad/m}$$

(i) If R, G, and C are kept constant, then the value of L for which attenuation will be minimum is as follows:

$$L = \frac{CR}{G} = \frac{1.033 \times 10^{-10} \times 3.242}{4.861 \times 10^{-15}} = 68.895\ \text{kH/m}$$

(ii) If R, G, and L are kept constant then the value of C for which attenuation will be minimum is as follows:

$$C = \frac{LG}{R} = \frac{2.585 \times 10^{-7} \times 4.861 \times 10^{-15}}{3.242} = 3.876 \times 10^{-22}\ \text{pF/m}$$

(iii) The minimum attenuation that can be achieved is as follows:

$$\alpha_{min} = \sqrt{RG} = \sqrt{3.242 \times 4.861 \times 10^{-15}} = 0.126\ \mu\text{Np/m}$$

Practice Problems

2.11 For the transmission line system described in Practice Problem 2.3, calculate the following parameters, using Eqs (2.206) and (2.207):
(a) Attenuation constant   (b) Phase constant

Calculate (i) the value of L for which the attenuation will be minimum if R, G, and C are kept constant and (ii) the value of C for which the attenuation will be minimum if R, G, and L are kept constant. What will be the minimum attainable attenuation with these modifications?

[0.0098 Np/m, 30.3716 rad/m, 6.2857×10^{11} H/m, 3.717×10^{-29} F/m, 14.712 pNp/m]

2.12 For the transmission line system described in Practice Problem 2.4, calculate the following factors using Eqs (2.206) and (2.207):
(a) Attenuation constant   (b) Phase constant

Calculate (i) the value of L for which the attenuation will be minimum if R, G, and C are kept constant and (ii) the value of C for which the attenuation will be minimum if R, G, and L are kept constant. What will be the minimum attainable attenuation with these modifications?

[0.005 Np/m, 20.9607 rad/m, 1.5569 kH/m, 7.148×10^{-21} F/m, 0.1467 μNp/m]

2.10 Power Transmission and Loss Characterization

Assuming a time harmonic variation, the instantaneous forward travelling voltage and current in a lossy transmission line can be expressed, based on Eqs (2.59) and (2.102), as follows:

$$V_f(z,t) = \text{Re}\left\{V_+ e^{-\alpha z} e^{j(\omega t - \beta z)}\right\} = |V_+| e^{-\alpha z} \cos(\omega t - \beta z) \qquad (2.222)$$

$$I_f(z,t) = \text{Re}\left\{I_+ e^{-\alpha z} e^{j(\omega t - \beta z + \phi)}\right\} = |I_+| e^{-\alpha z} \cos(\omega t - \beta z + \phi) \qquad (2.223)$$

where ϕ is the phase difference between the instantaneous voltage and current, if any. Therefore, the instantaneous power can be expressed by the following form:

$$P(z,t) = V_f(z,t) I_f(z,t) = |V_+||I_+| e^{-2\alpha z} \cos(\omega t - \beta z)\cos(\omega t - \beta z + \phi) \qquad (2.224)$$

The time average power can be expressed as follows:

$$\langle P(z)\rangle = \frac{1}{T}\int_0^T |V_+||I_+| e^{-2\alpha z} \cos(\omega t - \beta z)\cos(\omega t - \beta z + \phi)\,dt$$

or $\qquad \langle P(z)\rangle = \frac{1}{2T}|V_+||I_+| e^{-2\alpha z}\int_0^T [\cos(\phi) + \cos(2\omega t - 2\beta z + \phi)]\,dt$

or $\qquad \langle P(z)\rangle = \frac{1}{2}|V_+||I_+| e^{-2\alpha z} \cos(\phi)$

or $\qquad \langle P(z)\rangle = \langle P(0)\rangle e^{-2\alpha z} \qquad\qquad\qquad (2.225)$

where T is the time period, and

$$\langle P(0)\rangle = \frac{1}{2}|V_+||I_+|\cos(\phi) \qquad\qquad\qquad (2.226)$$

is the time average power at $z = 0$.

Equation (2.225) can be written as follows:

$$\frac{\langle P(z)\rangle}{\langle P(0)\rangle} = e^{-2\alpha z} = 10^{-\kappa\alpha z} \qquad\qquad\qquad (2.227)$$

κ can be calculated by setting $\alpha z = 1$ in Eq. (2.227), which yields the following relation:

$$e^{-2} = 10^{-\kappa}$$

or $\qquad \kappa = \log_{10}(e^2) = 0.869 \qquad\qquad\qquad (2.228)$

Now, by definition, the power loss in decibels (dB) is as follows:

$$\text{Power loss (dB)} = 10\log_{10}\left[\frac{\langle P(0)\rangle}{\langle P(z)\rangle}\right] = 8.69\alpha z \qquad\qquad (2.229)$$

The preceding equations are valid for a lossy line in the absence of a reflected wave, that is, when the line is match terminated. If the transmission line is lossless and is terminated in an unmatched load, the time average power flow in the $+z$ direction is given by the following equation:

$$\langle P(z)\rangle = \frac{1}{2}\mathrm{Re}\left[V(z)I(z)^*\right] \tag{2.230}$$

Now $\quad V(z) = V_+ e^{-j\beta z}\left[1 + \Gamma(z)\right]$ (2.231)

and $\quad I(z) = \frac{V_+}{Z_0} e^{-j\beta z}\left[1 - \Gamma(z)\right]$ (2.232)

Substituting the expressions for voltage and current from Eqs (2.231) and (2.232) in Eq. (2.230), we get the following relation:

$$\langle P(z)\rangle = \frac{1}{2}\mathrm{Re}\left[\frac{V_+ V_+^*}{Z_0}\left\{1 + \Gamma(z)\right\}\left\{1 - \Gamma(z)^*\right\}\right]$$

or $\quad \langle P(z)\rangle = \frac{1}{2}\frac{|V_+|^2}{Z_0}\mathrm{Re}\left[1 + \Gamma(z) - \Gamma(z)^* - \Gamma(z)\Gamma(z)^*\right] = \frac{1}{2}\frac{|V_+|^2}{Z_0}\left[1 - |\Gamma(z)|^2\right]$

or $\quad \langle P(z)\rangle = \frac{1}{2}\frac{|V_+|^2}{Z_0}\left[1 - |\Gamma_L|^2\right]$ (2.233)

as $\Gamma(z) - \Gamma(z)^*$ is purely imaginary and $|\Gamma(z)| = |\Gamma_L|$.

Expressions for the voltage and currents in the forward and reverse directions are as follows:

$$V_+(z) = V_+ e^{-j\beta z} \tag{2.234}$$

$$I_+(z) = \frac{V_+}{Z_0} e^{-j\beta z} \tag{2.235}$$

$$V_-(z) = V_- e^{+j\beta z} \tag{2.236}$$

and $\quad I_-(z) = -\frac{V_-}{Z_0} e^{+j\beta z}$ (2.237)

where the '+' sign indicates a positive z-directed wave and a '−' sign indicates a negative z-directed wave.

Using Eqs (2.234)–(2.237), the expressions for forward and reverse time average power flow are given as follows:

$$\langle P_+(z)\rangle = \frac{1}{2}\frac{|V_+|^2}{Z_0} \tag{2.238}$$

and $\quad \langle P_-(z)\rangle = -\frac{1}{2}\frac{|V_-|^2}{Z_0} = -\frac{1}{2}\frac{|V_+|^2|\Gamma(z)|^2}{Z_0} = -\frac{1}{2}\frac{|V_+|^2|\Gamma_L|^2}{Z_0}$ (2.239)

The expression for net power flow is the sum of Eqs (2.238) and (2.239), which is the same as obtained in Eq. (2.233).

The ratio of the incident power to the reflected power at any point on the line is given by the following equation:

$$\Gamma_P = -\frac{\langle P_-(z) \rangle}{\langle P_+(z) \rangle}$$

or $\qquad \Gamma_P = |\Gamma_L|^2$ (2.240)

2.11 Other Transmission Line Models

In addition to the equivalent lumped circuit model of transmission line, shown in Fig. 2.13, several other lumped circuit equivalent models, namely, lumped backward Γ, lumped π, lumped T, and lumped Γ models, also exist and are shown in Fig. 2.22. All these models would give good results provided the line section is electrically short. However, depending on the value of the load impedance, one model will extend the accuracy of the result further in frequency than another model. For example, if the load impedance is low, compared to the characteristic impedance of the line, lumped Γ and lumped T models would extend the frequency range of accuracy slightly higher than the rest. This is because a low impedance load would be in parallel with the right-most parallel capacitance elements of lumped backward Γ and lumped π models, thus making them ineffective. The converse applies to high impedance loads, where lumped backward Γ and lumped π models will extend the frequency range of accuracy slightly higher than lumped Γ and lumped T models.

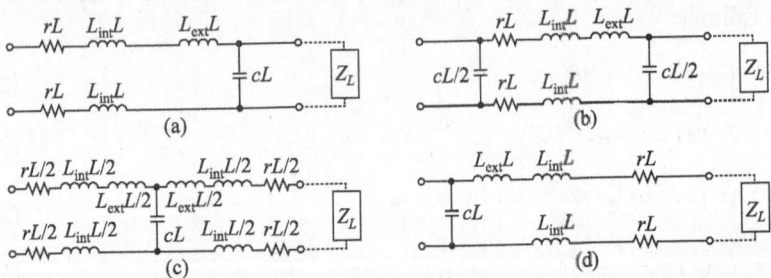

Fig. 2.22 Lumped equivalent circuits for a pair of parallel wires (a) Lumped backward Γ model (b) Lumped π model (c) Lumped T model (d) Lumped Γ model

2.12 Field Theoretic Analysis of Coaxial Transmission Lines

In this section, we will re-derive the telegrapher's equation (Eqs 2.55 and 2.56) starting from Maxwell's equation for the specific case of a coaxial transmission line, shown in Fig. 2.23.

The field within the transmission line will satisfy the following Maxwell's curl equations:

$$\vec{\nabla} \times \vec{E} = -j\omega\mu\vec{H}$$ (2.241)

$$\vec{\nabla} \times \vec{H} = j\omega\varepsilon\vec{E}$$ (2.242)

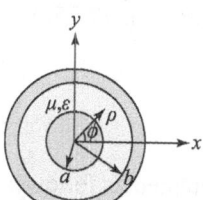

Fig. 2.23 Cross-sectional view of a coaxial transmission line

A TEM wave on the coaxial line is characterized by $E_z = 0$ and $H_z = 0$. Furthermore, azimuthal symmetry of the line ensures that for the TEM wave no ϕ variation of the fields will occur. Therefore, in a cylindrical coordinate system, Eqs (2.241) and (2.242) reduce to the following forms:

$$-\frac{\partial E_\phi}{\partial z}\hat{\rho}+\frac{\partial E_\rho}{\partial z}\hat{\phi}+\frac{1}{\rho}\frac{\partial}{\partial\rho}\left(\rho E_\phi\right)\hat{z}=-j\omega\mu\left(H_\rho\hat{\rho}+H_\phi\hat{\phi}\right)$$

(2.243)

$$-\frac{\partial H_\phi}{\partial z}\hat{\rho}+\frac{\partial H_\rho}{\partial z}\hat{\phi}+\frac{1}{\rho}\frac{\partial}{\partial\rho}\left(\rho H_\phi\right)\hat{z}=j\omega\varepsilon\left(E_\rho\hat{\rho}+E_\phi\hat{\phi}\right)$$

(2.244)

Since the RHS of Eqs (2.243) and (2.244) have no z-component of fields, we can write the following equations:

$$\frac{1}{\rho}\frac{\partial}{\partial\rho}\left(\rho E_\phi\right)=0$$

(2.245)

and $$\frac{1}{\rho}\frac{\partial}{\partial\rho}\left(\rho H_\phi\right)=0$$

(2.246)

The solution of Eqs (2.245) and (2.246) can be expressed as follows:

$$E_\phi=\frac{f(z)}{\rho}$$

(2.247)

and $$H_\phi=\frac{g(z)}{\rho}$$

(2.248)

Since the tangential component of the electric field must vanish at the boundary, $E_\phi = 0$ at $\rho = a$ and b. This boundary condition, when substituted in Eq. (2.247), ensures that $E_\phi = 0$ everywhere. Therefore, Eqs (2.243) and (2.244) reduce to the following forms:

$$\frac{\partial E_\rho}{\partial z}=-j\omega\mu H_\phi$$

(2.249)

and $$\frac{\partial H_\phi}{\partial z}=-j\omega\varepsilon E_\rho$$

(2.250)

Substituting Eq. (2.248) into Eq. (2.249), we can write that the following equation:

$$\frac{\partial E_\rho}{\partial z}=-j\omega\mu\frac{g(z)}{\rho}$$

or $$E_\rho=-\frac{j\omega\mu}{\rho}\int g(z)dz=\frac{h(z)}{\rho}$$

(2.251)

Substituting Eqs (2.248) and (2.251) back into Eq. (2.249), we get the following relation:

$$\frac{\partial h(z)}{\partial z}=-j\omega\mu g(z)$$

(2.252)

Similarly substituting Eqs (2.248) and (2.251) back into Eq. (2.250), the following equation is obtained:

$$\frac{\partial g(z)}{\partial z} = -j\omega\varepsilon h(z) \tag{2.253}$$

Now, the voltage between the two conductors can be calculated as follows:

$$V(z) = \int_{\rho=a}^{b} E_\rho(\rho, z)\,d\rho = h(z)\int_{\rho=a}^{b}\frac{d\rho}{\rho} = h(z)\ln\left(\frac{b}{a}\right) \tag{2.254}$$

or $\quad h(z) = V(z)\Big/\ln\left(\frac{b}{a}\right)$ $\tag{2.255}$

Similarly, current on the inner conductor at $\rho = a$ can be calculated from the following equation:

$$I(z) = \int_{\phi=0}^{2\pi} H_\phi(a, z)a\,d\phi = g(z)\int_{\phi=0}^{2\pi} d\phi = 2\pi g(z) \tag{2.256}$$

or $\quad g(z) = \frac{I(z)}{2\pi}$ $\tag{2.257}$

Substituting Eqs (2.255) and (2.257) in Eq. (2.252), we get the following equation:

$$\frac{1}{\ln(b/a)}\frac{\partial V(z)}{\partial z} = -j\omega\mu\frac{I(z)}{2\pi}$$

or $\quad \frac{\partial V(z)}{\partial z} = -j\omega\left[\frac{\mu}{2\pi}\ln\left(\frac{b}{a}\right)\right]I(z)$ $\tag{2.258}$

Since for a coaxial transmission line, as given in Eq. (2.14),

$$L_{\text{coaxial}} = \frac{\mu}{2\pi}\ln\left(\frac{b}{a}\right) \tag{2.259}$$

Eq. (2.258) reduces to the following form:

$$\frac{\partial V(z)}{\partial z} = -j\omega L_{\text{coaxial}}I(z) \tag{2.260}$$

which is the same as Eq. (2.55).

Similarly, substitution of Eqs (2.255) and (2.257) into Eq. (2.253) results in the following expression:

$$\frac{1}{2\pi}\frac{\partial I(z)}{\partial z} = -j\omega\varepsilon\frac{V(z)}{\ln(b/a)}$$

or $\quad \frac{\partial I(z)}{\partial z} = -j\omega\left[\frac{2\pi\varepsilon}{\ln(b/a)}\right]V(z)$ $\tag{2.261}$

Since for a coaxial transmission line, as given in Eq. (2.26),

$$C_{\text{coaxial}} = \frac{2\pi\varepsilon}{\ln(b/a)} \tag{2.262}$$

Eq. (2.261) reduces to the following form:

$$\frac{\partial I(z)}{\partial z} = -j\omega C_{\text{coaxial}} V(z) \tag{2.263}$$

which is same as equation (2.56).

2.13 Field Theoretic Analysis of Parallel-plate Transmission Lines

Let us consider an infinitely extended parallel-plate transmission line, shown in Fig. 2.12. We will also assume that $w \gg h$ so that fringing fields can be neglected. This assumption also ensures that there is no variation in the field vectors along the width of the plates or $\dfrac{\partial}{\partial y} \to 0$.

Assuming TEM mode propagation, that is, $E_z = 0$ and $H_z = 0$, the Maxwell's curl equations for electric field can be written as follows:

$$\vec{\nabla} \times \vec{E} = -\mu \frac{\partial \vec{H}}{\partial t}$$

or $\quad \left(-\dfrac{\partial E_y}{\partial z}\right)\hat{a}_x + \left(\dfrac{\partial E_x}{\partial z}\right)\hat{a}_y + \left(\dfrac{\partial E_y}{\partial x}\right)\hat{a}_z = -\mu\left(\dfrac{\partial H_x}{\partial t}\hat{a}_x + \dfrac{\partial H_y}{\partial t}\hat{a}_y\right)$

Equating the vector components on each side, we get the following equations:

$$\frac{\partial E_y}{\partial z} = \mu \frac{\partial H_x}{\partial t} \tag{2.264}$$

$$\frac{\partial E_x}{\partial z} = -\mu \frac{\partial H_y}{\partial t} \tag{2.265}$$

$$\frac{\partial E_y}{\partial x} = 0 \tag{2.266}$$

Equation (2.266) implies that E_y is independent of the variable x. Simultaneously, being the tangential component, it must also satisfy the following boundary condition:

$$E_y = 0 \text{ at } x = 0 \text{ and } x = h \tag{2.267}$$

However, E_y cannot be simultaneously independent of x and satisfy the boundary condition in Eq. (2.267) unless it is zero. Therefore,

$$E_y = 0 \tag{2.268}$$

and the only remaining component of the electric field is E_x.

Now from the Maxwell's curl equations for a magnetic field, we can write the following relations:

$$\vec{\nabla} \times \vec{H} = \varepsilon \frac{\partial \vec{E}}{\partial t}$$

or
$$\left(-\frac{\partial H_y}{\partial z} \right) \hat{a}_x + \left(\frac{\partial H_x}{\partial z} \right) \hat{a}_y + \left(\frac{\partial H_y}{\partial x} \right) \hat{a}_z = \varepsilon \left(\frac{\partial E_x}{\partial t} \hat{a}_x + \frac{\partial E_y}{\partial t} \hat{a}_y \right)$$

Equating the vector components on each side, we get the following equations:

$$\frac{\partial H_y}{\partial z} = -\varepsilon \frac{\partial E_x}{\partial t} \tag{2.269}$$

$$\frac{\partial H_x}{\partial z} = \varepsilon \frac{\partial E_y}{\partial t} \tag{2.270}$$

$$\frac{\partial H_y}{\partial x} = 0 \tag{2.271}$$

Since $E_y = 0$, as given in Eq. (2.268), H_x must also be zero, and therefore we are left with only the H_y component, which is independent of x, as evident from Eq. (2.271).

Thus, the total field vectors can be written as follows:

$$\vec{E} = E_x(z,t) \hat{a}_x \tag{2.272}$$

and
$$\vec{H} = H_y(z,t) \hat{a}_y \tag{2.273}$$

Integrating Eq. (2.265) with respect to x we get the following equation:

$$\frac{\partial}{\partial z} \int_{x=0}^{h} E_x dx = -\mu \frac{\partial}{\partial t} \int_{x=0}^{h} H_y dx \tag{2.274}$$

Now,

$$V(z,t) = -\int_{x=0}^{h} E_x dx \tag{2.275}$$

Substituting Eq. (2.275) into Eq. (2.274), we can write the following equation:

$$\frac{\partial V(z,t)}{\partial z} = \mu \frac{\partial}{\partial t} \int_{x=0}^{h} H_y dx \tag{2.276}$$

Since H_y is independent of x, Eq. (2.276) can be written as follows:

$$\frac{\partial V(z,t)}{\partial z} = \mu h \frac{\partial H_y}{\partial t} \tag{2.277}$$

Now,

$$I(z,t) = -\int_{y=0}^{w} H_y dy \tag{2.278}$$

Since we have assumed that the field components do not vary along the y-direction as $w \gg h$, Eq. (2.278) can be written in the following form:

$$I(z,t) = -wH_y \qquad (2.279)$$

Substituting Eq. (2.279) into Eq. (2.277), we get the following expression:

$$\frac{\partial V(z,t)}{\partial z} = \mu h \frac{\partial}{\partial t}\left(-\frac{I(z,t)}{w}\right)$$

or

$$\frac{\partial V(z,t)}{\partial z} = -\mu \frac{h}{w}\frac{\partial I(z,t)}{\partial t} \qquad (2.280)$$

For a parallel-plate transmission line:

$$L_{\text{parallel plate}} = \frac{\psi_{\text{ext}}}{I} = \int\limits_{x=0}^{h} \mu H_y \, dx \bigg/ \int\limits_{y=0}^{w} H_y \, dy = \frac{\mu h}{w} \qquad (2.281)$$

Therefore, Eq. (2.280) can be written as follows:

$$\frac{\partial V(z,t)}{\partial z} = -L_{\text{parallel plate}} \frac{\partial I(z,t)}{\partial t} \qquad (2.282)$$

For a time harmonic variation of the fields, Eq. (2.282) modifies to the following form:

$$\frac{\partial V(z,t)}{\partial z} = -j\omega L_{\text{parallel plate}} I(z) \qquad (2.283)$$

which has the same form as Eq. (2.55).

Similarly, integrating Eq. (2.269) with respect to y, we get the following relation:

$$\frac{\partial}{\partial z}\int\limits_{y=0}^{w} H_y \, dy = -\varepsilon \frac{\partial}{\partial t}\int\limits_{y=0}^{w} E_x \, dy \qquad (2.284)$$

Substituting Eq. (2.278) into Eq. (2.284), we get the following equation:

$$\frac{\partial I(z,t)}{\partial z} = \varepsilon \frac{\partial}{\partial t}\int\limits_{y=0}^{w} E_x \, dy \qquad (2.285)$$

Since E_x is independent of y, Eq. (2.285) can be written as follows:

$$\frac{\partial I(z,t)}{\partial z} = \varepsilon \frac{\partial}{\partial t}(wE_x) \qquad (2.286)$$

Again since H_y is independent of x, as shown in Eq. (2.271), for the TEM wave E_x must also be independent of x. Therefore, Eq. (2.275) can be written as follows:

$$V(z,t) = -\int\limits_{x=0}^{h} E_x \, dx = -hE_x \qquad (2.287)$$

Substituting Eq. (2.287) into Eq. (2.286), we get the following expression:

$$\frac{\partial I(z,t)}{\partial z} = \varepsilon \frac{\partial}{\partial t}\left(-\frac{w}{h}V(z,t)\right)$$

or $\qquad \dfrac{\partial I(z,t)}{\partial z} = -\varepsilon \dfrac{w}{h}\dfrac{\partial V(z,t)}{\partial t}$ $\qquad\qquad$ (2.288)

For a parallel-plate transmission line,

$$C_{\text{parallel plate}} = \frac{q}{V} = \int\limits_{y=0}^{w} \varepsilon E_x dy \Big/ \int\limits_{x=0}^{h} E_x dx = \frac{\varepsilon w}{h}$$ (2.289)

Substituting Eq. (2.289) into Eq. (2.288), we get the following equation:

$$\frac{\partial I(z,t)}{\partial z} = -C_{\text{parallel plate}} \frac{\partial V(z,t)}{\partial t}$$ (2.290)

For a time harmonic variation of the field, Eq. (2.290) modifies to the following format:

$$\frac{\partial I(z,t)}{\partial z} = -j\omega C_{\text{parallel plate}} V(z,t)$$ (2.291)

which has the same form as Eq. (2.56).

2.14 Smith Chart

A Smith chart consists of a combined plot of the normalized impedance or admittance and the angle and magnitude of a generalized complex reflection coefficient in a unity circle. It is applicable to the analysis of a lossless as well as a lossy transmission line. By simple rotation of the chart, the effect of the position on the transmission line can be determined.

Based on Eq. (2.173), we can write the following equation:

$$Z_{\text{in}}(z) = Z_0 \frac{1+\Gamma_L e^{-2\gamma d}}{1-\Gamma_L e^{-2\gamma d}} = Z_0 \frac{1+|\Gamma_L|e^{j\theta_L}e^{-2\alpha d}e^{-j2\beta d}}{1-|\Gamma_L|e^{j\theta_L}e^{-2\alpha d}e^{-j2\beta d}}$$

or $\qquad Z_{\text{in}}(z) = Z_0 \dfrac{1+|\Gamma_L|e^{-2\alpha d}e^{-j(2\beta d-\theta_L)}}{1-|\Gamma_L|e^{-2\alpha d}e^{-j(2\beta d-\theta_L)}}$ $\qquad$ (2.292)

Without any loss of generality, the normalized input impedance can be written as follows:

$$z_{\text{in}}(z) = \frac{Z_{\text{in}}(z)}{Z_0} = r_{\text{in}} + jx_{\text{in}}$$ (2.293)

Now at any point z on the transmission line, $Z_{in}(z)$ will behave as a load and hence the reflection coefficient at this point can be written as follows:

$$\Gamma_{in} = \frac{Z_{in}(z) - Z_0}{Z_{in}(z) + Z_0} = \left[\frac{Z_{in}(z)}{Z_0} - 1\right] \bigg/ \left[\frac{Z_{in}(z)}{Z_0} + 1\right] = \frac{z_{in}(z) - 1}{z_{in}(z) + 1} = \Gamma_r + j\Gamma_i \quad (2.294)$$

or $\quad z_{in}(z) - 1 = \Gamma_r z_{in}(z) + \Gamma_r + j\Gamma_i z_{in}(z) + j\Gamma_i$

or $\quad z_{in}(z) = (1 + \Gamma_r + j\Gamma_i)/(1 - \Gamma_r - j\Gamma_i)$

or $\quad r_{in} + jx_{in} + 1 = \dfrac{1 + \Gamma_r + j\Gamma_i}{1 - \Gamma_r - j\Gamma_i} + 1 = \dfrac{2}{1 - \Gamma_r - j\Gamma_i} = \dfrac{2(1 - \Gamma_r + j\Gamma_i)}{(1 - \Gamma_r - j\Gamma_i)(1 - \Gamma_r + j\Gamma_i)}$

or $\quad r_{in} + jx_{in} + 1 = \dfrac{2(1 - \Gamma_r + j\Gamma_i)}{(1 - \Gamma_r)^2 + \Gamma_i^2}$

or $\quad r_{in} + jx_{in} + 1 = \dfrac{2(1 - \Gamma_r)}{(1 - \Gamma_r)^2 + \Gamma_i^2} + j\dfrac{2\Gamma_i}{(1 - \Gamma_r)^2 + \Gamma_i^2} \quad (2.295)$

Equating the real and imaginary parts on the both side of Eq. (2.295), we get the following expression:

$$r_{in} + 1 = \frac{2(1 - \Gamma_r)}{(1 - \Gamma_r)^2 + \Gamma_i^2} \quad (2.296)$$

or $\quad r_{in} = \dfrac{2(1 - \Gamma_r)}{(1 - \Gamma_r)^2 + \Gamma_i^2} - 1 = \dfrac{2(1 - \Gamma_r) - (1 - \Gamma_r)^2 - \Gamma_i^2}{(1 - \Gamma_r)^2 + \Gamma_i^2} = \dfrac{1 - \Gamma_r^2 - \Gamma_i^2}{(1 - \Gamma_r)^2 + \Gamma_i^2} \quad (2.297)$

and $\quad x_{in} = (2\Gamma_i)/\{(1 - \Gamma_r)^2 + \Gamma_i^2\} \quad (2.298)$

Taking the ratio of Eqs (2.297) and (2.296), we get the following expression:

$$\frac{r_{in}}{r_{in} + 1} = \left[\frac{1 - \Gamma_r^2 - \Gamma_i^2}{(1 - \Gamma_r)^2 + \Gamma_i^2}\right] \bigg/ \left[\frac{2(1 - \Gamma_r)}{(1 - \Gamma_r)^2 + \Gamma_i^2}\right] = \frac{1 - \Gamma_r^2 - \Gamma_i^2}{2(1 - \Gamma_r)}$$

or $\quad \dfrac{2r_{in}(1 - \Gamma_r)}{r_{in} + 1} = 1 - \Gamma_r^2 - \Gamma_i^2$

or $\quad \Gamma_r^2 - \dfrac{2r_{in}\Gamma_r}{r_{in} + 1} + \Gamma_i^2 = 1 - \dfrac{2r_{in}}{r_{in} + 1} \quad (2.299)$

Adding $\dfrac{r_{in}^2}{(r_{in} + 1)^2}$ to both sides of Eq. (2.299), we get the following expression:

$$\Gamma_r^2 - \frac{2r_{in}\Gamma_r}{r_{in} + 1} + \frac{r_{in}^2}{(r_{in} + 1)^2} + \Gamma_i^2 = 1 - \frac{2r_{in}}{r_{in} + 1} + \frac{r_{in}^2}{(r_{in} + 1)^2}$$

or $\qquad \left\{ \Gamma_r - \dfrac{r_{in}}{(r_{in}+1)} \right\}^2 + \Gamma_i^2 = \left(1 - \dfrac{r_{in}}{r_{in}+1} \right)^2$

or $\qquad \left\{ \Gamma_r - \dfrac{r_{in}}{(r_{in}+1)} \right\}^2 + \Gamma_i^2 = \left(\dfrac{1}{r_{in}+1} \right)^2 \qquad (2.300)$

Equation (2.300) represents a family of circles with a centre $\left\{ \dfrac{r_{in}}{(r_{in}+1)}, 0 \right\}$ and a radius $\dfrac{1}{r_{in}+1}$, as shown in Fig. 2.24.

Equation (2.298) can be written as follows:

$$(\Gamma_r - 1)^2 + \Gamma_i^2 = \dfrac{2\Gamma_i}{x_{in}} \qquad (2.301)$$

Adding $\dfrac{1}{x_{in}^2}$ to both sides of Eq. (2.301), we get the following equation:

$$(\Gamma_r - 1)^2 + \Gamma_i^2 + \dfrac{1}{x_{in}^2} = \dfrac{2\Gamma_i}{x_{in}} + \dfrac{1}{x_{in}^2}$$

or $\qquad (\Gamma_r - 1)^2 + \dfrac{1}{x_{in}^2} - \dfrac{2\Gamma_i}{x_{in}} + \Gamma_i^2 = \dfrac{1}{x_{in}^2}$

or $\qquad (\Gamma_r - 1)^2 + \left(\Gamma_i - \dfrac{1}{x_{in}} \right)^2 = \dfrac{1}{x_{in}^2} \qquad (2.302)$

Equation (2.302) represents a family of circles with a centre $\left\{ 1, \dfrac{1}{x_{in}} \right\}$ and a radius $\dfrac{1}{x_{in}}$ and is shown in Fig. 2.25.

To measure the distance along the line, a second scale is added on the circumference, which is in wavelength units. Since the properties of the transmission line are

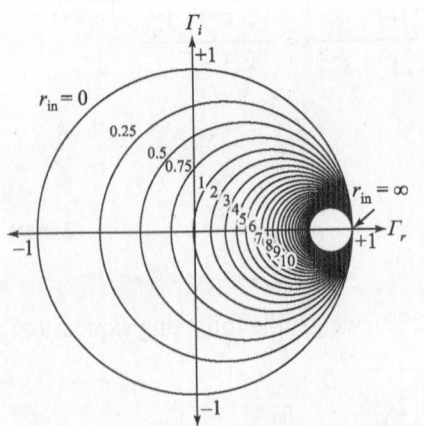

Fig. 2.24 Constant resistance circles

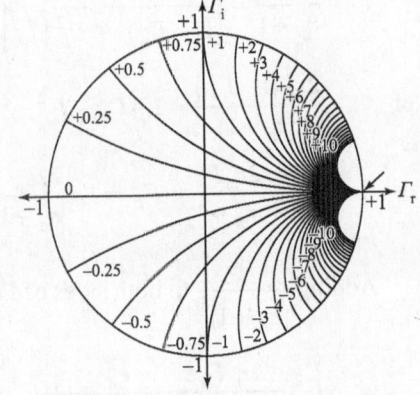

Fig. 2.25 Constant reactance circles

repeated at an interval of $\lambda/2$, the scale is adjusted so that a complete travel along the periphery of the Smith chart results in an equivalent travelling of $\lambda/2$ distance along the transmission line.

For a lossless transmission line, from Eq. (2.173), we can write as the following equation:

$$Z_{\text{in}}(z) = Z_0 \frac{1 + \Gamma_L e^{-j2\beta d}}{1 - \Gamma_L e^{-j2\beta d}} \tag{2.303}$$

Therefore, if we proceed from the load Z_L to the input impedance Z_{in} at any point $z = l - d$ on the transmission line, we actually move a distance d towards the generator. The negative sign in the angle $\angle - 2\beta d$ indicates that the corresponding motion in the Smith chart is in the clockwise direction. The reverse is true when we move from the source to the load along the transmission line.

Since for a passive load, $|\Gamma|_{\text{Max}} = 1$

or $\qquad |\Gamma_r^2 + \Gamma_i^2|_{\text{Max}} = 1 \tag{2.304}$

the circle on which the distance scale is plotted is an unit circle. For $|\Gamma| < 1$, the reflection coefficient circles are of radius $|\Gamma|$ and are centred at $(0, 0)$, as shown in Fig. 2.26. In practice, the constant reflection coefficient circles are not included in the main chart; instead, their radii are projected at the bottom of the chart. The corresponding transmission coefficients, SWRs, return losses, etc. are also plotted along with them, as shown in Fig. 2.27.

Information regarding the locations of the voltage maxima and minima can also be obtained from a Smith chart. We already know that the first voltage maximum/minimum occurs at the position of the load provided that Z_L is a pure resistance. Conversely, it can be said that at the points of voltage maxima and minima, the input impedance of the transmission line is purely resistive. Since the reactance part of the input impedance vanishes at these points, the voltage maxima and minima lie on the $\Gamma_i = 0$ axis. Further since for voltage maximum $R_{\text{in}} > Z_0$ or $r_{\text{in}} > 1$, the possible positions of the voltage maxima in the Smith chart are $(\Gamma_r > 0, \Gamma_i = 0)$. Similarly, the possible positions of the voltage minima in the Smith chart are $(\Gamma_r < 0, \Gamma_i = 0)$. At $(\Gamma_r = 0, \Gamma_i = 0)$, neither the voltage maxima nor the voltage minima occur, and this correspond to the position of the matched load in the Smith chart.

For an open-circuit termination, $r_{\text{in}} = \infty$ and $x_{\text{in}} = 0$. Therefore, the position of the open-circuit point in a Smith chart is at the intersection point of the $|\Gamma| = 1$ circle and the positive Γ_r axis. Similarly, the position of the short-circuit point in a Smith chart is at the intersection point of the $|\Gamma| = 1$ circle and the negative Γ_r axis.

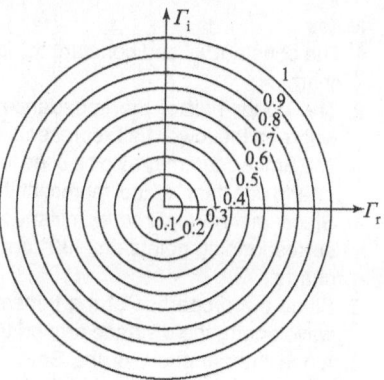

Fig. 2.26 Constant reflection coefficient circles

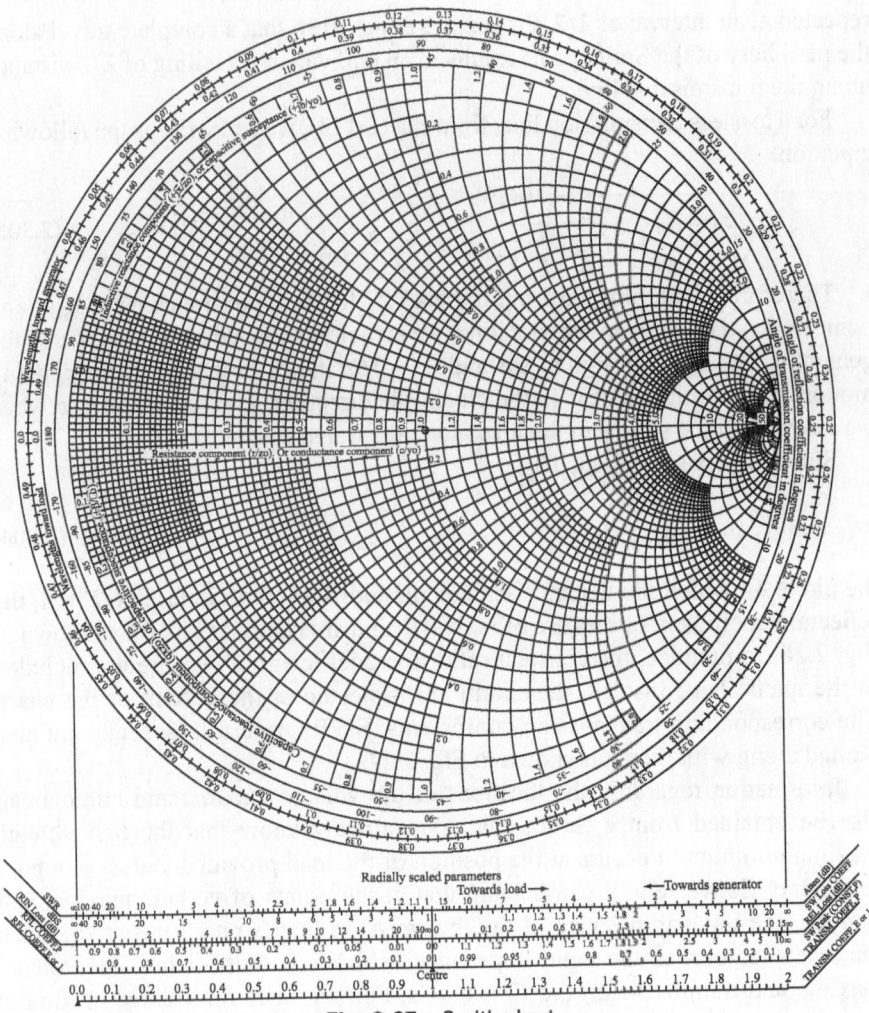

Fig. 2.27 Smith chart

Notes

1. The constant 'r_{in}' and constant 'x_{in}' loci form two orthogonal sets of circles in the Smith chart.

2. The upper half of the impedance Smith chart represents the inductive reactance, whereas the lower half represents the capacitive reactance.

3. The Smith chart may also be used for normalized admittance. In such cases, the constant 'r_{in}' and constant 'x_{in}' loci represent the constant 'g_{in}' and constant 'b_{in}' circles, respectively.

4. Since the normalized admittance is reciprocal to the normalized impedance, the corresponding points are 180° out of phase in the impedance and admittance Smith charts.

5. Since the properties of the transmission lines are repeated after a distance of half-wavelength, the distance around the Smith chart is $\lambda/2$.

6. Along the periphery of the Smith chart, distances in terms of wavelengths towards the generator are given in the clockwise direction. Similarly, distances in terms of wavelengths towards the load are given in the anticlockwise direction.

7. At a point $z_{min} = 1/\rho$, there is a V_{min} on the line. The horizontal radius in the left half of the Smith chart corresponds to the V_{min}, I_{max}, z_{min}, and $1/\rho$. Similarly, at a point $z_{max} = \rho$, there is a V_{max} on the line. The horizontal radius in the right half of the Smith chart corresponds to the V_{max}, I_{min}, z_{max}, and ρ.

EXAMPLE 2.7 For the transmission line system described in Example 2.3, find the position of the following factors using a Smith chart:

(a) First voltage maximum   (b) First voltage minimum   (c) VSWR

Assume the transmission line to be lossless. In addition, calculate the input impedance at a distance of 64 mm from the load. A description of the problem is given in Fig. 2.28.

Solution   Given: $Z_L = 78.27 + j60.93 \ \Omega$, $Z_0 = 50.026 \ \Omega$, $d = 64$ mm, and $f = 1$ GHz

1. The normalized load impedance is as follows:

$$z_L = \frac{Z_L}{Z_0} = \frac{78.27 + j60.93}{50.026}$$

$$= 1.565 + j1.218$$

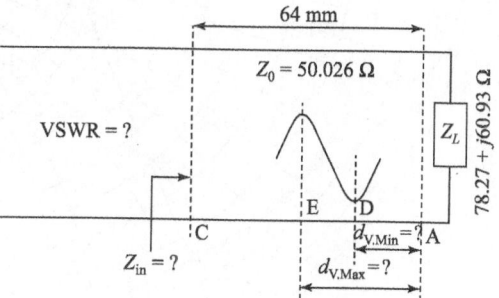

Fig. 2.28 Problem description

2. Enter the $z_l = 1.565 + j1.218$ point on the chart as shown by point A on Fig. 2.29. Point A represents the position of the load.

3. Read 0.194λ on the 'wavelength towards the generator scale' by drawing a straight line from the centre of the chart through the load point A and intersecting the distance scale.

4. Since the voltage maximum or minimum exists on the line in a direction from the load to the source, move a distance from the point at 0.194λ towards the generator and stop at the voltage maximum point at 0.25λ. Then the distance of the voltage maximum point from the load is $d_{V,Max} = 0.25\lambda - 0.194\lambda = 0.056\lambda = 1.68$ cm, as the frequency is 1 GHz or wavelength is 30 cm.

5. Similarly, move a distance from the point at 0.194λ towards the generator and stop at the voltage minimum point at 0.50λ. Then the distance of the voltage minimum point from the load is $d_{V,Min} = 0.50\lambda - 0.194\lambda = 0.306\lambda = 9.18$ cm.

6. Next, draw a standing wave circle with O as the centre and OA as a radius. The circle cuts the right-hand real axis at point B, which indicates the SWR. This is $\rho = 2.8$.

7. The wavelength of the voltage signal is $\lambda = 3 \times 10^{11}/10^9 = 300$ mm.

8. The distance of the measurement point in terms of wavelengths is $d = 64\lambda/300 = 0.213\lambda$.

9. Move a distance of 0.213λ towards the generator from the 0.194λ point; the new location is $d_{in} = 0.213\lambda + 0.194\lambda = 0.407\lambda$.

10. Draw a straight line connecting the 0.407λ point and the centre of the Smith chart O. The line cuts the $\rho = 2.8$ circle at a point C, where $z_l = 0.48 - j0.54$.

11. Therefore, the input impedance of the line at a distance 64 mm from the load is
$$Z_{in} = 50.026 \times (0.48 - j0.54) = 24.012 - j27.014 \ \Omega.$$

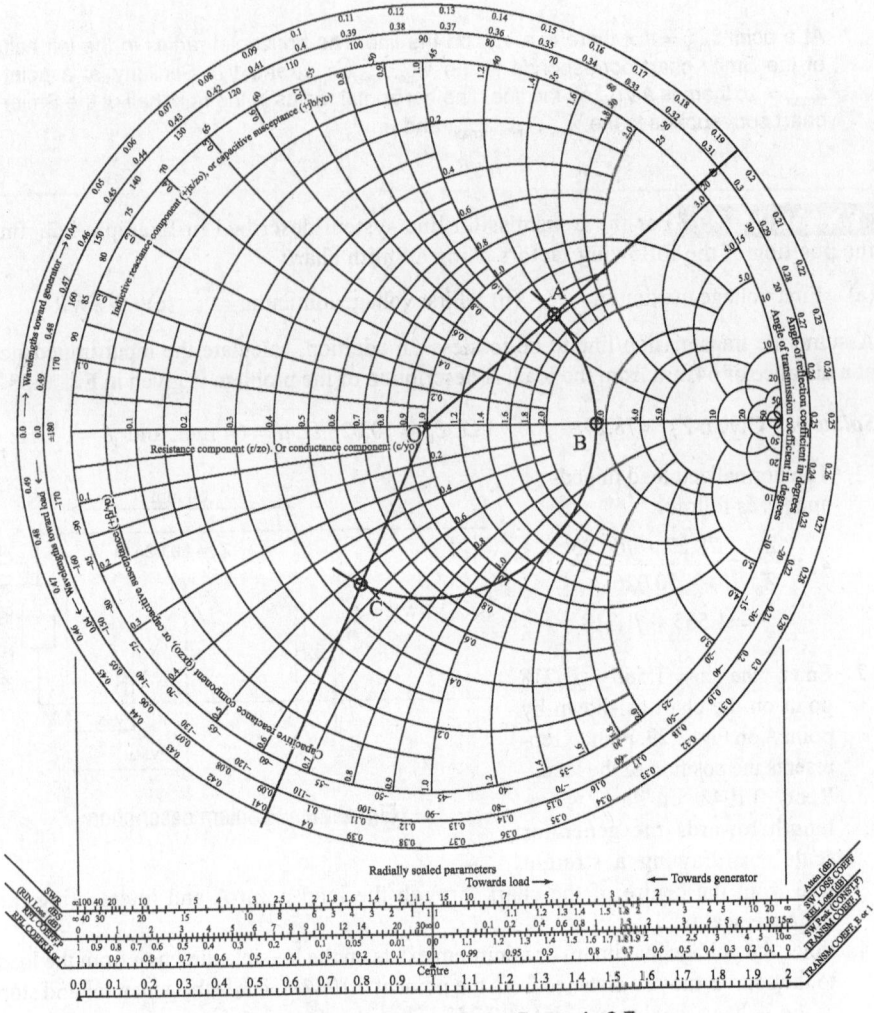

Fig. 2.29 Smith chart for Example 2.7

Practice Problems

2.13[@] Assume that the SWR of a 50 Ω transmission line is 1.8 when the line is loaded. When the load is shorted, the voltage minimum shifts by 6 cm towards the load from its previous position. If the distance between two successive voltage minima is 15 cm, then determine the load impedance.

2.14[@] A load of $60.57 + j32.87\ \Omega$ terminates a section of 50 Ω transmission line of length 0.167λ. The 50 Ω line is preceded by a 74 Ω line. Calculate the following parameters:

(a) Input impedance at a distance of 0.204λ from the junction in the second transmission line
(b) VSWR in the second line

@Solution in Online Resources

2.15@ A capacitive iris having susceptance of 0.014 ℧ is located 0.18λ from load admittance $Y_L = 0.012 + j0.008$ ℧. Find the VSWR on the generator side of the iris, provided that the characteristic admittance of the line is 0.02 ℧.

2.16@ Determine the admittance of the load that produces an SWR of 2.8 in a 0.02 ℧ line, provided that the first voltage minimum is at a distance of 0.13λ from the load.

2.17 For the transmission line system described in Practice Problem 2.5, find the position of the following factors using a Smith chart:

(a) First voltage maximum   (b) First voltage minimum   (c) VSWR

Assume the transmission line to be lossless. In addition, calculate the input impedance at a distance 108.6 mm from the load.   [3.06 cm, 10.56 cm, 4.5, 16.34 + i4.439 Ω]

2.18 For the transmission line system described in Practice Problem 2.6, find the position of the following parameters using a Smith chart:

(a) First voltage maximum   (b) First voltage minimum   (c) VSWR

Assume the transmission line to be lossless. In addition, calculate the input impedance at a distance of 72 mm from the load.   [0.21 cm, 7.71 cm, 1.65, 61.002 − j6.5 Ω]

2.19 Assume that the SWR of a 50 Ω transmission line is 2.4 when the line is loaded. When the load is shorted, the voltage minimum shifted by 10 cm towards the load from its previous position. If the distance between two successive voltage minima is 16 cm, then determine the load impedance.   [70.5 + j50 Ω]

2.20 A load of $74.32 + j21.93$ Ω terminates a section of 50 Ω transmission line of length 0.115λ. The 50 Ω line is preceded by a 60 Ω line. Calculate the following parameters:

(a) Input impedance at a distance 0.312λ from the junction in the second transmission line

(b) VSWR in the second line   [64.80 + j27 Ω, 1.538]

2.21 A capacitive iris having susceptance of 0.006 ℧ is located 0.11λ from the load admittance $Y_L = 0.032 + j0.012$ ℧. Find the VSWR on the generator side of the iris, provided that the characteristic admittance of the line is 0.02 ℧.   [1.59]

2.22 Determine the admittance of the load that produces an SWR of 1.7 in a 0.02 ℧ line, provided the first voltage minimum is at a distance of 0.21λ from the load.

[0.0124 + j0.0033 ℧]

2.15 Transient Analysis of Transmission Lines

So far, we have discussed the analysis of a transmission line under the steady-state excitation condition, that is, the voltage and currents are sinusoidal at a single frequency. Now, we will deviate from the steady-state condition and try to understand the response of the transmission line to voltage step function, pulses that will be grouped under the general heading of 'transients'. The voltage step function is particularly important to understand the behaviour of the transmission line at the instant when it is initially energized by switching on the source, whereas the pulse propagation is important for digital communication in which the signals are composed

---

@Solution in Online Resources

of a series of pulses. During the analysis, we will assume that the transmission line is lossless and there is no dispersion in it. In practice, no transmission line is purely lossless and transients are also composed of numerous frequencies. Since the phase constant of a propagating wave in a transmission line is a function of frequency, as given by Eq. (2.207), each frequency component of the transients will propagate at different group velocities and thus will reach the load at different instants, resulting in pulse broadening. If the load is not properly matched, then the incident frequency components will be reflected from it, following which they will be vectorically added to the forward travelling wave, thus changing the pulse shape further.

Let us consider that a transmission line of length L is terminated by load impedance Z_L, as shown in Fig. 2.30. Initially we will assume that $Z_L = Z_0$ and the line is energized at $z = 0$ with the help of a battery of voltage V_0 and a key. Let us assume that at $t = 0$ the switch is closed. This will instantly make the voltage of the line at the point $z = 0$ equal to V_0. However, the voltage of the transmission line at a point $z > 0$ at $t = 0$ will still be zero, as a finite time is required for the wave to travel from one point of the line to another due to its finite group velocity. The voltage that has been initiated at $z = 0$ and $t = 0$ will now start propagating along the transmission line with its finite group velocity v_g. The wave will reach the load at $z = L$ at a time $t = T = L/v_g$. Since the load is matched, reflection will not occur and the load voltage will remain at a value of $V_L = V_0$. The voltage at the load as a function of time t will resemble the pattern shown in Fig. 2.31. As for $t < T$, $V_L = 0$ and for $t > T$, $V_L = V_0$, the load voltage can be assumed to be a step function. The corresponding load current is $I_L = V_L/Z_L = V_0/Z_0$, where $Z_L = Z_0$ is the load impedance/characteristic impedance of the transmission line.

Now let us consider the more general case where the load impedance is not purely matched, that is, the transmission line is terminated by a impedance Z_L where $Z_L \neq Z_0$. If we further assume that the battery has some finite impedance Z_S, then

$$V_G = \frac{V_0 Z_0}{Z_S + Z_0} \tag{2.305}$$

The presence of the unmatched load will launch a reflected wave V_1^-, travelling along the negative z-direction. This reflected voltage wave will now be added to the voltage wave travelling towards the load, say V_1^+, where $V_1^+ = V_G$. Therefore, just after the forward travelling wave has reached the load and reflection has taken place, the voltage at the load will be $V_1^+ + V_1^-$. The following relation holds:

$$\frac{V_1^-}{V_1^+} = \Gamma_L = \frac{Z_L - Z_0}{Z_L + Z_0} \text{ or } V_1^- = \Gamma_L V_1^+ = V_G \frac{Z_L - Z_0}{Z_L + Z_0} \tag{2.306}$$

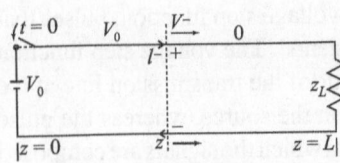

Fig. 2.30 Voltage propagation along the transmission line

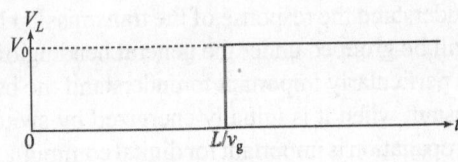

Fig. 2.31 Voltage at the load as a function of time

The total voltage at the load, just after the reflection, will be as follows:

$$V_L = V_G\left(1 + \frac{Z_L - Z_0}{Z_L + Z_0}\right) = V_G\frac{2Z_L}{Z_L + Z_0} \tag{2.307}$$

As the reflected voltage travels towards the battery, it leaves behind a total voltage of $V_1^+ + V_1^-$. The voltage ahead of the reflected voltage remains at a value of V_1^+, until and unless it reaches the battery at time $t = 2L / v_g$. Upon reaching the battery, the voltage reflected by the load, V_1^-, will be re-reflected by the battery and will start travelling towards the load in the +z-direction. Let this re-reflected voltage be V_2^+:

$$V_2^+ = \Gamma_S V_1^- = \Gamma_S V_G \frac{Z_L - Z_0}{Z_L + Z_0} = V_G\left(\frac{Z_S - Z_0}{Z_S + Z_0}\right)\left(\frac{Z_L - Z_0}{Z_L + Z_0}\right) \tag{2.308}$$

where Γ_S is the reflection coefficient at the source and is given in Eq. (2.100).

Therefore, after the voltage V_1^- has reached the battery and re-reflection by the battery taken place, the total voltage at the source will be as follows:

$$V_S = V_1^+ + V_1^- + V_2^+ = V_G\left\{1 + \left(\frac{Z_L - Z_0}{Z_L + Z_0}\right) + \left(\frac{Z_S - Z_0}{Z_S + Z_0}\right)\left(\frac{Z_L - Z_0}{Z_L + Z_0}\right)\right\} \tag{2.309}$$

The voltage wave V_2^+ will reach the load at a time $t = 3L / v_g$ and will suffer further reflection by the load. The whole process will be repeated infinitely. If the voltage reflected by the load and source for their nth reflection is denoted by V_n^- and V_n^+, respectively, then the final load voltage will be as follows:

$$V_L = V_1^+ + V_1^- + V_2^+ + V_2^- + \dots + V_n^+ + V_n^- + \dots\infty \tag{2.310}$$

In terms of load and source reflection coefficients, Eq. (2.310) can be written as follows:

$$V_L = V_G\left(1 + \Gamma_L + \Gamma_S\Gamma_L + \Gamma_S\Gamma_L^2 + \dots + \Gamma_S^{n-1}\Gamma_L^n + \Gamma_S^n\Gamma_L^n + \Gamma_S^n\Gamma_L^{n+1} + \dots\infty\right)$$

or $\quad V_L = V_G\left\{(1 + \Gamma_L) + \Gamma_S\Gamma_L(1 + \Gamma_L) + \dots\infty\right\}$

or $\quad V_L = V_G(1 + \Gamma_L)\left(1 + \Gamma_S\Gamma_L + \Gamma_S^2\Gamma_L^2 + \dots\infty\right) \tag{2.311}$

Since both $|\Gamma_S| < 1$ and $|\Gamma_L| < 1$ are true, Eq. (2.311) can be written as follows:

$$V_L = V_G\frac{1 + \Gamma_L}{1 - \Gamma_S\Gamma_L} \tag{2.312}$$

To find the voltage at a point z of the transmission line at a time t, the voltage reflection diagram, shown in Fig. 2.32, is often used. It is a 2D plot in which the position of the line, z, is measured along the horizontal axis, whereas the time, t, is measured along the vertical axis. Another vertical line parallel to the time axis is also drawn at $z = L$. This vertical line along with the time axis, drawn at $z = 0$, defines the boundaries of the transmission line. The left-hand time axis at $z = 0$ indicates the position of the battery, whereas the right-hand vertical line at $z = L$ indicates the position of the load. If the switch is located at $z = 0$ and is closed at $t = 0$, the initial wave, V_1^+, starts at the origin. Since the wave will reach the

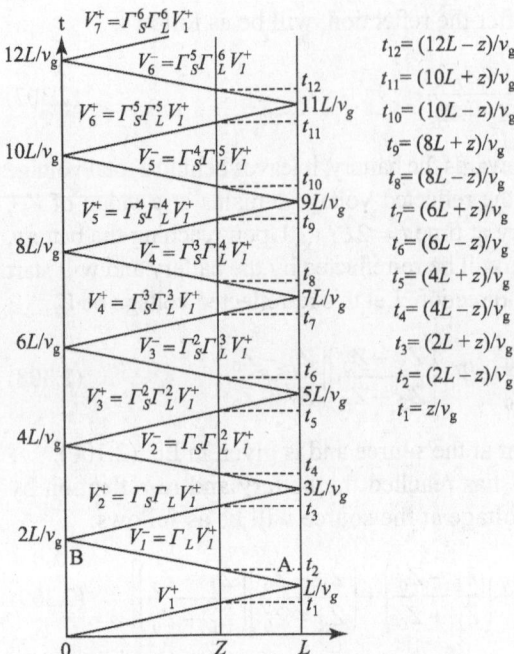

Fig. 2.32 Voltage reflection diagram for transmission line

load at $t = T = L/v_g$, for the first time, the point $\left(L, L/v_g\right)$ is located on the voltage reflection diagram. This point has been represented by A in Fig. 2.32. Next, the origin O and the point is connected by drawing a straight line. The location of the voltage wave V_1^+ as a function of time is represented by this straight line. From point A, the reflected voltage wave V_1^- will be generated, which will travel towards the battery and reach there at time $t = 2L/v_g$. This is represented by point B in Fig. 2.32. As before, points A and B are connected by a straight line that will represent the location of the reflected voltage wave V_1^- along the transmission line, as a function of time. Subsequent reflected waves and their positions can also be determined in a similar fashion from the voltage reflection diagram and is shown in the figure. To find the voltage at a given position z at a particular time t, initially the point (z, t) is located in the diagram and a straight line connecting the points (z, t) and $(z, 0)$ is drawn. This line represents the desired location of the transmission line. The voltage can now be found by adding the voltages of the waves that cut this constant-z line. The addition is performed starting at the bottom of the diagram and progressing upwards along the time axis. Whenever a voltage wave crosses this constant-z line, its value is added to the total at that time. This is shown in Fig. 2.33.

Line current can also be found in a similar way by drawing a current reflection diagram. The current reflection diagram can easily be drawn using the voltage reflection coefficient diagram, by determining the value of the current that is associated with each voltage wave. Values of the current waves reflected by the battery and the load can be obtained using, respectively,

$$I_n^+ = V_n^+ / Z_0 \tag{2.313}$$

and

$$I_n^- = -V_n^- / Z_0 \tag{2.314}$$

The negative sign in Eq. (2.314) arises because the current direction of I_n^+ and I_n^- are opposite to each other. Figure 2.34 shows the current reflection diagram that has been obtained from the voltage diagram of Fig. 2.33. The current at a point z as a function of time is shown in Fig. 2.35.

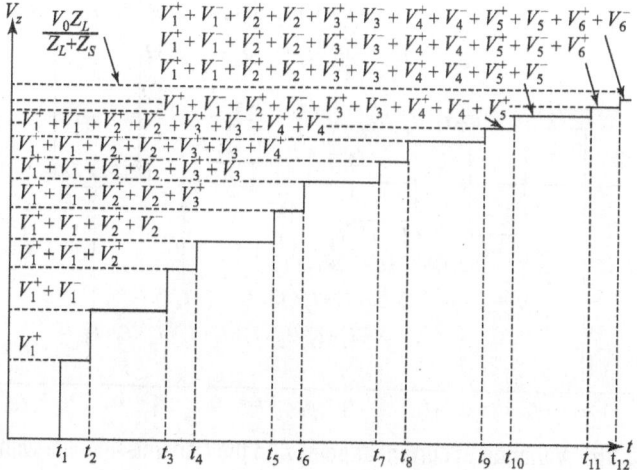

Fig. 2.33 Variation of voltage at point z on transmission line with time

---

EXAMPLE 2.8 Assume that a 50 Ω, 500 m long transmission line is terminated in a load impedance of 75 Ω and has been energized with a 30 V battery with zero source resistance at $t = 0$. If the velocity of propagation is 250 m/μs then plot the voltage and current as a function of time in the following cases:

(a) At the load

(b) At $z = 250$ m

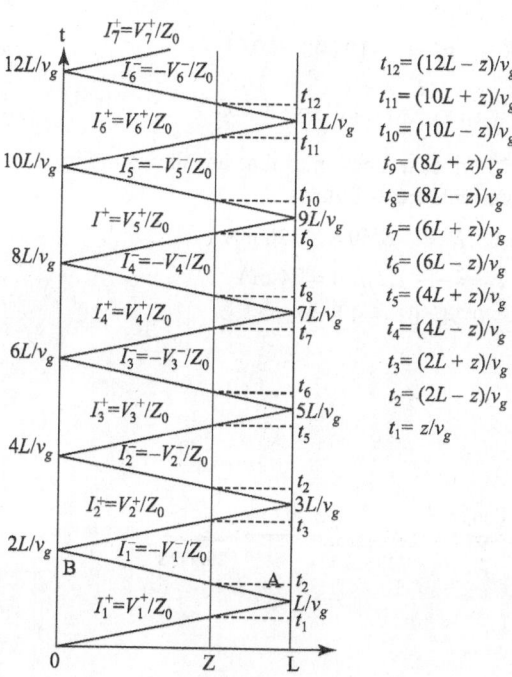

$$t_{12} = (12L - z)/v_g$$
$$t_{11} = (10L + z)/v_g$$
$$t_{10} = (10L - z)/v_g$$
$$t_9 = (8L + z)/v_g$$
$$t_8 = (8L - z)/v_g$$
$$t_7 = (6L + z)/v_g$$
$$t_6 = (6L - z)/v_g$$
$$t_5 = (4L + z)/v_g$$
$$t_4 = (4L - z)/v_g$$
$$t_3 = (2L + z)/v_g$$
$$t_2 = (2L - z)/v_g$$
$$t_1 = z/v_g$$

Fig. 2.34 Current reflection diagram for transmission line

Solution Since the length of the line is 500 m and the velocity of propagation is 250 m/μs, the time required by the wave to reach the load end from the source end is given by the following relation:

$$T = \frac{500}{250} = 2 \ \mu s$$

Therefore, the voltage waves from/after multiple reflections from the source and load will reach (a) the load at 2 μs, 6 μs, 10 μs, 14 μs, 18 μs, 22 μs, etc. and (b) at $z = 250$ m at 1 μs, 3 μs, 5 μs, 7 μs, 9 μs, 11 μs, 13 μs, 15 μs, 17 μs, 19 μs, 21 μs, 23 μs, etc. The reflection coefficients at the load (Γ_L) and the source (Γ_S) are as follows:

$$\Gamma_L = \frac{75 - 50}{75 + 50} = \frac{1}{5}$$

and $$\Gamma_s = \frac{0 - 50}{0 + 50} = -1$$

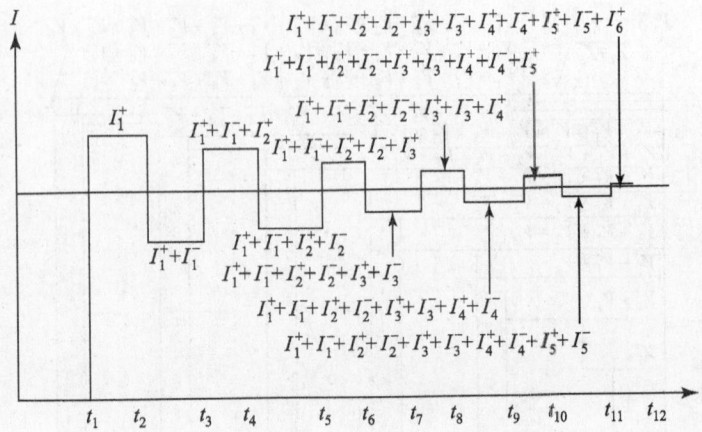

Fig. 2.35 Variation of current at point z on the transmission line with time

Therefore,

$$v_1^+ = \frac{V_0 Z_0}{R_s + Z_0} = \frac{30 \times 50}{0 + 50} = 30 \text{ V} \quad v_1^- = \frac{1}{5} \times 30 = 6 \text{ V}$$

$$v_2^+ = (-1) \times 6 = -6 \text{ V} \quad v_2^- = \frac{1}{5} \times (-6) = -1.2 \text{ V}$$

$$v_3^+ = (-1) \times (-1.2) = 1.2 \text{ V} \quad v_3^- = \frac{1}{5} \times 1.2 = 0.24 \text{ V}$$

$$v_4^+ = (-1) \times 0.24 = -0.24 \text{ V} \quad v_4^- = \frac{1}{5} \times (-0.24) = -0.048 \text{ V}$$

$$v_5^+ = (-1) \times (-0.048) = 0.048 \text{ V} \quad v_5^- = \frac{1}{5} \times (0.048) = 0.01 \text{ V}$$

$$v_6^+ = (-1) \times (0.01) = -0.01 \text{ V} \quad v_6^- = \frac{1}{5} \times (-0.01) = -0.002 \text{ V}$$

Plots of voltage at the load and at $z = 250$ m as a function of time are shown in Figs 2.36 and 2.37, respectively. The respective currents are as follows:

$$I_1^+ = 30/50 = 0.6 \text{ A} = 600 \text{ mA} \quad I_1^- = -6/50 = -120 \text{ mA}$$

$$I_2^+ = (-6)/50 = -120 \text{ mA} \quad I_2^- = -(-1.2)/50 = 24 \text{ mA}$$

$$I_3^+ = 1.2/50 = 24 \text{ mA} \quad I_3^- = -0.24/50 = -4.8 \text{ mA}$$

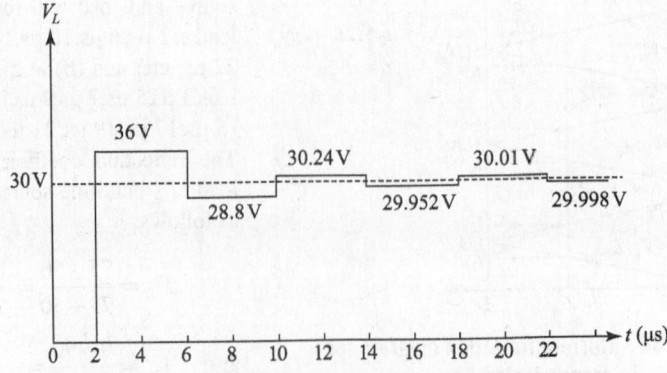

Fig. 2.36 Plot of voltage at the load as a function of time

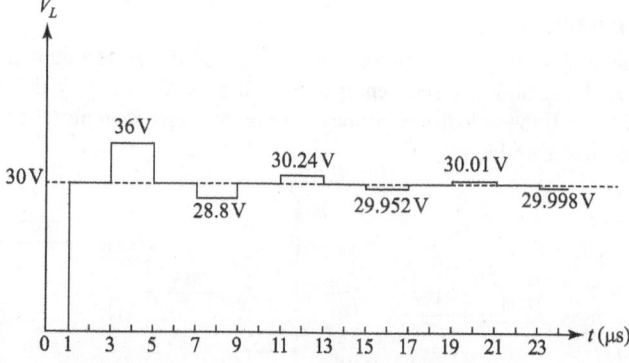

Fig. 2.37 Plot of voltage at $z = 250$ m as a function of time

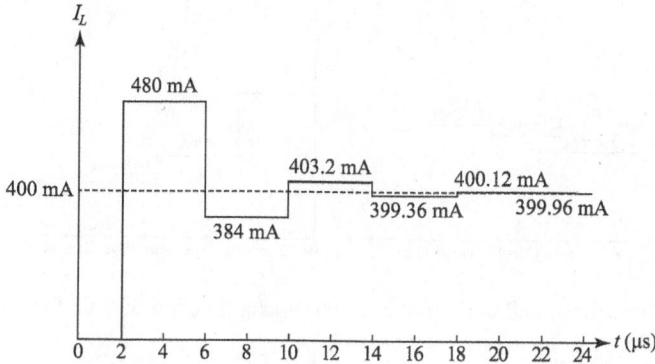

Fig. 2.38 Plot of current at the load as a function of time

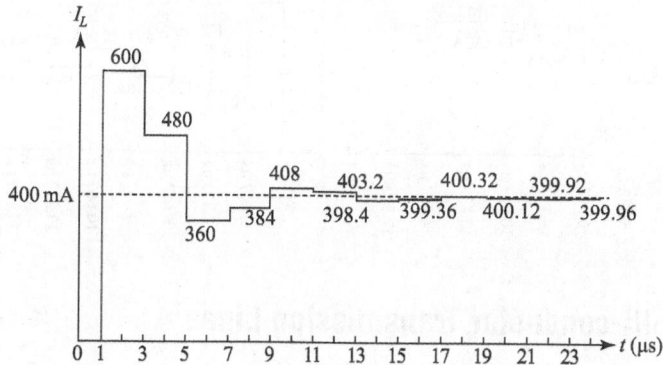

Fig. 2.39 Plot of current at $z = 250$ m as a function of time

$I_4^+ = (-0.24)/50 = -4.8$ mA $I_4^- = -(-0.048)/50 = 0.96$ mA

$I_5^+ = 0.048/50 = 0.96$ mA $I_5^- = -0.01/50 = -0.2$ mA

$I_6^+ = (-0.01)/50 = -0.2$ mA $I_6^- = -(-0.002)/50 = 0.04$ mA

Plots of current at the load and at $z = 250$ m as a function of time are shown in Figs 2.38 and 2.39, respectively.

Practice Problems

2.23 Assume that a 50 Ω, 300 m long transmission line is terminated in a load impedance of 100 Ω and has been energized with a 30 V battery with 75 Ω source resistance at $t = 0$. If the velocity of propagation is 150 m/µs, then plot the voltage and current at the source and load.

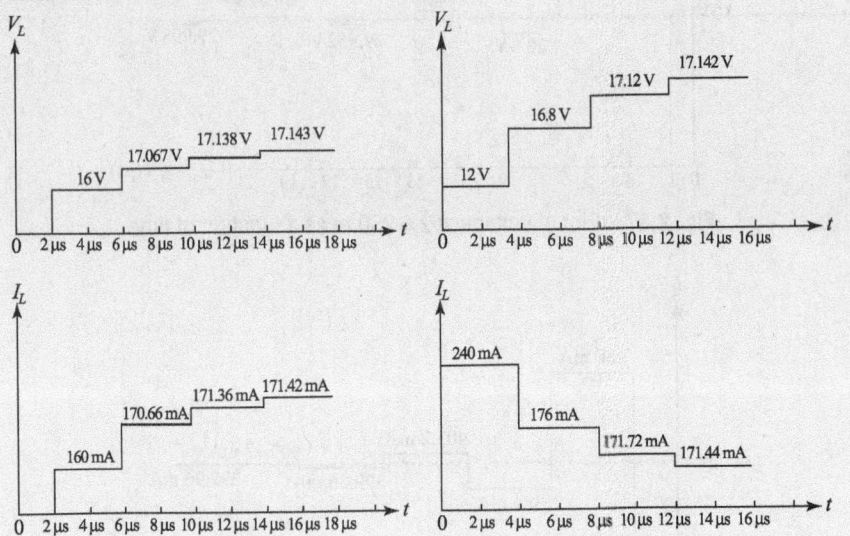

2.24 Plot the voltage and current at $z = 200$ m as a function of time for the Practice Problem 2.23.

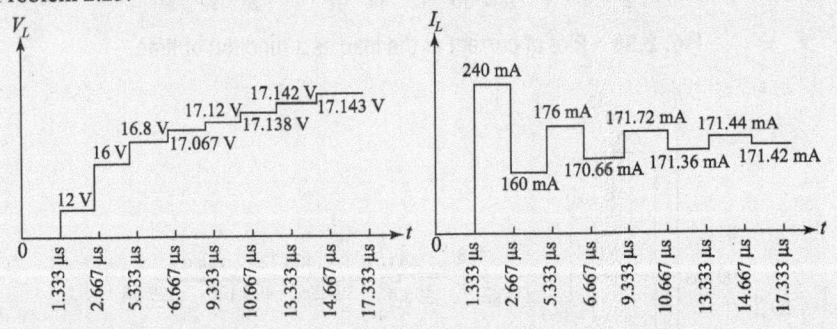

2.16 Multi-conductor Transmission Lines

Till now, we have discussed transmission lines having two conductors only. In practice, due to the existence of ground planes and shields, the problems get modified and result in multi-conductor transmission lines (MTLs). Here the term 'multi-conductor transmission lines' has been used to represent the transmission lines consisting of three or more conductors. For the time being, we will consider three-conductor transmission lines. MTLs having more than three conductors can be represented in a similar way. The only modification required is the change in the dimensions of the matrix representing the voltages, currents, and per-unit-length parameters.

The equivalent lumped element representation of a general three-conductor transmission line is shown in Fig. 2.40.

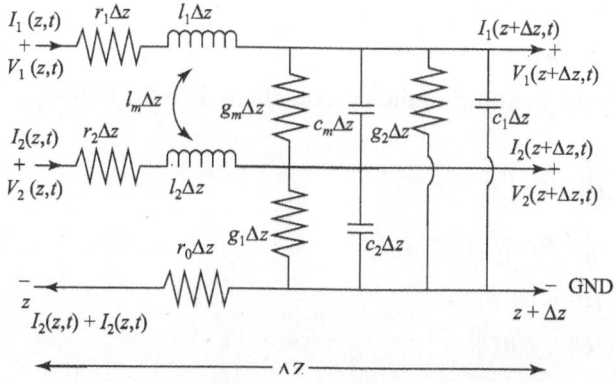

Fig. 2.40 Equivalent lumped element representation of a three-conductor transmission line

Applying the KVL and KCL, the following transmission line equations can be obtained:

$$\frac{\partial}{\partial z}[V(z,t)] = -[R][I(z,t)] - [L]\frac{\partial}{\partial t}[I(z,t)] \tag{2.315}$$

$$\frac{\partial}{\partial z}[I(z,t)] = -[G][V(z,t)] - [C]\frac{\partial}{\partial t}[V(z,t)] \tag{2.316}$$

Where $[V(z,t)] = \begin{bmatrix} V_1(z,t) \\ V_2(z,t) \end{bmatrix}$ $\hspace{2cm}$ (2.317)

$$[I(z,t)] = \begin{bmatrix} I_1(z,t) \\ I_2(z,t) \end{bmatrix} \tag{2.318}$$

$$[R] = \begin{bmatrix} (r_1 + r_0) & r_0 \\ r_0 & (r_2 + r_0) \end{bmatrix} \tag{2.319}$$

$$[L] = \begin{bmatrix} l_1 & l_m \\ l_m & l_2 \end{bmatrix} \tag{2.320}$$

$$[G] = \begin{bmatrix} (g_1 + g_m) & -g_m \\ -g_m & (g_2 + g_m) \end{bmatrix} \tag{2.321}$$

$$[C] = \begin{bmatrix} (c_1 + c_m) & -c_m \\ -c_m & (c_2 + c_m) \end{bmatrix} \tag{2.322}$$

The aforementioned MTL equations for the sinusoidal steady-state domain or frequency domain take the following forms:

$$\frac{d}{dz}[V(z)] = -[Z][I(z)] \tag{2.323}$$

$$\frac{d}{dz}[I(z)] = -[Y][V(z)] \tag{2.324}$$

Where $[V(z)] = \begin{bmatrix} V_1(z) \\ V_2(z) \end{bmatrix}$ $\hspace{2cm}$ (2.325)

$$[I(z)] = \begin{bmatrix} I_1(z) \\ I_2(z) \end{bmatrix} \qquad (2.326)$$

If we consider the time harmonic excitation, then the following relations are obtained:

$$[V(z,t)] = \text{Re}\left\{ [V(z) e^{j\omega t}] \right\} \qquad (2.327)$$

$$[I(z,t)] = \text{Re}\left\{ [I(z) e^{j\omega t}] \right\} \qquad (2.328)$$

$$|Z| = |R| + j\omega |L| \qquad (2.329)$$

$$|Y| = |G| + j\omega |C| \qquad (2.330)$$

The transmission line equations in the frequency domain are given by Eqs (2.323) and (2.324). These are coupled first-order equations. We can obtain uncoupled second-order equations by differentiating each equation with respect to z and substituting the other to yield the following expressions:

$$\frac{d^2}{dz^2} [V(z)] = [Z][Y][V(z)] \qquad (2.331)$$

$$\frac{d^2}{dz^2} [I(z)] = [Y][Z][I(z)] \qquad (2.332)$$

2.17 RF Coaxial Connectors and Adaptors

RF connectors are mostly used to connect a transmission line with a load or source and should have zero attenuation, dispersion, and reflection. RF connectors are mostly coaxial and hence also called coaxial RF connectors. The coaxial geometry enables the connectors to be used with the coaxial cables and to maintain the shielding that the coaxial cable offers. The mechanical design of coaxial connectors consists of a centre conductor shielded by another coaxial outer conductor, with a low loss dielectric separation and having fastening mechanisms such as thread, bayonet braces, and push–pull for a low Ohmic electric contact. Apart from the M-type (male) or F-type (female), coaxial connectors can also be of standard, miniature, micro-miniature, sub-miniature, precision, flange, and quick-lock types. Each class groups a number of connectors. Some of the commonly used RF connectors are as follows:

BNC connector   It was designed by Paul Neill of Bell Lab and Carl Concelman of Amphenol. It has two versions: 50 and 75 Ω, which found wide applications in radio, television, electronic test equipment, and computer networks. The 50 Ω version can work in the frequency range DC-3 GHz, whereas the 75 Ω version can work in the frequency range DC-2 GHz.

TNC connector   It is basically the threaded version of a BNC connector and was designed by Paul Neill of Bell Lab and Carl Concelman of Amphenol. Like a BNC connector, it also has two versions: 50 and 75 Ω, though the use of the 75 Ω is not very common. The 50 Ω version can work in the frequency range DC-11 GHz, whereas the 75 Ω version can work only in the frequency range DC-1 GHz.

N-connector It was initially designed by Paul Neill of Bell Lab. It also has two versions, namely, 50 and 75 Ω. The 50 Ω version finds wide use in mobile and cellular systems, while the 75 Ω version finds applications in cable television system. The common N-type connectors can operate in the DC-11 GHz bandwidth. Later, Julius Botka of Hewlett Packard has modified this connector to operate at a bandwidth of up to 18 GHz.

F-connector It was designed by Eric E. Winston of Jerold Electronics. It has an input impedance of 75 Ω and found wide applications in terrestrial, cable, and satellite television systems. The bandwidth of an F-connector is DC-1 GHz.

Miniature UHF connector It is the miniature version of the standard UHF connector. The standard UHF connector can operate only up to 300 MHz, whereas the miniature version can operate up to 2.5 GHz. Miniature UHF connectors found wide applications in mobile communication.

UFL connector It was developed by Hirose Electronic group and it finds applications in laptop and mini PCI cards. It also has an input impedance of 50 Ω and can operate in the DC-6 GHz bandwidth.

SMA connector It was initially designed by Bendix Scintilla Corporation. Later, it has been manufactured by Omni-Spectra Inc. as an OSM connector. An SMA connector uses polytetraflouroethylene or PTFE as the dielectric and can operate typically in the frequency range DC-18 GHz. However, some proprietary versions of SMA can also work at a frequency of up to 26.5 GHz. These connectors have an input impedance of 50 Ω and are widely used in microwave systems.

SMB connector As compared to SMA connectors that use screw-type coupling mechanism, an SMB connector uses snap on coupling and is available both in 50 Ω and 75 Ω versions. It can operate in the bandwidth range of DC-4 GHz. A 50 Ω smaller version of an SMB connector, known as an SSMB connector, is also available that can operate up to a frequency of 12.4 GHz.

SMC connector It was originally designed by Sealectro Corporation and is available both in 50 Ω and 75 Ω versions. They can operate in the frequency range DC-10 GHz.

MCX connector It usually has 50 Ω input impedance and can operate in the DC-6 GHz frequency range. These connectors have a snap on interface and find applications in USB DVB-T tuners for computers and laptops.

MMCX connector It is a miniature version of an MCX connector and has a lock–snap mechanism that allows its 360° rotation. It can work in the DC-6 GHz bandwidth range and find applications in GPS antenna, GPS receivers, and Wi-Fi PCMCIA cards. The input impedance of an MMCX connector is 50 Ω.

FME connector It is a miniature 50 Ω connector and finds applications in mobile equipment like vehicles. It can operate in the DC-2 GHz frequency range.

APC-7 connector This connector was originally developed by Hewlett Packard. Later, Amphenol improved the design for better performance. Unlike others, this connector has no male–female distinction and is the most repeatable connecting device.

It has 50 Ω input impedance and can work in the DC-18 GHz frequency bandwidth range. Though it can produce a VSWR of as low as 1.025 at 18 GHz, because of its high cost it is seldom used outside the laboratory where extreme accuracy is required.

APC-3.5 connector Like an APC-7 connector, an APC-3.5 connector was also originally developed by Hewlett Packard and later improved by Amphenol for better performance. However, it is not genderless and can mate with the opposite type of SMA connector. It has an input impedance of 50 Ω and can work in the DC-34 GHz bandwidth.

QMA and QN connectors These connectors were developed by Quick Lock Formula Alliance, consisting of Amphenol, Huber + Suhner, Radiall, and Rosenberger Hochfrequenztechnik, to replace the traditional threaded 50 Ω SMA and N-type connectors with a snap fastening and to allow a simultaneous 360° cable rotation after installation to make the cable routing easier. The QMA and QN connectors can operate in the DC-18 and DC-11 GHz frequency ranges, respectively, and are applied in defence and cellular base station applications. A miniature version of QMA, known as mini-QMA, is also available to operate in the DC-18 GHz bandwidth range. A mini-QMA is panel mount and finds wide applications in PCBs.

Apart from these, numerous other types of connectors, such as MMPX, SK, BMA, MMBX, 7/16, SMS, 10/2.3, BNT, QLA, MHV, SHV, BNO, SSMA, GPO, GMS, 2.9 mm, 1 mm, and 7 mm are also available in the market. Different types of coaxial RF connectors are shown in Fig. 2.41, whereas Fig. 2.42 shows the bandwidths of some of the coaxial connectors available in the market.

In RF and microwave applications, often two or more similar or different types of connectors required to be connected. Adaptors are used for this purpose. Depending on the connectors that are being connected, the adaptors are called

Fig. 2.41 Different types of coaxial RF connectors

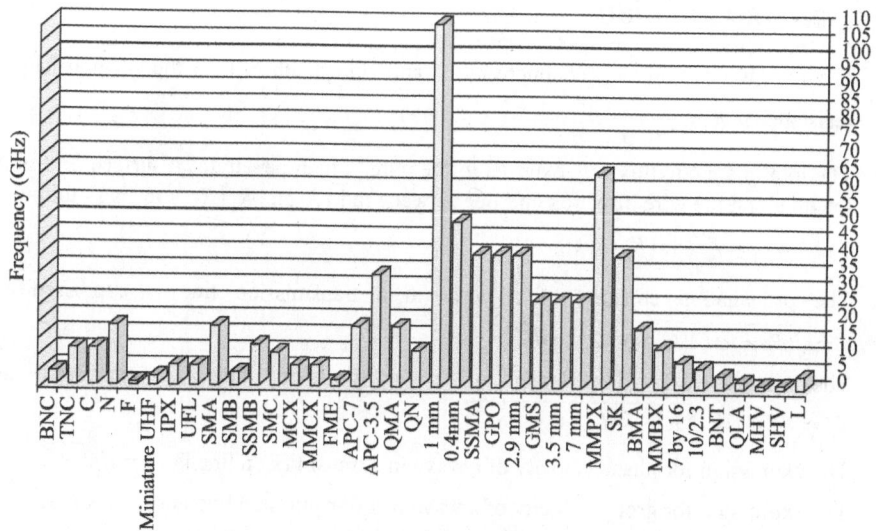

Fig. 2.42 Bandwidths of different coaxial connectors

intra- or inter-series adapters. Inter-series adaptors connect connectors of two different series, whereas intra-series adaptors connect those of the same series. Examples of inter-series adaptors are SMA to N-type adaptors, while SMA to SMA adaptors are the examples of intra-series adaptors. Figure 2.43 shows some intra- and inter-series adaptors.

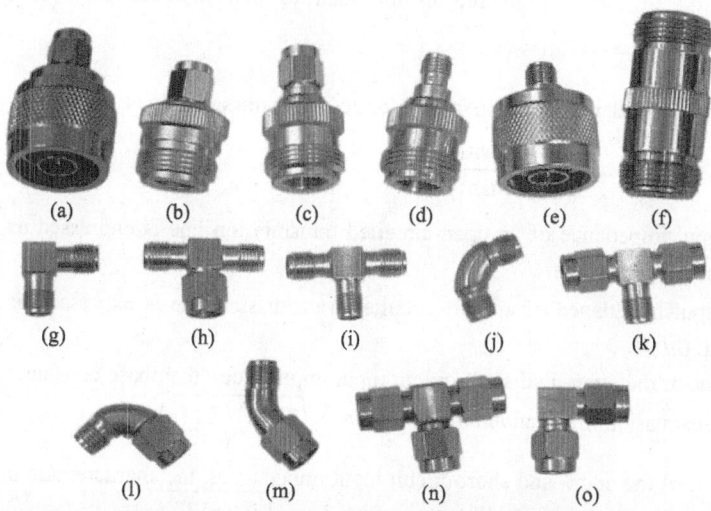

Fig. 2.43 Different intra- and inter-series adapters (a) SMA male to N male (b) and (c) SMA male to N female (d) SMA female to N female (e) SMA female to N male (f) N female to N female (g) Right-angle bend: SMA female to SMA female (h) SMA T: female–male–female (i) SMA T: female–female–female (j) Smooth 90° SMA bend: SMA female to SMA female (k) SMA T: male–female–male (l) Smooth 90° SMA bend: SMA male to SMA female (m) Smooth 45° SMA bend: SMA male to SMA female (n) SMA T: male–male–male (o) Right-angle bend: SMA male to SMA male

Important Formulae

- If a wire has a radius r_w, conductivity σ, and skin depth δ, then the per-unit-length resistance is $R = \dfrac{1}{2\pi r_w \sigma \delta}$

- The total magnetic flux ψ_{ext} external to the wire, which penetrates a surface S of unit length along the wire direction and lies between radii R_1 and R_2 $(R_2 \geq R_1)$ is as follows:

$$\psi_{\text{ext}} = \frac{\mu I}{2\pi r} \ln(R_2/R_1).$$

- The propagation constant of a wave in a transmission line is expressed as
$$\gamma = \sqrt{(R + j\omega L)(G + j\omega C)} = \sqrt{ZY}$$

- The characteristic impedance of a transmission line is expressed as
$$Z_0 = \sqrt{Z/Y} = \sqrt{(R + j\omega L)/(G + j\omega C)}$$

- The expression for phase velocity of a wave in a transmission line is $v_p = 1/\sqrt{LC}$.

- The expression for group velocity of a wave in a transmission line is $v_g = 1/\sqrt{LC}$.

- The reflection coefficient at the load is given by $\Gamma_L = (Z_L - Z_0)/(Z_L + Z_0)$.

- The transmission coefficient at the load is given by $T_L = 2Z_L/(Z_L + Z_0)$.

- The transmission and reflection coefficients in a lossless transmission line are related by the formula $Z_L\left(1 - \Gamma_L^2\right) = Z_0 T_L^2$.

- The VSWR in a transmission line is expressed as $\rho = \left(1 + |\Gamma_L|\right)/\left(1 - |\Gamma_L|\right)$.

- The maximum normalized impedance is given by $|z_{\max}| = \rho$.

- The minimum normalized impedance is given by $|z_{\min}| = 1/\rho$.

- The ratio of power delivered in the load to that incident on it is given by
$$\frac{P_L}{P_i} = 1 - |\Gamma_L|^2 = 4\rho/(\rho + 1)^2.$$

- The input impedance at a point z on a lossless transmission line is as follows:

$$Z_{\text{in}}(z) = Z_0 \frac{Z_L + jZ_0 \tan(\beta d)}{Z_0 + jZ_L \tan(\beta d)}.$$

- The input impedance of an open-circuited transmission line is expressed as $Z_{\text{in}}^{\text{oc}}(z) = -jZ_0 \cot(\beta d)$.

- The input impedance of a short-circuited transmission line is expressed as $Z_{\text{in}}^{\text{sc}}(z) = jZ_0 \tan(\beta d)$.

- In terms of the open- and short-circuit input impedances, the phase constant of a wave in a transmission line is given by $\beta = \dfrac{1}{d} \tan^{-1} \sqrt{Z_{\text{in}}^{\text{sc}}(z)/Z_{\text{in}}^{\text{oc}}(z)}$.

- In terms of the open- and short-circuit input impedances, the characteristic impedance of a wave in a transmission line is given by $Z_0 = \sqrt{Z_{\text{in}}^{\text{sc}}(z) Z_{\text{in}}^{\text{oc}}(z)}$.

- Condition for removing both amplitude and phase distortion is $RC = LG$.

- The minimum attenuation that can be achieved by varying L and keeping other parameters constant is $\alpha_{\min, L} = \sqrt{RG}$, whereas the minimum attenuation that can be achieved by varying C and keeping other parameters constant is $\alpha_{\min, C} = \sqrt{RG}$.

Exercises

Objective-type Questions

2.1 At a high frequency, the per-unit-length external inductance of a coaxial cable is approximately expressed as

(a) $\dfrac{\mu}{2\pi}\log\left[\dfrac{r_s}{r_w}\right]$

(c) $\dfrac{\mu}{2\pi}\ln\left[\dfrac{r_s}{r_w}\right]$

(b) $\dfrac{\mu}{4\pi}\log\left[\dfrac{r_s}{r_w}\right]$

(d) $\dfrac{\mu}{4\pi}\ln\left[\dfrac{r_s}{r_w}\right]$

2.2 The characteristic impedance of a lossless transmission line is

(a) $\sqrt{\dfrac{R+j\omega L}{G+j\omega C}}$ (b) $\sqrt{\dfrac{L}{C}}$ (c) $\sqrt{\dfrac{R+j\omega L}{G+j\omega C}}$ (d) $\sqrt{\dfrac{C}{L}}$

2.3 A transmission line has the primary constants R, L, G, and C and secondary constants Z_0 and γ. If the line is lossless then
(a) $R = 0$, $G \neq 0$ and $\alpha = 0$
(b) $R = 0$, $G = \alpha$ and $\beta = |\gamma|$
(c) $G = 0$ and $\alpha = \beta$
(d) $R = G = \alpha = 0$ and $\beta = |\gamma|$

2.4 For a lossless two-wire transmission line, the propagation constant is given by
(a) $\sqrt{ZY}$ (b) $j\omega\sqrt{ZY}$ (c) $\omega\sqrt{ZY}$ (d) $-j\omega\sqrt{ZY}$

2.5 The characteristic impedance of a lossless two-wire line is given by

(a) $\dfrac{59.96}{\sqrt{\varepsilon_r}}\log\left[\dfrac{s^2}{r_{w1}r_{w2}}\right]$

(c) $\dfrac{59.96}{\sqrt{\varepsilon_r}}\ln\left[\dfrac{s^2}{r_{w1}r_{w2}}\right]$

(b) $\dfrac{59.96}{\varepsilon_r}\log\left[\dfrac{s^2}{r_{w1}r_{w2}}\right]$

(d) $\dfrac{59.96}{\varepsilon_r}\ln\left[\dfrac{s^2}{r_{w1}r_{w2}}\right]$

2.6 The phase constant of a lossless coaxial cable is given by
(a) $\omega\sqrt{\mu\varepsilon}$ (b) $j\omega\sqrt{\mu\varepsilon}$ (c) $-j\omega\sqrt{\mu\varepsilon}$ (d) $-\omega\sqrt{\mu\varepsilon}$

2.7 In terms of the characteristic impedance and reflection coefficient, the load impedance can be expressed as

(a) $Z_L = Z_0\left[\dfrac{1-\Gamma_L}{1+\Gamma_L}\right]$

(c) $Z_L = Z_0\left[\dfrac{1-|\Gamma_L|}{1+|\Gamma_L|}\right]$

(b) $Z_L = Z_0\left[\dfrac{1+\Gamma_L}{1-\Gamma_L}\right]$

(d) $Z_L = Z_0\left[\dfrac{1+|\Gamma_L|}{1-|\Gamma_L|}\right]$

2.8 In a transmission line, the reflection coefficient at the load end is given by $0.25\angle-27°$. What is the reflection coefficient at a distance 0.15λ towards the source?
(a) $0.25\angle81°$ (b) $0.25\angle135°$ (c) $0.25\angle-81°$ (d) $0.25\angle225°$

2.9 The relation between reflection and the transmission coefficient is
(a) $T_L = 1 + \Gamma_L$ (b) $T_L = 1 - \Gamma_L$ (c) $|T_L| = 1 + |\Gamma_L|$ (d) $|T_L| = 1 - |\Gamma_L|$

2.10 Percentage of incident power reflected from 100 Ω load of a 75 Ω transmission line is
(a) 2% (b) 1.43% (c) 1% (d) 0.75%

2.11   The distance of the voltage nodal point from the load for open-circuit termination is

(a) $\dfrac{n\lambda}{2}$

(c) ∞

(b) $\dfrac{(2n+1)\lambda}{4}$

(d) At the source end

2.12   The relation between the SWR (ρ) and (Γ) is

(a) $\rho = \dfrac{1-|\Gamma_L|}{1+|\Gamma_L|}$

(c) $\rho = \dfrac{1-\Gamma_L}{1+\Gamma_L}$

(b) $\rho = \dfrac{1+|\Gamma_L|}{1-|\Gamma_L|}$

(d) $\rho = \dfrac{1+\Gamma_L}{1-\Gamma_L}$

2.13   The VSWR can have any value between

(a) 0 and 1   (b) -1 and 1   (c) 1 and ∞ (d) 0 and ∞

2.14   In a lossless transmission line with matched generator and load, the SWR and reflection coefficient are

(a) Unequal in magnitude

(c) Both equal to ∞

(b) Both equal to 1

(d) Both equal to 0

2.15   A 75 Ω transmission line is terminated in a load impedance of 100 Ω. The minimum impedance measured on the line is

(a) 75 Ω (b) 100 Ω (c) 56.23 Ω (d) 0 Ω

2.16   In a two-wire transmission line in air, the adjacent voltage minimum are at 2.7 and 17.7 cm. The operating frequency is

(a) 2 GHz   (b) 1 GHz   (c) 500 MHz   (d) 4 GHz

2.17   A 50 Ω transmission line is to be terminated in two resistive loads R_1 and R_2 such that the standing wave patterns in the two cases have the same SWR ρ. To obtain the desired result, the values of R_1 and R_2 in Ohms should be

(a) 50ρ and $50\sqrt{\rho}$

(c) $50/\rho$ and $50\sqrt{\rho}$

(b) $50/\sqrt{\rho}$ and 50ρ

(d) $50/\rho$ and 50ρ

2.18   For a short-circuited lossless transmission line, the input impedance is given by

(a) $-jZ_0 \cot(\beta d)$

(c) $Z_0 \coth(\beta d)$

(b) $jZ_0 \tan(\beta d)$

(d) $Z_0 \tanh(\beta d)$

2.19   A short-circuit transmission line behaves as an inductive load provided its length d is

(a) $2n\lambda/4 < d < (2n+1)\lambda/4$

(c) An even multiple of $\lambda/4$

(b) $(2n+1)\lambda/4 < d < (n+1)\lambda/2$

(d) An odd multiple of $\lambda/4$

2.20   The equivalent inductance of an open-circuited transmission line is

(a) $\dfrac{Z_0}{\omega} \cot(\beta d)$

(c) $\dfrac{Z_0}{\omega} \tan(\beta d)$

(b) $-\dfrac{Z_0}{\omega} \cot(\beta d)$

(d) $-\dfrac{Z_0}{\omega} \tan(\beta d)$

2.21   If $Z_{in}^{sc}(z)$ and $Z_{in}^{oc}(z)$ be the short- and open-circuit input impedances, respectively, of a transmission line of length d, then the phase constant can be expressed as

(a) $\dfrac{1}{d}\tan^{-1}\sqrt{\dfrac{|Z_{in}^{sc}(z)|}{|Z_{in}^{oc}(z)|}}$

(c) $\dfrac{1}{d}\cot^{-1}\sqrt{\dfrac{Z_{in}^{sc}(z)}{Z_{in}^{oc}(z)}}$

(b) $\sqrt{Z_{in}^{sc}(z)Z_{in}^{oc}(z)}$

(d) $\sqrt{Z_{in}^{sc}(z)/Z_{in}^{oc}(z)}$

2.22 A transmission line with the characteristic impedance Z_0 is terminated in load impedance Z_0. The input impedance of the line will be

(a) Z_0 (b) $2Z_0$ (c) $Z_0/2$ (d) $\sqrt{2}Z_0$

2.23 The variation of attenuation constant with frequency is known as

(a) Frequency distortion   (c) Modulation

(b) Phase distortion   (d) Fading

2.24 The condition for a line with no phase distortion is

(a) $RL = CG$ (b) $RG = LC$ (c) $RC = LG$ (d) $RLCG = 1$

2.25 If R, G, and C are constant, then the attenuation will be minimum if the value of L is

(a) RG/C (b) $\sqrt{CR/G}$ (c) CG/R (d) CR/G

2.26 The minimum attenuation that can be achieved by varying L and/or C, if R and G are not zero, is

(a) $\sqrt{LG}$ (b) $\sqrt{RG}$ (c) $\sqrt{RL}$ (d) $\sqrt{LC}$

2.27 A complete travel along the periphery of a Smith chart results an equivalent travelling of

(a) λ along the line   (c) $\lambda/4$ along the line

(b) $\lambda/2$ along the line   (d) 2λ along the line

2.28 The possible positions of the voltage maximum in a Smith chart are

(a) 0.25λ point   (c) 0.2λ point

(b) 0.5λ point   (d) At the centre of the Smith chart

2.29 In a Smith chart, the normalized impedance and its corresponding normalized admittance are

(a) $0°$ out of phase   (c) $180°$ out of phase

(b) $90°$ out of phase   (d) $270°$ out of phase

2.30 If the length of a transmission line is 400 m and the velocity of wave propagation is 2×10^8 m/s, then the equivalent length of the line is

(a) 2 µs   (b) 2 ns   (c) 20 µs   (d) 20 ns

Review Questions

2.1 How do voltage and current waves propagate through a transmission line?

2.2 With suitable assumptions, derive the expressions for per-unit-length inductance, capacitance, and conductance of a two-wire transmission line.

2.3 Using the uniform distributed circuit theory, derive the coupled and uncoupled transmission line equations for a two-conductor transmission line.

2.4 Solve the uncoupled transmission line equations and find the general expressions for the following parameters in a two-conductor transmission line:

(a) Characteristic impedance

(b) Propagation constant

(c) Phase velocity

(d) Group velocity of a wave

2.5 Starting from the generalized expressions of the characteristic impedance and propagation constant, find their expressions for a coaxial cable.

2.6 Derive the expressions for the (a) reflection coefficient and (b) transmission coefficient at the load in terms of the load impedance and characteristic impedance. Hence, show that $T_L = 1 + \Gamma_L$. Also prove that $Z_L\left(1 - \Gamma_L^2\right) = Z_0 T_L^2$.

2.7 Derive the expressions for the voltage and current standing waves in a two-conductor transmission line terminated in an open-circuit load. Find the positions of voltage and current nodal and antinodal points on it. Plot the standing wave pattern.

2.8 Derive the relations between the SWR and (a) reflection coefficient at the load and (b) load impedance.

2.9 Prove that the ratio of the power delivered at the load to that incident on it can be expressed as $4\rho/(\rho+1)^2$.

2.10 Derive the generalized expression for the input impedance of a transmission line at any point z on it, in terms of the load impedance and characteristic impedance.

2.11 Starting from the generalized expression for the input impedance of a two-conductor transmission line, show that a short-circuited transmission line can be used as an inductor/capacitor/parallel resonant circuit/series resonant circuit depending on its length. Hence, find the expressions for the equivalent inductance and capacitance.

2.12 How can you calculate the (a) characteristic impedance and (b) propagation constant of a transmission line if the input impedances of that line for open- and short-circuit terminations are known?

2.13 Show that a one-eighth-wavelength transmission line can be used to transform any load impedance to the characteristic impedance.

2.14 Starting from the generalized expression of the propagation constant, derive the exact expressions for (a) attenuation constant and (b) phase constant.

2.15 Derive the condition(s) that must be satisfied to remove the amplitude/frequency distortion and phase distortion in a transmission line.

2.16 Show that the attenuation in a transmission line cannot be zero until and unless the per-unit-length resistance and conductance of the line is zero.

2.17 What is the minimum value of attainable attenuation in a two-conductor transmission line if both the per-unit-length resistance and conductance of the line are non-zero? Find an expression for the (a) per unit length inductance and (b) per unit length conductance of the line to achieve the minimum attainable attenuation, provided that other per-unit-length parameters remain constant.

2.18 Using the field theoretic approach, find the uncoupled transmission line equations for a coaxial cable.

2.19 Derive the equations for the constant resistance and constant reactance circles of a Smith chart.

2.20 How can you find the voltage and current at any point z on a transmission line at any instant t by performing transient analysis of the line?

Problems

2.1 A line in air with a single conductor of 1 cm radius over a ground plane has an impedance of 50 Ω. Find the height. *[Hint: Use Eq. (2.66)]*

2.2 For a distortionless line with $\gamma = 0.035 + j27.75$ /m, $Z_0 = 75\ \Omega$, and $f = 1$ GHz, determine the primary constants R, L, G, and C. *[Hint: Use Eqs (2.64), (2.73), (2.208), and $\alpha = \sqrt{RG}$]*

2.3 A lossless transmission line has a characteristic impedance of 50 Ω and phase constant of 31.21 rad/m at 3 GHz. Find the per-unit-length inductance and capacitance of the line. *[Hint: Use Eqs (2.64) and (2.73)]*

2.4 A coaxial line has an open circuit impedance of $150\angle 24.8°$ and a short-circuit impedance of $16.667\angle -22.6°$. Calculate the characteristic impedance of the line. *[Hint:
Use Eq. (2.193)]*

2.5 A telephone cable has the following electrical parameters: $R = 37.5\ \Omega/\text{mile}$,
$L = 1.2\ \text{mH/mile}$, $G = 0.0012\ \mho/\text{mile}$, and $C = 0.075\ \mu\text{F/mile}$. Loading coils are added, which provide an additional inductance of 23.2 mH/mile as well as an additional
resistance of 8.4 Ω/mile. Obtain the (a) characteristic impedance, (b) attenuation constant, and (c) phase velocity at 918 MHz. *[Hint: Use Eqs (2.54), (2.63), and (2.80)]*

2.6 A lossless transmission line has $C = 73.7\ \text{pF/m}$, $Z_0 = 125\ \Omega$. If the operating frequency is 2.7 GHz, calculate the (a) phase constant, (b) group velocity, and (c) phase
velocity. *[Hint: Use Eqs (2.64), (2.73), (2.80), and (2.81)]*

2.7 A 50 Ω distortionless line has a capacitance of 0.10 nF/m and attenuation of
0.00278 dB/m. Calculate the per-unit-length line parameters. *[Hint: Use Eqs (2.64),
(2.72), and (2.208)]*

2.8 A signal generator has an internal resistance of 45 Ω and an open-circuit voltage
$v(t) = 7\cos(3\pi \times 10^9 t)$. It is connected to a 50 Ω lossless line that is 6 m long and
terminated by a matched load at the other end. If the signal propagates on this line at
a velocity of 2.5×10^8 m/s, find the instantaneous voltage and current at a point z on
the line. *[Hint: Use the voltage and current division rule and Eq. (2.80)]*

2.9 A 50 Ω coaxial line is connected to a 100 Ω load through a 70.71 Ω coaxial section
of length 0.75λ at 2 GHz. What is the VSWR in the 50 Ω line if it is fed by a 1 GHz
source? *[Hint: Use Eqs (2.175), (2.177), and (2.184)]*

2.10 A 75 Ω lossless line is terminated by a series LCR circuit, as shown in Fig. 2.44. Find
the source frequency f_0 for which the line will be matched with the load. *[Hint: Load
impedance should be matched with the characteristic impedance]*

2.11 Calculate the input impedance for the transmission line network shown in Fig. 2.45.
[Hint: Use Eqs (2.175), (2.198), and (2.200)]

2.12 A lossless line has a characteristic impedance of 50 Ω and is terminated in a
load impedance of 75 Ω. The line is energized by a generator that has an output impedance of 50 Ω and an open-circuit output voltage of 25 V. If the line
is 1.25-wavelength long, then determine the (a) input impedance, (b) instantaneous load voltage, and (c) instantaneous power delivered to the load. *[Hint: Use
Eqs (2.89), (2.106), and (2.175)]*

2.13 A line of characteristic impedance 75 Ω is terminated in a load impedance Z_L. The
VSWR measured on the line is 2.3 and the first voltage minimum occurs at a distance 6 cm from the load. If the frequency of the signal is 980 MHz, find the load
impedance. *[Hint: Use a Smith chart]*

2.14 Using an impedance bridge, the input impedance of a transmission line, terminated
by a load impedance Z_L, is measured to be $Z_{in} = 18.669 - j43.302\ \Omega$. The experiment is repeated twice, with the load being replaced first by an open circuit and then

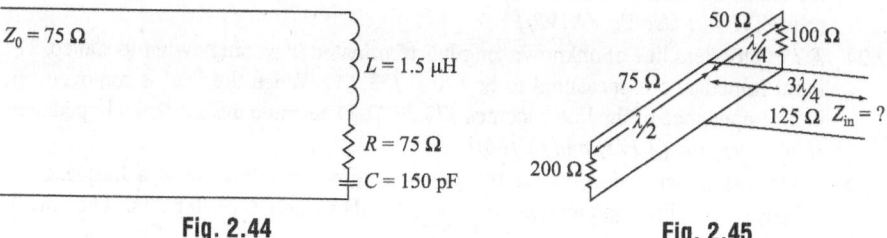

Fig. 2.44   Fig. 2.45

by a short circuit. If these impedances are measured to be $Z_{in}^{oc}(z) = -j125.79\ \Omega$ and $Z_{in}^{sc}(z) = j11.24\ \Omega$, respectively, then find the load impedance. *[Hint: Use Eqs (2.175), (2.186), and (2.193)]*

2.15 Find the length of a short-circuited $50\ \Omega$ lossless transmission line so that its input impedance at 1.37 GHz is equivalent to the reactance of an inductor $L_{eq}^{sc} = 2.3$ nH. The wave velocity on the line is $0.65c$ where c is the velocity of light in free space. *[Hint: Use Eq. (2.190)]*

2.16 A $50\ \Omega$ line is terminated by a $98.8 + j52.3\ \Omega$ load. How far from the load will the line impedance be $130.41 + j16.97\ \Omega$? *[Hint: Use Eq. (2.175)]*

2.17 A $50\ \Omega$ lossless line is 4 m long. At the operating frequency of 938 MHz, the input impedance at the middle of the line is $46.7 + j17.23\ \Omega$. Find the input impedance at the generator side. Assume that the velocity of propagation is 75% of that in free space. *[Hint: Use Eq. (2.175)]*

2.18 A 21.925 m long lossless transmission line having the per-unit-length parameters $L = 250$ nH/m and $C = 25$ pF/m is terminated in a load impedance $Z_L = 89 + j1.2\ \Omega$. Find the series impedance of the generator so that a match can be obtained at 1.37 GHz. *[Hint: Use Eqs (2.64), (2.73), and (2.175)]*

2.19 A 0.67λ, $50\ \Omega$ transmission line is connected to a matched $75\ \Omega$ line. Determine the reactance that, when connected across the sending end of the line, will make the input impedance a pure resistance. *[Hint: Use Eq. (2.175)]*

2.20 Two quarter-wave transmission line with the characteristic impedances Z_{01} and Z_{02} are in series and have been used to match a $50\ \Omega$ line with a $100\ \Omega$ load. The quarter-wave line with the characteristic impedance Z_{02} is on the load side, whereas that with the characteristic impedance Z_{01} is on the $50\ \Omega$ line side. Now if $Z_{02} = 40\ \Omega$, calculate Z_{01}. *[Hint: Use Eq. (2.198)]*

2.21 Two antennas with input impedances $72.5 + j1.2\ \Omega$ and $74.3 - j2.1\ \Omega$ are fed with three $50\ \Omega$ coaxial cables, as shown in Fig. 2.46. Calculate the impedance of the generator so that a match can be obtained at the source end. *[Hint: Use Eq. (2.175)]*

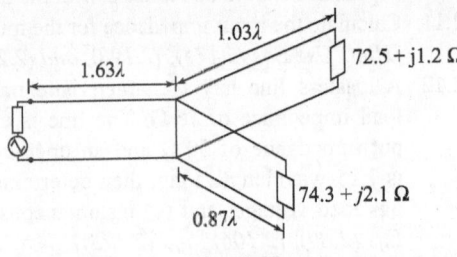

Fig. 2.46

2.22 A voltage source drives the series combination of impedance $Z_g = 75 + j60\ \Omega$ and a lossless transmission line of length L open circuited at the load end. The characteristic impedance of the line is $75\ \Omega$. Determine the shortest length of the line that will result in the voltage source driving a total impedance of $75\ \Omega$. *[Hint: Use Eq. (2.186)]*

2.23 A 1 GHz voltage source drives the series combination of an impedance $Z_g = 35 + j27\ \Omega$ and a lossless transmission line of length $\lambda/4$, terminated by a load impedance Z_L. If the characteristic impedance of the line is $50\ \Omega$, then determine the load impedance value required to achieve a net impedance of $50\ \Omega$ seen by the source. *[Hint: Use Eq. (2.198)]*

2.24 A $75\ \Omega$ lossless line of unknown length is terminated in an unknown impedance. The input impedance is measured to be $170 + j33.7\ \Omega$. When the load is removed, the input impedance of the line becomes $j73.57\ \Omega$. Determine the unknown impedance. *[Hint: Use Eqs (2.175) and (2.186)]*

2.25 A lossless transmission line is 100 cm in length and operating at a frequency of 2.3 GHz. The line parameters are $L = 0.15\ \mu$H/m and $C = 90$ pF/m. The line is shorted at $z = 100$ cm and there is a load $Z_L = 60.23 + j17.23$ Ω at $z = 67$ cm. Calculate the input impedance. *[Hint: Use Eqs (2.64), (2.73), (2.175), and (2.189)]*

2.26   A 25 m long transmission line is producing a 1.5 dB drop in power from end to end. What fraction of the input power reaches the output? Determine the attenuation constant of the line. *[Hint: Use Eq. (2.227)]*

2.27   Three lossy lines are joined end to end. The first line is 15.7 m long and has a loss rating of 0.21 dB/m, the second line is 25.1 m long and has a loss rating of 0.13 dB/m, and the third line is 37.3 m long and has a loss rating of 0.17 dB/m. The reflection coefficient at the junction of the first and second lines is 0.12, while that at the junction of the second and third lines is 0.23. If the input power to line 1 is 137.5 mW, then determine the power delivered to a matched load at the output end of line 3. *[Hint: Add individual losses]*

2.28   A 250 m long 100 Ω transmission line is terminated in a load impedance of $137\angle 21.4°$. It operates at 950 MHz. Find the reflection coefficient, positions of voltage maximum and minimum on the line, and input impedance. *[Hint: Use a Smith chart]*

2.29   A lossless transmission line has a characteristic impedance of 75 Ω and is terminated in a load of 100 Ω. The magnitude of the incident voltage to the line is 13 V (r.m.s.). Determine the (a) SWR on the line, (b) maximum and minimum voltages on the line, (c) maximum and minimum currents on the line, and (d) power transmitted to the load. *[Hint: Use Eqs (2.90) and (2.142)–(2.146)]*

2.30   A 75 Ω lossless line connects a signal source of 938 MHz to a 100 Ω load. The load power is 87.35 mW. Calculate the (a) position of the first V_{min} and V_{max}, (b) values of V_{min} and V_{max}, and (c) input impedance of the line at the position of the V_{min} and V_{max}. *[Hint: Use Eqs (2.90), (2.142), and (2.144)]*

2.31   A lossless transmission line has a characteristic impedance of 50 Ω and is terminated with an unknown load. The SWR on the line is 2.7. Successive voltage minima are 1.5 cm apart and the first minimum is 0.37 cm from the load. Now determine the following factors
   (a)   Admittance of the terminating load
   (b)   Shortest length of the line from the load for which the admittance is purely conductive

2.32   A lossless 50 Ω transmission line is connected to a load that produces the reflection coefficient $0.27\angle - 45°$ at a distance 1.37λ from the load. Find the (a) load resistance and (b) SWR in the line.

2.33   An unknown load attached to a 50 Ω line produces $V_{max} = 7.3$ V and $V_{min} = 1.3$ V and adjacent minima are 12.2 and 19.7 cm from the load. Calculate the frequency of operation, SWR, and load impedance.

2.34   At an operating frequency of 918 MHz, an equivalent load reactance of $j27.9$ Ω is to be constructed using a section of a short-circuited, lossless 75 Ω line. If the phase velocity of the wave in the line is 80% of that in free space, what is the shortest possible length that would exhibit the desired reactance at its inputs? Recalculate the same if the line is terminated in an open circuit. *[Hint: Use Eqs (2.186) and (2.189)]*

2.35   A lossless 100 Ω transmission line of 0.37λ in length is terminated in an unknown impedance (Z_L). If the input impedance is $Z_{in} = j9.4$ Ω use a Smith chart to find Z_L. What lengths of open- and short-circuit lines could be used to replace Z_L?

2.36   For a lossless 50 Ω transmission line terminated in an unknown load produces an SWR of 3. If the first voltage maximum is 0.17λ from the load, then determine the load impedance.

2.37 A 50 Ω lossless line is 0.23λ long. If the SWR is 1.5 and angle of reflection coefficient at the load is $\angle -30°$, use a Smith chart to find $Z_L, |\Gamma|$ and Z_{in}.

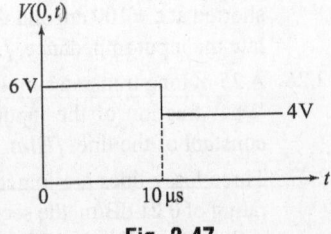

Fig. 2.47

2.38 In response to a step voltage, the voltage waveform shown in Fig. 2.47 was observed at the sending end of a lossless transmission line with $R_g = Z_0 = 60\ \Omega$ and $\varepsilon_r = 2.4$. Determine the (a) length of the line, (b) load impedance, and (c) generator voltage. *[Hint: Use Eqs (2.90) and (2.305)]*

2.39 In response to a step voltage, the voltage waveform shown Fig. 2.48 was observed at the sending end of a shorted line of with characteristic impedance 50 Ω and $\varepsilon_r = 2.4$. Determine the (a) length of the line, (b) source voltage, and (c) source resistance. *[Hint: Use Eqs (2.100) and (2.305)]*

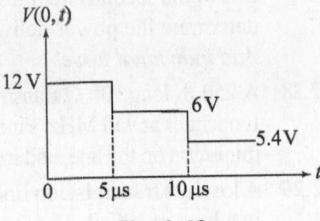

Fig. 2.48

2.40 Suppose the voltage waveform, shown in Fig. 2.49, was observed at the sending end of a 50 Ω underwater transmission line in response to a step voltage introduced by a generator with an open loop voltage 30 V and an unknown series impedance R_g. The line is 3000 km long and terminated with $Z_L = 60\ \Omega$. Determine R_g. If the velocity of propagation in the cable is 2×10^8 m/s then explain why the drop in the level of $V(0,t)$ at $t = 5$ ms cannot be due to the reflection in the load. Determine the shunt resistance and

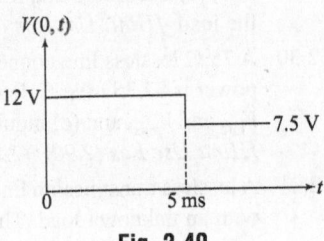

Fig. 2.49

location of the fault responsible for the observed waveform. *[Hint: Use Eqs (2.90) and (2.305)]*

3

Transmission Line Matching Networks

3.1 Introduction

It has been shown in Chapter 2 that when a transmission line of characteristic impedance Z_0 is terminated in load impedance Z_L, then reflection occurs at the load. This reflected wave travels towards the source and couples with it, which may lead to system malfunction. In addition, reflection minimizes the power transferred to the load, thereby causing power loss. A voltage reflection coefficient of 0.4 corresponds to a power reflection coefficient of 0.16, resulting in a 16% power loss. To design an efficient transmission line system, therefore, the voltage/power reflection coefficient should be as small as possible so that the maximum amount of power can be transferred to the load. Since in transmission line systems Z_L and Z_0 may not be matched, an efficient matching network should be designed to match Z_L with Z_0. In this chapter, we will describe how this can be accomplished using different types of matching networks.

3.2 Mismatch Losses in Transmission Lines

The different types of mismatch losses that occur in an unmatched transmission line system are defined in the following subsections.

Attenuation loss Attenuation loss in a line is defined as follows:

$$\text{Attenuation loss (dB)} = 8.686\alpha l \tag{3.1}$$

where l is the distance and α is the attenuation constant.

Reflection loss Reflection loss in a line is defined as follows:

$$\text{Reflection loss (dB)} = 10\log_{10}\left(\frac{1}{1-|\Gamma|^2}\right) \tag{3.2}$$

where Γ is the reflection coefficient.

Transmission loss Transmission loss in a line is defined as follows:

$$\text{Transmission loss (dB)} = 0.868\alpha l + 10\log_{10}\left(\frac{1}{1-|\Gamma|^2}\right) \tag{3.3}$$

Return loss Return loss in a line is defined as follows:

$$\text{Return loss (dB)} = 10\log_{10}\left(1/|\Gamma|^2\right) = -20\log_{10}\left(|\Gamma|\right) \tag{3.4}$$

Insertion loss If P_i and P_r be the powers received by a load when it is connected to the source with and without a transmission line, respectively, then insertion loss is defined as follows:

$$\text{Insertion loss (dB)} = 10\log_{10}\left(P_i/P_r\right) \tag{3.5}$$

The total insertion loss is the summation of losses due to source mismatch, attenuation in the line, and load mismatch.

3.3 Matching with Lumped Elements

As discussed earlier, to get rid of the losses due to reflections, we need to insert a matching network between the load and the line, as shown in Fig. 3.1 The simplest way to obtain a load matching is to use lumped elements to form the matching network. Depending on the arrangement of the lumped elements in the matching network, it can be classified as L-network, pi-network, or T-network.

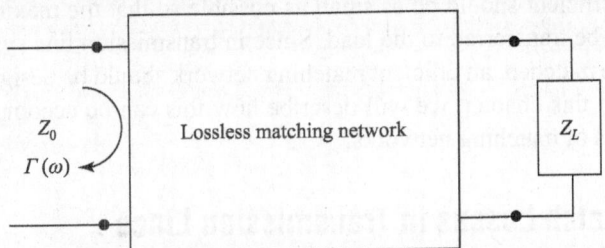

Z_0

$\Gamma(\omega)$

Lossless matching network

Z_L

Fig. 3.1 Matching load Z_L with a line of characteristic impedance Z_0 using lossless matching network

3.3.1 L-network

An L-section matching network is probably the simplest type of matching network and uses two reactive elements to match an arbitrary complex load to a transmission line. This type of network has two possible configurations, as shown in Figs 3.2(a) and (b). The matching network shown in Fig. 3.2(a) is used to match the load having the normalized impedance inside the $1 + jx$ circle of the Smith chart, whereas that shown in Fig. 3.2(b) is used to match the normalized load impedance lying outside the $1 + jx$ circle of the chart.

Let us first consider that the normalized load impedance lies inside the $1 + jx$ circle on the Smith chart. Therefore, we will proceed by considering the matching network shown in Fig. 3.2(a). The input impedance at 'AB', seen looking towards the load, can be expressed as follows:

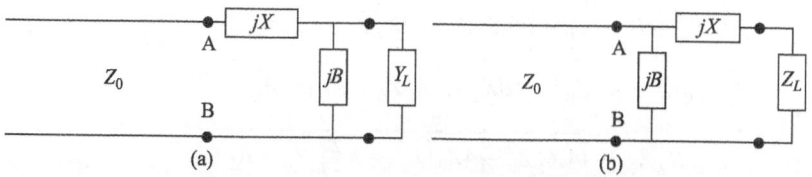

Fig. 3.2 *L*-section matching networks (a) Networks with normalized load impedance inside $1 + jx$ circle (b) Networks with normalized load impedance outside $1 + jx$ circle

$$Z_{\text{in,AB}} = jX + \frac{1}{(jB \| Y_L)} = jX + \frac{1}{jB + 1/(R_L + jX_L)} = jX + \frac{(R_L + jX_L)}{(1 - BX_L) + jBR_L}$$

or
$$Z_{\text{in,AB}} = \frac{jX\{(1 - BX_L) + jBR_L\} + (R_L + jX_L)}{(1 - BX_L) + jBR_L}$$

or
$$Z_{\text{in,AB}} = \frac{(R_L - BXR_L) + j(X_L + X - BXX_L)}{(1 - BX_L) + jBR_L} \tag{3.6}$$

Now matching requires that $Z_{\text{in,AB}} = Z_0$. Therefore, Eq. (3.6) can be rewritten as follows:

$$Z_0 = \frac{(R_L - BXR_L) + j(X_L + X - BXX_L)}{(1 - BX_L) + jBR_L} \tag{3.7}$$

or
$$Z_0\{(1 - BX_L) + jBR_L\} = (R_L - BXR_L) + j(X_L + X - BXX_L) \tag{3.8}$$

Equating the real parts on both sides of Eq. (3.8), we get the following expression:

$$Z_0(1 - BX_L) = (R_L - BXR_L)$$

or
$$B(XR_L - Z_0X_L) = R_L - Z_0 \tag{3.9}$$

Similarly, equating the imaginary parts on both sides of Eq. (3.8) the following relation is obtained:

$$X(1 - BX_L) = Z_0BR_L - X_L \tag{3.10}$$

From Eq. (3.9), we can write that

$$X = \frac{1}{B} + \frac{Z_0X_L}{R_L} - \frac{Z_0}{BR_L} \tag{3.11}$$

Substituting the expression of X from Eq. (3.11) into Eq. (3.10), we get the following equation:

$$(1 - BX_L)\left(\frac{1}{B} + \frac{Z_0X_L}{R_L} - \frac{Z_0}{BR_L}\right) = Z_0BR_L - X_L$$

or
$$(1 - BX_L)\frac{BZ_0X_L + R_L - Z_0}{BR_L} = Z_0BR_L - X_L$$

or $\quad BZ_0 X_L + R_L - Z_0 - B^2 X_L^2 Z_0 + BX_L Z_0 = Z_0 B^2 R_L^2$

or $\quad B^2 \left(Z_0 R_L^2 + X_L^2 Z_0 \right) - 2BX_L Z_0 + \left(Z_0 - R_L \right) = 0$

or $\quad B = \dfrac{2X_L Z_0 \pm \sqrt{4X_L^2 Z_0^2 - 4Z_0 \left(R_L^2 + X_L^2 \right) \left(Z_0 - R_L \right)}}{2Z_0 \left(R_L^2 + X_L^2 \right)}$

or $\quad B = \dfrac{2X_L Z_0 \pm \sqrt{-4Z_0^2 R_L^2 + 4Z_0 R_L^3 + 4Z_0 R_L X_L^2}}{2Z_0 \left(R_L^2 + X_L^2 \right)}$

or $\quad B = \dfrac{2X_L Z_0 \pm 2Z_0 \sqrt{R_L/Z_0} \sqrt{-Z_0 R_L + R_L^2 + X_L^2}}{2Z_0 \left(R_L^2 + X_L^2 \right)}$

or $\quad B = \dfrac{X_L \pm \sqrt{R_L/Z_0} \sqrt{R_L^2 + X_L^2 - Z_0 R_L}}{\left(R_L^2 + X_L^2 \right)} \qquad (3.12)$

Since the normalized load impedance lies inside the $1 + jx$ circle on the Smith chart, $R_L > Z_0$ and hence the argument of the second square root is always positive. Equations 3.11 and 3.12 indicate that two solutions are possible for B and X, as these may have both positive and negative values. However, one solution may lead to significantly smaller values for the reactive components and may be the preferred solution if the bandwidth of the match is better. Since $R_L > Z_0$, this matching circuit is also known as a downward impedance transformer.

Now let us consider the second case, that is, the normalized load impedance lies outside the $1 + jx$ circle on the Smith chart. Therefore, we will proceed by considering the matching network shown in Fig. 3.2(b). The input admittance at 'AB', seen looking towards the load, can be expressed as follows:

$$Y_{\text{in,AB}} = jB + \frac{1}{Z_L + jX} = jB + \frac{1}{R_L + j(X + X_L)} \qquad (3.13)$$

Now, matching requires that $Y_{\text{in,AB}} = Y_0$. Therefore, Eq. (3.13) can be rewritten as follows:

$$Y_0 = jB + \frac{1}{R_L + j(X + X_L)} \quad \text{or} \quad \frac{1}{Z_0} = \frac{1 - B(X + X_L) + jBR_L}{R_L + j(X + X_L)}$$

or $\quad R_L + j(X + X_L) = Z_0 - BZ_0(X + X_L) + jBZ_0 R_L \qquad (3.14)$

Equating the real parts on both sides of Eq. (3.14), we get the following relation:

$$BZ_0(X + X_L) = Z_0 - R_L \qquad (3.15)$$

Similarly, equating the imaginary parts on both sides of Eq. (3.14), the following equation is obtained: $(X + X_L) = BZ_0 R_L$

or $\quad B = \dfrac{(X + X_L)}{Z_0 R_L} \qquad (3.16)$

Substitution of the expression of B from Eq. (3.16) into Eq. (3.15) yields the following expression: $(X + X_L)^2 = Z_0 R_L - R_L^2$

or $\quad X = -X_L \pm \sqrt{R_L (Z_0 - R_L)}$ \hfill (3.17)

Substituting the expression of X from Eq. (3.17) into Eq. (3.16), we get the following relation:

$$B = \frac{\pm\sqrt{R_L (Z_0 - R_L)}}{Z_0 R_L} = \pm\frac{\sqrt{(Z_0 - R_L)/R_L}}{Z_0} \hspace{2cm} (3.18)$$

Since the normalized load impedance lies outside the $1 + jx$ circle of the Smith chart, $R_L < Z_0$, and hence the argument of the square root is always positive. As discussed earlier, Eqs (3.17) and (3.18) indicate that two solutions are possible for B and X, as both positive and negative values exist for B and X. However, again, one solution may lead to significantly smaller values for the reactive components and may be the preferred solution if the bandwidth of the match is better. Since $R_L < Z_0$, this matching circuit is also known as an upward impedance transformer.

For a frequency of up to about 1 GHz, actual lumped element inductors and capacitors are realizable and can be used. However, lumped element inductors and capacitors are not realizable for a larger frequency, because the component length becomes comparable with the wavelength. This limits the use of L-section matching technique at higher frequencies. Another disadvantage of this matching network is that here we have only two degrees of freedom, that is, we can choose only X and B. Hence, once the frequency of matching and R_L/Z_0 have been specified, the network Q (defined as an inverse of the fractional bandwidth of the network) is determined automatically. If we want a different Q, we must use a network that offers a third degree of freedom.

EXAMPLE 3.1 Consider the transmission line example given in Example 2.1 in Chapter 2 for a lossless case. The characteristic impedance of the line is 50.026 Ω and the load impedance is $78.27 + j60.93\ \Omega$. Design the possible L-section matching networks for the system using and without using a Smith chart.

Solution Given: $Z_0 = 50.026\ \Omega$ and $Z_L = 78.27 + j60.93\ \Omega$

1. The normalized load impedance is as follows:

$$z_L = \frac{Z_L}{Z_0} = \frac{78.27 + j60.93}{50.026} = 1.5646 + j1.218$$

2. Enter the $1.5646 + j1.218$ point in the Smith chart as point 'A' (Fig. 3.3). Since z_L is inside the $1 + jx$ circle, we need to use the circuit shown in Fig. 3.2(a).
3. Since the first element of Fig. 3.2(a) is a shunt susceptance, we need to convert the normalized load impedance to the normalized load admittance. The normalized load admittance, y_L, is exactly opposite to point A on the constant SWR circle, that is, they are 180° apart. Let this point be 'B' and the value of the normalized load admittance $y_L = 0.398 - j0.31$.
4. Now, we need to add a shunt susceptance jB to the y_L so that, after addition, the equivalent normalized impedance z_L' lies on the $1 + jx$ circle. To achieve this, we have drawn an inverted $1 + jx$ circle as shown in Fig. 3.3.

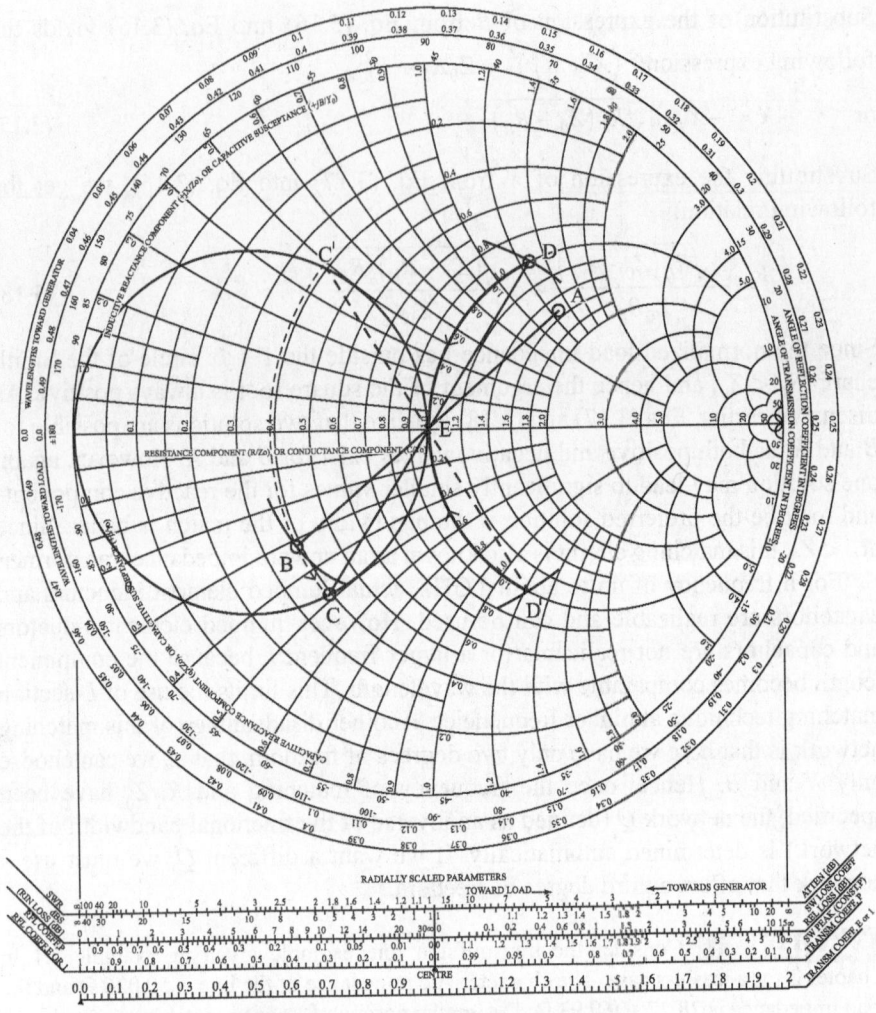

Fig. 3.3 Smith chart for Example 3.1

5. Since addition of jB changes only the susceptive part of the normalized load admittance, we should move along the constant conductance circle to reach the nearest $1 + jb$ circle. Consequently, we will reach a normalized admittance point $y'_L = 0.398 - j0.49$, shown by point 'C'. Therefore, the normalized susceptance that must be added with the normalized load admittance is $-j0.49 - (-j0.31) = -j0.18$.

6. The corresponding value of the normalized impedance is $z'_L = 1 + j1.23$ and is denoted by point 'D' on the Smith chart.

7. This implies that we need to insert a normalized series impedance of $-j1.23$ to reach the matched point 'E'.

8. Instead of adding the normalized load admittance $-j1.18$ in Step 5, we can also add a reactance of $j0.49 - (-j0.31) = j0.80$ to reach a normalized admittance point $y''_L = 0.398 + j0.49$, shown by point 'C''.

9. The corresponding value of normalized impedance is $z''_L = 1 - j1.23$ and is denoted by point 'D''' on the Smith chart.

10. This implies that we need to insert a normalized series impedance of $j1.23$ to reach the matched point E.

The problem can also be solved without using a Smith chart, as follows:

$$B = \frac{X_L \pm \sqrt{R_L/Z_0}\sqrt{R_L^2 + X_L^2 - Z_0 R_L}}{\left(R_L^2 + X_L^2\right)}$$

or $$B = \frac{60.93 \pm \sqrt{78.27/50.026}\sqrt{(78.27)^2 + (60.93)^2 - 50.026 \times 78.27}}{(78.27)^2 + (60.93)^2}$$

or $$B = \frac{60.93 \pm 96.266}{9838.658}$$

or $$b = \frac{1}{Y_0}\frac{60.93 \pm 96.266}{9838.658} = Z_0\frac{60.93 \pm 96.266}{9838.658} = 50.026 \times \frac{60.93 \pm 96.266}{9838.658} = \begin{cases} -0.18 \\ +0.80 \end{cases}$$

These are similar to those obtained using a Smith chart. Now, for $b = -0.18$,

$$X = \frac{1}{B} + \frac{Z_0 X_L}{R_L} - \frac{Z_0}{BR_L} = \frac{50.026}{-0.18} + \frac{50.026 \times 60.93}{78.27} + \frac{(50.026)^2}{0.18 \times 78.27} = -61.346$$

or $x = -1.23$

and for $b = 0.80$,

$$X = \frac{1}{B} + \frac{Z_0 X_L}{R_L} - \frac{Z_0}{BR_L} = \frac{50.026}{0.80} + \frac{50.026 \times 60.93}{78.27} - \frac{(50.026)^2}{0.80 \times 78.27} = 61.508$$

or $x = 1.23$

These are again similar to those obtained using a Smith chart.

Practice Problems

3.1 Consider the transmission line example given in Practice Problem 2.1 in Chapter 2 for a lossless case. The characteristic impedance of the line is 73.986 Ω and the load impedance is $43.72 + j87.63$ Ω. Design the possible L-section matching networks for the system using and without using a Smith chart.
[Smith chart: $x = 0.704/1.664$, $b = \pm0.86$; formula: $x = -0.693/1.676$, $b = \pm0.832$]

3.2 Consider the transmission line example given in Practice Problem 2.2 in Chapter 2 for a lossless case. The characteristic impedance of the line is 100.004 Ω and the load impedance is $165.21 + j6.97\,\Omega$. Design the possible L-section matching networks for the system using and without using a Smith chart.
[Smith chart: $b = -0.45/0.51$, $b = \pm0.86$; formula: $b = -0.464/0.514$, $b = \pm0.81$]

3.3.2 Pi-network

As we have already discussed, one of the major limitations of an L-network is that we have no control on the network Q[1] and the network should have a third degree of

---

[1]For details, see *Foundations for Microwave Engineering* by Robert E. Collin, McGraw-Hill International Edition, Second edition, Singapore, pp. 325–330.

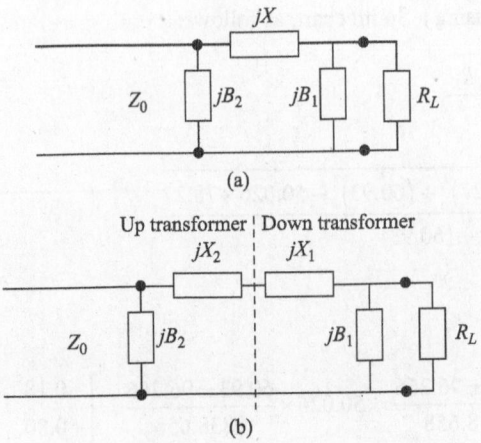

(a)

(b)

Fig. 3.4 Pi-network (a) Lumped pi-matching network (b) Equivalent pi-matching network constructed using two L-matching networks where $X_1 + X_2 = X$

freedom to have a control on Q. One of the circuits that offer the third degree of freedom is a pi-network, shown in Fig. 3.4(a). The pi-matching network decouples Q from the transformation ratio R_L/Z_0 by introducing an intermediate resistance R_I. This allows the user to obtain a much higher Q than that obtained with an L-section matching network.

The best way to understand the function of a pi-network is to consider it as a cascade connection of two L-networks, as shown in Fig. 3.4(b); here, the first L-network is a downward impedance transformer and the second one an upward impedance transformer.

The downward impedance transformer transforms the load resistance R_L to a lower intermediate or image resistance R_I. The image resistance R_I is then transformed to the line characteristic impedance Z_0 by the upward impedance transformer.

3.3.3 *T*-network

Another way to obtain the third degree of freedom to achieve a control on Q is to use a *T*-matching network, as shown in Fig. 3.5(a). Like a pi-matching network, a *T*-network also decouples Q from the transformation ratio R_L/Z_0 by introducing an intermediate resistance R_I. This again allows the user to obtain a much higher Q than that obtained with an L-section matching network.

The equivalent representation of a *T*-network as a cascade connection of two L-networks is shown in Fig. 3.5(b), where the first network is an upward impedance transformer and the second one a downward impedance transformer.

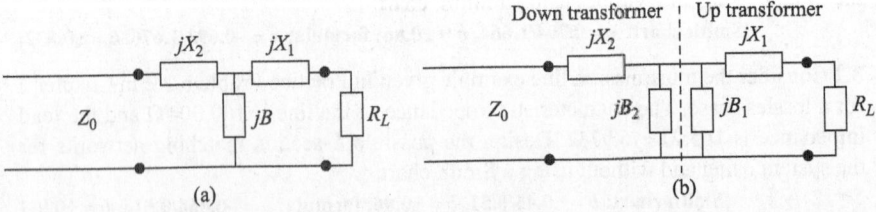

(a)

(b)

Fig. 3.5 T-network (a) Lumped *T*-matching network (b) Equivalent *T*-matching network constructed using two L-matching networks where $B_1 + B_2 = B$

3.4 Single-stub Matching

A stub is basically a piece of transmission line that is terminated, normally, in a short circuit at one end. It may also be terminated in an open circuit; however, this practice is not common as an open-circuited stub has an inherent tendency to radiate from its open end. Furthermore, it is also more difficult to obtain a good open-circuit termination than a good short-circuit termination.

Irrespective of whether a stub is short or open circuited, it is always true that the input impedance of the stub is purely reactive and any reactive input impedance can be obtained by a proper choice of the electrical length of the stub, as described in Section 2.8.

Because of the pure reactive/susceptive behaviour of the input impedance/admittance of a stub, it can be used to tune out the reactive/susceptive component of the line impedance/admittance at any desired point on the line. For example, if the line impedance at any point z on the line is $R_{in} + jX_{in}$, then insertion of a series stub with an input impedance $-jX_{in}$ at z will tune out the reactive part of the line; thus, the modified input impedance will be $R_{in} + jX_{in} - jX_{in} = R_{in}$. Now, if the point z is so chosen that at this point $R_{in} = Z_0$, where Z_0 is the characteristic impedance of the line, then the line will be matched at z. Theoretical determination of the stub location and stub length is described in the following paragraphs.

Let us consider the series stub tuning circuit shown in Fig. 3.6. The input admittance of the line at a'b' will be as follows:

$$Y_{in} = Y_0 \frac{Y_L + jY_0 \tan(\beta d)}{Y_0 + jY_L \tan(\beta d)} \tag{3.19}$$

Now assuming that $Y_L = G_L + jB_L$, Eq. (3.19) modifies to the following form:

$$Y_{in} = Y_0 \frac{(G_L + jB_L) + jY_0 \tan(\beta d)}{Y_0 + j(G_L + jB_L)\tan(\beta d)}$$

or $$Z_{in} = \frac{1}{Y_0} \frac{Y_0 + j(G_L + jB_L)\tan(\beta d)}{(G_L + jB_L) + jY_0 \tan(\beta d)}$$

or $$Z_{in} = \frac{1}{Y_0} \frac{\left[Y_0 + j(G_L + jB_L)\tan(\beta d)\right]\left[G_L - j\{B_L + Y_0 \tan(\beta d)\}\right]}{\left[G_L + j\{B_L + Y_0 \tan(\beta d)\}\right]\left[G_L - j\{B_L + Y_0 \tan(\beta d)\}\right]}$$

or $$Z_{in} = \frac{Y_0 G_L - jY_0 B_L - jY_0^2 \tan(\beta d) + jG_L^2 \tan(\beta d) + G_L Y_0 \tan^2(\beta d)}{+ jB_L^2 \tan(\beta d) + jB_L Y_0 \tan^2(\beta d)} \Big/ Y_0\left[G_L^2 + \{B_L + Y_0 \tan(\beta d)\}^2\right] \tag{3.20}$$

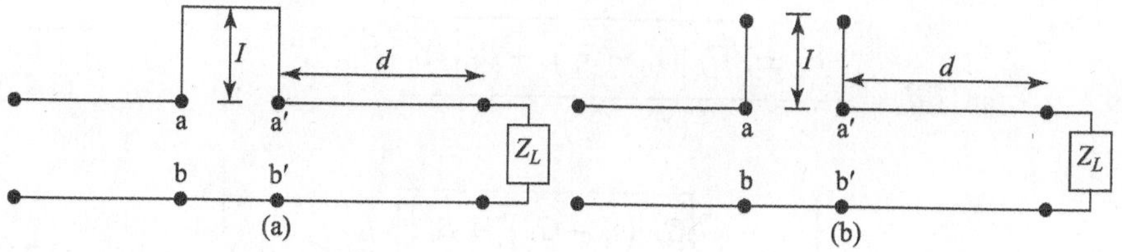

Fig. 3.6 Transmission line series stub matching (a) Using short-circuited stub (b) Using open-circuited stub

Substituting $Z_{in} = R_{in} + jX_{in}$ and equating the real and imaginary parts of Eq. (3.20), we get the following expressions:

$$R_{in} = \frac{1}{Y_0} \frac{Y_0 G_L + G_L Y_0 \tan^2(\beta d)}{G_L^2 + \{B_L + Y_0 \tan(\beta d)\}^2} = \frac{G_L \{1 + \tan^2(\beta d)\}}{G_L^2 + \{B_L + Y_0 \tan(\beta d)\}^2} \tag{3.21}$$

and $\quad X_{in} = \dfrac{1}{Y_0} \dfrac{-Y_0 B_L - Y_0^2 \tan(\beta d) + G_L^2 \tan(\beta d) + B_L^2 \tan(\beta d) + B_L Y_0 \tan^2(\beta d)}{G_L^2 + \{B_L + Y_0 \tan(\beta d)\}^2}$

or $\quad X_{in} = \dfrac{1}{Y_0} \dfrac{G_L^2 \tan(\beta d) - \{Y_0 - B_L \tan(\beta d)\}\{B_L + Y_0 \tan(\beta d)\}}{G_L^2 + \{B_L + Y_0 \tan(\beta d)\}^2} \tag{3.22}$

Now, since for matching $R_{in} = Z_0 = 1/Y_0$, we can rewrite Eq. (3.21) as follows:

$$\frac{1}{Y_0} = \frac{G_L \{1 + \tan^2(\beta d)\}}{G_L^2 + \{B_L + Y_0 \tan(\beta d)\}^2}$$

or $\quad Y_0^2 \tan^2(\beta d) + 2Y_0 B_L \tan(\beta d) + G_L^2 + B_L^2 = Y_0 G_L + Y_0 G_L \tan^2(\beta d)$

or $\quad \left(Y_0^2 - Y_0 G_L\right)\tan^2(\beta d) + 2Y_0 B_L \tan(\beta d) + \left(G_L^2 + B_L^2 - Y_0 G_L\right) = 0$

or $\quad \tan(\beta d) = \dfrac{-2Y_0 B_L \pm \sqrt{4 Y_0^2 B_L^2 - 4\left(Y_0^2 - Y_0 G_L\right)\left(G_L^2 + B_L^2 - Y_0 G_L\right)}}{2\left(Y_0^2 - Y_0 G_L\right)}$

or $\quad \tan(\beta d) = \dfrac{-2Y_0 B_L \pm \sqrt{4 Y_0^3 G_L + 4Y_0 G_L^3 + 4Y_0 G_L B_L^2 - 8 Y_0^2 G_L^2}}{2Y_0 \left(Y_0 - G_L\right)}$

or $\quad \tan(\beta d) = \dfrac{-2Y_0 B_L \pm 2\sqrt{Y_0 G_L \left[\left(Y_0^2 - 2Y_0 G_L + G_L^2\right) + B_L^2\right]}}{2Y_0 \left(Y_0 - G_L\right)}$

or $\quad \tan(\beta d) = \dfrac{-Y_0 B_L \pm Y_0 \sqrt{G_L \left[\left(Y_0 - G_L\right)^2 + B_L^2\right]/Y_0}}{Y_0 \left(Y_0 - G_L\right)}$

or $\quad \tan(\beta d) = \dfrac{-B_L \pm \sqrt{G_L \left[\left(Y_0 - G_L\right)^2 + B_L^2\right]/Y_0}}{\left(Y_0 - G_L\right)} \tag{3.23}$

or $\quad d = \dfrac{1}{\beta} \tan^{-1}\left[\left\{-B_L \pm \sqrt{\dfrac{G_L \left[\left(Y_0 - G_L\right)^2 + B_L^2\right]}{Y_0}}\right\} \middle/ \left(Y_0 - G_L\right)\right] \tag{3.24}$

We can determine the position of the stub using Eq. (3.24), provided that $G_L \neq Y_0$. For $G_L = Y_0$, we can insert the stub at the position of the load. Once the value of $\tan(\beta d)$ is obtained using Eq. (3.23), we can easily calculate the value of X_{in} from Eq. (3.22).

Now, if we use a short-circuited stub, as shown in Fig. 3.6(a),

$$j\frac{\tan(\beta l_{ss})}{Y_{OS}} = jX_{ss} \tag{3.25}$$

where Y_{OS} is the characteristic stub admittance, and X_{ss} is the reactance of the stub and is equal to $-X_{in}$. Thus, $\tan(\beta l_{ss}) = -Y_{OS}X_{in}$

or $\qquad l_{ss} = \frac{1}{\beta}\tan^{-1}\left[-Y_{OS}X_{in}\right] \tag{3.26}$

Substituting the expression of X_{in} from Eq. (3.22) into Eq. (3.26), we get the following relation:

$$l_{ss} = \frac{1}{\beta}\tan^{-1}\left[-\frac{Y_{OS}}{Y_0}\frac{G_L^2\tan(\beta d) - \{Y_0 - B_L\tan(\beta d)\}\{B_L + Y_0\tan(\beta d)\}}{G_L^2 + \{B_L + Y_0\tan(\beta d)\}^2}\right] \tag{3.27}$$

Once the value of $\tan(\beta d)$ is obtained from Eq. (3.23), we can easily calculate the value of l_{ss} from Eq. (3.27).

Now if we use an open-circuited stub, as shown in Fig. 3.6(b), $-j\frac{\cot(\beta l_{os})}{Y_{OS}} = jX_{so}$

where X_{so} is the reactance of the stub and is equal to $-X_{in}$. Thus,

$$\cot(\beta l_{os}) = Y_{OS}X_{in} \quad \text{or} \quad l_{os} = \frac{1}{\beta}\cot^{-1}\left[Y_{OS}X_{in}\right] \tag{3.28}$$

Substituting the expression of X_{in} from Eq. (3.22) into Eq. (3.28), we get the following equation:

$$l_{os} = \frac{1}{\beta}\cot^{-1}\left[\frac{Y_{OS}}{Y_0}\frac{G_L^2\tan(\beta d) - (Y_0 - B_L\tan(\beta d))(B_L + Y_0\tan(\beta d))}{G_L^2 + \{B_L + Y_0\tan(\beta d)\}^2}\right] \tag{3.29}$$

Once the value of $\tan(\beta d)$ is obtained from Eq. (3.23), we can easily calculate the value of l_{os} from Eq. (3.29).

If the length given by Eqs (3.26)–Eq.(3.29) is negative, then $\lambda/2$ must be added to give a positive result.

Like a series stub, a shunt stub can also be used for matching purpose. For the shunt stub case, the basic concept is to select a point z on the line such that the admittance Y_{in}, seen looking into the line from z is of the form $Y_{in} = G_0 + jB_{in}$, where G_0 is the characteristic admittance of the line. Now, if the input susceptance of the shunt stub, inserted at z, is $-jB_{in}$, then the resultant input admittance of the line is $G_0 + jB_{in} - jB_{in} = G_0$, which is the matched condition. To derive the necessary equations, let us consider the shunt stub tuning circuit shown in Fig. 3.7.

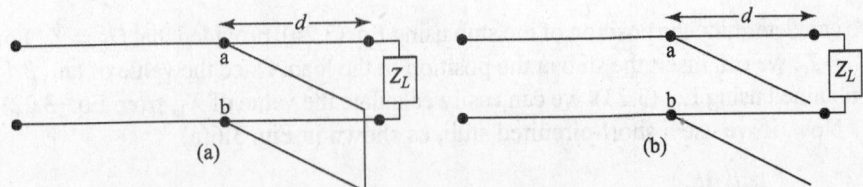

Fig. 3.7 Transmission line shunt stub matching (a) Using short-circuited stub
(b) Using open-circuited stub

The input admittance of the line at ab will be as follows:

$$Z_{in} = Z_0 \frac{Z_L + jZ_0 \tan(\beta d)}{Z_0 + jZ_L \tan(\beta d)} \tag{3.30}$$

Now, assuming that $Z_L = R_L + jX_L$, Eq. (3.30) modifies to the following form:

$$Z_{in} = Z_0 \frac{(R_L + jX_L) + jZ_0 \tan(\beta d)}{Z_0 + j(R_L + jX_L)\tan(\beta d)}$$

or $$Y_{in} = \frac{1}{Z_0} \frac{Z_0 + j(R_L + jX_L)\tan(\beta d)}{(R_L + jX_L) + jZ_0 \tan(\beta d)}$$

or $$Y_{in} = \frac{\left[\{Z_0 - X_L \tan(\beta d)\} + jR_L \tan(\beta d)\right]\left[R_L - j\{Z_0 \tan(\beta d) + X_L\}\right]}{Z_0\left[R_L + j\{Z_0 \tan(\beta d) + X_L\}\right]\left[R_L - j\{Z_0 \tan(\beta d) + X_L\}\right]}$$

$$Z_0 R_L - jZ_0^2 \tan(\beta d) - jZ_0 X_L + jZ_0 X_L \tan^2(\beta d)$$

or $$Y_{in} = \frac{+jX_L^2 \tan(\beta d) + jR_L^2 \tan(\beta d) + R_L Z_0 \tan^2(\beta d)}{Z_0\left[R_L^2 + \{Z_0 \tan(\beta d) + X_L\}^2\right]} \tag{3.31}$$

Substituting $Y_{in} = G_{in} + jB_{in}$ and equating the real and imaginary parts of Eq. (3.31), we get the following expression:

$$G_{in} = \frac{1}{Z_0} \frac{Z_0 R_L + R_L Z_0 \tan^2(\beta d)}{R_L^2 + \{Z_0 \tan(\beta d) + X_L\}^2} = \frac{R_L\{1 + \tan^2(\beta d)\}}{\left[R_L^2 + \{Z_0 \tan(\beta d) + X_L\}^2\right]} \tag{3.32}$$

and $$B_{in} = \frac{1}{Z_0} \frac{-Z_0^2 \tan(\beta d) - Z_0 X_L + Z_0 X_L \tan^2(\beta d) + X_L^2 \tan(\beta d) + R_L^2 \tan(\beta d)}{R_L^2 + \{Z_0 \tan(\beta d) + X_L\}^2}$$

or $$B_{in} = \frac{R_L^2 \tan(\beta d) - \{Z_0 - X_L \tan(\beta d)\}\{X_L + Z_0 \tan(\beta d)\}}{Z_0\left[R_L^2 + \{Z_0 \tan(\beta d) + X_L\}^2\right]} \tag{3.33}$$

Since $G_{in} = Y_0 = \dfrac{1}{Z_0}$, we can write Eq. (3.32) as follows:

$$\frac{1}{Z_0} = \frac{R_L\{1+\tan^2(\beta d)\}}{\left[R_L^2+\{Z_0\tan(\beta d)+X_L\}^2\right]}$$

or $\quad Z_0^2\tan^2(\beta d)+2Z_0X_L\tan(\beta d)+R_L^2+X_L^2 = Z_0R_L+Z_0R_L\tan^2(\beta d)$

or $\quad \left(Z_0^2-Z_0R_L\right)\tan^2(\beta d)+2Z_0X_L\tan(\beta d)+\left(R_L^2+X_L^2-Z_0R_L\right)=0$

or $\quad \tan(\beta d)=\dfrac{-2Z_0X_L\pm\sqrt{4Z_0^2X_L^2-4\left(Z_0^2-Z_0R_L\right)\left(R_L^2+X_L^2-Z_0R_L\right)}}{2\left(Z_0^2-Z_0R_L\right)}$

or $\quad \tan(\beta d)=\dfrac{-2Z_0X_L\pm\sqrt{4Z_0^3R_L+4Z_0R_L^3+4Z_0R_LX_L^2-8Z_0^2R_L^2}}{2\left(Z_0^2-Z_0R_L\right)}$

or $\quad \tan(\beta d)=\dfrac{-2Z_0X_L\pm\sqrt{4Z_0R_L\left[Z_0^2-2Z_0R_L+R_L^2+X_L^2\right]}}{2Z_0\left(Z_0-R_L\right)}$

or $\quad \tan(\beta d)=\dfrac{-2Z_0X_L\pm2Z_0\sqrt{R_L\left[\left(Z_0-R_L\right)^2+X_L^2\right]/Z_0}}{2Z_0\left(Z_0-R_L\right)}$

or $\quad \tan(\beta d)=\dfrac{-X_L\pm\sqrt{R_L\left[\left(Z_0-R_L\right)^2+X_L^2\right]/Z_0}}{\left(Z_0-R_L\right)}$ $\qquad$ (3.34)

or $\quad d=\dfrac{1}{\beta}\tan^{-1}\left[\left[-X_L\pm\sqrt{\dfrac{R_L\left[\left(Z_0-R_L\right)^2+X_L^2\right]}{Z_0}}\right]\Big/\left(Z_0-R_L\right)\right]$ $\qquad$ (3.35)

Using Eq. (3.35) we can determine the position of the stub, provided that $R_L\neq Z_0$. For $R_L=Z_0$, we can insert the stub at the position of the load. Once the value of $\tan(\beta d)$ is obtained from Eq. (3.34), we can easily calculate the value of B_{in} from Eq. (3.33).

If we use a short-circuited stub, as shown in Fig. 3.7(a), we get the following expression:

$$\frac{-jY_{0S}}{\tan(\beta l_{ss})}=jB_{ss} \qquad (3.36)$$

where Y_{0S} is the characteristic admittance of the stub, and B_{ss} is its susceptance and is equal to $-B_{\text{in}}$. Thus,

$$\tan(\beta l_{ss})=\frac{Y_{0S}}{B_{\text{in}}} \quad \text{or} \quad l_{ss}=\frac{1}{\beta}\tan^{-1}\left(\frac{Y_{0S}}{B_{\text{in}}}\right) \qquad (3.37)$$

Substituting the expression of B_{in} from Eq. (3.33) into Eq. (3.37), we get the following equation:

$$l_{ss} = \frac{1}{\beta} \tan^{-1} \left(\frac{Z_0}{Z_{0S}} \frac{\left[R_L^2 + \{Z_0 \tan(\beta d) + X_L\}^2 \right]}{R_L^2 \tan(\beta d) - \{Z_0 - X_L \tan(\beta d)\}\{X_L + Z_0 \tan(\beta d)\}} \right) \quad (3.38)$$

Here, Z_{0S} is the characteristic impedance of the stub. Once the value of $\tan(\beta d)$ is obtained from Eq. (3.34), we can easily calculate the value of l_{ss} from Eq. (3.38).

If we use an open-circuited stub, as shown in Fig. 3.7(b), the following equation is obtained:

$$\frac{jY_{0S}}{\cot(\beta l_{os})} = jB_{so} \quad (3.39)$$

where B_{so} is the susceptance of the stub and is equal to $-B_{in}$. Thus,

$$\cot(\beta l_{os}) = -\frac{Y_{0S}}{B_{in}} \quad \text{or} \quad l_{os} = -\frac{1}{\beta} \cot^{-1}\left(\frac{Y_{0S}}{B_{in}} \right) \quad (3.40)$$

Substituting the expression of B_{in} from Eq. (3.33) into Eq. (3.40), we get the following expression:

$$l_{os} = -\frac{1}{\beta} \cot^{-1} \left(\frac{Z_0}{Z_{0S}} \frac{\left[R_L^2 + \{Z_0 \tan(\beta d) + X_L\}^2 \right]}{R_L^2 \tan(\beta d) - \{Z_0 - X_L \tan(\beta d)\}\{X_L + Z_0 \tan(\beta d)\}} \right) \quad (3.41)$$

Once the value of $\tan(\beta d)$ is obtained from Eq. (3.34), we can easily calculate the value of l_{os} from Eq. (3.41).

If the length given by Eqs (3.37), (3.38), (3.40), and (3.41) is negative, then $\lambda/2$ must be added to give a positive result.

Since only a single stub is being used for matching a complex load, this method of matching is called single-stub matching. It requires a very precise location and electrical length of the matching stub. Any deviation from this will lead to a mismatch. Since both the location and the electrical length of a stub is a function of the wavelength and hence the frequency of the signal, single-stub matching is very narrow band in nature.

EXAMPLE 3.2 Design single shunt stub matching network(s) for the transmission line of Example 3.1 using a Smith chart. Assume that the characteristic impedance of the matching stub is 74 Ω. Verify the result using direct formulas.

Solution Given: $Z_0 = 50.026 \ \Omega$ and $Z_L = 78.27 + j60.93 \ \Omega$

1. The normalized load impedance is as follows:

$$z_L = \frac{Z_L}{Z_0} = \frac{78.27 + j60.93}{50.026} = 1.5646 + j1.218$$

2. Enter the $1.5646 + j1.218$ point in the Smith chart as point A (Fig. 3.8).

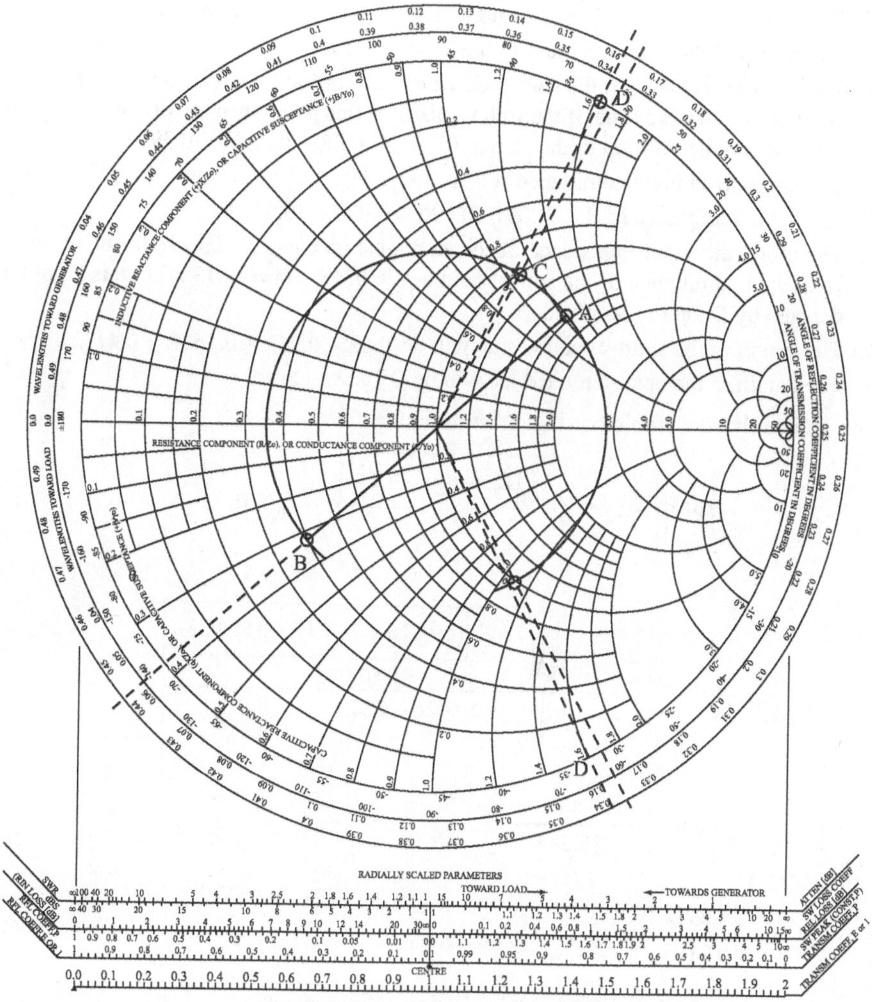

Fig. 3.8 Smith chart for Example 3.2

3. Now we need to convert the normalized load impedance to the normalized load admittance. The normalized load admittance, y_L, is exactly opposite to point A on the constant SWR circle, that is, they are 180° apart. Let this point be B and the value of normalized load admittance $y_L = 0.398 - j0.31$.

4. Since the stub must be inserted at a point where the normalized conductance is 1, we need to move along the constant SWR circle towards the generator till we reach the $1 + jb$ circle. Let us assume that the constant SWR circle cuts the $1 + jb$ circle at point C for the first time. The normalized admittance at point C is $1 + j1.09$.

5. Therefore, the distance between the position of the stub and the load is $0.5\lambda - 0.443\lambda + 0.165\lambda = 0.222\lambda$.

6. The input admittance of the line at point C is $(1 + j1.09)/50.026 = 0.020 + j0.0218$.

7. Therefore, the input admittance of the stub must be equal to $-j0.0218$ or the normalized input admittance of the stub should be $-j0.0218 \times 74 = -1.6132$. Let this point be denoted by D on the Smith chart.

8. The length of the short-circuited stub will be $0.338\lambda - 0.25\lambda = 0.088\lambda$.

9. The length of the open-circuited stub will be 0.338λ.

10. The alternate position of the stub is C', where the constant SWR circle cuts the $1 + jb$ circle for the second time. If the stub is placed at this point, then the distance between the load and the position of the stub is $0.5\lambda - 0.443\lambda + 0.335\lambda = 0.392\lambda$.

11. The normalized input admittance at C' is $1 - j1.09$.

12. The input admittance is $(1 - j1.09)/50.026 = 0.020 - j0.0218$.

13. Therefore, the input impedance of the stub must be equal to $j0.0218$ or the normalized input admittance of the stub should be $j0.0218 \times 74 = 1.6132$. Let this point be denoted by D' on the Smith chart.

14. The length of the short-circuited stub will be $0.5\lambda - 0.25\lambda + 0.162\lambda = 0.412\lambda$.

15. The length of the open-circuited lab will be 0.162λ.

Now, let us solve the problem using direct formulas:

$$d = \frac{1}{\beta}\tan^{-1}\left[\left[-X_L \pm \sqrt{\frac{R_L\left[(Z_0 - R_L)^2 + X_L^2\right]}{Z_0}}\right]\bigg/(Z_0 - R_L)\right]$$

$$\frac{d}{\lambda} = \frac{1}{2\pi}\tan^{-1}\left[\frac{-60.93 \pm \sqrt{\dfrac{78.27\left[(50.026 - 78.27)^2 + (60.93)^2\right]}{50.026}}}{(50.026 - 78.27)}\right]$$

$$= \frac{1}{2\pi}\tan^{-1}\left[\frac{60.93 \mp 84.003}{28.244}\right]$$

or $\qquad d = \begin{cases} 0.391\lambda \\ 0.219\lambda \end{cases}$

These are almost similar to those obtained using a Smith chart.

For $d = 0.219\lambda$ and a short-circuited stub:

$$l_{ss} = \frac{1}{\beta}\tan^{-1}\left(\frac{Z_0}{Z_{0S}}\frac{\left[R_L^2 + \{Z_0\tan(\beta d) + X_L\}^2\right]}{R_L^2\tan(\beta d) - \{Z_0 - X_L\tan(\beta d)\}\{X_L + Z_0\tan(\beta d)\}}\right)$$

or $\qquad \dfrac{l_{ss}}{\lambda} = \dfrac{1}{2\pi}\tan^{-1}\left(\dfrac{50.026}{74}\dfrac{\left[(78.27)^2 + \{50.026 \times 5.0689 + 60.93\}^2\right]}{(78.27)^2 \times 5.0689 - \{50.026 - 60.93 \times 5.0689\}}\right)$
$$\{60.93 + 50.026 \times 5.0689\}$$

or $\qquad \dfrac{l_{ss}}{\lambda} = \dfrac{1}{2\pi}\tan^{-1}\left(\dfrac{50.026}{74}\dfrac{1,05,040.715}{31,053.0592 + 81,401.301}\right) = \dfrac{1}{2\pi}\tan^{-1}\left(\dfrac{52,54,766.809}{83,21,622.655}\right)$

or $\qquad l_{ss} = 0.090\lambda$

For $d = 0.219\lambda$ and an open-circuited stub:

$$\frac{l_{os}}{\lambda} = -\frac{1}{2\pi}\cot^{-1}\left(\frac{Z_0}{Z_{0S}}\frac{\left[R_L^2 + \{Z_0\tan(\beta d) + X_L\}^2\right]}{R_L^2\tan(\beta d) - \{Z_0 - X_L\tan(\beta d)\}\{X_L + Z_0\tan(\beta d)\}}\right)$$

$$= -\frac{1}{2\pi}\cot^{-1}\left(\frac{52,54,766.809}{83,21,622.655}\right)$$

or $\qquad l_{os} = 0.340\lambda$

These are again almost similar to those obtained using a Smith chart.
   For $d = 0.391\lambda$ and a short-circuited stub:

$$l_{ss} = \frac{1}{\beta}\tan^{-1}\left(\frac{Z_0}{Z_{0S}}\frac{\left[R_L^2 + \{Z_0\tan(\beta d) + X_L\}^2\right]}{R_L^2\tan(\beta d) - \{Z_0 - X_L\tan(\beta d)\}\{X_L + Z_0\tan(\beta d)\}}\right)$$

or $\qquad \frac{l_{ss}}{\lambda} = \frac{1}{2\pi}\tan^{-1}\left(\frac{50.026}{74}\frac{\left[(78.27)^2 + \{-50.026\times0.8167 + 60.93\}^2\right]}{(78.27)^2(-0.8167) - (50.026 + 60.93\times0.8167)}\right)$
$$(60.93 - 50.026\times0.8167)$$

Practice Problems

3.3@ Design single-series stub matching network(s) for the transmission line of Example 3.1 using a Smith chart. Assume that the characteristic impedance of the matching stub is 74 Ω. Verify the result using direct formulas.

3.4 Design single shunt stub matching network(s) for the transmission line of Practice Problem 3.1 using a Smith chart. Assume that the characteristic impedance of the matching stub is 100 Ω. Verify the result using direct formulas. [Chart: 0.283λ, 0.067λ, 0.317λ, 0.424λ, 0.434λ, 0.184λ; formula: 0.280λ, 0.068λ, 0.317λ, 0.422λ, 0.432λ, 0.182λ]

3.5 Design single shunt stub matching network(s) for the transmission line of Practice Problem 3.2 using a Smith chart. Assume that the characteristic impedance of the matching stub is 50Ω. Verify the result using direct formulas. [Chart: 0.152λ, 0.209λ, 0.459λ, 0.361λ, 0.292λ, 0.042λ; formula: 0.210λ, 0.460λ, 0.362λ, 0.29λ, 0.04λ]

3.6 Design single series stub matching network(s) for the transmission line of Practice Problem 3.1 using a Smith chart. Assume that the characteristic impedance of the matching stub is 100Ω. Verify the result using direct formulas. [Chart: 0.032λ, 0.362λ, 0.112λ, 0.172λ, 0.138λ, 0.388λ; formula: 0.030λ, 0.361λ, 0.111λ, 0.171λ, 0.141λ, 0.391λ]

3.7 Design single series stub matching network(s) for the transmission line of Practice Problem 3.2 using a Smith chart. Assume that the characteristic impedance of the matching stub is 50Ω. Verify the result using direct formulas. [Chart: 0.111λ, 0.125λ, 0.375λ, 0.401λ, 0.375λ, 0.125λ; formula: 0.112λ, 0.128λ, 0.378λ, 0.403λ, 0.372λ, 0.122λ]

@Solution in Online Resources

or $\dfrac{l_{ss}}{\lambda} = \dfrac{1}{2\pi}\tan^{-1}\left(\dfrac{50.026}{74} - \dfrac{6,529.149}{-5,003.2617 - 2,003.1115}\right) = \dfrac{1}{2\pi}\tan^{-1}\left(-\dfrac{3,26,627.2079}{5,18,471.6168}\right)$

or $l_{ss} = 0.411\lambda$

For $d = 0.219\lambda$ and an open-circuited stub:

$$\dfrac{l_{os}}{\lambda} = -\dfrac{1}{2\pi}\cot^{-1}\left(\dfrac{Z_0}{Z_{0S}}\dfrac{\left[R_L^2 + \{Z_0\tan(\beta d)+X_L\}^2\right]}{R_L^2\tan(\beta d)-\{Z_0 - X_L\tan(\beta d)\}\{X_L + Z_0\tan(\beta d)\}}\right)$$

$$= -\dfrac{1}{2\pi}\cot^{-1}\left(-\dfrac{3,26,627.2079}{5,18,471.6168}\right)$$

or $l_{os} = 0.161\lambda$

These are again almost similar to those obtained using a Smith chart.

3.5 Double-stub Matching

The single-stub matching technique, just described, is able to match any complex load but the difficulty comes from the placement of the stub at the exact location. For a satisfactory operation, it must be placed at a point where the normalized input resistance/conductance is equal to 1. Since a stub cannot always be placed physically at the ideal location, double-stub matching is used. The double-stub matching technique uses two short- or open-circuited stubs in parallel with a fixed length between them, as shown in Fig. 3.9.

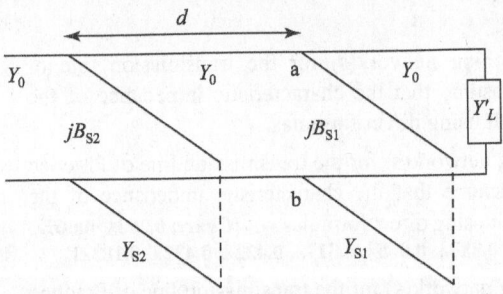

Fig. 3.9 General double-stub matching circuit

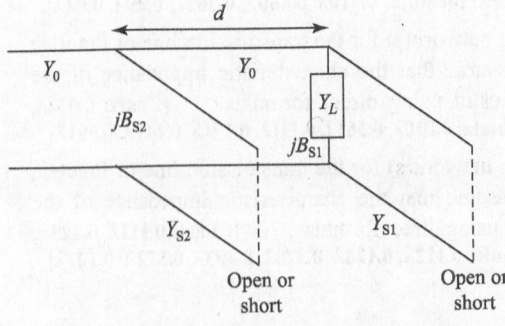

Fig. 3.10 Equivalent circuit of Fig. 3.9

Without any loss of generality, the circuit shown in Fig. 3.9 can be modified as that in Fig. 3.10. Here, Y_L is the input admittance of the line at position 'ab' due to the load Y_L', jB_{S1} is the input admittance of stub 1, jB_{S2} is the input admittance of stub 2, Y_{S1} is the characteristic admittance of stub 1, and Y_{S2} is the characteristic admittance of stub 2.

The equivalent admittance, on the generator side of the first stub, is as follows:

$$Y_1 = G_L + j(B_L + B_{s1}) \tag{3.42}$$

where $Y_L = G_L + jB_L$. The input admittance on the load side of the second stub is as follows:

$$Y_2 = Y_0 \frac{Y_1 + jY_0 \tan(\beta d)}{Y_0 + jY_1 \tan(\beta d)} \tag{3.43}$$

Substituting Eq. (3.42) into Eq. (3.43), we get the following relation:

$$Y_2 = Y_0 \left(\frac{G_L + j\{B_L + B_{s1} + Y_0 \tan(\beta d)\}}{\{Y_0 - (B_L + B_{s1}) \tan(\beta d)\} + jG_L \tan(\beta d)} \right) \tag{3.44}$$

$$Y_2 = Y_0 \frac{\left[G_L + j\{B_L + B_{s1} + Y_0 \tan(\beta d)\}\right]\left[\{Y_0 - (B_L + B_{s1}) \tan(\beta d)\} - jG_L \tan(\beta d)\right]}{\left[\{Y_0 - (B_L + B_{s1}) \tan(\beta d)\} + jG_L \tan(\beta d)\right]\left[\{Y_0 - (B_L + B_{s1}) \tan(\beta d)\} - jG_L \tan(\beta d)\right]}$$

$$Y_2 = Y_0 \frac{G_L\{Y_0 - (B_L + B_{s1}) \tan(\beta d)\} + G_L\{B_L + B_{s1} + Y_0 \tan(\beta d)\} \tan(\beta d)}{\{Y_0 - (B_L + B_{s1}) \tan(\beta d)\}^2 + G_L^2 \tan^2(\beta d)}$$

$$+ jY_0 \frac{\{B_L + B_{s1} + Y_0 \tan(\beta d)\}\{Y_0 - (B_L + B_{s1}) \tan(\beta d)\} - G_L^2 \tan(\beta d)}{\{Y_0 - (B_L + B_{s1}) \tan(\beta d)\}^2 + G_L^2 \tan^2(\beta d)} \tag{3.45}$$

Since the real part of Eq. (3.45) must be equal to Y_0 to achieve matching, we can write the following equation:

$$Y_0 = Y_0 \frac{G_L\{Y_0 - (B_L + B_{s1}) \tan(\beta d)\} + G_L\{B_L + B_{s1} + Y_0 \tan(\beta d)\} \tan(\beta d)}{\{Y_0 - (B_L + B_{s1}) \tan(\beta d)\}^2 + G_L^2 \tan^2(\beta d)}$$

or

$$G_L\{Y_0 - (B_L + B_{s1}) \tan(\beta d)\} + G_L\{B_L + B_{s1} + Y_0 \tan(\beta d)\} \tan(\beta d)$$
$$= \{Y_0 - (B_L + B_{s1}) \tan(\beta d)\}^2 + G_L^2 \tan^2(\beta d)$$

or

$$G_L^2 \tan^2(\beta d) - G_L Y_0 \{1 + \tan^2(\beta d)\}$$
$$+ \{Y_0 - (B_L + B_{s1}) \tan(\beta d)\}^2 = 0 \tag{3.46}$$

$$G_L = \frac{Y_0\{1 + \tan^2(\beta d)\} \pm \sqrt{Y_0^2\{1 + \tan^2(\beta d)\}^2 - 4\{Y_0 - (B_L + B_{s1}) \tan(\beta d)\}^2 \tan^2(\beta d)}}{2 \tan^2(\beta d)}$$

$$G_L = \frac{Y_0\{1 + \tan^2(\beta d)\} \pm Y_0\{1 + \tan^2(\beta d)\} \sqrt{1 - \dfrac{4 \tan^2(\beta d)\{Y_0 - (B_L + B_{s1}) \tan(\beta d)\}^2}{Y_0^2\{1 + \tan^2(\beta d)\}^2}}}{2 \tan^2(\beta d)}$$

or $\quad G_L = Y_0 \dfrac{\left\{1+\tan^2\left(\beta d\right)\right\}}{2\tan^2\left(\beta d\right)}\left[1\pm\sqrt{1-\dfrac{4\tan^2\left(\beta d\right)\left\{Y_0-\left(B_L+B_{s1}\right)\tan\left(\beta d\right)\right\}^2}{Y_0^2\left\{1+\tan^2\left(\beta d\right)\right\}^2}}\;\right]$ (3.47)

Since G_L is real and positive, the quantity within the square brackets must be positive. Therefore,

$$0 \le \frac{4\tan^2\left(\beta d\right)\left\{Y_0-\left(B_L+B_{s1}\right)\tan\left(\beta d\right)\right\}^2}{Y_0^2\left\{1+\tan^2\left(\beta d\right)\right\}^2} \le 1 \qquad (3.48)$$

For the regions outside this limit, no solution exists. Such a region in a Smith chart is called a *forbidden region*.

In the lower limit of Eq. (3.48), $0 \le G_L \le Y_0\left\{1+\tan^2\left(\beta d\right)\right\}\big/\tan^2\left(\beta d\right)$

On the other hand, in the higher limit of Eq. (3.48),

$$G_L = Y_0 \frac{\left\{1+\tan^2\left(\beta d\right)\right\}}{2\tan^2\left(\beta d\right)}$$

Therefore, the range of G_L that can be matched for a given stub spacing d is given by the following equation:

$$0 \le G_L \le Y_0\left\{1+\tan^2\left(\beta d\right)\right\}\big/\tan^2\left(\beta d\right)$$

or $\qquad 0 \le G_L \le Y_0\big/\sin^2\left(\beta d\right)$ (3.49)

Since G_L is known, the designer can choose a value of d that satisfies the limit in Eq. (3.49). Once the distance d has been fixed, the designer can find the input susceptance of the first stub from Eq. (3.46) as follows:

$$\left\{Y_0-\left(B_L+B_{s1}\right)\tan\left(\beta d\right)\right\}^2 = G_L Y_0\left\{1+\tan^2\left(\beta d\right)\right\}-G_L^2\tan^2\left(\beta d\right)$$

or $\qquad B_{s1} = -B_L + \dfrac{Y_0 \pm \sqrt{G_L Y_0\left\{1+\tan^2\left(\beta d\right)\right\}-G_L^2\tan^2\left(\beta d\right)}}{\tan\left(\beta d\right)}$ (3.50)

The real part of the input admittance of the line, at the position of the second stub, is equal to the characteristic admittance of the line; hence, we must adjust the input susceptance of the second stub in such a way that it cancels out the input susceptance of the line at the position of the stub. Therefore, the input susceptance of the stub must be equal in magnitude but opposite in sign to the input susceptance of the line at the position of the stub. The input susceptance of the line at the position of the second stub can be calculated from Eq. (3.45) as follows:

$$B_{s2} = -Y_0 \frac{\left\{B_L+B_{s1}+Y_0\tan\left(\beta d\right)\right\}\left\{Y_0-\left(B_L+B_{s1}\right)\tan\left(\beta d\right)\right\}-G_L^2\tan\left(\beta d\right)}{\left\{Y_0-\left(B_L+B_{s1}\right)\tan\left(\beta d\right)\right\}^2+G_L^2\tan^2\left(\beta d\right)} \qquad (3.51)$$

Based on Eq. (3.50), we can write the following expression:

$$Y_0 - (B_L + B_{s1}) \tan(\beta d) = \mp \sqrt{G_L Y_0 \left\{1 + \tan^2(\beta d)\right\} - G_L^2 \tan^2(\beta d)} \quad (3.52)$$

Now substituting Eq. (3.52) into Eq. (3.51), we get the following formula:

$$B_{s2} = -Y_0 \frac{\left| \mp \left\{B_L + B_{s1} + Y_0 \tan(\beta d)\right\} \sqrt{G_L Y_0 \left\{1 + \tan^2(\beta d)\right\} - G_L^2 \tan^2(\beta d)} - G_L^2 \tan(\beta d) \right|}{G_L Y_0 \left\{1 + \tan^2(\beta d)\right\} - G_L^2 \tan^2(\beta d) + G_L^2 \tan^2(\beta d)}$$

$$B_{s2} = \frac{\left| \pm \left\{B_L + B_{s1} + Y_0 \tan(\beta d)\right\} \sqrt{G_L Y_0 \left\{1 + \tan^2(\beta d)\right\} - G_L^2 \tan^2(\beta d)} + G_L^2 \tan(\beta d) \right|}{G_L \left\{1 + \tan^2(\beta d)\right\}}$$

$$(3.53)$$

Again from Eq. (3.50), we get the following expression:

$$B_L + B_{s1} = \frac{Y_0 \pm \sqrt{G_L Y_0 \left\{1 + \tan^2(\beta d)\right\} - G_L^2 \tan^2(\beta d)}}{\tan(\beta d)} \quad (3.54)$$

Substituting Eq. (3.54) into Eq. (3.53) and assuming the $p = \tan(\beta d)$ for the compactness of the derivation, the following equation can be obtained:

$$B_{s2} = \frac{\left| \pm \left\{ \dfrac{Y_0 \pm \sqrt{G_L Y_0 \left(1 + p^2\right) - G_L^2 p^2}}{p} + Y_0 p \right\} \sqrt{G_L Y_0 \left(1 + p^2\right) - G_L^2 p^2} + G_L^2 p \right|}{G_L \left(1 + p^2\right)}$$

$$\text{or} \quad B_{s2} = \frac{\left| \pm \left\{ \dfrac{Y_0 \left(1 + p^2\right) \pm \sqrt{G_L Y_0 \left(1 + p^2\right) - G_L^2 p^2}}{p} \right\} \sqrt{G_L Y_0 \left(1 + p^2\right) - G_L^2 p^2} + G_L^2 p \right|}{G_L \left(1 + p^2\right)}$$

$$\text{or} \quad B_{s2} = \frac{\left| \pm \left\{ \dfrac{Y_0 \left(1 + p^2\right) \sqrt{G_L Y_0 \left(1 + p^2\right) - G_L^2 p^2} \pm \left\{G_L Y_0 \left(1 + p^2\right) - G_L^2 p^2\right\}}{p} \right\} + G_L^2 p \right|}{G_L \left(1 + p^2\right)}$$

$$\text{or} \quad B_{s2} = \frac{\pm Y_0 \left(1 + p^2\right) \sqrt{G_L Y_0 \left(1 + p^2\right) - G_L^2 p^2} + G_L Y_0 \left(1 + p^2\right)}{G_L p \left(1 + p^2\right)}$$

$$\text{or} \quad B_{s2} = \frac{\pm Y_0 \sqrt{G_L Y_0 \left(1 + p^2\right) - G_L^2 p^2} + G_L Y_0}{G_L p} \quad (3.55)$$

Substituting back $p = \tan(\beta d)$ into Eq. (3.55), we get the following relation:

$$B_{s2} = \frac{\pm Y_0 \sqrt{G_L Y_0 \left\{1 + \tan^2(\beta d)\right\} - G_L^2 \tan^2(\beta d)} + G_L Y_0}{G_L \tan(\beta d)} \tag{3.56}$$

The positive and negative signs in Eqs (3.50) and (3.56) correspond to the same solution. Once the values of B_{s1} and B_{s2} are obtained, the corresponding stub lengths can be estimated in the same manner as described in Section 3.4.

Note Equation (3.49) reveals that if $G_L > Y_0/\{\sin^2(\beta d)\}$, then it cannot be matched using the double-stub matching technique. The triple-stub matching technique can be used in such cases. A triple-stub matching network consists of three stubs, with the third stub being at the position of the load. The third stub adds a susceptance to the load admittance and results in an equivalent load admittance that provides an input conductance on the load side of the first stub that satisfies this condition.

EXAMPLE 3.3 Design double-shunt-stub matching network(s) for the transmission line of Example 3.1 using a Smith chart. Assume that the distance between the load and the first stub is 0.158λ and that between the two stubs is $\lambda/8$. Verify the result using direct formulas. The characteristic impedance of the line and that of the stubs are identical.

Solution Given: $Z_0 = 50.026\ \Omega$ and $Z_L = 78.27 + j60.93\ \Omega$

(a) The normalized load impedance is as follows:

$$z_L = \frac{Z_L}{Z_0} = \frac{78.27 + j60.93}{50.026} = 1.5646 + j1.218$$

(b) Enter the $1.5646 + j1.218$ point on the Smith chart as A (Fig. 3.11).

(c) Now we need to convert the normalized load impedance into the normalized load admittance. The normalized load admittance, Y_L, is exactly opposite to point A on the constant SWR circle, that is, they are $180°$ apart; the normalized load admittance is obtained as $y_L = 0.398 - j0.32$. Let this point be B.

(d) The input admittance at the load side of the first stub can be obtained by moving a distance of 0.158λ towards the generator on the constant SWR circle. The new input admittance is $Y_d = 0.5 + j0.6$. The corresponding point is denoted by C on the Smith chart.

(e) The distance between the two stubs is $\lambda/8$ and the input conductance on the generator side of the second stub must be unity; hence, we must draw a spacing circle at a distance $\lambda/8$ towards the load from the unity conductance circle so that once we move $\lambda/8$ towards the generator from this spacing circle we reach at the unity conductance circle. The spacing circle touches the $g = 2$ circle tangentially.

(f) Since the input admittance lies outside the $g = 2$ circle, the load can be matched using the double-stub matching technique.

(g) The input admittance on the generator side of the first stub (y_1) is equal to the sum of the input admittance on the load side of the first stub (y_d) and that of the first stub (y_{s1}).

(h) The input admittance y_1, when transferred to the position of the second stub, must lie on the unit conductance circle. This implies that y_1 must lie on the spacing circle.

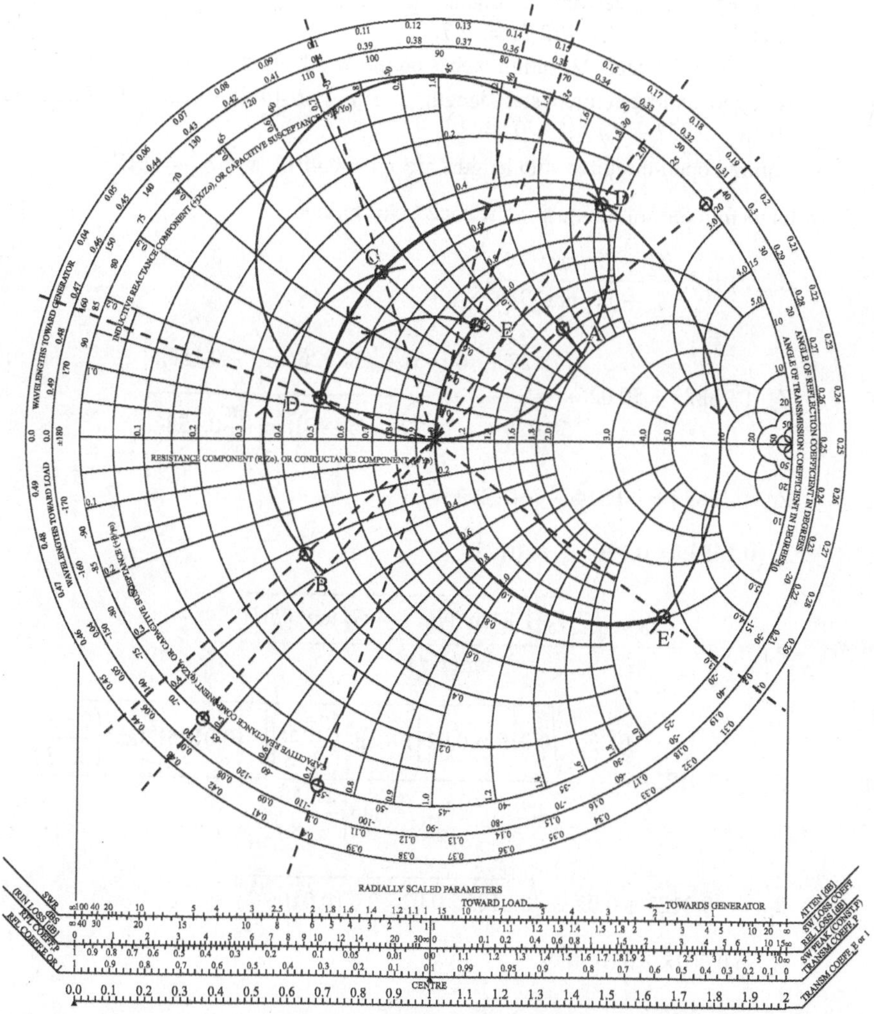

Fig. 3.11   Smith chart for Example 3.3

(i)   Since y_1 lies on the spacing circle, two solutions are possible, as denoted by points D and D'. Solutions are obtained by moving along the constant g circle on which y_{d1} exists (as the stub produces only the susceptive admittance, the g value does not change). The solutions are $y_{1,D} = 0.5 + j0.14$ and $y_{1,D'} = 0.5 + j1.85$.

(j)   The required stub susceptance to obtain these solutions are $j_{s1,D} = 0.5 + j0.14 - (0.5 + j0.6) = -j0.46$ and $y_{s1,D'} = (0.5 + j1.85) - (0.5 + j0.6) = j1.25$.

(k)   Locate the $-j0.46$ and $j1.25$ admittances on the Smith chart.

(l)   The required short-circuited stub lengths are $l_{s1} = 0.431\lambda - 0.25\lambda = 0.181\lambda$ and $l_{s1'} = 0.5\lambda - 0.25\lambda + 0.143\lambda = 0.393\lambda$.

(m)   The required open-circuited stub lengths are $l_{o1} = 0.431\lambda$ and $l_{o1'} = 0.141\lambda$.

(n)   The corresponding input admittance on the load side of stub 2 is obtained by moving towards the generator on the respective SWR circles. The respective input admittances are $y_{d2} = 1 + j0.72$ and $y_{d2'} = 1 - j2.7$.

(o) The required stub susceptance to obtain these solutions are $y_{s2, E} = 1-(1+j0.72) = -j0.72$ and $y_{s2,E'} = 1-(1-j2.7) = j2.7$.

(p) Locate the $-j0.72$ and $j2.7$ admittances on the Smith chart.

(q) The required short-circuited stub lengths are $l_{s2} = 0.401\lambda - 0.25\lambda = 0.151\lambda$ and $l_{s2'} = 0.5\lambda - 0.25\lambda + 0.194\lambda = 0.444\lambda$.

(r) The required open-circuited stub lengths are $l_{o2} = 0.401\lambda$ and $l_{o2'} = 0.194\lambda$.

Now let us find the solution using the analytic technique:

$$Z_{in}(z) = Z_0 \frac{Z_L + jZ_0 \tan(\beta d)}{Z_0 + jZ_L \tan(\beta d)}$$

or $\quad Z_{in}(0.158\lambda) = 50.026 \dfrac{(78.27 + j60.93) + j50.026 \tan\left(\dfrac{2\pi}{\lambda}0.158\lambda\right)}{50.026 + j(78.27 + j60.93)\tan\left(\dfrac{2\pi}{\lambda}0.158\lambda\right)}$

or $\quad Z_{in}(0.158\lambda) = 40.3144 - j47.2096$

or $\quad Y_{in}(0.158\lambda) = 0.0105 + j0.0122$

Now, $\quad B_{s1} = -B_L + \dfrac{Y_0 \pm \sqrt{G_L Y_0 \left\{1 + \tan^2(\beta d)\right\} - G_L^2 \tan^2(\beta d)}}{\tan(\beta d)}$

or $\quad B_{s1} = -0.0122 + \dfrac{0.02 \pm \sqrt{0.0105 \times 0.02\left\{1 + \tan^2\left(\dfrac{2\pi}{\lambda}\dfrac{\lambda}{8}\right)\right\} - (0.0105)^2 \tan^2\left(\dfrac{2\pi}{\lambda}\dfrac{\lambda}{8}\right)}}{\tan\left(\dfrac{2\pi}{\lambda}\dfrac{\lambda}{8}\right)}$

or $\quad B_{s1} = -0.0122 + 0.02 \pm \sqrt{0.0105 \times 0.02 \times 2 - (0.0105)^2}$

or $\quad B_{s1} = 0.0078 \pm 0.0176$

or $\quad b_{s1} = \begin{cases} -0.49 \\ 1.27 \end{cases}$

For a short-circuited stub: $j\tan(\beta l_{s1}) = \dfrac{1}{jb_{s1}}$

or $\quad l_{s1} = \dfrac{\lambda}{2\pi}\tan^{-1}\left(-\dfrac{1}{b_{s1}}\right) = \dfrac{\lambda}{2\pi}\tan^{-1}\left(\dfrac{1}{0.49}\right) = 0.1775\lambda$

Similarly, $l_{s1'} = 0.3938\lambda$

For an open-circuited stub: $-j\cot(\beta l_{o1}) = \dfrac{1}{jb_{s1}}$

or $\quad l_{o1} = \dfrac{\lambda}{2\pi}\cot^{-1}\left(\dfrac{1}{b_{s1}}\right) = \dfrac{\lambda}{2\pi}\cot^{-1}\left(-\dfrac{1}{0.49}\right) = 0.4275\lambda$

Similarly, $l_{o1'} = 0.1438\lambda$

Again
$$B_{s2} = \frac{\pm Y_0 \sqrt{G_L Y_0 \left\{ 1 + \tan^2 (\beta d) \right\} - G_L^2 \tan^2 (\beta d)} + G_L Y_0}{G_L \tan (\beta d)}$$

or
$$B_{s2} = \frac{\pm 0.02 \sqrt{0.0105 \times 0.02 \times 2 - (0.0105)^2} + 0.0105 \times 0.02}{0.0105}$$

or
$$B_{s2} = \frac{(\pm 3.5199 + 2.1) \times 10^{-4}}{0.0105}$$

or
$$b_{s2} = \begin{cases} -0.676 \\ 2.677 \end{cases}$$

Now for a short-circuited stub: $j \tan (\beta l_{s2}) = \dfrac{1}{j b_{s2}}$

or
$$l_{s2} = \frac{\lambda}{2\pi} \tan^{-1} \left(-\frac{1}{b_{s2}} \right) = \frac{\lambda}{2\pi} \tan^{-1} \left(\frac{1}{0.952} \right) = 0.1289\lambda$$

Similarly, $l_{s2'} = 0.4464\lambda$

For an open-circuited stub: $-j \cot (\beta l_{o2}) = \dfrac{1}{j b_{s2}}$

or
$$l_{o2} = \frac{\lambda}{2\pi} \cot^{-1} \left(\frac{1}{b_{s2}} \right) = \frac{\lambda}{2\pi} \cot^{-1} \left(-\frac{1}{0.952} \right) = 0.3789\lambda$$

Similarly, $l_{o2'} = 0.1964\lambda$

These are again almost similar to those obtained using a Smith chart.

Practice Problems

3.8 Design double-shunt-stub matching network(s) for the transmission line of Practice Problem 3.1 using a Smith chart. Assume that the distance between the load and the first stub is 0.19λ and that between the two stubs is $\lambda/8$. Verify the result using direct formulas. The characteristic impedance of the line and that of the stubs are identical. [Chart: $-j0.33, j1.14, 0.199\lambda, 0.386\lambda, 0.449\lambda, 0.136\lambda, -j1.22, j3.19, 0.109\lambda, 0.451\lambda, 0.359\lambda, 0.201\lambda$; formula: $-j0.311, j1.141, 0.202\lambda, 0.385\lambda, 0.452\lambda, 0.135\lambda, -j1.326, j3.326, 0.103\lambda, 0.453\lambda, 0.353\lambda, 0.203\lambda$]

3.9 Design double-shunt-stub matching network(s) for the transmission line of Practice Problem 3.2 using a Smith chart. Assume that the distance between the load and the first stub is 0.19λ and that between the two stubs is $\lambda/8$. Verify the result using direct formulas. The characteristic impedance of the line and that of the stubs are identical. [Chart: $-j0.42, j1.4, 0.186\lambda, 0.401\lambda, 0.436\lambda, 0.151\lambda, j0.28, j1.69, 0.294\lambda, 0.413\lambda, 0.044\lambda, 0.163\lambda$; formula: $-j0.44, j1.46, 0.184\lambda, 0.404\lambda, 0.434\lambda, 0.154\lambda, j0.28, j1.72, 0.294\lambda, 0.416\lambda, 0.044\lambda, 0.166\lambda$]

3.6 Quarter-wave Transformers

In Section 2.8, it has already been demonstrated that a transmission line having a length of $\lambda/4$ and a characteristic impedance of $Z_0' = \sqrt{Z_{in}(z) Z_L}$ can transform a real load Z_L into another load $Z_{in}(z)$. Now if we assume that $Z_{in}(z) = Z_0$, where Z_0 is

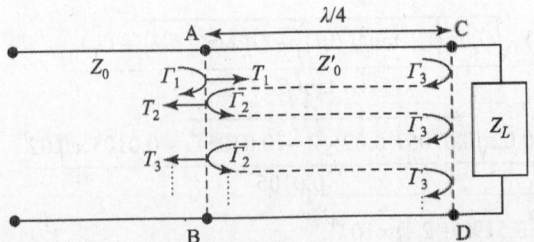

Fig. 3.12 Multiple reflections in quarter-wave transformer

the characteristic impedance of the line, then a quarter-wavelength line transforms the real load Z_L into the characteristic impedance of the line Z_0, and behaves like a matching network between the load Z_L and the line.

To interpret the matching technique, let us consider Fig. 3.12. When the incident wave arrives at the junction 'AB' for the first time, it sees only an impedance Z_0' as it has not yet reached the load Z_L. On incidence, a part of the wave is reflected back with a reflection coefficient Γ_1, while the remaining part is transmitted into the quarter-wave line with a transmission coefficient T_1

where $\quad \Gamma_1 = \dfrac{Z_0' - Z_0}{Z_0' + Z_0}$ (3.57)

and $\quad T_1 = \dfrac{2Z_0'}{Z_0' + Z_0}$ (3.58)

The transmitted wave travels through the quarter-wave line, arrives at the load, and is reflected back from the load to the quarter-wave line with a reflection coefficient of Γ_3, as follows:

$$\Gamma_3 = \frac{Z_L - Z_0'}{Z_L + Z_0'} \qquad (3.59)$$

This reflected wave travels through the quarter-wave line towards the generator and arrives at the Z_0 impedance line. A part of this wave is re-reflected back from the from the Z_0 impedance line to the quarter-wave line with a reflection coefficient Γ_2 and the remaining part is transmitted into the Z_0 line with a transmission coefficient T_2

where $\quad \Gamma_2 = \dfrac{Z_0 - Z_0'}{Z_0 + Z_0'} = -\Gamma_1$ (3.60)

and $\quad T_2 = \dfrac{2Z_0}{Z_0' + Z_0}$ (3.61)

This process is continued an infinite number of times. The total reflection coefficient Γ in the line is represented by the vector sum of these reflected and transmitted waves on the generator side of the boundary AB.

$$\Gamma = \Gamma_1 - T_1 T_2 \Gamma_3 + T_1 T_2 \Gamma_2 \Gamma_3^2 - T_1 T_2 \Gamma_2^2 \Gamma_3^3 + \ldots = \Gamma_1 - T_1 T_2 \Gamma_3 \sum_{n=0}^{\infty} (-\Gamma_2 \Gamma_3)^n \quad (3.62)$$

In this equation, the alternate negative signs indicate that each round trip in the quarter-wave line contributes to a total phase change of 180°.

Since $|\Gamma_2| < 1$ and $|\Gamma_3| < 1$, the infinite series in Eq. (3.62) can be written as follows:

$$\Gamma = \Gamma_1 - \frac{T_1 T_2 \Gamma_3}{1 + \Gamma_2 \Gamma_3} = \frac{\Gamma_1 + \Gamma_1 \Gamma_2 \Gamma_3 - T_1 T_2 \Gamma_3}{1 + \Gamma_2 \Gamma_3} \tag{3.63}$$

Here, we have used the relations $|x| < 1$, $\displaystyle\sum_{n=0}^{\infty} x^n = \frac{1}{1-x}$, and $x = -\Gamma_2 \Gamma_3$.

Based on Eqs (3.57)–(3.61), the numerator of Eq. (3.63) can be written as follows:

$$\Gamma_1 + \Gamma_1 \Gamma_2 \Gamma_3 - T_1 T_2 \Gamma_3 = \Gamma_1 - \Gamma_3 \left(\Gamma_1^2 + T_1 T_2 \right)$$

or $\quad \Gamma_1 + \Gamma_1 \Gamma_2 \Gamma_3 - T_1 T_2 \Gamma_3 = \left(\dfrac{Z_0' - Z_0}{Z_0' + Z_0} \right) - \left(\dfrac{Z_L - Z_0'}{Z_L + Z_0'} \right) \left[\left(\dfrac{Z_0' - Z_0}{Z_0' + Z_0} \right)^2 + \dfrac{4 Z_0' Z_0}{\left(Z_0' + Z_0 \right)^2} \right]$

or $\quad \Gamma_1 + \Gamma_1 \Gamma_2 \Gamma_3 - T_1 T_2 \Gamma_3 = \left(\dfrac{Z_0' - Z_0}{Z_0' + Z_0} \right) - \left(\dfrac{Z_L - Z_0'}{Z_L + Z_0'} \right) \left[\dfrac{\left(Z_0' - Z_0 \right)^2 + 4 Z_0' Z_0}{\left(Z_0' + Z_0 \right)^2} \right]$

or $\quad \Gamma_1 + \Gamma_1 \Gamma_2 \Gamma_3 - T_1 T_2 \Gamma_3 = \left(\dfrac{Z_0' - Z_0}{Z_0' + Z_0} \right) - \left(\dfrac{Z_L - Z_0'}{Z_L + Z_0'} \right)$

or $\quad \Gamma_1 + \Gamma_1 \Gamma_2 \Gamma_3 - T_1 T_2 \Gamma_3 = \dfrac{\left(Z_0' - Z_0 \right) \left(Z_L + Z_0' \right) - \left(Z_0' + Z_0 \right) \left(Z_L - Z_0' \right)}{\left(Z_0' + Z_0 \right) \left(Z_L + Z_0' \right)}$

or $\quad \Gamma_1 + \Gamma_1 \Gamma_2 \Gamma_3 - T_1 T_2 \Gamma_3 = \dfrac{2 \left\{ \left(Z_0' \right)^2 - Z_0 Z_L \right\}}{\left(Z_0' + Z_0 \right) \left(Z_L + Z_0' \right)} \tag{3.64}$

If we choose $Z_0' = \sqrt{Z_0 Z_L}$ $\tag{3.65}$

then from Eq. (3.64) we get that $\Gamma_1 + \Gamma_1 \Gamma_2 \Gamma_3 - T_1 T_2 \Gamma_3 = 0$ and hence $\Gamma = 0$. Therefore, the line will be matched.

Like other matching networks, a quarter-wave line or quarter-wave transformer is also a narrow-band matching network. This can be explained by the fact that, as the frequency deviates from the design frequency, the corresponding wavelength also changes and the length of the line no more remains $\lambda/4$.

At a frequency $f' \neq f_0$, where f_0 is the design frequency, the input impedance of the line is given by the following equation:

$$Z_{\text{in}} = Z_0' \frac{Z_L + j Z_0' \tan \left(\beta' d \right)}{Z_0' + j Z_L \tan \left(\beta' d \right)} \tag{3.66}$$

where β' is the phase constant of the quarter-wave line at $f = f'$ and d is the length of the line. Since we are considering a frequency $f' \neq f_0$, the length d

is no more a quarter wavelength. The resultant reflection coefficient will be as follows:

$$\Gamma = \frac{Z_{in} - Z_0}{Z_{in} + Z_0} \tag{3.67}$$

Substituting Eq. (3.66) into Eq. (3.67), we get the following relation:

$$\Gamma = \frac{Z_0' \dfrac{Z_L + jZ_0' \tan(\beta'd)}{Z_0' + jZ_L \tan(\beta'd)} - Z_0}{Z_0' \dfrac{Z_L + jZ_0' \tan(\beta'd)}{Z_0' + jZ_L \tan(\beta'd)} + Z_0} = \frac{Z_0'\{Z_L + jZ_0' \tan(\beta'd)\} - Z_0\{Z_0' + jZ_L \tan(\beta'd)\}}{Z_0'\{Z_L + jZ_0' \tan(\beta'd)\} + Z_0\{Z_0' + jZ_L \tan(\beta'd)\}}$$

or

$$\Gamma = \frac{Z_0'(Z_L - Z_0) + j\{(Z_0')^2 - Z_0 Z_L\} \tan(\beta'd)}{Z_0'(Z_L + Z_0) + j\{(Z_0')^2 + Z_0 Z_L\} \tan(\beta'd)} \tag{3.68}$$

Substituting Eq. (3.65) into Eq. (3.68), we get the following expression:

$$\Gamma = \frac{\sqrt{Z_0 Z_L}(Z_L - Z_0)}{\sqrt{Z_0 Z_L}(Z_L + Z_0) + j2Z_0 Z_L \tan(\beta'd)}$$

or

$$\Gamma = \frac{(Z_L - Z_0)}{(Z_L + Z_0) + j2\sqrt{Z_0 Z_L} \tan(\beta'd)} \tag{3.69}$$

Taking the magnitude, the following equation is obtained:

$$|\Gamma| = \frac{|Z_L - Z_0|}{\sqrt{(Z_L + Z_0)^2 + 4Z_0 Z_L \tan^2(\beta'd)}} = \frac{1}{\sqrt{\left(\dfrac{Z_L + Z_0}{Z_L - Z_0}\right)^2 + \dfrac{4Z_0 Z_L}{(Z_L - Z_0)^2} \tan^2(\beta'd)}}$$

or

$$|\Gamma| = \frac{1}{\sqrt{1 + \dfrac{4Z_0 Z_L}{(Z_L - Z_0)^2} + \dfrac{4Z_0 Z_L}{(Z_L - Z_0)^2} \tan^2(\beta'd)}} = \frac{1}{\sqrt{1 + \dfrac{4Z_0 Z_L}{(Z_L - Z_0)^2}\{1 + \tan^2(\beta'd)\}}}$$

or

$$|\Gamma| = \frac{1}{\sqrt{1 + \dfrac{4Z_0 Z_L}{(Z_L - Z_0)^2} \sec^2(\beta'd)}} \tag{3.70}$$

Now if $f' \simeq f_0$, then $d \simeq \lambda_0/4$ and $\beta'd \simeq \pi/2$. Since $\sec^2(\pi/2) \gg 1$, the second term under the square root in the denominator of Eq. (3.70) is significantly larger than the first term, and we can write that $1 + \dfrac{4Z_0 Z_L}{(Z_L - Z_0)^2} \sec^2(\beta'd) \approx \dfrac{4Z_0 Z_L}{(Z_L - Z_0)^2} \sec^2(\beta'd)$. Therefore, Eq. (3.70) reduces to the following form:

$$|\Gamma| \approx \frac{|Z_L - Z_0|}{2\sqrt{Z_0 Z_L}}\left|\cos\left(\beta'd\right)\right| \qquad (3.71)$$

If we restrict the maximum allowable reflection coefficient to Γ_m and it occurs at $(\beta'd) = \theta_m$, then the bandwidth of the matching network will be as follows:

$$\Delta\theta = 2\left(\frac{\pi}{2} - \theta_m\right) \qquad (3.72)$$

This is shown in Fig. 3.13. Since $\Gamma = \Gamma_m$ at $(\beta'd) = \theta_m$, Eq. (3.70) can be written as follows:

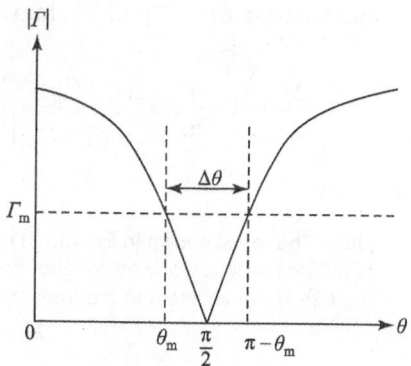

Fig. 3.13 Frequency response of quarter-wave transformer operating near design frequency

$$\Gamma_m = \frac{1}{\sqrt{1 + \dfrac{4Z_0 Z_L}{\left(Z_L - Z_0\right)^2}\sec^2\left(\theta_m\right)}}$$

or $$1 + \frac{4Z_0 Z_L}{\left(Z_L - Z_0\right)^2}\sec^2\left(\theta_m\right) = \frac{1}{\Gamma_m^2}$$

or $$\sec^2\left(\theta_m\right) = \frac{\left(Z_L - Z_0\right)^2}{4Z_0 Z_L}\left[\frac{1}{\Gamma_m^2} - 1\right]$$

or $$\sec\left(\theta_m\right) = \sqrt{\frac{\left(Z_L - Z_0\right)^2}{4Z_0 Z_L}\left[\frac{1 - \Gamma_m^2}{\Gamma_m^2}\right]} = \frac{|Z_L - Z_0|}{2\sqrt{Z_0 Z_L}}\frac{\sqrt{1 - \Gamma_m^2}}{\Gamma_m}$$

or $$\cos\left(\theta_m\right) = \frac{\Gamma_m}{\sqrt{1 - \Gamma_m^2}}\frac{2\sqrt{Z_0 Z_L}}{|Z_L - Z_0|} \qquad (3.73)$$

or $$\theta_m = \cos^{-1}\left[\frac{\Gamma_m}{\sqrt{1 - \Gamma_m^2}}\frac{2\sqrt{Z_0 Z_L}}{|Z_L - Z_0|}\right] \qquad (3.74)$$

Now, $$\theta = \beta d = \frac{2\pi}{\lambda}\frac{\lambda_0}{4} = \frac{2\pi f}{c}\frac{c}{4f_0} = \frac{\pi f}{2f_0} \qquad (3.75)$$

Therefore, $$\theta_m = \frac{\pi f_m}{2f_0}$$

or $$f_m = \frac{2\theta_m f_0}{\pi} \qquad (3.76)$$

Hence, the fractional bandwidth is as follows:

$$\frac{\Delta f}{f_0} = \frac{2(f_0 - f_m)}{f_0} = 2 - \frac{2f_m}{f_0} \tag{3.77}$$

Substituting from Eq. (3.76) into Eq. (3.77), we get the following relation:

$$\frac{\Delta f}{f_0} = 2 - \frac{4\theta_m}{\pi} \tag{3.78}$$

Substituting from Eq. (3.74) into Eq. (3.78), we get the following expression:

$$\frac{\Delta f}{f_0} = 2 - \frac{4}{\pi}\cos^{-1}\left[\frac{\Gamma_m}{\sqrt{1-\Gamma_m^2}}\frac{2\sqrt{Z_0 Z_L}}{|Z_L - Z_0|}\right] \tag{3.79}$$

Note The result shown in Eq. (3.79) is valid, provided that the propagation constant is a linear function of wavelengths and the reactance developed at the junction of the quarter-wave line and the transmission line is negligibly small.

The theory described so far assumes a real load. In practice, a complex load can also be matched using a quarter-wave transmission line, because the reactance part of the load admittance can easily be cancelled out using a short/open-circuited stub at the position of the load. Alternatively, we can also insert the quarter-wave line at a point on the line where the input impedance is purely real, that is, at the voltage maximum or voltage minimum point. These have been shown in Figs 3.14 and 3.15.

EXAMPLE 3.4 Design a quarter-wave transformer to match a 74 Ω load with a line of characteristic impedance 50 Ω at 1 GHz. If the maximum allowable reflection coefficient is 0.05, then calculate the percentage of bandwidth.

Solution Given: $Z_0 = 50\ \Omega$, $Z_L = 74\ \Omega$, and $\Gamma_m = 0.05$
The characteristic impedance of the quarter-wave line will be as follows:

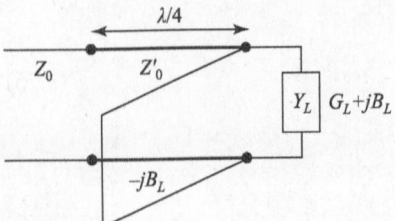

Fig. 3.14 Matching of complex load using quarter-wave line and stub at the position of load

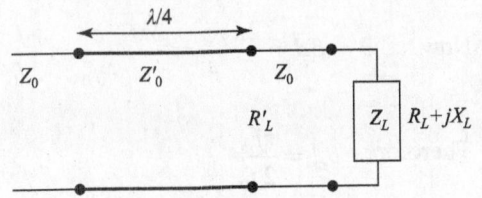

Fig. 3.15 Matching of complex load inserting quarter-wave line at voltage maximum/minimum point

$$Z_t = \sqrt{Z_L Z_0} = \sqrt{74 \times 50} = 60.8276 \; \Omega$$

The length of the line is given the following equation:

$$l_t = \frac{\lambda}{4} = \frac{c}{4f} = \frac{3 \times 10^{10}}{4 \times 10^9} = 7.5 \text{ cm}$$

The fractional bandwidth is as follows:

$$\frac{\Delta f}{f_0} = 2 - \frac{4}{\pi} \cos^{-1} \left[\frac{\Gamma_m}{\sqrt{1 - \Gamma_m^2}} \frac{2\sqrt{Z_0 Z_L}}{|Z_L - Z_0|} \right] = 2 - \frac{4}{\pi} \cos^{-1} \left[\frac{0.05}{\sqrt{1 - (0.05)^2}} \frac{2\sqrt{50 \times 74}}{|74 - 50|} \right]$$

or $$\frac{\Delta f}{f_0} = 2 - \frac{4}{\pi} \cos^{-1} [0.2538] = 0.3267$$

Therefore, the percentage of the bandwidth is 32.67%.

Practice Problem

3.10 Design a quarter-wave transformer to match a 100 Ω load with a line of characteristic impedance 74 Ω at 3 GHz. If the maximum allowable reflection coefficient is 0.05, then calculate the percentage of bandwidth.　　　　　[**86.023 Ω, 2.5 cm, 21.19%**]

3.7 Theory of Small Reflections

Let us consider the quarter-wave line shown in Fig. 3.12. If θ be the phase shift at any frequency f, then $\theta = \beta d$ and the reflection coefficient Γ can be written as follows:

$$\Gamma = \Gamma_1 + T_1 T_2 \Gamma_3 e^{-j2\theta} + T_1 T_2 \Gamma_2 \Gamma_3^2 e^{-j4\theta} + \ldots + T_1 T_2 \Gamma_2^n \Gamma_3^{n+1} e^{-j2(n+1)\theta} + \ldots \infty$$

or $$\Gamma = \Gamma_1 + T_1 T_2 \Gamma_3 e^{-2j\theta} \sum_{n=0}^{\infty} \Gamma_2^n \Gamma_3^n e^{-j2n\theta} \tag{3.80}$$

The expression of reflection coefficient given in Eq. (3.62) is valid only for $f = f_0$, whereas that given in Eq. (3.80) represents a general expression and valid for any frequency of excitation. As before, since for $|x| < 1$, $\sum_{n=0}^{\infty} x^n = \frac{1}{1-x}$ and $|\Gamma_2| < 1$, $|\Gamma_3| < 1$, the infinite series in Eq. (3.80) can be written as follows:

$$\Gamma = \Gamma_1 + \frac{T_1 T_2 \Gamma_3 e^{-2j\theta}}{1 - \Gamma_2 \Gamma_3 e^{-2j\theta}} \tag{3.81}$$

Again since $\Gamma_2 = -\Gamma_1$, Eq. (3.81) modifies to the following form:

$$\Gamma = \Gamma_1 + \frac{T_1 T_2 \Gamma_3 e^{-2j\theta}}{1 + \Gamma_1 \Gamma_3 e^{-2j\theta}} \tag{3.82}$$

Further substituting $T_1 = 1 + \Gamma_1$ and $T_2 = 1 + \Gamma_2 = 1 - \Gamma_1$ into Eq. (3.82), we get the following expression:

$$\Gamma = \Gamma_1 + \frac{(1+\Gamma_1)(1-\Gamma_1)\Gamma_3 e^{-2j\theta}}{1+\Gamma_1\Gamma_3 e^{-2j\theta}} = \frac{\Gamma_1\left(1+\Gamma_1\Gamma_3 e^{-2j\theta}\right) + \left(1-\Gamma_1^2\right)\Gamma_3 e^{-2j\theta}}{1+\Gamma_1\Gamma_3 e^{-2j\theta}}$$

or $$\Gamma = \frac{\Gamma_1 + \Gamma_3 e^{-2j\theta}}{1+\Gamma_1\Gamma_3 e^{-2j\theta}} \qquad (3.83)$$

Now if we assume that the discontinuities between Z_0 and Z_0' and between Z_0' and Z_L are small, then $|\Gamma_1\Gamma_3| \ll 1$ and Eq. (3.83) modifies to the following form:

$$\Gamma \simeq \Gamma_1 + \Gamma_3 e^{-2j\theta} \qquad (3.84)$$

Equation (3.84) reveals that if the discontinuities between Z_0 and Z_0' and between Z_0' and Z_L are small, then the total reflection in the line is dominated by the reflection from the discontinuity between Z_0 and Z_0' and the first reflection from the discontinuity between Z_0' and Z_L.

3.8 Multisection Transformers

In Section 3.7, a quarter-wave transformer has been illustrated as a narrow-band matching network. However, if the bandwidth requirement is more than that can be provided by a single quarter-wave transformer, a multisection quarter-wave transformer can be used. A multisection quarter-wave transformer consists of a number of cascaded quarter-wave transmission lines with predefined individual characteristic impedance, as shown in Fig. 3.16.

Now, if we assume that the discontinuities between any two successive quarter-wave section impedances and between Z_N and Z_L are small, then using Eq. (3.84) the total reflection coefficient can be written as follows:

$$\Gamma(\theta) = \Gamma_0 + \Gamma_1 e^{-j2\theta} + \Gamma_2 e^{-j4\theta} + \Gamma_3 e^{-j6\theta} + \ldots$$
$$+ \Gamma_n e^{-j2n\theta} + \ldots + \Gamma_{N-1} e^{-j2(N-1)\theta} + \Gamma_N e^{-j2N\theta} \qquad (3.85)$$

If we further assume that the transformer is symmetrical from both ends, then $\Gamma_0 = \Gamma_N$, $\Gamma_1 = \Gamma_{N-1}$, $\Gamma_2 = \Gamma_{N-2}$, $\Gamma_3 = \Gamma_{N-3}$, ... Therefore, Eq. (3.85) can be written as follows:

$$\Gamma(\theta) = \Gamma_0 + \Gamma_1 e^{-j2\theta} + \Gamma_2 e^{-j4\theta} + \Gamma_3 e^{-j6\theta} + \ldots + \Gamma_n e^{-j2n\theta} + \ldots$$
$$+ \Gamma_3 e^{-j2(N-3)\theta} + \Gamma_2 e^{-j2(N-2)\theta} + \Gamma_1 e^{-j2(N-1)\theta} + \Gamma_0 e^{-j2N\theta}$$

or $$\Gamma(\theta) = \Gamma_0\left\{1+e^{-j2N\theta}\right\} + \Gamma_1\left\{e^{-j2\theta} + e^{-j2(N-1)\theta}\right\} + \Gamma_2\left\{e^{-j4\theta} + e^{-j2(N-2)\theta}\right\}$$
$$+ \Gamma_3\left\{e^{-j6\theta} + e^{-j2(N-3)\theta}\right\} + \ldots$$

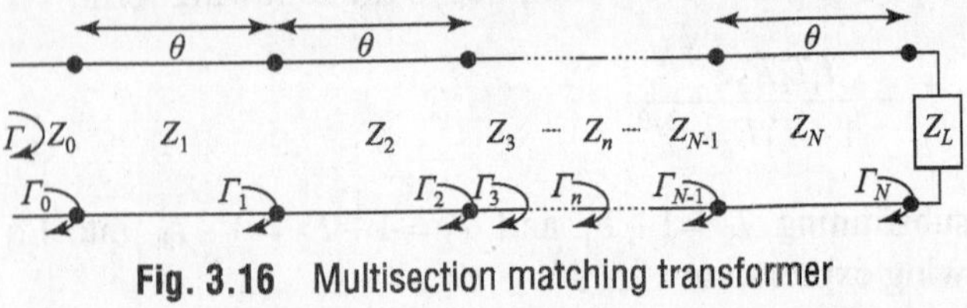

Fig. 3.16 Multisection matching transformer

or
$$\Gamma(\theta)=e^{-jN\theta}\left[\Gamma_0\left\{e^{jN\theta}+e^{-jN\theta}\right\}+\Gamma_1\left\{e^{j(N-2)\theta}+e^{-j(N-2)\theta}\right\}\right.$$
$$\left.+\Gamma_2\left\{e^{j(N-4)\theta}+e^{-j(N-4)\theta}\right\}+\Gamma_3\left\{e^{j(N-6)\theta}+e^{-j(N-6)\theta}\right\}+...\right] \quad (3.86)$$

Now, if N is even, then the last term will be $\Gamma_{N/2}$ and Eq. (3.86) can be written as follows:

$$\Gamma(\theta)=2e^{-jN\theta}\left[\Gamma_0\cos(N\theta)+\Gamma_1\cos\left\{(N-2)\theta\right\}+\Gamma_2\cos\left\{(N-4)\theta\right\}\right.$$
$$\left.+\Gamma_3\cos\left\{(N-6)\theta\right\}+...+\Gamma_n\cos\left\{(N-2n)\theta\right\}+...+\frac{1}{2}\Gamma_{N/2}\right] \quad (3.87)$$

If N is odd, then the last term of Eq. (3.86) will be $\Gamma_{(N-1)/2}\left\{e^{j\theta}+e^{-j\theta}\right\}$, and we can write the following equation:

$$\Gamma(\theta)=2e^{-jN\theta}\left[\Gamma_0\cos(N\theta)+\Gamma_1\cos\left\{(N-2)\theta\right\}+\Gamma_2\cos\left\{(N-4)\theta\right\}\right.$$
$$\left.+\Gamma_3\cos\left\{(N-6)\theta\right\}+...+\Gamma_n\cos\left\{(N-2n)\theta\right\}+...+\Gamma_{(N-1)/2}\cos(\theta)\right]$$

$$(3.88)$$

It can be noted that Eq. (3.86) represents a form of finite Fourier cosine series in θ. Since a Fourier series, consisting of a sufficient number of terms, can represent an arbitrary continuous function, with a proper choice of Γ_n and N we can also synthesize a desired reflection coefficient response of the network. This has been demonstrated in the following examples.

3.8.1 Binomial Multisection Matching Transformers

A binomial matching transformer exhibits a reflection coefficient response that is as flat as possible in the pass band. That is why it is also called a *maximally flat matching transformer*. For an N-element multisection matching transformer, a maximally flat reflection coefficient response can be obtained by setting the first $(N-1)$ derivatives of $|\Gamma(\theta)|$ to zero at the centre frequency. Therefore, we can write the following equation:

$$\Gamma(\theta)=A\left(1+e^{-j2\theta}\right)^N \quad (3.89)$$

Note At the centre frequency, $d=\lambda/4$ and $\theta=\pi/2$. Therefore, both $|\Gamma(\theta)|=0$ and $d^n\Gamma(\theta)/d\theta^n=0$, where $n=1, 2, 3, ..., N-1$.

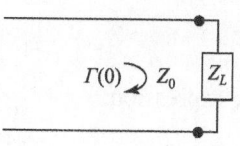

Fig. 3.17 Multisection transmission line at limit $f=0$

If we assume that $f=0$, then $\theta=0$ and Fig. 3.16 modifies to Fig. 3.17.

From Fig. 3.17, we can write the following equation:

$$\Gamma(0)=\frac{Z_L-Z_0}{Z_L+Z_0} \quad (3.90)$$

Substituting Eq. (3.90) into Eq. (3.89) we get the following expression:

$$\Gamma(0) = A\left(1 + e^0\right)^N = A2^N = \frac{Z_L - Z_0}{Z_L + Z_0}$$

or $\qquad A = 2^{-N} \dfrac{Z_L - Z_0}{Z_L + Z_0}$ (3.91)

Since $\Gamma(\theta)$ shows a binomial response,

$$\Gamma(\theta) = A\sum_{n=0}^{N} C_n^N e^{-j2n\theta}$$ (3.92)

where $\qquad C_n^N = \dfrac{N!}{(N-n)!n!}$ (3.93)

are the binomial coefficients.

Equating Eqs (3.92) and (3.85), we get the following relation:

$$A\sum_{n=0}^{N} C_n^N e^{-j2n\theta} = \Gamma_0 + \Gamma_1 e^{-j2\theta} + \Gamma_2 e^{-j4\theta} + \Gamma_3 e^{-j6\theta} + \ldots + \Gamma_n e^{-j2n\theta}$$

$$+ \ldots + \Gamma_{N-1} e^{-j2(N-1)\theta} + \Gamma_N e^{-j2N\theta}$$

or $\qquad A\{C_0^N + C_1^N e^{-j2\theta} + C_2^N e^{-j4\theta} + C_3^N e^{-j6\theta} + \ldots + C_n^N e^{-j2n\theta}$

$$+ \ldots + C_{N-1}^N e^{-j2(N-1)\theta} + C_N^N e^{-j2N\theta}\}$$

$$= \Gamma_0 + \Gamma_1 e^{-j2\theta} + \Gamma_2 e^{-j4\theta} + \Gamma_3 e^{-j6\theta} + \ldots + \Gamma_n e^{-j2n\theta}$$

$$+ \ldots + \Gamma_{N-1} e^{-j2(N-1)\theta} + \Gamma_N e^{-j2N\theta}$$ (3.94)

Equating the coefficients of the similar exponent terms of both sides of Eq. (3.94), we can write the following formula:

$$\Gamma_n = AC_n^N$$ (3.95)

Substituting Eq. (3.91) into Eq. (3.95), the following relation is obtained:

$$\Gamma_n = 2^{-N} C_n^N \frac{Z_L - Z_0}{Z_L + Z_0}$$ (3.96)

Now, $\qquad \Gamma_n = \dfrac{Z_{n+1} - Z_n}{Z_{n+1} + Z_n} = \left(\dfrac{Z_{n+1}}{Z_n} - 1\right) \bigg/ \left(\dfrac{Z_{n+1}}{Z_n} + 1\right) \approx \dfrac{1}{2}\ln\left(\dfrac{Z_{n+1}}{Z_n}\right)$ (3.97)

since $\quad \ln(x) \approx 2(x-1)/(x+1).$

Equating Eqs (3.96) and (3.97), we get the following expression:

$$\frac{1}{2}\ln\left(\frac{Z_{n+1}}{Z_n}\right) = 2^{-N} C_n^N \frac{Z_L - Z_0}{Z_L + Z_0} \approx 2^{-N} C_n^N \frac{1}{2}\ln\left(\frac{Z_L}{Z_0}\right)$$

or $$\ln\left(\frac{Z_{n+1}}{Z_n}\right) = 2^{-N} C_n^N \ln\left(\frac{Z_L}{Z_0}\right)$$ (3.98)

Note From Eq. (3.98) we get $Z_{N+1} = Z_L$, that is as expected.

Now, as before, if Γ_m be the maximum allowable reflection coefficient and the lower cut-off frequency occurs at $\theta = \theta_m$, then using Eq. (3.89) we can write the following expression:

$$\Gamma_m = A\left(1 + e^{-j2\theta_m}\right)^N = A\left(e^{-j\theta_m}\right)^N\left(e^{j\theta_m} + e^{-j\theta_m}\right)^N = A\left(e^{-j\theta_m}\right)^N 2^N \cos^N(\theta_m)$$

or $$|\Gamma_m| = |A|\left|e^{-j\theta_m}\right|^N 2^N \left|\cos(\theta_m)\right|^N = 2^N |A|\left|\cos(\theta_m)\right|^N$$

or $$\left|\cos(\theta_m)\right|^N = 2^{-N}|\Gamma_m|/|A|$$

or $$\left|\cos(\theta_m)\right| = \frac{1}{2}(|\Gamma_m|/|A|)^{\frac{1}{N}}$$ (3.99)

Since θ_m corresponds to the lower cut-off frequency, $\theta_m < \pi/2$ and $|\cos(\theta_m)| = \cos(\theta_m)$. Hence, Eq. (3.99) can be written as follows:

$$\cos(\theta_m) = \frac{1}{2}(|\Gamma_m|/|A|)^{\frac{1}{N}} \text{ or } \theta_m = \cos^{-1}\left[\frac{1}{2}(|\Gamma_m|/|A|)^{\frac{1}{N}}\right]$$ (3.100)

Substituting Eq. (3.100) into Eq. (3.78), we get the following equation:

$$\frac{\Delta f}{f_0} = 2 - \frac{4\theta_m}{\pi} = 2 - \frac{4}{\pi}\cos^{-1}\left[\frac{1}{2}(|\Gamma_m|/|A|)^{\frac{1}{N}}\right]$$ (3.101)

If N is large, then the calculation of the characteristic impedance of each individual section is time consuming. A simple program can be useful in such a case.[1]

EXAMPLE 3.5   Design a three-section binomial quarter-wave transformer for the transmission line system of Example 3.4. Calculate the percentage of bandwidth. Write a MATLAB program to plot the reflection coefficient as a function of θ for $N = 1$–5.

Solution   Given: $Z_0 = 50\ \Omega$, $Z_L = 74\ \Omega$, $N = 3$, and $\Gamma_m = 0.05$

The necessary binomial coefficients are as follows:

$$C_0^3 = \frac{3!}{3!0!} = 1, \ C_1^3 = \frac{3!}{2!1!} = 3, \text{ and } C_2^3 = \frac{3!}{1!2!} = 3$$

Now, $$\ln(Z_{n+1}) = \ln(Z_n) + 2^{-N} C_n^N \ln\left(\frac{Z_L}{Z_0}\right)$$

For $n = 0$: $$\ln(Z_1) = \ln(Z_0) + 2^{-N} C_0^N \ln\left(\frac{Z_L}{Z_0}\right)$$

---

[1]Program 3.1 in Online Resources

or
$$\ln(Z_1) = \ln(50) + 2^{-3}C_0^3 \ln\left(\frac{74}{50}\right) = 3.9120 + 0.0490 = 3.961$$

or
$$Z_1 = 52.5098 \ \Omega$$

For $n = 1$:
$$\ln(Z_2) = \ln(Z_1) + 2^{-N}C_1^N \ln\left(\frac{Z_L}{Z_0}\right)$$

or
$$\ln(Z_2) = \ln(52.5098) + 2^{-3}C_1^3 \ln\left(\frac{74}{50}\right) = 3.961 + 0.1470 = 4.108$$

or
$$Z_2 = 60.8249 \ \Omega$$

For $n = 2$:
$$\ln(Z_3) = \ln(Z_2) + 2^{-N}C_2^N \ln\left(\frac{Z_L}{Z_0}\right)$$

or
$$\ln(Z_3) = \ln(60.8249) + 2^{-3}C_2^3 \ln\left(\frac{74}{50}\right) = 4.108 + 0.147 = 4.255$$

or
$$Z_3 = 70.4568 \ \Omega$$

Now,
$$A = 2^{-N}\frac{Z_L - Z_0}{Z_L + Z_0} = 2^{-3}\frac{74 - 50}{74 + 50} = 0.0242$$

Therefore, the fractional bandwidth is as follows:

$$\frac{\Delta f}{f_0} = 2 - \frac{4}{\pi}\cos^{-1}\left[\frac{1}{2}\left(\frac{|\Gamma_m|}{|A|}\right)^{\frac{1}{N}}\right] = 2 - \frac{4}{\pi}\cos^{-1}\left[\frac{1}{2}\left(\frac{0.05}{0.0242}\right)^{\frac{1}{3}}\right] = 2 - \frac{4}{\pi}\cos^{-1}[0.6368]$$

or
$$\frac{\Delta f}{f_0} = 0.8790$$

The percentage bandwidth is therefore 87.90%.

A plot of the magnitudes of reflection coefficient with θ for a binomial multisection quarter-wave transformer of Example 3.5 is illustrated in Fig. 3.18.

MATLAB Program

```
% Program for Studying and Designing Binomial Transformer
clear all;
clf;
ZL=input('Enter the Load Impedance:- ');
Z0=input('Enter the Characteristic Impedance:- ');
NMAX=input('Enter the Highest Order:- ');
Theta_L=input('Enter the Start Value of Theta Axis:- ');
Theta_H=input('Enter the End Value of Theta Axis:- ');
d_theta=(Theta_H-Theta_L)/100;
for N=1:NMAX
A=2^-N*abs((ZL-Z0)/(ZL+Z0));
n=0;
```

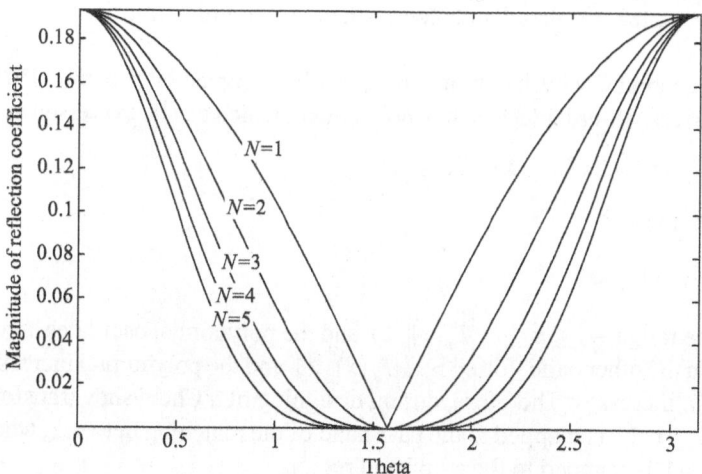

Fig. 3.18 Plot of magnitude of reflection coefficient with θ for binomial multisection quarter-wave transformer of Example 3.5

```
for theta_x=Theta_L:d_theta:Theta_H
    n=n+1;
    Theta(n)=theta_x;
    Gama(n)=abs(A*(1+exp(-i*2*theta_x))^N);
end
Gama_min=min(Gama);
Gama_max=max(Gama);
plot(Theta, Gama)
hold on
axis([Theta_L Theta_H Gama_min Gama_max]);
xlabel ('Theta');
ylabel('Magnitude of Reflection Coefficient');
end
```

Practice Problem

3.11 Design a three-section quarter-wave transformer for the transmission line system of Practice Problem 3.10. Calculate the percentage of bandwidth.

[76.838 Ω, 86.0217 Ω, 96.3031 Ω, 97.66%]

3.8.2 Chebyshev Multisection Matching Transformers

In contrast to a binomial multisection matching transformer, which is characterized by a maximally flat response in the pass band and a slow pass-band-to-stop-band transition characteristic, a Chebyshev matching transformer has a comparatively sharp pass-band-to-stop-band transition, but at the expense of a pass band ripple. For a given bandwidth, the number of elements required in a Chebyshev transformer is also less than that required in a binomial transformer. Therefore, if pass band ripples can be tolerated, then a Chebyshev multisection transformer would be much more advantageous than a binomial multisection matching transformer.

An N-section Chebyshev transformer can be designed by equating $\Gamma(\theta)$ of the network to an Nth-order Chebyshev polynomial, which can be expressed as follows:

$$T_N(x) = 2xT_{N-1}(x) - T_{N-2}(x) \tag{3.102}$$

where $\quad T_1(x) = x \tag{3.103}$

and $\quad T_2(x) = 2x^2 - 1 \tag{3.104}$

For the range $-1 \le x \le 1$, $|T_n(x)| < 1$ and the polynomial oscillates between $+1$ and -1; on the other hand, for $|x| > 1$, $|T_n(x)| > 1$ and the polynomial increases faster with x as n increases. Therefore, during designing of a Chebyshev transformer the range $-1 \le x \le 1$ is mapped to the pass band of the matching network, whereas the range $|x| > 1$ is mapped to the stop band region.

Now, if we assume that $x = \cos(\theta)$ or $\theta = \cos^{-1}(x)$ for $|x| < 1$, then it can be shown that the following relation holds:

$$T_n\{\cos(\theta)\} = \cos(n\theta) \quad \text{or} \quad T_n(x) = \cos\{n\cos^{-1}(x)\} \tag{3.105}$$

Similarly for $|x| > 1$, we can write the following equation:

$$T_n(x) = \cosh\{n\cosh^{-1}(x)\} \tag{3.106}$$

Once the Chebyshev polynomial has been defined, θ_m is required to be mapped to $x = 1$ and $\pi - \theta_m$ to $x = -1$. This can be achieved if we replace $\cos(\theta)$ with $\cos(\theta)/\cos(\theta_m)$, that is, if $x = \sec(\theta_m)\cos(\theta)$. As a consequence, Eq. (3.105) modifies to the following form:

$$T_n\{\sec(\theta_m)\cos(\theta)\} = \cos\left[n\cos^{-1}\{\sec(\theta_m)\cos(\theta)\}\right] \tag{3.107}$$

Note Since $\left|\sec(\theta_m)\cos(\theta)\right| \le 1$ for $\theta_m \le \theta \le \pi - \theta_m$, $\left|T_n\{\sec(\theta_m)\cos(\theta)\}\right| \le 1$ for $\theta_m \le \theta \le \pi - \theta_m$.

The pass band reflection coefficient response of an N-section Chebyshev transformer can be expressed as follows:

$$\Gamma(\theta) = Ae^{-jN\theta} T_N\{\sec(\theta_m)\cos(\theta)\} \tag{3.108}$$

Equating Eq. (3.108) with Eq. (3.87)/(3.88), depending on whether the number of elements is even or odd, we can write the following equation:

$$Ae^{-jN\theta} T_N\{\sec(\theta_m)\cos(\theta)\} = 2e^{-jN\theta}\left[\Gamma_0\cos(N\theta) + \Gamma_1\cos\{(N-2)\theta\}\right.$$

$$\left. + \Gamma_2\cos\{(N-4)\theta\} + \Gamma_3\cos\{(N-6)\theta\} + \dots + \Gamma_n\cos\{(N-2n)\theta\} + \dots\right]$$

or $\quad AT_N\{\sec(\theta_m)\cos(\theta)\} = 2\left[\Gamma_0\cos(N\theta) + \Gamma_1\cos\{(N-2)\theta\}\right.$

$$\left. + \Gamma_2\cos\{(N-4)\theta\} + \Gamma_3\cos\{(N-6)\theta\} + \dots + \Gamma_n\cos\{(N-2n)\theta\} + \dots\right] \tag{3.109}$$

The value of Γ_n can be obtained by rewriting the LHS of Eq. (3.109) in a polynomial form and equating them to similar terms on the RHS. However, if the values of Z_n and Z_{n+1} are known, the value of Γ_n can be obtained using the following formula:

$$\Gamma_n = \frac{Z_{n+1} - Z_n}{Z_{n+1} - Z_n} \simeq \frac{1}{2} \ln\left(\frac{Z_{n+1}}{Z_n}\right) \tag{3.110}$$

Again, as before, if we assume that $f = 0$, then $\theta = 0$ and Eq. (3.108) can be written as follows: $\Gamma(0) = Ae^0 T_N \{\sec(\theta_m)\cos(0)\}$

or $\qquad \Gamma(0) = AT_N \{\sec(\theta_m)\}$ $\tag{3.111}$

Equating Eq. (3.111) and Eq. (3.90), we get the following relation:

$$AT_N \{\sec(\theta_m)\} = \frac{Z_L - Z_0}{Z_L + Z_0} \quad \text{or} \quad A = \frac{1}{T_N \{\sec(\theta_m)\}} \frac{Z_L - Z_0}{Z_L + Z_0} \tag{3.112}$$

Since the maximum value of $|T_n \{\sec(\theta_m)\cos(\theta)\}|$ in the pass band is 1, using Eq. (3.108), the maximum allowable reflection coefficient magnitude in the pass band can be expressed as follows:

$$|\Gamma_m| = A \tag{3.113}$$

Substituting Eq. (3.113) into Eq. (3.112), we get the following expression:

$$|\Gamma_m| = \frac{1}{T_N \{\sec(\theta_m)\}} \left|\frac{Z_L - Z_0}{Z_L + Z_0}\right|$$

or $\qquad T_N \{\sec(\theta_m)\} = \frac{1}{|\Gamma_m|} \left|\frac{Z_L - Z_0}{Z_L + Z_0}\right| \approx \frac{1}{2|\Gamma_m|} |\ln(Z_L/Z_0)| \approx \left|\frac{\ln(Z_L/Z_0)}{2\Gamma_m}\right|$ $\tag{3.114}$

Now for $n = N$ and $x = \sec(\theta_m)$, Eq. (3.106) can be written as follows:

$$\cosh\{N\cosh^{-1}(\sec(\theta_m))\} = T_N \{\sec(\theta_m)\}$$

or $\qquad N\cosh^{-1}(\sec(\theta_m)) = \cosh^{-1}[T_N \{\sec(\theta_m)\}]$

or $\qquad \cosh^{-1}(\sec(\theta_m)) = \frac{1}{N}\cosh^{-1}[T_N \{\sec(\theta_m)\}]$

or $\qquad \sec(\theta_m) = \cosh\left[\frac{1}{N}\cosh^{-1}[T_N \{\sec(\theta_m)\}]\right]$ $\tag{3.115}$

Substituting Eq. (3.114) into Eq. (3.115), we get the following relation:

$$\sec(\theta_m) = \cosh\left[\frac{1}{N}\cosh^{-1}\left\{\left|\frac{\ln(Z_L/Z_0)}{2\Gamma_m}\right|\right\}\right] \tag{3.116}$$

or $\qquad \theta_m = \sec^{-1}\left[\cosh\left[\frac{1}{N}\cosh^{-1}\left\{\left|\frac{\ln(Z_L/Z_0)}{2\Gamma_m}\right|\right\}\right]\right]$ $\tag{3.117}$

Substituting Eq. (3.117) into Eq. (3.78), the fractional bandwidth can be obtained as follows:

$$\frac{\Delta f}{f_0} = 2 - \frac{4}{\pi} \sec^{-1}\left[\cosh\left[\frac{1}{N}\cosh^{-1}\left[T_N\left\{\sec(\theta_m)\right\}\right]\right]\right] \tag{3.118}$$

Unlike a binomial transformer, no straightforward relation exists between Z_{n+1}, Z_n, Z_L, and Z_0 in a Chebyshev transformer. Therefore, it is not possible to use a simple MATLAB program to calculate the characteristic impedance of each section of the Chebyshev transformer from the given value of N, Z_L, Z_0, and $|\Gamma_m|$. However, once the values of all Γ_ns are obtained, we can use a MATLAB program to calculate the characteristic impedances of each section of the network.[1]

EXAMPLE 3.6 Design a three-section Chebyshev quarter-wave transformer for the transmission line system of Example 3.4. Calculate the percentage of bandwidth. Write a MATLAB program to plot the reflection coefficient as a function of for $N = 1$–5.

Solution Given: $Z_0 = 50\ \Omega$, $Z_L = 74\ \Omega$, $N = 3$, and $\Gamma_m = 0.05$

For $N = 3$:

$$\Gamma(\theta) = Ae^{-j3\theta}T_3\left\{\sec(\theta_m)\cos(\theta)\right\} = 2e^{-j3\theta}\left[\Gamma_0\cos(3\theta) + \Gamma_1\cos(\theta)\right]$$

or

$$A\left[\sec^3(\theta_m)\left\{\cos(3\theta) + 3\cos(\theta)\right\} - 3\sec(\theta_m)\cos(\theta)\right] = 2\left[\Gamma_0\cos(3\theta) + \Gamma_1\cos(\theta)\right]$$

Equating the $\cos(3\theta)$ terms on both sides, we get the following relation:

$$A\sec^3(\theta_m) = 2\Gamma_0 \quad \text{or} \quad \Gamma_0 = A\sec^3(\theta_m)/2$$

Similarly, equating the $\cos(\theta)$ terms on both sides, we get the following expression:

$$3A\sec(\theta_m)\left[\sec^2(\theta_m) - 1\right] = 2\Gamma_1 \quad \text{or} \quad \Gamma_1 = 3A\sec(\theta_m)\left[\sec^2(\theta_m) - 1\right]/2$$

Now, $\sec(\theta_m) = \cosh\left[\frac{1}{N}\cosh^{-1}\left\{\left|\frac{\ln(Z_L/Z_0)}{2\Gamma_m}\right|\right\}\right] = \cosh\left[\frac{1}{3}\cosh^{-1}\left\{\left|\frac{\ln(74/50)}{2\times0.05}\right|\right\}\right] = 1.2409$

and $A = \Gamma_m = 0.05$

Therefore,

$$\Gamma_0 = \frac{A\sec^3(\theta_m)}{2} = \frac{0.05\times(1.2409)^3}{2} = 0.0478$$

$$\Gamma_1 = \frac{3A\sec(\theta_m)\left[\sec^2(\theta_m) - 1\right]}{2} = \frac{3\times0.05\times1.2409\left[(1.2409)^2 - 1\right]}{2} = 0.0502$$

On the basis of symmetry, we can write the following equations:

$$\Gamma_2 = \Gamma_1 = 0.0502 \quad \text{and} \quad \Gamma_3 = \Gamma_0 = 0.0478$$

Now, $\ln(Z_{n+1}) = \ln(Z_n) + 2\Gamma_n$

---

[1] Program 3.2 in Online Resources

For $n = 0$: $\ln(Z_1) = \ln(Z_0) + 2\Gamma_0 = \ln(50) + 2 \times 0.0478 = 4.0076$

or $Z_1 = 55.0159 \ \Omega$

For $n = 1$: $\ln(Z_2) = \ln(Z_1) + 2\Gamma_1 = \ln(55.0159) + 2 \times 0.0502 = 4.1080$

or $Z_2 = 60.8263 \ \Omega$

For $n = 2$: $\ln(Z_3) = \ln(Z_2) + 2\Gamma_2 = \ln(60.8263) + 2 \times 0.0502 = 4.2084$

or $Z_3 = 67.2504 \ \Omega$

The fractional bandwidth is as follows:

$$\frac{\Delta f}{f_0} = 2 - \frac{4}{\pi}\sec^{-1}\left[\cosh\left[\frac{1}{N}\cosh^{-1}\left[4\sec^3(\theta_m) - 3\sec(\theta_m)\right]\right]\right]$$

or $$\frac{\Delta f}{f_0} = 2 - \frac{4}{\pi}\sec^{-1}\left[\cosh\left[\frac{1}{3}\cosh^{-1}[3.9204]\right]\right] = 2 - \frac{4}{\pi}\sec^{-1}\left[\cosh[0.6809]\right]$$

or $$\frac{\Delta f}{f_0} = 2 - \frac{4}{\pi}\sec^{-1}[1.2409] = 2 - \frac{4}{\pi} \times 0.6337 = 1.1932$$

Therefore, the percentage bandwidth is 119.32%.

A plot of the magnitudes of reflection coefficient with θ for a Chebyshev multisection quarter-wave transformer of Example 3.6 is illustrated in Fig. 3.19.

MATLAB Program

```
% Program for Studying and Designing Chebyshev Transformer
clear all;
clf;
NMAX=input('Enter the Highest Order of Filter:- ');
```

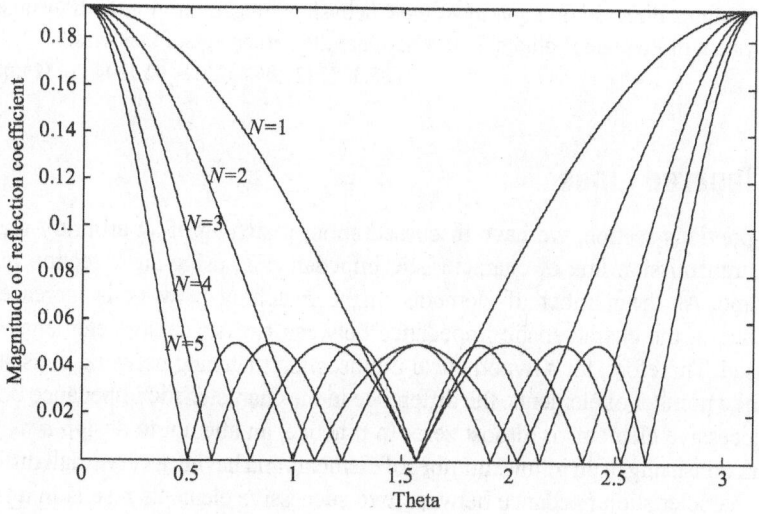

Fig. 3.19   Plot of magnitude of reflection coefficient with θ for Chebyshev multisection quarter-wave transformer of Example 3.6

```
ZL=input('Enter the Load Impedance:- ');
Z0=input('Enter the Characteristic Impedance:- ');
gm=input('Enter the Magnitude of Maximum Tolerable Reflection Coefficient:-');
Theta_L=input('Enter the Start Value of Theta Axis:- ');
Theta_H=input('Enter the End Value of Theta Axis:- ');
A=gm;
d_theta=(Theta_H-Theta_L)/1000;
for N=1:NMAX
    sec_thetam=cosh(acosh((abs((ZL-Z0)/(ZL+Z0)))/gm)/N);
n=0;
for theta_x=Theta_L:d_theta:Theta_H
    n=n+1;
    Theta(n)=theta_x;
    T(1)=sec_thetam*cos(theta_x);
    T(2)=2*(sec_thetam*cos(theta_x))^2-1;
    for k=3:NMAX
        T(k)=2*sec_thetam*cos(theta_x)*T(k-1)-T(k-2);
    end
    Gama(n)=A*exp(-i*N*theta_x)*T(N);
end
Gama_min=min(Gama);
Gama_max=max(Gama);
plot(Theta, abs(Gama))
hold on
axis([Theta_L Theta_H Gama_min Gama_max]);
xlabel ('Theta');
ylabel('Magnitude of Reflection Coefficient');
end
```

Practice Problem

3.12 Design a three-section quarter-wave Chebyshev transformer for the transmission line system of Practice Problem 3.10. Calculate the percentage of bandwidth.

[80.3077 Ω, 86.0273 Ω, 92.1544 Ω, 129.01%]

3.9 Tapered Lines

In the previous section, we have discussed about matching of an arbitrary real load with a transmission line of characteristic impedance Z_0 using an N-section quarter-wave line. As the number of elements in the matching network is increased, the difference in the characteristic impedance between two successive elements is also decreased. Therefore, for a hypothetical multisection matching network, consisting of an infinite number of elements, the difference in the characteristic impedance between two successive elements is almost zero. In practice, an attempt to design a matching network consisting of an infinite number of elements and having a very small difference in the characteristic impedance between two successive elements results in a tapered line, as shown in Fig. 3.20. Depending on the nature of tapering, such lines can be classified as triangular tapers, exponential tapers, Chebyshev/Klopfenstein tapers, etc.

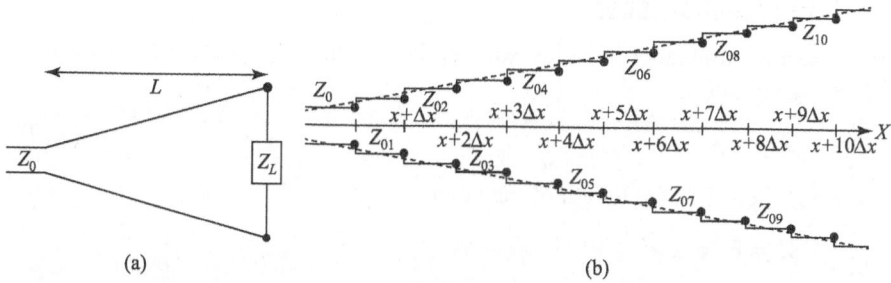

Fig. 3.20 Schematic diagram (a) Tapered line (b) Incremental step model of tapered line

Let us consider a section of the tapered line shown in Fig. 3.21. The incremental reflection coefficient at the step $\Delta\Gamma$ can be expressed as follows:

$$\Delta\Gamma = \frac{(Z+\Delta Z)-Z}{(Z+\Delta Z)+Z} \approx \frac{\Delta Z}{2Z} \qquad (3.119)$$

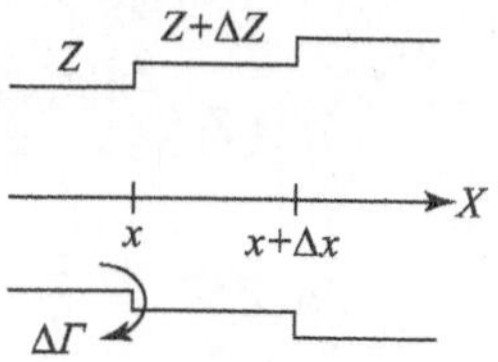

Fig. 3.21 Incremental step change in tapered line

If we assume that $\Delta x \to 0$, then Eq. (3.119) transforms to an exact differential and can be written as follows:

$$d\Gamma = \frac{1}{2}\frac{dZ}{Z} = \frac{1}{2}\frac{Z_0}{Z}\frac{dZ}{Z_0} = \frac{1}{2}\frac{d(Z/Z_0)}{Z/Z_0} \qquad (3.120)$$

Now using the substitution $f(x) = Z/Z_0$ in the following mathematical identity

$$\frac{d\{f(x)\}}{f(x)} = \frac{d}{dx}\big[\ln\{f(x)\}\big]dx,$$

we get $\quad \dfrac{d(Z/Z_0)}{(Z/Z_0)} = \dfrac{d}{dx}\big[\ln(Z/Z_0)\big]dx \qquad (3.121)$

Substituting Eq. (3.121) into Eq. (3.120), we get the following expression:

$$d\Gamma = \frac{1}{2}\frac{d}{dx}\big[\ln(Z/Z_0)\big]dx \qquad (3.122)$$

Using the theory of small reflections, the total reflection coefficient at $x = 0$ can be found by integrating Eq. (3.122) over the complete taper length and is given by the following equation:

$$\Gamma(\theta) = \frac{1}{2}\int_{x=0}^{L} \frac{d}{dx}\big[\ln(Z/Z_0)\big]e^{-j2\beta x}dx \qquad (3.123)$$

If Z is known, Eq. (3.123) can be used to find the variation of $\Gamma(\theta)$ as a function of frequency.

Note Equation (3.123) neglects the multiple reflections between each section of the taper.

3.9.1 Exponential Taper

For an exponential taper, Z varies exponentially with x; therefore, we can write the following equation:

$$Z = Z_0 e^{\alpha x} \tag{3.124}$$

at $x = L$, where L is the length of the taper,

$$Z = Z_L = Z_0 e^{\alpha L}$$

or $\quad \ln(Z_L) = \ln(Z_0) + \alpha L \quad$ or $\quad \alpha = \dfrac{1}{L}\ln(Z_L/Z_0) \tag{3.125}$

Substituting Eq. (3.125) into Eq. (3.124), we get the following relation:

$$Z = Z_0 e^{\frac{x}{L}\ln(Z_L/Z_0)} \quad \text{or} \quad \ln(Z) = \ln(Z_0) + \ln\left[e^{\frac{x}{L}\ln(Z_L/Z_0)}\right]$$

or $\quad \ln(Z/Z_0) = \dfrac{x}{L}\ln(Z_L/Z_0) \tag{3.126}$

Substitution of Eq. (3.126) into Eq. (3.123) produces the following equation:

$$\Gamma(\theta) = \frac{1}{2}\int\limits_{x=0}^{L} \frac{d}{dx}\left[\frac{x}{L}\ln(Z_L/Z_0)\right]e^{-j2\beta x}dx = \frac{1}{2}\int\limits_{x=0}^{L}\left[\frac{1}{L}\ln(Z_L/Z_0)\right]e^{-j2\beta x}dx$$

or $\quad \Gamma(\theta) = \dfrac{1}{2L}\ln(Z_L/Z_0)\left[\dfrac{e^{-j2\beta x}}{-j2\beta}\right]_0^L$

or $\quad \Gamma(\theta) = \dfrac{1}{2L}\ln(Z_L/Z_0)\left[\dfrac{e^{-j2\beta L}-1}{-j2\beta}\right] = \dfrac{1}{2\beta L}\ln(Z_L/Z_0)e^{-j\beta L}\left[\dfrac{e^{j\beta L}-e^{-j\beta L}}{2j}\right]$

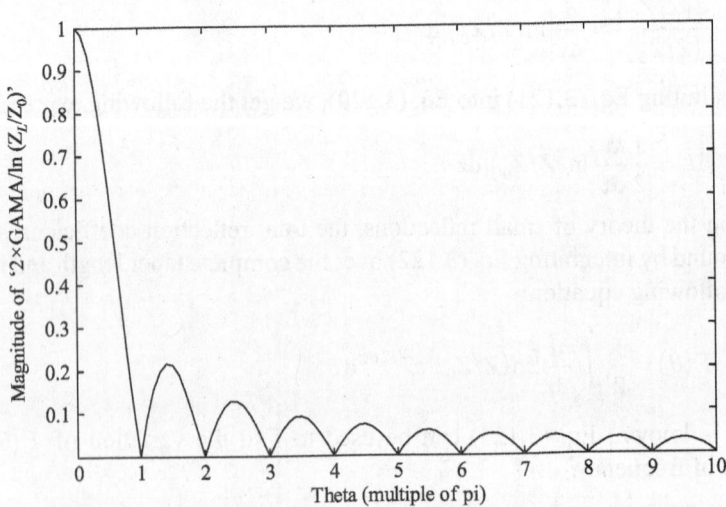

Fig. 3.22 Plot of input reflection coefficient as a function of theta for exponential taper

or $\qquad \Gamma(\theta) = \frac{1}{2}\ln(Z_L/Z_0)e^{-j\beta L}\frac{\sin(\beta L)}{(\beta L)} = \frac{1}{2}\ln(Z_L/Z_0)e^{-j\theta}\text{sinc}(\theta)$ (3.127)

This analysis assumes that β is independent of x. A plot of $|\Gamma(\theta)|$ as a function of θ is shown in Fig. 3.22. The figure reveals that when $L > \lambda/2$, the reflection coefficient is quite small, the first minor lobe being about 22% of the maximum. Using a simple program[1], we can plot $|\Gamma(\theta)|$ as a function of θ.

Note If the length is fixed then $\beta L = \frac{2\pi}{\lambda}L = \frac{2\pi L}{c}f = Kf$, where $K = 2\pi L/c$ is a constant. Figure 3.22 therefore illustrates a plot of $|\Gamma(\theta)|$ as a function of frequency.

3.9.2 Triangular Taper

For a triangular taper, the impedance is defined as follows:

$$Z = \begin{cases} Z_0 e^{2(x/L)^2 \ln(Z_L/Z_0)} & \text{for } 0 \le x \le L/2 \\ Z_0 e^{(4x/L - 2x^2/L^2 - 1)\ln(Z_L/Z_0)} & \text{for } L/2 \le x \le L \end{cases}$$ (3.128)

or $\qquad \dfrac{Z}{Z_0} = \begin{cases} e^{2(x/L)^2 \ln(Z_L/Z_0)} & \text{for } 0 \le x \le L/2 \\ e^{(4x/L - 2x^2/L^2 - 1)\ln(Z_L/Z_0)} & \text{for } L/2 \le x \le L \end{cases}$

or $\qquad \ln\!\left(\dfrac{Z}{Z_0}\right) = \begin{cases} \left(2x^2/L^2\right)\ln(Z_L/Z_0) & \text{for } 0 \le x \le L/2 \\ \left(4x/L - 2x^2/L^2 - 1\right)\ln(Z_L/Z_0) & \text{for } L/2 \le x \le L \end{cases}$ (3.129)

Note Since $d\{\ln(z)\}/dx$ is a triangular function, such tapers are called 'triangular tapers'.

Substituting Eq. (3.129) into Eq. (3.123), we get the following relation:

$$\Gamma(\theta) = \frac{1}{2}\int_{x=0}^{L}\frac{d}{dx}\left[\begin{cases}\left(2x^2/L^2\right)\ln(Z_L/Z_0) & \text{for } 0 \le x \le L/2 \\ \left(4x/L - 2x^2/L^2 - 1\right)\ln(Z_L/Z_0) & \text{for } L/2 \le x \le L\end{cases}\right]e^{-j2\beta x}dx$$

or $\qquad \Gamma(\theta) = \dfrac{1}{2}\int_{x=0}^{L}\left[\begin{cases}\left(4x/L^2\right)\ln(Z_L/Z_0) & \text{for } 0 \le x \le L/2 \\ \left(4/L - 4x/L^2\right)\ln(Z_L/Z_0) & \text{for } L/2 \le x \le L\end{cases}\right]e^{-j2\beta x}dx$

or $\qquad \Gamma(\theta) = \dfrac{2}{L^2}\ln(Z_L/Z_0)\int_{x=0}^{L}\left[\begin{cases}xe^{-j2\beta x} & \text{for } 0 \le x \le L/2 \\ (L-x)e^{-j2\beta x} & \text{for } L/2 \le x \le L\end{cases}\right]dx$

---

[1]Program 3.3 in Online Resources

or $\quad \Gamma(\theta) = \dfrac{2}{L^2} \ln(Z_L/Z_0) \left[\int\limits_{x=0}^{L/2} x e^{-j2\beta x} dx + \int\limits_{L/2}^{L} (L-x) e^{-j2\beta x} dx \right]$ $\qquad$ (3.130)

Now using the identity $\int x e^{ax} dx = \dfrac{e^{ax}}{a}\left(x - \dfrac{1}{a}\right)$ and assuming that $a = -j2\beta$, Eq. (3.130) can be written as follows:

$$\Gamma(\theta) = \frac{2}{L^2} \ln(Z_L/Z_0) \left[\left\{ \frac{e^{-j2\beta x}}{-j2\beta}\left(x - \frac{1}{-j2\beta}\right)\right\}_0^{L/2} + L\left(\frac{e^{-j2\beta x}}{-j2\beta}\right)_{L/2}^{L} \right.$$
$$\left. - \left\{ \frac{e^{-j2\beta x}}{-j2\beta}\left(x - \frac{1}{-j2\beta}\right)\right\}_{L/2}^{L} \right]$$

or $\quad \Gamma(\theta) = \dfrac{2}{L^2} \ln(Z_L/Z_0) \left[\left\{ j\dfrac{e^{-j2\beta x}}{2\beta}\left(x - j\dfrac{1}{2\beta}\right)\right\}_0^{L/2} + jL\left(\dfrac{e^{-j2\beta x}}{2\beta}\right)_{L/2}^{L} \right.$

$\qquad\qquad\qquad\left. - j\left\{ \dfrac{e^{-j2\beta x}}{2\beta}\left(x - j\dfrac{1}{2\beta}\right)\right\}_{L/2}^{L} \right]$

or $\quad \Gamma(\theta) = \dfrac{2}{L^2} \ln(Z_L/Z_0) \left[\left\{\left(j\dfrac{x e^{-j2\beta x}}{2\beta} + \dfrac{e^{-j2\beta x}}{4\beta^2}\right)\right\}_0^{L/2} + jL\left(\dfrac{e^{-j2\beta x}}{2\beta}\right)_{L/2}^{L} \right.$

$\qquad\qquad\qquad\left. - \left\{\left(j\dfrac{x e^{-j2\beta x}}{2\beta} + \dfrac{e^{-j2\beta x}}{4\beta^2}\right)\right\}_{L/2}^{L} \right]$

or $\quad \Gamma(\theta) = \dfrac{2}{L^2} \ln(Z_L/Z_0) \left[j\dfrac{L}{2}\dfrac{e^{-j\beta L}}{2\beta} + \dfrac{e^{-j\beta L}}{4\beta^2} - \dfrac{1}{4\beta^2} + jL\dfrac{e^{-j2\beta L}}{2\beta} \right.$

$\qquad\qquad\qquad\left. - jL\dfrac{e^{-j\beta L}}{2\beta} - jL\dfrac{e^{-j2\beta L}}{2\beta} + j\dfrac{L}{2}\dfrac{e^{-j\beta L}}{2\beta} - \dfrac{e^{-j2\beta L}}{4\beta^2} + \dfrac{e^{-j\beta L}}{4\beta^2} \right]$

or $\quad \Gamma(\theta) = \dfrac{2}{L^2} \ln(Z_L/Z_0) \left[\dfrac{e^{-j\beta L}}{2\beta^2} - \dfrac{1}{4\beta^2} - \dfrac{e^{-j2\beta L}}{4\beta^2} \right]$

or $\quad \Gamma(\theta) = -\dfrac{1}{2\beta^2 L^2} \ln(Z_L/Z_0) e^{-j\beta L}\left[e^{j\beta L} - 2 + e^{-j\beta L} \right]$

or $\quad \Gamma(\theta) = -\dfrac{1}{2\beta^2 L^2} \ln(Z_L/Z_0) e^{-j\beta L}\left(e^{j\beta L/2} - e^{-j\beta L/2} \right)^2$

or $\quad \Gamma(\theta) = -\dfrac{1}{2\beta^2 L^2} \ln(Z_L/Z_0) e^{-j\beta L}\left\{ 2j\sin(\beta L/2)\right\}^2$

or $\quad \Gamma(\theta) = \dfrac{1}{2} e^{-j\beta L} \ln(Z_L/Z_0)\left\{ \sin(\beta L/2)/(\beta L/2)\right\}^2$

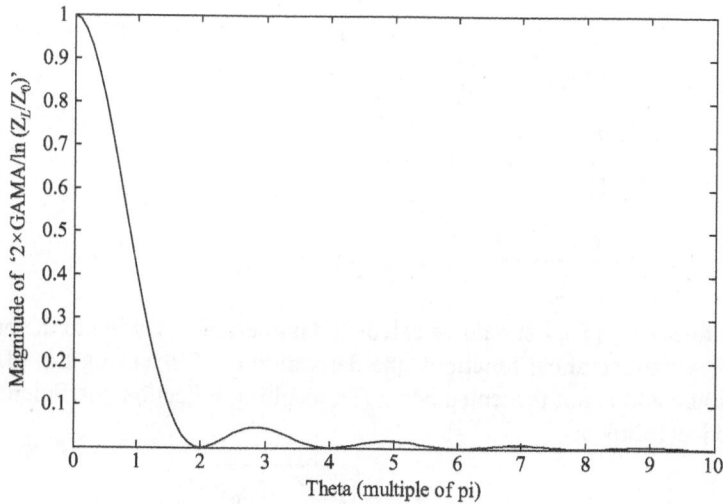

Fig. 3.23 Plot of input reflection coefficient as a function of theta for triangular taper

or $\Gamma(\theta) = \frac{1}{2} e^{-j\beta L} \ln(Z_L/Z_0) \mathrm{sinc}^2(\beta L/2)$

or $\Gamma(\theta) = \frac{1}{2} e^{-j\theta} \ln(Z_L/Z_0) \mathrm{sinc}^2(\theta/2)$ (3.131)

This analysis again assumes that β is independent of x. A plot of $|\Gamma(\theta)|$ as a function of θ is shown in Fig. 3.23. The figure reveals that when $L > \lambda$, the reflection coefficient is quite small.

Using a simple MATLAB program we can plot $|\Gamma(\theta)|$ as a function of θ.[1]

3.9.3 Chebyshev/Klopfenstein Taper

A Chebyshev/Klopfenstein taper can be obtained if the number of elements in a Chebyshev transformer is increased while keeping its overall length fixed. This taper is characterized by its equal-amplitude minor lobes and has the advantage of providing the smallest reflection coefficient for a fixed taper length as compared to others. In other words, for a given pass band reflection coefficient, it yields the shortest matching section compared to others. For a Chebyshev transformer, the following relation holds:

$$\ln(Z) = \frac{1}{2}\ln(Z_0 Z_L) + \frac{1}{2}\ln(Z_L/Z_0)\frac{(\beta_0 L)^2}{\cosh(\beta_0 L)}\varphi\left(\frac{2x}{L} - 1, \beta_0 L\right) \qquad (3.132)$$

where β_0 is the value of β at the cut-off and

$$\varphi(\xi,\varsigma) = -\varphi(-\xi,\varsigma) = \int_0^\xi \frac{J_1\left(\varsigma\sqrt{1-\psi^2}\right)}{\varsigma\sqrt{1-\psi^2}}\,d\psi \quad \text{for } |\xi| \le 1 \qquad (3.133)$$

---

[1]Program 3.4 in Online Resources

where $J_1\left(\varsigma\sqrt{1-\psi^2}\right)$ is the modified Bessel's function. Equation (3.133) takes the following special values:

$$\varphi(0,\varsigma)=0 \tag{3.134}$$

$$\varphi(\xi,0)=\xi/2 \tag{3.135}$$

$$\varphi(1,\varsigma)=\frac{\cosh(\varsigma)-1}{\varsigma^2} \tag{3.136}$$

Others values of $\varphi(\xi,\varsigma)$ should be calculated numerically. Owing to the presence of complex mathematical functions, the derivation of $\Gamma(\theta)$ using Eq. (3.123) is cumbersome and is not presented here. The resulting reflection coefficient can be expressed as follows:

$$\Gamma(\theta)=\frac{1}{2}e^{-j\beta L}\ln(Z_L/Z_0)\frac{\cos\left\{\sqrt{(\beta L)^2-(\beta_0 L)^2}\right\}}{\cosh(\beta_0 L)}$$

or $\quad\Gamma(\theta)=\frac{1}{2}e^{-j\theta}\ln(Z_L/Z_0)\dfrac{\cos\left(\sqrt{\theta^2-\theta_0^2}\right)}{\cosh(\theta_0)} \tag{3.137}$

The maximum pass band reflection coefficient occurs at $\theta=\theta_0$ and is given by the following expression:

$$\Gamma(\theta_0)=\frac{1}{2}\frac{\ln(Z_L/Z_0)}{\cosh(\theta_0)}e^{-j\theta_0} \tag{3.138}$$

or $\quad|\Gamma(\theta_0)|=\dfrac{1}{2}\dfrac{|\ln(Z_L/Z_0)|}{\cosh(\theta_0)} \tag{3.139}$

Substituting Eq. (3.138) into Eq. (3.137), we get the following relation:

$$\Gamma(\theta)=\Gamma(\theta_0)\cos\left(\sqrt{\theta^2-\theta_0^2}\right) \tag{3.140}$$

Within the range $0\le\theta\le\theta_0$, $\cos\left(\sqrt{\theta^2-\theta_0^2}\right)=\cosh\left(\sqrt{\theta_0^2-\theta^2}\right)$ and $\Gamma(\theta)$ is a hyperbolic function of θ, whereas for $\theta>\theta_0$, $\Gamma(\theta)$ is a cosine function of θ. Since the cosine function oscillates between ±1, the pass band reflection coefficient $\Gamma(\theta)$ oscillates between $\pm|\Gamma(\theta_0)|$. If $|\Gamma(\theta_0)|$ is specified, then the taper length can be calculated from Eq. (3.139) as follows:

$$\cosh(\theta_0)=\frac{1}{2}\frac{|\ln(Z_L/Z_0)|}{|\Gamma(\theta_0)|}\quad\text{or}\quad\theta_0=\beta_0 L=\cosh^{-1}\left[\frac{1}{2}\frac{|\ln(Z_L/Z_0)|}{|\Gamma(\theta_0)|}\right]$$

or $\quad L=\dfrac{1}{\beta_0}\cosh^{-1}\left[\dfrac{1}{2}\dfrac{|\ln(Z_L/Z_0)|}{|\Gamma(\theta_0)|}\right] \tag{3.141}$

Conversely, if β_0 and L are given, then $|\Gamma(\theta_0)|$ is fixed.

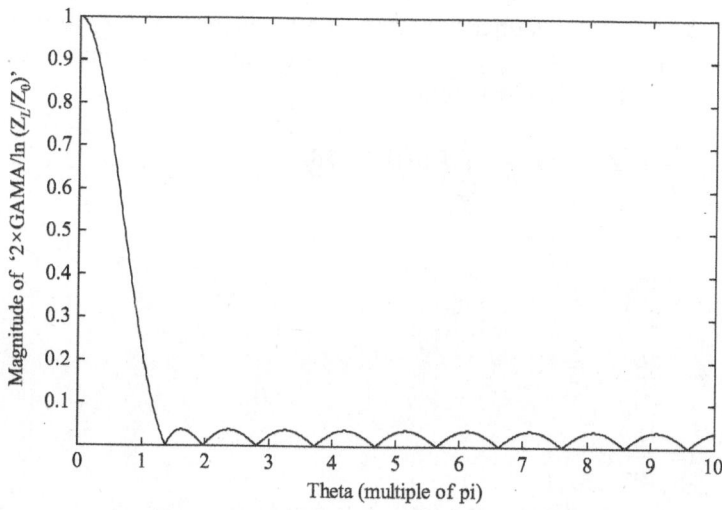

Fig. 3.24 Plot of input reflection coefficient as a function of theta for Klopfenstein/Chebyshev taper

At $\theta = 0$, the reflection coefficient becomes maximum in the stop band and it is represented by the following equation:

$$\Gamma(0) = \frac{1}{2}\ln(Z_L/Z_0)\frac{\cos\left\{\sqrt{-(\theta_0)^2}\right\}}{\cosh(\theta_0)} = \frac{1}{2}\ln(Z_L/Z_0)\frac{\cosh(\theta_0)}{\cosh(\theta_0)} = \frac{1}{2}\ln(Z_L/Z_0)$$

or $\qquad |\Gamma(0)| = \frac{1}{2}|\ln(Z_L/Z_0)| \qquad\qquad\qquad\qquad$ (3.142)

Taking the ratio of Eq. (3.142) to Eq. (3.139), we get the following expression:

$$|\Gamma(0)|/|\Gamma(\theta_0)| = \cosh(\theta_0) \qquad\qquad\qquad\qquad \text{(3.143)}$$

> **Note** A Chebyshev taper has steps at $x = 0$ and $x = L$.

Using a simple program[1], we can plot $|\Gamma(\theta)|$ as a function of θ. A typical plot of input reflection coefficient as a function of θ for Klopfenstein/Chebyshev taper is shown in Fig. 3.24.

3.10 Synthesis of Transmission Line Tapers

The reflection coefficient at a point x on the taper can be calculated from Eq. (3.123). For convenience it is rewritten as follows:

$$\Gamma(\theta) = \frac{1}{2}\int_{x=0}^{L}\frac{\mathrm{d}}{\mathrm{d}x}\left[\ln(Z/Z_0)\right]e^{-j2\beta x}\mathrm{d}x \qquad\qquad \text{(3.144)}$$

---

[1]Program 3.5 in Online Resources

Equation (3.144) is the Fourier transform of a function $d[\ln(Z/Z_0)]/dx$, which is zero outside the range $0 \le x \le L$. Therefore, using the inverse Fourier transform, we can write the following equation:

$$\frac{1}{2}\frac{d}{dx}[\ln(Z/Z_0)] = \frac{1}{2\pi}\int_{-\infty}^{\infty}\Gamma(\theta)e^{j2\beta x}2d\beta \qquad (3.145)$$

Now let us assume that

$$p = \frac{2\pi}{L}(x - L/2) \qquad (3.146)$$

or $\qquad x = \frac{pL}{2\pi} + L/2 \qquad (3.147)$

or $\qquad dx = \frac{L}{2\pi}dp \qquad (3.148)$

Substituting Eqs (3.147) and (3.148) into Eq. (3.144), we get

$$\Gamma(\theta) = \frac{1}{2}\int_{-\pi}^{\pi}\frac{2\pi}{L}\frac{d}{dp}[\ln(Z/Z_0)]e^{-j2\beta\left(\frac{pL}{2\pi}+L/2\right)}\frac{L}{2\pi}dp$$

or $\qquad \Gamma(\theta) = \frac{1}{2}\int_{-\pi}^{\pi}\frac{d}{dp}[\ln(Z/Z_0)]e^{-j\frac{\beta L}{\pi}p}e^{-j\beta L}dp \qquad (3.149)$

Again if we substitute

$$u = \beta L/\pi \qquad (3.150)$$

or $\qquad \beta = \frac{u\pi}{L} \qquad (3.151)$

or $\qquad d\beta = \frac{\pi}{L}du \qquad (3.152)$

in Eq. (3.149), we get the following relation:

$$\Gamma(\theta) = \frac{1}{2}e^{-j\beta L}\int_{-\pi}^{\pi}\frac{d}{dp}[\ln(Z/Z_0)]e^{-jpu}dp \qquad (3.153)$$

Now if we assume that

$$g(p) = \frac{d}{dp}[\ln(Z/Z_0)] \qquad (3.154)$$

then Eq. (3.153) modifies to the following form:

$$\Gamma(\theta) = \frac{1}{2}e^{-j\beta L}\int_{-\pi}^{\pi}g(p)e^{-jpu}dp \qquad (3.155)$$

A further substitution of

$$F(u) = \int_{-\pi}^{\pi}g(p)e^{-jpu}dp \qquad (3.156)$$

into Eq. (3.155) yields the following equation:

$$\Gamma(\theta) = \frac{1}{2} e^{-j\beta L} F(u) \tag{3.157}$$

Substitution of Eqs (3.147) and (3.148) into Eq. (3.145), produces the following expression:

$$\frac{1}{2}\frac{2\pi}{L}\frac{d}{dp}[\ln(Z/Z_0)] = \frac{1}{2\pi}\int_{-\infty}^{\infty} \Gamma(\theta) e^{j2\beta\left(\frac{pL}{2\pi}+L/2\right)} 2d\beta$$

or

$$\frac{1}{2}\frac{2\pi}{L}\frac{d}{dp}[\ln(Z/Z_0)] = \frac{1}{2\pi}\int_{-\infty}^{\infty} \Gamma(\theta) e^{j\frac{\beta L}{\pi}p} e^{j\beta L} 2d\beta \tag{3.158}$$

Substituting Eqs (3.150), (3.152), and (3.154) into Eq. (3.158), we get the following expression:

$$\frac{1}{2}\frac{2\pi}{L}g(p) = \frac{1}{2\pi}\int_{-\infty}^{\infty} e^{j\beta L}\Gamma(\theta)e^{jup} 2\frac{\pi}{L}du$$

or

$$\frac{1}{2}g(p) = \frac{1}{2\pi}\int_{-\infty}^{\infty} e^{j\beta L}\Gamma(\theta)e^{jup}du \tag{3.159}$$

Again substituting Eq. (3.157) into Eq. (3.159), we get the following relation:

$$\frac{1}{2}g(p) = \frac{1}{2\pi}\int_{-\infty}^{\infty} e^{j\beta L}\frac{1}{2}e^{-j\beta L}F(u)e^{jup}du$$

or

$$g(p) = \frac{1}{2\pi}\int_{-\infty}^{\infty} F(u)e^{jup}du \tag{3.160}$$

Now let us define $g(p)$ as follows:

$$g(p) = \begin{cases} \displaystyle\sum_{n=-\infty}^{\infty} a_n e^{jnp} & \text{for } -\pi \le p \le \pi \\ 0 & \text{for } |p| \end{cases} \tag{3.161}$$

Substituting Eq. (3.161) into Eq. (3.156), we get the following equation:

$$F(u) = \int_{-\pi}^{\pi}\sum_{n=-\infty}^{\infty} a_n e^{jnp}e^{-jpu}dp = \sum_{n=-\infty}^{\infty} a_n\int_{-\pi}^{\pi} e^{-jp(u-n)}dp$$

or

$$F(u) = \sum_{n=-\infty}^{\infty} a_n\left[\frac{e^{-j(u-n)p}}{-j(u-n)}\right]_{-\pi}^{\pi}$$

or

$$F(u) = \sum_{n=-\infty}^{\infty} a_n\left[\frac{e^{-j(u-n)\pi}-e^{j(u-n)\pi}}{-j(u-n)}\right] = 2\pi\sum_{n=-\infty}^{\infty} a_n\left[\frac{\sin\{\pi(u-n)\}}{\pi(u-n)}\right] \tag{3.162}$$

Now,
$$\frac{\sin\{\pi(u-n)\}}{\{\pi(u-n)\}} = \begin{cases} 1 & \text{if } u=n \\ 0 & \text{if } u\neq n \end{cases} \tag{3.163}$$

Substituting Eq. (3.163) into Eq. (3.162), the following relation is obtained:

$$F(n) = 2\pi a_n \tag{3.164}$$

or
$$a_n = F(n)/2\pi \tag{3.165}$$

Substituting Eq. (3.165) back into Eq. (3.162), we get the following relation:

$$F(u) = 2\pi \sum_{n=-\infty}^{\infty} \frac{F(n)}{2\pi}\left[\frac{\sin\{\pi(u-n)\}}{\pi(u-n)}\right] = \sum_{n=-\infty}^{\infty} F(n)\left[\frac{\sin\{\pi(u-n)\}}{\pi(u-n)}\right] \tag{3.166}$$

Equation (3.166) basically represents the sampling theorem in communication theory and states that $F(u)$ can be uniquely reconstructed from its sampled values at $u=n$, where $n=0, \pm1, \pm2, \pm3,\dots$. Equation (3.166) guarantees the desired performance of the taper only at the integer values of u, but not at its other values. A better method for obtaining $F(u)$ is as follows:

If we assume that the values of a_n for $|n| > N$ are zero, then Eq. (3.162) can be written as follows:

$$F(u) = 2\pi \sum_{n=-N}^{N} a_n\left[\frac{\sin\{\pi(u-n)\}}{\pi(u-n)}\right] \tag{3.167}$$

Since n is an integer, Eq. (3.167) can be written as follows:

$$F(u) = 2\pi \sum_{n=-N}^{N} a_n(-1)^n\left[\frac{\sin(\pi u)}{\pi(u-n)}\right] = 2\pi\frac{\sin(\pi u)}{(\pi u)} \sum_{n=-N}^{N} a_n(-1)^n \frac{u}{(u-n)} \tag{3.168}$$

The summation in Eq. (3.168) can be represented as the partial-fraction expansion of a function $P(u)\big/\prod_{n=1}^{N}(u^2-n^2)$, where $P(u)$ is a polynomial in u of degree $2N$ and satisfies the condition $P(-u) = P^*(u)$. Therefore, Eq. (3.168) can be written as follows:

$$F(u) = 2\pi\frac{\sin(\pi u)}{(\pi u)} \frac{P(u)}{\prod_{n=1}^{N}(u^2-n^2)} \tag{3.169}$$

With a proper choice of the function $P(u)$, we can synthesize a transmission line taper using Eq. (3.169). This has been explained with the following examples.

3.10.1 Transmission Line Tapers with Double Zeros at $u=\pm2, \pm4, \pm6, \dots$ or Triangular Tapers

Since we have double zeros at $u = \pm2,\pm4,\pm6, \dots$, we can represent $P(u)$ as follows:

$$P(u) = \prod_{n=1}^{N}(u^2-4n^2)^2 \tag{3.170}$$

Substituting Eq. (3.170) into Eq. (3.169), we get the following expression:

$$F(u) = 2\pi\frac{\sin(\pi u)}{(\pi u)} \frac{\prod_{n=1}^{N}(u^2-4n^2)^2}{\prod_{n=1}^{N}(u^2-n^2)} \tag{3.171}$$

However, the product in Eq. (3.171) does not converge even when $N \to \infty$. Therefore, $F(u)$ must be modified. Let the modified form be the following:

$$F(u) = C \frac{\sin(\pi u)}{(\pi u)} \frac{\prod_{n=1}^{N} \left\{1 - u^2 / (4n^2)\right\}^2}{\prod_{n=1}^{N} \left(1 - u^2 / n^2\right)} \tag{3.172}$$

Now

$$\frac{\sin(\pi u)}{(\pi u)} = \prod_{n=1}^{N} \left(1 - u^2 / n^2\right) \tag{3.173}$$

and

$$\left[\frac{\sin(\pi u/2)}{(\pi u/2)}\right]^2 = \prod_{n=1}^{N} \left\{1 - u^2 / (4n^2)\right\}^2 \tag{3.174}$$

Using Eqs (3.173) and (3.174) in Eq. (3.172), we get the following relation:

$$F(u) = C \frac{\sin(\pi u)}{(\pi u)} \left[\frac{\sin(\pi u/2)}{(\pi u/2)}\right]^2 \bigg/ \left[\frac{\sin(\pi u)}{(\pi u)}\right] = C \left[\frac{\sin(\pi u/2)}{(\pi u/2)}\right]^2 \tag{3.175}$$

Now at $u = n$:

$$F(n) = C \left[\frac{\sin(\pi n/2)}{(\pi n/2)}\right]^2 \tag{3.176}$$

Substituting Eq. (3.164) into Eq. (3.176), we get the following relation:

$$2\pi a_n = C \left[\frac{\sin(\pi n/2)}{(\pi n/2)}\right]^2 \quad \text{or} \quad a_n = \frac{C}{2\pi} \left[\frac{\sin(\pi n/2)}{(\pi n/2)}\right]^2 \tag{3.177}$$

For $n = 0$: $a_0 = \dfrac{C}{2\pi}$ $\tag{3.178}$

Now integrating Eq. (3.154) with respect to p, we get the following expression:

$$\int_{-\pi}^{\pi} g(p)\,dp = \int_{-\pi}^{\pi} \frac{d}{dp}[\ln(Z/Z_0)]\,dp = \int_{-\pi}^{\pi} d[\ln(Z/Z_0)] = [\ln(Z/Z_0)]_{-\pi}^{\pi}$$

or

$$\int_{-\pi}^{\pi} g(p)\,dp = \ln(Z_L/Z_0) - \ln(Z_0/Z_0)$$

or

$$\int_{-\pi}^{\pi} g(p)\,dp = \ln(Z_L/Z_0) - \ln(1) = \ln(Z_L/Z_0) \tag{3.179}$$

Repeating the same for Eq. (3.161) produces the following equation:

$$\int_{-\pi}^{\pi} g(p)\,dp = \int_{-\pi}^{\pi} \sum_{n=-\infty}^{\infty} a_n e^{jnp}\,dp = \sum_{n=-\infty}^{\infty} a_n \int_{-\pi}^{\pi} e^{jnp}\,dp = \sum_{n=-\infty}^{\infty} a_n \left[\frac{e^{jnp}}{jn}\right]_{-\pi}^{\pi}$$

or

$$\int_{-\pi}^{\pi} g(p)\,dp = \sum_{n=-\infty}^{\infty} a_n \left[\frac{e^{jn\pi} - e^{-jn\pi}}{jn}\right]$$

or

$$\int_{-\pi}^{\pi} g(p)\,dp = \pi \sum_{n=-\infty}^{\infty} a_n \left[\frac{2j\sin(n\pi)}{jn\pi}\right] = 2\pi \sum_{n=-\infty}^{\infty} a_n \left[\frac{\sin(n\pi)}{n\pi}\right] \tag{3.180}$$

Since $\dfrac{\sin(n\pi)}{n\pi} = \begin{cases} 1 & \text{for } n = 0 \\ 0 & \text{for } n \neq 0 \end{cases}$, Eq. (3.180) is modified to the following form:

$$\int_{-\pi}^{\pi} g(p)\,dp = 2\pi a_0 \tag{3.181}$$

Equating Eqs (3.179) and (3.181), we get the following relation:

$$2\pi a_0 = \ln(Z_L/Z_0) \text{ or } a_0 = \frac{\ln(Z_L/Z_0)}{2\pi} \tag{3.182}$$

Further equating Eqs (3.178) and (3.182), we get the following expression:

$$\frac{C}{2\pi} = \frac{\ln(Z_L/Z_0)}{2\pi} \text{ or } C = \ln(Z_L/Z_0) \tag{3.183}$$

Substitution of the expression of C from Eq. (3.183) into Eq. (3.175) yields the following equation:

$$F(u) = \ln(Z_L/Z_0)\left[\frac{\sin(\pi u/2)}{(\pi u/2)}\right]^2 \tag{3.184}$$

A further substitution of Eq. (3.184) into Eq. (3.157) gives the following relation:

$$\Gamma(\theta) = \frac{1}{2}e^{-j\beta L}\ln(Z_L/Z_0)\left[\frac{\sin(\pi u/2)}{(\pi u/2)}\right]^2 \tag{3.185}$$

Substituting Eq. (3.150) in Eq. (3.185), we get the following expression:

$$\Gamma(\theta) = \frac{1}{2}e^{-j\beta L}\ln(Z_L/Z_0)\left[\frac{\sin(\pi\beta L/2\pi)}{(\pi\beta L/2\pi)}\right]^2 = \frac{1}{2}e^{-j\beta L}\ln(Z_L/Z_0)\left[\frac{\sin(\beta L/2)}{(\beta L/2)}\right]^2$$

or $$\Gamma(\theta) = \frac{1}{2}e^{-j\theta}\ln(Z_L/Z_0)\left[\frac{\sin(\theta/2)}{(\theta/2)}\right]^2 = \frac{1}{2}e^{-j\theta}\ln(Z_L/Z_0)\,\text{sinc}^2(\theta/2) \tag{3.186}$$

Equation (3.186) is similar to Eq. (3.131), which validates the analysis.

3.10.2 Transmission Line Tapers with Zeros at $u = -1, -2, -3, \ldots$ or Exponential Tapers

Since we have zeros at $u = \pm 1, \pm 2, \pm 3, \ldots$, we can represent $P(u)$ as follows:

$$P(u) = \prod_{n=1}^{N}\left(u^2 - n^2\right) \tag{3.187}$$

Substituting Eq. (3.187) into Eq. (3.169), we get the following relation:

$$F(u) = 2\pi\frac{\sin(\pi u)}{(\pi u)}\frac{\prod_{n=1}^{N}\left(u^2 - n^2\right)}{\prod_{n=1}^{N}\left(u^2 - n^2\right)} \tag{3.188}$$

However, the products in Eq. (3.187) does not converge even when $N \to \infty$. Therefore, $F(u)$ must be modified. Let the modified form be the following:

$$F(u) = C\frac{\sin(\pi u)}{(\pi u)}\frac{\prod_{n=1}^{N}\left(1 - u^2/n^2\right)}{\prod_{n=1}^{N}\left(1 - u^2/n^2\right)} \tag{3.189}$$

Using Eq. (3.173) in Eq. (3.189), we get the following formula:

$$F(u) = C\frac{\sin(\pi u)}{(\pi u)}\left[\frac{\sin(\pi u)}{(\pi u)}\right] \Bigg/ \left[\frac{\sin(\pi u)}{(\pi u)}\right] = C\frac{\sin(\pi u)}{(\pi u)} \tag{3.190}$$

Now at $u = n$:

$$F(n) = C\left[\frac{\sin(\pi n)}{(\pi n)}\right] \tag{3.191}$$

Substituting Eq. (3.164) into Eq. (3.191), we get the following expression:

$$2\pi a_n = C\left[\frac{\sin(\pi n)}{(\pi n)}\right] \text{ or } a_n = \frac{C}{2\pi}\left[\frac{\sin(\pi n)}{(\pi n)}\right] \tag{3.192}$$

For $n = 0$: $a_0 = \dfrac{C}{2\pi}$ (3.193)

Equating Eq. (3.193) to Eq. (3.182), we get the following expression:

$$\frac{C}{2\pi} = \frac{\ln(Z_L/Z_0)}{2\pi}$$

or $\quad C = \ln(Z_L/Z_0)$ (3.194)

Substituting the expression of C in Eq. (3.194) into Eq. (3.190), we get the following expression:

$$F(u) = \ln(Z_L/Z_0)\left[\frac{\sin(\pi u)}{(\pi u)}\right] \tag{3.195}$$

A further substitution of Eq. (3.195) into Eq. (3.157) gives the following expression:

$$\Gamma(\theta) = \frac{1}{2}e^{-j\beta L}\ln(Z_L/Z_0)\left[\frac{\sin(\pi u)}{(\pi u)}\right] \tag{3.196}$$

Substituting Eq. (3.150) into Eq. (3.196), we get the following relation:

$$\Gamma(\theta) = \frac{1}{2}e^{-j\beta L}\ln(Z_L/Z_0)\left[\frac{\sin(\pi \beta L/\pi)}{(\pi \beta L/\pi)}\right] = \frac{1}{2}e^{-j\beta L}\ln(Z_L/Z_0)\left[\frac{\sin(\beta L)}{(\beta L)}\right]$$

or $\quad \Gamma(\theta) = \frac{1}{2}e^{-j\theta}\ln(Z_L/Z_0)\left[\frac{\sin(\theta)}{(\theta)}\right] = \frac{1}{2}e^{-j\theta}\ln(Z_L/Z_0)\mathrm{sinc}(\theta)$ (3.197)

Equation (3.197) is similar to Eq. (3.127), which validates the analysis.

EXAMPLE 3.7 Design a transmission line taper that has double zeros at $\beta L = \pm 2\pi$ and $\pm 3\pi$ by moving the zeros at $\beta L = \pm \pi$ and $\pm 4\pi$ at $\beta L = \pm 2\pi$ and $\pm 3\pi$, respectively.

Solution In the present case, $N = 4$.

Since the taper has double zeros at $\beta L = \pm 2\pi$ and $\beta L = \pm 3\pi$, we can choose the following equation:

$$P(u) = C\left(u^2 - 4\right)^2\left(u^2 - 9\right)^2$$

Thus $F(u) = 2\pi C \dfrac{\sin(\pi u)}{(\pi u)} \dfrac{\left(u^2 - 4\right)^2 \left(u^2 - 9\right)^2}{\left(u^2 - 1\right)\left(u^2 - 4\right)\left(u^2 - 9\right)\left(u^2 - 16\right)}$

The coefficients of expansion for $g(p)$ are as follows:

$$a_0 = \frac{1}{2\pi} F(0) = \frac{1}{2\pi} 2\pi C \frac{\sin(0)}{(0)} \frac{\left(0^2 - 4\right)^2 \left(0^2 - 9\right)^2}{\left(0^2 - 1\right)\left(0^2 - 4\right)\left(0^2 - 9\right)\left(0^2 - 16\right)}$$

or $a_0 = C \times 1 \times \dfrac{16 \times 81}{(-1) \times (-4) \times (-9) \times (-16)} = \dfrac{9}{4} C$

Again, $a_0 = \dfrac{1}{2\pi} \ln(z_L)$ or $\dfrac{9}{4} C = \dfrac{1}{2\pi} \ln(z_L)$ or $C = \dfrac{4}{18\pi} \ln(z_L)$

Now, $a_n = \dfrac{1}{2\pi} F(n) = \dfrac{1}{2\pi} 2\pi C \dfrac{\sin(\pi u)}{(\pi u)} \dfrac{\left(u^2 - 4\right)^2 \left(u^2 - 9\right)^2}{\left(u^2 - 1\right)\left(u^2 - 4\right)\left(u^2 - 9\right)\left(u^2 - 16\right)}$

or $a_n = \dfrac{4}{18\pi} \ln(z_L) \dfrac{\sin(\pi u)}{(\pi u)} \dfrac{\left(u^2 - 4\right)^2 \left(u^2 - 9\right)^2}{\left(u^2 - 1\right)\left(u^2 - 4\right)\left(u^2 - 9\right)\left(u^2 - 16\right)}$

Therefore,

$$a_1 = a_{-1} = \frac{4}{18\pi} \frac{\ln(z_L)}{\pi} \left[\frac{\sin(\pi u)}{\left(u^2 - 1\right)}\right]_{u=1} \frac{\left(1^2 - 4\right)^2 \left(1^2 - 9\right)^2}{\left(1^2 - 4\right)\left(1^2 - 9\right)\left(1^2 - 16\right)}$$

or $a_1 = a_{-1} = \dfrac{4}{18\pi} \dfrac{\ln(z_L)}{\pi} \times \left(-\dfrac{\pi}{2}\right) \times \dfrac{(-9) \times (-64)}{(-3) \times (-8) \times (-15)} = \dfrac{0.3556}{2\pi} \ln(z_L)$

$$a_2 = a_{-2} = \frac{4}{18\pi} \frac{\ln(z_L)}{2\pi} \left[\frac{\sin(\pi u)}{\left(u^2 - 4\right)}\right]_{u=2} \frac{\left(2^2 - 4\right)^2 \left(2^2 - 9\right)^2}{\left(2^2 - 1\right)\left(2^2 - 9\right)\left(2^2 - 16\right)}$$

or $a_2 = a_{-2} = \dfrac{4}{18\pi} \dfrac{\ln(z_L)}{2\pi} \left(\dfrac{\pi}{4}\right) \dfrac{0 \times 25}{3 \times (-5) \times (-12)} = 0$

$$a_3 = a_{-3} = \frac{4}{18\pi} \frac{\ln(z_L)}{3\pi} \left[\frac{\sin(\pi u)}{\left(u^2 - 9\right)}\right]_{u=3} \frac{\left(3^2 - 4\right)^2 \left(3^2 - 9\right)^2}{\left(3^2 - 1\right)\left(3^2 - 4\right)\left(3^2 - 16\right)}$$

or $a_3 = a_{-3} = \dfrac{4}{18\pi} \dfrac{\ln(z_L)}{3\pi} \left(\dfrac{-\pi}{6}\right) \dfrac{25 \times 0}{8 \times 5 \times (-7)} = 0$

$$a_4 = a_{-4} = \frac{4}{18\pi} \frac{\ln(z_L)}{4\pi} \left[\frac{\sin(\pi u)}{\left(u^2 - 16\right)}\right]_{u=4} \frac{\left(4^2 - 4\right)^2 \left(4^2 - 9\right)^2}{\left(4^2 - 1\right)\left(4^2 - 4\right)\left(4^2 - 9\right)}$$

or $a_4 = a_{-4} = \dfrac{4}{18\pi} \dfrac{\ln(z_L)}{4\pi} \left(\dfrac{\pi}{8}\right) \dfrac{144 \times 49}{15 \times 12 \times 7} = \dfrac{0.0778}{2\pi} \ln(z_L)$

$$g(p) = \frac{d\{\ln(z)\}}{dp} = \sum_{n=-3}^{3} a_n \cos(np)$$

or $$g(p) = \frac{d\{\ln(z)\}}{dp} = \frac{\ln(z_L)}{2\pi}\{1 + 0.7112\cos(p) + 0.1556\cos(4p)\}$$

or $$\ln(z) = \int \frac{\ln(z_L)}{2\pi}\{1 + 0.7112\cos(p) + 0.1556\cos(4p)\}dp$$

or $$\ln(z) = \frac{\ln(z_L)}{2\pi}\{p + 0.7112\sin(p) + 0.0389\sin(4p)\} + A \text{ (constant)}$$

Now $\ln(z) = 0$ at $p = -\pi$; therefore,

$$0 = \frac{\ln(z_L)}{2\pi}\{-\pi + 0.7112\sin(-\pi) + 0.0389\sin(-4\pi)\} + A$$

or $$-\frac{\ln(z_L)}{2} + A = 0 \text{ or } A = \ln(z_L)/2$$

Therefore, $$\ln(z) = \frac{\ln(z_L)}{2\pi}\{\pi + p + 0.7112\sin(p) + 0.0389\sin(4p)\}$$

where $$p = \frac{2\pi}{L}\left(x - \frac{L}{2}\right)$$

Practice Problem

3.13 Design a transmission line taper that has triple zeros at $\beta L = \pm 2\pi$ by moving the zeros at $\beta L = \pm \pi$ and $\beta L = \pm 3\pi$ at $\beta L = \pm 2\pi$.

$$\left[\ln(z) = \frac{\ln(z_L)}{2\pi}\{\pi + p + 0.632\sin(p) - 0.0653\sin(3p)\}\right]$$

3.11 Bode–Fano Criterion

If a lossless network is used to match an arbitrary load impedance Z_L with a line of characteristic impedance Z_0, as shown in Fig. 3.1, then the Bode–Fano criterion states that

$$\int_0^\infty \ln\left(\frac{1}{|\Gamma_\omega|}\right)d\omega \le \frac{\pi}{\tau} \tag{3.198}$$

where τ is the time constant of the network. The expression of π/τ depends on the characteristics of the load impedance and is given in Table 3.1. Since $\ln(1/|\Gamma|)$ is proportional to the return loss at the input of the matching network, the LHS of Eq. (3.198) represents the total area between the return loss curve and the return loss = 0 axis. Therefore, according to the Bode–Fano criterion, the total area between the return loss curve and the return loss = 0 axis must be less than or equal to π/τ. It reveals that an attempt to improve matching over a given frequency band will result in a poorer matching on the other part of the spectrum.

Table 3.1 Load characteristics and corresponding expression of $1/\tau$

Load Characteristics	$1/\tau$
Series R–L	R/L
Series R–C	$\omega_0^2 RC$
Parallel R–L	$\omega_0^2 L/R$
Parallel R–C	$1/(RC)$

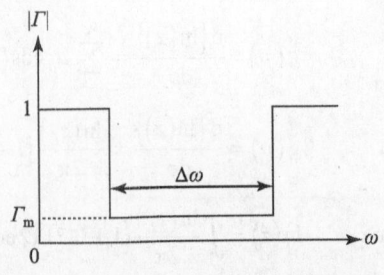

Fig. 3.25 Optimum frequency response graph according to Bode–Fano criterion

Now let us assume that a hypothetical matching network produces a uniform reflection coefficient Γ_m over a bandwidth $\Delta\omega$ and a value of unity outside the bandwidth, as shown in Fig. 3.25.

Since $\ln(1) = 0$, we can write the following equation:

$$\int_0^{\Delta\omega} \ln\left(1/|\Gamma_\omega|\right) d\omega \le \frac{\pi}{\tau} \quad \text{or} \quad -\int_0^{\Delta\omega} \ln\left(|\Gamma_\omega|\right) d\omega \le \frac{\pi}{\tau}$$

or $\qquad -\Delta\omega \ln\left(|\Gamma_m|\right) \le \frac{\pi}{\tau}$ $\qquad\qquad$ (3.199)

Based on Eq. (3.199), we can write the following relation:

$$|\Gamma_m| \le e^{-\frac{\pi}{\tau(\Delta\omega)}}$$ $\qquad\qquad$ (3.200)

and $\qquad \Delta\omega \le -\dfrac{\pi}{\tau \ln\left(|\Gamma_m|\right)}$

or $\qquad \Delta f \le -\dfrac{1}{2\tau \ln\left(|\Gamma_m|\right)}$ $\qquad\qquad$ (3.201)

Equation (3.201) reveals that a larger permissible mismatch corresponds to a greater bandwidth of matching. Since $\ln|\Gamma_m| \to -\infty$ as $|\Gamma_m| \to 0$, the inequality of Eq. (3.201) also represents that it is impossible to obtain a perfect match over a non-zero bandwidth. At best, we can obtain a perfect match at finite discrete frequencies within any finite-frequency interval.

Note Practical realization of a matching network having the reflection coefficient response, as shown in Fig. 3.25, requires an infinite number of elements and hence cannot be realized in practice. However, it can be shown that if the pass band ripple of a Chebyshev matching transformer is made equal to Γ_m, then it closely approximates the pass band of Fig. 3.25.

Important Formulae

- Attenuation loss (dB) $= 8.686\alpha l$.
- Reflection loss (dB) $= 10\log_{10}\left\{1/\left(1 - |\Gamma|^2\right)\right\}$.

- Transmission loss (dB) $= 0.868\alpha l + 10\log_{10}\left\{1/\left(1-|\Gamma|^2\right)\right\}$.
- Return loss (dB) $= -20\log_{10}\left(|\Gamma|\right)$.
- The characteristic impedance of a quarter-wave transformer used to match a load Z_L with a line of characteristic impedance Z_0 is $\sqrt{Z_0 Z_L}$.
- The reflection coefficient out of a single-section quarter-wave transformer-matched line at a frequency f', other than the designed frequency f_0, is

$$\Gamma = \frac{(Z_L - Z_0)}{(Z_L + Z_0) + j2\sqrt{Z_0 Z_L}\,\tan\left(\beta' d\right)}.$$

- The fractional bandwidth of a single-section quarter-wave transformer is

$$\frac{\Delta f}{f_0} = 2 - \frac{4}{\pi}\cos^{-1}\left|\frac{\Gamma_m}{\sqrt{1-\Gamma_m^2}}\frac{2\sqrt{Z_0 Z_L}}{|Z_L - Z_0|}\right|.$$

- The reflection coefficient out of an N-section binomial quarter-wave transformer can be expressed as $\Gamma(\theta) = A\left(1 + e^{-j2\theta}\right)^N$, where $A = 2^{-N}\dfrac{Z_L - Z_0}{Z_L + Z_0}$.
- The fractional bandwidth of an N-section binomial quarter-wave transformer is

$$\frac{\Delta f}{f_0} = 2 - \frac{4\theta_m}{\pi} = 2 - \frac{4}{\pi}\cos^{-1}\left[\frac{1}{2}\left(\frac{|\Gamma_m|}{|A|}\right)^{\frac{1}{N}}\right].$$

- For binomial quarter-wave transformers, the characteristic impedance of two successive sections is related by the relation $\ln\left(\dfrac{Z_{n+1}}{Z_n}\right) = 2^{-N}C_n^N\ln\left(\dfrac{Z_L}{Z_0}\right)$.
- The reflection coefficient out of an N-section Chebyshev quarter-wave transformer can be expressed as $\Gamma(\theta) = Ae^{-jN\theta}T_n\left\{\sec\left(\theta_m\right)\cos(\theta)\right\}$, where $|\Gamma_m| = A$.
- The fractional bandwidth of an N-section Chebyshev quarter-wave transformer is

$$\frac{\Delta f}{f_0} = 2 - \frac{4}{\pi}\sec^{-1}\left[\cosh\left[\frac{1}{N}\cosh^{-1}\left[T_N\left\{\sec\left(\theta_m\right)\right\}\right]\right]\right].$$

- For an exponential taper, Z varies exponentially with x; therefore, we can write $Z = Z_0 e^{\alpha x}$
- For a triangular taper, we can write

$$Z = \begin{cases} Z_0 e^{2(x/L)^2 \ln(Z_L/Z_0)} & \text{for } 0 \le x \le L/2 \\ Z_0 e^{(4x/L - 2x^2/L^2 - 1)\ln(Z_L/Z_0)} & \text{for } L/2 \le x \le L \end{cases}.$$

Exercises

Objective-type Questions

3.1 Attenuation loss in transmission line is defined as
  (a) $8.686/\alpha l$ (b) $8.686\alpha l$ (c) $8.686^{\alpha l}$ (d) $1/8.686\alpha l$

3.2 Basic design geometry of a lumped element matching network depends on whether the normalized load is
  (a) inductive or capacitive in nature
  (b) real or complex in nature
  (c) inside or outside $1 + jx$ circle in the Smith chart
  (d) larger or lower than normalized source impedance

3.3 For a T-matching network, if the image resistance is R_I, then
(a) $Z_0 < R_I < R_L$
(c) $Z_0 < R_I = R_L$
(b) $Z_0 = R_I < R_L$
(d) $Z_0 < R_I$ and $R_I > R_L$

3.4 Short-circuited stubs are preferred to open-circuited stubs because the latter
(a) are liable to radiate
(b) are incapable of giving full range of reactance
(c) are difficult to fabricate
(d) require larger stub length

3.5 In single-stub matching, the stub should be inserted at a point on the line where the normalized conductance is
(a) 0
(b) ∞
(c) 1
(d) 0.5

3.6 For double-stub matching, the conductance of the line at the position of the first stub should be
(a) less than or equal to $Y_0 \sin^2(\beta d)$
(c) less than or equal to $Y_0 \cosec^2(\beta d)$
(b) less than or equal to $Y_0 \sec^2(\beta d)$
(d) greater than $Y_0 \cosec^2(\beta d)$

3.7 A quarter-wave transformer of characteristic impedance $60\,\Omega$ has been used to match a transmission line of characteristic impedance Z_0 with a load of $72\,\Omega$. The characteristic impedance of the transformer, when the load of $72\,\Omega$ is replaced by that of $98\,\Omega$, is
(a) $60\,\Omega$
(b) $70\,\Omega$
(c) $50\,\Omega$
(d) $98\,\Omega$

3.8 A $50\,\Omega$ lossless transmission line is terminated in a load of $100\,\Omega$. The line has been matched using the single-stub matching technique. If the distance of the stub from the load and the length of the stub are d and l, respectively, then the SWR in the line sections of length d and length l are, respectively,
(a) ∞ and 2
(b) 2 and ∞
(c) 1 and 1
(d) 2 and 2

3.9 A quarter-wave transformer is used for
(a) low impedance loads
(b) high impedance loads
(c) connecting higher impedance loads to a lower-impedance line and vice versa
(d) provide extra space in the system

3.10 A transmission line of a characteristic impedance of $50\,\Omega$ has been matched with a load of $100\,\Omega$ using a single-section quarter-wave transmission line. If the maximum allowable reflection coefficient is 0.05, then the percentage of fractional bandwidth will be
(a) 38.19
(b) 18.09
(c) 0.1809
(d) 0.3820

Review Questions

3.1 Describe different lumped section matching networks.
3.2 How can matching be obtained using a single stub? Describe the theory in detail.
3.3 Prove that, for a given stub spacing, the double-stub matching technique cannot be used to match any arbitrary load admittances. What are the ranges of load admittance that cannot be matched using double-stub matching for a given stub spacing d?
3.4 Describe the principle of quarter-wave transformer matching. Can you use a single quarter-wave transformer section to match any arbitrary load? If no, what modifications are required?

3.5 Derive the expression of the reflection coefficient out of a single-section quarter-wave transformer matching section at a frequency f' other than the matching frequency f_0. Next, derive the expression of the bandwidth of the single-section quarter-wave transformer for a maximum allowable reflection coefficient magnitude Γ_m.

3.6 Describe the detailed design procedure of a binomial multisection matching transformer. In addition, derive the expression for its fractional bandwidth.

3.7 Describe the detailed design procedure of a Chebyshev multisection matching transformer and derive the expression for its fractional bandwidth.

3.8 Derive the expression of the reflection coefficient out of an exponential taper line.

3.9 Derive the expression of the reflection coefficient out of a triangular taper line.

3.10 Prove that a transmission line taper, exhibiting double zeros at $u = \pm 2, \pm 4, \pm 6,...$ in its frequency response, is basically a triangular taper.

3.11 Prove that a transmission line taper, exhibiting zeros at $u = \pm 1, \pm 2, \pm 3,...$ in its frequency response, is basically an exponential taper.

3.12 Describe the Bode–Fano criterion.

Problems

3.1 Design a pi-section matching network to match a 90 Ω line with a $100 + j63$ Ω load. Assume the image impedance to be 74 Ω. *[Hint: Use the Smith chart]*

3.2 Design a T-section matching network to match a 74 Ω line with a $69 + j100$ Ω load. Assume the image impedance to be 90 Ω. *[Hint: Use the Smith chart]*

3.3 What is the minimum stub spacing required to match a normalized load admittance $2 + j3.7$ using a double-stub matching network, if spacing is restricted to greater than $2\lambda/13$? If spacing is restricted to less than $2\lambda/13$, then determine the maximum stub spacing. Can you match a normalized load admittance $0.87 + j1.43$ using a double-stub matching network, with a stub spacing $\lambda/8$? *[Hint: Use Eq. (3.49)]*

3.4 An exponential taper has been used to match a 40 Ω line with a 74 Ω line. Find out the reflection coefficient at a distance 2.3λ. *[Hint: Use Eq. (3.127)]*

3.5 Design an exponential taper to match a 64 Ω line with a 130 Ω load at 3 GHz. Assume that the line is 0.73λ long. *[Hint: Use Eqs (3.124) and (3.125)]*

3.6 The characteristic impedance of an exponential taper is defined as $Z = Z_0 e^{0.0367x}$. If the taper has been used to match a 100 Ω load with a 70 Ω line, then calculate the length of the taper. *[Hint: Use Eqs (3.124) and (3.125)]*

3.7 A triangular taper has been used to match a 60 Ω line with a 90 Ω line. Find out the reflection coefficient at a distance 3.7λ. *[Hint: Use Eq. (3.131)]*

3.8 Design a triangular taper to match a 45 Ω line with a 63 Ω load at 3 GHz. Assume that the line is 1.23λ long. *[Hint: Use Eq. (3.128)]*

3.9 The characteristic impedance of a triangular taper is defined as $Z = Z_0 e^{0.4578x^2}$ in the range $0 \le x \le L/2$. If the taper has been used to match a 123 Ω load with a 87 Ω line, then calculate the length of the taper. In addition, find the expression of Z in the range $L/2 \le x \le L$. *[Hint: Use Eq. (3.128)]*

3.10 A Chebyshev taper has been used to match a 50 Ω line with a 74 Ω line. If the cut-off occurs at $\beta L = 1.7$, then calculate the maximum tolerable reflection coefficient. In addition, calculate the reflection coefficient at $\beta L = 5.7$. *[Hint: Use Eq. (3.139)]*

3.11 A Chebyshev taper has been used to match a 74 Ω line with a 100 Ω line. If the maximum tolerable reflection coefficient is 0.05, then calculate the electrical length of the line at the cut-off. In addition, calculate the value of βL at the cut-off. *[Hint: Use Eq. (3.141)]*

4

Planar Transmission Lines

4.1 Introduction

A planar transmission line is characterized by two or more conducting metal strips that lie entirely in parallel planes. In its most common form, a planar transmission line consists of one or more conducting strips that are supported by a low-loss dielectric substrate and adjacent to a ground plane. In planar transmission line structures, more than one ground plane is also common. Depending on the arrangement of the conducting strip(s) and the ground plane(s), planar transmission lines can be classified as strip lines, microstrip lines, coplanar lines, slot lines, etc.

Out of these different types of planar transmission lines, microstrip lines are most common and used most widely. These have the advantages of low cost, small size, easiness of fabrication, mass production, good repeatability, and reproducibility, and mostly have the provision for integration of active devices such as diodes and transistors. The last advantage has opened the gateway for the design and development of monolithic microwave integrated circuits (MMICs).

Unlike the two- or multi-conductor transmission lines, described in the previous chapters, it is very difficult to obtain closed-form expressions for the secondary constants (propagation constant and characteristic impedance) of a planar transmission line. The analytical procedures for the analysis of planar transmission lines are also very complicated and cumbersome. Most of the available closed-form expressions are based on either the curve-fitting method or some assumptions. Therefore, neither the expressions are 100% accurate nor they can be used for all types of substrates. This shortcoming of a planar line has put a hurdle in the designing of planar transmission line circuits. However, in present days, with the availability of CAD software with optimization techniques, designing of a planar circuit has become easier than before.

4.2 General Analysis of Transverse Electric and Transverse Magnetic Modes

The following are the Maxwell's curl equations in a source-free region:

$$\vec{\nabla} \times \vec{H} = \frac{\partial \vec{D}}{\partial t} + \sigma \vec{E} \tag{4.1}$$

$$\vec{\nabla} \times \vec{E} = -\frac{\partial \vec{B}}{\partial t} \qquad (4.2)$$

For the frequency-domain analysis $\frac{\partial}{\partial t} \rightarrow j\omega$ and Eqs (4.1)–(4.2) can be written as follows:

$$\vec{\nabla} \times \vec{H} = j\omega\varepsilon\vec{E} + \sigma\vec{E} \qquad (4.3)$$

and $\qquad \vec{\nabla} \times \vec{E} = -j\omega\mu\vec{H} \qquad (4.4)$

Now from Eq. (4.3), the following expression is obtained:

$$\vec{\nabla} \times \vec{H} = \begin{vmatrix} \hat{a}_x & \hat{a}_y & \hat{a}_z \\ \dfrac{\partial}{\partial x} & \dfrac{\partial}{\partial y} & \dfrac{\partial}{\partial z} \\ H_x & H_y & H_z \end{vmatrix} = (j\omega\varepsilon + \sigma)(E_x\hat{a}_x + E_y\hat{a}_y + E_z\hat{a}_z) \qquad (4.5)$$

Equating each component on the RHS to the corresponding component on the LHS, we get the following relations:

$$\frac{\partial H_z}{\partial y} - \frac{\partial H_y}{\partial z} = (j\omega\varepsilon + \sigma)E_x \qquad (4.6)$$

$$\frac{\partial H_x}{\partial z} - \frac{\partial H_z}{\partial x} = (j\omega\varepsilon + \sigma)E_y \qquad (4.7)$$

$$\frac{\partial H_y}{\partial x} - \frac{\partial H_x}{\partial y} = (j\omega\varepsilon + \sigma)E_z \qquad (4.8)$$

Similarly, from Eq. (4.4), we get the following relations:

$$\frac{\partial E_z}{\partial y} - \frac{\partial E_y}{\partial z} = -j\omega\mu H_x \qquad (4.9)$$

$$\frac{\partial E_x}{\partial z} - \frac{\partial E_z}{\partial x} = -j\omega\mu H_y \qquad (4.10)$$

$$\frac{\partial E_y}{\partial x} - \frac{\partial E_x}{\partial y} = -j\omega\mu H_z \qquad (4.11)$$

For a uniform plane wave propagation along the z-direction, the electric and magnetic fields can be written as follows:

$$\vec{E}(x,y,z) = E_x\hat{a}_x + E_y\hat{a}_y + E_z\hat{a}_z = (E_{0x}\hat{a}_x + E_{0y}\hat{a}_y + E_{0z}\hat{a}_z)e^{-\gamma z}$$

or $\qquad \vec{E}(x,y,z) = (\vec{e}_t + \vec{e}_z)e^{-\gamma z} \qquad (4.12)$

$$\vec{H}(x,y,z) = H_x\hat{a}_x + H_y\hat{a}_y + H_z\hat{a}_z = \left(H_{0x}\hat{a}_x + H_{0y}\hat{a}_y + H_{0z}\hat{a}_z\right)e^{-\gamma z}$$

or $\qquad \vec{H}(x,y,z) = \left(\vec{h_t} + \vec{h_z}\right)e^{-\gamma z}$ $\qquad\qquad$ (4.13)

where E_{0x}, E_{0y}, E_{0z}, H_{0x}, H_{0y}, and H_{0z} are all functions of (x, y); $\vec{e_t}$, and $\vec{h_t}$, are the transverse components of the electric and magnetic field intensity, respectively; and $\vec{e_z}$, and $\vec{h_z}$, are the longitudinal components of the electric and magnetic field intensity, respectively. Substituting Eqs (4.12) and (4.13) into Eqs (4.6)–(4.11), we get the following expressions:

$$\frac{\partial H_{0z}}{\partial y} + \gamma H_{0y} = (j\omega\varepsilon + \sigma)E_{0x} \qquad\qquad (4.14)$$

$$-\gamma H_{0x} - \frac{\partial H_{0z}}{\partial x} = (j\omega\varepsilon + \sigma)E_{0y} \qquad\qquad (4.15)$$

$$\frac{\partial H_{0y}}{\partial x} - \frac{\partial H_{0x}}{\partial y} = (j\omega\varepsilon + \sigma)E_{0z} \qquad\qquad (4.16)$$

and $\qquad \dfrac{\partial E_{0z}}{\partial y} + \gamma E_{0y} = -j\omega\mu H_{0x}$ $\qquad\qquad$ (4.17)

$$-\gamma E_{0x} - \frac{\partial E_{0z}}{\partial x} = -j\omega\mu H_{0y} \qquad\qquad (4.18)$$

$$\frac{\partial E_{0y}}{\partial x} - \frac{\partial E_{0x}}{\partial y} = -j\omega\mu H_{0z} \qquad\qquad (4.19)$$

From Eq. (4.18), we can write the following relation:

$$H_{0y} = \frac{1}{j\omega\mu}\frac{\partial E_{0z}}{\partial x} + \frac{\gamma}{j\omega\mu}E_{0x} \qquad\qquad (4.20)$$

Substituting Eq. (4.20) into Eq. (4.14) and simplifying, the following expression is obtained:

$$E_{0x} = -\frac{\gamma}{\left[\gamma^2 - j\omega\mu(\sigma + j\omega\varepsilon)\right]}\frac{\partial E_{0z}}{\partial x} - \frac{j\omega\mu}{\left[\gamma^2 - j\omega\mu(\sigma + j\omega\varepsilon)\right]}\frac{\partial H_{0z}}{\partial y} \qquad (4.21)$$

Substitution of

$$k_c^2 = \gamma^2 - j\omega\mu(\sigma + j\omega\varepsilon) \qquad\qquad (4.22)$$

into Eq. (4.21) yields the following expression:

$$E_{0x} = -\frac{\gamma}{k_c^2}\frac{\partial E_{0z}}{\partial x} - \frac{j\omega\mu}{k_c^2}\frac{\partial H_{0z}}{\partial y} \qquad\qquad (4.23)$$

Based on Eq. (4.17), we can write the following equation:

$$H_{0x} = -\frac{1}{j\omega\mu}\frac{\partial E_{0z}}{\partial y} - \frac{\gamma}{j\omega\mu}E_{0y} \tag{4.24}$$

Substituting Eq. (4.24) into Eq. (4.15) and simplifying, we get the following relation:

$$E_{0y} = -\frac{\gamma}{\left[\gamma^2 - j\omega\mu\left(j\omega\varepsilon + \sigma\right)\right]}\frac{\partial E_{0z}}{\partial y} + \frac{j\omega\mu}{\left[\gamma^2 - j\omega\mu\left(j\omega\varepsilon + \sigma\right)\right]}\frac{\partial H_{0z}}{\partial x} \tag{4.25}$$

Using Eq. (4.22) in Eq. (4.25), we get the following relation:

$$E_{0y} = -\frac{\gamma}{k_c^2}\frac{\partial E_{0z}}{\partial y} + \frac{j\omega\mu}{k_c^2}\frac{\partial H_{0z}}{\partial x} \tag{4.26}$$

Again from Eq. (4.15), we can write the following expression:

$$E_{0y} = -\frac{\gamma}{\left(j\omega\varepsilon + \sigma\right)}H_{0x} - \frac{1}{\left(j\omega\varepsilon + \sigma\right)}\frac{\partial H_{0z}}{\partial x} \tag{4.27}$$

Substituting Eq. (4.27) into Eq. (4.17) and simplifying, we get the following relation:

$$\frac{\partial E_{0z}}{\partial y} - \frac{\gamma^2}{\left(j\omega\varepsilon + \sigma\right)}H_{0x} - \frac{\gamma}{\left(j\omega\varepsilon + \sigma\right)}\frac{\partial H_{0z}}{\partial x} = -j\omega\mu H_{0x}$$

or $$H_{0x} = -\frac{\gamma}{\left[\gamma^2 - j\omega\mu\left(j\omega\varepsilon + \sigma\right)\right]}\frac{\partial H_{0z}}{\partial x} + \frac{\left(j\omega\varepsilon + \sigma\right)}{\left[\gamma^2 - j\omega\mu\left(j\omega\varepsilon + \sigma\right)\right]}\frac{\partial E_{0z}}{\partial y} \tag{4.28}$$

Using Eq. (4.22) in Eq. (4.28), we get the following expression:

$$H_{0x} = -\frac{\gamma}{k_c^2}\frac{\partial H_{0z}}{\partial x} + \frac{\left(\sigma + j\omega\varepsilon\right)}{k_c^2}\frac{\partial E_{0z}}{\partial y} \tag{4.29}$$

Finally, the following equation is obtained from Eq. (4.14):

$$E_{0x} = \frac{1}{\left(j\omega\varepsilon + \sigma\right)}\frac{\partial H_{0z}}{\partial y} + \frac{\gamma}{\left(j\omega\varepsilon + \sigma\right)}H_{0y} \tag{4.30}$$

Substituting Eq. (4.30) into Eq. (4.18) and simplifying, we get the following relation:

$$-\frac{\gamma}{\left(j\omega\varepsilon + \sigma\right)}\frac{\partial H_{0z}}{\partial y} - \frac{\gamma^2}{\left(j\omega\varepsilon + \sigma\right)}H_{0y} - \frac{\partial E_{0z}}{\partial x} = -j\omega\mu H_{0y}$$

or $$H_{0y} = -\frac{\gamma}{\left[\gamma^2 - j\omega\mu\left(j\omega\varepsilon + \sigma\right)\right]}\frac{\partial H_{0z}}{\partial y} - \frac{\left(j\omega\varepsilon + \sigma\right)}{\left[\gamma^2 - j\omega\mu\left(j\omega\varepsilon + \sigma\right)\right]}\frac{\partial E_{0z}}{\partial x} \tag{4.31}$$

Using Eq. (4.22) in Eq. (4.31), the following equation is derived:

$$H_{0y} = -\frac{\gamma}{k_c^2}\frac{\partial H_{0z}}{\partial y} - \frac{(\sigma + j\omega\varepsilon)}{k_c^2}\frac{\partial E_{0z}}{\partial x} \tag{4.32}$$

Equations (4.23), (4.26), (4.29), and (4.32) give general relationships for field components for a plane wave.

If the medium is lossless then $\gamma \to j\beta$ and $\sigma = 0$, and Eqs (4.14)–(4.19) can be written as follows:

$$\frac{\partial H_{0z}}{\partial y} + j\beta H_{0y} = j\omega\varepsilon E_{0x} \tag{4.33}$$

$$-j\beta H_{0x} - \frac{\partial H_{0z}}{\partial x} = j\omega\varepsilon E_{0y} \tag{4.34}$$

$$\frac{\partial H_{0y}}{\partial x} - \frac{\partial H_{0x}}{\partial y} = j\omega\varepsilon E_{0z} \tag{4.35}$$

$$\frac{\partial E_{0z}}{\partial y} + j\beta E_{0y} = -j\omega\mu H_{0x} \tag{4.36}$$

$$-j\beta E_{0x} - \frac{\partial E_{0z}}{\partial x} = -j\omega\mu H_{0y} \tag{4.37}$$

$$\frac{\partial E_{0y}}{\partial x} - \frac{\partial E_{0x}}{\partial y} = -j\omega\mu H_{0z} \tag{4.38}$$

The transverse field components, given in Eqs (4.29), (4.32), (4.23), and (4.26) can be written as follows:

$$H_{0x} = -\frac{j\beta}{k_c^2}\frac{\partial H_{0z}}{\partial x} + \frac{j\omega\varepsilon}{k_c^2}\frac{\partial E_{0z}}{\partial y} = \frac{j}{k_c^2}\left[\omega\varepsilon\frac{\partial E_{0z}}{\partial y} - \beta\frac{\partial H_{0z}}{\partial x}\right] \tag{4.39}$$

$$H_{0y} = -\frac{j\beta}{k_c^2}\frac{\partial H_{0z}}{\partial y} - \frac{j\omega\varepsilon}{k_c^2}\frac{\partial E_{0z}}{\partial x} = -\frac{j}{k_c^2}\left[\omega\varepsilon\frac{\partial E_{0z}}{\partial x} + \beta\frac{\partial H_{0z}}{\partial y}\right] \tag{4.40}$$

$$E_{0x} = -\frac{j\beta}{k_c^2}\frac{\partial E_{0z}}{\partial x} - \frac{j\omega\mu}{k_c^2}\frac{\partial H_{0z}}{\partial y} = -\frac{j}{k_c^2}\left[\beta\frac{\partial E_{0z}}{\partial x} + \omega\mu\frac{\partial H_{0z}}{\partial y}\right] \tag{4.41}$$

$$E_{0y} = -\frac{j\beta}{k_c^2}\frac{\partial E_{0z}}{\partial y} + \frac{j\omega\mu}{k_c^2}\frac{\partial H_{0z}}{\partial x} = \frac{j}{k_c^2}\left[-\beta\frac{\partial E_{0z}}{\partial y} + \omega\mu\frac{\partial H_{0z}}{\partial x}\right] \tag{4.42}$$

where $\quad k_c^2 = (j\beta)^2 - j^2\omega^2\mu\varepsilon = \omega^2\mu\varepsilon - \beta^2 = k^2 - \beta^2 \tag{4.43}$

as $\quad \omega^2\mu\varepsilon = k^2$

Equations (4.39)–(4.42) represent the transverse components of electric and magnetic fields in terms of their longitudinal components. Since the transverse electric (TE), transverse magnetic (TM), and transverse electro-magnetic (TEM) modes of waves are characterized by their longitudinal components, these expressions help in simplifying the analysis of these modes.

4.2.1 TE Mode

A TE or H-wave has zero electric field component along the propagation direction (here z-direction). Therefore, the only existing longitudinal component is H_z. The z-component of magnetic field intensity must satisfy the Helmholtz equation:[1]

$$\left(\nabla^2 + k^2\right)H_z = 0 \tag{4.44}$$

For a lossless case, we can write the following expression based on Eq. (4.13):

$$H_z = h_z e^{-j\beta z} \tag{4.45}$$

Substituting Eq. (4.45) into Eq. (4.44), we get the following equation:

$$\left(\frac{\partial^2}{\partial x^2} + \frac{\partial^2}{\partial y^2} + \frac{\partial^2}{\partial z^2} + k^2\right)h_z e^{-j\beta z} = 0$$

or

$$\left(\frac{\partial^2}{\partial x^2} + \frac{\partial^2}{\partial y^2} + (-j\beta)^2 + k^2\right)h_z e^{-j\beta z} = 0$$

or

$$\left(\frac{\partial^2}{\partial x^2} + \frac{\partial^2}{\partial y^2} + k^2 - \beta^2\right)h_z e^{-j\beta z} = 0 \tag{4.46}$$

Since $e^{-j\beta z} \neq 0$ and $k_c^2 = k^2 - \omega^2 \mu \varepsilon = k^2 - \beta^2$, we can write Eq. (4.46) as follows:

$$\left(\frac{\partial^2}{\partial x^2} + \frac{\partial^2}{\partial y^2} + k_c^2\right)h_z = 0 \tag{4.47}$$

Equation (4.47) can be solved, using proper boundary conditions, to find h_z and hence H_{0z}. Once H_{0z} is obtained, we can find the transverse field components using Eqs (4.39)–(4.42) for TE modes. For TE modes, Eqs (4.39)–(4.42) are modified as follows:

$$H_{0x} = -\frac{j\beta}{k_c^2}\frac{\partial H_{0z}}{\partial x} \tag{4.48}$$

---

[1]For detailed derivation and explanation of Helmholtz equation, consult *Principles of Electromagnetics* by M.N.O. Sadiku, Oxford University Press, Fourth Edition, New Delhi, pp. 372–373.

$$H_{0y} = -\frac{j\beta}{k_c^2}\frac{\partial H_{0z}}{\partial y} \tag{4.49}$$

$$E_{0x} = -\frac{j\omega\mu}{k_c^2}\frac{\partial H_{0z}}{\partial y} \tag{4.50}$$

$$E_{0y} = \frac{j\omega\mu}{k_c^2}\frac{\partial H_{0z}}{\partial x} \tag{4.51}$$

4.2.2 TM Mode

A TM or E-wave has zero magnetic field component along the propagation direction (here z-direction). Therefore, the only existing longitudinal component is E_z. The z-component of electric field intensity must satisfy the Helmholtz equation:

$$\left(\nabla^2 + k^2\right)E_z = 0 \tag{4.52}$$

For a lossless case, we can write the following formula based on Eq. (4.12):

$$E_z = e_z e^{-j\beta z} \tag{4.53}$$

Substituting Eq. (4.53) into Eq. (4.52) we get the following relation:

$$\left(\frac{\partial^2}{\partial x^2} + \frac{\partial^2}{\partial y^2} + \frac{\partial^2}{\partial z^2} + k^2\right)e_z e^{-j\beta z} = 0$$

or $$\left(\frac{\partial^2}{\partial x^2} + \frac{\partial^2}{\partial y^2} + k_c^2\right)e_z = 0 \tag{4.54}$$

Equation (4.54) can be solved, with proper boundary conditions, to find e_z and hence E_{0z}. Once E_{0z} is obtained, we can find the transverse field components using Eqs (4.39)–(4.42) for TM modes. For TM modes, Eqs (4.39)–(4.42) are modified as follows:

$$H_{0x} = \frac{j\omega\varepsilon}{k_c^2}\frac{\partial E_{0z}}{\partial y} \tag{4.55}$$

$$H_{0y} = -\frac{j\omega\varepsilon}{k_c^2}\frac{\partial E_{0z}}{\partial x} \tag{4.56}$$

$$E_{0x} = -\frac{j\beta}{k_c^2}\frac{\partial E_{0z}}{\partial x} \tag{4.57}$$

$$E_{0y} = -\frac{j\beta}{k_c^2}\frac{\partial E_{0z}}{\partial y} \tag{4.58}$$

Solutions of Eqs (4.47) and (4.54) are subjected to boundary conditions, which are different for different structures. In the following sections, we will solve these equations for different cases.

4.3 Surface Waves on Grounded Dielectric Slab

Surface waves, in the dielectric interface of a planar transmission line, are characterized by a field that is mostly concentrated in or near the dielectric and decays exponentially away from the interface. As the frequency increases, the field becomes more tightly bound to the dielectric. Detailed analyses of TE and TM mode surface waves in a grounded dielectric slab are given in the subsequent sections. For the analysis of surface waves, we will assume that the grounded dielectric slab has thickness h and relative permittivity ε_r, and is infinitely extended in the y- and z-direction, as shown in Fig. 4.1. We will further assume that the wave is propagating along the z-direction with an $e^{-j\beta z}$ propagation factor and is uniform along the y-direction, that is, $\partial/\partial y = 0$. Since there are two distinct regions, the dielectric and free-space regions, the fields in these regions will satisfy different wave equations.

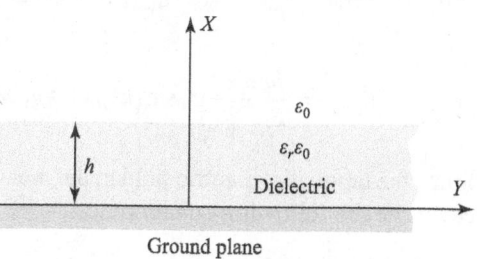

Fig. 4.1 Infinitely extended grounded dielectric slab

4.3.1 TE-mode Propagation

From Eq. (4.13), the z-component of magnetic field intensity in the dielectric and free-space regions can be written as follows:

$$H_{0z,d} = h_{z,d} e^{-j\beta z} \tag{4.59}$$

and $\qquad H_{0z,a} = h_{z,a} e^{-j\beta z} \tag{4.60}$

These fields will satisfy the wave equations

$$\left(\frac{\partial^2}{\partial x^2} + k_{c,d}^2\right) h_{z,d}(x,y) = 0 \quad \text{for } 0 \le x \le h \tag{4.61}$$

within dielectric region and

$$\left(\frac{\partial^2}{\partial x^2} - k_{c,a}^2\right) h_{z,a}(x,y) = 0 \quad \text{for } h \le x \le \infty \tag{4.62}$$

in the air. In Eqs (4.61) and (4.62), we have substituted the following relations:

$$k_{c,d}^2 = \varepsilon_r k_0^2 - \beta^2 \tag{4.63}$$

$$k_{c,a}^2 = \beta^2 - k_0^2 \tag{4.64}$$

This ensures that both $k_{c,d}$ and $k_{c,a}$ are real. The well-known general solution of Eqs (4.61) and (4.62) are expressed as follows:

$$h_{z,d}(x,y) = g_0 \cos(k_{c,d}x) + g_1 \sin(k_{c,d}x) \tag{4.65}$$

and

$$h_{z,a}(x,y) = h_0 e^{k_{c,a}x} + h_1 e^{-k_{c,a}x} \tag{4.66}$$

Using Eqs (4.51), (4.59), and (4.65), we can write the following expression:

$$E_{0y,d} = \frac{j\omega\mu}{k_{c,d}^2}\frac{\partial H_{z,d}}{\partial x} = \frac{j\omega\mu}{k_{c,d}^2}\frac{\partial}{\partial x}\left(h_{z,d}e^{-j\beta z}\right) = \frac{j\omega\mu}{k_{c,d}^2}\frac{\partial h_{z,d}}{\partial x}e^{-j\beta z}$$

or

$$E_{0y,d} = \frac{j\omega\mu}{k_{c,d}^2}e^{-j\beta z}\frac{\partial}{\partial x}\left\{g_0 \cos\left(k_{c,d}x\right) + g_1 \sin\left(k_{c,d}x\right)\right\}$$

or

$$E_{0y,d} = \frac{j\omega\mu}{k_{c,d}}\left\{-g_0 \sin\left(k_{c,d}x\right) + g_1 \cos\left(k_{c,d}x\right)\right\}e^{-j\beta z} \tag{4.67}$$

Since the tangential electric field at the metal–dielectric interface must be zero, we can write the following equation:

$$E_{0y,d} = 0 \text{ at } x = 0 \tag{4.68}$$

Substituting Eq. (4.68) into Eq. (4.67), we get the following expression:

$$\frac{j\omega\mu}{k_{c,d}}\left\{-g_0 \sin\left(0\right) + g_1 \cos\left(0\right)\right\}e^{-j\beta z} = 0$$

or

$$g_1 = 0 \quad \left(\text{as } \frac{j\omega\mu}{k_{c,d}}e^{-j\beta z} \neq 0\right) \tag{4.69}$$

Substitution of Eq. (4.69) into Eq. (4.67) yields the following relation:

$$E_{0y,d} = -\frac{j\omega\mu}{k_{c,d}}g_0 \sin\left(k_{c,d}x\right)e^{-j\beta z} \tag{4.70}$$

and substitution of Eqs (4.69) and (4.65) into Eq. (4.59) gives the following expression:

$$H_{0z,d} = g_0 \cos\left(k_{c,d}x\right)e^{-j\beta z} \tag{4.71}$$

For air region, the electric field can be found using the following equation:

$$E_{0y,a} = \frac{j\omega\mu}{\left(-k_{c,a}^2\right)}\frac{\partial H_{0z,a}}{\partial x} = -\frac{j\omega\mu}{k_{c,a}^2}\frac{\partial H_{0z,a}}{\partial x} \tag{4.72}$$

The appearance of the negative sign in Eq. (4.72), in contrast to Eq. (4.51), is a result of our representation of $k_{c,a}$ in Eq. (4.64) (i.e., $k_{c,a}^2 = \beta^2 - k_0^2$ has been used in place of $k_{c,a}^2 = k_0^2 - \beta^2$). Using Eqs (4.72), (4.60), and (4.66) we can write that the following expressions:

$$E_{0y,a} = -\frac{j\omega\mu}{k_{c,a}^2}\frac{\partial}{\partial x}\left(h_{z,a}e^{-j\beta z}\right) = -\frac{j\omega\mu}{k_{c,a}^2}\frac{\partial h_{z,a}}{\partial x}e^{-j\beta z}$$

or

$$E_{0y,a} = -\frac{j\omega\mu}{k_{c,a}^2}e^{-j\beta z}\frac{\partial}{\partial x}\left(h_0 e^{k_{c,a}x} + h_1 e^{-k_{c,a}x}\right)$$

or

$$E_{0y,a} = -\frac{j\omega\mu}{k_{c,a}}e^{-j\beta z}\left(h_0 e^{k_{c,a}x} - h_1 e^{-k_{c,a}x}\right) \tag{4.73}$$

Since the electric field must vanish at $x \to \infty$, we can write the following equation:

$$E_{0y,a} \to 0 \text{ at } x \to \infty \tag{4.74}$$

Substituting Eq. (4.74) into Eq. (4.73) gives the following expression:

$$\frac{j\omega\mu}{k_{c,a}}e^{-j\beta z}\left(h_0 e^{\infty} - h_1 e^{-\infty}\right) = 0$$

or

$$h_0 = 0 \quad \left(\text{as } \frac{j\omega\mu}{k_{c,a}}e^{-j\beta z} \neq 0\right) \tag{4.75}$$

Substitution of Eq. (4.75) into Eq. (4.73) leads to the following relation:

$$E_{0y,a} = \frac{j\omega\mu}{k_{c,a}}h_1 e^{-k_{c,a}x}e^{-j\beta z} \tag{4.76}$$

and substitution of Eqs (4.75) and (4.66) into Eq. (4.60) leads to the following formula:

$$H_{0z,a} = h_1 e^{-k_{c,a}x}e^{-j\beta z} \tag{4.77}$$

Since the tangential component of electric field is continuous at the dielectric–dielectric boundary, the following relation holds at $x = h$:

$$E_{0y,d}\big|_{x=h} = E_{0y,a}\big|_{x=h} \tag{4.78}$$

Using Eqs (4.70) and (4.76) in Eq. (4.78), we can write the following relation:

$$-\frac{j\omega\mu}{k_{c,d}}g_0 \sin\left(k_{c,d}h\right)e^{-j\beta z} = \frac{j\omega\mu}{k_{c,a}}h_1 e^{-k_{c,a}h}e^{-j\beta z}$$

or $\quad -\dfrac{g_0}{k_{c,d}}\sin\left(k_{c,d}h\right)=\dfrac{h_1}{k_{c,a}}e^{-k_{c,a}h}$ $\qquad$ (4.79)

Similarly, since the tangential component of magnetic field is continuous at the dielectric–dielectric boundary, the following relation holds at $x = h$:

$$H_{0z,d}\big|_{x=h} = H_{0z,a}\big|_{x=h} \qquad (4.80)$$

Using Eqs (4.71) and (4.77) in Eq. (4.80), we can write the following relation:

$$g_0\cos\left(k_{c,d}h\right)e^{-j\beta z} = h_1 e^{-k_{c,a}h}e^{-j\beta z}$$

or $\quad g_0\cos\left(k_{c,d}h\right)=h_1 e^{-k_{c,a}h}$ $\qquad$ (4.81)

Taking the ratio of Eq. (4.79) to Eq. (4.81), the following expression can be obtained:

$$-k_{c,d}\cot\left(k_{c,d}h\right)=k_{c,a} \qquad (4.82)$$

or $\quad -\left(k_{c,d}h\right)\cot\left(k_{c,d}h\right)=\left(k_{c,a}h\right)$ $\qquad$ (4.83)

Adding Eq. (4.63) to Eq. (4.64), we get the following relation:

$$k_{c,d}^2 + k_{c,a}^2 = \left(\varepsilon_r - 1\right)k_0^2 \qquad (4.84)$$

or $\quad (k_{c,d}h)^2 + (k_{c,a}h)^2 = (\varepsilon_r - 1)(k_0 h)^2$ $\qquad$ (4.85)

Equations (4.83) and (4.85) constitute a set of simultaneous transcendental equations that must be solved graphically or numerically to find the expressions of $k_{c,d}$ and $k_{c,a}$ in terms of ε_r and k_0. A graphical method of solution is shown in Fig. 4.2. The figure consists of simultaneous plots of the family of curves described by Eqs (4.83) and (4.85) in the $k_{c,d}h$ and $k_{c,a}h$ plane. The intersection point of these curves corresponds to the solution. Now, since $k_{c,a}$ cannot be negative for an exponentially decaying field, any negative solution obtained using the aforementioned method must be ignored. A plot of Eq. (4.83), as shown in Fig. 4.2, depicts that $k_{c,a}h$ is negative for $-\pi/2 < k_{c,d}h < \pi/2$ and attains a zero value at $|k_{c,d}h| = \pm\pi/2$. Therefore, for an acceptable solution, the radius of the circle, described by Eq. (4.85), must be greater than $\pi/2$ or

$$k_0 h\sqrt{\varepsilon_r - 1} > \pi/2 \qquad (4.86)$$

Equation (4.86) sets the condition for the cut-off frequency, which is given by the following formula:

$$f_{c,\mathrm{TE}_n} = \dfrac{(2n-1)c}{4h\sqrt{\varepsilon_r - 1}} \quad \text{for } n = 1, 2, 3, \ldots \qquad (4.87)$$

where each n corresponds to a particular TE_n mode.

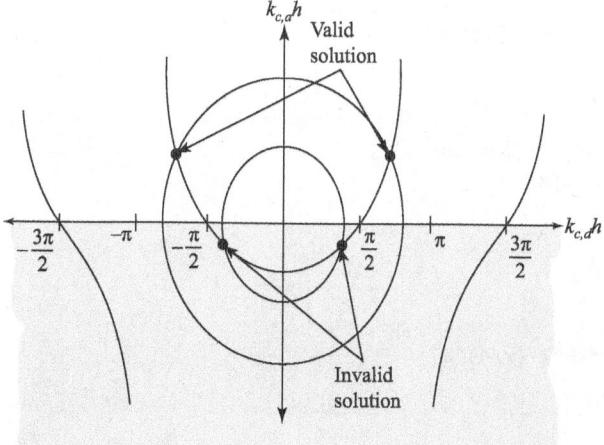

Fig. 4.2 Graphical solution of transcendental equations
for the cut-off frequency of TE surface wave mode in
grounded dielectric slabs

To find a numerical solution, let us substitute Eq. (4.82) into Eq. (4.84). This will result the following relation:

$$k_{c,d}\operatorname{cosec}\left(k_{c,d}h\right) = k_0\sqrt{\varepsilon_r - 1} \tag{4.88}$$

Multiplying both sides of Eq. (4.88) by h, we get the following expression:

$$\left(k_{c,d}h\right)\operatorname{cosec}\left(k_{c,d}h\right) - k_0 h\sqrt{\varepsilon_r - 1} = 0 \tag{4.89}$$

The following MATLAB code, which solves Eq. (4.89) numerically, can be used to find the propagation constants.

MATLAB Program

```
% Program for finding the propagation constants for TE mode surface
wave
% Frequency must be greater than the cut-off frequency for the lowest
propagating mode
clear all;
er=input('Enter the Relative Dielectric Permittivity:- ');
h=input('Enter the Thickness of the Dielectric Substrate in mm:- ');
n=input('Enter the TE Mode:- ');
c=3*10^11;
fc=(2*n-1)*c*10^-9/(4*h*sqrt(er-1))
freq=input('Enter the Frequency in GHz (Greater than fc):- ');
f=freq*10^9;
la=c/f;
k0=2*pi/la;
xin=input('Guess value (In multiple of Pi):- ');
x=xin*pi;
ep=1*10^-10;
```

```
for p=1:1000;
    fx=x*csc(x)-k0*h*sqrt(er-1);
    if abs(fx)<ep
        break;
    else
        dfx=csc(x)*(1-x*cot(x));
        x=x-fx/dfx;
    end
end
y=-x*cot(x);
kcdh=x
kcah=y
Beta=sqrt(er*k0^2-(x/h)^2)
Error=fx
```

The field components can be calculated using Eqs (4.48)–(4.51) and can be given as follows:

$$H_{0x} = -\frac{j\beta}{k_{c,d}^2}\frac{\partial H_{0z}}{\partial x} = -\frac{j\beta}{k_{c,d}^2}\frac{\partial}{\partial x}\left\{g_0 \cos\left(k_{c,d}x\right)e^{-j\beta z}\right\}$$

or $\qquad H_{0x} = \frac{j\beta g_0}{k_{c,d}}\sin\left(k_{c,d}x\right)e^{-j\beta z}$ for $0 \leq x \leq h$ $\qquad$ (4.90)

$$H_{0x} = \frac{j\beta}{k_{c,a}^2}\frac{\partial H_{0z}}{\partial x} = \frac{j\beta}{k_{c,a}^2}\frac{\partial}{\partial x}\left(h_1 e^{-k_{c,a}x}e^{-j\beta z}\right)$$

or $\qquad H_{0x} = -\frac{j\beta h_1}{k_{c,a}}e^{-k_{c,a}x}e^{-j\beta z}$ for $h \leq x \leq \infty$ $\qquad$ (4.91)

Using Eq. (4.81) in Eq. (4.91), we get the following relations:

$$H_{0x} = -\frac{j\beta}{k_{c,a}}g_0 \cos\left(k_{c,d}h\right)e^{k_{c,a}h}e^{-k_{c,a}x}e^{-j\beta z}$$

or $\qquad H_{0x} = -\frac{j\beta g_0}{k_{c,a}}\cos\left(k_{c,d}h\right)e^{-k_{c,a}(x-h)}e^{-j\beta z}$ for $h \leq x \leq \infty$ $\qquad$ (4.92)

$$H_{0y} = -\frac{j\beta}{k_{c,d}^2}\frac{\partial H_{0z}}{\partial y} = -\frac{j\beta}{k_{c,d}^2}\frac{\partial}{\partial y}\left\{g_0 \cos\left(k_{c,d}x\right)e^{-j\beta z}\right\}$$

or $\qquad H_{0y} = 0$ for $0 \leq x \leq h$ $\qquad$ (4.93)

$$H_{0y} = \frac{j\beta}{k_{c,a}^2}\frac{\partial H_{0z}}{\partial y} = \frac{j\beta}{k_{c,a}^2}\frac{\partial}{\partial y}\left(h_1 e^{-k_{c,a}x}e^{-j\beta z}\right) = 0$$ for $h \leq x \leq \infty$ $\qquad$ (4.94)

$$H_{0z} = g_0 \cos\left(k_{c,d}x\right)e^{-j\beta z}$$ for $0 \leq x \leq h$ $\qquad$ (4.95)

$$H_{0z} = h_1 e^{-k_{c,a}x}$$ for $h \leq x \leq \infty$ $\qquad$ (4.96)

Substituting Eq. (4.81) into Eq. (4.96), we get the following expressions:

$$H_{0z} = g_0 \cos\left(k_{c,d}h\right)e^{k_{c,a}h}e^{-k_{c,a}x}$$

or $\qquad H_{0z} = g_0 \cos\left(k_{c,d}h\right)e^{-k_{c,a}(x-h)}$ for $h \le x \le \infty$ (4.97)

$$E_{0x} = -\frac{j\omega\mu}{k_{c,d}^2}\frac{\partial H_{0z}}{\partial y} = -\frac{j\omega\mu}{k_{c,d}^2}\frac{\partial}{\partial y}\left\{g_0 \cos\left(k_{c,d}x\right)e^{-j\beta z}\right\}$$

or $\qquad E_{0x} = 0$ for $0 \le x \le h$ (4.98)

$$E_{0x} = \frac{j\omega\mu}{k_{c,a}^2}\frac{\partial H_{0z}}{\partial y} = \frac{j\omega\mu}{k_{c,a}^2}\frac{\partial}{\partial y}\left(h_1 e^{-k_{c,a}x}e^{-j\beta z}\right) = 0$ for $h \le x \le \infty$ (4.99)

$$E_{0y} = -\frac{j\omega\mu}{k_{c,d}}g_0 \sin\left(k_{c,d}x\right)e^{-j\beta z} \quad \text{for } 0 \le x \le h$$ (4.100)

Substitution of Eq. (4.81) into Eq. (4.76) results in the following relation:

$$E_{0y} = \frac{j\omega\mu}{k_{c,a}}g_0 \cos\left(k_{c,d}h\right)e^{k_{c,a}h}e^{-k_{c,a}x}e^{-j\beta z}$$

or $\qquad E_{0y} = \frac{j\omega\mu}{k_{c,a}}g_0 \cos\left(k_{c,d}h\right)e^{-k_{c,a}(x-h)}e^{-j\beta z}$ for $h \le x \le \infty$ (4.101)

and $\qquad E_{0z} = 0$ for $0 \le x \le h$ (4.102)

$\qquad E_{0z} = 0$ for $h \le x \le \infty$ (4.103)

4.3.2 TM-mode Propagation

From Eq. (4.12), the z-component of electric field intensity in the dielectric and free-space regions can be written as follows:

$$E_{0z,d}(x,y,z) = e_{z,d}(x,y)e^{-j\beta z}$$ (4.104)

and $\qquad E_{0z,a}(x,y,z) = e_{z,a}(x,y)e^{-j\beta z}$ (4.105)

These fields will satisfy the wave equations

$$\left[\frac{\partial^2}{\partial x^2} + k_{c,d}^2\right]e_{z,d}(x,y) = 0 \quad \text{for } 0 \le x \le h$$ (4.106)

within the dielectric region and

$$\left[\frac{\partial^2}{\partial x^2} - k_{c,a}^2\right]e_{z,a}(x,y) = 0 \quad \text{for } h \le x \le \infty$$ (4.107)

in the air. In Eqs (4.106) and (4.107), we have substituted the following equations:

$$k_{c,d}^2 = \varepsilon_r k_0^2 - \beta^2$$ (4.108)

$$k_{c,a}^2 = \beta^2 - k_0^2 \tag{4.109}$$

This ensures that both $k_{c,d}$ and $k_{c,a}$ are real. The well-known general solution of Eqs (4.106) and (4.107) can be written as follows:

$$e_{z,d}(x,y) = g_0 \cos\left(k_{c,d}x\right) + g_1 \sin\left(k_{c,d}x\right) \tag{4.110}$$

and $\quad e_{z,a}(x,y) = h_0 e^{k_{c,a}x} + h_1 e^{-k_{c,a}x} \tag{4.111}$

Since the tangential electric field at the metal–dielectric interface must be zero, we can write the following relation:

$$E_{0z,d} = 0 \text{ at } x = 0 \tag{4.112}$$

Substitution of Eqs (4.112) and (4.110) into Eq. (4.104) gives the following expression:

$$\left\{ g_0 \cos(0) + g_1 \sin(0) \right\} e^{-j\beta z} = 0$$

or $\quad g_0 = 0 \tag{4.113}$

Substitution of Eq. (4.113) into Eq. (4.110) results in the following relation:

$$e_{z,d}(x,y) = g_1 \sin\left(k_{c,d}x\right) \tag{4.114}$$

The z-component of electric field in the dielectric region can be found by substituting Eq. (4.114) into Eq. (4.104):

$$E_{0z,d}(x,y,z) = g_1 \sin\left(k_{c,d}x\right) e^{-j\beta z} \tag{4.115}$$

Substituting Eq. (4.115) into Eq. (4.56), we get the following expression:

$$H_{0y,d} = -\frac{j\omega\varepsilon_0\varepsilon_r}{k_{c,d}^2} \frac{\partial E_{0z,d}}{\partial x} = -\frac{j\omega\varepsilon_0\varepsilon_r}{k_{c,d}^2} \frac{\partial}{\partial x} \left\{ g_1 \sin\left(k_{c,d}x\right) e^{-j\beta z} \right\}$$

or $\quad H_{0y,d} = -\frac{j\omega\varepsilon_0\varepsilon_r}{k_{c,d}} g_1 \cos\left(k_{c,d}x\right) e^{-j\beta z} \tag{4.116}$

Since the electric field must vanish at $x \to \infty$, we can write the following relation:

$$E_{0y,a} \to 0 \text{ at } x \to \infty \tag{4.117}$$

Substitution of Eqs (4.117) and (4.111) into Eq. (4.105) gives the following formulae:

$$\left(h_0 e^\infty + h_1 e^{-\infty} \right) e^{-j\beta z} = 0 \quad \text{or} \quad h_0 = 0 \tag{4.118}$$

Substitution of Eq. (4.118) into Eq. (4.111) yields the following expression:

$$e_{z,a}(x,y) = h_1 e^{-k_{c,a}x} \tag{4.119}$$

The z-component of electric field in the air region can be found by substituting Eq. (4.119) into Eq. (4.105):

$$E_{0z,a}(x,y,z) = h_1 e^{-k_{c,a}x} e^{-j\beta z} \tag{4.120}$$

Further substituting Eq. (4.120) into Eq. (4.56), we get the following expression:

$$H_{0y,a} = \frac{j\omega\varepsilon_0}{k_{c,a}^2}\frac{\partial E_{0z,a}}{\partial x} = \frac{j\omega\varepsilon_0}{k_{c,a}^2}\frac{\partial}{\partial x}\left(h_1 e^{-k_{c,a}x}e^{-j\beta z}\right) = -\frac{j\omega\varepsilon_0}{k_{c,a}}h_1 e^{-k_{c,a}x}e^{-j\beta z} \tag{4.121}$$

Since the tangential component of electric field is continuous at the dielectric–dielectric boundary, the following relation holds at $x = h$:

$$E_{0z,d}\big|_{x=h} = E_{0z,a}\big|_{x=h} \tag{4.122}$$

Using Eqs (4.115) and (4.120) in Eq. (4.122), we can write the following equation:

$$g_1 \sin\left(k_{c,d}h\right) = h_1 e^{-k_{c,a}h} \tag{4.123}$$

Similarly, since the tangential component of magnetic field is continuous at the dielectric–dielectric boundary, the following relation holds at $x = h$:

$$H_{0y,d}\big|_{x=h} = H_{0y,a}\big|_{x=h} \tag{4.124}$$

Using Eqs (4.116) and (4.121) in Eq. (4.124), we can write the following expression:

$$\frac{\varepsilon_r g_1}{k_{c,d}}\cos\left(k_{c,d}h\right) = \frac{h_1}{k_{c,a}}e^{-k_{c,a}h} \tag{4.125}$$

Taking the ratio of Eq. (4.123) to Eq. (4.125), the following equations are obtained:

$$k_{c,d}\tan\left(k_{c,d}h\right) = \varepsilon_r k_{c,a} \tag{4.126}$$

or $$\left(k_{c,d}h\right)\tan\left(k_{c,d}h\right) = \varepsilon_r\left(k_{c,a}h\right) \tag{4.127}$$

Adding Eq. (4.108) to Eq. (4.109), we get the following relation:

$$k_{c,d}^2 + k_{c,a}^2 = (\varepsilon_r - 1)k_0^2 \tag{4.128}$$

or $$\left(k_{c,d}h\right)^2 + \left(k_{c,a}h\right)^2 = (\varepsilon_r - 1)(k_0 h)^2 \tag{4.129}$$

Equations (4.127) and (4.129) constitute a set of simultaneous transcendental equations that must be solved graphically or numerically for finding the expressions of $k_{c,d}$ and $k_{c,a}$ in terms of ε_r and k_0. A graphical method of solution is shown in Fig. 4.3. The figure consists of simultaneous plots of the family of curves described by Eqs (4.127) and (4.129) in the $k_{c,d}h$ and $k_{c,a}h$ plane. The intersection point of these curves corresponds to the solution. Now, since $k_{c,a}$ cannot be negative for an exponentially decaying field, any negative

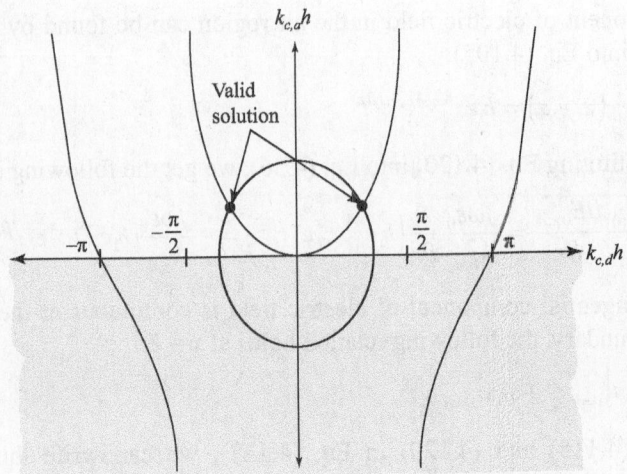

Fig. 4.3   Graphical solution of transcendental equations
for cut-off frequency of TM surface wave mode in
grounded dielectric slabs

solution obtained using the aforementioned method must be ignored. A plot of Eq. (4.127), as shown in Fig. 4.3, depicts that $k_{c,d}h$ is always positive for $-\pi/2 < k_{c,d}h < \pi/2$. Since slab thickness can never be zero, at least one propagating TM mode, generally called TM_0 mode, exists in a grounded dielectric slab. It is also the dominant mode of the dielectric slab waveguide and has a zero cut-off frequency. Since for $\pi > k_{c,d}h > \pi/2$ and $-\pi/2 > -k_{c,d}h > -\pi$, $k_{c,a}h$ attains a negative value, if the radius of the circle, represented by Eq. (4.129), lies within this range, then no further TM-mode propagation is possible. The next TM mode propagates when $k_0h\sqrt{\varepsilon_r - 1} > \pi$. As the value of $k_0h\sqrt{\varepsilon_r - 1}$ becomes larger, the circle, described by Eq. (4.129), cuts more and more branches of the tangent curves in the positive $k_{c,a}h$ plane of Fig. 4.3, which represents the propagation of higher-order TM modes. The cut-off frequency is expressed as follows:

$$f_{c,TM_n} = \frac{nc}{2h\sqrt{\varepsilon_r - 1}} \quad \text{for } n = 0,1,2,\ldots \tag{4.130}$$

To find a numrical solution, let us substitute Eq. (4.126) into Eq. (4.128). This results in the following relation:

$$k_{c,d}^2 + \left[\frac{k_{c,d}\tan\left(k_{c,d}h\right)}{\varepsilon_r}\right]^2 = (\varepsilon_r - 1)k_0^2$$

or $$\varepsilon_r^2 k_{c,d}^2 + \left\{k_{c,d}\tan\left(k_{c,d}h\right)\right\}^2 = \varepsilon_r^2(\varepsilon_r - 1)k_0^2$$

or $$k_{c,d}^2\left[\varepsilon_r^2 + \tan^2\left(k_{c,d}h\right)\right] = \varepsilon_r^2(\varepsilon_r - 1)k_0^2$$

or $$k_{c,d}\sqrt{\varepsilon_r^2 + \tan^2\left(k_{c,d}h\right)} = k_0\varepsilon_r\sqrt{\varepsilon_r - 1} \tag{4.131}$$

Multiplying both sides of Eq. (4.131) by h we get the following equation:

$$\left(k_{c,d}h\right)\sqrt{\varepsilon_r^2 + \tan^2\left(k_{c,d}h\right)} = k_0 h \varepsilon_r \sqrt{\varepsilon_r - 1}$$

or $\qquad \left(k_{c,d}h\right)\sqrt{\varepsilon_r^2 + \tan^2\left(k_{c,d}h\right)} - k_0 h \varepsilon_r \sqrt{\varepsilon_r - 1} = 0$ (4.132)

As described earlier, a MATLAB code can be written to solve Eq. (4.132) to find the propagation constants.[1]

The other field components can be calculated using Eqs (4.55) – (4.58) and can be given as follows:

$$H_{0x} = \frac{j\omega\varepsilon_0\varepsilon_r}{k_{c,d}^2} \frac{\partial E_{0z,d}}{\partial y} = \frac{j\omega\varepsilon_0\varepsilon_r}{k_{c,d}^2} \frac{\partial}{\partial y}\left\{g_1 \sin\left(k_{c,d}x\right)e^{-j\beta z}\right\}$$

or $\qquad H_{0x} = 0 \quad \text{for } 0 \le x \le h$ (4.133)

$$H_{0x} = -\frac{j\omega\varepsilon_0}{k_{c,a}^2} \frac{\partial E_{0z,a}}{\partial y} = \frac{j\omega\varepsilon_0}{k_{c,a}^2} \frac{\partial}{\partial y}\left(h_1 e^{-k_{c,a}x}e^{-j\beta z}\right)$$

or $\qquad H_{0x} = 0 \quad \text{for } h \le x \le \infty$ (4.134)

$$H_{0y} = -\frac{j\omega\varepsilon_0\varepsilon_r}{k_{c,d}^2} \frac{\partial E_{0z,d}}{\partial x}$$

or $\qquad H_{0y} = -\frac{j\omega\varepsilon_0\varepsilon_r}{k_{c,d}^2} \frac{\partial}{\partial x}\left\{g_1 \sin\left(k_{c,d}x\right)e^{-j\beta z}\right\} \quad \text{for } 0 \le x \le h$

or $\qquad H_{0y} = -\frac{j\omega\varepsilon_0\varepsilon_r}{k_{c,d}} g_1 \cos\left(k_{c,d}x\right)e^{-j\beta z} \quad \text{for } 0 \le x \le h$ (4.135)

$$H_{0y} = \frac{j\omega\varepsilon_0}{k_{c,a}^2} \frac{\partial E_{0z,a}}{\partial x} = \frac{j\omega\varepsilon_0}{k_{c,a}^2} \frac{\partial}{\partial x}\left(h_1 e^{-k_{c,a}x}e^{-j\beta z}\right)$$

or $\qquad H_{0y} = -\frac{j\omega\varepsilon_0}{k_{c,a}} h_1 e^{-k_{c,a}x}e^{-j\beta z} \quad \text{for } h \le x \le \infty$ (4.136)

Substituting Eq. (4.123) into Eq. (4.136), we get the following relations:

$$H_{0y} = -\frac{j\omega\varepsilon_0}{k_{c,a}} g_1 \sin\left(k_{c,d}h\right)e^{k_{c,a}h}e^{-k_{c,a}x}e^{-j\beta z}$$

or $\qquad H_{0y} = -\frac{j\omega\varepsilon_0}{k_{c,a}} g_1 \sin\left(k_{c,d}h\right)e^{-k_{c,a}(x-h)}e^{-j\beta z} \quad \text{for } h \le x \le \infty$ (4.137)

$$H_{0z} = 0 \quad \text{for } 0 \le x \le h$$ (4.138)

---

[1]Program 4.1 in Online Resources

$$H_{0z} = 0 \quad \text{for } h \leq x \leq \infty \tag{4.139}$$

$$E_{0x} = -\frac{j\beta}{k_{c,d}^2}\frac{\partial E_{0z,d}}{\partial x} = -\frac{j\beta}{k_{c,d}^2}\frac{\partial}{\partial x}\left\{g_1 \sin\left(k_{c,d}x\right)e^{-j\beta z}\right\}$$

or
$$E_{0x} = -\frac{j\beta}{k_{c,d}}g_1 \cos\left(k_{c,d}x\right)e^{-j\beta z} \quad \text{for } 0 \leq x \leq h \tag{4.140}$$

$$E_{0x} = \frac{j\beta}{k_{c,a}^2}\frac{\partial E_{0z,a}}{\partial x} = \frac{j\beta}{k_{c,a}^2}\frac{\partial}{\partial x}\left(h_1 e^{-k_{c,a}x}e^{-j\beta z}\right)$$

or
$$E_{0x} = -\frac{j\beta}{k_{c,a}}h_1 e^{-k_{c,a}x}e^{-j\beta z} \quad \text{for } h \leq x \leq \infty \tag{4.141}$$

Substituting Eq. (4.123) into Eq. (4.141), we get the following expressions:

$$E_{0x} = -\frac{j\beta}{k_{c,a}}g_1 \sin\left(k_{c,d}h\right)e^{k_{c,a}h}e^{-k_{c,a}x}e^{-j\beta z}$$

or
$$E_{0x} = -\frac{j\beta}{k_{c,a}}g_1 \sin\left(k_{c,d}h\right)e^{-k_{c,a}(x-h)}e^{-j\beta z} \quad \text{for } h \leq x \leq \infty \tag{4.142}$$

$$E_{0y} = -\frac{j\beta}{k_{c,d}^2}\frac{\partial E_{0z,d}}{\partial y} = -\frac{j\beta}{k_{c,d}^2}\frac{\partial}{\partial y}\left\{g_1 \sin\left(k_{c,d}x\right)e^{-j\beta z}\right\}$$

or
$$E_{0y} = 0 \quad \text{for } 0 \leq x \leq h \tag{4.143}$$

$$E_{0y} = \frac{j\beta}{k_{c,a}^2}\frac{\partial E_{0z,a}}{\partial y} = \frac{j\beta}{k_{c,a}^2}\frac{\partial}{\partial y}\left(h_1 e^{-k_{c,a}x}e^{-j\beta z}\right)$$

or
$$E_{0y} = 0 \quad \text{for } h \leq x \leq \infty \tag{4.144}$$

and
$$E_{0z} = g_1 \sin\left(k_{c,d}x\right)e^{-j\beta z} \quad \text{for } 0 \leq x \leq h \tag{4.145}$$

$$E_{0z} = h_1 e^{-k_{c,a}x}e^{-j\beta z} \quad \text{for } h \leq x \leq \infty \tag{4.146}$$

Substitution of Eq. (4.123) into Eq. (4.146) provides the following relation:

$$E_{0z}\left(x,y,z\right) = g_1 \sin\left(k_{c,d}h\right)e^{k_{c,a}h}e^{-k_{c,a}x}e^{-j\beta z}$$

or
$$E_{0z} = g_1 \sin\left(k_{c,d}h\right)e^{-k_{c,a}(x-h)}e^{-j\beta z} \quad \text{for } h \leq x \leq \infty \tag{4.147}$$

4.4 Strip Lines

A strip line structure basically consists of a flat, thin metal strip of width W that is sandwiched between two parallel ground planes separated by a distance h, and the entire region between the ground planes is filled with lossless dielectric(s). Though, in general, the metal strip may be placed at any height in-between the ground conductors, and the dielectric material covering the strip from the top may be different from that covering it from the bottom, a common practice is to place it at a distance $h/2$ from each ground plane with identical dielectric materials on both sides, as shown in Fig. 4.4(a). Since the strip line structure has two conductors (centre conductor and ground plane) and a homogeneous dielectric, it can support TEM waves. However, it can also support higher-order TE and TM modes. These higher-order modes can be suppressed by using vias running parallel to the strip on both sides and restricting the ground plane spacing to less than $\lambda/4$. The TEM-mode field configuration within a strip line is shown in Fig. 4.4(b).

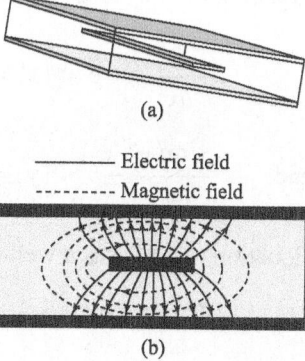

(a)

——— Electric field
- - - - - Magnetic field

(b)

Fig. 4.4 Strip line (a) General structure (b) Field configuration within strip line

For a simplified analysis of a strip line, let us assume that the field does not extend very far from the conducting strip. With this assumption we can truncate the ground plates beyond some distance, say $|x| > a/2$ where $a \gg h$, and place metal walls on the sides, as shown in Fig. 4.5. Since $a \gg h$, fields near the strips are not disturbed by sidewalls. The potential $V(x,y)$, at any point (x, y) inside the closed region, must satisfy the Laplace equation,[1] and hence we can write the following equation:

$$\nabla_t^2 V(x, y) = 0 \quad \text{for } |x| \leq a/2$$
$$\text{and } 0 \leq y \leq h \qquad (4.148)$$

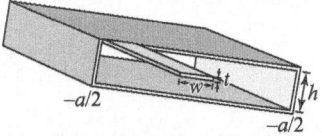

Fig. 4.5 Approximate model of strip line structure

where
$$\nabla_t^2 = \frac{\partial^2}{\partial x^2} + \frac{\partial^2}{\partial y^2} \qquad (4.149)$$

The boundary conditions are as follows:

$$V(x, y) = 0 \quad \text{at } x = \pm a/2, y = 0 \text{ and } y = h \qquad (4.150)$$

Equation (4.148) can be solved using the method of separation of variables, and hence we can assume the following relation:

$$V(x, y) = \Phi(x)\Phi(y) \qquad (4.151)$$

Substituting Eq. (4.151) into Eq. (4.148), we get the following equation:

$$-\frac{1}{\Phi(x)}\frac{d^2\Phi(x)}{dx^2} = \frac{1}{\Phi(y)}\frac{d^2\Phi(y)}{dy^2} \qquad (4.152)$$

---

[1]For detailed derivation and explanation of Laplace equation, consult *Principles of Electromagnetics* by M.N.O. Sadiku, Oxford University Press, Fourth Edition, New Delhi, pp. 169–170.

Since the LHS of Eq. (4.152) is a function of x and the RHS is a function of y, both sides must be equal to a constant, say β^2, and Eq. (4.152) can be written as follows:

$$-\frac{1}{\Phi(x)}\frac{d^2\Phi(x)}{dx^2} = \frac{1}{\Phi(y)}\frac{d^2\Phi(y)}{dy^2} = \beta^2 \tag{4.153}$$

Equation (4.153) leads to the following two differential equations:

$$\frac{d^2\Phi(x)}{dx^2} + \beta^2\Phi(x) = 0 \tag{4.154}$$

and $\quad \dfrac{d^2\Phi(y)}{dy^2} - \beta^2\Phi(y) = 0 \tag{4.155}$

Equation (4.154) has a well-known solution of the following form:

$$\Phi(x) = g_0\cos(\beta x) + g_1\sin(\beta x) \tag{4.156}$$

Now, since $\Phi(x) = 0$ at $x = -a/2$ Eq. (4.156) gives the following equation:

$$g_0 \cos(\beta a/2) - g_1 \sin(\beta a/2) = 0 \tag{4.157}$$

Again since $\Phi(x) = 0$ at $x = a/2$, Eq. (4.156) yields the following expression:

$$g_0 \cos(\beta a/2) + g_1 \sin(\beta a/2) = 0 \tag{4.158}$$

Adding Eq. (4.157) to Eq. (4.158), we get the following equation:

$$g_0 \cos(\beta a/2) = 0 \tag{4.159}$$

If we assume that $g_0 \neq 0$, then Eq. (4.159) implies that

$$\cos(\beta a/2) = 0 = \cos(n\pi/2), \text{ where } n = 1, 3, 5, 7, \ldots$$

or $\quad \beta = n\pi/a \tag{4.160}$

Subtraction of Eq. (4.157) from (4.158) results in the following equation:

$$g_1 \sin(\beta a/2) = 0 \tag{4.161}$$

Substituting the expression of β from Eq. (4.160) into Eq. (4.161), we get the following relation:

$$g_1 \sin(n\pi/2) = 0 \tag{4.162}$$

Since $\quad \sin(n\pi/2) \neq 0, \quad g_1 = 0 \tag{4.163}$

Substituting Eqs (4.160) and (4.163) into Eq. (4.156) and keeping in mind that an infinite number of solutions may be obtained for each odd integer n, we can write the following equation:

$$\Phi_n(x) = g_n \cos\left(\frac{n\pi x}{a}\right), \text{ where } n \text{ is an odd integer} \qquad (4.164)$$

Now let us find the solution for $\Phi(y)$. The solution of Eq. (4.155) has the following form:

$$\Phi(y) = h_0 \cosh(\beta y) + h_1 \sinh(\beta y) \qquad (4.165)$$

Since the strip at $y = h/2$ will contain a surface charge density, the potential $V(x,y)$ will have a slope discontinuity there. Therefore, separate solutions exist for $\Phi(y)$ above and below the strip. For $y \le h/2$, we have $\Phi(y) = 0$ for $y = 0$; consequently, we can write Eq. (4.165) as follows:

$$\Phi(y) = h_0 \cosh(0) + h_1 \sinh(0) = 0$$

or $\qquad h_0 = 0 \qquad (4.166)$

Using Eqs (4.166) and (4.160) in Eq. (4.165) and keeping in mind that β can have an infinite number of values for each odd integer n, we can write the following equation:

$$\Phi_n(y) = h_n \sinh\left(\frac{n\pi y}{a}\right), \text{ where } n \text{ is an odd integer} \qquad (4.167)$$

Substituting Eqs (4.164) and (4.167) into Eq. (4.151), we obtain the following relation:

$$V_n(x,y) = g_n h_n \cos\left(\frac{n\pi x}{a}\right) \sinh\left(\frac{n\pi y}{a}\right), \text{ where } n \text{ is an odd integer} \qquad (4.168)$$

Equation (4.168) shows that an infinite number of solutions $V_1, V_3, V_5, \dots$ are possible for odd integer values of n. Now by the superposition theorem, the following linear combination of $V_1, V_3, V_5, \dots$

$$V = c_1 V_1 + c_3 V_3 + c_5 V_5 + \dots \infty$$

where $c_1, c_3, c_5, \dots$ are constants, is also a solution of Laplace equation. Thus, the solution of Eq. (4.148) for $y \le h/2$ is as follows:

$$V(x,y) = \sum_{\substack{n=1 \\ \text{odd}}}^{\infty} P_n \cos\left(\frac{n\pi x}{a}\right) \sinh\left(\frac{n\pi y}{a}\right) \qquad (4.169)$$

For $y \geq h/2$, we have $\Phi(y) = 0$ for $y = h$, and we can write Eq. (4.165) as follows:

$$\Phi(y) = h_0 \cosh(\beta h) + h_1 \sinh(\beta h) = 0$$

or $\quad h_0 = -h_1 \tanh(\beta h)$ $\qquad\qquad$ (4.170)

Substituting Eq. (4.170) into Eq. (4.165), we get the following relation:

$$\Phi(y) = h_1 \sinh(\beta y) - h_1 \tanh(\beta h) \cosh(\beta y)$$

or $\quad \Phi(y) = \dfrac{h_1}{\cosh(\beta h)} \left[\cosh(\beta h) \sinh(\beta y) - \sinh(\beta h) \cosh(\beta y) \right]$

or $\quad \Phi(y) = u_1 \sinh\{\beta(h - y)\}$ $\qquad\qquad$ (4.171)

Using Eq. (4.160) in Eq. (4.171) and keeping in mind that an infinite number of values can exist for β for each odd integer n, we can write the following equation:

$$\Phi_n(y) = u_n \sinh\left\{ \frac{n\pi}{a}(h - y) \right\}$$ $\qquad\qquad$ (4.172)

Substituting Eqs (4.164) and (4.172) in Eq. (4.151), we obtain the following expression:

$$V_n(x, y) = g_n u_n \cos\left(\frac{n\pi x}{a} \right) \sinh\left\{ \frac{n\pi}{a}(h - y) \right\}$$ $\qquad\qquad$ (4.173)

where n is an odd integer.

Equation (4.173) again shows that there are an infinite number of possible solutions $V_1, V_3, V_5, \ldots$ for odd integer values of n. Now by the superposition theorem, a linear combination of $V_1, V_3, V_5, \ldots$,

$$V = d_1 V_1 + d_3 V_3 + d_5 V_5 + \ldots \infty$$

where $d_1, d_3, d_5, \ldots$ are constants, is also a solution of Laplace equation. Thus, the solution of Eq. (4.148) for $y \geq h/2$ is as follows:

$$V(x, y) = \sum_{\substack{n=1 \\ \text{odd}}}^{\infty} q_n \cos\left(\frac{n\pi x}{a} \right) \sinh\left\{ \frac{n\pi}{a}(h - y) \right\}$$ $\qquad\qquad$ (4.174)

Combining Eqs (4.169) and (4.174), we can write the following expression:

$$V(x, y) = \begin{cases} \displaystyle\sum_{\substack{n=1 \\ \text{odd}}}^{\infty} p_n \cos\left(\frac{n\pi x}{a} \right) \sinh\left(\frac{n\pi y}{a} \right) & \text{for } |x| < a/2, 0 < y < h/2 \\[4mm] \displaystyle\sum_{\substack{n=1 \\ \text{odd}}}^{\infty} q_n \cos\left(\frac{n\pi x}{a} \right) \sinh\left\{ \frac{n\pi}{a}(h - y) \right\} & \text{for } |x| < a/2, h/2 < y < h \end{cases}$$ $\qquad$ (4.175)

Since potential must be continuous at $y = h/2$,

$$\sum_{\substack{n=1 \\ \text{odd}}}^{\infty} P_n \cos\left(\frac{n\pi x}{a}\right)\sinh\left(\frac{n\pi h}{2a}\right) = \sum_{\substack{n=1 \\ \text{odd}}}^{\infty} q_n \cos\left(\frac{n\pi x}{a}\right)\sinh\left(\frac{n\pi h}{2a}\right)$$

or $\quad P_n = q_n$ $\hfill$ (4.176)

Therefore, Eq. (4.176) can be written in the following form:

$$V(x,y) = \begin{cases} \displaystyle\sum_{\substack{n=1 \\ \text{odd}}}^{\infty} P_n \cos\left(\frac{n\pi x}{a}\right)\sinh\left(\frac{n\pi y}{a}\right) & \text{for } |x| \le a/2, \quad 0 \le y \le h/2 \\[2em] \displaystyle\sum_{\substack{n=1 \\ \text{odd}}}^{\infty} P_n \cos\left(\frac{n\pi x}{a}\right)\sinh\left(\frac{n\pi}{a}(h-y)\right) & \text{for } |x| \le a/2, \quad h/2 \le y \le h \end{cases}$$

$\hfill$ (4.177)

The electric field inside the dielectric can be calculated from Eq. (4.177) as follows:

$$E(x,y) = -\frac{\partial V(x,y)}{\partial y}$$

or $\quad E(x,y) = \begin{cases} \displaystyle -\sum_{\substack{n=1 \\ \text{odd}}}^{\infty} P_n \left(\frac{n\pi}{a}\right)\cos\left(\frac{n\pi x}{a}\right)\cosh\left(\frac{n\pi y}{a}\right) & \text{for } |x| \le a/2, 0 \le y \le h/2 \\[2em] \displaystyle \sum_{\substack{n=1 \\ \text{odd}}}^{\infty} P_n \left(\frac{n\pi}{a}\right)\cos\left(\frac{n\pi x}{a}\right)\cosh\left\{\frac{n\pi}{a}(h-y)\right\} & \text{for } |x| \le a/2, h/2 \le y \le h \end{cases}$

$\hfill$ (4.178)

The surface charge density on the strip at $y = h/2$ is as follows:

$$\rho_s = D_y(x, y = h/2_+) - D_y(x, y = h/2_-)$$

or $\quad \rho_s = \varepsilon_0 \varepsilon_r E_y(x, y = h/2_+) - \varepsilon_0 \varepsilon_r E_y(x, y = h/2_-)$ $\hfill$ (4.179)

Substituting Eq. (4.178) into Eq. (4.179), we get the following relation:

$$\rho_s = \varepsilon_0 \varepsilon_r \left[\sum_{\substack{n=1 \\ \text{odd}}}^{\infty} P_n \left(\frac{n\pi}{a}\right)\cos\left(\frac{n\pi x}{a}\right)\cosh\left(\frac{n\pi h}{2a}\right) \right.$$

$$\left. + \sum_{\substack{n=1 \\ \text{odd}}}^{\infty} P_n \left(\frac{n\pi}{a}\right)\cos\left(\frac{n\pi x}{a}\right)\cosh\left(\frac{n\pi h}{2a}\right) \right]$$

or $\quad \rho_s = 2\varepsilon_0 \varepsilon_r \displaystyle\sum_{\substack{n=1 \\ \text{odd}}}^{\infty} P_n \left(\frac{n\pi}{a}\right)\cos\left(\frac{n\pi x}{a}\right)\cosh\left(\frac{n\pi h}{2a}\right)$ $\hfill$ (4.180)

Equation (4.180) implies that to find the unknown constant p_n, the surface charge density at $y = h/2$ must be known. Let us approximate that

$$\rho_s = \begin{cases} 1 \text{ for } |x| < W/2 \\ 0 \text{ for } |x| > W/2 \end{cases}$$

(4.181)

Multiplying both sides of Eq. (4.180) by $\cos\left(\dfrac{m\pi x}{a}\right)$ and integrating in the range $-a/2$ to $a/2$, we get the following equation:

$$\int_{-a/2}^{a/2} \rho_s \cos\left(\frac{m\pi x}{a}\right) dx = 2\varepsilon_0 \varepsilon_r \times$$

$$\sum_{\substack{n=1 \\ \text{odd}}}^{\infty} P_n\left(\frac{n\pi}{a}\right) \cosh\left(\frac{n\pi h}{2a}\right) \int_{-a/2}^{a/2} \cos\left(\frac{n\pi x}{a}\right) \cos\left(\frac{m\pi x}{a}\right) dx$$

(4.182)

Using Eq. (4.181) and the orthogonality property of cosine function, which states that

$$\int_{-a/2}^{a/2} \cos\left(\frac{n\pi x}{a}\right) \cos\left(\frac{m\pi x}{a}\right) dx = \begin{cases} a/2 \text{ for } m = n \\ 0 \quad \text{ for } m \neq n \end{cases}$$

(4.183)

we can write Eq. (4.182) as follows:

$$\int_{-W/2}^{W/2} \cos\left(\frac{n\pi x}{a}\right) dx = a\varepsilon_0 \varepsilon_r P_n\left(\frac{n\pi}{a}\right) \cosh\left(\frac{n\pi h}{2a}\right)$$

or

$$\left(\frac{2}{n\pi}\right) \sin\left(\frac{n\pi W}{2a}\right) = \varepsilon_0 \varepsilon_r P_n\left(\frac{n\pi}{a}\right) \cosh\left(\frac{n\pi h}{2a}\right)$$

or

$$P_n = \frac{2a \sin\left(\dfrac{n\pi W}{2a}\right)}{\varepsilon_0 \varepsilon_r (n\pi)^2 \cosh\left(\dfrac{n\pi h}{2a}\right)}$$

(4.184)

Therefore, the voltage at the strip with respect to the bottom ground plane can be written as follows:

$$V = -\int_0^{h/2} E_y(x = 0, y) dy$$

(4.185)

Substituting Eq. (4.178) into Eq. (4.185), we get the following relation:

$$V = \sum_{\substack{n=1 \\ \text{odd}}}^{\infty} P_n\left(\frac{n\pi}{a}\right) \int_0^{h/2} \cosh\left(\frac{n\pi y}{a}\right) dy = \sum_{\substack{n=1 \\ \text{odd}}}^{\infty} P_n \sinh\left(\frac{n\pi h}{2a}\right)$$

(4.186)

The per-unit-length total charge on the strip is as follows:

$$Q = \int_{-W/2}^{W/2} \rho_s \, dx \tag{4.187}$$

Substituting Eq. (4.181) into Eq. (4.187), we get the following formula:

$$Q = \int_{-W/2}^{W/2} dx = x\big|_{-W/2}^{W/2} = W \tag{4.188}$$

Using Eqs (4.186) and (4.188), the per-unit-length capacitance of the strip line can be written as follows:

$$C = \frac{Q}{V} = W \Big/ \sum_{\substack{n=1 \\ \text{odd}}}^{\infty} P_n \sinh\left(\frac{n\pi h}{2a}\right) \tag{4.189}$$

Substituting Eq. (4.184) into Eq. (4.189), the following expression can be obtained:

$$C = \frac{W}{\displaystyle\sum_{\substack{n=1 \\ \text{odd}}}^{\infty} \frac{2a \sin\left(\dfrac{n\pi W}{2a}\right) \sinh\left(\dfrac{n\pi h}{2a}\right)}{\varepsilon_0 \varepsilon_r \, (n\pi)^2 \cosh\left(\dfrac{n\pi h}{2a}\right)}} \tag{4.190}$$

The characteristic impedance of the line is as follows:

$$Z_0 = \sqrt{\frac{L}{C}} = \frac{1}{C}\sqrt{LC} = \frac{1}{v_p C} = \frac{\sqrt{\varepsilon_r}}{cC} \tag{4.191}$$

Substituting Eq. (4.190) into Eq. (4.191), we get the following relation:

$$Z_0 = \frac{\sqrt{\varepsilon_r}}{cW} \sum_{\substack{n=1 \\ \text{odd}}}^{\infty} \frac{2a \sin\left(\dfrac{n\pi W}{2a}\right) \sinh\left(\dfrac{n\pi h}{2a}\right)}{\varepsilon_0 \varepsilon_r \, (n\pi)^2 \cosh\left(\dfrac{n\pi h}{2a}\right)} \tag{4.192}$$

Equation (4.190) and hence Eq. (4.192) are approximate solutions. For an accurate expression of the per-unit-length capacitance, the Laplace equation must be solved using the conformal mapping technique. The resulting solution, however, contains a complicated special function. A simpler formula can also be developed by curve fitting to the exact solution. Using this simpler expression of the per-unit-length capacitance, the characteristic impedance can be expressed as follows:

$$Z_0 = \frac{30\pi}{\sqrt{\varepsilon_r}} \frac{1}{0.441 + \dfrac{W_e}{h}} \tag{4.193}$$

where W_e is the effective width of the strip and can be expressed as follows:

$$\frac{W_e}{h} = \begin{cases} \dfrac{W}{h} & \text{for } \dfrac{W}{h} > 0.35 \\[2ex] \dfrac{W}{h} - \left(0.35 - W/h\right)^2 & \text{for } \dfrac{W}{h} < 0.35 \end{cases} \tag{4.194}$$

Note Equation (4.193) assumes zero strip thickness and is almost 99% accurate.

The following simple MATLAB program can be used to plot Z_0 as a function of W/h:

MATLAB Program

```
% Program to plot Z0 of strip line as a function of W/h for different er
% It assumes a zero strip thickness
% It neglects dispersion
% wbh=W/h;
clear all
clf
er=input('Enter relative permittivity of the dielectric medium:- ');
n=0;
for wbh=0.1:0.1:10
    n=n+1;
    wh(n)=wbh;
    if wbh>0.35
        webh=wbh;
    else
        webh=wbh-(0.35-wbh)^2;
    end
    Z0(n)=30*pi/(sqrt(er)*(0.441+webh));
end
plot(wh, Z0)
```

Equation (4.193) can be used to find the characteristic impedance of a line for a given strip width. However, for designing strip line circuits, the width of the strip for a given characteristic impedance is required to be calculated. This can be achieved by using the inverse formula of Eq. (4.193), which is given by the following equation:

$$\frac{W}{h} = \begin{cases} \kappa & \text{for } \sqrt{\varepsilon_r}\, Z_0 < 120 \\[1ex] 0.85 - \sqrt{0.6 - \kappa} & \text{for } \sqrt{\varepsilon_r}\, Z_0 > 120 \end{cases} \tag{4.195}$$

where $\quad \kappa = \dfrac{30\pi}{\sqrt{\varepsilon_r}\, Z_0} - 0.441 \tag{4.196}$

Like other transmission lines, strip lines are also subjected to losses. The total per-unit-length loss of the dominant mode in a strip line consists of two types of losses: dielectric loss in the substrate and Ohmic loss.

Dielectric loss The dielectric loss in a strip line circuit can be calculated from the expression of the propagation constant. The complex propagation constant in a strip line circuit can be written as follows:

$$\gamma = \alpha_d + j\beta_d = \sqrt{k_c^2 - \omega^2 \mu_0 \varepsilon_0 \varepsilon_r \left\{1 - j \tan(\delta)\right\}}$$

or $\quad \gamma = \sqrt{k_c^2 - k^2 + jk^2 \tan(\delta)}$ (4.197)

Since for most of the dielectric materials $\tan(\delta) \ll 1$, Eq. (4.197) can be simplified by using the first two terms of Taylor's series expansion, as follows:

$$\sqrt{a^2 + x^2} \simeq a + \frac{1}{2}\left(\frac{x^2}{a}\right) \quad \text{for } x \ll a$$

Assuming that $a = \sqrt{k_c^2 - k^2}$ and $x = \sqrt{jk^2 \tan(\delta)}$, and using Eq. (4.197), we can write the following expression:

$$\gamma = \sqrt{k_c^2 - k^2 + jk^2 \tan(\delta)} \simeq \sqrt{k_c^2 - k^2} + \frac{jk^2 \tan(\delta)}{2\sqrt{k_c^2 - k^2}}$$ (4.198)

Now since $\sqrt{k_c^2 - k^2} = j\beta$, from Eq. (4.198), the following expression can be obtained:

$$\gamma = \alpha_d + j\beta_d = j\beta + \frac{jk^2 \tan(\delta)}{2j\beta} = \frac{k^2 \tan(\delta)}{2\beta} + j\beta$$

or $\quad \alpha_d = \dfrac{k^2 \tan(\delta)}{2\beta}$ (4.199)

Conductor loss The conductor loss can be calculated by the perturbation method or Wheeler's incremental rule.[1] An approximate expression for the per-unit-length attenuation constant is as follows:

$$\alpha_c = \begin{cases} \dfrac{2.7 \times 10^{-3} R_s \varepsilon_r \chi Z_0}{30\pi (h-t)} & \text{for } \sqrt{\varepsilon_r} Z_0 < 120 \\[4mm] \dfrac{0.16 R_s \varsigma}{Z_0 h} & \text{for } \sqrt{\varepsilon_r} Z_0 > 120 \end{cases}$$ (4.200)

where the surface resistance $R_s = \sqrt{\pi f \mu / \sigma}$ (4.201)

---

[1]For details, see Appendix A.

$$\chi = 1 + \frac{2W}{h-t} + \frac{1}{\pi}\left(\frac{h+t}{h-t}\right)\ln\left(\frac{2h-t}{t}\right) \tag{4.202}$$

$$\zeta = 1 + \left(\frac{h}{0.5W + 0.7t}\right)\left\{0.5 + \frac{0.414t}{W} + \frac{1}{2\pi}\ln\left(\frac{4\pi W}{t}\right)\right\} \tag{4.203}$$

Here, t is the strip thickness.

EXAMPLE 4.1 A strip line has been designed with a dielectric substrate having relative permittivity 10.5, thickness 3.1 mm, and strip width 2.5 mm. Calculate the characteristic impedance of the line.

Solution Given: $\varepsilon_r = 10.5$, $h = 3.1$ mm, and $W = 2.5$ mm

Now, $\dfrac{W}{h} = \dfrac{2.5}{3.1} = 0.806$

Hence, $\dfrac{W_e}{h} = \dfrac{W}{h} = 0.806$

$$Z_0 = \frac{30\pi}{\sqrt{\varepsilon_r}}\frac{1}{0.441 + \dfrac{W_e}{h}} = \frac{30\pi}{\sqrt{10.5}}\frac{1}{0.441 + 0.806} = 23.324\ \Omega$$

Practice Problems

4.1[@] If the thickness of the strip line, described in Example 4.1, is 0.07 mm and conductivity of the strip material is 5.8×10^7 S/m, then calculate the conductor loss at 3 GHz.

4.2 A strip line has been designed with a dielectric substrate having relative permittivity 9.7, thickness 3.7 mm, and strip width 2 mm. Calculate the characteristic impedance of the line. **[30.821 Ω]**

4.3 If the thickness of the strip line, described in Practice Problem 4.2, is 0.03 mm and conductivity of the strip material is 5.8×10^7 S/m, then calculate the conductor loss at 5 GHz. **[0.016 Np/m]**

4.5 Microstrip Lines

A microstrip line consists of a conducting strip of width W printed on a thin, grounded dielectric substrate of relative permittivity ε_r and thickness h. The region above the strip is air, as shown in Fig. 4.6. Owing to the presence of an air–dielectric interface, some fields also exist in the air region above the strip or dielectric substrate. Existence of fields both in the air and in the dielectric region imposes the condition that a microstrip structure cannot support a pure TEM

@Solution in Online Resources

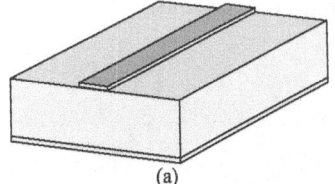

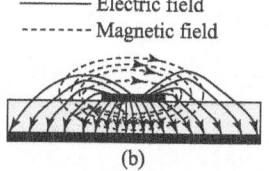

────── Electric field
------- Magnetic field

(a) (b)

Fig. 4.6 Microstrip line (a) General microstrip structure
(b) Field distribution

wave. This can be explained in the following way. The phase velocity of a pure
TEM wave in air is c, whereas that in a dielectric is $c/\sqrt{\varepsilon_r}$. Therefore, the existence
of a pure TEM wave in a microstrip structure will result in a phase mismatch
at the air–dielectric interface, which is not possible. Practically, the exact field
of a microstrip line consists of a hybrid TE–TM wave, which requires a very
complicated analysis to study the microstrip structures. However, for most of the
practical applications $\lambda \gg h$; hence, we can approximate the field as quasi-TEM
and can find an approximate electrostatic solution for it.

> **Note** Unlike a pure TEM wave, the longitudinal component of field is not zero
> in a quasi-TEM mode. This longitudinal component, however, is small and can
> be neglected during analysis. In a microstrip line, the quasi-TEM mode arises
> because the wave is not completely confined in the dielectric and a portion of the
> wave propagates in the free space.

For a simplified analysis of a
microstrip line, we will again assume that
the field does not extend very far from the
conducting strip. With this assumption,
we can truncate the ground plane beyond
some distance, say $|x| > a/2$ where
$a \gg h$, and place metal walls on the sides,
as shown in Fig. 4.7. Since $a \gg h$, the
fields near the strips are not disturbed by
the sidewalls. The potential $V(x, y)$, at any
point (x, y) inside the closed region, must
satisfy the Laplace equation, and hence
we can write the following relation:

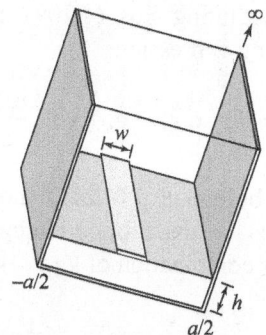

Fig. 4.7 Approximate
microstrip model

$$\nabla_t^2 V(x,y) = 0 \quad \text{for } |x| \leq a/2 \text{ and } 0 \leq y \leq \infty \tag{4.204}$$

The boundary conditions are as follows:

$$V(x,y) = 0 \quad \text{at } x = \pm a/2, y = 0 \text{ and } y = \infty \tag{4.205}$$

Equation (4.204) can be solved using the same procedure as described for the
strip line structures. Thus, the potential $V(x,y)$ at any point (x,y) can be written
as follows:

$$V(x,y) = \Phi(x)\Phi(y) \tag{4.206}$$

which will lead to the following two differential equations:

$$\frac{d^2\Phi(x)}{dx^2} + \beta^2\Phi(x) = 0 \tag{4.207}$$

and $$\frac{d^2\Phi(y)}{dy^2} - \beta^2\Phi(y) = 0 \tag{4.208}$$

where β is a constant to be determined.

Since both differential equations and the boundary condition for $\Phi(x)$ are similar to those of a strip line structure, its expression for a microstrip line structure will also be the same as that of a strip line structure, and hence we can write the following equation:

$$\Phi_n(x) = g_n \cos\left(\frac{n\pi x}{a}\right), \text{ where } n \text{ is an odd integer} \tag{4.209}$$

Now let us find the solution for $\Phi(y)$. For the region below the conducting strip (i.e., for $y < h$), $\Phi(y)$ also satisfies the same differential equation and boundary condition as those of a strip line structure; hence, its solution for the region $y < h$ can be written as follows:

$$\Phi_n(y) = h_n \sinh\left(\frac{n\pi y}{a}\right), \text{ where } n \text{ is an odd integer} \tag{4.210}$$

Substituting Eqs (4.209) and (4.210) into Eq. (4.206), we can write the following expression:

$$V_n(x,y) = g_n h_n \cos\left(\frac{n\pi x}{a}\right) \sinh\left(\frac{n\pi y}{a}\right) \text{ where } n \text{ is an odd integer} \tag{4.211}$$

As before, Eq. (4.211) shows that an infinite number of possible solutions $V_1, V_3, V_5, \ldots$ exist for odd integer values of n. Now by the superposition theorem, a linear combination of $V_1, V_3, V_5, \ldots,$

$$V = c_1 V_1 + c_3 V_3 + c_5 V_5 + \ldots \infty$$

where $c_1, c_3, c_5, \ldots$ are constants, is also a solution of Laplace equation. Thus, the solution of Eq. (4.204) for $y < h$ is as follows:

$$V(x,y) = \sum_{\substack{n=1 \\ \text{odd}}}^{\infty} p_n \cos\left(\frac{n\pi x}{a}\right) \sinh\left(\frac{n\pi y}{a}\right) \tag{4.212}$$

When $y \geq h$, we have $\Phi(y) = 0$ for $y = \infty$, and we can solve Eq. (4.208) assuming that

$$\Phi(y) = e^{my} \tag{4.213}$$

Substituting Eq. (4.213) into Eq. (4.208), we get the following relation:

$$\left(m^2 - \beta^2\right)e^{my} = 0 \tag{4.214}$$

Since $e^{my} \neq 0$, we can write the following equation: $m = \pm\beta$ (4.215)

Therefore, $\Phi(y) = h_0 e^{\beta y} + h_1 e^{-\beta y}$ (4.216)

Now imposing the boundary condition $\Phi(y) = 0$ for $y = \infty$ in Eq. (4.216), we get the following equation:

$$\Phi(y) = h_0 e^{\infty} + h_1 e^{-\infty} = 0$$

or $\quad h_0 = 0$ (4.217)

Substituting Eqs (4.217) and (4.160) into Eq. (4.216) and keeping in mind that an infinite number of solutions may be obtained for each odd integer n, we can write the following expression:

$$\Phi_n(y) = h_n e^{-\left(\frac{n\pi y}{a}\right)} \tag{4.218}$$

Substituting Eqs (4.209) and (4.218) into Eq. (4.206), the following relation is obtained:

$$V_n(x,y) = g_n h_n \cos\left(\frac{n\pi x}{a}\right) e^{-\left(\frac{n\pi y}{a}\right)}, \text{ where } n \text{ is an odd integer} \tag{4.219}$$

Equation (4.219) again shows that an infinite number of solutions $V_1, V_3, V_5, \dots$ are possible for odd integer values of n. Now by the superposition theorem, a linear combination of $V_1, V_3, V_5, \dots,$

$$V = d_1 V_1 + d_3 V_3 + d_5 V_5 + \dots\infty$$

where $d_1, d_3, d_5, \dots$ are constants, is also a solution of the Laplace equation. Thus, the solution of Eq. (4.204) for $y \geq h$ can be written as follows:

$$V(x,y) = \sum_{\substack{n=1 \\ \text{odd}}}^{\infty} q_n \cos\left(\frac{n\pi x}{a}\right) e^{-\left(\frac{n\pi y}{a}\right)} \tag{4.220}$$

Combining Eqs (4.212) and (4.220), we can write the following expression:

$$V(x,y) = \begin{cases} \displaystyle\sum_{\substack{n=1 \\ \text{odd}}}^{\infty} p_n \cos\left(\frac{n\pi x}{a}\right) \sinh\left(\frac{n\pi y}{a}\right) & \text{for } |x| < a/2, 0 < y < h \\[4mm] \displaystyle\sum_{\substack{n=1 \\ \text{odd}}}^{\infty} q_n \cos\left(\frac{n\pi x}{a}\right) e^{-\left(\frac{n\pi y}{a}\right)} & \text{for } |x| < a/2, h < y < \infty \end{cases} \tag{4.221}$$

Since potential must be continuous at $y = h$,

$$\sum_{\substack{n=1 \\ \text{odd}}}^{\infty} P_n \cos\left(\frac{n\pi x}{a}\right) \sinh\left(\frac{n\pi h}{a}\right) = \sum_{\substack{n=1 \\ \text{odd}}}^{\infty} q_n \cos\left(\frac{n\pi x}{a}\right) e^{-\left(\frac{n\pi h}{a}\right)}$$

or $\quad P_n \sinh\left(\frac{n\pi h}{a}\right) = q_n e^{-\left(\frac{n\pi h}{a}\right)} \quad$ or $\quad q_n = P_n \sinh\left(\frac{n\pi h}{a}\right) e^{\left(\frac{n\pi h}{a}\right)}$ (4.222)

Therefore, Eq. (4.221) can be written in the following form:

$$V(x,y) = \begin{cases} \displaystyle\sum_{\substack{n=1 \\ \text{odd}}}^{\infty} P_n \cos\left(\frac{n\pi x}{a}\right) \sinh\left(\frac{n\pi y}{a}\right) & \text{for } |x| < a/2, 0 < y < h \\[4mm] \displaystyle\sum_{\substack{n=1 \\ \text{odd}}}^{\infty} P_n \sinh\left(\frac{n\pi h}{a}\right) \cos\left(\frac{n\pi x}{a}\right) e^{-\left[\frac{n\pi(y-h)}{a}\right]} & \text{for } |x| < a/2, h < y < \infty \end{cases}$$

(4.223)

The electric field inside the dielectric can be calculated from Eq. (4.223), as follows:

$$E(x,y) = -\frac{\partial V(x,y)}{\partial y}$$

or $E(x,y) = \begin{cases} \displaystyle -\sum_{\substack{n=1 \\ \text{odd}}}^{\infty} P_n \left(\frac{n\pi}{a}\right) \cos\left(\frac{n\pi x}{a}\right) \cosh\left(\frac{n\pi y}{a}\right) & \text{for } |x| < a/2, 0 < y < h \\[4mm] \displaystyle \sum_{\substack{n=1 \\ \text{odd}}}^{\infty} P_n \left(\frac{n\pi}{a}\right) \sinh\left(\frac{n\pi h}{a}\right) \cos\left(\frac{n\pi x}{a}\right) e^{-\left[\frac{n\pi(y-h)}{a}\right]} & \text{for } |x| < a/2, h < y < \infty \end{cases}$

(4.224)

The surface charge density on the strip at $y = h$ is as follows:

$$\rho_s = D_y(x, y = h_+) - D_y(x, y = h_-)$$

or $\quad \rho_s = \varepsilon_0 E_y(x, y = h_+) - \varepsilon_0 \varepsilon_r E_y(x, y = h_-)$ (4.225)

Substituting Eq. (4.224) into Eq. (4.225), we get the following relation:

$$\rho_s = \varepsilon_0 \sum_{\substack{n=1 \\ \text{odd}}}^{\infty} P_n \left(\frac{n\pi}{a}\right) \cos\left(\frac{n\pi x}{a}\right) \left[\sinh\left(\frac{n\pi h}{a}\right) + \varepsilon_r \cosh\left(\frac{n\pi h}{a}\right)\right]$$ (4.226)

Equation (4.226) implies that to find the unknown constant p_n, the surface charge density at $y = h$ must be known. As in the earlier case, let us approximate that

$$
\rho_s = \begin{cases} 1 \text{ for } |x| < W/2 \\ 0 \text{ for } |x| > W/2 \end{cases}
\tag{4.227}
$$

Multiplying both sides of Eq. (4.226) by $\cos\left(\dfrac{m\pi x}{a}\right)$ and integrating in the range $-a/2$ to $a/2$, we get the following equation:

$$
\int_{-a/2}^{a/2} \rho_s \cos\left(\frac{m\pi x}{a}\right) dx = \varepsilon_0 \sum_{\substack{n=1 \\ \text{odd}}}^{\infty} P_n \left(\frac{n\pi}{a}\right) \left[\frac{\sinh\left(\dfrac{n\pi h}{a}\right)}{+\varepsilon_r \cosh\left(\dfrac{n\pi h}{a}\right)} \right] \int_{-a/2}^{a/2} \cos\left(\frac{n\pi x}{a}\right) \cos\left(\frac{m\pi x}{a}\right) dx
\tag{4.228}
$$

Using Eq. (4.227) and the orthogonality property of Eq. (4.183), we can write Eq. (4.228) in the following form:

$$
\int_{-W/2}^{W/2} \cos\left(\frac{n\pi x}{a}\right) dx = \varepsilon_0 P_n \left(\frac{n\pi}{a}\right) \left[\sinh\left(\frac{n\pi h}{a}\right) + \varepsilon_r \cosh\left(\frac{n\pi h}{a}\right) \right] \frac{a}{2}
$$

or
$$
\left(\frac{2}{n\pi}\right) \sin\left(\frac{n\pi W}{2a}\right) = \varepsilon_0 P_n \left(\frac{n\pi}{2a}\right) \left[\sinh\left(\frac{n\pi h}{a}\right) + \varepsilon_r \cosh\left(\frac{n\pi h}{a}\right) \right]
$$

or
$$
P_n = \frac{4a \sin\left(\dfrac{n\pi W}{2a}\right)}{\varepsilon_0 (n\pi)^2 \left[\sinh\left(\dfrac{n\pi h}{a}\right) + \varepsilon_r \cosh\left(\dfrac{n\pi h}{a}\right) \right]}
\tag{4.229}
$$

Therefore, the voltage at the strip with respect to the bottom ground plane can be written as follows:

$$
V = -\int_0^h E_y (x = 0, y) \, dy
\tag{4.230}
$$

Substituting Eq. (4.224) into Eq. (4.230), we get the following expression:

$$
V = \sum_{\substack{n=1 \\ \text{odd}}}^{\infty} P_n \left(\frac{n\pi}{a}\right) \int_0^h \cosh\left(\frac{n\pi y}{a}\right) dy = \sum_{\substack{n=1 \\ \text{odd}}}^{\infty} P_n \sinh\left(\frac{n\pi h}{a}\right)
\tag{4.231}
$$

The per-unit-length total charge on the strip is expressed by the following equation:

$$
Q = \int_{-W/2}^{W/2} \rho_s \, dx
\tag{4.232}
$$

Substituting Eq. (4.227) into Eq. (4.232), we get the following relation:

$$Q = \int_{W/2}^{W/2} dx = x\Big|_{-W/2}^{W/2} = W \tag{4.233}$$

Using Eqs (4.231) and (4.233), the per-unit-length capacitance of the strip line can be written as follows:

$$C = \frac{Q}{V} = \frac{W}{\sum\limits_{\substack{n=1 \\ \text{odd}}}^{\infty} P_n \sinh\left(\frac{n\pi h}{a}\right)} \tag{4.234}$$

Substituting Eq. (4.229) into Eq. (4.234), the following equation is obtained:

$$C = \frac{W}{\sum\limits_{\substack{n=1 \\ \text{odd}}}^{\infty} \dfrac{4a \sin\left(\dfrac{n\pi W}{2a}\right) \sinh\left(\dfrac{n\pi h}{a}\right)}{\varepsilon_0 (n\pi)^2 \left[\sinh\left(\dfrac{n\pi h}{a}\right) + \varepsilon_r \cosh\left(\dfrac{n\pi h}{a}\right)\right]}} \tag{4.235}$$

If C and C_0 are the per-unit-length capacitance of the structure for $\varepsilon_r \neq 1$ and $\varepsilon_r = 1$ (air dielectric), respectively, then we can define the *effective dielectric constant* of the structure as follows:

$$\varepsilon_{\text{eff}} = \frac{C}{C_0} \tag{4.236}$$

The effective dielectric constant can be defined as the dielectric constant of an equivalent homogeneous medium that can replace the air and dielectric regions of the microstrip structure without changing the properties of the structure.

The characteristic impedance of the line then can be written as follows:

$$Z_0 = \sqrt{\frac{L}{C}} = \frac{1}{C}\sqrt{LC} = \frac{1}{v_p C} = \frac{\sqrt{\varepsilon_{\text{eff}}}}{cC} \tag{4.237}$$

Substituting Eq. (4.235) into Eq. (4.237), we get the following relation:

$$Z_0 = \frac{\sqrt{\varepsilon_{\text{eff}}}}{cW} \sum\limits_{\substack{n=1 \\ \text{odd}}}^{\infty} \frac{4a \sin\left(\dfrac{n\pi W}{2a}\right) \sinh\left(\dfrac{n\pi h}{a}\right)}{\varepsilon_0 (n\pi)^2 \left[\sinh\left(\dfrac{n\pi h}{a}\right) + \varepsilon_r \cosh\left(\dfrac{n\pi h}{a}\right)\right]} \tag{4.238}$$

Equation (4.235) and hence Eq. (4.238) are approximate solutions. A more accurate expression for the characteristic impedance can be given as follows:

$$Z_0 = \begin{cases} \dfrac{60}{\sqrt{\varepsilon_{\text{eff}}}} \ln\left(\dfrac{8h}{W} + \dfrac{W}{4h}\right) & \text{for } \dfrac{W}{h} \leq 1 \\[4mm] \dfrac{120\pi}{\sqrt{\varepsilon_{\text{eff}}}\left[\dfrac{W}{h} + 1.393 + 0.667\ln\left(\dfrac{W}{h} + 1.444\right)\right]} & \text{for } \dfrac{W}{h} \geq 1 \end{cases} \tag{4.239}$$

$$
\text{where} \quad \varepsilon_{\text{eff}} = \begin{cases} \dfrac{\varepsilon_r+1}{2}+\dfrac{\varepsilon_r-1}{2}\left\{\left(1+12\dfrac{h}{W}\right)^{-\frac{1}{2}}+0.04\left(1-\dfrac{W}{h}\right)^2\right\} & \text{for } \dfrac{W}{h}\leq 1 \\[4mm] \dfrac{\varepsilon_r+1}{2}+\dfrac{\varepsilon_r-1}{2}\left(1+12\dfrac{h}{W}\right)^{-\frac{1}{2}} & \text{for } \dfrac{W}{h}\geq 1 \end{cases}
$$
(4.240)

A simple MATLAB program can be used to plot Z_0 as a function of W/h.[1]

Equation (4.239) can be used to find the characteristic impedance of a line for a given strip width. However, for designing microstrip line circuits, the width of the strip for a given characteristic impedance is required to be calculated. This can be achieved by using the following equations:

$$
\frac{W}{h}=\begin{cases} \left[\left(\dfrac{e^\kappa}{8}-\dfrac{1}{4e^\kappa}\right)^{-1}\right] & \text{for } Z_0 > 44-2\varepsilon_r\ \Omega \\[4mm] \dfrac{2}{\pi}\left\{(v-1)-\ln(2v-1)\right\}+\dfrac{\varepsilon_r-1}{\pi\varepsilon_r}\left\{\ln(v-1)+0.293-\dfrac{0.517}{\varepsilon_r}\right\} & \text{for } Z_0 < 44-2\varepsilon_r\ \Omega \end{cases}
$$
(4.241)

$$
\text{where} \quad \kappa=\frac{Z_0\sqrt{2(\varepsilon_r+1)}}{119.9}+\frac{1}{2}\left(\frac{\varepsilon_r-1}{\varepsilon_r+1}\right)\left[\ln\left(\frac{\pi}{2}\right)+\frac{1}{\varepsilon_r}\ln\left(\frac{4}{\pi}\right)\right]
$$
(4.242)

$$
v=\frac{59.95\pi^2}{Z_0\sqrt{\varepsilon_r}}
$$
(4.243)

If Z_0 is known, then ε_{eff} can be calculated using the following equation:

$$
\varepsilon_{\text{eff}}=\frac{\varepsilon_r}{0.96+\varepsilon_r\left(0.109-0.004\varepsilon_r\right)\left\{\log(10+Z_0)-1\right\}}
$$
(4.244)

Like strip lines, microstrip lines are also subjected to loss. However, the total per-unit-length loss of the dominant mode in a microstrip line consists of three types of losses: dielectric loss in the substrate, Ohmic loss, and radiation loss.

Dielectric loss   If we consider a microstrip line as a quasi-TEM line, then the attenuation constant for the dielectric loss can be expressed as follows:

$$
\alpha_d=\frac{k_0\varepsilon_r\left(\varepsilon_{\text{eff}}-1\right)}{2\sqrt{\varepsilon_{\text{eff}}}\left(\varepsilon_r-1\right)}\tan(\delta)\ \text{Np/m}
$$
(4.245)

where $\tan(\delta)$ is the loss tangent of the dielectric material of the substrate.

---

[1]Program 4.2 in Online Resources

Conductor loss For a microstrip line on a low-loss dielectric substrate, imperfect conductors are the dominant contributors to the total microstrip loss at a microwave frequency. The conductor loss can be calculated by the perturbation method or Wheeler's incremental rule.

Radiation loss In addition to dielectric and conductor losses, a microstrip line is also subjected to radiation loss that depends on the substrate thickness, dielectric structure, and line geometry. If we assume that the line is a TEM line having a uniform dielectric with an effective dielectric constant ε_{eff} and the substrate thickness is much less than the free-space wavelength λ, then the ratio of radiated power to the total dissipated power for an open-air microstrip line is given by the following equation:

$$\frac{P_r}{P_d} = 240\pi^2 \left(\frac{h}{\lambda_0}\right)^2 \frac{F\left(\varepsilon_{\text{eff}}\right)}{Z_0} \tag{4.246}$$

where $F\left(\varepsilon_{\text{eff}}\right)$ is called the radiation factor and can be expressed as follows:

$$F\left(\varepsilon_{\text{eff}}\right) = \frac{\varepsilon_{\text{eff}} + 1}{\varepsilon_{\text{eff}}} - \frac{\left(\varepsilon_{\text{eff}} - 1\right)^2}{2\left(\varepsilon_{\text{eff}}\right)^{3/2}} \ln\left(\frac{\sqrt{\varepsilon_{\text{eff}}} + 1}{\sqrt{\varepsilon_{\text{eff}}} - 1}\right) \tag{4.247}$$

The total quality factor $\left(Q_T\right)$ of a microstrip line can be defined as follows:

$$\frac{1}{Q_T} = \frac{1}{Q_c} + \frac{1}{Q_d} + \frac{1}{Q_r} \tag{4.248}$$

where

$$\frac{1}{Q_c} + \frac{1}{Q_d} = \frac{\lambda_0 \left(\alpha_c + \alpha_d\right)}{\pi\sqrt{\varepsilon_{\text{eff}}\left(f\right)}} \tag{4.249}$$

and

$$Q_r = \frac{Z_0}{480\pi \left(h/\lambda_0\right)^2 F\left(\varepsilon_{\text{eff}}\right)} \tag{4.250}$$

EXAMPLE 4.2 A microstrip line has been designed with a dielectric substrate having relative permittivity 10.5, thickness 3.1 mm, and strip width 2.5 mm. Calculate the characteristic impedance of the line.

Solution Given: $\varepsilon_r = 10.5$, $h = 3.1$ mm, and $W = 2.5$ mm

Now, $\dfrac{W}{h} = \dfrac{2.5}{3.1} = 0.806$

$$\varepsilon_{\text{eff}} = \frac{\varepsilon_r + 1}{2} + \frac{\varepsilon_r - 1}{2}\left\{\left(1 + 12\frac{h}{W}\right)^{-\frac{1}{2}} + 0.04\left(1 - \frac{W}{h}\right)^2\right\}$$

or $\qquad \varepsilon_{\text{eff}} = \dfrac{10.5+1}{2} + \dfrac{10.5-1}{2}\left\{\left(1+\dfrac{12}{0.806}\right)^{-\frac{1}{2}} + 0.04(1-0.806)\right\}$

or $\qquad \varepsilon_{\text{eff}} = 5.75 + 4.75(0.251+0.008) = 6.979$

$$Z_0 = \frac{60}{\sqrt{\varepsilon_{\text{eff}}}}\ln\left(\frac{8h}{W}+\frac{W}{4h}\right) = \frac{60}{\sqrt{6.979}}\ln\left(\frac{8}{0.806}+\frac{0.806}{4}\right) = 52.583\Omega$$

Practice Problem

4.4 A microstrip line has been designed with a dielectric substrate having relative permittivity 9.7, thickness 3.7 mm, and strip width 2 mm. Calculate the characteristic impedance of the line. [64.354Ω]

4.6 Coupled Microstrip Lines

A coupled microstrip line consists of two parallel microstrip lines in close proximity, as shown in Fig. 4.8. When the two strips are identical, we call them *symmetric lines*; otherwise, they are called *asymmetric lines*. Since the two lines are in close proximity, a continuous coupling exists between them, which makes them useful as basic elements for filters, directional couplers, and such other circuits and also enables them to support two modes of propagation.

Since the dielectric medium of a microstrip line is not homogeneous, as discussed earlier, a fraction of field exists in the air. This fraction is different for the two existing modes and results in different effective dielectric constants, phase velocities, and characteristic impedances of the line for the two modes. Different analytical techniques, such as even- and odd-mode techniques, coupled mode formulation, congruent transformation technique, and graph

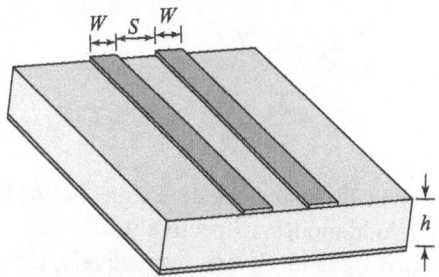

Fig. 4.8 Coupled microstrip line

transformation technique, exist for determining the propagation characteristic of a coupled line. Out of these, the even- and odd-mode technique is most convenient for symmetrical coupled lines. In this method, the symmetry plane is replaced by a magnetic or electric wall, and the wave propagating along the line is expressed in terms of two modes— an even- and an odd-mode symmetry about the wall. The even- and odd-mode field distributions of coupled microstrip lines are shown in Fig. 4.9.

In terms of the line constants, the line impedances and phase constant of a coupled transmission line can be expressed as follows:

$$Z_{0,e} = \frac{\omega}{\beta_e}(L_o + L_m) = \frac{\beta_e}{\omega}\left(\frac{1}{C_o - C_m}\right) \qquad (4.251)$$

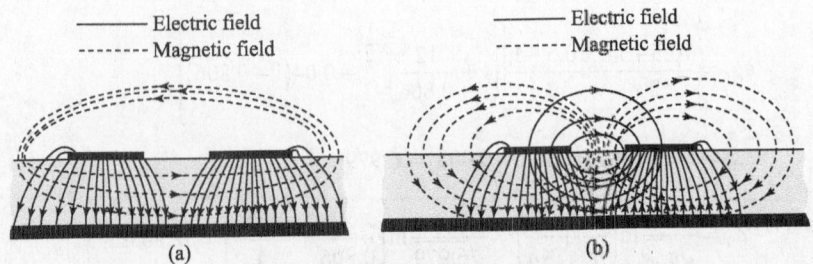

Fig. 4.9 Coupled microstrip line (a) Even mode field configuration
(b) Odd mode field configuration

$$Z_{0,o} = \frac{\omega}{\beta_o}(L_o - L_m) = \frac{\beta_o}{\omega}\left(\frac{1}{C_o + C_m}\right) \tag{4.252}$$

$$\beta_{e,o} = \omega\sqrt{L_oC_o - L_mC_m \pm (L_mC_o - L_oC_m)} \tag{4.253}$$

The phase constant can also be expressed as follows:

$$\beta_{e,o} = \beta_{\text{eff}}\sqrt{1 \pm \delta} \tag{4.254}$$

where
$$\beta_{\text{eff}} = \beta_0\sqrt{1 - k_L k_C} \tag{4.255}$$

$$\beta_0 = \omega\sqrt{L_0 C_0} \tag{4.256}$$

$$k_L = L_m/L_0 \tag{4.257}$$

$$k_C = C_m/C_0 \tag{4.258}$$

$$\delta = \frac{k_L - k_C}{1 - k_L k_C} \tag{4.259}$$

The positive and negative signs in Eqs (4.253) and (4.254) correspond to the even
and odd modes, respectively.

If C_e^a and C_o^a are, respectively, the even- and odd-mode capacitances with the
air dielectric and C_e^d, C_o^d are those with the substrate, then

$$L_o = \frac{\mu_o \varepsilon_0}{2}\left(\frac{1}{C_e^a} + \frac{1}{C_o^a}\right) \tag{4.260}$$

$$L_m = \frac{\mu_o \varepsilon_0}{2}\left(\frac{1}{C_e^a} - \frac{1}{C_o^a}\right) \tag{4.261}$$

$$C_0 = \frac{1}{2}\left(C_e^d + C_o^d\right) \tag{4.262}$$

$$C_m = \frac{1}{2}\left(C_o^d - C_e^d\right) \tag{4.263}$$

Substituting Eqs (4.260)–(4.263) into Eqs (4.256)–(4.258), we get the following equations:

$$\beta_0 = \omega\sqrt{L_0 C_0} = \omega\sqrt{\frac{\mu_0\varepsilon_0}{2}\left(\frac{1}{C_e^a}+\frac{1}{C_o^a}\right)\frac{1}{2}\left(C_e^d+C_o^d\right)} = \frac{\omega}{2c}\sqrt{\frac{\left(C_e^a+C_o^a\right)\left(C_e^d+C_o^d\right)}{C_e^a C_o^a}}$$

(4.264)

$$k_L = \frac{L_m}{L_0} = \frac{\dfrac{\mu_0\varepsilon_0}{2}\left(\dfrac{1}{C_e^a}-\dfrac{1}{C_o^a}\right)}{\dfrac{\mu_0\varepsilon_0}{2}\left(\dfrac{1}{C_e^a}+\dfrac{1}{C_o^a}\right)} = \frac{C_o^a-C_e^a}{C_e^a C_o^a}\frac{C_e^a C_o^a}{C_e^a+C_o^a} = \frac{C_o^a-C_e^a}{C_o^a+C_e^a}$$

(4.265)

$$k_C = \frac{C_m}{C_0} = \frac{\dfrac{1}{2}\left(C_o^d-C_e^d\right)}{\dfrac{1}{2}\left(C_e^d+C_o^d\right)} = \frac{C_o^d-C_e^d}{C_o^d+C_e^d}$$

(4.266)

Substituting Eqs (4.264)–(4.266) into Eqs (4.255) and (4.259), we get the following relation:

$$\beta_{\text{eff}} = \frac{\omega}{2c}\sqrt{\frac{\left(C_e^a+C_o^a\right)\left(C_e^d+C_o^d\right)}{C_e^a C_o^a}}\sqrt{1-\frac{\left(C_o^a-C_e^a\right)\left(C_o^d-C_e^d\right)}{\left(C_o^a+C_e^a\right)\left(C_o^d+C_e^d\right)}}$$

or $$\beta_{\text{eff}} = \frac{\omega}{2c}\sqrt{\frac{\left(C_e^a+C_o^a\right)\left(C_e^d+C_o^d\right)}{C_e^a C_o^a}}\sqrt{\frac{\left(C_o^a+C_e^a\right)\left(C_o^d+C_e^d\right)-\left(C_o^a-C_e^a\right)\left(C_o^d-C_e^d\right)}{\left(C_o^a+C_e^a\right)\left(C_o^d+C_e^d\right)}}$$

or $$\beta_{\text{eff}} = \frac{\omega}{2c}\sqrt{\frac{2\left(C_o^a C_e^d+C_e^a C_o^d\right)}{C_e^a C_o^a}} = \frac{\omega}{c}\sqrt{\frac{C_o^a C_e^d+C_e^a C_o^d}{2C_e^a C_o^a}}$$

(4.267)

and $$\delta = \frac{k_L-k_C}{1-k_L k_C} = \frac{\left(\dfrac{C_o^a-C_e^a}{C_o^a+C_e^a}\right)-\left(\dfrac{C_o^d-C_e^d}{C_o^d+C_e^d}\right)}{1-\left(\dfrac{C_o^a-C_e^a}{C_o^a+C_e^a}\right)\left(\dfrac{C_o^d-C_e^d}{C_o^d+C_e^d}\right)}$$

or $$\delta = \frac{\dfrac{\left(C_o^a-C_e^a\right)\left(C_o^d+C_e^d\right)-\left(C_o^d-C_e^d\right)\left(C_o^a+C_e^a\right)}{\left(C_o^a+C_e^a\right)\left(C_o^d+C_e^d\right)}}{\dfrac{\left(C_o^a+C_e^a\right)\left(C_o^d+C_e^d\right)-\left(C_o^a-C_e^a\right)\left(C_o^d-C_e^d\right)}{\left(C_o^a+C_e^a\right)\left(C_o^d+C_e^d\right)}}$$

or
$$\delta = \frac{\left(C_o^a - C_e^a\right)\left(C_o^d + C_e^d\right) - \left(C_o^a + C_e^a\right)\left(C_o^d - C_e^d\right)}{\left(C_o^a + C_e^a\right)\left(C_o^d + C_e^d\right) - \left(C_o^a - C_e^a\right)\left(C_o^d - C_e^d\right)}$$

or
$$\delta = \frac{2\left(C_o^a C_e^d - C_e^a C_o^d\right)}{2\left(C_o^a C_e^d + C_e^a C_o^d\right)} = \frac{C_o^a C_e^d - C_e^a C_o^d}{C_o^a C_e^d + C_e^a C_o^d} \tag{4.268}$$

Substituting Eqs (4.267) and (4.268) into Eq. (4.254), the following relation is obtained:

$$\beta_{e,o} = \beta_{\text{eff}} \sqrt{1 \pm \delta} = \frac{\omega}{c} \sqrt{\frac{\left(C_o^a C_e^d + C_e^a C_o^d\right)}{2 C_e^a C_o^a}} \sqrt{1 \pm \frac{C_o^a C_e^d - C_e^a C_o^d}{C_o^a C_e^d + C_e^a C_o^d}}$$

or
$$\beta_{e,o} = \begin{cases} \dfrac{\omega}{c} \sqrt{\dfrac{\left(C_o^a C_e^d + C_e^a C_o^d\right)}{2 C_e^a C_o^a}} \sqrt{\dfrac{\left(C_o^a C_e^d + C_e^a C_o^d\right) + \left(C_o^a C_e^d - C_e^a C_o^d\right)}{\left(C_o^a C_e^d + C_e^a C_o^d\right)}} & \text{for } \beta_e \\[2em] \dfrac{\omega}{c} \sqrt{\dfrac{\left(C_o^a C_e^d + C_e^a C_o^d\right)}{2 C_e^a C_o^a}} \sqrt{\dfrac{\left(C_o^a C_e^d + C_e^a C_o^d\right) - \left(C_o^a C_e^d - C_e^a C_o^d\right)}{\left(C_o^a C_e^d + C_e^a C_o^d\right)}} & \text{for } \beta_o \end{cases}$$

or
$$\beta_{e,o} = \begin{cases} \dfrac{\omega}{c} \sqrt{\dfrac{2 C_o^a C_e^d}{2 C_e^a C_o^a}} & \text{for } \beta_e \\[1.5em] \dfrac{\omega}{c} \sqrt{\dfrac{2 C_e^a C_o^d}{2 C_e^a C_o^a}} & \text{for } \beta_o \end{cases} = \begin{cases} \dfrac{\omega}{c} \sqrt{\dfrac{C_e^d}{C_e^a}} & \text{for } \beta_e \\[1.5em] \dfrac{\omega}{c} \sqrt{\dfrac{C_o^d}{C_o^a}} & \text{for } \beta_o \end{cases} \tag{4.269}$$

Substitution of Eqs (4.262), (4.263), and (4.269) into Eqs (4.251) and (4.252) yields the following expressions:

$$Z_{0,e} = \frac{1}{\omega} \frac{\omega}{c} \sqrt{\frac{C_e^d}{C_e^a}} \left(\frac{1}{\frac{1}{2}\left(C_e^d + C_o^d\right) - \frac{1}{2}\left(C_o^d - C_e^d\right)} \right) = \frac{2}{c} \sqrt{\frac{C_e^d}{C_e^a}} \left(\frac{1}{2 C_e^d} \right)$$

or
$$Z_{0,e} = \frac{1}{c} \sqrt{\frac{1}{C_e^a C_e^d}} = \left(c \sqrt{C_e^a C_e^d} \right)^{-1} \tag{4.270}$$

$$Z_{0,o} = \frac{1}{\omega} \frac{\omega}{c} \sqrt{\frac{C_o^d}{C_o^a}} \left(\frac{1}{\frac{1}{2}\left(C_e^d + C_o^d\right) + \frac{1}{2}\left(C_o^d - C_e^d\right)} \right) = \frac{2}{c} \sqrt{\frac{C_o^d}{C_o^a}} \left(\frac{1}{2 C_o^d} \right)$$

or
$$Z_{0,o} = \frac{1}{c} \sqrt{\frac{1}{C_o^a C_o^d}} = \left(c \sqrt{C_o^a C_o^d} \right)^{-1} \tag{4.271}$$

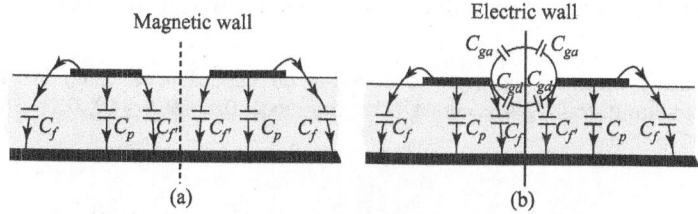

Fig. 4.10 Coupled microstrip line (a) Even mode capacitances
(b) Odd mode capacitances

Equations (4.269)–(4.271) imply that if the even- and odd-mode capacitances are known, then we can calculate the line impedances and phase constant. The even- and odd-mode capacitances along with the symmetry planes are shown in Fig. 4.10.

The even-mode per-unit-length capacitance can be expressed as follows:

$$C_e = C_p + C_f + C_{f'} \tag{4.272}$$

where

$$C_p = \frac{\varepsilon_0 \varepsilon_r W}{h} \tag{4.273}$$

$$C_f = \frac{1}{2}\left[\frac{\sqrt{\varepsilon_{\text{eff}}}}{cZ_0} - \frac{\varepsilon_0 \varepsilon_r W}{h}\right] \tag{4.274}$$

$$C_{f'} = \frac{C_f}{1 + A(h/S)\tanh(10S/h)}\left(\frac{\varepsilon_r}{\varepsilon_{\text{eff}}}\right)^{\frac{1}{4}} \tag{4.275}$$

$$A = e^{-0.1e^{(2.33-1.5W/h)}} \tag{4.276}$$

where Z_0 is the characteristic impedance of the microstrip line of width W.

The odd-mode capacitance can be expressed as follows:

$$C_0 = C_p + C_f + C_{f'} + C_{gd} + C_{ga} \tag{4.277}$$

It can be shown that Eq. (4.277) can be rewritten in the following form:

$$C_0 = 0.5C_{os} + C_{CPS} \tag{4.278}$$

where

$$C_{os} = 4\varepsilon_0 \varepsilon_r \frac{K(k_1)}{K(k_1')} \tag{4.279}$$

$$C_{CPS} = \varepsilon_0 \frac{K(k_2')}{K(k_2)} \tag{4.280}$$

$$k_1 = \tanh\left(\frac{\pi W}{4h}\right)\coth\left[\frac{\pi}{4}\left(\frac{W+S}{h}\right)\right] \tag{4.281}$$

$$k_2 = \frac{S}{S+2W} \tag{4.282}$$

and $K(m)$ is the complete elliptic integral of the first kind with modulus m.

> **Note** Values obtained for even- and odd-mode capacitances using the above equations are almost 97% accurate within the range $0.1 \leq W/h \leq 10$, $0.1 \leq S/h \leq 4$, and $2 \leq \varepsilon_r \leq 18$.

Once the even- and odd-mode capacitances have been found, the even- and odd-mode effective dielectric constants can be calculated using the following equations:

$$\varepsilon_{\mathrm{eff}}^e = C_e^d / C_e^a \tag{4.283}$$

and

$$\varepsilon_{\mathrm{eff}}^o = C_o^d / C_o^a \tag{4.284}$$

4.7 Suspended and Inverted Microstrip Lines

Microstrip lines are used in various other forms: thin-film, multilayered, valley, suspended, and inverted microstrips. Out of these, suspended and inverted microstrip structures are common and are shown in Fig. 4.11. Such structures provide a higher Q than the conventional microstrip structures and also a wide range of achievable characteristic impedances. The achievable Q value from these structures lies in the range 500–1500. The potential of these structures to achieve a wide range of characteristic impedances make them suitable for filter designing.

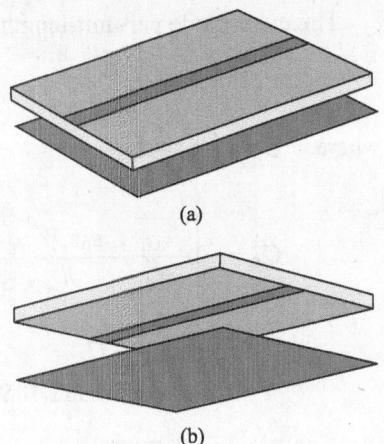

Fig. 4.11 Different forms of microstrip line (a) Suspended microstrip line (b) Inverted microstrip line

4.8 Coplanar Waveguide and Coplanar Strip Line Structures

Unlike a strip or microstrip line, the coplanar waveguide (CPW) or coplanar strip line (CPS) structures are uniplanar, that is, both the strip conductor and the ground plane are located on the same plane of the dielectric substrate, as shown in Fig. 4.12. Such structures have unique advantages such as small dispersion, easier implementation of short circuits at the ends, easier integration of lumped elements and active circuits, and circumventing the need for via holes. The existence of parasitic modes, non-confinement of fields, and lower

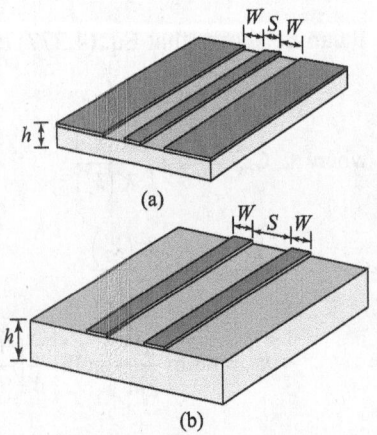

Fig. 4.12 3D figures (a) CPW (b) CPS

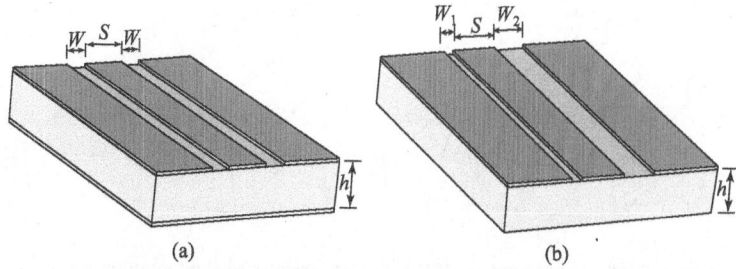

Fig. 4.13 Different CPW structures (a) Symmetric CPW with ground plane (b) Asymmetric CPW

power handling capability are the disadvantages of such structures. Different types of CPW configurations are shown in Fig. 4.13.

The electric and magnetic field configurations, under a quasistatic approximation, of CPW structures are shown in Fig. 4.14. Because of the existence of longitudinal components of the magnetic field, the mode of propagation becomes non-TEM at higher frequencies. In such cases, the magnetic field becomes elliptically polarized at the dielectric–air interface and hence the CPW structures become suitable for non-reciprocal ferrite devices. The electric and magnetic field configurations, under a quasistatic approximation, of CPS structures are shown in Fig. 4.15.

Sometimes, a conductor backing is required to be introduced to improve the mechanical strength as well as power handling capability of a CPW line. This also allows easy

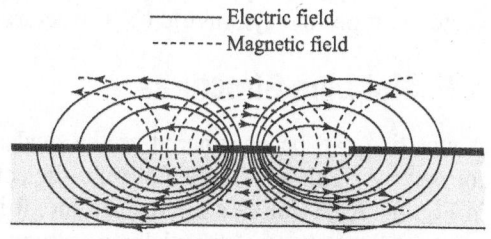

Fig. 4.14 Electric and magnetic field distributions in CPW

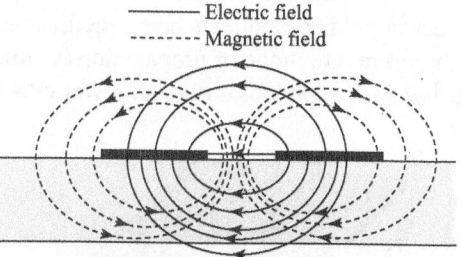

Fig. 4.15 Electric and magnetic field distributions in CPS

integration of the CPW and a microstrip circuit for MIC and MMIC applications. A conductor-backed CPW (CBCPW) is practically a mixed coplanar–microstrip structure, and, depending on the geometry, one structure may dominate the other. The microstrip behaviour is dominant when $W/h > 2$, whereas the coplanar behaviour is dominant when $W/h \ll 1$. For a moderate aspect ratio, a CBCPW is less dispersive than its microstrip counterpart. A CBCPW has the disadvantage of supporting leaky modes and parasitic microstrip modes.

Notes 1. Fields of a CPW are less confined in the dielectric than a microstrip line. This makes a CPW structure more sensitive to the covers or shields placed above it.
2. The asymmetry in CPW structures gives rise to a decrease in the characteristic impedance and an increase in the effective dielectric constant.

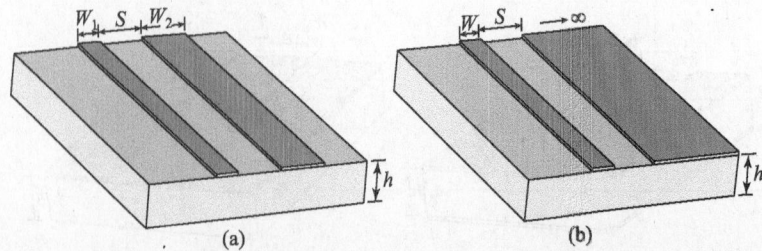

Fig. 4.16 Asymmetric CPS (a) Finite strip (b) Infinitely wide strip

Like CPW, CPS also offers the flexibility of mounting devices in series as well as in shunt configurations. It is also easier to achieve higher characteristic impedances with a CPS than with a CPW. In addition, being a balanced line in nature, CPS finds wide applications in mixers, modulators, and high-speed digital circuits. Different types of asymmetric CPS structures are shown in Fig. 4.16.

4.9 Slot Line Structures

A slot line can be considered as the dual structure of a microstrip line and is formed by etching a single slot on a dielectric-supported metal layer, as shown in Fig. 4.17(a). Unlike a microstrip line, it is uniplanar, that is, no ground plane exists at the bottom of the dielectric structure, supporting the slot line. Slot line structures are useful in circuits requiring high impedance lines, short-circuited ends, series stubs, etc. and can also be combined with CPW or microstrip lines.

In slot line circuits, a wave propagates along the slot, with most of the electric field components being inclined across the slot, as shown in Fig. 4.17(b). Therefore, the mode of propagation is almost TE in nature. However, the slot line being a two-conductor line, no lower cut-off frequency exists for such modes.

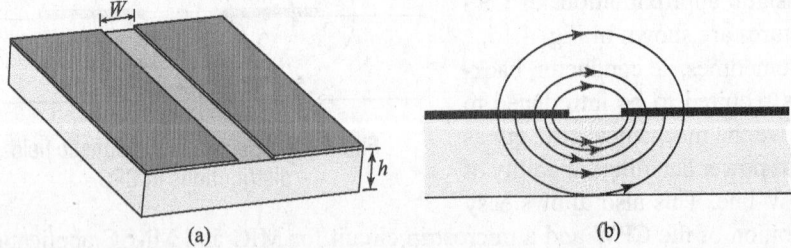

Fig. 4.17 Slot line (a) General slot line structure (b) Electric field distribution in slot line

4.10 Lumped Elements in Microstrips

In this section, various types of inductors, capacitors, and resistors have been discussed.

4.10.1 Inductors

Lumped inductor circuits can be realized by using a high-impedance microstrip line, loop, or spiral conductor, as shown in Fig. 4.18. A straight section of a microstrip line can provide an inductance of up to 2 or 3 nH, whereas for MMICs

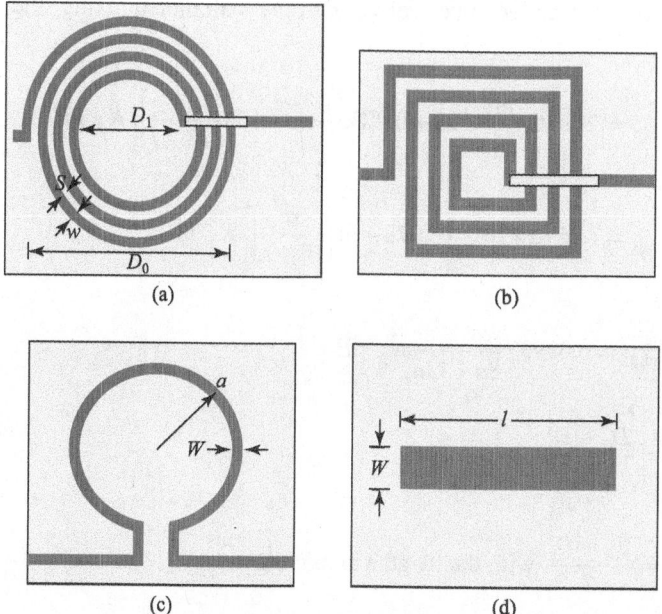

Fig. 4.18 Lumped inductor circuits (a) and (b) Spiral inductors
(c) Loop inductor (d) Strip inductor

operating above the L-band, the achievable inductance values ranges from 0.5 to 10 nH.

In general, spiral inductors that find wide applications in high-density circuits have higher Q values and can also provide higher impedance values. However, when a spiral inductor is placed adjacent to a ground plane, the presence of the ground plane changes its inductance value. In general, the inductance decreases as the ground plane is brought closer. This decrease in inductance values can be taken into account with the introduction of a correction factor K_g, which relates the modified inductance (L) with the free-space inductance (L_0) by the following relation:

$$L = K_g L_0 \tag{4.285}$$

where $\quad K_g = 0.57 - 0.145 \ln\left(\dfrac{W}{h}\right) \quad$ for $\dfrac{W}{h} > 0.05 \tag{4.286}$

Equivalent circuits of a strip, loop, and spiral inductor are illustrated in Fig. 4.19.

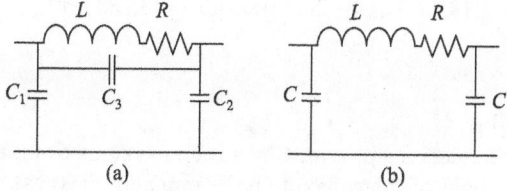

Fig. 4.19 Equivalent circuits (a) Spiral inductor
(b) Strip and loop inductors

The respective inductance values can be calculated using the following expressions:

$$L_{strip} = 2l \times 10^{-4} \left\{ 1.193 + 0.2235 \frac{W+t}{l} + \ln\left(\frac{l}{W+t}\right) \right\} K_g \ nH \qquad (4.287)$$

$$L_{loop} = 1.257a \times 10^{-3} \left\{ 0.078 + \ln\left(\frac{a}{W+t}\right) \right\} K_g \ nH \qquad (4.288)$$

and $\quad L_{spiral} = 0.03937 \dfrac{a^2 n^2}{8a + 11c} K_g \ nH \qquad (4.289)$

where $\quad a = \dfrac{D_0 + D_1}{4} \qquad (4.290)$

$$c = \frac{D_0 - D_1}{2} \text{ with the lengths in micrometres.} \qquad (4.291)$$

EXAMPLE 4.3 For a 3-cm-long microstrip line, described in Example 4.2, calculate the inductance.

Solution Given: $\varepsilon_r = 10.5$, $h = 3.1\,mm$, $W = 2.5\,mm$, $t = 0.03\,mm$, and $L = 3\,cm$

Now, $\quad \dfrac{W}{h} = \dfrac{2.5}{3.1} = 0.806$

$$K_g = 0.57 - 0.145 \ln\left(\frac{W}{h}\right) = 0.57 - 0.145 \ln(0.806) = 0.601$$

$$L_{strip} = 2l \times 10^{-4} \left\{ 1.193 + 0.2235 \frac{W+t}{l} + \ln\left(\frac{l}{W+t}\right) \right\} K_g \ nH$$

or $\quad L_{strip} = 2 \times 3 \times 10^4 \times 10^{-4} \left\{ \begin{array}{l} 1.193 + 0.2235\left(\dfrac{2.5 + 0.03}{30}\right) \\[2mm] + \ln\left(\dfrac{30}{2.5 + 0.03}\right) \end{array} \right\} 0.601$

or $\quad L_{strip} = 6 \times (1.193 + 0.019 + 2.473) \times 0.601 = 13.288 \ nH$

Practice Problem

4.5 Calculate the inductance provided by a loop of radius 6 mm and width 2 mm. Assume that the height of the substrate is 3.7 mm and thickness of the conductor is 0.05 mm. **[5.465 nH]**

4.10.2 Capacitors

In general, two types of capacitors, namely metal–insulator–metal (MIM) and interdigital capacitors, are mostly used in microwave and millimetre-wave circuits. A choice of one between the two depends on the required capacitance value, frequency of operation, size, and available processing technology. Usually, an interdigital capacitor, with an acceptable size, can provide a capacitance of up to 1 pF. If higher capacitance values are required while maintaining the overall size, MIM capacitors can be used. The basic diagrams of interdigital and MIM capacitors are shown in Fig. 4.20.

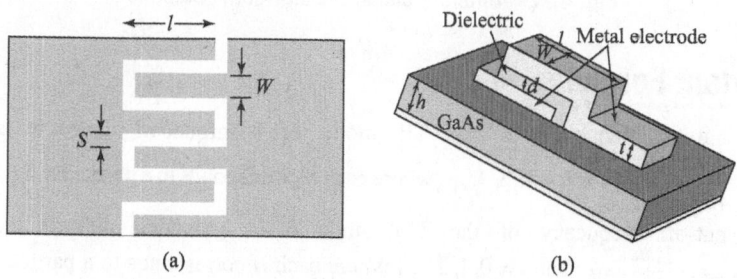

(a)   (b)

Fig. 4.20 Different capacitors (a) Interdigital capacitor (b) MIM capacitor

An interdigital capacitor is fabricated using the technique used for the fabrication of MICs, whereas an MIM capacitor is constructed using a low-loss dielectric, like silicon nitride, between two metal planes. Generally, the top metal of an MIM capacitor is made very thick, to reduce the loss.

4.10.3 Thin-film Resistors

Resistors are generally realized by depositing a semiconductor film on a semi-insulating substrate or by depositing thin films of a lossy material on a dielectric substrate. For the latter configuration, the use of Nichrome or tantalum nitride is most common. The basic diagram of a thin-film resistor is shown in Fig. 4.21.

If ρ_s, l, w, and t are the surface resistivity, length, width, and thickness of the thin film, respectively, then

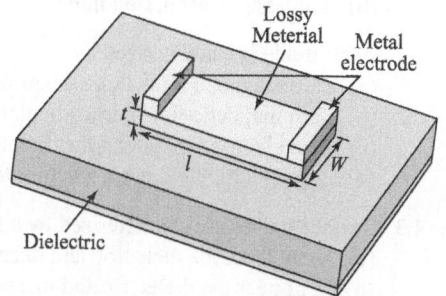

Fig. 4.21 General structure of thin-film resistor

$$R = \frac{\rho_s l}{wt} \ \Omega \qquad\qquad (4.292)$$

Semiconductor-based planar resistors can be fabricated by forming an isolated band of conducting epitaxial films on the substrate. Alternatively, a high-resistivity region can also be implemented within the semi-insulating substrate. Several configurations of planar resistors are shown in Fig. 4.22.

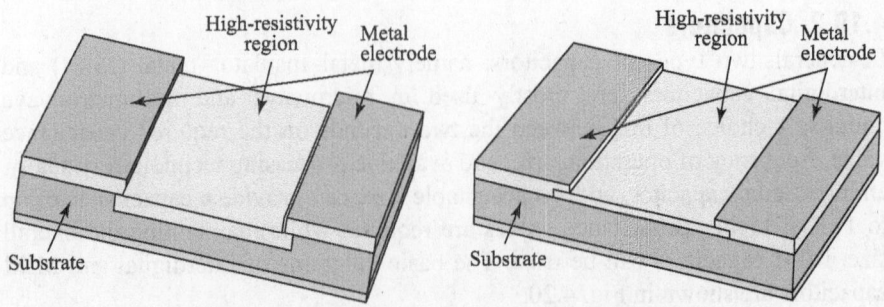

Fig. 4.22 Different planar resistor configurations

Important Formulae

- The cut-off frequency of the TE mode of a grounded dielectric slab is
$$f_c = \frac{(2n-1)c}{4h\sqrt{\varepsilon_r - 1}}$$
for $n = 1, 2, 3, ...$, where each n corresponds to a particular TE_n mode.
- The cut-off frequency of the TM mode of a grounded dielectric slab is
$$f_{c,TMn} = \frac{nc}{2h\sqrt{\varepsilon_r - 1}}$$
for $n = 0, 1, 2, ...$, where each n corresponds to a particular TM_n mode.

Exercises

Objective-type Questions

4.1 A planar transmission line is a
    (a) two-conductor transmission line
    (b) uniplanar transmission line
    (c) non-planar transmission line
    (d) one-conductor transmission line

4.2 A TE mode is characterized by
    (a) Zero magnetic field along the direction of propagation
    (b) Zero magnetic field perpendicular to the direction of propagation
    (c) Zero electric field along the direction of propagation
    (d) Zero electric field perpendicular to the direction of propagation

4.3 Surface waves are characterized by a field that is mostly concentrated
    (a) Away from the dielectric and decays exponentially away from the interface
    (b) In or near the dielectric and increases exponentially away from the interface
    (c) Away from the dielectric and increases exponentially away from the interface
    (d) In or near the dielectric and decays exponentially away from the interface

4.4 The cut-off frequency of the TE mode of a surface wave of a grounded dielectric slab is
    (a) $\dfrac{(n-1)c}{4h\sqrt{\varepsilon_r - 1}}$
    (b) $\dfrac{(2n-1)c}{4h\sqrt{\varepsilon_r - 1}}$
    (c) $\dfrac{(n+1)c}{4h\sqrt{\varepsilon_r - 1}}$
    (d) $\dfrac{(2n+1)c}{4h\sqrt{\varepsilon_r - 1}}$

4.5 A strip line structure supports
    (a) only TEM mode
    (b) only TE mode
    (c) only TM mode
    (d) TE, TM, and TEM modes

4.6 In a CPW line, microstrip behaviour is dominant when
   (a) $W/h < 2$ (b) $W/h > 0$ (c) $W/h > 2$ (d) $W/h < 5$

4.7 In slot line structures, propagation is almost
   (a) TE mode with the cut-off frequency
   (b) TM mode with the cut-off frequency
   (c) TE mode with no cut-off frequency
   (d) TM mode with no cut-off frequency

4.8 If the ground plane is brought closer, inductance of a spiral inductor in general
   (a) decreases   (c) remains constant
   (b) increases   (d) becomes negative

Review Questions

4.1 Describe the propagation of TE-mode surface waves on a grounded dielectric slab.
4.2 Describe the propagation of TM-mode surface waves on a grounded dielectric slab.
4.3 Assuming that the sides of a strip line structure can be replaced by perfect conductor, derive the expressions for its capacitance and characteristic impedance.
4.4 Assuming that the sides of a microstrip line structure can be replaced by perfect conductor, derive the expressions for its capacitance and characteristic impedance.
4.5 Write a short note on CPW, CPS, and slot line structures.
4.6 Describe different types of lumped elements in microstrip lines.

Problems

4.1 Find the first five cut-off frequencies for the possible TE- and TM-mode surface waves in a grounded dielectric slab with relative permittivity 2.55 and thickness 1.4 mm. *[Hint: Use Eqs (4.87) and (4.130)]*

4.2 It is required to design a strip line with characteristic impedance 50 Ω. If the thickness and relative permittivity of the substrate are 1.4 mm and 2.55, respectively, then calculate the strip width assuming zero conductor thickness. *[Hint: Use Eqs (4.195) and (4.196)]*

4.3 It is required to design a microstrip line with characteristic impedance 50 Ω. The thickness and the relative permittivity of the substrate are 1.4 mm and 2.55, respectively.
   (a) Calculate the strip width assuming zero conductor thickness. Also calculate the effective dielectric constant of the line. *[Hint: Use Eqs (4.241), (4.242), and (4.244)]*
   (b) Find the attenuation constant due to dielectric loss for the microstrip line, provided that the loss tangent of the dielectric is 4.7×10^{-4}. Assume the frequency of excitation to be 10 GHz. *[Hint: Use Eq. (4.245)]*

4.4 A microstrip spiral inductor having five turns has been designed on a dielectric substrate of thickness 1.4 mm. If the width of the strip is 2 mm, and inner and outer diameters of the spiral structure are 8 and 40 mm, respectively, then calculate the inductance of the structure. *[Hint: Use Eqs (4.286), (4.289)–(4.291)]*

5

Waveguides and Finlines

5.1 Introduction

In the last few chapters, we have discussed about different types of transmission lines. However, all these transmission lines consist of two or more conductors. Unlike them, a waveguide is a single-conductor transmission line. A waveguide is a hollow metallic conductor that is used for guided wave propagation. Depending on the cross-sectional geometry, waveguides can be classified as rectangular, circular, elliptical, radial, spherical, ridge, etc. However, the most common are the rectangular and circular waveguides.

The first structure for guiding waves was proposed by J.J. Thompson in the year 1893. The structure was experimentally tested one year later by Oliver Lodge in 1894. Theoretical modelling of the propagation of electromagnetic waves in a waveguide, however, was established three years later, in 1897, based on the work of Lord Rayleigh, which finally opened up a new area of microwave engineering. Surprisingly, waveguides were almost forgotten in the next four decades; these were rediscovered in 1936 by George C. Southworth of the AT&T Company and W. L. Barrow of MIT independently.

Probably the main boost to the research and applications of waveguides came during World War II. A huge verity of waveguide-based circuits and antennas were developed at that time, which still continue to be in use. These have got wide applications in ground-based, air-borne, and ship-borne radars, as well as in onboard satellite applications across frequency bands ranging from 1 to 1000 GHz. The main reason for this is their high power handling capability, ability to sustain high environmental variations, ruggedness, ability to achieve high accuracy in fabrication, and low loss.

In a waveguide, electric and magnetic fields are confined within the hollow space of the guide and thus no power is lost through radiation. Due to air being the dielectric that fills the hollow region within the guide, the dielectric loss is also negligible. Further, if we choose a metal having good conductivity, like copper, for constructing waveguide walls, then the conductor losses or Ohmic losses at the guide walls are also negligible. Thus, a waveguide acts as a lossless medium of transferring microwave signals from one point to another.

The possible existence of higher-order *modes* in a waveguide, however, may cause power loss during the transmission of electromagnetic waves. These modes basically correspond to the different possible solutions of the Maxwell's equation inside the waveguide region and can exist if the frequency of the signal is above a certain value, called the *cut-off frequency*. If the frequency of a signal is below the cut-off frequency, then the corresponding mode will be attenuated within a short distance. The cut-off frequency depends on the order of the mode and the dimensions of waveguides. Thus, each mode has its individual field pattern and cut-off frequency. The mode that has the lowest cut-off frequency among all the possible modes is called a *dominant mode*. It is a common practice to operate a waveguide in the dominant mode for reducing mode loss. This can be achieved by a proper choice of the waveguide dimensions, such that the frequency of the input signal is greater than the cut-off frequency of the dominant mode but less than the cut-off frequency of the next higher-order mode.

5.2 Formation of Waveguides from Two-wire Transmission Lines

In a two-wire transmission line system, support is often required to keep the wires in place. One of the ways to support wires is an insulator stand. The mandatory requirement for such a support is that the insulator must have very high impedance at the junction. However, the wire, insulator, and ground together form a capacitive system that often decreases the overall system impedance. The effect becomes prominent at high frequencies at which the capacitor behaves like a short circuit. A better high-frequency insulator option, as have already been discussed in Chapter 2, is a short-circuited quarter-wave transmission line. This line is also called a *copper insulator* and is shown in Fig. 5.1. Copper insulators can be placed at any point on a transmission line. Moreover, a large number of such insulators can be placed at different points on a transmission line without affecting the system performance significantly, as shown in Fig. 5.2.

If we go on increasing the number of such sections on each side of the line, then after a certain time, the sections will touch each other and form a rectangular waveguide, as shown in Fig. 5.3.

Insertion of short-circuited quarter-wave sections has a prominent impact on the field configurations of a two-wire line system. As a result, field configurations of an isolated two-wire transmission line undergo many

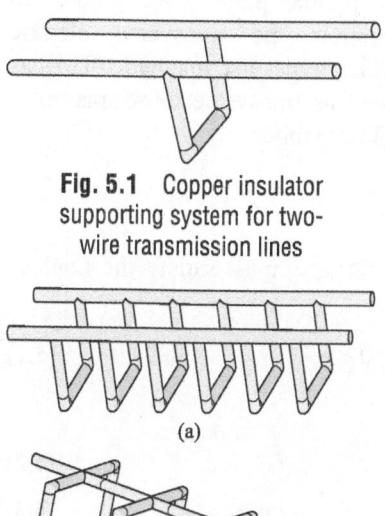

Fig. 5.1 Copper insulator supporting system for two-wire transmission lines

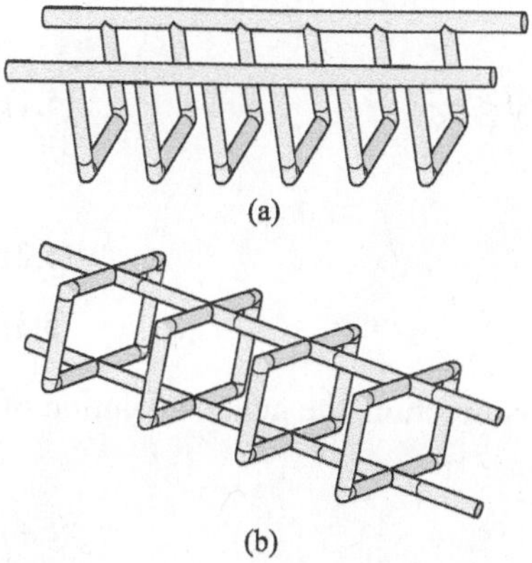

(a)

(b)

Fig. 5.2 Array of copper insulators for supporting two-wire transmission lines (a) Single side (b) Both sides

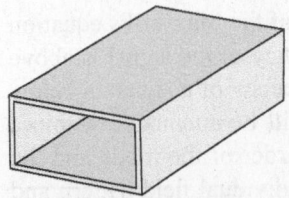

Fig. 5.3 Typical
rectangular waveguide

changes, and finally energy is confined within the hollow space of a waveguide rather than travelling along the length of the guide. In practice, short-circuited quarter-wave sections become a part of the rectangular waveguide system.

A waveguide appears to operate satisfactorily only at a particular frequency because of the existence of short-circuited quarter-wave sections. However, this is not true. In fact, waves with frequencies, for which the sections are $\lambda/4$ long or smaller, do not propagate efficiently through waveguides, but those with frequencies higher than this may propagate efficiently. Thus, a waveguide shows high-pass response characteristics. In subsequent sections, we will discuss more about this.

5.3 Parallel-plate Waveguides

A parallel-plate waveguide consists of two flat parallel plates of width a that are separated by a distance b such that $a > b$, as shown in Fig. 5.4. The space

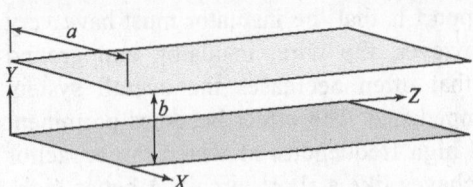

Fig. 5.4 Parallel-plate waveguide

between the plates is filled with a lossless dielectric. The condition $a > b$ ensures that we can neglect the fringing field at the edges. A parallel-plate wave guide can support the transverse electric (TE), transverse magnetic (TM), as well as transverse electromagnetic (TEM) modes.

5.3.1 Propagation of TEM Waves

For the TEM mode, the electrostatic potential $\Phi(x,y)$ must satisfy the Laplace equation:

$$\nabla_t^2 \Phi(x,y) = 0 \quad \text{for } 0 \le x \le a \text{ and } 0 \le y \le b \tag{5.1}$$

and the boundary conditions

$$\Phi(x,0) = 0 \tag{5.2}$$

$$\Phi(x,b) = V_0 \tag{5.3}$$

Since there is no variation of $\Phi(x,y)$ along the x-direction, the general solution of Eq. (5.1) can be written as follows:

$$\Phi(x,y) = A + By \tag{5.4}$$

Substituting the boundary condition of Eq. (5.2), we get the following relation:

$$A = 0 \tag{5.5}$$

Further substituting the boundary condition of Eq. (5.3), we get the following expression:

$$V_0 = Bb$$

or $$B = V_0/b \qquad (5.6)$$

Substitution of Eqs (5.5) and (5.6) into Eq. (5.4) produces the following equation:

$$\Phi(x, y) = V_0 y/b \qquad (5.7)$$

The transverse electric field is therefore given by the following equation:

$$\vec{e}_t(x, y) = -\vec{\nabla}_t \Phi(x, y) = -\frac{V_0}{b} \hat{a}_y \qquad (5.8)$$

Therefore, the complete electric field can be expressed as follows:

$$\vec{E}(x, y, z) = \vec{e}_t(x, y) e^{-jkz} = -\frac{V_0}{b} e^{-jkz} \hat{a}_y \qquad (5.9)$$

where $k = \omega\sqrt{\mu\varepsilon} \qquad (5.10)$

is the propagation constant of the TEM wave. The magnetic field can be expressed as follows:

$$\vec{H}(x, y, z) = \hat{a}_z \times \frac{1}{\eta} \vec{E}(x, y, z) = \frac{V_0}{\eta b} e^{-jkz} \hat{a}_x \qquad (5.11)$$

where $\eta = \sqrt{\mu/\varepsilon} \qquad (5.12)$

is the *intrinsic impedance* of the medium between the parallel plates.

The voltage at the top plate with respect to the bottom plate can be expressed as follows:

$$V = -\int_0^b E_y \, dy = -\int_0^b \left(-\frac{V_0}{b} e^{-jkz} \right) dy = \frac{V_0}{b} e^{-jkz} \int_0^b dy = V_0 e^{-jkz} \qquad (5.13)$$

Similarly, the current at the top plate can be expressed as follows:

$$I = \int_0^a H_x \, dx = \int_0^a \frac{V_0}{\eta b} e^{-jkz} \, dx = \frac{V_0}{\eta b} e^{-jkz} \int_0^a dx = \frac{V_0 a}{\eta b} e^{-jkz} \qquad (5.14)$$

Therefore, the characteristic impedance of the line is given by the following expression:

$$Z_0 = \frac{V}{I} = \frac{\eta b}{a} \qquad (5.15)$$

The phase velocity of the wave in the line is as follows:

$$v_p = \frac{\omega}{k} = \frac{1}{\sqrt{\mu\varepsilon}} \qquad (5.16)$$

Attenuation due to dielectric loss is given by the following equation:

$$\alpha_d = \frac{k \tan(\delta)}{2} \text{ Np/m} \tag{5.17}$$

EXAMPLE 5.1 A parallel-plate transmission line consists of plates of width 22.86 mm that are separated by a distance of 10.16 mm. If the medium between the plates is air, then calculate the characteristic impedance of the line for TEM-mode propagation.

Solution Given: $\eta = 376.73 \, \Omega$, $b = 10.16$ mm, and $a = 22.86$ mm

$$Z_0 = \frac{\eta b}{a} = \frac{376.73 \times 10.16}{22.86} = 167.436 \, \Omega$$

Practice Problem

5.1 A parallel-plate transmission line consists of plates of width 12.95 mm that are separated by a distance of 6.48 mm. If the medium between the plates is air, then calculate the characteristic impedance of the line for TEM-mode propagation.   **[188.51 Ω]**

5.3.2 Propagation of TE Waves

The TE mode has already been discussed in detail in Chapter 4. Since there is no x-variation, $\frac{\partial}{\partial x} \to 0$ and the TE wave must satisfy the following wave equation:

$$\left(\frac{\partial^2}{\partial y^2} + k_c^2 \right) h_z(x, y) = 0 \tag{5.18}$$

where $\quad k_c^2 = \varepsilon_r k_0^2 - \beta^2 = k^2 - \beta^2 \tag{5.19}$

The general solution of Eq. (5.18) is as follows:

$$h_z(x, y) = g_0 \cos(k_c y) + g_1 \sin(k_c y) \tag{5.20}$$

Therefore, $\quad E_x = -\frac{j\omega\mu}{k_c^2} \frac{\partial H_z}{\partial y} = -\frac{j\omega\mu}{k_c^2} \frac{\partial}{\partial y} \left\{ h_z(x, y) e^{-j\beta z} \right\}$

or $\quad E_x = -\frac{j\omega\mu}{k_c} \left[\left\{ -g_0 \sin(k_c y) + g_1 \cos(k_c y) \right\} e^{-j\beta z} \right] \tag{5.21}$

where g_0 and g_1 are unknown constants and can be found from the following boundary conditions:

$$E_x = 0 \text{ at } y = 0, b \tag{5.22}$$

Applying the first boundary condition, that is, $E_x = 0$ at $y = 0$ in Eq. (5.21), we get the following relation:

$$0 = -\frac{j\omega\mu}{k_c} \left[\left\{ -g_0 \sin(0) + g_1 \cos(0) \right\} e^{-j\beta z} \right] = g_1 \frac{j\omega\mu}{k_c} e^{-j\beta z}$$

or $\quad g_1 = 0 \text{ as } \frac{j\omega\mu}{k_c} e^{-j\beta z} \neq 0 \tag{5.23}$

Now applying the second boundary condition, that is, $E_x = 0$ at $y = b$, we get the following equation:

$$0 = g_0 \frac{j\omega\mu}{k_c} \sin(k_c b) e^{-j\beta z}$$

or $\qquad \sin(k_c b) = 0$ (5.24)

As $g_0 \dfrac{j\omega\mu}{k_c} e^{-j\beta z} \neq 0$ Eq. (5.24) implies that

$$\sin(k_c b) = \sin(n\pi) \quad \text{for } n = 1, 2, 3, \ldots$$

or $\qquad k_c = \dfrac{n\pi}{b} \quad \text{for } n = 1, 2, 3, \ldots$ (5.25)

In Eq. (5.25), each n corresponds to a particular mode called the TE_n mode. Therefore, we can write the following equation:

$$E_x = g_n \frac{j\omega\mu}{k_c} \sin\left(\frac{n\pi y}{b}\right) e^{-j\beta z}$$ (5.26)

Substituting Eqs (5.23) and (5.25) into Eq. (5.20), we get the following equation:

$$h_z(x, y) = g_n \cos\left(\frac{n\pi y}{b}\right)$$

or $\qquad H_z = h_z(x, y) e^{-j\beta z} = g_n \cos\left(\dfrac{n\pi y}{b}\right) e^{-j\beta z}$ (5.27)

$$H_x = -\frac{j\beta}{k_c^2} \frac{\partial H_z}{\partial x} = -\frac{j\beta}{k_c^2} \frac{\partial}{\partial x}\left\{ g_n \cos\left(\frac{n\pi y}{b}\right) e^{-j\beta z}\right\} = 0$$ (5.28)

$$H_y = -\frac{j\beta}{k_c^2} \frac{\partial H_z}{\partial y} = -\frac{j\beta}{k_c^2} \frac{\partial}{\partial y}\left\{ g_n \cos\left(\frac{n\pi y}{b}\right) e^{-j\beta z}\right\}$$

or $\qquad H_y = g_n \dfrac{j\beta}{k_c^2}\left(\dfrac{n\pi}{b}\right) \sin\left(\dfrac{n\pi y}{b}\right) e^{-j\beta z}$

or $\qquad H_y = g_n \dfrac{j\beta}{k_c^2}(k_c) \sin\left(\dfrac{n\pi y}{b}\right) e^{-j\beta z} = g_n \dfrac{j\beta}{k_c} \sin\left(\dfrac{n\pi y}{b}\right) e^{-j\beta z}$ (5.29)

$$E_y = \frac{j\omega\mu}{k_c^2} \frac{\partial H_z}{\partial x} = \frac{j\omega\mu}{k_c^2} \frac{\partial}{\partial x}\left\{ g_n \cos\left(\frac{n\pi y}{b}\right) e^{-j\beta z}\right\} = 0$$ (5.30)

and $\qquad E_z = 0$ (5.31)

Equations (5.26) and (5.28)–(5.30) show that for $n = 0$, all transverse field components vanish and hence the TE_0 mode cannot exist.

Note TE modes in parallel-plate waveguides are orthogonal.

The propagation constant for the TE_n mode can be found from Eq. (5.19) as follows:

$$\beta = \sqrt{k^2 - k_c^2} \qquad (5.32)$$

Using Eq. (5.25) in Eq. (5.32), we can write the following relation:

$$\beta = \sqrt{k^2 - \left(\frac{n\pi}{b}\right)^2} \qquad (5.33)$$

Equation (5.33) leads to two cases.

Case 1 $k^2 < \left(\frac{n\pi}{b}\right)^2$. In this case, β is imaginary; hence, the wave will not propagate and will be attenuated.

Case 2 $k^2 > \left(\frac{n\pi}{b}\right)^2$. In this case, β is real and positive; hence, the wave will propagate.

These two cases imply that there must be a cut-off frequency above which the wave will propagate. This reveals the high-pass filter nature of a parallel-plate waveguide. The expression for the cut-off frequency can be obtained from the boundary of these two cases. At the cut-off frequency, the following relation holds:

$$k^2 = k_c^2 = \left(\frac{n\pi}{b}\right)^2 \text{ or } k_c = \omega_c \sqrt{\mu\varepsilon} = \frac{n\pi}{b}$$

or $\quad f_c = \dfrac{n}{2b\sqrt{\mu\varepsilon}} \qquad (5.34)$

The guided wavelength can be defined as follows:

$$\lambda_g = \frac{2\pi}{\beta} = \frac{2\pi}{\sqrt{k^2 - k_c^2}} = \frac{2\pi}{k\sqrt{1 - \left(k_c^2/k^2\right)}} = \frac{\lambda}{\sqrt{1 - \left(\lambda/\lambda_c\right)^2}} \qquad (5.35)$$

The phase velocity is expressed as follows:

$$v_p = \frac{\omega}{\beta} = \frac{\omega}{\sqrt{k^2 - k_c^2}} = \frac{\omega}{\omega\sqrt{\mu\varepsilon}\sqrt{1 - \left(\omega_c^2/\omega^2\right)}} = \frac{v}{\sqrt{1 - \left(f_c^2/f^2\right)}}$$

or $\quad v_p = \dfrac{v}{\sqrt{1 - \left(\lambda/\lambda_c\right)^2}} \qquad (5.36)$

The group velocity is given by the following relation:

$$v_g = \frac{d\omega}{d\beta} \qquad (5.37)$$

Now, $\dfrac{d\beta}{d\omega} = \dfrac{d}{d\omega}\left(\sqrt{k^2 - k_c^2}\right) = \dfrac{d}{d\omega}\left(\sqrt{\omega^2 \mu\varepsilon - \omega_c^2 \mu\varepsilon}\right) = \dfrac{\omega\mu\varepsilon}{\sqrt{\omega^2\mu\varepsilon - \omega_c^2\mu\varepsilon}}$

or $\qquad \dfrac{d\beta}{d\omega} = \dfrac{\sqrt{\mu\varepsilon}}{\sqrt{1 - \left(\omega_c^2/\omega^2\right)}} = \dfrac{\sqrt{\mu\varepsilon}}{\sqrt{1 - \left(f_c^2/f^2\right)}} = \dfrac{1}{v\sqrt{1 - \left(\lambda/\lambda_c\right)^2}}$ (5.38)

Therefore, $v_g = \dfrac{d\omega}{d\beta} = v\sqrt{1 - \left(\lambda/\lambda_c\right)^2}$ (5.39)

and $\qquad v_p v_g = \dfrac{v}{\sqrt{1 - \left(\lambda/\lambda_c\right)^2}} \times v\sqrt{1 - \left(\lambda/\lambda_c\right)^2} = v^2$ (5.40)

From Eqs (5.26) and (5.29), $Z_{0,\text{TE}} = \dfrac{E_x}{H_y} = \dfrac{\omega\mu}{\beta}$ (5.41)

The power carried by the TE_n mode is given by the following equation:

$$P_{\text{TE}_n} = \frac{1}{2}\operatorname{Re}\int_{x=0}^{a}\int_{y=0}^{b}\left(\vec{E}\times\vec{H^*}\right)\cdot\hat{a}_z\,dxdy = \frac{1}{2}\operatorname{Re}\int_{x=0}^{a}\int_{y=0}^{b} E_x H_y^* dxdy$$

or $\qquad P_{\text{TE}_n} = \dfrac{\omega\mu}{2k_c^2}\left|g_n\right|^2 \operatorname{Re}(\beta)\int_{x=0}^{a}\int_{y=0}^{b}\sin^2\left(\dfrac{n\pi y}{b}\right)dxdy$

or $\qquad P_{\text{TE}_n} = \dfrac{\omega\mu}{2k_c^2}\left|g_n\right|^2 \operatorname{Re}(\beta)\left(\dfrac{ab}{2}\right) = \dfrac{\omega\mu ab}{4k_c^2}\left|g_n\right|^2 \operatorname{Re}(\beta)$ (5.42)

The attenuation constant due to dielectric loss can be expressed as follows:

$$\alpha_d = \frac{k^2 \tan(\delta)}{2\beta} \text{ Np/m}$$ (5.43)

The attenuation constant due to conductor loss can be expressed as follows:

$$\alpha_c = \frac{2k_c^2 R_s}{\omega\mu\beta b} \text{ Np/m}$$ (5.44)

EXAMPLE 5.2 For the parallel-plate transmission line described in Example 5.1, calculate (a) wave impedance, (b) cut-off frequency, (c) phase velocity, (d) group velocity, and (e) guided wavelength for the TE_1 wave propagation at 20 GHz.

Solution Given: $\eta = 376.73\ \Omega$, $b = 10.16\,\text{mm}$, and $f = 20\,\text{GHz}$

(a) $\lambda = \dfrac{c}{f} = \dfrac{3\times10^{11}}{20\times10^9} = 15\ \text{mm}$

$k = \dfrac{2\pi}{\lambda} = \dfrac{2\pi}{0.015} = 418.879\ \text{rad/m}$

$$\beta = \sqrt{k^2 - \left(\frac{n\pi}{b}\right)^2} = \sqrt{418.879^2 - \left(\frac{\pi}{10.16 \times 10^{-3}}\right)^2} = 282.573 \text{ rad/m}$$

$$Z_{0,\text{TE}_1} = \frac{\omega\mu}{\beta} = \frac{2\pi \times 20 \times 10^9 \times 4\pi \times 10^{-7}}{282.573} = 558.842 \ \Omega$$

(b) $f_c = \dfrac{n}{2b\sqrt{\mu\varepsilon}} = \dfrac{3 \times 10^8}{2 \times 10.16 \times 10^{-3}} = 14.764 \text{ GHz}$

(c) $\lambda_c = \dfrac{c}{f_c} = \dfrac{3 \times 10^{11}}{14.764 \times 10^9} = 20.32 \text{ mm}$

$$v_p = \frac{v}{\sqrt{1 - (\lambda/\lambda_c)^2}} = \frac{3 \times 10^8}{\sqrt{1 - (15/20.32)^2}} = \frac{3 \times 10^8}{\sqrt{0.455}} = 4.447 \times 10^8 \text{ m/s}$$

(d) $v_g = v\sqrt{1 - (\lambda/\lambda_2)^2} = 3 \times 10^8 \sqrt{0.455} = 2.024 \times 10^8 \text{ m/s}$

(e) $\lambda_g = \dfrac{\lambda}{\sqrt{1 - (\lambda/\lambda_c)^2}} = \dfrac{15}{\sqrt{0.455}} = 22.237 \text{ mm}$

Practice Problem

5.2 For the parallel-plate transmission line described in Practice Problem 5.1, calculate (a) wave impedance, (b) cut-off frequency, (c) phase velocity, (d) group velocity, and (e) guided wavelength for TE_1 wave propagation at 30 GHz.

$$\left[592.659 \ \Omega, \ 23.148 \text{ GHz}, \ 12.96 \text{ mm}, \ 4.714 \times 10^8 \text{ m/s}, \ 1.909 \times 10^8 \text{ m/s}, \ 15.713 \text{ mm}\right]$$

5.3.3 Propagation of TM Waves

A detailed discussion on the TM mode has been carried out in Chapter 4. Since there is no x-variation, $\dfrac{\partial}{\partial x} \to 0$ and the TM wave must satisfy the following wave equation:

$$\left(\frac{\partial^2}{\partial y^2} + k_c^2\right) e_z(x, y) = 0 \tag{5.45}$$

where $k_c^2 = \varepsilon_r k_0^2 - \beta^2 = k^2 - \beta^2$ $\hspace{2cm}$ (5.46)

The general solution of Eq. (5.45) is as follows:

$$e_z(x, y) = g_0 \cos(k_c y) + g_1 \sin(k_c y)$$

or $\quad E_z(x, y, z) = e_z(x, y)e^{-j\beta z} = \{g_0 \cos(k_c y) + g_1 \sin(k_c y)\} e^{-j\beta z}$ $\hspace{1cm}$ (5.47)

where g_0 and g_1 are unknown constants and can be found from the following boundary conditions:

$$E_z = 0 \text{ at } y = 0, \ b \tag{5.48}$$

Applying the first boundary condition, that is, $E_z = 0$ at $y = 0$, in Eq. (5.47), we get the following expression:

$$0 = \{g_0 \cos(0) + g_1 \sin(0)\} e^{-j\beta z} = g_0 e^{-j\beta z}$$

or   . $g_0 = 0$ as $e^{-j\beta z} \neq 0$ (5.49)

Therefore, Eq. (5.47) is modified to the following form:

$$E_z(x, y, z) = g_1 \sin(k_c y) e^{-j\beta z} \tag{5.50}$$

Now applying the second boundary condition, that is, $E_z = 0$ at $y = b$, in Eq. (5.50), we get the following relation:

$$0 = g_1 \sin(k_c b) e^{-j\beta z}$$

or $\sin(k_c b) = 0$ (5.51)

As $g_1 e^{-j\beta z} \neq 0$, Eq. (5.51) implies that,

$$\sin(k_c b) = \sin(n\pi) \quad \text{for } n = 1, 2, 3, \ldots$$

or $k_c = \dfrac{n\pi}{b}$ for $n = 1, 2, 3, \ldots$ (5.52)

In Eq. (5.52), each n corresponds to a particular mode, called the TM_n mode.

Substituting Eq. (5.52) into Eq. (5.50) and considering the fact that n number of solutions can exist, we can write the following expression:

$$E_z = g_n \sin\left(\frac{n\pi y}{b}\right) e^{-j\beta z} \tag{5.53}$$

Therefore,

$$H_x = \frac{j\omega\varepsilon}{k_c^2} \frac{\partial E_z}{\partial y} = \frac{j\omega\varepsilon}{k_c^2} \frac{\partial}{\partial y}\left\{g_n \sin\left(\frac{n\pi y}{b}\right) e^{-j\beta z}\right\}$$

or $H_x = g_n \dfrac{j\omega\varepsilon}{k_c^2}\left(\dfrac{n\pi}{b}\right)\cos\left(\dfrac{n\pi y}{b}\right) e^{-j\beta z} = g_n \dfrac{j\omega\varepsilon}{k_c}\cos\left(\dfrac{n\pi y}{b}\right) e^{-j\beta z}$ (5.54)

$$H_y = -\frac{j\omega\varepsilon}{k_c^2} \frac{\partial E_z}{\partial x} = -\frac{j\omega\varepsilon}{k_c^2} \frac{\partial}{\partial x}\left\{g_n \sin\left(\frac{n\pi y}{b}\right) e^{-j\beta z}\right\} = 0 \tag{5.55}$$

$$E_x = -\frac{j\beta}{k_c^2} \frac{\partial E_z}{\partial x} = -\frac{j\beta}{k_c^2} \frac{\partial}{\partial x}\left\{g_n \sin\left(\frac{n\pi y}{b}\right) e^{-j\beta z}\right\} = 0 \tag{5.56}$$

$$E_y = -\frac{j\beta}{k_c^2} \frac{\partial E_z}{\partial y} = -\frac{j\beta}{k_c^2} \frac{\partial}{\partial y}\left\{g_n \sin\left(\frac{n\pi y}{b}\right) e^{-j\beta z}\right\}$$

or $E_y = -g_n \dfrac{j\beta}{k_c^2}\left(\dfrac{n\pi}{b}\right)\cos\left(\dfrac{n\pi y}{b}\right) e^{-j\beta z} = -g_n \dfrac{j\beta}{k_c}\cos\left(\dfrac{n\pi y}{b}\right) e^{-j\beta z}$ (5.57)

Equations (5.52), (5.46), and (5.53) show that for $n = 0$, $\beta = k = \omega\sqrt{\mu\varepsilon}$ and $E_z = 0$. Therefore, the TM_0 mode is basically the TEM mode.

Note Like the TE modes, the TM modes in parallel-plate waveguides are also orthogonal.

The propagation constant for the TM_n mode can be found from Eq. (5.46), as follows:

$$\beta = \sqrt{k^2 - k_c^2} \tag{5.58}$$

Using Eq. (5.52) in Eq. (5.58), we can write the following relation:

$$\beta = \sqrt{k^2 - \left(\frac{n\pi}{b}\right)^2} \tag{5.59}$$

Equation (5.59) leads to the following two cases:

Case 1 $k^2 < \left(\frac{n\pi}{b}\right)^2$. In this case, β is imaginary; hence, wave will not propagate and will be attenuated.

Case 2 $k^2 > \left(\frac{n\pi}{b}\right)^2$. In this case, β is real and positive, and hence wave will propagate.

At the cut-off frequency:

$$k^2 = k_c^2 = \left(\frac{n\pi}{b}\right)^2 \quad \text{or} \quad k_c = \frac{2\pi f_c}{v_p} = 2\pi f_c\sqrt{\mu\varepsilon} = \frac{n\pi}{b}$$

or $$f_c = \frac{n}{2b\sqrt{\mu\varepsilon}} \tag{5.60}$$

This is same as Eq. (5.34). Using Eqs (5.54) and (5.57), we get the relation

$$Z_{0,TM} = -\frac{E_y}{H_x} = \frac{\beta}{\omega\varepsilon} \tag{5.61}$$

Multiplying Eq. (5.41) by Eq. (5.61), we get the following relation:

$$Z_{0,TE}Z_{0,TM} = \frac{\omega\mu}{\beta}\frac{\beta}{\omega\varepsilon} = \frac{\mu}{\varepsilon} = \eta^2 \tag{5.62}$$

Since the expression of β is the same as that of the TE mode, expressions for λ_g, v_p, and v_g are also same.

The power carried by the TM_n mode is given by the following expression:

$$P_{TM_n} = \frac{1}{2}\text{Re}\int_{x=0}^{a}\int_{y=0}^{b}\left(\vec{E}\times\vec{H^*}\right).\hat{a}_z dxdy = -\frac{1}{2}\text{Re}\int_{x=0}^{a}\int_{y=0}^{b} E_y H_x^* dxdy$$

or $$P_{TM_n} = \frac{\omega\varepsilon}{2k_c^2}|g_n|^2\,\text{Re}(\beta)\int_{x=0}^{a}\int_{y=0}^{b}\cos^2\left(\frac{n\pi y}{b}\right)dxdy$$

or $\quad P_{\mathrm{TM}_n} = \dfrac{\omega \varepsilon a}{2k_c^2} |g_n|^2 \operatorname{Re}(\beta) \displaystyle\int_{y=0}^{b} \cos^2\!\left(\dfrac{n\pi y}{b}\right) dy$

or $\quad P_{\mathrm{TM}_n} = \begin{cases} \dfrac{\omega \varepsilon a b}{4k_c^2} |g_n|^2 \operatorname{Re}(\beta) & \text{for } n > 0 \\[3mm] \dfrac{\omega \varepsilon a b}{2k_c^2} |g_n|^2 \operatorname{Re}(\beta) & \text{for } n = 0 \end{cases}$ $\qquad$ (5.63)

The attenuation constant due to conductor loss can be expressed as follows:

$$\alpha_c = \frac{P_l}{2P_0} \ \text{Np/m} \qquad\qquad (5.64)$$

where $\quad P_l = R_s \displaystyle\int_{x=0}^{a} |J_s|^2 \, dx = |g_n|^2 \dfrac{\omega^2 \varepsilon^2 R_s a}{k_c^2}$ $\qquad\qquad$ (5.65)

Substituting Eqs (5.63) and (5.65) into Eq. (5.64), we get the following relations:

$$\alpha_c = \frac{P_l}{2P_0} = \frac{|g_n|^2 \dfrac{\omega^2 \varepsilon^2 R_s a}{k_c^2}}{2\dfrac{\omega \varepsilon a b}{4k_c^2} |g_n|^2 \operatorname{Re}(\beta)} = \frac{2\omega \varepsilon R_s}{b\beta} = \frac{2\omega\sqrt{\mu\varepsilon}\,R_s}{b\beta}\sqrt{\frac{\varepsilon}{\mu}}$$

or $\quad \alpha_c = \dfrac{2kR_s}{b\eta\beta} \ \text{for } n > 0 \ \text{Np/m}$ $\qquad\qquad$ (5.66)

and $\quad \alpha_c = \dfrac{P_l}{2P_0} = \dfrac{|g_n|^2 \dfrac{\omega^2 \varepsilon^2 R_s a}{k_c^2}}{2\dfrac{\omega \varepsilon a b}{2k_c^2} |g_n|^2 \operatorname{Re}(\beta)} = \dfrac{\omega \varepsilon R_s}{b\beta} = \dfrac{\omega\sqrt{\mu\varepsilon}\,R_s}{b\beta}\sqrt{\dfrac{\varepsilon}{\mu}}$

or $\quad \alpha_c = \dfrac{kR_s}{b\eta\beta} \ \text{for } n = 0 \ \text{Np/m}$ $\qquad\qquad$ (5.67)

Since the TM_0 mode is basically the TEM mode, Eq. (5.67) is also valid for a TEM wave.

Let us now consider the TM_1 mode propagation inside a parallel-plate waveguide, as shown in Fig. 5.5. The phase constant and the z-component of the electric field can be written as follows:

$$\beta_1 = \sqrt{k^2 - (\pi/b)^2} \qquad\qquad (5.68)$$

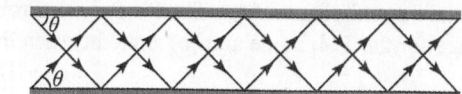

Fig. 5.5 Propagation of TM_1 mode inside parallel plate waveguide

and $\quad E_z = g_1 \sin\left(\dfrac{\pi y}{b}\right)e^{-j\beta_1 z} = g_1\left[\dfrac{e^{j(\pi y/b)} - e^{-j(\pi y/b)}}{2j}\right]e^{-j\beta_1 z}$

or $\quad E_z = \dfrac{g_1}{2j}\left[e^{j\{(\pi y/b)-\beta_1 z\}} - e^{-j\{(\pi y/b)+\beta_1 z\}}\right]$ $\qquad\qquad$ (5.69)

Equation (5.69) represents two waves propagating obliquely in the $(-y, +z)$- and $(+y, +z)$-direction, respectively. The angle θ that each plane wave makes with the z-axis satisfies the following relations:

$$k\sin(\theta) = \pi/b \qquad\qquad (5.70)$$

and $\quad k\cos(\theta) = \beta_1 \qquad\qquad (5.71)$

Taking the ratio of Eq. (5.70) to Eq. (5.71), we get the following equation:

$$\tan(\theta) = \dfrac{\pi}{b\beta_1} = \dfrac{\pi/b}{\sqrt{k^2 - (\pi/b)^2}} > 0 \text{ for propagating modes} \qquad (5.72)$$

Therefore, the range of θ is $0° \le \theta \le 90°$. As $f \to f_c$, $\beta_1 \to 0$ and $\theta \to 90°$. That is, the wave bounces up and down between the two plates and there is no motion in the $+z$-direction.

The phase velocity of the wave inside the waveguide is as follows:

$$v_p = \dfrac{\omega}{\beta_1} = \dfrac{k}{k\cos(\theta)\sqrt{\mu\varepsilon}} = \dfrac{1}{\cos(\theta)\sqrt{\mu\varepsilon}} \qquad (5.73)$$

Since $\cos(\theta) \le 1$, $v_p \ge c$, where c is the velocity of the wave in the free space.

Therefore, we can write the following relation:

$$\dfrac{v_p}{c} = \dfrac{f\lambda_g}{f\lambda} \ge 1$$

or $\quad \lambda_g \ge \lambda \qquad\qquad (5.74)$

where λ_g is the wavelength inside the guided structure, called a *guided wavelength*, and λ is the wavelength in the unguided medium. Therefore, Eq. (5.74) implies that a guided wavelength, in general, is larger than the wavelength in the unguided medium.

Practice Problem

5.3@ For the parallel-plate transmission line described in Example 5.2, calculate the (a) wave impedance for the TM_1 mode and (b) angle between the guide walls and the incident wave.

5.4 For the parallel-plate transmission line described in Practice Problem 5.2, calculate the (a) wave impedance for the TM_1 mode and (b) angle between the guide walls and the incident wave. $\qquad$ [239.472 Ω, 50.498°]

@ Solutions in Online Resources

5.4 Introduction to Rectangular Waveguides

A rectangular waveguide is basically a hollow metallic tube that is made of a good conductor and has a rectangular cross-section. A typical rectangular waveguide is illustrated in Fig. 5.3. Like a parallel-plate waveguide, a rectangular waveguide can also support a number of modes, each of which will have individual cut-off frequencies. The cut-off frequency of a mode depends on the dimension of the waveguide. Though there are few applications of a rectangular waveguide for supporting higher-order modes, a common practice is to design a waveguide in such a way that only the dominant mode can propagate.

Like parallel-plate waveguides, here also a wave travels down the guide through multiple reflections from the guide walls, as shown in Fig. 5.5, and hence is characterized by a guided wavelength that is longer than the free-space wavelength. This procedure of propagation down the guide ensures that either an electric or a magnetic field will develop in the direction of propagation. Therefore, a rectangular waveguide supports only the TE or the TM mode. Since the guide has four metal walls, these modes are denoted by TE_{mn} or TM_{mn}, where m and n are the number of half-cycles of the electric or magnetic field in the directions of width and height, respectively.

5.5 Propagation of TE Modes in Rectangular Waveguides

The wave that propagates inside a waveguide is a plane wave in nature and hence obeys the following equations, as derived in Chapter 4:

$$\frac{\partial H_{0z}}{\partial y} + \gamma H_{0y} = (j\omega\varepsilon + \sigma)E_{0x} \tag{5.75}$$

$$-\gamma H_{0x} - \frac{\partial H_{0z}}{\partial x} = (j\omega\varepsilon + \sigma)E_{0y} \tag{5.76}$$

$$\frac{\partial H_{0y}}{\partial x} - \frac{\partial H_{0x}}{\partial y} = (j\omega\varepsilon + \sigma)E_{0z} \tag{5.77}$$

$$\frac{\partial E_{0z}}{\partial y} + \gamma E_{0y} = -j\omega\mu H_{0x} \tag{5.78}$$

$$-\gamma E_{0x} - \frac{\partial E_{0z}}{\partial x} = -j\omega\mu H_{0y} \tag{5.79}$$

$$\frac{\partial E_{0y}}{\partial x} - \frac{\partial E_{0x}}{\partial y} = -j\omega\mu H_{0z} \tag{5.80}$$

Since the TE modes in a rectangular waveguide are characterized by $E_{0z} = 0$, Eqs (5.75)–(5.80) are reduced to the following forms:

$$\frac{\partial H_{0z}}{\partial y} + \gamma H_{0y} = (j\omega\varepsilon + \sigma)E_{0x} = j\omega\varepsilon_c E_{0x} \tag{5.81}$$

$$-\gamma H_{0x} - \frac{\partial H_{0z}}{\partial x} = (j\omega\varepsilon + \sigma)E_{0y} = j\omega\varepsilon_c E_{0y} \tag{5.82}$$

$$\frac{\partial H_{0y}}{\partial x} - \frac{\partial H_{0x}}{\partial y} = 0 \tag{5.83}$$

$$\gamma E_{0y} = -j\omega\mu H_{0x} \tag{5.84}$$

$$\gamma E_{0x} = j\omega\mu H_{0y} \tag{5.85}$$

$$\frac{\partial E_{0y}}{\partial x} - \frac{\partial E_{0x}}{\partial y} = -j\omega\mu H_{0z} \tag{5.86}$$

Differentiating Eq. (5.81) with respect to y and Eq. (5.82) with respect to x and then subtracting the latter from the former, we get the following relation:

$$\frac{\partial^2 H_{0z}}{\partial x^2} + \frac{\partial^2 H_{0z}}{\partial y^2} + \gamma\left(\frac{\partial H_{0x}}{\partial x} + \frac{\partial H_{0y}}{\partial y}\right) = -j\omega\varepsilon_c\left(\frac{\partial E_{0y}}{\partial x} - \frac{\partial E_{0x}}{\partial y}\right) \tag{5.87}$$

Differentiating Eq. (5.85) with respect to y and Eq. (5.84) with respect to x and then subtracting the latter from the former, we get the following expression:

$$\left(\frac{\partial H_{0x}}{\partial x} + \frac{\partial H_{0y}}{\partial y}\right) = -\frac{\gamma}{j\omega\mu}\left(\frac{\partial E_{0y}}{\partial x} - \frac{E_{0x}}{\partial y}\right) \tag{5.88}$$

Substituting Eq. (5.88) into Eq. (5.87), we get the following relation:

$$\frac{\partial^2 H_{0z}}{\partial x^2} + \frac{\partial^2 H_{0z}}{\partial y^2} = \left(\frac{\omega^2\mu\varepsilon_c + \gamma^2}{j\omega\mu}\right)\left(\frac{\partial E_{0y}}{\partial x} - \frac{\partial E_{0x}}{\partial y}\right) \tag{5.89}$$

Now substituting Eq. (5.86) into (5.89), the following equation is obtained:

$$\frac{\partial^2 H_{0z}}{\partial x^2} + \frac{\partial^2 H_{0z}}{\partial y^2} = -\left(\omega^2\mu\varepsilon_c + \gamma^2\right)H_{0z} = -k_c^2 H_{0z} \tag{5.90}$$

where $\quad \gamma^2 + \omega^2\mu\varepsilon_c = \gamma^2 - j\omega\mu(\sigma + j\omega\varepsilon) = k_c^2 \tag{5.91}$

Equation (5.90) is a two-dimensional wave equation and can be solved using the method of separation of variable. For this, we assume that

$$H_{0z}(x,y) = H_{0z}(x)H_{0z}(y) \tag{5.92}$$

Substitution of Eq. (5.92) into Eq. (5.90) results in the following expression:

$$\frac{1}{H_{0z}(x)}\frac{\partial^2 H_{0z}(x)}{\partial x^2} = -k_c^2 - \frac{1}{H_{0z}(y)}\frac{\partial^2 H_{0z}(y)}{\partial y^2} \tag{5.93}$$

The RHS of Eq. (5.93) is a function of y, whereas the LHS is a function of x. This implies that both sides of Eq. (5.93) must be a constant (say $-k_x^2$) and we can write Eq. (5.93) as follows:

$$\frac{\partial^2 H_{0z}(x)}{\partial x^2} + k_x^2 H_{0z}(x) = 0 \tag{5.94}$$

and $\quad \dfrac{1}{H_{0z}(y)}\dfrac{\partial^2 H_{0z}(y)}{\partial y^2} = k_x^2 - k_c^2 = -k_y^2$

or $\quad \dfrac{\partial^2 H_{0z}(y)}{\partial y^2} + k_y^2 H_{0z}(y) = 0$ $\hspace{2cm}$ (5.95)

where $\quad k_x^2 + k_y^2 = k_c^2$ $\hspace{4cm}$ (5.96)

Equations (5.94) and (5.95) have well-known solutions of the following forms:

$$H_{0z}(x) = A\cos(k_x x) + B\sin(k_x x)$$ (5.97)

and $\quad H_{0z}(y) = C\cos(k_x x) + D\sin(k_x x)$ $\hspace{2cm}$ (5.98)

Substituting Eqs (5.97) and (5.98) into Eq. (5.92), we get the following relation:

$$H_{0z} = \left[A\cos(k_x x) + B\sin(k_x x)\right]\left[C\cos(k_y y) + D\sin(k_y y)\right]$$ (5.99)

As shown in Chapter 4, for TE modes we can write the following equation:

$$E_{0x} = -\frac{j\omega\mu}{k_c^2}\frac{\partial H_{0z}}{\partial y}$$ (5.100)

$$E_{0y} = \frac{j\omega\mu}{k_c^2}\frac{\partial H_{0z}}{\partial x}$$ (5.101)

$$H_{0x} = -\frac{\gamma}{k_c^2}\frac{\partial H_{0z}}{\partial x}$$ (5.102)

and $\quad H_{0y} = -\dfrac{\gamma}{k_c^2}\dfrac{\partial H_{0z}}{\partial y}$ $\hspace{2cm}$ (5.103)

Therefore, substituting Eq. (5.99) into Eqs (5.100)–(5.103), we get the following relations:

$$E_{0x} = -\frac{j\omega\mu}{k_c^2}\left[A\cos(k_x x) + B\sin(k_x x)\right]\left[-Ck_y\sin(k_y y) + Dk_y\cos(k_y y)\right] \quad (5.104)$$

$$E_{0y} = \frac{j\omega\mu}{k_c^2}\left[-Ak_x\sin(k_x x) + Bk_x\cos(k_x x)\right]\left[C\cos(k_y y) + D\sin(k_y y)\right] \quad (5.105)$$

$$H_{0x} = -\frac{\gamma}{k_c^2}\left[-Ak_x\sin(k_x x) + Bk_x\cos(k_x x)\right]\left[C\cos(k_y y) + D\sin(k_y y)\right] \quad (5.106)$$

and $H_{0y} = -\dfrac{\gamma}{k_c^2}\left[A\cos(k_x x) + B\sin(k_x x)\right]\left[-Ck_y\sin(k_y y) + Dk_y\cos(k_y y)\right]$ (5.107)

Equations (5.104)–(5.107) are subjected to the boundary conditions that the tangential components of electric fields must vanish at the conductor walls. Considering Fig. 5.6, the following relations can be obtained:

at $\quad x = 0$ and a, $E_{0y} = 0$ or $E_y = 0$ $\hspace{2cm}$ (5.108)

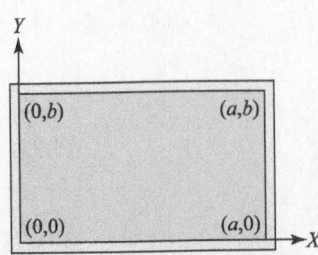

Y

(0,b)   (a,b)

(0,0)   (a,0)

→X

• **Fig. 5.6** Coordinate system
for deriving field expressions
inside the waveguide for
TE and TM modes

and at $y = 0$ and b, $E_{0x} = 0$ or $E_x = 0$ (5.109)

Substituting the boundary condition, at $x = 0$, $E_{0y} = 0$, in Eq. (5.105), we get the following relation:

$$E_{0y} = \frac{j\omega\mu}{k_c^2}\big[-Ak_x \sin(0) + Bk_x \cos(0)\big]$$

$$\big[C\cos(k_y y) + D\sin(k_y y)\big] = 0$$

or $B = 0$ (5.110)

as $\dfrac{j\omega\mu}{k_c^2}k_x\big[C\cos(k_y y) + D\sin(k_y y)\big] \neq 0$

Similarly, substituting the boundary condition, at $y = 0$, $E_{0x} = 0$, in Eq. (5.104), we get the following equation:

$$E_{0x} = -\frac{j\omega\mu}{k_c^2}\big[A\cos(k_x x) + B\sin(k_x x)\big]\big[-Ck_y \sin(0) + Dk_y \cos(0)\big] = 0$$

or $D = 0$ (5.111)

as $\dfrac{j\omega\mu}{k_c^2}k_y\big[A\cos(k_x x) + B\sin(k_x x)\big] \neq 0$

Substitution of Eqs (5.110) and (5.111) into Eqs (5.104)–(5.107) results in the following expressions:

$$E_{0x} = \frac{j\omega\mu}{k_c^2}ACk_y \cos(k_x x)\sin(k_y y) = \frac{j\omega\mu}{k_c^2}H_0 k_y \cos(k_x x)\sin(k_y y) \quad (5.112)$$

$$E_{0y} = -\frac{j\omega\mu}{k_c^2}ACk_x \sin(k_x x)\cos(k_y y)$$

or $E_{0y} = -\dfrac{j\omega\mu}{k_c^2}H_0 k_x \sin(k_x x)\cos(k_y y)$ (5.113)

$$H_{0x} = \frac{\gamma}{k_c^2}ACk_x \sin(k_x x)\cos(k_y y) = \frac{\gamma}{k_c^2}H_0 k_x \sin(k_x x)\cos(k_y y) \quad (5.114)$$

$$H_{0y} = \frac{\gamma}{k_c^2}k_y AC \cos(k_x x)\sin(k_y y) = \frac{\gamma}{k_c^2}H_0 k_y \cos(k_x x)\sin(k_y y) \quad (5.115)$$

where $AC = H_0$ (5.116)

Substituting the boundary condition, at $x = a$, $E_{0y} = 0$, in Eq. (5.113), we get the following relation:

$$E_{0y} = -\frac{j\omega\mu}{k_c^2}H_0 k_x \sin(k_x a)\cos(k_y y) = 0$$

Since $\dfrac{j\omega\mu}{k_c^2}H_0 k_x \cos(k_y y) \neq 0$

$\sin(k_x a) = 0 = \sin(m\pi)$, where $m = 0, 1, 2, 3, \ldots$

or $\quad k_x = m\pi/a$ (5.117)

Substituting the boundary condition, at $y = b$, $E_{0x} = 0$, in Eq. (5.112), we get the following relation:

$$E_{0x} = \frac{j\omega\mu}{k_c^2} H_0 k_y \cos(k_x x) \sin(k_y b) = 0$$

Since $\dfrac{j\omega\mu}{k_c^2} H_0 k_y \cos(k_x x) \neq 0$, $\sin(k_y b) = 0 = \sin(n\pi)$, where $n = 0, 1, 2, 3, \dots$

or $\quad k_y = n\pi/b$ (5.118)

Therefore, the complete field components are as follows:

$$E_x = E_{0x} e^{j\omega t - \gamma z} = \frac{j\omega\mu}{k_c^2} H_0 \frac{n\pi}{b} \cos\left(\frac{m\pi x}{a}\right) \sin\left(\frac{n\pi y}{b}\right) e^{j\omega t - \gamma z} \tag{5.119}$$

$$E_y = E_{0y} e^{j\omega t - \gamma z} = -\frac{j\omega\mu}{k_c^2} H_0 \frac{m\pi}{a} \sin\left(\frac{m\pi x}{a}\right) \cos\left(\frac{n\pi y}{b}\right) e^{j\omega t - \gamma z} \tag{5.120}$$

$$E_z = E_{0z} e^{j\omega t - \gamma z} = 0 \tag{5.121}$$

$$H_x = H_{0x} e^{j\omega t - \gamma z} = \frac{\gamma}{k_c^2} H_0 \frac{m\pi}{a} \sin\left(\frac{m\pi x}{a}\right) \cos\left(\frac{n\pi y}{b}\right) e^{j\omega t - \gamma z} \tag{5.122}$$

$$H_y = H_{0y} e^{j\omega t - \gamma z} = \frac{\gamma}{k_c^2} H_0 \frac{n\pi}{b} \cos\left(\frac{m\pi x}{a}\right) \sin\left(\frac{n\pi y}{b}\right) e^{j\omega t - \gamma z} \tag{5.123}$$

$$H_z = H_{0z} e^{j\omega t - \gamma z} = H_0 \cos\left(\frac{m\pi x}{a}\right) \cos\left(\frac{n\pi y}{b}\right) e^{j\omega t - \gamma z} \tag{5.124}$$

Note TE modes in rectangular waveguides are orthogonal.

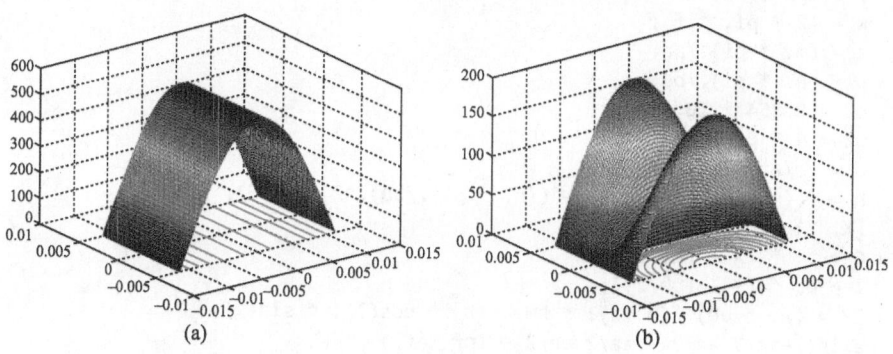

Fig. 5.7 Magnitude of Y-component of electric field distribution at the aperture for different modes (a) TE_{10} (b) TE_{11}

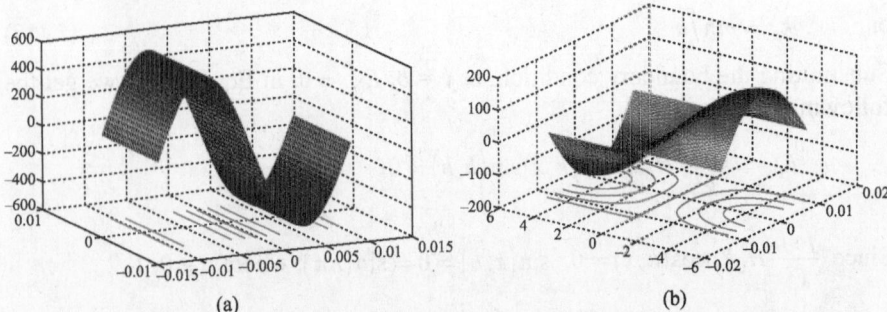

Fig. 5.8 Y-component of electric field distribution with phase at the aperture for different modes (a) TE_{20} (b) TE_{11}

Three-dimensional aperture field distributions for different TE modes are shown in Figs 5.7 and 5.8.

The following is a simple MATLAB program to plot the Y-component of the electric fields for different TE modes.

Similar programs can be written for the other components also.

MATLAB PROGRAM

```
% Program for finding distribution of Y component of electric field
(Ey) for TE mode
clear all;
clf;
aa = input('Enter the Width of the Guide(in Meter):- ');
bb = input('Enter the Height of the Guide(in Meter):- ');
a = linspace(-aa/2,aa/2,100);
b = linspace(-bb/2,bb/2,100);
h0 = 1;
ff = input('Enter the Frequency(in GHz):- ');
f = ff * 10^9;
m = input('Enter "m"(Mode No):- ');
n = input('Enter "n"(Mode No):- ');
w = (2 * pi. * f);
t = (m. * pi)./aa;
p = (n. * pi)./bb;
e0 = 8.854 * 10^(-12);
u0 = 4 * pi * 10^(-7);
d = (m/aa)^2 + (n/bb)^2;
h = (((m. * pi)./aa).^2 + ((n. * pi)./bb).^2).^(0.5);
[X, Y] = meshgrid(a, b);
G = t. * (X + (aa/2));
l = p. * (Y + (bb/2));
EY = (w. * u0)./(h.^2). * h0. * t. * cos(l). * sin(g);
axis([-aa/2 aa/2 - aa/2 aa/2 - Inf Inf]);
surfc(X, Y, EY)
```

TE$_{10}$ TE$_{11}$

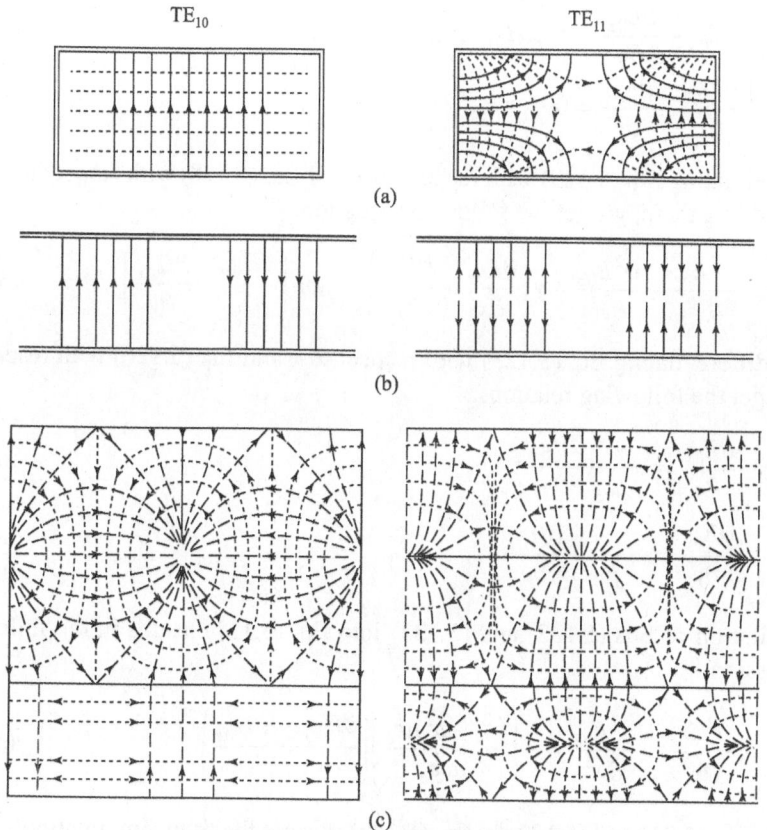

(a)

(b)

(c)

Fig. 5.9 Different TE modes in rectangular waveguide
(a) Cross-sectional view (b) Longitudinal view (c) Surface view

Contour plots of the TE$_{10}$ and TE$_{11}$ modes at different surfaces are shown in Fig. 5.9.

5.6 Propagation of TM Modes in Rectangular Waveguides

For TM-mode wave propagation, $H_{0z} = 0$ and Eqs (5.75)–(5.80) reduce to the following forms:

$$\gamma H_{0y} = (j\omega\varepsilon + \sigma)E_{0x} = j\omega\varepsilon_c E_{0x} \tag{5.125}$$

$$-\gamma H_{0x} = (j\omega\varepsilon + \sigma)E_{0y} = j\omega\varepsilon_c E_{0y} \tag{5.126}$$

$$\frac{\partial H_{0y}}{\partial x} - \frac{\partial H_{0x}}{\partial y} = (j\omega\varepsilon + \sigma)E_{0z} = j\omega\varepsilon_c E_{0z} \tag{5.127}$$

$$\frac{\partial E_{0z}}{\partial y} + \gamma E_{0y} = -j\omega\mu H_{0x} \tag{5.128}$$

$$\gamma E_{0x} + \frac{\partial E_{0z}}{\partial x} = j\omega\mu H_{0y} \tag{5.129}$$

$$\frac{\partial E_{0y}}{\partial x} - \frac{\partial E_{0x}}{\partial y} = 0 \tag{5.130}$$

Differentiating Eq. (5.129) with respect to x and Eq. (5.128) with respect to y, and then adding them, we get the following equation:

$$\frac{\partial^2 E_{0z}}{\partial x^2} + \frac{\partial^2 E_{0z}}{\partial y^2} + \gamma\left(\frac{\partial E_{0x}}{\partial x} + \frac{\partial E_{0y}}{\partial y}\right) = j\omega\mu\left(\frac{\partial H_{0y}}{\partial x} - \frac{\partial H_{0x}}{\partial y}\right) \tag{5.131}$$

Now differentiating Eq. (5.125) with respect to x and Eq. (5.126) with respect to y, we get the following relations:

$$\frac{\partial E_{0x}}{\partial x} = \frac{\gamma}{j\omega\varepsilon_c}\frac{\partial H_{0y}}{\partial x} \tag{5.132}$$

and

$$\frac{\partial E_{0y}}{\partial y} = -\frac{\gamma}{j\omega\varepsilon_c}\frac{\partial H_{0x}}{\partial y} \tag{5.133}$$

Substitution of Eqs (5.132) and (5.133) into Eq. (5.131) results in the following expression:

$$\frac{\partial^2 E_{0z}}{\partial x^2} + \frac{\partial^2 E_{0z}}{\partial y^2} = -\left(\frac{\gamma^2 + \omega^2\mu\varepsilon_c}{j\omega\varepsilon_c}\right)\left(\frac{\partial H_{0y}}{\partial x} - \frac{\partial H_{0x}}{\partial y}\right) \tag{5.134}$$

Using Eqs (5.91) and (5.127) in Eq. (5.134), we get the following relation:

$$\frac{\partial^2 E_{0z}}{\partial x^2} + \frac{\partial^2 E_{0z}}{\partial y^2} = -k_c^2 E_{0z} \tag{5.135}$$

To solve Eq. (5.135) for E_{0z}, let us assume that

$$E_{0z} = E_{0z}(x)E_{0z}(y) \tag{5.136}$$

Substituting Eq. (5.136) into Eq. (5.135), the following equation can be obtained:

$$\frac{1}{E_{0z}(x)}\frac{\partial^2 E_{0z}(x)}{\partial x^2} = -k_c^2 - \frac{1}{E_{0z}(y)}\frac{\partial^2 E_{0z}(y)}{\partial y^2} \tag{5.137}$$

The RHS of Eq. (5.137) is a function of y only, whereas the LHS is a function of x only. This implies that both sides of Eq. (5.137) must be a constant (say $-k_x^2$). Therefore, proceeding as before, we can write the following relations:

$$\frac{\partial^2 E_{0z}(x)}{\partial x^2} + k_x^2 E_{0z}(x) = 0 \tag{5.138}$$

and

$$\frac{\partial^2 E_{0z}(y)}{\partial y^2} + k_y^2 E_{0z}(y) = 0 \text{ where } k_x^2 + k_y^2 = k_c^2 \tag{5.139}$$

Equations (5.138) and (5.139) have well-known solutions of the following forms:

$$E_{0z}(x) = P\cos(k_x x) + Q\sin(k_x x) \tag{5.140}$$

and
$$E_{0z}(y) = R\cos(k_y y) + S\sin(k_y y) \tag{5.141}$$

Substituting Eqs (5.140) and (5.141) into Eq. (5.136), we get the following relation:

$$E_{0z} = \left[P\cos(k_x x) + Q\sin(k_x x) \right] \left[R\cos(k_y y) + S\sin(k_y y) \right] \tag{5.142}$$

Using Eqs (4.55)–(4.58), we can write the following expressions for the TM mode:

$$E_{0x} = -\frac{\gamma}{k_c^2} \frac{\partial E_{0z}}{\partial x} \tag{5.143}$$

$$E_{0y} = -\frac{\gamma}{k_c^2} \frac{\partial E_{0z}}{\partial y} \tag{5.144}$$

$$H_{0x} = \frac{j\omega\varepsilon_c}{k_c^2} \frac{\partial E_{0z}}{\partial y} \tag{5.145}$$

and
$$H_{0y} = -\frac{j\omega\varepsilon_c}{k_c^2} \frac{\partial E_{0z}}{\partial x} \tag{5.146}$$

Substituting Eq. (5.142) into Eqs (5.143)–(5.146), we get the following relation:

$$E_{0x} = -\frac{\gamma}{k_c^2} \left[-Pk_x \sin(k_x x) + Qk_x \cos(k_x x) \right] \left[R\cos(k_y y) + S\sin(k_y y) \right] \tag{5.147}$$

$$E_{0y} = -\frac{\gamma}{k_c^2} \left[P\cos(k_x x) + Q\sin(k_x x) \right] \left[-Rk_y \sin(k_y y) + Sk_y \cos(k_y y) \right] \tag{5.148}$$

$$H_{0x} = \frac{j\omega\varepsilon_c}{k_c^2} \left[P\cos(k_x x) + Q\sin(k_x x) \right] \left[-Rk_y \sin(k_y y) + Sk_y \cos(k_y y) \right] \tag{5.149}$$

$$H_{0y} = -\frac{j\omega\varepsilon_c}{k_c^2} \left[-Pk_x \sin(k_x x) + Qk_x \cos(k_x x) \right] \left[R\cos(k_y y) + S\sin(k_y y) \right] \tag{5.150}$$

These equations are subjected to boundary conditions that the tangential components of electric fields must vanish at the walls. Considering Fig. 5.6, this gives the following relations:

at $x = 0$ and a, $E_{0z} = 0$ or $E_z = 0$ $\tag{5.151}$

and $y = 0$ and b, $E_{0z} = 0$ or $E_z = 0$ $\tag{5.152}$

Substituting the boundary condition, at $x = 0$, $E_{0z} = 0$, in Eq. (5.142), we get the following relation:

$$E_{0z} = \left[P\cos(0) + Q\sin(0) \right] \left[R\cos(k_y y) + S\sin(k_y y) \right] = 0$$

or $\quad P = 0 \ \left(\text{as } R\cos\left(k_y y\right) + S\sin\left(k_y y\right) \neq 0\right)$ $\qquad$ (5.153)

Similarly, substituting the condition, at $y = 0$, $E_{0z} = 0$, in Eq. (5.142), we get the following relation:

$$E_{0z} = Q\sin\left(k_x x\right)\left[R\cos\left(0\right) + S\sin\left(0\right)\right] = 0$$

or $\quad R = 0 \ \left(\text{as } Q\sin\left(k_x x\right) \neq 0\right)$ $\qquad$ (5.154)

Using Eqs (5.153) and (5.154) in Eq. (5.142), we get the following expression:

$$E_{0z} = QS\sin\left(k_x x\right)\sin\left(k_y y\right) = E_0\sin\left(k_x x\right)\sin\left(k_y y\right) \qquad (5.155)$$

where $\quad E_0 = QS$ $\qquad$ (5.156)

Now substituting the boundary condition that at $x = a$, $E_{0z} = 0$, in Eq. (5.155), we get the following relation:

$$E_{0z} = E_0\sin\left(k_x a\right)\sin\left(k_y y\right) = 0$$

Since $\quad E_0\sin\left(k_y y\right) \neq 0, \ \sin\left(k_x a\right) = 0 = \sin\left(m\pi\right)$ where $m = 0, 1, 2, 3, \ldots$

or $\quad k_x = \dfrac{m\pi}{a}$ $\qquad$ (5.157)

Similarly, substituting the boundary condition that at $y = b$, $E_{0z} = 0$, in Eq. (5.155), we get the following relation:

$$E_{0z} = E_0\sin\left(k_x x\right)\sin\left(k_y b\right) = 0$$

Since $\quad E_0\sin\left(k_x x\right) \neq 0, \ \sin\left(k_y b\right) = 0 = \sin\left(n\pi\right)$

or $\quad k_y = \dfrac{n\pi}{b}$ $\qquad$ (5.158)

First substituting Eqs (5.157) and (5.158) in Eq. (5.155) and then substituting the resultant expression of E_{0z} into Eqs (5.143)–(5.146), the following expressions are obtained:

$$E_{0x} = -\frac{\gamma}{k_c^2} E_0 \frac{m\pi}{a}\cos\left(\frac{m\pi x}{a}\right)\sin\left(\frac{n\pi y}{b}\right) \qquad (5.159)$$

$$E_{0y} = -\frac{\gamma}{k_c^2} E_0 \frac{n\pi}{b}\sin\left(\frac{m\pi x}{a}\right)\cos\left(\frac{n\pi y}{b}\right) \qquad (5.160)$$

$$H_{0x} = \frac{j\omega\varepsilon_c}{k_c^2} E_0 \frac{n\pi}{b}\sin\left(\frac{m\pi x}{a}\right)\cos\left(\frac{n\pi y}{b}\right) \qquad (5.161)$$

$$H_{0y} = -\frac{j\omega\varepsilon_c}{k_c^2} E_0 \frac{m\pi}{a}\cos\left(\frac{m\pi x}{a}\right)\sin\left(\frac{n\pi y}{b}\right) \qquad (5.162)$$

Therefore, the complete field components are as follows:

$$E_x = E_{0x}e^{j\omega t - \gamma z} = -\frac{\gamma}{k_c^2}E_0\frac{m\pi}{a}\cos\left(\frac{m\pi x}{a}\right)\sin\left(\frac{n\pi y}{b}\right)e^{j\omega t - \gamma z} \qquad (5.163)$$

$$E_y = E_{0y}e^{j\omega t - \gamma z} = -\frac{\gamma}{k_c^2}E_0\frac{n\pi}{b}\sin\left(\frac{m\pi x}{a}\right)\cos\left(\frac{n\pi y}{b}\right)e^{j\omega t - \gamma z} \qquad (5.164)$$

$$E_z = E_{0z}e^{j\omega t - \gamma z} = E_0\sin\left(\frac{m\pi x}{a}\right)\sin\left(\frac{n\pi y}{b}\right)e^{j\omega t - \gamma z} \qquad (5.165)$$

$$H_x = H_{0x}e^{j\omega t - \gamma z} = \frac{j\omega\varepsilon_c}{k_c^2}E_0\frac{n\pi}{b}\sin\left(\frac{m\pi x}{a}\right)\cos\left(\frac{n\pi y}{b}\right)e^{j\omega t - \gamma z} \qquad (5.166)$$

$$H_y = H_{0y}e^{j\omega t - \gamma z} = -\frac{j\omega\varepsilon_c}{k_c^2}E_0\frac{m\pi}{a}\cos\left(\frac{m\pi x}{a}\right)\sin\left(\frac{n\pi y}{b}\right)e^{j\omega t - \gamma z} \qquad (5.167)$$

$$H_z = H_{0z}e^{j\omega t - \gamma z} = 0 \qquad (5.168)$$

Note TM modes in rectangular waveguides are also orthogonal.

Three-dimensional aperture field distributions for different TM modes are shown in Figs 5.10 and 5.11.

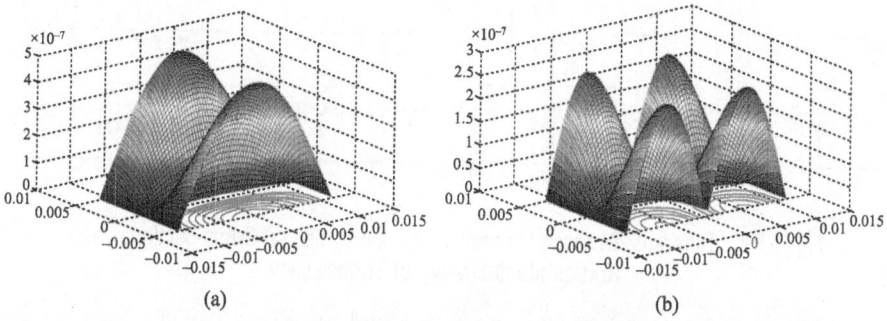

(a)   (b)

Fig. 5.10 Magnitude of Y-component of electric field distribution at the aperture for different modes (a) TM_{11} (b) TM_{21}

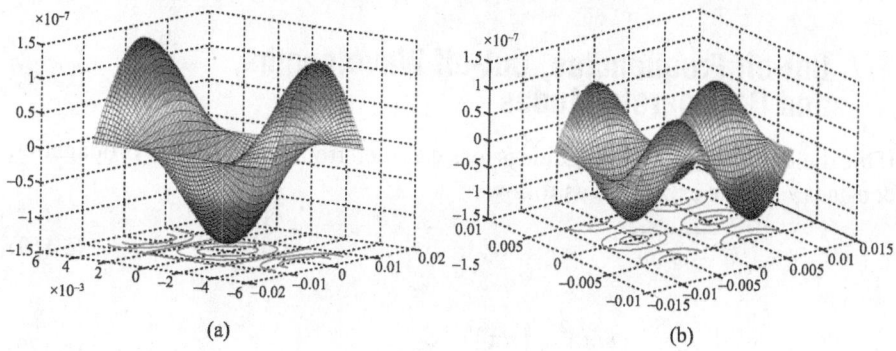

(a)   (b)

Fig. 5.11 Y-component of electric field distribution at the aperture with phase for different modes (a) TM_{11} (b) TM_{21}

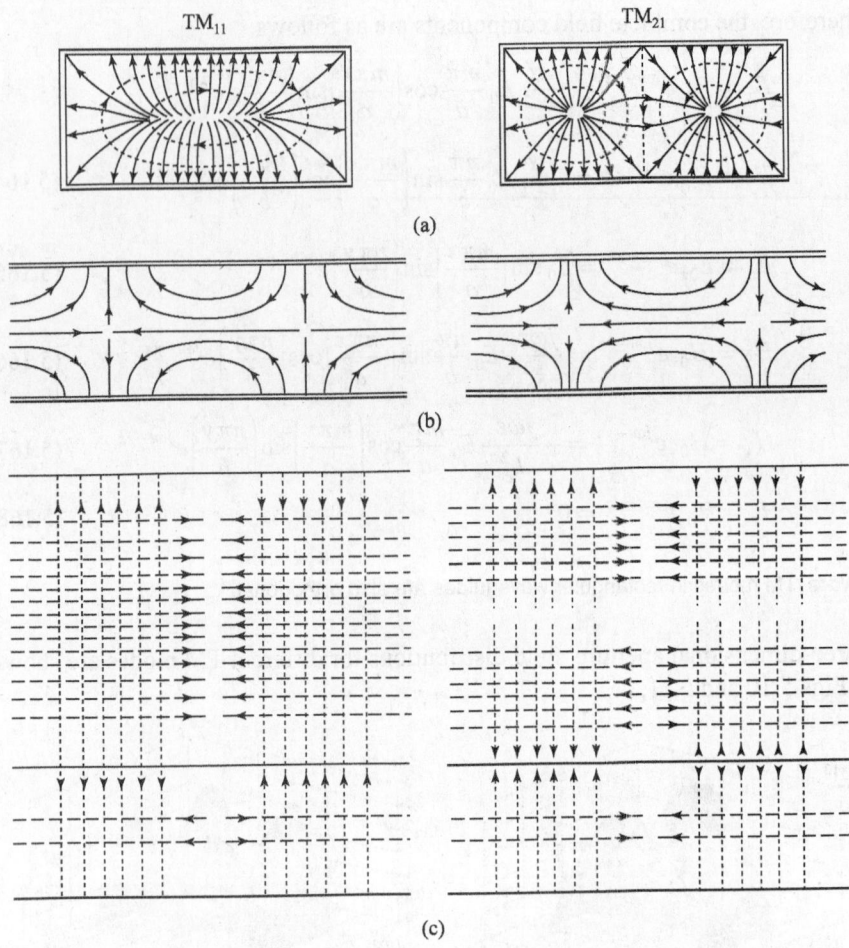

Fig. 5.12 Different TM modes in rectangular waveguide (a) Cross-sectional view (b) Longitudinal view (c) Surface view

Contour plots of the TM_{10} and TM_{11} modes at different surfaces are shown in Fig. 5.12.

5.7 Cut-off Frequencies, Cut-off Wavelengths, and Degenerate Modes

If the dielectric within a waveguide is lossless, then from Eq. (5.91) the propagation constant γ can be expressed as follows:

$$\gamma = \sqrt{k_c^2 - \omega^2 \mu \varepsilon} \tag{5.169}$$

Now, $\qquad k_c^2 = k_x^2 + k_y^2 = \left(\dfrac{m\pi}{a}\right)^2 + \left(\dfrac{n\pi}{b}\right)^2 \tag{5.170}$

Substituting the expression of k_c^2 from Eq. (5.170) into Eq. (5.169), we get the following relation:

$$\gamma = \sqrt{\left(\frac{m\pi}{a}\right)^2 + \left(\frac{n\pi}{b}\right)^2 - \omega^2 \mu\varepsilon} \tag{5.171}$$

Equation (5.171) leads to the following two cases:

Case 1 $\quad \left(\frac{m\pi}{a}\right)^2 + \left(\frac{n\pi}{b}\right)^2 > \omega^2 \mu\varepsilon$

In this case, γ is real and positive; hence, the wave will not propagate and will be attenuated.

Case 2 $\quad \left(\frac{m\pi}{a}\right)^2 + \left(\frac{n\pi}{b}\right)^2 < \omega^2 \mu\varepsilon$

In this case, γ is purely imaginary and hence the wave will propagate.

These two cases imply that there is a cut-off frequency above which a wave will propagate. This reveals the high-pass-filter nature of a waveguide. The expression for the cut-off frequency can be obtained from the boundary of these two cases. At the cut-off frequency, the following relation holds:

$$\omega_c^2 \mu\varepsilon = \left(\frac{m\pi}{a}\right)^2 + \left(\frac{n\pi}{b}\right)^2 \quad \text{or} \quad \omega_c = 2\pi f_c = \frac{1}{\sqrt{\mu\varepsilon}} \left[\left(\frac{m\pi}{a}\right)^2 + \left(\frac{n\pi}{b}\right)^2\right]^{\frac{1}{2}}$$

or $\qquad f_c = \frac{c}{2}\left[\left(\frac{m}{a}\right)^2 + \left(\frac{n}{b}\right)^2\right]^{\frac{1}{2}}$ $\hfill$ (5.172)

where $c = 1/\sqrt{\mu\varepsilon}$ is the velocity of a wave in the free space.

The corresponding cut-off wavelength is expresses as follows:

$$\lambda_c = \frac{c}{f_c} = \frac{c}{\frac{c}{2}\left[\left(\frac{m}{a}\right)^2 + \left(\frac{n}{b}\right)^2\right]^{\frac{1}{2}}} = \frac{2}{\left[\left(\frac{m}{a}\right)^2 + \left(\frac{n}{b}\right)^2\right]^{\frac{1}{2}}} = \frac{2ab}{\sqrt{m^2 b^2 + n^2 a^2}} \tag{5.173}$$

The wavelengths greater than λ_c are attenuated during propagation through the waveguide, whereas those less than λ_c are allowed to propagate through it.

The TE and TM modes that have the same cut-off frequency or wavelength are called degenerate modes. Since the cut-off frequencies of a pair of TE_{mn} and TM_{mn} modes (e.g., TE_{11} and TM_{11}) are identical, these modes are always degenerate.

Note Since for TM_{mn} modes neither m nor n can be zero, as in such cases the entire TM field components will vanish, there are no degenerate TM_{m0} and TM_{0n} modes corresponding to the TE_{m0} and TE_{0n} modes.

Degenerate modes have found wide applications in multimode circuits like dual band filters. To obtain a degenerate mode of the TE_{10} mode, we often take a square waveguide for which the width and height are equal. In such case, the TE_{10} and TE_{01} modes have the same cut-off frequency and hence these become degenerate modes. In a square waveguide, the TE_{mn}, TE_{nm}, TM_{mn}, and TM_{nm} modes form a foursome of degeneracy.

> **Note** Waveguides are used in a shielded room to restrict the passing of electromagnetic waves through the ventilators of the room. Waveguides having cut-off frequencies above the maximum operating frequency are arranged in a honey-comb fashion. Since the frequencies of the interfering electromagnetic waves are below the cut-off frequency of the guide, they cannot propagate through it while air can easily pass.
>
> Another important application of a waveguide that takes advantage of the cut-off frequency is found in the front door of a microwave oven, where a number of thin waveguides are arranged in a net pattern. The frequency of the microwave radiation in the oven cavity is much lower than the cut-off frequencies of these waveguides and hence the radiation cannot pass through the door and come outside. On the other hand, light waves, having much higher frequencies than the cut-off frequencies of these waveguides, can pass through it. Thus, the user can easily see the condition of the food inside the oven and simultaneously stays protected from the high-power microwave radiation.

EXAMPLE 5.3   A rectangular waveguide has width $a = 22.86$ mm and height $b = 10.16$ mm.

(a) Calculate the cut-off frequency and cut-off wavelength of the first four modes.
(b) If the waveguide is operated at a frequency of 20 GHz, then identify the propagating modes. In addition, determine the degenerate modes.

Solution   Given: $a = 22.86$ mm, $b = 10.16$ mm, and $f = 20$ GHz

(a) The first four propagating modes are TE_{10}, TE_{20}, TE_{01}, and TE_{11}/TM_{11}.
The cut-off frequency for TE_{10} mode is as follows:

$$f_{c,TE_{10}} = \frac{c}{2}\left[\left(\frac{m}{a}\right)^2 + \left(\frac{n}{b}\right)^2\right]^{\frac{1}{2}} = \frac{3\times10^8}{2}\left[\left(\frac{1}{22.86\times10^{-3}}\right)^2 + \left(\frac{0}{10.16\times10^{-3}}\right)^2\right]^{\frac{1}{2}}$$

or $f_{c,TE_{10}} = 6.562$ GHz

The corresponding cut-off wavelength is as follows:

$$\lambda_{c,TE_{10}} = \frac{2ab}{\sqrt{m^2b^2 + n^2a^2}} = \frac{2\times22.86\times10.16}{\sqrt{10.16^2}} = 45.72 \text{ mm}$$

The cut-off frequency for the TE_{20} mode is given by the following relation:

$$f_{c,TE_{20}} = \frac{c}{2}\left[\left(\frac{m}{a}\right)^2 + \left(\frac{n}{b}\right)^2\right]^{\frac{1}{2}} = \frac{3\times10^8}{2}\left[\left(\frac{2}{22.86\times10^{-3}}\right)^2 + \left(\frac{0}{10.16\times10^{-3}}\right)^2\right]^{\frac{1}{2}}$$

or $f_{c,TE_{20}} = 13.123$ GHz

The corresponding cut-off wavelength is as follows:

$$\lambda_{c,\mathrm{TE}_{20}} = \frac{2ab}{\sqrt{m^2b^2 + n^2a^2}} = \frac{2 \times 22.86 \times 10.16}{\sqrt{4 \times 10.16^2}} = 22.86 \text{ mm}$$

The cut-off frequency for the TE_{01} mode is given by the following equation:

$$f_{c,\mathrm{TE}_{01}} = \frac{c}{2}\left[\left(\frac{m}{a}\right)^2 + \left(\frac{n}{b}\right)^2\right]^{\frac{1}{2}} = \frac{3 \times 10^8}{2}\left[\left(\frac{0}{22.86 \times 10^{-3}}\right)^2 + \left(\frac{1}{10.16 \times 10^{-3}}\right)^2\right]^{\frac{1}{2}}$$

or $\quad f_{c,\mathrm{TE}_{01}} = 14.764 \text{ GHz}$

The corresponding cut-off wavelength is as follows:

$$\lambda_{c,\mathrm{TE}_{01}} = \frac{2ab}{\sqrt{m^2b^2 + n^2a^2}} = \frac{2 \times 22.86 \times 10.16}{\sqrt{22.86^2}} = 20.32 \text{ mm}$$

The cut-off frequency for the $\mathrm{TE}_{11}/\mathrm{TM}_{11}$ mode is represented by the following equation:

$$f_{c,\mathrm{TE}_{11}/\mathrm{TM}_{11}} = \frac{c}{2}\left[\left(\frac{m}{a}\right)^2 + \left(\frac{n}{b}\right)^2\right]^{\frac{1}{2}} = \frac{3 \times 10^8}{2}\left[\left(\frac{1}{22.86 \times 10^{-3}}\right)^2 + \left(\frac{1}{10.16 \times 10^{-3}}\right)^2\right]^{\frac{1}{2}}$$

or $\quad f_{c,\mathrm{TE}_{11}/\mathrm{TM}_{11}} = 16.156 \text{ GHz}$

The corresponding cut-off wavelength is

$$\lambda_{c,\mathrm{TE}_{11}/\mathrm{TM}_{11}} = \frac{2ab}{\sqrt{m^2b^2 + n^2a^2}} = \frac{2 \times 22.86 \times 10.16}{\sqrt{10.16^2 + 22.86^2}} = 18.569 \text{ mm}$$

(b) Other modes that have cut-off frequencies below 20 GHz are TE_{30} and $\mathrm{TE}_{21}/\mathrm{TM}_{21}$. The corresponding cut-off frequencies are as follows:

$$f_{c,\mathrm{TE}_{30}} = \frac{c}{2}\left[\left(\frac{m}{a}\right)^2 + \left(\frac{n}{b}\right)^2\right]^{\frac{1}{2}} = \frac{3 \times 10^8}{2}\left[\left(\frac{3}{22.86 \times 10^{-3}}\right)^2 + \left(\frac{0}{10.16 \times 10^{-3}}\right)^2\right]^{\frac{1}{2}}$$

or $\quad f_{c,\mathrm{TE}_{30}} = 19.685 \text{ GHz}$

$$f_{c,\mathrm{TE}_{21}/\mathrm{TM}_{21}} = \frac{c}{2}\left[\left(\frac{m}{a}\right)^2 + \left(\frac{n}{b}\right)^2\right]^{\frac{1}{2}} = \frac{3 \times 10^8}{2}\left[\left(\frac{2}{22.86 \times 10^{-3}}\right)^2 + \left(\frac{1}{10.16 \times 10^{-3}}\right)^2\right]^{\frac{1}{2}}$$

or $\quad f_{c,\mathrm{TE}_{21}/\mathrm{TM}_{21}} = 19.753 \text{ GHz}$

Therefore, the modes TE_{10}, TE_{20}, TE_{01}, $\mathrm{TE}_{11}/\mathrm{TM}_{11}$, TE_{30}, and $\mathrm{TE}_{21}/\mathrm{TM}_{21}$ will exist when the waveguide is operated at 20 GHz.
The degenerate modes are $(\mathrm{TE}_{11}, \mathrm{TM}_{11})$ and $(\mathrm{TE}_{21}, \mathrm{TM}_{21})$.

Practice Problem

5.5 A rectangular waveguide has width $a = 12.95$ mm and height $b = 6.48$ mm.
(a) Calculate the cut-off frequency and cut-off wavelength of the first four modes.
(b) If the waveguide is operated at a frequency of 30 GHz, then identify the propagating modes. In addition, determine the degenerate modes.

[11.583 GHz, 23166 GHz, 23.148 GHz, 25.884 GHz, 25.90 mm, 12.95 mm, 12.96 mm, 11.59 mm, (TE_{10}, TE_{20}, TE_{01}, $\mathrm{TE}_{11}/\mathrm{TM}_{11}$), ($\mathrm{TE}_{11}/\mathrm{TM}_{11}$)]

5.8 Dominant Mode in Rectangular Waveguides

The *dominant mode* corresponds to a particular combination of m and n, as shown in Eq. (5.172) or (5.173), which results in the minimum cut-off frequency or the maximum cut-off wavelength.

5.8.1 TE Mode

In this section, the dominant TE mode has been selected after comparing various TE modes and their cut-off frequencies.

TE$_{00}$ mode For $m = 0$ and $n = 0$, all the transverse field components in Eqs (5.119), (5.120), (5.122), and (5.123) vanish. Therefore, the TE$_{00}$ mode cannot exist inside a waveguide.

TE$_{01}$ mode For $m = 0$ and $n = 1$, E_y and H_x vanish, but E_x and H_y exist. Therefore, the TE$_{01}$ mode exists. From Eq. (5.173), the cut-off wavelength for the TE$_{01}$ mode is given by the following relation:

$$\lambda_{c,01} = \frac{2ab}{\sqrt{a^2}} = 2b \tag{5.174}$$

TE$_{10}$ mode For $m = 1$ and $n = 0$, E_x and H_y vanish, but E_y and H_x exist. Therefore, the TE$_{10}$ mode also exists. From Eq. (5.173), the cut-off wavelength for the TE$_{10}$ mode is given by the following relation:

$$\lambda_{c,10} = \frac{2ab}{\sqrt{b^2}} = 2a \tag{5.175}$$

TE$_{11}$ mode For $m = 1$ and $n = 1$, all the field components exist and therefore the TE$_{11}$ mode exists. Based on Eq. (5.173), the cut-off wavelength for the TE$_{11}$ mode is given by the following relation:

$$\lambda_{c,11} = \frac{2ab}{\sqrt{a^2 + b^2}} \tag{5.176}$$

Since in a waveguide $a > b$, the cut-off wavelength corresponding to $m = 1$ and $n = 0$ has the maximum value. Therefore, the dominant TE mode is TE$_{10}$.

5.8.2 TM Mode

In this section, the dominant TM mode has been selected after comparing various TM modes and their cut-off frequencies.

TM$_{00}$ mode For $m = 0$ and $n = 0$, all the field components in Eqs (5.163)–(5.168) vanish. Therefore, the TM$_{00}$ mode cannot exist inside a waveguide.

TM$_{01}$ mode For $m = 0$ and $n = 1$, all the field components in Eqs (5.163)–(5.168) vanish. Therefore, the TM$_{01}$ mode cannot exist inside a waveguide.

TM$_{10}$ mode For $m = 1$ and $n = 0$, all the field components in Eqs (5.163)–(5.168) vanish. Therefore, the TM$_{10}$ mode cannot exist inside a waveguide.

TM$_{11}$ mode   For $m = 1$ and $n = 1$, all the field components exist and therefore the TM$_{11}$ mode exists. From Eq. (5.173), the cut-off wavelength for the TM$_{11}$ mode is given by the following relation:

$$\lambda_{c,11} = \frac{2ab}{\sqrt{a^2 + b^2}} \tag{5.177}$$

For the TM modes, the cut-off wavelength corresponding to $m = 1$ and $n = 1$ has the maximum cut-off wavelength. Thus, the dominant TM mode is TM$_{11}$.

> **Note** Waveguides are operated mainly in the dominant mode because of the easiness in coupling and extracting power into and from it. The dominant mode also has the simplest field configuration.

5.9 Physical Explanation for Wave Propagation in Rectangular Waveguides

Equations (5.119)–(5.124) and (5.163)–(5.168) show that the expressions for the TE- and TM-mode field components involve sine or cosine function of the terms $(m\pi x/a)$ and $(n\pi y/b)$. Since sine and cosine functions can be represented as a linear combination of the terms $e^{j\theta}$ and $e^{-j\theta}$, the waves, at a point within a waveguide, can also be resolved into two components propagating along the $+\theta$ and $-\theta$ directions with respect to the waveguide axis, as shown in Fig. 5.13.

To demonstrate, this let us consider the propagation of the dominant TE$_{10}$ mode inside a waveguide. For this mode in a lossless case, using Eq. (5.120), we can write the following expression:

$$E_y = -\frac{j\omega\mu}{k_c^2} H_0 \frac{\pi}{a} \sin\left(\frac{\pi x}{a}\right) e^{j(\omega t - \beta z)} \tag{5.178}$$

Using Eq. (5.170), for $m = 1$ and $n = 0$, we can write the following relation:

$$k_c^2 = k_x^2 + k_y^2 = (\pi/a)^2 \tag{5.179}$$

Substituting Eq. (5.179) into Eq. (5.178), we get the following equation:

$$E_y = -\frac{j\omega\mu}{(\pi/a)^2} H_0 \frac{\pi}{a} \sin\left(\frac{\pi x}{a}\right) e^{j(\omega t - \beta z)}$$

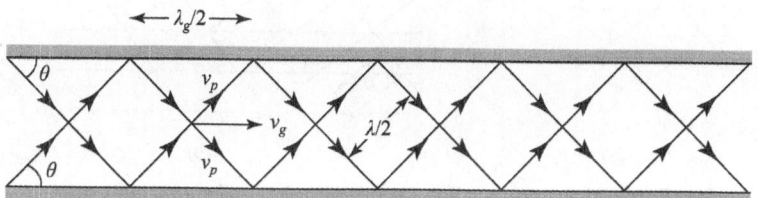

Fig. 5.13   Propagation of waves inside rectangular waveguide

or $\qquad E_y = -\dfrac{j\omega\mu a}{\pi} H_0 \left(\dfrac{e^{\,j\frac{\pi x}{a} - j\beta z} - e^{-j\frac{\pi x}{a} - j\beta z}}{2j} \right) e^{j\omega t}$

or $\qquad E_y = \dfrac{\omega\mu a}{2\pi} H_0 \left\{ e^{-j\beta\left(z + \frac{\pi x}{\beta a} \right)} - e^{-j\beta\left(z - \frac{\pi x}{\beta a} \right)} \right\} e^{j\omega t}$ $\qquad\qquad$ (5.180)

Equation (5.180) represents two TEM waves travelling along the positive Z-axis at an angle θ such that

$$\theta = \pm \tan^{-1}\left(\frac{\pi}{\beta a} \right) \qquad\qquad (5.181)$$

Therefore, the TE_{10} mode can be considered as a combination of two TEM modes propagating along a zigzag path between the guide walls $x = 0$ and $x = a$.

The concepts of wave propagation in a rectangular wave, described earlier, can also be established by the following explanations.

Let us consider the propagation of a plane wave in free space. The wavefront of such a wave is shown in Fig. 5.14(a). If a second wavefront, propagating in a different direction, is present and overlaps with the first, then a combined wavefront results, as shown in Fig. 5.14(b). Figure 5.14(b) reveals that the two wavefronts add on the reference axis, and alternate cancellation (intersection point of a dashed- and a solid-lined wavefront) and addition (intersection point of two dashed- or two solid-lined wavefronts) of the two wavefronts occur at progressive half-wavelength apart from it. That is the wavefront addition takes places along the axes A, C, F, and H, whereas wavefront cancellation occurs along the axes B, D, E, and G. Now, if we place two metallic plates along the cancellation lines D and E or B and G, the boundary condition (i.e., the tangential component of the electric field on a metal–dielectric boundary must be zero) will be satisfied. Satisfaction of the first boundary condition automatically satisfies the other boundary conditions.

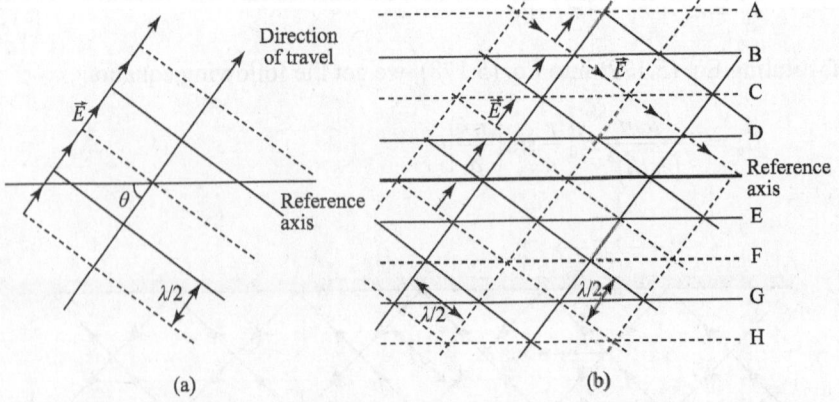

Fig. 5.14 Wave propagation (a) Wavefront of a plane wave in space
(b) Combined wavefront of two plane waves propagating at different directions

$$f > f_1 > f_2 > f_3 > f_4 > f_5 > f_6 > f_7 > f_8 > f_9 > f_c$$

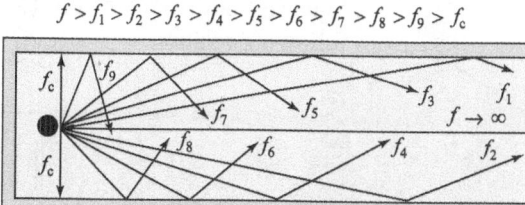

Fig. 5.15 Angle of incidence for different frequencies of excitation

Placing of the metal plates thus ensures that the wavefront propagates along the reference axis, which is basically along the length of the guide, with the individual wavefront moving in a zigzag fashion after complete reflection from the metallic walls at the top and bottom (or from side walls).

Equation (5.181) reveals that the angle of incidence depends on β, which in turn depends on frequency. The value of β decreases with a decrease in frequency, and for a particular frequency it will become zero. At this value of β, the angle of incidence becomes 90° and the wavefront bounces back and forth across the guide walls. Therefore no energy propagates down the waveguide. The frequency for which β becomes zero represents the cut-off frequency. At a very high frequency, β tends to infinity and the angle of incidence tends to zero. This has been illustrated in Fig. 5.15.

5.10 Guided Wavelengths

A guided wavelength is defined as the distance travelled by a wave in order to undergo a phase shift of 2π radians. In a guided medium it is less than the cut-off wavelength. Mathematically, a guided wavelength is expressed as follows:

$$\lambda_g = \frac{2\pi}{\beta} \tag{5.182}$$

Now from Eq. (5.171),

$$\gamma = j\beta = \sqrt{\left(\frac{m\pi}{a}\right)^2 + \left(\frac{n\pi}{b}\right)^2 - \omega^2 \mu\varepsilon} = \sqrt{\omega_c^2 \mu\varepsilon - \omega^2 \mu\varepsilon}$$

or $$\beta = \sqrt{\omega^2 \mu\varepsilon - \omega_c^2 \mu\varepsilon} = \omega\sqrt{\mu\varepsilon}\sqrt{1 - (\omega_c/\omega)^2} \tag{5.183}$$

or $$\beta = \omega\sqrt{\mu\varepsilon}\sqrt{1 - (\lambda/\lambda_c)^2} \tag{5.184}$$

Substituting Eq. (5.184) into Eq. (5.182), we get the following expression:

$$\lambda_g = \frac{2\pi}{\omega\sqrt{\mu\varepsilon}\sqrt{1 - (\lambda/\lambda_c)^2}} = \frac{2\pi c}{2\pi f \sqrt{1 - (\lambda/\lambda_c)^2}}$$

or $$\lambda_g = \frac{2\pi c\lambda}{2\pi c\sqrt{1 - (\lambda/\lambda_c)^2}} = \frac{\lambda}{\sqrt{1 - (\lambda/\lambda_c)^2}} \tag{5.185}$$

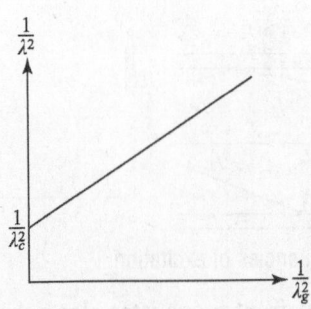

Fig. 5.16 Plot of $1/\lambda_g^2$ vs $1/\lambda^2$

Squaring both sides of Eq. (5.185), we get the following relation:

$$\lambda_g^2 = \frac{\lambda^2}{\left[1-\left(\dfrac{\lambda}{\lambda_c}\right)^2\right]} = \frac{1}{\left[\left(\dfrac{1}{\lambda}\right)^2 - \left(\dfrac{1}{\lambda_c}\right)^2\right]}$$

or $\quad \dfrac{1}{\lambda_g^2} = \dfrac{1}{\lambda^2} - \dfrac{1}{\lambda_c^2}$ \hfill (5.186)

A plot of $\dfrac{1}{\lambda_g^2}$ versus $\dfrac{1}{\lambda^2}$, shown in Fig. 5.16, is often used in experimental determination of λ_c.

EXAMPLE 5.4 If the rectangular waveguide described in Example 5.3 is operated at 20 GHz, then calculate the guided wavelength of the dominant mode.

Solution Given: $a = 22.86$ mm, $b = 10.16$ mm, $f = 20$ GHz

$$\lambda_{c,\text{TE}_{10}} = 45.72 \text{ mm}; \quad \lambda = \frac{c}{f} = \frac{3 \times 10^{11}}{20 \times 10^9} = 15 \text{ mm}$$

$$\lambda_g = \frac{\lambda}{\sqrt{1-(\lambda/\lambda_c)^2}} = \frac{15}{\sqrt{1-(15/45.72)^2}} = 15.878 \text{ mm}$$

Practice Problem

5.6 If the rectangular waveguide described in Practice Problem 5.5 is operated at 30 GHz, then calculate the guided wavelength of the dominant mode. **[10.841 mm]**

5.11 Characteristic Impedances of TE and TM Modes

The TE- and/or TM-mode wave impedances are defined as follows:

$$\eta = E_x/H_y = -E_y/H_x \hfill (5.187)$$

TE mode For the TE mode propagating in a lossless waveguide,

$$\eta_{\text{TE}} = \frac{E_x}{H_y} = \frac{\dfrac{j\omega\mu}{k_c^2} H_0 \left(\dfrac{n\pi}{b}\right)\cos\left(\dfrac{m\pi x}{a}\right)\sin\left(\dfrac{n\pi y}{b}\right)e^{j\omega t - \gamma z}}{\dfrac{j\beta}{k_c^2} H_0 \left(\dfrac{n\pi}{b}\right)\cos\left(\dfrac{m\pi x}{a}\right)\sin\left(\dfrac{n\pi y}{b}\right)e^{j\omega t - \gamma z}} = \frac{\omega\mu}{\beta} \hfill (5.188)$$

Substituting the expression of β from Eq. (5.184) into Eq. (5.188), we get the following equation:

$$\eta_{\text{TE}} = \frac{\omega\mu}{\omega\sqrt{\mu\varepsilon}\sqrt{1-(\lambda/\lambda_c)^2}} = \sqrt{\frac{\mu}{\varepsilon}}\frac{1}{\sqrt{1-(\lambda/\lambda_c)^2}} = \frac{\eta}{\sqrt{1-(\lambda/\lambda_c)^2}} \hfill (5.189)$$

where $\eta = \sqrt{\mu/\varepsilon}$ is the intrinsic wave impedance in a medium characterized by permeability μ and permittivity ε. Similarly, it can be shown that if we start with the expression $\eta = -E_y/H_x$, we will again arrive at Eq. (5.189).

TM mode For the TM mode propagating in a lossless waveguide,

$$\eta_{TM} = \frac{E_x}{H_y} = \frac{-\dfrac{j\beta}{k_c^2} E_0 \left(\dfrac{m\pi}{a}\right)\cos\left(\dfrac{m\pi x}{a}\right)\sin\left(\dfrac{n\pi y}{b}\right)e^{j\omega t - \gamma z}}{-\dfrac{j\omega\varepsilon}{k_c^2} E_0 \left(\dfrac{m\pi}{a}\right)\cos\left(\dfrac{m\pi x}{a}\right)\sin\left(\dfrac{n\pi y}{b}\right)e^{j\omega t - \gamma z}} = \frac{\beta}{\omega\varepsilon} \qquad (5.190)$$

Substituting the expression of β from Eq. (5.184) into Eq. (5.190), we get the following relation:

$$\eta_{TM} = \frac{\omega\sqrt{\mu\varepsilon}\sqrt{1-(\lambda/\lambda_c)^2}}{\omega\varepsilon} = \sqrt{\frac{\mu}{\varepsilon}}\sqrt{1-(\lambda/\lambda_c)^2} = \eta\sqrt{1-(\lambda/\lambda_c)^2} \qquad (5.191)$$

Similarly, it can be shown that if we start with the expression $\eta = -E_y/H_x$, we will again arrive at Eq. (5.191).

Multiplying Eq. (5.189) by Eq. (5.191), we get the following relation:

$$\eta_{TE}\eta_{TM} = \eta^2 \qquad (5.192)$$

A plot of η_{TE} and η_{TM} as a function of frequency is shown in Fig. 5.17.

EXAMPLE 5.5 Calculate the wave impedances of the dominant TE and TM modes at 20 GHz for the rectangular waveguide described in Example 5.3.

Solution Given: $a = 22.86$ mm, $b = 10.16$ mm, and $f = 20$ GHz

The dominant TE and TM modes are TE_{10} and TM_{11}, respectively and the corresponding cut-off wavelengths 45.72 and 18.569 mm, respectively.

The wavelength at 20 GHz is as follows:

$$\lambda = \frac{c}{f} = \frac{3\times10^{11}}{20\times10^9} = 15 \text{ mm}$$

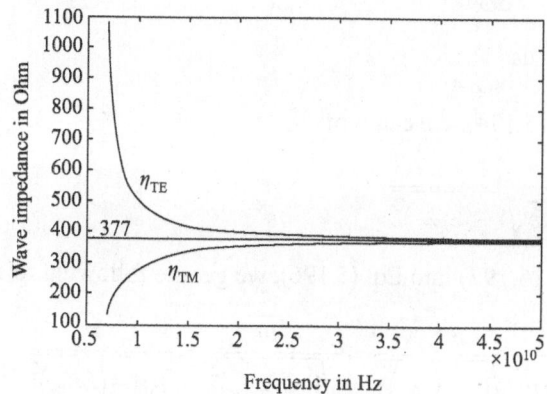

Fig. 5.17 Plot of η_{TE} and η_{TM} as function of frequency

The dominant TE mode wave impedance is, therefore, given by the following equation:

$$\eta_{TE_{10}} = \frac{\eta}{\sqrt{1-(\lambda/\lambda_c)^2}} = \frac{376.73}{\sqrt{1-(15/45.72)^2}} = \frac{376.73}{0.945} = 398.804\ \Omega$$

The dominant TM mode wave impedance is as follows:

$$\eta_{TM_{11}} = \eta\sqrt{1-(\lambda/\lambda_c)^2} = 376.73\sqrt{1-(15/45.72)^2} = 355.877\ \Omega$$

Practice Problem

5.7 Calculate the wave impedances of the dominant TE and TM modes at 30 GHz for the rectangular waveguide described in Practice Problem 5.5.   [408.398 Ω, 347.517 Ω]

5.12 Phase Velocity and Group Velocity

The phase velocity of a wave inside a waveguide is expressed as follows:

$$v_p = \frac{\omega}{\beta} = \lambda_g f = \frac{\lambda_g c}{\lambda} \tag{5.193}$$

Substituting the expression of λ_g from Eq. (5.185) into Eq. (5.193), we get the following relation:

$$v_p = \frac{\omega}{\beta} = \lambda_g f = \frac{\lambda}{\sqrt{1-(\lambda/\lambda_c)^2}}\frac{c}{\lambda} = \frac{c}{\sqrt{1-(\lambda/\lambda_c)^2}} \tag{5.194}$$

Since $\lambda_c > \lambda$, Eq. (5.194) implies that $v_p > c$.

On the other hand, group velocity is defined as follows:

$$v_g = \frac{d\omega}{d\beta} \tag{5.195}$$

Squaring and then differentiating Eq. (5.183) with respect to ω, we get the following equation:

$$\frac{d}{d\omega}\left(\beta^2\right) = \frac{d}{d\omega}\left(\omega^2\mu\varepsilon - \omega_c^2\mu\varepsilon\right)$$

or $$\frac{d\beta}{d\omega} = \frac{\omega}{\beta}\mu\varepsilon \tag{5.196}$$

Again from Eq. (5.194), we can write that

$$\frac{\omega}{\beta} = \frac{1}{\sqrt{\mu\varepsilon}\sqrt{1-(\lambda/\lambda_c)^2}} \tag{5.197}$$

Substituting Eq. (5.197) into Eq. (5.196), we get the following relation:

$$\frac{d\beta}{d\omega} = \frac{\mu\varepsilon}{\sqrt{\mu\varepsilon}\sqrt{1-(\lambda/\lambda_c)^2}} = \frac{\sqrt{\mu\varepsilon}}{\sqrt{1-(\lambda/\lambda_c)^2}} = \frac{1}{c\sqrt{1-(\lambda/\lambda_c)^2}}$$

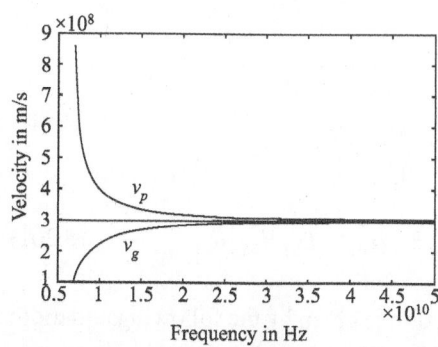

Fig. 5.18 Plot of v_p and v_g as function of frequency

or $\quad v_g = \dfrac{d\omega}{d\beta} = c\sqrt{1-(\lambda/\lambda_c)^2}$ (5.198)

Multiplying Eq. (5.194) by Eq. (5.198), we get the relation

$$v_p v_g = c^2 \qquad (5.199)$$

A plot of v_p and v_g as a function of frequency is shown in Fig. 5.18.

EXAMPLE 5.6 Calculate the phase velocity and group velocity of the dominant TE mode at 20 GHz for the rectangular waveguide described in Example 5.3.

Solution Given: $a = 22.86\,\text{mm}$, $b = 10.16\,\text{mm}$, and $f = 20\,\text{GHz}$, $\lambda_{c,\text{TE}_{10}} = 45.72$ mm, and $\lambda = 15$ mm

The phase velocity is as follows:

$$v_p = \frac{c}{\sqrt{1-\left(\dfrac{\lambda}{\lambda_c}\right)^2}} = \frac{3\times10^8}{\sqrt{1-\left(\dfrac{15}{45.72}\right)^2}} = 3.175\times10^8 \text{ m/s}$$

The group velocity is as follows:

$$v_g = \frac{d\omega}{d\beta} = c\sqrt{1-\left(\frac{\lambda}{\lambda_c}\right)^2} = 3\times10^8\sqrt{1-\left(\frac{15}{45.72}\right)^2} = 2.835\times10^8 \text{ m/s}$$

Practice Problem

5.8 Calculate the phase velocity and group velocity of the dominant TE mode at 30 GHz for the rectangular waveguide described in Example 5.3.

[3.254×10^8 m/s, 2.766×10^8 m/s]

5.13 Power Flow in Rectangular Waveguides

Power flow through a waveguide can be calculated by means of the complex Poynting theorem. Throughout the analysis, it will be assumed that both ends of the waveguide are matched terminated or the waveguide is infinitely extended so that there are no reflections from the ends. According to the Poynting theorem, the power transmitted through a guide is given by the following equation:

$$P_{tr} = \oint \vec{P}\cdot\overline{dS} = \frac{1}{2}\oint\left(\vec{E}\times\overline{H^*}\right)\cdot\overline{dS} \qquad (5.200)$$

TE mode

$$P_{TE} = \frac{1}{2}\left(\vec{E} \times \overrightarrow{H^*}\right) = \frac{1}{2}\begin{vmatrix} \hat{a}_x & \hat{a}_y & \hat{a}_z \\ E_{0x} & E_{0y} & 0 \\ H_{0x}^* & H_{0y}^* & H_{0z}^* \end{vmatrix}$$

or $\qquad P_{TE} = \frac{1}{2}\left[E_{0y}H_{0z}^*\hat{a}_x - E_{0x}H_{0z}^*\hat{a}_y + \left(E_{0x}H_{0y}^* - E_{0y}H_{0x}^*\right)\hat{a}_z\right]$ (5.201)

Therefore, the power flowing in the z-direction is given by the following equation:

$$\left(p_z\right)_{TE} = \frac{1}{2}\left(E_{0x}H_{0y}^* - E_{0y}H_{0x}^*\right) \tag{5.202}$$

Substituting the field components from Eqs (5.119), (5.120), (5.122), and (5.123), we get the following expression:

$$\left(p_z\right)_{TE} = \frac{1}{2}\frac{\omega\mu\beta}{k_c^4}H_0^2\left[\begin{array}{l}\left(\dfrac{n\pi}{b}\right)^2\cos^2\left(\dfrac{m\pi x}{a}\right)\sin^2\left(\dfrac{n\pi y}{b}\right) + \\[2ex] \left(\dfrac{m\pi}{a}\right)^2\sin^2\left(\dfrac{m\pi x}{a}\right)\cos^2\left(\dfrac{n\pi y}{b}\right)\end{array}\right] \tag{5.203}$$

The total power passing through the cross-section of the waveguide is as follows:

$$\left(P_z\right)_{TE} = \int_0^a\int_0^b \left(p_z\right)_{TE}dxdy \tag{5.204}$$

Substituting Eq. (5.203) into Eq. (5.204) and integrating, we get the following relation:

$$\left(P_z\right)_{TE} = \frac{1}{2}\frac{\omega\mu\beta}{k_c^4}H_0^2\left[\left(\frac{n\pi}{b}\right)^2\int_0^a\cos^2\left(\frac{m\pi x}{a}\right)dx\int_0^b\sin^2\left(\frac{n\pi y}{b}\right)dy\right.$$

$$\left. + \left(\frac{m\pi}{a}\right)^2\int_0^a\sin^2\left(\frac{m\pi x}{a}\right)dx\int_0^b\cos^2\left(\frac{n\pi y}{b}\right)dy\right]$$

or $\qquad \left(P_z\right)_{TE} = \begin{cases} \dfrac{\omega\mu\beta}{8k_c^4}H_0^2ab\left[\left(\dfrac{m\pi}{a}\right)^2 + \left(\dfrac{n\pi}{b}\right)^2\right] & \text{for } m \neq 0 \text{ and } n \neq 0 \\[3ex] \dfrac{\omega\mu\beta}{4k_c^4}H_0^2ab\left(\dfrac{n\pi}{b}\right)^2 & \text{for } m = 0 \text{ and } n \neq 0 \\[3ex] \dfrac{\omega\mu\beta}{4k_c^4}H_0^2ab\left(\dfrac{m\pi}{a}\right)^2 & \text{for } m \neq 0 \text{ and } n = 0 \end{cases}$ (5.205) where $\displaystyle\int_0^a \cos^2\left(\frac{m\pi x}{a}\right)dx \int_0^b \sin^2\left(\frac{n\pi y}{b}\right)dy = \int_0^a \sin^2\left(\frac{m\pi x}{a}\right)dx \int_0^b \cos^2\left(\frac{n\pi y}{b}\right)dy$ (5.206)

$$
= \begin{cases} \dfrac{ab}{4} & \text{for } m \neq 0 \text{ and } n \neq 0 \\[2mm] \dfrac{ab}{2} & \text{for } m = 0 \text{ and } n \neq 0 \\[2mm] \dfrac{ab}{2} & \text{for } m \neq 0 \text{ and } n = 0 \end{cases}
$$

TM mode

$$
P_{\text{TM}} = \frac{1}{2}\left(\vec{E} \times \overrightarrow{H^*}\right) = \frac{1}{2}\begin{vmatrix} \hat{a}_x & \hat{a}_y & \hat{a}_z \\ E_{0x} & E_{0y} & E_{0z} \\ H_{0x}^* & H_{0y}^* & 0 \end{vmatrix}
$$

or $\qquad P_{\text{TM}} = \dfrac{1}{2}\left[-E_{0z}H_{0y}^*\hat{a}_x + E_{0z}H_{0x}^*\hat{a}_y + \left(E_{0x}H_{0y}^* - E_{0y}H_{0x}^*\right)\hat{a}_z\right]$ (5.207)

Therefore, the power flowing in the z-direction is given by the following equation:

$$
\left(p_z\right)_{\text{TM}} = \frac{1}{2}\left(E_{0x}H_{0y}^* - E_{0y}H_{0x}^*\right)
$$
(5.208)

Substituting the field components from Eqs (5.159)–(5.162), we get the following expression:

$$
\left(p_z\right)_{\text{TM}} = \frac{1}{2}\frac{\omega\varepsilon\beta}{k_c^4}E_0^2 \left[\begin{aligned} &\left(\frac{m\pi}{a}\right)^2 \cos^2\left(\frac{m\pi x}{a}\right)\sin^2\left(\frac{n\pi y}{b}\right) + \\ &\left(\frac{n\pi}{b}\right)^2 \sin^2\left(\frac{m\pi x}{a}\right)\cos^2\left(\frac{n\pi y}{b}\right) \end{aligned} \right]
$$
(5.209)

The total power passing through the cross-section of a waveguide is as follows:

$$
\left(P_z\right)_{\text{TM}} = \int_0^a \int_0^b \left(p_z\right)_{\text{TM}} dx\,dy
$$
(5.210)

Substituting Eq. (5.209) into Eq. (5.210) and integrating, we get the following relation:

$$
\left(P_z\right)_{\text{TM}} = \frac{1}{2}\frac{\omega\varepsilon\beta}{k_c^4}E_0^2 \times
$$

$$
\left[\left(\frac{m\pi}{a}\right)^2 \int_0^a \cos^2\left(\frac{m\pi x}{a}\right)dx \int_0^b \sin^2\left(\frac{n\pi y}{b}\right)dy + \left(\frac{n\pi}{b}\right)^2 \int_0^a \sin^2\left(\frac{m\pi x}{a}\right)dx \int_0^b \cos^2\left(\frac{n\pi y}{b}\right)dy\right]
$$

$$\text{or} \quad (P_z)_{\text{TM}} = \begin{cases} \dfrac{\omega\varepsilon\beta}{8k_c^4} E_0^2 ab \left[\left(\dfrac{m\pi}{a}\right)^2 + \left(\dfrac{n\pi}{b}\right)^2 \right] & \text{for } m \neq 0 \text{ and } n \neq 0 \\[2mm] \dfrac{\omega\varepsilon\beta}{4k_c^4} E_0^2 ab \left(\dfrac{n\pi}{b}\right)^2 & \text{for } m = 0 \text{ and } n \neq 0 \\[2mm] \dfrac{\omega\varepsilon\beta}{4k_c^4} E_0^2 ab \left(\dfrac{m\pi}{a}\right)^2 & \text{for } m \neq 0 \text{ and } n = 0 \end{cases} \quad (5.211)$$

5.14 Power-handling Capability of Waveguides

The normalized dominant-mode fields inside a rectangular waveguide can be expressed as follows:

$$E_{0x} = 0 \tag{5.212}$$

$$E_{0y} = A \sin\left(\frac{\pi x}{a}\right) \tag{5.213}$$

$$H_{0x} = \frac{A}{\eta_{\text{TE}}} \sin\left(\frac{\pi x}{a}\right) \tag{5.214}$$

$$H_{0y} = 0 \tag{5.215}$$

Therefore,

$$p_{\text{TE},10} = \frac{1}{2}\left(E_{0x}H_{0y}^* - E_{0y}H_{0x}^*\right) = \frac{1}{2}\frac{A^2}{\eta_{\text{TE}}}\sin^2\left(\frac{\pi x}{a}\right)$$

$$P_{\text{TE},10} = \int_0^a \int_0^b p_{\text{TE},10}\, dx\, dy = \int_0^a \int_0^b \frac{1}{2}\frac{A^2}{\eta_{\text{TE}}}\sin^2\left(\frac{\pi x}{a}\right) dx\, dy$$

$$\text{or} \quad P_{\text{TE},10} = \frac{1}{2}\frac{A^2}{\eta_{\text{TE}}}\int_0^a \sin^2\left(\frac{\pi x}{a}\right) dx \int_0^b dy = \frac{A^2 ab}{4\eta_{\text{TE}}} \tag{5.216}$$

Substituting Eq. (5.189) into Eq. (5.216), we get the following relation:

$$P_{\text{TE},10} = \frac{A^2 ab}{4\eta}\sqrt{1 - \left(\frac{f_{c,10}}{f}\right)^2} \tag{5.217}$$

Assuming break down voltage of dry air to be 30 kV/cm and $\eta = 377\ \Omega$, we get from Eq. (5.217) the following expression:

$$P_{\text{TE},10} = \frac{\left(30\times10^3\right)^2 ab}{4\times377}\sqrt{1 - \left(\frac{f_{c,10}}{f}\right)^2} = 597\times10^3\, ab\sqrt{1 - \left(\frac{f_c}{f}\right)^2}$$

$$\text{or} \quad P_{\text{TE},10} = 597\, ab\sqrt{1 - \left(\frac{f_{c,10}}{f}\right)^2}\ \text{kW} \tag{5.218}$$

where a and b are expressed in cm units.

EXAMPLE 5.7 For the air-core rectangular waveguide described in Example 5.3, calculate the maximum power that can be handled if it is operated at 12 GHz.

Solution Given: $a = 22.86\,\text{mm}$, $b = 10.16\,\text{mm}$, $f = 12\,\text{GHz}$, and $f_{c,\text{TE}_{10}} = 6.562$ GHz

Since the operating frequency is 12 GHz, only the dominant TE mode will exist inside the guide. Therefore, the power handling capability of the guide is as follows:

$$P_{\text{TE},10} = 597ab\sqrt{1 - \left(\frac{f_{c,10}}{f}\right)^2} = 597 \times 2.286 \times 1.016\sqrt{1 - \left(\frac{6.562 \times 10^9}{12 \times 10^9}\right)^2} = 1160.9 \text{ kW}$$

Practice Problem

5.9 For the air-core rectangular waveguide described in Example 5.3, calculate the maximum power that can be handled if it is operated at 20 GHz. **[408.408 kW]**

5.15 Waveguide Attenuation

Due to finite conductivity of both the conducting walls and the dielectric filling, two types of power losses occur in a rectangular waveguide: losses in guide walls and losses in dielectric.

First, let us consider the power losses caused by dielectric attenuation. Attenuation caused by the low-loss dielectric in a rectangular waveguide for the TE mode and that for the TM mode are, respectively, given by

$$\alpha_{d,\text{TE}} = \frac{\sigma\eta}{2\sqrt{1 - (f_c/f)^2}} \tag{5.219}$$

and $\quad \alpha_{d,\text{TM}} = \dfrac{\sigma\eta}{2}\sqrt{1 - (f_c/f)^2} \tag{5.220}$

In most of the practical applications, air is considered as the dielectric material for which the attenuation loss is very small. In such cases, losses in the guide walls play a major role. We can calculate the power losses caused by the guide walls by assuming a uniform current density through the guide walls. To do this, let us consider a thin filament of width dx, depth δ, and length dz, as shown in Fig. 5.19, where δ is the skin depth and is expressed as follows:

$$\delta = \sqrt{\frac{2}{\omega\mu_m\sigma_m}} \tag{5.221}$$

Here, μ_m and σ_m are, respectively, the permeability and conductivity of the medium.

From the Ampere's law:

$$i_x = H_{0z}dz \tag{5.222}$$

and $i_z = H_{0x}dx \tag{5.223}$

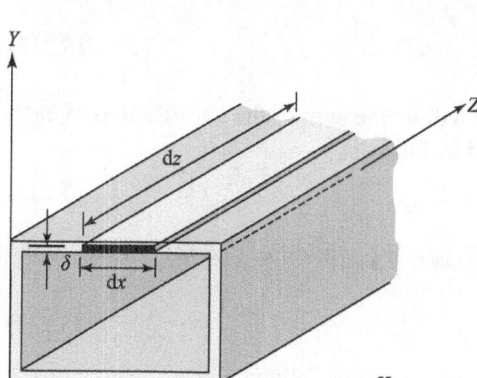

Fig. 5.19 Calculation of losses in waveguide

Again, $R_x = \dfrac{1}{\sigma_m}\dfrac{dx}{\delta dz}$ (5.224)

and $R_z = \dfrac{1}{\sigma_m}\dfrac{dz}{\delta dx}$ (5.225)

The total power loss for the element is given by the following equation:

$$dP_z = \frac{1}{2}\left(i_x^2 R_x + i_z^2 R_z\right)$$ (5.226)

Substituting Eqs (5.222)–(5.225) into Eq. (5.226), we get the following relation:

$$dP_z = \frac{1}{2}\left[H_{0z}^2\,(dz)^2\,\frac{1}{\sigma_m}\frac{dx}{\delta dz} + H_{0x}^2\,(dx)^2\,\frac{1}{\sigma_m}\frac{dz}{\delta dx}\right]$$

or $\dfrac{dP_z}{dz} = \dfrac{dx}{2\sigma_m \delta}\left(H_{0x}^2 + H_{0z}^2\right)$ (5.227)

Again substituting the expression for δ from Eq. (5.221) into Eq. (5.227), the following expression is obtained:

$$\frac{dP_z}{dz} = \frac{dx}{2}\sqrt{\frac{\omega\mu_m}{2\sigma_m}}\left(H_{0x}^2 + H_{0z}^2\right)$$ (5.228)

Therefore, the total per-unit-length power loss in the x–z plane is given by the following equation:

$$W_L(x, z) = \frac{1}{2}\sqrt{\frac{\omega\mu_m}{2\sigma_m}}\iint\left(H_{0x}^2 + H_{0z}^2\right)dxdz$$

or $W_L(x, z) = \dfrac{R_s}{2}\displaystyle\iint\left(H_{0x}^2 + H_{0z}^2\right)dxdz$ (5.229)

where $R_s = \sqrt{\dfrac{\omega\mu_m}{2\sigma_m}}$ (5.230)

and is called the surface resistance of the waveguide.

Similarly for the x–y plane,

$$W_L(y, z) = \frac{R_s}{2}\iint\left(H_{0y}^2 + H_{0z}^2\right)dydz$$ (5.231)

The total power loss in guide walls will be the sum of the contribution of each wall of the waveguide and is expressed as follows:

$$W_{\text{Total}} = W_{\text{Bottom}} + W_{\text{Top}} + W_{\text{Left}} + W_{\text{Right}}$$ (5.232)

Due to symmetry, $W_{\text{Bottom}} = W_{\text{Top}}$ and $W_{\text{Left}} = W_{\text{Right}}$ and Eq. (5.232) reduces to the following form:

$$W_{\text{Total}} = 2\left(W_{\text{Bottom}} + W_{\text{Left}}\right)$$ (5.233)

The power flow in a waveguide can be expressed as follows:

$$P_z = P_0 e^{-2\alpha z}$$ (5.234)

Differentiating Eq. (5.234) with respect to z, we get the following relation:

$$\frac{\partial P_z}{\partial z} = \frac{\partial}{\partial z}\left[P_0 e^{-2\alpha z}\right] = -2\alpha P_0 e^{-2\alpha z} \tag{5.235}$$

Substituting Eq. (5.234) into Eq. (5.235), we get the following expression:

$$\frac{\partial P_z}{\partial z} = -2\alpha P_z \text{ or } \alpha = -\frac{\partial P_z/\partial z}{2P_z} \tag{5.236}$$

The negative sign in Eq. (5.236) indicates a power loss at any point z.

TE mode Substituting Eqs (5.122) and (5.124) into Eq. (5.229), the power loss at any point z can be expressed as follows:

$$W_{\text{Bottom,TE}}(x,z)\Big|_{y=0} = \frac{R_s}{2}\int_0^a \frac{|\gamma|^2}{k_c^4} H_0^2 \left(\frac{m\pi}{a}\right)^2 \sin^2\left(\frac{m\pi x}{a}\right)\cos^2\left(\frac{n\pi y}{b}\right) dx\Bigg|_{y=0}$$

$$+ \frac{R_s}{2}\int_0^a H_0^2 \cos^2\left(\frac{m\pi x}{a}\right)\cos^2\left(\frac{n\pi y}{b}\right) dx\Bigg|_{y=0}$$

or $\quad W_{\text{Bottom,TE}}\Big|_{y=0} = \frac{R_s}{2} H_0^2 \int_0^a \left[\frac{|\gamma|^2}{k_c^4}\left(\frac{m\pi}{a}\right)^2 \sin^2\left(\frac{m\pi x}{a}\right) + \cos^2\left(\frac{m\pi x}{a}\right)\right] dx$

or $\quad W_{\text{Bottom,TE}}\Big|_{y=0} = \begin{cases} \dfrac{1}{2} R_s H_0^2 \dfrac{|\gamma|^2}{k_c^4}\left(\dfrac{m\pi}{a}\right)^2 \dfrac{a}{2} + \dfrac{a}{2} & \text{for } m \neq 0 \\[3mm] \dfrac{1}{2} R_s H_0^2 a & \text{for } m = 0 \end{cases} \tag{5.237}$

Similarly, substituting Eqs (5.123) and (5.124) in Eq. (5.231), the power loss at any point z can be written as follows:

$$W_{\text{Left,TE}}(y,z)\Big|_{x=0} = \frac{R_s}{2}\int_0^b \frac{|\gamma|^2}{k_c^4} H_0^2 \left(\frac{n\pi}{b}\right)^2 \cos^2\left(\frac{m\pi x}{a}\right)\sin^2\left(\frac{n\pi y}{b}\right) dy\Bigg|_{x=0}$$

$$+ \frac{R_s}{2}\int_0^b H_0^2 \cos^2\left(\frac{m\pi x}{a}\right)\cos^2\left(\frac{n\pi y}{b}\right) dy\Bigg|_{x=0}$$

or $\quad W_{\text{Left,TE}}\Big|_{x=0} = \frac{R_s}{2} H_0^2 \int_0^b \left[\frac{|\gamma|^2}{k_c^4}\left(\frac{n\pi}{b}\right)^2 \sin^2\left(\frac{n\pi y}{b}\right) + \cos^2\left(\frac{n\pi y}{b}\right)\right] dy$

or $\quad W_{\text{Left,TE}}\Big|_{x=0} = \begin{cases} \dfrac{1}{2} R_s H_0^2 \dfrac{|\gamma|^2}{k_c^4}\left(\dfrac{n\pi}{b}\right)^2 \dfrac{b}{2} + \dfrac{b}{2} & \text{for } n \neq 0 \\[3mm] \dfrac{1}{2} R_s H_0^2 b & \text{for } n = 0 \end{cases} \tag{5.238}$

Therefore, the total cross-sectional loss (for $m \neq 0$ and $n \neq 0$) is given by the following equation:

$$W_{z,\text{TE}} = 2\left(W_{\text{Bottom,TE}}\big|_z + W_{\text{Left,TE}}\big|_z\right)$$

or

$$W_{z,\text{TE}} = R_s H_0^2 \left[\frac{|\gamma|^2}{k_c^4}\left(\frac{m\pi}{a}\right)^2 \frac{a}{2} + \frac{a}{2}\right] + R_s H_0^2 \left[\frac{|\gamma|^2}{k_c^4}\left(\frac{n\pi}{b}\right)^2 \frac{b}{2} + \frac{b}{2}\right]$$

or

$$W_{z,\text{TE}} = \frac{1}{2} R_s H_0^2 \left[(a+b) + \frac{|\gamma|^2 \pi^2}{k_c^4}\left(\frac{m^2}{a} + \frac{n^2}{b}\right)\right] \tag{5.239}$$

Substituting Eqs (5.205) and (5.239) into Eq. (5.236), we get the following expression:

$$\alpha = \frac{\dfrac{1}{2} R_s H_0^2 \left[(a+b) + \dfrac{|\gamma|^2 \pi^2}{k_c^4}\left(\dfrac{m^2}{a} + \dfrac{n^2}{b}\right)\right]}{2\dfrac{\omega\mu\beta}{8k_c^4} H_0^2 ab \left[\left(\dfrac{m\pi}{a}\right)^2 + \left(\dfrac{n\pi}{b}\right)^2\right]}$$

or

$$\alpha = \frac{2k_c^4 R_s}{\omega\mu\beta} \frac{\left[(a+b) + \dfrac{|\gamma|^2 \pi^2}{k_c^4}\left(\dfrac{m^2}{a} + \dfrac{n^2}{b}\right)\right]}{ab\left[\left(\dfrac{m\pi}{a}\right)^2 + \left(\dfrac{n\pi}{b}\right)^2\right]} \tag{5.240}$$

If attenuation is small, then $|\gamma|^2 \approx \beta^2$ and Eq. (5.240) modifies to the following form:

$$\alpha = \frac{2k_c^4 R_s}{\omega\mu\beta} \frac{\left[(a+b) + \dfrac{\beta^2 \pi^2}{k_c^4}\left(\dfrac{m^2}{a} + \dfrac{n^2}{b}\right)\right]}{ab\left[\left(\dfrac{m\pi}{a}\right)^2 + \left(\dfrac{n\pi}{b}\right)^2\right]} \tag{5.241}$$

For the dominant-mode excitation, $m = 1$ and $n = 0$. Therefore, substituting Eqs (5.237) and (5.238) for $m = 1$ and $n = 0$ in Eq. (5.233), we obtain the following expression:

$$W_{z,\text{TE}} = 2\left(W_{\text{Bottom,TE}}\big|_z + W_{\text{Left,TE}}\big|_z\right)$$

or

$$W_{z,\text{TE}} = R_s H_0^2 \left[\frac{\beta^2}{k_c^4}\left(\frac{\pi}{a}\right)^2 \frac{a}{2} + \frac{a}{2} + b\right] = R_s H_0^2 \left[b + \frac{a}{2}\left\{1 + \frac{\beta^2}{(\pi/a)^4}(\pi/a)^2\right\}\right]$$

or

$$W_{z,\text{TE}} = R_s H_0^2 \left[b + \frac{a}{2}\left(1 + \frac{\beta^2 a^2}{\pi^2}\right)\right] \tag{5.242}$$

Substituting Eqs (5.205) and (5.242) into Eq. (5.236), we get the following relation:

$$\alpha = \frac{R_s H_0^2 \left[b + \dfrac{a}{2}\left(1 + \dfrac{\beta^2 a^2}{\pi^2}\right)\right]}{2\dfrac{\omega\mu\beta}{4k_c^4} H_0^2 ab \left(\dfrac{\pi}{a}\right)^2} = \frac{R_s H_0^2 \left[b + \dfrac{a}{2}\left(1 + \dfrac{\beta^2 a^2}{\pi^2}\right)\right]}{2\dfrac{\omega\mu\beta}{4(\pi/a)^4} H_0^2 ab \left(\pi/a\right)^2}$$

or $\qquad \alpha = \dfrac{2\pi^2 R_s \left[b + \dfrac{a}{2}\left(1 + \dfrac{\beta^2 a^2}{\pi^2}\right)\right]}{\omega\mu\beta a^3 b}$ $\qquad\qquad$ (5.243)

TM mode Substituting Eqs (5.166) and (5.168) into Eq. (5.229), the power loss at any point z can be written as follows:

$$W_{\text{Bottom,TM}}\left(x,z\right)\Big|_{y=0} = \frac{R_s}{2}\int_0^a \frac{\omega^2 |\varepsilon_c|^2}{k_c^4} E_0^2 \left(\frac{n\pi}{b}\right)^2 \sin^2\left(\frac{m\pi x}{a}\right)\cos^2\left(\frac{n\pi y}{b}\right) dx \Bigg|_{y=0}$$

or $\qquad W_{\text{Bottom,TM}}\Big|_{y=0} = \dfrac{R_s}{2} E_0^2 \displaystyle\int_0^a \dfrac{\omega^2 |\varepsilon_c|^2}{k_c^4}\left(\dfrac{n\pi}{b}\right)^2 \sin^2\left(\dfrac{m\pi x}{a}\right) dx$

or $\qquad W_{\text{Bottom,TM}}\Big|_{y=0} = \dfrac{1}{2} R_s E_0^2 \dfrac{\omega^2 |\varepsilon_c|^2}{k_c^4}\left(\dfrac{n\pi}{b}\right)^2 \dfrac{a}{2}$ $\qquad\qquad$ (5.244)

Similarly, substituting Eqs (5.167) and (5.168) into Eq. (5.231), the power loss at any point z can be expressed as follows:

$$W_{\text{Left,TM}}\left(y,z\right)\Big|_{x=0} = \frac{R_s}{2}\int_0^b \frac{\omega^2 |\varepsilon_c|^2}{k_c^4} E_0^2 \left(\frac{m\pi}{a}\right)^2 \cos^2\left(\frac{m\pi x}{a}\right)\sin^2\left(\frac{n\pi y}{b}\right) dy \Bigg|_{x=0}$$

or $\qquad W_{\text{Left,TM}}\Big|_{x=0} = \dfrac{R_s}{2} E_0^2 \displaystyle\int_0^a \dfrac{\omega^2 |\varepsilon_c|^2}{k_c^4}\left(\dfrac{m\pi}{a}\right)^2 \sin^2\left(\dfrac{n\pi y}{b}\right) dy$

or $\qquad W_{\text{Left,TM}}\Big|_{x=0} = \dfrac{1}{2} R_s E_0^2 \dfrac{\omega^2 |\varepsilon_c|^2}{k_c^4}\left(\dfrac{m\pi}{a}\right)^2 \dfrac{b}{2}$ $\qquad\qquad$ (5.245)

Therefore, the total cross-sectional loss (for $m \neq 0$ and $n \neq 0$) is given by the following equation:

$$W_{z,\text{TM}} = 2\left(W_{\text{Bottom,TM}}\Big|_z + W_{\text{Left,TM}}\Big|_z\right)$$

or $\qquad W_{z,\text{TM}} = \dfrac{1}{2} R_s E_0^2 \dfrac{\omega^2 |\varepsilon_c|^2 \pi^2}{k_c^4}\left[\left(\dfrac{m}{a}\right)^2 b + \left(\dfrac{n}{b}\right)^2 a\right]$ $\qquad\qquad$ (5.246)

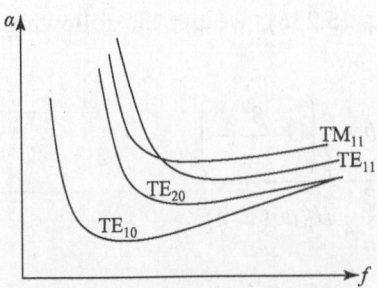

Fig. 5.20 Plot of attenuation for different TE and TM modes cross-sectional view

Substituting Eqs (5.211) and (5.246) into Eq. (5.236), the following expression is obtained:

$$\alpha = \frac{\dfrac{1}{2}R_s E_0^2 \dfrac{\omega^2 |\varepsilon_c|^2 \pi^2}{k_c^4}\left[\left(\dfrac{m}{a}\right)^2 b + \left(\dfrac{n}{b}\right)^2 a\right]}{2\dfrac{\omega\varepsilon_c\beta}{8k_c^4}E_0^2 ab\left[\left(\dfrac{m\pi}{a}\right)^2 + \left(\dfrac{n\pi}{b}\right)^2\right]}$$

$$\text{or } \alpha = \frac{2R_s\omega\varepsilon_c\left[\left(\dfrac{m}{a}\right)^2 b + \left(\dfrac{n}{b}\right)^2 a\right]}{\beta ab\left[\left(\dfrac{m}{a}\right)^2 + \left(\dfrac{n}{b}\right)^2\right]} \qquad (5.247)$$

A plot of attenuation for different TE and TM modes are shown in Fig. 5.20.

5.16 Quality Factor *Q* of Waveguides

The quality factor of a waveguide is defined as follows:

$$Q = \omega\,\frac{\text{Energy stored per unit length}}{\text{Energy lost per unit length per second}}$$

$$\text{or} \qquad Q = \omega\,\frac{\text{Energy stored per unit length}}{\text{Power lost per unit length}} \qquad (5.248)$$

For a waveguide,

$$\text{Energy stored per unit length} = \frac{\text{Power transmitted}}{v_g} \qquad (5.249)$$

Substituting Eq. (5.249) into Eq. (5.248), the following relation is obtained:

$$Q = \frac{\omega}{v_g}\,\frac{\text{Power transmitted}}{\text{Power lost per unit length}} \qquad (5.250)$$

$$\text{Now} \qquad \frac{\text{Power lost per unit length}}{\text{Power transmitted}} = 2\alpha \qquad (5.251)$$

Substituting Eq. (5.251) into Eq. (5.250), we get the following expression:

$$Q = \frac{\omega}{2\alpha v_g} \qquad (5.252)$$

Using Eq. (5.199), the following relation is obtained:

$$Q = \frac{\omega v_p}{2\alpha c^2} \qquad (5.253)$$

Using Eq. (5.194), we get the following expression:

$$Q = \frac{\omega c}{2\alpha c^2 \sqrt{1-(\lambda/\lambda_c)^2}} = \frac{\omega}{2\alpha c \sqrt{1-(\lambda/\lambda_c)^2}} = \frac{\omega}{2\alpha c \sqrt{1-(\omega_c/\omega)^2}} \qquad (5.254)$$

Note Equation (5.254) indicates that for a lossless waveguide, Q should be ∞.

5.17 Non-existence of TEM Modes in Hollow Waveguides

Let us first consider that the TEM mode exists within a waveguide. Now, the Maxwell's equation, $\vec{\nabla}.\vec{B} = 0$, ensures that $\vec{\nabla}.\vec{H} = 0$ or $\vec{H}$ forms a closed loop. Since for a TEM wave $\vec{H}$ must be perpendicular to the direction of propagation, the closed loop of $\vec{H}$ lines must also be perpendicular to the direction of propagation or must be in the cross-sectional plane of the waveguide.

Now, since inside a waveguide $\vec{J} = 0$, the Maxwell's equation, $\oint \vec{H}.\vec{dl} = \int_S (\vec{D}+\vec{J}).\vec{dS}$, implies that $\vec{D} \neq 0$ and must be along the axis of the waveguide. The presence of $\vec{D}$ along the axis of the waveguide ensures the existence of $\vec{E}$ along the same axis, that is, in the direction of propagation. The last conclusion contradicts our initial assumption of the existence of a TEM wave within a waveguide. Therefore, a TEM wave cannot exist inside a hollow waveguide. Practically, propagation of an axial current requires a centre conductor. For this reason, the TEM-mode propagation requires a two-conductor system.

5.18 Transmission Line Analogy of Waveguides

In Chapter 4, we have already discussed about the TE and TM modes and the inter-relations between different field components for these modes. Using these relations, we can find transmission line equivalent models of a waveguide for such modes.

5.18.1 TE Equivalent Model

For a TE wave, using the Maxwell's curl equations, we can write that

$$\frac{\partial E_y}{\partial z} = j\omega\mu H_x \qquad (5.255)$$

and $\quad \dfrac{\partial H_z}{\partial y} - \dfrac{\partial H_y}{\partial z} = j\omega\varepsilon E_x \qquad (5.256)$

Since $E_z = 0$ for the TE mode, we can write the following expression:

$$\left(\vec{\nabla} \times \vec{H}\right)_z = 0 \qquad (5.257)$$

Equation (5.257) implies that in the X–Y plane, we can define a magnetic scalar potential satisfying the following relations:

$$H_x = -\frac{\partial U}{\partial x} \qquad (5.258)$$

$$H_y = -\frac{\partial U}{\partial y} \tag{5.259}$$

Again, for the TE mode: $E_y = \dfrac{j\omega\mu}{k_c^2} \dfrac{\partial H_z}{\partial x}$ (5.260)

Differentiating both sides with respect to z, we get the following equation:

$$\frac{\partial E_y}{\partial z} = \frac{\partial}{\partial z}\left(\frac{j\omega\mu}{k_c^2} \frac{\partial H_z}{\partial x} \right) \tag{5.261}$$

Equating Eq. (5.255) to Eq. (5.261), we get the following relation:

$$\frac{\partial}{\partial z}\left(\frac{j\omega\mu}{k_c^2} \frac{\partial H_z}{\partial x} \right) = j\omega\mu H_x \tag{5.262}$$

Substituting Eq. (5.258) into Eq. (5.262), the following relation is obtained:

$$\frac{j\omega\mu}{k_c^2} \frac{\partial^2 H_z}{\partial x \partial z} = -j\omega\mu \frac{\partial U}{\partial x} \quad \text{or} \quad \frac{\partial}{\partial z}\left(\frac{j\omega\mu}{k_c^2} H_z \right) = -j\omega\mu U \tag{5.263}$$

Further, for the TE mode:

$$E_x = -\frac{j\omega\mu}{k_c^2} \frac{\partial H_z}{\partial y} \tag{5.264}$$

Substituting Eq. (5.264) into Eq. (5.256), we get the following relation:

$$\frac{\partial H_z}{\partial y} - \frac{\partial H_y}{\partial z} = j\omega\varepsilon\left(-\frac{j\omega\mu}{k_c^2} \frac{\partial H_z}{\partial y} \right) = \frac{\omega^2 \mu\varepsilon}{k_c^2} \frac{\partial H_z}{\partial y}$$

or $\quad -\dfrac{\partial H_y}{\partial z} = -\dfrac{\partial H_z}{\partial y} + \dfrac{\omega^2 \mu\varepsilon}{k_c^2} \dfrac{\partial H_z}{\partial y} = \left(-1 + \dfrac{\omega^2 \mu\varepsilon}{k_c^2} \right)\dfrac{\partial H_z}{\partial y}$ (5.265)

Substituting Eq. (5.259) into Eq. (5.265), we get the following expression:

$$\frac{\partial^2 U}{\partial y \partial z} = \frac{\partial}{\partial y}\left\{ \left(-1 + \frac{\omega^2 \mu\varepsilon}{k_c^2} \right) H_z \right\}$$

or $\quad \dfrac{\partial U}{\partial z} = -\left(\dfrac{k_c^2}{j\omega\mu} + j\omega\varepsilon \right)\left(\dfrac{j\omega\mu}{k_c^2} H_z \right)$ (5.266)

Since the parameter $j\omega\mu H_z/k_c^2$ has a unit of voltage and U has a unit of current, Eqs (5.263) and (5.266) may be written as follows:

$$\frac{\partial V_{TE}}{\partial z} = -Z_{TE} I_{TE} \tag{5.267}$$

and $\quad \dfrac{\partial I_{TE}}{\partial z} = -Y_{TE} V_{TE}$ (5.268)

where $\quad Z_{TE} = j\omega\mu$ (5.269)

$$Y_{TE} = \frac{k_c^2}{j\omega\mu} + j\omega\varepsilon \tag{5.270}$$

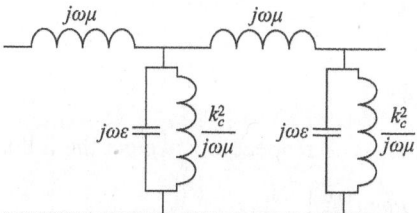

Fig. 5.21 Equivalent transmission line representation of waveguide for TE mode

The equivalent circuit of a waveguide for the TE mode is shown in Fig. 5.21. The characteristic impedance is expressed as follows:

$$\eta_{TE} = \sqrt{\frac{Z}{Y}} = \sqrt{\frac{j\omega\mu}{\dfrac{k_c^2}{j\omega\mu} + j\omega\varepsilon}} = \sqrt{\frac{\mu}{\dfrac{k_c^2}{j^2\omega^2\mu} + \varepsilon}} = \sqrt{\frac{\mu}{\varepsilon - \dfrac{k_c^2}{\omega^2\mu}}} \tag{5.271}$$

For a rectangular waveguide, $k_c^2 = \omega_c^2\mu\varepsilon$, and we get from Eq. (5.271)

$$\eta_{TE} = \sqrt{\frac{\mu}{\varepsilon - \dfrac{\omega_c^2\mu\varepsilon}{\omega^2\mu}}} = \sqrt{\frac{\mu}{\varepsilon - \dfrac{\omega_c^2\varepsilon}{\omega^2}}} = \sqrt{\frac{\mu/\varepsilon}{1 - \dfrac{\omega_c^2}{\omega^2}}}$$

or

$$\eta_{TE} = \frac{\sqrt{\mu/\varepsilon}}{\sqrt{1 - \dfrac{f_c^2}{f^2}}} = \frac{\eta}{\sqrt{1 - \left(\dfrac{f_c}{f}\right)^2}} = \frac{\eta}{\sqrt{1 - \left(\dfrac{\lambda}{\lambda_c}\right)^2}} \tag{5.272}$$

Equation (5.272) is identical to Eq. (5.189).

5.18.2 TM Equivalent Model

For a TM wave, using the Maxwell's curl equations, we can write that

$$\frac{\partial H_y}{\partial z} = -j\omega\varepsilon E_x \tag{5.273}$$

and

$$\frac{\partial E_x}{\partial z} - \frac{\partial E_z}{\partial x} = -j\omega\mu H_y \tag{5.274}$$

Since for the TM mode $H_z = 0$, we can write the following expression:

$$\left(\vec{\nabla} \times \vec{E}\right)_z = 0 \tag{5.275}$$

Equation (5.275) implies that, in the X–Y plane, we can define an electric scalar potential satisfying the following relations:

$$E_x = -\frac{\partial V}{\partial x} \tag{5.276}$$

$$E_y = -\frac{\partial V}{\partial y} \tag{5.277}$$

Again, for the TM mode:

$$H_y = -\frac{j\omega\varepsilon}{k_c^2}\frac{\partial E_z}{\partial x} \tag{5.278}$$

Differentiating both sides with respect to z, we get the following relation:

$$\frac{\partial H_y}{\partial z} = \frac{\partial}{\partial z}\left(-\frac{j\omega\varepsilon}{k_c^2}\frac{\partial E_z}{\partial x}\right) \tag{5.279}$$

Equating Eq. (5.273) to Eq. (5.279), we get the following expression:

$$\frac{\partial}{\partial z}\left(\frac{j\omega\varepsilon}{k_c^2}\frac{\partial E_z}{\partial x}\right) = j\omega\varepsilon E_x \tag{5.280}$$

Substituting Eq. (5.276) into Eq. (5.280), we get the following relation:

$$\frac{j\omega\varepsilon}{k_c^2}\frac{\partial^2 E_z}{\partial x \partial z} = -j\omega\varepsilon\frac{\partial V}{\partial x} \quad \text{or} \quad \frac{\partial}{\partial z}\left(\frac{j\omega\varepsilon}{k_c^2}E_z\right) = -j\omega\varepsilon V \tag{5.281}$$

Substituting Eq. (5.278) into Eq. (5.274), the following expression is obtained:

$$\frac{\partial E_x}{\partial z} - \frac{\partial E_z}{\partial x} = -j\omega\mu H_y = -j\omega\mu\left(-\frac{j\omega\varepsilon}{k_c^2}\frac{\partial E_z}{\partial x}\right) = -\frac{\omega^2\mu\varepsilon}{k_c^2}\frac{\partial E_z}{\partial x}$$

or $\quad \dfrac{\partial E_x}{\partial z} = \dfrac{\partial E_z}{\partial x} - \dfrac{\omega^2\mu\varepsilon}{k_c^2}\dfrac{\partial E_z}{\partial x} = \left(1 - \dfrac{\omega^2\mu\varepsilon}{k_c^2}\right)\dfrac{\partial E_z}{\partial x} \tag{5.282}$

Substitution of Eq. (5.276) into Eq. (5.282) results in the following relation:

$$-\frac{\partial^2 V}{\partial x \partial z} = \left(1 - \frac{\omega^2\mu\varepsilon}{k_c^2}\right)\frac{\partial E_z}{\partial x}$$

or $\quad \dfrac{\partial V}{\partial z} = \left(\dfrac{\omega^2\mu\varepsilon}{k_c^2} - 1\right)E_z = -\left(\dfrac{k_c^2}{j\omega\varepsilon} + j\omega\mu\right)\left(\dfrac{j\omega\varepsilon}{k_c^2}E_z\right) \tag{5.283}$

Since the parameter $j\omega\varepsilon E_z/k_c^2$ has a unit of current and V has a unit of voltage, Eqs (5.281) and (5.283) may be written as follows:

$$\frac{\partial I_{\text{TM}}}{\partial z} = -Y_{\text{TM}}V_{\text{TM}} \tag{5.284}$$

and $\quad \dfrac{\partial V_{\text{TM}}}{\partial z} = -Z_{\text{TM}}I_{\text{TM}} \tag{5.285}$

where $\quad Y_{\text{TM}} = j\omega\varepsilon \tag{5.286}$

$$Z_{\text{TM}} = \frac{k_c^2}{j\omega\varepsilon} + j\omega\mu \tag{5.287}$$

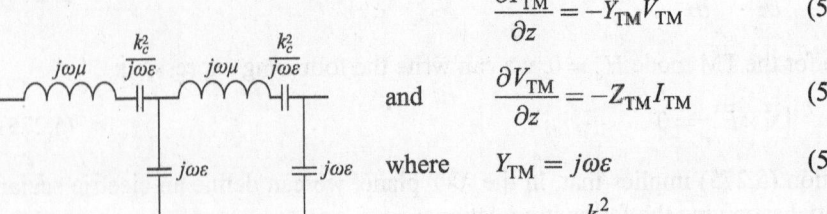

Fig. 5.22 Equivalent transmission line representation of waveguide for TM mode

The equivalent circuit of a waveguide for the TM mode is shown in Fig. 5.22.

The characteristic impedance is expressed as follows:

$$\eta_{TM} = \sqrt{\frac{Z}{Y}} = \sqrt{\frac{\dfrac{k_c^2}{j\omega\varepsilon} + j\omega\mu}{j\omega\varepsilon}}$$

or

$$\eta_{TM} = \sqrt{\frac{\dfrac{k_c^2}{j^2\omega^2\varepsilon} + \mu}{\varepsilon}} = \sqrt{\frac{\mu - \dfrac{k_c^2}{\omega^2\varepsilon}}{\varepsilon}} \tag{5.288}$$

For a rectangular waveguide, $k_c^2 = \omega_c^2 \mu\varepsilon$, and we get from Eq. (5.288) the following expression:

$$\eta_{TM} = \sqrt{\frac{\mu - \dfrac{\omega_c^2 \mu\varepsilon}{\omega^2\varepsilon}}{\varepsilon}} = \sqrt{\frac{\mu - \dfrac{\omega_c^2 \mu}{\omega^2}}{\varepsilon}} = \sqrt{\frac{\mu}{\varepsilon}}\sqrt{1 - \frac{\omega_c^2}{\omega^2}}$$

or

$$\eta_{TM} = \eta\sqrt{1 - \left(\frac{f_c}{f}\right)^2} = \eta\sqrt{1 - \left(\frac{\lambda}{\lambda_c}\right)^2} \tag{5.289}$$

Equation (5.289) is identical to Eq. (5.191).

5.19 Excitation of Modes in Rectangular Waveguides

Now it is known to us that a rectangular waveguide can support different TE and TM modes. The desired field intensities of a particular mode can be established by means of a probe or loop coupling. The probe and loop are basically a monopole and a loop antenna, respectively. The probe is used to excite the electric field of the mode, while the loop is used to generate the magnetic field of the desired mode. This can best be described with the help the following examples.

The aperture electric field distribution of the TE_{10} mode is shown in Fig. 5.23(a). For this mode, the electric field intensity is directed towards Y-direction and is maximum at the centre of the aperture. No X-component of the electric field exists for this mode. Therefore, the best position of the probe to generate the TE_{10} mode is at the top/bottom wall of the guide and at the centre of the aperture, as shown in

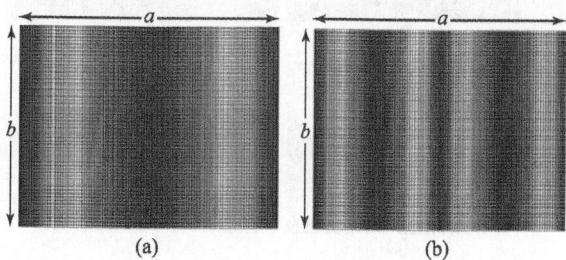

Fig. 5.23 *Y*-component of aperture electric field distribution for different modes (a) TE_{10} (b) TE_{20}

Fig. 5.24(a). In practice, a short circuit is placed at a distance of $\lambda/4$ at the source end. Therefore, the wave travelling in the negative z-direction will be reflected back by the short circuit and will travel a distance of $\lambda/2$ before adding to the negative z-directed wave in an opposite phase. The probes for exciting the TE_{20} mode can be placed as shown in Fig. 5.24(b). In this case, we must take into account the phase relation between the currents in the two probes. Since the directions of the electric fields at positions $a/4$ and $3a/4$ are opposite to each other, the currents at the respective probes should also have a phase difference of $180°$. This condition can be satisfied by inserting an extra cable of $\lambda/2$ length between the probes. The aperture electric field distribution of the TE_{20} mode is shown in Fig. 5.23(b).

Now let us consider the TM modes. Electric fields for the TM modes have both X- and Y-component. Therefore, the right place for inserting the feed probe is through the shorting wall. Positions of the probes for the TM_{11} and TM_{21} modes are shown in Figs 5.24(c) and (d), respectively. The corresponding aperture electric field distributions are shown in Figs 5.25(a) and (b), respectively.

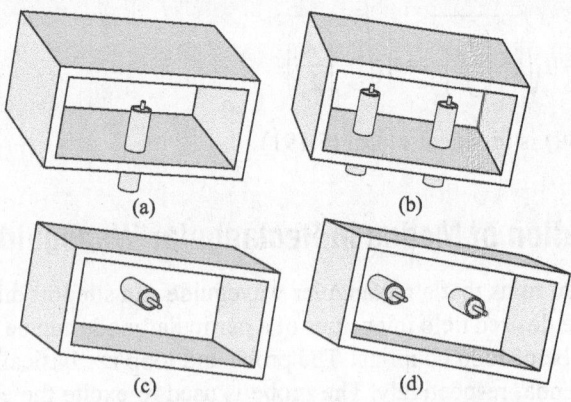

Fig. 5.24 Excitation of modes in rectangular waveguide (a) TE_{10} (b) TE_{20} (c) TM_{11} (d) TM_{21}

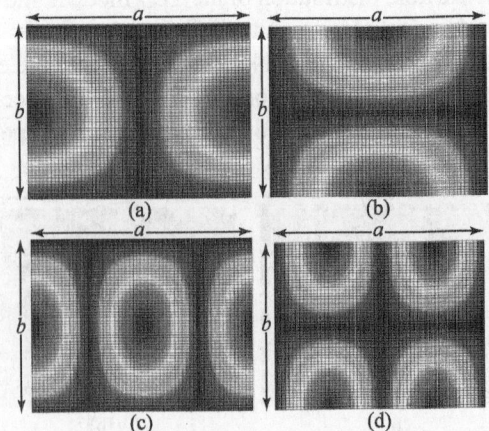

Fig. 5.25 X- and Y-component of aperture electric field distribution (a) X-component of TM_{11} (b) Y-component of TM_{11} (c) X-component of TM_{21} (d) Y-component of TM_{21}

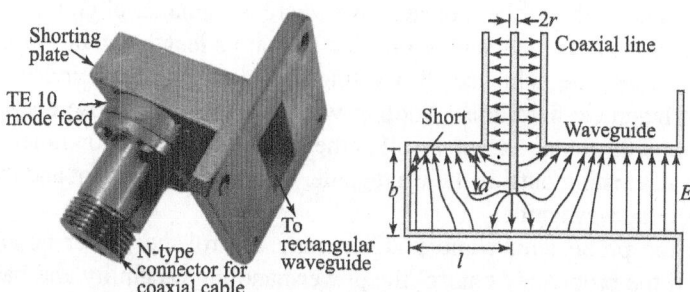

Fig. 5.26 Coaxial to waveguide adaptor (a) Basic TE_{10} mode adaptor (b) Detailed view with electric field distribution

It should be noted that in addition to the intended modes, some higher-order modes are also generated at the position of the probe. However, these higher-order modes are attenuated within a short distance and only the dominant mode propagates down the guide. A proper impedance matching reduces these higher-order modes at the feed point. Such impedance matching is, in general, obtained by varying the position and depth of the probe in the guide, or by using impedance matching stubs or transformers in the coaxial line/guide. It should also be remembered that a probe that excites a given mode in the guide also serves as a receptor of energy for that mode. For example, the probe that excites the TE_{10} mode in the guide can also be used to measure the energy/power of that mode.

A basic probe feeding mechanism to launch the dominant TE_{10} mode in a rectangular waveguide is shown in Fig. 5.26(a), while a detailed view of such a coupling is shown in Fig. 5.26(b).

To achieve a satisfactory result, the depth of the probe inside a waveguide should be small so that a constant current on the probe generates an electromagnetic field inside the guide. A short-circuit plate is also placed at a distance l (approximately quarter wavelength) from the probe to ensure propagation in one direction. The distances l and d are adjusted to achieve pure resistive impedance equal to the characteristic impedance of the waveguide. The reactance of the probe, resulting from higher-order modes, can be made negligible by making the probe diameter very small ($< 0.15a$).

Electromagnetic energy can also be excited inside a waveguide with the help of a coupling loop, as shown in Fig. 5.27. For exciting the dominant TE_{10} mode inside a waveguide, the loop is placed at the middle of the two broad walls of the guide, with its plane being transverse to the waveguide. A short circuit is placed at a distance l from the loop to ensure propagation in one direction only. The input impedance is made resistive and equal to the characteristic impedance of the guide by adjusting the loop diameter and length L.

In microwave circuits, sometimes transfer of a lesser amount of energy is required. This can be achieved through loose coupling. Loose coupling can be achieved by decreasing the length of the probe or moving it out of the centre of the E-field. It is also possible to achieve loose coupling by

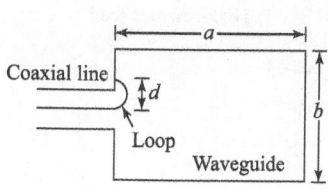

Fig. 5.27 Coupling loops

partially shielding the probe. For the case where the amount of coupling must be varied, the probe must be made retractable so that its length can be varied. Loose coupling can also be achieved by rotating the loop until it encircles a smaller number of magnetic field lines. Another way of achieving a loose coupling is to mill a slot/aperture at the shorting wall of the guide. For proper slot dimensions, the electric field, existing outside the waveguide, can penetrate the slot and excite EM fields inside the waveguide.

While the probe length and position may control the power coupling, the diameter of the probe may control the power-handling capability and bandwidth. A *door knob* probe (similar in shape to a door knob) is capable of handling much higher power and bandwidth than a conventional probe. In general, as the diameter of the probe is increased, the bandwidth is also increased.

5.20 Waveguide Terminations

Different types of waveguide terminations are discussed in the following subsections:

5.20.1 Matched Termination

Waveguide circuits often require a matched termination. The basic concept behind the design of a matched termination is to use a resistive or absorptive material that will absorb the field incident on it. Since the fields are absorbed, no reflection occurs from the termination and hence the standing wave ratio (SWR) of the line remains 1. The absorbed energy is dissipated as heat. One method of obtaining matched termination using this technique is to fill the end of the guide with a graphite and sand mixture. Instead of the sand and graphite mixture, a highly resistive rod can also be placed at the end. The electric field incident on it induces a current in the rod to flow, but the high resistance of the rod dissipates the energy in the form of heat. Another common form of a matched load uses a wedge of highly resistive material. Again, as before, the electric field gets absorbed by the wedge and the energy is dissipated as heat. Figure 5.28 shows the schematic diagrams of the matched loads.

The efficiency of a matched load, using wedges of highly resistive material as absorbents, depends on the position of the wedges and the existing mode pattern inside

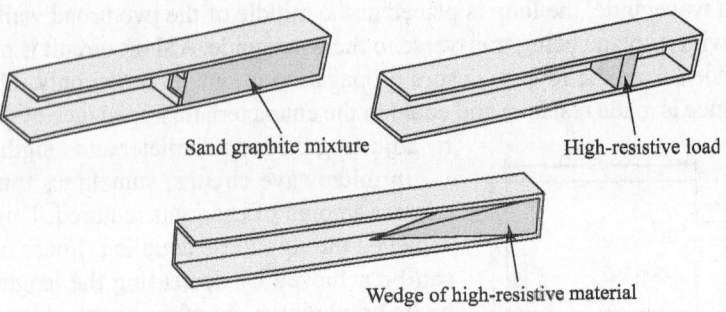

Sand graphite mixture   High-resistive load

Wedge of high-resistive material

Fig. 5.28 Cross-sectional view of different types of matched loads

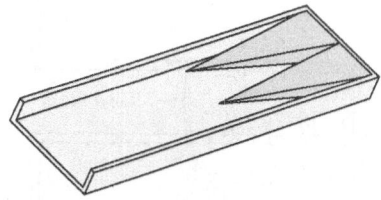

Fig. 5.29 Position of wedges for matching TE_{20} mode (cross-sectional view)

the guide. The basic rule is to align the tip of the wedge in the direction of the maximum electric field. Therefore, for the TE_{10} mode, the wedge should be placed as shown in Fig. 5.28. With such an arrangement, the maximum electric field at the centre will travel a longer path through the resistive wedge than a weaker electric field at the sides. Therefore, at the end of the wedge, the stronger electric field will be absorbed more than a weaker

electric field. Even if reflection occurs at the end, the reflected field, corresponding to the stronger electric field, will again travel a longer distance than that corresponding to a weaker field. Since the TE_{20} mode has zero electric field at the centre of the guide cross-section, the matched load geometry for the TE_{10} mode will not be suitable for matching the TE_{20} mode. For the TE_{20} mode, two such wedges, with their tips at $a/4$ and $3a/4$, should be used. This has been described in Fig. 5.29.

It is to be remembered that waveguide-matched loads provide matching by absorbing the incident and reflected power, and hence get heated easily. Therefore, for high-power applications, separate cooling mechanisms, such as the use of fans and running cold water jackets, are required to cool the matched load. For low-power applications, the generated heat is automatically lost by radiation and hence no separate cooling mechanism is required.

5.20.2 Short Circuit

A waveguide short circuit can be obtained by simply inserting a metal plate at the end of the guide, as shown in Fig. 5.30.

5.20.3 Short-circuit Plunger

A short-circuit plunger is basically a variable short circuit whose distance from the reference plane can be adjusted to achieve any reactive impedance in the range from $-j\infty$ to $j\infty$, as shown in Fig. 5.31(a). To obtain a better result, the contact resistance at the guide walls should be as low as possible and also be constant

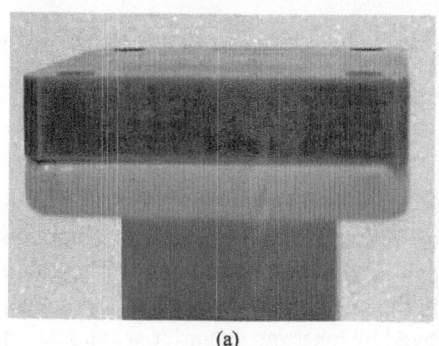
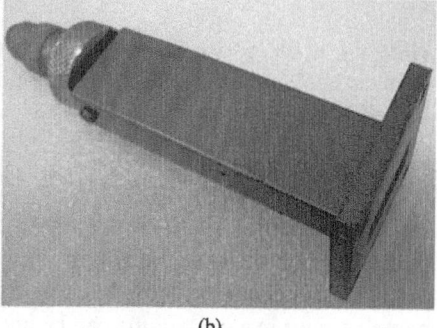

(a) (b)

Fig. 5.30 Waveguide short circuit (a) Fixed short (b) Variable short

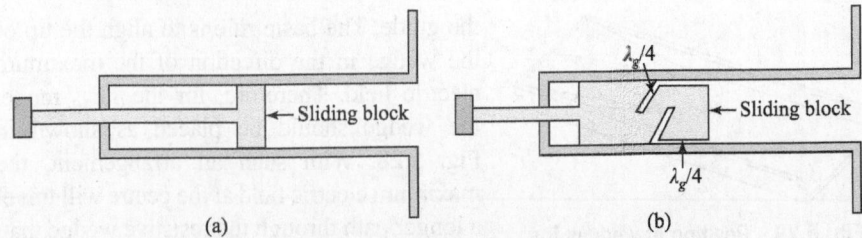

Fig. 5.31 Short-circuit plunger (a) Ordinary (b) Modified

along the line. These requirements can be met by modifying the shorting section as shown in Fig. 5.31(b). There are two cascaded $\lambda_g/4$ sections. The short circuit at the end of the second $\lambda_g/4$ section is transferred to an open circuit at the joint of the two $\lambda_g/4$ sections, which is further transferred to a short circuit at the reference plane of the short circuit.

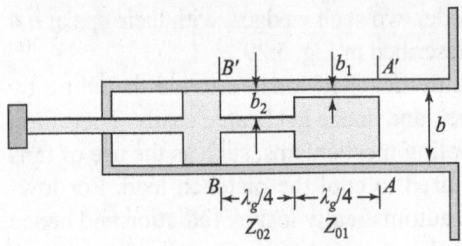

Fig. 5.32 Choke plunger

Another better-performing short-circuit plunger is a choke plunger. A choke plunger consists of two cascaded $\lambda_g/4$ sections, as shown in Fig. 5.32. The widths of the plungers are uniform and are slightly less than the internal guide width of the broad wall, whereas the height of the plungers are different—$b - 2b_1$ and $b - 2b_2$. For a satisfactory performance, b_1 should as small as possible, while b_2 be as large as possible. The extreme back section makes a sliding fit in the guide with almost zero air gap. If Z_{01} and Z_{02} are the characteristic impedances of the respective quarter-wave sections, then the input impedance seen at the plane AA' is given by the following relation:

$$Z_{in} = (Z_{01}/Z_{02})^2 Z_{sc} = (b_1/b_2)^2 Z_{sc} \qquad (5.290)$$

where Z_{sc} is the input impedance at the plane BB' and is almost zero. Since $b_2 \gg b_1$, that is, $Z_{02} \gg Z_{01}$, $Z_{in} \ll Z_{sc}$.

The main disadvantage of a choke plunger is its bandwidth limitation, which is only 20–30% at the mid frequency.

Note For a circular waveguide operating in the TE_{01} mode, choke plungers are not required because of the absence of longitudinal currents.

5.20.4 Open Circuit

Unlike in the case of a two-wire transmission line, it is not possible to obtain an open circuit for a waveguide simply by leaving the terminating end of the guide open. This is because radiation occurs from the open end of the guide. In practice, an open circuit in a waveguide can be achieved by inserting a quarter-wave section between the guide and a shorting plate.

5.21 Impedance Matching in Waveguides

Similar to the two-wire or other transmission line systems, waveguides are also subjected to load mismatch. The amount of mismatch depends on the load impedance and characteristic impedance of a guide. The concepts that are used to match a transmission line with a load are also applicable to waveguides and therefore are not discussed here. Some of the waveguide components used for load matching are described in the following subsections.

5.21.1 Irises

Irises are basically metallic obstacles inside a waveguide with an opening through which electromagnetic energy can pass. Depending on the geometry, waveguide irises may be classified as inductive, capacitive, or resonant irises, as shown in Fig. 5.33. When the dominant TE_{10} mode faces such discontinuities, higher-order TE and TM modes are generated to satisfy the boundary condition. These higher-order modes are evanescent in nature and die down within a quarter-wavelength distance from the junction. If the irises are inductive in nature, they will store magnetic energy from these higher-order modes; on the other hand, if the irises are capacitive in nature, they will store electric energy.

An inductive iris has an opening of cross-sectional dimension $l \times b$, where l is the width of the opening/aperture along the width of the guide, and b is the height of the opening/aperture along the height of the guide and equal to the guide height. That is, edges of inductive irises are perpendicular to the magnetic plane. As compared to an inductive iris, a capacitive iris has an opening of cross-sectional dimension $a \times d$, where a is the width of the opening/aperture along the width of the guide and equal to the guide width and d is the height of the opening/ aperture along the height of the guide. Therefore, the edges of a capacitive iris are perpendicular to the electric plane. Finally, a resonant iris has an opening of cross-sectional dimension $l \times d$, where l is the width of the opening/aperture along the width of the guide and d is the height of the opening/aperture along the height of the guide. It behaves as a shunt resistance at the resonance point, and as an inductor or a capacitor at other points. The amount of inductance and capacitance produced by such a circuit depends on its dimensions (width, height, and thickness). Normalized admittance produced by some commonly used irises, shown in Fig. 5.34, is as follows:

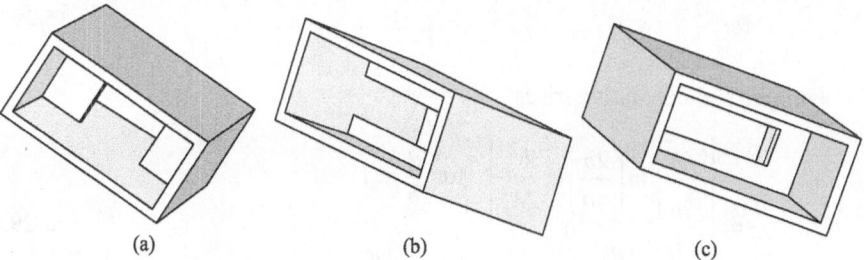

(a)	(b)	(c)

Fig. 5.33 Irises/Diaphragms inside rectangular waveguide (a) Inductive (b) Capacitive (c) Resonant

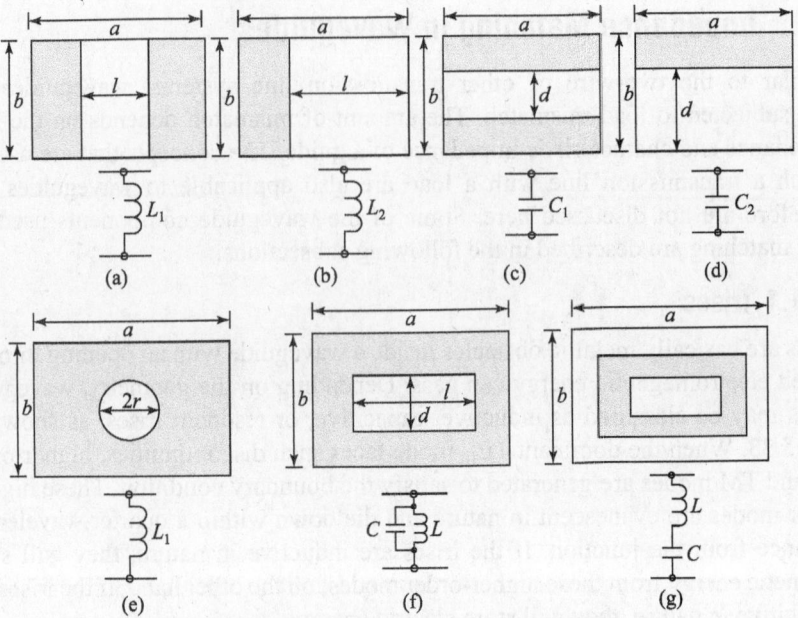

Fig. 5.34 Irises commonly used in rectangular waveguides
(a) Symmetrical inductive (b) Asymmetrical inductive
(c) Symmetrical capacitive (d) Asymmetrical capacitive
(e) Inductive (f) Parallel resonant (g) Series resonant

For symmetrical inductive irises:

$$\frac{B}{Y_0} = \begin{cases} -\dfrac{\lambda_g}{a}\cot^2\left(\dfrac{\pi l}{2a}\right)\Big/\left[1+\dfrac{1}{6}\left(\dfrac{\pi l}{\lambda_g}\right)^2\right] & \text{for } \dfrac{l}{a}\ll 1 \\[4mm] -\dfrac{\lambda_g}{a}\cot^2\left\{\dfrac{2\pi(a-l)}{a}\right\}\Big/\left[1+\dfrac{8}{3}\left\{\dfrac{\pi^2(a-l)^2}{\lambda_g^2}\right\}\right] & \text{for } \dfrac{a-l}{a}\ll 1 \end{cases}$$

$$(5.291)$$

For asymmetrical inductive irises:

$$\frac{B}{Y_0} = -\left[\lambda_g\pi^2(a-l)^2\Big/(2a^3)\right]\left[1+\frac{\pi^2}{\lambda_g^2}(a-l)^2\ln\left(\frac{\pi}{2}\right)\left(\frac{a-l}{a}\right)\right]$$

$$\text{for } \frac{a-l}{a}\ll 1 \qquad\qquad (5.292)$$

For symmetrical capacitive irises:

$$\frac{B}{Y_0} = \begin{cases} \left(\dfrac{4b}{\lambda_g}\right)\left[\ln\left(\dfrac{2b}{\pi d}\right)+\dfrac{b^2}{2\lambda_g^2}\right] & \text{for } \dfrac{d}{b}\ll 1 \\[4mm] \dfrac{\pi^2(b-d)^2}{2b\lambda_g} & \text{for } \dfrac{b-d}{b}\ll 1 \end{cases}$$

$$(5.293)$$

For asymmetrical capacitive irises:

$$\frac{B}{Y_0} = \begin{cases} \left[\left(\dfrac{8b}{\lambda_g}\right)\left|\ln\left(\dfrac{2b}{\pi d}\right)+\dfrac{2b^2}{\lambda_g^2}\right] & \text{for } \dfrac{d}{b} \ll 1 \\[3mm] \dfrac{\pi^2(b-d)^2}{b\lambda_g} & \text{for } \dfrac{b-d}{b} \ll 1 \end{cases} \tag{5.294}$$

For centre hole inductive irises:

$$\frac{B}{Y_0} = \frac{3ab\lambda_g}{16\pi r_0^3} \quad \text{for } r_0 \ll b \tag{5.295}$$

5.21.2 Posts and Screws

Waveguide posts and screws are often used for matching loads. A thin screw of diameter $\ll \lambda_g/4$ possesses capacitive reactance if $h < \lambda_g/4$, infinite reactance if $h = \lambda_g/4$, and inductive reactance if $h > \lambda_g/4$, where h is the depth of penetration inside a waveguide. However, the exact amount of capacitance/inductance produced by the screws depends on the radius and length of the post. Screws can also be inserted inside a waveguide through a longitudinal slot cut along the broad wall of a waveguide. Such an arrangement provides more flexibility of varying both the penetration and the position of the screw for better matching.

Thin cylindrical posts inserted through the broad wall and extending completely across the narrow width of the guide provide inductive susceptance for the dominant TE_{10} mode, whereas those inserted through the narrow wall and extending completely across the broad width of the guide provide capacitive susceptance. However, again, the exact value of the inductive/capacitive reactance depends on the diameter of the post. Figure 5.35 shows the structures of some waveguide posts and screws.

A combination of two screws/posts, one having length $h < \lambda_g/4$ and the other $h > \lambda_g/4$, and being separated by a distance of $3\lambda_g/8$, produces a parallel L–C combination and is shown in Fig. 5.36.

In practice, if a post or screw is parallel to the electric field, it will have more effect than when it is perpendicular to the electric field. Furthermore, posts or

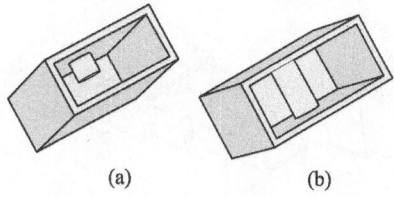

Fig. 5.35 Posts and screws in waveguides (a) Capacitive (b) Inductive

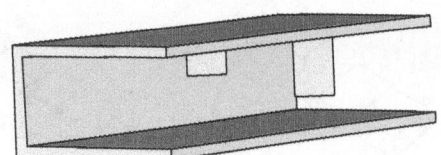

Fig. 5.36 Parallel L–C circuit (cross-sectional view)

screws that are nearer to the side walls produce more reflection than those at the centre.

5.21.3 Waveguide Stubs

In rectangular waveguides, stubs are connected in the E- or H-plane by placing a waveguide short-circuit plunger, as shown in Fig. 5.37. For the E-plane stub, the electric field lines penetrate into the stub guide and thus offer reactance in series with the main line, whereas for the H-plane stub, the magnetic field lines penetrate into the stub guide and thus present susceptance in parallel with the main guide.

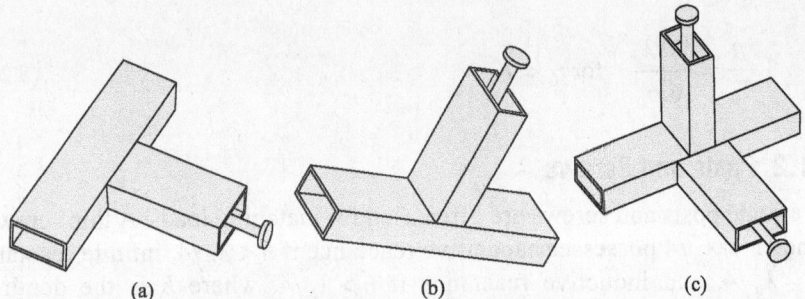

Fig. 5.37 Waveguide stubs (a) H-plane (b) E-plane (c) E–H plane

Such stubs are also called E- and H-plane tuners. A combination of an E- and an H-plane tuner, called an E–H-plane tuner, is also used to match a wide range of load impedances.

5.21.4 Waveguide Quarter-wave Impedance Transformers

Waveguide quarter-wave impedance transformers are designed by making steps either in the E- or in the H-plane, as shown in Fig. 5.38. The corresponding step sizes can be calculated from the following relations:

$$b_2 = \sqrt{b_1/b_3} \qquad (5.296)$$

and
$$\frac{a_2}{\lambda_{g2}} = \sqrt{\frac{a_1 a_3}{\lambda_{g1}\lambda_{g3}}} \qquad (5.297)$$

where λ_{g1}, λ_{g2} and λ_{g3}, are the guided wavelengths in guides 1, 2, and 3, respectively.

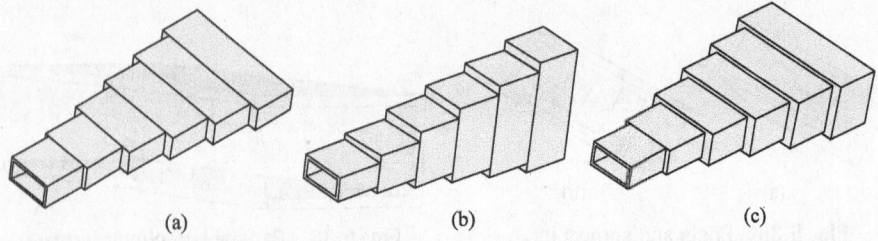

Fig. 5.38 Quarter-wave waveguide transformers (a) H-plane (b) E-plane (c) E–H plane

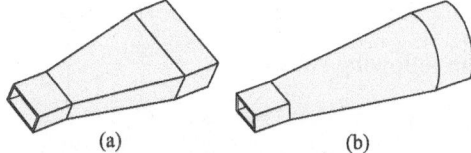

Fig. 5.39 Waveguide tapers between two dissimilar waveguides (a) H-plane (b) Rectangular to circular

5.21.5 Waveguide Tapers

A waveguide taper results from the gradual connection of two similar or dissimilar waveguides of different cross-sections, as shown in Fig. 5.39. For a smooth transition and lower reflection coefficient, the taper length must be at least $2\lambda_g$ at the operating frequency. Depending on the flaring dimensions, tapers can be classified as E-, H-, or E–H-plane tapers.

5.22 Waveguide Bends and Twists

The following subsections discuss about waveguide bends and twists.

Bends In microwave circuits, often a waveguide is required to be bent to achieve desired flexibility for connection with adapters, terminators, or other loads. However, any abrupt change in the shape of the guide will cause reflection of waves from the discontinuity. Therefore, special care must be taken to form the bends. Depending on the geometry, waveguide bends may be classified as E-plane, H-plane, or sharp bends, as shown in Fig. 5.40. Out of these, E- and H-plane bends are gradual bends. E-plane bends distort only the electric field distribution, whereas H-plane bends distort only the magnetic field distribution. To achieve a satisfactory performance, the bending radius of both the bends must be greater than two wavelengths, and the mean length of the bend must be an odd multiple of a quarter wavelength, to cancel reflections from both ends. In contrast to smooth bends, sharp bends can also be used. For the sharp bend shown in Fig. 5.40(c), the two 45° bends are quarter-wavelength apart, and therefore reflections that occur at each of the bends cancel each other, leaving the fields of the main guide as if no reflections have occurred. The bend shown in Fig. 5.40(c) is also called a *mittered corner*.

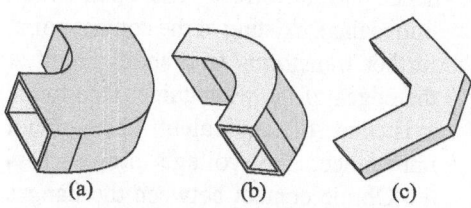

Fig. 5.40 Waveguide bends (a) H-plane (b) E-plane (c) 45°

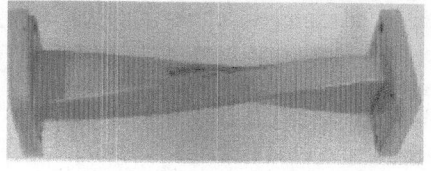

Fig. 5.41 Twist in waveguide

Twists Twists are sometimes used to achieve proper phase matching with the load. As before, care must be taken to form the twist so that the reflection becomes minimum. This can be achieved by twisting the guide gradually over a length of greater than two wavelengths, as shown in Fig. 5.41.

5.23 Waveguide Flanges

Two types of flanges are discussed in the following subsections.

5.23.1 Cover Flanges

A cover flange is basically a square-shaped metal block having a central slot of dimensions equal to the outer dimensions of the waveguide and four holes at each corner, as shown in Fig. 5.42(a). The surface of a flange must be smooth and clean to make a proper contact. Since RF current flows across it, the metal should also be highly conducting in order to reduce the Ohmic loss. A properly designed flange produces an SWR of less than 1.03 and has the following advantages: simpler structure, ease of fabrication, and low cost. However, there is often a fair chance of the existence of an air gap between the two flanges connected together, which may cause voltage breakdown in high-power applications. A schematic diagram of the connection between two waveguides using a cover flange is shown in Fig. 5.42(b).

5.23.2 Choke Flanges

A good alternative of a cover flange, for use in high-power applications, is a choke flange. It is machined to form a radial transmission line of length $\lambda_g/4$ between the guide and the point of contact of the two flanges, as shown in Fig. 5.43. At the point of contact between two flanges, another line of length $\lambda_g/4$ is formed by a circular groove, as shown in the same figure. The short circuit at the end of this groove is transformed into an open circuit at the contact point of the flanges. Any resistance that exists at the contact point is in series with this open-circuit impedance and hence has no effect. The open-circuit impedance, existing at the contact point, further transforms to a short circuit at the edges of the waveguide. Due to the existence of equivalent short-circuit impedance, the voltage drop across the Ohmic contact between the flanges is very small and thus eliminates the probability of voltage breakdown in high-power applications. In spite of these advantages, choke flanges have one major disadvantage: these are more frequency dependent than cover flanges due to the existence of the $\lambda_g/4$ sections. A choke flange can produce an SWR as low as 1.05.

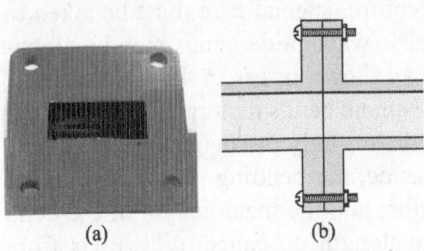

(a)   (b)

Fig. 5.42 Waveguide cover flange (a) Typical cover flange (b) Connection between two waveguides using cover flange

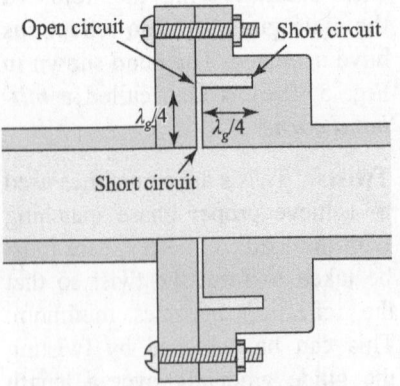

Fig. 5.43 Connection of two waveguides using choke flange

5.24 Introduction to Circular Cylindrical Waveguides

A circular cylindrical waveguide is a hollow metallic pipe having a circular cross-section, as shown in Fig. 5.44. Like a rectangular waveguide, the dimension of a circular waveguide is also determined by the cut-off frequency of the lowest order or the dominant mode and the next higher-order mode.

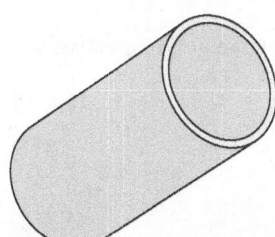

The Maxwell's curl equations inside a waveguide at steady state are as follows:

$$\vec{\nabla} \times \vec{E} = -j\omega\mu\vec{H} \tag{5.298}$$

Fig. 5.44 Hollow circular cylindrical waveguide

and

$$\vec{\nabla} \times \vec{H} = j\omega\varepsilon_c\vec{E} \tag{5.299}$$

Using the definition of curl of a vector $\vec{A}$ in a cylindrical coordinate, that is,

$$\vec{\nabla} \times \vec{A} = \frac{1}{r}\begin{vmatrix} \hat{a}_r & r\hat{a}_\phi & \hat{a}_z \\ \dfrac{\partial}{\partial r} & \dfrac{\partial}{\partial \phi} & \dfrac{\partial}{\partial z} \\ A_r & rA_\phi & A_z \end{vmatrix} \tag{5.300}$$

in Eqs (5.298) and (5.299), we get the following relations:

$$\frac{\partial E_z}{\partial \phi} - r\frac{\partial E_\phi}{\partial z} = -j\omega\mu r H_r \tag{5.301}$$

$$\frac{\partial E_r}{\partial z} - \frac{\partial E_z}{\partial r} = -j\omega\mu H_\phi \tag{5.302}$$

$$\frac{\partial}{\partial r}\left(rE_\phi\right) - \frac{\partial E_r}{\partial \phi} = -j\omega\mu r H_z \tag{5.303}$$

$$\frac{\partial H_z}{\partial \phi} - r\frac{\partial H_\phi}{\partial z} = j\omega\varepsilon_c r E_r \tag{5.304}$$

$$\frac{\partial H_r}{\partial z} - \frac{\partial H_z}{\partial r} = j\omega\varepsilon_c E_\phi \tag{5.305}$$

and

$$\frac{\partial}{\partial r}\left(rH_\phi\right) - \frac{\partial H_r}{\partial \phi} = j\omega\varepsilon_c r E_z \tag{5.306}$$

where the electric and magnetic fields are expressed as follows:

$$\vec{E}(r,\phi,z) = E_r\hat{a}_r + E_\phi\hat{a}_\phi + E_z\hat{a}_z = \left(E_{0r}\hat{a}_r + E_{0\phi}\hat{a}_\phi + E_{0z}\hat{a}_z\right)e^{-\gamma z} \tag{5.307}$$

and

$$\vec{H}(r,\phi,z) = H\hat{a}_r + H_\phi\hat{a}_\phi + H_z\hat{a}_z = \left(H_{0r}\hat{a}_r + H_{0\phi}\hat{a}_\phi + H_{0z}\hat{a}_z\right)e^{-\gamma z} \tag{5.308}$$

5.25 Propagation of TE Mode in Circular Waveguides

The TE mode is characterized by the following relation:

$$E_{0z} = 0 \tag{5.309}$$

Substituting Eqs (5.307)–(5.309) into Eqs (5.301)–(5.306), we get, respectively,

$$\gamma E_{0\phi} = -j\omega\mu H_{0r} \tag{5.310}$$

$$\gamma E_{0r} = j\omega\mu H_{0\phi} \tag{5.311}$$

$$\frac{\partial}{\partial r}\left(r E_{0\phi}\right) - \frac{\partial}{\partial \phi}\left(E_{0r}\right) = -j\omega\mu r H_{0z} \tag{5.312}$$

$$\frac{\partial}{\partial \phi}\left(H_{0z}\right) + r\gamma H_{0\phi} = j\omega\varepsilon_c r E_{0r} \tag{5.313}$$

$$\gamma H_{0r} + \frac{\partial}{\partial r}\left(H_{0z}\right) = -j\omega\varepsilon_c E_{0\phi} \tag{5.314}$$

$$\frac{\partial}{\partial r}\left(r H_{0\phi}\right) - \frac{\partial}{\partial \phi}\left(H_{0r}\right) = 0 \tag{5.315}$$

From Eq. (5.310), we get the following expression:

$$H_{0r} = -\frac{\gamma}{j\omega\mu} E_{0\phi} \tag{5.316}$$

Substituting the expression of H_{0r} from Eq. (5.316) into Eq. (5.314), we get the following relation:

$$\left(\gamma^2 + \omega^2 \mu\varepsilon_c\right) E_{0\phi} = j\omega\mu \frac{\partial}{\partial r}\left(H_{0z}\right) \tag{5.317}$$

With the help of Eq. (5.91), Eq. (5.317) reduces to the following form:

$$E_{0\phi} = \frac{j\omega\mu}{k_c^2} \frac{\partial}{\partial r}\left(H_{0z}\right) \tag{5.318}$$

Substituting the expression of $E_{0\phi}$ from Eq. (5.318) back into Eq. (5.316), we get the following relation:

$$H_{0r} = -\frac{\gamma}{k_c^2} \frac{\partial}{\partial r}\left(H_{0z}\right) \tag{5.319}$$

From Eq. (5.311), we get the following relation:

$$H_{0\phi} = \frac{\gamma}{j\omega\mu} E_{0r} \tag{5.320}$$

Substituting the expression of $H_{0\phi}$ from Eq. (5.320) into Eq. (5.313), the following expression is obtained:

$$\left(\gamma^2 + \omega^2 \mu\varepsilon_c\right) r E_{0r} = -j\omega\mu \frac{\partial}{\partial \phi}\left(H_{0z}\right) \tag{5.321}$$

With the help of Eq. (5.91), Eq. (5.321) reduces to the following form:

$$E_{0r} = -\frac{j\omega\mu^*}{k_c^2}\frac{1}{r}\frac{\partial}{\partial\phi}(H_{0z})$$

(5.322)

Substituting the expression of E_{0r} from Eq. (5.322) back into Eq. (5.320), we get the following relation:

$$H_{0\phi} = -\frac{\gamma}{k_c^2}\frac{1}{r}\frac{\partial}{\partial\phi}(H_{0z})$$

(5.323)

Multiplying Eq. (5.314) by r and then differentiating with respect to r, the following equation is obtained:

$$\gamma\frac{\partial}{\partial r}(rH_{0r}) + \frac{\partial}{\partial r}\left\{r\frac{\partial}{\partial r}(H_{0z})\right\} = -j\omega\varepsilon_c\frac{\partial}{\partial r}(rE_{0\phi})$$

(5.324)

Again differentiating Eq. (5.313) with respect to ϕ and dividing through by r, we get the following relation:

$$\frac{1}{r}\frac{\partial^2 H_{0z}}{\partial\phi^2} + \gamma\frac{\partial H_{0\phi}}{\partial\phi} = j\omega\varepsilon_c\frac{\partial E_{0r}}{\partial\phi}$$

(5.325)

Adding Eqs (5.324) and (5.325), we get the following expression:

$$\frac{1}{r}\frac{\partial^2 H_{0z}}{\partial\phi^2} + \gamma\frac{\partial H_{0\phi}}{\partial\phi} + \gamma\frac{\partial}{\partial r}(rH_{0r}) + \frac{\partial}{\partial r}\left(r\frac{\partial H_{0z}}{\partial r}\right)$$
$$= j\omega\varepsilon_c\left\{\frac{\partial E_{0r}}{\partial\phi} - \frac{\partial}{\partial r}(rE_{0\phi})\right\}$$

(5.326)

With the help of Eq. (5.312), Eq. (5.326) reduces to the following form:

$$\frac{\gamma}{r}\left[\frac{\partial H_{0\phi}}{\partial\phi} + \frac{\partial}{\partial r}(rH_{0r})\right] + \frac{1}{r}\frac{\partial}{\partial r}\left(r\frac{\partial H_{0z}}{\partial r}\right) + \frac{1}{r^2}\frac{\partial^2 H_{0z}}{\partial\phi^2}$$
$$= -\omega^2\mu\varepsilon_c H_{0z}$$

(5.327)

Substituting the expression of H_{0r} from Eq. (5.316) and the expression of $H_{0\phi}$ from Eq. (5.320) into Eq. (5.327), we get the following expression:

$$-\frac{\gamma^2}{j\omega\mu r}\left[\frac{\partial}{\partial r}(rE_{0\phi}) - \frac{\partial}{\partial\phi}(E_{0r})\right] + \frac{1}{r}\frac{\partial}{\partial r}\left(r\frac{\partial H_{0z}}{\partial r}\right) + \frac{1}{r^2}\frac{\partial^2 H_{0z}}{\partial\phi^2}$$
$$= -\omega^2\mu\varepsilon_c H_{0z}$$

(5.328)

Using Eq. (5.312) for the first term on the LHS in Eq. (5.328), we get the following equation:

$$\frac{1}{r}\frac{\partial}{\partial r}\left(r\frac{\partial H_{0z}}{\partial r}\right) + \frac{1}{r^2}\frac{\partial^2 H_{0z}}{\partial\phi^2} + \left(\gamma^2 + \omega^2\mu\varepsilon_c\right)H_{0z} = 0$$

(5.329)

Again using Eq. (5.91) in Eq. (5.329), we get the following relation:

$$\frac{1}{r}\frac{\partial}{\partial r}\left(r\frac{\partial H_{0z}}{\partial r}\right) + \frac{1}{r^2}\frac{\partial^2 H_{0z}}{\partial\phi^2} + k_c^2 H_{0z} = 0$$

(5.330)

To solve Eq. (5.330), let us start with the following trial solution:

$$H_{0z}(r,\phi) = H_{0z}(r)H_{0z}(\phi) \qquad (5.331)$$

Substituting Eq. (5.331) into Eq. (5.330) and then dividing by $H_{0z}(r)H_{0z}(\phi)/r^2$, the following equation is obtained:

$$\frac{r}{H_{0z}(r)}\frac{\partial}{\partial r}\left(r\frac{\partial H_{0z}(r)}{\partial r}\right) + \frac{1}{H_{0z}(\phi)}\frac{\partial^2 H_{0z}(\phi)}{\partial \phi^2} + k_c^2 r^2 = 0$$

$$r^2 \frac{1}{H_{0z}(r)}\frac{\partial^2 H_{0z}(r)}{\partial r^2} + r\frac{1}{H_{0z}(r)}\frac{\partial H_{0z}(r)}{\partial r} + k_c^2 r^2$$

or
$$= -\frac{1}{H_{0z}(\phi)}\frac{\partial^2 H_{0z}(\phi)}{\partial \phi^2} \qquad (5.332)$$

The LHS of Eq. (5.332) is a function of r, whereas its RHS is a function of ϕ only. This implies that Eq. (5.332) may be separated in the following two equations:

$$r^2 \frac{1}{H_{0z}(r)}\frac{\partial^2 H_{0z}(r)}{\partial r^2} + r\frac{1}{H_{0z}(r)}\frac{\partial H_{0z}(r)}{\partial r} + k_c^2 r^2 = n^2$$

or
$$r^2 \frac{\partial^2 H_{0z}(r)}{\partial r^2} + r\frac{\partial H_{0z}(r)}{\partial r} + \left(k_c^2 r^2 - n^2\right)H_{0z}(r) = 0 \qquad (5.333)$$

and
$$\frac{\partial^2 H_{0z}(\phi)}{\partial \phi^2} + n^2 H_{0z}(\phi) = 0 \qquad (5.334)$$

where n^2 is a constant. The general solution of Eq. (5.333) is expressed as follows:

$$H_{0z}(r) = AJ_n(k_c r) + BN_n(k_c r) \qquad (5.335)$$

where $J_n(k_c r)$ is the Bessel's function of order n of argument $(k_c r)$ and N_n is the Numen's function of order n of argument $(k_c r)$.

Now
$$|N_n(k_c r)| \to \infty \text{ at } r = 0 \qquad (5.336)$$

which implies that $B = 0$ (5.337)

and we get $H_{0z}(r) = AJ_n(k_c r)$ (5.338)

The solution of Eq. (5.334) is expressed as follows:

$$H_{0z}(\phi) = P\cos(n\phi) + Q\sin(n\phi) \qquad (5.339)$$

Substituting Eqs (5.338) and (5.339) into Eq. (5.331), we get the following expression:

$$H_{0z} = AJ_n(k_c r)\{P\cos(n\phi) + Q\sin(n\phi)\}$$

or
$$H_{0z} = J_n(k_c r)\{P'\cos(n\phi) + Q'\sin(n\phi)\} \qquad (5.340)$$

where $P' = AP$ (5.341)

and $Q' = AQ$ (5.342)

In Eq. (5.340), there are two arbitrary amplitude constants, P' and Q', which control the amplitudes of $\cos(n\phi)$ and $\sin(n\phi)$ terms, respectively. Because of the azimuthal symmetry of the waveguide, both these terms are valid and can be present. The actual amplitudes of these two terms depend on the excitation of the waveguide. However, by rotating the waveguide about the z-axis, any one of these two terms can be made equal to zero. Under such a circumstance, if we assume that the value of the other amplitude term is H_0, then Eq. (5.340) can be written as:

$$H_{0z} = H_0 J_n(k_c r) \begin{bmatrix} \cos(n\phi) \\ \sin(n\phi) \end{bmatrix} \tag{5.343}$$

Substituting Eq. (5.343) into Eqs (5.322), (5.318), (5.319), and (5.323), we get the following expressions:

$$E_{0r} = -\frac{j\omega\mu}{k_c^2}\frac{1}{r} H_0 n J_n(k_c r) \begin{bmatrix} -\sin(n\phi) \\ \cos(n\phi) \end{bmatrix} \tag{5.344}$$

$$E_{0\phi} = \frac{j\omega\mu}{k_c} H_0 J_n'(k_c r) \begin{bmatrix} \cos(n\phi) \\ \sin(n\phi) \end{bmatrix} \tag{5.345}$$

$$H_{0r} = -\frac{\gamma}{k_c} H_0 J_n'(k_c r) \begin{bmatrix} \cos(n\phi) \\ \sin(n\phi) \end{bmatrix} \tag{5.346}$$

$$H_{0\phi} = -\frac{\gamma}{k_c^2}\frac{1}{r} H_0 n J_n(k_c r) \begin{bmatrix} -\sin(n\phi) \\ \cos(n\phi) \end{bmatrix} \tag{5.347}$$

where $\quad \dfrac{\partial}{\partial r}\{J_n(k_c r)\} = J_n'(k_c r) \tag{5.348}$

Now at the boundaries, H_r should vanish, that is,

$$H_{0r} = 0 \text{ at } r = R \tag{5.349}$$

Applying the boundary condition of Eq. (5.349) into Eq. (5.346), we get the following relation:

$$J_n'(k_c R) = 0 \tag{5.350}$$

as $\quad \dfrac{\gamma}{k_c} H_0 \begin{bmatrix} \cos(n\phi) \\ \sin(n\phi) \end{bmatrix} \neq 0$

For $n = 0$, Eq. (5.350) gives the following relations:

$$k_{c01}R = 3.83 \text{ or } k_{c01} = \frac{3.83}{R} \tag{5.351}$$

and $\quad k_{c02}R = 7.02 \text{ or } k_{c02} = \dfrac{7.02}{R} \tag{5.352}$

Similarly, for $n = 1$, we get the following expressions:

$$k_{c11} = \frac{1.84}{R} \tag{5.353}$$

and $\quad k_{c12} = \dfrac{5.33}{R}$ $\qquad\qquad\qquad\qquad$ (5.354)

The zeros of $J_n'(k_c r)$ for the TE_{nm} modes are given in Table 5.1.
For a lossless dielectric:

$$k_{c,nm}^2 = -\beta_{nm}^2 + \omega^2 \mu\varepsilon \quad \text{or} \quad \beta_{nm}^2 = \omega^2 \mu\varepsilon - k_{c,nm}^2 \qquad (5.355)$$

Table 5.1   Zeros of $J_n'(k_c r)$ for TE_{nm} modes

	$n=0$	$n=1$	$n=2$	$n=3$	$n=4$	$n=5$
$m=1$	3.832	1.841	3.054	4.201	5.317	6.416
$m=2$	7.016	5.331	6.706	8.015	9.282	10.520
$m=3$	10.173	8.536	9.969	11.346	12.682	13.987

At the cut-off frequency:

$$\omega_c^2 \mu\varepsilon - k_{c,nm}^2 = 0 \quad \text{or} \quad f_c = \frac{k_{c,nm}}{2\pi\sqrt{\mu\varepsilon}} \qquad (5.356)$$

Since for a given R, k_{c11} has the minimum value, the dominant TE mode in a circular waveguide is TE_{11}. Figure 5.45 shows the field patterns of different TE modes in a circular waveguide.

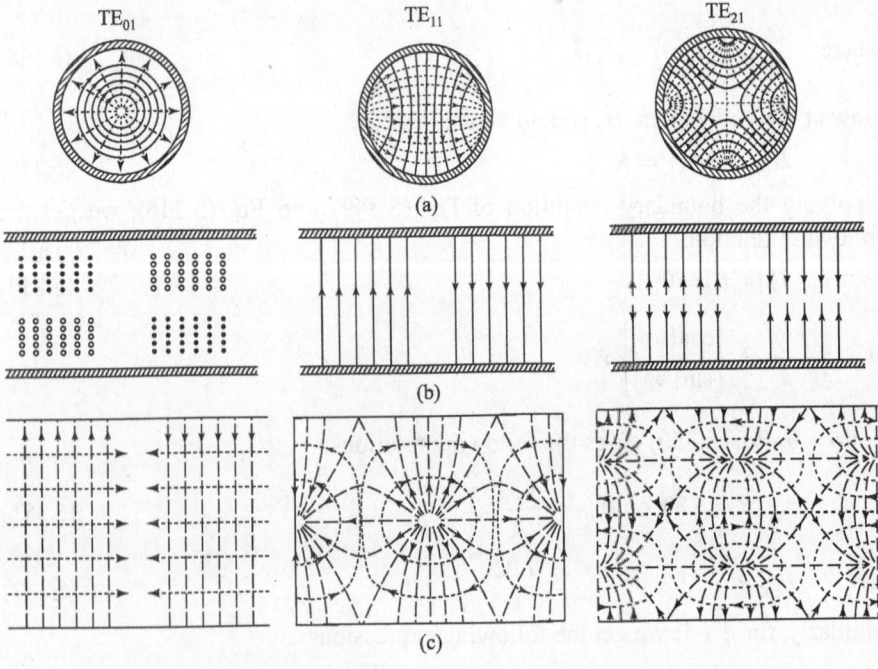

Fig. 5.45   Field patterns of different TE modes in circular waveguides
(a) Cross-sectional view (b) Longitudinal view (c) Surface view

EXAMPLE 5.8 A circular cylindrical waveguide has an inner diameter of 2.383 cm. Calculate the cut-off frequencies for the (a) TE_{01}, (b) TE_{02}, (c) TE_{11}, and (d) TE_{12} modes.

Solution   Given: $R = 2.383/2 = 1.192$ cm

(a) $k_{c01} = \dfrac{3.83}{R} = \dfrac{3.83}{1.192} = 3.213$

$f_{c,TE_{01}} = \dfrac{k_{cmn}}{2\pi\sqrt{\mu\varepsilon}} = \dfrac{3.213 \times 3 \times 10^{10}}{2\pi} = 15.34 \text{ GHz}$

(b) $k_{c02} = \dfrac{7.02}{R} = \dfrac{7.02}{1.192} = 5.889$

$f_{c,TE_{02}} = \dfrac{k_{cmn}}{2\pi\sqrt{\mu\varepsilon}} = \dfrac{5.889 \times 3 \times 10^{10}}{2\pi} = 28.118 \text{ GHz}$

(c) $k_{c11} = \dfrac{1.84}{R} = \dfrac{1.84}{1.192} = 1.544$

$f_{c,TE_{11}} = \dfrac{k_{cmn}}{2\pi\sqrt{\mu\varepsilon}} = \dfrac{1.544 \times 3 \times 10^{10}}{2\pi} = 7.370 \text{ GHz}$

(d) $k_{c12} = \dfrac{5.33}{R} = \dfrac{5.33}{1.192} = 4.471$

$f_{c,TE_{12}} = \dfrac{k_{cmn}}{2\pi\sqrt{\mu\varepsilon}} = \dfrac{4.471 \times 3 \times 10^{10}}{2\pi} = 21.347 \text{ GHz}$

Practice Problem

5.10 A circular cylindrical waveguide has an inner diameter of 1.27 cm. Calculate the cut-off frequencies for the (a) TE_{01}, (b) TE_{02}, (c) TE_{11}, and (d) TE_{12} modes.

[28.796 GHz, 52.784 GHz, 13.837 GHz, 40.078 GHz]

5.26 Propagation of TM Mode in Circular Waveguides

For the TM mode: $H_z = 0$ (5.357)

Substituting Eqs (5.307), (5.308) and (5.357) into Eqs (5.301)–(5.306), we get, respectively, the following equations:

$$\frac{\partial E_{0z}}{\partial \phi} + \gamma r E_{0\phi} = -j\omega\mu r H_{0r} \quad (5.358)$$

$$\gamma E_{0r} + \frac{\partial E_{0z}}{\partial r} = j\omega\mu H_{0\phi} \quad (5.359)$$

$$\frac{\partial}{\partial r}\left(r E_{0\phi}\right) - \frac{\partial E_{0r}}{\partial \phi} = 0 \quad (5.360)$$

$$\gamma H_{0\phi} = j\omega\varepsilon_c E_{0r} \quad (5.361)$$

$$\gamma H_{0r} = -j\omega\varepsilon_c E_{0\phi} \quad (5.362)$$

and
$$\frac{\partial}{\partial r}\left(rH_{0\phi}\right)-\frac{\partial H_{0r}}{\partial \phi}=j\omega\varepsilon_c rE_{0z} \tag{5.363}$$

From Eq. (5.362), we get the following relation:

$$E_{0\phi}=-\frac{\gamma}{j\omega\varepsilon_c}H_{0r} \tag{5.364}$$

Substituting the expression of $E_{0\phi}$ from Eq. (5.364) into Eq. (5.358), the following relation is obtained:

$$\left(\gamma^2+\omega^2\mu\varepsilon_c\right)rH_{0r}=j\omega\varepsilon_c\frac{\partial E_{0z}}{\partial \phi} \tag{5.365}$$

Using Eq. (5.91), Eq. (5.365) reduces to the following form:

$$H_{0r}=\frac{j\omega\varepsilon_c}{k_c^2}\frac{1}{r}\frac{\partial E_{0z}}{\partial \phi} \tag{5.366}$$

Substituting the expression of H_{0r} from Eq. (5.366) back into Eq. (5.364), we get the following relation:

$$E_{0\phi}=-\frac{\gamma}{k_c^2}\frac{1}{r}\frac{\partial E_{0z}}{\partial \phi} \tag{5.367}$$

Again from Eq. (5.361), the following relation is obtained:

$$E_{0r}=\frac{\gamma}{j\omega\varepsilon_c}H_{0\phi} \tag{5.368}$$

Substituting the expression of E_{0r} from Eq. (5.368) into Eq. (5.359), we get the following equation:

$$\left(\gamma^2+\omega^2\mu\varepsilon_c\right)H_{0\phi}=-j\omega\varepsilon_c\frac{\partial E_{0z}}{\partial r} \tag{5.369}$$

Using Eq. (5.91), Eq. (5.369) reduces to the following form:

$$H_{0\phi}=-\frac{j\omega\varepsilon_c}{k_c^2}\frac{\partial E_{0z}}{\partial r} \tag{5.370}$$

Substituting the expression of $H_{0\phi}$ from Eq. (5.370) back into Eq. (5.368), we get the following relation:

$$E_{0r}=-\frac{\gamma}{k_c^2}\frac{\partial E_{0z}}{\partial r} \tag{5.371}$$

Now multiplying Eq. (5.359) by r and then differentiating with respect to r, we get the following equation:

$$\gamma\frac{\partial}{\partial r}\left(rE_{0r}\right)+\frac{\partial}{\partial r}\left(r\frac{\partial E_{0z}}{\partial r}\right)=j\omega\mu\frac{\partial}{\partial r}\left(rH_{0\phi}\right) \tag{5.372}$$

Differentiating Eq. (5.358) with respect to ϕ and then dividing through by r, we get the following expression:

$$\frac{1}{r}\frac{\partial^2 E_{0z}}{\partial \phi^2} + \gamma \frac{\partial E_{0\phi}}{\partial \phi} = -j\omega\mu \frac{\partial H_{0r}}{\partial \phi} \tag{5.373}$$

Adding Eqs (5.372) and (5.373), we get the following expression:

$$\gamma \frac{\partial}{\partial r}(rE_{0r}) + \frac{\partial}{\partial r}\left(r\frac{\partial E_{0z}}{\partial r}\right) + \frac{1}{r}\frac{\partial^2 E_{0z}}{\partial \phi^2} + \gamma \frac{\partial E_{0\phi}}{\partial \phi}$$

$$= j\omega\mu \left[\frac{\partial}{\partial r}(rH_{0\phi}) - \frac{\partial H_{0r}}{\partial \phi}\right] \tag{5.374}$$

Using Eq. (5.363) in Eq. (5.374), the following equation is obtained:

$$\gamma\left[\frac{\partial}{\partial r}(rE_{0r}) + \frac{\partial E_{0\phi}}{\partial \phi}\right] + \frac{\partial}{\partial r}\left(r\frac{\partial E_{0z}}{\partial r}\right) + \frac{1}{r}\frac{\partial^2 E_{0z}}{\partial \phi^2} = -\omega^2 \mu\varepsilon_c rE_{0z} \tag{5.375}$$

Substituting the expressions of E_{0r} and $E_{0\phi}$ from Eqs (5.368) and (5.364) into Eq. (5.375), we get the following relation:

$$\frac{\gamma^2}{j\omega\varepsilon_c}\left[\frac{\partial}{\partial r}(rH_{0\phi}) - \frac{\partial H_{0r}}{\partial \phi}\right] + \frac{\partial}{\partial r}\left(r\frac{\partial E_{0z}}{\partial r}\right) + \frac{1}{r}\frac{\partial^2 E_{0z}}{\partial \phi^2} = -\omega^2 \mu\varepsilon_c rE_{0z} \tag{5.376}$$

Again using Eq. (5.363) in Eq. (5.376), we get the following equation:

$$\frac{\partial}{\partial r}\left(r\frac{\partial E_{0z}}{\partial r}\right) + \frac{1}{r}\frac{\partial^2 E_{0z}}{\partial \phi^2} + \left(\gamma^2 + \omega^2\mu\varepsilon_c\right)rE_{0z} = 0 \tag{5.377}$$

Using Eq. (5.91) in Eq. (5.377), the following expression is obtained:

$$\frac{1}{r}\frac{\partial}{\partial r}\left(r\frac{\partial E_{0z}}{\partial r}\right) + \frac{1}{r^2}\frac{\partial^2 E_{0z}}{\partial \phi^2} + k_c^2 E_{0z} = 0 \tag{5.378}$$

To solve Eq. (5.378), let us assume that

$$E_{0z}(r,\phi) = E_{0z}(r)E_{0z}(\phi) \tag{5.379}$$

Substitution of Eq. (5.379) into Eq. (5.378) yields the following relation:

$$\frac{1}{r}E_{0z}(\phi)\frac{\partial}{\partial r}\left(r\frac{\partial E_{0z}(r)}{\partial r}\right) + \frac{1}{r^2}E_{0z}(r)\frac{\partial^2 E_{0z}(\phi)}{\partial \phi^2} + k_c^2 E_{0z}(r)E_{0z}(\phi) = 0$$

or $\quad \dfrac{1}{r}\dfrac{1}{E_{0z}(r)}\left[\dfrac{\partial E_{0z}(r)}{\partial r} + r\dfrac{\partial^2 E_{0z}(r)}{\partial r^2}\right] + \dfrac{1}{r^2}\dfrac{1}{E_{0z}(\phi)}\dfrac{\partial^2 E_{0z}(\phi)}{\partial \phi^2} + k_c^2 = 0$

or $\quad r^2 \dfrac{1}{E_{0z}(r)} \dfrac{\partial^2 E_{0z}(r)}{\partial r^2} + r \dfrac{1}{E_{0z}(r)} \dfrac{\partial E_{0z}(r)}{\partial r} + k_c^2 r^2 = -\dfrac{1}{E_{0z}(\phi)} \dfrac{\partial^2 E_{0z}(\phi)}{\partial \phi^2}$ (5.380)

The LHS of Eq. (5.380) is a function of r, whereas its RHS is a function of only ϕ. This implies that Eq. (5.380) may be separated in the following two equations:

$$r^2 \dfrac{1}{E_{0z}(r)} \dfrac{\partial^2 E_{0z}(r)}{\partial r^2} + r \dfrac{1}{E_{0z}(r)} \dfrac{\partial E_{0z}(r)}{\partial r} + k_c^2 r^2 = n^2$$

or $\quad r^2 \dfrac{\partial^2 E_{0z}(r)}{\partial r^2} + r \dfrac{\partial E_{0z}(r)}{\partial r} + \left(k_c^2 r^2 - n^2\right) E_{0z}(r) = 0$ (5.381)

and $\quad \dfrac{\partial^2 E_{0z}(\phi)}{\partial \phi^2} + n^2 E_{0z}(\phi) = 0$ (5.382)

where n^2 is a constant. The general solution of Eq. (5.381) is expressed as follows:

$$E_{0z}(r) = A J_n(k_c r) + B N_n(k_c r)$$ (5.383)

where $J_n(k_c r)$ is the Bessel's function of order n of argument $(k_c r)$ and N_n is the Numen's function of order n of argument $(k_c r)$.

Now $\quad |N_n(k_c r)| \to \infty$ at $r = 0$ (5.384)

which implies that $B = 0$ (5.385)

and we get the following expression:

$$E_{0z}(r) = A J_n(k_c r)$$ (5.386)

The solution of Eq. (5.382) is given by the following equation:

$$E_{0z}(\phi) = P \cos(n\phi) + Q \sin(n\phi)$$ (5.387)

Substituting Eqs (5.386) and (5.387) into Eq. (5.379), we get the following equality:

$$E_{0z} = A J_n(k_c r)\{P \cos(n\phi) + Q \sin(n\phi)\}$$

or $\quad E_{0z} = J_n(k_c r)\{P' \cos(n\phi) + Q' \sin(n\phi)\}$ (5.388)

where $\quad P' = AP$ (5.389)

and $\quad Q' = AQ$ (5.390)

Following previous arguments, Eq. (5.388) can also be written as follows:

$$E_{0z} = E_0 J_n(k_c r) \begin{bmatrix} \cos(n\phi) \\ \sin(n\phi) \end{bmatrix}$$ (5.391)

Substituting Eq. (5.391) into Eqs (5.371), (5.367), (5.366), and (5.370), we get the following relations:

$$E_{0r} = -\frac{\gamma}{k_c} E_0 J_n'(k_c r) \begin{bmatrix} \cos(n\phi) \\ \sin(n\phi) \end{bmatrix} \tag{5.392}$$

$$E_{0\phi} = -\frac{\gamma}{k_c^2} \frac{1}{r} E_0 n J_n(k_c r) \begin{bmatrix} -\sin(n\phi) \\ \cos(n\phi) \end{bmatrix} \tag{5.393}$$

$$H_{0r} = \frac{j\omega\varepsilon_c}{k_c^2} \frac{1}{r} E_0 n J_n(k_c r) \begin{bmatrix} -\sin(n\phi) \\ \cos(n\phi) \end{bmatrix} \tag{5.394}$$

$$H_{0\phi} = -\frac{j\omega\varepsilon_c}{k_c} E_0 J_n'(k_c r) \begin{bmatrix} \cos(n\phi) \\ \sin(n\phi) \end{bmatrix} \tag{5.395}$$

Now at the boundaries, E_z should vanish, that is, $E_{0z} = 0$ at $r = R$ (5.396)

Applying the boundary condition of Eq. (5.396) to Eq. (5.391), we get the following expression:

$$J_n(k_c R) = 0 \tag{5.397}$$

which results in the following expression for $n = 0$:

$$k_{c01} R = 2.405 \text{ or } k_{c01} = \frac{2.405}{R} \tag{5.398}$$

and $\quad k_{c02} = \frac{5.52}{R}$ (5.399)

Similarly, for $n = 1$:

$$k_{c11} = \frac{3.85}{R} \tag{5.400}$$

and $\quad k_{c12} = \frac{7.02}{R}$ (5.401)

The zeros of $J_n(k_c r)$ for the TM$_{nm}$ modes are given in Table 5.2.

Table 5.2 Zeros of $J_n(k_c r)$ for TM$_{nm}$ modes

	$n = 0$	$n = 1$	$n = 2$	$n = 3$	$n = 4$
$m = 1$	2.405	3.832	5.136	6.380	7.588
$m = 2$	5.520	7.106	8.417	9.761	11.065
$m = 3$	8.645	10.173	11.620	13.015	14.372

The expression of the cut-off frequency is the same as that expressed in Eq. (5.356). Since for a given R, $k_{c,01}$ has the minimum value, the dominant TM mode in a circular waveguide is TM$_{01}$. Figure 5.46 shows the field distribution of different TM modes.

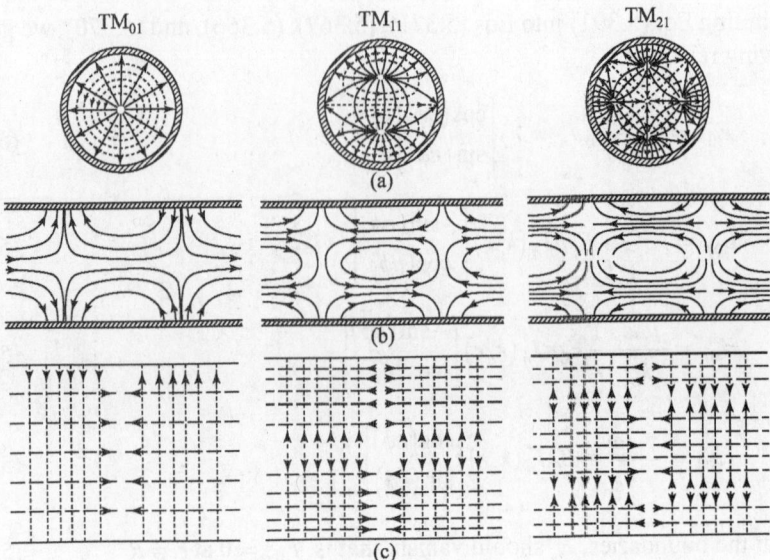

TM$_{01}$ TM$_{11}$ TM$_{21}$

(a)

(b)

(c)

Fig. 5.46 Field distribution of different TM modes in circular waveguide
(a) Cross-sectional view (b) Longitudinal view (c) Surface view

Note The TM$_{01}$ mode is preferred to the TE$_{01}$ mode, because the former requires a smaller waveguide diameter for the same cut-off wavelength.

EXAMPLE 5.9 A circular cylindrical waveguide has an inner diameter of 2.383 cm. Calculate the cut-off frequencies for the (a) TM$_{01}$, (b) TM$_{02}$, (c) TM$_{11}$, and (d) TM$_{12}$ modes.

Solution Given: $R = 2.383/2 = 1.192$ cm

(a) $k_{c01} = \dfrac{2.405}{R} = \dfrac{2.405}{1.192} = 2.018$

$f_{c,\text{TM}_{01}} = \dfrac{k_{cmn}}{2\pi\sqrt{\mu\varepsilon}} = \dfrac{2.018\times3\times10^{10}}{2\pi} = 9.635$ GHz

(b) $k_{c02} = \dfrac{5.52}{R} = \dfrac{5.52}{1.192} = 4.631$

$f_{c,\text{TM}_{02}} = \dfrac{k_{cmn}}{2\pi\sqrt{\mu\varepsilon}} = \dfrac{4.631\times3\times10^{10}}{2\pi} = 22.111$ GHz

(c) $k_{c11} = \dfrac{3.85}{R} = \dfrac{3.85}{1.192} = 3.23$

$f_{c,\text{TM}_{11}} = \dfrac{k_{cmn}}{2\pi\sqrt{\mu\varepsilon}} = \dfrac{3.23\times3\times10^{10}}{2\pi} = 15.422$ GHz

(d) $k_{c12} = \dfrac{7.02}{R} = \dfrac{7.02}{1.192} = 5.889$

$$f_{c,TM_{12}} = \dfrac{k_{cmn}}{2\pi\sqrt{\mu\varepsilon}} = \dfrac{5.889 \times 3 \times 10^{10}}{2\pi} = 28.118 \text{ GHz}$$

Practice Problem

5.11 A circular cylindrical waveguide has an inner diameter of 1.27 cm. Calculate the cut-off frequencies for the (a) TM_{01}, (b) TM_{02}, (c) TM_{11}, and (d) TM_{12} modes.

[18.082 GHz, 41.506 GHz, 28.949 GHz, 52.784 GHz]

5.27 Mode Numbering System in Circular Waveguides

As has already been described, for a rectangular waveguide, the first subscript in the mode numbering denotes the number of half-wave patterns along the guide width, whereas the second subscript denotes that along the guide height. However, for a circular waveguide, the subscripts have different meaning. Here, the first subscript represents the number of full-wave patterns around the circumference of the waveguide, whereas the second subscript indicates the number of half-wave patterns across the diameter. These can best be illustrated with the example

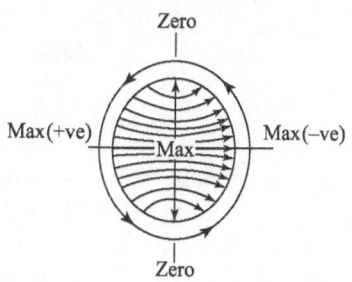

Fig. 5.47 TE_{11}-mode field distribution in circular waveguide

shown in Fig. 5.47. If we follow the electric field pattern in the counter-clockwise direction, the electric field lines start from zero at the top, pass through maximum positive, zero, and maximum negative, and finally come back to zero. This is one full cycle, and therefore the first subscript is 1. Along the diameter, the electric field lines go from zero through maximum and then returns to zero, and thus complete one half-cycle. Therefore, the second subscript is also 1.

5.28 Power Transmission in Circular Waveguides

Power transmitted through a circular waveguide can be calculated from the Poynting theorem. If the dielectric is lossless, then the time-average power transmitted through a circular waveguide is given by the following equation:

$$P_{tr} = \frac{1}{2Z} \int_0^{2\pi} \int_0^R \left[|E_r|^2 + |E_\phi|^2 \right] d\phi r dr = \frac{Z}{2} \int_0^{2\pi} \int_0^R \left[|H_r|^2 + |H_\phi|^2 \right] d\phi r dr \quad (5.402)$$

where Z is the wave impedance and is expressed as follows:

$$Z = \frac{E_r}{H_\phi} = -\frac{E_\phi}{H_r} \quad (5.403)$$

For the TE_{mn} mode, the average power transmitted through a circular waveguide is expressed as follows:

$$P_{\text{tr,TE}_{mn}} = \frac{\sqrt{1-(f_c/f)^2}}{2\eta} \int_0^{2\pi} \int_0^R \left[|E_r|^2 + |E_\phi|^2 \right] d\phi r dr \qquad (5.404)$$

where η is the wave impedance in the unbounded medium and is given by the following relation:

$$\eta = \sqrt{\mu/\varepsilon} \qquad (5.405)$$

Similarly, for the TM_{mn} mode, the average power transmitted through a circular waveguide is written as follows:

$$P_{\text{tr,TM}_{mn}} = \frac{1}{2\eta\sqrt{1-(f_c/f)^2}} \int_0^{2\pi} \int_0^R \left[|E_r|^2 + |E_\phi|^2 \right] d\phi r dr \qquad (5.406)$$

Now, let us calculate the attenuation constant for the TE_{11} mode. The TE_{11} mode fields can be written as follows:

$$H_z = H_0 J_1(k_c r)\sin(\phi)e^{-j\beta z} \qquad (5.407)$$

$$E_r = -\frac{j\omega\mu}{k_c^2 r} H_0 J_1(k_c r)\cos(\phi)e^{-j\beta z} \qquad (5.408)$$

$$E_\phi = \frac{j\omega\mu}{k_c} H_0 J_1'(k_c r)\sin(\phi)e^{-j\beta z} \qquad (5.409)$$

$$H_r = -\frac{j\beta}{k_c} H_0 J_1'(k_c r)\sin(\phi)e^{-j\beta z} \qquad (5.410)$$

$$H_\phi = -\frac{j\beta}{k_c^2 r} H_0 J_1(k_c r)\cos(\phi)e^{-j\beta z} \qquad (5.411)$$

$$E_z = 0 \qquad (5.412)$$

Therefore, power carried by the guide can be calculated from the following equations:

$$P_0 = \frac{1}{2}\text{Re}\int_{r=0}^{R}\int_{\phi=0}^{2\pi}\left(\vec{E}\times\vec{H^*}\right).\hat{z} r dr d\phi = \frac{1}{2}\text{Re}\int_{r=0}^{R}\int_{\phi=0}^{2\pi}\left(E_r H_\phi^* - E_\phi H_r^*\right)d\phi r dr$$

or $$P_0 = \frac{\omega\mu|H_0|^2\text{Re}(\beta)}{2k_c^4}\int_{r=0}^{R}\int_{\phi=0}^{2\pi}\left\{\frac{1}{r^2}\cos^2(\phi)J_1^2(k_c r) + k_c^2\sin^2(\phi)J_1'^2(k_c r)\right\}d\phi r dr$$

or $$P_0 = \frac{\pi\omega\mu|H_0|^2\text{Re}(\beta)}{2k_c^4}\int_{r=0}^{R}\left\{\frac{1}{r}J_1^2(k_c r) + rk_c^2 J_1'^2(k_c r)\right\}dr$$

or $$P_0 = \frac{\pi\omega\mu|H_0|^2\text{Re}(\beta)}{4k_c^4}\left(p_{11}'^2 - 1\right)J_1^2(k_c R) \qquad (5.413)$$

where p'_{11} is the first root of J'_1 and can be expressed as follows:

$$k_{c,11} = p'_{11}/R \tag{5.414}$$

Attenuation due to dielectric loss is given by the following equality:

$$\alpha_d = \frac{k^2 \tan(\delta)}{2\beta} \text{ Np/m} \tag{5.415}$$

Now $\quad P_l = \dfrac{R_s}{2} \displaystyle\int\limits_{\phi=0}^{2\pi} |J_s|^2 \, Rd\phi = \dfrac{R_s}{2} \displaystyle\int\limits_{\phi=0}^{2\pi} \left(|H_\phi|^2 + |H_z|^2 \right) Rd\phi$

or $\quad P_l = \dfrac{|H_0|^2 R_s}{2} \displaystyle\int\limits_{\phi=0}^{2\pi} \left[\dfrac{\beta^2}{k_c^4 R^2} \cos^2(\phi) + \sin^2(\phi) \right] J_1^2 (k_c R) \, Rd\phi$

or $\quad P_l = \dfrac{\pi |H_0|^2 R_s R}{2} \left(1 + \dfrac{\beta^2}{k_c^4 R^2} \right) J_1^2 (k_c R) \tag{5.416}$

Therefore, the attenuation constant due to conductor loss can be expressed by the following equation:

$$\alpha_c = \frac{P_l}{2P_0} = \frac{\dfrac{\pi |H_0|^2 R_s R}{2} \left(1 + \dfrac{\beta^2}{k_c^4 R^2} \right) J_1^2 (k_c R)}{2 \dfrac{\pi \omega \mu |H_0|^2 \beta}{4 k_c^4} \left({p'_{11}}^2 - 1 \right) J_1^2 (k_c R)}$$

or $\quad \alpha_c = \dfrac{R_s \left(k_c^4 R^2 + \beta^2 \right)}{\omega \mu \beta R \left({p'_{11}}^2 - 1 \right)} = \dfrac{R_s \left(k_c^4 R^2 + \beta^2 \right)}{\eta k \beta R \left({p'_{11}}^2 - 1 \right)} = \dfrac{R_s \left(k_c^4 R^2 + k^2 - k_c^2 \right)}{\eta k \beta R \left({p'_{11}}^2 - 1 \right)}$

or $\quad \alpha_c = \dfrac{R_s \left\{ k_c^2 \left(k_c^2 R^2 - 1 \right) + k^2 \right\}}{\eta k \beta R \left({p'_{11}}^2 - 1 \right)} = \dfrac{R_s \left\{ k_c^2 \left({p'_{11}}^2 - 1 \right) + k^2 \right\}}{\eta k \beta R \left({p'_{11}}^2 - 1 \right)}$

or $\quad \alpha_c = \dfrac{R_s}{\eta k \beta R} \left(k_c^2 + \dfrac{k^2}{{p'_{11}}^2 - 1} \right) \text{ Np/m} \tag{5.417}$

Power-handling capabilities of a circular waveguide for the dominant TE_{11} and TE_{01} modes are given by the following equations:

$$P_{TE_{11}} \approx 1790 R^2 \sqrt{1 - (f_{c,11}/f)^2} \text{ kW} \tag{5.418}$$

and $\quad P_{TE_{01}} \approx 1805 R^2 \sqrt{1 - (f_{c,01}/f)^2} \text{ kW} \tag{5.419}$

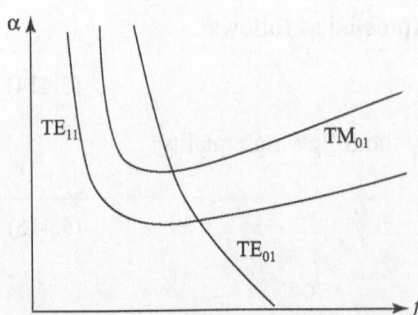

Fig. 5.48 Plot of attenuation constant vs frequency for different modes in circular waveguide

where $f_{c,11}$ and $f_{c,01}$ are the cut-off frequencies for the TE_{11} and TE_{01} modes, respectively. A plot of the attenuation constant against frequency for different modes in a circular waveguide is shown in Fig. 5.48.

Note The TE_{01} mode has the lowest attenuation per unit length for frequencies greater than 10 GHz.

5.29 Excitation of Modes in Circular Waveguides

Since for the TE modes, no electric field component exists along the direction of propagation, these modes can be excited if we place a probe/loop in the guide in such a way that it does not excite any component of the electric field along the axis of the guide. Similarly, if we place the probe/loop in such a way that it does not excite any component of the magnetic field along the axis of the guide, then the propagating mode will be a TM mode. For both the cases, however, the position and number of probes and phase of currents in the probe are determined by the field configurations of the intended modes. A basic methodology for exciting the TE_{11} and TM_{01} modes is shown in Fig. 5.49.

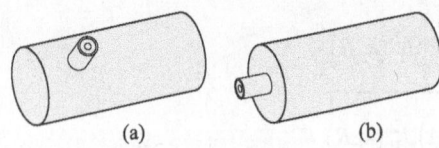

Fig. 5.49 Excitation of modes in circular waveguide (a) TE_{11} (b) TM_{01}

As before introduction of the probe to the guide creates an impedance discontinuity and excites higher order modes. The higher order modes decay within a very short distance and we are left only with the desired modes. The impedance mismatch however often requires tuning device around the probe to minimize the reflection.

Note The TE_{01} mode has no practical application.

5.30 Waveguide Mode Filters

We already know that the propagation of a particular waveguide mode depends upon the dimension of the guide and the operating frequency. In practice, waveguides are operated in the dominant mode only, and hence its dimensions and operating frequency are adjusted accordingly. If we want to operate the waveguide at a higher-order mode, then either the operating frequency should be increased or its dimensions should be readjusted. However, in both cases, by virtue of having

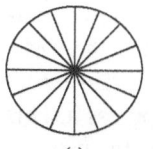 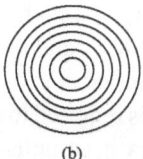 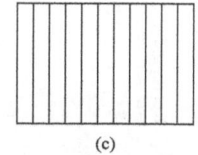

(a)   (b)   (c)

Fig. 5.50 Mode filters (a) TE_{01} transmission (circular) (b) TM_{01} transmission (circular) (c) TE_{01} transmission (rectangular)

a lower cut-off frequency, the lower-order modes will also exist along with the desired higher-order mode. Therefore, to operate the waveguide only at a particular higher-order mode, we need to use a mode filter.

A mode filter is a series of conducting strips or wires inserted in a guide to reflect unwanted modes, without interfering with the desired modes. The strips, in general, are arranged parallel to the electric field lines of the modes to be filtered out. The conducting strips being parallel to the electric field will reflect the modes strongly, thus filtering them out. Several mode filters are shown in Fig. 5.50.

5.31 Waveguide Transition

In a microwave circuit, often a transition between a rectangular and a circular waveguide is required, and vice versa. The simplest example is the transition from a rectangular to a circular waveguide taper, as shown in Fig. 5.39(b). In this circuit, the dominant TE_{10} mode in a rectangular waveguide is converted to the dominant TE_{11} mode in a circular waveguide. Another way to accomplish this job is shown in Fig. 5.51. If we choose the circular waveguide dimension such that the excitation frequency is above the cut-off frequency of the TE_{11} mode but below the cut-off frequency of the next higher-order modes, then all the higher-order modes, generated at the discontinuity, will be localized around the discontinuity. Therefore, at the output we will get only the TE_{11} mode.

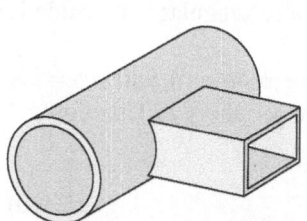

Fig. 5.51 Transition from rectangular to circular waveguide

The transition shown in Fig. 5.51 can also be used to convert the dominant TE_{10} in a rectangular waveguide to the TE_{01} mode in a circular waveguide. However, for this case, the guide dimension should be such that it can support the TE_{01} mode. Since the TE_{11} mode is the dominant mode, it will also exist with the desired TE_{01} mode. Now if a TE_{11} mode filter is used, then the TE_{11} mode will be suppressed and at the output we will get only the desired TE_{01} mode.

5.32 Waveguide Rotary Joints

In a waveguide rotary joint, two rectangular waveguides, operating in the TE_{10} mode, are connected by a circular waveguide that is specially designed to operate at the ϕ-*symmetric* TM_{01} mode, as shown in Fig. 5.52. The diameter of the circular waveguide is chosen in such a way that the modes higher than TM_{01} are not supported by the guide. Since the dominant mode for the circular waveguide is

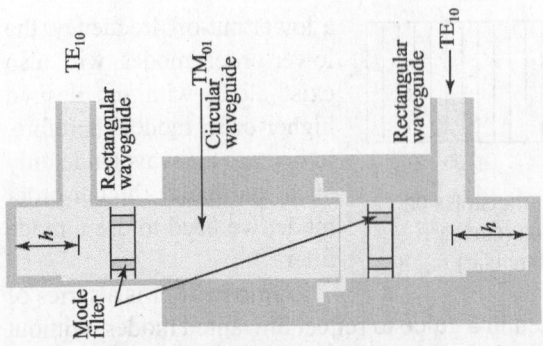

Fig. 5.52 Waveguide rotary junction

TE_{11}, which has a lower cut-off frequency than the TM_{01} mode, a special arrangement is required to suppress it. This is accomplished by using a ring filter. In addition, the end lengths of the circular waveguide are chosen such that these are odd multiples of a quarter guided wavelength for the TE_{11} mode, but even multiples of a quarter guided wavelength for the TM_{01} mode. This particular length ensures infinite input impedance at the junction between the rectangular and circular waveguides for the TE_{11} mode and zero input impedance at the junction for the TM_{01} mode. Therefore, inside the guide, the TE_{11} mode is weakly excited, whereas the TM_{01} mode is strongly excited. The distance between the input and output rectangular waveguides is also selected in such a way that resonance of the TE_{11} mode can be avoided in this section. The impedance at the junction of rectangular and circular waveguides is further matched by using receiving tuners such as irises. Such rotary joints can produce an SWR of less than 1.1.

5.33 Comparison of Rectangular and Circular Waveguides

In this section, a comparison between rectangular and circular waveguides is presented.

1. The dominant TE_{mn} mode is TE_{10} in a rectangular waveguide but TE_{11} in a circular waveguide. The dominant TM_{mn} mode in a rectangular waveguide is TM_{11}, whereas it is TM_{01} in a circular waveguide.
2. Let us consider that a rectangular waveguide of dimension $a \times b$, where $a = 2b$, and a circular waveguide of radius R have the same periphery L. Therefore,

$$2(a+b) = L$$

or $\qquad a = L/3$ $\hfill (5.420)$

and $\qquad 2\pi R = L$ or $R = \dfrac{L}{2\pi}$ $\hfill (5.421)$

The corresponding dominant-mode cut-off frequencies are as follows:

$$\lambda_{c,\text{Rectangular}} = 2a = 2 \times \frac{L}{3} = 0.667L \hfill (5.422)$$

and $\qquad \lambda_{c,\text{Circular}} = \dfrac{2\pi}{k_{c11}} = \dfrac{2\pi R}{k_{c11}R} = \dfrac{L}{1.84} = 0.543L$ $\hfill (5.423)$

Therefore, the cut-off frequency is lower in a circular waveguide, provided the periphery is the same for both the waveguides.

3. Attenuation is larger in a rectangular waveguide than a circular waveguide.

4. The volume of the material used in a circular waveguide is more than that used in a rectangular waveguide. Therefore, a circular waveguide is costlier than a rectangular waveguide.

5. A circular waveguide is much more difficult to bend than a rectangular waveguide.

6. Since α is lower for a circular waveguide than that for a rectangular waveguide, the Q value for a circular waveguide is higher than that for a rectangular waveguide.

5.34 Other Waveguides

In addition to the waveguides described so far, there are few more waveguides. These are described in the subsequent sections.

5.34.1 Ridge Waveguides

A ridge waveguide is formed by developing rectangular ridges inside a rectangular waveguide, as shown in Fig. 5.53(a).

Ridges are developed typically at the position of the maximum electric field. Therefore, for the dominant mode of operation, the preferred position of the ridges is the centre of the wide walls. It has the effect of increasing the capacitance between the wide walls of the guide, thereby reducing the effective impedance and cut-off frequency of the guide. The cut-off frequency of a ridge waveguide can be expressed as follows:

$$f_c = \frac{1}{2\pi}\sqrt{\frac{4g}{\mu\varepsilon bd\left(a-d\right)}} \tag{5.424}$$

The different parameters used in Eq. (5.424) have been explained in Fig. 5.53(b). The ridges also increase the frequency bandwidth of the guide.

Ridge waveguides also have the following disadvantages: higher attenuation and reduction in power-handling capability.

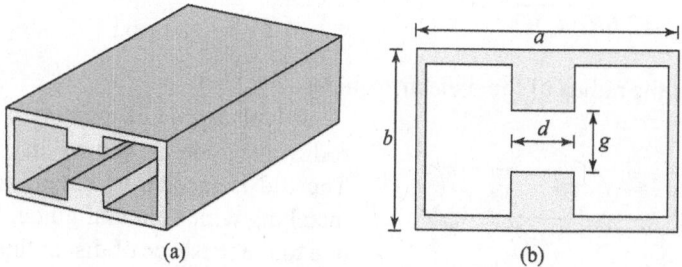

(a)   (b)

Fig. 5.53   Ridge waveguide (a) Typical example (b) Dimensions

Notes 1. Since for the TE_{20} mode, the electric field is either zero or very weak at/near the centre of the broad walls of the guide, the ridge shown in Fig. 5.53 does not disturb the TE_{20} mode. 2. A ridged section behaves almost like a parallel-plate waveguide, and consequently, a ridge waveguide has a much lower cut-off frequency than a hollow metallic waveguide.

EXAMPLE 5.10 Ridges of width 4.39 mm have been made in a rectangular waveguide of dimension 17.55 mm × 8.15 mm. If the gap between the ridges is 3.45 mm, then calculate the cut-off frequency.

Solution Given: $a = 17.55$ mm, $b = 8.15$ mm, $d = 4.39$ mm, and $g = 3.45$ mm

$$f_c = \frac{1}{2\pi}\sqrt{\frac{4g}{\mu\varepsilon bd(a-d)}} = \frac{3\times10^{11}}{2\pi}\sqrt{\frac{4\times3.45}{8.15\times4.39\times(17.55-4.39)}}$$

$$= \frac{3\times10^{11}}{2\pi}\sqrt{\frac{13.8}{470.845}} = 8.174 \text{ GHz}$$

Practice Problem

5.12 Ridges of width 1.829 mm have been made in a rectangular waveguide of dimension 7.32 mm × 3.4 mm. If the gap between the ridges is 1.448 mm, then calculate the cut-off frequency.   **[19.656 GHz]**

5.34.2 Dielectric Rod Waveguides

A uniform cylindrical low-loss, high-permittivity dielectric rod can guide waves through it by the process of total internal reflection from the dielectric–air boundary. The modes that are supported by such waveguides are both pure TE and pure TM modes, as well as hybrid HE or EH modes. The axi-symmetric modes are pure TE and TM modes with non-zero cut-off frequencies, whereas all other modes with angular dependence are a combination of both TE and TM modes and called hybrid EH or HE modes. Hybrid modes contain all the components of electric and magnetic fields. The dominant mode of a dielectric rod waveguide is the HE_{11} mode and has zero cut-off frequency. The cut-off wavelength for the first higher-order mode is expressed as follows:

$$\lambda_c \cong 2.6R\sqrt{\varepsilon_r - 1} \tag{5.425}$$

where R is the radius of the dielectric guide.

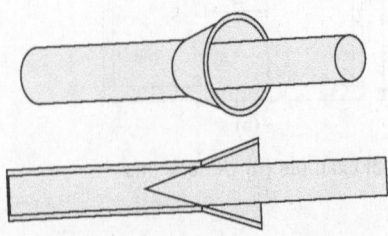

Fig. 5.54 Dielectric rod waveguide

A typical way of exciting a dielectric rod waveguide is shown in Fig. 5.54. The rod is tapered to obtain impedance matching with a circular guide. However, due to the presence of discontinuity at the aperture, a reflected wave is launched in the circular waveguide. This decreases the efficiency of launching, which can be expressed as follows:

$$\eta_L = \frac{\text{Surface wave power}}{\text{Input power} - \text{Reflected power}} = \frac{|S_{12}|^2}{1-|S_{11}|^2} \tag{5.426}$$

EXAMPLE 5.11 Calculate the cut-off wavelength of the first higher-order mode of a dielectric rod waveguide with a dielectric constant of 10 and radius of 4.8 mm.

Solution   Given: $R = 4.8$ mm and $\varepsilon_r = 10$

$$\lambda_c = 2.6 \times 4.8\sqrt{10-1} = 37.44 \text{ mm}$$

Practice Problem

5.13 Calculate the cut-off wavelength of the first higher-order mode of a dielectric rod waveguide having a dielectric constant of 7 and radius of 2.36 mm.   [15.03 mm]

5.35 Advantages and Disadvantages of Waveguides

The following are the advantages of a waveguide:

Lower Ohmic loss   Due to a larger surface area of the conductor, the Ohmic loss is lower.

Lower dielectric loss   A waveguide uses air as the dielectric. Since air has a low dielectric loss, the dielectric loss of the waveguide is also small.

High power-handling capability   Since air has higher breakdown voltages, a waveguide is suitable for carrying high power. If necessary, the hollow space can be vacuumed to handle more power.

Ease of fabrication   Due to its simpler structure, a waveguide can easily be fabricated.

Rugged   The single-conductor metallic structure of a waveguide makes it rugged.

The following are the disadvantages of a waveguide:

Bulky size at lower frequencies   Since the width of a waveguide is approximately half-wavelength, the waveguide is bulky at lower frequencies.

Heavy   The complete metallic structure makes a waveguide heavy compared to others.

Difficult to install   Because of its rigid and hollow pipe shape, installation of a waveguide is difficult and special care is required for its bending, twisting, and coupling.

Costly   Due to the requirement of a highly conducting metal such as copper, brass, or aluminium for its fabrication, a waveguide is costly as compared to others. Further, to reduce the skin effect, often the inner surface of a waveguide requires silver/gold plating, which increases the cost.

Not amendable to integrated circuit fabrication   Since the waveguides are not a planar structure, they are not amendable in integrated circuit fabrication

5.36 Finlines

A finline is a quasi-planar transmission line structure that was proposed to achieve a large bandwidth, compatibility with planar circuit technology in absence of radiation. A basic finline can be thought as a shielded slot line mounted on the E-plane of a rectangular waveguide. Therefore, for a specified frequency band, the dimensions of a finline are identical to that of a waveguide operating at that band. In addition, a finline may also be considered as a printed version of a ridge waveguide. Figure 5.55 shows some commonly used finline structures. Unilateral finlines are the simplest and best choice for finline component fabrication, whereas bilateral finlines provide greater flexibility for biasing of active devices due to metallization on both sides. Bilateral finlines also have lower losses and can provide characteristic impedance as low as 100 Ω. Antipodal finlines can provide characteristic impedance of the order of 10 Ω, and are suitable for transition between a microstrip line and a waveguide.

In finline structures, fins concentrate the electric field in the fin-gap region, as shown in Fig. 5.56. Such a field configuration leads to a capacitive loading to the propagation of the dominant HE mode in the slab waveguide. The capacitive loading, in turn, lowers the cut-off frequency of the fundamental mode by a considerable amount and that of the next higher-order modes by a negligible amount. Therefore the overall bandwidth of the fundamental mode of operation is increased.

In addition to larger bandwidths, the devices and components that are integrated in a finline for integrated circuit applications are subjected to high power densities, resulting in better matching. However, higher field densities also result in higher conduction and dielectric loss, due to the presence of a dielectric slab, than in case of a waveguide. It also results in high current density at the fin edges. Attenuation of a finline is high and of the order of 0.1 dB/wavelength. Therefore, finlines are not suitable for long-distance power transmission.

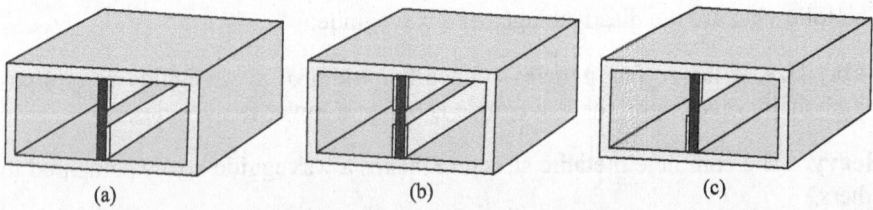

    (a)   (b)   (c)

Fig. 5.55 Finline structures (a) Unilateral (b) Bilateral (c) Antipodal

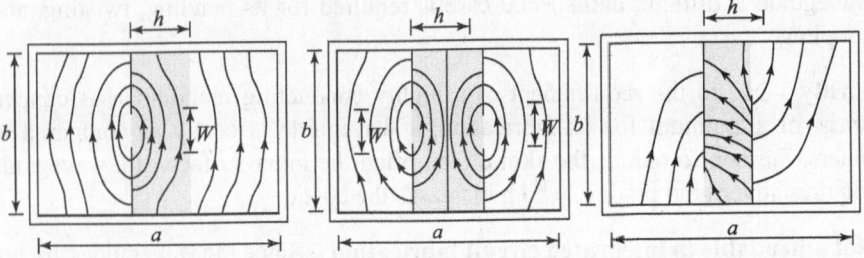

Fig. 5.56 Transverse electric field configuration in different finline structures

Similar to the case of a planar transmission line, discussed in Chapter 4, the analysis of finline structures is very difficult and cumbersome, because both the cases involve an inhomogeneous dielectric and sharp metal edges. A finline has the HE and EH modes with dominant H_z and E_z fields. Such fields, on cut-off, reduce to the TE and TM modes, respectively.

Important Formulae

- The characteristic impedance of a parallel-plate transmission line for the TEM mode is $Z_0 = \eta b/a$.
- The phase constant of a parallel-plate transmission line for the TE and TM modes is

$$\beta = \sqrt{k^2 - \left(\frac{n\pi}{b}\right)^2}.$$

- The cut-off frequency of a parallel-plate transmission line for the TE and TM modes is

$$f_c = \frac{n}{2b\sqrt{\mu\varepsilon}}.$$

- The guided wavelength of a parallel-plate transmission line for the TE and TM modes is $\lambda_g = \lambda / \sqrt{1 - (\lambda/\lambda_c)^2}$.
- The phase velocity of a parallel-plate transmission line for the TE and TM modes is

$$v_p = \frac{v}{\sqrt{1 - (\lambda/\lambda_c)^2}}.$$

- The group velocity of a parallel-plate transmission line for the TE and TM modes is

$$v_g = v\sqrt{1 - (\lambda/\lambda_c)^2}.$$

- For a parallel-plate transmission line, $v_p v_g = v^2$.
- The TE-mode characteristic impedance of a parallel-plate transmission line is $Z_{0,\text{TE}} = \omega\mu/\beta$.
- The TM-mode characteristic impedance of a parallel-plate transmission line is

$$Z_{0,\text{TM}} = \frac{\beta}{\omega\varepsilon}.$$

- Propagation constant inside a rectangular waveguide is expressed as

$$\gamma = \sqrt{\left(\frac{m\pi}{a}\right)^2 + \left(\frac{n\pi}{b}\right)^2 - \omega^2\mu\varepsilon}.$$

- The cut-off frequency of a rectangular waveguide is expressed as

$$f_c = \frac{c}{2}\left[\left(\frac{m}{a}\right)^2 + \left(\frac{n}{b}\right)^2\right]^{\frac{1}{2}}.$$

- The cut-off wavelength of a rectangular waveguide is expressed as $\lambda_c = \dfrac{2ab}{\sqrt{m^2 b^2 + n^2 a^2}}$.
- The TE-mode intrinsic impedance of a rectangular waveguide is expressed as

$$\eta_{\text{TE}} = \eta \left/ \sqrt{1 - \left(\frac{\lambda}{\lambda_c}\right)^2}\right..$$

- The TM-mode intrinsic impedance of a rectangular waveguide is expressed as
$$\eta_{TM} = \eta\sqrt{1-\left(\frac{\lambda}{\lambda_c}\right)^2}.$$

- The guided wavelength of a rectangular waveguide is expressed as
$$\lambda_g = \lambda\left/\sqrt{1-\left(\frac{\lambda}{\lambda_c}\right)^2}\right..$$

- The phase velocity of a wave in a rectangular waveguide is expressed as
$$v_p = c\left/\sqrt{1-\left(\frac{\lambda}{\lambda_c}\right)^2}\right..$$

- The group velocity of a wave in a rectangular waveguide is expressed as
$$v_g = c\sqrt{1-\left(\frac{\lambda}{\lambda_c}\right)^2}.$$

Exercises

Objective-type Questions

5.1 A waveguide behaves as a
   (a) low-pass filter
   (b) high-pass filter
   (c) band-pass filter
   (d) all-pass filter

5.2 The dominant mode is defined as the mode having
   (a) the lowest cut-off frequency
   (b) the highest cut-off frequency
   (c) the cut-off frequency equal to the frequency of the propagating signal
   (d) frequency-independent characteristics

5.3 When a particular mode is excited in a waveguide, an extra electric component appears in the direction of propagation. The resulting mode is
   (a) transverse electric
   (b) transverse magnetic
   (c) hybrid electric
   (d) hybrid magnetic

5.4 A guided wavelength is defined as the distance travelled by a wave in order to undergo a phase shift of 2π radians within the guide, and it is
   (a) greater than the cut-off wavelength but less than the free-space wavelength
   (b) less than both the cut-off and free-space wavelengths
   (c) less than the cut-off wavelength but greater than the free-space wavelength
   (d) greater than both the cut-off and free-space wavelengths

5.5 The guided wavelength of a parallel-plate transmission line for the TE and TM modes is

(a) $\dfrac{\lambda_c}{\sqrt{1-(\lambda/\lambda_c)^2}}$

(c) $\dfrac{\lambda}{\sqrt{1-(\lambda/\lambda_c)^2}}$

(b) $\dfrac{\lambda_c}{\sqrt{1+(\lambda/\lambda_c)^2}}$

(d) $\dfrac{\lambda}{\sqrt{1-(\lambda_c/\lambda)^2}}$

5.6 For a rectangular waveguide to support only the dominant TE mode, which of the following pairs of inequalities has to be satisfied?

(a) $b<\lambda<2b;\ \lambda>2a$

(c) $a<\lambda<2b;\ \lambda>2a$

(b) $b<\lambda<2b;\ \lambda<2a$

(d) $a<\lambda<2b;\ \lambda<2a$

5.7 In rectangular waveguide modes are denoted by TE_{mn} or TM_{mn}, where m and n are the number of

(a) full cycles of the electric or magnetic field in the direction of width and height, respectively

(b) half cycles of the electric or magnetic field in the direction of width and height, respectively

(c) quarter cycles of the electric or magnetic field in the direction of width and height, respectively

(d) half cycles of the electric or magnetic field in the direction of propagation

5.8 Which of the following mode of transmissions will not be supported by a rectangular waveguide?

(a) TE_{10} (b) TM_{10} (c) TM_{11} (d) TE_{11}

5.9 The dominant TM mode in a rectangular waveguide is

(a) TM_{21} (b) TM_{12} (c) TM_{11} (d) TM_{32}

5.10 The cut-off frequency of a rectangular waveguide is expressed as

(a) $\dfrac{c}{2}\left[\left(\dfrac{m}{a}\right)^2+\left(\dfrac{n}{b}\right)^2\right]^{\frac{1}{2}}$

(c) $2c\left[\left(\dfrac{m}{a}\right)^2+\left(\dfrac{n}{b}\right)^2\right]^{\frac{1}{2}}$

(b) $c\left[\left(\dfrac{m}{a}\right)^2+\left(\dfrac{n}{b}\right)^2\right]^{\frac{1}{2}}$

(d) $3c\left[\left(\dfrac{m}{a}\right)^2+\left(\dfrac{n}{b}\right)^2\right]^{\frac{1}{2}}$

5.11 The cut-off wavelength of the dominant mode propagating down a rectangular waveguide of dimension $a \times b$ will be

(a) $2a$ (b) a (c) $2b$ (d) b

5.12 When for a rectangular waveguide the free-space wavelength of a signal equals the cut-off wavelength, the TE mode characteristic impedance becomes

(a) zero

(c) 377 Ω

(b) less than 377 Ω but greater than zero

(d) infinite

5.13 The best position of a probe to generate the TE_{10} mode is

(a) at any point on the side wall of the guide

(b) at any point on the broad wall of the guide

    (c) at the narrow wall of the guide and at the centre of the aperture

    (d) at the broad wall of the guide and at the centre of the aperture

5.14 Which of the following modes is most difficult to excite in a waveguide?

    (a) TE_{10} (b) TM_{11} (c) TE_{11} (d) TE_{35}

5.15 H-plane stubs provide

    (a) reactance in series   (c) resistance in series

    (b) susceptance in parallel   (d) conductance in parallel

5.16 A waveguide twist must be at least

    (a) $\lambda_g/2$ at the operating frequency   (c) $3\lambda_g/2$ at the operating frequency

    (b) λ_g at the operating frequency   (d) $2\lambda_g$ at the operating frequency

5.17 The attenuation constant of the TE_{01} mode in a circular waveguide

    (a) is independent of frequency

    (b) increases with an increase in frequency

    (c) decreases with an increase in frequency

    (d) none of these

5.18 Which of the following modes is not possible in a circular waveguide?

    (a) TE_{10} (b) TE_{01} (c) TE_{11} (d) TE_{12}

5.19 In certain applications, circular waveguides are preferred to rectangular waveguides because of

    (a) lower attenuation   (c) small-cross section

    (b) freedom of spurious mode   (d) none of these

5.20 To reduce the cross-sectional area, the shape chosen for a waveguide is

    (a) rectangular   (b) circular   (c) ridge   (d) elliptical

Review Questions

5.1 List the advantages and disadvantages of waveguides.

5.2 Derive the expressions for the field components inside a parallel-plate waveguide for the TE-mode propagation.

5.3 Derive the expressions for the cut-off frequency, guided wavelength, phase velocity, and group velocity of the waves inside a parallel-plate waveguide.

5.4 Find out the characteristic impedance of a parallel-plate waveguide for the propagation of the TE and TM modes and hence prove that $Z_{0,TE} Z_{0,TM} = \eta^2$.

5.5 Derive the expressions for the field components inside a rectangular waveguide for the propagation of the TE and TM modes.

5.6 Derive the expressions for the cut-off frequency, guided wavelength, phase velocity, and group velocity of the TE and TM modes inside a rectangular waveguide. Hence, prove that $v_p v_g = c^2$.

5.7 Find out the characteristic impedance of a rectangular waveguide for the propagation of the TE and TM modes, and hence prove that $\eta_{TE} \eta_{TM} = \eta^2$.

5.8 For a rectangular waveguide, prove that $\dfrac{1}{\lambda_g^2} = \dfrac{1}{\lambda^2} - \dfrac{1}{\lambda_c^2}$.

5.9 Find out an expression for the power carried by the TE_{mn} mode inside a rectangular waveguide.

5.10 Find out an expression for the power carried by the TM_{mn} mode inside a rectangular waveguide.

5.11 Find out an expression for the attenuation constant of a rectangular waveguide for the propagation of the TE_{mn} mode. Hence, find out an expression for the attenuation constant for the dominant TE mode.

5.12 Why can the TEM mode not exist inside a waveguide? Justify your answer with a proper explanation.

5.13 Derive the expressions for the field components inside a circular waveguide for the propagation of the TE and TM modes, and find out the dominant TE mode.

5.14 Describe waveguide rotary joints.

5.15 Compare rectangular and circular waveguides.

Problems

5.1 A rectangular air-filled copper waveguide of length 15 cm and cross-sectional dimension 22.86 mm × 10.16 mm is operated at 9 GHz. Calculate the loss in the guide. The conductivity of copper is 5.8×10^7 S/m. *[Hint: Use Eqs (5.230) and (5.243)]*

5.2 A circular waveguide has a dominant TE_{11} mode cut-off frequency of 8.685 GHz. Calculate the (a) phase constant, (b) guided wavelength, (c) phase velocity, (d) group velocity, and (e) characteristic impedance at 12 GHz.
[Hint: Use Eqs (5.171), (5.185), (5.189), (5.194), and (5.198)]

5.3 A rectangular waveguide of dimension 22.86 mm × 10.16 mm is filled with a dielectric of dielectric constant 4.3. Calculate the cut-off frequency. *[Hint: Use the relation*
$$\lambda_{c,dielectric} = \lambda_{c,air} / \sqrt{\varepsilon_r}]$$

5.4 An air-filled rectangular waveguide of dimension 22.86 mm × 10.16 mm is excited with a frequency of 5 GHz. Calculate the attenuation constant. *[Hint: Use the relation*
$$\alpha = \sqrt{(\pi/a)^2 - \omega^2 \mu \varepsilon}]$$

5.5 A rectangular waveguide is to be designed to guide a 12 GHz signal under the dominant mode. Calculate the cross-sectional dimension of the waveguide. *[Hint: Use Eq. (5.175)]*

5.6 An air-filled circular waveguide of radius R_{air} is operated in the TE_{11} mode. Now, if the guide is filled with a lossless dielectric of relative permittivity 2.8, by what value must its radius be changed in order to maintain the cut-off frequency at its initial value? *[Hint: Use Eqs (5.353) and (5.356)]*

5.7 Design a rectangular waveguide that, at 8 GHz, will operate in the TE_{10} mode at 22% safety factor when the guide is filled with air. It is also required that the next two higher-order TE modes should be degenerate. *[Hint: Use the relation $f = 1.22 f_c$ and Eq. (5.175)]*

5.8 A dominant TE_{10} mode is propagating down a rectangular waveguide of dimension 22.86 mm × 10.16 mm. The distance between two successive minima is 20 mm. Find the signal frequency. *[Hint: Use Eqs (5.175) and (5.186)]*

5.9 A rectangular waveguide with $a = 2b$ has a TE_{01} mode cut-off frequency of 15 GHz. Find out the cut-off frequency of the dominant mode. *[Hint: Use Eq. (5.174) and the relation $a = 2b]$*

5.10 A waveguide, having a cross-sectional dimension of 22.86 mm × 10.16 mm, is filled with a dielectric of dielectric constant 6.3. If the waveguide operates at 5 GHz, calculate the phase constant, guided wavelength, phase velocity, and group velocity for the dominant-mode propagation. *[Hint: Use Eqs (5.171), (5.175), (5.182), (5.194), and (5.198)]*

5.11 Design an air-core circular cylindrical waveguide that will be able to guide only the dominant TE_{11} mode at 10 GHz, with a safety margin of 25%. What is the cut-off frequency of the next higher mode? *[Hint: Use Eqs (5.353), (5.356), and (5.398)]*

5.12 An air-filled rectangular waveguide of dimensions $a = 22.86$ mm and $b = 10.16$ mm transports energy at a rate of 3 mW, using the dominant-mode propagation. If the frequency of operation is 9 GHz, determine the peak value of the electric field in the waveguide. *[Hint: Use Eqs (5.175) and (5.189), and the relation*

$$E_0 = \sqrt{4P\eta_{TE_{10}}/(ab)}]$$

5.13 A rectangular waveguide has dimensions $a = 22.86$ mm and $b = 10.16$ mm and is operating at 14 GHz. Calculate the group delay between the propagating modes after a distance of 1 km. *[Hint: Use Eqs (5.172) and (5.198), and the relation*

$$\Delta t_g = L\left\{\left(v_{g,TE_{20}}\right)^{-1} - \left(v_{g,TE_{10}}\right)^{-1}\right\}]$$

5.14 It is required to achieve 450 Ω wave impedance at 12 GHz for the dominant TE_{10} mode propagation in a rectangular waveguide. Calculate the required guide width. *[Hint: Use Eqs (5.175) and (5.189)]*

5.15 A rectangular waveguide having a cross-section of 22.86 mm × 10.16 mm is filled with a lossless dielectric of relative permittivity 4.7. The x-component of the magnetic field and z-component of the electric field are expressed as

$$H_x = 2\sin\left(\frac{\pi x}{a}\right)\cos\left(\frac{2\pi y}{b}\right)\sin\left(30\pi \times 10^9 t - \beta z\right) \text{ A/m and } E_z = 0 \text{ V/m, respectively.}$$

Calculate the (a) mode of operation, (b) cut-off frequency, (c) phase constant, (d) propagation constant, and (e) intrinsic impedance.

What are the other modes that propagate through it at 7.5 GHz? *[Hint: Use Eqs (5.171), (5.172), and (5.189)]*

5.16 A rectangular waveguide with dimensions $a = 22.86$ mm and $b = 10.16$ mm has been designed for a single mode operation. Find the possible frequency range of operation such that the lowest frequency is 10% above the cut-off of the dominant mode and the highest frequency is 10% below the cut-off frequency of the next higher-order mode. In addition, calculate the frequency at which the next higher-order mode will have an attenuation of 3 dB/m. *[Hint: Use Eqs (5.171) and (5.172)]*

5.17 For a rectangular waveguide of dimension 22.86 mm × 10.16 mm, calculate the frequency at which the guided wavelength will be twice the free-space wavelength. *[Hint: Use Eqs (5.173) and (5.185)]*

5.18 A rectangular waveguide has dimension $a = 22.86$ mm. Calculate the b dimension such that the cut-off wavelength of the TE_{11} mode is half of that of the TE_{10} mode. *[Hint: Use Eq. (5.173)]*

5.19 An air-filled waveguide operating in the dominant mode has dimensions $a = 22.86$ mm and $b = 10.16$ mm. The guide is attached with another dielectric-filled guide of the same dimensions. The dielectric constant is 2.1. Calculate the dominant-mode reflection coefficient at the junction, at 10 GHz. *[Hint: Use Eq. (5.189) and the relations $\lambda_{di} = \lambda_{air}/\sqrt{\varepsilon_r}$, $\eta_{di} = \eta_{air}/\sqrt{\varepsilon_r}$]*

5.20 In an air-filled rectangular waveguide, the vector electric field is given by the relation $E = \cos(30\pi y)\exp\{-j(20\pi z - \omega t)\}\hat{a}_x$ V/m. Find the vector magnetic field. *[Hint: Start with Maxwell's curl equation for the electric field]*

5.21 A rectangular waveguide has an angular cut-off frequency of 41.23 radian/s and phase shift per unit length of 150 radian/s. Calculate the operating frequency. *[Hint: Use Eqs (5.182) and (5.186)]*

5.22 A rectangular waveguide is to be designed for operation at 8–10 GHz and the band centre is 1.5 times the cut-off frequency. What should be the dimension of the waveguide? *[Hint use Eq. (5.175) and the relations $f_0 = (f_l + f_h)/2$ and $f_c = f_0/1.5$]*

5.23 A rectangular waveguide has dimension $a = 22.86$ mm and is operating in the dominant TE_{10} mode. If the wave impedance is 600 Ω, calculate the operating frequency. *[Hint: Use Eqs (5.175) and (5.189)]*

5.24 A waveguide has a cut-off wavelength of 45.72 mm when it is air filled. Find the dielectric constant of the material that will bring down the cut-off frequency to 2 GHz. *[Hint: Use the relation $v = c/\sqrt{\varepsilon_r}$]*

5.25 Calculate the ratio of the area of a circular waveguide to that of a rectangular waveguide, provided that each has the same cut-off frequency for its dominant mode. Assume that $a = 2b$ for the rectangular waveguide. *[Hint: Use Eqs (5.175), (5.353), and (5.356)]*

5.26 A rectangular waveguide has the dimension $a \times b$. Study the cut-off frequencies for the TE_{10}, TE_{20}, TE_{01}, and TE_{11} modes for (a) $b = 0.25a$, (b) $b = 0.5a$, (c) $b = 0.75a$, (d) $b = a$, and (e) $b = 1.25a$. *[Hint: Use Eq. (5.172)]*

5.27 A square waveguide operates at 5 GHz in the dominant mode. If the group velocity is 1.5×10^8 m/s, calculate the largest dimension of the waveguide. Assume that the waveguide is filled with a dielectric of relative permittivity 2.7. *[Hint: Use Eqs (5.175) and (5.198)]*

5.28 Two rectangular waveguides are joined end to end and have the same dimension $a \times b$, where $a = 2b$. If the first guide is air filled and the second is filled with a dielectric of relative permittivity ε_r, then find the limit of ε_r such that a single dominant-mode operation is possible in both the guides. *[Hint: Use Eq. (5.172)]*

5.29 Two parallel-plate waveguides, filled with dielectrics of relative permittivity ε_1 and ε_2 ($\varepsilon_1 > \varepsilon_2$), are joined end to end. Determine the frequency range for which the TM_1 mode, flowing in waveguide 1, will not be able to reach waveguide 2. Assume the plate separation of both the guides to be equal to b. *[Hint: Use Eq. (5.70) and the relation $\sin\theta_c = (\varepsilon_2/\varepsilon_1)$]*

5.30 Two parallel-plate waveguides, filled with dielectrics of relative permittivity ε_1 and ε_2 ($\varepsilon_1 > \varepsilon_2$), are joined end to end. At which frequency, will the TM_1 mode cross the junction without suffering any reflective loss at the interface? Assume the plate separation of both the guides to be equal to b. *[Hint: Use Eq. (5.70) and the relation $\sin(\theta_i) = \sin(\theta_B) = \varepsilon_2/\sqrt{\varepsilon_1^2 + \varepsilon_2^2}$]*

5.31 A dielectric-filled rectangular waveguide has a cut-off frequency of 4 GHz for the TE_{10} mode and 7.5 GHz for the TE_{11} mode. If the relative permittivity of the dielectric material is 3.7, calculate the dimensions of the guide. *[Hint: Use Eq. (5.172)]*

6 Microwave Resonators

6.1 Introduction

A resonator is a device that exhibits resonance at a particular frequency, called the resonance frequency. Such structures are widely used in amplifiers, oscillators, filters, as well as frequency meters. The resonance frequency of a resonator can be tuned with the help of a tuner circuit.

A resonator is often characterized by a factor, called the Q-factor or quality factor. The Q-factor basically measures the frequency selectivity of a cavity and is inversely proportional to the bandwidth of a resonator. It also relates the capacity of storing electromagnetic energy within the cavity with energy dissipation through heat and dielectric loss. In practice, the Q-value should be very high for the cavity to have very high frequency selectivity and low loss. A typical Q-value of 10,000 can easily be achieved with a transmission line resonator formed using half- or quarter-wavelength short- or open-circuited transmission lines. However, their losses are very high for a frequency greater than 1000 MHz and hence are not much useful above this frequency. At a higher frequency, cavity resonators, formed by shorting both the ends of a waveguide, are used.

At resonance, the average energy stored in the electric field is equal to the average energy stored in the magnetic field, and hence the total energy stored within a resonator is twice of either of these energies. Energy stored in the electric field contributes a capacitive loading, whereas that stored in the magnetic field produces an inductive loading. Since both the energies are equal at resonance, they cancel each other, and therefore the effective circuit behaves like a series resistance or shunt conductance.

6.2 Designing High-frequency Resonators

Let us consider the parallel lumped element resonator circuit shown in Fig. 6.1. The resonant frequency of the circuit can be expressed as follows:

$$f_r = \frac{1}{2\pi\sqrt{LC}} \tag{6.1}$$

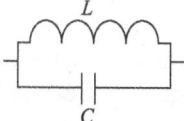

Fig. 6.1 Parallel resonant circuit

For a lumped circuit, the value of L and C are reasonably high, and hence the resonance frequency of the circuit is low. If we want to increase the resonance frequency, we have to decrease either C or L, or both. C can be decreased by reducing the plate area and increasing the distance between the plates. However, for a proper operation of the circuit, C cannot be reduced beyond a certain level. Therefore to obtain higher resonance frequencies, we need to reduce L. The inductance L can be reduced by reducing the number of turns, and the minimum value of L is obtained when the number of turns is zero. Under such a circumstance, the inductance becomes a plane strip across the capacitor plate, as shown in Fig. 6.2(a). The total L can further be reduced by adding more such inductance in parallel with it, which is equivalent to the addition of more strips across the capacitor plates. If we go on adding such strips across the capacitor plate, then a situation will arise when all the four sides of the capacitor plate will totally be covered with such strips, and the situation will be equivalent to a closed metallic box, which is basically a rectangular waveguide cavity, as shown in Fig. 6.2(b).

If we consider the short-circuited transmission line equivalent circuit of Fig. 6.1, then we can write the following equation:

$$B_1 = -jY_0\cot(\beta L_1) \tag{6.2}$$

and
$$B_2 = -jY_0\cot(\beta L_2) \tag{6.3}$$

At resonance,

$$B_1 + B_2 = -jY_0\cot(\beta L_1) - jY_0\cot(\beta L_2) = 0$$

or
$$\cot(\beta L_1) = -\cot(\beta L_2) = \cot(\pi - \beta L_2) \text{ or } \beta L_1 + \beta L_2 = \beta(L_1 + L_2) = \pi$$

or
$$L_1 + L_2 = \pi/\beta = \lambda_g/2 \tag{6.4}$$

This can be explained as follows. When one end of a waveguide is terminated in a short circuit, then total reflection will occur and a standing wave will be formed, with a minimum at the position of the short circuit. Now, if we place another short circuit at a distance of $\lambda_g/2$ or an integral multiple of $\lambda_g/2$, that is, at a position of another voltage minimum, then a hollow space will be formed, which will support a signal that bounces back and forth between the two shorting plates. This will result in resonance and form a cavity.

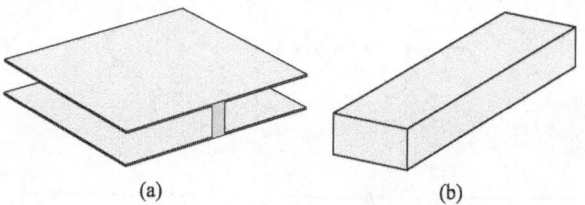

(a) (b)

Fig. 6.2 Formation of cavity from parallel plate waveguide
(a) Equivalent circuit of Fig. 6.1 with minimum achievable
L and C (b) Rectangular waveguide cavity

6.3 *Q*-factor of Cavity Resonators

Cavity resonators are often characterized by its Q-factor. The Q-factor is a measure of frequency selectivity of a resonant or anti-resonant cavity, and is defined as follows:

$$Q \equiv 2\pi \frac{\text{Maximum energy stored}}{\text{Energy dissipated per cycle}} = \frac{\omega W}{P} \tag{6.5}$$

At the resonance frequency, both the electric and the magnetic energies will be equal and in time quadrature, that is, when the electric field will be maximum the magnetic field will be zero, and vice versa. Therefore, we can write that

$$W_e = \frac{1}{2} \int_v \varepsilon |E|^2 \, dv = W_m = \frac{1}{2} \int_v \mu |H|^2 \, dv = W \tag{6.6}$$

Further, the average power loss in the resonator is given by the following equation:

$$P = \frac{R_s}{2} \int_s |H_t|^2 \, ds \tag{6.7}$$

where H_t is the tangential magnetic field intensity.

Substituting Eqs (6.6) and (6.7) into Eq. (6.5), we get the following relation:

$$Q = \frac{\omega \mu \int_v |H|^2 \, dv}{R_s \int_s |H_t|^2 \, ds} \tag{6.8}$$

An unloaded resonator can be represented by either a series or a parallel RLC circuit. If the cavity is coupled with a generator of internal impedance Z_g by means of a $N{:}1$ transformer and a series inductance L_s, as shown in Fig. 6.3, then the loaded and unloaded Q of the systems are given by the following equations:

$$Q_{\text{Unloaded}} = \omega_o L / R \tag{6.9}$$

and $$Q_{\text{Loaded}} = \frac{\omega_o L}{R + N^2 Z_g} = \frac{\omega_o L}{R(1+K)} \text{ for } \left| N^2 L_s \right| \ll \left| R + N^2 Z_g \right| \tag{6.10}$$

where $$K = N^2 Z_g / R \tag{6.11}$$

is the coupling coefficient.

Substituting Eq. (6.9) into Eq. (6.10), we get the following relation:

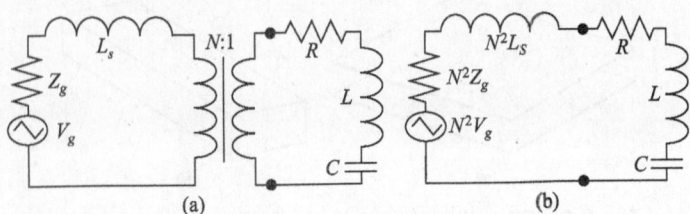

Fig. 6.3 Equivalent circuits of cavity resonator (a) Cavity coupled with generator (b) Equivalent circuit

or $\quad Q_{\text{Loaded}} = \dfrac{Q_{\text{Unloaded}}}{1+K}$ $\hspace{3cm}$ (6.12)

$$\frac{1}{Q_{\text{Loaded}}} = \frac{1+K}{Q_{\text{Unloaded}}} = \frac{1}{Q_{\text{Unloaded}}} + \frac{K}{Q_{\text{Unloaded}}}$$

or $\quad \dfrac{1}{Q_{\text{Loaded}}} = \dfrac{1}{Q_{\text{Unloaded}}} + \dfrac{1}{(Q_{\text{Unloaded}}/K)} = \dfrac{1}{Q_{\text{Unloaded}}} + \dfrac{1}{Q_{\text{Ext}}}$ $\hspace{1cm}$ (6.13)

where $\quad Q_{\text{Ext}} = Q_{\text{Unloaded}}/K$ $\hspace{4cm}$ (6.14)

is the external Q.

Depending on the value of Q, coupling may of the following three types:

Under coupling For under coupling, $K < 1$ and can be expressed as follows:

$\quad K = 1/\rho$ $\hspace{6cm}$ (6.15)

Substituting Eq. (6.15) into Eq. (6.12), we get $Q_{\text{Loaded}} = \dfrac{\rho}{\rho+1} Q_{\text{Unloaded}}$ $\hspace{0.5cm}$ (6.16)

Critical coupling For critical coupling, $K = 1$ $\hspace{2.5cm}$ (6.17)

Substituting Eq. (6.17) into Eq. (6.12), we get $Q_{\text{Loaded}} = Q_{\text{Unloaded}}/2$ $\hspace{0.5cm}$ (6.18)

Over coupling For over coupling, $K > 1$ and can be expressed as $K = \rho$ (6.19)

Substituting Eq. (6.19) into Eq. (6.12), we get $Q_{\text{Loaded}} = \dfrac{Q_{\text{Unloaded}}}{\rho+1}$ $\hspace{1cm}$ (6.20)

A plot of the standing wave ratio as a function of coupling coefficients is shown in Fig. 6.4.

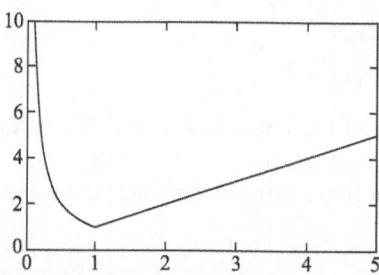

Fig. 6.4 Plot of standing wave ratio as function of coupling coefficients

EXAMPLE 6.1 A cavity has a loaded and an unloaded Q of 10,000 and 7327, respectively. Calculate the coupling coefficient and external Q. In addition, calculate the loaded Q for (a) an under coupling, (b) a critical coupling, and (c) an over coupling case, if we want to maintain an SWR of 1.25.

Solution Given: $Q_{\text{Unloaded}} = 10,000$, $Q_{\text{Loaded}} = 7327$, and $\rho = 1.25$

Now $\quad Q_{\text{Loaded}} = \dfrac{Q_{\text{Unloaded}}}{1+K}$

or $\quad 7327 = \dfrac{10,000}{1+K}$ or $1+K = \dfrac{10,000}{7327} = 1.365$ or $K = 0.365$.

And $Q_{\text{Ext}} = Q_{\text{Unloaded}}/K = 10,000/0.365 = 27,397.26$

   (a)   For under coupling: $Q_{\text{Loaded}} = \dfrac{\rho}{\rho+1} Q_{\text{Unloaded}} = \dfrac{1.25}{1.25+1} \times 10,000 = 5555.556$

   (b)   For critical coupling: $Q_{\text{Loaded}} = \dfrac{Q_{\text{Unloaded}}}{2} = \dfrac{10,000}{2} = 5000$

   (c)   For over coupling: $Q_{\text{Loaded}} = \dfrac{Q_{\text{Unloaded}}}{\rho+1} = \dfrac{10,000}{1.25+1} = 4444.444$

Practice Problem

6.1 A cavity has a loaded and an unloaded Q of 13,327 and 8975, respectively. Calculate the coupling coefficient and external Q. In addition, calculate the loaded Q for (a) an under coupling, (b) a critical coupling, and (c) an over coupling case, if we want to maintain an SWR of 1.2. **[0.485, 27,478.35, 7269.273, 6663.5, 6057.727]**

6.4 Transmission Line Resonators

Different types of transmission line resonators are discussed in detail in the following subsections.

6.4.1 Short-circuited Half-wavelength Transmission Line Resonators

Let us consider that a lossy short-circuited transmission line of length L has been excited with a signal of frequency f_0, which satisfies the following condition:

$$l = \frac{\lambda_0}{2} = \frac{v_p}{2f_0} = \frac{v_p \pi}{\omega_0} \tag{6.21}$$

where v_p is the velocity of the wave in the line. We further assume that the line is characterized by attenuation constant α, phase constant β, and characteristic impedance Z_0. The input impedance of the line can be expressed as follows:

$$Z_{\text{in}} = Z_0 \tanh\{(\alpha+j\beta)l\} \tag{6.22}$$

Using the trigonometric identity

$$\tanh\{(\alpha+j\beta)l\} = \frac{\tanh(\alpha l)+j\tan(\beta l)}{1+j\tanh(\alpha l)\tan(\beta l)} \tag{6.23}$$

in Eq. (6.22), we get the following relation:

$$Z_{\text{in}} = Z_0 \frac{\tanh(\alpha l)+j\tan(\beta l)}{1+j\tanh(\alpha l)\tan(\beta l)} \tag{6.24}$$

If the transmission line is characterized by a low α, then we can write that $\alpha L \ll 1$. Hence, Eq. (6.24) modifies to the following form:

$$Z_{\text{in}} = Z_0 \frac{\alpha l+j\tan(\beta l)}{1+j\alpha l\tan(\beta l)} \tag{6.25}$$

At a frequency $\omega = \omega_0 + \Delta\omega$, where $\Delta\omega$ is small,

$$\beta l = \frac{2\pi l}{\lambda} = \frac{2\pi f l}{v_p} = \frac{\omega l}{v_p} = \frac{(\omega_0 + \Delta\omega)l}{v_p} = \frac{\omega_0 l}{v_p} + \frac{\Delta\omega l}{v_p} \tag{6.26}$$

Substituting Eq. (6.21) into Eq. (6.26), we get the following expression:

$$\beta l = \frac{\omega_0 v_p \pi}{\omega_0 v_p} + \frac{\Delta\omega v_p \pi}{\omega_0 v_p} = \pi + \pi \frac{\Delta\omega}{\omega_0} \tag{6.27}$$

or $\qquad \tan(\beta l) = \tan\left[\pi + \pi \frac{\Delta\omega}{\omega_0}\right] = \tan\left(\pi \frac{\Delta\omega}{\omega_0}\right) \cong \pi \frac{\Delta\omega}{\omega_0} \tag{6.28}$

Substituting Eq. (6.28) into Eq. (6.25), we get the following relation:

$$Z_{in} \cong Z_0 \frac{\alpha l + j\pi \dfrac{\Delta\omega}{\omega_0}}{1 + j\alpha l \pi \dfrac{\Delta\omega}{\omega_0}} \tag{6.29}$$

Since both $\Delta\omega$ and α are small, the product $\pi \dfrac{\Delta\omega \alpha l}{\omega_0} \ll 1$; hence; Eq. (6.29) modifies to the following form:

$$Z_{in} \cong Z_0 \left(\alpha l + j\pi \frac{\Delta\omega}{\omega_0}\right) \tag{6.30}$$

The input impedance shown in Eq. (6.30) is similar to that of a series RLC circuit, which can be expressed as follows:

$$Z_{in} = R + j2L\Delta\omega \tag{6.31}$$

Comparing Eqs (6.30) and (6.31), we get the following relations:

$$R = Z_0 \alpha l \tag{6.32}$$

and $\qquad 2L\Delta\omega = Z_0 \pi \dfrac{\Delta\omega}{\omega_0}$ or $L = Z_0 \dfrac{\pi}{2\omega_0} \tag{6.33}$

Further, for a series LCR circuit, we have the following equation:

$$\omega_0 = \frac{1}{\sqrt{LC}} \tag{6.34}$$

or $\qquad LC = \dfrac{1}{\omega_0^2}$ or $C = \dfrac{1}{\omega_0^2 L} \tag{6.35}$

Substitution of Eq. (6.33) into Eq. (6.35) results in the following expression:

$$C = \frac{1}{\omega_0^2 L} = \frac{2}{\pi Z_0 \omega_0} \tag{6.36}$$

The Q of the resonator is given by the following equation:

$$Q = \frac{\omega_0 L}{R} = \frac{\pi Z_0 \omega_0}{2\omega_0 Z_0 \alpha l} = \frac{\pi}{2\alpha l} = \frac{2\pi}{2\alpha \lambda_0} = \frac{\beta}{2\alpha} \tag{6.37}$$

6.4.2 Short-circuited Quarter-wavelength Transmission Line Resonators

In contrast to a half-wavelength transmission line transformer that acts as a series RLC circuit, a quarter-wave transmission line resonator behaves like a parallel resonance (anti-resonance) circuit. For this case,

$$l = \frac{\lambda_0}{4} = \frac{v_p}{4f_0} = \frac{v_p \pi}{2\omega_0} \tag{6.38}$$

Dividing both the numerator and the denominator of Eq. (6.24) by $j\tan(\beta l)$, we get the following relation:

$$Z_{in} = Z_0 \frac{1 - j\tanh(\alpha l)\cot(\beta l)}{\tanh(\alpha l) - j\cot(\beta l)} \tag{6.39}$$

If the transmission line is characterized by a low α, then we can write that $\alpha L \ll 1$. Hence, Eq. (6.39) modifies to the following form:

$$Z_{in} = Z_0 \frac{1 - j\alpha l\cot(\beta l)}{\alpha l - j\cot(\beta l)} \tag{6.40}$$

Now substituting Eq. (6.38) into Eq. (6.26), we get the following expression for $\omega = \omega_0 + \Delta\omega$:

$$\beta l = \frac{\omega_0 v_p \pi}{2\omega_0 v_p} + \frac{\Delta\omega v_p \pi}{2\omega_0 v_p} = \frac{\pi}{2} + \pi\frac{\Delta\omega}{2\omega_0} \tag{6.41}$$

or $\quad \cot(\beta l) = \cot\left(\frac{\pi}{2} + \pi\frac{\Delta\omega}{2\omega_0}\right) = -\tan\left(\pi\frac{\Delta\omega}{2\omega_0}\right) \cong -\pi\frac{\Delta\omega}{2\omega_0} \tag{6.42}$

Substituting Eq. (6.42) into Eq. (6.40), the following relation is obtained:

$$Z_{in} = Z_0 \frac{1 + j\alpha l\pi\dfrac{\Delta\omega}{2\omega_0}}{\alpha l + j\pi\dfrac{\Delta\omega}{2\omega_0}} \tag{6.43}$$

Since both $\Delta\omega$ and α are small, the product $\alpha l\pi\dfrac{\Delta\omega}{2\omega_0} \ll 1$ and Eq. (6.43) modifies to the following form:

$$Z_{in} \cong 1 \bigg/ \left(\frac{\alpha l}{Z_0} + j\pi\frac{\Delta\omega}{2Z_0\omega_0}\right) \tag{6.44}$$

Equation (6.44) represents the input impedance that is similar to that of a parallel RLC circuit. The input impedance of a parallel RLC circuit can be written as follows:

$$Z_{in} = \frac{1}{\dfrac{1}{R} + j2C\Delta\omega} \tag{6.45}$$

Comparing Eqs (6.44) and (6.45), we get the following relation:

$$R = \frac{Z_0}{\alpha l} \tag{6.46}$$

and $\quad 2C\Delta\omega = \pi \dfrac{\Delta\omega}{2Z_0\omega_0}$ or $C = \dfrac{\pi}{4Z_0\omega_0}$ (6.47)

Further, for a parallel resonant circuit, we have $L = \dfrac{1}{\omega_0^2 C}$ (6.48)

Substituting Eq. (6.47) into Eq. (6.48), we get the following relation:

$$L = \dfrac{1}{\omega_0^2 C} = \dfrac{1}{\omega_0^2 \dfrac{\pi}{4Z_0\omega_0}} = \dfrac{4Z_0}{\pi\omega_0}$$ (6.49)

The Q of the resonator is expressed as follows:

$$Q = \omega_0 RC = \dfrac{\pi\omega_0 Z_0}{4\alpha l Z_0\omega_0} = \dfrac{\pi}{4\alpha l} = \dfrac{4\pi}{4\alpha\lambda_0} = \dfrac{\beta}{2\alpha}$$ (6.50)

6.4.3 Open-circuited Half-wavelength Transmission Line Resonators

Similar to a short-circuited quarter-wavelength transmission line resonator, an open-circuited half-wavelength transmission line resonator also behaves like a parallel resonant circuit. For this case, the input impedance is given by the following equality:

$$Z_{in} = Z_0 \coth\{(\alpha+j\beta)l\}$$ (6.51)

Using the trigonometric identity (6.23) in Eq. (6.51), we get the following relation:

$$Z_{in} = Z_0 \dfrac{1 + j\tan(\beta l)\tanh(\alpha l)}{\tanh(\alpha l) + j\tan(\beta l)}$$ (6.52)

If the transmission line is characterized by a low α, then Eq. (6.52) modifies to the following form:

$$Z_{in} = Z_0 \dfrac{1 + j\alpha l \tan(\beta l)}{\alpha l + j\tan(\beta l)}$$ (6.53)

Now substituting Eq. (6.28) into Eq. (6.53), we get the following expression:

$$Z_{in} = Z_0 \dfrac{1 + j\alpha l\pi\dfrac{\Delta\omega}{\omega_0}}{\alpha l + j\pi\dfrac{\Delta\omega}{\omega_0}}$$ (6.54)

Since both $\Delta\omega$ and α are small, the product $\pi\dfrac{\Delta\omega\alpha l}{\omega_0} \ll 1$; hence, Eq. (6.54) modifies to the following form:

$$Z_{in} = 1 \Big/ \left(\dfrac{\alpha l}{Z_0} + j\pi\dfrac{\Delta\omega}{Z_0\omega_0}\right)$$ (6.55)

The input impedance represented by Eq. (6.55) is similar to that of a parallel RLC circuit. Comparing Eqs (6.55) and (6.45), we get the following relations:

$$R = \dfrac{Z_0}{\alpha l}$$ (6.56)

and $\quad 2C\Delta\omega = \pi\dfrac{\Delta\omega}{Z_0\omega_0}$ or $C = \dfrac{\pi}{2Z_0\omega_0}$ $\qquad\qquad$ (6.57)

Further, for a parallel resonant circuit, we have $L = \dfrac{1}{\omega_0^2 C}$ $\qquad\qquad$ (6.58)

Substituting Eq. (6.57) into Eq. (6.58), we get $L = \dfrac{1}{\omega_0^2 C} = \dfrac{2Z_0}{\pi\omega_0}$ $\qquad\qquad$ (6.59)

The Q of the resonator is expressed as follows:

$$Q = \omega_0 RC = \frac{\pi\omega_0 Z_0}{2\alpha l Z_0 \omega_0} = \frac{\pi}{2\alpha l} = \frac{2\pi}{2\alpha\lambda_0} = \frac{\beta}{2\alpha} \qquad\qquad (6.60)$$

EXAMPLE 6.2 A 50 Ω short-circuited half-wavelength transmission line resonator has a resonant frequency of 1 GHz and a Q of 7984. Calculate R, L, and C.

Solution Given: $Z_0 = 50\,\Omega$, $f_0 = 1\,\text{GHz}$, and $Q = 7984$

Therefore,

$$L = Z_0\frac{\pi}{2\omega_0} = 50 \times \frac{\pi}{2 \times 2\pi \times 10^9} = 12.5 \times 10^{-9} = 12.5\,\text{nH}$$

$$C = \frac{2}{\pi Z_0 \omega_0} = \frac{2}{\pi \times 50 \times 2\pi \times 10^9} = 2.026 \times 10^{-12} = 2.026\,\text{pF}$$

$$R = \frac{\omega_0 L}{Q} = \frac{2\pi \times 10^9 \times 12.5 \times 10^{-9}}{7984} = 9.837 \times 10^{-3} = 9.837\,\text{m}\Omega$$

Practice Problem

6.2 A 75 Ω short-circuited quarter-wavelength transmission line resonator has a resonant frequency of 1 GHz and a Q of 8375. Calculate R, L, and C.

[**15.198 nH, 1.667 pF, 799.594 kΩ**]

6.5 Waveguide Cavities

Waveguide resonators are constructed by shorting both the ends of a guide. When the cavity is excited through a probe, a loop, or an aperture with a signal frequency equal to the resonant frequency of the guide, it bounces back and forth from the shorting walls, thus sustaining resonance. Due to resonance, the electric and magnetic energies are stored within the cavity. However, the power can be dissipated in the conducting walls of the cavity, as well as in the dielectric filling of the cavity.

6.5.1 Rectangular Waveguide Cavities

Let us consider a rectangular waveguide cavity of cross-sectional dimension $a \times b$ and length d, as shown in Fig. 6.2(b).

To find the resonance frequency, let us first consider the field components in a rectangular waveguide. Without any loss of generality, the TE- and TM-mode tangential electric field components in a lossless rectangular waveguide can be written as follows:

$$E_t^+ = e_t(x,y)e^{j(\omega t - \beta z)} \tag{6.61}$$

where $e_t(x,y)$ represents the product of the magnitude, x and y variation of the tangential electric field component. These tangential electric field components will travel along the positive z-direction, face a shorting plate at $z = d$, and are reflected back. The reflected field components can be expressed as follows:

$$E_t^- = \Gamma_L e_t(x,y)e^{j(\omega t + \beta z)} = -e_t(x,y)e^{j(\omega t + \beta z)} \tag{6.62}$$

where $\Gamma_L = -1$ is the reflection coefficient.

Therefore the resultant tangential electric and magnetic fields at any point z in the cavity is given by the following relation:

$$E_t = E_t^+ + E_t^- = e_t(x,y)e^{j(\omega t - \beta z)} - e_t(x,y)e^{j(\omega t + \beta z)}$$

or

$$E_t = e_t(x,y)\left(e^{-j\beta z} - e^{j\beta z}\right)e^{j\omega t}$$

or

$$E_t = -2je_t(x,y)\sin(\beta z)e^{j\omega t} \tag{6.63}$$

Since the tangential components of the electric field must vanish at the shorting wall at $z = d$, we can write the following equation:

$$E_t\big|_{z=d} = -2je_t(x,y)\sin(\beta d)e^{j\omega t} = 0 \tag{6.64}$$

Since $e_t(x,y)$ and $e^{j\omega t}$ cannot be equal to zero, as this would imply that $E_t = 0$ at all z, we can conclude that

$$\sin(\beta d) = 0 \text{ or } \sin(\beta d) = \sin(p\pi), \text{ where } p \text{ is an integer}$$

or

$$\beta = \frac{p\pi}{d} \tag{6.65}$$

Therefore, using Eqs (5.119)–(5.124), the TE$_{mnp}$-mode electric field components inside the rectangular waveguide cavity can be expressed as follows:

$$E_x = \frac{2\omega\mu}{k_c^2} H_0 \frac{n\pi}{b}\cos\left(\frac{m\pi x}{a}\right)\sin\left(\frac{n\pi y}{b}\right)\sin\left(\frac{p\pi z}{d}\right)e^{j\omega t} \tag{6.66}$$

$$E_y = -\frac{2\omega\mu}{k_c^2} H_0 \frac{m\pi}{a}\sin\left(\frac{m\pi x}{a}\right)\cos\left(\frac{n\pi y}{b}\right)\sin\left(\frac{p\pi z}{d}\right)e^{j\omega t} \tag{6.67}$$

$$E_z = 0 \tag{6.68}$$

$$H_x = \frac{j2\beta}{k_c^2} H_0 \frac{m\pi}{a}\sin\left(\frac{m\pi x}{a}\right)\cos\left(\frac{n\pi y}{b}\right)\cos\left(\frac{p\pi z}{d}\right)e^{j\omega t} \tag{6.69}$$

$$H_y = \frac{j2\beta}{k_c^2} H_0 \frac{n\pi}{b}\cos\left(\frac{m\pi x}{a}\right)\sin\left(\frac{n\pi y}{b}\right)\cos\left(\frac{p\pi z}{d}\right)e^{j\omega t} \tag{6.70}$$

$$H_z = -j2H_0\cos\left(\frac{m\pi x}{a}\right)\cos\left(\frac{n\pi y}{b}\right)\sin\left(\frac{p\pi z}{d}\right)e^{j\omega t} \tag{6.71}$$

During the derivation of Eqs (6.69) and (6.70), we should remember that, since the reflected wave is propagating in the $-z$-direction and the direction of the reflected electric field is the same as that of the incident field, the direction of the reflected magnetic field will be opposite to that of the incident field.

Similarly, using Eqs (5.163)–(5.168), the TM_{mnp}-mode field components inside the rectangular waveguide cavity can be expressed as follows:

$$E_x = -\frac{2\beta}{k_c^2} E_0 \frac{m\pi}{a} \cos\left(\frac{m\pi x}{a}\right)\sin\left(\frac{n\pi y}{b}\right)\sin\left(\frac{p\pi z}{d}\right)e^{j\omega t} \qquad (6.72)$$

$$E_y = -\frac{2\beta}{k_c^2} E_0 \frac{n\pi}{b} \sin\left(\frac{m\pi x}{a}\right)\cos\left(\frac{n\pi y}{b}\right)\sin\left(\frac{p\pi z}{d}\right)e^{j\omega t} \qquad (6.73)$$

$$E_z = 2E_0\sin\left(\frac{m\pi x}{a}\right)\sin\left(\frac{n\pi y}{b}\right)\cos\left(\frac{p\pi z}{d}\right)e^{j\omega t} \qquad (6.74)$$

$$H_x = \frac{j2\omega\varepsilon}{k_c^2} E_0 \frac{n\pi}{b} \sin\left(\frac{m\pi x}{a}\right)\cos\left(\frac{n\pi y}{b}\right)\cos\left(\frac{p\pi z}{d}\right)e^{j\omega t} \qquad (6.75)$$

$$H_y = -\frac{j2\omega\varepsilon}{k_c^2} E_0 \frac{m\pi}{a} \cos\left(\frac{m\pi x}{a}\right)\sin\left(\frac{n\pi y}{b}\right)\cos\left(\frac{p\pi z}{d}\right)e^{j\omega t} \qquad (6.76)$$

$$H_z = 0 \qquad (6.77)$$

Equation (6.65) can be written as follows: $\beta = 2\pi/\lambda_g = p\pi/d$

or $\qquad d = p\lambda_g/2 \qquad (6.78)$

Equation (6.78) reveals that the length of the cavity must be an integer multiple of the half-guided wavelength at the resonance frequency.

Using Eq. (5.91) for a lossless cavity, we can write the following expression:

$$k_c^2 = \omega^2\mu\varepsilon - \beta^2 = \omega^2\mu\varepsilon - \left(\frac{p\pi}{d}\right)^2 \qquad (6.79)$$

Further, from Eq. (5.170),

$$k_c^2 = k_x^2 + k_y^2 = \left(\frac{m\pi}{a}\right)^2 + \left(\frac{n\pi}{b}\right)^2 \qquad (6.80)$$

Equating Eq. (6.79) to Eq. (6.80), we get the following relation:

$$\omega^2\mu\varepsilon = \left(\frac{m\pi}{a}\right)^2 + \left(\frac{n\pi}{b}\right)^2 + \left(\frac{p\pi}{d}\right)^2$$

or $\qquad 2\pi f_{mnp}/v_p = \sqrt{\left(\frac{m\pi}{a}\right)^2 + \left(\frac{n\pi}{b}\right)^2 + \left(\frac{p\pi}{d}\right)^2}$

or $\qquad f_{mnp} = \frac{v_p}{2}\sqrt{\left(\frac{m}{a}\right)^2 + \left(\frac{n}{b}\right)^2 + \left(\frac{p}{d}\right)^2} \qquad (6.81)$

where f_{mnp} is the resonance frequency of the corresponding TE_{mnp} or TM_{mnp} mode. If $d > a > b$, then the TE_{101} mode has the lowest resonance frequency and hence is the dominant mode. The TE_{101} mode basically corresponds to a TE_{10} mode in a short-circuited waveguide of length $\lambda_g/2$. The dominant TM_{mnp} mode in a rectangular waveguide cavity is TM_{110}.

Time-average power loss at the cavity walls is given by the following equation:

$$P_c = R_S \left[\int_{S_1} |J_{S1}|^2 \, dS_1 + \int_{S_2} |J_{S2}|^2 \, dS_2 + \int_{S_3} |J_{S3}|^2 \, dS_3 \right] \tag{6.82}$$

For the TE mode, we can write the following equation:

$$P_c = R_S \left[\int_{S_1} \left\{ |H_{0x}|^2 + |H_{0y}|^2 \right\} \bigg|_{z=0} dxdy + \int_{S_2} \left\{ |H_{0y}|^2 + |H_{0z}|^2 \right\} \bigg|_{x=0} dydz \right.$$
$$\left. + \int_{S_3} \left\{ |H_{0x}|^2 + |H_{0z}|^2 \right\} \bigg|_{y=0} dxdz \right] \tag{6.83}$$

Now, the field components can be expressed as follows:

$$H_{0x} = \frac{j2\beta}{k_c^2} H_0 \frac{m\pi}{a} \sin\left(\frac{m\pi x}{a}\right) \cos\left(\frac{n\pi y}{b}\right) \cos\left(\frac{p\pi z}{d}\right)$$

or $\quad H_{0x} = \frac{2E_{yo}}{\eta_{TE}} \sin\left(\frac{m\pi x}{a}\right) \cos\left(\frac{n\pi y}{b}\right) \cos\left(\frac{p\pi z}{d}\right) \tag{6.84}$

where

$$E_{yo} = \frac{j\beta}{k_c^2} H_0 \eta_{TE} \frac{m\pi}{a} = \frac{j\beta}{k_c^2} H_0 \frac{\omega\mu}{\beta} \frac{m\pi}{a} = \frac{j\omega\mu}{k_c^2} H_0 \frac{m\pi}{a} \tag{6.85}$$

$$H_{0y} = \frac{j2\beta}{k_c^2} H_0 \frac{n\pi}{b} \cos\left(\frac{m\pi x}{a}\right) \sin\left(\frac{n\pi y}{b}\right) \cos\left(\frac{p\pi z}{d}\right)$$

or $\quad H_{0y} = \frac{2E_{xo}}{\eta_{TE}} \cos\left(\frac{m\pi x}{a}\right) \sin\left(\frac{n\pi y}{b}\right) \cos\left(\frac{p\pi z}{d}\right) \tag{6.86}$

where

$$E_{xo} = \frac{j\beta}{k_c^2} \eta_{TE} H_0 \frac{n\pi}{b} = \frac{j\beta}{k_c^2} \frac{\omega\mu}{\beta} H_0 \frac{n\pi}{b} = \frac{j\omega\mu}{k_c^2} H_0 \frac{n\pi}{b} \tag{6.87}$$

$$H_{0z} = -j2H_0 \cos\left(\frac{m\pi x}{a}\right) \cos\left(\frac{n\pi y}{b}\right) \sin\left(\frac{p\pi z}{d}\right)$$

or $\quad H_{0z} = 2H_{zo} \cos\left(\frac{m\pi x}{a}\right) \cos\left(\frac{n\pi y}{b}\right) \sin\left(\frac{p\pi z}{d}\right) \tag{6.88}$

where $\quad H_{zo} = -jH_0 \tag{6.89}$

Using these field equations, we can write the following expression:

$$P_1 = R_S \int_{S_1} \left\{ |H_{0x}|^2 + |H_{0y}|^2 \right\}\Big|_{z=0} dxdy$$

or
$$P_1 = \frac{4R_S}{\eta_{TE}^2} \left[(E_{yo})^2 \int_0^a \sin^2\left(\frac{m\pi x}{a}\right) dx \int_0^b \cos^2\left(\frac{n\pi y}{b}\right) dy \right.$$
$$\left. + (E_{xo})^2 \int_0^a \cos^2\left(\frac{m\pi x}{a}\right) dx \int_0^b \sin^2\left(\frac{n\pi y}{b}\right) dy \right]$$

or
$$P_1 = \frac{4R_S}{\eta_{TE}^2} \left[(E_{yo})^2 \frac{ab}{4} + (E_{xo})^2 \frac{ab}{4} \right] = \frac{R_S ab}{\eta_{TE}^2} \left[(E_{yo})^2 + (E_{xo})^2 \right] \tag{6.90}$$

$$P_2 = R_S \int_{S_2} \left\{ |H_{0y}|^2 + |H_{0z}|^2 \right\}\Big|_{x=0} dydz$$

or
$$P_2 = 4R_S \left[\left(\frac{E_{xo}}{\eta_{TE}}\right)^2 \int_0^b \sin^2\left(\frac{n\pi y}{b}\right) dy \int_0^d \cos^2\left(\frac{p\pi z}{d}\right) dz \right.$$
$$\left. + (H_{zo})^2 \int_0^b \cos^2\left(\frac{n\pi y}{b}\right) dy \int_0^d \sin^2\left(\frac{p\pi z}{d}\right) dz \right]$$

or
$$P_2 = 4R_S \left[\left(\frac{E_{xo}}{\eta_{TE}}\right)^2 \frac{bd}{4} + (H_{zo})^2 \frac{bd}{4} \right] = R_S bd \left[\left(\frac{E_{xo}}{\eta_{TE}}\right)^2 + (H_{zo})^2 \right] \tag{6.91}$$

$$P_3 = R_S \int_{S_3} \left\{ |H_{0x}|^2 + |H_{0z}|^2 \right\}\Big|_{y=0} dxdz$$

or
$$P_3 = 4R_S \left[\left(\frac{E_{yo}}{\eta_{TE}}\right)^2 \int_0^a \sin^2\left(\frac{m\pi x}{a}\right) dx \int_0^d \cos^2\left(\frac{p\pi z}{d}\right) dz \right.$$
$$\left. + (H_{zo})^2 \int_0^a \cos^2\left(\frac{m\pi x}{a}\right) dx \int_0^d \sin^2\left(\frac{p\pi z}{d}\right) dz \right]$$

or
$$P_3 = 4R_S \left[\left(\frac{E_{yo}}{\eta_{TE}}\right)^2 \frac{ad}{4} + (H_{zo})^2 \frac{ad}{4} \right] = R_S ad \left[\left(\frac{E_{yo}}{\eta_{TE}}\right)^2 + (H_{zo})^2 \right] \tag{6.92}$$

The total loss is, therefore,

$$P_c = (P_1 + P_2 + P_3) = \frac{R_S ab}{\eta_{TE}^2} \left[(E_{yo})^2 + (E_{xo})^2 \right]$$
$$+ R_S bd \left[\left(\frac{E_{xo}}{\eta_{TE}}\right)^2 + (H_{zo})^2 \right] + R_S ad \left[\left(\frac{E_{yo}}{\eta_{TE}}\right)^2 + (H_{zo})^2 \right]$$

Using Eqs (6.85), (6.87), and (6.89) in this equation, we get the following relations:

$$P_c = \frac{R_S ab}{\eta_{TE}^2}\left[\left(\frac{\omega\mu}{k_c^2}H_0\frac{m\pi}{a}\right)^2 + \left(\frac{\omega\mu}{k_c^2}H_0\frac{n\pi}{b}\right)^2\right]$$

$$+ R_S bd\left[\left(\frac{1}{\eta_{TE}}\frac{\omega\mu}{k_c^2}H_0\frac{n\pi}{b}\right)^2 + (H_0)^2\right] + R_S ad\left[\left(\frac{1}{\eta_{TE}}\frac{\omega\mu}{k_c^2}H_0\frac{m\pi}{a}\right)^2 + (H_0)^2\right]$$

or

$$P_c = R_S ab\left(\frac{\beta}{\omega\mu}\right)^2\left[\left(\frac{\omega\mu}{k_c^2}H_0\frac{m\pi}{a}\right)^2 + \left(\frac{\omega\mu}{k_c^2}H_0\frac{n\pi}{b}\right)^2\right]$$

$$+ R_S bd\left[\left(\frac{\beta}{\omega\mu}\frac{\omega\mu}{k_c^2}H_0\frac{n\pi}{b}\right)^2 + (H_0)^2\right] + R_S ad\left[\left(\frac{\beta}{\omega\mu}\frac{\omega\mu}{k_c^2}H_0\frac{m\pi}{a}\right)^2 + (H_0)^2\right]$$

or

$$P_c = R_S H_0^2\left[ab\left\{\left(\frac{\beta}{k_c^2}\frac{m\pi}{a}\right)^2 + \left(\frac{\beta}{k_c^2}\frac{n\pi}{b}\right)^2\right\} + bd\left\{\left(\frac{\beta}{k_c^2}\frac{n\pi}{b}\right)^2 + 1\right\}\right.$$

$$\left. + ad\left\{\left(\frac{\beta}{k_c^2}\frac{m\pi}{a}\right)^2 + 1\right\}\right]$$

or

$$P_c = R_S H_0^2\left[(ab+ad)\left(\frac{m\pi}{a}\right)^2\frac{\beta^2}{k_c^4} + (ab+bd)\left(\frac{n\pi}{b}\right)^2\frac{\beta^2}{k_c^4} + (ad+bd)\right] \quad (6.93)$$

Now,

$$E_{0x} = \frac{2\omega\mu}{k_c^2}H_0\frac{n\pi}{b}\cos\left(\frac{m\pi x}{a}\right)\sin\left(\frac{n\pi y}{b}\right)\sin\left(\frac{p\pi z}{d}\right) \quad (6.94)$$

and

$$E_{0y} = -\frac{2\omega\mu}{k_c^2}H_0\frac{m\pi}{a}\sin\left(\frac{m\pi x}{a}\right)\cos\left(\frac{n\pi y}{b}\right)\sin\left(\frac{p\pi z}{d}\right) \quad (6.95)$$

The time-average stored electric energy in the cavity is expressed as follows:

$$W_e = \frac{\varepsilon}{4}\int_V\left[|E_x|^2 + |E_y|^2\right]dv$$

or

$$W_e = \frac{\varepsilon}{4}\left[\left(\frac{2\omega\mu}{k_c^2}H_0\frac{n\pi}{b}\right)^2\int_0^a\int_0^b\int_0^d\cos^2\left(\frac{m\pi x}{a}\right)\sin^2\left(\frac{n\pi y}{b}\right)\sin^2\left(\frac{p\pi z}{d}\right)dxdydz\right.$$

$$\left. + \left(-\frac{2\omega\mu}{k_c^2}H_0\frac{m\pi}{a}\right)^2\int_0^a\int_0^b\int_0^d\sin^2\left(\frac{m\pi x}{a}\right)\cos^2\left(\frac{n\pi y}{b}\right)\sin^2\left(\frac{p\pi z}{d}\right)dxdydz\right]$$

or

$$W_e = \frac{\varepsilon}{4}\left[\left(\frac{2\omega\mu}{k_c^2}H_0\frac{n\pi}{b}\right)^2\frac{abd}{8} + \left(\frac{2\omega\mu}{k_c^2}H_0\frac{m\pi}{a}\right)^2\frac{abd}{8}\right]$$

or
$$W_e = \frac{\varepsilon abd}{8}\left(\frac{\omega\mu}{k_c^2}H_0\right)^2\left[\left(\frac{m\pi}{a}\right)^2+\left(\frac{n\pi}{b}\right)^2\right]$$

or
$$W_e = \frac{\varepsilon abd}{8}\frac{\omega^2\mu^2}{k_c^4}H_0^2k_c^2 = \frac{\varepsilon abd\omega^2\mu^2 H_0^2}{8k_c^2} \tag{6.96}$$

The following equation represents the time-average stored magnetic energy in the cavity:

$$W_m = \frac{\mu}{4}\int_V\left[|H_{0x}|^2+|H_{0y}|^2+|H_{0z}|^2\right]dxdydz$$

or
$$W_m = \frac{\mu}{4}\left[\int_0^a\int_0^b\int_0^d\left(\frac{2\beta}{k_c^2}H_0\frac{m\pi}{a}\right)^2\sin^2\left(\frac{m\pi x}{a}\right)\cos^2\left(\frac{n\pi y}{b}\right)\cos^2\left(\frac{p\pi z}{d}\right)dxdydz\right.$$

$$+\int_0^a\int_0^b\int_0^d\left(\frac{2\beta}{k_c^2}H_0\frac{n\pi}{b}\right)^2\cos^2\left(\frac{m\pi x}{a}\right)\sin^2\left(\frac{n\pi y}{b}\right)\cos^2\left(\frac{p\pi z}{d}\right)dxdydz$$

$$\left.+\int_0^a\int_0^b\int_0^d(2H_0)^2\cos^2\left(\frac{m\pi x}{a}\right)\cos^2\left(\frac{n\pi y}{b}\right)\sin^2\left(\frac{p\pi z}{d}\right)dxdydz\right]$$

or
$$W_m = \frac{\mu}{4}\left[\left(\frac{2\beta}{k_c^2}H_0\frac{m\pi}{a}\right)^2\frac{abd}{8}+\left(\frac{2\beta}{k_c^2}H_0\frac{n\pi}{b}\right)^2\frac{abd}{8}+(2H_0)^2\frac{abd}{8}\right]$$

or
$$W_m = \frac{\mu abd}{8}H_0^2\left[\frac{\beta^2}{k_c^4}\left\{\left(\frac{m\pi}{a}\right)^2+\left(\frac{n\pi}{b}\right)^2\right\}+1\right] = \frac{\mu abd}{8}H_0^2\left[\frac{\beta^2}{k_c^4}k_c^2+1\right]$$

or
$$W_m = \frac{\mu abd}{8}H_0^2\left[\frac{\beta^2+k_c^2}{k_c^2}\right] = \frac{\mu abdk^2}{8k_c^2}H_0^2$$

or
$$W_m = \frac{\mu abd\omega^2\mu\varepsilon}{8k_c^2}H_0^2 = \frac{\varepsilon abd\omega^2\mu^2}{8k_c^2}H_0^2 \tag{6.97}$$

Therefore, the total energy stored is expressed as follows:

$$W = W_e+W_m = \frac{\varepsilon abd\omega^2\mu^2}{8k_c^2}H_0^2+\frac{\varepsilon abd\omega^2\mu^2}{8k_c^2}H_0^2 = \frac{\varepsilon abd\omega^2\mu^2}{4k_c^2}H_0^2 \tag{6.98}$$

Consequently, the Q-factor for the TE mode is given by the following relation:

$$Q_c = \frac{\omega W}{P_c} = \frac{\omega\varepsilon abd\omega^2\mu^2 H_0^2}{4R_S H_0^2 k_c^2\left[(ab+ad)\left(\frac{m\pi}{a}\right)^2\frac{\beta^2}{k_c^4}+(ab+bd)\left(\frac{n\pi}{b}\right)^2\frac{\beta^2}{k_c^4}+(ad+bd)\right]}$$

or
$$Q_c = \frac{\omega\varepsilon abd\omega^2\mu^2 k_c^2}{4R_S\left[(ab+ad)\left(\frac{m\pi}{a}\right)^2\beta^2+(ab+bd)\left(\frac{n\pi}{b}\right)^2\beta^2+(ad+bd)k_c^4\right]} \tag{6.99}$$

For the TM mode, the time-average power loss at the cavity walls is given by the following equation:

$$P_c = R_S \left[\int_{S_1} \left\{ |H_{0x}|^2 + |H_{0y}|^2 \right\} \Big|_{z=0} dxdy + \int_{S_2} |H_{0y}|^2 \Big|_{x=0} dydz \right.$$

$$\left. + \int_{S_3} |H_{0x}|^2 \Big|_{y=0} dxdz \right] \qquad (6.100)$$

whereas the time-average stored electric energy in the cavity is expressed as follows:

$$W_e = \frac{\varepsilon}{4} \int_V \left[|E_x|^2 + |E_y|^2 + |E_z|^2 \right] dv \qquad (6.101)$$

and the time-average stored magnetic energy in the cavity is represented by the following equation:

$$W_m = \frac{\mu}{4} \int_V \left[|H_{0x}|^2 + |H_{0y}|^2 \right] dxdydz \qquad (6.102)$$

Proceeding as before, it can be shown that

$$P_c = \frac{R_S \omega^2 \varepsilon^2 E_0^2}{k_c^4} \left[(ab + bd) \left(\frac{m\pi}{a} \right)^2 + (ab + ad) \left(\frac{n\pi}{b} \right)^2 \right] \qquad (6.103)$$

$$W_e = \frac{\mu abd\omega^2 \varepsilon^2}{8k_c^2} E_0^2 \qquad (6.104)$$

and $\qquad W_m = \frac{\mu abd\omega^2 \varepsilon^2}{8k_c^2} E_0^2 \qquad (6.105)$

Therefore, the total energy stored is as follows:

$$W = W_e + W_m = \frac{\mu abd\omega^2 \varepsilon^2}{8k_c^2} E_0^2 + \frac{\mu abd\omega^2 \varepsilon^2}{8k_c^2} E_0^2 = \frac{\mu abd\omega^2 \varepsilon^2}{4k_c^2} E_0^2 \quad (6.106)$$

The Q-factor for the TM mode is represented by the following equation:

$$Q_c = \frac{\omega W}{P_c} = \frac{\omega \mu abd\omega^2 \varepsilon^2 E_0^2 k_c^4}{4R_S \omega^2 \varepsilon^2 E_0^2 k_c^2 \left[(ab + bd) \left(\frac{m\pi}{a} \right)^2 + (ab + ad) \left(\frac{n\pi}{b} \right)^2 \right]}$$

or $\qquad Q_c = \dfrac{\omega \mu abd k_c^2}{4R_S \left[(ab + bd) \left(\dfrac{m\pi}{a} \right)^2 + (ab + ad) \left(\dfrac{n\pi}{b} \right)^2 \right]} \qquad (6.107)$

EXAMPLE 6.3 A rectangular waveguide cavity has the dimension $22.86 \text{ mm} \times 10.16 \text{ mm} \times 15 \text{ mm}$. Calculate the resonant frequency for the TE_{101} mode.

Solution Given: $a = 22.86$ mm, $b = 10.16$ mm, and $d = 15$ mm

The resonant frequency is expressed as follows:

$$f_{101} = \frac{v_p}{2}\sqrt{\left(\frac{m}{a}\right)^2 + \left(\frac{n}{b}\right)^2 + \left(\frac{p}{d}\right)^2}$$

$$= \frac{3 \times 10^{11}}{2}\sqrt{\left(\frac{1}{22.86}\right)^2 + \left(\frac{0}{10.16}\right)^2 + \left(\frac{1}{15}\right)^2} = 11.96 \text{ GHz}$$

6.5.2 Circular Waveguide Cavities

Similar to rectangular waveguide cavities, circular waveguide cavities are also formed by shorting both the ends of a circular waveguide, as shown in Fig. 6.5.

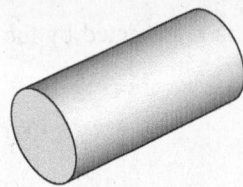

The effects of the shorting plate on the incident tangential field components are also the same for both cases, that is, the incident field gets reflected back with a reflection coefficient of -1. Further, since for both the cases, the field components have a $e^{-j\beta z}$ dependence on z, Eq. (6.65) also holds for a circular cavity. However, for a circular waveguide, Eq. (6.63) is modified to the following form:

Fig. 6.5 Circular waveguide cavity

$$E_t = -2je_t(r,\phi)\sin(\beta z)e^{j\omega t} \tag{6.108}$$

Using Eq. (6.108), the TE field components in a circular waveguide cavity can be written as follows:

$$E_r = -\frac{2\omega\mu n}{k_c^2 r}\{Q'\cos(n\phi) - P'\sin(n\phi)\}J_n(k_c r)\sin\left(\frac{p\pi z}{d}\right)e^{j\omega t} \tag{6.109}$$

$$E_\phi = \frac{2\omega\mu}{k_c}\{Q'\sin(n\phi) + P'\cos(n\phi)\}J_n'(k_c r)\sin\left(\frac{p\pi z}{d}\right)e^{j\omega t} \tag{6.110}$$

$$E_z = 0 \tag{6.111}$$

$$H_r = -\frac{j2\beta}{k_c}\{Q'\sin(n\phi) + P'\cos(n\phi)\}J_n'(k_c r)\cos\left(\frac{p\pi z}{d}\right)e^{j\omega t} \tag{6.112}$$

$$H_\phi = -\frac{j2\beta n}{k_c^2 r}\{Q'\cos(n\phi) - P'\sin(n\phi)\}J_n(k_c r)\cos\left(\frac{p\pi z}{d}\right)e^{j\omega t} \tag{6.113}$$

$$H_z = -j2\{Q'\sin(n\phi) + P'\cos(n\phi)\}J_n(k_c r)\sin\left(\frac{p\pi z}{d}\right)e^{j\omega t} \tag{6.114}$$

Similarly, the TM-mode field components can be written as follows:

$$E_r = -\frac{2\beta}{k_c}\{Q'\sin(n\phi) + P'\cos(n\phi)\}J_n'(k_c r)\sin\left(\frac{p\pi z}{d}\right)e^{j\omega t} \tag{6.115}$$

$$E_\phi = -\frac{2\beta n}{k_c^2 r}\left\{Q'\cos(n\phi) - P'\sin(n\phi)\right\}J_n(k_c r)\sin\left(\frac{p\pi z}{d}\right)e^{j\omega t} \tag{6.116}$$

$$E_z = 2\left\{Q'\sin(n\phi) + P'\cos(n\phi)\right\}J_n(k_c r)\cos\left(\frac{p\pi z}{d}\right)e^{j\omega t} \tag{6.117}$$

$$H_r = \frac{j2\omega\varepsilon n}{k_c^2 r}\left\{Q'\cos(n\phi) - P'\sin(n\phi)\right\}J_n(k_c r)\cos\left(\frac{p\pi z}{d}\right)e^{j\omega t} \tag{6.118}$$

$$H_\phi = -\frac{j2\omega\varepsilon}{k_c}\left\{Q'\sin(n\phi) + P'\cos(n\phi)\right\}J_n'(k_c r)\cos\left(\frac{p\pi z}{d}\right)e^{j\omega t} \tag{6.119}$$

$$H_z = 0 \tag{6.120}$$

The cut-off frequency for the TE mode in a circular waveguide is defined as $J_n'(k_c R) = 0$. If p_{mn}' is the root of $J_n'(k_c R)$, then

$$J_n'(p_{nm}') = 0 \tag{6.121}$$

Therefore, $k_c R = p_{nm}'$ or $k_c = p_{nm}'/R$ \hfill (6.122)

Again, $k_c^2 = \omega^2 \mu\varepsilon - \beta^2$ \hfill (6.123)

Substituting Eqs (6.65) and (6.122) into Eq. (6.123), we get the following relation:

$$\omega^2 \mu\varepsilon = \left(\frac{p_{nm}'}{R}\right)^2 + \left(\frac{p\pi}{d}\right)^2 \text{ or } \omega\sqrt{\mu\varepsilon} = \sqrt{\left(\frac{p_{nm}'}{R}\right)^2 + \left(\frac{p\pi}{d}\right)^2}$$

or $\quad \dfrac{2\pi f}{v_p} = \sqrt{\left(\dfrac{p_{nm}'}{R}\right)^2 + \left(\dfrac{p\pi}{d}\right)^2}$

or $\quad f = \dfrac{v_p}{2\pi}\sqrt{\left(\dfrac{p_{nm}'}{R}\right)^2 + \left(\dfrac{p\pi}{d}\right)^2}$ \hfill (6.124)

For the TM mode, the cut-off frequency in a circular waveguide is defined as $J_n(k_c R) = 0$. If p_{nm} is the root of $J_n(k_c R)$, then $J_n(p_{nm}) = 0$ \hfill (6.125)

Therefore, $k_c R = p_{nm}$ or $k_c = p_{nm}/R$ \hfill (6.126)

Substituting Eqs (6.65) and (6.126) into Eq. (6.123) and proceeding in a similar way, we get the following equation:

$$f = \frac{v_p}{2\pi}\sqrt{\left(\frac{p_{nm}}{R}\right)^2 + \left(\frac{p\pi}{d}\right)^2} \tag{6.127}$$

The dominant TE and TM modes in a circular waveguide cavity can, respectively, be calculated as TE_{111} and TM_{110}.

Without any loss of generality, the TE_{nmp} mode field components can be written as follows:

$$H_{0z} = -j2P'J_n(k_cr)\cos(n\phi)\sin\left(\frac{p\pi z}{d}\right) = H_A J_n\left(\frac{p'_{nm}r}{R}\right)\cos(n\phi)\sin\left(\frac{p\pi z}{d}\right) \quad (6.128)$$

where $H_A = -j2P'$ $\quad (6.129)$

$$E_{0r} = \frac{2P'\omega\mu n}{k_c^2 r}\sin(n\phi)J_n(k_cr)\sin\left(\frac{p\pi z}{d}\right)$$

or $\quad E_{0r} = j\frac{(-j2P')\omega\mu nR^2}{(p'_{mn})^2 r}J_n\left(\frac{p'_{nm}r}{R}\right)\sin(n\phi)\sin\left(\frac{p\pi z}{d}\right)$

or $\quad E_{0r} = \frac{jH_A k\eta nR^2}{(p'_{mn})^2 r}J_n\left(\frac{p'_{nm}r}{R}\right)\sin(n\phi)\sin\left(\frac{p\pi z}{d}\right) \quad (6.130)$

$$E_{0\phi} = \frac{2P'\omega\mu}{k_c}J'_n(k_cr)\cos(n\phi)\sin\left(\frac{p\pi z}{d}\right)$$

or $\quad E_{0\phi} = j\frac{(-j2P')\omega\mu R}{p'_{nm}}J'_n\left(\frac{p'_{nm}r}{R}\right)\cos(n\phi)\sin\left(\frac{p\pi z}{d}\right)$

or $\quad E_{0\phi} = \frac{jH_A k\eta R}{p'_{nm}}J'_n\left(\frac{p'_{nm}r}{R}\right)\cos(n\phi)\sin\left(\frac{p\pi z}{d}\right) \quad (6.131)$

$$E_{0z} = 0 \quad (6.132)$$

$$H_{0r} = -\frac{jP'2\beta}{k_c}J'_n(k_cr)\cos(n\phi)\cos\left(\frac{p\pi z}{d}\right)$$

or $\quad H_{0r} = \frac{(-2jP')\beta R}{p'_{nm}}J'_n\left(\frac{p'_{nm}r}{R}\right)\cos(n\phi)\cos\left(\frac{p\pi z}{d}\right)$

or $\quad H_{0r} = \frac{H_A\beta R}{p'_{nm}}J'_n\left(\frac{p'_{nm}r}{R}\right)\cos(n\phi)\cos\left(\frac{p\pi z}{d}\right) \quad (6.133)$

$$H_{0\phi} = \frac{j2P'\beta n}{k_c^2 r}J_n(k_cr)\sin(n\phi)\cos\left(\frac{p\pi z}{d}\right)$$

or $\quad H_{0\phi} = -\frac{(-j2P')\beta R^2 n}{(p'_{nm})^2 r}J_n\left(\frac{p'_{nm}r}{R}\right)\sin(n\phi)\cos\left(\frac{p\pi z}{d}\right)$

or $\quad H_{0\phi} = -\frac{H_A\beta R^2 n}{(p'_{nm})^2 r}J_n\left(\frac{p'_{nm}r}{R}\right)\sin(n\phi)\cos\left(\frac{p\pi z}{d}\right) \quad (6.134)$

Since the time-average stored electric and magnetic energies are equal, we can write the following equation:

$$W = 2W_e = \frac{\varepsilon}{2}\int_0^R\int_0^{2\pi}\int_0^d\left(|E_{0r}|^2 + |E_{0\phi}|^2\right)r\,dr\,d\phi\,dz \quad (6.135)$$

Substituting Eqs (6.130) and (6.131) into Eq. (6.135), we get the following relation:

$$W = \frac{\varepsilon}{2} \frac{H_A^2 k^2 \eta^2 R^2}{\left(p'_{nm}\right)^2} \int_0^R \int_0^{2\pi} \int_0^d \left\{ \left[J'_n \left(\frac{p'_{nm}r}{R} \right) \right]^2 \cos^2 (n\phi) \sin^2 \left(\frac{p\pi z}{d} \right) \right.$$

$$\left. + \frac{n^2 R^2}{\left(p'_{nm}\right)^2 r^2} J_n^2 \left(\frac{p'_{nm}r}{R} \right) \sin^2 (n\phi) \sin^2 \left(\frac{p\pi z}{d} \right) \right\} r\, dr\, d\phi\, dz$$

or $\qquad W = \dfrac{\varepsilon k^2 \eta^2 R^2 H_A^2}{2\left(p'_{nm}\right)^2} \pi \dfrac{d}{2} \int_0^R \left[\left\{ J'_n \left(\dfrac{p'_{nm}r}{R} \right) \right\}^2 + \left(\dfrac{nR}{p'_{nm}r} \right)^2 J_n^2 \left(\dfrac{p'_{nm}r}{R} \right) \right] r\, dr$ $\qquad$ (6.136)

Now, let us consider the following identity:

$$\int_0^{p'_{mn}} \left[\left\{ J'_n (x) \right\}^2 + \left(\frac{n}{x} \right)^2 J_n^2 (x) \right] x\, dx = \frac{\left(p'_{nm}\right)^2}{2} \left[1 - \left(\frac{n}{p'_{nm}} \right)^2 \right] J_n^2 \left(p'_{nm} \right) \qquad (6.137)$$

If we assume that

$$x = p'_{nm} r / R \qquad (6.138)$$

Then $\qquad dx = p'_{nm} dr / R \qquad (6.139)$

and $\qquad x\, dx = \left(p'_{nm}/R \right)^2 r\, dr \qquad (6.140)$

Further, at $x = 0$, $r = 0$ and at $x = p'_{nm}$, $r = R$.

Using Eqs (6.138)–(6.140) in Eq. (6.137), we get the following expression:

$$\int_0^R \left[\left\{ J'_n \left(\frac{p'_{nm}r}{R} \right) \right\}^2 + \left(\frac{nR}{p'_{nm}r} \right)^2 J_n^2 \left(\frac{p'_{nm}r}{R} \right) \right] \left(\frac{p'_{nm}}{R} \right)^2 r\, dr$$

$$= \frac{\left(p'_{nm}\right)^2}{2} \left[1 - \left(\frac{n}{p'_{nm}} \right)^2 \right] J_n^2 \left(p'_{nm} \right)$$

or $\qquad \displaystyle\int_0^R \left[\left\{ J'_n \left(\frac{p'_{nm}r}{R} \right) \right\}^2 + \left(\frac{nR}{p'_{nm}r} \right)^2 J_n^2 \left(\frac{p'_{nm}r}{R} \right) \right] r\, dr = \frac{R^2}{2} \left[1 - \left(\frac{n}{p'_{nm}} \right)^2 \right] J_n^2 \left(p'_{nm} \right)$

$$\qquad (6.141)$$

Substituting Eq. (6.141) into Eq. (6.136), we get the following relation:

$$W = \frac{\varepsilon k^2 \eta^2 R^2 H_A^2 \pi d}{4\left(p'_{nm}\right)^2} \frac{R^2}{2} \left[1 - \left(\frac{n}{p'_{nm}} \right) \right] J_n^2 \left(p'_{nm} \right)$$

or $\qquad W = \dfrac{\varepsilon k^2 \eta^2 R^4 H_A^2 \pi d}{8\left(p'_{nm}\right)^2} \left[1 - \left(\dfrac{n}{p'_{nm}} \right)^2 \right] J_n^2 \left(p'_{nm} \right) \qquad (6.142)$

Power loss in the conducting walls of the cavity can be expressed as follows:

$$P_c = \frac{R_s}{2} \int_S |H_t|^2 \, ds$$

or

$$P_c = \frac{R_s}{2} \Bigg[\int_0^{2\pi} \int_0^d \left\{ \left| H_{0\phi}(r=R) \right|^2 + \left| H_{0z}(r=R) \right|^2 \right\} R \, d\phi \, dz$$

$$+ 2 \int_0^R \int_0^{2\pi} \left\{ \left| H_{0r}(z=0) \right|^2 + \left| H_{0\phi}(z=0) \right|^2 \right\} r \, dr \, d\phi \Bigg] \tag{6.143}$$

Using Eqs (6.128), (6.133), and (6.134) in Eq. (6.143), we get the following relation:

$$P_c = \frac{R_s}{2} \Bigg[\frac{H_A^2 \beta^2 R^2 n^2}{(p'_{nm})^4} J_n^2(p'_{nm}) R \int_0^{2\pi} \sin^2(n\phi) \, d\phi \int_0^d \cos^2\left(\frac{p\pi z}{d}\right) dz$$

$$+ H_A^2 J_n^2(p'_{nm}) R \int_0^{2\pi} \cos^2(n\phi) \, d\phi \int_0^d \sin^2\left(\frac{p\pi z}{d}\right) dz$$

$$+ 2 \frac{H_A^2 \beta^2 R^2}{(p'_{nm})^2} \int_0^R \int_0^{2\pi} \left\{ \left[J_n'\left(\frac{p'_{nm} r}{R}\right) \right]^2 \cos^2(n\phi) \right.$$

$$\left. + \frac{R^2 n^2}{(p'_{nm})^2 r^2} J_n^2\left(\frac{p'_{nm} r}{R}\right) \sin^2(n\phi) \right\} r \, dr \, d\phi \Bigg]$$

$$P_c = \frac{R_s}{2} \Bigg[\frac{H_A^2 \beta^2 R^2 n^2}{(p'_{nm})^4} J_n^2(p'_{nm}) R \frac{\pi d}{2} + H_A^2 J_n^2(p'_{nm}) R \frac{\pi d}{2}$$

or

$$+ 2 \frac{H_A^2 \beta^2 R^2}{(p'_{nm})^2} \pi \int_0^R \left\{ \left[J_n'\left(\frac{p'_{nm} r}{R}\right) \right]^2 + \frac{R^2 n^2}{(p'_{nm})^2 r^2} J_n^2\left(\frac{p'_{nm} r}{R}\right) \right\} r \, dr \Bigg] \tag{6.144}$$

Substituting Eq. (6.141) into Eq. (6.144), we get the following expression:

$$P_c = \frac{R_s}{2} \Bigg[\frac{H_A^2 \beta^2 R^2 n^2}{(p'_{nm})^4} J_n^2(p'_{nm}) R \frac{\pi d}{2} + H_A^2 J_n^2(p'_{nm}) R \frac{\pi d}{2}$$

$$+ 2 \frac{H_A^2 \beta^2 R^2}{(p'_{nm})^2} \pi \frac{R^2}{2} \left\{ 1 - \left[\frac{n}{p'_{nm}} \right] \right\} J_n^2(p'_{nm}) \Bigg]$$

or

$$P_c = \frac{\pi R_s H_A^2 J_n^2(p'_{nm})}{2} \Bigg[\frac{Rd}{2} \left\{ 1 + \left[\frac{\beta R n}{(p'_{nm})^2} \right]^2 \right\} + \left(\frac{\beta R^2}{p'_{nm}} \right)^2 \left\{ 1 - \left(\frac{n}{p'_{nm}} \right)^2 \right\} \Bigg] \tag{6.145}$$

Using Eqs (6.142) and (6.145), the Q of the cavity, due to conductor loss, can be expressed as follows:

$$Q_c = \omega \frac{\varepsilon k^2 \eta^2 R^4 d}{4 R_s \left(p'_{nm}\right)^2} \frac{\left\{1 - \left(\dfrac{n}{p'_{nm}}\right)^2\right\}}{\dfrac{Rd}{2}\left\{1 + \left[\dfrac{\beta R n}{\left(p'_{nm}\right)^2}\right]^2\right\} + \left(\dfrac{\beta R^2}{p'_{nm}}\right)^2 \left\{1 - \left(\dfrac{n}{p'_{nm}}\right)^2\right\}} \tag{6.146}$$

Since $\omega\varepsilon = k/\eta$, Eq. (6.146) can be written as follows:

$$Q_c = \frac{\eta\, Rd\, (kR)^3}{4 R_s \left(p'_{nm}\right)^2} \frac{\left\{1 - \left(\dfrac{n}{p'_{nm}}\right)^2\right\}}{\dfrac{Rd}{2}\left\{1 + \left[\dfrac{\beta R n}{\left(p'_{nm}\right)^2}\right]^2\right\} + \left(\dfrac{\beta R^2}{p'_{nm}}\right)^2 \left\{1 - \left(\dfrac{n}{p'_{nm}}\right)^2\right\}} \tag{6.147}$$

Equation (6.65), which also holds for a circular waveguide cavity resonator, shows that β is a constant. Therefore, using Eqs (6.123) and (6.122), we can write the following expression:

$$\beta = \sqrt{\omega^2 \mu\varepsilon - k_c^2} = \sqrt{k^2 - k_c^2} = \sqrt{k^2 - \left(\frac{p'_{nm}}{R}\right)^2}$$

or $\qquad \beta = \dfrac{1}{R}\sqrt{(kR)^2 - \left(p'_{nm}\right)^2} = \text{constant} \tag{6.148}$

Since for given the given values of m and n $p'_{nm} = $ constant, the product kR is also a constant. This implies that Q of a circular cavity varies as follows:

$$Q_c \infty \frac{1}{R_s} \infty \frac{1}{\sqrt{f}} \tag{6.149}$$

Power dissipated in the dielectric filling is given by the following equation:

$$P_d = \frac{\sigma}{2}\int_V \left[|E_{0r}|^2 + |E_{0\phi}|^2\right] dv \tag{6.150}$$

Substituting Eqs (6.130) and (6.131) into Eq. (6.150), we get the following expression:

$$P_d = \frac{\sigma}{2}\frac{H_A^2 k^2 \eta^2 R^2}{\left(p'_{nm}\right)^2}\left[\int_0^R \int_0^{2\pi} \int_0^d \frac{n^2 R^2}{\left(p'_{nm}\right)^2 r^2} J_n^2\left(\frac{p'_{nm} r}{R}\right)\sin^2(n\phi)\sin^2\left(\frac{p\pi z}{d}\right) r\, dr\, d\phi\, dz \right.$$

$$\left. + \int_0^R \int_0^{2\pi} \int_0^d \left\{J_n'\left(\frac{p'_{nm} r}{R}\right)\right\}^2 \cos^2(n\phi)\sin^2\left(\frac{p\pi z}{d}\right) r\, dr\, d\phi\, dz\right]$$

or $\qquad P_d = \dfrac{\sigma}{2}\dfrac{H_A^2 k^2 \eta^2 R^2}{\left(p'_{nm}\right)^2}\dfrac{\pi d}{2}\displaystyle\int_0^R \left[\left[J'_n\left(\dfrac{p'_{nm}r}{R}\right)\right]^2 + \left(\dfrac{nR}{p'_{nm}r}\right)^2 J_n^2\left(\dfrac{p'_{nm}r}{R}\right)\right] r\, dr$ (6.151)

Substituting Eq. (6.141) into Eq. (6.151), we get the following relation:

$$P_d = \dfrac{\sigma \pi H_A^2 k^2 \eta^2 R^2 d}{4\left(p'_{nm}\right)^2}\dfrac{R^2}{2}\left\{1-\left(\dfrac{n}{p'_{nm}}\right)^2\right\} J_n^2\left(p'_{nm}\right)$$

or $\qquad P_d = \dfrac{\sigma \pi H_A^2 k^2 \eta^2 R^4 d}{8\left(p'_{nm}\right)^2}\left\{1-\left(\dfrac{n}{p'_{nm}}\right)^2\right\} J_n^2\left(p'_{nm}\right)$ (6.152)

Therefore, the Q of the cavity due to dielectric loss can be expressed as follows:

$$Q_d = \dfrac{\omega W}{P_d} = \omega\, \dfrac{\dfrac{\varepsilon k^2 \eta^2 R^4 H_A^2 \pi d}{8\left(p'_{nm}\right)^2}\left\{1-\left(\dfrac{n}{p'_{nm}}\right)^2\right\} J_n^2\left(p'_{nm}\right)}{\dfrac{\sigma \pi H_A^2 k^2 \eta^2 R^4 d}{8\left(p'_{nm}\right)^2}\left\{1-\left(\dfrac{n}{p'_{nm}}\right)^2\right\} J_n^2\left(p'_{nm}\right)}$$

or $\qquad Q_d = \dfrac{\omega\varepsilon}{\sigma} = \dfrac{\omega\varepsilon}{\omega\varepsilon''} = \dfrac{\varepsilon}{\varepsilon''}$ (6.153)

EXAMPLE 6.4 A circular waveguide cavity has a diameter of 2.383 cm and length of 15 mm. Calculate the resonant frequency for the TE_{111} mode.

Solution Given: $D = 2.383$ cm, $d = 15$ mm

For the TE_{111} mode $p'_{nm} = 1.841$. Therefore,

$$f_{111} = \dfrac{v_p}{2\pi}\sqrt{\left(\dfrac{p'_{nm}}{R}\right)^2 + \left(\dfrac{p\pi}{d}\right)^2} = \dfrac{3\times10^{11}}{2\pi}\sqrt{\left(\dfrac{2\times1.841}{23.83}\right)^2 + \left(\dfrac{\pi}{15}\right)^2} = 12.427\,\text{GHz}$$

Practice Problems

6.3[@] Derive the expression for the quality factor of a rectangular waveguide for the dominant TE_{110} mode.

6.4 A circular waveguide cavity has a diameter of 17.48 m and length of 10 mm. Calculate the resonant frequency for the TM_{110} mode. [20.934 GHz]

6.6 Dielectric Resonators

A cube or disc made of a low-loss, high-dielectric-constant material, shown in Fig. 6.6, can also be used as a microwave resonator. The dielectric constant preferred for this purpose is generally in the range $10 \le \varepsilon_r \le 100$. Due to the high dielectric constant of the material, most of the fields are contained within the

---

[@]Solution in Online Resources

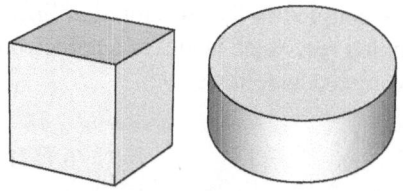

Fig. 6.6 Dielectric resonator
(a) Cubical (b) Cylindrical

resonator. However, unlike metal cavities, some fringing and leakage fields also exist outside it. Operation of a dielectric resonator is based on the following facts: because of the high permittivity of the resonator, propagating modes will be developed inside the resonator structure; however, at the outside, because of low dielectric constants, the modes will be turned into evanescent modes. For a circular cylindrical dielectric resonator of radius r, height h, and dielectric permittivity ε_r, the resonant frequency can be expressed as follows:

$$f_r = \frac{34}{r\varepsilon_r}\left(3.45 + \frac{r}{h}\right) \text{ GHz} \qquad (6.154)$$

where r and h are expressed in millimetres. The preceding expression is 98% accurate for $2 > r/h > 0.5$ and $50 > \varepsilon_r > 30$.

Equation (6.154) reveals that the resonant frequency of a dielectric resonator depends on the dimensions and relative permittivity of the dielectric medium. However, these parameters change with the variation in temperature. Therefore, dielectric materials with relatively small temperature dependence, such as barium tetratitanate and titanium dioxide, are mostly preferred. Q-values of several thousands can be achieved with such materials.

It may be noted that isolated dielectric resonators are rarely used. Instead, they are coupled with microstrip lines, as shown in Fig. 6.7, where the separation between the microstrip line and the resonator is adjusted to achieve the required coupling. As a result of coupling, an oscillating resonator launches a propagating wave on the microstrip line. Such a wave causes a flow of power and applies an external load on the resonator. In practice, the resonator can be represented by a parallel resonant circuit that is coupled to a microstrip line by means of an ideal transformer, as shown in Fig. 6.8. During the designing of a resonator, the values of R and L are chosen such that

$$Q_{\text{Unloaded}} = \omega_0 CR \qquad (6.155)$$

and $\quad \omega_0^2 LC = 1 \qquad (6.156)$

at the resonance frequency.

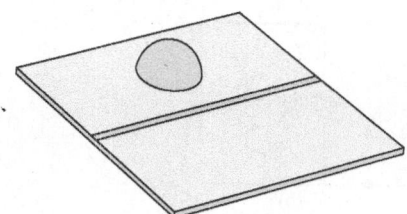

Fig. 6.7 Coupling of dielectric resonator and microstrip line

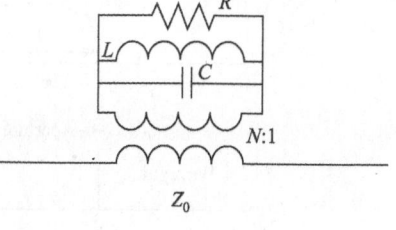

Fig. 6.8 Equivalent circuit of dielectric resonator coupled to microstrip line

At resonance, a series resistance of N^2R is coupled to the transmission line. When the line is matched terminated, this coupled resistance launches a reflected wave with a reflection coefficient, which is expressed as follows:

$$\Gamma = \frac{\left(N^2R + Z_0\right) - Z_0}{\left(N^2R + Z_0\right) + Z_0} = \frac{N^2R}{N^2R + 2Z_0} \tag{6.157}$$

or $$\frac{N^2R}{Z_0} = \frac{2\Gamma}{1 - \Gamma} \tag{6.158}$$

Therefore, from the required values of Γ and N, which decide the coupling, the value of R can be fixed. Again from the calculated value of R, required value of Q_{Unloaded}, and resonance frequency, we can calculate L and C using Eqs (6.155) and (6.156).

Dielectric resonators are lower in cost, size, and weight than metal cavities, and can be incorporated in planar transmission lines and microwave integrated circuits. Furthermore, due to the absence of metallic conductors, conductor loss of a dielectric resonator is zero. However, dielectric loss of a dielectric resonator is usually higher than the metal cavities and increases with an increase in the dielectric constants of the materials. Q-values of the order of several hundreds can easily be achieved with a dielectric resonator.

6.7 Coupled Cavities

Various types of coupled cavities are discussed in the subsequent sections.

6.7.1 Reflection Cavities

A reflection cavity is coupled only to the feeder line. A waveguide cavity fed by a waveguide through an aperture is an example of a reflection cavity and is shown in Fig. 6.9, along with its equivalent transmission line model. The unloaded and external Q of such a cavity can be expressed as follows:

$$Q_{\text{Unloaded}} = \frac{R_{\text{in}}}{\omega_r L Z_0} \tag{6.159}$$

and $$Q_{\text{Ext}} = \frac{1}{\omega_r L} \tag{6.160}$$

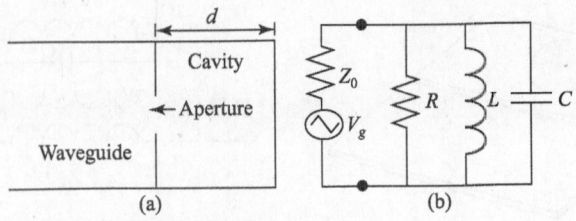

Fig. 6.9 Reflection cavity (a) Reflection-type cavity resonator (b) Equivalent circuits

Therefore, the loaded Q is expressed as follows:

$$\frac{1}{Q_{\text{Loaded}}} = \frac{1}{Q_{\text{Unloaded}}} + \frac{1}{Q_{\text{Ext}}} = \frac{\omega_r L Z_0}{R_{\text{in}}} + \omega_r L$$

or

$$\frac{1}{Q_{\text{Loaded}}} = \frac{\omega_r L Z_0 + \omega_r L R_{\text{in}}}{R_{\text{in}}} = \frac{\omega_r L (Z_0 + R_{\text{in}})}{R_{\text{in}}}$$

or

$$Q_{\text{Loaded}} = \frac{R_{\text{in}}}{\omega_r L (Z_0 + R_{\text{in}})} \qquad (6.161)$$

Now from Eq. (6.14),

$$K = \frac{Q_{\text{Unloaded}}}{Q_{\text{Ext}}} = \frac{R_{\text{in}} \omega_r L}{\omega_r L Z_0} = \frac{R_{\text{in}}}{Z_0} \qquad (6.162)$$

Over coupling will occur when $R_{\text{in}} > Z_0$ for parallel resonance and $R_{\text{in}} < Z_0$ for series resonance. The converse is true for under coupling. Critical coupling occurs when $R_{\text{in}} = Z_0$ for both types of resonance.

6.7.2 Transmission Cavities

Unlike a reflection cavity, a transmission cavity is coupled both to a generator and to a load. A transmission cavity along with its equivalent circuits is shown in Fig. 6.10. The unloaded, loaded, and external Q of the circuit can be expressed as follows:

$$Q_{\text{Unloaded}} = \omega_r L / R \qquad (6.163)$$

$$Q_{\text{Loaded}} = \frac{\omega_r L}{R + n_1^2 R_g + n_2^2 R_L} \qquad (6.164)$$

and

$$Q_{\text{Ext}} = \omega_r L / Z_0 \qquad (6.165)$$

From Eq. (6.13),

$$\frac{1}{Q_{\text{Loaded}}} = \frac{1}{Q_{\text{Unloaded}}} + \frac{1}{Q_{\text{Ext}}}$$

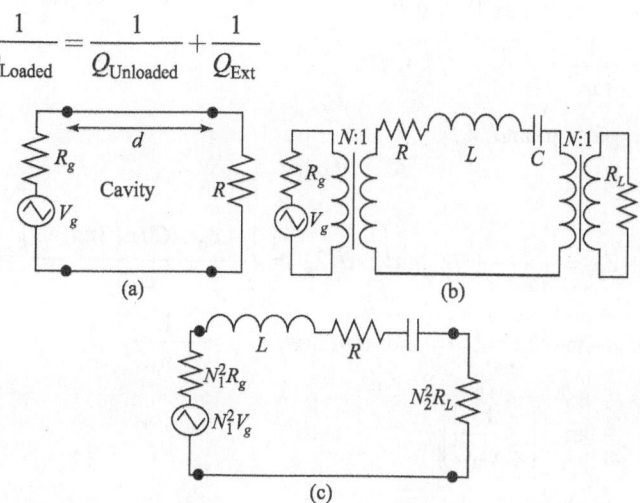

Fig. 6.10 Transmission-type cavity (a) Transmission-line equivalent circuit (b) Coupled with generator and load using *N*:1 transformer (c) Final equivalent circuit

or $$\frac{R+n_1^2 R_g + n_2^2 R_L}{\omega_r L} = \frac{R}{\omega_r L} + \frac{Z_0}{\omega_r L}$$

or $$R + Z_0 = R + n_1^2 R_g + n_2^2 R_L \text{ or } Z_0 = n_1^2 R_g + n_2^2 R_L \tag{6.166}$$

Now, from Eq. (6.14), $K = \dfrac{Q_{\text{Unloaded}}}{Q_{\text{Ext}}} = \dfrac{\omega_r L}{R}\dfrac{Z_0}{\omega_r L} = \dfrac{Z_0}{R}$ (6.167)

Using Eq. (6.166) in Eq. (6.167), we get the following expression:

$$K = \frac{n_1^2 R_g + n_2^2 R_L}{R} = \frac{n_1^2 R_g}{R} + \frac{n_2^2 R_L}{R} = K_1 + K_2 \tag{6.168}$$

where $K_1 = n_1^2 R_g / R$ (6.169)

$$K_2 = n_2^2 R_L / R \tag{6.170}$$

are the input and output coupling coefficients, respectively.

6.8 Re-entrant Cavities

In microwave tubes it is required to transfer maximum energy from an electron beam to a high-Q cavity. To achieve this, a special type of cavity, called a re-entrant cavity, is used. A schematic diagram of a re-entrant cavity is shown in Fig. 6.11.

A re-entrant cavity can be considered as a coaxial cavity with an inner and an outer radius of a and b, respectively. Therefore,

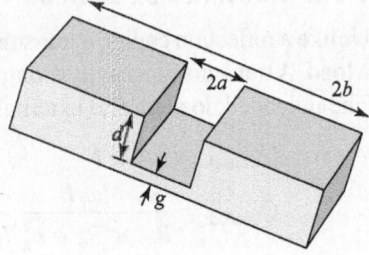

Fig. 6.11 Re-entrant cavity

the impedance provided by the gap capacitance C and the coaxial line below the gap can be written as follows:

$$Z_C = \frac{1}{j\omega C} \tag{6.171}$$

and $$Z_{\text{in}} = jZ_0 \tan\left(2\pi d/\lambda_g\right) \tag{6.172}$$

At resonance,

$$Z_C + Z_{\text{in}} = \frac{1}{j\omega_0 C} + jZ_0 \tan\left(2\pi d/\lambda_g\right) = \frac{1 - Z_0\omega_0 C \tan\left(2\pi d/\lambda_g\right)}{j\omega_0 C} = 0$$

or $$1 - Z_0\omega C \tan\left(2\pi d/\lambda_g\right) = 0 \text{ or } \tan\left(2\pi d/\lambda_g\right) = \frac{1}{Z_0\omega_0 C}$$

or $$d = \frac{\lambda_g}{2\pi}\tan^{-1}\left(\frac{1}{Z_0\omega_0 C}\right) \tag{6.173}$$

where $$C = \varepsilon_o \pi a^2 / g \tag{6.174}$$

and $$Z_0 = 60\ln\left(b/a\right) \tag{6.175}$$

If $C \ll 1$, then from Eq. (6.173), we get the following relation:

$$d \simeq \frac{\lambda_g}{2\pi} \tan^{-1}(\infty) \simeq \frac{\lambda_g}{2\pi} \frac{\pi}{2} \simeq \frac{\lambda_g}{4} \qquad (6.176)$$

that is, the line is quarter-wavelength long. On the other hand, when $C \gg 1$,

$$d \simeq \frac{\lambda_g}{2\pi} \tan^{-1}(0) \simeq 0 \qquad (6.177)$$

that is, the line is shortened. Equations (6.176) and (6.177) reveal that by changing the gap thickness g, we can vary the cavity length from 0 to $\lambda_g/4$. If C is increased, the energy stored in the electric field also increases. As per resonance criteria, the corresponding energy in the magnetic field also increases. An increase in energy in the magnetic field corresponds to large microwave currents in the cavity walls, which in turn increase the conductor loss and hence lower the Q-value.

Note If $d > (b-a)$ and $b < \lambda_g/4$ then the cavity is of TEM type. On the other hand, when $d < (b-a)$ and $b = \lambda_g/4$ the cavity is of TM type.

Figure 6.12 shows cross-sectional views of different re-entrant cavities.

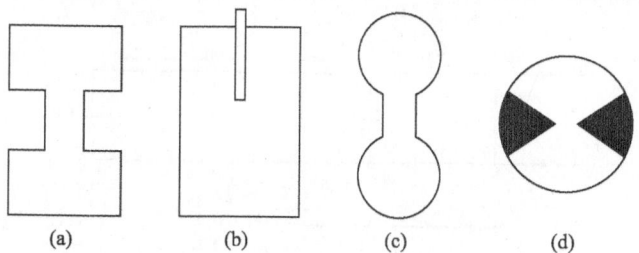

(a)   (b)   (c)   (d)

Fig. 6.12 Different types of re-entrant cavities (a) Coaxial (b) Tuneable (c) Toroidal (d) Butterfly

6.9 Hole and Slot Cavities

Similar to re-entrant cavities, hole and slot cavities, shown in Fig. 6.13, are also used in microwave tubes. A hole and slot cavity consists of a hole, equivalent to a loop of metallic band, and a slot. The holes provide an equivalent inductance L, whereas the slots provide an equivalent capacitance C. At resonance, the reactance provided by both of them is equal; therefore,

$$\omega_0 = \frac{1}{\sqrt{LC}} \qquad (6.178)$$

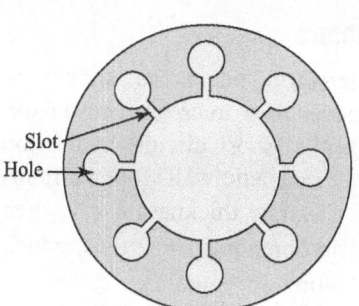

Fig. 6.13 Hole and slot cavity

where ω_0 is the resonant frequency.

6.10 Microstrip Resonators

Microstrip resonators are formed by developing a metallic patch, closed loop, or a line left open or short circuited at both ends on a dielectric substrate. In the following sections, we will discuss about few of the transmission line resonators.

6.10.1 Half-wavelength Gap-coupled Microstrip Line Resonators

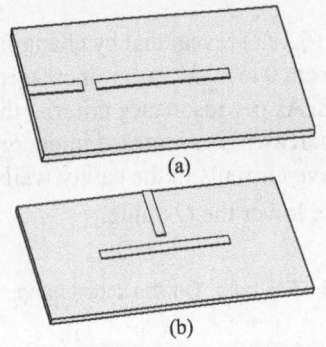

Fig. 6.14 Gap-coupled microstrip line resonator (a) Coupled at the end (b) Coupled at the side

A microstrip line that is half-wavelength long and open circuited at both ends behaves like a resonator. Such resonators are usually fed by another feeder line, as shown in Fig. 6.14. The equivalent circuits are shown in Fig. 6.15. The open end of the resonator can be represented as an extra line of length Δl_{oc}. Similarly, the gap region between the resonator and the feeder line can also be represented by an effective length Δl_g. Therefore, the resonance condition can be expressed as follows:

$$l + \Delta l_{oc} + \Delta l_g = \lambda_g / 2 \qquad (6.179)$$

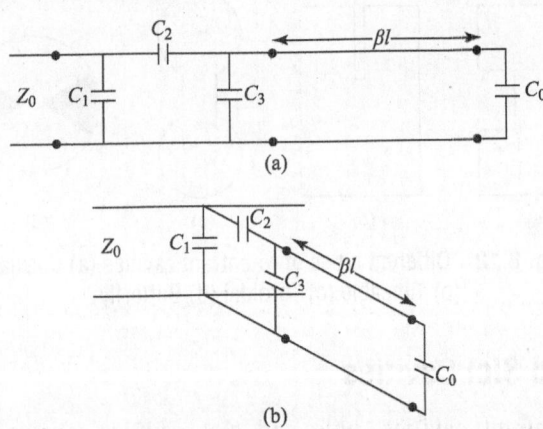

Fig. 6.15 Equivalent circuit representation (a) Fig. 6.14(a) (b) Fig. 6.14(b)

6.10.2 Rectangular Microstrip Patch Resonators

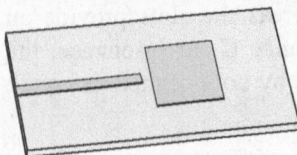

Fig. 6.16 Rectangular microstrip patch resonator

A rectangular microstrip patch, as shown in Fig. 6.16, also behaves as a microstrip resonator, with two electric walls (short circuit) at the top and bottom, and four magnetic walls (open circuit) at the sides. If the substrate thickness $h \ll \lambda$, then only one electric field component E_z exists, which satisfies the scalar Helmholtz equation:

$$\left(\nabla^2 + k^2\right) E_z(x, y) = 0 \qquad (6.180)$$

Equation (6.180) is subjected to the following boundary conditions:

$$H_x(y=0) = H_x(y=W) = 0$$

or $$\left.\frac{\partial E_z}{\partial y}\right|_{y=0} = \left.\frac{\partial E_z}{\partial y}\right|_{y=W} = 0 \qquad (6.181)$$

and $$H_y(x=0) = H_y(x=L) = 0$$

or $$\left.\frac{\partial E_z}{\partial x}\right|_{x=0} = \left.\frac{\partial E_z}{\partial x}\right|_{x=L} = 0 \qquad (6.182)$$

Solving Eq. (6.180) with the given boundary conditions, the following field equations are obtained:

$$E_z = E_{mn}\cos\left(\frac{m\pi x}{L}\right)\cos\left(\frac{n\pi y}{W}\right) \qquad (6.183)$$

$$H_x = \frac{j}{\omega\mu}\frac{\partial E_z}{\partial y} = -j\frac{E_{mn}}{\omega\mu}\frac{n\pi}{W}\cos\left(\frac{m\pi x}{L}\right)\sin\left(\frac{n\pi y}{W}\right) \qquad (6.184)$$

$$H_y = \frac{j}{\omega\mu}\frac{\partial E_z}{\partial y} = -j\frac{E_{mn}}{\omega\mu}\frac{m\pi}{L}\sin\left(\frac{m\pi x}{L}\right)\cos\left(\frac{n\pi y}{W}\right) \qquad (6.185)$$

Resonance occurs at a frequency for which

$$k_{0,mn}^2 = \omega_{0,mn}^2\mu\varepsilon = \left(\frac{m\pi}{L}\right)^2 + \left(\frac{n\pi}{W}\right)^2$$

or $$f_{0,mn} = \frac{1}{2\pi\sqrt{\mu\varepsilon}}\sqrt{\left(\frac{m\pi}{L}\right)^2 + \left(\frac{n\pi}{W}\right)^2} \qquad (6.186)$$

Physically, resonance occurs when either L or W becomes equal to or an integral multiple of $\lambda/2$. If $L > W$, then lowest-order resonance occurs for $m = 1$ and $n = 0$; hence, the dominant mode will be TM_{100}.

The total energy stored in the resonator for the dominant mode is given by the following equation:

$$W_0 = 2W_e = \frac{\varepsilon'}{2}\int_0^h\int_0^W\int_0^L |E_z|^2\,dxdydz$$

or $$W_0 = 2W_e = \frac{\varepsilon'}{2}\int_0^h\int_0^W\int_0^L |E_{10}|^2\cos^2\left(\frac{\pi x}{L}\right)dxdydz = \frac{\varepsilon'hWL}{4}|E_{10}|^2 \qquad (6.187)$$

Power loss in the dielectric is given by the following expression:

$$P_d = \frac{\omega\varepsilon''}{2}\int_0^h\int_0^W\int_0^L |E_z|^2\,dxdydz = \frac{\omega\varepsilon''}{2}\int_0^h\int_0^W\int_0^L |E_{10}|^2\cos^2\left(\frac{\pi x}{L}\right)dxdydz$$

or $$P_d = \frac{\omega\varepsilon''hWL}{4}|E_{10}|^2 \qquad (6.188)$$

Power loss in the top and bottom conducting walls are expressed as follows:

$$P_c = 2\frac{R_s}{2}\int_0^L\int_0^W \left(\left|H_x^2\right|+\left|H_y^2\right|\right)dxdy = R_s\int_0^L\int_0^W \left(\frac{\left|E_{10}\right|}{\omega\mu}\frac{\pi}{L}\right)^2 \sin^2\left(\frac{\pi x}{L}\right)dxdy$$

$$\text{or}\qquad P_c = \frac{R_sLW}{2}\left(\frac{\pi\left|E_{10}\right|}{\omega\mu L}\right)^2 \tag{6.189}$$

If we neglect the radiation loss, then the Q of the cavity is given by the following expression:

$$Q = \frac{\omega W_0}{(P_c + P_d)} = \frac{\omega\dfrac{\varepsilon' hWL}{4}\left|E_{10}\right|^2}{\dfrac{R_sLW}{2}\left(\dfrac{\pi\left|E_{10}\right|}{\omega\mu L}\right)^2 + \dfrac{\omega\varepsilon'' hWL}{4}\left|E_{10}\right|^2}$$

$$\text{or}\qquad Q = \frac{\omega\varepsilon' h}{\left[2R_s\left(\dfrac{\pi}{\omega\mu L}\right)^2 + \omega\varepsilon'' h\right]} \tag{6.190}$$

EXAMPLE 6.5 Calculate the resonance frequency of a microstrip patch of length 1.5 cm and width 1.0 cm for the TM_{100} mode.

Solution Given: $L = 1.5$ cm and $W = 1$ cm

Therefore,

$$f_{100} = \frac{c}{2}\sqrt{\left(\frac{m}{L}\right)^2 + \left(\frac{n}{W}\right)^2} = \frac{3\times10^{10}}{2}\sqrt{\left(\frac{1}{1.5}\right)^2 + \left(\frac{0}{1}\right)^2} = 10\,\text{GHz}$$

6.10.3 Circular Microstrip Patch Resonators

The circular microstrip patch resonator, shown in Fig. 6.17, can be excited by a probe from the ground plane or by an input microstrip line. If we assume that $h \ll \lambda$, then only one electric field component E_z exists, which satisfies the scalar Helmholtz equation:

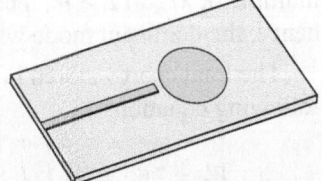

Fig. 6.17 Microstrip circular patch resonator

$$\left(\nabla^2 + k^2\right)E_z = 0 \tag{6.191}$$

subject to the following boundary condition:

$$H_\phi = 0 \text{ at } r = R \tag{6.192}$$

Solving Eq. (6.191), field components can be written as follows:

$$E_z = E_{mn}J_n(kr)\cos(n\phi) \tag{6.193}$$

$$H_r = \frac{j}{\omega\mu}\frac{\partial E_z}{\partial \phi} = -\frac{jE_{mn}}{\omega\mu} n J_n\left(kr\right)\sin\left(n\phi\right) \tag{6.194}$$

and $$H_\phi = \frac{j}{\omega\mu}\frac{\partial E_z}{\partial r} = \frac{jkE_{mn}}{\omega\mu} J_n'\left(kr\right)\cos\left(n\phi\right) \tag{6.195}$$

Again from Eq. (6.192), $\dfrac{jkE_{mn}}{\omega\mu} J_n'\left(kR\right)\cos\left(n\phi\right) = 0$

or $$J_n'\left(kR\right) = 0 \tag{6.196}$$

or $$k = p_{nm}'/R \tag{6.197}$$

where p_{nm}' are the roots of Eq. (6.196). The minimum value of p_{nm}' is 1.841 for $n = 1$ and $m = 1$. Therefore, the dominant mode is TM_{110}. The resonance frequency can be calculated from Eq. (6.197), as follows:

$$\frac{2\pi f_0}{v_p} = \frac{p_{nm}'}{R}$$

or $$f_0 = \frac{v_p}{2\pi}\frac{p_{nm}'}{R} = \frac{p_{nm}'}{2\pi R\sqrt{\mu\varepsilon}} \tag{6.198}$$

The total energy stored in the resonator in the dominant mode is expressed as follows:

$$W_0 = 2W_e = \frac{\varepsilon'}{2}\int_0^h\int_0^{2\pi}\int_0^R |E_z|^2\, r\,dr\,d\phi\,dz$$

or $$W_0 = \frac{\varepsilon'}{2}|E_{mn}|^2 \int_0^h\int_0^{2\pi}\int_0^R J_n^2\left(kr\right)\cos^2\left(n\phi\right) r\,dr\,d\phi\,dz$$

or $$W_0 = \frac{\pi\varepsilon' hR^2}{4}|E_{mn}|^2 J_1^2\left(kR\right)\left(1 - \frac{1}{k^2R^2}\right) \tag{6.199}$$

Power loss in the dielectric is given by the following expression:

$$P_d = \frac{\omega\varepsilon''}{2}\int_0^h\int_0^{2\pi}\int_0^R |E_z|^2\, r\,dr\,d\phi\,dz$$

or $$P_d = \frac{\omega\varepsilon''}{2}|E_{mn}|^2 \int_0^h\int_0^{2\pi}\int_0^R J_n^2\left(kr\right)\cos^2\left(n\phi\right) r\,dr\,d\phi\,dz$$

or $$P_d = \frac{\pi\omega\varepsilon'' hR^2}{4}|E_{mn}|^2 J_1^2\left(kR\right)\left(1 - \frac{1}{k^2R^2}\right) \tag{6.200}$$

Power loss in the conductor is expressed as follows:

$$P_c = 2\frac{R_s}{2}\int_0^{2\pi}\int_0^R \left(|H_r|^2 + |H_\phi|^2\right) r\,dr\,d\phi$$

or $$P_c = R_s\frac{|E_{mn}|^2}{(\omega\mu)^2}\int_0^{2\pi}\int_0^R \left[n^2 J_n^2\left(kr\right)\sin^2\left(n\phi\right) + k^2\left\{J_n'\left(kr\right)\right\}^2 \cos^2\left(n\phi\right)\right] r\,dr\,d\phi$$

or $$P_c = \frac{\pi R_s k^2 R^2}{2k_o^2 \eta^2} |E_{mn}|^2 J_1^2 (kR) \left(1 - \frac{1}{k^2 R^2}\right)$$ (6.201)

The cavity Q is, therefore, given by the following relation:

$$Q = \frac{\omega W_0}{P_c + P_d} = \frac{\omega \frac{\pi \varepsilon' h R^2}{4} |E_{mn}|^2 J_1^2 (kR) \left(1 - \frac{1}{k^2 R^2}\right)}{\frac{\pi \omega \varepsilon'' h R^2}{4} |E_{mn}|^2 J_1^2 (kR) \left(1 - \frac{1}{k^2 R^2}\right) + \frac{\pi R_s k^2 R^2}{2k_o^2 \eta^2} |E_{mn}|^2 J_1^2 (kR) \left(1 - \frac{1}{k^2 R^2}\right)}$$

or $$Q = \frac{\omega \varepsilon' h}{\omega \varepsilon'' h + \frac{2R_s k^2}{k_o^2 \eta^2}} = \frac{\omega \varepsilon_r' \varepsilon_0 h}{\omega \varepsilon_r'' \varepsilon_0 h + \frac{2R_s k_0^2 \varepsilon_r'}{k_o^2 \eta^2}} = \frac{\omega h}{\omega h \frac{\varepsilon_r''}{\varepsilon'} + \frac{2R_s}{\eta^2 \varepsilon_0}}$$

or $$Q = \frac{\omega h}{\omega h \frac{\varepsilon_r''}{\varepsilon'} + \frac{2R_s}{\eta \sqrt{\mu_0 \varepsilon_0}}} = \frac{h \omega \sqrt{\mu_0 \varepsilon_0}}{h \omega \sqrt{\mu_0 \varepsilon_0} \frac{\varepsilon_r''}{\varepsilon'} + \frac{2R_s}{\eta}}$$

or $$Q = \frac{k_0 h}{k_0 h \frac{\varepsilon_r''}{\varepsilon'} + \frac{2R_s}{\eta}}$$ (6.202)

As before, Eq. (6.202) neglects the radiation loss.

Practice Problem

6.5 For a circular patch of radius $1.5\,\mathrm{cm}$, calculate the resonant frequency for the TM_{110} mode.   **[5.86 GHz]**

6.10.4 Microstrip Ring Resonators

A microstrip ring resonator with an inner and outer radius of a and b, respectively, is shown in Fig. 6.18. In principle, a ring resonator resonates at a frequency for which

$$\frac{a+b}{2} = \frac{nc}{2f_r \sqrt{\varepsilon_{\mathrm{eff}}}} = \frac{nv_p}{2f_r} = n\frac{\lambda}{2}$$ (6.203)

that is, the mean circumference equals an integral number of half wavelengths.

Like other microstrip resonators, a ring resonator can also be assumed as a cavity with electric walls at the top and bottom, and magnetic walls at the edges. The fields within it obeys the scalar Helmholtz equation, subject to the following boundary condition:

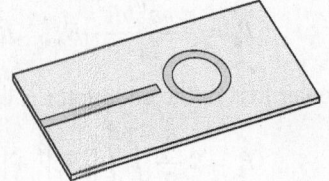

Fig. 6.18 Microstrip ring resonator

$$H_\phi = 0 \text{ at } r = a \text{ and } b$$ (6.204)

After solving the scalar Helmholtz equation, the electric fields can be expressed as follows:

$$E_z = E_{nm} \left\{ J_n(kr) - \frac{J_n'(kr)}{N_n'(kr)} N_n(kr) \right\} \cos(n\phi) \qquad (6.205)$$

$$H_r = -\frac{jn}{\omega\mu_0} E_{nm} \left\{ J_n(kr) - \frac{J_n'(kr)}{N_n'(kr)} N_n(kr) \right\} \sin(n\phi) \qquad (6.206)$$

$$H_\phi = -\frac{jk}{\omega\mu_0} E_{nm} \left\{ J_n(kr) - \frac{J_n'(kr)}{N_n'(kr)} N_n(kr) \right\} \cos(n\phi) \qquad (6.207)$$

Once the field equations are known, the Q of the cavity can be found following a similar procedure that has been described earlier. The detailed derivations are much more complex and beyond the scope of this book, and hence are not presented here.

6.11 Excitation of Resonators

In the previous section, we have discussed the different types of resonator cavities. Since most of the resonators are closed structures, their excitation, without affecting their normal operations, is a challenging task. In practice, different coupling mechanisms are used for this purpose. In the later part of this section, we will discuss microstrip gap coupling and aperture coupling; however, initially we will discuss the critical coupling of a resonator.

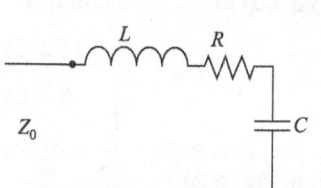

Fig. 6.19 Series resonant circuit coupled with two-wire transmission line

In order to transfer maximum power to the cavity from the feed line, these must be matched at the resonant frequency. Under such a circumstance, the cavity is said to be critically coupled to its feed. A basic example of achieving critical coupling is shown in Fig. 6.19, where a series LCR network is coupled to a two-wire transmission line.

The input impedance of a series resonant circuit, at a frequency close to the resonance, is given by the following equation:

$$Z_{in} = R + j2L\Delta\omega \qquad (6.208)$$

Further, the Q of a series resonant circuit is given by the following relation:

$$Q_{Unloaded} = Q = \omega_0 L / R \qquad (6.209)$$

Substituting Eq. (6.208) into Eq. (6.209), we get the following expression:

$$Z_{in} = R + j\frac{2QR\Delta\omega}{\omega_0} \qquad (6.210)$$

At the resonance point, $\Delta\omega = 0$ and Eq. (6.210) reduces to the following form:

$$Z_{in} = R \qquad (6.211)$$

Therefore, if the characteristic impedance of the feed line is chosen to be $Z_0 = R$, then the resonator will be matched critically with the feed line.

6.11.1 Gap-coupled Microstrip Resonators

Let us consider that a half-wavelength open-circuited transmission line resonator has been excited by a feed line using the gap coupling methodology, as shown in Fig. 6.14. The equivalent circuit is shown in Fig. 6.20, where the gap has been replaced by a *gap capacitance*.

The input impedance, seen by the feed line, can approximately be expressed as follows:

$$Z_{in} = \frac{1}{j\omega C_{gap}} - jZ_0 \cot(\beta l) \tag{6.212}$$

Therefore, the normalized input impedance is expressed as follows:

$$z_{in} = \frac{Z_{in}}{Z_0} = -j\left[\frac{1}{\omega C_{gap} Z_0} + \cot(\beta l)\right] \text{ or } z_{in} = -j\left[\frac{1}{b_c} + \frac{1}{\tan(\beta l)}\right]$$

$$\text{or} \qquad z_{in} = -j\left[\frac{b_c + \tan(\beta l)}{b_c \tan(\beta l)}\right] \tag{6.213}$$

$$\text{where} \quad b_c = Z_0 \omega C_{gap} \tag{6.214}$$

is the normalized susceptance of the gap capacitor C_{gap}.

At resonance, the imaginary part of the input impedance must vanish, that is,

$$z_{in} = 0 \tag{6.215}$$

$$\text{or} \qquad \tan(\beta l) = -b_c \tag{6.216}$$

Therefore, the resonance frequency corresponds to the intersection of $\tan(\beta l)$ curve and the straight line $-b_c = -Z_0 \omega C_{gap}$, as shown in Fig. 6.21.

Now, expanding $z_{in}(\omega)$ in a Taylor series around the angular resonance frequency ω_0 and assuming that b_c is very small, we get the following equation:

$$z_{in}(\omega) = z_{in}(\omega_0) + (\omega - \omega_0)\frac{dz_{in}(\omega)}{d\omega}\bigg|_{\omega_0} + \dots \tag{6.217}$$

Differentiating Eq. (6.213) with respect to ω, we get the following relation at $\omega = \omega_0$:

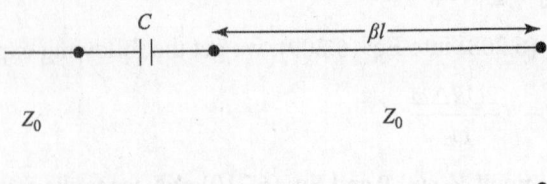

Fig. 6.20 Equivalent circuit of gap-coupled microstrip resonator circuit

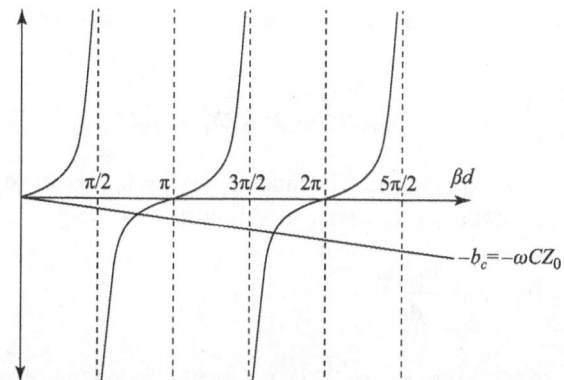

Fig. 6.21 Graphical solution of Eq. (6.216) for resonance frequency

$$\frac{\mathrm{d}z_{\text{in}}(\omega)}{\mathrm{d}\omega}\bigg|_{\omega_0} = -j\left[\frac{1}{b_c \tan(\beta l)}\bigg|_{\omega_0} \frac{\mathrm{d}}{\mathrm{d}\omega}\{b_c + \tan(\beta l)\}\bigg|_{\omega_0}\right.$$

$$\left. + \{b_c + \tan(\beta l)\}\bigg|_{\omega_0} \frac{\mathrm{d}}{\mathrm{d}\omega}\left\{\frac{1}{b_c \tan(\beta l)}\right\}\bigg|_{\omega_0}\right] \qquad (6.218)$$

Using Eq. (6.216) in Eq. (6.218), we get the following equation:

$$\frac{\mathrm{d}z_{\text{in}}(\omega)}{\mathrm{d}\omega}\bigg|_{\omega_0} = -j\left[\frac{1}{b_c \tan(\beta l)} \frac{\mathrm{d}}{\mathrm{d}\omega}\{b_c + \tan(\beta l)\}\right]\bigg|_{\omega_1}$$

or $\qquad \dfrac{\mathrm{d}z_{\text{in}}(\omega)}{\mathrm{d}\omega}\bigg|_{\omega_0} = \dfrac{-j\sec^2(\beta l)}{b_c \tan(\beta l)} \dfrac{\mathrm{d}(\beta l)}{\mathrm{d}\omega}\bigg|_{\omega_0}$ $\qquad (6.219)$

where we have assumed that b_c is constant or $\dfrac{\mathrm{d}b_c}{\mathrm{d}\omega} = 0$. Equation (6.219) can further be simplified as follows:

$$\frac{\mathrm{d}z_{\text{in}}(\omega)}{\mathrm{d}\omega}\bigg|_{\omega_0} = \frac{-j\{1 + \tan^2(\beta l)\}}{b_c \tan(\beta l)} \frac{\mathrm{d}(\omega l/v_p)}{\mathrm{d}\omega}\bigg|_{\omega_0}$$

or $\qquad \dfrac{\mathrm{d}z_{\text{in}}(\omega)}{\mathrm{d}\omega}\bigg|_{\omega_0} = \dfrac{-j\{1 + \tan^2(\beta l)\}}{b_c \tan(\beta l)} \dfrac{l}{v_p}\bigg|_{\omega_0}$ $\qquad (6.220)$

Using Eq. (6.216) in Eq. (6.220), we get the following relation:

$$\frac{\mathrm{d}z_{\text{in}}(\omega)}{\mathrm{d}\omega}\bigg|_{\omega_0} = \frac{-j\left(1 + b_c^2\right)}{b_c\left(-b_c\right)} \frac{l}{v_p} = \frac{j\left(1 + b_c^2\right)}{b_c^2} \frac{l}{v_p} \qquad (6.221)$$

Since $b_c \ll 1, 1 + b_c^2 \approx 1$ and Eq. (6.221) reduces to the following form:

$$\left. \frac{dz_{in}(\omega)}{d\omega} \right|_{\omega_0} \approx \frac{jl}{v_p b_c^2} = \frac{j\lambda_0}{f_0 \lambda_0 2b_c^2} = \frac{j2\pi}{2\pi f_0 2b_c^2} = \frac{j\pi}{\omega_0 b_c^2} \qquad (6.222)$$

Now, let us concentrate on Eq. (6.217). Since $z_{in}(\omega_0) = 0$, we can write Eq. (6.217) after neglecting the higher-order terms, as follows:

$$z_{in}(\omega) = (\omega - \omega_0) \left. \frac{dz_{in}(\omega)}{d\omega} \right|_{\omega_0} \qquad (6.223)$$

Substituting Eq. (6.222) into Eq. (6.223), we get the following relation:

$$z_{in}(\omega) = \frac{j\pi(\omega - \omega_0)}{\omega_0 b_c^2} \qquad (6.224)$$

So far, we have assumed that the resonator is lossless. If the resonator incorporates a small loss, then the aforementioned analysis must be modified. Since a lossy resonator can be treated as a lossless resonator by replacing its resonant frequency ω_0 with $\omega_0 \left(1 + \dfrac{j}{2Q}\right)$, Eq. (6.224) can approximately be written as follows:

$$z_{in}(\omega) = \frac{j\pi \left\{ \omega - \omega_0 \left(1 + \dfrac{j}{2Q} \right) \right\}}{\omega_0 b_c^2} = \frac{\pi}{2Q b_c^2} + j \frac{\pi(\omega - \omega_0)}{\omega_0 b_c^2} \qquad (6.225)$$

Equation (6.225) reveals that though an uncoupled half-wave open-circuited resonator behaves like a parallel resonant circuit, the coupled resonator circuit behaves like a series LCR network, because of the inclusion of the coupling gap capacitor. Therefore, a series coupling capacitor has the effect of inverting the driving impedance of a resonator.

At resonance, the imaginary part vanishes and we get the following relation:

$$Z_{in} = R = \frac{\pi Z_0}{2Q b_c^2} \qquad (6.226)$$

For critical coupling, the characteristic impedance of the feed line must be equal to Z_0. Therefore, from Eq. (6.226),

$$Z_0 = \frac{\pi Z_0}{2Q b_c^2} \text{ or } b_c = \sqrt{\frac{\pi}{2Q}} \qquad (6.227)$$

Since the resultant circuit is a series LCR network, the coefficient of coupling is given by the following equation:

$$K = \frac{Q_{Unloaded}}{Q_{ext}} = \frac{\omega_o L}{Z_o} \frac{R}{\omega_o L} = \frac{R}{Z_o} = \frac{\pi Z_0}{2Q Z_0 b_c^2} = \frac{\pi}{2Q b_c^2} \qquad (6.228)$$

If $b_c > \sqrt{\dfrac{\pi}{2Q}}$, then $K < 1$ and the resonator is under coupled, whereas if $b_c < \sqrt{\dfrac{\pi}{2Q}}$, then $K > 1$ and the resonator is over coupled

6.11.2 Aperture-coupled Cavities

Let us consider the aperture coupled cavity shown in Fig. 6.22(a). The aperture will act as a shunt inductance, and hence the equivalent circuit of Fig. 6.22(a) can be represented as a short-circuited transmission line loaded with a shunt inductor, as shown in Fig. 6.22(b).

The input admittance, seen by the feed line, is given by the following equation:

$$Y_{in} = \frac{1}{j\omega L} + \frac{1}{jZ_0 \tan(\beta l)} = -j \left\{ \frac{1}{X_L} + \frac{1}{Z_0 \tan(\beta l)} \right\} \tag{6.229}$$

The normalized input admittance is, therefore,

$$y_{in} = \frac{Y_{in}}{Y_0} = -j \left\{ \frac{Z_0}{X_L} + \frac{1}{\tan(\beta l)} \right\} = -j \left\{ \frac{X_L + Z_0 \tan(\beta l)}{X_L \tan(\beta l)} \right\}$$

or
$$y_{in} = -j \left\{ \frac{\dfrac{X_L}{Z_0} + \tan(\beta l)}{\dfrac{X_L}{Z_0} \tan(\beta l)} \right\} = -j \left\{ \frac{x_L + \tan(\beta l)}{x_L \tan(\beta l)} \right\} \tag{6.230}$$

Anti-resonance will occur when

$$x_L + \tan(\beta l) = 0 \tag{6.231}$$

Therefore, the anti-resonance frequency correspondence to the intersection of $\tan(\beta l)$ curve and the straight line $-x_L = -\omega L / Z_0$, as shown in Fig. 6.23.

Now expanding $y_{in}(\omega)$ in a Taylor series around the angular resonance frequency ω_0 and assuming that b_c is very small, we get the following equation:

$$y_{in}(\omega) = y_{in}(\omega_0) + (\omega - \omega_0) \frac{dy_{in}(\omega)}{d\omega} \bigg|_{\omega_0} + \cdots \tag{6.232}$$

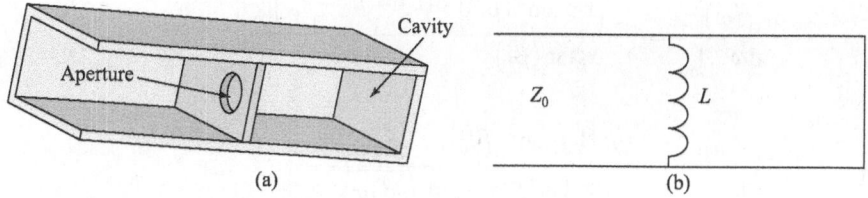

Fig. 6.22   Aperture-coupled cavity resonator (a) Cut-plane view (b) Equivalent circuit

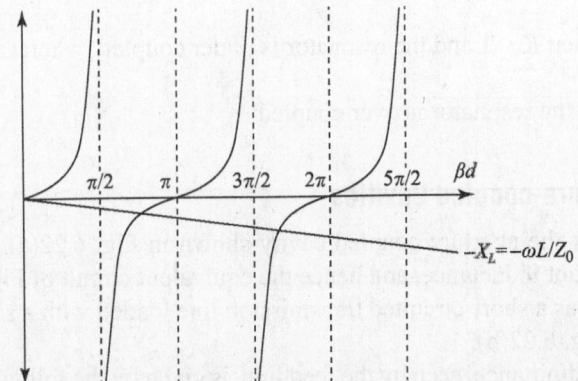

Fig. 6.23 Graphical solution of Eq. (6.231) for resonance frequency

Differentiating Eq. (6.230) with respect to ω, we get the following relation at $\omega = \omega_0$:

$$\left.\frac{dy_{in}(\omega)}{d\omega}\right|_{\omega_0} = -j\left[\left.\frac{1}{x_L\tan(\beta l)}\right|_{\omega_0}\left.\frac{d}{d\omega}\{x_L+\tan(\beta l)\}\right|_{\omega_0}\right.$$
$$\left.+\{x_L+\tan(\beta l)\}\right|_{\omega_0}\left.\frac{d}{d\omega}\left\{\frac{1}{x_L\tan(\beta l)}\right\}\right|_{\omega_0}\right] \tag{6.233}$$

Using Eq. (6.231) in Eq. (6.233), we get the following relation:

$$\left.\frac{dy_{in}(\omega)}{d\omega}\right|_{\omega_0} = -j\left[\frac{1}{x_L\tan(\beta l)}\frac{d}{d\omega}\{x_L+\tan(\beta l)\}\right]_{\omega_0}$$

or

$$\left.\frac{dy_{in}(\omega)}{d\omega}\right|_{\omega_0} = \left.\frac{-j\sec^2(\beta l)}{x_L\tan(\beta l)}\frac{d(\beta l)}{d\omega}\right|_{\omega_0} \tag{6.234}$$

where we have assumed that x_L is constant or $\frac{dx_L}{d\omega} = 0$. Equation (6.234) can be further simplified as follows:

$$\left.\frac{dy_{in}(\omega)}{d\omega}\right|_{\omega_0} = \left.\frac{-j\{1+\tan^2(\beta l)\}}{x_L\tan(\beta l)}\frac{d\left(l\sqrt{\omega^2\mu\varepsilon-k_c^2}\right)}{d\omega}\right|_{\omega_0}$$

or

$$\left.\frac{dy_{in}(\omega)}{d\omega}\right|_{\omega_0} = \left.\frac{-jl\{1+\tan^2(\beta l)\}}{x_L\tan(\beta l)}\frac{2\omega\mu\varepsilon}{2\sqrt{\omega^2\mu\varepsilon-k_c^2}}\right|_{\omega_0}$$

or $\quad\left.\dfrac{dy_{in}\left(\omega\right)}{d\omega}\right|_{\omega_0}=\dfrac{-jl\left\{1+\tan^2\left(\beta l\right)\right\}}{x_L\tan\left(\beta l\right)}\left.\dfrac{\omega}{\beta v_p^2}\right|_{\omega_0}=\dfrac{-j\lambda_g\left\{1+\tan^2\left(\beta l\right)\right\}}{2x_L\tan\left(\beta l\right)}\dfrac{k_0 v_p}{\beta v_p^2}$

or $\quad\left.\dfrac{dy_{in}\left(\omega\right)}{d\omega}\right|_{\omega_0}=\dfrac{-j\lambda_g\left\{1+\tan^2\left(\beta l\right)\right\}}{2x_L\tan\left(\beta l\right)}\dfrac{k_0}{\beta v_p}$ (6.235)

Using Eq. (6.231) in Eq. (6.235), we get the following expression:

$$\left.\frac{dy_{in}\left(\omega\right)}{d\omega}\right|_{\omega_0}=\frac{-j\lambda_g\left(1+x_L^2\right)}{2x_L\left(-x_L\right)}\frac{k_0}{\beta v_p}=\frac{j\lambda_g\left(1+x_L^2\right)}{2x_L^2}\frac{k_0}{\beta v_p}$$ (6.236)

Since $x_L^2\ll1,1+x_L^2\approx1$ and Eq. (6.236) reduces to the following form:

$$\left.\frac{dy_{in}\left(\omega\right)}{d\omega}\right|_{\omega_0}\approx\frac{j\lambda_g k_0}{2x_L^2\beta v_p}=\frac{jk_0}{2x_L^2\beta v_p}\frac{2\pi}{\beta}=\frac{j\pi k_0}{x_L^2\beta^2 v_p}$$ (6.237)

Now let us concentrate on Eq. (6.232). Since $y_{in}\left(\omega_0\right)=0$, we can write Eq. (6.232) after neglecting the higher-order terms, as follows:

$$y_{in}\left(\omega\right)=\left(\omega-\omega_0\right)\left.\frac{dy_{in}\left(\omega\right)}{d\omega}\right|_{\omega_0}$$ (6.238)

Substituting Eq. (6.237) into Eq. (6.238), we get the following relation:

$$y_{in}\left(\omega\right)=\frac{j\pi k_0\left(\omega-\omega_0\right)}{x_L^2\beta^2 v_p}$$ (6.239)

As before, since a lossy resonator can be treated as a lossless resonator by replacing its resonant frequency ω_0 with $\omega_0\left(1+\dfrac{j}{2Q}\right)$, Eq. (6.239) can approximately be written as follows:

$$y_{in}\left(\omega\right)=\frac{j\pi k_0\left\{\omega-\omega_0\left(1+\dfrac{j}{2Q}\right)\right\}}{x_L^2\beta^2 v_p}=\frac{\pi\omega_0 k_0}{2Qx_L^2\beta^2 v_p}+\frac{j\pi k_0\left(\omega-\omega_0\right)}{x_L^2\beta^2 v_p}$$ (6.240)

At resonance, the imaginary part of the input admittance vanishes, whereas the real part becomes equal to the characteristic admittance of the line. Therefore, we get the following equation:

$$y_{in}\left(\omega\right)=\frac{\pi\omega_0 k_0}{2Qx_L^2\beta^2 v_p}$$

or $\quad Y_{in}\left(\omega\right)=Y_0\dfrac{\pi\omega_0 k_0}{2Qx_L^2\beta^2 v_p}=\dfrac{\pi\omega_0 k_0}{2Z_0 Qx_L^2\beta^2 v_p}=Y_0$

or $\qquad x_L^2 = \dfrac{\pi \omega_0 k_0}{2Z_0 Y_0 Q \beta^2 v_p} = \dfrac{\pi \omega_0 k_0}{2Q\beta^2 v_p}$

or $\qquad X_L = Z_0 \sqrt{\dfrac{\pi \omega_0 k_0}{2Q\beta^2 v_p}}$ (6.241)

Equation (6.241) provides the necessary inductance value required for critical coupling.

Important Formulae

- The relation between loaded, unloaded and external Q can be expressed as

$$\frac{1}{Q_{\text{Loaded}}} = \frac{1}{Q_{\text{Unloaded}}} + \frac{1}{Q_{\text{Ext}}}$$

- The coupling coefficient is expressed by the relation $K = Q_{\text{Unloaded}} / Q_{\text{Ext}}$.
- The resonance frequency of a rectangular cavity waveguide is

$$f_{mnp} = \frac{v_p}{2} \sqrt{\left(\frac{m}{a}\right)^2 + \left(\frac{n}{b}\right)^2 + \left(\frac{p}{d}\right)^2}$$

- The resonance frequency of a circular cavity waveguide operating in the TE mode is

$$f = \frac{v_p}{2\pi} \sqrt{\left(\frac{p'_{nm}}{R}\right)^2 + \left(\frac{p\pi}{d}\right)^2}.$$

- The resonance frequency of a circular cavity waveguide operating in the TM mode is

$$f = \frac{v_p}{2\pi} \sqrt{\left(\frac{p_{nm}}{R}\right)^2 + \left(\frac{p\pi}{d}\right)^2}.$$

- When both conductor and dielectric losses are present, the total Q of a cavity will be

$$\frac{1}{Q_T} = \frac{1}{Q_c} + \frac{1}{Q_d}.$$

- The resonance frequency of a circular cylindrical dielectric can be expressed as

$$f_r = \frac{34}{r\varepsilon_r}\left(3.45 + \frac{r}{h}\right) \text{ GHz, where } 2 > \frac{r}{h} > 0.5 \text{ and } 50 > \varepsilon_r > 30, \text{ and } r \text{ and } h \text{ are}$$

expressed in millimetres.

- For a rectangular microstrip patch resonator, $f_{0,mn} = \dfrac{1}{2\pi\sqrt{\mu\varepsilon}} \sqrt{\left(\dfrac{m\pi}{L}\right)^2 + \left(\dfrac{n\pi}{W}\right)^2}.$

- For a rectangular microstrip patch resonator, $Q = \dfrac{\omega\varepsilon' h}{\left[2R_s\left(\dfrac{\pi}{\omega\mu L}\right)^2 + \omega\varepsilon'' h\right]}.$

- For a circular microstrip patch resonator, $f_0 = \dfrac{p'_{nm}}{2\pi R\sqrt{\mu\varepsilon}}.$

- For a circular microstrip patch resonator, $Q = \dfrac{k_0 h}{k_0 h \dfrac{\varepsilon''_r}{\varepsilon'} + \dfrac{2R_s}{\eta}}.$

Exercises

Objective-type Questions

6.1 The Q-factor measures
(a) frequency selectivity
(b) energy dissipation
(c) energy stored in the cavity
(d) all of these

6.2 A transmission line resonator cannot be used above 1 GHz because
(a) the Q-value is very high
(b) losses are very high
(c) dimensions are very large
(d) none of these

6.3 The dominant TE mode in a circular waveguide cavity is
(a) TE_{101}
(b) TM_{111}
(c) TE_{111}
(d) TM_{101}

6.4 Waveguide cavities cannot be used at very high frequencies because of
(a) small dimensions
(b) low Q
(c) both (a) and (b)
(d) none of these

6.5 In microwave frequencies, a cavity may be considered as a
(a) low-pass filter
(b) high-pass filter
(c) band-pass filter
(d) band-stop filter

6.6 A cylindrical cavity resonator has a diameter R and length d. What is the dominant resonant mode if (i) $d > R$ and $d < R$?
(a) TE_{111} and TM_{111}
(b) TE_{010} and TM_{111}
(c) TE_{111} and TM_{010}
(d) TE_{111} and TM_{011}

6.7 A cavity resonator can be represented by
(a) an LC circuit
(b) an LCR circuit
(c) a lossy inductor
(d) a lossy capacitor

6.8 In a cylindrical cavity resonator, the two modes that are degenerate include
(a) TE_{111} and TM_{111}
(b) TE_{011} and TM_{011}
(c) TE_{022} and TM_{111}
(d) TE_{111} and TM_{011}

6.9 For a critical coupling, the loaded and unloaded Q of a cavity resonator, having an SWR (ρ) are related by
(a) $Q_{\text{Loaded}} = \dfrac{\rho}{\rho+1} Q_{\text{Unloaded}}$
(b) $Q_{\text{Loaded}} = \dfrac{Q_{\text{Unloaded}}}{2}$
(c) $Q_{\text{Loaded}} = \dfrac{Q_{\text{Unloaded}}}{\rho+1}$
(d) $Q_{\text{Loaded}} = Q_{\text{Unloaded}}$

6.10 The resonance frequency of a rectangular cavity waveguide is
(a) $\dfrac{v_p}{2} \sqrt{\left(\dfrac{m\pi}{a}\right)^2 + \left(\dfrac{n\pi}{b}\right)^2 + \left(\dfrac{p\pi}{d}\right)^2}$
(b) $\dfrac{v_p}{2} \sqrt{\left(\dfrac{m}{a}\right)^2 + \left(\dfrac{n}{b}\right)^2 + \left(\dfrac{p}{d}\right)^2}$
(c) $v_p \sqrt{\left(\dfrac{m\pi}{a}\right)^2 + \left(\dfrac{n\pi}{b}\right)^2 + \left(\dfrac{p\pi}{d}\right)^2}$
(d) $v_p \sqrt{\left(\dfrac{m}{a}\right)^2 + \left(\dfrac{n}{b}\right)^2 + \left(\dfrac{p}{d}\right)^2}$

Review Questions

6.1 What is meant by the quality factor of a cavity resonator? Show that the unloaded, load-
ed, and external Q of a cavity are related by the equation $\dfrac{1}{Q_{\text{Loaded}}} = \dfrac{1}{Q_{\text{Unloaded}}} + \dfrac{1}{Q_{\text{Ext}}}$.
In addition, describe coupling coefficient, under coupling, over coupling, and critical
coupling.

6.2 Show that for a short-circuited quarter-wavelength transmission line resonator, the
Q-factor can be expressed as $Q = \dfrac{\beta}{2\alpha}$.

6.3 Find the expressions for the resonance frequency of a rectangular waveguide cavity
resonator for the TE and TM modes.

6.4 Find an expression for the Q-factor of a rectangular waveguide cavity resonator for
TE-mode propagation in the presence of conductor loss.

6.5 Find an expression for the resonance frequency of a circular waveguide cavity resonator.

6.6 Find an expression for the Q-factor of a rectangular waveguide cavity resonator for
TM-mode propagation in the presence of conductor loss.

6.7 Find the resonance frequency and Q-factor of a rectangular microstrip patch resonator.

6.8 Find the resonance frequency and Q-factor of a circular microstrip patch resonator.

6.9 Find out the condition of critical coupling in a gap-coupled microstrip patch resonator.

6.10 Find out the condition of critical coupling in an aperture-coupled cavity resonator.

Problems

6.1 A 50 Ω open-circuited half-wavelength transmission line resonator has a resonance
frequency of 1 GHz and Q of 9737. Calculate R, L, and C. *[Hint: Use Eqs 6.57,
(6.59), and (6.60)]*

6.2 A dielectric resonator has a radius of 15 mm and height of 10 mm, and is made of a
dielectric with relative permittivity 37.9. Calculate the resonance frequency. *[Hint:
Use Eq. (6.154)]*

6.3 A cylindrical cavity resonator resonates at 10 GHz in the TE_{111} mode. The 3 dB bandwidth has been measured to be 2.5 MHz. Calculate the Q of the cavity. *[Use the relation $Q = f_0 / \Delta f_0$]*

6.4 A rectangular waveguide cavity, filled with a dielectric of dielectric constant 3.7, has
the dimension 22.86 mm × 10.16 mm × 20 mm. Calculate the resonance frequency
for the TE_{101} mode. *[Hint: Use Eq. (6.81)]*

6.5 A circular cavity resonator of diameter 23.83 mm has been designed to operate at
18 GHz in the TE_{011} mode. Determine the length of the cavity. *[Hint: Use Eq. (6.124)]*

6.6 Design a circular disk resonator to operate at 10 GHz in the TM_{110} mode. If the
substrate has a dielectric constant of $\varepsilon_r' - j\varepsilon_r' = 4.7 - j0.007$ and is 1.4 mm thick,
calculate the Q of the resonator, provided that $\sigma = 5.8 \times 10^7$ S/m. *[Hint: Use
Eqs (6.194) and (6.202)]*

6.7 Determine the length of a rectangular waveguide cavity that will operate at 10 GHz
in the TE_{101} mode, provided that $a = 22.86$ mm and $b = 10.16$ mm. Calculate the Q
of the cavity if it is made of copper. The conductivity of copper is $\sigma = 5.8 \times 10^7$ S/m
[Hint: Use Eq. (6.81)]

6.8 A half-wavelength coaxial resonator having an inner conductor radius of 0.75 mm
and outer conductor radius of 3 mm has been formed to operate at 1 GHz. Assume
that the conductors are made of copper, having conductivity $\sigma = 5.8 \times 10^7$ S/m, and
the dielectric is Teflon. Calculate the Q of the resonator. Teflon has a dielectric constant of 2.1 and loss tangent of 0.0004. *[Use the relation $Q = \beta / (2\alpha)$]*

6.9 An aperture-coupled waveguide cavity has a resonance frequency of 10 GHz and Q of 10,000. If the width and height of the waveguide are 22.86 and 10.16 mm, respectively, then find the normalized reactance of the aperture required for critical coupling. *[Hint: Use Eq. (6.241)]*

6.10 A quarter-wavelength transmission line of characteristic impedance Z_0 is terminated in a series RLC circuit. Show that at the vicinity of the resonance, the input imped-ance behaves like a parallel RLC circuit *[Hint: Start with the expression of input impedance of a transmission line]*

6.11 A resonator, constructed of a 5 cm, 75 Ω line, terminated at a short circuit at the load end and a capacitive reactance at the source end. Find the capacitance value to obtain the lowest-order resonance at 10 GHz. *[Hint: Start with the expression of input impedance of a transmission line]*

6.12 An air-filled rectangular waveguide cavity resonator has dimensions $a = 22.86$ mm, $b = 10.16$ mm, and $d = 30$ mm. Calculate the resonance frequency of the first five lowest-order modes. *[Hint: Use Eq. (6.81)]*

6.13 An air-filled cubical cavity operates at 10 GHz when excited at the TE_{101} mode. Determine the cavity dimension. *[Hint: Use Eq. (6.81)]*

6.14 A rectangular waveguide cavity resonator, having dimensions $a = 22.86$ mm and $b = 10.16$ m and operating in the TE_{101} mode, has a resonance frequency of 10 GHz. Now, the resonator is required to operate at 12 GHz in the same mode. Find the change in the distance between the shorting plates required to achieve it. *[Hint: Use Eq. (6.81)]*

6.15 A circular waveguide cavity having diameter 23.83 mm can be tuned with a plunger that can be varied from 3 to 7 cm. Determine the range of resonance frequency for the TE_{111} mode. *[Hint: Use Eq. (6.124)]*

6.16 Show that the Q-factor of a cubical cavity resonator having dimension a and operat-ing in the TM_{111}-mode can be given by $\dfrac{\pi \eta \sqrt{3}}{8R_S}$.

6.17 Show that the TE_{111}-mode Q-factor of a cubical cavity resonator having dimension a can be given by $\dfrac{a}{4\delta}$.

7

Microwave Network Representations

7.1 Introduction

In the previous few chapters, we have discussed different types of transmission lines, their matching networks, terminations, junctions, etc. For microwave applications, such transmission lines are often connected together to perform a specific task, such as power division and filtering. Such types of circuits, formed by connecting same or different types of transmission lines in a pre-defined manner, are called microwave networks. T-junctions, hybrid rings, directional couplers, ortho-modal couplers, turnstile couplers, monopulse comparators, filters, isolators, phase shifters, gyrators, etc. are a few examples of such microwave networks. The procedures used to analyse low-frequency networks cannot be applied directly to analyse these high-frequency networks, but these concepts can be extended to analyse and design these microwave circuits.

Solutions of the Maxwell's equations in these networks provide a complete solution of the problem, as it gives the fields at each point within the network. However, in most of the cases, we do not require all the information about these networks; we are interested only in voltages, currents, or powers at the terminals or ports (ports are defined as any type of transmission line or transmission line equivalent circuit of a single propagating waveguide or transmission line mode). For example, during the use of a power divider, we are interested only in the input power and power division ratios at different ports, and not in the details of the fields at each point within the network. In such cases, the application of microwave network analysis is much more convenient than the application of the Maxwell's equation. Further, since most of these microwave networks are extremely complex in nature, solving the Maxwell's equations for these networks is also very difficult.

In the following sections, we will discuss different microwave network representations such as Z, Y, S, and $ABCD$ parameters. These network theories were originally developed at the MIT radiation laboratory to study the radar system and its components in 1940. Later, in 1942, these theories were extended at the Microwave Research Institute.

7.2 Impedance and Equivalent Voltage and Current

At microwave frequencies, measurement of the equivalent voltage and current in a circuit is difficult until and unless a terminal pair is available. Such terminal pairs exist in TEM-type lines, such as coaxial cables, microstrips, or strip lines, but not in non-TEM-type lines like waveguides. In fact, there is no correct voltage, in the sense of being unique, in these lines. To explain this, let us consider the TE_{10} mode propagating in a rectangular waveguide. The TE_{10}-mode field components in a rectangular waveguide can be written as follows:

$$E_y(x,y,z) = -\frac{j\omega\mu a}{\pi} H_0 \sin\left(\frac{\pi x}{a}\right) e^{-j\beta z} = H_0 e_y(x,y) e^{-j\beta z} \tag{7.1}$$

and

$$H_x(x,y,z) = \frac{j\beta a}{\pi} H_0 \sin\left(\frac{\pi x}{a}\right) e^{-j\beta z} = H_0 h_x(x,y) e^{-j\beta z} \tag{7.2}$$

Therefore, the equivalent voltage is given by the following equation:

$$V = \int E_y dy = -\frac{j\omega\mu a}{\pi} H_0 \sin\left(\frac{\pi x}{a}\right) e^{-j\beta z} \int_y dy \tag{7.3}$$

Equation (7.3) shows that the voltage depends on x, as well as on the length of the integration contour along the y-direction. The same is true for current also. To avoid this, we generally define equivalent voltage and current, and hence impedance. Interestingly, the definition of equivalent voltage and current is also not unique, and there are many ways to define them. However, the definition that gives the most useful result is based on the following considerations:

1. Equivalent voltage and current are defined for a particular waveguide mode. For different modes, equivalent voltages and currents are also different.
2. Equivalent voltage is proportional to the transverse electric field, whereas equivalent current is proportional to the transverse magnetic field.
3. The product of the equivalent voltage and current for a particular mode results in the power carried by the mode.
4. The ratio of the equivalent voltage to the equivalent current for a particular mode results in the characteristic impedance of the line for the mode.

For an arbitrary waveguide mode with both positive and negative z-directed travelling waves, the transverse field components are given by the following equations:

$$\vec{E}_t(x,y,z) = \vec{e}_t(x,y)\left(A_0^+ e^{-j\beta z} + A_0^- e^{j\beta z}\right)$$

or

$$\vec{E}_t(x,y,z) = \frac{\vec{e}_t(x,y)}{C_1}\left(V^+ e^{-j\beta z} + V^- e^{j\beta z}\right) = \frac{\vec{e}_t(x,y)V}{C_1} \tag{7.4}$$

and

$$\vec{H}_t(x,y,z) = \vec{h}_t(x,y)\left(A_0^+ e^{-j\beta z} - A_0^- e^{j\beta z}\right)$$

or

$$\vec{H}_t(x,y,z) = \frac{\vec{h}_t(x,y)}{C_2}\left(I^+ e^{-j\beta z} + I^- e^{j\beta z}\right) = \frac{\vec{h}_t(x,y)I}{C_2} \tag{7.5}$$

where $\vec{e_t}(x,y)$ and $\vec{h_t}(x,y)$ are transverse field variations and A_0^+ and A_0^- are the field amplitudes of positive and negative z-directed waves. The $\vec{e_t}(x,y)$ and $\vec{h_t}(x,y)$ are related by the following equation:

$$\vec{h_t}(x,y) = \left(\hat{a}_z \times \vec{e_t}(x,y)\right)/z_w \tag{7.6}$$

where z_w is the wave impedance. In Eqs (7.4) and (7.5),

$$V^+/I^+ = V^-/I^- = Z_0 \tag{7.7}$$

and
$$C_1 = V^+/A_0^+ = V^-/A_0^- \tag{7.8}$$

$$C_2 = I^+/A_0^+ = I^-/A_0^- \tag{7.9}$$

The power carried by the z-directed wave is expressed as follows:

$$P^+ = \frac{1}{2}\left|A_0^+\right|^2 \iint_S \left(\vec{e} \times \vec{h^*}\right).\hat{a}_z ds = \frac{V^+ I^+}{2 C_1 C_2^*} \iint_S \left(\vec{e} \times \vec{h^*}\right).\hat{a}_z ds \tag{7.10}$$

Since power should be equal to $V^+ I^{+^*}/2$, from Eq. (7.10), we can write that

$$C_1 C_2^* = \iint_S \left(\vec{e} \times \vec{h^*}\right).\hat{a}_z ds \tag{7.11}$$

Further, $Z_0 = V^+/I^+ = V^-/I^- = C_1/C_2$ \hfill (7.12)

Equations (7.11) and (7.12) can be solved to find C_1 and C_2. This has been demonstrated in the following example:

TE$_{10}$ mode propagating in rectangular waveguide   Without any loss in generality, the TE$_{10}$-mode field components in a rectangular waveguide can be written as follows:

$$E_y = \left(A^+ e^{-j\beta z} + A^- e^{j\beta z}\right)\sin(\pi x/a) \tag{7.13}$$

and
$$H_x = -\frac{1}{\eta_{TE}}\left(A^+ e^{-j\beta z} + A^- e^{j\beta z}\right)\sin\left(\frac{\pi x}{a}\right) \tag{7.14}$$

The power carried by the z-directed wave is expressed as follows:

$$P^+ = \frac{1}{2}V^+ I^{+^*} = \frac{1}{2}C_1 A^+ \left(C_2 A^+\right)^* = \frac{1}{2}\left|A^+\right|^2 C_1 C_2^* \tag{7.15}$$

Again,

$$P^+ = -\frac{1}{2}\iint_S E_y H_x^* dxdy$$

or
$$P^+ = \frac{1}{2\eta_{TE}}\int_0^a \int_0^b \left(A^+ e^{-j\beta z}\right)\left(A^+ e^{-j\beta z}\right)^* \sin^2\left(\frac{\pi x}{a}\right) dxdy = \frac{\left|A^+\right|^2 ab}{4\eta_{TE}} \tag{7.16}$$

Equating Eqs. (7.15) and (7.16), we get the following expression:

$$\frac{1}{2}\left|A^+\right|^2 C_1 C_2^* = \frac{\left|A^+\right|^2 ab}{4\eta_{TE}}$$

or $\qquad C_1 C_2^* = \dfrac{ab}{2\eta_{TE}}$ (7.17)

Again $\quad C_1/C_2 = \eta_{TE}$ or $C_1 = \eta_{TE} C_2$ (7.18)

Substituting Eq. (7.18) into Eq. (7.17), we get the following relation:

$$\eta_{TE} C_2 C_2^* = \dfrac{ab}{2\eta_{TE}}$$

or $\qquad |C_2|^2 = \dfrac{ab}{2\eta_{TE}^2}$ or $C_2 = \dfrac{1}{\eta_{TE}} \sqrt{\dfrac{ab}{2}}$ (7.19)

Substituting Eq. (7.19) back into Eq. (7.18), we get the following expression:

$$C_1 = \eta_{TE} C_2 = \sqrt{ab/2}$$ (7.20)

7.3 Impedance and Admittance Parameters

Let us consider the N-port microwave network shown in Fig. 7.1. The voltage at any port i of the network depends on the currents existing at all the ports and can be written as follows:

$$V_i = Z_{i1}I_1 + Z_{i2}I_2 + Z_{i3}I_3 + \ldots + Z_{ij}I_j + \ldots + Z_{i(N-1)}I_N + Z_{iN}I_N$$ (7.21)

where $\quad Z_{ij} = \dfrac{V_i}{I_j}\bigg|_{I_k=0 \text{ for } k \neq j}$ (7.22)

In other words, Z_{ij} is the ratio of the open-circuit voltage at port i to the current at port j, with all the ports, except j, being left open circuited. The complete port voltages can, therefore, be written as follows:

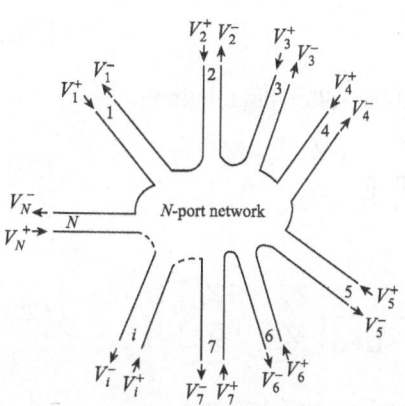

Fig. 7.1 Arbitrary N-port microwave network

$$\begin{bmatrix} V_1 \\ V_2 \\ \vdots \\ V_i \\ \vdots \\ V_N \end{bmatrix} = \begin{bmatrix} Z_{11} & Z_{12} & \cdots & Z_{1j} & \cdots & Z_{1N} \\ Z_{21} & Z_{22} & \cdots & Z_{2j} & \cdots & Z_{2N} \\ \vdots & \vdots & \cdots & \vdots & \cdots & \vdots \\ Z_{i1} & Z_{i2} & \cdots & Z_{ij} & \cdots & Z_{iN} \\ \vdots & \vdots & \cdots & \vdots & \cdots & \vdots \\ Z_{N1} & Z_{N2} & \cdots & Z_{Nj} & \cdots & Z_{NN} \end{bmatrix} \begin{bmatrix} I_1 \\ I_2 \\ \vdots \\ I_i \\ \vdots \\ I_N \end{bmatrix}$$

or $\qquad |V| = |Z||I|$ (7.23)

where $|Z|$ is known as the *impedance matrix* of the network.

Similarly, the current at any port i of the network depends on the voltages existing at all the ports and can be written as follows:

$$I_i = Y_{i1}V_1 + Y_{i2}V_2 + Y_{i3}V_3 + \ldots + Y_{ij}V_j + \ldots + Y_{i(N-1)}V_N + Y_{iN}V_N$$ (7.24)

where $\quad Y_{ij} = \dfrac{I_i}{V_j}\bigg|_{V_k=0 \text{ for } k \neq j}$ (7.25)

In other words, Y_{ij} is the ratio of the short-circuit current at port i to the voltage at port j, with all the ports, except j, being left short circuited. The complete port currents can, therefore, be written as follows:

$$
\begin{bmatrix} I_1 \\ I_2 \\ \vdots \\ I_i \\ \vdots \\ I_N \end{bmatrix} =
\begin{bmatrix}
Y_{11} & Y_{12} & \cdots & Y_{1j} & \cdots & Y_{1N} \\
Y_{21} & Y_{22} & \cdots & Y_{2j} & \cdots & Y_{2N} \\
\vdots & \vdots & \cdots & \vdots & \cdots & \vdots \\
Y_{i1} & Y_{i2} & \cdots & Y_{ij} & \cdots & Y_{iN} \\
\vdots & \vdots & \cdots & \vdots & \cdots & \vdots \\
Y_{N1} & Y_{N2} & \cdots & Y_{Nj} & \cdots & Y_{NN}
\end{bmatrix}
\begin{bmatrix} V_1 \\ V_2 \\ \vdots \\ V_i \\ \vdots \\ V_N \end{bmatrix}
$$

or $\qquad |I| = |Y||V| \qquad\qquad\qquad (7.26)$

where $|Y|$ is known as the *admittance* matrix of the network.

For an N-port network, [Z] or [Y] has a dimension of $N \times N$, and therefore each network representation has N^2 independent quantities. In practice, many of the microwave networks are reciprocal for some particular combinations of (i, j), that is, $Z_{ij} = Z_{ji}$ or $Y_{ij} = Y_{ji}$, and hence the total number of unknowns is fewer.

For a two-port network:

$$
\begin{bmatrix} I_1 \\ I_2 \end{bmatrix} = \begin{bmatrix} Y_{11} & Y_{12} \\ Y_{21} & Y_{22} \end{bmatrix} \begin{bmatrix} V_1 \\ V_2 \end{bmatrix}
$$

or $\qquad \begin{bmatrix} V_1 \\ V_2 \end{bmatrix} = \begin{bmatrix} Y_{11} & Y_{12} \\ Y_{21} & Y_{22} \end{bmatrix}^{-1} \begin{bmatrix} I_1 \\ I_2 \end{bmatrix} \qquad\qquad (7.27)$

Comparing Eq. (7.27) with Eq. (7.23), we get the following relation:

$$
\begin{bmatrix} Z_{11} & Z_{12} \\ Z_{21} & Z_{22} \end{bmatrix} = \begin{bmatrix} Y_{11} & Y_{12} \\ Y_{21} & Y_{22} \end{bmatrix}^{-1} = \frac{1}{Y_{11}Y_{22} - Y_{12}Y_{21}} \begin{bmatrix} Y_{22} & -Y_{12} \\ -Y_{21} & Y_{11} \end{bmatrix} \qquad (7.28)
$$

Alternatively,

$$
\begin{bmatrix} Y_{11} & Y_{12} \\ Y_{21} & Y_{22} \end{bmatrix} = \begin{bmatrix} Z_{11} & Z_{12} \\ Z_{21} & Z_{22} \end{bmatrix}^{-1} = \frac{1}{Z_{11}Z_{22} - Z_{12}Z_{21}} \begin{bmatrix} Z_{22} & -Z_{12} \\ -Z_{21} & Z_{11} \end{bmatrix} \qquad (7.29)
$$

EXAMPLE 7.1 Calculate the Z-matrix of the following T-network:

Solution The calculations are as follows:

Z_{11} can be calculated by making $I_2 = 0$, that is, open circuiting port 2:

$$
Z_{11} = \frac{V_1}{I_1}\bigg|_{I_2=0} = \frac{(Z_A + Z_C)I_1}{I_1} = (Z_A + Z_C)
$$

Z_{12} can be calculated by making $I_1 = 0$, that is, open circuiting port 1:

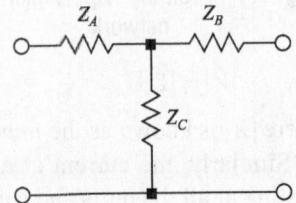

Fig. 7.2 T-network of Example 7.1

$$
Z_{12} = \frac{V_1}{I_2}\bigg|_{I_1=0} = \frac{V_2 Z_C}{(Z_B + Z_C)I_2} = \frac{(Z_B + Z_C)I_2 Z_C}{(Z_B + Z_C)I_2} = Z_C
$$

Z_{21} can be calculated by making $I_2 = 0$, that is, open circuiting port 2:

$$Z_{21} = \frac{V_2}{I_1}\bigg|_{I_2=0} = \frac{V_1 Z_C}{(Z_A + Z_C)I_1} = \frac{(Z_A + Z_C)I_1 Z_C}{(Z_A + Z_C)I_1} = Z_C$$

Z_{22} can be calculated by making $I_1 = 0$, that is, open circuiting port 1:

$$Z_{22} = \frac{V_2}{I_2}\bigg|_{I_1=0} = \frac{(Z_B + Z_C)I_2}{I_2} = (Z_B + Z_C)$$

Practice Problems

7.1@ Starting with the Z parameters, obtained in Example 7.1, draw the equivalent network for a symmetrical lossless network.

7.2 Calculate the Y-matrix of the following pi-network:

$$\begin{bmatrix} Y_A + Y_B & -Y_B \\ -Y_B & Y_B + Y_C \end{bmatrix}$$

7.3 Starting with the Y parameters, obtained in the Practice Problem 7.2, draw the equivalent network for a symmetrical lossless network. $|Y_A = Y_{11} + Y_{12},\ Y_B = -Y_{12},\ Y_C = Y_{22} + Y_{12}|$

7.4 Scattering Matrix

In the previous section, we have discussed the impedance and admittance matrix of a network. However, at microwave frequencies, measurement of voltages and currents is very difficult, because it involves the measurement of magnitude from power and phase from a travelling or a standing wave. On the other hand, direct measurements of the impedance and admittance matrix generally require the open-circuit voltage and short-circuit currents to be measured, which is a very challenging task since achieving a perfect open and short circuit at microwave frequencies is very difficult. In addition, for some single-conductor non-TEM lines, like waveguides, defining voltages and currents is also not straightforward. Therefore, for an efficient representation of a microwave network, some additional network parameters must be defined.

Since at microwave frequencies, measurement of the waves scattered from ports are much easier than that of an open-circuit voltage or a short-circuit current, a network representation of a microwave circuit in terms of these scattered waves is much more convenient. Such a network matrix is called a *scattering matrix* or an *S*-matrix. The direct measurement of a scattering matrix generally requires matched termination, which is much easier to achieve than a perfect open and short circuit.

While the impedance and admittance matrices relate the voltages with the currents at the ports, the scattering matrix relates the wave reflected from the port

---

@Solution in Online Resources

with that incident at the port. To demonstrate, let us consider the N-port network shown in Fig. 7.1. We can represent the reflected waves in terms of the incident waves, as follows:

$$V_i^- = S_{i1}V_1^+ + S_{i2}V_2^+ + S_{i3}V_3^+ + \ldots + S_{ii}V_i^+ + \ldots + S_{i(N-1)}V_{(N-1)}^+ + S_{iN}V_N^+ \quad (7.30)$$

where

$$S_{ij} = \left. \frac{V_i^-}{V_j^+} \right|_{V_k^+ = 0 \text{ for } k \neq j} \quad (7.31)$$

In other words, S_{ij} is the ratio of the voltage coming out of port i, when port j is fed with a voltage V_j^+, provided all other ports are matched terminated. Alternatively, we can say that S_{ij} is the ratio of the transmitted wave at port i to the incident wave at port j, with all other ports being terminated by a matched load. Since the ratio of the transmitted wave to the incident wave defines the transmission coefficient, S_{ij} is basically the transmission coefficient at port i due to feed at port j. Similarly, S_{ii} is the ratio of the reflected wave to the incident wave at port i, with all other ports being terminated by a matched load, and hence represents the reflection coefficient at port i. The complete network representation is expressed as follows:

$$\begin{bmatrix} V_1^- \\ V_2^- \\ \vdots \\ V_i^- \\ \vdots \\ V_N^- \end{bmatrix} = \begin{bmatrix} S_{11} & S_{12} & \cdots & S_{1j} & \cdots & S_{1N} \\ S_{21} & S_{22} & \cdots & S_{2j} & \cdots & S_{2N} \\ \vdots & \vdots & \cdots & \vdots & \cdots & \vdots \\ S_{i1} & S_{i2} & \cdots & S_{ij} & \cdots & S_{iN} \\ \vdots & \vdots & \cdots & \vdots & \cdots & \vdots \\ S_{N1} & S_{N2} & \cdots & S_{Nj} & \cdots & S_{NN} \end{bmatrix} \begin{bmatrix} V_1^+ \\ V_2^+ \\ \vdots \\ V_i^+ \\ \vdots \\ V_N^+ \end{bmatrix}$$

or $\quad \left[V^- \right] = \left[S \right] \left[V^+ \right] \quad (7.32)$

To relate the S matrix to the normalized Z matrix, let us assume that the characteristic impedance of all the ports is identical and equal to unity. Therefore, port voltages and currents can be expressed as follows:

$$V_i = V_i^+ + V_i^- \quad (7.33)$$

and $\quad I_i = I_i^+ - I_i^- = V_i^+ - V_i^- \quad (7.34)$

Therefore, using Eq. (7.34), Eq. (7.23) can be written in the following form:

$$[V] = [Z][I] = [Z]\left\{ \left[V^+ \right] - \left[V^- \right] \right\} \quad (7.35)$$

Substituting Eq. (7.33) into Eq. (7.35), we get the following relation:

$$\left[V^+ \right] + \left[V^- \right] = [Z]\left\{ \left[V^+ \right] - \left[V^- \right] \right\} = [Z]\left[V^+ \right] - [Z]\left[V^- \right]$$

or $\quad ([Z] + [U])\left[V^- \right] = ([Z] - [U])\left[V^+ \right]$

or $\quad \left[V^- \right] = \left\{ ([Z] + [U])^{-1} ([Z] - [U]) \right\}\left[V^+ \right] \quad (7.36)$

where $[U]$ is a unit or an identity matrix.

Comparing Eq. (7.32) with Eq. (7.36), we can write the following relation:

$$[S] = ([Z]+[U])^{-1}([Z]-[U]) \tag{7.37}$$

Alternatively,

$$([Z]+[U])[S] = [Z]-[U] \text{ or } [Z][S]+[U][S] = [Z]-[U]$$

or $\quad [Z]([U]-[S]) = [U]+[S] \text{ or } [Z] = ([U]+[S])([U]-[S])^{-1}$ $\qquad$ (7.38)

Equation (7.38) shows that if $[Z]$ is symmetrical, then $[S]$ is also symmetrical.

Note An $[S]$ matrix is defined for a given set of reference plane. If the reference planes are changed, the scattering coefficients vary only in phase. This is not possible with a $[Z]$ or $[Y]$ matrix. Both magnitude and phase of impedance or admittance parameters are changed, if there is a change in the reference planes.

Practice Problem

7.4[@] Find the S parameter of the following circuit when it is terminated with a matched load of 50 Ω.

EXAMPLE 7.2 Find the modified S-matrix of an N-port lossless network for an arbitrary shift in the reference plane.

Solution Let us consider that the network has an S-matrix $[S]$ when its terminal ports are defined at $z_i = 0$. Therefore,

$$[V^-] = [S][V^+] \tag{7.39}$$

Defining new reference planes at $z_i = l_i$ will change its S-matrix. Let us assume that the modified network has an S-matrix $[S']$. Therefore,

$$[V'^-] = [S'][V'^+] \tag{7.40}$$

Using the theory of travelling waves in a lossless line, we can write the following equations:

$$V_i'^+ = V_i^+ e^{j\theta_i} \tag{7.41}$$

and $\quad V_i'^- = V_i^- e^{-j\theta_i}$ $\qquad$ (7.42)

where $\quad \theta_i = \beta_i l_i$.

Using Eqs (7.41) and (7.42) in Eq. (7.39), we can write the following relation:

[@]Solution in Online Resources

$$\begin{bmatrix} e^{j\theta_1} & 0 & \cdots & 0 \\ 0 & e^{j\theta_2} & \cdots & 0 \\ \vdots & \vdots & \ddots & \vdots \\ 0 & 0 & \cdots & e^{j\theta_N} \end{bmatrix} [V'^-] = [S] \begin{bmatrix} e^{-j\theta_1} & 0 & \cdots & 0 \\ 0 & e^{-j\theta_2} & \cdots & 0 \\ \vdots & \vdots & \ddots & \vdots \\ 0 & 0 & \cdots & e^{-j\theta_N} \end{bmatrix} [V'^+]$$

or
$$[V'^-] = \begin{bmatrix} e^{j\theta_1} & 0 & \cdots & 0 \\ 0 & e^{j\theta_2} & \cdots & 0 \\ \vdots & \vdots & \ddots & \vdots \\ 0 & 0 & \cdots & e^{j\theta_N} \end{bmatrix}^{-1} [S] \begin{bmatrix} e^{-j\theta_1} & 0 & \cdots & 0 \\ 0 & e^{-j\theta_2} & \cdots & 0 \\ \vdots & \vdots & \ddots & \vdots \\ 0 & 0 & \cdots & e^{-j\theta_N} \end{bmatrix} [V'^+] \qquad (7.43)$$

Comparing Eq. (7.40) with Eq. (7.43), we get the following relation:

$$[S'] = \begin{bmatrix} e^{j\theta_1} & 0 & \cdots & 0 \\ 0 & e^{j\theta_2} & \cdots & 0 \\ \vdots & \vdots & \ddots & \vdots \\ 0 & 0 & \cdots & e^{j\theta_N} \end{bmatrix}^{-1} [S] \begin{bmatrix} e^{-j\theta_1} & 0 & \cdots & 0 \\ 0 & e^{-j\theta_2} & \cdots & 0 \\ \vdots & \vdots & \ddots & \vdots \\ 0 & 0 & \cdots & e^{-j\theta_N} \end{bmatrix}$$

Practice Problem

7.5 Two transmission lines of characteristic impedance Z_{01} and Z_{02} are joined face to face. Find the scattering matrix of the network.
$$\begin{bmatrix} \dfrac{Z_{02} - Z_{01}}{Z_{02} + Z_{01}} & \dfrac{2Z_{01}}{Z_{01} + Z_{02}} \\ \dfrac{2Z_{01}}{Z_{01} + Z_{02}} & \dfrac{Z_{01} - Z_{02}}{Z_{01} + Z_{02}} \end{bmatrix}$$

7.5 Properties of Scattering Parameters for Reciprocal and Lossless Networks

In this section, we will find the properties of scattering parameters for reciprocal and lossless networks. Adding Eqs. (7.34) and (7.33), we get the following relation:

$$V_i^+ = (V_i + I_i)/2$$

or $\qquad [V^+] = \dfrac{1}{2}([Z][I] + [I]) = \dfrac{1}{2}([Z] + [U])[I] \qquad (7.44)$

Subtracting Eq. (7.34) from Eq. (7.33), we get the following expression:

$$V_i^- = (V_i - I_i)/2$$

or $\qquad [V^-] = \dfrac{1}{2}([Z][I] - [I]) = \dfrac{1}{2}([Z] - [U])[I] \qquad (7.45)$

From Eq. (7.44), we can write the following relation:

$$[I] = 2([Z] + [U])^{-1}[V^+] \qquad (7.46)$$

Substituting Eq. (7.46) into Eq. (7.45), the following equation is obtained:

$$[V^-] = \dfrac{1}{2}([Z] - [U])2([Z] + [U])^{-1}[V^+] = ([Z] - [U])([Z] + [U])^{-1}[V^+] \qquad (7.47)$$

Comparing Eq. (7.47) with Eq. (7.32), we can write the following relation:

$$[S] = ([Z]-[U])([Z]+[U])^{-1}$$ (7.48)

or $\quad [S]^t = \left\{([Z]-[U])([Z]+[U])^{-1}\right\}^t = \left\{([Z]+[U])^{-1}\right\}^t ([Z]-[U])^t$ (7.49)

Since $[U]^t = [U]$ and for a reciprocal network $[Z]^t = [Z]$:

$$[S]^t = ([Z]+[U])^{-1}([Z]-[U])$$ (7.50)

Comparing Eq. (7.50) with Eq. (7.37), we get the following relation:

$$[S]^t = [S]$$ (7.51)

The average power delivered to the network is expressed as follows:

$$P_{av} = \frac{1}{2}\text{Re}\left\{[V]^t[I]^*\right\} = \frac{1}{2}\text{Re}\left\{\left([V^+]^t + [V^-]^t\right)\left([V^+]^* - [V^-]^*\right)\right\}$$

or $\quad P_{av} = \frac{1}{2}\text{Re}\left\{[V^+]^t[V^+]^* - [V^+]^t[V^-]^* + [V^-]^t[V^+]^* - [V^-]^t[V^-]^*\right\}$ (7.52)

The term $-[V^+]^t[V^-]^* + [V^-]^t[V^+]^*$ has the form $A - A^*$ and hence is purely imaginary. Since the average power is a real quantity,

$$[V^-]^t[V^+]^* - [V^+]^t[V^-]^* = 0$$

and $\quad P_{av} = \frac{1}{2}[V^+]^t[V^+]^* - \frac{1}{2}[V^-]^t[V^-]^*$ (7.53)

where the first term represents the total incident power and the second term represents the total reflected power. Since for a lossless network, the total power delivered to the network must be zero, Eq. (7.53) reveals that

$$[V^+]^t[V^+]^* = [V^-]^t[V^-]^*$$ (7.54)

Substituting Eq. (7.32) into Eq. (7.54), we get the following relation:

$$[V^+]^t[V^+]^* = \left\{[S][V^+]\right\}^t\left\{[S][V^+]\right\}^* = [V^+]^t[S]^t[S]^*[V^+]^*$$ (7.55)

Equation (7.55) reveals that for a lossless network:

$$[S]^t[S]^* = [U]$$

or $\quad [S]^* = \left\{[S]^t\right\}^{-1}$ (7.56)

which represents that [S] must be a unitary matrix.

For a lossless network, the total power leaving all the N ports must be equal to the sum of the incident power. Therefore, if a_i and b_i are, respectively, the incident and reflected power at port i, then we can write the following equation:

$$\sum_{i=1}^{N}\left(a_i a_i^* - b_i b_i^*\right) = 0 \text{ or } \sum_{i=1}^{N} a_i a_i^* = \sum_{i=1}^{N} b_i b_i^*$$ (7.57)

Again from the definition of S parameters, we can write the following relation:

$$b_i = \sum_{j=1}^{N} S_{ij} a_j \qquad (7.58)$$

Substituting Eq. (7.58) into Eq. (7.57), we get the following expression:

$$\sum_{i=1}^{N} \left| \sum_{j=1}^{N} S_{ij} a_j \right|^2 = \sum_{i=1}^{N} a_i a_i^* \qquad (7.59)$$

Now since a_i's are independent, we can choose all a_i's except a_p as zero. Therefore, Eq. (7.59) modifies to the following form:

$$\sum_{i=1}^{N} \left| S_{ip} a_p \right|^2 = a_p a_p^*$$

or $\qquad \displaystyle \sum_{i=1}^{N} \left| S_{ip} \right|^2 = \sum_{i=1}^{N} S_{ip} S_{ip}^* = 1 \qquad (7.60)$

Since p is arbitrary, Eq. (7.60) is valid for all values of p. Therefore, for a lossless network, the product of any column of S-matrix and the conjugate of this column equals unity.

Now, let us assume that all a_i's, except a_q and a_r, are zero. Therefore, from Eq. (7.59), the following relation is obtained:

$$\sum_{i=1}^{N} \left| S_{iq} a_q + S_{ir} a_r \right|^2 = a_q a_q^* + a_r a_r^*$$

or $\qquad \displaystyle \sum_{i=1}^{N} \left(S_{iq} a_q + S_{ir} a_r \right) \left(S_{iq} a_q + S_{ir} a_r \right)^* = \left| a_q \right|^2 + \left| a_r \right|^2$

or $\qquad \displaystyle \sum_{i=1}^{N} \left(S_{iq} a_q S_{iq}^* a_q^* + S_{ir} a_r S_{iq}^* a_q^* + S_{iq} a_q S_{ir}^* a_r^* + S_{ir} a_r S_{ir}^* a_r^* \right) = \left| a_q \right|^2 + \left| a_r \right|^2$

or $\qquad \displaystyle \sum_{i=1}^{N} \left(S_{iq} S_{iq}^* \left| a_q \right|^2 + S_{ir} S_{iq}^* a_r a_q^* + S_{iq} S_{ir}^* a_q a_r^* + S_{ir} S_{ir}^* \left| a_r \right|^2 \right) = \left| a_q \right|^2 + \left| a_r \right|^2 \qquad (7.61)$

Now, from Eq. (7.60):

$$\sum_{i=1}^{N} S_{iq} S_{iq}^* = \sum_{i=1}^{N} S_{ir} S_{ir}^* = 1 \qquad (7.62)$$

Substituting Eq. (7.62) into Eq. (7.61), we get the following relation:

$$\sum_{i=1}^{N} \left(S_{ir} S_{iq}^* a_r a_q^* + S_{iq} S_{ir}^* a_q a_r^* \right) + \left| a_q \right|^2 + \left| a_r \right|^2 = \left| a_q \right|^2 + \left| a_r \right|^2$$

or $\qquad \displaystyle \sum_{i=1}^{N} \left(S_{ir} S_{iq}^* a_r a_q^* + S_{iq} S_{ir}^* a_q a_r^* \right) = 0 \qquad (7.63)$

If we assume that $a_r = a_q$, then Eq. (7.63) modifies to the following form:

$$\left| a_r \right|^2 \sum_{i=1}^{N} \left(S_{ir} S_{iq}^* + S_{iq} S_{ir}^* \right) = 0 \qquad (7.64)$$

Since $|a_r|^2 \neq 0$, $\displaystyle\sum_{i=1}^{N}\left(S_{ir}S_{iq}^* + S_{iq}S_{ir}^*\right) = 0$ (7.65)

Further, if we assume that $a_r = ja_q$, then Eq. (7.63) modifies to the following form:

$$j|a_q|^2 \sum_{i=1}^{N}\left(S_{ir}S_{iq}^* - S_{iq}S_{ir}^*\right) = 0 \qquad (7.66)$$

Since $j|a_q|^2 \neq 0$, $\displaystyle\sum_{i=1}^{N}\left(S_{ir}S_{iq}^* - S_{iq}S_{ir}^*\right) = 0$ (7.67)

Since both a_q and a_r are arbitrary, and the S parameters of a network are independent of the incident field strengths, Eqs. (7.65) and (7.67) should hold simultaneously; this occurs only when the following condition is satisfied:

$$\sum_{i=1}^{N} S_{ir}S_{iq}^* = 0 \text{ for } r \neq q \qquad (7.68)$$

That is, the product of any column of the scattering matrix and the complex conjugate of another column is equal to zero.

Notes 1. The unitary property of an [S] matrix can be used to check the power balance of a lossless microwave network. No such check is possible with a [Z] or [Y] matrix.
2. For a symmetrical lossless network, the total number of unknown parameters in the scattering matrix is $N(N + 1)/2$.

EXAMPLE 7.3 Prove that for a reciprocal lossless N-port network, the impedance matrix is purely imaginary.

Solution

The average power delivered to the network is as follows:

$$P_{av} = \frac{1}{2}[V]^t[I]^* = \frac{1}{2}[[Z][I]]^t[I]^* = \frac{1}{2}[I]^t[Z]^t[I]^* = \frac{1}{2}[I]^t[Z][I]^*$$

or $\qquad P_{av} = \frac{1}{2}\left(I_1 Z_{11} I_1^* + I_1 Z_{12} I_2^* + I_2 Z_{21} I_1^* + \ldots\right)$

or $\qquad P_{av} = \frac{1}{2}\sum_{n=1}^{N}\sum_{m=1}^{N} I_m Z_{mn} I_n^*$

Since I_n's are independent and the network is lossless, the real part of the self-terms $I_n Z_{nn} I_n^*$ must be zero, that is, $\text{Re}\{I_n Z_{nn} I_n^*\} = |I_n|^2 \text{Re}\{Z_{nn}\} = 0$.

Further, since we can set all port currents equal to zero, except for the nth current, that is, $|I_n|^2 \neq 0$,

$$\text{Re}\{Z_{nn}\} = 0$$

Further, let us assume that all port currents except m and n are zero. Now, to satisfy the lossless condition of the network, that is, $\text{Re}\{P_{av}\} = 0$, we must have

$$\text{Re}\{I_m Z_{mn} I_n^* + I_n Z_{nm} I_m^*\} = \text{Re}\left\{\left(I_m I_n^* + I_n I_m^*\right)Z_{mn}\right\} = 0$$

Since $\left(I_m I_n^* + I_n I_m^*\right)$ is a purely real quantity and in general not equal to zero, we have the following relation:

$$\text{Re}\{Z_{mn}\} = 0$$

Therefore, $\text{Re}\{Z_{mn}\}$ is equal to zero for all m and n.

7.6 Generalized Scattering Parameters

Till now, we have assumed that the characteristic impedance of each port is same. However, in practical cases, this condition may not hold, especially in the case of different waveguide junctions and steps where the characteristic impedance of each port is different. In such cases, we need to define *generalized scattering parameters*. For this, we need to define new sets of incident and reflected wave amplitudes, as follows:

$$a_i = V_i^+ / \sqrt{Z_{0i}} \tag{7.69}$$

and

$$b_i = V_i^- / \sqrt{Z_{0i}} \tag{7.70}$$

where Z_{0i} is the characteristic impedance of port i.
Substituting Eqs. (7.69) and (7.70) into Eqs. (7.33) and (7.34), we get the following relations:

$$V_i = V_i^+ + V_i^- = \sqrt{Z_{0i}}\,(a_i + b_i) \tag{7.71}$$

and

$$I_i = \left(V_i^+ - V_i^-\right) / Z_{0i} = (a_i - b_i) / \sqrt{Z_{0i}} \tag{7.72}$$

The average power delivered to the ith port is as follows:

$$P_{\text{av},i} = \frac{1}{2}\text{Re}\{V_i I_i^*\} = \frac{1}{2}\text{Re}\left\{\sqrt{Z_{0i}}\,(a_i + b_i)\frac{(a_i - b_i)^*}{\sqrt{Z_{0i}}}\right\}$$

or

$$P_{\text{av},i} = \frac{1}{2}\text{Re}\left\{(a_i + b_i)(a_i^* - b_i^*)\right\}$$

or

$$P_{\text{av},i} = \frac{1}{2}\text{Re}\left\{a_i a_i^* - a_i b_i^* + b_i a_i^* - b_i b_i^*\right\} = \frac{1}{2}|a_i|^2 - \frac{1}{2}|b_i|^2 \tag{7.73}$$

As $b_i a_i^* - a_i b_i^*$ is purely imaginary. Equation (7.73) reveals that the power delivered to the ith port is equal to the power incident minus the power reflected from this port.

The general form of the scattering matrix equation is as follows:

$$|b| = |S||a| \tag{7.74}$$

where

$$S_{ij} = \left.\frac{b_i}{a_j}\right|_{a_k = 0 \text{ for } k \neq j} \tag{7.75}$$

Using Eqs. (7.69) and (7.70) in Eq. (7.75), we get the following expression:

$$S_{ij} = \frac{V_i^- \sqrt{Z_{0j}}}{V_j^+ \sqrt{Z_{0i}}}\bigg|_{V_k^+ = 0 \text{ for } k \neq j} \tag{7.76}$$

The relation between the generalized scattering matrix and the impedance matrix is given by the following equations:

$$[Z] = \left[\sqrt{z}\right]\left([U] + [S]\right)\left([U] - [S]\right)^{-1}\left[\sqrt{z}\right]$$
$$= \left[\sqrt{z}\right]\left([U] - [S]\right)^{-1}\left([U] + [S]\right)\left[\sqrt{z}\right] \tag{7.77}$$

and
$$[S] = \left(\left[\sqrt{y}\right][Z]\left[\sqrt{y}\right] - [U]\right)\left(\left[\sqrt{y}\right][Z]\left[\sqrt{y}\right] + [U]\right)^{-1}$$
$$= \left(\left[\sqrt{y}\right][Z]\left[\sqrt{y}\right] + [U]\right)^{-1}\left(\left[\sqrt{y}\right][Z]\left[\sqrt{y}\right] - [U]\right) \tag{7.78}$$

where
$$\left[\sqrt{z}\right] = \begin{vmatrix} \sqrt{z_{01}} & 0 & \cdots & 0 \\ 0 & \sqrt{z_{02}} & \cdots & 0 \\ 0 & 0 & \ddots & 0 \\ 0 & 0 & \cdots & \sqrt{z_{0N}} \end{vmatrix} \tag{7.79}$$

and
$$\left[\sqrt{y}\right] = \left[\sqrt{z}\right]^{-1} \tag{7.80}$$

7.7 *S* Parameters of Two-port Networks with Mismatched Loads

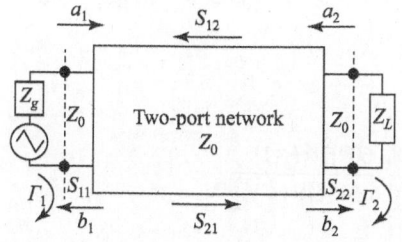

Fig. 7.3 Mismatched two-port network

Let us consider that a two-port network has been terminated by a mismatched load at both ends, as shown in Fig. 7.3. Therefore, the reflection coefficient at the load will be the following:

$$\Gamma_2 = \frac{a_2}{b_2} = \frac{Z_L - Z_0}{Z_L + Z_0} \tag{7.81}$$

Now
$$b_1 = S_{11}a_1 + S_{12}a_2 = S_{11}a_1 + S_{12}\frac{a_2}{b_2}b_2 = S_{11}a_1 + \Gamma_2 S_{12}b_2 \tag{7.82}$$

$$b_2 = S_{21}a_1 + S_{22}a_2 = S_{21}a_1 + S_{22}\frac{a_2}{b_2}b_2 = S_{21}a_1 + \Gamma_2 S_{22}b_2 \tag{7.83}$$

From Eq. (7.82),

$$\Gamma_1 = \frac{b_1}{a_1} = S_{11} + \Gamma_2 S_{12}\frac{b_2}{a_1} \tag{7.84}$$

and from Eq. (7.83),

$$\frac{b_2}{a_1} = S_{21} + \Gamma_2 S_{22} \frac{b_2}{a_1}$$

or $\quad (1 - \Gamma_2 S_{22}) \dfrac{b_2}{a_1} = S_{21}$ or $\dfrac{b_2}{a_1} = \dfrac{S_{21}}{1 - \Gamma_2 S_{22}}$ $\hfill (7.85)$

Substituting Eq. (7.85) into Eq. (7.84), we get the following expression:

$$\Gamma_1 = S_{11} + \frac{S_{12} S_{21} \Gamma_2}{1 - S_{22} \Gamma_2} \hfill (7.86)$$

Therefore, for a mismatched load $\Gamma_1 \neq S_{11}$. Now, if the network is reciprocal, then $S_{12} = S_{21}$ and Eq. (7.86) reduces to the following form:

$$\Gamma_1 = S_{11} + \frac{S_{12}^2 \Gamma_2}{1 - S_{22} \Gamma_2} \hfill (7.87)$$

In a similar fashion, it can be shown that

$$\Gamma_2 = S_{22} + \frac{S_{21}^2 \Gamma_1}{1 - S_{11} \Gamma_1} \hfill (7.88)$$

where $\quad \Gamma_1 = \dfrac{a_1}{b_1} = \dfrac{Z_g - Z_0}{Z_g + Z_0} \hfill (7.89)$

EXAMPLE 7.4 A two-port network has the following scattering matrix:

$$\begin{bmatrix} 0.6\angle 45° & 0.8\angle 60° \\ 0.8\angle 60° & 0.6\angle 45° \end{bmatrix}$$

What is the reflection coefficient seen at port 1 if a short circuit is placed at port 2.

Solution Given: $[S] = \begin{bmatrix} 0.6\angle 45° & 0.8\angle 60° \\ 0.8\angle 60° & 0.6\angle 45° \end{bmatrix}$

For short-circuit termination, $\Gamma_2 = -1$.
Since $S_{12} = S_{21}$,

$$\Gamma_1 = S_{11} + \frac{S_{12}^2 \Gamma_2}{1 - S_{22} \Gamma_2} = 0.6\angle 45° + \frac{(0.8\angle 60°)^2 \times (-1)}{1 - \{(0.6\angle 45°) \times (-1)\}}$$

or $\quad \Gamma_1 = 0.6\angle 45° - \dfrac{0.64\angle 120°}{1 + 0.6\angle 45°}$

or $\quad \Gamma_1 = 0.6\angle 45° - \dfrac{0.64\angle 120°}{1.4862\angle 16.5879°} = 0.5242\angle 0.5854°$

7.8 Transmission or *ABCD* matrix

The Z-, Y-, or S-parameter matrices are used widely to characterize N-port networks. However, in many practical applications, cascaded connections of a number of two-port networks, as shown in Fig. 7.4, are used. In such cases, it is

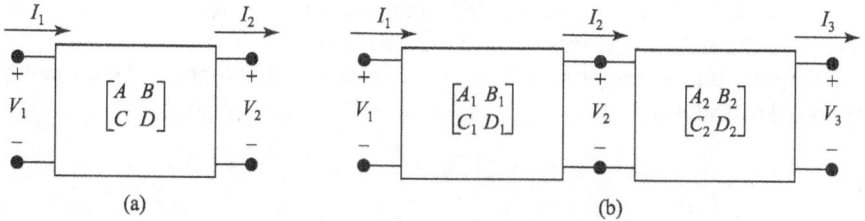

Fig. 7.4 Two-port networks (a) Isolated (b) Cascaded

more convenient to define a transmission or *ABCD matrix*. In the *ABCD* matrix representation of a two-port network, the voltages and currents at the ports are related by the following relations:

$$V_1 = AV_2 + BI_2 \tag{7.90}$$

$$I_1 = CV_2 + DI_2 \tag{7.91}$$

where V_1 and V_2 are the voltages at ports 1 and 2, respectively; I_1 is the current flowing in at port 1; and I_2 is the current flowing out of port 2.

In a matrix form, Eqs. (7.90) and (7.91) can be written as follows:

$$\begin{bmatrix} V_1 \\ I_1 \end{bmatrix} = \begin{bmatrix} A & B \\ C & D \end{bmatrix} \begin{bmatrix} V_2 \\ I_2 \end{bmatrix} \tag{7.92}$$

For a cascaded connection, I_2 will be the current flowing into the adjacent network, as shown in Fig. 7.4(b). For each of the two networks of Fig. 7.4(b), we can write the following equations:

$$\begin{bmatrix} V_1 \\ I_1 \end{bmatrix} = \begin{bmatrix} A_1 & B_1 \\ C_1 & D_1 \end{bmatrix} \begin{bmatrix} V_2 \\ I_2 \end{bmatrix} \tag{7.93}$$

and $$\begin{bmatrix} V_2 \\ I_2 \end{bmatrix} = \begin{bmatrix} A_2 & B_2 \\ C_2 & D_2 \end{bmatrix} \begin{bmatrix} V_3 \\ I_3 \end{bmatrix} \tag{7.94}$$

Substituting Eq. (7.94) into Eq. (7.93), we get the following relation:

$$\begin{bmatrix} V_1 \\ I_1 \end{bmatrix} = \begin{bmatrix} A_1 & B_1 \\ C_1 & D_1 \end{bmatrix} \begin{bmatrix} A_2 & B_2 \\ C_2 & D_2 \end{bmatrix} \begin{bmatrix} V_3 \\ I_3 \end{bmatrix} \tag{7.95}$$

If the transmission matrix of the complete cascaded network is represented as follows:

$$\begin{bmatrix} V_1 \\ I_1 \end{bmatrix} = \begin{bmatrix} A & B \\ C & D \end{bmatrix} \begin{bmatrix} V_3 \\ I_3 \end{bmatrix} \tag{7.96}$$

then comparing Eq. (7.95) with Eq. (7.96), we can write the following equality:

$$\begin{bmatrix} A & B \\ C & D \end{bmatrix} = \begin{bmatrix} A_1 & B_1 \\ C_1 & D_1 \end{bmatrix} \begin{bmatrix} A_2 & B_2 \\ C_2 & D_2 \end{bmatrix} \tag{7.97}$$

That is, the *ABCD* matrix of a cascaded connection of two networks is equal to the product of the *ABCD* matrices of individual networks.

The individual parameters of the *ABCD* matrix of a two-port network can be expressed as follows:

$$A = \frac{V_1}{V_2}\bigg|_{I_2=0} \tag{7.98}$$

$$B = \frac{V_1}{I_2}\bigg|_{V_2=0} \tag{7.99}$$

$$C = \frac{I_1}{V_2}\bigg|_{I_2=0} \tag{7.100}$$

and $\qquad D = \frac{I_1}{I_2}\bigg|_{V_2=0} \tag{7.101}$

Using the current convention, used to define an *ABCD* matrix, the *Z*-parameter matrix of the network can be defined based on the following relations:

$$V_1 = Z_{11}I_1 - Z_{12}I_2 \tag{7.102}$$

and $\qquad V_2 = Z_{21}I_1 - Z_{22}I_2 \tag{7.103}$

Using Eqs. (7.102) and (7.103) in the definition of *A*, *B*, *C*, and *D* in Eqs (7.98)–(7.101), the relation between the transmission matrix and the impedance matrix parameters can be expressed as follows:

$$A = \frac{V_1}{V_2}\bigg|_{I_2=0} = \frac{Z_{11}I_1}{Z_{21}I_1} = \frac{Z_{11}}{Z_{21}} \tag{7.104}$$

$$B = \frac{V_1}{I_2}\bigg|_{V_2=0} = \frac{Z_{11}I_1 - Z_{12}I_2}{I_2} = Z_{11}\frac{I_1}{I_2}\bigg|_{V_2=0} - Z_{12}$$

or $\qquad B = \frac{Z_{11}Z_{22}}{Z_{21}} - Z_{12} = \frac{Z_{11}Z_{22} - Z_{12}Z_{21}}{Z_{21}} \tag{7.105}$

$$C = \frac{I_1}{V_2}\bigg|_{I_2=0} = \frac{I_1}{Z_{21}I_1} = \frac{1}{Z_{21}} \tag{7.106}$$

and $\qquad D = \frac{I_1}{I_2}\bigg|_{V_2=0} = \frac{Z_{22}}{Z_{21}} \tag{7.107}$

For a reciprocal network, $Z_{11} = Z_{22}$ and $Z_{21} = Z_{12}$. Therefore,

$$AD - BC = \frac{Z_{11}}{Z_{21}}\frac{Z_{22}}{Z_{21}} - \frac{Z_{11}Z_{22} - Z_{12}Z_{21}}{Z_{21}}\frac{1}{Z_{21}}$$

or $\qquad AD - BC = \left(\frac{Z_{11}}{Z_{21}}\right)^2 - \left(\frac{Z_{11}}{Z_{21}}\right)^2 + 1 = 1 \tag{7.108}$

From Eq. (7.91),

$$V_2 = \frac{1}{C}I_1 - \frac{D}{C}I_2 = \frac{1}{C}(I_1) + \frac{D}{C}(-I_2) \tag{7.109}$$

Substituting Eq. (7.109) into (7.90), we get the following relation:

$$V_1 = AV_2 + BI_2 = \frac{A}{C}I_1 - \frac{AD}{C}I_2 + BI_2 = \frac{A}{C}(I_1) + \frac{AD-BC}{C}(-I_2) \tag{7.110}$$

Equations (7.109) and (7.110) can be written in a matrix equation of the following form:

$$\begin{bmatrix} V_1 \\ V_2 \end{bmatrix} = \begin{bmatrix} \dfrac{A}{C} & \dfrac{AD-BC}{C} \\ \dfrac{1}{C} & \dfrac{D}{C} \end{bmatrix} \begin{bmatrix} I_1 \\ -I_2 \end{bmatrix} \tag{7.111}$$

Comparing Eq. (7.111) with Eq. (7.23), and taking care of the proper current direction used to define the [Z] and ABCD parameters, we can write the following equation:

$$[Z] = \begin{bmatrix} Z_{11} & Z_{12} \\ Z_{21} & Z_{22} \end{bmatrix} = \begin{bmatrix} \dfrac{A}{C} & \dfrac{AD-BC}{C} \\ \dfrac{1}{C} & \dfrac{D}{C} \end{bmatrix} \tag{7.112}$$

Similarly, from Eq. (7.90),

$$-I_2 = -\frac{1}{B}V_1 + \frac{A}{B}V_2 \tag{7.113}$$

Substituting Eq. (7.113) into Eq. (7.91), we get the following relation:

$$I_1 = CV_2 + DI_2 = CV_2 + \frac{D}{B}V_1 - \frac{AD}{B}V_2 = \frac{D}{B}V_1 - \frac{AD-BC}{B}V_2 \tag{7.114}$$

Equations (7.113) and (7.114) can be written in a matrix equation of the following form:

$$\begin{bmatrix} I_1 \\ -I_2 \end{bmatrix} = \begin{bmatrix} \dfrac{D}{B} & -\dfrac{AD-BC}{B} \\ -\dfrac{1}{B} & \dfrac{A}{B} \end{bmatrix} \begin{bmatrix} V_1 \\ V_2 \end{bmatrix} \tag{7.115}$$

Comparing between Eqs. (7.115) and (7.26), and taking care of the proper current direction used to define the [Y] and ABCD parameters, we can write the following equation:

$$[Y] = \begin{bmatrix} Y_{11} & Y_{12} \\ Y_{21} & Y_{22} \end{bmatrix} = \begin{bmatrix} \dfrac{D}{B} & -\dfrac{AD-BC}{B} \\ -\dfrac{1}{B} & \dfrac{A}{B} \end{bmatrix} \tag{7.116}$$

EXAMPLE 7.5 Find the relation between the ABCD and Z parameters of a two-port network shown in Fig. 7.5.

Solution

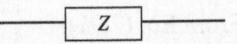

$$A = \frac{V_1}{V_2}\bigg|_{I_2=0} = 1 \qquad B = \frac{V_1}{I_2}\bigg|_{V_2=0} = \frac{ZV_1}{V_1} = Z$$

$$C = \frac{I_1}{V_2}\bigg|_{I_2=0} = 0 \qquad D = \frac{I_1}{I_2}\bigg|_{V_2=0} = \frac{I_1}{I_1} = 1$$

Fig. 7.5 Two-port network
of Example 7.5

Practice Problem

7.6 Find the relation between the *ABCD* and *Z* parameters of a two-port network, shown in the following figure:

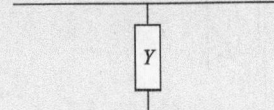

$$\begin{bmatrix} 1 & 0 \\ Y & 1 \end{bmatrix}$$

Important Formulae

- $[Z] = \dfrac{1}{Y_{11}Y_{22} - Y_{12}Y_{21}} \begin{bmatrix} Y_{22} & -Y_{12} \\ -Y_{21} & Y_{22} \end{bmatrix}$

- $[Y] = \dfrac{1}{Z_{11}Z_{22} - Z_{12}Z_{21}} \begin{bmatrix} Z_{22} & -Z_{12} \\ -Z_{21} & Z_{22} \end{bmatrix}$

- $[S] = ([z] + [U])^{-1}([z] - [U])$

- $[z] = ([U] + [S])([U] - [S])^{-1}$

- The reflection coefficient at port 1 of a two-port unmatched network can be expressed as follows: $\Gamma_1 = S_{11} + \dfrac{S_{12}S_{21}\Gamma_2}{1 - S_{22}\Gamma_2}$

- $\begin{bmatrix} A & B \\ C & D \end{bmatrix} = \begin{bmatrix} Z_{11}/Z_{21} & (Z_{11}Z_{22} - Z_{12}Z_{21})/Z_{21} \\ 1/Z_{21} & Z_{22}/Z_{21} \end{bmatrix}$

- $AD - BC = 1$

- $[Z] = \begin{bmatrix} A/C & (AD - BC)/C \\ 1/C & D/C \end{bmatrix}$

- $[Y] = \begin{bmatrix} D/B & -(AD - BC)/B \\ -1/B & A/B \end{bmatrix}$

Exercises

Objective-type Questions

7.1 The admittance parameter of a network relates
  (a) Port voltages with port currents
  (b) Input voltage and current with output voltage and current
  (c) Input voltages with output voltages
  (d) Input current with output current

7.2 If only the position of the port of a network is changed, then its scattering parameter undergoes a
(a) Change in the magnitude
(b) Change in the phase
(c) Change in both magnitude and phase
(d) No change

7.3 If the $[S]$ parameter of a network is known, then its normalized $[Z]$ parameter can be found using the formula
(a) $[z]=([U]+[S])([U]-[S])$
(b) $[z]=([U]+[S])([U]-[S])^*$
(c) $[z]=([U]+[S])([U]-[S])^{-1}$
(d) $[z]=([U]+[S])([U]+[S])^{-1}$

7.4 For a reciprocal network
(a) $[S]^t=[S]$
(b) $|S|=|U|+|S|$
(c) $[S]^t=-[S]$
(d) $[S]=[S]^*$

7.5 When the port of a network is mismatched, its reflection coefficient is
(a) Still equal to S_{11}
(b) Not equal to S_{11} but independent of Γ_2
(c) Not equal to S_{11} but dependent of Γ_2
(d) None of these

7.6 For the cascaded two-port network analysis, the most suitable network representation is
(a) $[Z]$ parameter
(b) $[Y]$ parameter
(c) $[S]$ parameter
(d) $[ABCD]$ parameter

7.7 For a two-port network to be reciprocal, which of the following relations holds?
(a) $z_{11}=z_{22}$
(b) $y_{21}=y_{12}$
(c) $z_{11}=z_{12}$
(d) $AD-BC=0$

7.8 The $[Z]$ parameters of a network are $\begin{bmatrix} 10 & 5 \\ 5 & 20 \end{bmatrix}$. Which one is not true?
(a) $A=2$ (b) $B=35$ (c) $C=0.2$ (d) $D=9$

7.9 The voltage transfer ratio of two-port networks connected in a cascade may conveniently be obtained from the
(a) Product of voltage transfer ratios of the two individual networks
(b) Sum of the individual Y matrices of the two networks
(c) Product of the $ABCD$ matrices of each network
(d) Subtraction of the S matrices of the two networks

7.10 The relation $AD-BC=1$, where A, B, C, and D are the elements of a transmission line matrix of a network, is valid for
(a) A passive and reciprocal network
(b) A passive and non-reciprocal network
(c) All passive networks
(d) All passive and active networks

7.11 An ideal transformer has a turn ratio of 2:1. Considering the high-voltage side as port 1 and low-voltage side as port 2, the $ABCD$ matrix parameter of the transformer is
(a) $\begin{bmatrix} 2 & 0 \\ 0 & 0.5 \end{bmatrix}$ (b) $\begin{bmatrix} 0.5 & 0 \\ 0 & 2 \end{bmatrix}$ (c) $\begin{bmatrix} 0 & 2 \\ 0.5 & 0 \end{bmatrix}$ (d) $\begin{bmatrix} 0 & 0.5 \\ 2 & 0 \end{bmatrix}$

7.12 In a two-port network containing linear bidirectional passive circuit elements, which one of the following conditions for $[Z]$ parameters will hold?
  (a) $Z_{11} = Z_{22}$
  (b) $Z_{11}Z_{22} = Z_{12}Z_{21}$
  (c) $Z_{11}Z_{12} = Z_{22}Z_{21}$
  (d) $Z_{12} = Z_{21}$

7.13 The impedance matrices of two two-port networks are given by $\begin{bmatrix} 3 & 2 \\ 2 & 4 \end{bmatrix}$ and $\begin{bmatrix} 1 & 3 \\ 3 & 2 \end{bmatrix}$.

If these two networks are connected in series, the impedance matrix of the resulting two-port network will be

  (a) $\begin{bmatrix} 4 & 5 \\ 5 & 6 \end{bmatrix}$
  (b) $\begin{bmatrix} 2 & -1 \\ -1 & 2 \end{bmatrix}$
  (c) $\begin{bmatrix} -2 & 2 \\ 2 & -2 \end{bmatrix}$
  (d) $\begin{bmatrix} 3 & 6 \\ 6 & 8 \end{bmatrix}$

7.14 For an ideal step-down transformer of turn ratio $N{:}1$, the $ABCD$ matrix is given as

  (a) $\begin{bmatrix} N & 1 \\ 1 & N \end{bmatrix}$
  (b) $\begin{bmatrix} N & 0 \\ 0 & N \end{bmatrix}$
  (c) $\begin{bmatrix} N & 0 \\ 0 & 1/N \end{bmatrix}$
  (d) $\begin{bmatrix} N & 1 \\ 1 & 1/N \end{bmatrix}$

7.15 Two two-port networks are connected in parallel. The combination is to be represented as a single two-port network. The parameters of this network are obtained by addition of the individual
  (a) Z parameters
  (b) Y parameters
  (c) S parameters
  (d) $ABCD$ parameters

Review Questions

7.1 Define the equivalent voltage, current, and impedance for a non-TEM line.

7.2 Define the impedance, admittance, scattering, and $ABCD$ parameters of an N-port network.

7.3 Find the properties of the scattering matrix for a reciprocal and lossless network.

7.4 Show that for a two-port network terminated with an unmatched load, the reflection coefficient is not the same as S_{11}.

7.5 Find the relation between $ABCD$ parameters and $[Z]$ and $[Y]$ parameters.

Problems

7.1 If $Z_{\text{SC}}^{(1)}$, $Z_{\text{SC}}^{(2)}$, $Z_{\text{OC}}^{(1)}$, and $Z_{\text{OC}}^{(2)}$ be the input impedance of a T-network when terminal 2 is short circuited, terminal 1 is short circuited, terminal 2 is open circuited, and terminal 1 is open circuited, respectively, then show that $Z_{12}^{(2)} = \left(Z_{\text{OC}}^{(1)} - Z_{\text{SC}}^{(1)} \right) Z_{\text{OC}}^{(2)} = \left(Z_{\text{OC}}^{(2)} - Z_{\text{SC}}^{(2)} \right) Z_{\text{OC}}^{(1)}$, $Z_{11} = Z_{\text{OC}}^{(1)}$, $Z_{22} = Z_{\text{OC}}^{(2)}$. *[Hint: Use the definition of Z parameters]*

7.2 Derive the $[Z]$ and $[Y]$ matrices for the networks in Fig. 7.6.

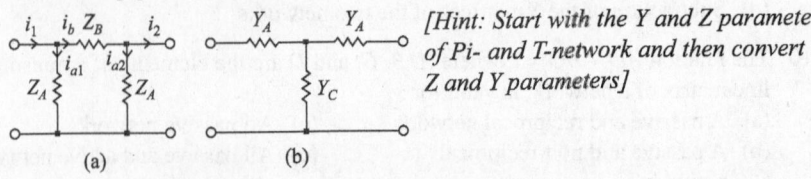

[Hint: Start with the Y and Z parameters of Pi- and T-network and then convert to Z and Y parameters]

(a)   (b)

Fig. 7.6

7.3 Two microwave networks having scattering matrices $\left[S^A \right]$ and $\left[S^B \right]$ are cascaded. Prove that for the cascaded network, the overall $S_{21} = S_{21}^A S_{21}^B \big/ \left(1 - S_{22}^A S_{11}^B \right)$. *[Hint: Start with the definition of S parameters]*

7.4 A four-port network has the following scattering matrix:

$$\begin{bmatrix} 0.3\angle 45° & 0.7\angle 20° & 0.6\angle 75° & 0.25\angle 45° \\ 0.7\angle 20° & 0.4\angle 30° & 0.5\angle 70° & 0.32\angle 60° \\ 0.6\angle 75° & 0.5\angle 70° & 0.25\angle 35° & 0.57\angle 25° \\ 0.25\angle 45° & 0.32\angle 60° & 0.57\angle 25° & 0.72\angle 60° \end{bmatrix}$$

Now if ports 1 and 3 are connected with a lossless matched transmission line of length 75°, find the modified scattering matrix. *[Hint: Use the result of Example 7.2]*

7.5 The scattering parameters of a two-port network is given as

$$\begin{bmatrix} 0.6\angle 45° & 0.8\angle 60° \\ 0.8\angle 60° & 0.6\angle 45° \end{bmatrix}$$

Find the normalized impedance matrix. *[Hint: Use Eq. (7.38)]*

7.6 Find the S parameters for the series and shunt load case in Fig. 7.7. Show that $S_{12} = 1 - S_{11}$ for the series case, whereas $S_{12} = 1 + S_{11}$ for the shunt case.

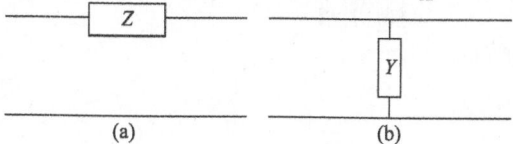

[Hint: Start with the ABCD matrix of the network]

Fig. 7.7

7.7 The scattering parameter of a two-port network is given as

$$\begin{bmatrix} 0.6\angle 45° & 0.8\angle 60° \\ 0.8\angle 60° & 0.6\angle 45° \end{bmatrix}$$

Find the angle by which port 1 must be shifted to make S_{12} or S_{21} real. *[Hint: Use the result of Example 7.2]*

7.8 A reciprocal transmission line network has a return loss of 4.64 dB and an insertion loss of 1.93 dB. Calculate the S parameters. *[Hint: S_{11} is the reflection coefficient, whereas S_{21} is the transmission coefficient]*

7.9 Find the $[Z]$ parameter of the networks shown in Fig. 7.8.

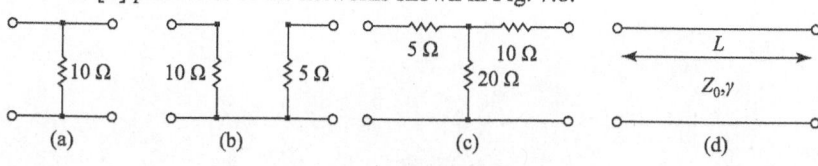

Fig. 7.8

[Hint: Use the definition of Z parameters]

7.10 Find the $[Y]$ parameter of the networks shown in Fig. 7.9.

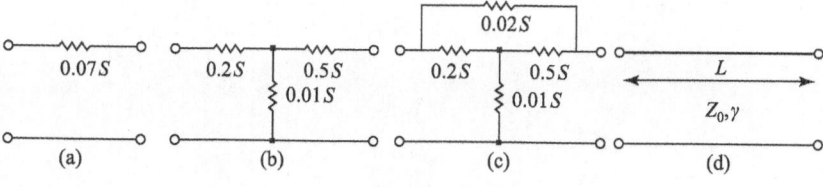

Fig. 7.9

[Hint: Use the definition of Y parameters]

7.11 Find the *ABCD* parameters of the networks shown in Fig. 7.10.

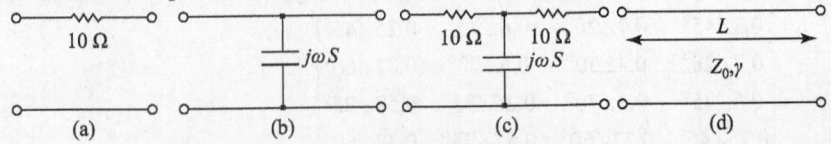

Fig. 7.10

[Hint: Use the definition of ABCD parameters]

7.12 Find the [*S*] parameter of the networks shown in Fig. 7.11.

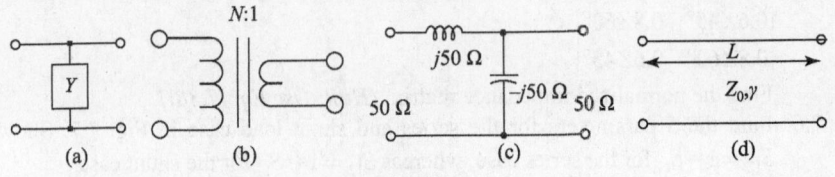

Fig. 7.11

[Hint: Use the definition of S parameters]

Microwave Power Dividers and Couplers

8.1 Introduction

In the previous chapter, we got familiarized with different types of microwave network representations. In this chapter, we will discuss about different types of microwave power dividers, hybrids, and couplers, which comprise an important part of a microwave system and are widely used to split or couple a required amount of electromagnetic energy in a transmission line. They are also widely used in antenna beam forming, connecting feeds to an antenna, monitoring and measuring signals, power feedback, etc. Different types of power dividers and couplers were invented to operate in the microwave range. However, out of them, T-junctions, hybrid junctions, rat-race junction, Wilkinson power divider, Bethe-hole directional couplers, Lange couplers, branch-line coupler, etc., are most well known and will be described in the following sections.

A large variety of waveguide power dividers, hybrids, and couplers, such as E- and H-plane T-junction, magic T-junction, Bethe-hole coupler, multi-hole directional coupler, and Schwinger coupler were invented and characterized during the Second World War at the MIT radiation laboratory and many of them are still used. Later during the development of planar circuit technology, many of these couplers were re-fabricated using strip and microstrip line technologies. A few new power dividers and couplers such as Wilkinson power dividers, branch-line hybrids, and coupled-line directional couplers were also designed.

Based on the number of ports available, power dividers and couplers can also be categorized as three-, four-, or multi-port networks. Waveguide E- and H-plane T-junctions have three ports and hence are called three-port networks, whereas magic T-junctions, rat-race junctions, Bethe-hole couplers, multi-hole directional couplers, Schwinger couplers, Wilkinson power dividers, branch-line hybrids, and coupled-line directional couplers are examples of four-port networks.

8.2 T-junctions

T-junctions are three-port power dividers or combiners that are used to split the power, fed into one designated port, between the other two ports or to combine the power fed into the two designated ports at the third port. Examples of different types of T-junctions are E- and H-plane T-junctions.

8.2.1 *E*-plane T-junctions

An E-plane T-junction, shown in Fig. 8.1(a), is a waveguide three port network having one main arm and one side arm. The side arm is parallel to the E-field of the main guide. Electric field distribution inside the network is shown in Fig. 8.1(b). The figure reveals that, when used as a power divider, the electric field fed into port 3 will be divided between ports 1 and 2 and will have opposite phases. Alternatively, when applied as a power combiner, the electric field fed into ports 1 and 2 will be subtracted at port 3. If port 3 is symmetric, then during power division the amplitude of the electric fields at ports 1 and 2 will be equal; hence, a matched E-plane T-junction can be used as a 3 dB power divider. Further, if the fields, fed into ports 1 and 2, are of the same amplitude and phase, then the output power at port 3 will be zero due to phase cancellation. The equivalent circuit of an E-plane T-junction is shown in Fig. 8.2. For a symmetric three-port E-plane T-junction $S_{11} = S_{22}$ and $S_{ij} = S_{ji}$ for $i \neq j$. Therefore,

(a)

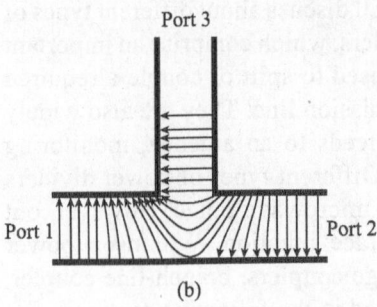

(b)

Fig. 8.1 *E*-plane T-junction network (a) Typical structure (b) Electric field distribution inside the network

$$[S] = \begin{bmatrix} S_{11} & S_{12} & S_{13} \\ S_{21} & S_{22} & S_{23} \\ S_{31} & S_{32} & S_{33} \end{bmatrix} = \begin{bmatrix} S_{11} & S_{12} & S_{13} \\ S_{12} & S_{11} & S_{23} \\ S_{13} & S_{23} & S_{33} \end{bmatrix}$$

Further, since $S_{13} = -S_{23}$,

$$[S] = \begin{bmatrix} S_{11} & S_{12} & S_{13} \\ S_{12} & S_{11} & S_{23} \\ S_{13} & S_{23} & S_{33} \end{bmatrix} = \begin{bmatrix} S_{11} & S_{12} & S_{13} \\ S_{12} & S_{11} & -S_{13} \\ S_{13} & -S_{13} & S_{33} \end{bmatrix}$$

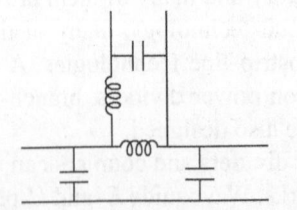

Fig. 8.2 Equivalent circuit of *E*-plane T-junction

(8.1)

Therefore, the S-matrix of an E-plane T-junction consists of nine elements, out of which four are unknown.

Note An E-plane T-junction behaves as a voltage series junction.

(a)

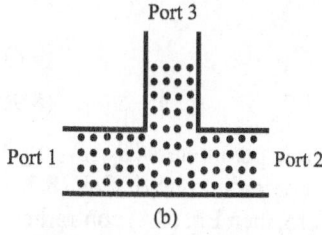

(b)

Fig. 8.3 *H*-plane T-junction network (a) Typical structure (b) Electric field distribution inside the network

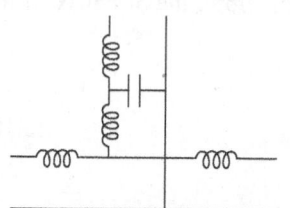

Fig. 8.4 Equivalent circuit of *H*-plane T-junction

8.2.2 *H*-plane T-junction

An *H*-plane T-junction is another type of waveguide T, with its side arm being parallel to the *H*-field of the main guide and is shown in Fig. 8.3(a). Electric field distribution inside the network is shown in Fig. 8.3(b). The figure reveals that, when used as a power divider, the electric field fed into port 3 will be divided between ports 1 and 2 and will have equal phases. Alternatively, when applied as a power combiner, the electric field fed into ports 1 and 2 will be combined at port 3. If port 3 is symmetric, then during power division the amplitude of the electric fields at ports 1 and 2 will be equal, and hence a matched *H*-plane T-junction can also be used as a 3 dB power divider. Further, if the fields fed into ports 1 and 2 are of the same amplitude but opposite phase, then the output power at port 3 will be zero due to phase cancellation. The equivalent circuit of an *H*-plane T-junction is shown in Fig. 8.4. For a symmetric three-port *H*-plane T-junction, $S_{11} = S_{22}$, $S_{ij} = S_{ji}$ for $i \neq j$ and $S_{13} = S_{23}$. Therefore,

$$[S] = \begin{bmatrix} S_{11} & S_{12} & S_{13} \\ S_{12} & S_{11} & S_{13} \\ S_{13} & S_{13} & S_{33} \end{bmatrix} \qquad (8.2)$$

Therefore, like the *S*-matrix of an *E*-plane T-junction, an *H*-plane T-junction also consists of nine elements, out of which four are unknown.

The characteristics of a T-junction can be described by three theorems, as follows:

1. A short circuit may be placed in one of the arms of a T-junction in such a way that no power can be transferred through the other two arms.
2. If the junction is symmetric about one of its arms, a short circuit can be placed in that arm so that no reflections occur in power transmission between the other two arms.
3. It is not possible to design a lossless, reciprocal T-junction having all its ports matched simultaneously.

Theorem 3 can be proved in the following way. If all the ports of a reciprocal *T*-network are matched, then

$$[S] = \begin{bmatrix} 0 & S_{12} & S_{13} \\ S_{12} & 0 & S_{23} \\ S_{13} & S_{23} & 0 \end{bmatrix} \qquad (8.3)$$

Further, if the network is lossless then the scattering matrix must be unitary. Therefore,

$$S_{13}^* S_{23} = 0 \tag{8.4}$$

$$S_{23}^* S_{12} = 0 \tag{8.5}$$

and $\quad S_{12}^* S_{13} = 0 \tag{8.6}$

In addition, conservation of energy in a lossless network establishes the following conditions:

$$|S_{12}|^2 + |S_{13}|^2 = 1 \tag{8.7}$$

$$|S_{12}|^2 + |S_{23}|^2 = 1 \tag{8.8}$$

$$|S_{13}|^2 + |S_{23}|^2 = 1 \tag{8.9}$$

Equations (8.4)–(8.6) show that at least two of the three parameters (S_{12}, S_{13}, S_{23}) must be zero. However, this conclusion will always contradict one of Eqs (8.7)–(8.9). For example, if we assume that S_{12} and S_{23} are zero, then Eq. (8.8) contradicts the conclusion. Therefore, a three-port, lossless, reciprocal network can never be designed with all its ports matched.

Now, we will check whether it is possible to design a lossless and reciprocal T-junction with two of its ports being matched, while the third is not matched. For this, let us assume that ports 1 and 2 are matched. Therefore, the S-matrix of the network can be represented as follows:

$$[S] = \begin{bmatrix} 0 & S_{12} & S_{13} \\ S_{12} & 0 & S_{23} \\ S_{13} & S_{23} & S_{33} \end{bmatrix} \tag{8.10}$$

Since the network is lossless, we can write the following equations:

$$S_{13}^* S_{23} = 0 \tag{8.11}$$

$$S_{12}^* S_{13} + S_{23}^* S_{33} = 0 \tag{8.12}$$

$$S_{23}^* S_{12} + S_{33}^* S_{13} = 0 \tag{8.13}$$

$$|S_{12}|^2 + |S_{13}|^2 = 1 \tag{8.14}$$

$$|S_{12}|^2 + |S_{23}|^2 = 1 \tag{8.15}$$

$$|S_{13}|^2 + |S_{23}|^2 + |S_{33}|^2 = 1 \tag{8.16}$$

Equations (8.14) and (8.15) reveal that

$$|S_{13}| = |S_{23}| \tag{8.17}$$

Furthermore, application of the condition of Eq. (8.11) in Eq. (8.17) provides the following relation:

$$S_{13} = S_{23} = 0 \tag{8.18}$$

Substituting Eq. (8.18) back into Eqs (8.14)–(8.16) gives the following relation:

$$|S_{12}| = |S_{33}| = 1 \tag{8.19}$$

Equations (8.17)–(8.19) ensure that the network will have a scattering matrix of the following form:

$$[S] = \begin{bmatrix} 0 & e^{j\theta} & 0 \\ e^{j\theta} & 0 & 0 \\ 0 & 0 & e^{j\phi} \end{bmatrix} \tag{8.20}$$

Therefore, it is possible to design a three-port reciprocal and lossless network with any of its two ports matched.

It is also possible to design a three-port network with all its ports matched, provided the network is lossless but not reciprocal. For such a case, the S-matrix can be represented as follows:

$$[S] = \begin{bmatrix} 0 & S_{12} & S_{13} \\ S_{21} & 0 & S_{23} \\ S_{31} & S_{32} & 0 \end{bmatrix} \tag{8.21}$$

Since the network is still lossless, we can write the following equations:

$$S_{31}^{*}S_{32} = 0 \tag{8.22}$$
$$S_{21}^{*}S_{23} = 0 \tag{8.23}$$
$$S_{12}^{*}S_{13} = 0 \tag{8.24}$$
$$|S_{12}|^2 + |S_{13}|^2 = 1 \tag{8.25}$$
$$|S_{21}|^2 + |S_{23}|^2 = 1 \tag{8.26}$$
$$|S_{31}|^2 + |S_{32}|^2 = 1 \tag{8.27}$$

Equations (8.22)–(8.27) can be satisfied simultaneously if any one of the following two conditions is satisfied:

$$S_{12} = S_{23} = S_{31} = 0, \quad |S_{21}| = |S_{32}| = |S_{13}| = 1 \tag{8.28}$$
$$S_{21} = S_{32} = S_{13} = 0, \quad |S_{12}| = |S_{23}| = |S_{31}| = 1 \tag{8.29}$$

Equations (8.28) and (8.29) result in the following scattering matrices:

$$[S] = \begin{bmatrix} 0 & 0 & 1 \\ 1 & 0 & 0 \\ 0 & 1 & 0 \end{bmatrix} \tag{8.30}$$

$$[S] = \begin{bmatrix} 0 & 1 & 0 \\ 0 & 0 & 1 \\ 1 & 0 & 0 \end{bmatrix} \tag{8.31}$$

Equations (8.30) and (8.31) reveal the $[S]$ matrix of a three-port circulator.

It can be proved that a three-port network with all its ports matched can be designed if the network is lossy. To demonstrate this, let us consider the three-port resistive power divider network shown in Fig. 8.5.

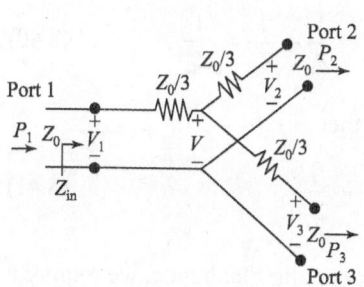

Fig. 8.5 Three-port resistive power divider

The input impedance of each of the output lines is as follows:

$$Z = \frac{Z_0}{3} + Z_0 = \frac{4Z_0}{3} \tag{8.32}$$

Therefore, the input impedance of the input line is given as follows:

$$Z_{in} = \frac{Z_0}{3} + \frac{2Z_0}{3} = Z_0 \tag{8.33}$$

Hence, $S_{11} = 0$ (8.34)

Since the network is symmetric from all the three ports,

$$S_{11} = S_{22} = S_{33} = 0 \tag{8.35}$$

If the voltage at port 1 is V_1, then using the voltage division rule, the following relation is obtained:

$$V = V_1 \frac{(2Z_0/3)}{(Z_0/3) + (2Z_0/3)} = \frac{2V_1}{3} \tag{8.36}$$

Again by the voltage division rule:

$$V_2 = V_3 = V \frac{Z_0}{Z_0 + (Z_0/3)} = \frac{3V}{4} = \frac{3}{4} \times \frac{2V_1}{3} = \frac{V_1}{2} \tag{8.37}$$

Therefore, $S_{21} = S_{31} = S_{23} = 1/2$ (8.38)

Thus, the complete S-matrix is as follows:

$$[S] = \frac{1}{2} \begin{bmatrix} 0 & 1 & 1 \\ 1 & 0 & 1 \\ 1 & 1 & 0 \end{bmatrix} \tag{8.39}$$

Equation (8.39) represents the S-matrix of a 6-dB power divider.

In contrast to a lossy transmission line, which incorporates resistances, a lossless T-junction can be modelled as a junction of three transmission lines with a lumped susceptance B at the junction, as shown in Fig. 8.6. The lumped susceptance incorporates the net energy stored in the higher-order modes, generated due to the discontinuity at the junction.

If the input port of the divider is matched, then

$$Y_{in} = jB + \frac{1}{Z_1} + \frac{1}{Z_2} = \frac{1}{Z_0} \tag{8.40}$$

If B is negligible, then

$$\frac{1}{Z_0} = \frac{1}{Z_1} + \frac{1}{Z_2} \tag{8.41}$$

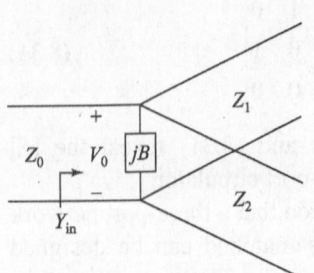

Fig. 8.6 Transmission line model of lossless T-network

In practice, B is not negligible; hence, we require a reactive tuning element, such as an iris, a post, or a screw, to match the input port.

Note An *H*-plane T-junction is also called a shunt junction.

EXAMPLE 8.1 A lossless T-junction has an input impedance of 50 Ω. Find the output characteristic impedances for an input power division ratio of 2:3. Compute the reflection coefficient at the ports.

Solution Given: $Z_0 = 50\ \Omega$ and $P_1 : P_2 = 2 : 3$

If the input voltage is V_0, then the input power is expressed as follows:

$$P_{in} = \frac{V_0^2}{2Z_0}$$

Output powers are as follows:

$$P_1 = \frac{V_0^2}{2Z_1} = \frac{2}{5}P_{in} = \frac{2}{5}\frac{V_0^2}{2Z_0} = \frac{V_0^2}{5Z_0}$$

or
$$Z_1 = \frac{5}{2}Z_0 = \frac{5}{2} \times 50 = 125\ \Omega \text{ and } P_2 = \frac{V_0^2}{2Z_2} = \frac{3}{5}P_{in} = \frac{3V_0^2}{10Z_0}$$

or
$$Z_2 = \frac{5Z_0}{3} = \frac{5}{3} \times 50 = 83.3333\ \Omega$$

The input impedance of the junction is as follows:

$$Z_{in} = 125\ \|\ 83.3333 = \frac{125 \times 83.3333}{125 + 83.3333} = 50\ \Omega$$

Therefore, the junction is matched with the input port, and hence the reflection coefficient is zero. Looking towards the 125 Ω line, the input impedance is expressed as follows:

$$Z_{in1} = 83.33\ \|\ 50 = \frac{83.33 \times 50}{83.33 + 50} = 31.25\ \Omega$$

Therefore, the reflection coefficient is as follows:

$$\Gamma_1 = \frac{31.25 - 125}{31.25 + 125} = -0.6$$

Looking towards the 83.3333 Ω line, the input impedance can be written as follows:

$$Z_{in2} = 125\ \|\ 50 = \frac{125 \times 50}{125 + 50} = 35.7143\ \Omega$$

Therefore, the reflection coefficient is expressed as follows:

$$\Gamma_1 = \frac{35.7143 - 83.3333}{35.7143 + 83.3333} = -0.4$$

Practice Problem

8.1 A lossless T-junction has an input impedance of 100 Ω. Find the output characteristic impedances for an input power division ratio of 1:2. Compute the reflection coefficient at the output ports. $[300\ \Omega,\ 150\ \Omega,\ 0,\ -0.6667,\ -0.3333]$

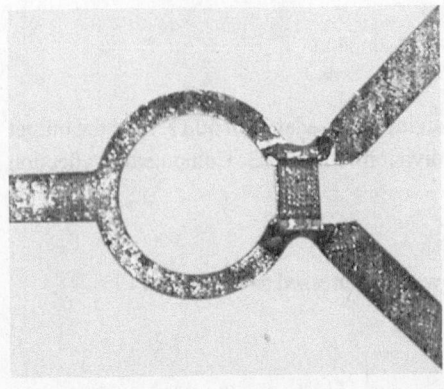

(a)

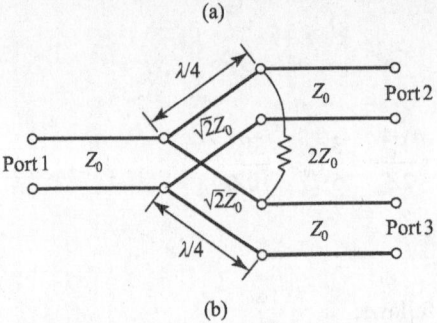

(b)

Fig. 8.7 Wilkinson power divider (a) Typical structure (b) Its transmission line equivalent circuit

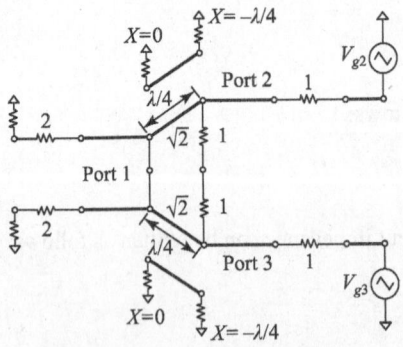

Fig. 8.8 Equivalent circuit of Wilkinson power divider with voltage generators at output ports and normalized impedances

8.3 Wilkinson Power Dividers

Previous discussion shows that a general three-port reciprocal, lossless T-junction cannot be matched at all its ports. This problem of matching can be avoided by using a resistive power divider network, however at the cost of poor isolation between its output ports. A Wilkinson power divider addresses the problems of both isolation and matching, and provides isolation between the output ports as well as matching. A Wilkinson power divider is generally designed in a microstrip or a strip line form for both equal and unequal power division. An equal-split Wilkinson power divider is shown in Fig. 8.7(a). Its equivalent transmission line model is shown in Fig. 8.7(b).

The best way to analyse a Wilkinson power divider is through the even mode–odd mode analysis. For this, we will first redraw Fig. 8.7(b), with voltage generators at the output ports and all impedances normalized to the characteristic impedance, as shown in Fig. 8.8. In the figure, two normalized source resistors of value 2 have been kept in parallel to give an equivalent normalized source resistance of value 1.

For the even mode $V_{g2} = V_{g3} = 2V$ and hence $V_{e2} = V_{e3}$. The last condition implies that no current flows through the shunt resistor, and hence Fig. 8.8 modifies to Fig. 8.9(a). Figure 8.9(a) reveals that for an even mode both the output ports are matched to the input port and thus results in a 3 dB power division. Further, output ports are also isolated.

The voltage at any point x on the transmission line is given by the following equation:

$$V(x) = V_+ \left(e^{-j\beta x} + \Gamma e^{j\beta x} \right) \qquad (8.42)$$

If we assume that port 1 is located at $x = 0$ and port 2/3 at $x = -\lambda/4$, then

$$V_{e2} = V\left(\frac{-\lambda}{4}\right) = V_+\left(e^{-j\left(\frac{2\pi}{\lambda}\right)\left(\frac{-\lambda}{4}\right)} + \Gamma e^{j\left(\frac{2\pi}{\lambda}\right)\left(\frac{-\lambda}{4}\right)}\right)$$

or $\qquad V_{e2} = V_+\left(e^{j\pi/2} + \Gamma e^{-j\pi/2}\right) = jV_+\left(1 - \Gamma\right)$ (8.43)

and $\qquad V_{e1} = V(0) = V_+\left(e^0 + \Gamma e^0\right) = V_+\left(1 + \Gamma\right)$ (8.44)

Furthermore, since port 2 is matched with port 1, $V_{e2} = V$ (8.45)

Equating Eqs (8.43) and (8.45), we get $jV_+\left(1 - \Gamma\right) = V$ (8.46)

Substituting the expression of V_+ from Eq. (8.46) into Eq. (8.44), we get the following relation:

$$V_{e1} = V_+\left(1 + \Gamma\right) = \frac{V}{j(1-\Gamma)}\left(1 + \Gamma\right) = -jV\frac{1+\Gamma}{1-\Gamma} = jV\frac{\Gamma+1}{\Gamma-1}$$ (8.47)

where $\qquad \Gamma = \dfrac{2 - \sqrt{2}}{2 + \sqrt{2}}$ (8.48)

Substituting Γ from Eq. (8.48) into Eq. (8.47), we get the following equation:

$$V_{e1} = jV\frac{\Gamma+1}{\Gamma-1} = jV\left(\frac{2-\sqrt{2}}{2+\sqrt{2}} + 1\right)\bigg/\left(\frac{2-\sqrt{2}}{2+\sqrt{2}} - 1\right)$$

or $\qquad V_{e1} = -jV\dfrac{4}{2\sqrt{2}} = -jV\sqrt{2}$ (8.49)

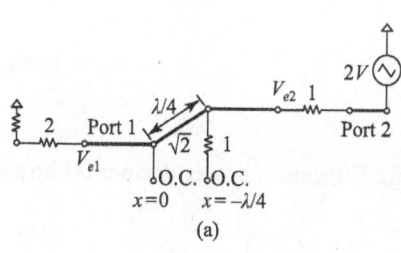

(a)

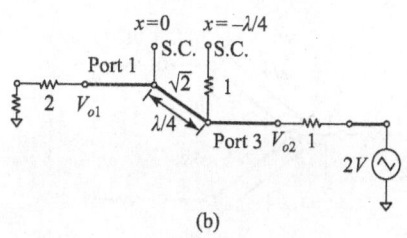

(b)

Fig. 8.9 Equivalent excitation circuit of Wilkinson power divider (a) Even mode (b) Odd mode

Now let us consider the odd-mode excitation. For this, $V_{g2} = -V_{g3} = 2V$ and $V_{o2} = -V_{o3}$. Therefore, a voltage null exists at the middle of the circuit. For the odd mode, Fig. 8.8, therefore, modifies to Fig. 8.9(b). The short circuit, existing at port 1, will be transformed into an open circuit at port 2 or 3. Therefore, no current will pass from port 2 or 3 to port 1. This implies that ports 2 and 3 are also matched. Due to the existence of the short circuit, they are also isolated. Using the voltage division rule, we can write the following equality:

$$V_{o2} = V$$ (8.50)

Further, since port 1 is short circuited, $V_{o1} = 0$ (8.51)

Therefore, using the property of a reciprocal network and the superposition principle, we can write the following relation:

$$S_{12} = S_{21} = \frac{V_{e1} + V_{o1}}{V_{e2} + V_{o2}} = \frac{-jV\sqrt{2} + 0}{V + V} = -j/\sqrt{2} \qquad (8.52)$$

Further due to symmetry of ports 2 and 3, we can write the following equation:

$$S_{13} = S_{31} = -j/\sqrt{2} \qquad (8.53)$$

Again since ports 2 and 3 are matched, for both even and odd modes, the following relation holds: $S_{22} = S_{33} = 0$ $\qquad (8.54)$

To find S_{11}, let us match both port 2 and port 3. Figure 8.8, therefore, modifies to Fig. 8.10(a). Due to symmetry, $V_2 = V_3$ and hence no current will flow through the shunt resistor. This implies that we can remove the shunt resistor from the equivalent circuit of Fig. 8.10(a). The resultant circuit modifies to Fig. 8.10(b). The final figure shows that port 1 is connected with two parallel quarter-wave transformers terminated in unity normalized load impedance. The input impedance at port 1, looking towards the outputs, is therefore expressed as follows:

$$\frac{1}{Z_{in1}} = \frac{2}{\dfrac{\sqrt{2}Z_0^2 + j2Z_0^2 \tan(\pi/2)}{\sqrt{2}Z_0 + jZ_0 \tan(\pi/2)}} \approx \frac{2}{\dfrac{j2Z_0^2 \tan(\pi/2)}{jZ_0 \tan(\pi/2)}} = \frac{j2Z_0 \tan(\pi/2)}{j2Z_0^2 \tan(\pi/2)} = \frac{1}{Z_0}$$

or $\qquad Z_{in1} = Z_0$ $\qquad\qquad (8.55)$

Therefore, $S_{11} = 0$ $\qquad\qquad (8.56)$

Thus, the complete S-matrix can be written as follows:

$$[S] = \begin{bmatrix} 0 & -j/\sqrt{2} & -j/\sqrt{2} \\ -j/\sqrt{2} & 0 & -j/\sqrt{2} \\ -j/\sqrt{2} & -j/\sqrt{2} & 0 \end{bmatrix} \qquad (8.57)$$

The lumped element equivalent circuit of a Wilkinson power divider is shown in Fig. 8.11.

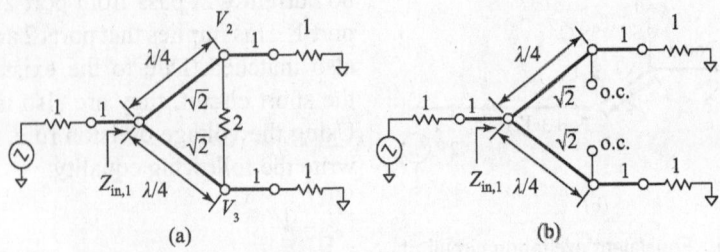

Fig. 8.10 Equivalent circuit of Wilkinson power divider
(a) Considering the shunt resistor (b) Removing the shunt resistor

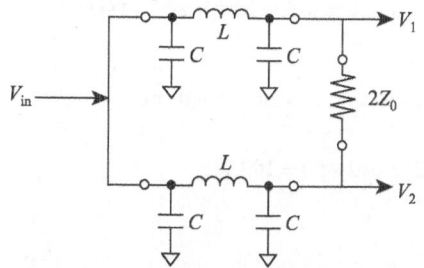

Fig. 8.11 Lumped element equivalent circuit of a Wilkinson power divider

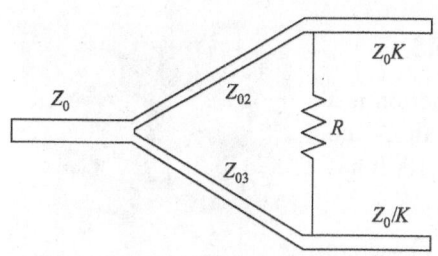

Fig. 8.12 Wilkinson power divider for unequal power division

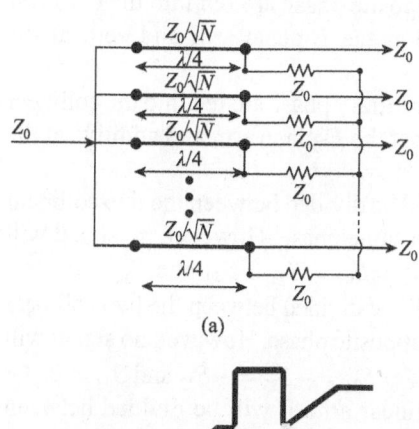

(a)

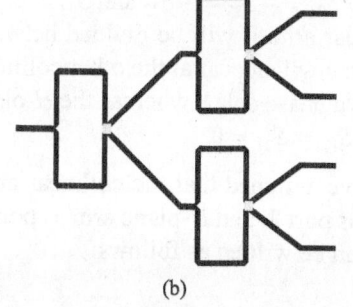

(b)

Fig. 8.13 *N*-port Wilkinson power divider
(a) *N*-way equal-split power divider
(b) Four-port corporate Wilkinson power divider using three Wilkinson power dividers

For unequal power division, circuit parameters of a Wilkinson power divider must be modified. A modified circuit is shown in Fig. 8.12.

If P_2 and P_3 are the powers at ports 2 and 3, respectively, then

$$K = \sqrt{P_3/P_2} \qquad (8.58)$$

$$Z_{03} = Z_0 \sqrt{\left(1+K^2\right)\big/K^3} \qquad (8.59)$$

$$Z_{02} = K^2 Z_{03} = Z_0 \sqrt{K\left(1+K^2\right)} \quad (8.60)$$

$$\text{and } R = Z_0 \left(1+K^2\right)\big/K \qquad (8.61)$$

Wilkinson power dividers can also be used for *N*-port power division, with all ports being matched and all output ports isolated from each other, as shown in Fig. 8.13(a). However, if $N > 10$, the required quarter-wave matching line impedance grows to a value that is difficult to implement. In addition, for $N > 2$, connection of the bridging resistors gets complex. A better procedure for *N*-port power division is shown in Fig. 8.13(b).

Wilkinson power dividers of wider bandwidths can also be designed using multi-section quarter-wavelength matching transformers and bridge resistors, as shown in Fig. 8.14. In practice, however, the matching sections are neither straight nor uniform.

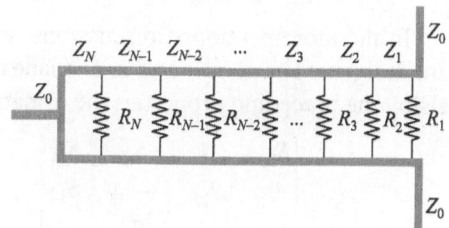

Fig. 8.14 Broad-bandwidth Wilkinson power divider

EXAMPLE 8.2   Design an equal-split Wilkinson power divider for a 50 Ω system.

Solution   Given: $Z_0 = 50\ \Omega$

The characteristic impedance of the quarter-wave transformer will be $Z = \sqrt{2}Z_0 = 50\sqrt{2} = 70.7107\ \Omega.$

The value of the shunt resistor will be $R = 2Z_0 = 2 \times 50 = 100\ \Omega.$

Practice Problem

8.2 Design an equal-split Wilkinson power divider for a 100 Ω system.

$$[141.4214\ \Omega, 200\ \Omega]$$

8.4 Waveguide Magic T-junctions

A waveguide magic, or *E–H* Plane, T-junction is a four-port device that is a combination of an *E*- and an *H*-plane T-junction, as shown in Fig. 8.15. It has several interesting characteristics:

Fig. 8.15   Magic T-junction

1. If two waves of equal magnitude and phase are fed into the collinear arms, then the output will be zero at the *E*-plane arm and additive at the *H*-plane arm.
2. If two waves of equal magnitude but opposite phase are fed into the collinear arms, then the output will be additive at the *E*-plane arm and zero at the *H*-plane arm.
3. If two waves of unequal magnitude and arbitrary phase are fed into the collinear arms, then the output will be subtractive at the *E*-plane arm and additive at the *H*-plane arm.
4. If signal is fed into the *H*-plane arm, it will be divided between the two collinear arms and will have equal magnitude and same phase. However, no signal will be coupled in the *E*-plane arm, that is, $S_{14} = S_{41} = S_{24} = S_{42}$ and $S_{34} = 0.$
5. If signal is fed into the *E*-plane arm, it will be divided between the two collinear arms and will have equal magnitude and opposite phase. However, no signal will be coupled in the *H*-plane arm, that is, $S_{13} = S_{31} = -S_{23} = -S_{32}$ and $S_{43} = 0.$
6. If signal is fed into only one of its collinear arm, it will be divided between the *E*- and *H*-plane arms. However, no signal will appear at the other collinear arm, because the *E*-plane arm will produce a phase delay, whereas the *H*-plane arm will produce a phase advance, that is, $S_{12} = S_{21} = 0.$

In the aforementioned discussions, we have assumed that the collinear arms are designated as ports 1 and 2, *E*- plane arm as port 3, and *H*-plane arm as port 4. Using the preceding properties the *S*-matrix can be written as follows:

$$[S] = \begin{bmatrix} S_{11} & 0 & S_{13} & S_{14} \\ 0 & S_{22} & -S_{13} & S_{14} \\ S_{13} & -S_{13} & S_{33} & 0 \\ S_{14} & S_{14} & 0 & S_{44} \end{bmatrix} \qquad (8.62)$$

The E- and H-plane arms of a magic T-junction can easily be matched by placing suitable matching elements in the respective arms, without destroying the symmetry of the junction. In such cases, $S_{33} = S_{44} = 0$ and Eq. (8.62) modifies to the following form:

$$[S] = \begin{bmatrix} S_{11} & 0 & S_{13} & S_{14} \\ 0 & S_{22} & -S_{13} & S_{14} \\ S_{13} & -S_{13} & 0 & 0 \\ S_{14} & S_{14} & 0 & 0 \end{bmatrix} \tag{8.63}$$

Further, using the lossless property of the network, we get the following equality:

$$2|S_{13}|^2 = 1 \text{ or } |S_{13}| = 1/\sqrt{2} \tag{8.64}$$

Similarly, $|S_{14}| = 1/\sqrt{2}$ (8.65)

Furthermore, $|S_{11}|^2 + |S_{13}|^2 + |S_{14}|^2 = |S_{11}|^2 + 1 = 1$

or $\qquad S_{11} = 0$ (8.66)

Similarly, it can be shown that $S_{22} = 0$ (8.67)

By a proper choice of the reference plane, S_{13} and S_{14} can be made real. In such a case, the S-matrix in Eq. (8.63) modifies to the following form:

$$[S] = \frac{1}{\sqrt{2}} \begin{bmatrix} 0 & 0 & 1 & 1 \\ 0 & 0 & -1 & 1 \\ 1 & -1 & 0 & 0 \\ 1 & 1 & 0 & 0 \end{bmatrix} \tag{8.68}$$

Magic T-junctions are widely used for impedance measurement, as duplexers and mixers, for doubling the feed power of antenna circulators, and as monopulse comparators.

Duplexer The basic setup for the use of a magic T-junction as a duplexer is shown in Fig. 8.16. Here, the transmitter and the receiver are connected at the collinear arms, whereas the antenna is connected at the E-plane arm. A matched load is connected at the H-plane arm. The wave from the transmitter will be coupled into the E- and H-plane arms. The power coupled to the H-plane arm will be absorbed by the matched load, and the power fed into the antenna will be radiated. During reception, the power received by the antenna will be divided equally between the two collinear arms, and hence half of it will be coupled to the receiver.

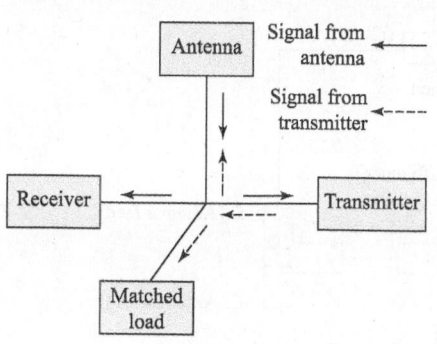

Fig. 8.16 Use of magic T-junction as duplexer

Mixer The basic setup for the use of a magic T-junction as a mixer is shown in Fig. 8.17. The antenna and the local

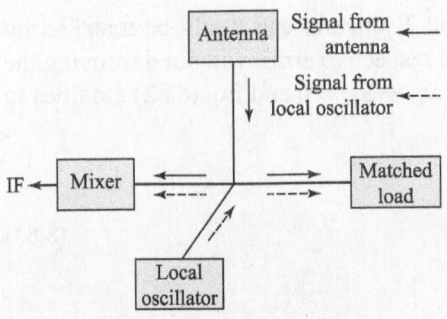

Fig. 8.17 Use of magic T-junction as mixer

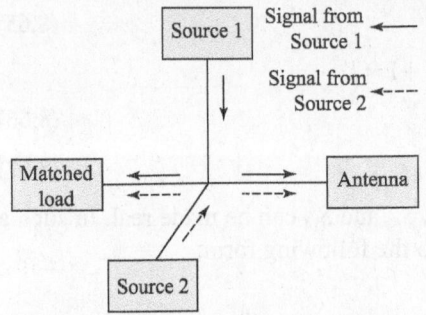

Fig. 8.18 Use of magic T-junction to double feed power

oscillators are connected at the E- and H-plane arms, respectively, while a matched load is connected to one of the collinear arms. The output is obtained from the other collinear arm. The field received by the antenna is mixed with the field from the local oscillator and produces an intermediate frequency.

Power enhancer The basic setup for the use of a magic T-junction to double the feed power of the antenna is shown in Fig. 8.18. Transmitters are connected to the E- and H-plane arms, while the antenna is connected to one of the collinear arms. The other collinear arm is matched terminated. Waves from the transmitters will be coupled either in phase or out of phase in the collinear arms. The collinear arm in which the waves are coupled in phase is connected with the antenna, and hence the feed power gets doubled. In the other arm, the waves are coupled out of phase and hence cancel each other. A matched load is placed in this arm to absorb the residue, if any.

Monopulse comparator A monopulse comparator is an eight-port junction with four input ports and four output ports, and consists of four magic T-junctions, as shown in Fig. 8.19. Waves from four antennas A, B, C, and D are fed into the

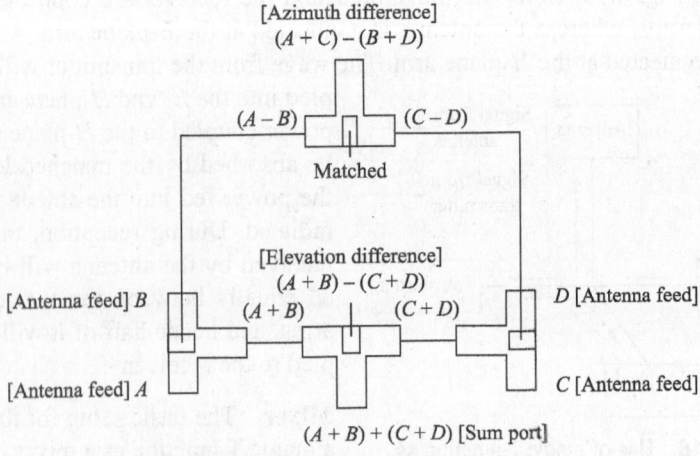

Fig. 8.19 Basic monopulse comparator

four collinear arms of two magic T-junctions. The E-plane arm of the first magic T-junction will produce an output of $(A-B)$, whereas the H-plane arm of the same magic T-junction will produce an output of $(A+B)$. Similarly, the E- and H-plane arms of the second magic T-junction will produce outputs $(C-D)$ and $(C+D)$, respectively. Outputs from both the E-plane arms are fed into the collinear arms of a third magic T-junction, whereas those from both the H-plane arms are fed into the collinear arms of a fourth magic T-junction. Thus, the H-plane arm of the third comparator produces an output of $[(A+C)-(B+D)]$, which gives the azimuth difference. Similarly, the E-plane arm of the fourth comparator will produce an output of $[(A+B)-(C+D)]$, which gives the elevation difference. The E-plane arm of the third magic T-junction is not used and usually kept matched terminated. The H-plane arm of the fourth magic T-junction provides the output of $[(A+B)+(C+D)]$, and is known as the sum port. During transmission, power is fed through this port so that it is divided equally among the antennas A, B, C, and D at the same phase.

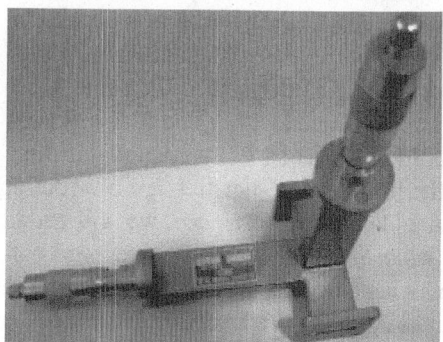

Fig. 8.20 *E–H* tuner

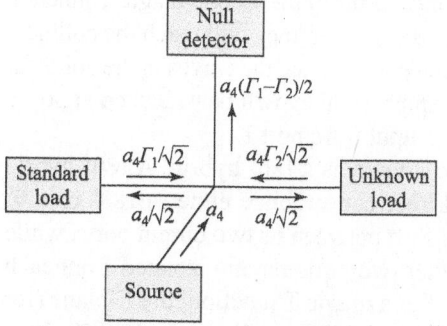

Fig. 8.21 Unknown impedance measurement using magic T-junction

E–H tuner   In an *E–H* tuner, both the E- and the H-plane arm are terminated by movable shorts, which act as E-and H-plane stubs. By adjusting the stub position, a wide range of load impedance, connected at the collinear arms, can be matched. A typical *E–H* plane tuner is shown in Fig. 8.20.

Impedance calculator   The basic set-up for impedance measurement using a magic T-junction is shown in Fig. 8.21. A load with unknown impedance is connected at one of the collinear arms, whereas the standard variable impedance is connected at the other collinear arm. The H-plane arm is connected to the source, and a null detector is placed at the E-plane arm. Since the source is in the H-plane arm, the field from it will be divided between both the collinear arms in the same phase; that is, if the amplitude of the wave in the H-plane arm is a_4, then the field travelling towards the respective loads at the collinear arms will be $a_4/\sqrt{2}$. These waves become the incident waves to the unknown load and the standard variable load. If the loads are different from the matched load, then a part of the incident waves will be reflected. If Γ_1 and Γ_2, respectively, be the reflection coefficients of the standard and the unknown load, then the fields reflected by them can, respectively, be expressed as $a_4\Gamma_1/\sqrt{2}$ and $a_4\Gamma_2/\sqrt{2}$. These reflected waves will travel towards the junction, and a part of them will be coupled to the E-plane arm. The wave in the E-plane arm will have an amplitude of $a_4 (\Gamma_1 - \Gamma_2)/2$. Now, if we adjust the standard load in such a way that $\Gamma_1 = \Gamma_2$, then the null detector in the E-plane arm will show a null. However, the relation $\Gamma_1 = \Gamma_2$ is true only when the standard load impedance is equal to the unknown load impedance. Since the value of the standard load for which a null has been obtained is known, we can find the value of the unknown load easily.

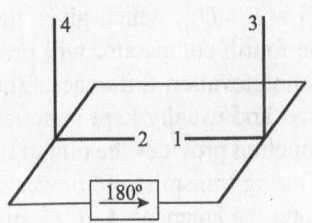

Fig. 8.22 Four-port circulator using magic T-junctions

Four-port circulator The basic configuration of a four-port circulator using magic T-junction is shown in Fig. 8.22. It requires two magic T-junctions and a gyrator. Wave from port 1 will be divided in the same phase between the collinear arms of the first magic T-junction. These waves will reach the collinear arms of the second magic T-junction at the same phase and will be added at port 2. Since they will be subtracted at port 4, no output will be obtained from port 4.

On the other hand, the power fed into port 2 will be divided in the same phase between the collinear arms of the second magic T-junction, but due to the gyrator, they will be out of phase on reaching the collinear arms of the first magic T-junction. Since the waves in the collinear arms of the first magic T-junction are now out of phase, they will be cancelled at port 1 and added at port 3. Hence, we will get an output from port 3. The wave fed at port 3 will be divided at out of phase in the collinear arms of the first magic T-junction and will also reach the collinear arms of the second magic T-junction at out of phase. Since the waves in the collinear arms of the second magic T-junction are now out of phase, they will be cancelled at port 2 and added at port 4. Hence, we will now get an output from port 4. Finally, the waves fed into port 4 will be divided between the collinear arms of the second magic T-junction and will be at out of phase; however, due to the gyrator, they will reach the collinear arms of the first magic T-junction at the same phase. Since the waves in the collinear arms of the first magic T-junction are now in phase, they will be cancelled at port 3 and added at port 1. Hence, we will get an output from port 1.

A waveguide magic T-junction is also known as a 180° hybrid. A 180° hybrid junction is basically a four-port network that can produce either a 180° or a 0°

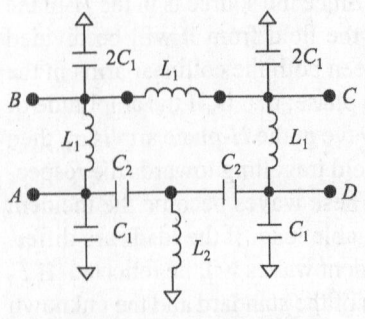

Fig. 8.23 Lumped element equivalent circuit of 180° hybrid

phase shift between its two output ports, while the other two ports remain isolated from each other. For a magic T-junction, the E-plane (for a 180° phase shift) and H-plane (for a 0° phase shift) arms are the input ports, whereas the collinear arms are the output ports. When used as a combiner, the E-plane arm is also called a Δ port, whereas the H-plane port is called a Σ port, as signals fed into the collinear arms subtract in the E-plane arm and add in the H-plane arm. The lumped element equivalent circuit of a 180° hybrid is shown in Fig. 8.23.

EXAMPLE 8.3 A magic T-junction is terminated at collinear ports 1 and 2 and difference port 3 by impedances of reflection coefficients $\Gamma_1 = 0.2$, $\Gamma_2 = 0.2$, and $\Gamma_3 = 0.3$, respectively. If 1 W power is fed into the sum port 4, calculate the power reflected from port 4 and that transmitted to the other three ports.

Solution   Given: $\Gamma_1 = 0.2$, $\Gamma_2 = 0.2$, $\Gamma_3 = 0.3$, and $P_4 = 1$ W

If a_1, a_2, a_3, and a_4 be the incident voltages and b_1, b_2, b_3, and b_4 be the corresponding output voltages at ports 1, 2, 3, and 4, respectively, then

$$a_1 = \Gamma_1 b_1 = 0.2b_1,\ a_2 = \Gamma_2 b_2 = 0.2b_2,$$
$$a_3 = \Gamma_3 b_3 = 0.3b_3,\ \text{and}\ a_4 = \sqrt{P_4} = 1$$

Therefore,

$$\begin{bmatrix} b_1 \\ b_2 \\ b_3 \\ b_4 \end{bmatrix} = \frac{1}{\sqrt{2}} \begin{bmatrix} 0 & 0 & 1 & 1 \\ 0 & 0 & -1 & 1 \\ 1 & -1 & 0 & 0 \\ 1 & 1 & 0 & 0 \end{bmatrix} \begin{bmatrix} a_1 \\ a_2 \\ a_3 \\ a_4 \end{bmatrix} = \frac{1}{\sqrt{2}} \begin{bmatrix} 0 & 0 & 1 & 1 \\ 0 & 0 & -1 & 1 \\ 1 & -1 & 0 & 0 \\ 1 & 1 & 0 & 0 \end{bmatrix} \begin{bmatrix} 0.2b_1 \\ 0.2b_2 \\ 0.3b_3 \\ 1.0 \end{bmatrix}$$

Hence,

$$b_1 = 0.7071 + 0.2121b_3 \tag{8.69}$$
$$b_2 = 0.7071 - 0.2121b_3 \tag{8.70}$$
$$b_3 = 0.1414b_1 - 0.1414b_2 \tag{8.71}$$

and $$b_4 = 0.1414b_1 + 0.1414b_2 \tag{8.72}$$

Substituting Eqs (8.69) and (8.70) into Eq. (8.71), we get the following expression:

$$b_3 = 0.1414b_1 - 0.1414b_2$$

or $$b_3 = 0.1414(0.7071 + 0.2121b_3) - 0.1414(0.7071 - 0.2121b_3)$$
or $$b_3 = 0 \tag{8.73}$$

Substituting Eq. (8.73) into Eqs (8.69) and (8.70), we get the following values:

$$b_1 = 0.7071 \tag{8.74}$$
$$b_2 = 0.7071 \tag{8.75}$$

Substituting Eqs (8.74) and (8.75) into Eq. (8.72), we get the following relation:

$$b_4 = 0.1414b_1 + 0.2121b_2 = 0.1414 \times 0.7071 + 0.2121 \times 0.7071 = 0.25$$

Therefore,

Power transmitted to port 1: $|b_1|^2 = 0.7071^2 = 0.5$ W

Power transmitted to port 2: $|b_2|^2 = 0.7071^2 = 0.5$ W

Power transmitted to port 3: $|b_3|^2 = 0$

Power reflected from port 4: $|b_4|^2 = 0.25^2 = 0.06$ W

Practice Problem

8.3 A magic T-junction is terminated at collinear ports 1 and 2 and difference port 3 by impedances of reflection coefficients $\Gamma_1 = 0.2$, $\Gamma_2 = 0.3$, and $\Gamma_3 = 0.4$, respectively. If 1 W power is fed into the sum port 4, calculate the power reflected from port 4 and that transmitted to the other three ports.   [0.478 W, 0.5224 W, 0.0003 W, 0.0631 W]

8.5 Rat-race Junctions

The basic structure of a rat-race junction is shown in Fig. 8.24(a). It consists of an annular ring of length $3\lambda_g/2$ and four ports connected at particular intervals. Theoretically, ports 1 and 4 are kept at a distance of $3\lambda_g/4$, whereas each pair of ports 1 and 2, 2 and 3, and 3 and 4 is separated by a distance of $\lambda_g/4$. This specific path difference ensures the following facts:

1. Signals fed into port 1 and travelling in clockwise and counterclockwise directions arrive at ports 2 and 4 in phase and port 3 out of phase.
2. Signals fed into port 2 and travelling in clockwise and counterclockwise directions arrive at ports 1 and 3 in phase and port 4 out of phase.
3. Signals fed into port 3 and travelling in clockwise and counterclockwise directions arrive at ports 2 and 4 in phase and port 1 out of phase.
4. Signals fed into port 4 and traveling in the clockwise and counterclockwise directions arrive at ports 1 and 3 in phase and port 2 out of phase.

Therefore, the scattering matrix of a rat-race junction can be expressed as follows:

$$[S] = \begin{bmatrix} S_{11} & S_{12} & 0 & -S_{14} \\ S_{21} & S_{22} & S_{23} & 0 \\ 0 & S_{32} & S_{33} & S_{34} \\ -S_{41} & 0 & S_{43} & S_{44} \end{bmatrix} \tag{8.76}$$

where the signs of the S-matrix elements represent the phase difference of the signal between the input and output ports (i.e., $e^{j\theta}$ where θ is 270° between ports 1 and 4 and 90° between other ports).

Since at least two ports are isolated from each other, we can match them independently without destroying the symmetry of the junction. Next, using the properties of the S-matrix of a reciprocal and a lossless network and choosing a proper port length, the final S-matrix of a rat-race junction can be obtained, which is as follows:

$$[S] = \frac{1}{\sqrt{2}} \begin{bmatrix} 0 & 1 & 0 & -1 \\ 1 & 0 & 1 & 0 \\ 0 & 1 & 0 & 1 \\ -1 & 0 & 1 & 0 \end{bmatrix} \tag{8.77}$$

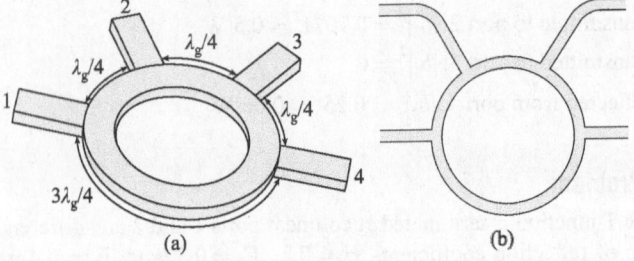

Fig. 8.24 Basic structure of rat-race junction (a) Waveguide rat-race junction (b) Planar rat-race junction

Like a magic T-junction, a rat-race junction can also be used as a 180° hybrid with ports 1 and 3 as Δ and Σ ports, respectively, and ports 2 and 4 as output ports. However, the main advantage of a rat-race junction over a magic T-junction is that, unlike a magic T-junction, a rat-race junction can also be constructed using planar technology, as shown in Fig. 8.24(b).

It should be noted that as the operating frequency changes, the ports are no longer the specific electrical length apart. Therefore, it is a very-narrow-band circuit. In addition, in a practical rat-race junction, small leakage couplings exist between ports, and hence the zero elements of the scattering matrix, given in Eq. (8.76), are also not perfectly zero.

> **Note** If the characteristic impedance of a port is Z_0, then the characteristic impedance of the ring should be $\sqrt{2}Z_0$.

8.6 Directional Couplers

A directional coupler is a four-port device having a main arm and a coupled or auxiliary arm, as shown in Fig. 8.25(a). The two commonly used symbols for directional couplers are shown in Fig. 8.25(b).

A part of the signal fed into port 1 of the main arm is coupled to port 3 of the auxiliary arm, and the remaining power is delivered to port 2 of the main arm. In such a case, port 3 is called a coupled port and port 2 a through port. For a directional coupler, theoretically, no power should be coupled at port 4; hence, port 4 is called an isolated port and is generally kept matched terminated. Following a similar logic when the network is fed through port 2, port 1 is a through port, port 4 is a coupled port, and port 3 is an isolated port. In practice, the signal can also be fed through port 3 or 4; however, in such a case, the main arm behaves as an auxiliary arm, while the auxiliary arm becomes the main arm. Coupled, through, and isolated ports can be defined in a similar manner. A directional coupler is generally characterized by three parameters, namely

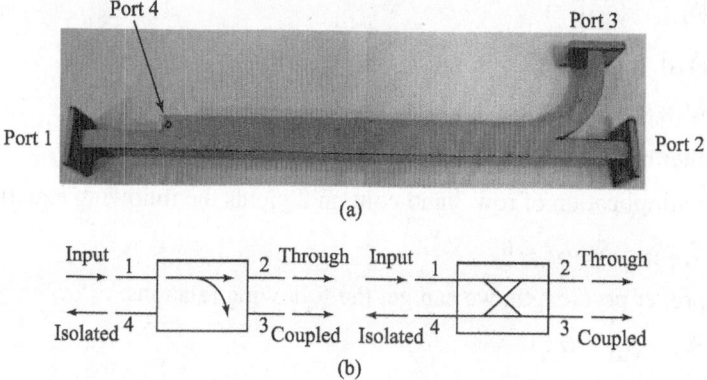

(a)

(b)

Fig. 8.25 Directional coupler (a) Typical structure (b) Two commonly used symbols

coupling, directivity, and isolation. If the power fed into port 1 is P_1 and the powers received at ports 2, 3, and 4 are P_2, P_3, and P_4, respectively, then

$$\text{Coupling}(C) = 10\log_{10}(P_1/P_3) \text{ dB} \tag{8.78}$$

$$\text{Directivity}(D) = 10\log_{10}(P_3/P_4) \text{ dB} \tag{8.79}$$

and $$\text{Isolation}(I) = 10\log_{10}(P_1/P_4) \text{ dB} \tag{8.80}$$

Therefore, $I = C + D$ dB $\tag{8.81}$

In general, the coupling factor represents the fraction of input power coupled to the output port, whereas directivity represents its efficiency to isolate forward and backward waves in the auxiliary arm. An ideal coupler should have infinite directivity and isolation.

The S-matrix of a reciprocal four-port network matched at all ports can be expressed as follows:

$$[S] = \begin{bmatrix} 0 & S_{12} & S_{13} & 0 \\ S_{12} & 0 & 0 & S_{24} \\ S_{13} & 0 & 0 & S_{34} \\ 0 & S_{24} & S_{34} & 0 \end{bmatrix} \tag{8.82}$$

Using the unitary property of the S-matrix, we can write the following equation:

$$[S]^t [S]^* = [U]$$

or

$$\begin{bmatrix} 0 & S_{12} & S_{13} & 0 \\ S_{12} & 0 & 0 & S_{24} \\ S_{13} & 0 & 0 & S_{34} \\ 0 & S_{24} & S_{34} & 0 \end{bmatrix} \begin{bmatrix} 0 & S_{12}^* & S_{13}^* & 0 \\ S_{12}^* & 0 & 0 & S_{24}^* \\ S_{13}^* & 0 & 0 & S_{34}^* \\ 0 & S_{24}^* & S_{34}^* & 0 \end{bmatrix} = \begin{bmatrix} 1 & 0 & 0 & 0 \\ 0 & 1 & 0 & 0 \\ 0 & 0 & 1 & 0 \\ 0 & 0 & 0 & 1 \end{bmatrix} \tag{8.83}$$

which gives

$$|S_{12}|^2 + |S_{13}|^2 = 1 \tag{8.84}$$

$$|S_{12}|^2 + |S_{24}|^2 = 1 \tag{8.85}$$

$$|S_{13}|^2 + |S_{34}|^2 = 1 \tag{8.86}$$

and $$|S_{24}|^2 + |S_{34}|^2 = 1 \tag{8.87}$$

Equations (8.84) and (8.85) imply $|S_{13}| = |S_{24}|$ $\tag{8.88}$

On the other hand, Eqs (8.85) and (8.87) imply $|S_{12}| = |S_{34}|$ $\tag{8.89}$

Further, multiplication of row 3 and column 2 yields the following equation:

$$S_{13}S_{12}^* + S_{34}S_{24}^* = 0 \tag{8.90}$$

Using a proper port length, we can get the following relations:

$$S_{12} = S_{34} = \alpha \tag{8.91}$$

$$S_{13} = \beta e^{j\theta} \tag{8.92}$$

and $$S_{24} = \beta e^{j\phi} \tag{8.93}$$

where α and β are real, and θ and ϕ are phase. Substituting Eqs (8.91)–(8.93) into Eq. (8.90), we get the following expressions:

$$\alpha\beta e^{j\theta} + \alpha\beta e^{-j\phi} = 0 \text{ or } e^{j\theta} = -e^{-j\phi} = e^{j\pi}e^{-j\phi}$$

or $\qquad \theta + \phi = \pi$ $\qquad\qquad\qquad\qquad\qquad\qquad\qquad\qquad$ (8.94)

Now, we have the following two choices:

1. $\theta = \phi = \pi/2$: Such couplers are called symmetrical couplers and have a scattering matrix of the following form:

$$[S] = \begin{bmatrix} 0 & \alpha & j\beta & 0 \\ \alpha & 0 & 0 & j\beta \\ j\beta & 0 & 0 & \alpha \\ 0 & j\beta & \alpha & 0 \end{bmatrix} \qquad\qquad (8.95)$$

2. $\theta = 0$, $\phi = \pi$: Such couplers are called asymmetrical couplers and have a scattering matrix of the following form:

$$[S] = \begin{bmatrix} 0 & \alpha & \beta & 0 \\ \alpha & 0 & 0 & -\beta \\ \beta & 0 & 0 & \alpha \\ 0 & -\beta & \alpha & 0 \end{bmatrix} \qquad\qquad (8.96)$$

It should be noted that these couplers differ only in the choice of the reference plane. The application of lossless condition to these matrices implies that

$$\alpha^2 + \beta^2 = 1 \qquad\qquad\qquad\qquad\qquad\qquad\qquad (8.97)$$

For a 3 dB coupler, $\alpha = \beta = 1/\sqrt{2}$; hence, Eqs (8.96) and (8.97) modify as follows:

$$[S] = \frac{1}{\sqrt{2}} \begin{bmatrix} 0 & 1 & j & 0 \\ 1 & 0 & 0 & j \\ j & 0 & 0 & 1 \\ 0 & j & 1 & 0 \end{bmatrix} \qquad\qquad (8.98)$$

and $\qquad [S] = \frac{1}{\sqrt{2}} \begin{bmatrix} 0 & 1 & 1 & 0 \\ 1 & 0 & 0 & -1 \\ 1 & 0 & 0 & 1 \\ 0 & -1 & 1 & 0 \end{bmatrix} \qquad\qquad (8.99)$

For Eq. (8.98) $\theta = \phi = \pi/2$, and hence such a coupler is called a 90° hybrid. For Eq. (8.99) $\theta = 0$, $\phi = \pi$, and such couplers are called 180° hybrids. It should be noted that magic T-junctions and rat-race junctions, described earlier, are also 3 dB 180° hybrids. The difference in the form of S-matrix, given in Eqs (8.68), (8.77), and (8.99) is only due to the difference in port numbering. The lumped element equivalent circuit of a directional coupler is shown in Fig. 8.26.

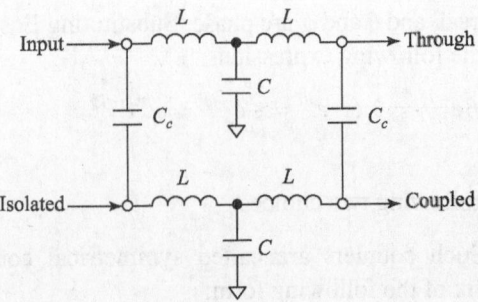

Fig. 8.26 Lumped element equivalent circuit of directional coupler

Note Directional couplers are widely used in power measurement and load power monitoring, without affecting the normal operation of the circuit.

8.7 Bethe-hole Couplers

The simplest ways to achieve the directional properties in a network is to couple two waveguides using a simple hole in the common broad wall of the waveguides, as shown in Fig. 8.27. Such types of couplers are known as Bethe-hole couplers. Bethe-hole couplers have two versions, as shown in the figure. In the first case, the guide axes are parallel, whereas in the second case, they are inclined at an angle θ. The required phase relationship is obtained by adjusting the aperture offset and aperture dimension in the parallel case and by adjusting the inclination angle θ in the second case.

According to the small aperture coupling theorem, the aperture in the common broad wall can be replaced by an equivalent source consisting of an electric dipole and a magnetic dipole. The normal electric dipole and the axial magnetic dipole radiate in the guides with an even symmetry and the normal electric dipole and the transverse magnetic dipole radiate in the guides with an odd symmetry. Adjusting the relative amplitudes of these two equivalent sources, by adjusting position and dimension of the coupling aperture or guide orientation, the radiation towards the isolated port can be cancelled while enhancing the radiation in the direction of coupled port. Thus, a Bethe-hole coupler behaves similar to a directional coupler. If the amplitude of the incident wave at port 1 is A, then the amplitude of the wave

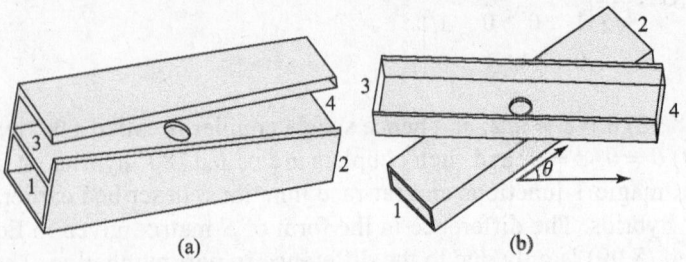

Fig. 8.27 Bethe-hole coupler (a) Parallel coupler (b) Skewed coupler

travelling in the forward and reverse directions in the guides can be expressed as follows:

$$A_{10}^{+} = -\frac{j\omega A}{P_{10}}\left[\varepsilon_0\alpha_e \sin^2\left(\frac{\pi s}{a}\right) - \frac{\mu_0\alpha_m}{\eta_{10}^2}\left\{\sin^2\left(\frac{\pi s}{a}\right) + \frac{\pi^2}{\beta^2 a^2}\cos^2\left(\frac{\pi s}{a}\right)\right\}\right] \quad (8.100)$$

and $\quad A_{10}^{-} = -\frac{j\omega A}{P_{10}}\left[\varepsilon_0\alpha_e \sin^2\left(\frac{\pi s}{a}\right) + \frac{\mu_0\alpha_m}{\eta_{10}^2}\left\{\sin^2\left(\frac{\pi s}{a}\right) - \frac{\pi^2}{\beta^2 a^2}\cos^2\left(\frac{\pi s}{a}\right)\right\}\right] \quad (8.101)$

where

$$P_{10} = ab/\eta_{10} \quad (8.102)$$
$$\alpha_e = 2r_0^3/3 \quad (8.103)$$

and $\quad \alpha_m = 4r_0^3/3 \quad (8.104)$

Since no wave should reach the isolated port, we can write the following equation:

$$A_{10}^{+} = -\frac{j\omega A}{P_{10}}\left[\varepsilon_0\alpha_e \sin^2\left(\frac{\pi s}{a}\right) - \frac{\mu_0\alpha_m}{\eta_{10}^2}\left\{\sin^2\left(\frac{\pi s}{a}\right) + \frac{\pi^2}{\beta^2 a^2}\cos^2\left(\frac{\pi s}{a}\right)\right\}\right] = 0$$

or $\quad \left(\varepsilon_0\alpha_e - \frac{\mu_0\alpha_m}{\eta_{10}^2}\right)\sin^2\left(\frac{\pi s}{a}\right) - \frac{\mu_0\alpha_m\pi^2}{\beta^2 a^2\eta_{10}^2}\cos^2\left(\frac{\pi s}{a}\right) = 0 \quad (8.105)$

Substituting Eqs (8.103) and (8.104) into Eq. (8.105), we get the following expression:

$$\left(\varepsilon_0 - \frac{2\mu_0}{\eta_{10}^2}\right)\sin^2\left(\frac{\pi s}{a}\right) - \frac{2\pi^2\mu_0}{\beta^2 a^2\eta_{10}^2}\cos^2\left(\frac{\pi s}{a}\right) = 0$$

Now, substituting $\eta_{TE} = \omega\mu/\beta$ in this equation, we get the following relation:

$$\left(\frac{\omega^2\mu_0\varepsilon_0 - 2\beta^2}{\omega^2\mu_0}\right)\sin^2\left(\frac{\pi s}{a}\right) - \frac{2\pi^2}{a^2\omega^2\mu_0}\cos^2\left(\frac{\pi s}{a}\right) = 0$$

or $\quad (k_0^2 - 2\beta^2)\sin^2\left(\frac{\pi s}{a}\right) - \frac{2\pi^2}{a^2}\cos^2\left(\frac{\pi s}{a}\right) = 0$

or $\quad \left[k_0^2 - 2\left\{k_0^2 - \left(\frac{\pi}{a}\right)^2\right\}\right]\sin^2\left(\frac{\pi s}{a}\right) = \frac{2\pi^2}{a^2}\left\{1 - \sin^2\left(\frac{\pi s}{a}\right)\right\}$

or $\quad \left(\frac{4\pi^2}{a^2} - k_0^2\right)\sin^2\left(\frac{\pi s}{a}\right) = \frac{2\pi^2}{a^2}$

or $\quad \sin\left(\frac{\pi s}{a}\right) = \sqrt{\frac{2\pi^2}{(4\pi^2 - k_0^2 a^2)}} = \sqrt{\frac{2\pi^2}{4\pi^2 - \frac{4\pi^2}{\lambda_0^2}a^2}}$

or $\quad \sin\left(\frac{\pi s}{a}\right) = \sqrt{\frac{\lambda_0^2}{2(\lambda_0^2 - a^2)}} = \frac{\lambda_0}{\sqrt{2(\lambda_0^2 - a^2)}} \quad (8.106)$

Equation (8.106) can be used to calculate the offset of the aperture.

The coupling factor and directivity can be expressed as follows:

$$C = 20 \log_{10} \left| A/A_{10}^- \right| \text{ dB} \tag{8.107}$$

$$D = 20 \log_{10} \left| A_{10}^-/A_{10}^+ \right| \text{ dB} \tag{8.108}$$

For a given coupling factor C, the ratio A/A_{10}^- can be found from Eq. (8.107). Once A/A_{10}^- is known, the value of r_0 can be calculated from Eqs (8.101)–(8.104).

For the skewed geometry, the amplitude of the wave travelling in the forward and reverse directions in the guides can be expressed as follows:

$$A_{10}^+ = -\frac{j\omega A}{P_{10}} \left\{ \varepsilon_0 \alpha_e - \frac{\mu_0 \alpha_m}{\eta_{10}^2} \cos(\theta) \right\} \tag{8.109}$$

and

$$A_{10}^- = -\frac{j\omega A}{P_{10}} \left\{ \varepsilon_0 \alpha_e + \frac{\mu_0 \alpha_m}{\eta_{10}^2} \cos(\theta) \right\} \tag{8.110}$$

For phase cancellation at the isolation port, the following relations must hold:

$$A_{10}^+ = -\frac{j\omega A}{P_{10}} \left\{ \varepsilon_0 \alpha_e - \frac{\mu_0 \alpha_m}{\eta_{10}^2} \cos(\theta) \right\} = 0$$

or

$$\varepsilon_0 \alpha_e - \frac{\mu_0 \alpha_m}{\eta_{10}^2} \cos(\theta) = 0 \tag{8.111}$$

Substituting Eqs (8.103) and (8.104) into Eq. (8.111), we get the following expression:

$$\cos(\theta) = \frac{\eta_{10}^2 \varepsilon_0}{2\mu_0} \tag{8.112}$$

Further substituting $\eta_{TE} = \omega\mu/\beta$ into Eq. (8.112), we get the following relation:

$$\cos(\theta) = \frac{\omega^2 \mu_0^2 \varepsilon_0}{2\beta^2 \mu_0} = \frac{\omega^2 \mu_0 \varepsilon_0}{2\beta^2} = \frac{k_0^2}{2\beta^2} \tag{8.113}$$

Substituting Eqs (8.102)–(8.104) and (8.113) into Eq. (8.110), the following equation is obtained:

$$A_{10}^- = -\frac{j\omega\eta_{10}A}{ab} \left(\frac{2r_0^3 \varepsilon_0}{3} + \frac{4r_0^3 k_0^2 \mu_0}{6\beta^2 \eta_{10}^2} \right) = -\frac{j2\omega\eta_{10}r_0^3 A}{3ab} \left(\varepsilon_0 + \frac{k_0^2 \mu_0}{\beta^2 \eta_{10}^2} \right) \tag{8.114}$$

Again, substituting $\eta_{10} = \omega\mu/\beta$ into Eq. (8.114), we get the following relation:

$$A_{10}^- = -\frac{j2\omega^2 \mu_0 r_0^3 A}{3ab\beta} \left(\varepsilon_0 + \frac{k_0^2}{\omega^2 \mu_0} \right) = -\frac{j2\omega^2 \mu_0 r_0^3 A}{3ab\beta} \frac{2k_0^2}{\omega^2 \mu_0} = -\frac{j4k_0^2 r_0^3 A}{3ab\beta}$$

or

$$\left| \frac{A_{10}^-}{A} \right| = \frac{4k_0^2 r_0^3}{3ab\beta} \tag{8.115}$$

Substitution of Eq. (8.115) into Eq. (8.107) produces the following relation:

$$C = 20 \log_{10} \left| A/A_{10}^- \right| = -20 \log_{10} \left(\frac{4k_0^2 r_0^3}{3ab\beta} \right) \text{ dB} \tag{8.116}$$

Equation (8.116) can be used to calculate r_0 for the given coupling coefficient.

A simple program to design and plot the frequency response of a parallel Bethe-hole coupler is as follows:

MATLAB Program

```
% Program to design a Bethe-hole coupler
% Also Plots Frequency Response over the Bandwidth
clf;
clear all;
a = input('Insert the Guide Width in mm:-  ');
a = a * 10^-3;
b = input('Insert the Guide Height in mm:-  ');
b = b * 10^-3;
f = input('Insert Frequency in GHz:-  ');
f = f * 10^9;
c = 3 * 10^8;
lambda = c/f;
k = 2*pi/lambda;
beta = sqrt(k^2 - (pi/a)^2);
eta = 376.73;
fc = c/(2*a);
eta10 = eta/sqrt(1 - (fc/f)^2);
p10 = a * b/eta10;
sinpsba = lambda/sqrt(2 * (lambda^2 - a^2));
s = a * asin(sinpsba)/pi
C = input('Insert coupling coefficient in dB:-  ');
Ratio = 1/(10^(C/20));
w = 2 * pi * f;
p1 = w/(3 * p10);
ep0 = 8.8542 * 10^-12;
p2 = 2 * ep0 * sinpsba^2;
cospsba = cos(pi * s/a);
p3 = pi^2 * cospsba^2/(beta * a)^2;
p4 = sinpsba^2 - p3^2;
mu0 = 4 * pi * 10^-7;
p5 = 4 * mu0 * p4/eta10^2;
p6 = p1 * (p2 + p5);
r0 = (ratio/p6)^(1/3)
alphae = 2 * r0^3/3;
alpham = 4 * r0^3/3;
stf = input('Insert Start Frequency in GHz:-  ');
stf = stf * 10^9;
spf = input('Insert Stop Frequency in GHz:-  ');
spf = spf * 10^9;
n = 0;
for f = stf:0.01 * 10^9:spf
    n = n + 1;
    Frequency(n) = f/10^9;
    w = 2 * pi * f;
    lambda = c/f;
```

```
    k = 2 * pi/lambda;
    beta = sqrt(k^2 - (pi/a)^2);
    eta10 = eta/sqrt(1 - (fc/f)^2);
    p10 = a * b/eta10;
    A10mbA = -i * w * (ep0 * alphae * sinpsba^2 + mu0 * alpham *
(sinpsba^2 - (pi/(beta * a))^2 * cospsba^2)/eta10^2)/p10;
    A10pbA = -i * w * (ep0 * alphae *sinpsba^2 - mu0 * alpham *
(sinpsba^2 + (pi/(beta * a))^2 * cospsba^2)/eta10^2)/p10;
    C(n) = -20 * log10(1/abs(A10mbA));
    D(n) = -20 * log10(abs(A10mbA)/abs(A10pbA));
end
plot(Frequency, C)
hold on
plot(Frequency, D)
xlabel('Frequency (GHz)');
ylabel('Coupling Coefficient & Directivity');
```

EXAMPLE 8.4   Design a 15 dB Bethe-hole coupler at 9 GHz for a WR-90 waveguide. Plot the coupling coefficient and directivity from 8.2 to 12.4 GHz.

Solution   Given: $C = 15$ dB, $f = 10$ GHz, $a = 22.86 \times 10^{-3}$ m, and $b = 10.16 \times 10^{-3}$ m

From these data, we can derive the following equations:

$$\lambda_0 = \frac{c}{f} = \frac{3 \times 10^8}{9 \times 10^9} = 0.0333 \text{ m}$$

$$k_0 = \frac{2\pi}{\lambda_0} = 188.6842 \text{ m}^{-1}$$

$$\beta = \sqrt{k_0^2 - (\pi/a)^2} = \sqrt{(188.6842)^2 - \left(\pi/(22.86 \times 10^{-3})\right)^2} = 129.288 \text{ m}^{-1}$$

$$f_c = \frac{c}{2a} = \frac{3 \times 10^8}{2 \times 22.86 \times 10^{-3}} = 6.5617 \times 10^9 \text{ Hz}$$

$$\eta_{10} = \frac{\eta}{\sqrt{1 - (f_c/f)^2}} = \frac{376.73}{\sqrt{1 - (6.5617/9)^2}} = 550.4281 \text{ } \Omega$$

$$P_{10} = \frac{ab}{\eta_{10}} = \frac{22.86 \times 10^{-3} \times 10.16 \times 10^{-3}}{550.4281} = 4.2196 \times 10^{-7} \text{ m}^2/\Omega$$

Therefore,

$$\sin\left(\frac{\pi s}{a}\right) = \frac{\lambda_0}{\sqrt{2(\lambda_0^2 - a^2)}} = \frac{0.0333}{\sqrt{2\left\{(0.0333)^2 - (22.86 \times 10^{-3})^2\right\}}} = 0.9724$$

or

$$s = \frac{a}{\pi} \sin^{-1}(0.9724) = \frac{22.86 \times 10^{-3}}{\pi} \sin^{-1}(0.9724)$$

$$= \frac{22.86 \times 10^{-3}}{\pi} \times 1.3353 = 9.7164 \times 10^{-3} \text{ m}$$

The coupling is 15 dB. Therefore,

$$20\log_{10}\left|\frac{A}{A_{10}^-}\right| = 15 \quad \text{or} \quad \left|\frac{A}{A_{10}^-}\right| = 10^{15/20} = 5.6234$$

or

$$\left|\frac{A_{10}^-}{A}\right| = \frac{1}{5.6234} = 0.1778$$

Now,

$$\frac{\omega r_0^3}{3P_{10}}\left[2\varepsilon_0 \sin^2\left(\frac{\pi s}{a}\right) + \frac{4\mu_0}{\eta_{10}^2}\left\{\sin^2\left(\frac{\pi s}{a}\right) - \frac{\pi^2}{\beta^2 a^2}\cos^2\left(\frac{\pi s}{a}\right)\right\}\right] = 0.1778$$

or

$$\frac{2\pi \times 9 \times 10^9 \times r_0^3}{3 \times 4.2196 \times 10^{-7}}\left[2\varepsilon_0 \times 0.9724^2 + \frac{4\mu_0}{550.4281^2}\left\{0.9724^2\right.\right.$$

$$\left.\left. - \frac{\pi^2 \times 0.0544}{(129.288 \times 22.86 \times 10^{-3})^2}\right\}\right] = 0.1778$$

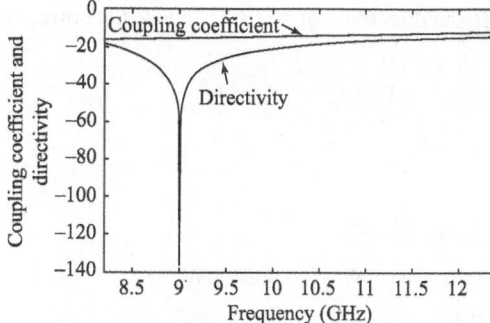

or $\quad 1.4032 r_0^3 \times 10^6 = 0.1778$

or $\quad r_0 = \left(\dfrac{0.1778}{1.4032 \times 10^6}\right)^{1/3}$

$$= 5.0227 \times 10^{-3} \text{ m}$$

The plot of coupling coefficient and directivity from 8.2 to 12.4 GHz is shown in Fig. 8.28.

Fig. 8.28 Plot of coupling coefficient and directivity for Example 8.4

Practice Problem

8.4 Design a 20 dB Bethe-hole coupler at 13 GHz for a WR-62 waveguide.

[6.6507 mm, 2.9923 mm]

8.8 Two-hole Directional Couplers

As compared to a Bethe-hole coupler, a two-hole directional coupler consists of two waveguides having two coupling apertures, separated by a distance $(2n+1)\lambda_g/4$ in the common broad wall. The basic structure is shown in Fig. 8.29.

A part of the signal, fed into port 1, passes through hole 1 and produces a forward and a reverse wave in the auxiliary guide. Similarly, another part of the signal, fed into port 1, passes through hole 2 and produces a forward and a reverse wave in the auxiliary guide. The two forward traveling waves are in phase at the position of hole 2 and hence they add, resulting in a net forward traveling wave in the auxiliary guide. However, the two backward traveling waves are

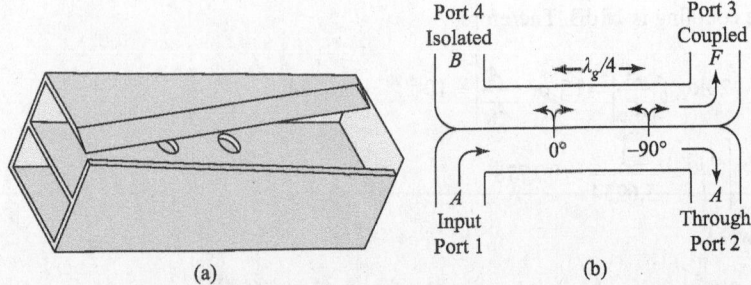

Fig. 8.29 Two-hole directional coupler (a) 3D view (b) Schematic view

out of phase by $(2n+1)\pi$ at the position of hole 1 and hence cancel each other, resulting in no net backward traveling wave in the auxiliary guide. The signal, which does not couple to the auxiliary guide via hole 1 or 2, arrives at port 2 of the main guide. Thus, port 2 is a through port, port 3 is a coupled port, and port 4 is an isolation port. Similarly, it can be shown that port 2 and port 3 are isolated.

EXAMPLE 8.5 The input power in a two-hole directional coupler is 1 mW. The coupler has a coupling coefficient of 15 dB and directivity of 50 dB. Calculate the power at all ports.

Solution   Given: $C = 15\,\text{dB}$, $D = 50\,\text{dB}$, and $C = 10\log_{10}(P_1/P_3) = 15$

or
$$P_3 = P_1 10^{-C/10} = 10^{-15/10} = 0.0316 \text{ mW}$$
$$D = 10\log_{10}(P_3/P_4) = 50$$

or
$$P_4 = P_3 10^{-D/10} = 0.0316 \times 10^{-50/10} = 0.316 \text{ nW}$$
$$P_2 = P_1 - (P_3 + P_4) = 1 - 0.0316 - (0.316 \times 10^{-6}) = 0.9684 \text{ mW}$$

Practice Problem

8.5 The input power in a two-hole directional coupler is 10 mW. The coupler has a coupling coefficient of 20 dB and directivity of 60 dB. Calculate the power at all ports.   [0.1 mW, 0.1 nW, 9.9 mW]

8.9 Multi-hole Directional Couplers

Let us now consider the multi-hole directional coupler shown in Fig. 8.30, where $(N+1)$ equally spaced apertures have been milled in the common broad wall of the two waveguide. As in the case of a Bethe-hole coupler, each of the apertures will excite a forward and a reverse travelling wave in the waveguides. If F_n and B_n be the coupling coefficients of the nth aperture in the forward and backward directions, respectively, then the respective wave amplitudes can be expressed as follows:

$$F = Ae^{-j\beta Nd} \sum_{n=0}^{N} F_n \tag{8.117}$$

and
$$B = A\sum_{n=0}^{N} B_n e^{-j2\beta nd} \tag{8.118}$$

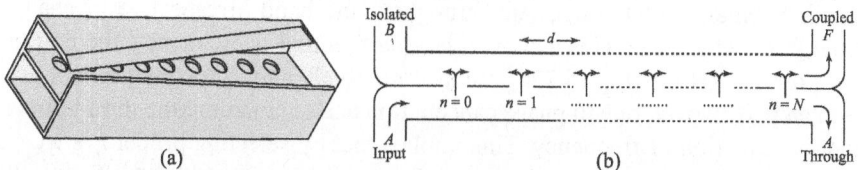

Fig. 8.30 Multi-hole directional coupler (a) 3D view (b) Schematic view

where d is the spacing between two successive apertures.

Since for small coupling the amplitude of the through wave is almost equal to that of the incident wave, A, we can write the following equations:

$$C = -20\log_{10}|F/A| = -20\log_{10}\left|\sum_{n=0}^{N} F_n\right| \text{ dB} \tag{8.119}$$

and $$D = -20\log_{10}|B/F| = -20\log_{10}\left|\sum_{n=0}^{N} B_n e^{-j2\beta nd}\bigg/\sum_{n=0}^{N} F_n\right|$$

or $$D = -C - 20\log_{10}\left|\sum_{n=0}^{N} B_n e^{-j2\beta nd}\right| \text{ dB} \tag{8.120}$$

Now if we assume that all the coupling apertures are placed at an offset s relative to the edge of guide and the nth aperture has a radius r_n, then Eqs (8.100)–(8.104) show that F_n and B_n can be written as follows:

$$F_n = K_f r_n^3 \tag{8.121}$$

and $$B_n = K_b r_n^3 \tag{8.122}$$

where K_f and K_b are constants for the forward and backward coupling coefficients and are same for all apertures. Substituting Eqs (8.121) and (8.122) into Eqs (8.119) and (8.120), we get the following relations:

$$C = -20\log_{10}\left|\sum_{n=0}^{N} K_f r_n^3\right| = -20\log_{10}|K_f| - 20\log_{10}\sum_{n=0}^{N} r_n^3 \text{ dB} \tag{8.123}$$

and $$D = -C - 20\log_{10}\left|\sum_{n=0}^{N} K_b r_n^3 e^{-j2\beta nd}\right|$$

or $$D = -C - 20\log_{10}|K_b| - 20\log_{10}\left|\sum_{n=0}^{N} r_n^3 e^{-j2\beta nd}\right| \text{ dB}$$

or $$D = -C - 20\log_{10}|K_b| - 20\log_{10}(S) \text{ dB} \tag{8.124}$$

where $$S = \left|\sum_{n=0}^{N} r_n^3 e^{-j2\beta nd}\right| \tag{8.125}$$

The first term on the RHS of Eq. (8.123) is a relatively slowly varying function of frequency, whereas the second term is independent of frequency. Therefore, coupling remains almost constant throughout the band, irrespective of changes in the frequency. Since C is a slowly varying function of frequency, the first two terms on the RHS of Eq. (8.124) are also relatively slowly varying functions of frequency. However, due to phase cancellation and summation, the third term is a sensitive function of frequency. This implies that by selecting proper r_n's we can achieve the desired frequency response for the directivity without affecting the coupling coefficient.

8.9.1 Binomial Response

To obtain a binomial response, let us assume that $r_n^3 = kC_n^N$ (8.126)

where k is an unknown constant and C_n^N is a binomial coefficient. Substituting Eq. (8.126) into Eq. (8.123), we get the following relation:

$$C = -20\log_{10}\left|K_f\right| - 20\log_{10} \sum_{n=0}^{N} kC_n^N$$

or $C = -20\log_{10}\left|K_f\right| - 20\log_{10}(k) - 20\log_{10} \sum_{n=0}^{N} C_n^N$ dB   (8.127)

Since C, K_f, and N are known, we can solve Eq. (8.127) to find k. Once k is known, we can find individual r_n's using Eq. (8.126). Throughout the analysis, d should be equal to $\lambda_g/4$ at the centre frequency.

8.9.2 Chebyshev Response

To obtain a Chebyshev response, let us first assume that N is even and the coupler is symmetric, that is, $r_0 = r_N$, $r_1 = r_{N-1}$, $r_2 = r_{N-2}$, ... Therefore, Eq. (8.125) can be written as follows:

$$S = \left|\sum_{n=0}^{N} r_n^3 e^{-j2\beta nd}\right| = \left|\sum_{n=0}^{N} r_n^3 e^{-j2n\theta}\right|$$

or $S = \left|r_0^3 + r_1^3 e^{-j2\theta} + r_2^3 e^{-j4\theta} + \ldots + r_{N-2}^3 e^{-j2(N-2)\theta} + r_{N-1}^3 e^{-j2(N-1)\theta} + r_N^3 e^{-j2N\theta}\right|$

or $S = \left|r_0^3 + r_1^3 e^{-j2\theta} + r_2^3 e^{-j4\theta} + \ldots + r_2^3 e^{-j2(N-2)\theta} + r_1^3 e^{-j2(N-1)\theta} + r_0^3 e^{-j2N\theta}\right|$

or $S = \left|r_0^3\left(1 + e^{-j2N\theta}\right) + r_1^3\left\{e^{-j2\theta} + e^{-j2(N-1)\theta}\right\} + r_2^3\left\{e^{-j4\theta} + e^{-j2(N-2)\theta}\right\} + \ldots\right|$

or $S = \left|e^{-jN\theta} r_0^3\left(e^{jN\theta} + e^{-jN\theta}\right) + e^{-jN\theta} r_1^3\left\{e^{j(N-2)\theta} + e^{-j(N-2)\theta}\right\}\right.$

$$\left. + e^{-jN\theta} r_2^3\left\{e^{-j(N-4)\theta} + e^{-j(N-4)\theta}\right\} + \ldots\right|$$

or $S = \left|e^{-jN\theta} 2\sum_{n=0}^{N/2} r_n^3 \cos\left\{(N-2n)\theta\right\}\right| = 2\sum_{n=0}^{N/2} r_n^3 \cos\left\{(N-2n)\theta\right\}$ (8.128)

where $\theta = \beta d$ (8.129)

To achieve a Chebyshev response, we can make the following substitution:

$$S = 2\sum_{n=0}^{N/2} r_n^3 \cos\left\{(N-2n)\theta\right\} = k\left|T_n\left\{\sec(\theta_m)\cos(\theta)\right\}\right|$$ (8.130)

where k and θ_m are unknown constants. For $\theta = 0$:

$$S\big|_{\theta=0} = 2\sum_{n=0}^{N/2} r_n^3 = \sum_{n=0}^{N} r_n^3 = k\left|T_n\left\{\sec\left(\theta_m\right)\right\}\right| \tag{8.131}$$

Using Eq. (8.131) in Eq. (8.123), we get the following relation:

$$C = -20\log_{10}\left|K_f\right| - 20\log_{10}\left(S\right)\big|_{\theta=0}$$

or $\qquad C = -20\log_{10}\left|K_f\right| - 20\log_{10}\left(k\left|T_n\left\{\sec\left(\theta_m\right)\right\}\right|\right)$ dB

or $\qquad C = -20\log_{10}\left|K_f\right| - 20\log_{10}\left(k\right) - 20\log_{10}\left|T_n\left\{\sec\left(\theta_m\right)\right\}\right|$ dB $\qquad$ (8.132)

Since the terms on the RHS of Eq. (8.132) are either independent or slowly varying functions of frequency, C is also a slowly varying function of frequency.

Substituting Eqs (8.130) and (8.132) into Eq. (8.124), we get the following expression:

$$D = 20\log_{10}\left|K_f\right| + 20\log_{10}\left(k\right) + 20\log_{10}\left|T_n\left\{\sec\left(\theta_m\right)\right\}\right| - 20\log_{10}\left|K_b\right|$$
$$\qquad - 20\log_{10}\left(k\right) - 20\log_{10}\left|T_n\left\{\sec\left(\theta_m\right)\cos\left(\theta\right)\right\}\right|$$

$$D = 20\log_{10}\left(\left|K_f\right|/\left|K_b\right|\right)$$

or
$$\qquad + 20\log_{10}\left|\left|T_n\left\{\sec\left(\theta_m\right)\right\}\right|/T_n\left\{\sec\left(\theta_m\right)\cos\left(\theta\right)\right\}\right|\ \text{dB} \tag{8.133}$$

The first term on the RHS of Eq. (8.133) is a slowly varying function of frequency, and hence the complete response of D is not purely Chebyshev. However, since the term $20\log_{10}\left(\left|K_f\right|/\left|K_b\right|\right)$ is usually very small, the deviation from the pure Chebyshev response is also very small. As $\left|T_n\left\{\sec\left(\theta_m\right)\right\}\right| \geq T_n\left\{\sec\left(\theta_m\right)\cos\left(\theta\right)\right\}$, D will have the lowest value when $T_n\left\{\sec\left(\theta_m\right)\cos\left(\theta\right)\right\} = 1$ and is given as follows:

$$D_{\min} = 20\log_{10}\left[T_n\left\{\sec\left(\theta_m\right)\right\}\right]\ \text{dB} \tag{8.134}$$

If we specify the minimum directivity $D_{\min}$, then θ_m can be calculated from the aforementioned equation. Since K_f is known, for a given coupling coefficient we can find k using Eq. (8.132). Once k has been calculated, we can find the r_n's using Eq. (8.131).

If N is odd, we need to assume that

$$S = 2\sum_{n=0}^{(N-1)/2} r_n^3 \cos\left\{(N-2n)\theta\right\} = k\left|T_n\left\{\sec\left(\theta_m\right)\cos\left(\theta\right)\right\}\right| \tag{8.135}$$

However, expressions for C, D, and $D_{\min}$, given in Eqs (8.132)–(8.134) still hold.

EXAMPLE 8.6 Design a 15 dB four-hole Chebyshev coupler at 9 GHz using a WR-90 waveguide. Assume that the minimum directivity is 50 dB and the coupling apertures are round and located at $s = a/4$. Write a MATLAB code to plot the frequency response from 8.2 to 12.4 GHz.

Solution Given: $C = 15$ dB, $f = 10$ GHz, $a = 22.86 \times 10^{-3}$ m,
$\qquad\qquad b = 10.16 \times 10^{-3}$ m, and $s = a/4$

From the given data, we get the following relations:

$$\lambda_0 = \frac{c}{f} = \frac{3\times10^8}{9\times10^9} = 0.0333 \text{ m and } k_0 = \frac{2\pi}{\lambda_0} = 188.6842 \text{ m}^{-1}$$

$$\beta = \sqrt{k_0^2 - (\pi/a)^2} = \sqrt{(188.6842)^2 - \left(\pi/(22.86\times10^{-3})\right)^2} = 129.288 \text{ m}^{-1}$$

$$f_c = \frac{c}{2a} = \frac{3\times10^8}{2\times22.86\times10^{-3}} = 6.5617\times10^9 \text{ Hz}$$

$$\eta_{10} = \frac{\eta}{\sqrt{1-(f_c/f)^2}} = \frac{376.73}{\sqrt{1-(6.5617/9)^2}} = 550.4281 \ \Omega$$

$$P_{10} = \frac{ab}{\eta_{10}} = \frac{22.86\times10^{-3}\times10.16\times10^{-3}}{550.4281} = 4.2196\times10^{-7} \text{ m}^2/\Omega$$

Now,

$$|K_f| = \frac{2\omega}{3P_{10}}\left[\varepsilon_0 \sin^2\left(\frac{\pi s}{a}\right) - \frac{2\mu_0}{\eta_{10}^2}\left\{\sin^2\left(\frac{\pi s}{a}\right) + \frac{\pi^2}{\beta^2 a^2}\cos^2\left(\frac{\pi s}{a}\right)\right\}\right]$$

or $\quad |K_f| = 3.953\times10^5$

and $\quad |K_b| = \frac{2\omega}{3P_{10}}\left[\varepsilon_0 \sin^2\left(\frac{\pi s}{a}\right) + \frac{2\mu_0}{\eta_{10}^2}\left\{\sin^2\left(\frac{\pi s}{a}\right) - \frac{\pi^2}{\beta^2 a^2}\cos^2\left(\frac{\pi s}{a}\right)\right\}\right]$

or $\quad |K_b| = 3.454\times10^5$

Now, for a four-hole coupler $N = 3$. Therefore,

$$20\log_{10}\{T_3 \sec(\theta_m)\} = 50$$

or $\quad T_3 \sec(\theta_m) = \cosh\left[3\cosh^{-1}\{\sec(\theta_m)\}\right] = 10^{50/20} = 316.2278$

or $\quad 3\cosh^{-1}\{\sec(\theta_m)\} = 6.4496$ or $\sec(\theta_m) = 4.3501$ or $\theta_m = 76.71°$

The other band edge occurs at $\sec(\theta_m) = -4.3501$ or $\theta_m = 103.29°$

Now,

$$C = -20\log_{10}|K_f| - 20\log_{10}(k) - 20\log_{10}|T_n\{\sec(\theta_m)\}| \text{ dB}$$

or $\quad 15 = -20\log_{10}(3.953\times10^5) - 20\log_{10}(k) - 50$

or $\quad 20\log_{10}(k) = -176.9385$

or $\quad k = 10^{-176.9385/20} = 1.4226\times10^{-9}$

Again,

$$S = k|T_n\{\sec(\theta_m)\cos(\theta)\}|$$

or $\quad 2\left[r_0^3\cos(3\theta) + r_1^3\cos(\theta)\right] = k\left[\sec^3(\theta_m)\{\cos(3\theta) + \cos(\theta)\} - \sec(\theta_m)\cos(\theta)\right]$

Comparing the equal terms on both sides of this equation, we can write the following relations:

$$2r_0^3 = k\sec^3(\theta_m) = 1.4226\times10^{-9}\sec^3(76.71) = 1.1711\times10^{-7}$$

or $\quad r_0 = r_3 = 3.8831\times10^{-3} \text{ m}$

and $\quad 2r_1^3 = 3k\left\{\sec^3\left(\theta_m\right) - \sec\left(\theta_m\right)\right\}$

$$= 3 \times 1.4226 \times 10^{-9}\left\{\sec^3\left(76.71\right) - \sec\left(76.71\right)\right\} = 3.3275 \times 10^{-7}$$

or $\quad r_1 = r_2 = 5.5 \times 10^{-3}$ m

The following is the MATLAB code for plotting the *coupling coefficient and directivity* over 8.2 to 12.4 GHz (Fig. 8.31).

MATLAB Program

```
% Program to design a 3rd Order Chebyshev Multi-hole Coupler
% Also Plots Frequency Response over the Bandwidth
clf;
clear all;
a = input('Insert the Guide Width in mm:-  ');
a = a * 10^-3;
b = input('Insert the Guide Height in mm:-  ');
b = b * 10^-3;
f = input('Insert Frequency in GHz:-  ');
f = f * 10^9;
w = 2 * pi * f;
c = 3 * 10^8;
lambda = c/f;
k0 = 2 * pi/lambda;
beta = sqrt(k0^2 - (pi/a)^2);
eta = 376.73;
fc = c/(2 * a);
eta10 = eta/sqrt(1 - (fc/f)^2);
lambdag = lambda/sqrt(1 - (fc/f)^2);
d = lambdag/4;
p10 = a * b/eta10;
s = input('Insert Distance of the Hole in Multiple of a:-  ');
s = a * s;
spsa = sin(pi * s/a);
cpsa = cos(pi * s/a);
ep0 = 8.8542 * 10^-12;
mu0 = 4 * pi * 10^-7;
Kf = abs(2 * w * (ep0 * spsa^2 - ((2 * mu0/eta10^2) * (spsa^2 + ((pi/
(beta * a))^2 * cpsa^2))))/(3 * p10));
Kb = abs(2 * w * (ep0 * spsa^2 + ((2 * mu0/eta10^2) * (spsa^2 - ((pi/
(beta * a))^2 * cpsa^2))))/(3 * p10));
Dmin = input('Insert the Minimum Directivity in dB:-  ');
T3_sec_thetam = 10^(Dmin/20);
sec_thetam = cosh(acosh(T3_sec_thetam)/3);
Cc = input('Insert the Coupling Coefficient in dB:-  ');
K = 10^-((Cc + (20 * log10(Kf))+Dmin)/20);
r0 = ((k * sec_thetam^3)/2)^(1/3);
r3 = r0;
```

```
r1 = ((3 * k * (sec_thetam^3 - sec_thetam))/2)^(1/3);
r2 = r1
stf = input('Insert Start Frequency in GHz:-  ');
stf = stf * 10^9;
spf = input('Insert Stop Frequency in GHz:-  ');
spf = spf * 10^9;
n = 0;
for f = stf:0.01 * 10^9:spf
    n = n + 1;
    Frequency(n) = f/10^9;
    Lambda = c/f;
    w = 2 * pi * f;
    eta10 = eta/sqrt(1 - (fc/f)^2);
    p10 = a * b/eta10;
    k0 = 2 * pi/lambda;
    beta = sqrt(k0^2 - (pi/a)^2);
    theta = beta * d;
    Kf = abs(2 * w * (ep0 * spsa^2 - ((2 * mu0/eta10^2) * (spsa^2 +
((pi/(beta * a))^2 * cpsa^2))))/(3 * p10));
    Kb = abs(2 * w * (ep0 * spsa^2+((2 * mu0/eta10^2) * (spsa^2 -
((pi/(beta * a))^2 * cpsa^2))))/(3 * p10));
    C(n) = 20 * log10(Kf) + 20 * log10(k) + 20 * log10(T3_sec_
thetam);
    T3 = sec_thetam^3 * (cos(3 * theta) + 3 * cos(theta)) - 3 * sec_
thetam * cos(theta);
    D(n) = -abs((20 * log10(Kf/Kb)) + (20 * log10(T3_sec_thetam/
T3)));
end
plot(Frequency, C)
hold on
plot(Frequency, D)
xlabel('Frequency (GHz)');
ylabel('Coupling Coefficient & Directivity');
```

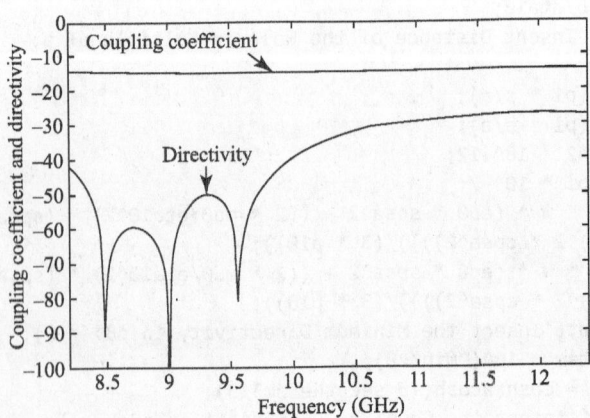

Fig. 8.31 Plot of coupling coefficient and directivity for Example 8.6

Practice Problem

8.6 Design a 20 dB four-hole Chebyshev coupler at 13 GHz using a WR-62 waveguide. Assume that the minimum directivity is 60 dB and the coupling apertures are round and located at $s = a/4$.   [2.2868 mm, 3.2706 mm]

8.10 Quadrature (90°) Hybrids

Quadrature hybrids are 3 dB directional couplers that produce a 90° phase difference between the output signals at the through and coupled arms. Thus, the S-matrix of a 90° hybrid has the following form:

(a)

$$[S] = \frac{1}{\sqrt{2}} \begin{bmatrix} 0 & 1 & j & 0 \\ 1 & 0 & 0 & j \\ j & 0 & 0 & 1 \\ 0 & j & 1 & 0 \end{bmatrix} \quad (8.136)$$

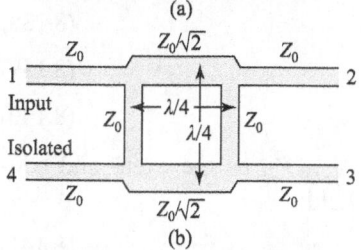

(b)

Fig. 8.32 Branch-line coupler (a) Typical view (b) Typical dimensions

Quadrature hybrids are often made in a microstrip or a strip line form, as shown in Fig. 8.32. When a signal is fed into port 1, port 2 behaves as a through port, whereas port 3 behaves as a coupled port. The remaining port 4 is the isolated port. Since a quadrature hybrid is symmetric, any port of the network can be used as the input port. The through, coupled, and isolated ports can then be defined accordingly.

Since the length of a branch-line coupler is a quarter wavelength, it is highly frequency dependent. A wideband frequency response can be obtained using a multi-section branch-line coupler, as shown in Fig. 8.33.

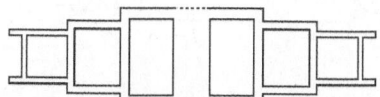

Fig. 8.33 Multi-section branch-line coupler

The lumped element equivalent circuit of a 90° hybrid is shown in Fig. 8.34.

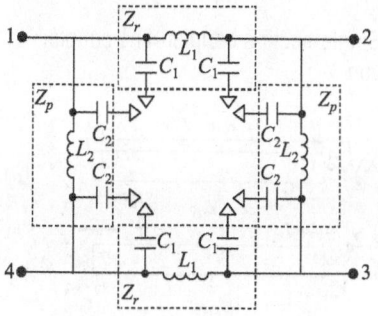

Fig. 8.34 Lumped element equivalent circuit of 90° hybrid

EXAMPLE 8.7 Design a 50 Ω branch-line quadrature hybrid junction.

Solution Given: $Z_0 = 50 \ \Omega$

Therefore, the impedances of the branch line will be as follows:

$$Z_0' = Z_0/\sqrt{2} = 50/\sqrt{2} = 35.3553 \ \Omega$$

The length of the branch line will be $\lambda/4$.

Practice Problem

8.7 Design a 100 Ω branch-line quadrature hybrid junction.   **[70.7107 Ω]**

8.11 Coupled-line Directional Couplers

A brief description of coupled microstrip lines has already been given in Section 4.6. It can be shown that such structures can effectively be used as directional couplers. To demonstrate this, let us consider the coupled-line structure shown in Fig. 8.35(a) and assume that a signal is fed into port 1, while rest of the ports are terminated with matched loads. The schematic circuit is shown in Fig. 8.35(b).

Using the superposition principle, the excitation of port 1 can be decomposed into an even- and an odd-mode excitation, as shown in Fig. 8.36. If I_{e1}, I_{e2}, I_{e3}, I_{e4} and I_{o1}, I_{o2}, I_{o3}, I_{o4} be the even- and odd-mode port currents, respectively, then the following relations hold:

$$I_{e1} = I_{e3} \tag{8.137}$$

$$I_{e2} = I_{e4} \tag{8.138}$$

$$I_{o1} = -I_{o3} \tag{8.139}$$

and $$I_{o2} = -I_{o4} \tag{8.140}$$

Similarly, if V_{e1}, V_{e2}, V_{e3}, V_{e4} and V_{o1}, V_{o2}, V_{o3}, V_{o4} be the even- and odd-mode port voltages, then the following relations hold:

$$V_{e1} = V_{e3} \tag{8.141}$$

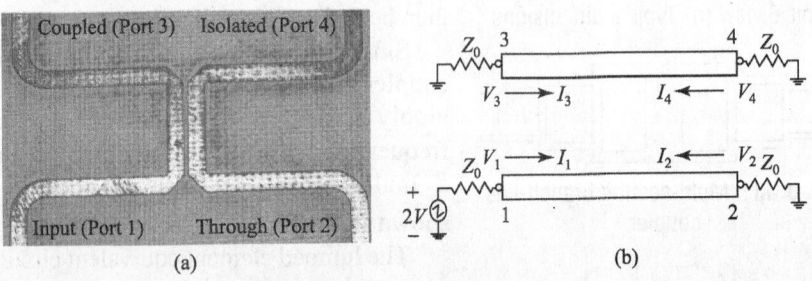

Fig. 8.35 Coupled microstrip line structure (a) Single-section coupled-line coupler (b) Schematic circuit

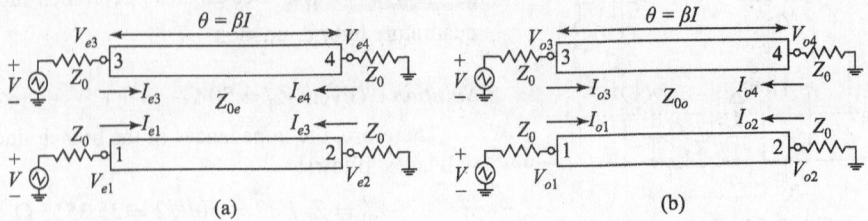

Fig. 8.36 Decomposed circuit (a) Even mode (b) Odd mode

$$V_{e2} = V_{e4} \tag{8.142}$$

$$V_{o1} = -V_{o3} \tag{8.143}$$

and $$V_{o2} = -V_{o4} \tag{8.144}$$

The input impedance at port 1 can be expressed as follows:

$$Z_{in} = \frac{V_1}{I_1} = \frac{V_{e1} + V_{o1}}{I_{e1} + I_{o1}} \tag{8.145}$$

Now, if Z_{in}^e and Z_{in}^o be the even- and odd-mode input impedances, respectively, then we can write the following relations:

$$Z_{in}^e = Z_{e0} \frac{Z_0 + jZ_{e0} \tan(\theta)}{Z_{e0} + jZ_0 \tan(\theta)} \tag{8.146}$$

and $$Z_{in}^o = Z_{o0} \frac{Z_0 + jZ_{o0} \tan(\theta)}{Z_{o0} + jZ_0 \tan(\theta)} \tag{8.147}$$

For each mode, the line resembles a transmission line of characteristic impedance Z_{in}^e or Z_{in}^o terminated by a load Z_0; hence, according to the voltage divider rule, the even- and odd-mode currents and voltages at port 1 can be written as follows:

$$I_{e1} = \frac{V}{Z_{in}^e + Z_0} \tag{8.148}$$

$$I_{o1} = \frac{V}{Z_{in}^o + Z_0} \tag{8.149}$$

$$V_{e1} = V \frac{Z_{in}^e}{Z_{in}^e + Z_0} \tag{8.150}$$

$$V_{o1} = V \frac{Z_{in}^o}{Z_{in}^o + Z_0} \tag{8.151}$$

where V is the source voltage. Substituting Eqs (8.148)–(8.151) into Eq. (8.145), we get the following expression:

$$Z_{in} = \frac{V_{e1} + V_{o1}}{I_{e1} + I_{o1}} = \frac{V \dfrac{Z_{in}^e}{Z_{in}^e + Z_0} + V \dfrac{Z_{in}^o}{Z_{in}^o + Z_0}}{\dfrac{V}{Z_{in}^e + Z_0} + \dfrac{V}{Z_{in}^o + Z_0}}$$

or $$Z_{in} = \frac{Z_{in}^e Z_0 + Z_{in}^o Z_0 + 2Z_{in}^e Z_{in}^o}{Z_{in}^e + Z_{in}^o + 2Z_0} = Z_0 + \frac{2\left(Z_{in}^e Z_{in}^o - Z_0^2\right)}{Z_{in}^e + Z_{in}^o + 2Z_0} \tag{8.152}$$

Now if we choose

$$Z_0 = \sqrt{Z_{e0} Z_{o0}} \tag{8.153}$$

then Eqs (8.146) and (8.147) modify to the following forms:

$$Z_{in}^e = Z_{e0} \frac{Z_0 + jZ_{e0} \tan(\theta)}{Z_{e0} + jZ_0 \tan(\theta)} = Z_{e0} \frac{\sqrt{Z_{e0} Z_{o0}} + jZ_{e0} \tan(\theta)}{Z_{e0} + j\sqrt{Z_{e0} Z_{o0}} \tan(\theta)}$$

or
$$Z_{in}^e = Z_{e0} \frac{\sqrt{Z_{o0}} + j\sqrt{Z_{e0}} \tan(\theta)}{\sqrt{Z_{e0}} + j\sqrt{Z_{o0}} \tan(\theta)} \tag{8.154}$$

$$Z_{in}^o = Z_{o0} \frac{Z_0 + jZ_{o0} \tan(\theta)}{Z_{o0} + jZ_0 \tan(\theta)} = Z_{o0} \frac{\sqrt{Z_{e0}Z_{o0}} + jZ_{o0} \tan(\theta)}{Z_{o0} + j\sqrt{Z_{e0}Z_{o0}} \tan(\theta)}$$

or
$$Z_{in}^o = Z_{o0} \frac{\sqrt{Z_{e0}} + j\sqrt{Z_{o0}} \tan(\theta)}{\sqrt{Z_{o0}} + j\sqrt{Z_{e0}} \tan(\theta)} \tag{8.155}$$

Multiplying Eq. (8.154) by Eq. (8.155), we get the following expression:

$$Z_{in}^e Z_{in}^o = Z_{e0}Z_{o0} = Z_0^2 \tag{8.156}$$

Therefore, Eq. (8.152) ensures matching at port 1. Further, when Eq. (8.156) is satisfied, addition of Eqs (8.150) and (8.151) produces the following relation:

$$V_1 = V_{e1} + V_{o1} = V\left[\frac{Z_{in}^e}{Z_{in}^e + Z_0} + \frac{Z_{in}^o}{Z_{in}^o + Z_0} \right]$$

or
$$V_1 = V\left[\frac{Z_{in}^e Z_0 + Z_{in}^o Z_0 + 2Z_{in}^o Z_{in}^e}{Z_{in}^e Z_{in}^o + Z_{in}^e Z_0 + Z_{in}^o Z_0 + Z_0^2} \right] = V\left[\frac{Z_{in}^e Z_0 + Z_{in}^o Z_0 + 2Z_0^2}{Z_{in}^e Z_0 + Z_{in}^o Z_0 + 2Z_0^2} \right] = V \tag{8.157}$$

Using Eqs (8.141) and (8.143), the voltage at port 3 can be calculated as follows:

$$V_3 = V_{e3} + V_{o3} = V_{e1} - V_{o1} \tag{8.158}$$

Substituting Eqs (8.150) and (8.151) into Eq. (8.158), we get the following relation:

$$V_3 = V\left[\frac{Z_{in}^e}{Z_{in}^e + Z_0} - \frac{Z_{in}^o}{Z_{in}^o + Z_0} \right] \tag{8.159}$$

Now using Eq. (8.154), we get the following expression:

$$\frac{Z_{in}^e}{Z_{in}^e + Z_0} = \frac{Z_{e0} \dfrac{\sqrt{Z_{o0}} + j\sqrt{Z_{e0}} \tan(\theta)}{\sqrt{Z_{e0}} + j\sqrt{Z_{o0}} \tan(\theta)}}{Z_{e0} \dfrac{\sqrt{Z_{o0}} + j\sqrt{Z_{e0}} \tan(\theta)}{\sqrt{Z_{e0}} + j\sqrt{Z_{o0}} \tan(\theta)} + Z_0}$$

or
$$\frac{Z_{in}^e}{Z_{in}^e + Z_{in}^o} = \frac{Z_{e0}\sqrt{Z_{o0}} + jZ_{e0}\sqrt{Z_{e0}} \tan(\theta)}{Z_{e0}\sqrt{Z_{o0}} + jZ_{e0}\sqrt{Z_{e0}} \tan(\theta) + Z_0\sqrt{Z_{e0}} + jZ_0\sqrt{Z_{o0}} \tan(\theta)}$$

or
$$\frac{Z_{in}^e}{Z_{in}^e + Z_0} = \frac{\sqrt{Z_{e0}Z_{o0}} + jZ_{e0} \tan(\theta)}{\sqrt{Z_{e0}Z_{o0}} + jZ_{e0} \tan(\theta) + Z_0 + jZ_0\sqrt{Z_{o0}/Z_{e0}} \tan(\theta)} \tag{8.160}$$

Substituting Eq. (8.156) into Eq. (8.160), the following relation is obtained:

$$\frac{Z_{in}^e}{Z_{in}^e + Z_{in}^o} = \frac{Z_0 + jZ_{e0} \tan(\theta)}{Z_0 + jZ_{e0} \tan(\theta) + Z_0 + j\sqrt{Z_{e0}Z_{o0}} \sqrt{Z_{o0}/Z_{e0}} \tan(\theta)}$$

or
$$\frac{Z_{in}^e}{Z_{in}^e + Z_0} = \frac{Z_0 + jZ_{e0}\tan(\theta)}{2Z_0 + j(Z_{e0} + Z_{o0})\tan(\theta)} \qquad (8.161)$$

Similarly, it can be shown that

$$\frac{Z_{in}^o}{Z_{in}^o + Z_0} = \frac{Z_0 + jZ_{o0}\tan(\theta)}{2Z_0 + j(Z_{e0} + Z_{o0})\tan(\theta)} \qquad (8.162)$$

Substituting Eqs (8.161) and (8.162) into Eq. (8.159), we get the following equation:

$$V_3 = V\left[\frac{Z_{in}^e}{Z_{in}^e + Z_0} - \frac{Z_{in}^o}{Z_{in}^o + Z_0}\right]$$

or
$$V_3 = V\left[\frac{Z_0 + jZ_{e0}\tan(\theta)}{2Z_0 + j(Z_{e0} + Z_{o0})\tan(\theta)} - \frac{Z_0 + jZ_{o0}\tan(\theta)}{2Z_0 + j(Z_{e0} + Z_{o0})\tan(\theta)}\right]$$

or
$$V_3 = V\frac{j(Z_{e0} - Z_{o0})\tan(\theta)}{2Z_0 + j(Z_{e0} + Z_{o0})\tan(\theta)} = V\frac{j\left(\dfrac{Z_{e0} - Z_{o0}}{Z_{e0} + Z_{o0}}\right)\tan(\theta)}{2\left(\dfrac{Z_0}{Z_{e0} + Z_{o0}}\right) + j\tan(\theta)} \qquad (8.163)$$

If the length of the line is $\lambda/4$ then $\theta = \pi/2$ or $\tan(\theta) \to \infty$, and Eq. (8.163) reduces to the following form:

$$V_3 = V\frac{j\left(\dfrac{Z_{e0} - Z_{o0}}{Z_{e0} + Z_{o0}}\right)\tan(\theta)}{j\tan(\theta)} = V\left(\frac{Z_{e0} - Z_{o0}}{Z_{e0} + Z_{o0}}\right) \qquad (8.164)$$

Therefore, the coupling coefficient is written as follows:

$$C = \frac{V_3}{V} = \frac{Z_{e0} - Z_{o0}}{Z_{e0} + Z_{o0}} \qquad (8.165)$$

Now,

$$1 - C^2 = 1 - \left(\frac{Z_{e0} - Z_{o0}}{Z_{e0} + Z_{o0}}\right)^2 = \frac{(Z_{e0} + Z_{o0})^2 - (Z_{e0} - Z_{o0})^2}{(Z_{e0} + Z_{o0})^2}$$

or
$$1 - C^2 = \frac{4Z_{e0}Z_{o0}}{(Z_{e0} + Z_{o0})^2} = \left(\frac{2Z_0}{Z_{e0} + Z_{o0}}\right)^2$$

or
$$\sqrt{1 - C^2} = \frac{2Z_0}{Z_{e0} + Z_{o0}} \qquad (8.166)$$

Substituting Eqs (8.165) and (8.166) into Eq. (8.163), we get the following expression:

$$V_3 = V\frac{jC\tan(\theta)}{\sqrt{1 - C^2} + j\tan(\theta)} \qquad (8.167)$$

Proceeding in a similar manner, the voltages at ports 4 and 2 can be expressed as follows:

$$V_4 = V_{e4} + V_{o4} = V_{e2} - V_{o2} = 0 \tag{8.168}$$

and

$$V_2 = V_{e2} + V_{o2} = V \frac{\sqrt{1-C^2}}{\sqrt{1-C^2}\cos(\theta) + j\sin(\theta)} \tag{8.169}$$

For $\theta = \pi/2$, Eq. (8.169) reduces to the following form:

$$V_2 = -jV\sqrt{1-C^2} \tag{8.170}$$

Equations (8.167) and (8.170) shows that there is a 90° phase shift between ports 2 and 3, and hence such structure can be used as a 90° or quadrature hybrid.

If C is known, then the even- and odd-mode characteristic impedances of a coupled-line directional coupler can be calculated from the following equations:

$$Z_{e0} = Z_0\sqrt{\frac{1+C}{1-C}} \tag{8.171}$$

and

$$Z_{o0} = Z_0\sqrt{\frac{1-C}{1+C}} \tag{8.172}$$

This analysis assumes that the line has the same electrical length for both the even- and the odd-mode excitation, which is not true because the even- and odd-mode phase velocities are different. This results in poor directivity for a coupled-line directional coupler. Directivity can be enhanced to a certain extent by the use of two capacitors, as shown in Fig. 8.37(a). Coupling between the lines is also small, as fringing fields only from one side of the line are used. Better coupling can be obtained using additional lines, as shown in Fig. 8.37(b).

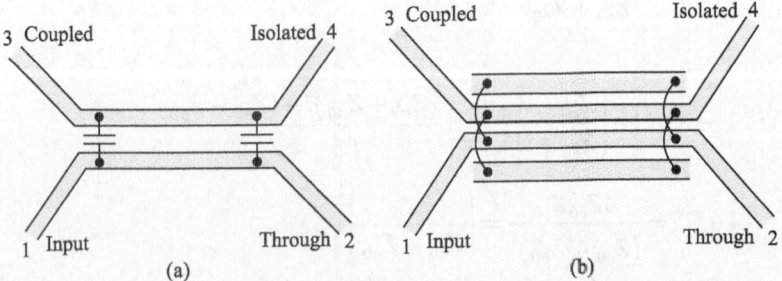

Fig. 8.37 Different directional couplers (a) Compensated microstrip coupled-line directional coupler (b) Coupled-line directional coupler with additional lines for tighter coupling

EXAMPLE 8.8 Calculate the even- and odd-mode impedances of a 15 dB, 50 Ω coupled-line coupler.

Solution Given: $C = 15$ dB and $Z_0 = 50\ \Omega$

Therefore,

$$C = 10^{-15/20} = 0.1778$$

Hence,

$$Z_{e0} = Z_0\sqrt{\frac{1+C}{1-C}} = 50\sqrt{\frac{1+0.1778}{1-0.1778}} = 59.8435\ \Omega$$

and $\quad Z_{o0} = Z_0\sqrt{\frac{1-C}{1-C}} = 50\sqrt{\frac{1-0.1778}{1+0.1778}} = 41.7756\ \Omega$

Practice Problem

8.8 Calculate the even- and odd-mode impedances of a 20 dB, 100 Ω coupled-line coupler. $\qquad$ [**110.5541 Ω, 90.4534 Ω**]

8.12 Multi-section Coupled-line Directional Couplers

The previous section describes a single-section coupled-line directional coupler. Such couplers are of low bandwidth as the section must be a quarter-wavelength long. To achieve a higher bandwidth, cascaded multi-section transmission lines can be used. Due to better phase characteristics, a multi-section transmission line is generally consists of an odd number of transmission line sections. If we assume that the coupling is weak, then from Eq. (8.167) we can write the following relation:

$$\frac{V_3}{V_1} = \frac{V_3}{V} = \frac{jC\tan(\theta)}{\sqrt{1-C^2}+j\tan(\theta)} \approx \frac{jC\tan(\theta)}{1+j\tan(\theta)}$$

or $\quad \dfrac{V_3}{V_1} = \dfrac{jC\sin(\theta)}{\cos(\theta)+j\sin(\theta)} = \dfrac{jC\sin(\theta)}{e^{j\theta}} = jC\sin(\theta)e^{-j\theta} \qquad (8.173)$

Similarly, Eq. (8.169) gives the following expression:

$$\frac{V_2}{V_1} = \frac{V_2}{V} = \frac{\sqrt{1-C^2}}{\sqrt{1-C^2}\cos(\theta)+j\sin(\theta)} = \frac{1}{\cos(\theta)+j\sin(\theta)} = \frac{1}{e^{j\theta}} = e^{-j\theta}$$

$$(8.174)$$

Using Eq. (8.173), the total coupled voltage at port 3 of the multi-section coupler, shown in Fig. 8.38, is given by the following relation:

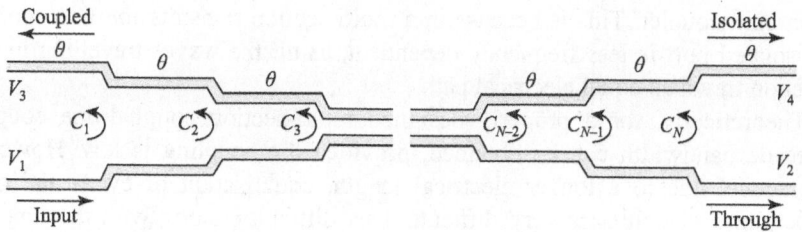

Fig. 8.38 *N*-section coupled-line coupler

$$\frac{V_3}{V_1} = jC_1 \sin(\theta) e^{-j\theta} + jC_2 \sin(\theta) e^{-j\theta} e^{-j2\theta}$$

$$+ jC_3 \sin(\theta) e^{-j\theta} e^{-j4\theta} + \ldots + jC_N \sin(\theta) e^{-j\theta} e^{-j2(N-1)\theta} \qquad (8.175)$$

where C_n is the coupling coefficient of the nth section. Now if the coupler is symmetric, then $C_1 = C_N$, $C_2 = C_{N-1}$, $C_3 = C_{N-2}$, ... and Eq. (8.175) modifies to the following form:

$$\frac{V_3}{V_1} = j\sin(\theta) e^{-j\theta} \left[\left\{ C_1 \left(1 + e^{-j2(N-1)\theta} \right) \right\} + \left\{ C_2 \left(e^{-j2\theta} + e^{-j2(N-2)\theta} \right) \right\} \right.$$

$$\left. + C_3 \left\{ e^{-j4\theta} + e^{-j2(N-3)\theta} \right\} + \ldots + C_{\frac{N+1}{2}} e^{-j(N-1)\theta} \right]$$

or

$$\frac{V_3}{V_1} = j\sin(\theta) e^{-j\theta} \left[\left\{ C_1 e^{-j(N-1)\theta} \left(e^{j(N-1)\theta} + e^{-j(N-1)\theta} \right) \right\} \right.$$

$$+ \left\{ C_2 e^{-j(N-1)\theta} \left(e^{j(N-3)\theta} + e^{-j(N-3)\theta} \right) \right\}$$

$$\left. + C_3 e^{-j(N-1)\theta} \left\{ e^{j(N-5)\theta} + e^{-j(N-5)\theta} \right\} + \ldots + C_{\frac{N+1}{2}} e^{-j(N-1)\theta} \right]$$

or

$$\frac{V_3}{V_1} = j2\sin(\theta) e^{-jN\theta} \left[C_1 \cos\{(N-1)\theta\} + C_2 \cos\{(N-3)\theta\} \right.$$

$$\left. + C_3 \cos\{(N-5)\theta\} + \ldots + \frac{1}{2} C_{\frac{N+1}{2}} \right] \qquad (8.176)$$

For $\theta = \pi/2$, we can write the following equation:

$$C_0 = \left| \frac{V_3}{V_1} \right|_{\theta=\pi/2} = 2 \left[C_1 \cos\left\{ \frac{(N-1)\pi}{2} \right\} + C_2 \cos\left\{ \frac{(N-3)\pi}{2} \right\} \right.$$

$$\left. + C_3 \cos\left\{ \frac{(N-5)\pi}{2} \right\} + \ldots + \frac{1}{2} C_{\frac{N+1}{2}} \right] \qquad (8.177)$$

Equation (8.177) can be used to synthesize the desired coupling response.

It should be noted that we synthesize the coupling coefficient in a multi-section coupled-line directional coupler, whereas directivity is synthesized in a multi-hole directional coupler. This is because in a multi-section transmission line coupler, the isolated port is less frequency dependent, as all the waves travelling in this direction travel an equal electrical path.

Theoretically, for a properly designed multi-section coupled-line coupler, a decade bandwidth can be obtained, provided the coupling is low. However, in practice, due to a longer electrical length, equalization of even- and odd-mode phase velocities is very difficult. This difficulty, along with the junction discontinuity, generally degrades the directivity of the coupler.

Note Since the equalization of even- and odd-mode phase velocities is easier in strip line circuits than in the microstrip lines, the former are usually preferred for the designing of multi-section coupled line transmission line couplers.

EXAMPLE 8.9 Calculate the even- and odd-mode impedances of a three-section, 15 dB, 50 Ω multi-section coupled-line directional coupler with maximally flat frequency response.

Solution Given: $C = 15\,\text{dB}$, $Z_0 = 50\,\Omega$, and $N = 3$

For a maximally flat response:

$$\left.\frac{d^n C(\theta)}{d\theta^n}\right|_{\theta=\pi/2} = 0 \quad \text{for } n = 1, 2$$

Therefore,

$$C = |V_3/V_1| = 2\sin(\theta)\left[C_1\cos(2\theta) + C_2/2\right]$$

or

$$C = C_1\{\sin(3\theta) - \sin(\theta)\} + C_2\sin(\theta) = C_1\sin(3\theta) + (C_2 - C_1)\sin(\theta)$$

or

$$\frac{dC(\theta)}{d\theta} = 3C_1\cos(3\theta) + (C_2 - C_1)\cos(\theta)$$

or

$$\frac{d^2C(\theta)}{d\theta^2} = -9C_1\sin(3\theta) - (C_2 - C_1)\sin(\theta)$$

Now,

$$\left.\frac{dC(\theta)}{d\theta}\right|_{\theta=\pi/2} = 3C_1\cos(3\theta) + (C_2 - C_1)\cos(\theta)\big|_{\theta=\pi/2} = 0$$

and

$$\left.\frac{d^2C(\theta)}{d\theta^2}\right|_{\theta=\pi/2} = -9C_1\sin(3\theta) - (C_2 - C_1)\sin(\theta)\big|_{\theta=\pi/2} = 10C_1 - C_2 = 0$$

or

$$C_2 = 10C_1$$

Again at $\theta = \pi/2$,

$$C = 15\,\text{dB} = 10^{-15/20} = 0.1778 = C_2 - 2C_1$$

Substituting $C_2 = 10C_1$ into the aforementioned equation, the following relation is obtained:

$$10C_1 - 2C_1 = 8C_1 = 0.1778$$

or

$$C_1 = 0.0222$$

and

$$C_2 = 10C_1 = 0.222$$

From symmetry,

$$C_3 = C_1 = 0.0222$$

The even- and odd-mode impedances are as follows:

$$Z_{e0}^1 = Z_{e0}^3 = Z_0\sqrt{\frac{1+C_1}{1-C_1}} = 50\sqrt{\frac{1+0.0222}{1-0.0222}} = 51.1226\,\Omega$$

$$Z_{e0}^2 = Z_0 \sqrt{\frac{1+C_2}{1-C_2}} = 50\sqrt{\frac{1+0.222}{1-0.222}} = 62.6637\ \Omega$$

$$Z_{o0}^1 = Z_{o0}^3 = Z_0 \sqrt{\frac{1-C_1}{1-C_1}} = 50\sqrt{\frac{1-0.0222}{1+0.0222}} = 48.9021\ \Omega$$

$$Z_{o0}^2 = Z_0 \sqrt{\frac{1-C_2}{1-C_2}} = 50\sqrt{\frac{1-0.198}{1+0.198}} = 39.8955\ \Omega$$

Practice Problem

8.9 Calculate the even- and odd-mode impedances of a three-section, 20 dB, 100 Ω multi-section coupled-line directional coupler with maximally flat frequency response.

$$[101.2579\ \Omega,\ 113.3893\ \Omega,\ 98.7577\ \Omega,\ 88.1917\ \Omega]$$

8.13 Lange Couplers

The fringing fields from only one side of a transmission line are used to provide coupling in a coupled-line coupler. This generally results in a loose coupling, as the fringing fields from the other side remain unutilized. Therefore, in order to obtain 3 or 6 dB coupling using a coupled-line coupler, the fringing fields from both sides of the line must be used. One of the practical implementation of this concept is the Lange coupler or inter-digital coupler, as shown in Fig. 8.39(a). In addition to the use of the fringing fields from both sides of the lines, interconnections between them are also used to achieve tighter coupling in a Lange coupler. Such a structure, when implemented, can provide a 3 dB coupling over an octave bandwidth. However, the main disadvantage of a Lange coupler is its practical implementation. Since the lines are very narrow and also close together, it is very difficult to fabricate the necessary bonding wires across the lines. To avoid this, unfolded Lange line couplers can be used, as shown in Fig. 8.39(b).

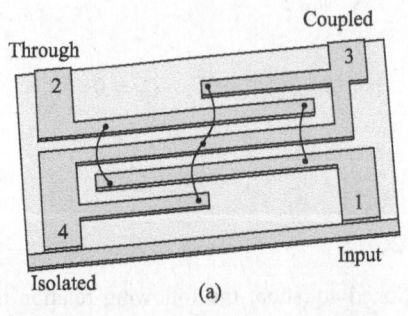

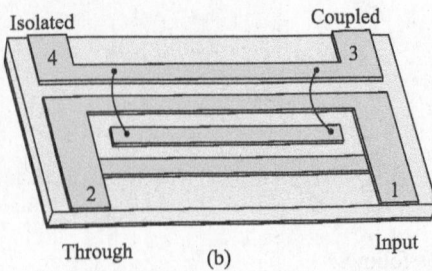

Fig. 8.39 Lange coupler (a) Conventional (b) Unfolded

The equivalent circuit of a Lange coupler is shown in Fig. 8.40(a), where the lines have equal width and spacing. If we further assume that each line is coupled only with its nearest neighbours, then we can effectively modify the equivalent circuit of Fig. 8.40(a) to that shown in Fig. 8.40(b). If Z_{e0} and Z_{o0}, respectively, be the even- and odd-mode characteristic impedances of

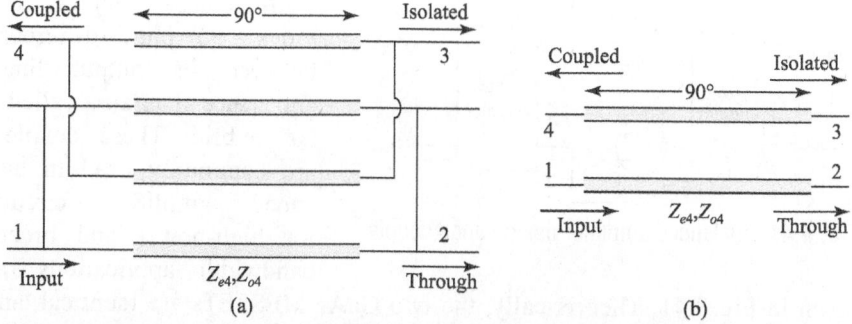

Fig. 8.40 Equivalent circuit of Lange couplers (a) Four-wire coupled-line model (b) Approximate two-wire coupled-line model

any two adjacent pairs, and Z_{e4} and Z_{o4} those of the equivalent two wire model, then the following relations hold:

$$Z_{e4} = Z_{e0} \frac{Z_{o0} + Z_{e0}}{3Z_{o0} + Z_{e0}} \tag{8.178}$$

and

$$Z_{o4} = Z_{o0} \frac{Z_{o0} + Z_{e0}}{3Z_{e0} + Z_{o0}} \tag{8.179}$$

Therefore,

$$Z_0 = \sqrt{Z_{e4}Z_{o4}} = \sqrt{\frac{Z_{e0}Z_{o0}(Z_{o0} + Z_{e0})^2}{(3Z_{e0} + Z_{o0})(3Z_{o0} + Z_{e0})}} \tag{8.180}$$

and

$$C = \frac{Z_{e4} - Z_{o4}}{Z_{e4} + Z_{o4}} = \frac{Z_{e0} \dfrac{Z_{o0} + Z_{e0}}{3Z_{o0} + Z_{e0}} - Z_{o0} \dfrac{Z_{o0} + Z_{e0}}{3Z_{e0} + Z_{o0}}}{Z_{e0} \dfrac{Z_{o0} + Z_{e0}}{3Z_{o0} + Z_{e0}} + Z_{o0} \dfrac{Z_{o0} + Z_{e0}}{3Z_{e0} + Z_{o0}}}$$

or

$$C = \frac{(Z_{o0} + Z_{e0})\{Z_{e0}(3Z_{e0} + Z_{o0}) - Z_{o0}(3Z_{o0} + Z_{e0})\}}{(Z_{o0} + Z_{e0})\{Z_{e0}(3Z_{e0} + Z_{o0}) + Z_{o0}(3Z_{o0} + Z_{e0})\}}$$

or

$$C = \frac{3Z_{e0}^2 + Z_{e0}Z_{o0} - 3Z_{o0}^2 - Z_{e0}Z_{o0}}{3Z_{e0}^2 + Z_{e0}Z_{o0} + 3Z_{o0}^2 + Z_{e0}Z_{o0}} = \frac{3(Z_{e0}^2 - Z_{o0}^2)}{3(Z_{e0}^2 + Z_{o0}^2) + 2Z_{e0}Z_{o0}} \tag{8.181}$$

If C and Z_0 are known, then Eqs (8.180) and (8.181) can be used to find the even- and odd-mode characteristic impedances of any two adjacent pairs, as follows:

$$Z_{e0} = Z_0 \frac{4C - 3 + \sqrt{9 - 8C^2}}{2C\sqrt{(1 - C)/(1 + C)}} \tag{8.182}$$

and

$$Z_{o0} = Z_0 \frac{4C + 3 - \sqrt{9 - 8C^2}}{2C\sqrt{(1 + C)/(1 - C)}} \tag{8.183}$$

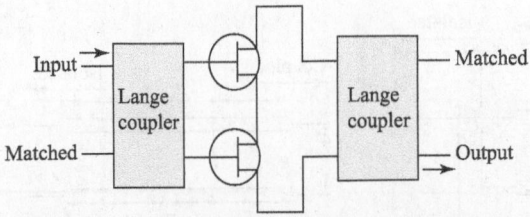

Fig. 8.41 Balanced amplifier using Lange coupler

A Lange coupler provides a 90° phase difference between its output lines, and hence it is also called a 90° hybrid. These couplers are commonly used in balanced amplifier circuits for high-power and broad-bandwidth applications, as shown in Fig. 8.41. Theoretically, the two GaAs MESFETs are identical, and the amplifier is balanced and provides a unity VSWR. Practically, however, the MESFETs are not identical, and fine tuning is required to balance it. Lange couplers are also used in balanced mixers, attenuators, phase shifters, modulators, and discriminators.

8.14 Other Couplers and Power Dividers

Some couplers and power dividers, in addition to those already discussed, are described in the following subsections.

8.14.1 Moreno Crossed Guide Coupler

This is a waveguide directional coupler, consisting of two crossed waveguides and two crossed coupling slots in the common broad wall, as shown in Fig. 8.42. The crossed slots provide tight coupling in both guides. By properly designing the coupling slots, the two waves excited at the auxiliary guide can be cancelled in one direction and added in the other direction.

8.14.2 Schwinger Reversed-phase Coupler

A Schwinger reversed-phase coupler consists of two waveguides, with the narrow wall of the auxiliary waveguide connected to the broad wall of the main waveguide, as shown in Fig. 8.43. Two coupling slots are milled on the common wall of both guides at a distance of $\lambda_g/4$ apart, as shown in the figure. As the slots are on the opposite side of the centreline of waveguide axis, phase cancellation occurs at the isolation port (forward port). The $\lambda_g/4$ spacing facilitates the in-phase addition of the waves at the coupled port (backward port). The main advantage of a Schwinger reversed-phase coupler is that its directivity is essentially independent of frequency,

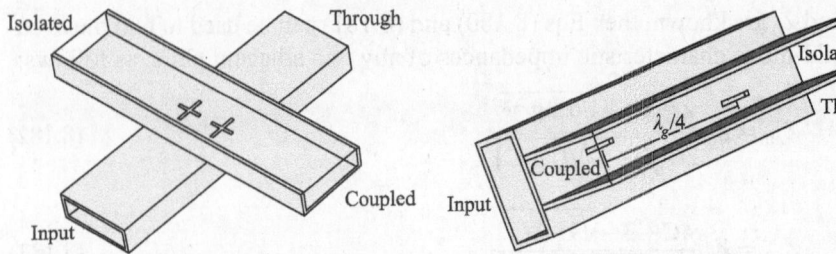

Fig. 8.42 Moreno crossed guide coupler   **Fig. 8.43** Schwinger reversed-phase coupler

whereas its main disadvantage is that its coupling coefficient is very sensitive to the excitation frequency.

8.14.3 Riblet Short-slot Coupler

A Riblet short-slot coupler consists of two parallel waveguides with a common side wall, as shown in Fig. 8.44. Coupling between the two waveguides takes place through a coupling aperture milled on the common side wall of the two waveguides. At the region of the aperture, both the even mode (TE_{10}) and the odd mode (TE_{20}) exist, which propagate through the coupled guide. With proper designing, these modes can be cancelled at the isolated port and added at the coupling port.

8.14.4 Ortho-modal Coupler

An ortho-modal coupler is used to either combine or separate two microwave signals and is shown in Fig. 8.45. The X-component of the electric field existing in waveguide 1 (a square waveguide) will be supported by waveguide 3 and will appear at port 3. Similarly, the Y-component of the electric field existing in waveguide 1 will be supported by waveguide 2 and will appear at port 2. No power can be coupled from port 2 to port 3 and vice versa. Due to this characteristic, this type of a coupler is also called a *polarization duplexer*.

In satellite applications, port 1 is connected to the antenna feed, either port 2 or 3 is connected to the transmitter, and the rest is connected to the receiver. During uplink, the signal follows the path from port 2 to port 1 (if the transmitter is connected to port 2) or from port 3 to port 1 (if the transmitter is connected to port 3).

On the contrary, during downlink, the signal follows the other path (from port 1 to port 3 or from port 1 to port 2), provided that polarization of the uplink and downlink signals is orthogonal.

8.14.5 Turnstile Junction

A turnstile junction is a symmetrical six-port device, as shown in Fig. 8.46. In the network, ports 1–4 correspond

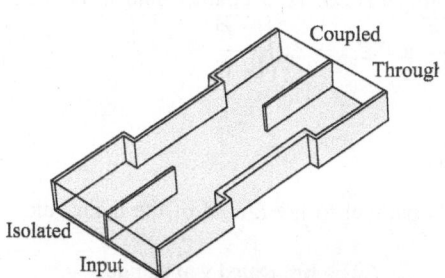

Fig. 8.44   Riblet short-slot coupler

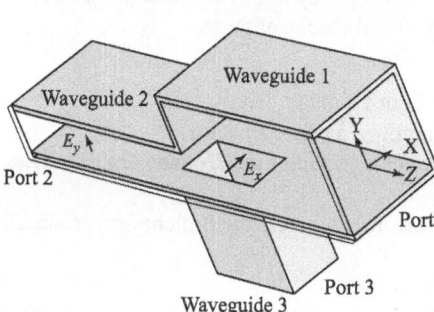

Fig. 8.45   Ortho-modal coupler

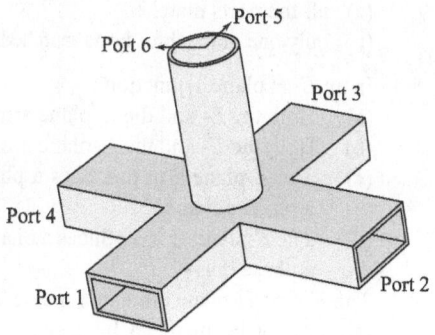

Fig. 8.46   Turnstile junction

to a rectangular waveguide operating in the dominant TE_{10} mode. On the other hand, the circular waveguide, operating in the dominant TE_{11} mode, supports two independent cross-polarized waves and thus exhibits two ports, ports 5 and 6, corresponding to each polarization.

Ports 1–4 can be matched by inserting a tuning post along the symmetry axis within the junction from the bottom, whereas ports 5 and 6 can be matched by inserting a tuning screw in the corresponding arm. Further, ports 5 and 6 are isolated. Thus, the scattering matrix of a turnstile junction can be expressed as follows:

$$[S] = \frac{1}{2} \begin{bmatrix} 0 & 1 & 0 & 1 & \sqrt{2} & 0 \\ 1 & 0 & 1 & 0 & 0 & \sqrt{2} \\ 0 & 1 & 0 & 1 & -\sqrt{2} & 0 \\ 1 & 0 & 1 & 0 & 0 & -\sqrt{2} \\ \sqrt{2} & 0 & -\sqrt{2} & 0 & 0 & 0 \\ 0 & \sqrt{2} & 0 & -\sqrt{2} & 0 & 0 \end{bmatrix} \tag{8.184}$$

Important Formulae

- Coupling$(C) = 10\log_{10}(P_1/P_3)$ dB
- Directivity$(D) = 10\log_{10}(P_3/P_4)$ dB
- Isolation$(I) = 10\log_{10}(P_1/P_4)$ dB
- The even- and odd-mode characteristic impedances of a coupled-line directional coupler can be expressed as $Z_{e0} = Z_0\sqrt{\dfrac{1+C}{1-C}}$ and $Z_{o0} = Z_0\sqrt{\dfrac{1-C}{1+C}}$.

Exercises

Objective-type Questions

8.1 In a three-port network, the side arm is parallel to the E-field of the main guide. The network can be classified as
   (a) an H-plane T-junction
   (b) an E-plane T-junction
   (c) a bifurcated waveguide
   (d) none of these

8.2 It is possible to design a three-port reciprocal and lossless network with
   (a) all the ports matched
   (b) only one of the three ports matched
   (c) any of its two ports matched
   (d) none of these

8.3 In an E–H plane T-junction,
   (a)   Both the E- and the H-plane arm produce a phase delay
   (b)   Both the E- and the H-plane arms produce a phase advance
   (c)   The E-plane arm produces a phase delay, whereas the H-plane arm produces a phase advance
   (d)   The E-plane arm produces a phase advance, whereas the E-plane arm produces a phase delay

8.4 The E- and H-plane arms of a magic T-junction
   (a) Cannot be matched by placing suitable matching elements in the respective arms without destroying the symmetry of the junction

   (b) Can be matched by placing suitable matching elements in the respective arms at the cost of destruction in the symmetry of the junction

   (c) Can be matched by placing suitable matching elements in the respective arms without destroying the symmetry of the junction

   (d) Can never be matched

8.5 When used as a mixer, the E-plane arm of an E–H plane junction is terminated in
   (a) a null detector   (c) a matched load
   (b) a local oscillator   (d) an antenna

8.6 The total length of a rat-race junction is
   (a) $3\lambda_g/2$ (b) $3\lambda_g/4$ (c) $2\lambda_g/3$ (d) $\lambda_g/2$

8.7 Directional couplers are mainly used
   (a) to measure power   (c) to combine power
   (b) to split power   (d) all of these

8.8 Quadrature hybrids are 3 dB directional couplers that produce a phase difference between the output signals of the through and coupled arms of
   (a) 45°   (b) 90°   (c) 180°   (d) 270°.

8.9 A turnstile junction is a
   (a) three-port junction   (c) five-port junction
   (b) four-port junction   (d) six-port junction

8.10 A three-port network has the following S-matrix: $[S] = \begin{bmatrix} S_{11} & S_{12} & S_{13} \\ S_{12} & S_{11} & -S_{13} \\ S_{13} & -S_{13} & S_{33} \end{bmatrix}$. The network is
   (a) An E-plane T-junction
   (b) An H-plane T-junction
   (c) A directional coupler with the isolated port short circuited
   (d) A magic T-junction with one of the collinear arms short circuited

8.11 The coupling coefficient of a directional coupler can be expressed as
   (a) $10\log_{10}(P_1/P_3)$ dB   (c) $10\log_{10}(P_1/P_4)$ dB
   (b) $10\log_{10}(P_3/P_4)$ dB   (d) $10\log_{10}(P_2/P_1)$ dB

8.12 A four-port microwave passive device has a scattering matrix of the following form:

$$\begin{bmatrix} 0 & p & 0 & jq \\ p & 0 & jq & 0 \\ 0 & jq & 0 & p \\ jq & 0 & p & 0 \end{bmatrix}$$

The device is
   (a) a hybrid T-junction   (c) an H-port circulator
   (b) a hybrid ring   (d) a directional coupler

8.13 The length of the E- and H-plane arms of a magic T-junction is $\lambda/2$. If two short circuits are placed at the end of these arms and one of the collinear arm is fed with a power of 10 mW, then the reflected power at the feed port will be
   (a) 0 mW   (b) 2.5 mW   (c) 5 mW   (d) 10 mW

8.14 A magic T-junction is a combination of
    (a) One H-plane T-junction and one E-plane T- junction
    (b) One H-plane T-junction and two E-plane T-junctions
    (c) Two H-plane T-junctions and one E-plane T-junction
    (d) None of these

8.15 A directional coupler with three or more holes is preferred to a two-hole coupler
    (a) To reduce spurious mode generation
    (b) To reduce coupling of the signal
    (c) To increase the bandwidth of the network
    (d) Because it is more efficient

Review Questions

8.1 Describe E- and H-plane T-junctions.
8.2 Show that it is impossible to design a three-port reciprocal and lossless network with all its ports matched.
8.3 Show that it is possible to obtain a three-port network with all its ports matched if it is lossless but not reciprocal.
8.4 Describe the operation of a rat-race junction.
8.5 Derive the S-matrix of a directional coupler.
8.6 Describe the basic operating principle of a two-hole directional coupler.

Problems

8.1 A 20 mW signal is fed into port 2 of a lossless directional coupler of coupling coefficient 20 dB and directivity 50 dB. Find the power at the output ports. *[Hint: Use the definitions of C and D in Eqs (8.78) and (8.79)]*

8.2 A power of 40 dB is incident on a directional coupler with infinite directivity. The power in the coupled port is 10 dB. The output power in the through arm is 35 dB. Find the coupling factor and the insertion loss. *[Hint: Use Eqs (8.78) and (8.80)]*

8.3 A directional coupler of 15 dB coupling and 50 dB directivity produces a transmission loss of 1 dB. For an input power of 10 W at the input port of the main arm, determine the power at the output ports. *[Hint: Use Eqs (8.78)–(8.80)]*

8.4 A magic T-junction is matched terminated at all its ports. A signal power of 1 W from a matched generator is fed into it through its H-arm. Find the output power amplitude at the output ports. *[Hint: Use the property of T-junction]*

8.5 A symmetrical directional coupler with infinite directivity and a forward attenuation of 20 dB is used to monitor the power delivered to a load. Bolometer 1 introduces an SWR of 1.5 on arm 3; bolometer 2 is attached to arm 4. If bolometer 1 reads 10 mW and bolometer 2 reads 1 mW, find the amount of power dissipated in the load and the SWR on arm 2. Assume coupling coefficient is 20 dB. *[Hint: Start with the relation between VSWR and reflection coefficient]*

8.6 Design a three-port resistive power divider for an equal power split and a 120 Ω system impedance. *[Hint: Use the relation $R = Z_0/3$]*

8.7 Design a Wilkinson power divider for a power division ratio of 2:3 and a source impedance of 50 Ω. *[Hint: Use Eqs (8.58)–(8.60)]*

8.8 Design a 20 dB four-hole binomial coupler at 9 GHz using a WR-90 waveguide. Assume that the minimum directivity is 50 dB, and the coupling apertures are round and located at $s = a/4$. Write a MATLAB code to plot the frequency response from 8.2 to 12.4 GHz. *[Hint: Use Eqs (8.126) and (8.127) and Example 8.6]*

Microwave Filters

9.1 Introduction

In Chapter 8, we have discussed different microwave power dividers and directional couplers that are used to divide or couple a part of the power in a specified arm. In a microwave network, sometimes the frequency response at a certain point of the network is also required to be controlled. Microwave filters are used for this purpose. A microwave filter is a two-port network that passes a certain part of a frequency, while blocking the remaining part. For example, a waveguide behaves as a high-pass filter by allowing the frequency above the cut-off to pass and blocking the frequency below the cut-off. Depending on the nature of the pass-band and stop-band, filters can be classified as follows: low-pass, high-pass, band-pass, and band-stop filters. From the point of view of the operating principle, filters can be classified as follows: reflective, absorptive, and lossy filters. A reflective filter consists of inductive and capacitive elements, and reflects the unwanted signal; an absorptive filter dissipates the unwanted signal internally and passes the wanted signal. A lossy filter uses a lossy material inside the filter that produces heavy loss in the unwanted signal, while producing low loss in the required signal. Absorptive and lossy filters are not commonly used, and hence we will describe only reflective filters.

Work on microwave filters started well before the World War II by Mason, Darlington, Fano, Richard, Lawson, etc. Initially, during the 1930s, the image parameter method was developed and was used to design low-frequency filters. However, today most of the filter design is carried out by the insertion loss method. This, added with CAD tools, provided great flexibility in designing more complex filters such as dielectric resonator, HTS, and LTCC.

The operation of a microwave filter is mainly based on the properties of periodic structures. Waveguides and transmission lines, loaded with identical obstacles at periodic intervals, form a periodic structure. Such periodic structures have two interesting properties: (a) pass-band-stop-band characteristics and (b) phase velocity much less than the velocity of light. The first property of periodic structures is used in filter designing, while the second property is used in designing microwave amplifiers, like a travelling wave tube (TWT).

In practice, an ideal filter should provide perfect transmission over the pass-band region and infinite attenuation over the stop-band region. However, such an ideal filter never exists, and we need to approximate the ideal behaviour within an acceptable tolerance. Next, we need to develop a network that shows this approximated frequency response. This procedure is called *filter synthesis*. Two filter synthesis techniques are known: (a) image parameter method (variations: constant-k and m-derived filters) and (b) insertion loss method, which are widely used. In the image parameter method, designing of a filter requires pass- and stop-band characteristics, but the exact frequency response over these regions is not specified. Cut-and-try procedures are often required in such synthesis procedures, to get an overall acceptable frequency response. In comparison, the insertion loss method begins with a complete specification of the filter frequency response and does not require any cut-and try-procedure. Hence, it is often preferred.

Efforts involved in the synthesis of a filter can largely be reduced by the use of impedance and frequency transformation and element normalization. These enable the synthesis of high-pass, band-pass, or band-stop filters with arbitrary frequency bandwidth and source impedance from a low-pass filter prototype with unit cut-off frequency and unit source impedance.

In this chapter, we will first discuss the theory of periodic structures, followed by discussion on various filter designing methods, such as image parameter and insertion loss methods. Both the image parameter and insertion loss methods result in a filter circuit that is based on lumped inductors and capacitors. Since these lumped elements are not useful at high frequencies, we need to replace them by distributed circuit elements. This can be achieved with the help of Richard transformation, Kuroda identities, and impedance and admittance inverters, the details of which will also be discussed. At the end, we will discuss a few widely used filters.

It may be noted that at some point, beyond the stop-band frequency, the response of a distributed circuit filter pops back even if the response of the equivalent lumped circuit filter monotonically decays. The periodic impedance behaviour of a transmission line produces these re-entrant modes. Tackling these re-entrant modes is a big challenge in microwave filter design.

9.2 Periodic Structures

In a transmission line, phase velocity can be expressed as follows:

$$v_p = 1/\sqrt{LC} = 1/\sqrt{\mu\varepsilon} = 1/\sqrt{\mu_0\varepsilon_r\varepsilon_0} \tag{9.1}$$

This equation implies that to reduce the phase velocity, ε_r should be increased. However, to attain this, we must also reduce the cross-sectional dimension of the guide to avoid propagation of higher-order modes. As an alternative, we can increase the value of the per-unit-length parameters L or C, to achieve a lower phase velocity. However, an attempt to increase one of the per-unit-length parameters will reduce the other by the same factor, in order to maintain the product $\mu\varepsilon$ constant. Thus, an effective reduction in phase velocity can be achieved by adding lumped capacitances in shunt. If the spacing between these shunt lumped capacitances is kept small compared to the wavelength, then they will behave as distributed capacitances. However, these additional capacitances will not change the per-unit-length inductance, as they are coming from lumped capacitances. The modified phase velocity can now be expressed as follows:

$$v_p = 1 \Big/ \sqrt{L\left(C + \frac{C_0}{d}\right)} \tag{9.2}$$

where C_0/d is the amount of lumped capacitance added per unit length.

Shunt capacitance can be added in the transmission line by incorporating reactive elements, such as stubs and diaphragms, in the main transmission line. The fringing electric fields in the vicinity of the diaphragm increase the local storage of electric energy and, hence, from a circuit viewpoint, introduce a capacitance.

An infinite transmission line that is periodically loaded with reactive elements, as shown in Fig. 9.1, is called a *periodic structure*. The loading elements, which often create discontinuities in the original transmission line, can generally be represented as lumped reactance, as shown in Fig. 9.2. Such structures support slow wave propagation and possess pass- and stop-band characteristics similar to a filter; hence, these are widely used in filter designing.

Periodic structures can be considered as cascaded connections of infinite unit cells of length d and normalized susceptance b, as shown in Fig. 9.2. Using the *ABCD* matrix representation, the nth unit element can be represented as follows:

$$\begin{bmatrix} V_n \\ I_n \end{bmatrix} = \begin{bmatrix} A & B \\ C & D \end{bmatrix} \begin{bmatrix} V_{n+1} \\ I_{n+1} \end{bmatrix} \tag{9.3}$$

where A, B, C, and D are the *ABCD* parameters of the cascaded connection of a transmission line of length $d/2$, shut susceptance b, and another transmission line of length $d/2$. Therefore,

$$\begin{bmatrix} A & B \\ C & D \end{bmatrix} = \begin{bmatrix} \cos(\theta/2) & j\sin(\theta/2) \\ j\sin(\theta/2) & \cos(\theta/2) \end{bmatrix} \begin{bmatrix} 1 & 0 \\ jb & 1 \end{bmatrix} \begin{bmatrix} \cos(\theta/2) & j\sin(\theta/2) \\ j\sin(\theta/2) & \cos(\theta/2) \end{bmatrix}$$

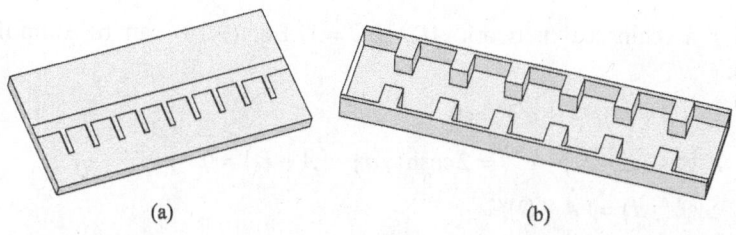

Fig. 9.1 Examples of periodic structures (a) in microstrip (b) in waveguide

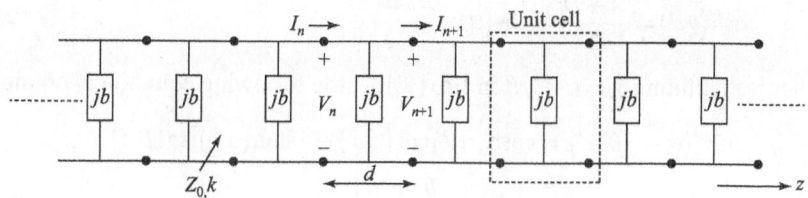

Fig. 9.2 Equivalent circuit of periodically loaded transmission line

or
$$\begin{bmatrix} A & B \\ C & D \end{bmatrix} = \begin{bmatrix} \left\{\cos(\theta) - \dfrac{b\sin(\theta)}{2}\right\} & j\left\{\sin(\theta) + \dfrac{b\cos(\theta)}{2} - \dfrac{b}{2}\right\} \\ j\left\{\sin(\theta) + \dfrac{b\cos(\theta)}{2} - \dfrac{b}{2}\right\} & \left\{\cos(\theta) - \dfrac{b\sin(\theta)}{2}\right\} \end{bmatrix} \quad (9.4)$$

where $\theta = kd$ and k is the propagation constant of the unloaded line.

Now, for the $+z$-direction propagation:

$$V(z) = V(0)e^{-\gamma z} \quad (9.5)$$

and
$$I(z) = I(0)e^{-\gamma z} \quad (9.6)$$

Therefore, in the nth unit cell,

$$V_{n+1} = V_n e^{-\gamma d} \quad (9.7)$$

and
$$I_{n+1} = I_n e^{-\gamma d} \quad (9.8)$$

Using Eqs (9.7) and (9.8) in the LHS of Eq. (9.3), we get the following relation:

$$\begin{bmatrix} V_{n+1}e^{\gamma d} \\ I_{n+1}e^{\gamma d} \end{bmatrix} = \begin{bmatrix} A & B \\ C & D \end{bmatrix}\begin{bmatrix} V_{n+1} \\ I_{n+1} \end{bmatrix}$$

or
$$\begin{bmatrix} AV_{n+1} + BI_{n+1} \\ CV_{n+1} + DI_{n+1} \end{bmatrix} - \begin{bmatrix} V_{n+1}e^{\gamma d} \\ I_{n+1}e^{\gamma d} \end{bmatrix} = \begin{bmatrix} V_{n+1}\left(A - e^{\gamma d}\right) + BI_{n+1} \\ CV_{n+1} + I_{n+1}\left(D - e^{\gamma d}\right) \end{bmatrix} = [0]$$

or
$$\begin{bmatrix} \left(A - e^{\gamma d}\right) & B \\ C & \left(D - e^{\gamma d}\right) \end{bmatrix}\begin{bmatrix} V_{n+1} \\ I_{n+1} \end{bmatrix} = [0] \quad (9.9)$$

For a nontrivial solution, the determinant of this matrix must vanish. Therefore,

$$\left(A - e^{\gamma d}\right)\left(D - e^{\gamma d}\right) - BC = AD - (A+D)e^{\gamma d} + e^{2\gamma d} - BC = 0 \quad (9.10)$$

Since for a reciprocal network $AD - BC = 1$, Eq. (9.10) can be simplified as follows:

$$1 - (A+D)e^{\gamma d} + e^{2\gamma d} = 0 \quad (9.11)$$

or
$$e^{-\gamma d} - (A+D) + e^{\gamma d} = 2\cosh(\gamma d) - (A+D) = 0$$

or
$$\cosh(\gamma d) = (A+D)/2 \quad (9.12)$$

Using Eq. (9.4) in Eq. (9.12), we get the following relation:

$$\cosh(\gamma d) = \frac{A+D}{2} = \cos(\theta) - \frac{b\sin(\theta)}{2} \quad (9.13)$$

Further, substituting $\gamma = \alpha + j\beta$ in Eq. (9.13), the following equation is obtained:

$$\cosh\{(\alpha + j\beta)d\} = \cosh(\alpha d)\cos(\beta d) + j\sinh(\alpha d)\sin(\beta d)$$

or
$$\cosh\{(\alpha + j\beta)d\} = \cos(\theta) - \frac{b\sin(\theta)}{2} \quad (9.14)$$

For $\alpha = 0$ and $\beta \neq 0$, Eq. (9.14) gives the following relation:

$$\cos(\beta d) = \cos(\theta) - \frac{b\sin(\theta)}{2} \tag{9.15}$$

and the case results in non-attenuating wave propagation in the line. This corresponds to pass-band characteristics. On the other hand, for $\alpha \neq 0$ and $\beta = 0$, Eq. (9.14) gives modifies to the following form:

$$\cosh(\alpha d) = \left|\cos(\theta) - \frac{b\sin(\theta)}{2}\right| \geq 1 \tag{9.16}$$

Equation (9.16) results in a non-propagating wave and hence represents stop-band characteristics. Thus, depending on the frequency and the value of the normalized susceptance, the periodically loaded line will exhibit either pass- or stop-band characteristics and will behave as a filter structure.

Note Waves, propagating in a periodic structure, are also called Bloch waves.

The characteristic impedance of the unit cell can be expressed as follows:

$$Z_B = Z_0 V_{n+1}/I_{n+1} \tag{9.17}$$

Here, V_{n+1} and I_{n+1} are both normalized quantities.
From Eq. (9.9), we get the following relation:

$$\left(A - e^{\gamma d}\right)V_{n+1} + BI_{n+1} = 0 \tag{9.18}$$

Using Eq. (9.18) in Eq. (9.17), we get the following relation:

$$Z_B = Z_0 \frac{V_{n+1}}{I_{n+1}} = -Z_0 \frac{BI_{n+1}}{\left(A - e^{\gamma d}\right)I_{n+1}} = -\frac{BZ_0}{A - e^{\gamma d}} \tag{9.19}$$

Further from Eq. (9.11), we get the following expression:

$$e^{\gamma d} = \frac{(A+D) \pm \sqrt{(A+D)^2 - 4}}{2} \tag{9.20}$$

Using Eq. (9.20) in Eq. (9.19), the following relation is obtained:

$$Z_B = -\frac{BZ_0}{A - e^{\gamma d}} = -\frac{BZ_0}{A - \dfrac{(A+D) \pm \sqrt{(A+D)^2 - 4}}{2}}$$

or $$Z_B = -\frac{2BZ_0}{A - D \mp \sqrt{(A+D)^2 - 4}} \tag{9.21}$$

Since the unit cell is symmetrical, $A = D$ and Eq. (9.21) simplifies to the following form:

$$Z_B = \pm \frac{2BZ_0}{\sqrt{4A^2 - 4}} = \pm \frac{BZ_0}{\sqrt{A^2 - 1}} \tag{9.22}$$

where '+' sign corresponds to the characteristic impedance of +z-directed wave and '−' sign corresponds to the characteristic impedance of −z-directed wave. The negative sign occurs because we have considered I_n as +z directed.

For the pass-band $\alpha = 0$, $\beta \neq 0$, and hence from Eq. (9.13) the following relation is obtained:

$$\cosh(\gamma d) = \cosh(j\beta d) = \cos(\beta d) = \frac{A+D}{2} = \frac{2A}{2} = A \tag{9.23}$$

Therefore, in the pass-band $A < 1$. Further, Eq. (9.4) reveals that B is imaginary. Therefore, from Eq. (9.22), we get that Z_B is purely real.

For the stop-band $\alpha \neq 0$, $\beta = 0$, and Eq. (9.13) can be written as follows:

$$\cosh(\gamma d) = \cosh(\alpha d) = \frac{A+D}{2} = \frac{2A}{2} = A \tag{9.24}$$

Therefore, in the stop-band $A > 1$. Thus, Eq. (9.22) reveals that Z_B is purely imaginary.

Now let us consider that the periodic structure is terminated in a load impedance Z_L, as shown in Fig. 9.3. At the terminals of the nth unit cell, pass-band voltages and currents can be represented as follows:

$$V_n = V_n^+ + V_n^- = V_0^+ e^{-j\beta nd} + V_0^- e^{j\beta nd} \tag{9.25}$$

and

$$I_n = I_n^+ + I_n^- = I_0^+ e^{-j\beta nd} + I_0^- e^{j\beta nd} = \frac{V_0^+}{Z_B^+} e^{-j\beta nd} + \frac{V_0^-}{Z_B^-} e^{j\beta nd} \tag{9.26}$$

For $n = N$,

$$V_N = V_N^+ + V_N^- = Z_L I_N \tag{9.27}$$

and

$$I_N = I_N^+ + I_N^- = \frac{V_N^+}{Z_B^+} + \frac{V_N^-}{Z_B^-} \tag{9.28}$$

Substituting Eq. (9.28) into Eq. (9.27), we get the following expression:

$$V_N^+ + V_N^- = Z_L \left(\frac{V_N^+}{Z_B^+} + \frac{V_N^-}{Z_B^-} \right) \text{ or } V_N^+ \left(1 - \frac{Z_L}{Z_B^+} \right) = \left(\frac{Z_L}{Z_B^-} - 1 \right) V_N^-$$

or

$$\Gamma = V_N^- / V_N^+ = -\left(\frac{Z_L}{Z_B^+} - 1 \right) \Big/ \left(\frac{Z_L}{Z_B^-} - 1 \right) \tag{9.29}$$

If the network is symmetrical, then $A = D$

and

$$Z_B^+ = -Z_B^- = Z_B \tag{9.30}$$

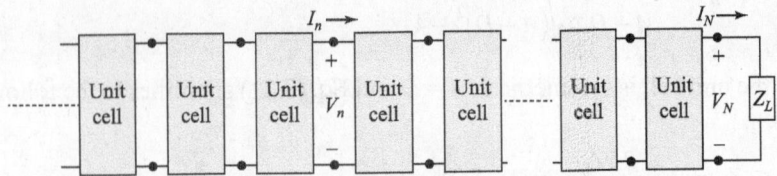

Fig. 9.3 Periodic structure terminated in load impedance

Substituting Eq. (9.30) into Eq. (9.29), we get the following relation:

$$\Gamma = -\left(\frac{Z_L}{Z_B}-1\right)\Bigg/\left(-\frac{Z_L}{Z_B}-1\right) = \left(\frac{Z_L}{Z_B}-1\right)\Bigg/\left(\frac{Z_L}{Z_B}+1\right) = \frac{Z_L - Z_B}{Z_L + Z_B} \qquad (9.31)$$

Therefore, to avoid any reflection from the end and to implement a finite terminated periodic structure as a virtually infinite periodic structure, we must set $Z_L = Z_B$. Z_B is the characteristic impedance of a unit cell and is also called Bloch impedance.

9.3 Filter Parameters

Let us consider the filter block shown in Fig. 9.4.
The basic filter parameters are as follows:

Insertion loss The insertion loss can be defined as follows:

$$\text{IL} = -10\log_{10}\left(P_L/P_{\text{in}}\right) = -10\log_{10}\left(1 - |\Gamma|^2\right) \qquad (9.32)$$

Return loss The return loss is defined as follows:

$$\text{RL} = -10\log_{10}\left(P_R/P_{\text{in}}\right) = -10\log_{10}\left(|\Gamma|^2\right) = -20\log_{10}|\Gamma| \qquad (9.33)$$

Group delay If ϕ_T be the transmission phase, that is, $\phi_T = \arg(\text{IL})$ $\qquad (9.34)$
then the group delay is defined as follows:

$$\tau_d = \frac{d\phi_T}{d\omega} = \frac{1}{2\pi}\frac{d\phi_T}{df} \qquad (9.35)$$

where the angle is in radians.

A group delay measures how long a signal takes to propagate through a filter. If the group delay is constant, then components of a multi-frequency signal will travel at the same velocity through the device and hence there will be no frequency dispersion. Alternatively, any deviation from the constant group delay will cause an FM signal to become distorted.

An ideal filter should have zero insertion loss and a constant group delay over the pass-band and infinite rejection everywhere else. In practice, such an ideal filter does not exist. In addition, no filter can operate over an unlimited frequency range. Sometimes, filters also exhibit spurious responses, where they have rejection in the pass-band and low loss in the stop-band.

Fig. 9.4 General two-port filter network

9.4 Lossless Ladder Network Synthesis

Let us consider that a ladder network, as shown in Fig. 9.5, is terminated in a load resistance R. To find the input

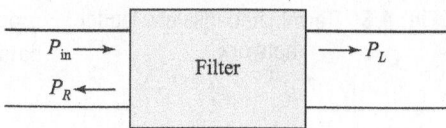

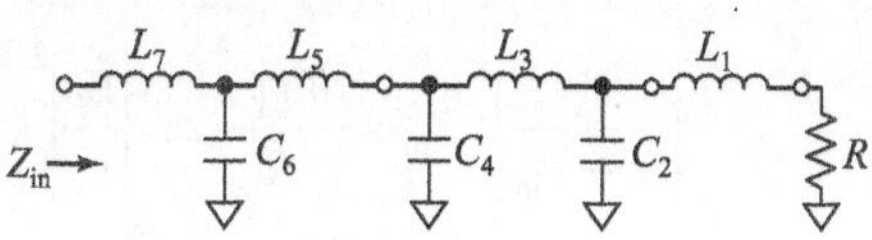

Fig. 9.5 Terminated LC ladder network

impedance of such a network, let us start from the load end. The input impedance at the left of L_1 is as follows: $Z_1 = R + j\omega L_1$ (9.36)

Similarly, the impedance at the left of the capacitor C_2 will be as follows:

$$Z_2 = \frac{1}{Y_2} = \frac{1}{j\omega C_2 + \frac{1}{Z_1}} = \frac{1}{j\omega C_2 + \frac{1}{R + j\omega L_1}} \qquad (9.37)$$

The impedance at the left of L_3 will be as follows:

$$Z_3 = j\omega L_3 + \frac{1}{j\omega C_2 + \frac{1}{j\omega L_1 + R}} \qquad (9.38)$$

Proceeding in a similar way, the impedance at the left of the capacitor C_4 will be as follows:

$$Z_4 = \frac{1}{Y_2} = \frac{1}{j\omega C_4 + \frac{1}{Z_3}} = \frac{1}{j\omega C_4 + \cfrac{1}{j\omega L_3 + \cfrac{1}{j\omega C_2 + \cfrac{1}{j\omega L_1 + R}}}} \qquad (9.39)$$

This process can be repeated to find the input impedance of the network, which will be of the following form:

$$Z_{in} = \cfrac{1}{\cdots + \cfrac{1}{j\omega C_4 + \cfrac{1}{j\omega L_3 + \cfrac{1}{j\omega C_2 + \cfrac{1}{j\omega L_1 + R}}}}} \qquad (9.40)$$

where the first new term of each successive step in the expansion will be the value of individual ladder elements.

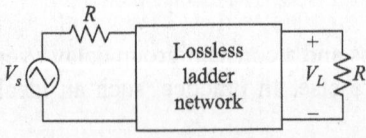

Now, we let us consider Fig. 9.6. Since the ladder network is lossless, the power delivered to the load must be equal to the power delivered to the input. Further, since power is delivered only to the real part of input impedance $Z_{in}(= R_{in} + jX_{in})$, the power balance criterion can be written as follows:

Fig. 9.6 Terminated lossless ladder network

$$|I_{in}|^2 R_{in} = |V_L|^2 / R \qquad (9.41)$$

where $I_{in} = \dfrac{V_s}{R + Z_{in}}$ (9.42)

Substituting Eq. (9.42) into Eq. (9.41), we get the following relation:

$$\left|\frac{V_s}{R + Z_{in}}\right|^2 R_{in} = \frac{|V_L|^2}{R} \quad \text{or} \quad \left|\frac{V_L}{V_s}\right|^2 = \frac{R R_{in}}{|R + Z_{in}|^2} \qquad (9.43)$$

Equation (9.43) thus relates the transfer function of the network with the known load impedance and input impedance of the network.

Now, let us consider an auxiliary function, which is defined as follows:

$$|A(j\omega)|^2 = 1 - 4\left|\frac{V_L}{V_s}\right|^2 \qquad (9.44)$$

Substituting Eq. (9.43) into Eq. (9.44), we get the following relation:

$$|A(j\omega)|^2 = 1 - \frac{4RR_{in}}{|R+Z_{in}|^2} = \frac{|R+Z_{in}|^2 - 4RR_{in}}{|R+Z_{in}|^2} = \frac{|R-Z_{in}|^2}{|R+Z_{in}|^2} \qquad (9.45)$$

Solution of Eq. (9.45) yields two results:

$$Z_{in} = R\frac{1-A(j\omega)}{1+A(j\omega)} \qquad (9.46)$$

and

$$Z_{in} = R\frac{1+A(j\omega)}{1-A(j\omega)} \qquad (9.47)$$

The two solutions will lead two different but equivalent networks. An observation of the two equations further reveals that if R is unity, then the expressions are reciprocal to each other.

If the normalized squared magnitude response can be expressed as

$$|H(j\omega)|^2 = \frac{1}{1+\varepsilon^2(\omega/\omega_c)^{2N}} \qquad (9.48)$$

then we can write the following equation:

$$4\left|\frac{V_L}{V_s}\right|^2 = |H(j\omega)|^2 = \frac{1}{1+\varepsilon^2(\omega/\omega_c)^{2N}} \qquad (9.49)$$

The presence of factor 4 in the LHS of Eq. (9.49) can be attributed to the fact that $|V_L/V_s|$ has a value of 0.5 at DC. Now, substituting Eq. (9.49) into Eq. (9.44), we get the following relation:

$$|A(j\omega)|^2 = 1 - 4\left|\frac{V_L}{V_s}\right|^2 = 1 - |H(j\omega)|^2$$

or

$$|A(j\omega)|^2 = 1 - \frac{1}{1+\varepsilon^2(\omega/\omega_c)^{2N}} = \frac{\varepsilon^2(\omega/\omega_c)^{2N}}{1+\varepsilon^2(\omega/\omega_c)^{2N}} \qquad (9.50)$$

If $\omega_c = 1$, this equation reduces to the following form:

$$|A(j\omega)|^2 = \frac{\varepsilon^2\omega^{2N}}{1+\varepsilon^2\omega^{2N}} \qquad (9.51)$$

If we assume that

$$s = j\omega \qquad (9.52)$$

then Eq. (9.51) can be rewritten as follows:

$$|A(s)|^2 = \frac{\varepsilon^2 s^{2N}}{1+\varepsilon^2 s^{2N}} \qquad (9.53)$$

Now to find $A(s)$, we need to find the poles of $|A(s)|^2$ lying on the left half plane. These poles can be found by forcing the following condition:

$$1+\varepsilon^2 s^{2N} = 0 \qquad (9.54)$$

or $\qquad s^{2N} = -\dfrac{1}{\varepsilon^2}$

or $\qquad s_k = \dfrac{1}{\varepsilon^{1/N}} e^{j\frac{(2k+1)\pi}{2N}}$ for $k = 0, 1, 2, \ldots, (2N-1)$ $\qquad$ (9.55)

For a fourth-order filter, we can find that s_0, s_1, s_6, and s_7 lie on the right half plane and hence can be neglected, whereas s_2–s_5 lie on the left half plane. Considering $\tau_k = 1/s_k$, we can write the following equations:

$$\tau_2 = \frac{1}{s_2} = \varepsilon^{1/N} e^{-j5\pi/8} \approx \varepsilon^{1/N}(-0.3827 - j0.9239) \qquad (9.56)$$

$$\tau_3 = \frac{1}{s_3} = \varepsilon^{1/N} e^{-j7\pi/8} \approx \varepsilon^{1/N}(-0.9239 - j0.3827) \qquad (9.57)$$

$$\tau_4 = \frac{1}{s_4} = \varepsilon^{1/N} e^{-j9\pi/8} \approx \varepsilon^{1/N}(-0.9239 + j0.3827) \qquad (9.58)$$

and $\qquad \tau_5 = \dfrac{1}{s_5} = \varepsilon^{1/N} e^{-j11\pi/8} \approx \varepsilon^{1/N}(-0.3827 + j0.9239) \qquad$ (9.59)

Using these values of τ's, $A(s)$ can be written as follows:

$$A(s) = \frac{\varepsilon s^4}{(\tau_2 s - 1)(\tau_3 s - 1)(\tau_4 s - 1)(\tau_5 s - 1)} \qquad (9.60)$$

Again substituting Eq. (9.60) into Eqs (9.46) and (9.47) and assuming that $R = 1$, the following relations are obtained:

$$Z_{in} = \frac{(\tau_2 s - 1)(\tau_3 s - 1)(\tau_4 s - 1)(\tau_5 s - 1) - \varepsilon s^4}{(\tau_2 s - 1)(\tau_3 s - 1)(\tau_4 s - 1)(\tau_5 s - 1) + \varepsilon s^4} \qquad (9.61)$$

and $\qquad Z_{in} = \dfrac{(\tau_2 s - 1)(\tau_3 s - 1)(\tau_4 s - 1)(\tau_5 s - 1) + \varepsilon s^4}{(\tau_2 s - 1)(\tau_3 s - 1)(\tau_4 s - 1)(\tau_5 s - 1) - \varepsilon s^4} \qquad$ (9.62)

Substituting the roots, we can write Eq. (9.61), after simplification, as follows:

$$Z_{in} = \frac{\left[\alpha^2 s^2 + (2\alpha a)s + 1\right]\left[\alpha^2 s^2 + (2\alpha b)s + 1\right] - \varepsilon s^4}{\left[\alpha^2 s^2 + (2\alpha a)s + 1\right]\left[\alpha^2 s^2 + (2\alpha b)s + 1\right] + \varepsilon s^4} \qquad (9.63)$$

where $\qquad \alpha = \varepsilon^{1/4}$ $\qquad\qquad\qquad\qquad\qquad\qquad\qquad\qquad\qquad\qquad\qquad$ (9.64)

$\qquad\qquad a \approx 0.3827$ $\qquad\qquad\qquad\qquad\qquad\qquad\qquad\qquad\qquad\qquad\qquad$ (9.65)

and $\qquad b \approx 0.9239$ $\qquad\qquad\qquad\qquad\qquad\qquad\qquad\qquad\qquad\qquad\qquad$ (9.66)

Equation (9.63) can further be simplified as follows:

$$Z_{in} = \frac{\varepsilon s^4 + 2\alpha^3(a+b)s^3 + 2\alpha^2(1+2ab)s^2 + 2\alpha(a+b)s + 1 - \varepsilon s^4}{\varepsilon s^4 + 2\alpha^3(a+b)s^3 + 2\alpha^2(1+2ab)s^2 + 2\alpha(a+b)s + 1 + \varepsilon s^4}$$

or $\qquad Z_{in} = \dfrac{2\alpha^3(a+b)s^3 + 2\alpha^2(1+2ab)s^2 + 2\alpha(a+b)s + 1}{2\varepsilon s^4 + 2\alpha^3(a+b)s^3 + 2\alpha^2(1+2ab)s^2 + 2\alpha(a+b)s + 1}$

or $\quad Z_{in} = \dfrac{2\alpha^3(a+b)s^3 + 2\alpha^2(1+2ab)s^2 + 2\alpha(a+b)s + 1}{2\alpha^4 s^4 + 2\alpha^3(a+b)s^3 + 2\alpha^2(1+2ab)s^2 + 2\alpha(a+b)s + 1}$ \qquad (9.67)

where we have used Eq. (9.64).
Similarly, if we proceed with Eq. (9.62), we will get the following expression:

$$Z_{in} = \dfrac{2\alpha^4 s^4 + 2\alpha^3(a+b)s^3 + 2\alpha^2(1+2ab)s^2 + 2\alpha(a+b)s + 1}{2\alpha^3(a+b)s^3 + 2\alpha^2(1+2ab)s^2 + 2\alpha(a+b)s + 1} \qquad (9.68)$$

Now, the expression of the input impedance requires to be rewritten in terms of continued fractions so that, by a comparison with Eq. (9.39), we can find the values of individual components. This can be accomplished using the synthetic division technique. After carrying out synthetic division on Eq. (9.67) and comparing the terms with Eq. (9.39), expressions for the inductance and capacitances can be written as follows:

$$L_1 = \frac{\alpha}{a+b} \qquad (9.69)$$

$$C_2 = \frac{\alpha(a+b)}{2ab} \qquad (9.70)$$

$$L_3 = \frac{8\alpha(ab)^2}{(a+b)(4ab-1)} \qquad (9.71)$$

and $\quad C_4 = \dfrac{\alpha(a+b)(4ab-1)}{2ab} \qquad (9.72)$

If we start with Eq. (9.68), we will also be able to find the component values in a similar way. However, the resulting filter for this case will be a 'C-first' type, as compared with the 'L-first' filter we obtained from Eq. (9.67).

9.5 Filter Design by Image Parameter Method

Let us consider the two-port network shown in Fig. 9.7. The network is terminated by impedances Z_{i1} on the input side and Z_{i2} on the output side, where Z_{i1} and Z_{i2} are the input and output impedances when the output and input are terminated by impedances Z_{i2} and Z_{i1}, respectively. Thus, both the input and output are matched. Impedances Z_{i1} and Z_{i2} are called the image impedances of Z_{i2} and Z_{i1}, respectively. Now for the network,

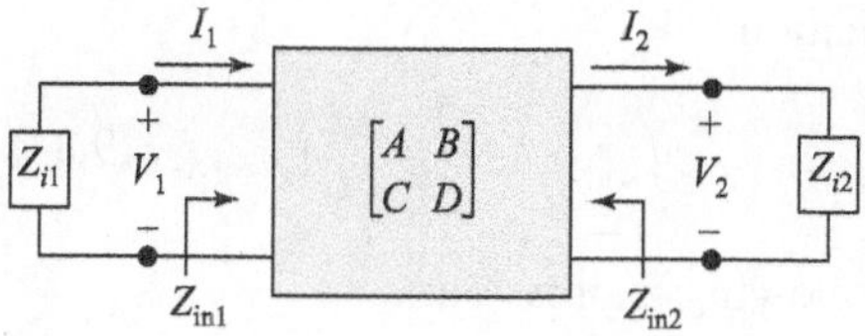

$$V_1 = AV_2 + BI_2 \qquad (9.73)$$

and $\quad I_1 = CV_2 + DI_2 \qquad (9.74)$

The input impedance at port 1 when port 2 is terminated by impedance Z_{i2} is written as follows:

Fig. 9.7 Two-port network terminated in its image impedances

$$Z_{in1} = \frac{V_1}{I_1} = \frac{AV_2 + BI_2}{CV_2 + DI_2} = \frac{AV_2/I_2 + B}{CV_2/I_2 + D} = \frac{AZ_{i2} + B}{CZ_{i2} + D} \tag{9.75}$$

Now,

$$\begin{bmatrix} V_1 \\ I_1 \end{bmatrix} = \begin{bmatrix} A & B \\ C & D \end{bmatrix}\begin{bmatrix} V_2 \\ I_2 \end{bmatrix}$$

or

$$\begin{bmatrix} V_2 \\ I_2 \end{bmatrix} = \begin{bmatrix} A & B \\ C & D \end{bmatrix}^{-1}\begin{bmatrix} V_1 \\ I_1 \end{bmatrix} = \frac{1}{AD - BC}\begin{bmatrix} D & -B \\ -C & A \end{bmatrix}\begin{bmatrix} V_1 \\ I_1 \end{bmatrix} = \begin{bmatrix} D & -B \\ -C & A \end{bmatrix}\begin{bmatrix} V_1 \\ I_1 \end{bmatrix} \tag{9.76}$$

as $AD - BC = 1$. Therefore,

$$V_2 = DV_1 - BI_1 \tag{9.77}$$

and

$$I_2 = -CV_1 + AI_1 \tag{9.78}$$

The input impedance at port 2 will be as follows:

$$Z_{in2} = -\frac{V_2}{I_2} = -\frac{DV_1 - BI_1}{-CV_1 + AI_1} = -\frac{DV_1/I_1 - B}{-CV_1/I_1 + A}$$

or

$$Z_{in2} = -\frac{D(-Z_{i1}) - B}{-C(-Z_{i1}) + A} = \frac{DZ_{i1} + B}{CZ_{i1} + A} \tag{9.79}$$

Now, as per definition:

$$Z_{i1} = Z_{in1} = \frac{AZ_{i2} + B}{CZ_{i2} + D} \tag{9.80}$$

and

$$Z_{i2} = Z_{in2} = \frac{DZ_{i1} + B}{CZ_{i1} + A} \tag{9.81}$$

Substituting Eq. (9.81) into Eq. (9.80), we get the following relation:

$$Z_{i1} = \frac{ADZ_{i1} + AB + BCZ_{i1} + AB}{CDZ_{i1} + BC + CDZ_{i1} + AD} = \frac{(AD + BC)Z_{i1} + 2AB}{2CDZ_{i1} + (AD + BC)}$$

or

$$2CDZ_{i1}^2 = 2AB \quad \text{or} \quad Z_{i1} = \sqrt{\frac{AB}{CD}} \tag{9.82}$$

Similarly, substituting Eq. (9.80) into Eq. (9.81), we get the following expression:

$$Z_{i2} = \sqrt{\frac{BD}{AC}} \tag{9.83}$$

If the network is symmetric, then $A = D$ and hence $Z_{i1} = Z_{i2}$. The impedance transfer ratio is given by the following equation:

$$\frac{Z_{i2}}{Z_{i1}} = \sqrt{\frac{BD}{AC}}\sqrt{\frac{CD}{AB}} = \frac{D}{A} \tag{9.84}$$

Again from Eq. (9.77), we can write the following expression:

$$V_2 = DV_1 - BI_1 = (D - BI_1/V_1)V_1 = (D - B/Z_{i1})V_1$$

or

$$V_2/V_1 = (D - B/Z_{i1}) \tag{9.85}$$

Substitution of Eq. (9.82) into Eq. (9.85) results in the following equation:

$$\frac{V_2}{V_1} = D - \frac{B}{Z_{i1}} = D - B\sqrt{\frac{CD}{AB}} = \sqrt{\frac{D}{A}}\left(\sqrt{AD} - \sqrt{BC}\right) \tag{9.86}$$

Similarly, from Eq. (9.78), we can write the following relation:

$$I_2 = (-CV_1/I_1 + A)I_1 = (-CZ_{i1} + A)I_1$$

or $\qquad I_2/I_1 = (-CZ_{i1} + A) \tag{9.87}$

Further substituting Eq. (9.82) into Eq. (9.87), we get the following relation:

$$\frac{I_2}{I_1} = -CZ_{i1} + A = -C\sqrt{\frac{AB}{CD}} + A = \sqrt{\frac{A}{D}}\left(\sqrt{AD} - \sqrt{BC}\right) \tag{9.88}$$

In Eqs (9.86) and (9.88), the factor $\sqrt{D/A}$ occurs in reciprocal positions and hence can be interpreted as the transfer turns ratio. We can also define that

$$e^{-\gamma} = \sqrt{AD} - \sqrt{BC} \text{ where } \gamma = \alpha + j\beta. \tag{9.89}$$

Now, from Eq. (9.89):

$$e^{\gamma} = \frac{1}{\sqrt{AD} - \sqrt{BC}} = \frac{AD - BC}{\sqrt{AD} - \sqrt{BC}} = \sqrt{AD} + \sqrt{BC} \tag{9.90}$$

Using Eqs (9.89) and (9.90), we can write the following relation:

$$\cosh(\gamma) = \frac{e^{-\gamma} + e^{\gamma}}{2} = \frac{\sqrt{AD} - \sqrt{BC} + \sqrt{AD} + \sqrt{BC}}{2} = \frac{2\sqrt{AD}}{2} = \sqrt{AD} \tag{9.91}$$

Similarly,

$$\sinh(\gamma) = \frac{e^{\gamma} - e^{-\gamma}}{2} = \frac{\sqrt{AD} + \sqrt{BC} - \sqrt{AD} + \sqrt{BC}}{2} = \frac{2\sqrt{BC}}{2} = \sqrt{BC} \tag{9.92}$$

In a pass-band, $\gamma = j\beta$. Therefore, $\cosh(\gamma) = \cosh(j\beta) = \cos(\beta) = \sqrt{AD}$

or $\qquad |AD| < 1 \tag{9.93}$

If the number of two-port networks cascaded is N, then from Eqs (9.86) and (9.89), voltage transfer function of the nth section can be written as follows:

$$\frac{V_{n+1}}{V_n} = \sqrt{\frac{D_n}{A_n}}\left(\sqrt{A_n D_n} - \sqrt{B_n C_n}\right) = T_n e^{-\gamma_n} \tag{9.94}$$

where $\qquad T_n = \sqrt{D_n/A_n} \tag{9.95}$

and $\qquad e^{-\gamma_n} = \sqrt{A_n D_n} - \sqrt{B_n C_n} \tag{9.96}$

The overall voltage transfer function can thus be written as follows:

$$\frac{V_{N+1}}{V_1} = \frac{V_{N+1}}{V_N}\frac{V_N}{V_{N-1}}\cdots\frac{V_{n+2}}{V_{n+1}}\frac{V_{n+1}}{V_n}\frac{V_n}{V_{n-1}}\cdots\frac{V_3}{V_2}\frac{V_2}{V_1}$$

$$= T_N e^{-\gamma_N} T_{N-1} e^{-\gamma_{N-1}} \cdots T_{n+1} e^{-\gamma_{n+1}} T_n e^{-\gamma_n} T_{n-1} e^{-\gamma_{n-1}} \cdots T_2 e^{-\gamma_2} T_1 e^{-\gamma_1}$$

or $\qquad \dfrac{V_{N+1}}{V_1} = \displaystyle\prod_{n=1}^{N} T_n e^{-\gamma_n} \tag{9.97}$

provided that the output image impedance of the nth section is equal to the input image impedance of $(n + 1)$th section.

Table 9.1 *ABCD* parameters, image impedances, and propagation factor of *T*- and pi-networks

Network	*ABCD* Parameters	Image Impedance	Propagation Factor
	$A = 1 + \dfrac{Z_1}{2Z_2}$ $B = Z_1 + \dfrac{Z_1^2}{4Z_2}$ $C = 1/Z_2$ $D = 1 + \dfrac{Z_1}{2Z_2}$	$Z_{iT} = \sqrt{Z_1 Z_2 \left(1 + \dfrac{Z_1}{4Z_2}\right)}$	$e^{\gamma} = 1 + \dfrac{Z_1}{2Z_2}$ $+ \sqrt{\dfrac{Z_1}{Z_2} + \dfrac{Z_1^2}{4Z_2^2}}$
	$A = 1 + \dfrac{Z_1}{2Z_2}$ $B = Z_1$ $C = \dfrac{1}{Z_2} + \dfrac{Z_1^2}{4Z_2}$ $D = 1 + \dfrac{Z_1}{2Z_2}$	$Z_{i\pi} = \sqrt{\dfrac{Z_1 Z_2}{1 + \dfrac{Z_1}{4Z_2}}}$	$e^{\gamma} = 1 + \dfrac{Z_1}{2Z_2}$ $+ \sqrt{\dfrac{Z_1}{Z_2} + \dfrac{Z_1^2}{4Z_2^2}}$

If the filter is terminated in a load Z_{iN}, equal to the image impedance of the output section, and in a source impedance Z_{i1}, equal to the image impedance of the input section, then the filter is matched and maximum power transfer takes place. The overall impedance transfer ratio is expressed as follows:

$$\frac{Z_{iN}}{Z_{i1}} = \frac{Z_{iN}}{Z_{i(N-1)}} \frac{Z_{i(N-1)}}{Z_{i(N-2)}} \cdots \frac{Z_{i(n+1)}}{Z_{in}} \frac{Z_{in}}{Z_{i(n-1)}} \cdots \frac{Z_{i3}}{Z_{i2}} \frac{Z_{i2}}{Z_{i1}}$$

$$= T_N^2 T_{N-1}^2 \cdots T_n^2 T_{n-1}^2 \cdots T_2^2 T_1^2 = \prod_{n=1}^{N} T_n^2 \tag{9.98}$$

For a symmetrical network, $Z_{iN} = Z_{i1}$, and hence no impedance transfer takes place.

In the following sections, we will discuss the two versions of image parameter method: (a) constant-*k* section and (b) *m*-derived filter. The basic understanding of these two versions requires the knowledge of the *ABCD* matrix of a *T*- and a pi-network. The *ABCD* parameters, image impedances, and propagation factors of a *T*- and a pi-network are given in Table 9.1.

9.5.1 Constant-*k* Filter Sections

Let us consider the basic lumped element low-pass filter prototype shown in Fig. 9.8. Comparing with Table 9.1, we can write that $Z_1 = j\omega L$ and $Z_2 = \dfrac{1}{j\omega C}$. Therefore, the image impedance of the network can be found as follows:

$$Z_{iT} = \sqrt{\frac{j\omega L}{j\omega C}\left(1 + \frac{j^2 \omega^2 LC}{4}\right)} = \sqrt{\frac{L}{C}\left(1 - \frac{\omega^2 LC}{4}\right)} \tag{9.99}$$

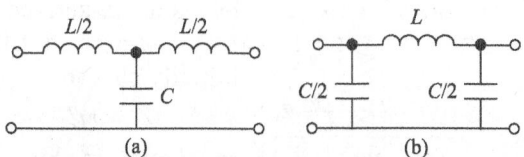

Fig. 9.8 Low-pass constant-k filter section (a) T-section
(b) Pi-section

The cut-off frequency and characteristic impedance of the network are as follows:

$$\omega_c = 2/\sqrt{LC} \tag{9.100}$$

$$R_0 = \sqrt{L/C} = k \tag{9.101}$$

Since the $\sqrt{L/C}$ ratio is constant throughout the circuit, it is called a constant-k filter section.

Substituting Eqs (9.100) and (9.101) into Eq. (9.99), we get the following relation:

$$Z_{iT} = \sqrt{\frac{L}{C}\left(1 - \frac{\omega^2 LC}{4}\right)} = R_0\sqrt{\left(1 - \frac{\omega^2}{\omega_c^2}\right)} \tag{9.102}$$

The propagation factor is expressed as follows:

$$e^\gamma = 1 + \frac{j^2\omega^2 LC}{2} + \sqrt{j^2\omega^2 LC + \frac{j^4\omega^4 L^2 C^2}{4}}$$

or
$$e^\gamma = 1 - \frac{2\omega^2}{\omega_c^2} + \sqrt{-\frac{4\omega^2}{\omega_c^2} + \frac{4\omega^4}{\omega_c^4}} = 1 - \frac{2\omega^2}{\omega_c^2} + \frac{2\omega}{\omega_c}\sqrt{\frac{\omega^2}{\omega_c^2} - 1} \tag{9.103}$$

If $\omega < \omega_c$, then Eqs (9.102) and (9.103) show that Z_{iT} is real and γ is imaginary, indicating that this is the pass-band of the filter. On the other hand, when $\omega > \omega_c$, Z_{iT} is imaginary and γ is real, which implies that this is the stop-band of the filter.

For the pi-network circuit, again we have $Z_1 = j\omega L$ and $Z_2 = 1/(j\omega C)$. Therefore, the propagation factor, characteristic impedance, and cut-off frequency of the pi-network are the same as those of the T-network. However, the image impedance $Z_{i\pi} = Z_{iT} = R_0$ only at $\omega = 0$, for other frequencies $Z_{i\pi} \neq Z_{iT}$.

The preceding results are based on the assumption that the network is terminated in its image impedances at both ends. This is one of the major disadvantages of the image parameter design method because, as evident from Eq. (9.102), image impedance is a function of frequency. Another disadvantage of this model is that its attenuation near the cut-off is very small.

9.5.2 *m*-Derived Filter Sections

Disadvantages of constant-k filter section, including dependence of image impedance on frequency and low attenuation near cut-off, can be overcome

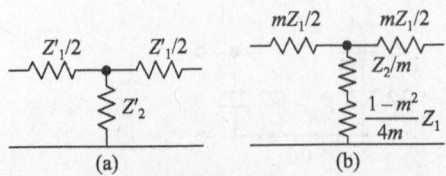

Fig. 9.9 Development of *m*-derived filter section (a) General *m*-derived section (b) Final *m*-derived section

by using *m*-derived filter sections. In an *m*-derived filter section, we initially choose

$$Z_1' = mZ_1 \qquad (9.104)$$

as shown in Fig. 9.9. Next, we choose Z_2' to obtain the same image impedance as for the constant-*k* section. Therefore,

$$Z_{iT} = \sqrt{Z_1 Z_2 \left(1 + \frac{Z_1}{4Z_2}\right)} = \sqrt{Z_1' Z_2' \left(1 + \frac{Z_1'}{4Z_2'}\right)} = \sqrt{mZ_1 Z_2' \left(1 + \frac{mZ_1}{4Z_2'}\right)}$$

or $$Z_1 Z_2 \left(1 + \frac{Z_1}{4Z_2}\right) = mZ_1 Z_2' \left(1 + \frac{mZ_1}{4Z_2'}\right)$$

or $$Z_2 + \frac{Z_1}{4} = mZ_2' + \frac{m^2 Z_1}{4}$$

or $$Z_2' = \frac{1}{m}\left(Z_2 + \frac{Z_1}{4} - \frac{m^2 Z_1}{4}\right) = \frac{1}{m} Z_2 + \frac{1-m^2}{4m} Z_1 \qquad (9.105)$$

Therefore, for an *m*-derived filter section, we have $Z_1' = j\omega mL$ $\qquad (9.106)$

and $$Z_2' = \frac{1}{j\omega mC} + \frac{1-m^2}{4m} j\omega L \qquad (9.107)$$

The propagation factor is as follows:

$$e^{\gamma} = 1 + \frac{Z_1'}{2Z_2'} + \sqrt{\frac{Z_1'}{Z_2'} + \frac{Z_1'^2}{4Z_2'^2}} = 1 + \frac{Z_1'}{2Z_2'} + \sqrt{\frac{Z_1'}{Z_2'}\left(1 + \frac{Z_1'}{4Z_2'}\right)} \qquad (9.108)$$

From Eqs (9.106) and (9.107), we get the following expression:

$$\frac{Z_1'}{Z_2'} = \frac{j\omega mL}{\dfrac{1}{j\omega mC} + \dfrac{1-m^2}{4m} j\omega L} = \frac{-4\omega^2 m^2 LC}{4 - \omega^2\left(1-m^2\right)LC}$$

or $$\frac{Z_1'}{Z_2'} = \frac{-16(\omega m/\omega_c)^2}{4 - 4\left(1-m^2\right)(\omega/\omega_c)^2} = \frac{-4(\omega m/\omega_c)^2}{1-\left(1-m^2\right)(\omega/\omega_c)^2} \qquad (9.109)$$

Therefore,

$$1 + \frac{Z_1'}{4Z_2'} = 1 - \frac{(\omega m/\omega_c)^2}{1-\left(1-m^2\right)(\omega/\omega_c)^2}$$

or $$1 + \frac{Z_1'}{4Z_2'} = \frac{1-\left(1-m^2\right)(\omega/\omega_c)^2 - (\omega m/\omega_c)^2}{1-\left(1-m^2\right)(\omega/\omega_c)^2} = \frac{1-(\omega/\omega_c)^2}{1-\left(1-m^2\right)(\omega/\omega_c)^2} \qquad (9.110)$$

If we restrict the values of m within the range $0 < m < 1$, then for $\omega < \omega_c$, $\left(1 + \dfrac{Z_1'}{4Z_2'}\right)$ is positive, whereas $\left(\dfrac{Z_1'}{Z_2'}\right)$ is negative. Thus, Eq. (9.108) reveals that e^γ is imaginary, and the pass-band is obtained. Similarly for $\omega > \omega_c$, if $(1 - m^2)(\omega/\omega_c)^2 > 1$, then both $\left(1 + \dfrac{Z_1'}{4Z_2'}\right)$ and $\left(\dfrac{Z_1'}{Z_2'}\right)$ are positive; else, for $(1 - m^2)(\omega/\omega_c)^2 < 1$, both $\left(1 + \dfrac{Z_1'}{4Z_2'}\right)$ and $\left(\dfrac{Z_1'}{Z_2'}\right)$ are negative. Therefore, for $\omega > \omega_c$, e^γ is real and stop-band is obtained.

Thus, the pass- to stop-band transfer takes place at $\omega = \omega_c$, and hence ω_c is the cut-off frequency.

If
$$\omega = \omega_\infty = \omega_c / \sqrt{1 - m^2} \qquad (9.111)$$

then the denominators of both $\left|1 + \dfrac{Z_1'}{4Z_2'}\right|$ and $\left(\dfrac{Z_1'}{Z_2'}\right)$ vanish, and we get $e^\gamma \to \infty$.

Therefore, infinite attenuation is achieved. Physically, this frequency corresponds to a pole of the attenuation characteristics of the filter and occurs due to resonance of the series LC circuit in the shunt arm of the T-network. Since we restrict the values of m within the range $0 < m < 1$, $\omega_\infty > \omega_c$.

The image impedance of an m-derived T-section is identical to that of a constant-k section and hence independent of m. Therefore, the problem of non-constant image impedance of a constant-k section persists for an m-derived T-section also. To avoid this shortcoming, a pi-section m-derived filter is used, as shown in Fig. 9.10. The image impedance of this network depends on m, and this extra degree of freedom can be used to design an optimum matching section. For a pi-section m-derived filter,

$$Z_{i\pi} = Z_1' Z_2' / Z_{iT} \qquad (9.112)$$

Using Eq. (9.102) in Eq. (9.112), we get the following relation:

$$Z_{i\pi} = \frac{Z_1' Z_2'}{R_0 \sqrt{\left|1 - \dfrac{\omega^2}{\omega_c^2}\right|}} \qquad (9.113)$$

Equation (9.113) shows that image impedance, $Z_{i\pi}$, still depends on frequency. However, the extra degree of freedom, provided by the parameter m, can be adjusted to design the optimum matching section.

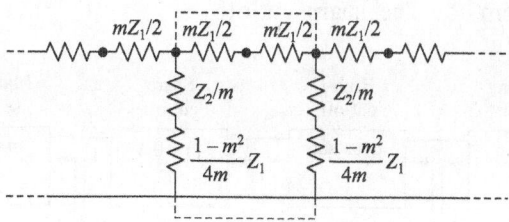

Fig. 9.10   Infinite cascaded m-derived section
pi-equivalent circuit

The bisected pi-section matching networks shown in Fig. 9.11 are often used at the input and output of a T-section section filter, to provide a nearly constant impedance match. The *ABCD* parameters of such a network are as follows:

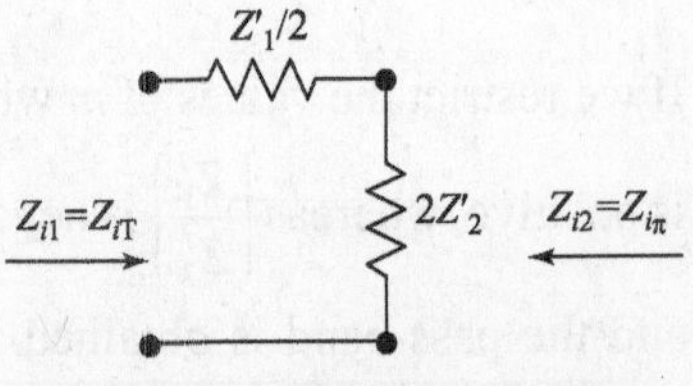

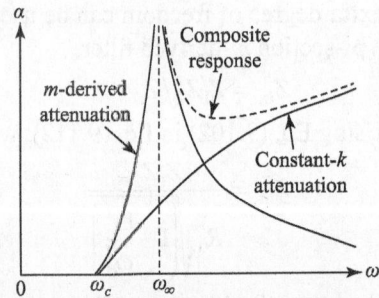

Fig. 9.11 Bisected pi-section matching section

$$\begin{bmatrix} A & B \\ C & D \end{bmatrix} = \begin{bmatrix} 1+\dfrac{Z_1'}{4Z_2'} & \dfrac{Z_1'}{2} \\ 1/(2Z_2') & 1 \end{bmatrix} \tag{9.114}$$

Using Eqs (9.82) and (9.83),

$$Z_{i1} = \sqrt{\dfrac{AB}{CD}} = \sqrt{Z_1'Z_2'\left(1+\dfrac{Z_1'}{4Z_2'}\right)} = Z_{iT} \tag{9.115}$$

and

$$Z_{i2} = \sqrt{\dfrac{BD}{AC}} = \sqrt{Z_1'Z_2' \Big/ \left(1+\dfrac{Z_1'}{4Z_2'}\right)} = Z_{i\pi} \tag{9.116}$$

Note The most chosen value of *m* is 0.6, because it yields a reasonably broad frequency range over which the transmission magnitude remains roughly constant.

9.5.3 Composite Filters

The main problem associated with an *m*-derived filter section is that its attenuation decreases for the frequency $\omega > \omega_\infty$. To overcome this shortcoming, the *m*-derived section is cascaded with a constant-*k* section. This results in a composite filter. The typical attenuation characteristics of a constant-*k* section, *m*-derived section, and composite response are shown in Fig. 9.12, whereas the typical diagram of a four-stage composite filter is shown in Fig. 9.13. At both ends, a matching section has been used to provide a nearly constant impedance match to and from R_0. The sharp cut-off

Fig. 9.12 Attenuation characteristics of constant-*k*- section, *m*-derived section, and composite response

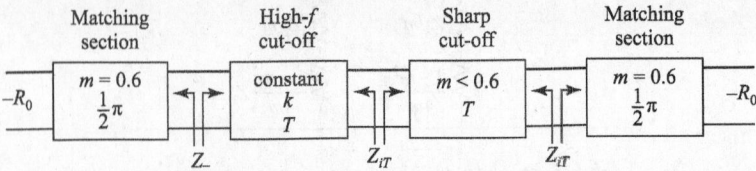

Fig. 9.13 Four-stage composite filter

section is used to place an attenuation pole near the cut-off, whereas the high-f cut-off section is used to provide even higher attenuation into the stop-band.

EXAMPLE 9.1   Design a low-pass composite filter with a cut-off frequency of 5 MHz and impedance 50 Ω, with an infinite attenuation pole at 5.1 MHz.

Solution   Given: $f_c = 5 \times 10^6$ Hz, $f_\infty = 5.1 \times 10^6$ Hz, and $Z_0 = 50 \Omega$

For the constant-k section:

$$L = \frac{2Z_0}{\omega_c} = \frac{2 \times 50}{2\pi \times 5 \times 10^6} = 3.1831 \ \mu H$$

or $\dfrac{L}{2} = 1.5916 \ \mu H$ and $C = \dfrac{2}{Z_0 \omega_c} = \dfrac{2}{50 \times 2\pi \times 5 \times 10^6} = 1.2732 \ nF$

Now, for the m-derived section:

$$m = \sqrt{1 - (f_c/f_\infty)^2} = \sqrt{1 - (5/5.1)^2} = 0.1971$$

Therefore,

$$\frac{mL}{2} = \frac{0.1971 \times 3.1831}{2} = 0.3137 \ \mu H$$

$$mC = 0.1971 \times 1.2732 = 0.2509 \ nF$$

$$\frac{1 - m^2}{4m} L = \frac{1 - 0.1971^2}{4 \times 0.1971} \times 3.1831 = 3.8806 \ \mu H$$

For the $m = 0.6$ section:

$$\frac{mL}{2} = \frac{0.6 \times 3.1831}{2} = 0.9549 \ \mu H \text{ and } \frac{mC}{2} = \frac{0.6 \times 1.2732}{2} = 0.382 \ nF$$

$$\frac{1 - m^2}{2m} L = \frac{1 - 0.6^2}{2 \times 0.6} \times 3.1831 = 1.6977 \ \mu H$$

Hence, the complete filter circuit will be similar to that shown in Fig. 9.14.

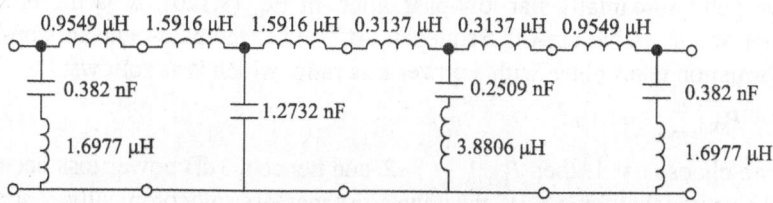

0.9549 μH   1.5916 μH   1.5916 μH   0.3137 μH   0.3137 μH   0.9549 μH

0.382 nF   1.2732 nF   0.2509 nF   0.382 nF

1.6977 μH   3.8806 μH   1.6977 μH

Fig. 9.14   Composite filter circuit of Example 9.1

Practice Problem

9.1 Design a low-pass composite filter with a cut-off frequency of 3 MHz and impedance 100 Ω with an infinite attenuation pole at 3.05 MHz.

[5.3052 μH, 1.061 nF, 0.9565 μH, 0.1913 nF, 14.2337 μH, 3.1831 μH, 0.3183 nF, 5.6588 μH]

9.6 Filter Design by Insertion Loss Method

The main disadvantage of the image parameter method is that an arbitrary frequency response cannot be incorporated into the design. In addition, there is no clear-cut way for design improvement. Therefore, the image parameter method is not suitable if we require a specific frequency response characteristic, and we need to switch to the insertion loss method. The insertion loss method allows control over pass- and stop-band characteristics and also offers a systematic way to generate a desired frequency response.

In the insertion loss method, the frequency response of filters is defined by the insertion loss, as follows:

$$\text{IL} = 10\log_{10}\left(P_{\text{LR}}\right) = 10\log_{10}\left(\frac{1}{1-\left|\Gamma\left(\omega\right)\right|^2}\right) \tag{9.117}$$

Since $\left|\Gamma\left(\omega\right)\right|^2$ is an even function of ω, it can be expressed as a polynomial of ω^2:

$$\left|\Gamma\left(\omega\right)\right|^2 = \frac{M\left(\omega^2\right)}{M\left(\omega^2\right)+N\left(\omega^2\right)} \tag{9.118}$$

Therefore,

$$P_{\text{LR}} = \frac{1}{1-\left|\Gamma\left(\omega\right)\right|^2} = \frac{M\left(\omega^2\right)+N\left(\omega^2\right)}{M\left(\omega^2\right)+N\left(\omega^2\right)-M\left(\omega^2\right)} = 1 + \frac{M\left(\omega^2\right)}{N\left(\omega^2\right)} \tag{9.119}$$

9.6.1 Maximally Flat Low-pass Filter Design

Equation (9.119) indicates that with a proper choice of $M\left(\omega^2\right)/N\left(\omega^2\right)$, we can obtain the desired frequency response of the filter. For example, if we choose

$$\frac{M\left(\omega^2\right)}{N\left(\omega^2\right)} = \varepsilon^2 \left(\frac{\omega}{\omega_c}\right)^{2N}$$

then $\qquad P_{\text{LR}} = 1 + \varepsilon^2 \left(\omega/\omega_c\right)^{2N} \tag{9.120}$

and we get a maximally flat low-pass filter. In Eq. (9.120), N is the order of the filter and ω_c is the cut-off frequency. At $\omega = \omega_c$, the pass-band to stop-band transformation takes place with a power loss ratio, which is as follows:

$$P_{\text{LR}}\big|_{\omega=\omega_c} = 1 + \varepsilon^2 \tag{9.121}$$

If we choose $\varepsilon = 1$, then $P_{\text{LR}}\big|_{\omega=\omega_c} = 2$, and hence a 3 dB power loss occurs. It should be noted that for $\omega > \omega_c$, the power loss increases monotonically. If $\omega \gg \omega_c$, then $P_{\text{LR}} = \varepsilon^2 \left(\omega/\omega_c\right)^{2N} \tag{9.122}$

and hence the insertion loss increases at a rate of $20N$ dB/decade.

Attenuation at the stop-band edge can be expressed as follows:

$$A_s^2 = 1 + \varepsilon^2 \left(\omega/\omega_c\right)^{2N} \tag{9.123}$$

or $\qquad \left(\omega/\omega_c\right)^{2N} = \left(A_s^2-1\right)\big/\varepsilon^2$ or $\left(\omega/\omega_c\right)^{N} = \sqrt{A_s^2-1}\big/\varepsilon$

or $\qquad N\ln\left(\omega/\omega_c\right) = \ln\left(\sqrt{A_s^2-1}\big/\varepsilon\right)$

or $\qquad N = \ln\left(\sqrt{A_s^2-1}\big/\varepsilon\right)\big/\ln\left(\omega/\omega_c\right) \approx \ln\left(A_s/\varepsilon\right)/\ln\left(\omega/\omega_c\right) \tag{9.124}$

Equation (9.124) can be used to find the order of the filter.

Now, let us demonstrate the synthesis of a maximally flat low-pass filter with an example. For this, we will consider the two-element filter prototype shown in Fig. 9.15, with a source impedance of 1, ω_c of 1, and 3 dB insertion loss at the cut-off. Therefore, $N = 2$ and $\varepsilon = 1$, and the power loss ratio can be written as follows:

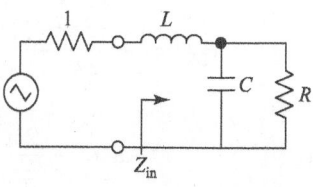

Fig. 9.15 Second-order low-pass filter

$$P_{LR} = 1 + \omega^4 \tag{9.125}$$

From the circuit theory, the input impedance of the filter is as follows:

$$Z_{in} = j\omega L + \left(R \parallel \frac{1}{j\omega C}\right) = j\omega L + \frac{R}{1 + j\omega CR} = j\omega L + \frac{R(1 - j\omega CR)}{\left(1 + \omega^2 R^2 C^2\right)} \tag{9.126}$$

The power loss ratio is given by

$$P_{LR} = \frac{1}{1 - |\Gamma|^2} = \frac{1}{1 - \left[\left\{(Z_{in} - 1)/(Z_{in} + 1)\right\}\left\{\left(Z_{in}^* - 1\right)/\left(Z_{in}^* + 1\right)\right\}\right]}$$

or

$$P_{LR} = \frac{(Z_{in} + 1)\left(Z_{in}^* + 1\right)}{(Z_{in} + 1)\left(Z_{in}^* + 1\right) - (Z_{in} - 1)\left(Z_{in}^* - 1\right)}$$

or

$$P_{LR} = \frac{|Z_{in} + 1|^2}{Z_{in}Z_{in}^* + Z_{in} + Z_{in}^* + 1 - Z_{in}Z_{in}^* + Z_{in} + Z_{in}^* - 1} = \frac{|Z_{in} + 1|^2}{2\left(Z_{in} + Z_{in}^*\right)} \tag{9.127}$$

Now from Eq. (9.126):

$$Z_{in} + Z_{in}^* = j\omega L + \frac{R(1 - j\omega CR)}{\left(1 + \omega^2 R^2 C^2\right)} - j\omega L + \frac{R(1 + j\omega CR)}{\left(1 + \omega^2 R^2 C^2\right)} = \frac{2R}{1 + \omega^2 R^2 C^2} \tag{9.128}$$

and

$$|Z_{in} + 1|^2 = \left|j\omega L + \frac{R(1 - j\omega CR)}{\left(1 + \omega^2 R^2 C^2\right)} + 1\right|^2$$

or

$$|Z_{in} + 1|^2 = \left|\frac{\left(R + 1 + \omega^2 R^2 C^2\right) + j\left\{\omega L\left(1 + \omega^2 R^2 C^2\right) - \omega CR^2\right\}}{\left(1 + \omega^2 R^2 C^2\right)}\right|^2$$

or

$$|Z_{in} + 1|^2 = \left(1 + \frac{R}{1 + \omega^2 R^2 C^2}\right)^2 + \left(\omega L - \frac{\omega CR^2}{1 + \omega^2 R^2 C^2}\right)^2 \tag{9.129}$$

Substituting Eqs (9.128) and (9.129) into Eq. (9.127), we get the following relation:

$$P_{LR} = \frac{|Z_{in} + 1|^2}{2\left(Z_{in} + Z_{in}^*\right)} = \frac{1 + \omega^2 R^2 C^2}{4R}\left[\left(1 + \frac{R}{1 + \omega^2 R^2 C^2}\right)^2 + \left(\omega L - \frac{\omega CR^2}{1 + \omega^2 R^2 C^2}\right)^2\right]$$

$$P_{LR} = \frac{1+\omega^2 R^2 C^2}{4R}\left[1+\frac{2R}{1+\omega^2 R^2 C^2}+\left(\frac{R}{1+\omega^2 R^2 C^2}\right)^2\right.$$

or

$$\left.+\omega^2 L^2 - \frac{2\omega^2 LCR^2}{1+\omega^2 R^2 C^2}+\left(\frac{\omega CR^2}{1+\omega^2 R^2 C^2}\right)^2\right]$$

$$P_{LR} = \frac{1}{4R}\left[\left(1+\omega^2 R^2 C^2\right)+2R+\frac{R^2}{1+\omega^2 R^2 C^2}+\omega^2 L^2 \left(1+\omega^2 R^2 C^2\right)\right.$$

or

$$\left.-2\omega^2 LCR^2 + \frac{\omega^2 C^2 R^4}{1+\omega^2 R^2 C^2}\right]$$

$$P_{LR} = \frac{1}{4R}\left[1+2R+\frac{R^2\left(1+\omega^2 C^2 R^2\right)}{1+\omega^2 R^2 C^2}+\omega^2 R^2 C^2 + \omega^2 L^2\right.$$

or

$$\left.+\omega^4 L^2 C^2 R^2 - 2\omega^2 LCR^2\right]$$

or

$$P_{LR} = \frac{1}{4R}\left[4R+\left(1-2R+R^2\right)+\omega^2\left(R^2 C^2 + L^2 - 2LCR^2\right)+\omega^4 L^2 C^2 R^2\right]$$

or

$$P_{LR} = 1+\frac{1}{4R}\left[\left(1-R\right)^2 + \omega^2\left(R^2 C^2 + L^2 - 2LCR^2\right)+\omega^4 L^2 C^2 R^2\right] \qquad (9.130)$$

Comparing Eq. (9.125) with Eq. (9.130), we can write the following relation:

$$R=1 \qquad (9.131)$$
$$R^2 C^2 + L^2 - 2LCR^2 = 0 \qquad (9.132)$$

and $\qquad L^2 C^2 R = 4 \qquad (9.133)$

From Eqs (9.131) and (9.132), we get the following relation:

$$R^2 C^2 + L^2 - 2LCR^2 = C^2 + L^2 - 2LC = (C-L)^2 = 0$$

or $\qquad C = L \qquad (9.134)$

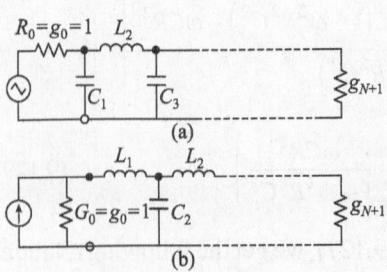

Fig. 9.16 *N*th-order low-pass filter geometry (a) Beginning with series element (b) Beginning with shunt element

Further, using Eqs (9.131) and (9.134) in Eq. (9.133), the following expression is obtained:

$$L^2 C^2 R = L^2 C^2 = L^4 = C^4 = 4$$

or $\qquad L = C = \sqrt{2} \qquad (9.135)$

The aforementioned methodology can be extended for any arbitrary value of *N*. A typical circuit diagram of the *N*th-order maximally flat low-pass filter is shown in Fig. 9.16.

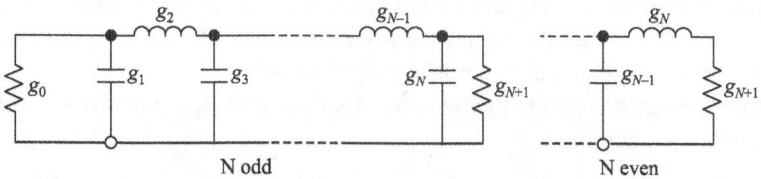

Fig. 9.17 *N*-section low-pass filter prototype with element values

For the maximally flat low-pass filter prototype shown in Fig. 9.17, element values can be calculated from the following equations:

$$g_0 = 1 \tag{9.136}$$

$$g_n = 2\varepsilon^{1/N} \sin\left\{\frac{(2n-1)\pi}{2N}\right\} \quad \text{for } n = 1, 2, 3, \ldots, N \tag{9.137}$$

$$g_{N+1} = 1 \tag{9.138}$$

The following MATLAB code calculates the values of *g*'s for a given *N*.

MATLAB Program

```
% Program to Calculate the Element Values of a Maximally Flat Low Pass
Filter Prototype
clear all;
N = input('Enter the Number of Sections (N):-  ')'
Ep = input('Enter the Ripple Parameter:-  ')'
g(1) = 1;
g(N + 2) = 1;
for k = 1:N
    g(k + 1) = 2 * ep^(1/N) * sin((2*k - 1) * pi/(2*N));
end
g
```

Note A maximally flat filter is also known as a Butterworth filter.

9.6.2 Chebyshev Low-pass Filters

To obtain a Chebyshev response, we choose the following equation:

$$P_{LR} = 1 + \varepsilon^2 T_N^2 (\omega/\omega_c) \tag{9.139}$$

where ε^2 determines the pass-band ripple level. Since at pass-band T_N oscillates between ± 1, the maximum pass-band power loss is $1 + \varepsilon^2$. For $\omega \gg \omega_c$, $T_N(\omega/\omega_c) \simeq \frac{1}{2}\left(\frac{2\omega}{\omega_c}\right)^N$, and we get the following equation:

$$P_{LR} \simeq 1 + \frac{\varepsilon^2}{4}\left(\frac{2\omega}{\omega_c}\right)^{2N} \tag{9.140}$$

Equation (9.140) shows that, similar to a maximally flat filter, the stop-band power loss in Chebyshev filter also increases at a rate of $20N$ dB/decade. However, it is $2^{2N}/4$ times larger than that in a maximally flat filter.

As before, we again consider the two-element filter with $\omega_c = 1$. For $N = 2$, we have the following equation:

$$T_N(\omega) = 2\omega^2 - 1 \tag{9.141}$$

Therefore, from Eq. (9.139) we get the following relation:

$$P_{LR} = 1 + \varepsilon^2 \left(4\omega^4 - 4\omega^2 + 1\right) \tag{9.142}$$

Equating Eq. (9.142) with Eq. (9.130), we get the following expression:

$$\varepsilon^2 \left(4\omega^4 - 4\omega^2 + 1\right) = \frac{1}{4R}\Big[(1-R)^2 + \omega^2 \left(R^2 C^2 + L^2 - 2LCR^2\right)$$
$$+ \omega^4 L^2 C^2 R^2 \Big] \tag{9.143}$$

Equating the power of ω from both sides, we get the following equations:

$$\varepsilon^2 = \frac{(1-R)^2}{4R} \tag{9.144}$$

$$-4\varepsilon^2 = \frac{R^2 C^2 + L^2 - 2LCR^2}{4R} \tag{9.145}$$

and $\qquad 4\varepsilon^2 = \dfrac{L^2 C^2 R^2}{4R} \tag{9.146}$

From Eq. (9.144), we get the following relation:

$$R^2 - \left(4\varepsilon^2 + 2\right)R + 1 = 0$$

or $\qquad R = \dfrac{\left(4\varepsilon^2 + 2\right) \pm \sqrt{\left(4\varepsilon^2 + 2\right)^2 - 4}}{2} = \dfrac{\left(4\varepsilon^2 + 2\right) \pm \sqrt{16\varepsilon^4 + 16\varepsilon^2 + 4 - 4}}{2}$

or $\qquad R = \dfrac{4\varepsilon^2 + 2 \pm 4\varepsilon\sqrt{1+\varepsilon^2}}{2}$

or $\qquad R = 1 + 2\varepsilon^2 \pm 2\varepsilon\sqrt{1+\varepsilon^2} \tag{9.147}$

Once R is known from Eq. (9.147), Eqs (9.145) and (9.146) can be solved for L and C.

It may be noted that since R is not unity, a mismatch exists at the output for unit load resistance. However, this can be avoided with the use of a quarter-wavelength matching section or an additional section to make N odd. It can be shown that if N is odd, then R is unity.

In a Chebyshev filter, if A_m be the maximum ripple amplitude in the pass-band, then the element values can be calculated using the following equations:

$$g_0 = 1 \tag{9.148}$$

$$g_1 = 2a_1/\gamma \tag{9.149}$$

$$g_n = \frac{4a_{n-1}a_n}{b_{n-1}g_{n-1}} \quad \text{for } n = 1, 2, 3, ..., N \tag{9.150}$$

$$g_{N+1} = \begin{cases} 1 & \text{for } N \text{ odd} \\ \coth^2(\beta/4) & \text{for } N \text{ even} \end{cases} \tag{9.151}$$

where $\quad a_n = \sin\left\{\dfrac{(2n-1)\pi}{2N}\right\} \quad \text{for } n = 1, 2, 3, ..., N \tag{9.152}$

$$b_n = \gamma^2 + \sin^2\left(\frac{n\pi}{N}\right) \quad \text{for } n = 1, 2, 3, ..., N \tag{9.153}$$

$$\gamma = \sinh\left(\frac{\beta}{2N}\right) \tag{9.154}$$

$$\beta = \ln\left\{\coth\left(A_m/17.37\right)\right\} \tag{9.155}$$

The order of a filter, required to achieve a predefined attenuation at the stop-band edge, can be calculated using the following formula:

$$N = \frac{\cosh^{-1}\left(\sqrt{A_s^2 - 1}/\varepsilon\right)}{\cosh^{-1}\left(\omega/\omega_c\right)} \simeq \frac{\cosh^{-1}\left(A_s/\varepsilon\right)}{\cosh^{-1}\left(\omega/\omega_c\right)} \tag{9.156}$$

A MATLAB code can be used to find the element values of a Chebyshev low-pass filter for a given order and pass-band ripple.[1]

> **Note** A Chebyshev filter is also known as an equal-ripple filter.

9.6.3 Inverse Chebyshev Filters

An inverse Chebyshev filter can be derived from an ordinary Chebyshev filter by two simple steps. In the first step, the ordinary Chebyshev response is subtracted from unity. This transforms a low-pass filter to a high-pass filter. Mathematically, it can be written as follows:

$$|H(j\omega)|^2 = 1 - \frac{1}{1 + \varepsilon^2 T_N^2(\omega/\omega_c)} = \frac{\varepsilon^2 T_N^2(\omega/\omega_c)}{1 + \varepsilon^2 T_N^2(\omega/\omega_c)} \tag{9.157}$$

In the next step, ω is replaced by $1/\omega$. Since this maps the high frequency into a low frequency, the second transformation converts the high-pass filter back into a low-pass filter. However, this also transforms the ripples, at low frequency, to high frequency, thus achieving an inverse Chebyshev response. The final transfer function can therefore be written as follows:

$$|H(j\omega)|^2 = \varepsilon^2 T_N^2(\omega_c/\omega)\Big/\left\{1 + \varepsilon^2 T_N^2(\omega_c/\omega)\right\} \tag{9.158}$$

9.6.4 Elliptic Filters

It has already been described that the transition from pass-band to stop-band can be made sharper by allowing ripple either in the pass-band or in the stop-band.

---

[1]Program 9.1 in Online Resources

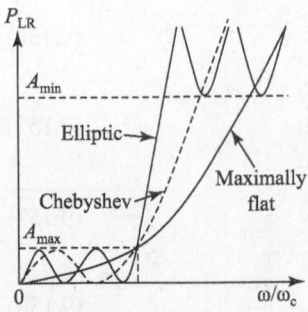

Fig. 9.18 Comparison of frequency responses of maximally flat, Chebyshev, and elliptic function

This concept leads to a Chebyshev or an inverse Chebyshev filter. However, in both these filters, ripples were allowed either in the pass-band or in the stop-band. Undoubtedly, the pass- to stop-band transition can be sharpened further by allowing ripple in both the bands. Such filters are known as elliptic filters, as elliptic functions are involved in its transfer function. The filter exploits the nulls provided by the finite zeros to create sharp pass-to stop-band transition. Frequency responses of a maximally flat, a Chebyshev, and an elliptic filter are shown in Fig. 9.18 for comparison.

The transfer function of an elliptic filter can be described by the following equation:

$$|H(j\omega)|^2 = \frac{1}{1+\varepsilon^2 F_n^2 (\omega/\omega_c)} \tag{9.159}$$

where $F_n (\omega/\omega_c)$, also known as a Chebyshev rational function, is a ratio of two polynomials rather than a single polynomial, as in the case of maximally flat and Chebyshev filters. Such rational polynomials have an additional degree of freedom that permits to meet the specifications with a lower number of elements. Similar to a Chebyshev polynomial, the magnitude of function $F_n (\omega/\omega_c)$ also oscillates between '0' and '1' for $\omega/\omega_c < 1$ and grows rapidly for $\omega/\omega_c > 1$. However, the difference between Chebyshev rational function and Chebyshev polynomial is that for the latter case , for $\omega/\omega_c > 1$, the magnitude grows monotonically whereas for the former case, it oscillates in some fashion between infinity and a specified finite value. This results in a stop-band ripple with a finite number of frequencies at which transmission is zero.

> **Note** Elliptic filters are also known as Cauer filters.

9.6.5 Linear Phase Filters

In all the previously mentioned filters, the amplitude response is generally specified. Some filters require to have a linear phase response in the pass-band to avoid signal distortion. Such filters are called linear-phase filter. Since a sharp cut-off generally has a worse phase response, amplitudes of linear-phase filters have inferior characteristics. The phase response in a linear-phase filter can be written as follows:

$$\phi(\omega) = A\omega\left[1+p(\omega/\omega_c)^{2N}\right] \tag{9.160}$$

where $\phi(\omega)$ is the phase of the voltage transfer function of the filter and p is a constant. The characteristic of a linear-phase filter is also described by the *group delay*. The group delay of a linear phase filter can be expressed as follows:

$$\tau_d = \frac{d}{d\omega}\phi(\omega) = A\left[1 + p(2N+1)\left(\frac{\omega}{\omega_c}\right)^{2N}\right] \qquad (9.161)$$

Equation (9.161) shows that the group delay of a linear phase filter has a maximally flat response.

9.7 Filter Transformations

In the last section, we have discussed different low-pass filters of different frequency characteristics. It is interesting to note that throughout all the cases, we have assumed a unity source impedance and cut-off frequency. However, in practical cases, the situation is completely different; neither the source impedance nor the cut-off frequency is unity. Therefore, a question arises whether the previous analysis is useful in practical cases or not. The answer is yes. The preceding designs can be scaled in terms of impedance and frequency. In addition, the aforementioned filter geometry can easily be converted to yield high-pass, band-pass, or band-stop characteristics. In the following sections, we will illustrate the required procedure.

9.7.1 Impedance Scaling

Impedance scaling is used to find the component values of a filter that has a non-unity source resistance. If the source resistance is R_S, then the new filter component values will be as follows:

$$L' = LR_S \qquad (9.162)$$

$$C' = C/R_S \qquad (9.163)$$

$$R'_S = R_S \qquad (9.164)$$

and $\qquad R'_L = R_S R_L \qquad (9.165)$

where L, C, and R_L are the component values in the original prototype.

9.7.2 Frequency Scaling

Frequency scaling is used to shift the angular cut-off frequency of the original prototype filter to an arbitrary value ω_c by substituting

$$\omega \leftarrow \omega/\omega_c \qquad (9.166)$$

Therefore,

$$P'_{LR}(\omega) = P_{LR}(\omega/\omega_c) \qquad (9.167)$$

$$jX_k = j\omega L_k/\omega_c = j\omega L'_k \text{ or } L'_k = L_k/\omega_c \qquad (9.168)$$

and $\qquad jB_k = j\omega C_k/\omega_c = j\omega C'_k \text{ or } C'_k = C_k/\omega_c \qquad (9.169)$

If both impedance and frequency scaling are required, then the preceding results may be combined to get the following equations:

$$L'_k = R_S L_k / \omega_c \tag{9.170}$$

and $\quad C'_k = C_k / (R_S \omega_c) \tag{9.171}$

9.7.3 Low- to High-pass Transformation

For low- to high-pass transformation, we require the following substitution:

$$\omega \leftarrow -\omega_c / \omega \tag{9.172}$$

This substitution modifies the frequency response as shown in Fig. 9.19. Such a substitution produces the following equations:

$$jX_k = -j\omega_c L_k / \omega \tag{9.173}$$

and $\quad jB_k = -j\omega_c C_k / \omega \tag{9.174}$

The negative reactance and susceptance in Eqs (9.173)–(9.174) imply that the inductor and capacitor should be replaced by a capacitor and an inductor, as the reactance of a capacitor and susceptance of an inductor are negative. Therefore,

$$-j\frac{\omega_c}{\omega} L_k = -\frac{j}{\omega C'_k} \text{ or } C'_k = \frac{1}{\omega_c L_k} \tag{9.175}$$

and $\quad -j\frac{\omega_c}{\omega} C_k = \frac{-j}{\omega L'_k} \text{ or } L'_k = \frac{1}{\omega_c C_k} \tag{9.176}$

Further, if we include impedance scaling, then $C'_k = \dfrac{1}{R_S \omega_c L_k} \tag{9.177}$

and $\quad L'_k = \dfrac{R_s}{\omega_c C_k} \tag{9.178}$

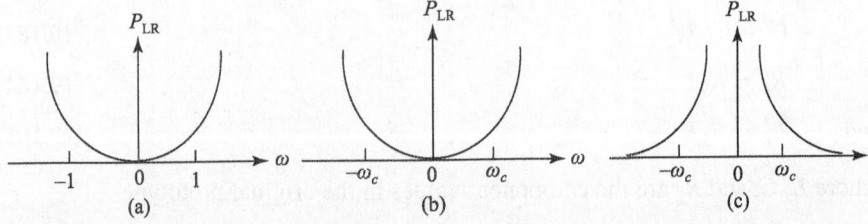

Fig. 9.19 Low- to high-pass transformation (a) Low-pass filter with unit cut-off frequency (b) Low-pass filter with arbitrary cut-off frequency (c) High-pass filter with arbitrary cut-off frequency

9.7.4 Low- to Band-pass Transformation

Let us consider that a band-pass filter is characterized by a lower and a higher angular cut-off frequency of ω_1 and ω_2, respectively. Then to transform a low-pass filter to a band-pass filter, we need to substitute

$$\omega \leftarrow \frac{\omega_0}{\omega_2 - \omega_1}\left(\frac{\omega}{\omega_0} - \frac{\omega_0}{\omega}\right) = \frac{1}{\Delta}\left(\frac{\omega}{\omega_0} - \frac{\omega_0}{\omega}\right) \tag{9.179}$$

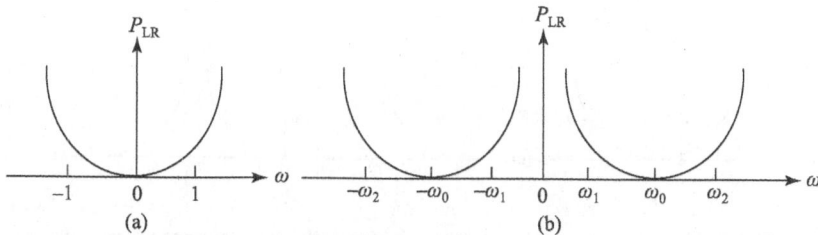

Fig. 9.20 Low- to band-pass transformation (a) Low-pass filter with unit cut-off frequency (b) Band-pass filter with arbitrary cut-off frequency

where Δ is the fractional bandwidth of the filter and can be expressed as follows:

$$\Delta = (\omega_2 - \omega_1)/\omega_0 \qquad (9.180)$$

and ω_0 is either the arithmetic or the geometric mean of ω_1 and ω_2. Since an equation will be simpler if we take the geometric mean, we will consider the following relation:

$$\omega_0 = \sqrt{\omega_1 \omega_2} \qquad (9.181)$$

The graphical representation of this transformation is shown in Fig. 9.20. With this substitution, we get the following equation:

$$jX_k = \frac{j}{\Delta}\left(\frac{\omega}{\omega_0} - \frac{\omega_0}{\omega}\right)L_k = j\frac{\omega L_k}{\Delta \omega_0} - j\frac{\omega_0 L_k}{\Delta \omega} = j\omega L_k' - \frac{j}{\omega C_k'}$$

or $\qquad L_k' = \dfrac{L_k}{\Delta \omega_0} \qquad (9.182)$

and $\qquad C_k' = \dfrac{\Delta}{\omega_0 L_k} \qquad (9.183)$

Similarly, $jB_k = \dfrac{j}{\Delta}\left(\dfrac{\omega}{\omega_0} - \dfrac{\omega_0}{\omega}\right)C_k = \dfrac{j\omega C_k}{\Delta \omega_0} - \dfrac{j\omega_0}{\Delta \omega}C_k = j\omega C_k' - \dfrac{j}{\omega L_k'}$

or $\qquad C_k' = \dfrac{C_k}{\Delta \omega_0} \qquad (9.184)$

and $\qquad L_k' = \dfrac{\Delta}{\omega_0 C_k} \qquad (9.185)$

Thus, the elements are modified by a series resonant circuit (low impedance at resonance) of resonance frequency ω_0 in the series arm and a parallel resonant circuit (high impedance at resonance) of resonance frequency ω_0 in the shunt arm.

9.7.5 Low-pass to Band-stop Transformation

For the low-pass to band-stop transformation, we need carry out the following substitution:

$$\omega \leftarrow \frac{\omega_2 - \omega_1}{\omega_0}\left(\frac{\omega}{\omega_0} - \frac{\omega_0}{\omega}\right)^{-1} = \Delta\left(\frac{\omega}{\omega_0} - \frac{\omega_0}{\omega}\right)^{-1} \qquad (9.186)$$

The graphical representation of this transformation is shown in Fig. 9.21.

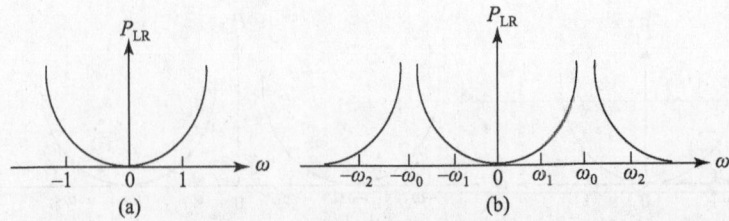

Fig. 9.21 Low-pass to band-stop transformation (a) Low-pass filter with unit cut-off frequency (b) Band-stop filter with arbitrary cut-off frequency

Table 9.2 Filter transformation

Low Pass	High Pass	Band Pass	Band Stop
L	$\dfrac{1}{\omega_c L}$	$\dfrac{L}{\omega_0 \Delta}$ $\dfrac{\Delta}{\omega_0 L}$	$\dfrac{1}{\omega_0 L \Delta}$ $\dfrac{L\Delta}{\omega_0}$
C	$\dfrac{1}{\omega_c C}$	$\dfrac{C}{\omega_0 \Delta}$ $\dfrac{\Delta}{\omega_0 C}$	$\dfrac{1}{\omega_0 C \Delta}$ $\dfrac{C\Delta}{\omega_0}$

With the substitution, the series inductor in the low-pass prototype is modified to a parallel LC circuit with

$$L'_k = \frac{\Delta L_k}{\omega_0} \tag{9.187}$$

and

$$C'_k = \frac{1}{\omega_0 \Delta L_k} \tag{9.188}$$

Similarly, the shunt capacitor in the low-pass prototype is modified to a series LC circuit with

$$C'_k = \frac{\Delta C_k}{\omega_0} \tag{9.189}$$

and

$$L'_k = \frac{1}{\omega_0 \Delta C_k} \tag{9.190}$$

Element transformations between different filter types are summarized in Table 9.2.

EXAMPLE 9.2 Design a maximally flat low-pass filter with a cut-off frequency 5 GHz, impedance 50 Ω, and 15 dB insertion loss at 7 GHz.

Solution Given: $f_c = 5$ GHz, $f = 7$ GHz, IL = 15 dB, and $Z_0 = 50\,\Omega$

$A_s = 15$ dB $= 10^{15/20} = 5.6234$

If we consider $\varepsilon = 1$, then the number of elements required is given by the following equation:

$$N \simeq \frac{\ln(A_s)}{\ln(\omega/\omega_c)} \simeq \frac{\ln(5.6234)}{\ln(7/5)} \simeq 5.1324$$

Hence, six elements will be sufficient.
Therefore,

$$g_0 = 1.0000 \qquad g_2 = 1.4142 \qquad g_4 = 1.9319 \qquad g_6 = 0.5176$$
$$g_1 = 0.5176 \qquad g_3 = 1.9319 \qquad g_5 = 1.4142 \qquad g_7 = 1.0000$$

Therefore, element values are as follows:

$$C_1' = \frac{g_1}{Z_0 \omega_c} = \frac{0.5176}{50 \times 2\pi \times 5 \times 10^9} = \frac{0.5176}{1570.7963} \times 10^{-9} = 0.3295 \text{ pF}$$

$$L_2' = \frac{Z_0 g_2}{\omega_c} = \frac{50 \times 1.4142}{2\pi \times 5 \times 10^9} = 1.5915 \times 1.4142 \times 10^{-9} = 2.2508 \text{ nH}$$

$$C_3' = \frac{g_3}{Z_0 \omega_c} = \frac{1.9319}{50 \times 2\pi \times 5 \times 10^9} = \frac{1.9319}{1570.7963} \times 10^{-9} = 1.2299 \text{ pF}$$

$$L_4' = \frac{Z_0 g_4}{\omega_c} = \frac{50 \times 1.9319}{2\pi \times 5 \times 10^9} = 1.5915 \times 1.9319 \times 10^{-9} = 3.0746 \text{ nH}$$

$$C_5' = \frac{g_5}{Z_0 \omega_c} = \frac{1.4142}{50 \times 2\pi \times 5 \times 10^9} = \frac{1.4142}{1570.7963} \times 10^{-9} = 0.9003 \text{ pF}$$

$$L_6' = \frac{Z_0 g_6}{\omega_c} = \frac{50 \times 0.5176}{2\pi \times 5 \times 10^9} = 1.5915 \times 0.5176 \times 10^{-9} = 0.8238 \text{ nH}$$

Practice Problems

9.2 @Design a three-element Chebyshev band-pass filter with 0.25 dB equal-ripple response and 2 GHz cut-off frequency, provided that bandwidth is 8% and impedance is 50 Ω.

9.3 Design a maximally flat low-pass filter with a cut-off frequency 3 GHz, impedance 50 Ω, and 15 dB insertion loss at 4.5 GHz.
$$[0.6557 \text{ pF}, 4.2919 \text{ nH}, 2.1221 \text{ pF}, 4.2919 \text{ nH}, 0.6557 \text{ pF}]$$
9.4 Design a three-element Chebyshev band-pass filter with 0.5 dB equal-ripple response and 3 GHz cut-off frequency. Provided bandwidth is 10% and impedance is 50 Ω.
$$[42.3432 \text{ nH}, 0.0665 \text{ pF}, 0.2419 \text{ nH}, 11.6363 \text{ pF}, 42.3432 \text{ nH}, 0.0665 \text{ pF}]$$

9.8 Filter Implementation

Till now, we have discussed the designing of a filter using lumped elements. However, at a microwave frequency there is limitation in the designable inductance and capacitance. Moreover, the electrical length of such circuit components is also not negligible. Therefore, we need to depend on the distributed inductance and capacitance of a structure. Richard transformation and Kuroda identity provide a

@Solution in Online Resources

flexible way to implement lumped element filters at microwave frequencies. While Richard transformation is used to convert lumped elements into transmission line sections, Kuroda identities are used to separate these filter elements using transmission line sections.

9.8.1 Richard Transformation

Richard transformation transforms the ω plane to a Ω plane with a period of $\omega l / v_p = 2\pi$ and can be described as follows:

$$\Omega = \tan(\beta l) = \tan(\omega l / v_p) \tag{9.191}$$

Therefore,

$$jX_L = j\Omega L = jL \tan(\beta l) \tag{9.192}$$

and

$$jB_L = j\Omega C = jC \tan(\beta l) \tag{9.193}$$

Equations (9.192) and (9.193) shows that an inductor or a capacitor can be replaced by a short- or an open- circuited transmission line of length l and characteristic impedance L or $1/C$, respectively, as shown in Fig. 9.22. Here, we have assumed unity filter impedance.

If we assume a unity cut-off frequency, then

$$\Omega = \tan(\beta l) = 1$$

or

$$l = \lambda/8 \tag{9.194}$$

where λ is the wavelength for $\omega = \omega_c$. For $\omega = 2\omega_c$, we will have $l = \lambda/4$ and obtain an attenuation pole. In practice, if ω deviates from ω_c, matching is lost and the filter response will not be the same as the lumped prototype response. In addition, the response will be periodic at an interval of $4\omega_c$.

9.8.2 Kuroda Identity

In the microwave filter design, Kuroda identities are widely used to physically separate different transmission line stubs, to transform series stubs into shunt stubs and vice versa, and also to change non-realizable characteristic impedances to more realizable ones. The additional transmission line sections that are used for this purpose are called *unit elements* and are $\lambda/8$ long at $\omega = \omega_c$.

Four Kuroda identities are used in filter design. In this section, we will prove only the first Kuroda identity. The proof of the rest three will be left to the readers.

Let us consider the first Kuroda identity, as shown in Fig. 9.23, where the box represents a unit element. Since from the transmission line theory we can

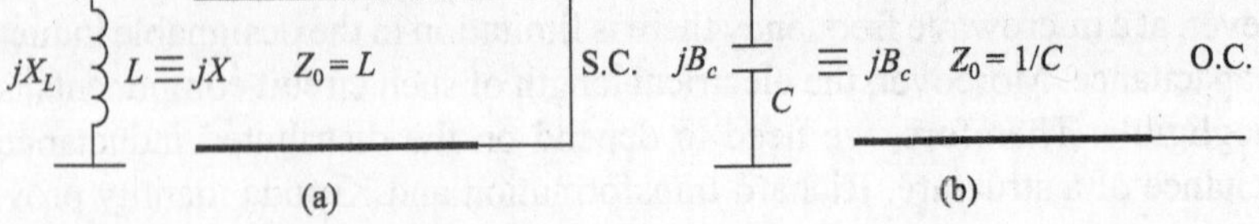

Fig. 9.22 Richard transformation (a) From inductor to short-circuited stub (b) From capacitor to open-circuited stub

Fig. 9.23 First Kuroda identity

represent a capacitor or an inductor
as an open-or a short-circuited stub,
we can redraw Fig. 9.23 as Fig. 9.24.

Using Eq. (9.191), the *ABCD*
matrix of a transmission line of length l
can be written as follows:

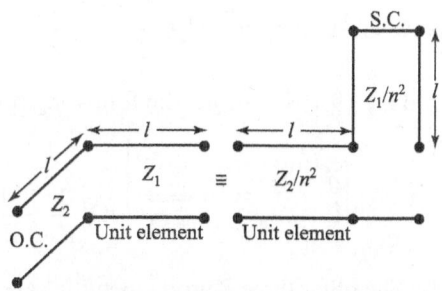

Fig. 9.24 Equivalent circuit of first Kuroda identity

$$\begin{bmatrix} A & B \\ C & D \end{bmatrix} = \begin{vmatrix} \cos(\beta l) & jZ_1 \sin(\beta l) \\ \dfrac{j}{Z_1} \sin(\beta l) & \cos(\beta l) \end{vmatrix} = \frac{1}{\sqrt{1+\Omega^2}} \begin{vmatrix} 1 & j\Omega Z_1 \\ \dfrac{j\Omega}{Z_1} & 1 \end{vmatrix} \tag{9.195}$$

Since the open-circuited stub has an input impedance of

$$Z_{in}^{OC} = -jZ_2 \cot(\beta l) = -jZ_2/\Omega \tag{9.196}$$

the *ABCD* matrix of the equivalent transmission line circuit, in the LHS of
Fig. 9.24, will be as follows:

$$\begin{bmatrix} A & B \\ C & D \end{bmatrix}_L = \frac{1}{\sqrt{1+\Omega^2}} \begin{bmatrix} 1 & 0 \\ \dfrac{j\Omega}{Z_2} & 1 \end{bmatrix} \begin{bmatrix} 1 & j\Omega Z_1 \\ \dfrac{j\Omega}{Z_1} & 1 \end{bmatrix}$$

or
$$\begin{bmatrix} A & B \\ C & D \end{bmatrix}_L = \frac{1}{\sqrt{1+\Omega^2}} \begin{bmatrix} 1 & j\Omega Z_1 \\ j\Omega\left(\dfrac{1}{Z_1} + \dfrac{1}{Z_2}\right) & 1 - \Omega^2 \dfrac{Z_1}{Z_2} \end{bmatrix} \tag{9.197}$$

Similarly, the short-circuited stub has an input impedance of

$$Z_{in}^{SC} = jZ_1 \tan(\beta l)/n^2 = j\Omega Z_1/n^2 \tag{9.198}$$

and the *ABCD* matrix of the equivalent transmission line circuit, in the RHS of
Fig. 9.24, will be as follows:

$$\begin{bmatrix} A & B \\ C & D \end{bmatrix}_R = \frac{1}{\sqrt{1+\Omega^2}} \begin{bmatrix} 1 & \dfrac{j\Omega Z_2}{n^2} \\ \dfrac{j\Omega n^2}{Z_2} & 1 \end{bmatrix} \begin{bmatrix} 1 & \dfrac{j\Omega Z_1}{n^2} \\ 0 & 1 \end{bmatrix}$$

or
$$\begin{bmatrix} A & B \\ C & D \end{bmatrix}_R = \frac{1}{\sqrt{1+\Omega^2}} \begin{bmatrix} 1 & \dfrac{j\Omega}{n^2}(Z_1 + Z_2) \\ \dfrac{j\Omega n^2}{Z_2} & 1 - \Omega^2 \dfrac{Z_1}{Z_2} \end{bmatrix} \tag{9.199}$$

Now substituting $n^2 = 1 + \dfrac{Z_2}{Z_1} = \dfrac{Z_1 + Z_2}{Z_1}$ (9.200)

into Eq. (9.199), we get the following relation:

$$\begin{bmatrix} A & B \\ C & D \end{bmatrix}_R = \frac{1}{\sqrt{1+\Omega^2}} \begin{bmatrix} 1 & j\Omega Z_1 \\ j\Omega\left(\dfrac{1}{Z_1} + \dfrac{1}{Z_2}\right) & 1 - \Omega^2 \dfrac{Z_1}{Z_2} \end{bmatrix} = \begin{bmatrix} A & B \\ C & D \end{bmatrix}_L$$ (9.201)

The other three Kuroda identities are shown in Fig. 9.25.

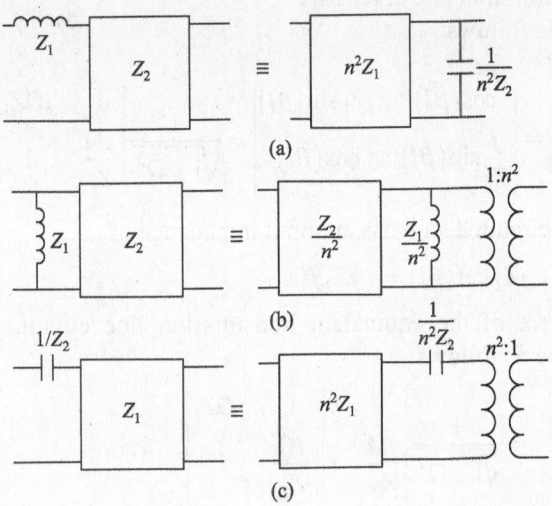

Fig. 9.25 Kuroda identities (a) Second (b) Third (c) Fourth

EXAMPLE 9.3 Design and implement a third-order low-pass microstrip filter with 5 GHz cut-off frequency, 50 Ω impedance, and 0.25 dB equiripple characteristics.

Solution Given: $f_c = 5$ GHz, $Z_0 = 50\,\Omega$, and $N = 3$

$\varepsilon = 0.25$ dB

The element values are as follows:

$g_0 = 1.0000$	$g_2 = 1.1463 = C_2$	$g_4 = 1.0000$
$g_1 = 1.3034 = L_1$	$g_3 = 1.3034 = L_3$	

The lumped element and its equivalent transmission line equivalent circuit are shown in Figs 9.26(a) and (b), respectively. It should be noted that, as per Eqs (9.192) and (9.193), the characteristic impedance of the series stub should be equal to L_1 and that of the shunt stub should be equal to $1/C_2$. For commensurate line synthesis, the stubs should be $\lambda/8$ long at $f = f_c$.

Now to convert the series stubs into shunt stubs, we will apply Kuroda identities. First, we will add unit elements at either end of the filter, as shown in Fig. 9.26(c). Since they are matched to the loads, they will not affect the performance of the filter. Applying Kuroda's first and second identities, the filter circuit modifies as shown in Fig. 9.26(d).

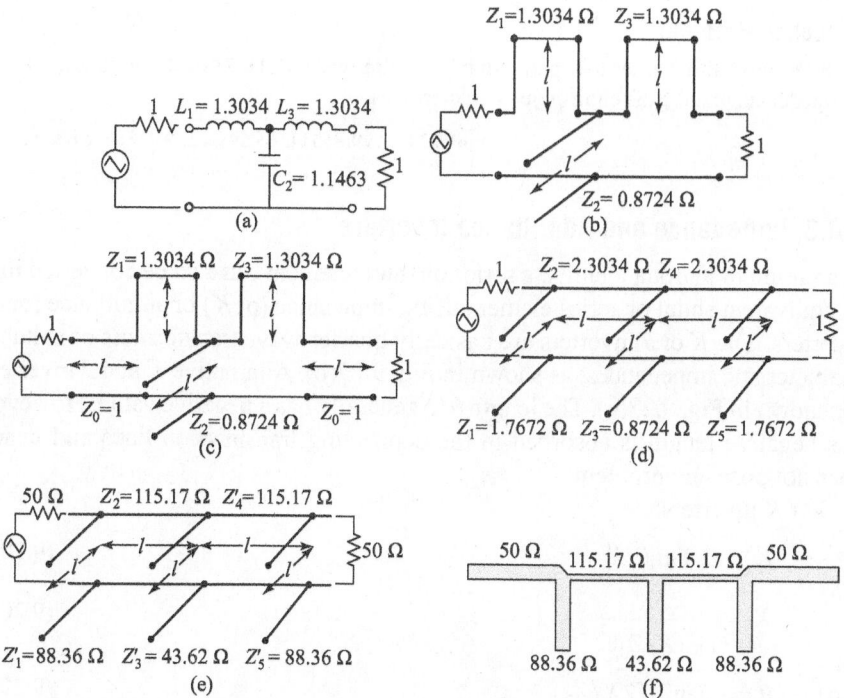

Fig. 9.26 Filter design procedure (a) Lumped element equivalent prototype
(b) Application of Richard transformation to convert lumped elements into transmission
line circuit (c) Addition of unit elements to facilitate application of Kuroda identity
(d) Filter prototype after application of Kuroda identity (e) Impedance
scaled filter circuit (f) Final filter circuit

$$n^2 = 1 + \frac{Z_2}{Z_1} = 1 + \frac{1}{1.3034} = 1.7672$$

$$Z_1 = 1.7672\,\Omega$$

$$Z_2 = 1.7672 \times 1.3034 = 2.3034\,\Omega$$

$$Z_3 = 0.8724\,\Omega$$

$$Z_4 = 1.7672 \times 1.3034 = 2.3034\,\Omega$$

$$Z_5 = 1.7672\,\Omega$$

As the final step, we will now need to perform impedance scaling. The impedance scaled values will be as follows:

$$Z_1' = 1.7672 \times 50 = 88.36\,\Omega$$

$$Z_2' = 2.3034 \times 50 = 115.17\,\Omega$$

$$Z_3' = 0.8724 \times 50 = 43.62\,\Omega$$

$$Z_4' = 2.3034 \times 50 = 115.17\,\Omega$$

$$Z_5' = 1.7672 \times 50 = 88.36\,\Omega$$

The final transmission line model and fabricated microstrip filter are shown in Figs 9.26(e) and (f), respectively.

Practice Problem

9.5 Design a third-order low-pass microstrip filter with 3 GHz cut-off frequency, 50 Ω impedance, and 0.5 dB equiripple characteristics.

$$[81.32\,\Omega,\ 129.815\,\Omega,\ 45.59\,\Omega,\ 129.815\,\Omega,\ 81.32\,\Omega]$$

9.8.3 Impedance and Admittance Inverters

In addition to Kuroda identity, a series or shunt element can also be converted into an equivalent shunt or series element using impedance (or K) or admittance (or J) inverters. The K or J inverters are basically quarter-wave transformers of suitable characteristic impedances, as shown in Fig. 9.27(b). Alternative K and J inverters are shown in Fig. 9.27(c). The length $\theta/2$ generally has a negative value. However, this negative length is absorbed in the connecting transmission lines and hence does not pose any problem.

For K inverters:

$$K = Z_0 \tan|\theta/2| \tag{9.202}$$

$$X = \frac{K}{1-(K/Z_0)^2} \tag{9.203}$$

and $\qquad \theta = -\tan^{-1}(2X/Z_0) \tag{9.204}$

For J inverters:

$$J = Y_0 \tan|\theta/2| \tag{9.205}$$

$$B = \frac{J}{1-(J/Y_0)^2} \tag{9.206}$$

and $\qquad \theta = -\tan^{-1}(2B/Y_0) \tag{9.207}$

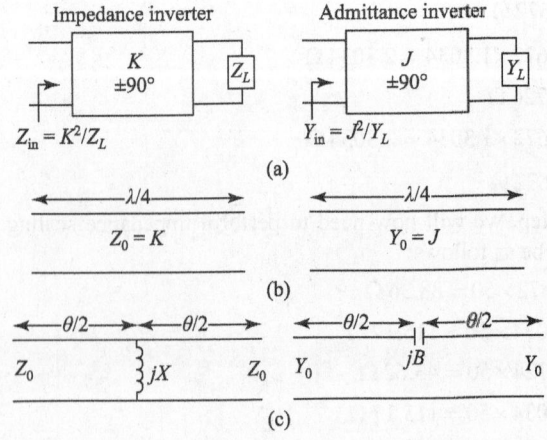

Fig. 9.27 Inverters (a) Impedance and admittance inverters (b) Implementation of inverters using quarter-wave transmission line (c) Alternative implementation of inverters

Inverters can also be used to convert series or parallel resonant circuits into equivalent parallel or series resonant circuit. If L_s and C_s be the series elements, and L_p and C_p be the parallel elements, then the impedance of the series and parallel networks will, respectively, be as follows:

$$Z_s = j\omega L_s + \frac{1}{j\omega C_s} = \frac{1 - \omega^2 L_s C_s}{j\omega C_s} \tag{9.208}$$

and $\quad Z_p = \dfrac{j\omega L_p}{1 - \omega^2 L_p C_p}$ (9.209)

The reciprocal of the series network impedance will be as follows:

$$Z_{eq} = \frac{K^2}{Z_s} = \frac{j\omega C_s K^2}{1 - \omega^2 L_s C_s} \tag{9.210}$$

where K has been introduced to avoid the unit mismatch on both sides. Comparing Eqs (9.209) and (9.210), we get the following relations:

$$L_p = C_s K^2 \text{ or } K = \sqrt{L_p / C_s} \tag{9.211}$$

and $\quad L_p C_p = L_s C_s$ (9.212)

Substituting Eq. (9.212) into Eq. (9.211), we get the following equation:

$$K = \sqrt{L_s / C_p} \tag{9.213}$$

Thus, a series resonant circuit can be transformed into a parallel resonant circuit provided that Eqs (9.211) and (9.212) are satisfied.

The following equation can be used to represent the relationship in terms of admittance: $Y_{eq} = J^2 / Y_s$ (9.214)

where J has been introduced to avoid the unit mismatch on both sides and can be expressed as follows: $J = \sqrt{C_s / L_p} = \sqrt{C_p / L_s}$ (9.215)

Since K has the dimension of impedance, we call it an impedance inverter, whereas J has the dimension of admittance and we call it an admittance inverter. Of course, a network that inverts impedance can also invert admittance. Therefore, the definition is not rigid and depends on the way we describe the network behaviour. Hence, instead of using only the term *impedance* or *admittance* inverter, we often use the term *immitance* inverter. Immitance inverters provide us an additional degree of freedom for the choice of element values within practical range.

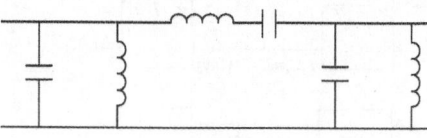

Fig. 9.28   Lumped band-pass filter

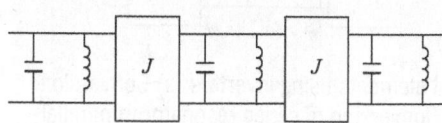

Fig. 9.29   Parallel-tank band-pass filter using immitance inverters

Impedance and admittance inverters found wide applications in designing narrow-band band pass and band-stop filters, using only series or parallel resonant circuits or by scaling a series or parallel resonant circuit into another series or parallel resonant circuit. To demonstrate this, let us consider the network shown in Fig. 9.28 and its equivalent circuit in Fig. 9.29. In Fig. 9.29, the series LC network of Fig. 9.28 has been replaced with a parallel LC network and two immitance inverters. Additional transformations are shown in Fig. 9.30.

Inverters can also be used to convert a conventional band-pass/band-stop filter network into another band-pass/band-stop network consisting of identical series/parallel tuned circuits and equal input and output terminating resistances. To illustrate this, let us consider the band-pass filter shown in Fig. 9.31(a). As a first step, we will convert the network into an all-series resonator network with immitance inverters, as shown in Fig. 9.31(b). In the next step, we will scale resonator 3 by changing inverters 2 and 3 so that the following equations are satisfied:

$$K_2 = K_2' \sqrt{\frac{L_0}{L_2}} = \omega_0 \sqrt{L_0 L_1} \sqrt{\frac{L_0}{L_2}} = \frac{1}{\sqrt{L_1 C_1}} \sqrt{\frac{L_0^2 L_1}{L_2}} = \sqrt{\frac{L_0^2}{L_2 C_1}} \qquad (9.216)$$

and

$$K_3 = K_3' \sqrt{\frac{L_0}{L_2}} = \omega_0 \sqrt{L_0 L_3} \sqrt{\frac{L_0}{L_2}} = \frac{1}{\sqrt{L_3 C_3}} \sqrt{\frac{L_0^2 L_3}{L_2}} = \sqrt{\frac{L_0^2}{L_2 C_3}} \qquad (9.217)$$

The resultant network is shown in Fig. 9.31(c). In order to concert R_L to R, we need to modify the first inverter as follows:

$$K_1 = K_1' \sqrt{\frac{R}{R_L}} = \omega_0 \sqrt{L_0 L_1} \sqrt{\frac{R}{R_L}} = \frac{1}{\sqrt{L_1 C_1}} \sqrt{\frac{R L_0 L_1}{R_L}} = \sqrt{\frac{R L_0}{R_L C_1}} \qquad (9.218)$$

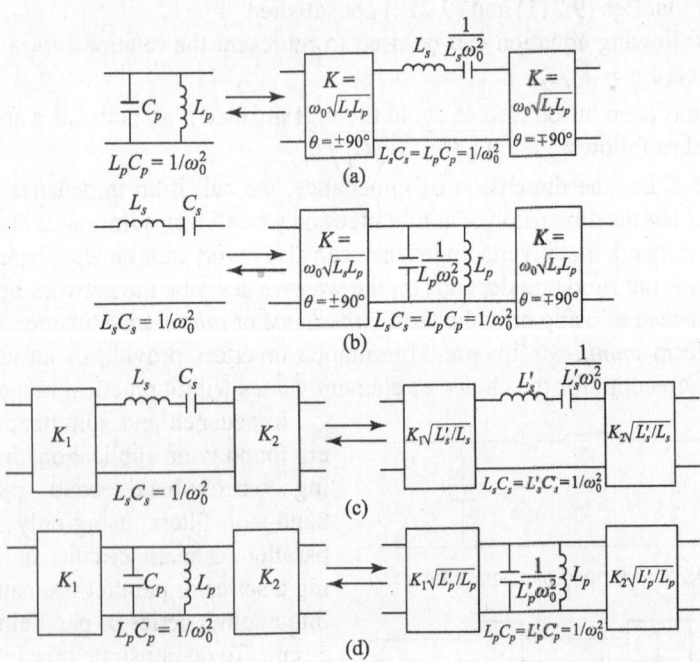

Fig. 9.30 Conversion and scaling of circuit elements using inverters (a) Conversion of parallel resonator to series resonator (b) Conversion of series resonator to parallel resonator (c) Scaling of series resonator to another series resonator (d) Scaling of parallel resonator to another parallel resonator

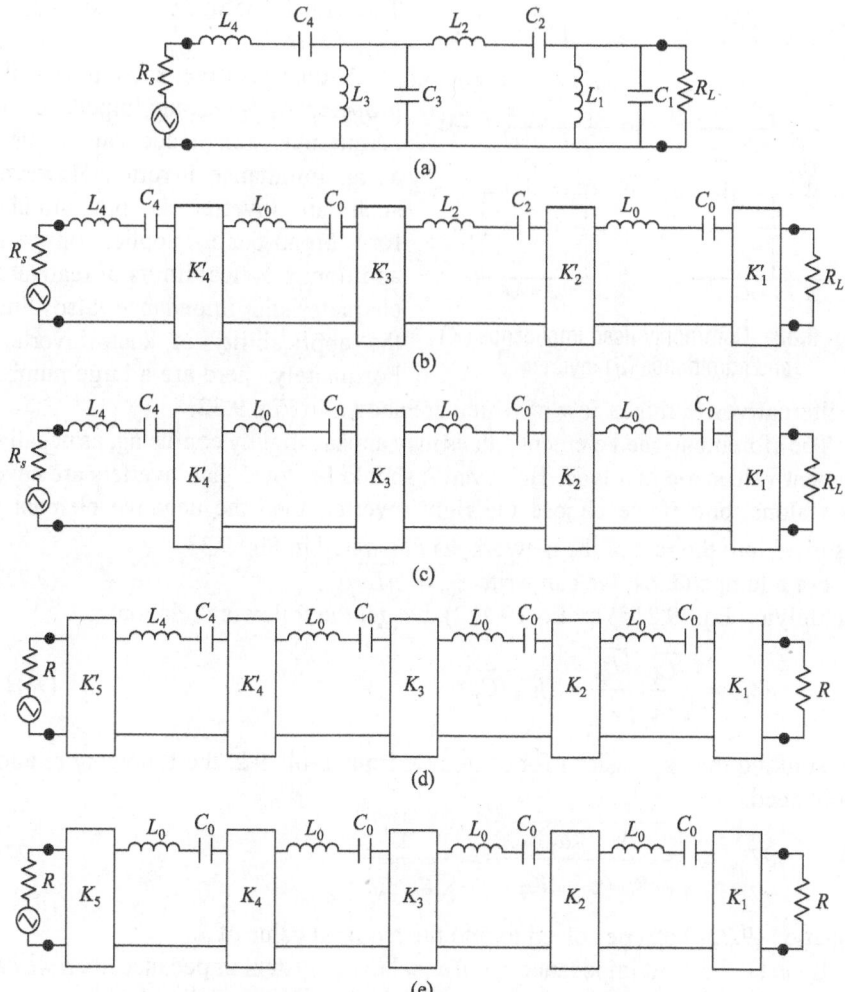

Fig. 9.31   Conversion of band-pass filter (a) Original filter prototype (b) Filter prototype with all-series element and immittance inverter (c) Filter prototype with all-series element after scaling resonator 2 to obtain same inductance and capacitance value as resonator 1 and 3 (d) Filter prototype with all-series element after scaling generator and load resistance to a value R (e) Final network with identical series tuned networks and equal input and output terminating resistance

Now, we need to change R_S to R with the help of another inverter, so that the following relation is satisfied: $K_5' = \sqrt{RR_S}$ (9.219)

The resultant network is shown in Fig. 9.31(d). Finally, we must change the fourth and fifth inverter so that it changes (L_4, C_4) to (L_0, C_0). Therefore, the new K values will be the following:

$$K_4 = K_4' \sqrt{\frac{L_0}{L_4}} = \omega_0 \sqrt{L_0 L_3} \sqrt{\frac{L_0}{L_4}} = \frac{1}{\sqrt{L_3 C_3}} \sqrt{\frac{L_0^2 L_3}{L_4}} = \sqrt{\frac{L_0^2}{L_4 C_3}}$$ (9.220)

and $$K_5 = K_5' \sqrt{L_0/L_4} = \sqrt{L_0 R R_S / L_4}$$ (9.221)

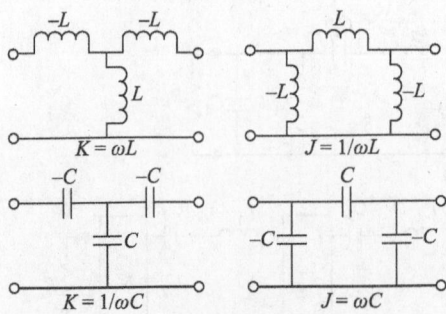

Fig. 9.32 Commonly used impedance (K) and admittance (J) inverters

The final network is shown in Fig. 9.31(e).

A quarter-wave section has the property of inverting impedance or admittance, and hence can be used as an immitance inverter. However, such an inverter is not suitable for broad-band applications. In addition, practical limits of realizable characteristic impedance also limit the applicability of such inverters. Fortunately, there are a large number of alternatives to this, a few of which are shown in Fig. 9.32.

The aforementioned element values may appear slightly confusing, as negative element values are involved. However, it should be noted that inverters are never used alone, and if we choose the right inverter, then the negative element is absorbed into the rest of the network, as illustrated in Fig. 9.33.

For a lumped tank, we can write $Z_{0r} = \sqrt{L_p/C_p}$ (9.222)

Multiplying Eq. (9.215) by Eq. (9.222), we get the following relation:

$$JZ_{0r} = \sqrt{\frac{C_s}{L_p}}\sqrt{\frac{L_p}{C_p}} = \sqrt{C_s/C_p} \tag{9.223}$$

Substituting the expressions for C_s and C_p from Table 9.2, the following relation is obtained:

$$JZ_{0r} = \sqrt{\frac{\Delta}{\omega_0 g_{ks} Z_{0r}} \frac{\omega_0 Z_{0r}\Delta}{g_{kp}}} = \frac{\Delta}{\sqrt{g_{ks}g_{kp}}} \tag{9.224}$$

Equation (9.224) can be solved to find the required value of J.

If the tank circuit impedance is not equal to the system impedance, then we can use immitance inverters to achieve the required matching. Following the analogy of the quarter-wave matching section, we can say that the desired impedance of an immitance inverter should be the geometric mean of the impedances to be matched. Therefore, if the prototype low-pass filter begins with an inductor, then

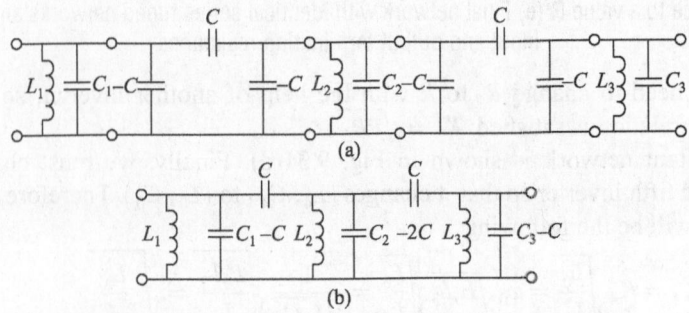

Fig. 9.33 All-parallel tank band-pass filter using J inverter (a) Before combining capacitor (b) After combining capacitor

$$J = \frac{1}{\sqrt{Z_0 \omega_0 L_1}} = \frac{1}{\sqrt{Z_0 \omega_0 \dfrac{g_1 Z_{0r}}{\omega_0 \Delta}}} = \frac{1}{Z_0}\sqrt{\frac{\Delta}{g_1 \dfrac{Z_{0r}}{Z_0}}} = \frac{1}{Z_0}\sqrt{\frac{\Delta}{m g_1}} \qquad (9.225)$$

where $\quad m = Z_{0r}/Z_0 \qquad\qquad\qquad\qquad\qquad\qquad\qquad\qquad\qquad (9.226)$

Equation (9.225) can be used to find the required value of J.

9.9 Coupled Line Filters

In the last chapter, we have shown that parallel coupled lines can be used as a directional coupler. Such a structure can also be used to design a band-pass or band-stop filter. Such a filter, when fabricated, can give a bandwidth of up to 20%. The bandwidth can further be increased with a tighter coupling. However, in such attempts, fabrication difficulties come into play. In the following discussion, we will first describe the filter properties of a coupled line section and then will show how to design a coupled line filter. For this, we will first consider Fig. 9.34.

Using even- and odd-mode excitations, the port currents can be described by the following equations:

$$I_1 = i_1 + i_2 \qquad\qquad\qquad\qquad\qquad\qquad\qquad (9.227)$$
$$I_2 = i_1 - i_2 \qquad\qquad\qquad\qquad\qquad\qquad\qquad (9.228)$$
$$I_3 = i_3 - i_4 \qquad\qquad\qquad\qquad\qquad\qquad\qquad (9.229)$$
and $\quad I_4 = i_3 + i_4 \qquad\qquad\qquad\qquad\qquad\qquad\qquad (9.230)$

Now, let us first consider that the structure has been excited with an even-mode current i_1, keeping the other port open circuited. The input impedance seen at port 1 or 2 is as follows: $Z_{in}^e = -jZ_{0e}\cot(\beta l) \qquad\qquad\qquad (9.231)$

The voltage at any point z on either conductor is as follows:

$$v_1^a(z) = v_1^b(z) = V_e^+\left[e^{-j\beta(z-l)} + e^{j\beta(z-l)}\right] = 2V_e^+\cos\{\beta(l-z)\} \qquad (9.232)$$

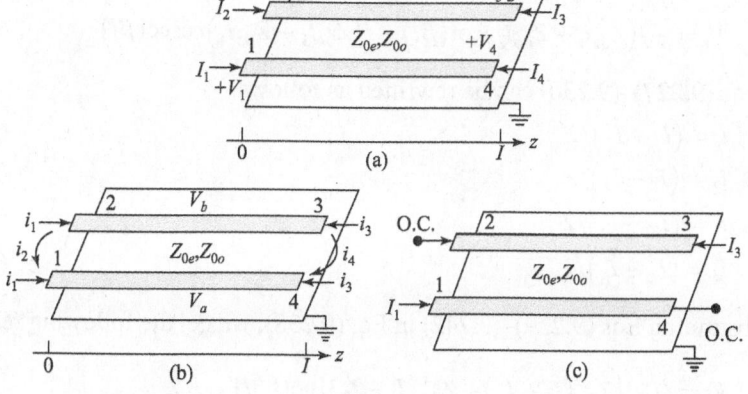

Fig. 9.34 Coupled line circuit (a) Parallel coupled line circuit with port voltage and currents (b) Parallel coupled line section with even- and odd-mode currents (c) Parallel coupled line section having band-pass response

Therefore, the port voltage is as follows:

$$v_1^a(0) = v_1^b(0) = 2V_e^+ \cos(\beta l) = i_1 Z_{in}^e$$

or $\qquad 2V_e^+ = i_1 Z_{in}^e / \cos(\beta l)$ (9.233)

Substituting Eqs (9.231) and (9.233) into Eq. (9.232), we get the following relation:

$$v_1^a(z) = v_1^b(z) = i_1 Z_{in}^e \frac{\cos\{\beta(l-z)\}}{\cos(\beta l)}$$

or $\qquad v_1^a(z) = v_1^b(z) = -ji_1 Z_{0e} \frac{\cot(\beta l)\cos\{\beta(l-z)\}}{\cos(\beta l)} = -jZ_{0e}\frac{\cos\{\beta(l-z)\}}{\sin(\beta l)}i_1$

(9.234)

Similarly, if the structure has been excited with an even-mode current i_3, keeping the other port open circuited, then we can write the following equation:

$$v_3^a(z) = v_3^b(z) = -jZ_{0e}\frac{\cos(\beta z)}{\sin(\beta l)}i_3$$ (9.235)

If the structure has been excited with an odd-mode current i_2, keeping the other port open circuited, then the following relation holds:

$$v_2^a(z) = -v_2^b(z) = -jZ_{0o}\frac{\cos\{\beta(l-z)\}}{\sin(\beta l)}i_2$$ (9.236)

Similarly, for odd-mode excitation with an odd-mode current i_4, with the other port being open circuited, we can write the following relation:

$$v_4^a(z) = -v_4^b(z) = -jZ_{0o}\frac{\cos(\beta z)}{\sin(\beta l)}i_4$$ (9.237)

Therefore, using the super position principle, the total voltage at port 1 is expressed as follows:

$$V_1 = v_1^a(0) + v_2^a(0) + v_3^a(0) + v_4^a(0)$$

or $\qquad V_1 = -j(Z_{0e}i_1 + Z_{0o}i_2)\cot(\beta l) - j(Z_{0e}i_3 + Z_{0o}i_4)\text{cosec}(\beta l)$ (9.238)

Now Eqs (9.227)–(9.230) can be rewritten as follows:

$$i_1 = (I_1 + I_2)/2$$ (9.239)
$$i_2 = (I_1 - I_2)/2$$ (9.240)
$$i_3 = (I_3 + I_4)/2$$ (9.241)

and $\qquad i_4 = (I_4 - I_3)/2$ (9.242)

Substituting Eqs (9.239)–(9.242) in Eq. (9.238), we get the following relation:

$$V_1 = -\frac{j}{2}\Big[\{Z_{0e}(I_1 + I_2) + Z_{0o}(I_1 - I_2)\}\cot(\beta l)$$
$$+ \{Z_{0e}(I_3 + I_4) + Z_{0o}(I_4 - I_3)\}\text{cosec}(\beta l)\Big]$$

$$V_1 = -\frac{j}{2}\Big[(Z_{0e}+Z_{0o})\cot(\beta l)I_1 + (Z_{0e}-Z_{0o})\cot(\beta l)I_2$$

or
$$+(Z_{0e}-Z_{0o})\operatorname{cosec}(\beta l)I_3 + (Z_{0e}+Z_{0o})\operatorname{cosec}(\beta l)I_4\Big] \tag{9.243}$$

Similarly, it can be shown that

$$V_2 = -\frac{j}{2}\Big[(Z_{0e}-Z_{0o})\cot(\beta l)I_1 + (Z_{0e}+Z_{0o})\cot(\beta l)I_2$$

$$+(Z_{0e}+Z_{0o})\operatorname{cosec}(\beta l)I_3 + (Z_{0e}-Z_{0o})\operatorname{cosec}(\beta l)I_4\Big] \tag{9.244}$$

$$V_3 = -\frac{j}{2}\Big[(Z_{0e}-Z_{0o})\operatorname{cosec}(\beta l)I_1 + (Z_{0e}+Z_{0o})\operatorname{cosec}(\beta l)I_2$$

$$+(Z_{0e}+Z_{0o})\cot(\beta l)I_3 + (Z_{0e}-Z_{0o})\cot(\beta l)I_4\Big] \tag{9.245}$$

$$V_4 = -\frac{j}{2}\Big[(Z_{0e}+Z_{0o})\operatorname{cosec}(\beta l)I_1 + (Z_{0e}-Z_{0o})\operatorname{cosec}(\beta l)I_2$$

and
$$+(Z_{0e}-Z_{0o})\cot(\beta l)I_3 + (Z_{0e}+Z_{0o})\cot(\beta l)I_4\Big] \tag{9.246}$$

Equations (9.243)–(9.246) can be written in the matrix form as follows:

$$\begin{bmatrix} V_1 \\ V_2 \\ V_3 \\ V_4 \end{bmatrix} = -\frac{j}{2}\begin{bmatrix} (Z_{0e}+Z_{0o})\cot(\beta l) & (Z_{0e}-Z_{0o})\cot(\beta l) & (Z_{0e}-Z_{0o})\operatorname{cosec}(\beta l) & (Z_{0e}+Z_{0o})\operatorname{cosec}(\beta l) \\ (Z_{0e}-Z_{0o})\cot(\beta l) & (Z_{0e}+Z_{0o})\cot(\beta l) & (Z_{0e}+Z_{0o})\operatorname{cosec}(\beta l) & (Z_{0e}-Z_{0o})\operatorname{cosec}(\beta l) \\ (Z_{0e}-Z_{0o})\operatorname{cosec}(\beta l) & (Z_{0e}+Z_{0o})\operatorname{cosec}(\beta l) & (Z_{0e}+Z_{0o})\cot(\beta l) & (Z_{0e}-Z_{0o})\cot(\beta l) \\ (Z_{0e}+Z_{0o})\operatorname{cosec}(\beta l) & (Z_{0e}-Z_{0o})\operatorname{cosec}(\beta l) & (Z_{0e}-Z_{0o})\cot(\beta l) & (Z_{0e}+Z_{0o})\cot(\beta l) \end{bmatrix}\begin{bmatrix} I_1 \\ I_2 \\ I_3 \\ I_4 \end{bmatrix} \tag{9.247}$$

or $\quad |V| = |Z||I|$

where

$$[Z] = -\frac{j}{2}\begin{bmatrix} (Z_{0e}+Z_{0o})\cot(\beta l) & (Z_{0e}-Z_{0o})\cot(\beta l) & (Z_{0e}-Z_{0o})\operatorname{cosec}(\beta l) & (Z_{0e}+Z_{0o})\operatorname{cosec}(\beta l) \\ (Z_{0e}-Z_{0o})\cot(\beta l) & (Z_{0e}+Z_{0o})\cot(\beta l) & (Z_{0e}+Z_{0o})\operatorname{cosec}(\beta l) & (Z_{0e}-Z_{0o})\operatorname{cosec}(\beta l) \\ (Z_{0e}-Z_{0o})\operatorname{cosec}(\beta l) & (Z_{0e}+Z_{0o})\operatorname{cosec}(\beta l) & (Z_{0e}+Z_{0o})\cot(\beta l) & (Z_{0e}-Z_{0o})\cot(\beta l) \\ (Z_{0e}+Z_{0o})\operatorname{cosec}(\beta l) & (Z_{0e}-Z_{0o})\operatorname{cosec}(\beta l) & (Z_{0e}-Z_{0o})\cot(\beta l) & (Z_{0e}+Z_{0o})\cot(\beta l) \end{bmatrix} \tag{9.248}$$

Now, since a four-terminal network can be reformed into a two-terminal network by either short or open circuiting the other two ports, 10 possible configurations exist. Depending on the conditions, they behave as low-pass, band-pass, all-pass or all-stop filters. Out of these 10 configurations, the image impedance and frequency response of four configurations are given in Table 9.3.

If we assume that ports 2 and 4 are open circuited, then the following relation can be written:

$$I_2 = I_4 = 0$$

Now, we can write the following equations:

$$V_1 = Z_{11}I_1 + Z_{12}I_2 \tag{9.249}$$

and $\quad V_2 = Z_{21}I_1 + Z_{22}I_2 \tag{9.250}$

Table 9.3 Canonical coupled line circuits

Circuit	Image Impedance	Response
	$Z_{i1} = \dfrac{2Z_{0e}Z_{0o}\cos(\theta)}{\sqrt{\left(Z_{0e}+Z_{0o}\right)^2\cos^2(\theta)-\left(Z_{0e}-Z_{0o}\right)^2}}$ $Z_{i2} = \dfrac{\sqrt{\left(Z_{0e}+Z_{0o}\right)^2\cos^2(\theta)-\left(Z_{0e}-Z_{0o}\right)^2}}{2\cos(\theta)}$	Low pass
	$Z_{i1} = \dfrac{2Z_{0e}Z_{0o}\sin(\theta)}{\sqrt{\left(Z_{0e}-Z_{0o}\right)^2-\left(Z_{0e}+Z_{0o}\right)^2\cos^2(\theta)}}$	Band pass
	$Z_{i1} = \dfrac{\sqrt{\left(Z_{0e}-Z_{0o}\right)^2-\left(Z_{0e}+Z_{0o}\right)^2\cos^2(\theta)}}{2\sin(\theta)}$	Band pass
	$Z_{i1} = \dfrac{\sqrt{Z_{0e}Z_{0o}}\sqrt{\left(Z_{0e}-Z_{0o}\right)^2-\left(Z_{0e}+Z_{0o}\right)^2\cos^2(\theta)}}{\left(Z_{0e}+Z_{0o}\right)\sin(\theta)}$ $Z_{i2} = \dfrac{\sqrt{Z_{0e}Z_{0o}}\left(Z_{0e}+Z_{0o}\right)\sin(\theta)}{\sqrt{\left(Z_{0e}-Z_{0o}\right)^2-\left(Z_{0e}+Z_{0o}\right)^2\cos^2(\theta)}}$	Band pass

Therefore, the image impedance is written as follows:

$$Z_{i1} = Z_{i2} = Z_i = \sqrt{Z_{11}^2 - \frac{Z_{11}Z_{13}^2}{Z_{33}}}$$

or

$$Z_i = \sqrt{\frac{j^2}{4}\left(Z_{0e}+Z_{0o}\right)^2\cot^2(\beta l) - \frac{2j^3\left(Z_{0e}+Z_{0o}\right)\cot(\beta l)\left(Z_{0e}-Z_{0o}\right)^2\mathrm{cosec}^2(\beta l)}{8j\left(Z_{0e}+Z_{0o}\right)\cot(\beta l)}}$$

or $\qquad Z_i = \dfrac{1}{2}\sqrt{\left(Z_{0e}-Z_{0o}\right)^2\mathrm{cosec}^2(\beta l) - \left(Z_{0e}+Z_{0o}\right)^2\cot^2(\beta l)}$ $\qquad$ (9.251)

Further, if the line is quarter-wavelength long, then

$$\beta l = \theta = \pi/2 \qquad (9.252)$$

and hence

$$Z_i = \frac{1}{2}\sqrt{\left(Z_{0e}-Z_{0o}\right)^2} = \frac{1}{2}\left(Z_{0e}-Z_{0o}\right) \qquad (9.253)$$

Since $Z_{0e} > Z_{0o}$, Eq. (9.253) reveals that Z_i is real, and a pass-band occurs. However, when $\beta l \to 0$ or π, $Z_i \to \pm\infty$, which indicates a stop-band. This is shown in Fig. 9.35. The cut-off occurs when $Z_i = 0$. Under such a condition, Eq. (9.251) can be written as follows:

$$\left(Z_{0e}-Z_{0o}\right)^2\mathrm{cosec}^2(\beta l) = \left(Z_{0e}+Z_{0o}\right)^2\cot^2(\beta l)$$

or $\qquad \left(Z_{0e}-Z_{0o}\right)\mathrm{cosec}(\beta l) = \pm\left(Z_{0e}+Z_{0o}\right)\cot(\beta l)$

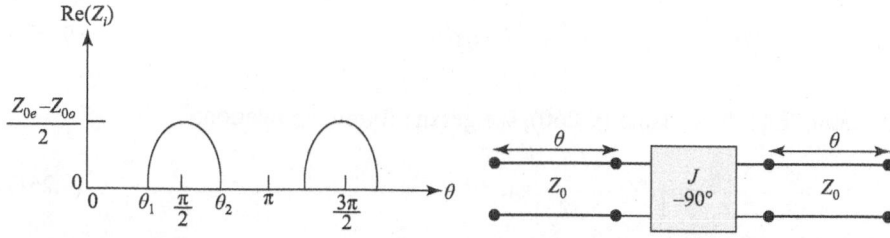

Fig. 9.35 Real part of image impedance of parallel coupled line section band-pass filter

Fig. 9.36 Equivalent circuit of coupled line section band-pass filter

or $$\cos(\beta l) = \pm \frac{Z_{0e} - Z_{0o}}{Z_{0e} + Z_{0o}} \qquad (9.254)$$

Equation (9.254) sets the upper and lower cut-off frequency. Again,

$$\cos(\beta) = \sqrt{\frac{Z_{11} Z_{33}}{Z_{13}^2}}$$

or $$\cos(\beta) = \sqrt{\frac{(Z_{0e} + Z_{0o})^2 \cot^2(\beta l)}{(Z_{0e} - Z_{0o})^2 \operatorname{cosec}^2(\beta l)}} = \frac{Z_{0e} + Z_{0o}}{Z_{0e} - Z_{0o}} \cos(\beta l) \qquad (9.255)$$

We will now show that an equivalent circuit of coupled line section band-pass filter can be modelled as shown in Fig. 9.36. *ABCD* parameters of the network, shown in Fig. 9.36, can be written as follows:

$$\begin{bmatrix} A & B \\ C & D \end{bmatrix} = \begin{bmatrix} \cos(\theta) & jZ_0 \sin(\theta) \\ \dfrac{j\sin(\theta)}{Z_0} & \cos(\theta) \end{bmatrix} \begin{bmatrix} 0 & -\dfrac{j}{J} \\ -jJ & 0 \end{bmatrix} \begin{bmatrix} \cos(\theta) & jZ_0 \sin(\theta) \\ \dfrac{j\sin(\theta)}{Z_0} & \cos(\theta) \end{bmatrix}$$

or $$\begin{bmatrix} A & B \\ C & D \end{bmatrix} = \begin{bmatrix} \left(JZ_0 + \dfrac{1}{JZ_0}\right)\sin\cos(\theta) & j\left\{JZ_0^2 \sin^2(\theta) - \dfrac{\cos^2(\theta)}{J}\right\} \\ j\left\{\dfrac{1}{JZ_0^2}\sin^2(\theta) - J\cos^2(\theta)\right\} & \left(JZ_0 + \dfrac{1}{JZ_0}\right)\sin\cos(\theta) \end{bmatrix} \qquad (9.256)$$

Therefore, $$Z_i = \sqrt{\frac{B}{C}} = \sqrt{Z_0^2 \frac{J^2 Z_0^2 \sin^2(\theta) - \cos^2(\theta)}{\sin^2(\theta) - J^2 Z_0^2 \cos^2(\theta)}} \qquad (9.257)$$

Further assuming that $\theta = \pi/2$, we get $Z_i = \sqrt{J^2 Z_0^4} = J Z_0^2 \qquad (9.258)$

Equating Eqs (9.253) and (9.258), we get $Z_{0e} - Z_{0o} = 2J Z_0^2 \qquad (9.259)$

Now from Eq. (9.91):

$$\cos(\beta) = \sqrt{AD} = \sqrt{\left(JZ_0 + \frac{1}{JZ_0}\right)^2 \sin^2(\theta) \cos^2(\theta)}$$

or
$$\cos(\beta) = \left(JZ_0 + \frac{1}{JZ_0}\right)\sin(\theta)\cos(\theta) \qquad (9.260)$$

Equating Eqs (9.255) and (9.260), we get the following relation:

$$\frac{Z_{0e} + Z_{0o}}{Z_{0e} - Z_{0o}} = \left(JZ_0 + \frac{1}{JZ_0}\right)\sin(\theta) \qquad (9.261)$$

Therefore, at $\theta = \pi/2$, the following relation is obtained:

$$\frac{Z_{0e} + Z_{0o}}{Z_{0e} - Z_{0o}} = JZ_0 + \frac{1}{JZ_0} \qquad (9.262)$$

Substitution of Eq. (9.259) into Eq. (9.262) results in the following relation:

$$\frac{Z_{0e} + Z_{0o}}{2JZ_0^2} = \frac{1 + J^2 Z_0^2}{JZ_0}$$

or
$$Z_{0e} + Z_{0o} = 2Z_0\left(1 + J^2 Z_0^2\right) \qquad (9.263)$$

Adding Eqs (9.259) and (9.263), we get the following relation:

$$Z_{0e} = Z_0\left[1 + JZ_0 + \left(JZ_0\right)^2\right] \qquad (9.264)$$

Subtracting Eq. (9.259) from Eq. (9.263), we get the following equation:

$$Z_{0o} = Z_0\left[1 - JZ_0 + \left(JZ_0\right)^2\right] \qquad (9.265)$$

Therefore, Fig. 9.36 can behave as an equivalent circuit of the coupled line filter element provided that Eqs (9.264) and (9.265) are satisfied.

Now, we will consider the band-pass filter composed of $(N + 1)$ cascaded unit, as shown in Fig. 9.37(a). Since each coupled line section can be represented as that shown in Fig. 9.36, the equivalent circuit of a multi-section filter can be represented as Fig. 9.37(b). Between any two consecutive inverters, we have a transmission line of effective length 2θ, which is equivalent to $\lambda/2$ at the centre frequency and has an equivalent circuit, as shown in Fig. 9.37(c). To show that the transmission line of length 2θ can behave as a parallel LC circuit, let us consider that the Z-matrix of the T-network, shown in Fig. 9.37(c), is $[Z]$. The $ABCD$ parameters of the T-network will be as follows:

$$\begin{bmatrix} A & B \\ C & D \end{bmatrix}_{TNW} = \begin{bmatrix} \dfrac{Z_{11}}{Z_{12}} & \dfrac{Z_{11}^2 - Z_{12}^2}{Z_{12}} \\ \dfrac{1}{Z_{12}} & \dfrac{Z_{11}}{Z_{12}} \end{bmatrix} \qquad (9.266)$$

The $ABCD$ parameters of the $1{:}{-}1$ transformer are as follows:

$$\begin{bmatrix} A & B \\ C & D \end{bmatrix}_{Tr} = \begin{bmatrix} -1 & 0 \\ 0 & -1 \end{bmatrix} \qquad (9.267)$$

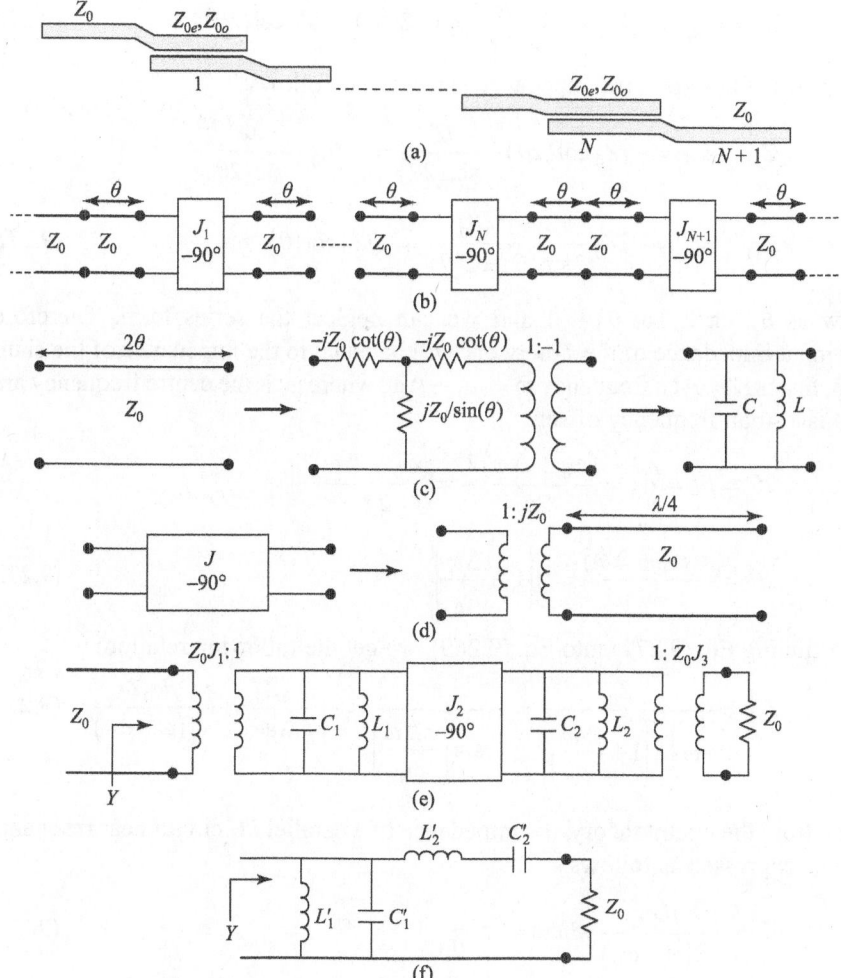

Fig. 9.37 Design and development of coupled line band-pass filter (a) $N + 1$ section coupled line band-pass filter (b) Equivalent circuit of $N + 1$-section coupled line band-pass filter (c) Equivalent circuit of transmission line of length 2θ (d) Equivalent circuit of admittance inverter (e) Equivalent circuit of coupled line band-pass filter for $n = 2$ (f) Lumped element circuit of band-pass filter for $N = 2$

Therefore, the complete ABCD parameters of the network are as follows:

$$
\begin{bmatrix} A & B \\ C & D \end{bmatrix} = \begin{bmatrix} \dfrac{Z_{11}}{Z_{12}} & \dfrac{Z_{11}^2 - Z_{12}^2}{Z_{12}} \\ \dfrac{1}{Z_{12}} & \dfrac{Z_{11}}{Z_{12}} \end{bmatrix} \begin{bmatrix} -1 & 0 \\ 0 & -1 \end{bmatrix} = \begin{bmatrix} -\dfrac{Z_{11}}{Z_{12}} & -\dfrac{Z_{11}^2 - Z_{12}^2}{Z_{12}} \\ -\dfrac{1}{Z_{12}} & -\dfrac{Z_{11}}{Z_{12}} \end{bmatrix}
\tag{9.268}
$$

Equating Eq. (9.268) with the *ABCD* parameters of a transmission line of length 2θ, we get the following relation:

$$
Z_{12} = Z_{21} = -\frac{1}{C} = \frac{jZ_0}{\sin(2\theta)}
\tag{9.269}
$$

and
$$Z_{11} = Z_{22} = -Z_{12}A = \frac{-jZ_0}{\sin(2\theta)}\cos(2\theta) = -jZ_0\cot(2\theta) \tag{9.270}$$

Therefore, the series arm impedance is written as follows:

$$Z_{11} - Z_{12} = -jZ_0\cot(2\theta) - \frac{jZ_0}{\sin(2\theta)} = -jZ_0\frac{1+\cos(2\theta)}{\sin(2\theta)}$$

or
$$Z_{11} - Z_{12} = -jZ_0\frac{2\cos^2(\theta)}{2\sin(\theta)\cos(\theta)} = -jZ_0\cot(\theta) \tag{9.271}$$

Now as $\theta \to \pi/2$, $\cot(\theta) \to 0$ and we can neglect the series term. Therefore, the input impedance of the T-network will be equal to the impedance of the shunt arm, that is, Z_{12}. At a frequency $\omega = \omega_0 + \Delta\omega$, where ω_0 is the centre frequency and $\Delta\omega$ is a small frequency offset:

$$2\theta = \beta l = \frac{\omega l}{v_p} = \frac{(\omega_0 + \Delta\omega)\lambda}{2v_p} = \frac{(\omega_0 + \Delta\omega)}{2f_0}$$

or
$$2\theta = \frac{\pi(\omega_0 + \Delta\omega)}{\omega_0} = \pi\left(1 + \frac{\Delta\omega}{\omega_0}\right) \tag{9.272}$$

Substituting Eq. (9.272) into Eq. (9.269), we get the following relation:

$$Z_{12} = \frac{jZ_0}{\sin\left\{\pi\left(1 + \frac{\Delta\omega}{\omega_0}\right)\right\}} = -\frac{jZ_0}{\sin\left(\frac{\pi\Delta\omega}{\omega_0}\right)} \approx -\frac{j\omega_0 Z_0}{\pi\Delta\omega} \approx -\frac{j\omega_0 Z_0}{\pi(\omega - \omega_0)} \tag{9.273}$$

Now from the circuit theory, the impedance of a parallel LC circuit near resonance can be expressed as follows:

$$Z = -\frac{jL\omega_0^2}{2(\omega - \omega_0)} \text{ where } \omega_0^2 = \frac{1}{LC} \tag{9.274}$$

or
$$C = \frac{1}{\omega_0^2 L} \tag{9.275}$$

Equating Eqs (9.273) and (9.274), the following equation is obtained:

$$-\frac{jL\omega_0^2}{2(\omega - \omega_0)} = -\frac{j\omega_0 Z_0}{\pi(\omega - \omega_0)}$$

or
$$L = \frac{2Z_0}{\pi\omega_0} \tag{9.276}$$

Substituting Eq. (9.276) into Eq. (9.275), we get the following expression:

$$C = \frac{1}{\omega_0^2 L} = \frac{1}{\omega_0^2}\frac{\pi\omega_0}{2Z_0} = \frac{\pi}{2\omega_0 Z_0} \tag{9.277}$$

The end section of the filters is a transmission line of length θ, which is matched with the feed line and hence can be ignored. The end inverters can be represented as a transformer followed by a quarter-wavelength line, as shown in Fig. 9.37(d). Therefore, the $ABCD$ matrix of J_1 and J_{N+1} will be as follows:

$$\begin{bmatrix} A & B \\ C & D \end{bmatrix} = \begin{bmatrix} 1/N & 0 \\ 0 & N \end{bmatrix} \begin{bmatrix} 0 & -jZ_0 \\ -j/Z_0 & 0 \end{bmatrix} = \begin{bmatrix} 0 & -jZ_0/N \\ -jN/Z_0 & 0 \end{bmatrix} \qquad (9.278)$$

where $N = JZ_0$.

The final equivalent circuit of Fig. 9.37(a), obtained using the equivalent circuits of different sections, is shown in Fig. 9.37(e) for $N = 2$.

The admittance just at the right of inverter J_2 is as follows:

$$Y' = j\omega C_2 + \frac{1}{j\omega L_2} + \frac{Z_0^2 J_3^2}{Z_0} = j\omega\sqrt{L_2 C_2}\sqrt{\frac{C_2}{L_2}} + \frac{1}{j\omega\sqrt{L_2 C_2}}\sqrt{\frac{C_2}{L_2}} + Z_0 J_3^2$$

or $\qquad Y' = j\sqrt{\dfrac{C_2}{L_2}}\left(\dfrac{\omega}{\omega_0} - \dfrac{\omega_0}{\omega}\right) + Z_0 J_3^2 \qquad\qquad (9.279)$

Hence,

$$Y = \frac{1}{J_1^2 Z_0^2}\left\{ j\omega C_1 + \frac{1}{j\omega L_1} + \frac{J_2^2}{j\sqrt{\dfrac{C_2}{L_2}}\left(\dfrac{\omega}{\omega_0} - \dfrac{\omega_0}{\omega}\right) + Z_0 J_3^2} \right\}$$

or $\qquad Y = \dfrac{1}{J_1^2 Z_0^2}\left\{ j\sqrt{\dfrac{C_1}{L_1}}\left(\dfrac{\omega}{\omega_0} - \dfrac{\omega_0}{\omega}\right) + \dfrac{J_2^2}{j\sqrt{\dfrac{C_2}{L_2}}\left(\dfrac{\omega}{\omega_0} - \dfrac{\omega_0}{\omega}\right) + Z_0 J_3^2} \right\} \qquad (9.280)$

Now let us consider the lumped element prototype of a band-pass filter (for $N = 2$), shown in Fig. 9.37(f). The input admittance of the network is expressed here:

$$Y = j\omega C_1' + \frac{1}{j\omega L_1'} + \frac{1}{j\omega L_2' + \dfrac{1}{j\omega C_2'} + Z_0}$$

or $\qquad Y = j\sqrt{\dfrac{C_1'}{L_1'}}\left(\dfrac{\omega}{\omega_0} - \dfrac{\omega_0}{\omega}\right) + \dfrac{1}{j\sqrt{\dfrac{L_2'}{C_2'}}\left(\dfrac{\omega}{\omega_0} - \dfrac{\omega_0}{\omega}\right) + Z_0} \qquad (9.281)$

which is identical in form to Eq. (9.280). Therefore, both the circuits will be identical if the following equations are satisfied:

$$\frac{1}{J_1^2 Z_0^2}\sqrt{\frac{C_1}{L_1}} = \sqrt{\frac{C_1'}{L_1'}}$$

or $\qquad J_1 Z_0 = \left(\dfrac{C_1 L_1'}{L_1 C_1'}\right)^{\frac{1}{4}} \qquad\qquad (9.282)$

$$\frac{J_1^2 Z_0^2}{J_2^2}\sqrt{\frac{C_2}{L_2}} = \sqrt{\frac{L_2'}{C_2'}}$$

or $\qquad J_2 Z_0 = J_1 Z_0^2\left(\dfrac{C_2 C_2'}{L_2 L_2'}\right)^{\frac{1}{4}} \qquad\qquad (9.283)$

and $\quad \dfrac{J_1^2 Z_0^3 J_3^2}{J_2^2} = Z_0$

or $\quad J_3 Z_0 = \dfrac{J_2}{J_1} = \dfrac{J_2 Z_0}{J_1 Z_0}$ $\qquad$ (9.284)

The value of L_n and C_n can be found from Eqs (9.276) and (9.277). After performing the impedance and frequency scaling, the component values of Fig. 9.37(e) can be written as follows:

$$L_1' = \frac{\Delta Z_0}{\omega_0 g_1} \qquad (9.285)$$

$$L_2' = \frac{g_2 Z_0}{\Delta \omega_0} \qquad (9.286)$$

$$C_1' = \frac{g_1}{\Delta Z_0 \omega_0} \qquad (9.287)$$

and $\quad C_2' = \dfrac{\Delta}{g_2 \omega_0 Z_0}$ $\qquad$ (9.288)

where $\quad \Delta = (\omega_2 - \omega_1)/\omega_0$ $\qquad$ (9.289)

Substituting Eqs (9.276), (9.277), and (9.285)–(9.288) into Eqs (9.282)–(9.284), inverter constants can be solved, as follows:

$$J_1 Z_0 = \left(\frac{C_1 L_1'}{L_1 C_1'}\right)^{\frac{1}{4}} = \left(\frac{\pi}{2\omega_0 Z_0} \frac{\Delta Z_0}{\omega_0 g_1} \frac{\pi \omega_0}{2 Z_0} \frac{\Delta Z_0 \omega_0}{g_1}\right)^{\frac{1}{4}} = \left(\frac{\pi^2 \Delta^2}{4 g_1^2}\right)^{\frac{1}{4}} = \sqrt{\frac{\pi \Delta}{2 g_1}} \qquad (9.290)$$

$$J_2 Z_0 = J_1 Z_0^2 \left(\frac{C_2 C_2'}{L_2 L_2'}\right)^{\frac{1}{4}} = Z_0 \sqrt{\frac{\pi \Delta}{2 g_1}} \left(\frac{\pi}{2\omega_0 Z_0} \frac{\Delta}{g_2 \omega_0 Z_0} \frac{\pi \omega_0}{2 Z_0} \frac{\Delta \omega_0}{g_2 Z_0}\right)^{\frac{1}{4}}$$

or $\quad J_2 Z_0 = Z_0 \sqrt{\dfrac{\pi \Delta}{2 g_1}} \left(\dfrac{\pi^2 \Delta^2}{4 g_2^2 Z_0^4}\right)^{\frac{1}{4}} = \dfrac{\pi \Delta}{2\sqrt{g_1 g_2}}$ $\qquad$ (9.291)

and $\quad J_3 Z_0 = \dfrac{J_2 Z_0}{J_1 Z_0} = \dfrac{\pi \Delta}{2\sqrt{g_1 g_2}} \sqrt{\dfrac{2 g_1}{\pi \Delta}} = \sqrt{\dfrac{\pi \Delta}{2 g_2}}$ $\qquad$ (9.292)

For an arbitrary value of N, the general solutions can be shown to be of the following forms:

$$Z_0 J_1 = \sqrt{\frac{\pi \Delta}{2 g_1}} \qquad (9.293)$$

$$Z_0 J_n = \frac{\pi \Delta}{2\sqrt{g_{n-1} g_n}} \quad \text{for } n = 2, 3, \ldots, N \qquad (9.294)$$

and $\quad Z_0 J_{N+1} = \sqrt{\dfrac{\pi \Delta}{2 g_N g_{N+1}}}$ $\qquad$ (9.295)

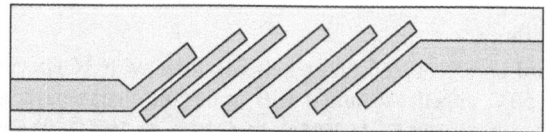

Fig. 9.38 More compact layout of coupled line band-pass filter

Coupled line filters are widely used because of their relatively straightforward design. However, the main disadvantage of such a filter is that it occupies a large area, especially when a large number of sections are used. To make it compact, the sections are often rotated, as shown in Fig. 9.38.

EXAMPLE 9.4 Design a third-order coupled line band-pass filter with 5 GHz centre frequency, 8% bandwidth, 50 Ω impedance, and 0.25 dB equiripple characteristics.

Solution Given: $\varepsilon = 0.25$ dB, $f_c = 5$ GHz, $Z_0 = 50\,\Omega$, $\Delta = 8\%$, and $N = 3$

Therefore,

$$g_0 = 1.0000 \qquad g_1 = 1.3034 \qquad g_2 = 1.1463 \qquad g_3 = 1.3034 \qquad g_4 = 1.0000$$

$$Z_0 J_1 = \sqrt{\frac{\pi\Delta}{2g_1}} = \sqrt{\frac{0.08\pi}{2\times1.3034}} = 0.3105$$

$$Z_0 J_2 = \frac{\pi\Delta}{2\sqrt{g_1 g_2}} = \frac{0.08\pi}{2\sqrt{1.3034\times1.1463}} = 0.1028$$

$$Z_0 J_3 = \frac{\pi\Delta}{2\sqrt{g_2 g_3}} = \frac{0.08\pi}{2\sqrt{1.1463\times1.3034}} = 0.1028$$

$$Z_0 J_4 = \sqrt{\frac{\pi\Delta}{2g_3 g_4}} = \sqrt{\frac{0.08\pi}{2\times1.3034}} = 0.3105$$

Therefore, the even- and odd-mode impedances are as follows:

$$Z_{0e,1} = Z_0\left[1 + Z_0 J_1 + \left(Z_0 J_1\right)^2\right] = 50\left[1 + 0.3105 + \left(0.3105\right)^2\right] = 70.3455\,\Omega$$

$$Z_{0o,1} = Z_0\left[1 - Z_0 J_1 + \left(Z_0 J_1\right)^2\right] = 50\left[1 - 0.3105 + \left(0.3105\right)^2\right] = 39.2955\,\Omega$$

$$Z_{0e,2} = Z_0\left[1 + Z_0 J_2 + \left(Z_0 J_2\right)^2\right] = 50\left[1 + 0.1028 + \left(0.1028\right)^2\right] = 55.6684\,\Omega$$

$$Z_{0o,2} = Z_0\left[1 - Z_0 J_2 + \left(Z_0 J_2\right)^2\right] = 50\left[1 - 0.1028 + \left(0.1028\right)^2\right] = 45.3884\,\Omega$$

$$Z_{0e,3} = Z_0\left[1 + Z_0 J_3 + \left(Z_0 J_3\right)^2\right] = 50\left[1 + 0.1028 + \left(0.1028\right)^2\right] = 55.6684\,\Omega$$

$$Z_{0o,3} = Z_0\left[1 - Z_0 J_3 + \left(Z_0 J_3\right)^2\right] = 50\left[1 - 0.1028 + \left(0.1028\right)^2\right] = 45.3884\,\Omega$$

$$Z_{0e,4} = Z_0\left[1 + Z_0 J_4 + \left(Z_0 J_4\right)^2\right] = 50\left[1 + 0.3105 + \left(0.3105\right)^2\right] = 70.3455\,\Omega$$

$$Z_{0o,4} = Z_0\left[1 - Z_0 J_4 + \left(Z_0 J_4\right)^2\right] = 50\left[1 - 0.3105 + \left(0.3105\right)^2\right] = 39.2955\,\Omega$$

Practice Problem

9.6 Design a third-order coupled line band-pass filter with 3 GHz centre frequency, 10% bandwidth, 50 Ω impedance, and 0.5 dB equiripple characteristics.

[70.6054 Ω, 39.2354 Ω, 56.6395 Ω, 44.7695 Ω, 56.6395 Ω, 44.7695 Ω, 70.6054 Ω, 39.2354 Ω]

9.10 Coupled Resonator Filters

It is now known that a band-pass or band-stop filter can be represented by a combination of different series and shunt LC resonant networks. One such filter design has already been described in Section 9.9. In this section we will discuss a few more types of coupled resonator filters.

9.10.1 Transmission Line Resonator Filters

The theory of transmission lines and resonators states that a short- or an open-circuited quarter-wave transmission line behaves as a parallel or a series resonator circuit, respectively. Therefore, such transmission lines can be arranged to form a band-pass or a band-stop response, as shown in Fig. 9.39. Since the internal impedance of such a stub filter is Z_0, no additional matching sections are required at the input and output ends. That is why, such filters are more compact. In addition, the design procedure is easy and straightforward. However, the main disadvantage of such filters is that stub resonators often require a characteristic impedance of the line, which is difficult to realize.

To demonstrate the design procedure, let us consider the N-section transmission line resonator band-stop filter shown in Fig. 9.40. Figure 9.40(a) shows an open-circuited quarter-wave transmission line and its equivalent circuit. The input impedance of the line is as follows: $Z = -jZ_{0n} \cot(\theta)$ (9.296)

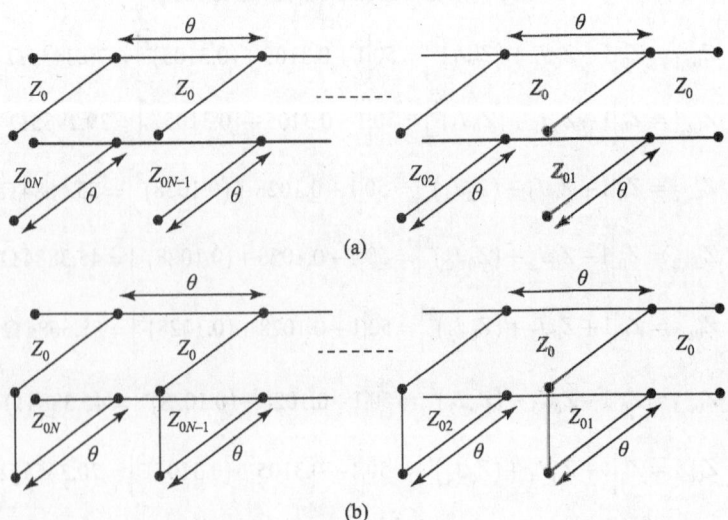

Fig. 9.39 Transmission line resonator filter (a) Band-stop (b) Band-pass

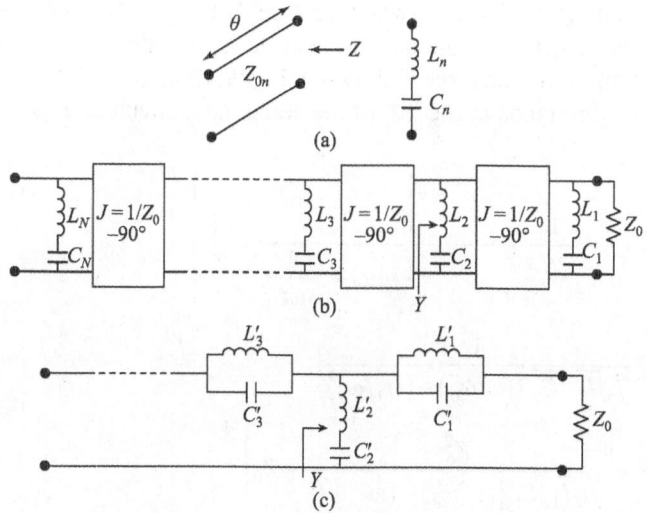

Fig. 9.40 Development of the equivalent circuit of Fig 9.39(a)
(a) Equivalent circuit of open-circuited stub (b) Equivalent filter
using resonator and admittance inverter (c) Equivalent lumped
circuit band-stop filter

where $\theta = \pi/2$ at $\omega = \omega_0$. At a frequency of $\omega = \omega_0 + \Delta\omega$, where $\Delta\omega \ll \omega_0$,

$$\theta = \beta l = \frac{2\pi}{\lambda}\frac{\lambda_0}{4} = \frac{\pi(\omega_0 + \Delta\omega)}{2\omega_0} = \frac{\pi}{2}\left(1 + \frac{\Delta\omega}{\omega_0}\right) \tag{9.297}$$

Thus, $Z = -jZ_{0n}\cot(\theta) = -jZ_{0n}\cot\left\{\frac{\pi}{2}\left(1 + \frac{\Delta\omega}{\omega_0}\right)\right\}$

or $Z = jZ_{0n}\tan\left(\frac{\pi\Delta\omega}{2\omega_0}\right) \approx \frac{jZ_{0n}\pi\Delta\omega}{2\omega_0} \approx \frac{jZ_{0n}\pi(\omega - \omega_0)}{2\omega_0} \tag{9.298}$

Now, let us consider the LC equivalent circuit. The input impedance of the line is
as follows:

$$Z = j\omega L_n + \frac{1}{j\omega C_n} = j\sqrt{\frac{L_n}{C_n}}\left(\frac{\omega}{\omega_0} - \frac{\omega_0}{\omega}\right)$$

or $Z = j\sqrt{\frac{L_n}{C_n}}\left(\frac{\omega^2 - \omega_0^2}{\omega\omega_0}\right) = j\sqrt{\frac{L_n}{C_n}}\frac{(\omega - \omega_0)(\omega + \omega_0)}{\omega\omega_0}$

or $Z \approx j\sqrt{\frac{L_n}{C_n}}\frac{2\omega(\omega - \omega_0)}{\omega\omega_0} \approx j2\sqrt{\frac{L_n}{C_n}}\sqrt{L_nC_n}(\omega - \omega_0) \approx j2L_n(\omega - \omega_0) \tag{9.299}$

Comparing Eqs (9.298) and (9.299), we get the following relation:

$$\frac{jZ_{0n}\pi(\omega - \omega_0)}{2\omega_0} = j2L_n(\omega - \omega_0)$$

or $Z_{0n} = 4\omega_0 L_n/\pi \tag{9.300}$

Now assuming the quarter-wave section between the stubs as an ideal admittance inverter, the band-stop filter can be represented as Fig. 9.40(b). The lumped element prototype filter is shown in Fig. 9.40(c).

The input admittance at the left of the series L_2C_2 circuit of Fig. 9.40(b) is as follows:

$$Y = \frac{1}{j\omega L_2 + \dfrac{1}{j\omega C_2}} + \frac{1}{Z_0^2}\left[\frac{1}{j\omega L_1 + \dfrac{1}{j\omega C_1}} + \frac{1}{Z_0}\right]^{-1}$$

$$Y = \frac{1}{j\sqrt{L_2/C_2}\left\{(\omega/\omega_0)-(\omega_0/\omega)\right\}}$$

or

$$+ \left[\frac{Z_0^2}{j\sqrt{L_1/C_1}\left\{(\omega/\omega_0)-(\omega_0/\omega)\right\}} + Z_0\right]^{-1} \tag{9.301}$$

The admittance at the corresponding point on the lumped prototype circuit is as follows:

$$Y = \frac{1}{j\omega L_2' + \dfrac{1}{j\omega C_2'}} + \left[\frac{1}{j\omega C_1' + \dfrac{1}{j\omega L_1'}} + Z_0\right]^{-1}$$

$$Y = \frac{1}{j\sqrt{L_2'/C_2'}\left\{(\omega/\omega_0)-(\omega_0/\omega)\right\}}$$

or

$$+ \left[\frac{1}{j\sqrt{C_1'/L_1'}\left\{(\omega/\omega_0)-(\omega_0/\omega)\right\}} + Z_0\right]^{-1} \tag{9.302}$$

Comparing these two results, we get the following equations:

$$Z_0^2\sqrt{\frac{C_1}{L_1}} = \sqrt{\frac{L_1'}{C_1'}} \quad \text{or} \quad \frac{Z_0^2}{L_1}\sqrt{L_1C_1} = L_1'\sqrt{\frac{1}{L_1'C_1'}} \quad \text{or} \quad \frac{Z_0^2}{L_1\omega_0} = L_1'\omega_0$$

or

$$L_1 = \frac{Z_0^2}{L_1'\omega_0^2} \tag{9.303}$$

and

$$\sqrt{\frac{L_2}{C_2}} = \sqrt{\frac{L_2'}{C_2'}} \quad \text{or} \quad L_2\sqrt{\frac{1}{L_2C_2}} = L_2'\sqrt{\frac{1}{L_2'C_2'}} \quad \text{or} \quad L_2\omega_0 = L_2'\omega_0$$

or

$$L_2 = L_2' \tag{9.304}$$

Now,

$$L_1' = \frac{g_1\Delta Z_0}{\omega_0} \tag{9.305}$$

$$C_1' = \frac{1}{\omega_0 g_1\Delta Z_0} \tag{9.306}$$

$$L_2' = \frac{Z_0}{\omega_0 g_2 \Delta} \tag{9.307}$$

and $\quad C_2' = \dfrac{g_2 \Delta}{\omega_0 Z_0} \tag{9.308}$

Using Eqs (9.300), (9.303), and (9.305), we get the following relation:

$$Z_{01} = \frac{4\omega_0 L_1}{\pi} = \frac{4\omega_0}{\pi}\frac{Z_0^2}{L_1'\omega_0^2} = \frac{4\omega_0 Z_0^2}{\pi\omega_0^2}\frac{\omega_0}{g_1\Delta Z_0} = \frac{4Z_0}{\pi g_1 \Delta} \tag{9.309}$$

Similarly, using Eqs (9.300), (9.304), and (9.307), we get the following expression:

$$Z_{02} = \frac{4\omega_0 L_2}{\pi} = \frac{4\omega_0 L_2'}{\pi} = \frac{4\omega_0}{\pi}\frac{Z_0}{\omega_0 g_2 \Delta} = \frac{4Z_0}{\pi g_2 \Delta} \tag{9.310}$$

The general equation for the characteristic impedance of the line can be shown to be the following:

$$Z_{0n} = \frac{4Z_0}{\pi g_n \Delta} \tag{9.311}$$

For a band-pass filter, the general equation modifies as, $Z_{0n} = \dfrac{\pi \Delta Z_0}{4 g_n} \tag{9.312}$

It should be noted that this analysis assumes the input and output impedances as Z_0 and hence cannot be used for a Chebyshev filter with an even N.

EXAMPLE 9.5 Design a third-order band-stop filter with 5 GHz centre frequency, 20% bandwidth, 50 Ω impedance, and 0.25 dB equiripple characteristics. Use a quarter-wave open-circuited stub to design the band-stop filter.

Solution Given: $\varepsilon = 0.25$ dB, $f_c = 5$ GHz, $Z_0 = 50\,\Omega$, $\Delta = 20\%$, and $N = 3$

Therefore, $g_0 = 1.0000$, $g_1 = 1.3034$, $g_2 = 1.1463$, $g_3 = 1.3034$, $g_4 = 1.0000$

$$Z_{01} = \frac{4Z_0}{\pi g_1 \Delta} = \frac{4 \times 50}{1.3034\pi \times 0.2} = 244.215\ \Omega$$

$$Z_{02} = \frac{4Z_0}{\pi g_2 \Delta} = \frac{4 \times 50}{1.1463\pi \times 0.2} = 277.6846\ \Omega$$

$$Z_{03} = \frac{4Z_0}{\pi g_3 \Delta} = \frac{4 \times 50}{1.3034\pi \times 0.2} = 244.215\ \Omega$$

Practice Problem

9.7 Design a third-order band-stop filter with 3 GHz centre frequency, 25% bandwidth, 50 Ω impedance, and 0.5 dB equiripple characteristics. Use a quarter-wave open-circuited stub to design the band-stop filter.

$$[199.4048\ \Omega,\ 290.2434\ \Omega,\ 290.2434\ \Omega,\ 199.4048\ \Omega]$$

9.10.2 Capacitively Coupled Microstrip Resonator Filters

A capacitively coupled microstrip resonator filter is shown in Fig. 9.41(a). The gap may be assumed as a series capacitor with a negative length of transmission line, and hence Fig. 9.41(a) can be redrawn as Fig. 9.41(b). The equivalent circuit

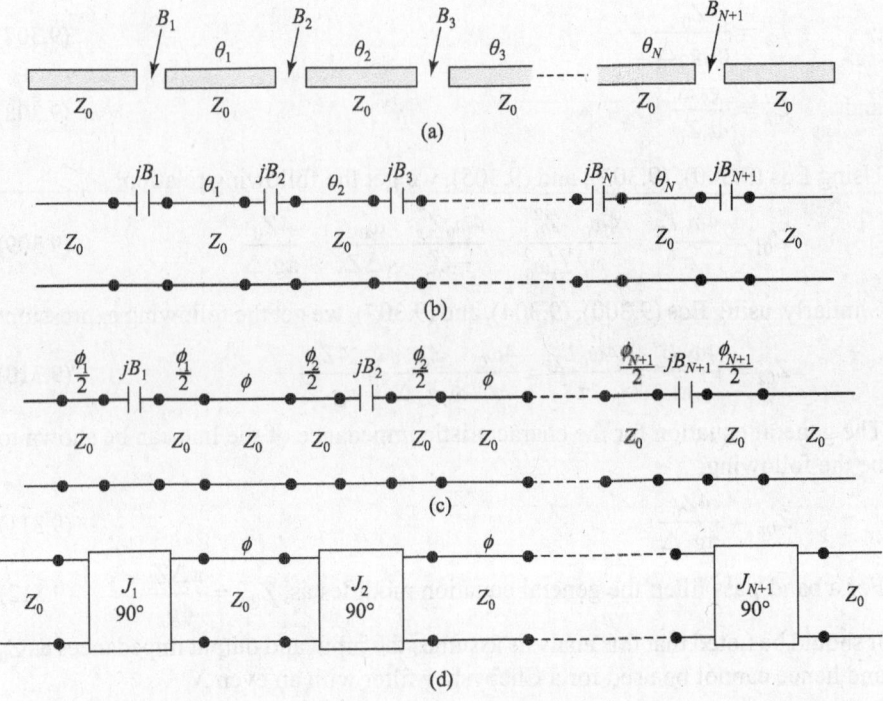

Fig. 9.41 Development of equivalent circuit (a) Capacitive gap coupled microstrip resonator filter (b) Equivalent transmission line model (c) Equivalent transmission line circuit with negative length section forming inverters (d) Equivalent circuit with admittance inverter and resonator

of Fig. 9.41(b) with a negative-length transmission line on both sides of the series capacitor is shown in Fig. 9.41(c). Since $\phi = \pi$ at $\omega = \omega_0$, we can write the following equation:

$$\theta_n = \pi + \frac{\phi_n}{2} + \frac{\phi_{n+1}}{2} \quad \text{for } n = 1, 2, 3, ..., N \tag{9.313}$$

where $\phi_n < 0$. Since the capacitor with a negative-length transmission line forms an admittance inverter, Fig. 9.41(c) can be modified as Fig. 9.41(d). However, to do this, we must have

$$\phi_n = -\tan^{-1}(2Z_0 B_n) \tag{9.314}$$

where

$$B_n = \frac{J_n}{1 - (Z_0 J_n)^2} \tag{9.315}$$

where J_n can be calculated using Eqs (9.293)–(9.295).

EXAMPLE 9.6 Design a third-order capacitively coupled resonator band-pass filter with 5 GHz centre off frequency, 8% bandwidth, 50 Ω impedance, and 0.25 dB equiripple characteristics.

Solution Given: $\varepsilon = 0.25$ dB, $f_c = 5$ GHz, $Z_0 = 50\,\Omega$, $\Delta = 8\%$, and $N = 3$.

Therefore,

$$g_0 = 1.0000 \qquad g_1 = 1.3034 \qquad g_2 = 1.1463 \qquad g_3 = 1.3034 \qquad g_4 = 1.0000$$

$$Z_0 J_1 = \sqrt{\frac{\pi\Delta}{2g_1}} = \sqrt{\frac{0.08\pi}{2\times1.3034}} = 0.3105$$

$$Z_0 J_2 = \frac{\pi\Delta}{2\sqrt{g_1 g_2}} = \frac{0.08\pi}{2\sqrt{1.3034\times1.1463}} = 0.1028$$

$$Z_0 J_3 = \frac{\pi\Delta}{2\sqrt{g_2 g_3}} = \frac{0.08\pi}{2\sqrt{1.1463\times1.3034}} = 0.1028$$

$$Z_0 J_4 = \sqrt{\frac{\pi\Delta}{2g_3 g_4}} = \sqrt{\frac{0.08\pi}{2\times1.3034}} = 0.3105$$

$$B_1 = \frac{1}{Z_0}\frac{Z_0 J_1}{1-(Z_0 J_1)^2} = \frac{0.3105}{50\left[1-(0.3105)^2\right]} = 6.8726\times10^{-3}$$

$$B_2 = \frac{1}{Z_0}\frac{Z_0 J_2}{1-(Z_0 J_2)^2} = \frac{0.1028}{50\left[1-(0.1028)^2\right]} = 2.078\times10^{-3}$$

$$B_3 = \frac{1}{Z_0}\frac{Z_0 J_3}{1-(Z_0 J_3)^2} = \frac{0.1028}{50\left[1-(0.1028)^2\right]} = 2.078\times10^{-3}$$

$$B_4 = \frac{1}{Z_0}\frac{Z_0 J_4}{1-(Z_0 J_4)^2} = \frac{0.3105}{50\left[1-(0.3105)^2\right]} = 6.8726\times10^{-3}$$

Hence,

$$C_1 = \frac{B_1}{\omega_0} = \frac{6.8726\times10^{-3}}{2\pi\times5\times10^9} = 0.2188\times10^{-12} = 0.2188 \text{ pF}$$

$$C_2 = \frac{B_2}{\omega_0} = \frac{2.078\times10^{-3}}{2\pi\times5\times10^9} = 0.0661\times10^{-12} = 0.0661 \text{ pF}$$

$$C_3 = \frac{B_3}{\omega_0} = \frac{2.078\times10^{-3}}{2\pi\times5\times10^9} = 0.0661\times10^{-12} = 0.0661 \text{ pF}$$

$$C_4 = \frac{B_4}{\omega_0} = \frac{6.8726\times10^{-3}}{2\pi\times5\times10^9} = 0.2188\times10^{-12} = 0.2188 \text{ pF}$$

$$\theta_1 = \pi - \frac{1}{2}\left[\tan^{-1}(2Z_0 B_1) + \tan^{-1}(2Z_0 B_1)\right]$$

or

$$\theta_1 = \pi - \frac{1}{2}\left[0.6021 + 0.2024\right] = 2.7393 \text{ rad} = 156.9528°$$

$$\theta_2 = \pi - \frac{1}{2}\left[\tan^{-1}(2Z_0 B_2) + \tan^{-1}(2Z_0 B_3)\right]$$

or

$$\theta_1 = \pi - \frac{1}{2}\left[0.2024 + 0.2024\right] = 2.9392 \text{ rad} = 168.4033°$$

$$\theta_3 = \pi - \frac{1}{2}\left[\tan^{-1}(2Z_0 B_3) + \tan^{-1}(2Z_0 B_4)\right]$$

or

$$\theta_3 = \pi - \frac{1}{2}\left[0.6021 + 0.2024\right] = 2.7393 \text{ rad} = 156.9528°$$

Practice Problem

9.8 Design a third-order capacitively coupled resonator band-pass filter with a centre off frequency of 3 GHz, a bandwidth of 10%, an impedance of 50 Ω, and equiripple characteristics of 0.5 dB.

[0.3692 pF, 0.1277 pF, 0.1277 pF, 0.3692 pF, 155.8155°, 166.461°, 155.8155°]

9.11 Other Filters

Some other commonly used filters are discussed in the following subsections.

9.11.1 Stepped Impedance Low-pass Filters

A stepped impedance filter consists of cascaded connections of low- and high-impedance lines to exhibit filter response and is shown in Fig. 9.42(a). To explain the structure, let us first consider the expression for input impedance of a loaded transmission line, given by the following equation:

$$Z_{in}(z) = Z_o \frac{Z_L + jZ_0 \tan(\beta z)}{Z_0 + jZ_L \tan(\beta z)}$$

or

$$\frac{Z_{in}(z)}{Z_o} = \frac{\dfrac{Z_L}{Z_0} + j \tan(\beta z)}{1 + j\dfrac{Z_L}{Z_0} \tan(\beta z)} \tag{9.316}$$

Now, if $Z_0 \gg Z_L$, Eq. (9.316) reduces to the following form:

$$Z_{in}(z) \approx jZ_o \tan(\beta z) \tag{9.317}$$

which is inductive for $\tan(\beta z) > 0$. Again if $Z_0 \ll Z_L$, then Eq. (9.316) reduces to the following form:

$$Z_{in}(z) \approx -jZ_o \cot(\beta z) \tag{9.318}$$

which is capacitive for $\tan(\beta z) < 0$.

These equations imply that we can choose Z_0 as low as possible to make a capacitor and Z_0 as high as possible to make an inductor. Alternatively, a thin microstrip line can be used as an inductor, whereas a thick microstrip line can be used as a capacitor. Thus, the equivalent circuit of the microstrip structure shown in Fig. 9.42(a) can be modified as the circuit shown in Fig. 9.42(b), which is basically a low-pass filter.

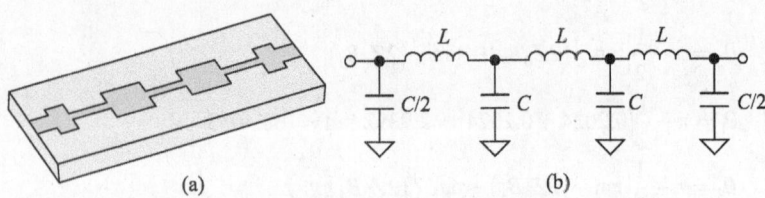

(a)   (b)

Fig. 9.42 Stepped impedance filter (a) Schematic diagram
(b) Lumped equivalent circuit

It should be noted that we cannot make arbitrarily high characteristic impedance because there is always a lower bound of strip width that can be fabricated reliably. Similarly, the maximum line width is also restricted, as all linear dimensions of a microstrip element must be small compared to the wavelength at the frequency of interest in order to validate the lumped element behaviour. In practice, the extreme range of characteristic impedance that can be implemented using microstrip lines lies between $10\,\Omega$ and $200\,\Omega$.

The transmission line behaviour of such structures becomes more prominent at higher frequency, especially after the cut-off frequency, and hence stepped impedance filters fail to roll off as quickly as its lumped equivalent prototype.

9.11.2 YIG Filters

Yttrium-iron-garnet or $Y_3Fe_2(FeO_4)_3$ spheres show resonance property when placed under a direct magnetic field and hence can be used as filters. A filter that uses such a structure is called a YIG filter and is shown in Fig. 9.43. The resonance frequency of such a filter is determined by the strength of the applied magnetic field. Therefore, by varying the current in the coil and thus varying the biasing magnetic field, the resonance frequency of such a filter can be tuned. Such filters have a very high Q and frequency stability.

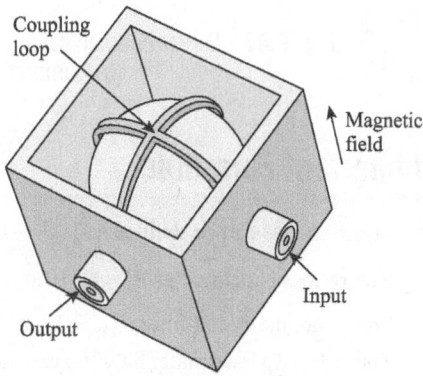

Fig. 9.43   YIG filter

9.11.3 Quarter-wave Coupled Cavity Band-pass Filters

Quarter-wave coupled cavity band-pass filters are realized by waveguides coupled through irises. The equivalent circuit of such a filter is shown in Fig. 9.44.

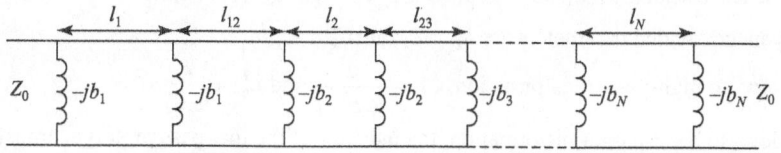

Fig. 9.44   Equivalent circuit of quarter-wave coupled cavity waveguide filter

9.11.4 Direct-coupled Cavity Waveguide Filters

A direct-coupled cavity waveguide filter is shown in Fig. 9.45(a). In such filters, inductive irises are placed along the waveguide. The equivalent circuit is shown in Fig. 9.45(b).

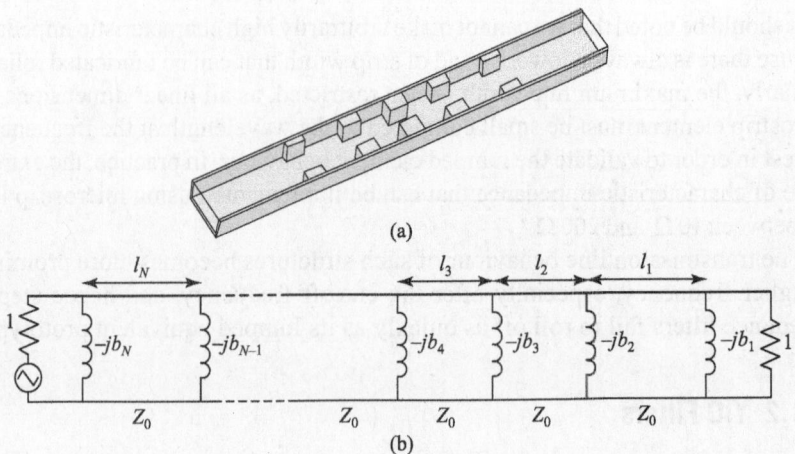

Fig. 9.45 Direct-coupled cavity waveguide filter (a) Schematic diagram
(b) Lumped equivalent circuit

Important Formulae

- Insertion loss can be defined as $\text{IL} = -10\log_{10}(P_L/P_{in}) = -10\log_{10}(1-|\Gamma|^2)$.

- Return loss is defined as $\text{RL} = -10\log_{10}(P_R/P_{in}) = -10\log_{10}(|\Gamma|^2) = -20\log_{10}|\Gamma|$.

- For a maximally flat filter, $P_{LR} = 1 + \varepsilon^2(\omega/\omega_c)^{2N}$.

- The order of a maximally flat filter can be calculated from $N = \ln\left(\sqrt{A_s^2-1}/\varepsilon\right)\Big/\ln(\omega/\omega_c)$.

- The phase response of a linear-phase filter is defined by $\phi(\omega) = A\omega\left[1 + p(\omega/\omega_c)^{2N}\right]$.

- For a Chebyshev filter, $P_{LR} = 1 + \varepsilon^2 T_N^2(\omega/\omega_c)$.

- The order of a Chebyshev filter can be calculated using the equation:
$$N = \frac{\cosh^{-1}\left(\sqrt{A_s^2-1}/\varepsilon\right)}{\cosh^{-1}(\omega/\omega_c)} \approx \frac{\cosh^{-1}(A_s/\varepsilon)}{\cosh^{-1}(\omega/\omega_c)}.$$

- For an inverse Chebyshev filter, $|H(j\omega)|^2 = 1 - \dfrac{1}{1+\varepsilon^2 T_N^2(\omega/\omega_c)} = \dfrac{\varepsilon^2 T_N^2(\omega/\omega_c)}{1+\varepsilon^2 T_N^2(\omega/\omega_c)}$.

- Impedance scaled parameters are $L' = LR_S$, $C' = C/R_S$, $R_S' = R_S$, $R_L' = R_S R_L$.

- Frequency scaled parameters are $L_k' = L_k/\omega_c$, $C_k' = C_k/\omega_c$.

- In low- to high-pass transformation, $C_k' = \dfrac{1}{R_S\omega_c L_k}$ and $L_k' = \dfrac{R}{\omega_c C_k}$.

- In low- to band-pass transformation, the inductor of the low-pass prototype transforms to the series connection of $L_k' = L_k/(\Delta\omega_0)$ and $C_k' = \Delta/(\omega_0 L_k)$, whereas the capacitor of the low-pass prototype transforms to the parallel combination of $C_k' = C_k/(\Delta\omega_0)$ and $L_k' = \Delta/(\omega_0 C_k)$.

- In low-pass to band-stop transformation, the inductor of the low-pass prototype transforms to the parallel connection of $L_k' = \Delta L_k/\omega_0$ and $C_k' = (\omega_0\Delta L_k)^{-1}$, whereas the capacitor of the low-pass prototype transforms to the series combination of $C_k' = \Delta C_k/\omega_0$ and $L_k' = (\omega_0\Delta C_k)^{-1}$.

Exercises

Objective-type Questions

9.1 Insertion loss is defined by
   (a) $-20\log_{10}\left(1-|\Gamma|^2\right)$

   (c) $\text{IL}=-10\log_{10}\left(1-|\Gamma|^2\right)$

   (b) $-10\log_{10}|\Gamma|$

   (d) $\text{IL}=-10\log_{10}\left(1+|\Gamma|^2\right)$

9.2 The image impedance of a network can be expressed as
   (a) $Z_{i1}=\sqrt{\dfrac{AB}{CD}}$ and $Z_{i2}=\sqrt{\dfrac{BD}{AC}}$

   (c) $Z_{i1}=\sqrt{\dfrac{BD}{AC}}$ and $Z_{i2}=\sqrt{\dfrac{BD}{AC}}$

   (b) $Z_{i1}=\sqrt{\dfrac{AB}{CD}}$ and $Z_{i2}=\sqrt{\dfrac{AD}{BC}}$

   (d) $Z_{i1}=\sqrt{\dfrac{BD}{AC}}$ and $Z_{i2}=\sqrt{\dfrac{AD}{BC}}$

9.3 The power loss ratio of a maximally flat low-pass filter can be expressed as
   (a) $1+\varepsilon^2\left(\dfrac{\omega}{\omega_c}\right)^{2N}$

   (c) $1+\varepsilon^2\left(\dfrac{\omega}{\omega_c}\right)^{2}$

   (b) $1-\varepsilon^2\left(\dfrac{\omega}{\omega_c}\right)^{2N}$

   (d) $1+\varepsilon^2\left(\dfrac{\omega_c}{\omega}\right)^{2N}$

9.4 The impedance scaled value of components are
   (a) $L'=LR_S$ and $C'=C/R_S$

   (c) $L'=LR_S$ and $C'=CR_S$

   (b) $L'=L/R_S$ and $C'=C/R_S$

   (d) $L'=L/R_S$ and $C'=CR_S$

9.5 Low-pass to band-stop transformation is done by
   (a) $\omega\leftarrow-\omega_c/\omega$

   (c) $\omega\leftarrow\dfrac{\omega_2-\omega_1}{\omega_0}\left[\dfrac{\omega}{\omega_0}-\dfrac{\omega_0}{\omega}\right]^{-1}$

   (b) $\omega\leftarrow\dfrac{\omega_0}{\omega_2-\omega_1}\left(\dfrac{\omega}{\omega_0}-\dfrac{\omega_0}{\omega}\right)$

   (d) $\omega\leftarrow-\omega/\omega_c$

9.6 For a K inverter
   (a) $K=L_pC_s$

   (c) $K=\sqrt{C_s/L_p}$

   (b) $K=\sqrt{1/L_pC_s}$

   (d) $K=\sqrt{L_p/C_s}$

9.7 For a J inverter
   (a) $J=C_sL_p$

   (c) $J=\sqrt{1/C_sL_p}$

   (b) $J=\sqrt{L_p/C_s}$

   (d) $J=\sqrt{C_s/L_p}=\sqrt{C_p/L_s}$

9.8 A microwave filter consists of two inductive susceptances placed L cm apart in a waveguide. The width of the inductive window is W cm. The centre frequency of resonance depends on
   (a) L

   (c) thickness

   (b) W

   (d) none of these

9.9 The pass-band of a typical filter network with Z_1 and Z_2 as the series and shunt arm impedances, respectively, is characterized by
   (a) $-1<\dfrac{Z_1}{4Z_2}<0$

   (c) $0<\dfrac{Z_1}{4Z_2}<1$

   (b) $-1<\dfrac{Z_1}{4Z_2}<1$

   (d) $0<\dfrac{Z_1}{4Z_2}<-1$

9.10 For an *m*-derived filter, the value of *m* is

(a) $\sqrt{1-\left(\dfrac{f_\infty}{f_c}\right)^2}$

(c) $\sqrt{1+\left(\dfrac{f_c}{f_\infty}\right)^2}$

(b) $\sqrt{1+\left(\dfrac{f_\infty}{f_c}\right)^2}$

(d) $\sqrt{1-\left(\dfrac{f_c}{f_\infty}\right)^2}$

Review Questions

9.1 Find an expression for the Bloch impedance of a periodic structure in terms of its *ABCD* parameters.

9.2 Show that a periodic structure can exhibit pass-band-stop-band characteristics.

9.3 Write short notes on (a) constant-*k* filter section and (b) *m*-derived filter.

9.4 With a suitable example, describe the insertion loss method for maximally flat low-pass filter design.

9.5 Describe impedance and frequency scaling in filter design.

9.6 How can the low-pass filter design procedure be used to design high-pass, band-pass, and band-stop filters?

9.7 Describe Richard transformation.

9.8 Describe and prove different Kuroda identities.

9.9 Write short notes on impedance and admittance inverters.

9.10 Write a short note on coupled line filters.

Problems

9.1 Calculate the image impedance and propagation factor for the networks shown in Fig. 9.46.
[*Hint: Use Table 9.1*]

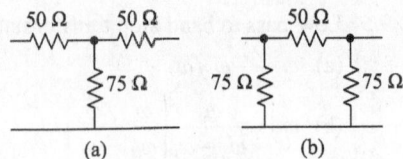

(a)   (b)

Fig. 9.46

9.2 For the constant-*k* filter shown in Fig. 9.47, calculate the image impedance and propagation factor at 8 MHz.
[*Hint: Use Eqs (9.100), (9.102), and (9.103)*]

9.3 Design a three-element maximally flat high-pass composite filter with a cut-off frequency of 5 MHz and impedance 50 Ω. [*Hint: Use Eqs (9.177) and (9.178)*]

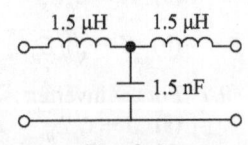

Fig. 9.47

9.4 Design a three-element Chebyshev band-stop filter with 0.25 dB equal-ripple response and 5 GHz cut-off frequency, provided that the bandwidth is 10% and impedance is 50 Ω. [*Hint: Use Eqs (9.187)–(9.190)*]

Microwave Non-reciprocal Networks

10.1 Introduction

In the last two chapters, we have discussed different types of reciprocal microwave networks. Here, the term *reciprocal* is used to imply that the response between any two ports of a network does not depend upon the direction of signal flow, that is, $S_{ij} = S_{ji}$. This condition is valid till the network consists of only isotropic material. If the network consists of an anisotropic material, for example, magnetically biased ferrites, then it will no longer be reciprocal. Such networks are called non-reciprocal networks. Ferrite phase shifters, gyrators, isolators, and circulators are a few examples of non-reciprocal microwave networks.

The main feature of an anisotropic material is that either its permittivity or its permeability is a tensor. If the permittivity of the material is a tensor, then it is called electrically anisotropic, whereas if the permeability of the material is a tensor, then it is called magnetically anisotropic. In microwave applications, the most commonly used anisotropic materials are yttrium–iron–garnet (or YIG) materials and ferrites. Ferrite is a generic name of ferromagnetic (ferrimagnetic) materials. It has the permeability of the order of 5×10^3, specific resistivity about 10^{14} times higher than that of the metal, and is mainly composed of iron oxides and various other materials. The general chemical composition of ferrite is $MOFe_2O_3$, where M is a divalent metal such as manganese, magnesium, iron, zinc, nickel, and cadmium or a mixture of these. A few examples of ferrites are $Mn_nZn_{(1-n)}Fe_2O_4$, $Ni_nZn_{(1-n)}Fe_2O_4$, etc.

Ferrites contain atoms with a large number of spinning electrons, which result in magnetic dipole moments. In the absence of magnetic biasing, the dipoles remain randomly oriented and hence the net dipole moment becomes zero. On the application of a bias field, the dipoles are aligned in a pre-ferred direction and the net magnetic dipole moment of the material becomes non-zero. These magnetic dipoles precess at a certain frequency, either in the clockwise or in the anti-clockwise direction. The frequency of this precession can be controlled by the strength of the bias field and the direction of rotation by the direction of the bias field. Now, if a circularly polarized magnetic field

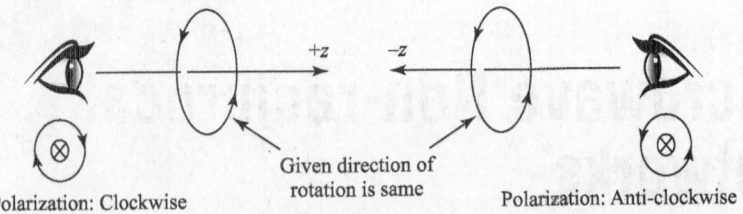

Fig. 10.1 Change in sense of polarization with direction of propagation

passes through such a biased ferrite medium, it will interact strongly with the dipole moment provided that its sense of polarization is the same as the direction of precession. On the other hand, this interaction will be weak if the polarization is opposite to the precession. Since for a given direction of rotation, the sense of polarization changes with the direction of propagation, as illustrated in Fig. 10.1, the same wave will interact differently when travelling in opposite directions. This non-reciprocal property of a biased ferrite is used to design several non-reciprocal microwave components such as phase shifter, gyrator, isolator, and circulator. In the later part of this chapter, we will discuss these components in detail.

Primarily, operations of ferrite-based microwave non-reciprocal devices can be described by the Faraday's principle, which states that 'If a circularly polarized wave (TE_{11} mode in circular waveguide) passes through a ferrite rod, biased with an axial magnetic field H_0, then the axis of polarization gets tilted in the clockwise direction and the amount of tilt depends on the strength of the magnetic field and geometry of the ferrite'. Such rotation of the axis of polarization is called *Faraday rotation*. This property of an electromagnetic wave was first observed by Faraday in 1845. He demonstrated that the plane of polarization of a linearly polarized wave is rotated along the direction of magnetic field when it passes through a magnetically biased ferrite material. Physically, when a linearly polarized wave passes through a biased ferrite, it is decomposed into two circularly polarized waves: right-handed circularly polarized (RHCP) and left-handed circularly polarized (LHCP). These two circularly polarized waves interact differently with the biased ferrite and travel through the ferrite with different speeds. As a result, the phase of one circularly polarized field leads the other. Therefore, when they are reinforced, the resultant plane of polarization suffers a tilt.

10.2 Plane Wave Propagation in Infinitely Extended Ferrite Medium

At the outset of this chapter we will first investigate the propagation of a plane wave in an infinitely extended ferrite medium. The property of the plane wave in a ferrite medium depends on the biasing of the ferrite medium and hence two cases—(a) biasing in the direction of propagation and (b) biasing perpendicular to the direction of propagation—will be discussed.

10.2.1 Propagation Along Direction of Bias

Let us consider that the following DC bias field

$$\overline{H_0^{DC}} = H_0^{DC} \hat{a}_z \tag{10.1}$$

has been applied in an infinitely extended ferrite field region in the direction of wave propagation. Under such a circumstance, permeability of the medium is a tensor and the Maxwell's equation is modified to the following form:

$$\vec{\nabla} \times \vec{E} = -j\omega [\mu] \vec{H} \tag{10.2}$$

$$\vec{\nabla} \times \vec{H} = j\omega \varepsilon \vec{E} \tag{10.3}$$

$$\vec{\nabla} \cdot \vec{D} = 0 \tag{10.4}$$

$$\vec{\nabla} \cdot \vec{B} = 0 \tag{10.5}$$

where $\quad [\mu] = \begin{bmatrix} \mu & j\kappa & 0 \\ -j\kappa & \mu & 0 \\ 0 & 0 & \mu_0 \end{bmatrix} \tag{10.6}$

for z-biasing. Otherwise,

$$[\mu] = \begin{bmatrix} \mu_0 & 0 & 0 \\ 0 & \mu & j\kappa \\ 0 & -j\kappa & \mu \end{bmatrix} \tag{10.7}$$

for x-biasing and

$$[\mu] = \begin{bmatrix} \mu & 0 & -j\kappa \\ 0 & \mu_0 & 0 \\ j\kappa & 0 & \mu \end{bmatrix} \tag{10.8}$$

for y-biasing.

Since the plane wave is propagating along the z-direction and $\partial/\partial x = \partial/\partial y = 0$, we can write the following equations:

$$\vec{E} = \vec{E}_0 e^{-j\beta z} \tag{10.9}$$

$$\vec{H} = \vec{H}_0 e^{-j\beta z} \tag{10.10}$$

Substituting Eqs (10.6), (10.9), and (10.10) into Eqs (10.2) and (10.3) we get the following relations:

$$j\beta E_y = -j\omega \left(\mu H_x + j\kappa H_y \right) \tag{10.11}$$

$$j\beta E_x = j\omega \left(-j\kappa H_x + \mu H_y \right) \tag{10.12}$$

$$0 = -j\omega \mu_0 H_z \tag{10.13}$$

$$j\beta H_y = j\omega \varepsilon E_x \tag{10.14}$$

$$-j\beta H_x = j\omega \varepsilon E_y \tag{10.15}$$

$$0 = j\omega \varepsilon E_z \tag{10.16}$$

Equations (10.13) and (10.16) reveal that the wave is TEM in nature, whereas Eqs (10.14) and (10.15) show that

$$Y = \frac{H_y}{E_x} = -\frac{H_x}{E_y} = \frac{\omega\varepsilon}{\beta} \tag{10.17}$$

where Y is the wave admittance. Substituting Eq. (10.17) into Eq. (10.11), we get the following expression:

$$j\beta E_y = -j\omega\left(\mu H_x + j\kappa H_y\right) = -j\omega\left\{\mu\left(-\frac{\omega\varepsilon}{\beta}\right)E_y + j\kappa\left(\frac{\omega\varepsilon}{\beta}\right)E_x\right\}$$

or $\quad j\beta E_y = -\dfrac{j\omega}{\beta}\left(-\omega\mu\varepsilon E_y + j\omega\kappa\varepsilon E_x\right)$

or $\quad j\omega^2\varepsilon\kappa E_x + \left(\beta^2 - \omega^2\mu\varepsilon\right)E_y = 0 \tag{10.18}$

Similarly, substituting Eq. (10.17) into Eq. (10.12), we get the following relation:

$$j\beta E_x = j\omega\left(-j\kappa H_x + \mu H_y\right) = j\omega\left\{-j\kappa\left(-\frac{\omega\varepsilon}{\beta}\right)E_y + \mu\left(\frac{\omega\varepsilon}{\beta}\right)E_x\right\}$$

or $\quad j\beta E_x = \dfrac{j\omega}{\beta}\left(j\omega\varepsilon\kappa E_y + \omega\mu\varepsilon E_x\right)$

or $\quad \left(\beta^2 - \omega^2\mu\varepsilon\right)E_x - j\omega^2\varepsilon\kappa E_y = 0 \tag{10.19}$

Equations (10.18) and (10.19) can be combined in the following matrix form:

$$\begin{bmatrix} j\omega^2\varepsilon\kappa & \left(\beta^2 - \omega^2\mu\varepsilon\right) \\ \left(\beta^2 - \omega^2\mu\varepsilon\right) & -j\omega^2\varepsilon\kappa \end{bmatrix}\begin{bmatrix} E_x \\ E_y \end{bmatrix} = \begin{bmatrix} 0 \\ 0 \end{bmatrix} \tag{10.20}$$

For a non-trivial solution of Eq. (10.20), the following relation must hold:

$$\begin{vmatrix} j\omega^2\varepsilon\kappa & \left(\beta^2 - \omega^2\mu\varepsilon\right) \\ \left(\beta^2 - \omega^2\mu\varepsilon\right) & -j\omega^2\varepsilon\kappa \end{vmatrix} = 0$$

or $\quad \omega^4\varepsilon^2\kappa^2 - \left(\beta^2 - \omega^2\mu\varepsilon\right)^2 = 0$

or $\quad \left(\beta^2 - \omega^2\mu\varepsilon\right) = \pm\omega^2\varepsilon\kappa$

or $\quad \beta = \sqrt{\omega^2\mu\varepsilon \pm \omega^2\varepsilon\kappa} = \omega\sqrt{\varepsilon(\mu\pm\kappa)} \tag{10.21}$

Since the wave is propagating along the $+z$-direction, we have neglected the negative sign in the final square root. Equation (10.21) reveals that two propagation constants are possible, namely β_+ and β_-. For proceeding further, let us first consider the constant β_+. Substituting Eq. (10.21) into Eq. (10.19), we get the following relation:

$$\left(\beta^2 - \omega^2\mu\varepsilon\right)E_x - j\omega^2\varepsilon\kappa E_y = \left(\omega^2\mu\varepsilon + \omega^2\varepsilon\kappa - \omega^2\mu\varepsilon\right)E_x - j\omega^2\varepsilon\kappa E_y$$

$$= \omega^2\varepsilon\kappa E_x - j\omega^2\varepsilon\kappa E_y = 0$$

or $\quad E_y = -jE_x$ $\hspace{5cm}$ (10.22)

Therefore, Eq. (10.9) can be written as $\vec{E}_+ = E_0\left(\hat{a}_x - j\hat{a}_y\right)e^{-j\beta_+ z}$ $\hspace{1cm}$ (10.23)

and hence $\vec{H}_+ = E_0 Y_+\left(j\hat{a}_x + \hat{a}_y\right)e^{-j\beta_+ z}$ $\hspace{2.5cm}$ (10.24)

Equations (10.23) and (10.24) reveal that the wave is RHCP. Y_+ is the wave impedance for this RHCP wave and can be expressed as follows:

$$Y_+ = \frac{\omega\varepsilon}{\beta_+} = \frac{\omega\varepsilon}{\omega\sqrt{\varepsilon(\mu+\kappa)}} = \sqrt{\frac{\varepsilon}{\mu+\kappa}} \hspace{3cm} (10.25)$$

Now, we will consider β_-. Substituting Eq. (10.21) into Eq. (10.19), we get the following relation:

$$\left(\beta^2 - \omega^2\mu\varepsilon\right)E_x - j\omega^2\varepsilon\kappa E_y = \left(\omega^2\mu\varepsilon - \omega^2\varepsilon\kappa - \omega^2\mu\varepsilon\right)E_x - j\omega^2\varepsilon\kappa E_y$$

$$= -\omega^2\varepsilon\kappa E_x - j\omega^2\varepsilon\kappa E_y = 0$$

or $\quad E_y = jE_x$ $\hspace{5cm}$ (10.26)

Therefore, Eq. (10.9) can be written as $\vec{E}_- = E_0\left(\hat{a}_x + j\hat{a}_y\right)e^{-j\beta_- z}$ $\hspace{0.5cm}$ (10.27)

and hence $\vec{H}_- = E_0 Y_-\left(-j\hat{a}_x + \hat{a}_y\right)e^{-j\beta_- z}$ $\hspace{2.5cm}$ (10.28)

Equations (10.27) and (10.28) reveals that the wave is LHCP. Y_- is the wave impedance for this LHCP wave and can be expressed as follows:

$$Y_- = \frac{\omega\varepsilon}{\beta_-} = \frac{\omega\varepsilon}{\omega\sqrt{\varepsilon(\mu-\kappa)}} = \sqrt{\frac{\varepsilon}{\mu-\kappa}} \hspace{3cm} (10.29)$$

Thus, in a biased ferrite medium an RHCP and an LHCP wave propagate with different propagation constants.

Since a linearly polarized wave can be assumed to be the sum of an RHCP and an LHCP wave, for $z = 0$ we can write the following equation:

$$\vec{E}\Big|_{z=0} = E_0\hat{a}_x = \frac{E_0}{2}\left(\hat{a}_x - j\hat{a}_y\right) + \frac{E_0}{2}\left(\hat{a}_x + j\hat{a}_y\right) \hspace{2cm} (10.30)$$

Now, as this wave propagates along the z-axis, we can write the following equation:

$$\vec{E} = \frac{E_0}{2}\left(\hat{a}_x - j\hat{a}_y\right)e^{-j\beta_+ z} + \frac{E_0}{2}\left(\hat{a}_x + j\hat{a}_y\right)e^{-j\beta_- z}$$

or $\quad \vec{E} = \frac{E_0}{2}\left(e^{-j\beta_+ z} + e^{-j\beta_- z}\right)\hat{a}_x - j\frac{E_0}{2}\left(e^{-j\beta_+ z} - e^{-j\beta_- z}\right)\hat{a}_y$

$$\vec{E} = \frac{E_0}{2}\left\{e^{-j(\beta_+ - \beta_-)z/2} + e^{j(\beta_+ - \beta_-)z/2}\right\}\hat{a}_x e^{-j(\beta_+ + \beta_-)z/2}$$

or $\quad -j\frac{E_0}{2}\left\{e^{-j(\beta_+ - \beta_-)z/2} - e^{j(\beta_+ - \beta_-)z/2}\right\}\hat{a}_y e^{-j(\beta_+ + \beta_-)z/2}$

or $$\vec{E} = E_0 \left[\cos\left\{ \left(\frac{\beta_+ - \beta_-}{2} \right) z \right\} \hat{a}_x - \sin\left\{ \left(\frac{\beta_+ - \beta_-}{2} \right) z \right\} \hat{a}_y \right] e^{-j(\beta_+ + \beta_-)z/2} \quad (10.31)$$

Thus, Eq. (10.31) still represents a linearly polarized wave; however, its polarization rotates as the wave propagates along the z-axis. At a given point on the z-axis, the angle of polarization with respect to the x-axis can be expressed as follows:

$$\phi = \tan^{-1}\left(\frac{E_y}{E_x} \right) = \tan^{-1}\left[-\frac{\sin\left\{ \left(\frac{\beta_+ - \beta_-}{2} \right) z \right\}}{\cos\left\{ \left(\frac{\beta_+ - \beta_-}{2} \right) z \right\}} \right]$$

or $$\phi = \tan^{-1}\left[-\tan\left\{ \left(\frac{\beta_+ - \beta_-}{2} \right) z \right\} \right] = -\left(\frac{\beta_+ - \beta_-}{2} \right) z \quad (10.32)$$

Such rotation of polarization direction, as the wave propagates along the z-axis, is called a Faraday rotation. The basic difference between a Faraday rotation and a circularly polarized wave is that, unlike in the latter case, in a Faraday rotation, the polarization direction at a fixed z-point is constant with time.

If $\omega < \omega_0$ then both μ and κ are positive and $\mu > \kappa$. Therefore, $\beta_+ > \beta_-$ and ϕ becomes more negative as the wave propagates along the z-axis. That is, the polarization rotates in the counter-clockwise direction, if we look into the +z-direction. Now, if we change the direction of biasing, then κ becomes negative. Therefore, $\beta_+ < \beta_-$ and ϕ becomes more positive as the wave propagates along the z-axis. That is, the polarization rotates in the clockwise direction, if we look into the +z-direction. Similarly, for +z biasing, the polarization of a wave, propagating along the −z-direction, rotates in the clockwise sense, if we look into the −z-direction and for −z biasing, the polarization of a wave, propagating along the −z-direction, rotates in the counter-clockwise sense if we look into the −z-direction. These are summarized in Table 10.1. The table shows that there will be a total polarization rotation of 2ϕ when the wave will travel from $z = 0$ to $z = L$ and back again to $z = 0$, or when the wave will travel from $z = 0$ to $z = -L$ and back again to $z = 0$ for a given biasing direction. This implies that a Faraday rotation is a non-reciprocal effect.

Table 10.1 Direction of rotation for different combinations of direction of propagation and direction of biasing

Direction of Biasing	Direction of Wave Propagation	Rotation, Looking into the Direction of Propagation
+z	+z	Counter-clockwise
−z	+z	Clockwise
+z	−z	Clockwise
−z	−z	Counter-clockwise

10.2.2 Propagation Transverse to Bias

In the last section, we assumed that both the direction of propagation and the direction of biasing are same. Now, we will consider that the biasing occurs along the x-direction, while the propagation direction remains the same, that is, in the z-direction. Under such a circumstance, the permeability tensor can be expressed by Eq. (10.7). The expression of the field remains the same as Eqs (10.9) and (10.10). Substituting Eqs (10.7), (10.9), and (10.10) in Eqs (10.2) and (10.3), we get the following relations:

$$j\beta E_y = -j\omega\mu_0 H_x \tag{10.33}$$

$$j\beta E_x = j\omega\left(\mu H_y + j\kappa H_z\right) \tag{10.34}$$

$$0 = -j\omega\left(-j\kappa H_y + \mu H_z\right) \tag{10.35}$$

$$j\beta H_y = j\omega\varepsilon E_x \tag{10.36}$$

$$-j\beta H_x = j\omega\varepsilon E_y \tag{10.37}$$

$$0 = j\omega\varepsilon E_z \tag{10.38}$$

Equations (10.36) and (10.37) show that

$$Y = \frac{H_y}{E_x} = -\frac{H_x}{E_y} = \frac{\omega\varepsilon}{\beta} \tag{10.39}$$

where Y is the wave admittance. Substituting Eq. (10.39) into Eq. (10.33), we get the following expression:

$$j\beta E_y = -j\omega\mu_0\left(-\frac{\omega\varepsilon}{\beta}E_y\right) = j\frac{\omega^2\mu_0\varepsilon}{\beta}E_y$$

or $\qquad \beta^2 E_y = \omega^2\mu_0\varepsilon E_y$ \hfill (10.40)

Similarly, substituting Eqs (10.35), (10.36), and (10.39) into Eq. (10.34), the following relation is obtained:

$$j\beta E_x = j\omega\left\{\mu\left(\frac{\omega\varepsilon E_x}{\beta}\right) + j\kappa\left(\frac{j\kappa H_y}{\mu}\right)\right\} = j\omega\left\{\mu\left(\frac{\omega\varepsilon E_x}{\beta}\right) - \frac{\kappa^2}{\mu}\left(\frac{\omega\varepsilon E_x}{\beta}\right)\right\}$$

or $\qquad j\beta E_x = j\frac{\omega^2\varepsilon}{\beta\mu}\left(\mu^2 - \kappa^2\right)E_x$

or $\qquad \beta^2\mu E_x = \omega^2\varepsilon\left(\mu^2 - \kappa^2\right)E_x$

or $\qquad \mu\left(\beta^2 - \omega^2\mu\varepsilon\right)E_x = -\omega^2\varepsilon\kappa^2 E_x$ \hfill (10.41)

If $E_x = 0$, from Eq. (10.40), the following expression is obtained:

$$\beta = \beta_0 = \omega\sqrt{\mu_0\varepsilon} \tag{10.42}$$

Such waves are called *ordinary waves*, and the field components for such waves can be expressed as follows:

$$\vec{E}_o = E_0 e^{-j\beta_0 z}\hat{a}_y \tag{10.43}$$

and $\qquad \vec{H}_o = -E_0 Y_o e^{-j\beta_0 z} \hat{a}_x$ (10.44)

where $\quad Y_o = \dfrac{\omega\varepsilon}{\beta_o} = \dfrac{\omega\varepsilon}{\omega\sqrt{\mu_0\varepsilon}} = \sqrt{\dfrac{\varepsilon}{\mu_0}}$ (10.45)

Such a wave propagates with the same propagation constant as in unbiased ferrite medium and is independent of the biasing field. The term $E_x = 0$ ensures that $H_y = 0$ and $H_z = 0$.

If $E_y = 0$, Eq. (10.41) gives the following relation:

$$\beta = \beta_e = \omega^2 \left(\mu^2 - \kappa^2\right)\varepsilon/\mu = \omega^2 \mu_e \varepsilon$$

or $\qquad \beta_e = \omega\sqrt{\mu_e\varepsilon}$ (10.46)

where $\quad \mu_e = \left(\mu^2 - \kappa^2\right)/\mu$ (10.47)

The term $E_y = 0$ ensures that $H_x = 0$. Such waves are called *extraordinary waves* and are affected by the biasing field. The field components for such waves can be expressed as follows:

$$\vec{E}_e = E_0 e^{-j\beta_e z} \hat{a}_x$$ (10.48)

and $\quad \vec{H}_e = H_y \hat{a}_y + H_z \hat{a}_z = H_y\left(\hat{a}_y + j\dfrac{\kappa}{\mu}\hat{a}_z\right) = E_0 Y_e e^{-j\beta_e z}\left(\hat{a}_y + j\dfrac{\kappa}{\mu}\hat{a}_z\right)$ (10.49)

where $\quad Y_e = \dfrac{\omega\varepsilon}{\beta_e} = \dfrac{\omega\varepsilon}{\omega\sqrt{\mu_e\varepsilon}} = \sqrt{\dfrac{\varepsilon}{\mu_e}}$ (10.50)

If $\kappa^2 > \mu^2$, Eq. (10.47) reveals that μ_e is negative, and hence β_e is imaginary. An imaginary β_e corresponds to an evanescent wave. Equation (10.48) reveals that the wave is still a linearly polarized wave, whereas Eq. (10.49) reveals that the magnetic field has a component along the propagation direction.

This analysis shows that the wave polarized in the y-direction propagates with a phase constant β_0, whereas that polarized in the x-direction propagates with a phase constant β_e. This effect is called *bifringence*.

10.3 Ferrite Isolators

Isolators are two-port non-reciprocal devices having unidirectional transmission characteristics. That is, if both ports are matched, transmission takes place only from port 1 to port 2 (or only from port 2 to port 1). The S-matrix of such networks can be expressed as follows:

$$[S] = \begin{bmatrix} 0 & 0 \\ 1 & 0 \end{bmatrix}$$ (10.51)

Since the device is non-reciprocal, the $[S]$ matrix is not symmetric. Further since $[S]$ is not a unitary matrix, the device must be lossy.

Isolators play an important role in microwave circuits and are often inserted after the source, to block the reflected wave from reaching the source and interfering with it, as shown in Fig. 10.2.

10.3.1 Resonance Isolator

It is known that a circularly polarized plane wave rotating in the same direction as the precessing magnetic dipoles interacts strongly with the material, whereas a circularly polarized plane wave rotating in

Fig. 10.2 Isolator (arrow represents direction of propagation)

the opposite direction of the precessing magnetic dipoles interacts weakly with the material. This property of ferrites is used to construct an isolator, known as resonance isolator. A resonance isolator consists of ferrite slabs or strips inside a waveguide and operates near gyromagnetic resonance. The ferrite slabs or strips are inserted at a point where the field is circularly polarized. Mathematically,

$$H_x/H_z = \pm j \tag{10.52}$$

In a rectangular waveguide,

$$H_x = \frac{j\beta_0}{k_c} A \sin(k_c x) e^{-j\beta_0 z} \tag{10.53}$$

and

$$H_z = A\cos(k_c x) e^{-j\beta_0 z} \tag{10.54}$$

Hence, $\dfrac{H_x}{H_z} = \dfrac{j\beta_0}{k_c} \tan(k_c x) = \pm j$

or

$$x = \tan^{-1}(\pm k_c/\beta_0)/k_c \tag{10.55}$$

Since the ferrite loading disturbs the field, Eq. (10.55) does not give the accurate location of the stub. However, it can provide a rough estimation of the position of the ferrite slab. Two basic resonance isolators are shown in Fig. 10.3.

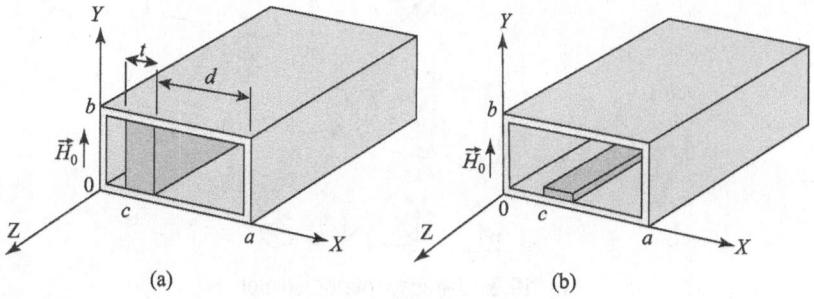

(a) (b)

Fig. 10.3 Basic resonance isolators consisting of (a) *E*-plane full-height ferrite slab (b) *H*-plane ferrite slab

10.3.2 Field Displacement Isolator

Ferrite resonance isolators rely on the different attenuation characteristics of the forward and reverse travelling waves in a ferrite slab-loaded waveguide, while field displacement isolators rely on the different field distributions of these waves inside the ferrite slab-loaded waveguide.[1] Field distribution of a forward and a reverse travelling wave is shown in Fig. 10.4. At $x = c + t$, electric field of the forward travelling wave is zero, while at the

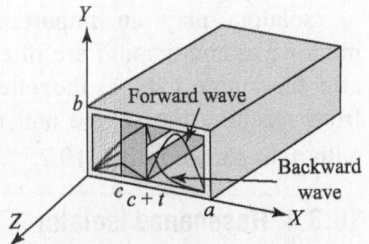

Fig. 10.4 Geometry of field displacement isolator and field distribution within it

same point that of the reverse travelling wave is quite large. Therefore, if we place a thin resistive plate at this point, it will not disturb the forward travelling wave but will attenuate the reverse travelling wave.

10.3.3 Faraday Rotator Isolator

A Faraday rotator isolator consists of two rectangular waveguide ports, one 45° rectangular waveguide twist, one fixed circular waveguide containing an attenuator pad at an angle of 45°, a Faraday rotator capable of providing 45° Faraday rotation, and two circular-to-rectangular waveguide transitions. The corresponding circuit is shown in Fig. 10.5. The input wave, fed at port 1, is rotated by an angle of 45° in the clockwise direction by the rectangular waveguide twist. The rotated field vector thus becomes perpendicular to the attenuator pad and passes it without any attenuation. It further suffers a 45° anti-clockwise rotation by the Faraday rotator, and thus the polarization becomes identical to the input wave. Finally, it appears at port 2 without any attenuation. When the wave is fed at port 2, its field vector suffers a 45° rotation in the clockwise direction by the Faraday rotator. The rotated field thus becomes parallel to the plane of the attenuator pad and is attenuated significantly when it passes through the attenuator pad. Therefore, the output becomes zero at port 1.

Practically, a Faraday rotator isolator has an insertion loss of the order of 0.5–1 dB and a reverse attenuation of the order of 30 dB. At high-power operations, the non-linearity effect comes into play and the angle of Faraday

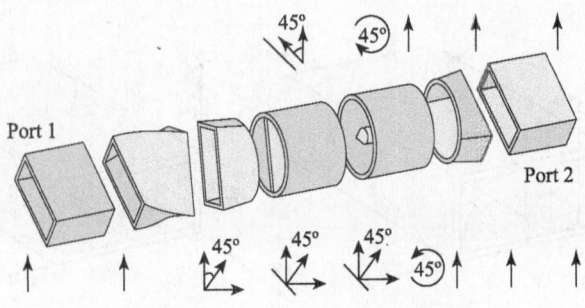

Fig. 10.5 Faraday rotator isolator

---

[1]For details, see *Microwave Engineering* by David M. Pozar, Fourth edition, Wiley, pp. 471–474

rotation deviates from the required 45°. This sets the maximum power-handling capability of the isolator to about 2 kW.

10.4 Ferrite Phase Shifters

Another important application of ferrites is found in designing phase shifters. A microwave phase shifter is an important two-port network that can produce fixed or variable phase shifts between its ports and find wide applications in phased array antenna. For a variable phase shifter, phase shift can be varied by changing the bias field of the ferrite.

10.4.1 Non-reciprocal Latching Phase Shifter

The diagram of a non-reciprocal latching phase shifter is shown in Fig. 10.6. It consists of a toroidal ferrite core symmetrically located in the waveguide. A bias wire passes through its centre. When magnetized, magnetization of the side walls of the toroidal ferrite core will be oppositely directed and perpendicular to the plane of circular polarization of the RF field. Further, since the sense of circular polarization is also opposite on opposite sidewalls of the waveguide, a strong interaction between the ferrite and the RF fields takes place. Such a network can provide a continuously variable phase shift by varying the bias current. Alternatively, hysteresis of the loop can also be used to provide a discrete phase shift between two values. Owing to hysteresis effect, magnetization of the loop can be set at two values, $\pm M_s$. The amount of phase shift produced this way can be controlled by the length of the ferrite toroid.

The main advantage of such a phase shifter is that we need not apply a continuous bias current. Only a pulsed current of proper polarity is required to set the required magnetization. The duration of the required pulse is a few microseconds only. Magnetization of the top and bottom walls of the ferrite core has little interaction with the RF field, because they are not perpendicular to the plane of polarization. They only provide the dielectric loading effect.

10.4.2 Reggia–Spencer Reciprocal Phase Shifter

Unlike a latching phase shifter, a Reggia–Spencer phase shifter is reciprocal in nature. In this phase shifter, a longitudinally biased ferrite rod is centred in a rectangular or circular waveguide, as shown in Fig. 10.7. If the diameter of the rod is greater than a critical size, the field becomes circularly polarized and tightly bound to the ferrite rod.

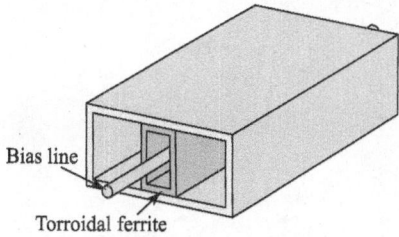

Fig. 10.6 Non-reciprocal latching phase shifter

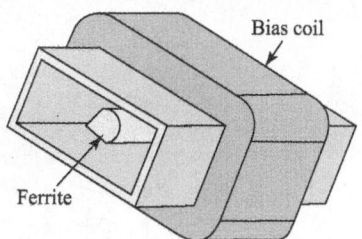

Fig. 10.7 Reggia–Spencer reciprocal phase shifter

The advantage of this phase shifter is that large phase shift can be obtained within a short length, while its main disadvantage is that it is frequency sensitive.

10.5 Ferrite Gyrators

A gyrator is a two-port non-reciprocal phase shifter that produces a differential phase shift of 180°. It is lossless, matched, and non-reciprocal. The S-matrix of an ideal gyrator is as follows:

$$[S] = \begin{bmatrix} 0 & 1 \\ -1 & 0 \end{bmatrix} \tag{10.56}$$

A ferrite gyrator is composed of two rectangular waveguide ports, one 90° twist, two rectangular-to-circular waveguide transitions, and a 90° Faraday rotator, as shown in Fig. 10.8. When the field moves from port 1 to port 2, it is first rotated by 90° in the anti-clockwise direction by the twist and then by another 90° in the anti-clockwise direction by the Faraday rotator. It finally exits from port 2 with a 180° phase difference. When it travels from port 2 to port 1, it is first rotated by 90° in the clockwise direction by the Faraday rotator and next by 90° in the anti-clockwise direction by the twist (the twist can be avoided by rotating the circular to rectangular waveguide transition by 45° with respect to its axis). Therefore, it exits port 2 with a 0° phase difference.

A gyrator can also be designed without using the 90° twist. However, in such a case, the port cross-sections must be orthogonal to each other, as shown in Fig. 10.9.

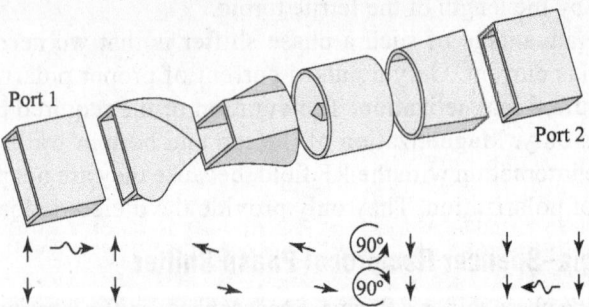

Fig. 10.8 Gyrator with twist

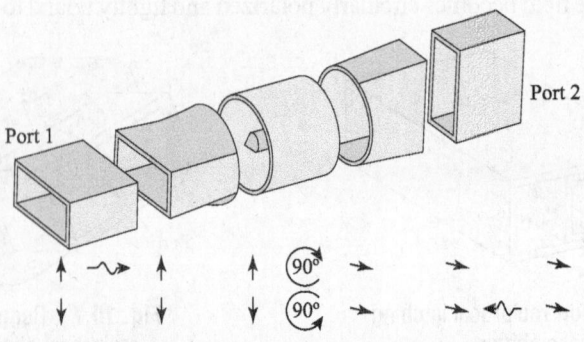

Fig. 10.9 Gyrator without twist

10.6 Ferrite Circulators

A circulator is an N-port non-reciprocal device that allows power flow from port 1 to 2, 2 to 3, 3 to 4, ..., N to 1, but not in the reverse direction. However, power flow in the reverse direction can be obtained by changing the polarity of the ferrite bias field. In general, a three-port circulator can be expressed by the following scattering matrix:

$$[S] = \begin{bmatrix} 0 & 0 & 1 \\ 1 & 0 & 0 \\ 0 & 1 & 0 \end{bmatrix} \tag{10.57}$$

Generally, a circulator uses a permanent magnet for its operation. However, electromagnets can also be used. In such cases, the circulator can operate in a latching mode as a single-pole double-throw switch.

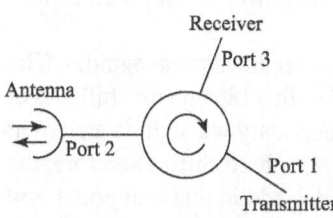

Fig. 10.10 Application of circulator as duplexer

Three-port circulators are widely used as duplexers in microwave systems. In such applications, the transmitter is connected at port 1, antenna at port 2, and receiver at port 3, as shown in Fig. 10.10.

The basic function of a four-port circulator can be understood with the help of the circuit shown in Fig. 10.11. It consists of two 3 dB side hole directional couplers and two non-reciprocal phase shifters. The first phase shifter provides an 180° phase shift in one direction and a 90° phase shift in the reverse direction, while the second phase shifter provides a 0° phase shift in one direction and a 90° phase shift in the reverse direction. An additional phase shift of 90° occurs when the wave passes from the primary to secondary or from the secondary to primary waveguide. A four-port circulator operates in the following way:

1. The wave fed at port 1 splits into the primary and secondary waveguides. The wave at the primary waveguide arrives at port 2 with a 180° phase shift and at port 4 with a 270° phase shift, while that at the secondary waveguide arrives at port 2 with a 180° phase shift and at port 4 with a 90° phase shift. Therefore, the waves from the primary and secondary waveguides add in phase at port 2 and cancel each other at port 4. Port 3 is isolated from port 1 and hence no wave arrives at port 3. Therefore, the power fed at port 1 arrives only at port 2.

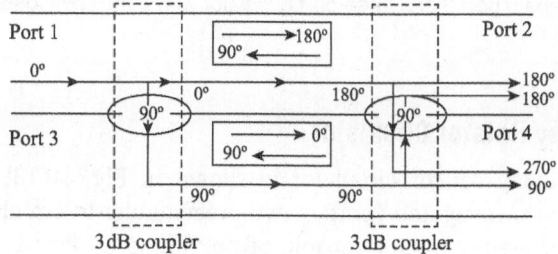

Fig. 10.11 Basic four-port circulator using 3 dB couplers and phase shifters

2. The wave fed at port 2 splits into the primary and secondary waveguides. The wave at the primary waveguide arrives at port 1 with a 90° phase shift and at port 3 with a 180° phase shift, while that at the secondary waveguide arrives at port 1 with a 270° phase shift and at port 3 with a 180° phase shift. Therefore, the waves from the primary and secondary waveguides add in phase at port 3 and cancel each other at port 1. Port 4 is isolated from port 2, and hence no wave arrives at port 4. Therefore, the power fed at port 2 arrives at port 3 only.

3. The wave fed at port 3 splits into the primary and secondary waveguides. The wave at the primary waveguide arrives at port 2 with a 270° phase shift and at port 4 with a 360° phase shift, while that wave at the secondary waveguide arrives at port 2 with a 90° phase shift and at port 4 with a 0° phase shift. Therefore, the waves from the primary and secondary waveguides add in phase at port 4 and cancel each other at port 2. Port 1 is isolated from port 3, and hence no wave arrives at port 1. Therefore, the power fed at port 3 arrives only at port 4.

4. The wave fed at port 4 splits into the primary and secondary waveguides. The wave at the primary waveguide arrives at port 1 with a 180° phase shift and at port 3 with a 270° phase shift, while that at the secondary waveguide arrives at port 1 with a 180° phase shift and at port 3 with a 90° phase shift. Therefore, the waves from the primary and secondary waveguides add in phase at port 1 and cancel each other at port 3. Port 2 is isolated from port 4, and hence no wave arrives at port 2. Therefore, the power fed at port 4 arrives only at port 1.

Circulators with more number of ports can be constructed by cascading the three/four-port circulators, as shown in Fig. 10.12. In general, a circulator constructed by cascading N number of four-port circulators has $2(N+1)$ ports.

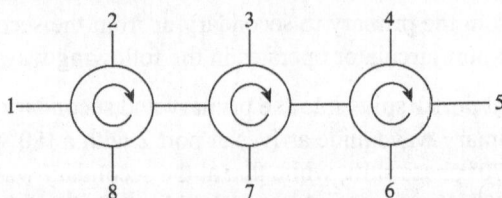

Fig. 10.12 Construction of eight-port circulator using three four-port circulators

> **Note** If we terminate port 3 of a three-port circulator with a matched load, then it results in an isolator.

10.6.1 Faraday Rotator Circulator

A four-port Faraday rotator circulator is shown in Fig. 10.13. It consists of four rectangular waveguide ports, two rectangular-to-circular waveguide transitions, one Faraday rotator capable of providing 45° Faraday rotation, and a 45° rectangular waveguide twist. Port 1 is connected to one of the inputs of

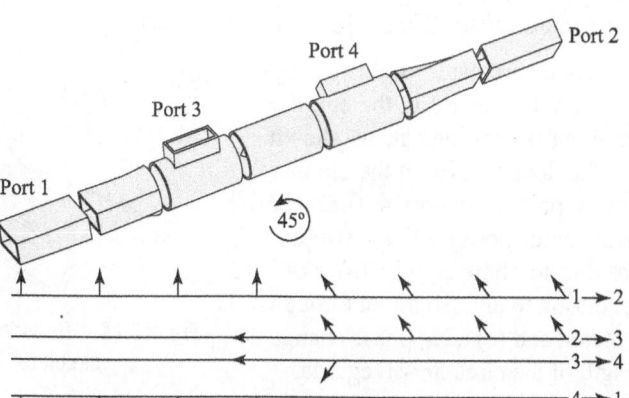

Fig. 10.13 Faraday rotator circulator

the circular waveguide through a rectangular-to-circular waveguide transition. Port 2 is connected to the other end of the circular waveguide through another circular-to-rectangular waveguide transition and a 45° rectangular waveguide twist. Therefore, ports 1 and 2 are inclined at an angle 45°, as shown in the figure. Port 3 is connected to the circular waveguide at the same side of the 45° Faraday rotator as port 1 and is perpendicular to port 1. Similarly, port 4 is connected to the circular waveguide at the same side of the 45° Faraday rotator as port 2 and is also perpendicular to port 2. Therefore, ports 3 and 4 are inclined at an angle 45°, as shown in the figure. The operation of such a circulator is as follows:

1. The field vector fed at port 1 is axial to port 3 and hence does not couple at port 3. When it passes through the Faraday rotator, it suffers a 45° rotation in the anti-clockwise direction and hence becomes axial to port 4. Therefore, it also does not couple to port 4. The rotated electric field, however, has the correct polarization for coupling with the 45° twist and thus exits from port 2.
2. The field vector fed at port 2 is axial to port 4 and hence does not couple at port 4. When it passes through the Faraday rotator, it suffers a 45° rotation in the clockwise direction and hence obtains correct polarization for coupling at port 3. Thus, it exits from port 3. The rotated electric field, however, has a wrong polarization for coupling with port 1 and hence does not couple at port 1.
3. The field vector fed at port 3 has a wrong polarization for coupling with port 1 and hence does not couple at port 1. When it passes through the Faraday rotator, it suffers a 45° rotation in the anti-clockwise direction and hence achieves the correct polarization for coupling with port 4. Thus, it exits from port 4. The rotated electric field, however, has a wrong polarization for coupling with the 45° twist and hence does not couple at port 2.
4. The field vector fed at port 4 has a wrong polarization for coupling at port 2 and hence does not couple at port 2. When it passes through the Faraday rotator, it suffers a 45° rotation in the clockwise direction and hence becomes axial to port 4. Therefore, it also does not couple to port 4. The rotated electric field, however, has the correct polarization for coupling with port 1 and thus exits from port 1.

10.6.2 Turnstile Junction Circulator

A turnstile junction circulator is formed by inserting a 45° Faraday rotator in the circular waveguide and short-circuiting it, as shown in Fig. 10.14. The field coupled at the circular waveguide from port 1 to port 4 (i.e., the rectangular waveguide ports) suffers (a) a 180° phase change due to short circuit, (b) a 90° phase change during to and fro travel through the faraday rotator, and (c) a $2\phi_0$ phase change due to the length of the circular waveguide.

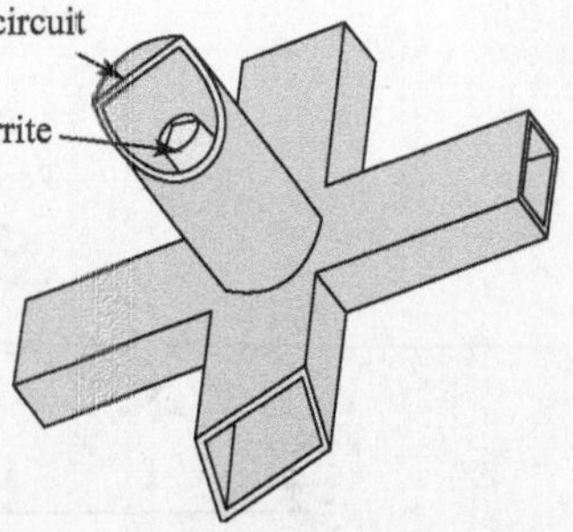

Fig. 10.14 Turnstile junction circulator

Thus, we can write the following equations:

$$a_6 = -b_5 e^{-j2\phi_0} \tag{10.58}$$

and

$$a_5 = b_6 e^{-j2\phi_0} \tag{10.59}$$

where ports 5 and 6 correspond to two orthogonal field vectors in a circular waveguide and is short circuited.

Therefore, if the circuit is fed at port 1, we can write the following equations:

$$
\begin{bmatrix} b_1 \\ b_2 \\ b_3 \\ b_4 \\ b_5 \\ b_6 \end{bmatrix} = \frac{1}{2}
\begin{bmatrix}
0 & 1 & 0 & 1 & \sqrt{2} & 0 \\
1 & 0 & 1 & 0 & 0 & \sqrt{2} \\
0 & 1 & 0 & 1 & -\sqrt{2} & 0 \\
1 & 0 & 1 & 0 & 0 & -\sqrt{2} \\
\sqrt{2} & 0 & -\sqrt{2} & 0 & 0 & 0 \\
0 & \sqrt{2} & 0 & -\sqrt{2} & 0 & 0
\end{bmatrix}
\begin{bmatrix} a_1 \\ 0 \\ 0 \\ 0 \\ b_6 e^{-j2\phi_0} \\ -b_5 e^{-j2\phi_0} \end{bmatrix}
$$

or

$$
\begin{bmatrix} b_1 \\ b_2 \\ b_3 \\ b_4 \\ b_5 \\ b_6 \end{bmatrix} = \frac{1}{2}
\begin{bmatrix}
\sqrt{2} b_6 e^{-j2\phi_0} \\
a_1 - \sqrt{2} b_5 e^{-j2\phi_0} \\
-\sqrt{2} b_6 e^{-j2\phi_0} \\
a_1 + \sqrt{2} b_5 e^{-j2\phi_0} \\
\sqrt{2} a_1 \\
0
\end{bmatrix}
\tag{10.60}
$$

or

$$b_6 = 0 \tag{10.61}$$

$$b_5 = \frac{1}{2}\sqrt{2} a_1 = \frac{a_1}{\sqrt{2}} \tag{10.62}$$

$$b_1 = \frac{1}{2}\sqrt{2} b_6 e^{-j2\phi_0} = 0 \tag{10.63}$$

$$b_3 = -\frac{1}{2}\sqrt{2} b_6 e^{-j2\phi_0} = 0 \tag{10.64}$$

$$b_2 = \frac{1}{2}\left(a_1 - \sqrt{2} b_5 e^{-j2\phi_0}\right) = \frac{a_1}{2} - \frac{a_1\sqrt{2}}{2\sqrt{2}} e^{-j2\phi_0} = \frac{a_1}{2}\left(1 - e^{-j2\phi_0}\right) \tag{10.65}$$

$$b_4 = \frac{1}{2}\left(a_1 + \sqrt{2}b_5 e^{-j2\phi_0}\right) = \frac{a_1}{2} + \frac{a_1\sqrt{2}}{2\sqrt{2}}e^{-j2\phi_0} = \frac{a_1}{2}\left(1 + e^{-j2\phi_0}\right) \qquad (10.66)$$

Now, if we choose $\phi_0 = \pi/2$ so that it satisfies the following condition:

$$e^{-j2\phi_0} = -1 \qquad (10.67)$$

then $b_2 = \frac{a_1}{2}\left(1 - e^{-j2\phi_0}\right) = \frac{a_1}{2}(1+1) = a_1$ \qquad (10.68)

and $b_4 = \frac{a_1}{2}\left(1 + e^{-j2\phi_0}\right) = \frac{a_1}{2}(1-1) = 0$ \qquad (10.69)

Therefore, if power is fed at port 1, it will exit at port 2. By carrying out similar exercises, it can be shown that power flows from port 2 to port 3, port 3 to port 4, and port 4 to port 1.

10.7 Quarter- and Half-wave Plates

This section discusses the quarter- and half-wave plates.

Quarter-wave plate A quarter-wave plate consists of a circular waveguide, operating in the TE_{11} mode, and a dielectric slab of finite length. The degenerate TE_{11} modes are polarized along and perpendicular to the plane of the slab, as shown in Fig. 10.15. The dielectric is tapered in such a way that the modes are matched. The length, thickness, and dielectric constant of the slab are chosen in such a way that the mode polarized along the slab is retarded by 90° more than the mode polarized perpendicular to the slab. That is why, this structure is known as a quarter-wave plate.

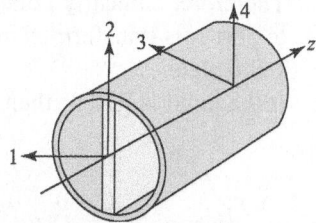

Fig. 10.15 Quarter-wave plate

Since a quarter-wave plate has one input and one output port, and two degenerate modes exist within it, the structure can be considered as a four-port device. We will assume that the wave whose electric vector is perpendicular to the plane of the dielectric slab enters at port 1, whereas the wave whose electric vector is parallel to the plane of the dielectric slab enters at port 2. The respective output ports are ports 3 and 4. The electric vector of a different polarization may be considered as the vector sum of two field vectors entering at ports 1 and 2. The scattering matrix therefore can be written as follows:

$$[S] = e^{-j\phi_1}\begin{bmatrix} 0 & 0 & 1 & 0 \\ 0 & 0 & 0 & -j \\ 1 & 0 & 0 & 0 \\ 0 & -j & 0 & 0 \end{bmatrix} \qquad (10.70)$$

The following are revealed by Eq. (10.70):

1. If $a_1 = a_2 = ae^{j\phi_0}$, then the output is as follows:

$$[b] = e^{-j\phi_1} \begin{bmatrix} 0 & 0 & 1 & 0 \\ 0 & 0 & 0 & -j \\ 1 & 0 & 0 & 0 \\ 0 & -j & 0 & 0 \end{bmatrix} e^{j\phi_0} \begin{bmatrix} a \\ a \\ 0 \\ 0 \end{bmatrix} = e^{-j(\phi_1 - \phi_0)} \begin{bmatrix} 0 \\ 0 \\ a \\ -ja \end{bmatrix} \tag{10.71}$$

Therefore, a linearly polarized wave, oriented in a plane at 45° with respect to port 1, is transformed into an RHCP wave while passing through a quarter-wave plate.

2. If $a_1 = -a_2 = ae^{j\phi_0}$, then the output is as follows:

$$[b] = e^{-j\phi_1} \begin{bmatrix} 0 & 0 & 1 & 0 \\ 0 & 0 & 0 & -j \\ 1 & 0 & 0 & 0 \\ 0 & -j & 0 & 0 \end{bmatrix} e^{j\phi_0} \begin{bmatrix} a \\ -a \\ 0 \\ 0 \end{bmatrix} = e^{-j(\phi_1 - \phi_0)} \begin{bmatrix} 0 \\ 0 \\ a \\ ja \end{bmatrix} \tag{10.72}$$

Therefore, a linearly polarized wave, oriented in a plane at −45° with respect to port 1, is transformed into an LHCP wave when it passes through a quarter-wave plate.

3. If the input is RHCP, then the output is as follows:

$$[b] = e^{-j\phi_1} \begin{bmatrix} 0 & 0 & 1 & 0 \\ 0 & 0 & 0 & -j \\ 1 & 0 & 0 & 0 \\ 0 & -j & 0 & 0 \end{bmatrix} e^{j\phi_0} \begin{bmatrix} a \\ -ja \\ 0 \\ 0 \end{bmatrix} = e^{-j(\phi_1 - \phi_0)} \begin{bmatrix} 0 \\ 0 \\ a \\ -a \end{bmatrix} \tag{10.73}$$

Therefore, an RHCP wave is transformed into a linearly polarized wave, oriented in a plane at −45° with respect to port 1, when it passes through a quarter-wave plate.

4. If the input is LHCP, then the output is as follows:

$$[b] = e^{-j\phi_1} \begin{bmatrix} 0 & 0 & 1 & 0 \\ 0 & 0 & 0 & -j \\ 1 & 0 & 0 & 0 \\ 0 & -j & 0 & 0 \end{bmatrix} e^{j\phi_0} \begin{bmatrix} a \\ ja \\ 0 \\ 0 \end{bmatrix} = e^{-j(\phi_1 - \phi_0)} \begin{bmatrix} 0 \\ 0 \\ a \\ a \end{bmatrix} \tag{10.74}$$

Therefore, an LHCP wave is transformed into a linearly polarized wave, oriented in a plane at 45° with respect to port 1, when it passes through a quarter-wave plate.

Half-wave plate By construction, a half-wave plate is similar to a quarter-wave plate, except that the dielectric slab is oriented at an angle θ, right-handed

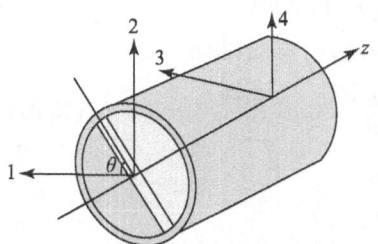

Fig. 10.16 Half-wave plate

positive, with respect to port 1, as shown in Fig. 10.16. A half-wave plate retards the component, polarized parallel to the plane of the slab, more than the component polarized perpendicular to the plane of the slab by an angle of 180°.

An input wave fed at port 1, a_1, may be considered as a vector sum of two orthogonal components, $a_1^\perp$ and $a_1^{\parallel}$, which are as follows:

$$a_1^\perp = a_1 \sin(\theta) \tag{10.75}$$

and $\quad a_1^{\parallel} = a_1 \cos(\theta) \tag{10.76}$

The respective outputs are as follows:

$$a_{1,\text{out}}^\perp = a_1 \sin(\theta) e^{-j\phi_2} \tag{10.77}$$

and $\quad a_{1,\text{out}}^{\parallel} = a_1 \cos(\theta) e^{-j(\pi+\phi_2)} = -a_1 \cos(\theta) e^{-j\phi_2} \tag{10.78}$

These outputs combine at ports 3 and 4 to give the following final outputs:

$$b_3 = a_{1,\text{out}}^\perp \sin(\theta) + a_{1,\text{out}}^{\parallel} \cos(\theta) = a_1 \left\{ \sin^2(\theta) - \cos^2(\theta) \right\} e^{-j\phi_2}$$

or $\quad b_3 = -a_1 \cos(2\theta) e^{-j\phi_2} \tag{10.79}$

and $\quad b_4 = -a_{1,\text{out}}^\perp \cos(\theta) + a_{1,\text{out}}^{\parallel} \sin(\theta) = -2a_1 \sin(\theta) \cos(\theta) e^{-j\phi_2}$

or $\quad b_4 = -a_1 \sin(2\theta) e^{-j\phi_2} \tag{10.80}$

Similarly, an input wave fed at port 2, a_2, may be considered as a vector sum of two orthogonal components, $a_2^\perp$ and $a_2^{\parallel}$, given by the following equations:

$$a_2^\perp = a_2 \cos(\theta) \tag{10.81}$$

and $\quad a_2^{\parallel} = a_2 \sin(\theta) \tag{10.82}$

The respective outputs are as follows:

$$a_{2,\text{out}}^\perp = a_2 \cos(\theta) e^{-j\phi_2} \tag{10.83}$$

and $\quad a_{2,\text{out}}^{\parallel} = a_2 \sin(\theta) e^{-j(\pi+\phi_2)} = -a_2 \sin(\theta) e^{-j\phi_2} \tag{10.84}$

These outputs combine at ports 3 and 4 to give the following final outputs:

$$b_3 = -a_{2,\text{out}}^\perp \sin(\theta) + a_{2,\text{out}}^{\parallel} \cos(\theta) = -2a_2 \sin(\theta) \cos(\theta) e^{-j\phi_2}$$

or $\quad b_3 = -a_2 \sin(2\theta) e^{-j\phi_2} \tag{10.85}$

and $\quad b_4 = a_{2,\text{out}}^\perp \cos(\theta) + a_{2,\text{out}}^{\parallel} \sin(\theta) = a_2 \left\{ \cos^2(\theta) - \sin^2(\theta) \right\} e^{-j\phi_2}$

or $\quad b_4 = a_2 \cos(2\theta) e^{-j\phi_2} \tag{10.86}$

The final scattering matrix therefore has the following form:

$$[S] = e^{-j\phi_2} \begin{bmatrix} 0 & 0 & -\cos(2\theta) & -\sin(2\theta) \\ 0 & 0 & -\sin(2\theta) & \cos(2\theta) \\ -\cos(2\theta) & -\sin(2\theta) & 0 & 0 \\ -\sin(2\theta) & \cos(2\theta) & 0 & 0 \end{bmatrix} \tag{10.87}$$

The following are implied by this scattering matrix:

1. An RHCP wave is transformed into an LHCP wave when it passes through a half-wave plate. Each portion of the circularly polarized wave suffers a delay, a part of which is fixed, while the other is twice the angle θ.

$$[b] = e^{-j\phi_2} \begin{bmatrix} 0 & 0 & -\cos(2\theta) & -\sin(2\theta) \\ 0 & 0 & -\sin(2\theta) & \cos(2\theta) \\ -\cos(2\theta) & -\sin(2\theta) & 0 & 0 \\ -\sin(2\theta) & \cos(2\theta) & 0 & 0 \end{bmatrix} e^{j\phi_0} \begin{bmatrix} a \\ -ja \\ 0 \\ 0 \end{bmatrix}$$

$$[b] = ae^{-j(\phi_2-\phi_0)} \begin{bmatrix} 0 \\ 0 \\ -\cos(2\theta)+j\sin(2\theta) \\ -\sin(2\theta)-j\cos(2\theta) \end{bmatrix} = ae^{-j(\phi_2-\phi_0)} \begin{bmatrix} 0 \\ 0 \\ -e^{-j2\theta} \\ -je^{-j2\theta} \end{bmatrix}$$

or $\quad [b] = e^{-j(\phi_2-\phi_0+\pi+2\theta)} \begin{bmatrix} 0 \\ 0 \\ a \\ ja \end{bmatrix} = e^{-j(\psi+2\theta)} \begin{bmatrix} 0 \\ 0 \\ a \\ ja \end{bmatrix}$ \hfill (10.88)

2. An LHCP wave is transformed into an RHCP wave when it passes through a half-wave plate. Each portion of the circularly polarized wave suffers both a fixed phase delay and a phase advance equal to twice the angle θ.

$$[b] = e^{-j\phi_2} \begin{bmatrix} 0 & 0 & -\cos(2\theta) & -\sin(2\theta) \\ 0 & 0 & -\sin(2\theta) & \cos(2\theta) \\ -\cos(2\theta) & -\sin(2\theta) & 0 & 0 \\ -\sin(2\theta) & \cos(2\theta) & 0 & 0 \end{bmatrix} e^{j\phi_0} \begin{bmatrix} a \\ ja \\ 0 \\ 0 \end{bmatrix}$$

$$[b] = ae^{-j(\phi_2-\phi_0)} \begin{bmatrix} 0 \\ 0 \\ -\cos(2\theta)-j\sin(2\theta) \\ -\sin(2\theta)+j\cos(2\theta) \end{bmatrix} = ae^{-j(\phi_2-\phi_0)} \begin{bmatrix} 0 \\ 0 \\ -e^{j2\theta} \\ je^{j2\theta} \end{bmatrix}$$

or $\quad [b] = e^{-j(\phi_2-\phi_0+\pi-2\theta)} \begin{bmatrix} 0 \\ 0 \\ a \\ -ja \end{bmatrix} = e^{-j(\psi-2\theta)} \begin{bmatrix} 0 \\ 0 \\ a \\ -ja \end{bmatrix}$ \hfill (10.89)

3. If a linearly polarized wave, polarized at angle θ_1, passes through a half-wave plate, then the output will also be a linearly polarized wave, but polarized at an angle of $(2\theta - \theta_1)$.

$$[b] = e^{-j\phi_2} \begin{bmatrix} 0 & 0 & -\cos(2\theta) & -\sin(2\theta) \\ 0 & 0 & -\sin(2\theta) & \cos(2\theta) \\ -\cos(2\theta) & -\sin(2\theta) & 0 & 0 \\ -\sin(2\theta) & \cos(2\theta) & 0 & 0 \end{bmatrix} \begin{bmatrix} a\cos(\theta_1) \\ a\sin(\theta_1) \\ 0 \\ 0 \end{bmatrix} e^{j\phi_0}$$

$$[b] = ae^{-j(\phi_2 - \phi_0)} \begin{bmatrix} 0 \\ 0 \\ -\{\cos(2\theta)\cos(\theta_1) + \sin(2\theta)\sin(\theta_1)\} \\ -\{\sin(2\theta)\cos(\theta_1) - \cos(2\theta)\sin(\theta_1)\} \end{bmatrix}$$

or $\quad [b] = e^{-j\psi} \begin{bmatrix} 0 \\ 0 \\ \cos(2\theta - \theta_1) \\ \sin(2\theta - \theta_1) \end{bmatrix}$ $\hfill (10.90)$

10.8 Precision Differential Phase Shifters

A precision differential phase shifter consists of two quarter-wave plates, one rotatable half-wave plate, and two rectangular-to-circular waveguide transitions, as shown in Fig. 10.17. The rectangular waveguide input is rotated at an angle of $45°$ with respect to port 1 of a quarter-wave plate. It is then connected to the quarter-wave plate through a rectangular-to-circular waveguide transition. Therefore, if we represent the input wave by $\sqrt{2}ae^{j\phi_0}$, then the wave entering ports 1 and 2 of the quarter-wave plate can be expressed as follows: $a_1 = a_2 = ae^{j\phi_0}$ $\hfill (10.91)$

Therefore, the output of the first quarter-wave plate will be an RHCP wave and can be represented as follows:

$$[b] = e^{-j(\phi_1 - \phi_0)} \begin{bmatrix} 0 \\ 0 \\ a \\ -ja \end{bmatrix} \hfill (10.92)$$

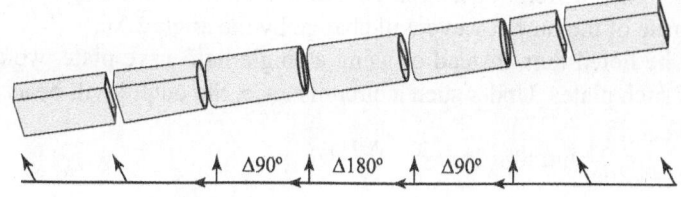

Fig. 10.17 Differential phase shifter

The wave emerging out of port 3 of the first quarter-wave plate now will be the input at port 1 of the half-wave plate. Similarly, the wave emerging out of port 4 of the first quarter-wave plate will now be the input at port 2 of the half-wave plate. Since the input of the half-wave plate is an RHCP wave, its output will be an LHCP wave of the following form:

$$[b] = e^{-j\phi_2} \begin{bmatrix} 0 & 0 & -\cos(2\theta) & -\sin(2\theta) \\ 0 & 0 & -\sin(2\theta) & \cos(2\theta) \\ -\cos(2\theta) & -\sin(2\theta) & 0 & 0 \\ -\sin(2\theta) & \cos(2\theta) & 0 & 0 \end{bmatrix} e^{-j(\phi_1-\phi_0)} \begin{bmatrix} a \\ -ja \\ 0 \\ 0 \end{bmatrix}$$

or

$$[b] = e^{-j(\psi_1+2\theta)} \begin{bmatrix} 0 \\ 0 \\ a \\ ja \end{bmatrix} \tag{10.93}$$

where $\psi_1 = \phi_1 + \phi_2 - \phi_0 + \pi$ \qquad (10.94)

The output of the half-wave plate will now be the input of the second quarter-wave plate. Since the input of the second quarter-wave plate is an LHCP wave, the output will be a linearly polarized wave and is as follows:

$$[b] = e^{-j\phi_1} \begin{bmatrix} 0 & 0 & 1 & 0 \\ 0 & 0 & 0 & -j \\ 1 & 0 & 0 & 0 \\ 0 & -j & 0 & 0 \end{bmatrix} e^{-j(\psi_1+2\theta)} \begin{bmatrix} a \\ ja \\ 0 \\ 0 \end{bmatrix}$$

or

$$[b] = e^{-j(\phi_1+\psi_1+2\theta)} \begin{bmatrix} 0 \\ 0 \\ a \\ a \end{bmatrix} = e^{-j(\psi_2+2\theta)} \begin{bmatrix} 0 \\ 0 \\ a \\ a \end{bmatrix} \tag{10.95}$$

where $\psi_2 = 2\phi_1 + \phi_2 - \phi_0 + \pi$ \qquad (10.96)

Outputs at the ports 3 and 4 of the second quarter-wave plate will now be combined during the second circular-to-rectangular waveguide transition. The final output will be of the following form: $b = \sqrt{2}ae^{-j(\psi_2+2\theta)}$ \qquad (10.97)

It may be noted that ψ_2 is fixed and cannot be changed. Therefore, by changing the angle θ we get different phase shifts. For example, if we change θ by an angle $\Delta\theta$, the phase of the output wave will change by the angle $2\Delta\theta$.

It may be noted that, instead of using a single half-wave plate, we can use a number of such plates. Under such a circumstance, the output will be as follows:

$$b = \sqrt{2}ae^{-j\{\psi_2+2\theta_1-2\theta_2+...+(-1)^{N-1}2\theta_N\}} \tag{10.98}$$

Precision differential phase shifters are broad band in nature. Furthermore, the calibration is also simple and accurate.

10.9 Precision Attenuators

A precision attenuator consists of two rectangular waveguide ports, two rectangular-to-circular waveguide transitions, two fixed circular waveguides with a horizontal attenuator strip, and one rotatable circular waveguide with an attenuator strip. The attenuator strip attenuates the wave component that is parallel to its plane, whereas the perpendicular component passes through it without any attenuation. The complete diagram is shown in Fig. 10.18.

If the rotatable circular waveguide is rotated by an angle θ, then the wave passing from left to right through it is attenuated by a factor of $\cos(\theta)$. The wave is further attenuated by a factor of $\cos(\theta)$ when it passes through the second fixed horizontal strip, and thus provides the total attenuation of $\cos^2(\theta)$. It may be noted that the first fixed attenuator strip does not provide any attenuation when the wave passes from left to right. However, it is effective when the wave passes from right to left.

Commercially available precision attenuators can provide attenuation in the range 0–50 dB. One major advantage of a precision attenuator is that the phase of the output signal is independent of attenuation. A precision attenuator is also known as a rotary vane attenuator.

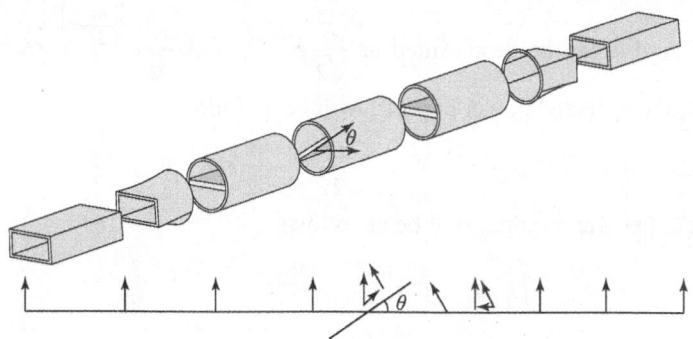

Fig. 10.18   Precision attenuator

10.10 Other Phase Shifters and Attenuators

In this section, we will discuss the dielectric phase shifter, hybrid phase shifter, and resistive card attenuator.

10.10.1 Dielectric Phase Shifter

A dielectric phase shifter consists of a rectangular waveguide with a centred broad-wall longitudinal non-radiating slot and a movable tapered low-loss dielectric slab, as shown in Fig. 10.19. The low-loss dielectric can be inserted or taken out of the waveguide through the broad-wall slot, and depending

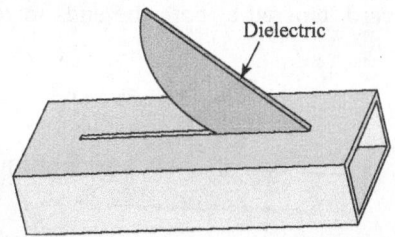

Fig. 10.19   Dielectric phase shifter

upon its position within the guide, a certain amount of phase shift is obtained. Insertion of the dielectric within the waveguide increases ε_{eff}, thereby decreasing λ_g and increasing β_g. The dielectric slab is tapered to reduce reflection within the guide.

10.10.2 Hybrid Phase Shifter

A hybrid phase shifter consists of a 3 dB quadrature coupler and a reflecting network, as shown in Fig. 10.20. The reflecting network may be a PIN diode or a movable short. Moving the short by a distance d, a phase shift of $\phi = 2\beta d$ can be obtained.

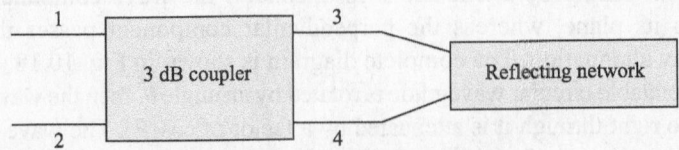

Fig. 10.20 Hybrid phase shifter

The 3 dB coupler divides the signal at port 1 equally between ports 3 and 4. The signal at port 3 is delayed by 90° relative to port 4. Therefore, the waves at ports 1, 3, and 4 can be written as E, $\dfrac{E}{\sqrt{2}}$, and $\dfrac{E}{\sqrt{2}}e^{j\pi/2}$, respectively. Port 2 is isolated from port 1 and hence no power is coupled to it. The waves at ports 3 and 4 are completely reflected by the short circuit, and hence the reflected waves at ports 3 and 4 can be represented as $\dfrac{E}{\sqrt{2}}e^{j(\pi+\phi)}$ and $\dfrac{E}{\sqrt{2}}e^{j\left(\frac{3\pi}{2}+\phi\right)}$, respectively.

Therefore, the reflected power at port 1 will be as follows:

$$E_{r1} = \frac{E}{2}e^{j(\pi+\phi)} + \frac{E}{2}e^{j(2\pi+\phi)} = -\frac{E}{2}e^{j\phi} + \frac{E}{2}e^{j\phi} = 0 \tag{10.99}$$

The reflected power at port 2 will be as follows:

$$E_{r4} = \frac{E}{2}e^{j\left(\frac{3\pi}{2}+\phi\right)} + \frac{E}{2}e^{j\left(\frac{3\pi}{2}+\phi\right)} = Ee^{j\left(\frac{3\pi}{2}+\phi\right)} \tag{10.100}$$

10.10.3 Resistive Card Attenuator

A resistive card attenuator has two versions: (a) fixed resistive card attenuator and (b) variable resistive card attenuator. A variable resistive card attenuator is also known as a flap attenuator.

A fixed resistive card attenuator consists of a ceramic-type absorbing resistive card, tapered at both the ends to reduce reflection in a rectangular waveguide. To achieve maximum attenuation, the card is placed at the position of maximum electric field. Attenuation of such an attenuator is a function of frequency and generally increases with frequency.

A flap attenuator, by construction, is similar to a dielectric phase shifter, except that the dielectric slab is replaced by a ceramic-type absorbing resistive card. The resistive card can be inserted or

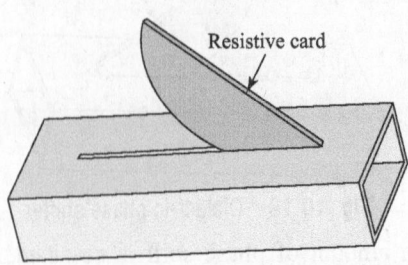

Fig. 10.21 Flap attenuator

taken out of the waveguide through the broad-wall slot, and depending upon its position within the guide, a certain amount of attenuation is obtained. The resistive card, as before, is tapered at both the ends to reduce reflection within the guide. The attenuation that can be obtained from such an arrangement is around 30 dB. The attenuation of a flap attenuator is a function of frequency, and the phase of the output wave is a function of attenuation, that is, of card position. A flap attenuator is shown in Fig. 10.21.

Important Formulae

- When a plane wave propagates through an infinite ferrite medium in the direction of bias, then the phase constants of the wave can be expressed as $\beta = \omega\sqrt{\varepsilon(\mu \pm \kappa)}$.
- When a plane wave propagates through an infinite ferrite medium in the direction of bias, then at a given point on the z-axis, the angle of polarization with respect to the x-axis can be expressed as $\phi = -(\beta_+ - \beta_-)z/2$.
- The propagation constant of an ordinary wave can be expressed as $\beta = \beta_0 = \omega\sqrt{\mu_0\varepsilon}$.
- The propagation constant of an extra-ordinary wave can be expressed as $\beta_e = \omega\sqrt{\mu_e\varepsilon}$.

Exercises

Objective-type Questions

10.1 When the ferrite is biased in the Y-direction, then the permeability tensor can be expressed as

(a) $\begin{bmatrix} \mu & 0 & -j\kappa \\ 0 & \mu_0 & 0 \\ j\kappa & 0 & \mu \end{bmatrix}$

(c) $\begin{bmatrix} \mu_0 & 0 & 0 \\ 0 & \mu & j\kappa \\ 0 & -j\kappa & \mu \end{bmatrix}$

(b) $\begin{bmatrix} \mu & j\kappa & 0 \\ -j\kappa & \mu & 0 \\ 0 & 0 & \mu_0 \end{bmatrix}$

(d) $\begin{bmatrix} \mu_0 & 0 & 0 \\ 0 & \mu_0 & j\kappa \\ 0 & -j\kappa & \mu_0 \end{bmatrix}$

10.2 When a plane wave propagates through an infinite ferrite medium biased in the direction of propagation, then the propagation constant can be expressed as

(a) $\beta = \sqrt{\varepsilon(\mu \pm \kappa)}$

(c) $\beta = \omega\sqrt{\varepsilon(\mu - \kappa)}$

(b) $\beta = \omega\sqrt{\varepsilon(\mu + \kappa)}$

(d) $\beta = \omega\sqrt{\varepsilon(\mu \pm \kappa)}$

10.3 When a plane wave propagates through an infinite ferrite medium biased in the direction of propagation, then the angle of polarization can be expressed as

(a) $-\left(\dfrac{\beta_+ - \beta_-}{2}\right)z$

(c) $-\left(\dfrac{\beta_+ + \beta_-}{2}\right)z$

(b) $\left(\dfrac{\beta_+ - \beta_-}{2}\right)z$

(d) $-(\beta_+ - \beta_-)z$

10.4 The S-matrix $[S] = \begin{bmatrix} 0 & 0 \\ 1 & 0 \end{bmatrix}$ corresponds to

    (a) an isolator   (b) a phase shifter   (c) a circulator   (d) an attenuator

10.5 A linearly polarized wave, oriented in a plane at 45° with respect to port 1, when passes through a quarter-wave plate is transformed into
    (a) An RHCP wave
    (b) An LHCP wave
    (c) A linearly polarized wave, oriented in a plane at −45° with respect to port 1
    (d) A linearly polarized wave, oriented in a plane at 45° with respect to port 1

10.6 If a linearly polarized wave, polarized at an angle θ_1, passes through a half-wave plate then the output will be
    (a) An RHCP wave
    (b) An LHCP wave
    (c) A linearly polarized wave, but polarized at an angle $(2\theta - \theta_1)$
    (d) A linearly polarized wave, but polarized at an angle $(2\theta + \theta_1)$

Review Questions

10.1 Find the expressions for propagation constant, wave impedances, and angle of polarization rotation when a plane wave propagates through an infinite ferrite medium biased in the direction of propagation.

10.2 Describe different types of ferrite isolators.

10.3 What is a gyrator? Describe its operating principle.

10.4 With a suitable figure, describe the basic operating principle of a four-port circulator.

10.5 Describe the operating principle of a Faraday rotator circulator and a turnstile circulator.

10.6 Describe the operating principle of a precision differential phase shifter.

Microwave Linear Beam Tubes

11.1 Introduction

In the last three chapters, we discussed the different types of microwave networks. In this and later chapters, we will discuss the different types of microwave sources and amplifiers. Like low-frequency oscillators and amplifiers, microwave oscillators and amplifiers are also of two types: (a) vacuum tubes and (b) semiconductor devices. In the current and next chapter we will discuss the different types of microwave tubes, after which we will discuss semiconductor devices.

As conventional tubes cannot work at frequencies greater than 1 GHz, because of lead inductance, inter-electrode capacitance, transit angle effects and gain–bandwidth product limitations, special types of tubes are required for high-frequency operations. High-frequency tubes are generally categorized into two classes: (a) linear beam tubes (or O-type) and (b) crossed field tubes (or M-type). In this chapter we will discuss linear beam tubes and in the next chapter, crossed field tubes.

In practice, there are different types of linear beam tubes, such as two-cavity klystron, reflex klystron, travelling-wave tubes (TWTs), forward wave amplifiers (FWAs), backward wave amplifiers and oscillators (BWAs and BWOs), and twystron. Out of these, klystron and reflex klystrons use resonant cavities, and hence are resonant structures. On the other hand, TWTs, FWAs, BWAs, and BWOs are non-resonant structures. A twystron is a hybrid structure and uses combinations of klystron and TWT components. Based on their operating principle and structure, microwave linear beam tubes can be classified as shown in Fig. 11.1.

In a linear beam tube, electrons emitted from the electron gun receive potential energy from the DC beam voltage and accelerate towards the anode. As a result, the potential energy is converted into kinetic energy before they arrive at the interaction region. In the interaction region, these electrons face the microwave field, and either accelerate or decelerate depending on the phase of that field. This acceleration or deceleration results in bunching of electrons,

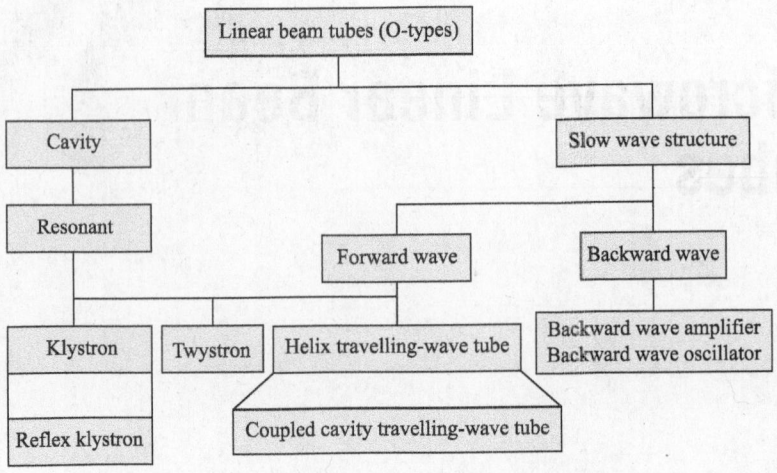

Fig. 11.1 Classification of linear beam tubes

which then drift down the tube and arrive at the output structure. At the output structure, these bunched electrons induce a current and give up their kinetic energy to the output microwave field. Finally, they are collected by the collector. Throughout the journey, electrons of the electron beam remain together with the help of a focusing magnetic field whose axis coincides with the axis of the electron beam.

> **Note** The name *O-type linear beam tubes* was derived either from their French name TPO (tubes á propagation des ondes) or from the word *original* (meaning original types of tubes).

11.2 High-frequency Limitation of Conventional Tubes

Conventional tubes, such as triodes, tetrodes, and pentodes, cannot work at frequencies greater than 1 GHz because of the following reasons:

11.2.1 Lead Inductance and Inter-electrode Capacitance Effect

Leads that are connected to electrodes in conventional tubes have an inductive effect, which, along with the inter-electrode capacitances between grid to plate and grid to cathode, sets a higher cut-off for the operating frequency. To elaborate, at microwave frequencies, the parasitic reactance of lead inductances and inter-electrode capacitances becomes very large compared to that of microwave resonant circuits, and hence such ordinary tubes cannot operate at microwave frequencies. This problem can be overcome by reducing the length and area of the leads. However, such attempts, in turn, minimize the power-handling capability of the tubes. Furthermore, at microwave frequencies, the input conductance of the tubes overloads the input circuit and thereby reduces their efficiency. To illustrate these facts, let us consider the triode circuits shown in Fig. 11.2.

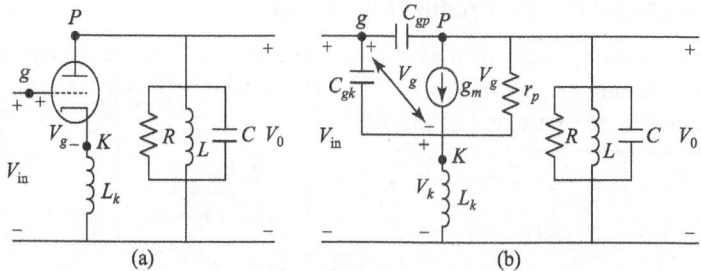

Fig. 11.2 Triode circuit (a) Without inter-electrode parasitics
(b) With inter-electrode parasitics

An equivalent circuit of the triode has been constructed assuming that inter-electrode capacitances and cathode inductances are the only parasitic parameters present. Since $C_{gp} \ll C_{gk}$ and $\omega L_k \ll 1/(\omega C_{gk})$, the input voltage and current can be written as follows:

$$V_{in} = V_g + V_k = V_g\left(1 + j\omega L_k g_m\right) \tag{11.1}$$

$$I_{in} = j\omega C_{gk} V_g \tag{11.2}$$

Therefore, the input admittance can be expressed as follows:

$$Y_{in} = \frac{I_{in}}{V_{in}} = \frac{j\omega C_{gk} V_g}{V_g\left(1 + j\omega L_k g_m\right)} = \frac{j\omega C_{gk}\left(1 - j\omega L_k g_m\right)}{1 + \omega^2 L_k^2 g_m^2}$$

or
$$Y_{in} = \frac{\omega^2 C_{gk} L_k g_m}{1 + \omega^2 L_k^2 g_m^2} + j\frac{\omega C_{gk}}{1 + \omega^2 L_k^2 g_m^2} \tag{11.3}$$

Since $1 \gg \omega^2 L_k^2 g_m^2$, Eq. (11.3) modifies to the following form:

$$Y_{in} = \omega^2 C_{gk} L_k g_m + j\omega C_{gk} \tag{11.4}$$

The input impedance can be written as follows:

$$Z_{in} = \frac{1}{\omega^2 L_k C_{gk} g_m} - j\frac{1}{\omega^3 L_k^2 C_{gk} g_m^2} \tag{11.5}$$

Equation (11.5) reveals that input resistance is inversely proportional to the square of the frequency, whereas input reactance is inversely proportional to the cube of frequency. Therefore, at high frequency the input impedance tends to zero or short and the output power decreases rapidly.

The input admittance and input impedance of a pentode can similarly be written in the following way:

$$Y_{in} = \omega^2 C_{gk} L_k g_m + j\omega\left(C_{gk} + C_{gs}\right) \tag{11.6}$$

and
$$Z_{in} = \frac{1}{\omega^2 L_k C_{gk} g_m} - j\frac{C_{gk} + C_{gs}}{\omega^3 L_k^2 C_{gk}^2 g_m^2} \tag{11.7}$$

where C_{gs} is the capacitance between the gate and the screen.

11.2.2 Gain–Bandwidth Product Limitation

In ordinary vacuum tubes, the maximum gain is generally achieved by resonating the output circuits, as shown in Fig. 11.3. If we assume that $r_p \gg \omega L_k$, then the load voltage can be expressed as follows:

$$V_l = \frac{g_m V_g}{G + j\left[\omega C - 1/(\omega L)\right]} \qquad (11.8)$$

$$\text{where} \quad G = \frac{1}{r_p} + \frac{1}{R} \qquad (11.9)$$

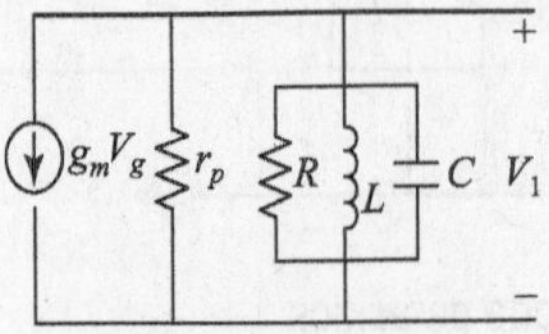

Fig. 11.3 Output tuned circuit of pentode

r_p is the plate resistance, R is the load resistance, and L and C are the tuning elements.

The corresponding resonance frequency can be expressed as follows:

$$f_r = \frac{1}{2\pi\sqrt{LC}} \qquad (11.10)$$

The maximum gain at resonance can be expressed as follows:

$$A_m = g_m / G \qquad (11.11)$$

Furthermore, $G = \omega C - \dfrac{1}{\omega L}$

$$\text{or} \qquad \omega^2 LC - \omega GL - 1 = 0 \qquad (11.12)$$

Since the bandwidth is between the half power points, the half power points can be expressed as follows:

$$\omega_{1,2} = \frac{GL \pm \sqrt{G^2 L^2 + 4LC}}{2LC} = \frac{G}{2C} \pm \sqrt{\frac{G^2}{4C^2} + \frac{1}{LC}}$$

$$\text{or} \qquad \omega_1 = \frac{G}{2C} - \sqrt{\left(\frac{G}{2C}\right)^2 + \frac{1}{LC}} \qquad (11.13)$$

$$\text{and} \qquad \omega_2 = \frac{G}{2C} + \sqrt{\left(\frac{G}{2C}\right)^2 + \frac{1}{LC}} \qquad (11.14)$$

The bandwidth can be expressed as follows:

$$BW = \omega_2 - \omega_1 = \frac{G}{2C} + \sqrt{\left(\frac{G}{2C}\right)^2 + \frac{1}{LC}} - \frac{G}{2C} + \sqrt{\left(\frac{G}{2C}\right)^2 + \frac{1}{LC}}$$

$$\text{or} \qquad BW = 2\sqrt{\left(\frac{G}{2C}\right)^2 + \frac{1}{LC}} \qquad (11.15)$$

Now since $\left(\dfrac{G}{2C}\right)^2 \gg \dfrac{1}{LC}$, the bandwidth becomes of the following form:

$$BW \approx 2\sqrt{\left(\frac{G}{2C}\right)^2} = 2\frac{G}{2C} = \frac{G}{C} \qquad (11.16)$$

Hence, the gain–bandwidth product can be written as follows:

$$A_m (BW) = g_m / C \qquad (11.17)$$

Equation (11.17) reveals that the gain–bandwidth product of a vacuum tube is independent of its frequency and is a constant. Therefore, an ordinary resonant circuit cannot be used with a microwave tube. In practice, microwave devices use either re-entrant cavities (in case of klystrons) or slow-wave structures (in case of TWTs) to obtain a high gain over a broad bandwidth.

11.2.3 RF Loss

RF loss at high frequencies can be of two types: (a) skin effect loss and (b) dielectric loss.

Skin effect loss   At a high frequency, current has a tendency to concentrate around the surface rather than being distributed throughout the cross section. This is known as skin effect. It reduces the effective surface area, which in turn increases the resistance and hence the loss of the device. Resistance loss is also proportional to the square of the frequency.

Losses due to skin effect can be reduced by increasing the current-carrying area, which, in turn, increases the inter-electrode capacitance and thus limits high-frequency operations.

Dielectric loss   Dielectric loss in a material is proportional to frequency, and hence plays an important role in the operations of high-frequency tubes. This loss can be avoided by eliminating the tube base and reducing the surface area of the dielectric materials, and can be reduced by placing insulating materials at the point of minimum electric field.

Radiation loss   At higher frequencies, the length of the leads approaches the operating wavelength, and as a result these start radiating. Radiation loss increases with the increase in frequency and hence is very severe at microwave frequencies. Proper shielding is required to avoid this loss. Radiation loss can be minimized by enclosing the tubes or using a concentric line construction.

11.2.4 Transit Angle Effect

Transit angles pose another limitation to the use of conventional tubes at high frequencies. The transit time of a tube is the time taken by an electron to travel the inter-electrode distance. Mathematically, an electron transit angle is defined as follows: $\theta_d \equiv \omega\tau_g = \omega d/v_0$ (11.18)
where

τ_g: transit time across the gap

d: distance between cathode and grid

$v_0 = 0.593\sqrt{V_0} \times 10^6$: velocity of electron   (11.19)

V_0: DC voltage

At low frequencies, an electron leaves the cathode and travels to the anode within a small fraction of the positive half-cycle of grid voltage, and hence the transit time effect is negligible. However, at a high frequency, transit time is large compared to the time period of the signal and so cannot be neglected. At these frequencies, even if the electron leaves the cathode during positive grid potential, the grid potential may become negative or even go several cycles before the electron passes through it. In practice, the potential between the cathode and grid may alternate 10–100 times during this time. In the positive half-cycle, grid potential attracts the electron beam and supplies energy to it, whereas in the negative half-cycle, it repels the electron beam and extracts energy from it. As a result, the electron beam oscillates back and forth in the region between the cathode and the grid, and may even return to the cathode. The overall result is a reduction of the operating frequency of the vacuum tube. Furthermore, since the transit time is not negligible, the transconductance of the device at microwave frequencies become complex with a relatively small magnitude. This, in turn, indicates a decrease in the output.

This analysis indicates that to use a vacuum device at microwave frequencies, transit angle must be reduced, either by increasing the anode voltage or by decreasing the inter-electrode spacing. However, the increase in anode voltage will increase the power dissipation, whereas the decrease in inter-electrode spacing will increase the inter-electrode capacitance. The increase in inter-electrode capacitance can be reduced by reducing the area of the electrodes, but this will reduce anode dissipation and hence the output power.

In microwave tubes, the transit angle can be reduced by first accelerating an electron beam with a very high DC voltage and then velocity modulating it. In fact, this is the basic principle of operation of a klystron tube, as discussed in the following section.

11.3 Klystron Amplifiers

A klystron is a widely used microwave amplifier that works on the principle of velocity and current modulation. It generally consists of an electron gun assembly, a buncher cavity, a catcher cavity, and a collector, as shown in Fig. 11.4. The microwave field to be amplified is fed into the buncher cavity and the amplified output is obtained from the catcher cavity. If electrons from the electron gun assembly pass the first cavity (or buncher cavity) at a zero cavity gap voltage (or microwave signal voltage), they remain unaffected and leave the cavity without any change in velocity. However, if they pass the cavity during

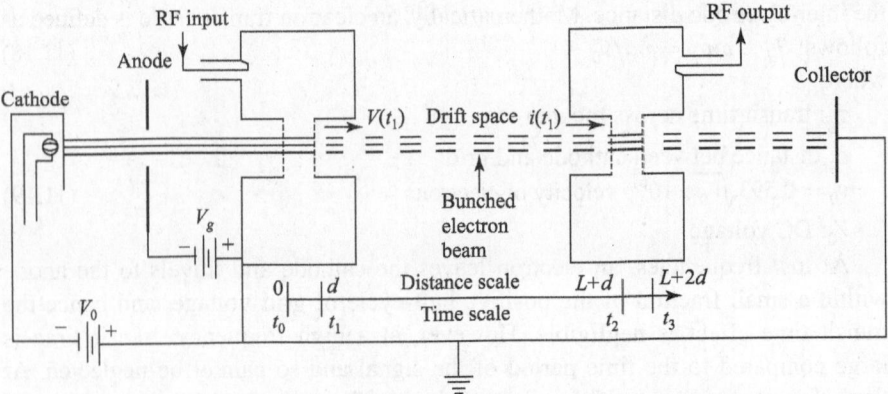

Fig. 11.4 Schematic diagram of two-cavity klystron amplifier

the positive half-cycle, they face an accelerating field and leave the cavity with a higher velocity. In contrast, if they pass the cavity during the negative half-cycle, they face a decelerating field and leave the cavity with a lower velocity. This process is known as velocity modulation, and results in the bunching of electrons. Bunched electrons then move towards the second cavity or the catcher cavity. Owing to bunching, the density of electrons in the catcher cavity varies cyclically with time.

Since the electron beam has been modulated by an RF field, it contains an AC component and is current modulated. If maximum bunching occurs approximately midway between the second cavity grids during its retarding phase, then the kinetic energy of the electron beam is transferred to the field of the second cavity, resulting in signal amplification. After giving up their energy in the catcher cavity, the electron beam emerges from it with a reduced velocity and is finally collected at the collector.

A reflex klystron can provide up to 500 kW CW power and 30 MW pulsed power at 10 GHz, with a power gain of about 30 dB and efficiency of about 40%. In the following section, we will perform a quantitative analysis of a two-cavity klystron under the following assumptions:

1. The electron beam has a uniform cross-sectional density.
2. Space charge and debunching effect are negligible.
3. Magnitude of the input microwave signal is much smaller than the DC accelerating voltage.
4. Electrons leave the cathode with a zero initial velocity.
5. The transit time of electrons across the cavity gap is very small in comparison with the time period of the input RF signal.
6. The cathode, anode, cavity grids, and collector of the tube are all parallel.
7. The cavity grids of the tube do not intercept any passing electron.
8. RF fields are totally confined within the cavities and are zero in the drift space.

The velocity of an electron, entering the buncher cavity under the influence of a high DC voltage, V_0, can be expressed as follows:

$$v_0 = \sqrt{\frac{2eV_0}{m}} = 0.593\sqrt{V_0} \times 10^6 \tag{11.20}$$

When the buncher cavity has been excited by an input microwave signal, the gap voltage can be written as follows: $V_s = V_1 \sin(\omega t)$ $\qquad(11.21)$ where V_1 ($V_1 \ll V_0$) is the amplitude of the input signal. Since $V_1 \ll V_0$, the average transit time through the buncher gap distance d is given by the following equation:

$$\tau \approx d/v_0 = t_1 - t_0 \tag{11.22}$$

Therefore, the average gap transit angle is as follows:

$$\theta_g = \omega\tau \approx \omega d/v_0 = \omega(t_1 - t_0) \tag{11.23}$$

Now, the average microwave voltage in the buncher gap can be expressed as follows:

$$\langle V_s \rangle = \frac{1}{\tau} \int_{t_0}^{t_1} V_1 \sin(\omega t) dt = -\frac{V_1}{\omega \tau} [\cos(\omega t_1) - \cos(\omega t_0)]$$

or $\quad \langle V_s \rangle = \frac{V_1}{\omega \tau} \left[\cos(\omega t_0) - \cos\left(\omega t_0 + \frac{\omega d}{v_0}\right) \right]$

or $\quad \langle V_s \rangle = V_1 \dfrac{\sin\left(\dfrac{\omega d}{2v_0}\right)}{\left(\dfrac{\omega d}{2v_0}\right)} \sin\left(\omega t_0 + \frac{\omega d}{2v_0}\right) = V_1 \frac{\sin(\theta_g/2)}{(\theta_g/2)} \sin\left(\omega t_0 + \frac{\theta_g}{2}\right)$ (11.24)

During the derivation of Eq. (11.24), we have used the trigonometric relation $\cos(A) - \cos(B) = -2\sin\left(\frac{A+B}{2}\right)\sin\left(\frac{A-B}{2}\right)$ where, $A = \omega t_0$ and $B = \omega t_0 + \frac{\omega d}{v_0}$. The term $\frac{\sin(\theta_g/2)}{(\theta_g/2)}$ is known as the beam coupling coefficient of the input cavity gap and can be denoted by the following equation:

$$\beta_i = \sin(\theta_g/2)/(\theta_g/2)$$ (11.25)

A simple program can be written to plot the beam coupling coefficient as a function of gap transit angle (Fig. 11.5).

At the buncher cavity, the input signal is superimposed on the DC voltage V_0, and hence the velocity of the electron, exiting from the buncher cavity, can be written as follows:

$$v(t_1) = \sqrt{\frac{2e}{m} \left\{ V_0 + \beta_i V_1 \sin\left(\omega t_0 + \frac{\theta_g}{2}\right) \right\}} = \sqrt{\frac{2eV_0}{m} \left\{ 1 + \frac{\beta_i V_1}{V_0} \sin\left(\omega t_0 + \frac{\theta_g}{2}\right) \right\}}$$

(11.26)

The term $\beta_i V_1/V_0$ is called the depth of velocity modulation. Since $\beta_i < 1$ [form Eq. (11.25)] and $V_1 \ll V_0$ (as per initial approximation), $\beta_i V_1 \ll V_0$ and the binomial expansion of Eq. (11.26) gives the following relation:

$$v(t_1) = \sqrt{\frac{2eV_0}{m}} \left\{ 1 + \frac{\beta_i V_1}{2V_0} \sin\left(\omega t_0 + \frac{\theta_g}{2}\right) \right\}$$ (11.27)

Substituting Eq. (11.20) into Eq. (11.27), we get the following relation:

$$v(t_1) = v_0 \left\{ 1 + \frac{\beta_i V_1}{2V_0} \sin\left(\omega t_0 + \frac{\theta_g}{2}\right) \right\}$$ (11.28)

Equation (11.28) is also known as the equation for velocity modulation and can be written as follows:

$$v(t_1) = v_0 \left\{ 1 + \frac{\beta_i V_1}{2V_0} \sin\left(\omega t_1 - \frac{\theta_g}{2}\right) \right\}$$ (11.29)

MATLAB Program

```
% Plot of beam coupling coefficient as a function of gap transit angle
clear all;
clf;
n = 0;
for th = 0:pi/1800:10 * pi
    n = n + 1;
    theta_g(n) = th;
    if th == 0
        beta(n) = 1;
    else
        beta(n) = sin(th/2)/(th/2);
    end
end
plot(theta_g, beta, 'k');
axis([0, 10 * pi, -0.5, 1.0]);
xlabel('Gap transit angle');
ylabel('Beam coupling coefficient');
```

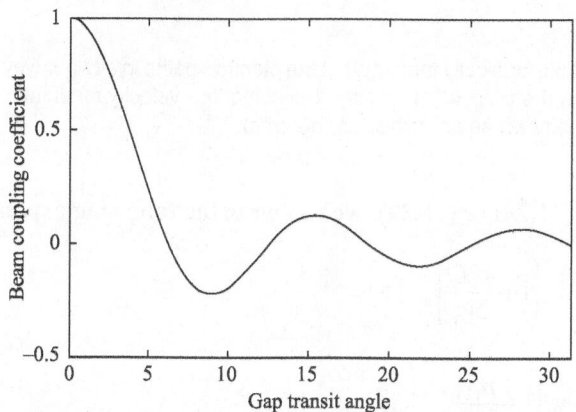

Fig. 11.5 Plot of beam coupling coefficient versus gap transit angle

Electrons leaving the buncher cavity at different instants of an RF cycle will move with different speeds and result in bunching (Fig. 11.6). Electrons leaving the cavity at t_a will move slower than electrons leaving the cavity at t_b, whereas those leaving the cavity at t_c will move faster. Therefore, to travel a distance ΔL, electrons leaving the cavity at t_a will take more time than those leaving the cavity at t_b, whereas those leaving the cavity at t_c will take lesser time. Since $t_c > t_b > t_a$, ΔL can be estimated such that it satisfies the following condition:

$$\Delta L = v_0 \left(t_d - t_b \right) = v_{min} \left(t_d - t_a \right) = v_{max} \left(t_d - t_c \right) \tag{11.30}$$

Under such circumstances, all the electrons leaving the cavity between t_a and t_c will arrive at a distance ΔL from the buncher cavity at the same time t_d and form a bunch. Equation (11.30) can also be written as follows:

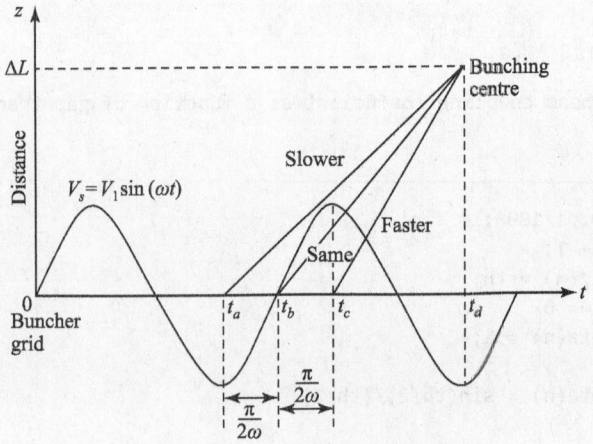

Fig. 11.6 Bunching of electrons

$$\Delta L = v_0 \left(t_d - t_b\right) = v_{\min}\left(t_d - t_b + \frac{\pi}{2\omega}\right) = v_{\max}\left(t_d - t_b - \frac{\pi}{2\omega}\right) \quad (11.31)$$

Note Figure 11.6 is basically the graph of the electron paths in a two-cavity klystron tube, and it describes the process of electron bunching in a velocity-modulated tube. Such a diagram is also known as an *Applegate diagram*.

Now from Eq. (11.28) or (11.29), we can write the following expressions:

$$v_{\max} = v_0 \left(1 + \frac{\beta_i V_1}{2V_0}\right) \quad (11.32)$$

and
$$v_{\min} = v_0 \left(1 - \frac{\beta_i V_1}{2V_0}\right) \quad (11.33)$$

Substituting Eqs (11.32) and (11.33) in Eq. (11.31), we can write the following relations:

$$\Delta L = v_0 \left(1 + \frac{\beta_i V_1}{2V_0}\right)\left(t_d - t_b - \frac{\pi}{2\omega}\right)$$

or
$$\Delta L = v_0 \left(t_d - t_b\right) + \left\{-v_0 \frac{\pi}{2\omega} + v_0 \frac{\beta_i V_1}{2V_0}\left(t_d - t_b\right) - v_0 \frac{\beta_i V_1}{2V_0}\frac{\pi}{2\omega}\right\} \quad (11.34)$$

$$\Delta L = v_0 \left(1 - \frac{\beta_i V_1}{2V_0}\right)\left(t_d - t_b + \frac{\pi}{2\omega}\right)$$

or
$$\Delta L = v_0 \left(t_d - t_b\right) + \left\{v_0 \frac{\pi}{2\omega} - v_0 \frac{\beta_i V_1}{2V_0}\left(t_d - t_b\right) - v_0 \frac{\beta_i V_1}{2V_0}\frac{\pi}{2\omega}\right\} \quad (11.35)$$

Comparing Eq. (11.31) with Eq. (11.34) or (11.35), we get the following expression:

$$v_0 \frac{\pi}{2\omega} - v_0 \frac{\beta_i V_1}{2V_0}(t_d - t_b) \pm v_0 \frac{\beta_i V_1}{2V_0} \frac{\pi}{2\omega} = 0$$

or $$\frac{\pi}{2\omega} - \frac{\beta_i V_1}{2V_0}(t_d - t_b) \pm \frac{\beta_i V_1}{2V_0} \frac{\pi}{2\omega} = 0$$

or $$\frac{\beta_i V_1}{2V_0}(t_d - t_b) = \frac{\pi}{2\omega} \pm \frac{\beta_i V_1}{2V_0} \frac{\pi}{2\omega}$$

or $$t_d - t_b = \frac{2\pi V_0}{2\omega \beta_i V_1} \pm \frac{\pi}{2\omega} = \frac{\pi V_0}{\omega \beta_i V_1} \pm \frac{\pi}{2\omega} \approx \frac{\pi V_0}{\omega \beta_i V_1} \text{ as } \omega \gg \pi \qquad (11.36)$$

Substituting Eq. (11.36) back in Eq. (11.31), the following relation is obtained:

$$\Delta L = v_0(t_d - t_b) = v_0 \frac{\pi V_0}{\omega \beta_i V_1} \qquad (11.37)$$

Note It may be noted that the distance given in Eq. (11.37) is not unique for maximum bunching.

Within the drift region, electrons move with a velocity $v(t_1)$. Now, if the catcher cavity is placed at a distance L from the buncher cavity, then the transit time of the electron can be expressed as follows:

$$T = t_2 - t_1 = \frac{L}{v(t_1)} = \frac{L}{v_0}\left[1 + \frac{\beta_i V_1}{2V_0}\sin\left(\omega t_1 - \frac{\theta_g}{2}\right)\right]^{-1}$$

or $$T = T_0\left[1 + \frac{\beta_i V_1}{2V_0}\sin\left(\omega t_1 - \frac{\theta_g}{2}\right)\right]^{-1} \qquad (11.38)$$

where $$T_0 = L/v_0 \qquad (11.39)$$

is the DC transit time. Since $\beta_i V_1 \ll V_0$, performing binomial expansion of the second bracketed term, we get the following equation:

$$T = T_0\left[1 - \frac{\beta_i V_1}{2V_0}\sin\left(\omega t_1 - \frac{\theta_g}{2}\right)\right] \qquad (11.40)$$

Therefore,

$$\omega T = \omega t_2 - \omega t_1 = \omega T_0 - \omega T_0 \frac{\beta_i V_1}{2V_0}\sin\left(\omega t_1 - \frac{\theta_g}{2}\right) = \theta_0 - X\sin\left(\omega t_1 - \frac{\theta_g}{2}\right) \qquad (11.41)$$

where $$\theta_0 = \omega T_0 = \frac{\omega L}{v_0} = 2\pi N \qquad (11.42)$$

is the DC transit angle between the cavities, N is the number of electron transit cycles in drift space, and $X \equiv \frac{\beta_i V_1}{2V_0}\theta_0 \qquad (11.43)$

is the bunching parameter of a klystron.

The instant at which an electron arrives at the catcher cavity is as follows:

$$t_2 = t_1 + T_0 \left\{ 1 - \frac{\beta_i V_1}{2V_0} \sin\left(\omega t_1 - \frac{\theta_g}{2}\right) \right\}$$

or
$$t_2 = t_0 + \tau + T_0 \left\{ 1 - \frac{\beta_i V_1}{2V_0} \sin\left(\omega t_0 + \omega\tau - \frac{\theta_g}{2}\right) \right\}$$

or
$$t_2 = t_0 + \tau + T_0 \left\{ 1 - \frac{\beta_i V_1}{2V_0} \sin\left(\omega t_0 + \theta_g - \frac{\theta_g}{2}\right) \right\}$$

or
$$t_2 = t_0 + \tau + T_0 \left\{ 1 - \frac{\beta_i V_1}{2V_0} \sin\left(\omega t_0 + \frac{\theta_g}{2}\right) \right\} \tag{11.44}$$

Alternatively, from Eq. (11.41), we can write the following expression:

$$\omega t_2 - \omega t_1 = \omega t_2 - \omega t_0 - \omega\tau = \theta_0 - X \sin\left(\omega t_1 - \frac{\theta_g}{2}\right)$$

or
$$\omega t_2 - \theta_0 - \omega\tau = \omega t_0 - X \sin\left(\omega t_0 + \omega\tau - \frac{\theta_g}{2}\right) \tag{11.45}$$

Substituting Eq. (11.23) into Eq. (11.45), we get the following relation:

$$\omega t_2 - \theta_0 - \theta_g = \omega t_0 - X \sin\left(\omega t_0 + \theta_g - \frac{\theta_g}{2}\right)$$

or
$$\omega t_2 - \left(\theta_0 + \frac{\theta_g}{2}\right) = \left(\omega t_0 + \frac{\theta_g}{2}\right) - X \sin\left(\omega t_0 + \frac{\theta_g}{2}\right) \tag{11.46}$$

where the term $\omega t_2 - \left(\theta_0 + \frac{\theta_g}{2}\right)$ represents a catcher cavity arrival angle and the term $\left(\omega t_0 + \frac{\theta_g}{2}\right)$ a buncher cavity departure angle. A simple program can be written to plot the catcher cavity arrival angle as a function of the buncher cavity departure angle (Fig. 11.7).

If we assume that dQ_0 amount of charge passes through the buncher cavity gap at a time interval dt_0, then we can write the following equation:

$$dQ_0 = I_0 dt_0 \tag{11.47}$$

where I_0 is the DC current. According to the law of conservation of charges, the same amount of charge will also pass the catcher cavity at a later time interval dt_2. Hence, we can write the following equation: $I_0 |dt_0| = i_2 |dt_2|$ (11.48)

The absolute value of time ratio is necessary in Eq. (11.48), because a negative value would indicate a negative resistance.

MATLAB Program

```
% Plot of catcher cavity arrival angle as a function of buncher cavity
departure angle
clear all;
clf;
for X = 0:0.5:1.5
    n = 0;
    for th = -pi:pi/180:pi
        n = n + 1;
        th_dep(n) = th;
        th_arr(n) = th - X * sin(th);
    end
    axis([-pi, pi, -pi, pi]);
    xlabel('Buncher cavity departure angle');
    ylabel('Catcher cavity arrival angle');
    plot(th_dep, th_arr, 'k');
    hold on;
end
```

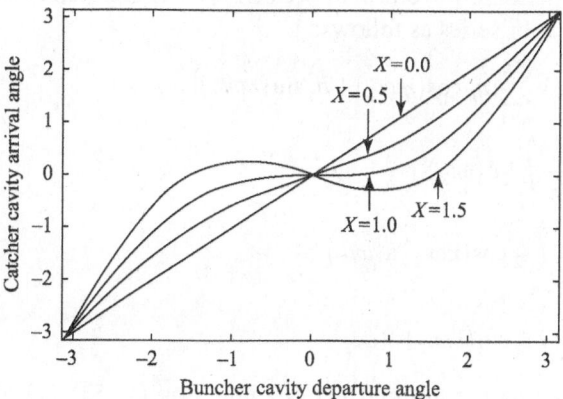

Fig. 11.7 Plots of catcher cavity arrival angle vs buncher cavity departure angle

Differentiating Eq. (11.46) with respect to t_0, we get the following relation:

$$\omega \frac{dt_2}{dt_0} = \omega - X\omega \cos\left(\omega t_0 + \frac{\theta_g}{2}\right)$$

or $\qquad dt_2 = dt_0\left\{1 - X\cos\left(\omega t_0 + \frac{\theta_g}{2}\right)\right\}$ $\qquad\qquad$ (11.49)

Substituting Eq. (11.49) into Eq. (11.48), we get the following expression:

$$I_0 dt_0 = i_2(t_0)dt_0\left\{1 - X\cos\left(\omega t_0 + \frac{\theta_g}{2}\right)\right\}$$

or $\qquad i_2\left(t_0\right)=I_0\Big/\left\{1-X\cos\left(\omega t_0+\dfrac{\theta_g}{2}\right)\right\}$ (11.50)

In terms of t_2, we get the following equation:

$$i_2\left(t_2\right)=\frac{I_0}{\left\{1-X\cos\left(\omega t_0+\dfrac{\theta_g}{2}\right)\right\}}=\frac{I_0}{\left\{1-X\cos\left(\omega t_2-\omega\tau-\omega T_0+\dfrac{\theta_g}{2}\right)\right\}}$$

or $\qquad i_2\left(t_2\right)=\dfrac{I_0}{\left\{1-X\cos\left(\omega t_2-\theta_0-\dfrac{\theta_g}{2}\right)\right\}}$ (11.51)

where we have used Eqs (11.23) and (11.42).

A simple program can be written to plot the normalized beam current as a function of catcher cavity arrival angle (see Fig. 11.8).

Since bunches are formed at periodic intervals, the beam current in the catcher cavity is also a periodic waveform of period $2\pi/\omega$. Therefore, the current i_2 can be expanded in Fourier series as follows:

$$i_2=a_0+\sum_{n=1}^{\infty}\left[a_n\cos\left(n\omega t_2\right)+b_n\sin\left(n\omega t_2\right)\right]$$ (11.52)

where $\qquad a_0=\dfrac{1}{2\pi}\displaystyle\int_{-\pi}^{\pi}i_2\mathrm{d}\left(\omega t_2\right)$ (11.53)

$$a_n=\dfrac{1}{\pi}\int_{-\pi}^{\pi}i_2\cos\left(n\omega t_2\right)\mathrm{d}\left(\omega t_2\right)$$ (11.54)

and $\qquad b_n=\dfrac{1}{\pi}\displaystyle\int_{-\pi}^{\pi}i_2\sin\left(n\omega t_2\right)\mathrm{d}\left(\omega t_2\right)$ (11.55)

Substituting Eqs (11.48) and (11.46) into Eqs (11.53)–(11.55), we get the following relations:

$$a_0=\frac{1}{2\pi}\int_{-\pi}^{\pi}i_2\mathrm{d}\left(\omega t_2\right)=\frac{1}{2\pi}\int_{-\pi}^{\pi}I_0\mathrm{d}\left(\omega t_0\right)=\frac{I_0}{2\pi}\left[\omega t_0\right]_{-\pi}^{\pi}=\frac{I_0}{2\pi}\left(\pi+\pi\right)=I_0$$

(11.56)

$$a_n=\frac{1}{\pi}\int_{-\pi}^{\pi}i_2\cos\left(n\omega t_2\right)\mathrm{d}\left(\omega t_2\right)$$

or $\qquad a_n=\dfrac{1}{\pi}\displaystyle\int_{-\pi}^{\pi}I_0\cos\left[n\left\{\theta_0+\theta_g+\omega t_0-X\sin\left(\omega t_0+\dfrac{\theta_g}{2}\right)\right\}\right]\mathrm{d}\left(\omega t_0\right)$

or $\qquad a_n=\dfrac{1}{\pi}\displaystyle\int_{-\pi}^{\pi}I_0\cos\left\{\left(n\omega t_0+n\theta_g+n\theta_0\right)-nX\sin\left(\omega t_0+\dfrac{\theta_g}{2}\right)\right\}\mathrm{d}\left(\omega t_0\right)$ (11.57)

and $\qquad b_n=\dfrac{1}{\pi}\displaystyle\int_{-\pi}^{\pi}I_0\sin\left\{\left(n\omega t_0+n\theta_g+n\theta_0\right)-nX\sin\left(\omega t_0+\dfrac{\theta_g}{2}\right)\right\}\mathrm{d}\left(\omega t_0\right)$ (11.58)

MATLAB Program

```
%Plot of normalized beam current of klystron as a function of catcher
cavity arrival angle
clear all;
clf;
for X = 0:0.5:1.5
    n = 0;
    for th = -pi:pi/180:pi
        n = n + 1;
        theta(n) = th;
        i2bI0(n) = 1/(1 - (X * cos(th)));
    end
    axis([-pi, pi, 0, 20])
    xlabel('Cather cavity arrival angle');
    ylabel('Normalized beam current')
    plot(theta, abs(i2bI0), 'k');
    hold on
end
```

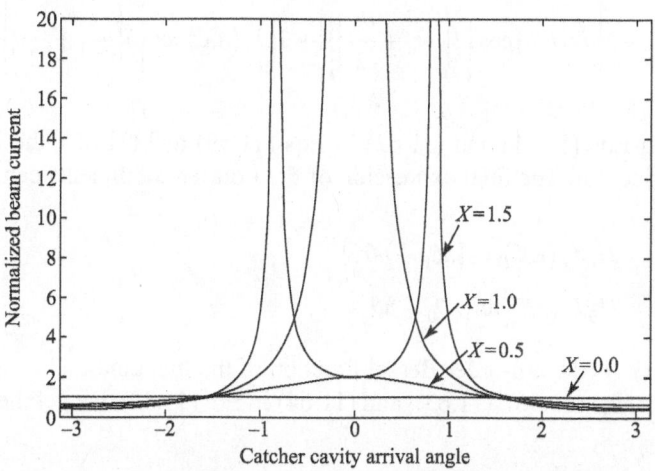

Fig. 11.8 Plot of beam current as function of catcher cavity arrival angle

To solve the integration given in Eqs (11.57) and (11.58), we need to split the cosines and sines of a sine function in the form of $\cos(A-B)$ and $\sin(A-B)$. This will result in the following equations:

$$a_n = \frac{1}{\pi}\int_{-\pi}^{\pi} I_0\left[\cos(n\omega t_0 + n\theta_g + n\theta_0)\cos\left\{nX\sin\left(\omega t_0 + \frac{\theta_g}{2}\right)\right\}\right.$$

$$\left. + \sin(n\omega t_0 + n\theta_g + n\theta_0)\sin\left\{nX\sin\left(\omega t_0 + \frac{\theta_g}{2}\right)\right\}\right]d(\omega t_0) \qquad (11.59)$$

and
$$b_n = \frac{1}{\pi}\int_{-\pi}^{\pi} I_0 \left[\sin\left(n\omega t_0 + n\theta_g + n\theta_0\right)\cos\left\{nX\sin\left(\omega t_0 + \frac{\theta_g}{2}\right)\right\} \right.$$
$$\left. - \cos\left(n\omega t_0 + n\theta_g + n\theta_0\right)\sin\left\{nX\sin\left(\omega t_0 + \frac{\theta_g}{2}\right)\right\} \right] d(\omega t_0) \qquad (11.60)$$

Cosines and sines of a sine function of Eqs (11.59) and (11.60) can be expressed as follows:

$$\cos\left\{nX\sin\left(\omega t_0 + \frac{\theta_g}{2}\right)\right\} = 2J_0(nX) + 2\left[J_2(nX)\cos\left\{2\left(\omega t_0 + \frac{\theta_g}{2}\right)\right\}\right]$$
$$+2\left[J_4(nX)\cos\left\{4\left(\omega t_0 + \frac{\theta_g}{2}\right)\right\}\right] + 2\left[J_6(nX)\cos\left\{6\left(\omega t_0 + \frac{\theta_g}{2}\right)\right\}\right] + \cdots \qquad (11.61)$$

and
$$\sin\left\{nX\sin\left(\omega t_0 + \frac{\theta_g}{2}\right)\right\} = 2\left[J_1(nX)\cos\left(\omega t_0 + \frac{\theta_g}{2}\right)\right]$$
$$+2\left[J_3(nX)\cos\left\{3\left(\omega t_0 + \frac{\theta_g}{2}\right)\right\}\right] + 2\left[J_5(nX)\cos\left\{5\left(\omega t_0 + \frac{\theta_g}{2}\right)\right\}\right] + \cdots \qquad (11.62)$$

Substituting Eqs (11.61) and (11.62) in Eqs (11.59) and (11.60), the integration can be carried out. The final expression of the Fourier coefficients can be written as follows:

$$a_n = 2I_0 J_n(nX)\cos(n\theta_0 + n\theta_g) \qquad (11.63)$$

and
$$b_n = 2I_0 J_n(nX)\sin(n\theta_0 + n\theta_g) \qquad (11.64)$$

where $J_n(nX)$ is the nth-order Bessel function of the first kind.

Substituting Eqs (11.56), (11.63), and (11.64) in Eq. (11.52), we get the following relation:

$$i_2 = I_0 + \sum_{n=1}^{\infty} 2I_0 J_n(nX)\left[\left\{\cos\left(n\theta_0 + n\theta_g\right)\cos\left(n\omega t_2\right)\right.\right.$$
$$\left.\left. + \sin\left(n\theta_0 + n\theta_g\right)\sin\left(n\omega t_2\right)\right\}\right]$$

or
$$i_2 = I_0 + \sum_{n=1}^{\infty} 2I_0 J_n(nX)\cos\left(n\theta_0 + n\theta_g - n\omega t_2\right) \qquad (11.65)$$

Substituting Eq. (11.23) in Eq. (11.65), we get the following expression:

$$i_2 = I_0 + \sum_{n=1}^{\infty} 2I_0 J_n(nX)\cos\left(n\omega T_0 + n\omega\tau - n\omega t_2\right)$$

or
$$i_2 = I_0 + \sum_{n=1}^{\infty} 2I_0 J_n(nX)\cos\left\{n\omega\left(t_2 - \tau - T_0\right)\right\} \qquad (11.66)$$

Equation (11.66) reveals that the magnitude of the fundamental component of beam current at the catcher cavity can be expressed as $I_f = 2I_0 J_1(X)$ (11.67)

This fundamental component has its maximum amplitude at $X = 1.841$ (11.68)

Substituting Eq. (11.43) in Eq. (11.68), we get $\dfrac{\beta_i V_1}{2V_0}\theta_0 = 1.841$ (11.69)

Further substituting Eq. (11.42) in Eq. (11.69), we get the following relation:

$$\frac{\beta_i V_1}{2V_0}\frac{\omega L_{optimum}}{v_0} = 1.841$$

or $\qquad L_{optimum} = 1.841\dfrac{2V_0 v_0}{\beta_i V_1 \omega} = \dfrac{3.682 v_0 V_0}{\omega \beta_i V_1}$ (11.70)

Comparing Eqs (11.37) and (11.70), we can find that the following relation holds:

$$\frac{\Delta L}{L_{optimum}} = \frac{\pi v_0 V_0}{\omega \beta_i V_1}\frac{\omega \beta_i V_1}{3.682 v_0 V_0} = \frac{\pi}{3.682} \approx 0.85$$ (11.71)

That is, ΔL is approximately 15% less than $L_{optimum}$. This is partly due to the approximations we have made for deriving Eq. (11.37) and partly due to the fact that the maximum fundamental component of current will not coincide with the maximum electron density along the beam because harmonic components also exist in the beam.

The current induced at the walls of the catcher cavity by the passing electron bunch is proportional to the input microwave voltage V_1. The fundamental component of induced microwave current in the catcher cavity is given by the following relation:

$$i_{2,ind} = \beta_o i_2 = 2\beta_o I_0 J_1(X)\cos\{\omega(t_2 - \tau - T_0)\}$$ (11.72)

where β_o is the beam coupling coefficient of the catcher gap. If the buncher and catcher cavities are identical, then $\beta_o = \beta_i$, and the magnitude of the fundamental component of current induced in the catcher cavity can be expressed as follows:

$$I_{2,ind} = \beta_o I_2 = 2\beta_o I_0 J_1(X)$$ (11.73)

The output equivalent circuit of a klystron is shown in Fig. 11.9, where R_{sho} is the wall resistance of the catcher cavity, R_B is the beam loading resistance, R_L is the external load, and R_{sh} is the effective shunt resistance. Therefore, the output power delivered to the load and the catcher cavity is given by the following equation:

$$P_{out} = (\beta_o I_2)^2\, R_{sh}/2 = \beta_o I_2 V_2 / 2$$ (11.74)

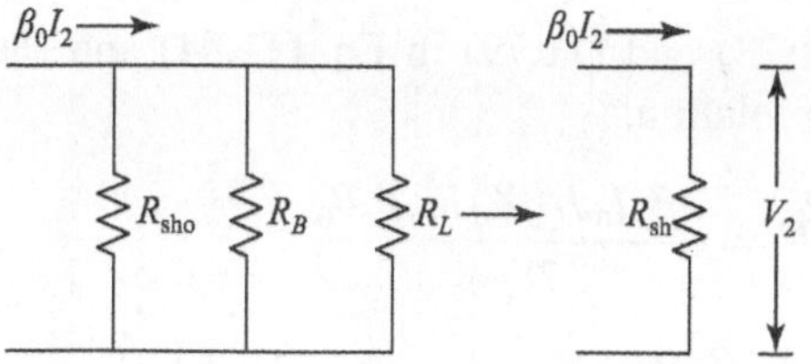

Fig. 11.9 Output equivalent circuit of klystron

where V_2 is the fundamental component of catcher gap voltage.

Therefore, the efficiency of the klystron amplifier can be expressed as follows:

$$\eta = \frac{P_{\text{out}}}{P_{\text{in}}} = \frac{\beta_o I_2 V_2}{2 I_0 V_0} \tag{11.75}$$

If coupling is perfect, then $\beta_0 = 1$, $V_2 = V_0$, and the maximum beam current approaches the following value:

$$I_{2,\text{max}} = 2 I_0 J_1 (1.841) = 2 \times 0.582 \times I_0 = 1.164 I_0 \tag{11.76}$$

Thus, the maximum electronic efficiency is expressed as follows:

$$\eta_{\text{max}} = \frac{\beta_o I_2 V_2}{2 I_0 V_0} = \frac{1.164 I_0 V_0}{2 I_0 V_0} \approx 0.58 \tag{11.77}$$

Therefore, the maximum electronic efficiency of a klystron is about 58%. In practice, the efficiency of a klystron is about 15–30%.

The equivalent mutual conductance of a klystron amplifier is as follows:

$$|G_m| \equiv \frac{i_{2,\text{ind}}}{V_1} = \frac{2 \beta_o I_0 J_1 (X)}{V_1} \tag{11.78}$$

From Eq. (11.43), we get the following expression:

$$V_1 = \frac{2 V_0 X}{\beta_i \theta_0} \tag{11.79}$$

Substituting Eq. (11.79) in Eq. (11.78), we get the following relation:

$$|G_m| = \frac{2 \beta_o \beta_i J_1 (X) \theta_0}{2 X} \frac{I_0}{V_0} = \frac{\beta_o \beta_i J_1 (X) \theta_0}{X} G_0 \tag{11.80}$$

where $\quad G_0 = I_0 / V_0 \tag{11.81}$

is the DC beam conductance.

If we assume that $\beta_o = \beta_i$, then Eq. (11.80) modifies as follows:

$$\frac{|G_m|}{G_0} = \beta_o^2 \theta_0 \frac{J_1 (X)}{X} \tag{11.82}$$

A simple program can be written to plot the normalized transconductance as a function of bunching parameter (Fig. 11.10).

For the maximum output $X = 1.841$ and Eq. (11.82) can be written as follows:

$$|G_m| / G_0 = \beta_o^2 J_1 (1.841) \theta_0 / 1.841 = 0.582 \beta_o^2 \theta_0 / 1.841 = 0.316 \beta_o^2 \theta_0 \tag{11.83}$$

The voltage gain of a klystron is as follows:

$$A_v = |V_2| / |V_1| = \beta_o I_2 R_{\text{sh}} / V_1 \tag{11.84}$$

Substituting Eqs (11.73) and (11.79) in Eq. (11.84) and assuming that $\beta_0 = \beta_i$, we get the following relation:

$$A_v = \frac{\beta_o I_2 R_{\text{sh}}}{V_1} = \frac{2 \beta_o I_0 J_1 (X) R_{\text{sh}} \beta_o \theta_0}{2 V_0 X}$$

or $\quad A_v = \frac{\beta_o^2 J_1 (X) R_{\text{sh}} \theta_0}{X} \frac{I_0}{V_0} = \beta_o^2 \theta_0 G_0 R_{\text{sh}} \frac{J_1 (X)}{X} \tag{11.85}$

MATLAB Program

```
%Program to plot the normalized transconductance of klystron as a
function of bunching parameter
clear all;
clf;
for beta0sqth0 = 5:5:30
    n = 0;
    for x = 0:0.01:4
        n = n + 1;
        GnbG0(n) = beta0sqth0 * besselj(1,x)/x;
        X(n) = x;
    end
    axis ([0, 4, 0, 15]);
    xlabel('Bunching Parameter');
    ylabel('Normalized Transconductance');
    plot(X, GnbG0, 'k')
    hold on
end
```

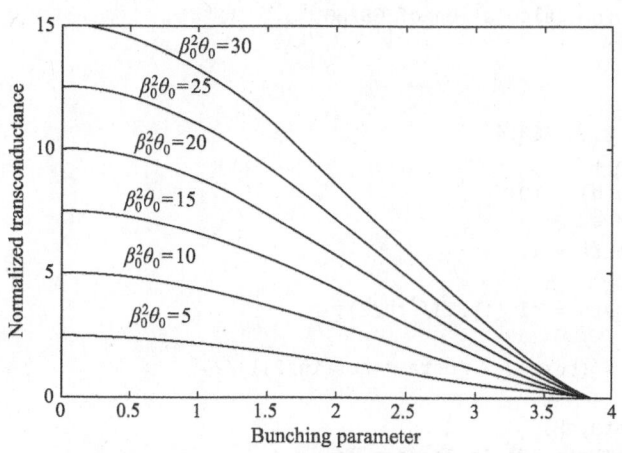

Fig. 11.10 Plot of normalized transconductance vs bunching parameter

Substituting Eq. (11.82) in Eq. (11.85), we get the following expression:

$$A_v = \frac{G_m}{G_0} G_0 R_{sh} = G_m R_{sh} \tag{11.86}$$

Power required to produce bunching can be expressed as follows:

$$P_B = \frac{P_0 V_1^2}{2V_0^2} \left[\frac{1}{2} \beta_i^2 - \frac{1}{2} \beta_i \cos\left(\frac{\theta_g}{2}\right) \right] = \frac{P_0 V_1^2}{2V_0^2} F(\theta_g) \tag{11.87}$$

where
$$P_0 = V_0^2 G_0 \tag{11.88}$$

is the DC power required to produce the electron beam and

$$F(\theta_g) = \frac{1}{2} \beta_i^2 - \frac{1}{2} \beta_i \cos\left(\frac{\theta_g}{2}\right) \tag{11.89}$$

The power required to produce bunching can also be expressed as follows:

$$P_B = V_1^2 G_B / 2 \tag{11.90}$$

where G_B is the bunching conductance.

Substituting Eqs (11.88) and (11.90) in Eq. (11.87), the following relation is obtained:

$$\frac{V_1^2}{2} G_B = V_0^2 G_0 F\left(\theta_g\right) \frac{V_1^2}{2V_0^2}$$

or $\qquad G_B / G_0 = F\left(\theta_g\right) \tag{11.91}$

A simple MATLAB program can be written to plot G_B / G_0 or $F\left(\theta_g\right)$ as a function of θ_g (Fig. 11.11). The figure shows that when $\theta_g = 3.5$, the equivalent bunching conductance is about one-fifth of the electron beam conductance, or alternatively bunching resistance is about five times the electron beam resistance.

MATLAB Program

```
%Program for calculation of GB/G0
clear all;
clf;
n = 0;
for th = 0:pi/360:4
    n = n + 1;
    Theta(n) = th;
    if th == 0;
        beta = 1;
    else
        beta = sin(th/2)/(th/2);
    end
    F(n) = (beta^2 - (beta * cos(th/2)))/2;
end
plot (Theta, F)
xlabel ('Theta (G) in Radians');
ylabel('Normalized Electronic Conductance');
```

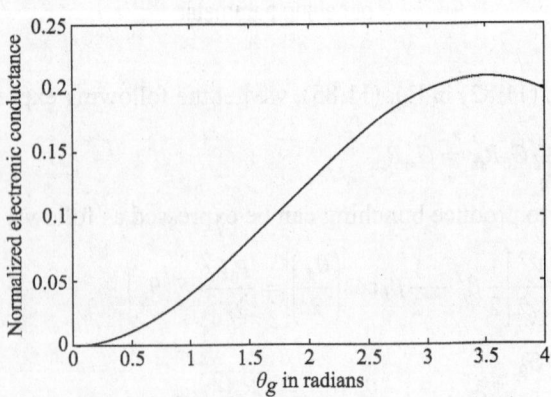

Fig. 11.11 Plot of normalized electronic conductance as function of θ_g

Power delivered by the electron beam to the catcher cavity can be written as follows:

$$\frac{V_2^2}{2R_{\text{sh}}} = \frac{V_2^2}{2R_{\text{sho}}} + \frac{V_2^2}{2R_B} + \frac{V_2^2}{2R_L} \tag{11.92}$$

or

$$\frac{1}{R_{\text{sh}}} = \frac{1}{R_{\text{sho}}} + \frac{1}{R_B} + \frac{1}{R_L} \tag{11.93}$$

The loaded Q-factor of the catcher cavity circuit, at the resonance frequency, can be expressed as follows:

$$\frac{1}{Q_L} = \frac{1}{Q_0} + \frac{1}{Q_B} + \frac{1}{Q_{\text{ext}}} \tag{11.94}$$

where Q_L is the loaded quality factor of the whole catcher cavity, Q_0 is the quality factor of the catcher cavity walls, Q_B is the quality factor of the beam loading, and Q_{ext} the quality factor of the external load.

> **Note** A klystron may have the following types of cavities: coaxial, radial, tunable, toroidal, and butterfly cavities.

An extended interaction in a klystron can be obtained by coupling two or more adjacent klystron cavities. A five-section extended interaction cavity is shown in Fig. 11.12.

The average energy of electrons leaving the buncher cavity over a complete cycle can be almost equal to that of the electrons entering the cavity during that period, provided that the buncher cavity gap is negligible. However, when this gap is not negligible, the average energy of the electrons leaving the buncher cavity over a cycle is larger than that of the electrons entering the cavity during that period. The difference in the average energy between the input and output electrons is a result of an interaction between the electrons and the RF field in the buncher cavity. This effect is known as *beam loading*.

Since two-cavity klystrons are associated with a considerable amount of noise, they are not generally used in receiver circuits. However, they find wide applications in troposcatter transmitters, UHF TV transmitter power amplifiers, and ground stations for satellite communication. Two-cavity klystrons can work in the frequency range starting from C-band up to 60 GHz and can provide an output power of 100–250 kW, with a possible power gain of 30–60 dB over a bandwidth of 10–60 MHz. The efficiency of such tubes is about 30–40%.

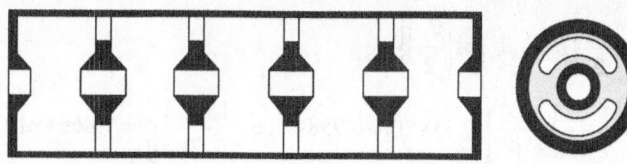 

Fig. 11.12 Coupling of cavities in klystron for extended interaction

EXAMPLE 11.1 A typical two-cavity klystron amplifier has the following parameters: $V_0 = 1$ kV, $R_0 = 50$ kΩ, $I_0 = 20$ mA, $f = 4$ GHz, gap spacing = 0.75 mm, spacing between the two cavities = 5 cm, and $R_{sh} = 25$ kΩ. Find the (a) input gap voltage to give the maximum voltage V_2; (b) voltage gain, neglecting beam loading in the output cavity; (c) efficiency of the amplifier, neglecting beam loading; and (d) beam loading conductance. It is provided that at $X = 1.841$, $J_1(X) = 0.582$.

Solution Given: $V_0 = 1$ kV, $R_0 = 50$ kΩ, $I_0 = 20$ mA, $f = 4$ GHz, $R_{sh} = 25$ kΩ, $d = 0.75$ mm, and $L = 5$ cm

(a) For maximum V_2, $J_1(X)$ must be maximum. Therefore, $J_1(X) = 0.582$ at $X = 1.841$. The velocity of the electron just leaving the cathode is as follows:

$$v_0 = 0.593 \times 10^6 \sqrt{V_0} = 0.593 \times 10^6 \sqrt{1000} = 1.8752 \times 10^7 \text{ m/s}$$

The gap transit angle is as follows:

$$\theta_g = \frac{\omega d}{v_0} = \frac{2\pi \times 4 \times 10^9 \times 0.75 \times 10^{-3}}{1.8752 \times 10^7} = 1.0052 \text{ rad}$$

The beam coupling coefficient is given by the following relation:

$$\beta_i = \beta_0 = \frac{\sin(\theta_g/2)}{(\theta_g/2)} = \frac{\sin(1.0052/2)}{(1.0052/2)} = 0.9584$$

The DC transit angle between the cavities is expressed as follows:

$$\theta_0 = \omega T_0 = \frac{\omega L}{v_0} = \frac{2\pi \times 4 \times 10^9 \times 5 \times 10^{-2}}{1.8752 \times 10^7} = 67.0135 \text{ rad}$$

The maximum input voltage is then given by the following equation:

$$V_{1,\max} = \frac{2V_0 X}{\beta_i \theta_0} = \frac{2 \times 1000 \times 1.841}{0.9584 \times 67.0135} = 57.329 \text{ V}$$

(b) The voltage gain is expressed as follows:

$$A_v = \frac{\beta_0^2 \theta_0}{R_0} \frac{J_1(X)}{X} R_{sh} = \frac{(0.9584)^2 \times 67.0135 \times 0.582 \times 25 \times 10^3}{50 \times 10^3 \times 1.841} = 9.7296$$

(c) Now,

$$I_2 = 2I_0 J_1(X) = 2 \times 20 \times 10^{-3} \times 0.582 = 23.28 \times 10^{-3} \text{ A}$$

$$V_2 = \beta_0 I_2 R_{sh} = 0.9584 \times 23.28 \times 10^{-3} \times 25 \times 10^3 = 557.7888 \text{ V}$$

and $\eta = \dfrac{\beta_0 I_2 V_2}{2 I_0 V_0} = \dfrac{0.9584 \times 23.28 \times 10^{-3} \times 557.7888}{2 \times 20 \times 10^{-3} \times 1000} = 0.3111 = 31.11\%$

(d) $G_B = \dfrac{G_0}{2}\left[\beta_0^2 - \beta_0 \cos\left(\dfrac{\theta_g}{2}\right)\right]$

or $G_B = \dfrac{1}{2 \times 50 \times 10^3}\left[(0.9584)^2 - 0.9584 \times \cos\left(\dfrac{1.0052}{2}\right)\right] = 7.8653 \times 10^{-7} \text{ ℧}$

The beam loading resistance is expressed as follows:

$$R_B = \frac{1}{G_B} = \frac{1}{7.8653 \times 10^{-7}} = 1.2714 \times 10^6 \ \Omega$$

Practice Problem

11.1 A typical two-cavity klystron amplifier has the following parameters: $V_0 = 1.2$ kV, $R_0 = 48$ kΩ, $I_0 = 25$ mA, $f = 3$ GHz, gap spacing = 1 mm, spacing between the two cavities = 4.5 cm, and $R_{sh} = 20$ kΩ. Find the (a) input gap voltage to give maximum voltage V_2; (b) voltage gain, neglecting beam loading in the output cavity; (c) efficiency of the amplifier, neglecting beam loading; and (d) beam loading conductance. It is given that at $X = 1.841$, $J_1(X) = 0.582$. $\left[110.8489 \text{ V}, 5.0682, 26.30\%, 7.3701 \times 10^{-7} \ \mho\right]$

11.4 Multi-cavity Klystrons

A multi-cavity klystron, shown in Fig. 11.13, is designed by cascading a number of cavities to achieve an enhanced gain. These intermediate cavities act as bunchers and are placed at a distance from the previous cavity for which the bunching parameter X equals to 1.841. Since velocity modulation increases as the beam progresses through various cavities, the requirement of $X = 1.841$ results in a decrease in the spacing between consecutive cavities. To keep the inter-cavity distance constant, and maintain $X = 1.841$, beam voltage (V_0) in consecutive cavities must be increased.

During the analysis of a two-cavity klystron, the space charge effect was assumed to be negligible because, in low-power operations, electron density of the beam is small. However, during high-power operations, electron density of the beam is high and therefore repulsion between electrons cannot be neglected. The space charge force within an electron bunch depends on the size and shape of the electron beam. For example, for an infinite electron beam, the electric field

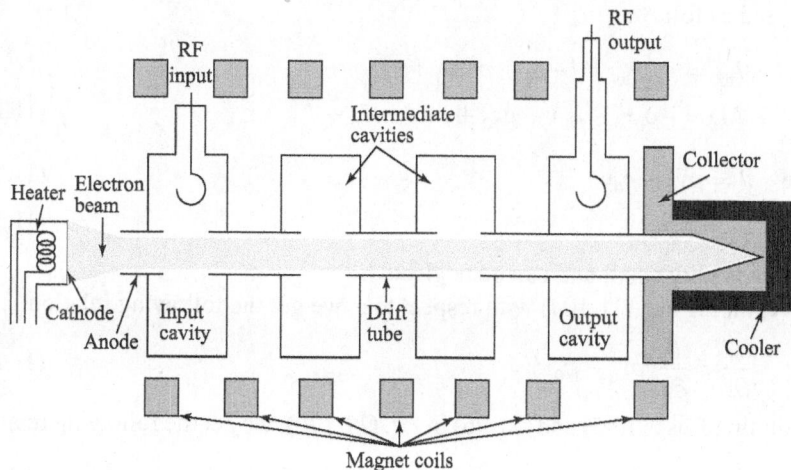

Fig. 11.13 Schematic diagram of multi-cavity klystron amplifier

acts only in the axial direction, whereas for a finite beam, the electric field is radial as well as axial. Thus, the axial component for a finite beam is less than that for an infinite beam. Due to a reduced axial space charge force, plasma frequency is reduced and plasma wavelength is increased.

If we assume that both charge density and velocity perturbation variations are simple sinusoidal in space and time, then we can write the following equations:

$$\rho = B\cos(\beta_e z - \omega t)\cos(\omega_q t + \theta) \tag{11.95}$$

and
$$v = -C\sin(\beta_e z - \omega t)\sin(\omega_q t + \theta) \tag{11.96}$$

where

B: constant of charge density perturbation
C: constant of velocity perturbation
$\beta_e = \omega/v_0$: DC phase constant of electron beam $\qquad$ (11.97)
$\omega_q = R\omega_p$: perturbation frequency or reduced plasma frequency $\qquad$ (11.98)
R: space charge reduction factor (varies from 0 to 1)
$\omega_p = \sqrt{\dfrac{e\rho_0}{m\varepsilon_0}}$: plasma frequency and a function of electron beam density

$$\tag{11.99}$$

θ: phase angle of oscillation

The total charge density, electron velocity, and beam current density can therefore be expressed as follows:

$$\rho_{tot} = -\rho_0 + \rho \tag{11.100}$$
$$v_{tot} = v_0 + v \tag{11.101}$$
and
$$J_{tot} = -J_0 + J \tag{11.102}$$

where ρ_0 is the DC electron charge density, v_0 is the DC electron velocity, J_0 is the DC beam current density, and J is the instantaneous RF beam current perturbation.

At any point in the beam, the instantaneous convection current density can be expressed as follows:

$$J_{tot} = \rho_{tot} v_{tot} = (-\rho_0 + \rho)(v_0 + v)$$
or
$$J_{tot} = -\rho_0 v_0 - \rho_0 v + \rho v_0 + \rho v = -J_0 + J \tag{11.103}$$

where $\quad J = \rho v_0 - \rho_0 v \tag{11.104}$
and $\quad J_0 = \rho_0 v_0 \tag{11.105}$

Here, ρv is very small and can be neglected.
Differentiating Eq. (11.104) with respect to z, we get the following relation:

$$\frac{\partial J}{\partial z} = \frac{\partial}{\partial z}(\rho v_0 - \rho_0 v) \tag{11.106}$$

Substituting Eqs (11.95) and (11.96) in Eq. (11.106), we get the following relation:

$$\frac{\partial J}{\partial z} = \frac{\partial}{\partial z}\{v_0 B\cos(\beta_e z - \omega t)\cos(\omega_q t + \theta) + \rho_0 C\sin(\beta_e z - \omega t)\sin(\omega_q t + \theta)\}$$

or $\quad \dfrac{\partial J}{\partial z} = -Bv_0\beta_e \sin(\beta_e z - \omega t)\cos(\omega_q t + \theta) + C\rho_0\beta_e \cos(\beta_e z - \omega t)\sin(\omega_q t + \theta)$

$$(11.107)$$

Further substituting Eq. (11.97) in Eq. (11.107), we get the following expression:

$$\dfrac{\partial J}{\partial z} = -B\omega \sin(\beta_e z - \omega t)\cos(\omega_q t + \theta) + C\rho_0\beta_e \cos(\beta_e z - \omega t)\sin(\omega_q t + \theta)$$

$$(11.108)$$

Differentiating Eq. (11.95) with respect to t, the following equation is obtained:

$$-\dfrac{\partial \rho}{\partial t} = -B\omega \sin(\beta_e z - \omega t)\cos(\omega_q t + \theta) + B\omega_q \cos(\beta_e z - \omega t)\sin(\omega_q t + \theta)$$

$$(11.109)$$

Now, from continuity condition:

$$\vec{\nabla}\cdot\vec{J} = -\dfrac{\partial \rho}{\partial t} \qquad (11.110)$$

Therefore, equating the RHS of Eqs (11.108) and (11.109), we get the following relation:

$$C\rho_0\beta_e = B\omega_q \qquad (11.111)$$

Substituting Eqs (11.95) and (11.96) in Eq. (11.104), we get the following expression:

$$J = v_0 B \cos(\beta_e z - \omega t)\cos(\omega_q t + \theta) + \rho_0 C \sin(\beta_e z - \omega t)\sin(\omega_q t + \theta) \quad (11.112)$$

Further, consecutive substitution of Eq. (11.111) and Eq. (11.97) in Eq. (11.112) gives the following expression:

$$J = v_0 B \cos(\beta_e z - \omega t)\cos(\omega_q t + \theta) + \dfrac{B\omega_q}{\beta_e} \sin(\beta_e z - \omega t)\sin(\omega_q t + \theta)$$

or $\quad J = v_0 B \cos(\beta_e z - \omega t)\cos(\omega_q t + \theta) + \dfrac{\omega_q}{\omega} v_0 B \sin(\beta_e z - \omega t)\sin(\omega_q t + \theta)$

$$(11.113)$$

Since $\omega_q/\omega \ll 1$, Eq. (11.113) can be approximated as follows:

$$J = v_0 B \cos(\beta_e z - \omega t)\cos(\omega_q t + \theta) \qquad (11.114)$$

It can be shown that Eq. (11.114) can be written as follows:

$$J = -\dfrac{1}{2}\dfrac{J_0\omega}{v_0\omega_q} \beta_i V_1 \sin(\beta_q z)\cos(\beta_e z - \omega t) \qquad (11.115)$$

where $\quad \beta_q = \omega_q/v_0 \qquad (11.116)$

is the plasma phase constant.
The modulated velocity can be expressed as follows:

$$v = -v_0 \dfrac{\beta_i V_1}{2V_0} \cos(\beta_q z)\sin(\beta_e z - \omega t) \qquad (11.117)$$

Electrons leaving the input gap of a klystron amplifier have a velocity given by the following relation:

$$v(t_1) = v_0 \left\{ 1 + \frac{\beta_i V_1}{2V_0} \sin(\omega\tau) \right\} \tag{11.118}$$

where V_1: magnitude of input signal

$$\tau = d/v_0 = t_1 - t_0 \tag{11.119}$$

D: dap distance

Since electrons under the influence of space charge forces exhibit a simple harmonic motion, velocity at a later time t can be expressed as follows:

$$v_{\text{tot}} = v_0 \left[1 + \frac{\beta_i V_1}{2V_0} \sin(\omega\tau) \cos\left\{ \omega_p(t-\tau) \right\} \right] \tag{11.120}$$

If the two cavities of a two-cavity klystron are identical and the second cavity is placed at a point where RF current modulation is maximum, then the magnitude of the RF convection current at the output cavity of the klystron can be written as follows:

$$|i_2| = \frac{1}{2} \frac{I_0 \omega}{V_0 \omega_q} \beta_i |V_1| \tag{11.121}$$

Therefore, magnitudes of induced current and voltage in the output cavity can be expressed as follows:

$$|I_2| = \beta_o |i_2| = \frac{1}{2} \frac{I_0 \omega}{V_0 \omega_q} \beta_o \beta_i |V_1| = \frac{1}{2} \frac{I_0 \omega}{V_0 \omega_q} \beta_o^2 |V_1| \tag{11.122}$$

and

$$|V_2| = |I_2| R_{\text{shl}} = \frac{1}{2} \frac{I_0 \omega}{V_0 \omega_q} \beta_o^2 |V_1| R_{\text{shl}} \tag{11.123}$$

Thus, the output power delivered to the load in a two-cavity klystron amplifier can be expressed as follows:

$$P_{\text{out}} = |I_2|^2 R_{\text{shl}} = \frac{1}{4} \left(\frac{I_0 \omega}{V_0 \omega_q} \right)^2 \beta_o^4 |V_1|^2 R_{\text{shl}} \tag{11.124}$$

Using Eq. (11.124), the power gain of the two-cavity klystron can be expressed as follows:

$$\frac{P_{\text{out}}}{P_{\text{in}}} = \frac{P_{\text{out}}}{|V_1^2|/R_{\text{sh}}} = \frac{1}{4} \left(\frac{I_0 \omega}{V_0 \omega_q} \right)^2 \beta_o^4 \frac{|V_1|^2 R_{\text{sh}} R_{\text{shl}}}{|V_1|^2} = \frac{1}{4} \left(\frac{I_0 \omega}{V_0 \omega_q} \right)^2 \beta_o^4 R_{\text{sh}} R_{\text{shl}} \tag{11.125}$$

The efficiency of a two-cavity klystron can be expressed as follows:

$$\eta = \frac{P_{\text{out}}}{P_{\text{in}}} = \frac{P_{\text{out}}}{I_0 V_0} = \frac{1}{4} \left(\frac{I_0 \omega}{V_0 \omega_q} \right)^2 \frac{\beta_o^4 |V_1|^2}{I_0 V_0} R_{\text{shl}} = \frac{1}{4} \left(\frac{I_0}{V_0} \right) \left(\frac{|V_1| \omega}{V_0 \omega_q} \right)^2 \beta_o^4 R_{\text{shl}} \tag{11.126}$$

This analysis assumes two cavities. Now we will perform a simple analysis of a four-cavity klystron under the following assumptions:

1. All the cavities are identical and have the same unloaded Q and coupling coefficient.
2. Two immediate cavities are not externally loaded.
3. Input and output cavities are matched.

If V_1 is the magnitude of the input cavity voltage, then the magnitude of the RF convection current density injected into the first immediate cavity gap can be expressed by Eq. (11.115), and the induced current and voltage in the first intermediate cavity by Eqs (11.122) and (11.123), respectively. The gap voltage of the first intermediate cavity produces a velocity modulation on the beam in the second intermediate cavity. The RF convection current in this second intermediate cavity can be expressed as follows:

$$|i_3| = \frac{1}{2}\frac{I_0\omega}{V_0\omega_q}\beta_i|V_2| = \frac{1}{2}\frac{I_0\omega}{V_0\omega_q}\beta_o|V_2| \tag{11.127}$$

Substituting Eq. (11.123) in Eq. (11.127) after replacing R_{shl} by R_{sh}, we get the following relation:

$$|i_3| = \frac{1}{4}\left(\frac{I_0\omega}{V_0\omega_q}\right)^2\beta_o^3|V_1|R_{\text{sh}} \tag{11.128}$$

The voltage of the second intermediate cavity will be as follows:

$$|V_3| = \beta_o|i_3|R_{\text{sh}} = \frac{1}{4}\left(\frac{I_0\omega}{V_0\omega_q}\right)^2\beta_o^4|V_1|R_{\text{sh}}^2 \tag{11.129}$$

This voltage produces further velocity modulation and is converted to an RF convection current at the final output of the four-cavity klystron. The output convection current density can be expressed as follows:

$$|i_4| = \frac{1}{2}\frac{I_0\omega}{V_0\omega_q}\beta_i|V_3| = \frac{1}{2}\frac{I_0\omega}{V_0\omega_q}\beta_o|V_3| \tag{11.130}$$

Substituting Eq. (11.129) into Eq. (11.130), the following equation is obtained:

$$|i_4| = \frac{1}{8}\left(\frac{I_0\omega}{V_0\omega_q}\right)^3\beta_o^5|V_1|R_{\text{sh}}^2 \tag{11.131}$$

Therefore,

$$|I_4| = \beta_o|i_4| = \frac{1}{8}\left(\frac{I_0\omega}{V_0\omega_q}\right)^3\beta_o^6|V_1|R_{\text{sh}}^2 \tag{11.132}$$

The output voltage can therefore be expressed as follows:

$$|V_4| = |I_4|R_{\text{shl}} = \frac{1}{8}\left(\frac{I_0\omega}{V_0\omega_q}\right)^3\beta_o^6|V_1|R_{\text{sh}}^2R_{\text{shl}} \tag{11.133}$$

The output power is, therefore,

$$P_{\text{out}} = |I_4|^2 R_{\text{shl}} = \frac{1}{64}\left(\frac{I_0\omega}{V_0\omega_q}\right)^6 \beta_o^{12}|V_1|^2 R_{\text{sh}}^4 R_{\text{shl}} \tag{11.134}$$

Multi-cavity klystrons are used as medium- or high-power amplifiers for both CW and pulsed applications. In general, such tubes can operate in the range from 250 MHz to 60 GHz, with a typical CW output power of 100 kW in the VHF range and 250 kW in the X-band. The pulsed output power may be as high as 25 MW. In addition, multi-cavity klystrons can provide a power gain of 30 dB at UHF up to 60 dB at X-band, in the bandwidth range of 8–60 MHz. The mechanical tuning range is about 300–600 MHz.

Multi-cavity klystrons are widely used in UHF television transmitters, troposphere scatter transmitters, and ground stations for satellite communication.

EXAMPLE 11.2 A typical four-cavity klystron has the following parameters: beam voltage = 15 kV, beam current = 1.5 A, operating frequency = 10 GHz, signal voltage = 10 V (rms), gap distance = 1 cm, input and output beam coupling coefficient = 1, DC electron charge density = 10^{-6} C/m², RF charge density = 10^{-8} C/m², and velocity perturbation = 10^5 m/s. Calculate (a) DC electron velocity, (b) DC phase constant, (c) plasma frequency, (d) reduced plasma frequency for $R = 0.5$, (e) DC beam current density, (f) instantaneous beam current density, (g) transit time across the input gap, and (h) electron velocity leaving the input gap.

Solution Given:

$$V_0 = 15 \text{ kV}, I_0 = 1.5 \text{ A}, f = 10 \text{ GHz}, V_1 = 10 \text{ V (rms)}, d = 1 \text{ cm}, \rho_0 = 10^{-6} \text{ C/m}^2,$$
$$\rho = 10^{-8} \text{ C/m}^2, v = 10^5 \text{ m/s, and } R = 0.5$$

(a) DC electron velocity is expressed as follows:

$$v_0 = 0.593 \times 10^6 \sqrt{15 \times 10^3} = 0.7263 \times 10^8 \text{ m/s}$$

(b) DC phase constant is given as follows:

$$\beta_e = \frac{\omega}{v_0} = \frac{2\pi \times 10 \times 10^9}{0.7263 \times 10^8} = 8.651 \times 10^2 \text{ rad/m}$$

(c) Plasma frequency is given by the following relation:

$$\omega_p = \sqrt{\frac{e}{m}\frac{\rho_0}{\varepsilon_0}} = \sqrt{\frac{1.759 \times 10^{11} \times 10^{-6}}{8.8542 \times 10^{-12}}} = 1.4095 \times 10^8 \text{ rad/s}$$

(d) Reduced plasma frequency at $R = 0.5$ is as follows:

$$\omega_q = R\omega_p = 0.5 \times 1.4095 \times 10^8 = 0.7047 \times 10^8 \text{ rad/s}$$

(e) DC beam current density is expressed as follows:

$$J_0 = \rho_0 v_0 = 10^{-6} \times 0.7268 \times 10^8 = 72.68 \text{ A/m}^2$$

(f) Instantaneous beam current density is as follows:

$$J = \rho v_0 - \rho_0 v = 10^{-8} \times 0.7268 \times 10^8 - 10^{-6} \times 10^5 = 0.6268 \text{ A/m}^2$$

(g) Transit time across the input gap is expressed as follows:

$$\tau = \frac{d}{v_0} = \frac{10^{-2}}{0.7268 \times 10^8} = 0.1376 \times 10^{-9} \text{ s} = 0.1376 \text{ ns}$$

(h) Electron velocity leaving the input gap is given by the following relation:

$$v(t_1) = v_0 \left[1 + \frac{\beta_i V_1}{2V_0} \sin(\omega\tau) \right]$$

or $\quad v(t_1) = 0.7263 \times 10^8 \left[1 + \frac{10}{2 \times 15 \times 10^3} \sin\left(2\pi \times 10 \times 10^9 \times 0.1376 \times 10^{-9}\right) \right]$

or $\quad v(t_1) = 0.727 \times 10^8 \, \text{m/s}$

Practice Problem

11.2 A typical four-cavity klystron has the following parameters: beam voltage = 12 kV, beam current = 1 A, operating frequency = 8 GHz, signal voltage = 9 V (rms), gap distance = 0.9 cm, input and output beam coupling coefficient = 1, DC electron charge density = 10^{-6} C/m^2, RF charge density = 10^{-8} C/m^2, and velocity perturbation = 10^5 m/s. Calculate (a) DC electron velocity, (b) DC phase constant, (c) plasma frequency, (d) reduced plasma frequency for $R = 0.4$, (e) DC beam current density, (f) instantaneous beam current density, (g) transit time across the input gap, and (h) electron velocity leaving the input gap. $\left[0.6496 \times 10^8 \text{ m/s}, 7.7379 \times 10^2 \text{ rad/m}, \right.$
$\left. 1.4095 \times 10^8 \text{ rad/s}, 0.5638 \times 10^8 \text{ rad/s}, 64.96 \text{ A/m}^2, 0.5496 \text{ A/m}^2, 0.1385 \text{ ns}, 0.6498 \times 10^8 \text{ m/s} \right]$

11.5 Two-cavity Klystron Oscillators

In the last two sections, we discussed two- and multi-cavity klystron amplifiers. If a fraction of the output of such an amplifier is fed back into the input with a unity feedback loop gain and a phase shift of an integral multiple of 2π (i.e., in positive feedback), then it will produce an oscillation, resulting in a klystron oscillator. Typical two-cavity klystron oscillators (Fig. 11.14) can produce an output power in the range 2–10 W, in the frequency band 5–50 GHz. Two-cavity oscillators with an output power of 200 W are also available, which find applications in CW Doppler radars, frequency modulators, and high-power microwave links, and as pump sources in parametric amplifiers.

A two-cavity klystron has the advantage of producing a relatively high CW power as compared to their size. However, it also suffers from some major disadvantages like frequency tuning. Cavities used in a two-cavity klystron have high Q values with narrow bandwidths, and thus, individual tuning is awkward.

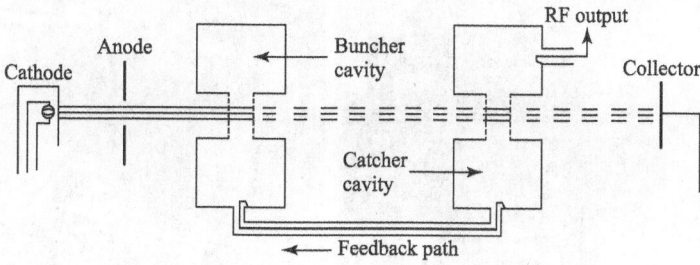

Fig. 11.14 Two-cavity klystron oscillator

In addition, maintaining the positive feedback is also difficult. Therefore, two-cavity klystrons are generally used for fixed-frequency applications.

11.6 Reflex Klystrons

The problem of frequency tuning of a two-cavity klystron can be solved using a special klystron structure, called reflex klystron, shown in Fig. 11.15. Fig. 11.16 shows a commercial packaged reflex klystron. A reflex klystron is a single-cavity structure and therefore its tuning is very easy.

In a reflex klystron, the DC voltage generates a wide band of RF noise in the cavity. However, the RF noise frequency that corresponds to the resonance frequency of a particular mode of the cavity, sustains and triggers the initial RF oscillation in the device. When the electron beam from the cathode enters the cavity, its velocity is modulated by this RF voltage or, more precisely, by the cavity gap voltage. Electrons entering the cavity gap at the positive half-cycle are accelerated and move with faster velocities, those entering the cavity gap at the negative half-cycle are decelerated and move with slower velocities, and those entering the cavity gap at the zero gap voltage move with an unchanged velocity. These result in velocity modulation. Velocity-modulated electrons then proceed to the repeller terminal and experience a repelling force. As a result, their velocities start decreasing and finally become zero before reaching the repeller terminal. The zero-velocity electrons still experience the repelling force and hence start moving, but towards the cavity. Therefore, after a certain time, electrons return to the cavity and are finally collected by the cavity walls or other grounded metal parts of the tube.

In practice, the total times taken by the individual electrons to get velocity-modulated and return back to the cavity are not same. This is because the electrons moving with a faster velocity, and hence with a higher kinetic energy, penetrate more distance towards the repeller than those moving with a slower velocity, as shown in

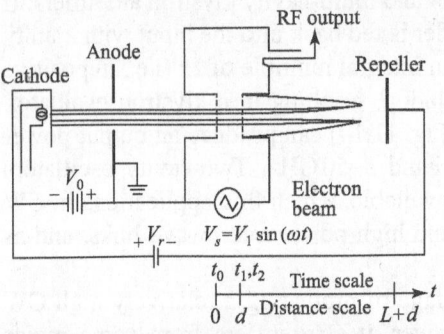

Fig. 11.15 Schematic diagram of reflex klystron

Fig. 11.16 Commercial packaged reflex klystron

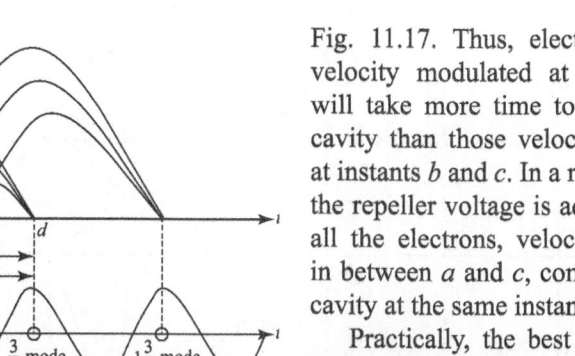

Fig. 11.17 Applegate diagram of reflex klystron

Fig. 11.17. Thus, electrons that are velocity modulated at an instant a will take more time to return to the cavity than those velocity modulated at instants b and c. In a reflex klystron, the repeller voltage is adjusted so that all the electrons, velocity modulated in between a and c, come back to the cavity at the same instant d.

Practically, the best possible time for electrons to return to the cavity gap is when the voltage existing across the gap will apply maximum retardation to them, that is, when the gap voltage is positive maximum. This causes the electrons to fall through the maximum negative voltage between the gap grids, thus transferring the maximum amount of energy to the gap. If the power delivered by the bunched electrons to the cavity is greater than the power loss in the cavity, then the RF field inside the cavity will increase in amplitude and the oscillation will be sustained. Figure 11.17 reveals that the electrons returning after $\frac{3}{4}$ and $1\frac{3}{4}$ cycles from b satisfy the condition for delivering maximum power to the RF field and hence for sustaining the oscillation. In practice, each of these numbers represents a mode of klystron. Theoretically, an infinite number of modes can exist. The transit time of the electrons corresponding to these modes can be written as follows:

$$ t_0 = \left(n - \frac{1}{4} \right) T = NT \text{ where } n \text{ is an integer} \tag{11.135} $$

The lowest-order mode (3/4) occurs for the maximum value of the repeller voltage and hence for the minimum transit time of electrons in the repeller space. On the other hand, higher-order modes occur at lower repeller voltages and hence for higher transit times of electrons in the repeller space. Since for the lowest-order mode, the repeller voltage and hence the acceleration of the bunched electrons on their return are maximum, the corresponding power output is also maximum.

The variation of the output power for different modes can be explained as follows. As the mode number increases, electron bunches are formed more slowly. As a result, electrons get more time for mutual repulsion and hence spread more. In addition, higher-order modes are associated with a long drift time, providing more time for the mutual repulsion and causing further spreading of electrons. Spreading of electrons from the bunch, also known as debunching, thus becomes more prominent as the mode number increases. Due to the debunching effect, the returning electron bunch is less populated, as compared with lower-order modes, and thus less power is delivered to the output cavity.

Plots of power output as a function of repeller voltage and operating frequency as a function of repeller voltage, for different modes of reflex klystron, are shown in Fig. 11.18. The figure reveals that lower-order modes are better for working on a fixed frequency, whereas higher-order modes are better when frequency tuning is required. In practice, two types of tuning mechanism can be used: (a) mechanical

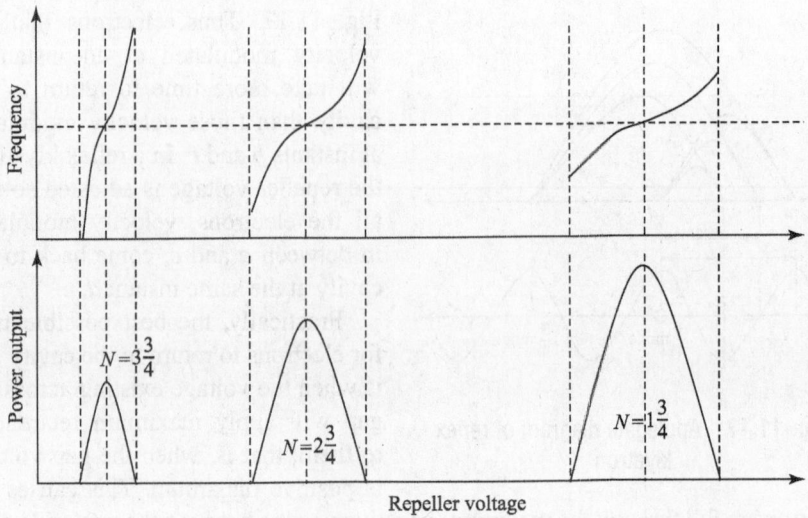

Fig. 11.18 Power output and frequency characteristics of reflex klystron

tuning and (b) electronic tuning. In mechanical tuning, dimensions of the cavity are varied either by flexing a portion of the cavity wall or by changing the space between the cavity grids. In electrical tuning, the repeller voltage is varied. Electronic tuning sensitivity (ETS) can be defined as follows:

$$ETS = \frac{f_2 - f_1}{V_2 - V_1} \text{ MHz/V} \tag{11.136}$$

where f_1 and f_2 are the frequencies in MHz at which mode power reduces to half of its value at the top. Figure 11.18 reveals that ETS is higher for higher-order modes, though the output is small.

The repeller can be overheated by the impact of high-velocity electrons and damaged very quickly. Thus, when operating a reflex klystron, electrons should be prevented from reaching the repeller terminal. This can be done by connecting a resistor to the cathode of the klystron so that the repeller does not get more positive than it. Alternatively, a protector diode can be used, with its anode connected to the repeller and cathode connected to the cathode of the klystron. With this arrangement, repeller voltage cannot become positive.

Note When a reflex klystron is switched on, a high negative voltage is first applied to the repeller and then a positive anode voltage is applied. This precaution prevents high-energy electrons from reaching the repeller terminal.

The analysis of a reflex klystron is similar to that of a two-cavity klystron to some extent and is subjected to the following approximations:

1. Cavity grids and repeller are plane and parallel, and also very large in extent.
2. RF field is absent in the repeller space.
3. Electrons are not intercepted by the cavity anode grid.

4. No debunching of electrons takes place in the repeller space.
5. RF gap voltage is small compared to the beam voltage.

Electrons entering the cavity gap from the cathode at time t_0 can be assumed to have a uniform velocity:

$$v_0 = 0.593\sqrt{V_0} \times 10^6 \tag{11.137}$$

When the electron leaves the cavity at $z = d$ at time t_1, it will have the following velocity:

$$v(t_1) = v_0 \left\{ 1 + \frac{\beta_i V_1}{2V_0} \sin\left(\omega t_1 - \frac{\theta_g}{2}\right) \right\} \tag{11.138}$$

These velocity-modulated electrons will experience a net retarding electric field:

$$E = \frac{V_r + V_0 + V_1 \sin(\omega t)}{L} \tag{11.139}$$

and return to the cavity at time t_2. The associated force equation can be written as follows:

$$m\frac{d^2 z}{dt^2} = -eE = -e\frac{V_r + V_0}{L} \tag{11.140}$$

where we have assumed that the electric field is only along the z-direction and $(V_r + V_0) \gg |V_1 \sin(\omega t)|$.

Integrating Eq. (11.140), we get the following equation:

$$\frac{dz}{dt} = \frac{-e(V_r + V_0)}{mL} \int_{t1}^{t} dt = \frac{-e(V_r + V_0)}{mL}(t - t_1) + C_1 \tag{11.141}$$

At $t = t_1$, $\dfrac{dz}{dt} = v(t_1)$. Therefore,

$$v(t_1) = \frac{-e(V_r + V_0)}{mL}(t_1 - t_1) + C_1 = C_1 \tag{11.142}$$

Substituting Eq. (11.142) into Eq. (11.141), we get the following relation:

$$\frac{dz}{dt} = \frac{-e(V_r + V_0)}{mL}(t - t_1) + v(t_1) \tag{11.143}$$

Further integrating Eq. (11.143), we get the following expression:

$$z = \frac{-e(V_r + V_0)}{mL} \int_{t_1}^{t}(t - t_1)dt + v(t_1)\int_{t_1}^{t} dt$$

or $$z = \frac{-e(V_r + V_0)}{mL}\left[\frac{t^2}{2} - tt_1\right]\Big|_{t_1}^{t} + v(t_1)t\Big|_{t_1}^{t} + C_2$$

or $$z = \frac{-e(V_r + V_0)}{2mL}(t^2 - 2tt_1 + t_1^2) + v(t_1)(t - t_1) + C_2$$

or
$$z = \frac{-e(V_r + V_0)}{2mL}(t - t_1)^2 + v(t_1)(t - t_1) + C_2 \tag{11.144}$$

Now at $t = t_1$, $z = d$. Therefore,

$$d = \frac{-e(V_r + V_0)}{2mL}(t_1 - t_1)^2 + v(t_1)(t_1 - t_1) + C_2 = C_2 \tag{11.145}$$

Substituting Eq. (11.145) back into Eq. (11.144), we get the following relation:

$$z = \frac{-e(V_r + V_0)}{2mL}(t - t_1)^2 + v(t_1)(t - t_1) + d \tag{11.146}$$

Since electrons return to the cavity gap at time t_2, at $t = t_2$, $z = d$. Thus, we can write that

$$d = \frac{-e(V_r + V_0)}{2mL}(t_2 - t_1)^2 + v(t_1)(t_2 - t_1) + d$$

or
$$\frac{-e(V_r + V_0)}{2mL}(t_2 - t_1) + v(t_1) = 0$$

or
$$T' = (t_2 - t_1) = \frac{2mL}{e(V_r + V_0)} v(t_1) \tag{11.147}$$

where T' is the round trip transit time. Substituting Eq. (11.138) in Eq. (11.147), we get the following expression:

$$T' = \frac{2mLv_0}{e(V_r + V_0)}\left[1 + \frac{\beta_i V_1}{2V_0} \sin\left(\omega t_1 - \frac{\theta_g}{2}\right)\right]$$

or
$$T' = T_0'\left\{1 + \frac{\beta_i V_1}{2V_0} \sin\left(\omega t_1 - \frac{\theta_g}{2}\right)\right\} \tag{11.148}$$

where
$$T_0' = \frac{2mLv_0}{e(V_r + V_0)} \tag{11.149}$$

is the round trip DC transit time of the centre of the bunch electron.

Multiplying Eq. (11.148) by ω, we get the following equation:

$$\omega T' = \omega(t_2 - t_1) = \omega T_0'\left\{1 + \frac{\beta_i V_1}{2V_0} \sin\left(\omega t_1 - \frac{\theta_g}{2}\right)\right\}$$

or
$$\omega T' = \theta_0' + X' \sin\left(\omega t_1 - \frac{\theta_g}{2}\right) \tag{11.150}$$

where
$$\theta_0' = \omega T_0' \tag{11.151}$$

is the round trip DC transit angle of the centre of the bunch electron and

$$X' = \frac{\beta_i V_1}{2V_0}\theta_0' \tag{11.152}$$

is the bunching parameter of the reflex klystron oscillator.

To transfer maximum energy to the oscillator, the returning electron must cross the cavity gap when the gap field is maximum retarding. Therefore, the round trip transit angle of the centre of the bunch is given as follows:

$$\omega T_0' = \left(n - \frac{1}{4}\right) 2\pi = 2\pi N \tag{11.153}$$

where $\quad N = \left(n - \frac{1}{4}\right) \tag{11.154}$

is the number of modes.

Current modulation of the electron beam when it returns to the cavity from the repeller region can be determined using the same procedure used in the analysis of two-cavity klystrons. However, since the beam current injected into the cavity is now in the negative z-direction, we can write the following relation:

$$i_{2t} = -I_0 - \sum_{n=1}^{\infty} 2I_0 J_n \left(nX'\right) \cos\left\{n\left(\omega t_2 - \theta_0' - \theta_g\right)\right\} \tag{11.155}$$

The fundamental component of the current induced in the cavity by the modulated electron beam is given by the following equation:

$$i_2 = -\beta_i I_2 = 2I_0 \beta_i J_1 \left(X'\right) \cos\left(\omega t_2 - \theta_0'\right) \tag{11.156}$$

where we have neglected θ_g as $\theta_0' \gg \theta_g$.

The magnitude of the fundamental component can be written as follows:

$$I_2 = 2I_0 \beta_i J_1 \left(X'\right) \tag{11.157}$$

The DC power supplied by the beam voltage is given by $P_{\text{dc}} = V_0 I_0 \tag{11.158}$

whereas the ac power delivered to the load is expressed as $P_{\text{ac}} = \dfrac{V_1 I_2}{2} \tag{11.159}$

Substituting Eq. (11.157) in Eq. (11.159), we get $P_{\text{ac}} = V_1 I_0 \beta_i J_1 \left(X'\right) \tag{11.160}$

Now first substituting Eq. (11.151) and then Eq. (11.153) in Eq. (11.152), the following relation is obtained:

$$X' = \frac{\beta_i V_1}{2V_0} \theta_0' = \frac{\beta_i V_1}{2V_0} \omega T_0' = \frac{\beta_i V_1}{2V_0}\left(n - \frac{1}{4}\right) 2\pi = \frac{\beta_i V_1}{2V_0}\left(2\pi n - \frac{\pi}{2}\right)$$

or $\quad V_1 = \dfrac{2V_0 X'}{\beta_i \left(2\pi n - \dfrac{\pi}{2}\right)} \tag{11.161}$

Substituting Eq. (11.161) in Eq. (11.160), we get the following relation:

$$P_{\text{ac}} = I_0 \beta_i J_1 \left(X'\right) \frac{2V_0 X'}{\beta_i\left(2\pi n - \dfrac{\pi}{2}\right)} = \frac{2V_0 I_0 X' J_1 \left(X'\right)}{\left(2\pi n - \dfrac{\pi}{2}\right)} \tag{11.162}$$

Therefore,

$$\eta = \frac{P_{\text{ac}}}{P_{\text{dc}}} = \frac{2V_0 I_0 X' J_1 \left(X'\right)}{V_0 I_0 \left(2\pi n - \dfrac{\pi}{2}\right)} = \frac{2X' J_1 \left(X'\right)}{\left(2\pi n - \dfrac{\pi}{2}\right)} \tag{11.163}$$

The factor $X'J_1(X')$ reaches a maximum value of 1.25 at $X' = 2.408$ and $J_1(X') = J_1(2.408) = 0.52$. In practice, the mode corresponding to $n = 2$ or $N = 1\frac{3}{4}$ has the most output power. Thus, the maximum efficiency of a reflex klystron is calculated as follows:

$$\eta = \frac{2 \times 2.408 \times 0.52}{\left(4\pi - \dfrac{\pi}{2}\right)} = 0.227 \tag{11.164}$$

Therefore, the maximum efficiency of reflex klystron is 22.7%.

Now, first multiplying Eq. (11.149) by ω and then substituting Eqs (11.153) and (11.137) in it, we get the following equation:

$$\omega T_0' = \frac{2m\omega L v_0}{e(V_r + V_0)}$$

or

$$2\pi\left(n - \frac{1}{4}\right) = \frac{2m\omega L \times 0.593\sqrt{V_0} \times 10^6}{e(V_r + V_0)}$$

or

$$\frac{\sqrt{V_0}}{(V_r + V_0)} = \frac{e\left(2\pi n - \dfrac{\pi}{2}\right)}{2m\omega L \times 0.593 \times 10^6}$$

or

$$\frac{V_0}{(V_r + V_0)^2} = \frac{e^2\left(2\pi n - \dfrac{\pi}{2}\right)^2}{4m^2\omega^2 L^2 \times \left(0.593 \times 10^6\right)^2} = \frac{\left(2\pi n - \dfrac{\pi}{2}\right)^2}{8\omega^2 L^2}\frac{e}{m} \tag{11.165}$$

Equation (11.165) reveals that for a given beam voltage V_0 and cycle number n or mode number N, the centre repeller voltage can be determined in terms of frequency.

Differentiating Eq. (11.165) with frequency, we get the following expression:

$$\frac{d}{d\omega}\left\{(V_r + V_0)^2\right\} = \frac{d}{d\omega}\left\{\frac{V_0 8\omega^2 L^2}{\left(2\pi n - \dfrac{\pi}{2}\right)^2}\frac{m}{e}\right\}$$

or

$$\frac{dV_r}{d\omega} = \frac{8V_0 m\omega L^2}{e\left(2\pi n - \dfrac{\pi}{2}\right)^2}\frac{1}{(V_r + V_0)} \tag{11.166}$$

Substituting Eq. (11.165) in Eq. (11.166), we get the following relation:

$$\frac{dV_r}{d\omega} = \frac{8V_0 m\omega L^2}{e\left(2\pi n - \dfrac{\pi}{2}\right)^2}\sqrt{\frac{e\left(2\pi n - \dfrac{\pi}{2}\right)^2}{V_0 8m\omega^2 L^2}} = \frac{L}{\left(2\pi n - \dfrac{\pi}{2}\right)}\sqrt{\frac{8mV_0}{e}}$$

or $\qquad \dfrac{dV_r}{df} = \dfrac{2\pi L}{\left(2\pi n - \dfrac{\pi}{2}\right)}\sqrt{\dfrac{8mV_0}{e}}$ (11.167)

Equation (11.167) establishes the relation between the repeller voltage and the frequency of operation of a reflex klystron.

The output power can be expressed in terms of repeller voltage V_r as follows:

$$P_{ac} = \dfrac{V_0 I_0 X' J_1(X')(V_r + V_0)}{\omega L}\sqrt{\dfrac{e}{2mV_0}}$$ (11.168)

Equation (11.168) can be used to calculate the power output at the centre frequency.

If an electron returns to the cavity a little before the time $\left(n - \frac{1}{4}\right)T$, the current lags behind the fields and an inductive reactance is presented to the circuit. On the other hand, if the electron returns to the cavity a little after the time $\left(n - \frac{1}{4}\right)T$, the current leads the fields and a capacitive is presented to the circuit. The electronic admittance can be written as follows:

$$Y_e = G_e + jB_e = \dfrac{I_0}{V_0}\dfrac{\beta_i^2\theta_0'}{2}\dfrac{2J_1(X')}{X'}e^{-j\left(\frac{\pi}{2}-\theta_0'\right)}$$ (11.169)

Equation (11.169) reveals that the phasor admittance is a function of DC beam admittance, DC transit angle, and the second transit of the electron beam through the cavity gap, and is non-linear. The plot of electronic admittance is shown in Fig. 11.19. Any value of θ_0' for which the spiral lies in the area at the left of the line $-G - jB$ will yield an oscillation, that is,

$$\theta_0' = \left(n - \dfrac{1}{4}\right)2\pi = 2\pi N$$

which is the same as Eq. (11.153).

The equivalent circuit of a reflex klystron is shown in Fig. 11.20; L and C are the energy storage elements of the cavity, G_c is the copper losses in the cavity, G_b is the beam loading conductance, and G_l is the load conductance. The necessary condition for oscillation is as follows:

$$|-G_e| \geq G$$ (11.170)

where $-G_e$ is the negative real part of the electronic admittance, given by Eq. (11.169), and

$$G = G_c + G_b + G_l = 1/R_{sh}$$ (11.171)

where R_{sh} is the effective shunt resistance.

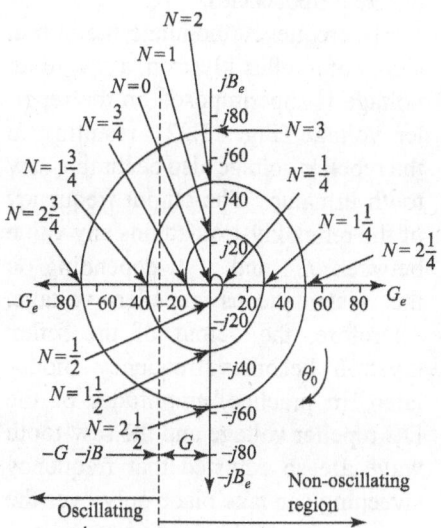

Fig. 11.19 Electronic admittance spiral of reflex klystron

Reflex klystrons are widely used for generation of low power in laboratories to find line and load characteristics. The most common measurable parameters are VSWR and the position of VSWR minimum. In practice, to avoid the use of costly microwave receivers and power meters, microwave signals are modulated with a low-frequency signal. This modulated signal is probed and detected with a crystal detector. The detected microwave signal carries the original amplitude and phase information of the microwave signal, and is measured using low-frequency receivers.

In practice, two basic modulation techniques are used for this purpose: (a) amplitude modulation and (b) frequency modulation. In amplitude modulation, the DC repeller voltage is adjusted at the left edge of the mode power curve, as shown in Fig. 11.21,

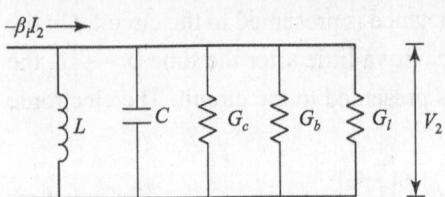

Fig. 11.20   Equivalent circuit of reflex klystron

and a low frequency (usually 1 kHz) square wave voltage is superimposed on it. The amplitude of this square wave is so adjusted that it attains the maximum power point, as shown in Fig. 11.21. The resultant repeller voltage is also a square wave with a frequency equal to the frequency of the modulating signal, and the output power of the reflex klystron can be either 0 or P_{max}, depending on the instantaneous repeller voltage. In amplitude modulation, care should be taken that the negative half-cycle of the modulating amplitude does not enter a higher mode or a part of the same mode, to avoid oscillation at two different frequencies.

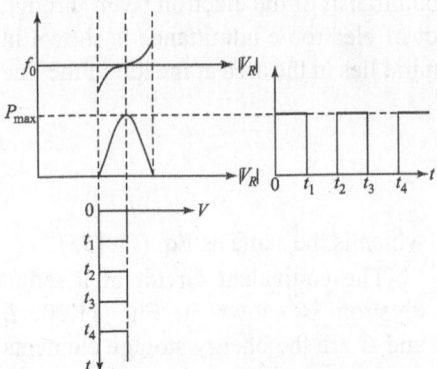

Fig. 11.21   Amplitude modulation

To frequency modulate the output signal of a reflex klystron, a saw tooth voltage is superimposed on the repeller voltage (Fig. 11.22), resulting in the repeller voltage also becoming saw tooth in nature. The output frequency of the reflex klystron attains any value between f_{min} and f_{max}, depending on the instantaneous repeller voltage. Therefore, the output of the reflex klystron becomes frequency modulated. In practice, amplitudes of the DC repeller voltage and the saw-tooth wave are so adjusted that frequency sweeping can take place only over the linear region of the frequency–voltage curve, ensuring almost constant power output and linear frequency sweeping.

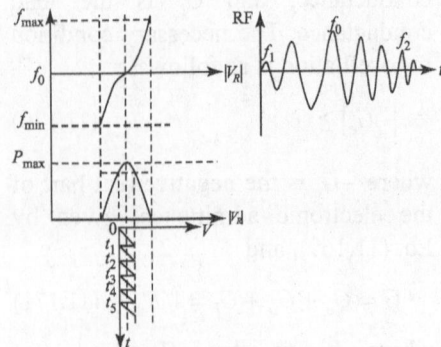

Fig. 11.22   Frequency modulation

In addition to their laboratory use, reflex klystrons are used in local oscillators employed in commercial, military, and airborne Doppler radars as well as in missiles. Such klystrons can produce an output power of 10–500 mW at a frequency range 1–25 GHz, with an η of about 20–30%.

Note Maximum power output of a reflex klystron is less than that of a two-cavity klystron.

EXAMPLE 11.3 A reflex klystron operates under the following conditions: $V_0 = 575$ V, $L = 1$ mm, $R_{sh} = 16$ kΩ, and $f = 10$ GHz. If the device is operating at the peak of $1\frac{3}{4}$ mode, calculate (a) repeller voltage, (b) direct current required to produce a gap voltage of 250 V, and (c) efficiency. Assume that $\beta_0 = 1$, given that at $X' = 1.841$, $J_1(X') = 0.582$.

Solution Given: $V_0 = 575$ V, $L = 1$ mm, $R_{sh} = 16$ kΩ, $f = 10$ GHz, $n = 2$, $\beta_0 = 1$, and $V_2 = 250$ V

(a) Now,

$$\frac{V_0}{(V_r + V_0)^2} = \left(\frac{e}{m}\right)\frac{\left(2n\pi - \frac{\pi}{2}\right)^2}{8\omega^2 L^2} = \frac{1.759 \times 10^{11} \times \left(4\pi - \frac{\pi}{2}\right)^2}{8 \times \left(2\pi \times 10 \times 10^9 \times 10^{-3}\right)^2} = 0.6734 \times 10^{-3}$$

or $\dfrac{575}{(V_r + 575)^2} = 0.6734 \times 10^{-3}$

or $V_r + 575 = \sqrt{\dfrac{575}{0.6734 \times 10^{-3}}} = 924.0765$

or $V_r = 924.0765 - 575 = 349.0765$ V

(b) $I_0 = \dfrac{V_2}{2J_1(X')R_{sh}} = \dfrac{250}{2 \times 0.582 \times 16 \times 10^3} = 0.0134$ A $= 13.4$ mA

(c) The efficiency is expressed as follows:

$$\eta = \frac{2X'J_1(X')}{2n\pi - \frac{\pi}{2}} = \frac{2 \times 1.841 \times 0.582}{4\pi - \frac{\pi}{2}} = 0.1949 = 19.49\%$$

Practice Problem

11.3 A reflex klystron operates under the following conditions: $V_0 = 625$ V, $L = 1$ mm, $R_{sh} = 14$ kΩ, and $f = 8$ GHz. If the device is operating at the peak of $1\frac{3}{4}$ mode, calculate (a) repeller voltage, (b) direct current required to produce a gap voltage of 175 V, and (c) efficiency. Assume that $\beta_0 = 1$, provided that at $X' = 1.841$, $J_1(X') = 0.582$.

[**145.7098 V, 1.07 mA, 19.49%**]

11.7 Helix Travelling-wave Tubes

Klystrons, as described in the last few sections, are resonant structures and hence narrow-band devices. In comparison, TWTs are non-resonant structures and hence wideband devices. A TWT incorporates a slow-wave structure within it, through which a wave propagates with a velocity almost equal to that of the electrons in the beam. As a result, the interaction time between the travelling RF field and the electrons in TWTs is much larger than that in a klystron and lasts over the entire length of the circuit. Due to the interaction between the RF field and electrons, a small amount of velocity modulation is introduced in the electron beam, which later transforms into current modulation. This current modulation, in turn, induces an RF current in the circuit, resulting in amplification. In general, two types of TWTs are available: (a) helix TWT and (b) coupled-cavity TWT. Helix TWTs are widely used in broadband applications, whereas coupled-cavity TWTs are widely used for high-power applications like in radar transmitters. It should be noted that the wave in a TWT is a travelling wave, which it is not true for klystrons. In addition, a coupling effect exists between the cavities of coupled-cavity TWTs, which is absent in case of klystrons.

A basic helix TWT, shown in Fig. 11.23, consists of an electron beam, focused by a constant magnetic field along the electron beam, and a slow-wave structure. A solenoid or permanent magnet is used for focusing the electron beam. The disadvantages of a solenoid are that it is relatively bulky and also consumes power. Therefore, this arrangement is suitable for high-power tubes where power output is more than a few kilowatts. For satellite communication and low-power applications, where the weight as well as power consumption should be minimized, permanent magnets are used. In satellite application, to reduce the bulk, the electron beam is focused using a periodic permanent magnet (PPM). In a PPM, a series of small magnets are located right along the tube, with gaps between successive magnets. The beam is slightly defocused in these gaps, but again refocused by the next magnet, as shown in Fig. 11.24. In a PPM, individual magnets are interconnected.

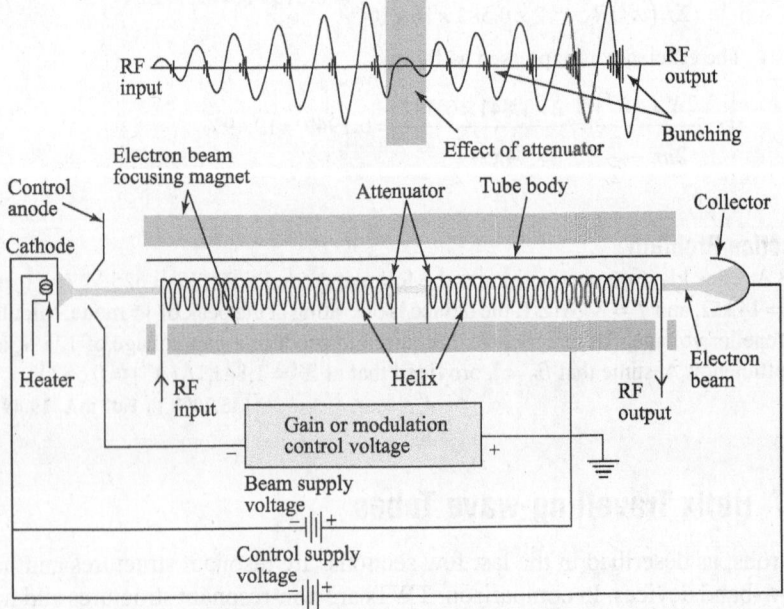

Fig. 11.23 Schematic diagram of helix TWT

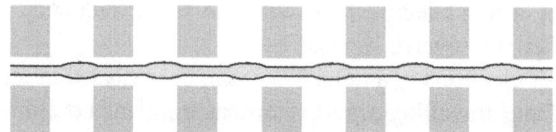

Fig. 11.24 Focusing of electron beam using PPM

The slow-wave structure used in a TWT is either a helix (more commonly used) or a folded back line. The helical slow-wave structure has both advantages and disadvantages. The main advantage is that it is inherently non-resonant and hence a large bandwidth can be obtained, whereas the main disadvantage is that the helical turns in a helix TWT are in close proximity and hence there is a potential chance of oscillation to set up due to feedback at high frequency. The helical structure also limits the use of the tube at high frequencies because the diameter of the helix must be small to allow a high RF field at the centre, which, in turn, presents focusing difficulties, especially under operating conditions where vibration is possible.

In a helix TWT, the applied RF signal propagates around the turn of the helix and results in an electric field at the centre of the helix along the helix axis. The axial electric field propagates with a velocity close to the product of the ratio of the helix pitch to helix circumference and the velocity of the light. In practice, the ratio is so adjusted that the velocity of the axial electric field becomes almost equal to that of electrons in free space. More precisely, DC velocity of electrons is maintained at a slightly higher value than the phase velocity of the travelling wave. When electrons enter the helix tube, an interaction takes place between them and the moving axial electric field. As a result, electrons transfer a net energy to the wave on the helix and a signal amplification takes place.

To understand the energy transfer process, let us assume that three electrons are entering the helix at three different instants. The first electron enters the helix when the RF field is retarding, and hence will move with a slower velocity; the second electron enters the helix when the RF field is zero, and hence will move with an unchanged velocity; and the third electron enters the helix when the RF field is accelerating, and hence will move with a faster velocity. Thus, the first electron will take more time than the second and third electrons and the third electron will take lesser time than the first and second electrons, to reach the collector. Therefore, if the first electron enters the helix, at a time before the others, and the third electron at a time later than the others, the length of the helix can be adjusted so that all the three electrons can reach the collector at the same time, thus forming a bunch at the collector end. Bunching shifts the phase by $\pi/2$. As a result, electrons in the bunch encounter a strong retarding field and energy is delivered to the RF field.

In TWTs, a mismatch exists between the input and output couplers over a wide frequency range, which causes a wave to be reflected from the output coupler and returned to the input. At the input, a part of the reflected signal is re-reflected, which now travels towards the load and is amplified by the tube. The total procedure results in an unwanted oscillation in the circuit, which can be avoided by placing an attenuator near the centre of the helix. Bunched electrons emerging from the attenuator induce a new electric field with the same frequency, which, in turn, results in amplified microwave signals.

The motion of electrons in a helix TWT can be analysed in terms of the axial electric field. If the travelling wave is propagating in the z-direction then the z-component of the electric field can be expressed as follows:

$$E_z = E_1 \sin\left(\omega t - \beta_p z\right) \tag{11.172}$$

where E_1 is the magnitude of the z-component of electric field, and β_p is the axial phase constant and is given by the following relation: $\beta_p = \omega/v_p$ (11.173)

Now, the force on the electrons exerted by the axial electric field can be expressed as follows:

$$m\frac{dv}{dt} = -eE_1 \sin\left(\omega t - \beta_p z\right) \tag{11.174}$$

Velocity of the velocity-modulated electrons can be assumed to be

$$v = v_0 + v_e \cos(\omega_e t + \theta_e) \tag{11.175}$$

where v_0 is the DC electron velocity, v_e is the magnitude of velocity fluctuation in the velocity-modulated electron beam, ω_e is the angular frequency of velocity fluctuation, and θ_e is the phase angle of fluctuation.

Substituting Eq. (11.175) in Eq. (11.174), we get the following relation:

$$m\omega_e v_e \sin\left(\omega_e t + \theta_e\right) = eE_1 \sin\left(\omega t - \beta_p z\right) \tag{11.176}$$

For interaction between electrons and the electric field, velocity of the velocity-modulated electron beam must be approximately equal to the DC electron beam velocity; therefore,

$$v \approx v_0 \tag{11.177}$$

and we can write

$$z = v_0\left(t - t_0\right) \tag{11.178}$$

Substituting Eq. (11.178) in Eq. (11.176), we get the following equation:

$$m\omega_e v_e \sin\left(\omega_e t + \theta_e\right) = eE_1 \sin\left\{\omega t - \beta_p v_0\left(t - t_0\right)\right\}$$

or $\quad m\omega_e v_e \sin\left(\omega_e t + \theta_e\right) = eE_1 \sin\left\{\left(\omega - \beta_p v_0\right)t + \beta_p v_0 t_0\right\}$ (11.179)

Substituting Eq. (11.173) in Eq. (11.179), the following expression is obtained:

$$m\omega_e v_e \sin\left(\omega_e t + \theta_e\right) = eE_1 \sin\left\{\left(\beta_p v_p - \beta_p v_0\right)t + \beta_p v_0 t_0\right\}$$

or $\quad m\omega_e v_e \sin\left(\omega_e t + \theta_e\right) = eE_1 \sin\left\{\beta_p\left(v_p - v_0\right)t + \beta_p v_0 t_0\right\}$ (11.180)

Comparing both sides of Eq. (11.180), we get the following relations:

$$m\omega_e v_e = eE_1$$

or $\quad v_e = \dfrac{eE_1}{m\omega_e}$ (11.181)

$$\omega_e = \beta_p\left(v_p - v_0\right) \tag{11.182}$$

and $\qquad \theta_e = \beta_p v_0 t_0$ (11.183)

Equation (11.181) reveals that the magnitude of velocity fluctuation of the electron beam is directly proportional to the magnitude of the axial electric field.

This analysis neglects the space charge effect. If the space charge effect is considered, then the electron velocity, charge density, current density, and axial electric field can be written as follows:

$$v = v_0 + v_1 e^{j\omega t - \gamma z}$$ (11.184)

$$\rho = \rho_0 + \rho_1 e^{j\omega t - \gamma}$$ (11.185)

$$J = -J_0 + J_1 e^{j\omega t - \gamma z}$$ (11.186)

and $\qquad E_z = E_1 e^{j\omega t - \gamma z}$ (11.187)

where $\gamma = \alpha_e + j\beta_e$ is the propagation constant of the axial waves. In Eq. (11.186), the negative sign added before J_0 ensures that J_0 is positive in the negative z-direction.

For a small signal,

$$J = \rho v = \left(\rho_0 + \rho_1 e^{j\omega t - \gamma z}\right)\left(v_0 + v_1 e^{j\omega t - \gamma z}\right)$$

or $\qquad J = \rho_0 v_0 + \left(\rho_0 v_1 + \rho_1 v_0\right)e^{j\omega t - \gamma z} + \rho_1 v_1 e^{2(j\omega t - \gamma z)}$

or $\qquad J \approx \rho_0 v_0 + \left(\rho_0 v_1 + \rho_1 v_0\right)e^{j\omega t - \gamma z}$ (11.188)

where we have neglected the term $\rho_1 v_1 e^{2(j\omega t - \gamma z)}$ as $\rho_1 v_1 \approx 0$.

Comparing Eqs (11.186) and (11.188), we get the following expressions:

$$-J_0 = \rho_0 v_0$$ (11.189)

and $\qquad J_1 = \rho_0 v_1 + \rho_1 v_0$ (11.190)

The force equation can be written as follows:

$$m\frac{dv}{dt} = -eE_1 e^{j\omega t - \gamma z}$$

or $\qquad m\left(\frac{\partial}{\partial t} + \frac{dz}{dt}\frac{\partial}{\partial z}\right)\left(v_0 + v_1 e^{j\omega t - \gamma z}\right) = -eE_1 e^{j\omega t - \gamma z}$

or $\qquad m\left(j\omega - \gamma v_0\right)v_1 e^{j\omega t - \gamma z} = -eE_1 e^{j\omega t - \gamma z}$

or $\qquad v_1 = \dfrac{-e/m}{j\omega - \gamma v_0} E_1$ (11.191)

where we have substituted $\dfrac{dz}{dt} = v_0$.

From the law of conservation of charges, we can write the following equation:

$$\vec{\nabla}\cdot\vec{J} + \frac{\partial \rho}{\partial t} = 0$$

or $\qquad \dfrac{\partial}{\partial z}\left(-J_0 + J_1 e^{j\omega t - \gamma z}\right) + \dfrac{\partial}{\partial t}\left(\rho_0 + \rho_1 e^{j\omega t - \gamma z}\right) = 0$

or $\qquad -\gamma J_1 e^{j\omega t - \gamma z} + j\omega \rho_1 e^{j\omega t - \gamma z} = 0$

or $\qquad \rho_1 = \dfrac{\gamma J_1}{j\omega} = -j\dfrac{\gamma J_1}{\omega}$ (11.192)

Substituting Eqs (11.191) and (11.192) in Eq. (11.190), we get the following relation:

$$J_1 = \rho_0 v_1 + \rho_1 v_0 = \rho_0 \left(\dfrac{-e/m}{j\omega - \gamma v_0} E_1\right) + \left(-j\dfrac{\gamma J_1}{\omega}\right) v_0$$

or $\qquad \left(1 + jv_0 \dfrac{\gamma}{\omega}\right) J_1 = -\dfrac{e\rho_0 E_1}{m(j\omega - \gamma v_0)}$

or $\qquad \dfrac{\omega + j\gamma v_0}{\omega} J_1 = -\dfrac{e\rho_0 E_1}{m(j\omega - \gamma v_0)}$

or $\qquad J_1 = -\dfrac{e\omega\rho_0 E_1}{m(j\omega - \gamma v_0)(\omega + j\gamma v_0)} = -\dfrac{e\omega\rho_0 E_1}{m(-j)(j\omega - \gamma v_0)^2}$

or $\qquad J_1 = -j\dfrac{e\omega\rho_0 E_1}{m(j\omega - \gamma v_0)^2} = -j\dfrac{e\omega\rho_0 v_0 E_1}{mv_0(j\omega - \gamma v_0)^2}$ (11.193)

Substitution of Eq. (11.189) in Eq. (11.193) results in the following expression:

$$J_1 = j\dfrac{\omega}{v_0}\dfrac{e}{m}\dfrac{J_0}{(j\omega - \gamma v_0)^2} E_1$$ (11.194)

If the magnitude of the axial electric field is uniform over the cross-sectional area of the electron beam then the spatial electric current i will be proportional to the DC current I_0 with the same proportionality constant for J_1 and J_0, and hence can be written as follows:

$$i = j\dfrac{\omega}{v_0}\dfrac{e}{m}\dfrac{I_0}{(j\omega - \gamma v_0)^2} E_1 = j\dfrac{\omega}{v_0}\dfrac{e}{m}\dfrac{I_0}{v_0^2\left(j\dfrac{\omega}{v_0} - \gamma\right)^2} E_1$$

or $\qquad i = j\dfrac{e}{mv_0^2}\dfrac{\beta_e I_0}{(j\beta_e - \gamma)^2} E_1$ (11.195)

where $\qquad \beta_e = \omega/v_0$ (11.196)

is the phase constant of the velocity-modulated electron beam.

Now from Eq. (11.20), we can write the following relation:

$$\dfrac{mv_0^2}{e} = 2V_0$$ (11.197)

Substituting Eq. (11.197) in Eq. (11.195), we get the following expression:

$$i = j\dfrac{\beta_e I_0}{2V_0(j\beta_e - \gamma)^2} E_1$$ (11.198)

Equation (11.198) is known as an *electronic equation*.

Convection current in the electron beam, given in Eq. (11.198), induces an electric field in the slow-wave circuit, which is added to the field already present in the circuit. As a result, the circuit power increases with distance.

To study the coupling, let us assume the slow-wave helix as a distributed lossless transmission line, as shown in Fig. 11.25. Therefore, using the transmission line equations of Chapter 2, we can write the following equations:

$$-\gamma I = -j\omega CV + \gamma i \qquad (11.199)$$

and $\quad -\gamma V = -j\omega LI \qquad (11.200)$

Substituting Eq. (11.199) in Eq. (11.200), we get the following relation:

$$-\gamma V = -j\omega LI = -j\omega L\left(\frac{-j\omega CV + \gamma i}{-\gamma}\right) = \frac{\omega^2 LCV + j\omega\gamma Li}{\gamma}$$

or $\quad \gamma^2 V = -\omega^2 LCV - j\omega\gamma Li \qquad (11.201)$

In the absence of the convection current i, the propagation constant and characteristic impedance of the line will be equal to that of a lossless transmission line, and can be expressed as follows:

$$\gamma_0 = j\omega\sqrt{LC} \qquad (11.202)$$

and $\quad Z_0 = \sqrt{L/C} \qquad (11.203)$

Substituting Eqs (11.202) and (11.203) in Eq. (11.201), we get the following relation:

$$\gamma^2 V = -\omega^2 LCV - j\omega\gamma Li = \gamma_0^2 V - \gamma j\omega\sqrt{LC}\sqrt{\frac{L}{C}}i = \gamma_0^2 V - \gamma\gamma_0 Z_0 i$$

or $\quad \left(\gamma^2 - \gamma_0^2\right)V = -\gamma\gamma_0 Z_0 i$

or $\quad V = -\dfrac{\gamma\gamma_0 Z_0 i}{\gamma^2 - \gamma_0^2} \qquad (11.204)$

Therefore,

$$E_1 = -\frac{\partial V}{\partial z} = -\frac{\partial}{\partial z}\left(-\frac{\gamma\gamma_0 Z_0 i}{\gamma^2 - \gamma_0^2}\right) = -\gamma\frac{\gamma\gamma_0 Z_0 i}{\gamma^2 - \gamma_0^2} = -\frac{\gamma^2\gamma_0 Z_0 i}{\gamma^2 - \gamma_0^2} \qquad (11.205)$$

Equation (11.205) is known as a circuit equation.

Substituting Eq. (11.198) in Eq. (11.205), we get the following relation:

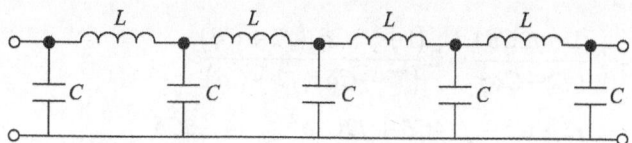

Fig. 11.25 Slow-wave helix as lossless transmission line

$$E_1 = -j \frac{\gamma^2 \gamma_0 Z_0}{\gamma^2 - \gamma_0^2} \frac{\beta_e I_0}{2V_0 (j\beta_e - \gamma)^2} E_1$$

or $\qquad (\gamma^2 - \gamma_0^2)(j\beta_e - \gamma)^2 = -j \dfrac{\gamma^2 \gamma_0 Z_0 \beta_e I_0}{2V_0}$ $\qquad\qquad$ (11.206)

Equation (11.206) is a fourth-order equation of the propagation constant γ and hence reveals that there are four distinct solutions for the propagation constant, each of which corresponds to a mode of the travelling wave in the tube.

The exact solution of Eq. (11.206) can be determined using numerical techniques. However, an approximate solution can be ascertained by equating the DC electron beam velocity to the axial phase velocity of the travelling wave. This assumption corresponds to

$$\gamma_0 = j\beta_e \qquad\qquad (11.207)$$

Substituting Eq. (11.207) in Eq. (11.206), we get the following expression:

$$\left(\gamma^2 - (j\beta_e)^2\right)(j\beta_e - \gamma)^2 = -j \frac{\gamma^2 j\beta_e Z_0 \beta_e I_0}{2V_0}$$

or $\qquad (\gamma - j\beta_e)^3 (\gamma + j\beta_e) = 2C^3 \gamma^2 \beta_e^2$ $\qquad\qquad$ (11.208)

where $\qquad C = \left(\dfrac{I_0 Z_0}{4V_0}\right)^{\frac{1}{3}}$ is the *gain parameter* of the circuit $\qquad\qquad$ (11.209)

Equation (11.208) reveals that there are three forward travelling waves that correspond to $e^{-j\beta_e z}$ and one backward travelling wave that corresponds to $e^{j\beta_e z}$. The propagation constant of the forward travelling waves can be approximated as follows:

$$\gamma = j\beta_e - \beta_e C\delta \qquad\qquad (11.210)$$

where $C\delta \ll 1$. Substituting Eq. (11.210) in Eq. (11.208), we get the following expression:

$$(j\beta_e - \beta_e C\delta - j\beta_e)^3 (j\beta_e - \beta_e C\delta + j\beta_e) = 2C^3 \beta_e^2 (j\beta_e - \beta_e C\delta)^2$$

or $\qquad (-\beta_e C\delta)^3 (j2\beta_e - \beta_e C\delta) = 2C^3 \beta_e^2 \left(-\beta_e^2 - j2\beta_e^2 C\delta + \beta_e^2 C^2 \delta^2\right)$ $\quad$ (11.211)

Since $C\delta \ll 1$, Eq. (11.211) simplifies as follows:

$$-\beta_e^4 C^3 \delta^3 (j2 - C\delta) \approx -2C^3 \beta_e^4 (1 + j2C\delta)$$

or $\qquad \delta^3 (j2 - C\delta) = 2(1 + j2C\delta)$

or $\qquad \delta^3 = 2\dfrac{(1 + j2C\delta)}{(j2 - C\delta)} = 2\dfrac{(1 + j2C\delta)(j2 + C\delta)}{(j2 - C\delta)(j2 + C\delta)}$

or $\qquad \delta^3 = 2\dfrac{j2 + C\delta + j^2 4C\delta + j2C^2\delta^2}{-4 - C^2\delta^2} \approx \dfrac{j4 - 6C\delta}{-4} \approx -j$

or $\qquad \delta = (-j)^{\frac{1}{3}} = \left\{ e^{-j\left(\frac{\pi}{2}+2n\pi\right)} \right\}^{\frac{1}{3}} = e^{-j\frac{\left(\frac{\pi}{2}+2n\pi\right)}{3}}$ where $n = 0, 1,$ and 2 $\qquad$ (11.212)

Therefore,

$$\delta_1 = e^{-j\frac{\pi}{6}} = \cos\left(\frac{\pi}{6}\right) - j\sin\left(\frac{\pi}{6}\right) = \frac{\sqrt{3}}{2} - j\frac{1}{2} \qquad (11.213)$$

$$\delta_2 = e^{-j\frac{5\pi}{6}} = \cos\left(\frac{5\pi}{6}\right) - j\sin\left(\frac{5\pi}{6}\right) = -\frac{\sqrt{3}}{2} - j\frac{1}{2} \qquad (11.214)$$

and $\qquad \delta_3 = e^{-j\frac{3\pi}{2}} = \cos\left(\frac{3\pi}{2}\right) - j\sin\left(\frac{3\pi}{2}\right) = j \qquad (11.215)$

To find the fourth root, we need to substitute

$$\gamma = -j\beta_e - \beta_e C\delta_4 \qquad (11.216)$$

in Eq. (11.208), which gives the following equation:

$$(-j2\beta_e - \beta_e C\delta_4)^3 (-\beta_e C\delta_4) = 2C^3 (-j\beta_e - \beta_e C\delta_4)^2 \beta_e^2$$

or $\qquad \delta_4 (j2\beta_e + \beta_e C\delta_4)^3 = 2C^2 (-j\beta_e - \beta_e C\delta_4)^2 \beta_e$

or $\qquad \delta_4 (j2 + C\delta_4)^3 = 2C^2 (j + C\delta_4)^2$

or $\qquad \delta_4 \left(-j8 - 12C\delta_4 + j6C^2\delta_4^2 + C^3\delta_4^3\right) = 2C^2 \left(-1 + j2C\delta_4 + C^2\delta_4^2\right)$

or $\qquad \delta_4 (j8 + 12C\delta_4) \approx 2C^2 (1 - j2C\delta_4)$ or $j8\delta_4 \approx 2C^2$

or $\qquad \delta_4 = -jC^2/4 \qquad (11.217)$

In deriving Eq. (11.217), we have again assumed that $C\delta \ll 1.$
$\qquad$ Therefore, the propagation constants can be written as follows:

$$\gamma_1 = j\beta_e - \beta_e C\delta_1 = j\beta_e - \beta_e C\left(\frac{\sqrt{3}}{2} - j\frac{1}{2}\right)$$

or $\qquad \gamma_1 = j\beta_e - \beta_e C\frac{\sqrt{3}}{2} + j\frac{1}{2}\beta_e C = -\beta_e C\frac{\sqrt{3}}{2} + j\beta_e\left(1 + \frac{C}{2}\right) \qquad (11.218)$

$$\gamma_2 = j\beta_e - \beta_e C\delta_2 = j\beta_e - \beta_e C\left(-\frac{\sqrt{3}}{2} - j\frac{1}{2}\right)$$

or $\qquad \gamma_2 = j\beta_e + \beta_e C\frac{\sqrt{3}}{2} + j\frac{1}{2}\beta_e C = \beta_e C\frac{\sqrt{3}}{2} + j\beta_e\left(1 + \frac{C}{2}\right) \qquad (11.219)$

$$\gamma_3 = j\beta_e - \beta_e C\delta_3 = j\beta_e - \beta_e Cj = j\beta_e (1 - C) \qquad (11.220)$$

and $\quad \gamma_4 = -j\beta_e - \beta_e C\delta_4 = -j\beta_e - \beta_e C\left(-j\dfrac{C^2}{4}\right) = -j\beta_e\left(1 - \dfrac{C^3}{4}\right)$ (11.221)

These equations reveal the following facts:

1. γ_1 is a forward wave and its amplitude grows exponentially with distance. This wave propagates at a phase velocity slightly lower than the electron beam velocity, and energy flows from the electron beam to the wave.
2. γ_2 is a forward wave and its amplitude decays exponentially with distance. This wave propagates at the same phase velocity as that of the growing wave and energy flows from the wave to the electron beam.
3. γ_3 is a forward wave and its amplitude remains constant with distance. This wave propagates at a phase velocity slightly higher than the electron beam velocity, and no energy transfer takes place between the electron beam and the wave.
4. γ_4 is a backward wave and its amplitude remains constant with distance. This wave propagates at a phase velocity slightly higher than the electron beam velocity, and no energy transfer takes place between the electron beam and the wave.

To calculate the gain of the amplifier, we will assume that the load is perfectly matched and there is no backward wave. Therefore, the total circuit voltage can be written as follows:

$$V(z) = V_1 e^{-\gamma_1 z} + V_2 e^{-\gamma_2 z} + V_3 e^{-\gamma_3 z} = \sum_{n=1}^{3} V_n e^{-\gamma_n z}$$ (11.222)

From Eq. (11.198), the input current can be written as follows:

$$i = j\frac{\beta_e I_0}{2V_0(j\beta_e - \gamma)^2} E_1 = j\frac{\beta_e I_0}{2V_0(j\beta_e - \gamma)^2}\left(-\frac{\partial V}{\partial z}\right) = j\frac{\beta_e I_0 \gamma V}{2V_0(j\beta_e - \gamma)^2}$$ (11.223)

Substituting Eq. (11.210) in Eq. (11.223), we get the following relation:

$$i = j\frac{\beta_e I_0(j\beta_e - \beta_e C\delta)V}{2V_0(j\beta_e - j\beta_e + \beta_e C\delta)^2} = j^2\frac{\beta_e^2 I_0(1 + jC\delta)V}{2V_0\beta_e^2 C^2\delta^2}$$

or $\quad i = -\dfrac{I_0(1 + jC\delta)V}{2V_0 C^2\delta^2} \approx -\dfrac{I_0 V}{2V_0 C^2\delta^2}$ (11.224)

Further substituting Eq. (11.222) in Eq. (11.224), we get the following expression:

$$i(z) = -\sum_{n=1}^{3} \frac{I_0}{2V_0 C^2}\frac{V_n}{\delta_n^2} e^{-\gamma_n z}$$ (11.225)

From Eq. (11.191), the input fluctuating component of the total wave velocity can be written as follows:

$$v_1 = \frac{-e/m}{j\omega - \gamma v_0} E_1 = \frac{-e/m}{j\omega - \gamma v_0}\left(-\frac{\partial V}{\partial z}\right) = -\frac{e}{m}\frac{\gamma V}{j\omega - \gamma v_0}$$ (11.226)

Substituting Eqs (11.20) and (11.210) in Eq. (11.226), the following relation is obtained:

$$v_1 = -\frac{v_0^2}{2V_0}\frac{(j\beta_e - \beta_e C\delta)V}{j\omega - (j\beta_e - \beta_e C\delta)v_0} = -\frac{v_0^2}{2V_0}\frac{(j\beta_e - \beta_e C\delta)V}{j\omega - j\beta_e v_0 + \beta_e C\delta v_0} \qquad (11.227)$$

Further, substituting Eq. (11.196) in Eq. (11.227), we get the following expression:

$$v_1 = -\frac{v_0^2}{2V_0}\frac{(j\beta_e - \beta_e C\delta)V}{j\omega - j\omega + \beta_e C\delta v_0} = -\frac{v_0^2}{2V_0}\frac{(j\beta_e - \beta_e C\delta)V}{\beta_e C\delta v_0}$$

or

$$v_1 = -\frac{v_0}{2V_0}\frac{(j - C\delta)V}{C\delta} \approx -j\frac{v_0}{2V_0}\frac{V}{C\delta} \qquad (11.228)$$

Finally, substitution of Eq. (11.222) in Eq. (11.228) gives the following relation:

$$v_1(z) = -\sum_{n=1}^{3} j\frac{v_0}{2V_0 C}\frac{V_n}{\delta_n}e^{-\gamma_n z} \qquad (11.229)$$

If we assume that the input is at $z = 0$, then the input voltage, input current, and fluctuating component of the velocity of the total wave can be written as follows:

$$V(0) = V_1 + V_2 + V_3 \qquad (11.230)$$

$$i(0) = -\frac{I_0}{2V_0 C^2}\left[\frac{V_1}{\delta_1^2} + \frac{V_2}{\delta_2^2} + \frac{V_3}{\delta_3^2}\right] \qquad (11.231)$$

and

$$v_1(0) = -j\frac{v_0}{2V_0 C}\left[\frac{V_1}{\delta_1} + \frac{V_2}{\delta_2} + \frac{V_3}{\delta_3}\right] \qquad (11.232)$$

Solving Eqs (11.230)–(11.232) using $i(0) = 0$ and $v_1(0) = 0$ we get the following relation:

$$V_1 = V_2 = V_3 = \frac{V(0)}{3} \qquad (11.233)$$

Since the growing wave increases exponentially with distance, at a sufficiently large distance $z = L$, it will dominate the other modes. Therefore, neglecting the other modes and substituting Eqs (11.218) and (11.233) in Eq. (11.222), the following equation is obtained:

$$V(L) = V_1 e^{-\gamma_1 L} = \frac{V(0)}{3}\exp\left[\frac{\sqrt{3}}{2}\beta_e CL\right]\exp\left\{-j\beta_e\left(1+\frac{C}{2}\right)L\right\} \qquad (11.234)$$

Now substituting $\beta_e L = 2\pi L_\lambda$ (11.235)
where L_λ is the circuit length in an electronic wavelength, that is,

$$L_\lambda = L/\lambda_e \qquad (11.236)$$

and

$$\beta_e = 2\pi/\lambda_e \qquad (11.237)$$

in Eq. (11.234), we get the following expression:

$$V(L) = \frac{V(0)}{3}\exp\left(\sqrt{3}\pi L_\lambda C\right)\exp\left\{j2\pi L_\lambda\left(1+\frac{C}{2}\right)\right\} \qquad (11.238)$$

The amplitude of the output voltage can be written as follows:

$$V(L) = \frac{|V(0)|}{3} \exp\left(\sqrt{3}\pi L_\lambda C\right) \tag{11.239}$$

Therefore, the output power gain is as follows:

$$A_p = 10\log_{10}\left|\frac{V(L)}{V(0)}\right|^2 = 20\log_{10}\left|\frac{\exp\left(\sqrt{3}\pi L_\lambda C\right)}{3}\right|$$

or

$$A_p = \frac{20}{2.3}\ln\left|\frac{\exp\left(\sqrt{3}\pi L_\lambda C\right)}{3}\right| = 8.6957\left\{\ln\left(\frac{1}{3}\right) + \sqrt{3}\pi L_\lambda C\right\}$$

or

$$A_p = -9.54 + 47.3 L_\lambda C \text{ dB} \tag{11.240}$$

Equation (11.240) reveals an initial loss of 9.54 dB at the circuit input, which results from the fact that the growing wave voltage is only one-third of the total input voltage. Equation (11.240) also reveals that the net power gain is proportional to the electrical length of the slow-wave structure and gain parameter (C) of the circuit.

The peak output power of a single-helix TWT is limited to about 3 kW, depending on the current-handling capability of the helix structure. For a low input power, the small signal gain of the helix TWT is almost constant. If the input power is increased, the output power does not increase in proportion, but instead, attains a maximum value and then starts decreasing, as shown in Fig. 11.26(a). The point at which the maximum output power is obtained is called a saturation point, and the corresponding gain is called saturation gain. In practice, the output remains in saturation over a range of input voltages. This range defines the overdrive capability of a TWT. A plot of gain as a function of input power is shown in Fig. 11.26(b).

From the performance point of view, there are two types of TWTs: (a) low-power low noise TWT and (b) high-power fairly noise TWT. Noise level of the former type ranges from 4 to 8 dB over the frequency range of 0.5–16 GHz. The typical output power for this tube ranges from 5 to 30 mW. Low-power CW TWTs can also provide output power in the range 5–25 mW for frequencies up to 40 GHz, with a noise level of around 25 dB. On the other hand, high-power TWTs can operate over the frequency range 0.5–95 GHz and produce a CW output power of 250 kW at 3 GHz or pulsed output power of 10 MW at 3 GHz. High-power TWTs are generally coupled cavity in nature and will be discussed in the next section.

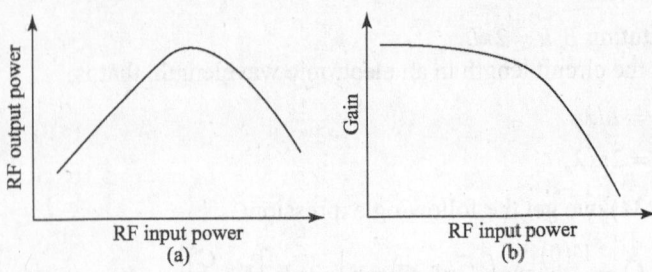

Fig. 11.26 Helix TWT (a) RF output power characteristics (b) Gain characteristics

Low-noise TWTs are widely used as RF amplifiers in broadband microwave receivers and repeater amplifiers, whereas CW high-power tubes are used in tropospheric scatter links due to their high power and large bandwidth. They can also be used in radar for jamming purpose. Pulsed TWTs are used in airborne and ship-borne radars, as well as in high-power ground-based radars. Due to their long tube life, TWTs are also used as power output tubes in communication satellites.

EXAMPLE 11.4 A typical helix TWT operates with the following conditions: beam voltage = 2.5 kV, beam current = 25 mA, helix characteristic impedance = 15 Ω, circuit length = 60, and operating frequency = 9 GHz. Calculate (a) gain parameter, (b) output power gain, and (c) all four propagation constants.

Solution Given: $V_0 = 2.5$ kV, $I_0 = 25$ mA, $Z_0 = 15 \Omega$, $L_\lambda = 60$, and $f = 9$ GHz

(a) $C = \left(\dfrac{I_0 Z_0}{4V_0}\right)^{\frac{1}{3}} = \left(\dfrac{25 \times 10^{-3} \times 15}{4 \times 25 \times 10^3}\right)^{\frac{1}{3}} = 0.0155$

(b) $A_p = -9.54 + 47.3 L_\lambda C = -9.54 + (47.3 \times 60 \times 0.0155) = 34.449$ dB

(c) $v_0 = 0.593 \times 10^6 \sqrt{V_0} = 0.593 \times 10^6 \times \sqrt{2.5 \times 10^3} = 2.965 \times 10^7$ m/s

$\beta_e = \dfrac{\omega}{v_0} = \dfrac{2\pi \times 9 \times 10^9}{2.965 \times 10^7} = 1.9072 \times 10^3$ rad/m

$\gamma_1 = -\beta_e C \dfrac{\sqrt{3}}{2} + j\beta_e \left(1 + \dfrac{C}{2}\right)$

or $\gamma_1 = -1.9072 \times 10^3 \times 0.0155 \times \dfrac{\sqrt{3}}{2} + j1.9072 \times 10^3 \left(1 + \dfrac{0.0155}{2}\right)$

or $\gamma_1 = -25.6011 + j1921.9808$

$\gamma_2 = \beta_e C \dfrac{\sqrt{3}}{2} + j\beta_e \left(1 + \dfrac{C}{2}\right)$

or $\gamma_2 = 1.9072 \times 10^3 \times 0.0155 \times \dfrac{\sqrt{3}}{2} + j1.9072 \times 10^3 \left(1 + \dfrac{0.0155}{2}\right)$

or $\gamma_2 = 25.6011 + j1921.9808$

$\gamma_3 = j\beta_e (1 - C) = j1.9072 \times 10^3 \times (1 - 0.0155) = j1877.6384$

$\gamma_4 = -j\beta_e \left(1 - \dfrac{C^3}{4}\right) = -j1.9072 \times 10^3 \times \left(1 - \dfrac{0.0155^3}{4}\right) = -j1907.1982$

Practice Problem

11.4 A typical helix TWT operates with the following conditions: beam voltage = 3.2 kV, beam current = 30 mA, helix characteristic impedance = 12 Ω, circuit length = 55, and operating frequency 10 GHz. Calculate (a) gain parameter, (b) output power gain, and (c) all four propagation constants.

$[0.0141, 27.1411 \,dB, -22.8723 + j1886.3054, 22.8723 + j1886.3054, j1846.6893, -j1873.0987]$

11.8 Coupled-cavity Travelling-wave Tubes

It has already been mentioned that in a helix TWT, the peak power is limited to a few kilowatts due to the limited current-handling capability of the helix structure. As the power level increases, the helix structure gets overheated, causing the helix geometry to wrap. Although the problem can partially be overcome by increasing the diameter of the helix wire, after a certain level it becomes impossible to obtain the required helix pitch using a thick wire. To overcome this problem, coupled-cavity TWTs were developed.

A coupled-cavity TWT consists of a series of coupled cavities, arranged axially along the beam to form a helical waveguide, as shown in Fig. 11.27. Such a structure has a serious drawback. The non-linear dispersion in a helical waveguide is very high and therefore such TWTs are narrowband, though larger than klystrons. In a couple-cavity TWT, amplification can occur via velocity modulation.

The term *coupled cavity* is used because a coupling is provided between the adjacent cavities by a long slot. The slot strongly couples the magnetic components of the field in adjacent cavities in such a manner that the pass band of the structure becomes mainly a function of this one variable. In practice, two types of coupled-cavity circuits can be used in TWTs. The first cavity structure consists of fundamentally forward wave circuits, whereas the second structure consists of first space harmonic circuits. The first cavity structure operates with fundamental space harmonics, and is used for pulse applications requiring at least 500 kW peak power, and exhibits negative inductive mutual coupling between the cavities. On the other hand, the second cavity structure operates with first spatial harmonics, and is used for pulse or continuous wave applications with output power from one to several hundred kilowatts and exhibits positive mutual coupling between the cavities.

From the theory of filter design it is known that any repetitive series or parallel lumped LC elements constitute a propagating filter-type circuit. Since cavities can be represented as an LC network, as shown in Fig. 11.28, coupled cavities can also be represented by such a circuit. In practice, cavities in coupled-cavity TWTs are usually kept over-coupled to obtain band-pass filter-type characteristics.

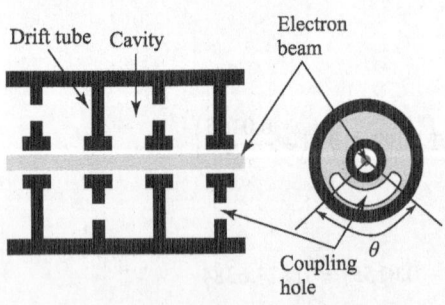

Fig. 11.27 Structure of coupled-cavity TWT

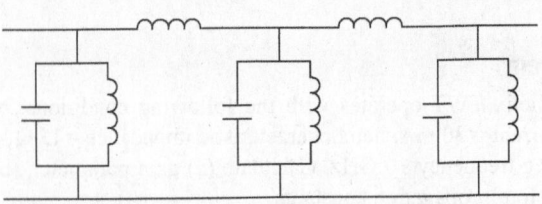

Fig. 11.28 Equivalent circuit of slot coupled cavity

When the slot angle θ (Fig. 11.27) is larger than 180°, the pass band is close to its practical limits. A drift tube is formed by the re-entrant part of the cavity, just like a klystron. During the interaction of the electron beam and the RF field, a phase change, as a function of frequency, occurs between the cavities. If the mutual inductive coupling of the slot is negative then an increasing phase characteristic is obtained. On the other hand, a decreasing phase characteristic is obtained if the mutual inductive coupling of the slot is positive.

To ensure stability, coupled-cavity TWTs are constructed with a limited amount of gain per section of cavities. Later, such cavity sections are cascaded to achieve higher tube gains. By this method, a stable gain of greater than 60 dB can be achieved over a 30% bandwidth. To reduce gain variation with frequency, each cavity section is terminated either by a matched load or by an input or output line.

The overall efficiency of a coupled-cavity TWT is determined by the amount of input energy converted into RF power and that dissipated in the collector. In practice, in a coupled-cavity TWT, an overall efficiency of 20–55% can be achieved.

The beam current and beam power of a coupled-cavity TWT can be expressed as follows:

$$I_{beam} = KV_0^{3/2} \qquad (11.241)$$

and

$$P_{beam} = I_{beam}V_0 = KV_0^{5/2} \qquad (11.242)$$

where V_0 is the beam voltage.

11.9 High-power Gridded Control Travelling-wave Tubes

High-power gridded control TWTs utilize a shadow grid technique to control the electron beam, and hence the device is also called a gridded TWT. In general, such tubes have four main sections: (a) electron gun structure, (b) slow-wave structure, (c) magnetic circuit, and (d) collector structure. Specifically, a gridded TWT consists of an electron emitter, a shadow grid, a control grid, a modulating anode, a coupled-cavity circuit, a solenoid magnetic circuit, and a collector depression structure.

In general, after electrons are emitted from the cathode, the electron beam has a tendency to spread due to its internal repulsions. This, in turn, reduces the interaction between the RF field and the electron beam, because for an effective interaction cross-section of the electron beam must be small. More precisely, the diameter of the electron beam should be smaller than 1/10th of the wavelength of the RF signal. To achieve this, the shadow grid technique is used, as shown in Fig. 11.29.

As mentioned before, the electron emitter of a gridded TWT has two control electrodes: (a) a shadow grid near the cathode and (b) a control grid. The shadow grid is kept at cathode potential and interposed between the cathode and the control grid, whereas the control grid is kept at a positive potential and placed slightly away from the cathode. The shadow grid suppresses electron emission from those portions of cathode that would give rise to interception at the control grid. The

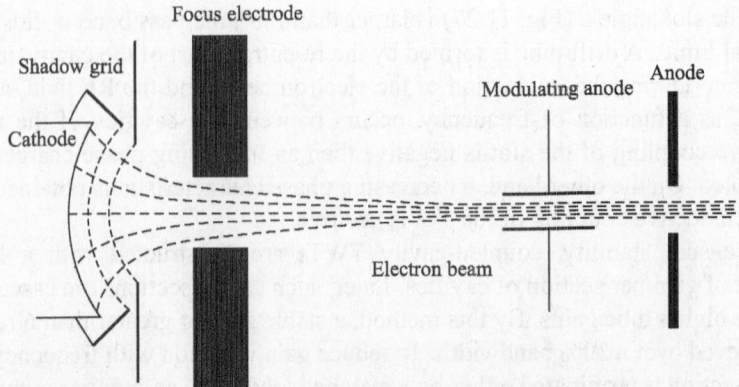

Fig. 11.29 Electron emitter with control electrodes for gridded
high-power TWTs

function of the control grid is to control the electron beam. These grids can control greater beam power than would otherwise be possible.

To reduce modulator power requirements for a high-pulsed output power and to eliminate the voltage pulsing through the lower unstable beam voltage, an anode modulation technique is frequently used in gridded TWTs. In this technique, a modulator applies a highly regulated positive grid voltage to turn on the electron beam or an unregulated negative grid voltage to turn it off. Alternatively, we can say that the anode modulator acts as a pulse switch.

In practice, the anode of an electron gun is kept at a higher potential than the slow-wave structure. This is required to prevent positive ions, formed by the electron beam in the slow-wave structure region, from reaching the cathode.

After passing the outer cavity, the electron beam strikes a collector electrode and is dissipated as heat. A cooling mechanism is provided at the collector to absorb the heat by thermal conduction to a cooler surface. In practice, a gridded TWT has a separate collector instead of a part of the slow wave structure being used as collector. This is due to the following two advantages:

1. The collector can be made as large as possible in order to collect the electron beam at a lower density. This arrangement minimizes localized heating. If the collector was a part of the slow-wave structure, its size would be limited by the maximum gap capacitance due to high-frequency operation.
2. The separate collector considerably reduces its potential below the beam voltage in the RF interaction region. This, in turn, reduces the power dissipated in the collector and hence increases the overall efficiency of the device.

The efficiency of a gridded TWT is defined as the ratio of the RF output power to the product of cathode voltage (or beam voltage) and cathode current (or beam current). Alternatively, it can also be expressed as the product of electronic efficiency and circuit efficiency. Electronic efficiency is expressed as the percentage of the DC or pulsed input power that is converted into the RF power on the slow-wave structure, whereas circuit efficiency is expressed as the percentage of the DC input power that is delivered to the external load. For a gridded TWT, the collector voltage is normally about 40% of the cathode voltage, and the overall efficiency of conversion of DC power to RF power is almost twice the electronic efficiency. Under this condition, the tube is operating with a collector voltage depression.

11.10 O-type Backward Wave Oscillators

The operating principle of O-type BWOs is the same as that of TWTs, except that in BWO the energy is extracted from the electron beam using a backward wave that travels from the collector end towards the electron gun. A typical BWO is constructed from a folded transmission line or waveguide that winds back and forth across the path of the electron beam, as shown in Fig. 11.30. Here, the folded transmission line or waveguide performs the same task as that of the helix in a TWT.

The operation of a BWO can be understood using a series of feedback loops, as shown in Fig. 11.31. Each of these regenerative loops can act as an amplifier or oscillator, and is designed in such a way that the total phase shift around the loop is a complete cycle. Conventionally, the forward wave or μ-circuit consists of a transmission line section and the backward wave or β-circuit consists of a unilateral amplifier. If the gain of these amplifiers is sufficiently high, then the transmission line losses can be overcome and positive feedback takes place. As a result, the $\mu-\beta$ loop oscillates at a frequency for which the total phase delay is one cycle. In a BWO, unilateral amplification is provided by the wave along the electron beam, and the μ-circuit corresponds to the backward wave on the helix. The electric field at a point on the helix, near the electron gun, velocity modulates the electron beam which moves towards the collector and forms an electron bunch. This is somehow similar to normal TWTs. At the collector end, the electron bunch supplies energy to the helix wave that travels towards the electron gun. The amplified wave comes back to the original position (i.e., electron gun end) and completes the feedback loop. If there is sufficient current in the electron beam, the feedback mechanism produces oscillations at a frequency for which the electron velocity is in step with the backward wave of the helix.

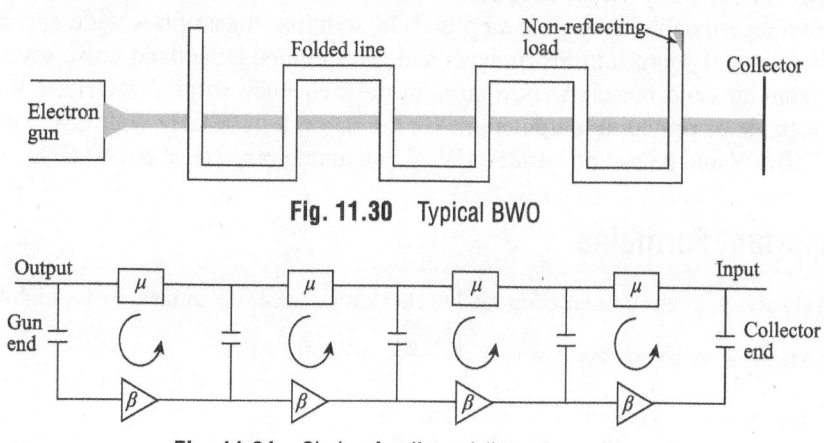

Fig. 11.30 Typical BWO

Fig. 11.31 Chain of unilateral (beam) amplifiers

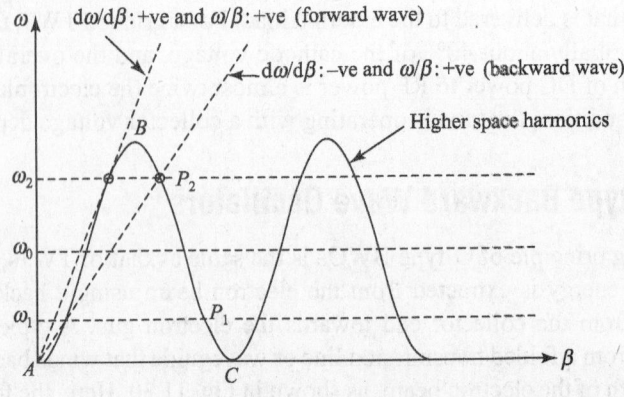

Fig. 11.32 ω–β diagram

A BWO is a self-oscillating TWT capable of delivering microwave signals over a wide frequency range. To establish this claim lets us recall that the helix and other periodic slow-wave structures can support a slow electromagnetic wave with phase velocity in one direction and group velocity in the opposite direction. Figure 11.32 shows a typical ω–β diagram of a helix structure. Both the phase velocity (ω/β) and the group velocity $(d\omega/d\beta)$ are positive in the portion AB of the ω–β curve, which corresponds to a forward wave and is used for the interaction in a conventional TWT amplifier. In the portion BC of the ω–β curve, the phase velocity is positive and group velocity is negative; this behaviour characterizes a backward wave and is used for the interaction in a BWO. The figure reveals that the frequency of interaction in a BWO can be varied from ω_1 to ω_2 when the electron beam velocity is varied in a way that the interaction point shifts from P_1 to P_2. This property makes a BWO tunable.

Since the oscillation frequency of a BWO can be changed by varying the beam voltage, it can be used as a sweep oscillator. In addition, in a BWO, the amplitude of oscillation can be decreased continuously to zero by changing the beam current. This feature of a BWO can be used to provide amplitude modulation or a constant output signal. A BWO operating below the oscillation threshold level can be used as a voltage tunable band-pass amplifier. In addition, it also finds wide applications as signal sources in instruments and transmitters, broadband noise sources for jamming, and noiseless oscillators in the frequency range 3–9 GHz. BWOs are capable of providing output signal in the range 1–1000 GHz with CW power 10–150 mW and pulsed power 250 kW, with a tuning range of about 40 GHz.

Important Formulae

- Velocity of an electron entering the buncher cavity under the influence of a high DC voltage V_0 can be expressed as $v_0 = \sqrt{\dfrac{2eV_0}{m}} = 0.593\sqrt{V_0} \times 10^6$.

- For a two-cavity klystron, $\dfrac{\Delta L}{L_{\text{optimum}}} = 0.85$.

- The DC phase constant of an electron beam can be expressed as $\beta_e = \omega/v_0$.
- For a helix TWT, the angular frequency of velocity fluctuation can be expressed as $\omega_e = \beta_p(v_p - v_0)$.
- For a helix TWT, the phase angle of velocity fluctuation $\theta_e = \beta_p v_0 t_0$.
- For a TWT, the gain parameter can be expressed as $C = \left(\dfrac{I_0 Z_0}{4 V_0}\right)^{\frac{1}{3}}$.
- The power gain of a TWT is $A_p = -9.5532 + 47.3165 L_\lambda C$ dB.

Exercises

Objective-type Questions

11.1 The major advantage of a TWT over a klystron lies in its
(a) compactness
(b) higher frequency
(c) higher gain
(d) all of these

11.2 In a reflex klystron, the velocity modulation occurs
(a) in the resonator gap
(b) near the reflector
(c) near the accelerating grid
(d) in between the resonator and the reflector

11.3 A backward wave oscillator is based on a
(a) travelling-wave tube
(b) coaxial magnetron
(c) rising sun magnetron
(d) crossed field amplifier

11.4 A multi-cavity klystron
(a) is an attenuator
(b) is not a good low-level amplifier because of noise
(c) is not suited for pulse operation
(d) has a higher repeller voltage to ensure small transit time

11.5 A reflex klystron is capable of generating such high frequencies as
(a) 0.1 GHz
(b) 1 GHz
(c) 10 GHz
(d) 100 GHz

11.6 The secondary cavity in a two-cavity klystron is called the
(a) buncher cavity
(b) catcher cavity
(c) velocity modulation cavity
(d) all of these

11.7 Klystron operates on the principle of
(a) velocity modulation
(b) frequency modulation
(c) phase modulation
(d) amplitude modulation

11.8 A reflex klystron is often preferred to a two-cavity klystron oscillator because the former
(a) is more efficient
(b) has a higher number of modes
(c) is easier to tune
(d) all of these

11.9 In a klystron, velocity modulation takes place in the
(a) drift tube
(b) collector
(c) catcher cavity resonator
(d) input cavity resonator

11.10 Which of the following microwave tubes has a repeller electrode?
   (a) Multi-cavity klystron
   (b) Coupled-cavity TWT
   (c) Helix travelling-wave tubes
   (d) Reflex klystron

11.11 A backward wave oscillator is based on a
   (a) coaxial magnetron
   (b) travelling-wave tube
   (c) rising sun magnetron
   (d) crossed field amplifier

11.12 The microwave amplifier characterized by a very high bandwidth is
   (a) a multi-cavity klystron
   (b) a travelling-wave tube
   (c) a reflex klystron
   (d) none of these

11.13 Which of the following is not a TWT slow-wave structure?
   (a) A helix
   (b) A ring bar
   (c) A periodic permanent magnet
   (d) A coupled cavity

11.14 TWT uses a helix to
   (a) reduce noise
   (b) ensure broad band operation
   (c) increase the efficiency
   (d) reduce the axial velocity of RF field

11.15 TWT is basically
   (a) an oscillator
   (b) a tuned amplifier
   (c) a low-gain amplifier
   (d) a wideband amplifier

11.16 Which of the following is basically a wideband amplifier?
   (a) A magnetron
   (b) A travelling-wave tube
   (c) A reflex klystron
   (d) A multi-cavity klystron

11.17 Which of the following is unlikely to be used as a pulsed device?
   (a) A travelling-wave tube
   (b) A crossed field amplifier
   (c) A multi-cavity klystron
   (d) A backward wave oscillator

11.18 Power gain of a klystron can be increased by
   (a) adding more cavities
   (b) using positive feedback
   (c) using negative feedback
   (d) using a reflex klystron

11.19 In a reflex klystron, electrons can be prevented from reaching the repeller terminal by
   (a) connecting a resistor to the cathode
   (b) using a protector diode
   (c) both (a) and (b)
   (d) none of these

11.20 The output power of a reflex klystron is
   (a) equal to that of a two-cavity klystron
   (b) less than that of a two-cavity klystron
   (c) more than that of a two-cavity klystron
   (d) more than that of a TWT

Review Questions

11.1 Describe the high-frequency operation limitations of conventional vacuum tubes.
11.2 With a suitable figure, describe the basic construction and operation of a two-cavity klystron.

11.3 Show that maximum achievable efficiency of a two-cavity klystron is 58%.

11.4 Write a short note on *multi-cavity klystron*.

11.5 With a suitable figure, describe the construction and operation of a two-cavity klystron oscillator.

11.6 With a suitable figure, describe the basic operation of a reflex klystron. Show that the maximum efficiency of such a tube is 22.70%.

11.7 How can the output of a reflex klystron be (a) amplitude and (b) frequency modulated?

11.8 With a suitable figure, describe the basic operation of a helix travelling-wave tube.

11.9 Write a short note on (a) coupled-cavity travelling-wave tubes, (b) gridded control travelling-wave tubes, and (c) O-type backward wave oscillators.

Problems

11.1 A typical two-cavity klystron has the following parameters: beam voltage = 18 kV, beam current = 1.75 A, frequency = 9 GHz, input and output beam coupling coefficient = 1, DC electron beam current density = 10^{-6} C/m², signal voltage = 10 V (rms), shunt resistance of the cavity = 12 kΩ, and total shunt resistance including load = 30 kΩ. Calculate (a) plasma frequency, (b) reduced plasma frequency for $R = 0.4$, (c) induced current in the output cavity, (d) induced voltage in the output cavity, (e) output power delivered to the load, (f) power gain, and (g) electronic efficiency. *[Hint: Use Eqs (11.99), (11.98), (11.122), (11.123), (11.124), (11.125), and (11.126)]*

11.2 A typical four-cavity klystron has the following parameters—beam voltage = 9 kV, beam current = 0.75 A, frequency = 4.5 GHz, input and output beam coupling coefficient = 1, DC electron beam current density = 4×10^{-5} C/m², signal voltage = 2.5 V (rms), shunt resistance of the cavity = 12 kΩ, and total shunt resistance including load = 15 kΩ. Calculate (a) plasma frequency, (b) reduced plasma frequency for $R = 0.7$, (c) induced current in the output cavity, (d) induced voltage in the output cavity, and (e) output power delivered to the load. *[Hint: Use Eqs (11.99), (11.98), (11.132), (11.133), and (11.134)]*

11.3 A reflex klystron is operating at 3 GHz with a repeller voltage of 2.5 kV and an accelerating voltage of 600 V. If the repeller voltage varies by 2.5%, then calculate the change in frequency. Assume that $n = 1$ and $L = 2.5$ cm. *[Hint: Use Eq. (11.167)]*

11.4 A typical two-cavity klystron amplifier operates with the following parameters: voltage gain = 18 dB, input power = 7.5 mW, shunt resistance of input cavity = 35 kΩ, shunt resistance of output cavity = 45 kΩ, and load resistance = 45 kΩ. Calculate (a) input rms voltage and (b) output rms voltage and power delivered to the load. *[Hint: Use the relations $V_{in} = \sqrt{P_{in} R_{sh,in}}$, $V_{out} = V_{in} 10^{A_v/20}$, and $P_{out} = V_{out}^2 / R_{sh,out}$]*

11.5 A typical reflex klystron operates with the following operating parameters: beam voltage = 500 V, beam current = 25 mA, and signal voltage = 35 V. Calculate (a) input power, (b) output power, and (c) efficiency. Assume that $n = 2$, provided that at $X' = 1.841$, $J_1(X') = 0.582$. *[Hint: Use Eqs (11.162) and (11.163)]*

11.6 A typical reflex klystron has the following parameters: DC accelerating voltage = 1.5 kV, repeller voltage = 150 V, frequency of operation = 9 GHz, and distance between cavity and repeller = 2.5 cm. Calculate the round trip DC transit time. *[Hint: Use Eqs (11.137) and (11.149)]*

11.7 A reflex klystron operates with the following conditions: acceleration potential = 600 V, repeller potential = 350 V, mode of operation = $1\frac{3}{4}$, and operating frequency = 9 GHz. Calculate the repeller to resonator grid spacing. *[Hint: Use Eq. (11.165)]*

11.8 If the useful frequency range of a vacuum device is limited to a point where the transit time becomes 1/10th of the time period, then calculate the maximum frequency at which the device can operate, provided the distance between the electrodes is 6 mm and anode voltage is 600 V. *[Hint: Use Eqs (11.20) and (11.22)]*

11.9 A helix TWT is operated with a beam current 250 mA, beam voltage 4.5 kV, and characteristic impedance 45 Ω. If a 50 dB gain is required at 9 GHz, calculate the helix length. *[Hint: Use Eqs (11.209) and (11.240)]*

11.10 A reflex klystron operates at the peak of the $n = 2$ mode. The DC input power is 50 mW and $V_1/V_0 = 0.3$. Calculate (a) efficiency of the reflex klystron, (b) output power in mW, and (c) power delivered to the load if 15% of the power delivered by the electron beam is dissipated in the cavity walls. It is given that at $X' = 1.841$, $J_1(X') = 0.582$. *[Hint: Use Eq. (11.163)]*

11.11 A TWT is operating under the following conditions: acceleration voltage = 4.5 kV, axial electric field = 3.8 V/m, and operating frequency = 3 GHz. If the phase velocity of the slow-wave structure is 1.08 times the average electron beam velocity, then determine the magnitude of velocity fluctuation. *[Hint: Use Eqs (11.20), (11.181), (11.182), and assume $v_p = 1.08\, v_0$]*

Microwave Crossed-field Tubes

12

12.1 Introduction

In the last chapter, we have discussed different types of microwave linear beam tubes, in which the DC magnetic field is parallel to the DC electric field and is used to focus the electron beam. In contrast, in crossed-field tubes, the DC magnetic and DC electric fields are perpendicular to each other. Unlike in linear beam tubes, the DC magnetic field plays a very important role in the operation of crossed-field tubes.

In a crossed-field tube, electrons facing a favourable electric field are accelerated by the electric field. However, as their velocity is increased, they are bent more towards the cathode by the magnetic field and subsequently return to the cathode. On the other hand, electrons facing an unfavourable field are decelerated and move with a slower velocity after giving up some energy to the field. Due to reduced velocity, they are bent less by the magnetic field and hence move towards the anode.

Crossed-field tubes are also known as M-type tubes, after the French term TPOM (tubes á propagation des ondes á champs magnétique or tubes for propagation of waves in a magnetic field), and can be classified as shown in Fig. 12.1.

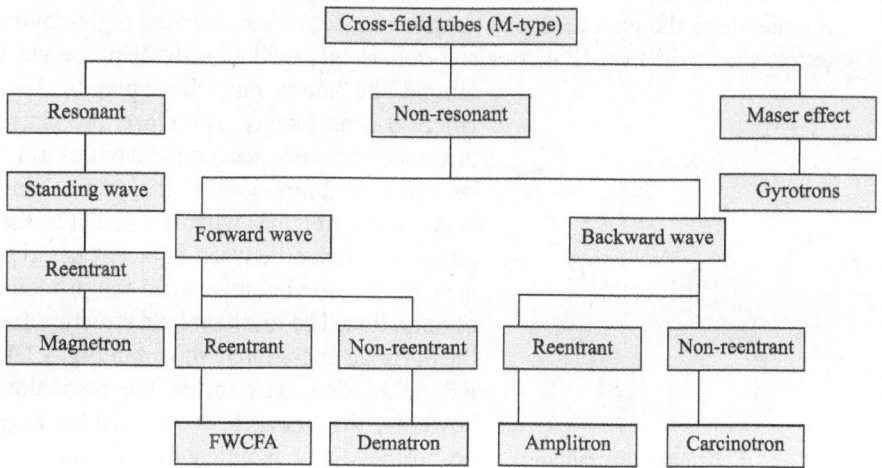

Fig. 12.1 Classification of crossed-field tubes

In this chapter, we will discuss different types of magnetrons (such as cylindrical magnetron, linear magnetron, coaxial magnetron, voltage tunable magnetron, inverted coaxial magnetron, and frequency agile coaxial magnetron), forward-wave crossed-field amplifiers (FWCFAs or CFAs), backward-wave crossed-field amplifiers (BWCFAs) or amplitrons, and backward-wave crossed-field oscillator (BWCFOs) or carcinotrons.

12.2 Magnetron Oscillators

Magnetron was invented by Hull in 1921. However, till 1940, it was used only in laboratories. During World War II there was an urgent need of high-power microwave generators and amplifiers, which led to the rapid development of magnetrons.

In general, magnetrons can be classified into three categories: (a) split anode magnetron, (b) cyclotron frequency magnetron, and (c) travelling-wave magnetrons. Split anode magnetrons use a static negative resistance between the anode segments and generally operate at frequencies below the microwave range, whereas cyclotron frequency magnetrons operate under the synchronization between the RF field and a periodic oscillation of electrons in a direction parallel to the field. Cyclotron frequency magnetrons can operate at microwave frequencies, however with low output power and low efficiency. Travelling-wave magnetrons use the interaction between electrons and a travelling RF field of linear velocity, and are customarily referred to as magnetrons. They can be classified as cylindrical, planar or linear, coaxial, voltage tunable, inverted coaxial, and frequency agile magnetrons.

12.2.1 Cylindrical Magnetron

In cylindrical magnetrons, several reentrant cavities are used, as shown in Fig. 12.2. A DC voltage V_0 is applied between the cathode and the anode, and a magnetic flux density B_0 is applied perpendicular to the electric field. Under such circumstances, electrons emitted from the cathode follow cycloidal paths, as shown in Fig. 12.3.

To understand the operation of a magnetron, let us consider the eight-cavity magnetron shown in Fig. 12.4. For self-consistent oscillation, total phase shift around the anode must be equal to $2n\pi$, where n is an integer. Therefore, the phase difference between successive cavities must be equal to $2n\pi/8 = n\pi/4$. In practice, the best result is obtained when $n = 4$, that is, the phase difference between adjacent cavities is π. Such a condition is called the pi-mode of operation. The resultant field structure for the pi-mode operation is shown in Fig. 12.4. RF fields also exist inside the resonator; however, they have been omitted as they do not have any significant contribution to magnetron operation.

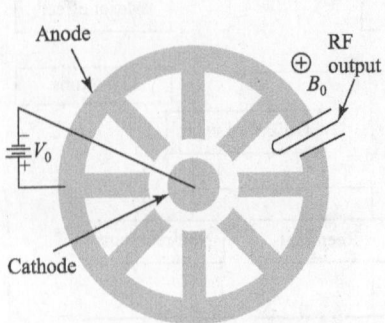

Fig. 12.2 Schematic diagram of cylindrical magnetron

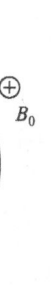

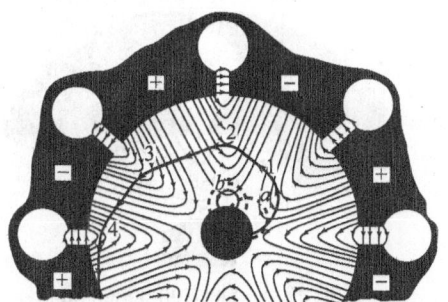

Fig. 12.3 Electron path in cylindrical magnetron

Fig. 12.4 Path traversed by electrons in magnetron under pi mode

The path followed by electrons '*a*' and '*b*' in the absence of an RF field is shown by the dotted line in Fig. 12.4. However, this path is modified in the presence of an RF field. For the formal operation of the tube, the anode voltage and magnetic field are adjusted so that electron *a* gains a particular tangential velocity with which it travels from position '1' to '2' at half the time period of the RF field. Therefore, by the time it reaches point 2 the electric field in the cavity has already reversed it polarity from that shown in Fig. 12.4. Thus, if electron *a* faces a retarding field at position 1, then it also faces a retarding field at position 2. As a result, electron *a* continues to be slowed down by transferring energy to the RF field and moves towards the anode. Finally, the electron strikes the anode surface after delivering the major part of its energy, gained from anode potential, to the RF field, thus amplifying it. If the electron starts from the cathode facing a favouring RF field, for example, electron *b*, then it is accelerated by the field. However, due to the presence of the magnetic field, it is deflected more sharply than in the absence of the RF field and returns quickly to the cathode, causing *back-heating* of the cathode. In practice, 5% of the anode power is lost in this manner. These types of electrons are undesirable as they also absorb energy from the RF field. However, such electrons do not stay long in the interaction region and hence do not have much time to absorb energy from the RF field. The path of electron *b*, in the presence of an RF field, is also shown in Fig. 12.4.

In addition to these, there is also a focusing mechanism which keeps the working electrons, that is, electron *a*, in step with the field in the interaction space. This focusing mechanism helps the working electrons to deliver the maximum possible energy to the RF field. To illustrate this, let us consider Fig. 12.5. The electron at position 'A' is in the vicinity of the 'positive' anode pole and hence the component of the RF fields aids the radial DC field, whereas the electron at position 'C' is in the vicinity of the 'negative' anode pole and hence the RF field opposes the radial DC field. Therefore, electrons at position A will face a favourable field and will be accelerated, while those at position C will face an unfavourable field and will be decelerated. However, if slower electrons are emitted before faster electrons, then there is a fair chance for the faster electrons to catch the slower electrons. This causes bunching of electrons around the electron whose relative position is indicated by 'B' in Fig. 12.5. The bunching procedure in magnetrons is also known as *phase focusing* and is similar to velocity modulation.

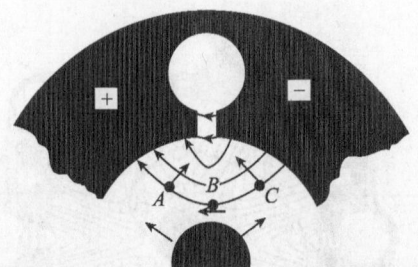

Fig. 12.5 Electron bunching in magnetron

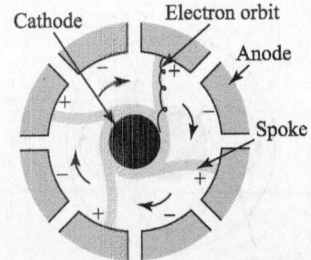

Fig. 12.6 Rotating space charge under oscillatory conditions

Selective grouping of the electrons, mentioned earlier, results in a spoke-shaped space charge cloud of electrons. In practice, there will be one spoke per cavity for the pi-mode operation. These spokes rotate at an angular velocity equivalent to two poles per cycle in the clockwise direction to keep up with the RF phase changes between adjoining anode poles. It is not difficult to imagine that the electric field itself is rotating in the clockwise direction at the same speed as that of the spokes (Fig. 12.6).

> **Notes** 1. Due to the rotating RF fields, such magnetrons are also called travelling-wave magnetrons. 2. It may be noted that the electric field existing in the interaction space of a magnetron corresponds to the rotating field in the air gap of a poly-phase electric machine. The pi-mode corresponds to a single-phase system, whereas other modes correspond to various poly-phase arrangements.

In a magnetron, the RF magnetic flux lines pass through the cavities and are parallel to the cathode axis, whereas the RF electric field lines are concentrated across the coupling slots and also fringe out to the interaction space in the transverse direction. In practice, a magnetron consisting of N resonant coupled cavities can support N resonant frequencies or modes. The equivalent resonant circuit is shown in Fig. 12.7.

If the magnetic flux density is assumed to be z-directed then the equation of motion of the electrons in a cylindrical magnetron can be written as follows:

$$\frac{d^2r}{dt^2} - r\left(\frac{d\phi}{dt}\right)^2 = \frac{e}{m}E_r - \frac{e}{m}rB_z\frac{d\phi}{dt} \tag{12.1}$$

and

$$\frac{1}{r}\frac{d}{dt}\left(r^2\frac{d\phi}{dt}\right) = \frac{e}{m}B_z\frac{dr}{dt} \tag{12.2}$$

or

$$\frac{d}{dt}\left(r^2\frac{d\phi}{dt}\right) = \frac{e}{m}rB_z\frac{dr}{dt} = \frac{1}{2}\left(\frac{e}{m}B_z\right)\frac{d}{dt}(r^2) = \frac{1}{2}\omega_c\frac{d}{dt}(r^2) \tag{12.3}$$

where

$$\omega_c = eB_z/m \tag{12.4}$$

and is called the cyclotron angular frequency.

Integrating Eq. (12.3), we get $r^2\dfrac{d\phi}{dt} = \dfrac{1}{2}\omega_c r^2 + C$ $\tag{12.5}$

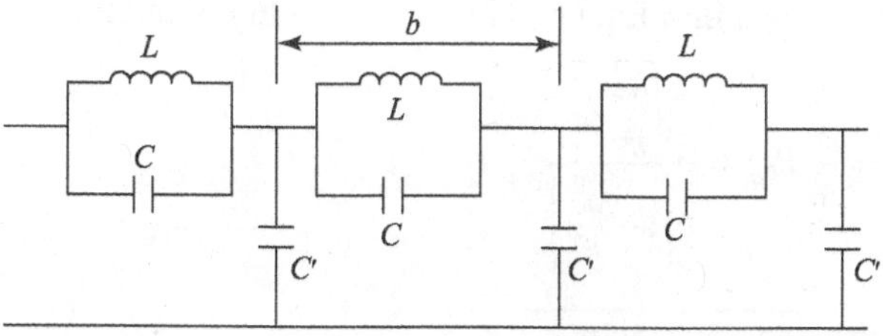

Fig. 12.7 Schematic band-pass equivalent circuit of magnetron

where C is an integration constant. At $r = a$, where a is the radius of cathode cylinder, $\dfrac{d\phi}{dt} = 0$. This gives the following equation: $C = -\dfrac{1}{2}\omega_c a^2$ (12.6)

Substituting Eq. (12.6) into Eq. (12.5), we get the following relation:

$$r^2 \frac{d\phi}{dt} = \frac{1}{2}\omega_c r^2 - \frac{1}{2}\omega_c a^2 = r^2 \frac{1}{2}\omega_c\left(1 - \frac{a^2}{r^2}\right)$$

or $\dfrac{d\phi}{dt} = \dfrac{1}{2}\omega_c\left(1 - \dfrac{a^2}{r^2}\right)$ (12.7)

Since the magnetic field does no work on the electrons, the kinetic energy of the electrons can be written as follows:

$$\frac{1}{2}mv^2 = eV$$

or $v^2 = v_r^2 + v_\phi^2 = \left(\dfrac{dr}{dt}\right)^2 + \left(r\dfrac{d\phi}{dt}\right)^2 = \dfrac{2e}{m}V$ (12.8)

Now at $r = b$, where b is the radius of edge of the anode, $V = V_0$ and $\dfrac{dr}{dt} = 0$. Therefore, when the electrons just graze the anode, Eqs (12.7) and (12.8) modify to the following forms:

$$\frac{d\phi}{dt} = \frac{1}{2}\omega_c\left(1 - \frac{a^2}{b^2}\right)$$ (12.9)

and $b^2\left(\dfrac{d\phi}{dt}\right)^2 = \dfrac{2e}{m}V_0$ (12.10)

Substituting Eq. (12.9) into Eq. (12.10), we get the following expression:

$$b^2\left(\frac{d\phi}{dt}\right)^2 = b^2\left\{\frac{1}{2}\omega_c\left(1 - \frac{a^2}{b^2}\right)\right\}^2 = \frac{b^2}{4}\omega_c^2\left(1 - \frac{a^2}{b^2}\right)^2 = \frac{2e}{m}V_0$$ (12.11)

or $\omega_c = \dfrac{\sqrt{\dfrac{8e}{m}V_0}}{b\left(1 - \dfrac{a^2}{b^2}\right)}$ (12.12)

Substituting Eq. (12.4) into Eq. (12.12), the following relation is obtained:

$$\frac{e}{m}B_z = \frac{e}{m}B_{0c} = \frac{\sqrt{\dfrac{8e}{m}}V_0}{b\left(1 - \dfrac{a^2}{b^2}\right)}$$

or $\qquad B_{0c} = \dfrac{\dfrac{m}{e}\sqrt{\dfrac{8e}{m}}V_0}{b\left(1 - \dfrac{a^2}{b^2}\right)} = \dfrac{\sqrt{\dfrac{8m}{e}}V_0}{b\left(1 - \dfrac{a^2}{b^2}\right)}$ \hfill (12.13)

Equation (12.13) is called the Hull cut-off magnetic flux density equation. If $B_0 > B_{0c}$ for a given V_0, then the electrons will not be able to reach the anode and will return to the cathode. Equation (12.13) can also be written as follows:

$$B_0 = \frac{\sqrt{\dfrac{8m}{e}}V_{0c}}{b\left(1 - \dfrac{a^2}{b^2}\right)}$$

or $\qquad V_{0c} = \dfrac{e}{8m}B_0^2 b^2 \left(1 - \dfrac{a^2}{b^2}\right)^2$ \hfill (12.14)

Equation (12.14) is called the Hull cut-off voltage equation. If $V_0 < V_{0c}$ for a given B_0, then the electrons will not be able to reach the anode and will return to the cathode.

Since the electrons follow a cycloidal path, the outward centrifugal force is equal to the pulling force, and we can write $mv^2/R = evB$ \hfill (12.15) where R is the radius of the cycloidal path and v is the tangential velocity of the electron. The cyclotron angular frequency can, therefore, be written as follows:

$$\omega_c = v/R = eB/m \tag{12.16}$$

The period of one complete revolution is expressed as $T = \dfrac{2\pi}{\omega_c} = \dfrac{2\pi m}{eB}$ \hfill (12.17)

Since the slow-wave structure is reentrant in nature, oscillation will occur when the total phase shift around the structure is an integral multiple of 2π. Thus, if there are N reentrant cavities in the anode structure, the phase shift between two anode cavities can be expressed as follows: $\phi_n = \dfrac{2\pi n}{N}$ \hfill (12.18)

where n is an integer. It has already been mentioned that the best result is obtained when $n = 4$, that is, in the pi-mode for which $\phi_n = \pi$ \hfill (12.19)
For the pi-mode operation, oscillation starts at a beam voltage

$$V_{0h} = \frac{\omega}{N}\left(b^2 - a^2\right)B_0 \tag{12.20}$$

known as the Hartree voltage. Plots of the Hull cut-off voltage and Hartree voltage as a function of B_0 are shown in Fig. 12.8.

The characteristic curves of a magnetron are shown in Fig. 12.9. The figure shows the combination of B_0 and V_0 for which oscillation will sustain. For the same B_0, if we move from $n = N/2$ mode to $n = \left(\dfrac{N}{2} - 1\right)$ mode, the voltage required for sustaining the oscillation also increases. If we now move to further lower n, at a certain time electrons emitted from the cathode will graze the anode without any oscillation, producing the region of steady anode current.

The performance of a magnetron can also be explained in terms of the characteristic curve. Let the magnetic flux density be kept sufficiently high and anode voltage increased. For low anode voltage, the anode current is cut-off and no oscillation is obtained. When $n = N/2$ line is reached, ϕ becomes equal to π and a pi-mode oscillation builds up. At this point, the space charge cloud encircles the cathode relatively slowly, since the electron has only $N/2$ oscillations to make a complete revolution. For a higher anode voltage, electrons rotate faster; the synchronization with the resonator field is lost and the oscillation stops. At further higher values of anode voltage, synchronization is established for the next higher-order mode $\left(\dfrac{N}{2} - 1\right)$ and oscillation restarts. In this mode, the space charge cloud revolves faster than before, since it must make a complete revolution in one less time period. As the anode voltage is further increased, alternate regions on no oscillation and oscillation of successively lower mode numbers appear until the region of steady anode current is reached, and all possibilities of oscillations are lost.

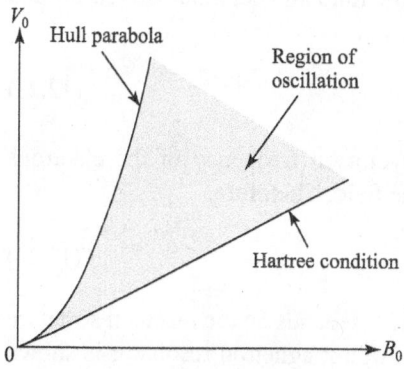

Fig. 12.8 Plot of Hull cut-off voltage and Hartree voltage as functions of magnetic flux density

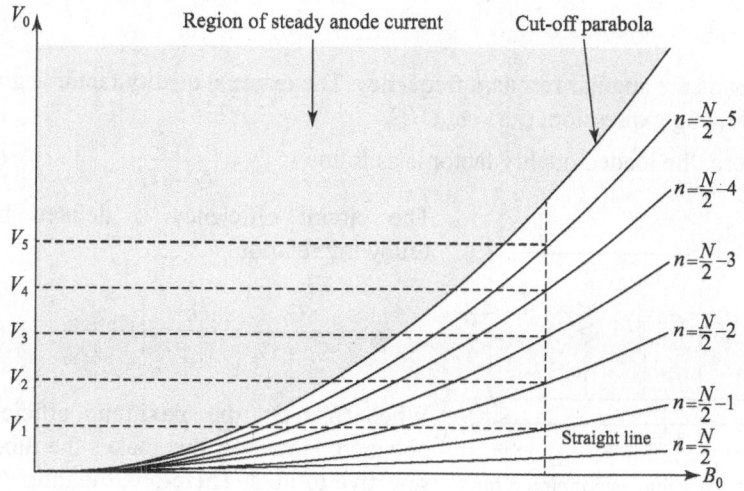

Fig. 12.9 Mode line diagram of magnetron

It should be noted that for low values of magnetic flux density or anode voltage the $V_0 - B_0$ combination reaches the cut-off parabola before any synchronization can occur and no oscillation takes place. The particular anode voltage or magnetic flux density for which the $V_0 - B_0$ combination reaches the cut-off parabola before any oscillation can happen, keeping the other parameter constant, is called Hull cut-off voltage or Hull cut-off magnetic flux density and has been expressed by Eqs (12.14) and (12.13) respectively.

If L is the mean separation between cavities, then the phase constant of the fundamental mode field is given by the equation $\beta_0 = \dfrac{2\pi n}{NL}$ (12.21)

The fundamental ϕ-component of the electric field can be obtained by solving the Maxwell equation with proper boundary conditions and is expressed as follows:

$$E_{\phi 0} = jE_1 e^{j(\omega t - \beta_0 \phi)}$$ (12.22)

where E_1 is a constant. The travelling field of the fundamental mode travels around the structure with an angular velocity, as follows:

$$\frac{d\phi}{dt} = \frac{\omega}{\beta_0}$$ (12.23)

The field and electrons interact when the cyclotron frequency of the electrons becomes equal to the angular frequency of the field. Therefore,

$$\omega_c = \beta_0 \frac{d\phi}{dt}$$ (12.24)

The efficiency and power output of a magnetron depends on the resonant structure and DC power supply. The equivalent circuit of a magnetron resonator is shown in Fig. 12.10, where Y_e is the electronic admittance, V is the RF voltage across the vane strips, C is the capacitance of the vane strips, L is the inductance of the resonator, G_r is the conductance of the resonator, and G is the load conductance per resonator. The unloaded quality factor of the resonator is given as follows:

$$Q_{un} = \omega_0 C / G_r$$ (12.25)

where ω_0 is the angular resonant frequency. The external quality factor is given by the following expression: $Q_{ex} = \omega_0 C / G_l$ (12.26)

Therefore, the loaded quality factor is as follows: $Q_l = \dfrac{\omega_0 C}{G_r + G_l}$ (12.27)

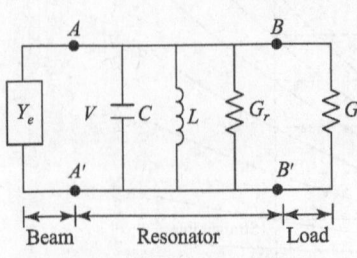

Beam   Resonator   Load

Fig. 12.10 Equivalent circuit for one resonator of a magnetron

The circuit efficiency is defined by the following relation:

$$\eta_c = \frac{G_l}{G_r + G_l} = \frac{G_l}{G_{ex}} = \left(1 + \frac{Q_{ex}}{Q_{un}}\right)^{-1}$$ (12.28)

When $G_l \gg G_r$, the maximum efficiency is obtained. However, this makes the tube quite sensitive to load. Therefore, the ratio G_l / G_{ex} is often chosen to compromise between high circuit efficiency and frequency stability. The electronic efficiency of a magnetron is defined as follows:

$$\eta_e = \frac{P_{\text{gen}}}{P_{\text{dc}}} = \frac{V_0 I_0 - P_{\text{lost}}}{V_0 I_0} \tag{12.29}$$

where P_{gen} is the RF power induced into the anode circuit, P_{dc} is the power obtained from the DC power supply, V_0 is the anode voltage, I_0 is the anode current, and P_{lost} is the power lost in the anode circuit. The electronic efficiency can also be written as follows:

$$\eta_e = \left(1 - \frac{m\omega_0^2}{2eV_0\beta^2}\right) \Big/ \left(1 + \frac{I_0 m M_1^2 Q_l}{\beta_z e N L^2 \omega_0 C}\right) \tag{12.30}$$

where $\quad M_1 = \dfrac{\sin(\beta_n \delta/2)}{(\beta_n \delta/2)} = 1$ for small δ $\qquad$ (12.31)

Here, δ is the gap factor for the pi-mode operation.

If the frequencies of different modes of a magnetron are very close, then the magnetron has a tendency of jumping from one mode to another during the operation. For example, a 3 cm π-mode oscillation, normal for a particular magnetron, can spuriously become a 3.05 cm $3\pi/4$-mode oscillation. This mode jumping can be prevented using straps, as shown in Fig. 12.11. Straps are basically two heavy-gauge rings that are connected only to the alternate anode poles. In the pi-mode operation, the rings are at a uniform potential, but in opposite polarity, and hence behave as a capacitor. This results in a capacitive loading to the cavities, which reduces the frequency of the π-mode. For other modes, the relative phase difference between two strapped points of the same strapping ring is not zero. This results in a non-uniform potential along the rings and a current flows through them. This flow of current results in an inductive loading to the cavity, which, in turn, raises the frequencies of these modes. The desired pi-mode is thus separated from other modes.

If the operating frequency of the magnetron is above 10 GHz, then the realization of the strapping technique becomes extremely difficult or almost impossible. This is because, at high frequencies, the cavities are small, and also a large number of cavities (16 or 32) are required to ensure a suitable RF field in the interaction space. These result in a large number of modes, which are supported by the magnetron, and strapping cannot prevent mode jumping. To avoid mode jumping, in such cases, a special anode block, having a pair of cavity systems that are quite dissimilar in shape and resonant frequencies, is designed. With such an anode geometry, the magnetron supports only one of the two resonant frequencies corresponding to the two sets of cavity, and hence mode

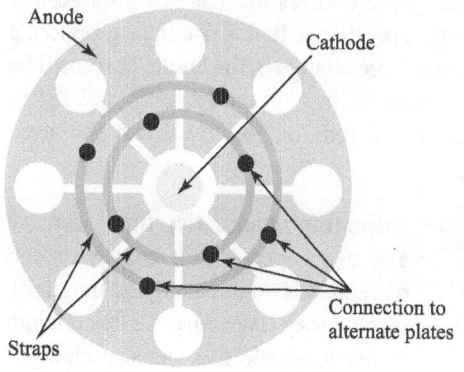

Fig. 12.11   Strapped magnetron

jumping can be prevented. Such magnetrons are called rising-sun magnetrons. Strapping is not required in rising-sun magnetron. In addition to high-frequency magnetrons, strapping may be unsatisfactory in very-high-power magnetrons due to the losses in the straps.

To tune the resonance frequency of a magnetron, a tunable resonant cavity can be coupled to one of the resonant cavities of the magnetron. Frequency tuning can also be achieved using a third strap or a tuning ring (also called a C-ring), as shown in Fig. 12.12. By changing the distance of the C-ring from the straps, the anode capacitive loading can be varied, which, in turn, changes the operating frequency. In contrast to the capacitive tuning, inductive tuning is also possible, where conducting rods are inserted lengthwise into the resonators at the region of high RF magnetic field.

A change in the anode voltage of the magnetron results in a change in the orbital velocity of electrons. This, in turn, alters the rate at which energy is given up to the anode resonator and the operating frequency. This is known as frequency pushing. Magnetrons are also susceptible to frequency variation due to a change in the load impedance. This phenomenon occurs regardless of whether these load variations are purely resistive or reactive, and is more severe in case of reactive load variations. Such a variation in operating frequency with variation in loads is known as frequency pulling. Frequency pushing in a magnetron is avoided using a stabilized power supply, whereas frequency pulling is avoided using a circulator, placed before the waveguide connection at the output of the magnetron.

During operation, often the output power of the magnetron is required to be adjusted. In practice, several methods can be used for this purpose:

Pulsing output power This method is common in most of the microwave ovens. In this method, a magnetron is pulsed very slowly, sometimes on for 2–5 seconds and then off for 2–5 seconds, so that the average power delivered to the load can be controlled. The timer circuit used for this purpose adds to the complexity and cost of the circuit.

Changing anode current This is based on the principle that a change in the anode current results in a change in the output power. The anode voltage–anode current graph, shown in Fig. 12.13(a), reveals that initially the anode current remains almost zero. However, after a certain anode voltage, known as the Hartree voltage, the anode current starts to flow and the magnetron oscillates. In this oscillating region, a small change in the anode voltage results in a large change in the anode current. The plot of output power/anode voltage versus anode current is shown in Fig. 12.13(b). The figure reveals that by changing the anode current, output power can be varied in the oscillation region.

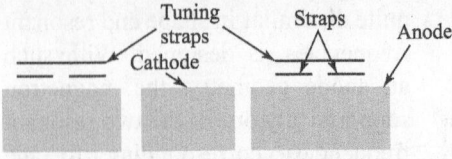

Fig. 12.12 Tuning of magnetron using C-ring

Adjusting magnetic flux density The output power of a magnetron can be reduced by increasing the magnetic flux density. Typically, a 10% change in the magnetic flux density changes the output power by 60%.

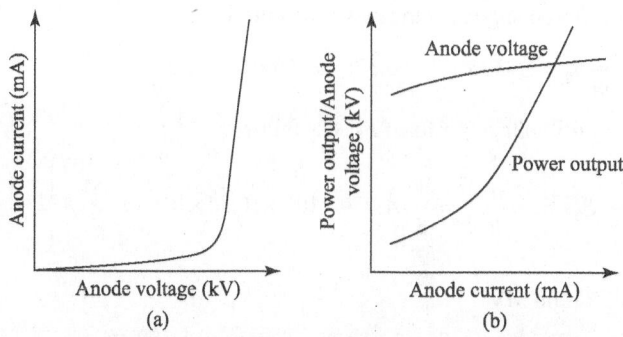

Fig. 12.13 Magnetron characteristics (a) Anode voltage versus anode current (b) Anode current versus output power

Attenuating microwave energy This method is based on the attenuation of microwave energy between the magnetron and the load, and therefore has no effect on the magnetron. However, it is a mechanical method, and is also complex and costly.

Magnetrons are widely used as a source in radar transmitters, in microwave ovens, and also for industrial heating. A domestic microwave oven requires a standard power of 600–900 W at a frequency of 915 or 2450 MHz. For industrial heating, the required power level is in kW level at MHz frequencies. A magnetron can deliver up to 40 MW of peak power at 10 GHz with just 50 kV input voltage. The average power output is about 800 kW. The efficiency of a magnetron ranges from 40% to 70%. In practice, no other microwave devices, having the same size, weight, operating voltage, and efficiency, can perform the same function. Conventional travelling-wave magnetrons are widely used to generate high-peak-power RF pulses of frequencies as high as 70 GHz. Beacon magnetrons—a miniaturized version of conventional magnetrons—can produce a peak output power of 3.5 kW, while exhibiting a negligible frequency shift and long-life performance under severe environmental and temperature conditions. The weight of such a magnetron is usually less than 2 pounds and hence it is ideal for applications that require very compact, low-voltage sources of pulsed power, such as, airborne missiles, satellites, and Doppler systems.

Note In the pi-mode, each cavity of the magnetron, along with its input gap, acts as a short-circuited transmission line of quarter wavelength and hence has a maximum electric field across the gap. The fields in two consecutive cavities also are oppositely directed.

EXAMPLE 12.1 A typical X-band pulsed cylindrical magnetron has the following operating parameters: anode voltage = 25 kV, beam current = 25 A, magnetic flux density = 0.35 Wb/m², radius of cathode cylinder = 5 cm, and radius of vane edge to centre = 10 cm. Calculate the (a) cyclotron angular frequency, (b) cut-off voltage for fixed B_0, and (c) cut-off magnetic flux density for fixed V_0.

Solution Given: $V_0 = 25$ kV, $I_0 = 25$ A, $B_0 = 0.35$ Wb/m², $a = 5$ cm, and $b = 10$ cm

(a) The cyclotron angular frequency is as follows:

$$\omega_c = \frac{e}{m} B_0 = 1.759 \times 10^{11} \times 0.35 = 6.1565 \times 10^{10} \text{ rad}$$

(b) The cut-off voltage for fixed B_0 is as follows:

$$V_{0c} = \frac{1}{8} \frac{e}{m} B_0^2 b^2 \left(1 - \frac{a^2}{b^2}\right)^2 = \frac{1}{8} \times 1.759 \times 10^{11} \times \left(0.35 \times 10 \times 10^{-2}\right)^2 \times \left\{1 - \left(\frac{5 \times 10^{-2}}{10 \times 10^{-2}}\right)^2\right\}^2$$

or $V_{0c} = 15.1508 \text{ MV}$

(c) The cut-off magnetic flux density for fixed V_0 is as follows:

$$B_{0c} = \frac{\sqrt{8V_0 \dfrac{m}{e}}}{b\left(1 - \dfrac{a^2}{b^2}\right)} = \frac{\sqrt{\dfrac{8 \times 25 \times 10^3}{1.759 \times 10^{11}}}}{10 \times 10^{-2} \times \left\{1 - \left(\dfrac{5 \times 10^{-2}}{10 \times 10^{-2}}\right)^2\right\}}$$

or $\qquad B_{0c} = \dfrac{1.0663 \times 10^{-3}}{0.075} = 14.2173 \times 10^{-3} \text{ Wb/m}^2$

or $\qquad B_{0c} = 14.2173 \text{ mWb/m}^2$

Practice Problem

12.1 A typical X-band pulsed cylindrical magnetron has the following operating parameters: anode voltage = 32 kV, beam current = 30 A, magnetic flux density = 0.3 Wb/m^2, radius of cathode cylinder = 6 cm, and radius of vane edge to centre = 12 cm. Calculate the (a) cyclotron angular frequency, (b) cut-off voltage for fixed B_0, and (c) cut-off magnetic flux density for fixed V_0.

$$\left[5.277 \times 10^{10} \text{ rad}, 16.0289 \text{ MV}, 13.4044 \text{ mWb} / \text{m}^2\right]$$

12.2.2 Linear Magnetron

The schematic diagram of a linear magnetron is shown in Fig. 12.14. If the electric field is along the x-axis and the magnetic field along the z-axis, then the motion of electrons can be written as follows:

$$\frac{d^2x}{dt^2} = -\frac{e}{m}\left(E_x + B_z \frac{dy}{dt}\right) \tag{12.32}$$

$$\frac{d^2y}{dt^2} = \frac{e}{m} B_z \frac{dx}{dt} \tag{12.33}$$

and $\qquad \dfrac{d^2z}{dt^2} = 0 \tag{12.34}$

In practice, due to the effect of space charges, the electric field is a non-linear function of the distance x and hence determining a complete solution of

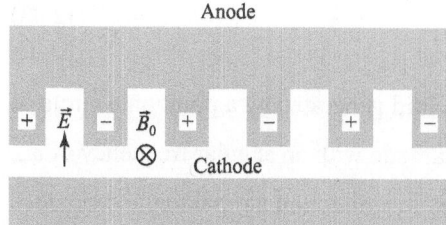

Fig. 12.14 Schematic diagram of linear magnetron

Eqs (12.32)–(12.34) is a very complex process. However, Eq. (12.33) can be integrated directly, which gives the following expression:

$$\frac{dy}{dt} = \frac{e}{m} B_z x + C \qquad (12.35)$$

where C is the integration constant. If we assume that the electron starts with zero initial velocity then at $x = 0$, $\frac{dy}{dt} = 0$ and Eq. (12.35) gives the following equality:

$$C = 0 \qquad (12.36)$$

Substituting Eq. (12.36) into Eq. (12.35), we get the following relation:

$$\frac{dy}{dt} = \frac{e}{m} B_z x \qquad (12.37)$$

Equation (12.37) reveals that the velocity of electron, parallel to the electrode surface, is proportional to the distance of the electron from the cathode and the magnetic flux density. In practice, the distance of the electron from the cathode depends on the magnetic flux density and the manner the potential V varies with x, which, in turn, depends on the space charge distribution, electrode spacing, and anode potential.

If we assume that the space charge is negligible, cathode potential is zero, and anode potential is V_0, then the differential electric field can be written as follows:

$$E_x = -\frac{dV}{dx} = -\frac{V_0}{d} \qquad (12.38)$$

where d is the distance between the cathode and the anode. Substituting Eqs (12.38) and (12.37) into Eq. (12.32), we get the following equation:

$$\frac{d^2x}{dt^2} = \frac{e}{m}\left(\frac{V_0}{d} - B_z \frac{e}{m} B_z x\right) = \frac{e}{m}\frac{V_0}{d} - \frac{e^2}{m^2} B_z^2 x$$

or

$$\frac{d^2x}{dt^2} + \left(\frac{e}{m} B_z\right)^2 x - \frac{e}{m}\frac{V_0}{d} = 0 \qquad (12.39)$$

The solution of Eq. (12.39) can be written as follows:

$$x = \frac{V_0}{B_z \omega_c d}\{1 - \cos(\omega_c t)\} \qquad (12.40)$$

Substituting Eq. (12.40) into Eq. (12.37) and then solving it for y, we get the following expression:

$$y = \frac{V_0}{B_z \omega_c d}[\omega_c t - \sin(\omega_c t)] \qquad (12.41)$$

Further solution of Eq. (12.34) gives $z = 0$ $\qquad (12.42)$

where $\omega_c = \dfrac{e}{m} B_z$ (12.43)

is the cyclotron angular frequency.

Equations (12.40)–(12.42) represent a cycloid generated by a point on a circle of radius $\dfrac{V_0}{B_z \omega_c d}$ rolling on the plane of the cathode with an angular frequency of ω_c. The maximum distance the electron moves in a direction normal to the cathode is given as follows:

$$\frac{2V_0 m}{B_z^2 e d} = d \tag{12.44}$$

Now, let us define a constant as $K = \dfrac{d^2 B_z^2}{V_0} = \dfrac{2m}{e} = 1.14 \times 10^{-11}$ (12.45)

If $K < 1.14 \times 10^{-11}$, then the electron strikes the anode, whereas if $K > 1.14 \times 10^{-11}$ the electron returns to the cathode. The path of the electron is shown in Fig. 12.15. The Hull cut-off voltage can be obtained from Eq. (12.44) as follows:

$$V_{0c} = \frac{1}{2} \frac{e}{m} B_0^2 d^2 \tag{12.46}$$

If $V_0 < V_{0c}$ for a given B_0, then the electron will not reach the anode. Similarly, the Hull cut-off magnetic flux density can be obtained from Eq. (12.44) as follows:

$$B_{0c} = \frac{1}{d} \sqrt{\frac{2m}{e}} V_0 \tag{12.47}$$

If $B_0 > B_{0c}$ for a given V_0, then the electron will not reach the anode.

In a linear magnetron tube, the electron beam extends up to a distance h from the cathode, called the hub thickness. If the velocity of the electron in the y-direction is v_y, then

$$v_y = -\frac{E_x}{B_0} = \frac{1}{B_0} \frac{dV}{dx} \tag{12.48}$$

Now the conservation of energy gives the following expression:

$$\frac{1}{2} m v_y^2 = eV \tag{12.49}$$

Substituting Eq. (12.48) into Eq. (12.49), we get the following relation:

$$\frac{1}{2} m \left(\frac{1}{B_0} \frac{dV}{dx} \right)^2 = eV \text{ or } \left(\frac{dV}{dx} \right)^2 = \frac{2eV}{m} B_0^2$$

or $$\sqrt{\frac{m}{2eB_0^2}} \frac{dV}{\sqrt{V}} = dx \tag{12.50}$$

Integrating Eq. (12.50), we get the following expression:

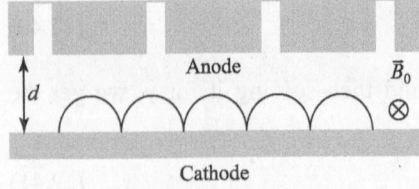

Fig. 12.15 Electron path in linear magnetron

$$\sqrt{\frac{m}{2eB_0^2}} 2\sqrt{V} = x + C_1 \tag{12.51}$$

where C_1 is the integration constant. If we assume that at $x = 0$, $V = 0$, then $C_1 = 0$ and Eq. (12.51) modifies as follows:

$$\sqrt{\frac{m}{2eB_0^2}}\,2\sqrt{V} = x \text{ or } V = \frac{eB_0^2}{2m}x^2 \tag{12.52}$$

The potential and electric fields at the hub surface, that is, at $x = h$, is given as follows:

$$V(h) = \frac{eB_0^2}{2m}h^2 \tag{12.53}$$

and
$$E_x(h) = -\frac{dV}{dx}\bigg|_{x=h} = -\frac{eB_0^2}{m}x\bigg|_{x=h} = -\frac{eB_0^2}{m}h \tag{12.54}$$

The potential at the anode is given by the following equation:

$$V_0 = -\int_0^d E_x dx = -\int_0^h E_x dx - \int_h^d E_x dx = V(h) - \int_h^d E_x dx \tag{12.55}$$

Substituting Eqs (12.53) and (12.54) into Eq. (12.55), we get the following relation:

$$V_0 = V(h) + \int_h^d \frac{eB_0^2}{m}h\,dx = \frac{eB_0^2}{2m}h^2 + \frac{eB_0^2}{m}h(d-h) = \frac{eB_0^2}{m}h\left(d - \frac{h}{2}\right) \tag{12.56}$$

Substituting Eq. (12.54) into Eq. (12.48), the following expression is obtained:

$$v_y(h) = \frac{1}{B_0}\frac{dV}{dx} = \frac{1}{B_0}\frac{eB_0^2}{m}h = \frac{eB_0}{m}h \tag{12.57}$$

For synchronization, this velocity must be equal to the phase velocity of the slow-wave structure. Therefore, $\dfrac{\omega}{\beta} = \dfrac{eB_0}{m}h$ $\qquad\qquad$ (12.58)

Substituting Eq. (12.58) into Eq. (12.56), the anode potential for the pi-mode of operation can be expressed as follows:

$$V_{0h} = \frac{eB_0^2}{m}h\left(d - \frac{h}{2}\right) = \frac{\omega}{\beta}B_0\left(d - \frac{h}{2}\right) = \frac{\omega}{\beta}B_0 d - \frac{\omega}{\beta}B_0\frac{h}{2}$$

or
$$V_{0h} = \frac{\omega B_0 d}{\beta} - \frac{\omega eB_0 h}{2m\beta}\frac{m}{e} = \frac{\omega B_0 d}{\beta} - \frac{m\omega^2}{2e\beta^2} \tag{12.59}$$

Equation (12.59) is called the Hartree anode voltage equation, which reveals that the Hartree anode voltage is a function of magnetic flux density and electrode spacing.

EXAMPLE 12.2 A typical linear magnetron has the following operating parameters: anode voltage = 10 kV, anode current = 1.2 A, magnetic flux density = 0.02 Wb/m², and distance between anode and cathode = 4.5 cm. Calculate the (a) Hull cut-off voltage for a fixed B_0 and (b) Hull cut-off magnetic flux density for a fixed V_0.

Solution   Given: $V_0 = 10$ kV, $I_0 = 1.2$ A, $B_0 = 0.02$ Wb/m², and $d = 4.5$ cm

(a) The Hull cut-off voltage for a fixed B_0 is expressed as follows:

$$V_{0c} = \frac{1}{2}\frac{e}{m}B_0^2 d^2 = \frac{1}{2}\times 1.759\times 10^{11}\times\left(0.02\times 4.5\times 10^{-2}\right)^2$$

or $V_{0c} = 71.2395\times 10^3 = 71.2395$ kV

(b) The following is the Hull cut-off magnetic flux density for a fixed V_0:

$$B_{0c} = \frac{1}{d}\sqrt{2V_0\frac{m}{e}} = \frac{1}{4.5\times 10^{-2}}\sqrt{\frac{2\times 10\times 10^3}{1.759\times 10^{11}}} = 7.4932\times 10^{-3} = 7.4932 \text{ mWb/m}^2$$

Practice Problem

12.2 A typical linear magnetron has the following operating parameters: anode voltage = 12 kV, anode current = 1.5 A, magnetic flux density = 0.015 Wb/m², and distance between anode and cathode = 6 cm. Calculate the (a) Hull cut-off voltage for a fixed B_0 and (b) Hull cut-off magnetic flux density for a fixed V_0.

$$\left[71.2395\,\text{kV},\ \ 6.1563\,\text{mWb} / \text{m}^2\right]$$

12.2.3 Coaxial Magnetron

A coaxial magnetron is composed of an unstrapped anode resonator structure that is tightly coupled to an inner, single, high-Q cavity, operating in the TE_{011} mode, through coupling slots in the back walls of alternate cavities, as shown in Fig. 12.16. An attenuator, present within the inner slotted cylinder and near the coupling slot ends, is used to attenuate the undesired modes. In the TE_{011} mode, the electric field follows a circular path within the cavity and reduces to zero at the cavity walls. As a result, the current in the cavity walls also follows a circular path about the axis of the tube. In the pi-mode operation, electric fields in every other cavity are in phase and hence they couple in the same direction into the surrounding cavity.

The tuning mechanism in a coaxial magnetron is very simple and reliable. Further, since coupling straps are not required, the anode resonator is larger and less complex than conventional strapped magnetrons. This results in lower cathode loading and reduced voltage gradients.

X-Band coaxial magnetrons can operate in the frequency range 8.9–9.6 GHz, with a minimum peak power of 400 kW. The required anode voltage is 32 kV and the corresponding peak anode current is 32 A.

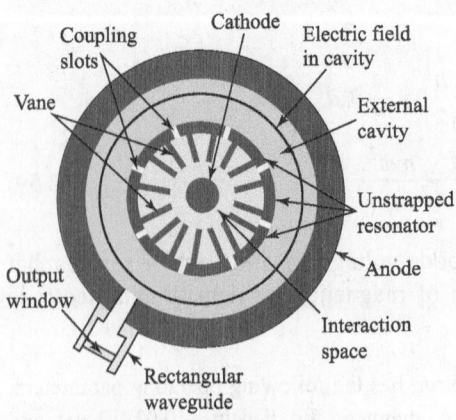

Fig. 12.16 Schematic diagram of coaxial cavity

12.2.4 Inverted Coaxial Magnetron

In an inverted coaxial magnetron (see Fig. 12.17), the anode and cathode are inverted, that is, the cathode is built as a ring around the anode and the cavity is located inside a slotted cylinder. A resonator vane array is arranged on the outside of the cavity.

The motion of electrons in an inverted coaxial magnetron can be written as follows:

$$\frac{d^2r}{dt^2} - r\left(\frac{d\phi}{dt}\right)^2 = \frac{e}{m}E_r - \frac{e}{m}rB_z\frac{d\phi}{dt} \qquad (12.60)$$

$$\frac{1}{r}\frac{d}{dt}\left(r^2\frac{d\phi}{dt}\right) = \frac{e}{m}B_z\frac{dr}{dt}$$

or $\quad \dfrac{d}{dt}\left(r^2\dfrac{d\phi}{dt}\right) = \dfrac{e}{m}B_z r\dfrac{dr}{dt} = \omega_c r\dfrac{dr}{dt} = \dfrac{1}{2}\omega_c\dfrac{d}{dt}\left(r^2\right) \qquad (12.61)$

where

$$\omega_c = \frac{e}{m}B_z = \frac{e}{m}B_0 \qquad (12.62)$$

is the cyclotron angular frequency.
Integrating Eq. (12.61) we get the following equation:

$$r^2\frac{d\phi}{dt} = \frac{1}{2}\omega_c r^2 + C \qquad (12.63)$$

where C is an integration constant. Now at $r = b$, $\dfrac{d\phi}{dt} = 0$. Therefore, Eq. (12.63) gives $C = -\dfrac{1}{2}\omega_c b^2$ $\qquad (12.64)$

Substituting Eq. (12.64) into Eq. (12.63), we get the following relation:

$$r^2\frac{d\phi}{dt} = \frac{1}{2}\omega_c\left(r^2 - b^2\right)$$

or $\quad \dfrac{d\phi}{dt} = \dfrac{1}{2r^2}\omega_c\left(r^2 - b^2\right) = \dfrac{1}{2}\omega_c\left(1 - \dfrac{b^2}{r^2}\right) \qquad (12.65)$

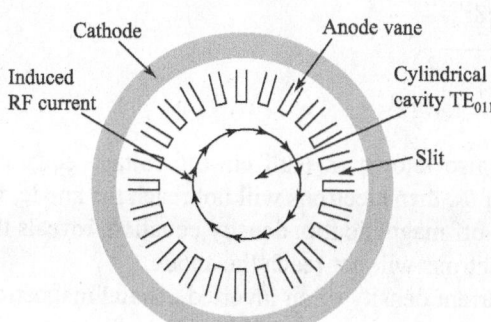

Fig. 12.17 Schematic diagram of inverted coaxial magnetron

Since the magnetic field does no work on the electrons, we can write the following equation:

$$\frac{1}{2}mv^2 = eV \tag{12.66}$$

where the square of the velocity of the electrons (v) can be written as follows:

$$v^2 = v_r^2 + v_\phi^2 = \left(\frac{dr}{dt}\right)^2 + \left(r\frac{d\phi}{dt}\right)^2 = \frac{2e}{m}V \tag{12.67}$$

Now at $r = a$, $V = V_0$ and $\frac{dr}{dt} = 0$. Therefore, when electrons just graze the anode, Eqs (12.65) and (12.67) modify as follows:

$$\frac{d\phi}{dt} = \frac{1}{2}\omega_c\left(1 - \frac{b^2}{a^2}\right) \tag{12.68}$$

and
$$a^2\left(\frac{d\phi}{dt}\right)^2 = \frac{2e}{m}V_0 \tag{12.69}$$

Substituting Eq. (12.68) into Eq. (12.69), we get the following expression:

$$a^2\left\{\frac{1}{2}\omega_c\left(1 - \frac{b^2}{a^2}\right)\right\}^2 = \frac{2e}{m}V_0 \tag{12.70}$$

Substituting Eq. (12.62) into Eq. (12.70), the following relation can be obtained:

$$a^2\left\{\frac{e}{2m}B_0\left(1 - \frac{b^2}{a^2}\right)\right\}^2 = \frac{2e}{m}V_0 \tag{12.71}$$

Whether the electron reaches the anode or returns back to the cathode depends on the relative magnitude of the anode voltage V_0 and the magnetic flux density B_0. The cut-off values can be obtained from Eq. (12.71) as follows:

$$V_{0c} = \frac{m}{2e}a^2\left\{\frac{e}{2m}B_0\left(1 - \frac{b^2}{a^2}\right)\right\}^2 = \frac{e}{8m}B_0^2 a^2\left(1 - \frac{b^2}{a^2}\right)^2 \tag{12.72}$$

and
$$B_{0c} = -\frac{\sqrt{8V_0\frac{m}{e}}}{a\left(1 - \frac{b^2}{a^2}\right)} \tag{12.73}$$

Equation (12.72), also known as Hull cut-off voltage equation, reveals that if $V_0 < V_{0c}$ for a given B_0, then electrons will not reach the anode, while Eq. (12.73), known as Hull cut-off magnetic flux density equation, reveals that if $B_0 > B_{0c}$ for a given V_0, then electrons will not reach the anode.

The cathode current density of an inverted coaxial magnetron is about 1/10th of that of a conventional magnetron. This makes the realization of a long-life millimetre magnetron practical. In addition, the output waveguide of an inverted coaxial magnetron can be in the circular electric mode that has extremely low transmission loss. Sometimes, the basic problem of mode suppression prevents the use of inverted coaxial magnetrons.

EXAMPLE 12.3 A typical inverted coaxial magnetron has the following parameters: anode voltage = 12 kV, anode current = 2 A, anode radius = 4 cm, cathode radius = 5 cm, and magnetic flux density = 0.02 Wb/m². Calculate the (a) cut-off voltage for a fixed B_0 and (b) cut-off magnetic flux density for a fixed V_0.

Solution   Given: $V_0 = 12$ kV, $I_0 = 2$ A, $a = 4$ cm, $b = 5$ cm, and $B_0 = 0.02$ Wb/m²

(a)   The cut-off voltage for a fixed B_0 is calculated as follows:

$$V_{0c} = \frac{1}{8}\frac{e}{m}B_0^2 a^2\left(1-\frac{b^2}{a^2}\right)^2 = \frac{1}{8}\times 1.759\times 10^{11}\times\left(0.02\times 4\times 10^{-2}\right)^2\left\{1-\left(\frac{5\times 10^{-2}}{4\times 10^{-2}}\right)^2\right\}^2$$

or $V_{0c} = 4452.4687 = 4.4525$ kV

(b)   The cut-off magnetic flux density for a fixed V_0 is calculated as follows:

$$B_{0c} = -\frac{\sqrt{8V_0\frac{m}{e}}}{a\left(1-\frac{b^2}{a^2}\right)} = -\frac{\sqrt{\dfrac{8\times 12\times 10^3}{1.759\times 10^{11}}}}{4\times 10^{-2}\times\left\{1-\left(\dfrac{5\times 10^{-2}}{4\times 10^{-2}}\right)^2\right\}}$$

or $B_{0c} = \dfrac{7.3876\times 10^{-4}}{0.0225} = 0.0328$ Wb/m²

Practice Problem

12.3 A typical inverted coaxial magnetron has the following parameters: anode voltage = 15 kV, anode current = 2.5 A, anode radius = 5 cm, cathode radius = 6 cm, and magnetic flux density = 0.01 Wb/m². Calculate the (a) cut-off voltage for a fixed B_0 and (b) cut-off magnetic flux density for a fixed V_0. $\left[1.0642\ \text{kV},\ 0.0375\ \text{Wb}/\text{m}^2\right]$

12.2.5 Voltage Tunable Magnetron

A voltage tunable magnetron consists of a short cylindrical cathode, a control electrode, a low Q-resonator, an anode, and a sole, arranged as shown in Fig. 12.18. Electrons are emitted from the cathode and form a hollow beam by the applied electric and magnetic fields. The electrons are then accelerated radially outward from the cathode and injected into the region between the anode and the sole. Here the beam rotates about the sole at a rate controlled by the axial magnetic field and the DC voltage applied between the anode and the sole. In the pi-mode operation, bunching of the hollow beam occurs in the resonator, and the oscillation frequency is determined by the rotational velocity of the electron. Since the rotational velocity of the electron beam can be varied by varying the voltage between the anode and the sole, the output frequency of a voltage tunable magnetron can also be varied by varying this voltage.

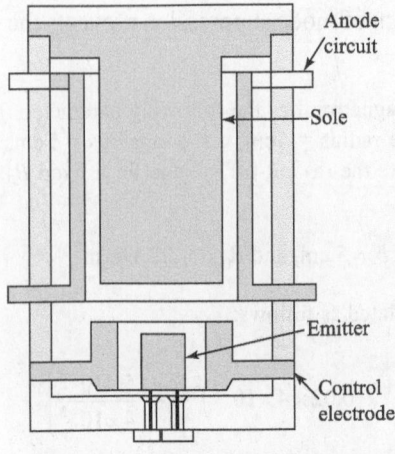

Fig. 12.18 Cross-sectional view of voltage tunable magnetron

A voltage variable magnetron is a broadband device and its output power can be adjusted up to some extent by the control electrode. For high-power and high-frequency applications, the percentage bandwidth of voltage tunable magnetron is limited. However, for low-power and low-frequency applications, the percentage bandwidth may approach 70%.

Since the output frequency of a voltage tunable magnetron can be varied by varying the voltage between the anode and the sole, such a magnetron can be used as a frequency modulator. In such applications, the modulating voltage is applied between the anode and the sole.

12.2.6 Frequency Agile Magnetron

In a frequency agile magnetron, the output frequency can also be varied; however, it differs from conventional tunable magnetrons. The frequency agility of such a magnetron is defined as the capability to tune the output frequency with sufficient high speed so that a pulse-to-pulse frequency change can be obtained. The frequency agile coaxial magnetrons can be classified as follows:

Dither magnetrons In dither magnetrons, the output RF frequency varies periodically with a constant excursion, constant rate, and has a fixed centre frequency. Earlier dither magnetrons used to have a piston that can be made to descend into the cavity, increasing and decreasing its volume and therefore tuning its operating frequency. The piston used to be operated by a processor-controlled servomotor, permitting very large frequency changes quickly. Dither tuning is also possible by electronic methods, yielding very rapid frequency changes during the transmission of pulse with a tuning range typically 1%.

Tunable/Dither magnetrons In tunable/dither magnetrons, the output RF frequency varies periodically with a constant excursion and constant rate; however, the centre frequency can be tuned manually by hand or mechanically by a servomotor.

Accutune magnetrons In accutune magnetrons, the output RF frequency variation is determined by the waveforms of an externally generated low-level voltage signal. If the tuning waveform is properly chosen, then an accutune magnetron can combine the features of dither and tunable/dither magnetrons, together with the capability of varying the excursion and rate.

The agile rate and agile excursion of a frequency agile magnetron is shown in Fig. 12.19. Agile rate is defined as the number of times per second that the transmitter frequency traverses the agile excursion and returns to its starting frequency. Agile excursion is defined as the total frequency variation of the transmitter during agile

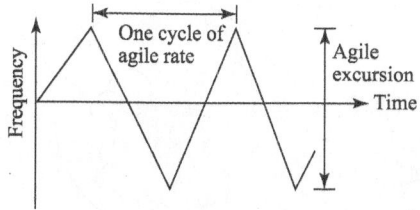

Fig. 12.19 Agile rate and agile excursion

operation. Mathematically, it can be expressed as follows:

$$\text{Agile excursion} = N/\tau \quad (12.74)$$

where N is the number of pulses placed on the target during one radar scan and τ is the shortest pulse duration used in the system.

The pulse repetition rate is given by the equation $f = \text{DC}/\tau$ (12.75)

where 'DC' is the duty cycle and is defined as $\text{DC} = \tau/T$ (12.76)

The agile rate can be expressed as Agile rate $= \dfrac{1}{2T}$ (12.77)

A typical frequency agile magnetron can provide 90 kW of peak output power with a centre frequency of 9.1 GHz and an accutune range of 1 GHz. Its pulse voltage is 15 kV and pulse current is 15 A.

12.2.7 Rising-sun Magnetron

In rising-sun magnetrons, two sets of resonators are used, and the operating frequency of the system lies in between the two resonant frequencies of these two sets of resonant structures. The number of resonators used in this magnetron is limited by practical mode separation requirement, while the size of the cathode is limited by the number of resonators. In practice, the size of the cathode decreases with an increase in the number of resonators for high-frequency operations. This, in turn, limits the input current and hence the output power. Frequencies of the different modes of a rising-sun magnetron are widely separated due to asymmetric cavity structures. Therefore, rising-sun magnetrons do not require strapping to prevent mode jumping. The schematic diagram of a rising-sun magnetron is shown in Fig. 12.20.

12.2.8 Negative Resistance Magnetron

A split anode negative resistance magnetron operates at a lower microwave frequency and is capable of providing higher output power than the basic magnetron. The general

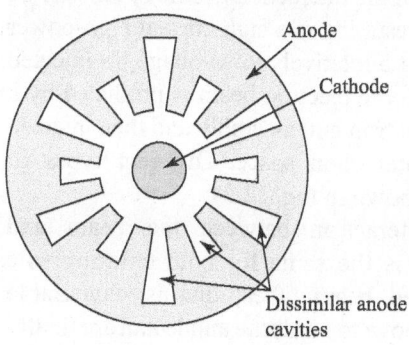

Fig. 12.20 Schematic diagram of rising-sun magnetron

construction of a split anode magnetron is shown in Fig. 12.21. The anodes of a split anode magnetron are operated at different potentials. Electrons leaving the cathode and moving towards the high-potential anode are deflected by the magnetic field and enter the electrostatic field generated by the low-potential anode after crossing the split. Here, the magnetic field has more effect on electrons and deflects them into a tighter curve. As a result, the electrons follow a path as shown in Fig. 12.22 and finally arrive at the low-potential anode.

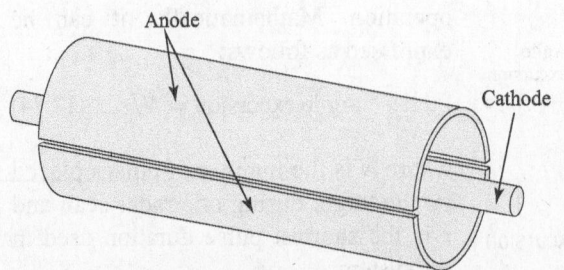
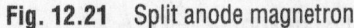

Fig. 12.21 Split anode magnetron

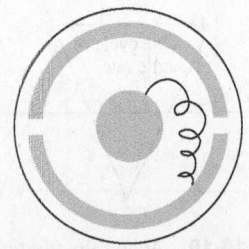

Fig. 12.22 Movement of electron in split anode magnetron

In a negative resistance magnetron, oscillations are initiated by applying a magnetic field that is slightly higher than the critical field required to force all the electrons to fail to reach the anode when its halves are operating at the same potential.

Since a very concentrated magnetic field is required for the operation of a negative resistance magnetron, the length of the anode of such a tube is limited to a few centimetres. In addition, the diameter of the anode is also limited due to its high operating frequencies. This overall reduction in the size makes the heating effect more severe in a split anode magnetron. In practice, a heavy-walled plate is used to increase the radiating property. Artificial cooling mechanisms, such as forced air- or water-cooled system, are also used to obtain higher dissipation of heat.

12.3 Forward-wave Crossed-field Amplifiers

Depending on the direction of the phase velocity, group velocity, and energy, cross-field amplifiers (CFAs) are classified as FWCFAs and BWCFAs. In FWCFAs, helix-type slow-wave structures are often used, whereas strapped bar lines are a satisfactory choice for BWCFAs. A schematic diagram of a strapped CFA is shown in Fig. 12.23.

Depending on the method by which the electrons reach the interaction space and the process used to control them, CFAs can also be classified as emitting sole type and injected beam type. In emitting sole tubes, current is a function of the applied voltage, dimensions, and emission property of the cathode. In these tubes, the

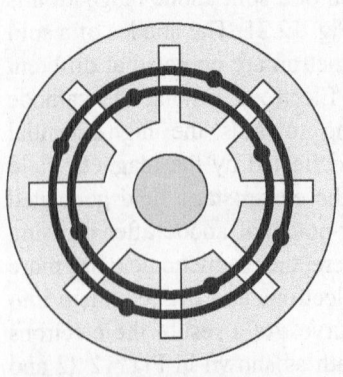

Fig. 12.23 Strapped FWCFA

perveance of the interaction geometry tends to be very high, resulting in a high-current high-power capability at a relatively low voltage. In injected beam tubes, the electron beam is produced by a separate electron gun assembly and then injected into the interaction space. Different types of CFAs are shown in Fig. 12.24.

The interaction between the beam and the circuit is the same for both emitting sole and injected beam tubes; that is, favourable electrons move towards the anode and are finally collected at the collector, while unfavourable electrons return to the cathode. In practice, in

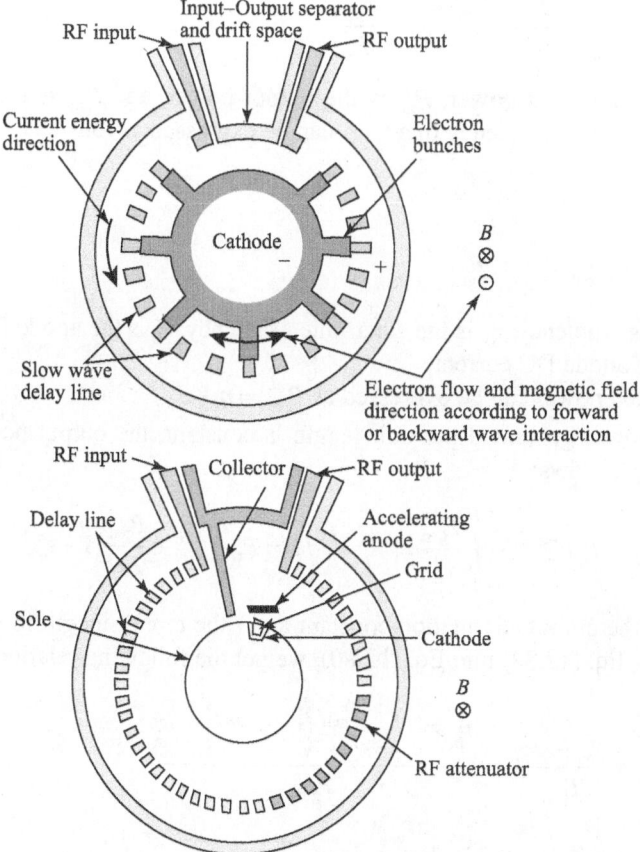

Fig. 12.24 Different types of FWCFAs

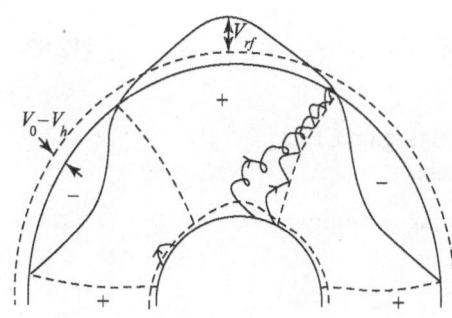

Fig. 12.25 Motion of electron in FWCFA

CFAs electrons face three forces: (a) DC electric field force, (b) magnetic field force, and (c) space charge force from other electrons. Under the influence of these forces, electrons move spirally along equipotentials, as shown in Fig. 12.25. This motion can be explained as follows. When the spoke is at positive potential, under the influence of an applied RF field, the electron speeds up to the anode, whereas when the spoke is at negative potential, under the influence of an applied RF field, the electron returns towards the cathode.

If the input power exceeds the threshold value for the spoke stability at the input, then the power generated by a CFA is independent of the input RF power and can be increased only by increasing the anode voltage and current. For negligible circuit attenuation, the circuit gain can be written as follows:

$$g = \frac{P_{\text{out}}}{P_{\text{in}}} = \frac{P_{\text{in}} + P_{\text{gen}}}{P_{\text{in}}} = 1 + \frac{P_{\text{gen}}}{P_{\text{in}}} \tag{12.78}$$

where P_{in} is the input power, P_{out} is the output power, and P_{gen} is the generated power. The overall efficiency of a CFA can be expressed as follows:

$$\eta = \eta_c \eta_e = \frac{P_{\text{out}} - P_{\text{in}}}{V_0 I_0}. \tag{12.79}$$

where

$$\eta_c = (P_{\text{out}} - P_{\text{in}})/P_{\text{gen}} \tag{12.80}$$

is the circuit efficiency, η_e is the electronic efficiency, V_0 is the anode DC voltage, and I_0 is the anode DC current.

The generated power can be expressed as $P_{\text{gen}} = \eta_e V_0 I_0$ (12.81)

Since the power generated per unit length is constant, the output power can be expressed as follows:

$$P_{\text{out}} = P_{\text{in}} e^{-2\alpha l} + \int_0^l \frac{P_{\text{gen}}}{l} e^{-2\alpha(l-\phi)} d\phi = P_{\text{in}} e^{-2\alpha l} + \frac{P_{\text{gen}}}{2\alpha l}\left(1 - e^{-2\alpha l}\right) \tag{12.82}$$

where α is the circuit attenuation constant and l the circuit length in ϕ direction. Substituting Eq. (12.82) into Eq. (12.80), we get the following relation:

$$\eta_c = \frac{P_{\text{out}} - P_{\text{in}}}{P_{\text{gen}}} = \frac{P_{\text{in}} e^{-2\alpha l} + \dfrac{P_{\text{gen}}}{2\alpha l}\left(1 - e^{-2\alpha l}\right) - P_{\text{in}}}{P_{\text{gen}}}$$

or

$$\eta_c = \frac{-P_{\text{in}}\left(1 - e^{-2\alpha l}\right) + \dfrac{P_{\text{gen}}}{2\alpha l}\left(1 - e^{-2\alpha l}\right)}{P_{\text{gen}}}$$

or

$$\eta_c = \left(\frac{1}{2\alpha l} - \frac{P_{\text{in}}}{P_{\text{gen}}}\right)\left(1 - e^{-2\alpha l}\right) \tag{12.83}$$

The term $P_{\text{in}}/P_{\text{gen}}$ becomes negligible for high-gain CFAs.

Now we assume the following conditions:

1. Input signal is sufficiently strong for spoke stability
2. DC current for the spoke is constant.
3. RF power grows linearly with distance.
4. Back bombardment loss is negligible.

Under these assumptions, the electronic efficiency can be expressed as follows:

$$\eta_e = \left(1 - \frac{m\omega^2}{2eV_0\beta^2}\right) \bigg/ \left[1 + \frac{I_0}{B^2}\frac{m\beta^2 Z_c}{2e}\left(\frac{g+1}{g-1}\right)\right] \tag{12.84}$$

where B is the magnetic flux density, Z_c is the beam coupling impedance, m is the mass of electron, e is the charge of electron, and β is the phase constant.

CFAs are characterized by high efficiency, low or moderate power gain, moderate bandwidth, high perveance, small size, low weight, and saturated amplification. Due to these features, CFAs are widely used in low-power and high-reliability space communications and high-average-power coherent pulsed radars. A typical CFA can produce 14 kW pulsed power with a 44–56 kV anode voltage and 58 A anode current over a frequency range 2.9–3.1 GHz.

EXAMPLE 12.4 A typical CFA operates with the following operating parameters: anode voltage = 2.5 kV, anode current = 1.8 A, electronic efficiency = 18%, and RF input power = 75 W. Calculate the (a) induced RF power, (b) total RF output power, and (c) power gain in dB.

Solution Given: $V_{a0} = 2.5$ kV, $I_{a0} = 1.8$ A, $\eta_e = 0.18$, and $P_{in} = 75$ W

(a) The induced RF power is calculated as follows:

$$P_{gen} = \eta_e V_{a0} I_{a0} = 0.18 \times 2.5 \times 10^3 \times 1.8 = 810 \text{ W}$$

(b) The following is the total RF output power:

$$P_{out} = P_{in} + P_{gen} = 75 + 810 = 885 \text{ W}$$

(c) The power gain is estimated to be as follows:

$$g = P_{out}/P_{in} = 885/75 = 11.8 \text{ dB}$$

Practice Problem

12.4 A typical CFA operates with the following operating parameters: anode voltage = 1.8 kV, anode current = 2 A, electronic efficiency = 22%, and RF input power = 100 W. Calculate the (a) induced RF power, (b) total RF output power, and (c) power gain in dB.   **[792 W, 892 W, 8.92 dB]**

12.4 Backward-wave Crossed-field Amplifiers

The schematic diagram of a BWCFA, or amplitron, is shown in Fig. 12.26. In this tube, the cavity and a pair of pins are excited in opposite phases using a strap line. The anode cavity and pins comprise the resonant circuit.

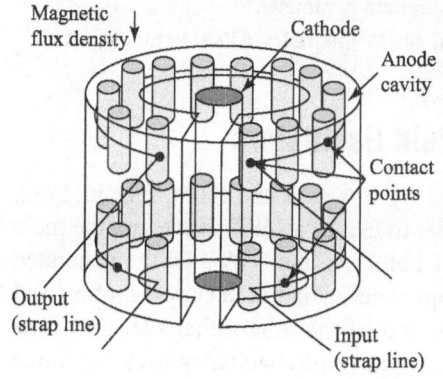

Fig. 12.26 Schematic diagram of BWCFA

A BWCFA can deliver 3 MW pulse power with a gain of about 8 dB in the S-band. Amplitron, having output power from a few hundred kilowatts to 3 MW with a gain as high as 16 dB, and efficiencies ranging from 60% for normal power to 76% for high power low gain operation has also been reported. In addition, two-stage super-power amplitrons with 425 kW of CW power, with a gain of 9 dB, an efficiency of 76%, and a bandwidth of 3% at a centre frequency of 3 GHz are also available.

EXAMPLE 12.5 A typical amplitron operates with the following operating parameters: anode voltage = 14 kV, anode current = 2.5 A, magnetic flux density = 0.18 Wb/m², frequency of operation = 9 GHz, and characteristic impedance = 50 Ω. Calculate the (a) DC electron beam velocity, (b) electron beam phase constant, (c) cyclotron angular frequency, (d) cyclotron phase constant, and (e) gain parameter.

Solution   Given: $V_0 = 14$ kV, $I_0 = 2.5$ A, $B_0 = 0.18$ Wb/m², $f = 9$ GHz, and $Z_0 = 50$ Ω

(a) The DC electron beam velocity is as follows:

$$v_0 = 0.593 \times 10^6 \sqrt{V_0} = 0.593 \times 10^6 \times \sqrt{14 \times 10^3} = 7.0165 \times 10^7 \text{ m/s}$$

(b) The following is the electron beam phase constant:

$$\beta_e = \frac{\omega}{v_0} = \frac{2\pi \times 9 \times 10^9}{7.0165 \times 10^7} = 805.9384 \text{ rad/m}$$

(c) The cyclotron angular frequency is estimated to be as follows:

$$\omega_c = \frac{e}{m} B_0 = 1.759 \times 10^{11} \times 0.18 = 31.662 \times 10^9 \text{ rad/s}$$

(d) The following is the cyclotron phase constant:

$$\beta_m = \frac{\omega_c}{v_0} = \frac{31.662 \times 10^9}{7.0165 \times 10^7} = 451.2506 \text{ rad/m}$$

(e) The gain parameter is calculated as follows:

$$C = \left(\frac{I_0 Z_0}{4 V_0}\right)^{\frac{1}{3}} = \left(\frac{2.5 \times 50}{4 \times 14 \times 10^3}\right)^{\frac{1}{3}} = 0.1307$$

Practice Problem

12.5 A typical amplitron operates with the following operating parameters: anode voltage = 16 kV, anode current = 2.5 A, magnetic flux density = 0.23 Wb/m², frequency of operation = 10 GHz, and characteristic impedance = 50 Ω. Calculate the (a) DC electron beam velocity, (b) electron beam phase constant, (c) cyclotron angular frequency, (d) cyclotron phase constant, and (e) gain parameter.

$$\left[7.5009 \times 10^7 \text{ m/s}, \; 837.6575 \text{ rad/m}, \; 40.457 \times 10^9 \text{ rad/s}, \; 539.3619 \text{ rad/m}, \; 0.125\right]$$

12.5 Backward-wave Crossed-field Oscillators

The BWCFO or carcinotron was invented by Epsztein and Kompfner in 1952, and its basic principle of operation is similar to that of a TWT. However, the main difference between a BWCFO and a TWT is that, in the BWCFO, the generated microwave signal moves in a direction opposite to that of electron motion and the output is taken from the electron gun end of the tube rather than from the collector end. In general, BWCFOs can be divided into two categories: (a) linear M-carcinotron and (b) circular M-carcinotron.

12.5.1 Linear M-carcinotron

The schematic diagram of a linear M-carcinotron is shown in Fig. 12.27. The slow-wave structure is in parallel with an electrode, known as the sole, and is terminated at the collector end. A DC electric field is provided between the grounded slow-wave structure and the sole, whereas a DC magnetic field is maintained in a direction perpendicular to it, that is, into the page. After emission from the electrode, electrons are bent at an angle of 90° by the magnetic field; thereafter, they enter the interaction region. Here, they interact with a backward-wave space harmonic of the circuit and results in a flow of energy in the circuit that is opposite in direction to that of electron motion. Therefore, the RF output signal is taken at the electron gun end. The efficiency of a linear M-carcinotron ranges from 30% to 60%.

12.5.2 Circular M-carcinotron

The schematic diagram of a circular M-carcinotron is shown in Fig. 12.28. Here, the slow-wave structure and the sole are circular and nearly reentrant to conserve the weight of the magnet. In this configuration, the delay line is terminated at the collector end using an attenuating material on the surface of the conductors. The output is taken from the gun end. The electron drift velocity in this tube must be synchronized with a backward space harmonic.

For a circular M-carcinotron, the oscillation condition can be written as follows:

$$2\beta_e Dl = (2n+1)\pi \qquad (12.85)$$

where n is any integer number. Now let us define N as follows:

$$\beta_e l = 2\pi N \qquad (12.86)$$

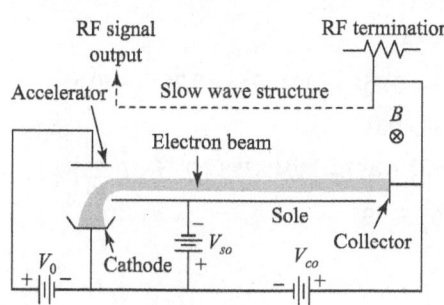

Fig. 12.27 Schematic diagram of linear M-carcinotron

Substituting Eq. (12.86) into Eq. (12.85), we get the following expression:

$$DN = (2n+1)/4 \qquad (12.87)$$

If we define

$$\delta_1 = j\frac{b-\sqrt{b^2+4}}{2} \qquad (12.88)$$

and

$$\delta_2 = j\frac{b+\sqrt{b^2+4}}{2} \qquad (12.89)$$

then the propagation constants can be written as follows:

$$\gamma_1 = j(\beta_e + b) + \beta_e D\delta_1 \qquad (12.90)$$

and

$$\gamma_2 = j(\beta_e + b) + \beta_e D\delta_2 \qquad (12.91)$$

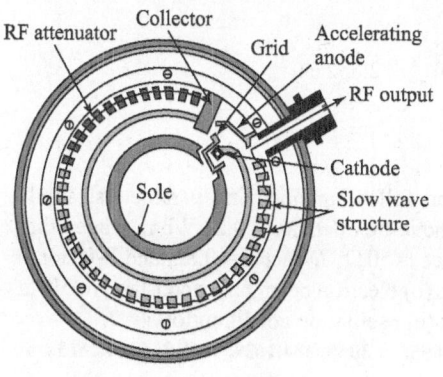

Fig. 12.28 Schematic diagram of circular M-carcinotron

EXAMPLE 12.6 A typical circular carcinotron has the following operating parameters: anode voltage = 18 kV, anode current = 3.3 A, magnetic flux density = 0.32 Wb/m², operating frequency = 5 GHz, characteristic impedance = 50 Ω, D-factor = 0.75, and b-factor = 0.45. Calculate the (a) DC electron velocity, (b) electron beam phase constant, (c) delta differentials, (d) propagation constants, and (e) and oscillation condition for $n = 2$.

Solution Given:

$$V_0 = 18 \text{ kV}, \ I_0 = 3.3 \text{ A}, \ B_0 = 0.32 \text{ Wb/m}^2, \ f = 5 \text{ GHz}, \ Z_0 = 50 \, \Omega, \ D = 0.75,$$
$$b = 0.45, \text{ and } n = 2$$

(a) The DC electron velocity is calculated to be the following:

$$v_0 = 0.593 \times 10^6 \sqrt{V_0} = 0.593 \times 10^6 \times \sqrt{18 \times 10^3} = 7.9559 \times 10^7 \text{ m/s}$$

(b) The following is the electron beam phase constant:

$$\beta_e = \frac{\omega}{v_0} = \frac{2\pi \times 5 \times 10^9}{7.9559 \times 10^7} = 394.8758 \text{ rad/m}$$

(c) The delta differentials are as follows:

$$\delta_1 = j\frac{b - \sqrt{b^2 + 4}}{2} = j\frac{0.45 - \sqrt{0.45^2 + 4}}{2} = -j0.8$$

$$\delta_2 = j\frac{b + \sqrt{b^2 + 4}}{2} = j\frac{0.45 + \sqrt{0.45^2 + 4}}{2} = j1.25$$

(d) The propagation constants are as follows:

$$\gamma_1 = j(\beta_e + b) + \beta_e D\delta_1 = j(394.8758 + 0.45) + \{394.8758 \times 0.75 \times (-j0.8)\}$$
$$\text{or } \gamma_1 = j395.3258 - j236.9255 = j158.4003$$
$$\gamma_2 = j(\beta_e + b) + \beta_e D\delta_2 = j(394.8758 + 0.45) + \{394.8758 \times 0.75 \times j1.25\}$$
$$\text{or } \gamma_1 = j395.3258 + j370.1961 = j765.5219$$

(e) For $n = 2$:

$$DN = \frac{2n + 1}{4} = \frac{5}{4} = 1.25$$

$$\text{or } N = \frac{1.25}{D} = \frac{1.25}{0.75} = 1.6667$$

Therefore,

$$l = \frac{2\pi N}{\beta_e} = \frac{2\pi \times 1.6667}{394.8758} = 2.652 \times 10^{-2} = 2.652 \text{ cm}$$

Practice Problem

12.6 A typical circular carcinotron has the following operating parameters: anode voltage = 20 kV, anode current = 3 A, magnetic flux density = 0.28 Wb/m², operating frequency = 3 GHz, characteristic impedance = 50 Ω, D-factor = 0.82, and b-factor = 0.54. Calculate the (a) DC electron velocity, (b) electron beam phase constant, (c) delta differentials, (d) propagation constants, and (e) oscillation condition for $n = 2$.

$$\left[8.3863 \times 10^7 \text{ m/s}, \ 224.7661 \text{ rad/m}, -j0.7658, \ j1.3058, \ j84.1629, \ j465.9758, \ 4.2613 \text{ cm} \right]$$

12.6 Gyrotrons

The tubes we have discussed so far use either a resonator cavity or a slow-wave structure to support the RF field for interaction with the electron beam. Dimensions of these structures are proportional to the operating wavelength of the device and hence become very small for millimetre-wave applications. Hence, at a high frequency, the available area for the interaction between the electron beam and the RF field decreases, which in turn, decreases the output power. Hence, conventional microwave tubes cannot be used above 100 GHz for high-power (of the order of kW) applications. To overcome this problem, gyrotrons have been developed.

A gyrotron does not rely on the use of a resonant cavity or slow-wave structure for its operation and hence the preceding problem does not arise. In this device, the RF field interacts with the electrons in a cyclotron motion in the presence of a strong static magnetic field. Under the influence of the static magnetic field, electrons move in a circular path, in a plane perpendicular to the magnetic field. Now if a sufficiently high magnetic field is used, then the frequency of rotation of the electrons, called the cyclotron frequency, can be of the order of the millimetre waves; hence, they can interact efficiently with the RF field having a frequency same as the cyclotron frequency. As a result of the interaction, the RF field grows in amplitude and amplification takes place.

In general, gyrotrons are of three types: (a) gyromonotron, (b) gyro-TWT, and (iii) gyroklystron. Gyromonotrons are oscillators, whereas Gyro-TWTs and gyroklystrons are amplifiers. Simplified drawings of these structures are shown in Fig. 12.29. Each of these tubes has a magnetron-type electron gun that imparts a high radial velocity to the electrons before they enter the high magnetic field region. The magnetic field is produced by a liquid-cooled or superconducting solenoid. In the gyromonotron, shown in Fig. 12.29(a), the interaction region is a circular waveguide, supporting a number of propagating modes, and the output is taken from an output waveguide through a transparent window. The transparent window also acts as a vacuum seal for the tube. In the case of gyro-TWTs, the input signal is coupled to the input of the interaction region through a waveguide. This input signal results in the initial bunching of the electron beam. As before, the interaction region is a circular waveguide that supports different higher-order modes. The radius of the electron beam, which is of the form of a hollow beam, is such that it can interact strongly with one, or at most only a few, of the modes supported by the waveguide and hence results in amplification. The schematic diagram of a gyro-TWT is shown in Fig. 12.29(b). The input and output cavity structures of a Gyroklystron are shown in Fig. 12.29(c).

Gyrotrons can produce an output power as high as 50 kW at 5 GHz with a gain of 24 dB. In addition, gyrotrons with 128 kW output power at 5.2 GHz operating with a 65 kV, 8 A electron beam and those with 200 kW output power operating at 28 GHz have also been reported. The pulsed power of gyrotrons is of the order of several hundred MW.

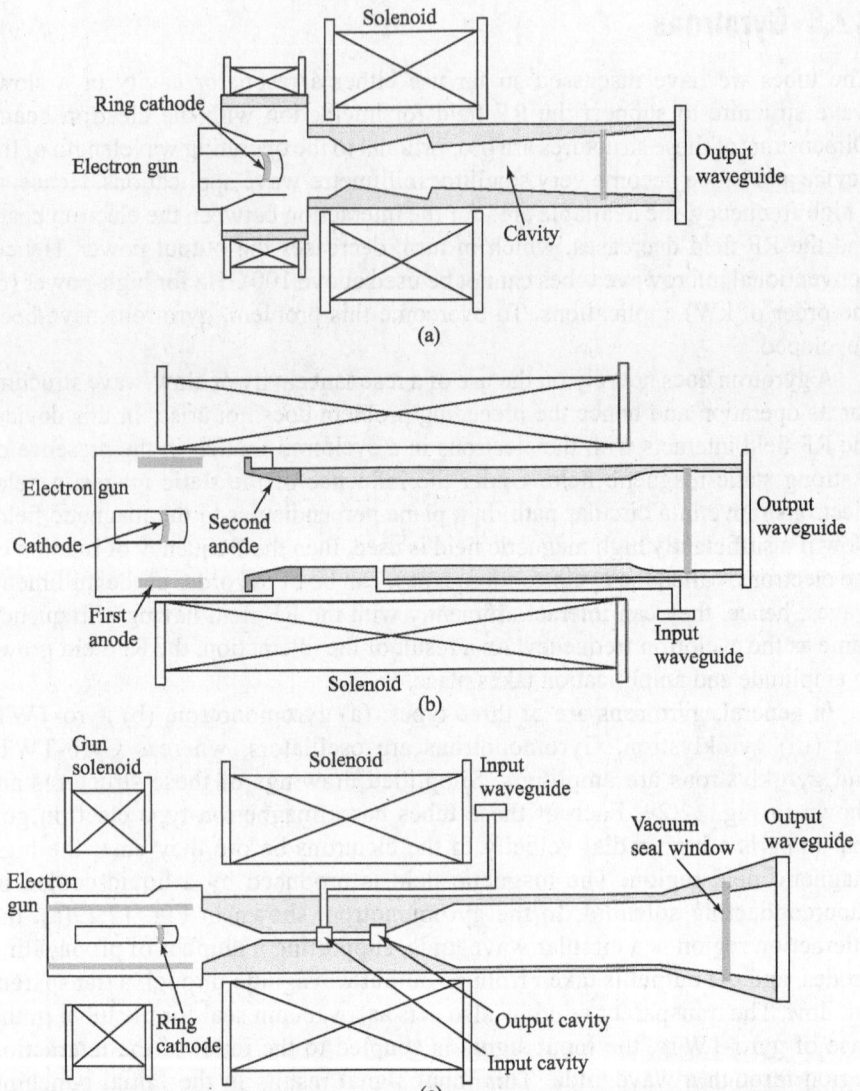

Fig. 12.29 Types of gyrotrons (a) Gyromonotron oscillator (b) Gyro-TWT amplifier (c) Gyroklystron amplifier

Important Formulae

- Cyclotron frequency of electrons in a cylindrical magnetron is expressed as

$$\omega_c = \frac{eB}{m} = \sqrt{\frac{8e}{m}V_0} \bigg/ \left[b\left(1 - \frac{a^2}{b^2}\right) \right] = \beta_0 \frac{d\phi}{dt}.$$

- For a cylindrical magnetron, the Hull cut-off magnetic flux density for a constant anode voltage can be expressed as $B_{0c} = \sqrt{\frac{8m}{e}V_0} \bigg/ \left[b\left(1 - \frac{a^2}{b^2}\right) \right]$.

- For a cylindrical magnetron, the Hull cut-off anode voltage for a constant magnetic flux density can be expressed as $V_{0c} = \dfrac{e}{8m} B_0^2 b^2 \left(1 - \dfrac{a^2}{b^2}\right)^2$.

- For a cylindrical magnetron, the Hartree voltage can be expressed as $V_{0h} = \dfrac{\omega}{N}\left(b^2 - a^2\right) B_0$.

- For a linear magnetron, the Hull cut-off magnetic flux density for a constant anode voltage can be expressed as $B_{0c} = \dfrac{1}{d}\sqrt{\dfrac{2m}{e}} V_0$.

- For a linear magnetron, the Hull cut-off anode voltage for a constant magnetic flux density can be expressed as $V_{0c} = \dfrac{1}{2}\dfrac{e}{m} B_0^2 d^2$.

- For a linear magnetron, the potential at the hub surface is expressed as $V(h) = \dfrac{eB_0^2}{2m} h^2$.

- For a linear magnetron, the electric field at the hub surface is expressed as $E_x(h) = -\dfrac{eB_0^2}{m} h$.

- For a linear magnetron, the Hartree voltage can be expressed as $V_{0h} = \dfrac{\omega B_0 d}{\beta} - \dfrac{m\omega^2}{2e\beta^2}$.

- For an inverted coaxial magnetron, the Hull cut-off magnetic flux density for a constant anode voltage can be expressed as $B_{0c} = \sqrt{8V_0 \dfrac{m}{e}} \left/ \left[a\left(1 - \dfrac{b^2}{a^2}\right)\right]\right.$.

- For an inverted coaxial magnetron, the Hull cut-off anode voltage for a constant magnetic flux density can be expressed as $V_{0c} = \dfrac{e}{8m} B_0^2 a^2 \left(1 - \dfrac{b^2}{a^2}\right)^2$.

Exercises

Objective-type Questions

12.1 In the pi-mode operation of magnetrons, due to the phase focusing effect, spokes rotate at an angular velocity corresponding to
(a) one pole/ cycle
(b) two poles/cycle
(c) four poles/cycle
(d) eight poles/cycle

12.2 In a cavity magnetron, strapping is used to
(a) improve the phase focusing effect
(b) prevent cathode back heating
(c) ensure bunching
(d) prevent mode jumping

12.3 A magnetron whose oscillating frequency is electronically adjustable over a wide range is called
(a) a coaxial magnetron
(b) a rising-sun magnetron
(c) an inverted coaxial magnetron
(d) a voltage tunable magnetron.

12.4 A dither tuned magnetron
(a) can be easily jammed
(b) can operate at a fixed frequency
(c) operates on a fixed centre frequency
(d) may be conventional or coaxial

12.5 Which of the flowing microwave tube amplifier uses an axial magnetic field and a radial electric field?
(a) Klystron
(b) Crossed-field amplifier
(c) Travelling-wave magnetron
(c) Reflex klystron

12.6 A magnetron is a
  (a) High-power microwave amplifier
  (b) low-power microwave amplifier
  (c) high-power microwave oscillator
  (d) low-power microwave oscillator

12.7 The cathode material used in a magnetron is
  (a) tungsten
  (b) silver
  (c) copper
  (d) barium and strontium oxides

12.8 To overcome the difficulties associated with strapping at high frequencies, the type of cavity structure desired for a magnetron is
  (a) slot
  (b) rising sun
  (c) vane
  (d) hole and slot

12.9 A cavity magnetron uses a magnetic field to
  (a) ensure that electrons orbit around the cathode
  (b) ensure broad band operation
  (c) ensure high-power operation
  (d) help in focusing the electron beam

12.10 A voltage tunable magnetron uses
  (a) a low Q-cavity
  (b) an extra injection electrode to help bunching
  (c) cold cathodes
  (d) all of these

12.11 Voltage tunable magnetrons are capable of generating frequencies in the range from
  (a) 50 kHz to 5 MHz
  (b) 5 to 50 MHz
  (c) 50 to 500 MHz,
  (d) 500 MHz to X-band.

12.12 In a magnetron, diode plate current is greatly influenced by the
  (a) plate voltage
  (b) operating frequency
  (c) magnetic field
  (d) electric field

12.13 In a frequency agile magnetron, the piston is operated by
  (a) a servomotor
  (b) the magnetic field
  (c) the electrostatic field
  (d) all of these

12.14 A travelling-wave magnetron having N resonators, with adjacent resonators tightly coupled, has
  (a) $(2N - 1)$ possible modes of resonance
  (b) N possible modes of resonance
  (c) $2N$ possible modes of resonance
  (d) $(2N + 1)$ possible modes of resonance

12.15 In a magnetron, electrons travel in a cycloidal path because
  (a) the anode is negative
  (b) the cathode is positive
  (c) permanent magnet generates a strong field
  (d) permanent magnet generates a low field

Review Questions

12.1 Describe the construction and operation of a cylindrical magnetron in the pi-mode.
12.2 With a suitable figure, describe the phase focusing effect of a cylindrical magnetron.

12.3 Starting from the basic equation of motion, find out an expression for the (a) cyclotron frequency, (b) Hull cut-off voltage for a fixed magnetic flux density, and (c) Hull cut-off magnetic flux density for a fixed beam voltage for a cylindrical magnetron.

12.4 What is strapping of a magnetron? Why is it used?

12.5 What is meant by *frequency pushing* and *frequency pulling* of a magnetron?

12.6 Starting from the basic equation of motion, find out an expression for the (a) cyclotron frequency, (b) Hull cut-off voltage for a fixed magnetic flux density, (c) Hull cut-off magnetic flux density for a fixed beam voltage, and (d) Hartree voltage for a linear magnetron.

12.7 Starting from basic equation of motion, find out an expression for the (a) Hull cut-off voltage for a fixed magnetic flux density and (b) Hull cut-off magnetic flux density for a fixed beam voltage for an inverted coaxial magnetron.

12.8 Write short notes on the following: (a) FWCFA, (b) BWCFO, and (c) gyrotron.

Problems

12.1 A typical X-band pulsed magnetron has the following operating parameters: anode voltage = 6 kV, beam current = 5 A, frequency of operation = 10 GHz, resonator conductance = 2.5×10^{-4} ℧, loaded conductance = 3×10^{-5} ℧, vane capacitance = 2 pF, duty cycle = 0.002, and power loss = 17 kW. Calculate the (a) unloaded quality factor, (b) loaded quality factor, (c) external quality factor, (d) circuit efficiency, and (e) electronic efficiency. *[Hint: Use Eqs (12.25), (12.27), (12.26), (12.28), and (12.29)]*

12.2 A typical linear magnetron has the following operating parameters: anode voltage = 16 kV, cathode current = 1.5 A, frequency = 10 GHz, magnetic flux density = 0.02 Wb/m^2, hub thickness = 2.5 cm, and distance between anode and cathode = 6 cm. Calculate the (a) electron velocity at the hub surface, (b) phase velocity, and (c) Hartree anode voltage. *[Hint: Use Eqs (12.57)–(12.59)]*

12.3 A typical frequency agile magnetron has the following operating parameters: pulse duration = 0.15, 0.3, and 0.6 μs; duty cycle = 0.001; and pulse rate on target 15 per scan. Calculate the (a) agile excursion, (b) pulse to pulse frequency separation, (c) signal frequency, (d) time for N pulses, and (e) agile rate. *[Hint: Use Eqs (12.74), (12.75) and (12.77)]*

13 Microwave Solid-state Diodes

13.1 Introduction

In the last two chapters, we discussed the different types of microwave vacuum tube devices. Various other solid-state devices, such as Gunn diode, Read diode, impact ionization avalanche transit time (IMPATT) diode, trapped plasma avalanche-triggered transit (TRAPATT) diode, barrier-injected transit time (BARITT) diode, and tunnel diode, are also widely used in microwave systems for generation, amplification, or detection of microwave signals. Such devices are compact and can also be integrated in a printed circuit board (PCB). However, they suffer from some major drawbacks, including low efficiency above 10 GHz, high noise, large dependence of frequency on temperature, and small tuning range.

Solid-state microwave sources can be classified as follows: (a) three-terminal devices (bipolar and field effect transistors), (b) transferred electron devices (Gunn diode, InP diode, CdTe diode), (c) avalanche transit-time devices (IMPATT, TRAPATT, BARITT diode, parametric devices), (d) varactor diodes, (e) quantum electronic devices (tunnel diodes, MASERS), (f) semiconductor lasers, and (g) infrared devices. Solid-state devices can also be categorized as follows, based on their electrical characteristics: (a) non-linear resistance type (varistors), (b) non-linear reactance type (varactors), (c) negative-resistance type (tunnel, IMPATT, and Gunn diodes), and (d) controllable impedance type (p–i–n junction diodes or PIN diodes). They can further be classified based on their contact type into the following four categories: (a) point-contact diode, (b) Schottky barrier diode, (c) metal–oxide–semiconductor devices, and (d) metal insulator devices.

Since the development of the semiconductor device theory, scientists have made numerous efforts to design a two-terminal negative-resistance device; tunnel diodes were the first such devices to be realized in practice in 1958. A tunnel diode is basically a forward-biased, heavily doped, p–n junction diode. Later transferred electron devices and avalanche transit-time devices were developed, which exhibit a negative resistance.

Tunnel diodes were discovered in 1957, when Leo Esaki was working on his Ph D thesis at Sony Corporation. He was investigating the breakdown voltage of

Ge diodes as a function of doping concentration. Since the result was almost well known and quite expected, the topic appeared unpromising initially. Esaki was probably aware of this and hence started employing extreme doping levels that no one had tried before. This study initially showed that heavier doping reduces breakdown voltage, which was expected. What was novel in his study was that with extremely heavy doping, the apparent breakdown voltage reduced to less than zero. With the increase in bias voltage, such a diode also exhibited a negative resistance over a certain period. If the forward bias was increased further, its characteristics eventually converged to those of a standard diode. Esaki explained this behaviour as a result of quantum mechanical tunnelling—a concept that physicists, then, found useful for considering various phenomena but no one had observed directly. Esaki's direct demonstration of tunnelling in such a simple structure was so dramatic that he was awarded a Nobel Prize in Physics in 1973.

When compared to tunnel diodes, transferred electron devices, also known as Gunn diodes, are basically bulk semiconductor samples that operate as negative-resistance devices on the application of a DC voltage above a threshold level. The oscillation frequency of such a diode depends on the natural resonance frequency of the circuit and also on the load.

The third type of negative-resistance device, avalanche transit-time device, uses the carrier impact ionization and drift in the high-field region of a semiconductor junction to produce a negative resistance. Such a device was originally proposed by Read in 1958 who found the negative-resistance property of an $n^+ - p - i - p^+$ diode. Named after him, such diodes are also called Read diodes. Such negative-resistance devices operate in two modes: IMPATT and TRAPATT. IMPATT diodes can operate at a frequency as high as 100 GHz with a DC to RF efficiency of 5–10%, whereas TRAPATT diodes can operate at a lower GHz frequency with a DC to RF efficiency of 20–60%.

Another semiconductor device, called BARITT diode, can also produce a negative resistance at microwave frequencies. These diodes can have several structures, such as p–n–p, p–n–v–p, p–n–metal, and metal–n–metal, and have a long drift region similar to IMPATT diodes. However, in contrast to IMPATT diodes, carriers, traversing the drift region of BARITT diodes, are generated by minority carrier injection from the forward-biased junction, rather than being extracted from the plasma of an avalanche region. BARITT diodes have low noise figures of 15 dB. They also have some major drawbacks such as low output power and narrow bandwidths.

During the end of the Second World War, researchers at MIT Radiation Laboratory observed a puzzling phenomenon: under certain circumstances, an ordinary point-contact diode provides power gain. No violation of energy was involved, as both a DC bias and a local oscillator were used as potential power sources. However, how energy from either or both of these sources could end up at the signal frequency was unclear. Later, it was revealed that the presence of non-linear junction capacitance was responsible for this effect. Today this phenomenon is known as parametric amplification. A parametric amplifier utilizes non-linear reactance or reactance that can be varied as a function of time by applying a suitable pump signal. The possibility of parametric amplification was first shown theoretically by Lord Rayleigh in 1831. However, the first analysis of non-linear capacitance was given by van der Ziel in 1948. He pointed out that since the diode essentially is a reactive device in which no thermal noise is generated, it can be useful as a low-noise amplifier. The first microwave parametric amplifier, however, was realized by Weiss following the earlier proposal of Suhl, suggesting non-linear effects in ferrites. Later, with the development of varactor diodes, semiconductor diode parametric amplifiers were also developed.

13.2 Tunnel and Backward Diodes

A tunnel diode is a heavily doped negative-resistance semiconductor p–n junction diode. The doping level in such a diode is about 10^{19}–10^{20} atoms/cm^3, which restricts the width of the depletion region at the junction to about 100 Å or 10^{-6} cm.

The negative resistance produced by a tunnel diode is due to *tunnel effect*. Tunnelling is a quantum mechanical effect and cannot be described simply by the classical theory. Classically, carriers can cross this potential barrier only if they have energy greater than or at least equal to the height of the barrier. However, quantum mechanically, if the barrier is less than a few Å there is an appreciable probability that the carriers will tunnel through it even though they do not have enough energy to pass over the same barrier. In addition to a thin barrier, tunnelling also requires a filled energy state on one side of the barrier, whereas an empty energy state on the other side. Under such circumstances, carriers from the filled energy state will tunnel through the barrier to the empty energy state. It may be noted that the tunnelling effect is a majority carrier effect and it is not governed by classical transit-time concept but rather by quantum transition probability per unit time.

To understand the tunnel effect fully, let us consider the energy-band diagram of a heavily doped p–n junction diode shown in Fig. 13.1. Under the open-circuit or zero-bias equilibrium condition, upper levels of electron energy of both n- and p-type semiconductors are lined up at the same Fermi level, as shown in Fig. 13.1. However, since the filled energy state of one side of the barrier is not at the same energy level as that of the empty state on the other side, no flow of carriers takes place across the junction, and thus the current flow is zero. This is shown by point A on the *I–V* characteristic graph of a tunnel diode shown in Fig. 13.2.

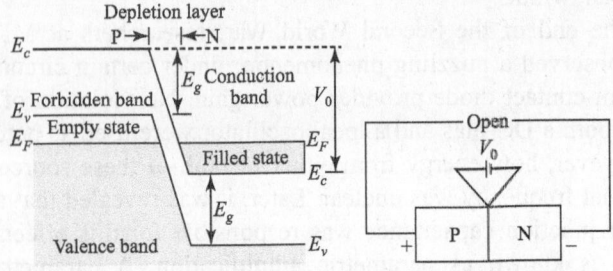

Fig. 13.1 Tunnel diode under zero-bias equilibrium

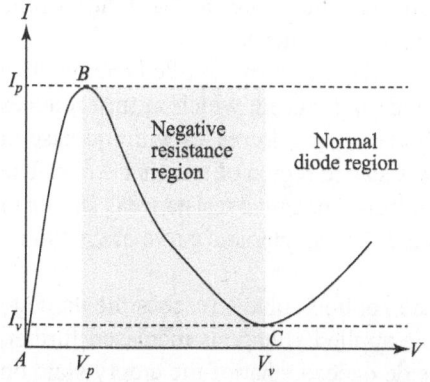

Fig. 13.2 Tunnel diode I–V characteristics

In contrast to ordinary diodes, where Fermi level exists in the forbidden region, in a heavily doped p–n junction diode, the Fermi level exists in the valence band in the p-type semiconductor and in the conduction band in the n-type semiconductor. Thus, if a forward bias of between 0 and V_p is applied to a tunnel diode, then the potential barrier will be decreased by the magnitude of the applied forward bias and a difference will be created between the Fermi levels of the two sides. The modified energy band diagram is shown in Fig. 13.3(a). Under such circumstances, a part of the filled energy state at one side of the barrier is at the same energy level as that of the empty state on the other side; hence, flow of carriers across the junction occurs through tunnelling. Thus, the current starts to flow. If the forward bias is increased gradually, the portion of overlapping also increases and so do the tunnelling rate and current. This is shown by AB on the I–V characteristic curve of a tunnel diode given in Fig. 13.2.

Tunnel current increases with applied voltage till V_p is reached, when the entire filled energy state of one side of the barrier is at the same energy level of the empty states on the other side, as shown in Fig. 13.3(b). In this stage, maximum number

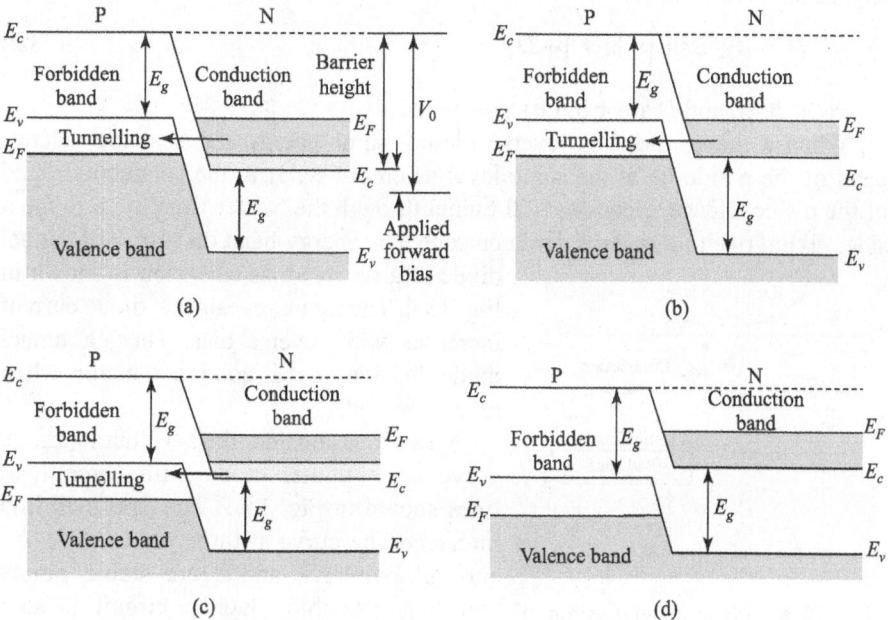

Fig. 13.3 Energy band diagram of tunnel diode with applied forward bias when (a) $0 < V < V_p$ (b) $V = V_p$ (c) $V_p < V < V_v$ (d) $V_v < V < \infty$

of electrons tunnel through the barrier from the filled state on the n side to the empty state on the p side, and a peak current, I_p, is achieved.

If the bias voltage is further increased, overlap between the filled energy state in the n side and the empty state in the p side is reduced, which in turn reduces tunnelling and diode current. Therefore, diode current decreases with increase in voltage. This corresponds to the negative-resistance region of a tunnel diode. The energy-band diagram of a tunnel diode exhibiting a negative resistance is shown in Fig. 13.3(c). The corresponding region in the I–V characteristic curve of the tunnel diode is marked by BC in Fig. 13.2.

A negative resistance is exhibited till the applied voltage reaches the value V_v and current I_v, as shown in Fig. 13.2. If the applied voltage is increased further, energy of the filled energy state on the n side exceeds that of the empty state on the p side, as shown in Fig. 13.3(d), and a carrier transfer takes place. Thus, the diode behaves like a normal p–n junction diode. Theoretically, for a tunnel diode the ratio I_p/I_v can be as high as 50–100. However, in practice, the achievable ratio is about 15.

A valley current, also called an excess current, is generated due to the presence of defect states within the semiconductor energy gap. The relation between the current and voltage of a tunnel diode can be explained by the following equation:

$$I = \frac{I_p - I_v}{\left(V_v - V_p\right)^5}\left[5\left(V - V_p\right)\left(V - V_v\right)^4 - \left(V_v - V_p\right)^5\right] + I_v \tag{13.1}$$

This relation has been obtained using curve fitting and does not yield $I = 0$ at $V = 0$ unless a specific relation exists between I_p/I_v and V_v/V_p. A simpler expression is as follows:

$$I = A + B\sin\left\{C\ln(V) + D\right\} \tag{13.2}$$

where A, B, C, and D are arbitrary constants.

When a tunnel diode is reverse biased, some energy states in the valence band of the p side lie at the same level as empty states in the conduction band of the n side. Hence, electrons will tunnel through the barrier from the p to the n side, giving rise to a reverse diode current. The energy-band diagram of a tunnel diode in a reverse-bias condition is shown in Fig. 13.4. The figure reveals that diode current increases with reverse bias. Thus, a tunnel diode behaves as a good conductor when reverse biased.

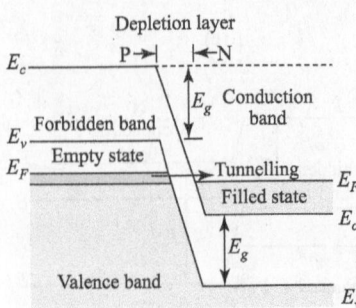

Fig. 13.4 Energy-band diagram of tunnel diode with applied reverse bias

Now let us consider the I–V characteristic curve of a tunnel diode with three load lines shown in Fig. 13.5. The first load line intersects the curve at three points, 'a', 'b', and 'c', where a and c are stable points and b is unstable. Such a circuit is said to be bistable and can be used as a binary device in switching circuits. The second load

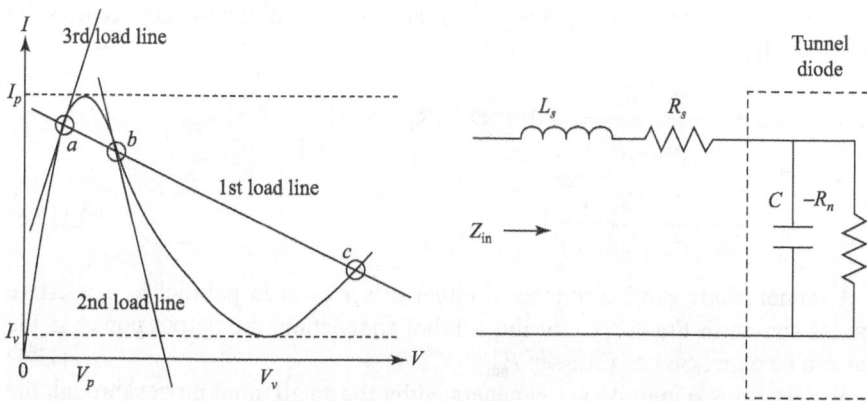

Fig. 13.5 Tunnel diode I–V characteristic curves with load lines

Fig. 13.6 Equivalent circuit of tunnel diode

line intersects the I–V characteristic curve at point b only, and such a circuit is called astable. The third load line crosses the curve at point a in the positive resistance region; such a circuit is called monostable.

A small signal equivalent circuit for the tunnel diode operating in the negative-resistance region is shown in Fig. 13.6. In the circuit, R_s and L_s are, respectively, the resistance and inductance of the packaging circuit of the tunnel diode; C_j is the junction capacitance; and $-R_n$ is the negative resistance. The junction capacitance can be expressed as follows:

$$C_j = \sqrt{\frac{enp\varepsilon_s}{2(n+p)(V_d - V)}} \tag{13.3}$$

where n and p are electron and hole concentrations and V_d is the diffusion voltage.

The input impedance of the circuit is expressed as follows:

$$Z_{in} = R_s + j\omega L_s + \left(-R_n \parallel \frac{-j}{\omega C_j}\right) = R_s + j\omega L_s + \frac{R_n(j/\omega C_j)}{-R_n - j/(\omega C_j)}$$

or

$$Z_{in} = R_s + j\omega L_s - \frac{R_n}{1 - jR_n\omega C_j}$$

or

$$Z_{in} = \left[R_s - \frac{R_n}{1 + (\omega C_j R_n)^2}\right] + j\left[\omega L_s - \frac{\omega C_j R_n^2}{1 + (\omega C_j R_n)^2}\right] \tag{13.4}$$

For the resistive cut-off frequency, the real part of the input impedance must be zero. Therefore,

$$R_s - \frac{R_n}{1 + (\omega_c C_j R_n)^2} = 0$$

or

$$f_c = \frac{1}{2\pi C_j R_n} \sqrt{\frac{R_n}{R_s} - 1} \tag{13.5}$$

For the self-resonance frequency, the imaginary part of the impedance must be zero. Therefore,

$$\omega_c L_s - \frac{\omega_c C_j R_n^2}{1+\left(\omega_c C_j R_n\right)^2} = 0 \text{ or } \left(\omega_c C_j R_n\right)^2 = \frac{C_j R_n^2}{L_s} - 1$$

or $\quad f_c = \dfrac{1}{2\pi C_j R_n}\sqrt{\dfrac{C_j R_n^2}{L_s}-1}$ $\qquad\qquad$ (13.6)

A tunnel diode can be connected either in series or in parallel to a resistive load, as shown in Fig. 13.7. For the parallel connection, the output power in the load can be expressed as follows: $P_{\text{out}} = V^2/R_L$ $\qquad$ (13.7)

A part of this output power is generated by the small input power through the tunnel diode amplifier of gain A. This part is represented as $P_{\text{in}} = \dfrac{V^2}{AR_L}$ $\quad$ (13.8)

The remaining part is generated by the negative resistance and is as follows:

$$P_n = V^2/R_n \qquad\qquad (13.9)$$

Thus, we can write that $\dfrac{V^2}{R_L} = \dfrac{V^2}{AR_L} + \dfrac{V^2}{R_n}$

or $\quad A = \dfrac{R_n}{R_n - R_L}$ $\qquad\qquad$ (13.10)

If $R_n \to R_L$, $A \to \infty$, and the system oscillates.
For the series circuit,

$$\frac{V^2}{R_L} = \frac{V^2}{R_n} - \frac{V^2}{AR_n}$$

or $\quad A = \dfrac{R_L}{R_L - R_n}$ $\qquad\qquad$ (13.11)

A tunnel diode can be connected to a circulator to make a negative-resistance amplifier, as shown in Fig. 13.8. The desired frequency signal is fed at port 1 of a circulator through a band-pass filter. The filter serves as both a bandwidth selector and an impedance matching device. This arrangement improves the gain of the amplifier. The input energy arrives at port 2 and is amplified by the tunnel diode. The amplified signal then enters port 3 and is delivered at the load. If there is any reflection at port 3, the reflected signal enters port 4, where it is absorbed by the matched load.

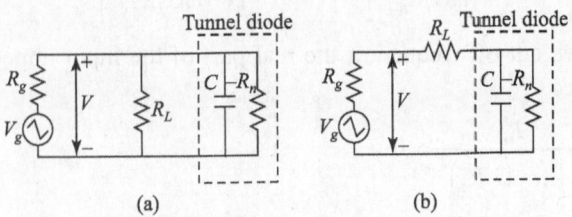

(a) $\qquad\qquad\qquad\qquad$ (b)

Fig. 13.7 Equivalent circuit of loaded tunnel diode
(a) Parallel loading (b) Series loading

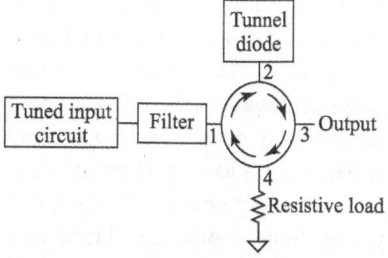

Fig. 13.8 Tuned diode amplifier

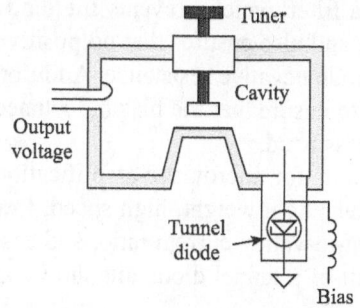

Fig. 13.9 Tunnel diode oscillator

A tunnel diode operating at the centre of the negative-resistance region and coupled to a tuning circuit or cavity behaves as a stable oscillator with an oscillation frequency same as the resonance frequency of the tuned circuit or cavity. Tunnel diode oscillators, operating at a microwave frequency, mainly use transmission lines as tune circuits. A simple tunnel diode oscillator circuit, which uses a high Q-tunable circuit, is shown in Fig. 13.9. Here, loose coupling is achieved using a short antenna feed probe placed off the centre in the cavity. Such an arrangement increases the stability of oscillations and the output power over a wide bandwidth. A few hundred microwatts of output power can be achieved with such an arrangement. The operating frequency of such an oscillator is determined by location and depth of the tuner screw in the cavity. Such tuning is called mechanical tuning. A tunnel diode oscillator can also be tuned electrically with the help of a changing bias voltage or using a varactor diode.

An equivalent tunnel diode oscillator circuit is shown in Fig. 13.10, which can generate microwave signals of up to about 100 GHz.

Tunnel diodes can also be used in mixer and converter circuits, as shown in Fig. 13.11. While other frequency converters usually have a conversion power loss, tunnel diode converters can have a conversion power gain. In addition,

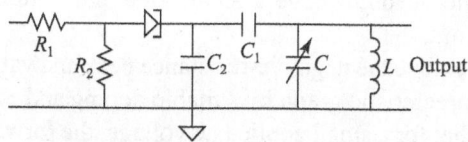

Fig. 13.10 Tunnel diode oscillator circuit

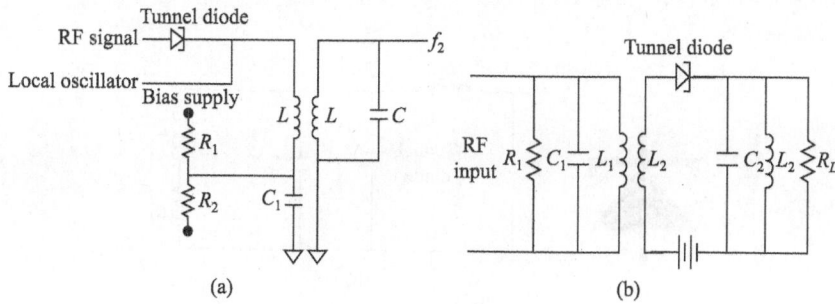

Fig. 13.11 Tunnel diode (a) Mixer (b) Converter

a single tunnel diode can also be designed to act as both a non-linear element in the converter and a negative-resistance element in the local oscillator simultaneously. Practical tunnel diodes usually have either a unity conversion gain or a small conversion loss. A conversion gain of 20 dB is also achievable if it is biased near or into the negative-resistance region, but its stability may be affected. One more attractive feature of a tunnel diode frequency converter is its low noise generation.

A tunnel diode, in all its application, should be loosely coupled to its tuned circuit. In addition, when mounted in a cavity, the diode should be placed at a point of significant coupling but not at maximum coupling. Another important concern is that its DC biasing should be done without interfering the tuning circuit. The simplest method to achieve this is to use a filter, which prevents the diode from being short-circuited by the supply source and also ensures that no positive resistance is added that can interfere with the diode negative resistance. Addition of capacitance across the diode is also avoided to ensure that the bias inductance does not introduce spurious frequencies in the pass band.

Tunnel diodes are widely used in many circuits for microwave amplification and oscillation, and as binary memory due to their light weight, high speed, low power requirement, low noise, high peak-current-to-valley-current ratio, and cost effectiveness. The symbol and equivalent circuit of a tunnel diode are shown in Fig. 13.12.

Tunnel diodes also suffer from some major disadvantages. Since it is a two-terminal element, construction of unilateral amplifiers using tunnel diodes is very difficult, as a change in the load impedance also affects the input. This property makes cascading very difficult. Another disadvantage of tunnel diodes is that its negative-resistance region is quite small, restricting full signal levels to only a fraction of a volt. Therefore, the output power lies below 1 mW. In addition, stabilizing the gain against variation in bias and temperature is also necessary in a tunnel diode. Finally, such a diode has a very narrow depletion region that results in high capacitance per unit area. Therefore, for high-frequency operations, the diode needs to have a small area and volume, which, in turn, results in low burn-out.

It is possible to remove the negative-resistance peak and valley region from the tunnel diode I–V characteristic graph by suitable doping and etching (Fig. 13.13). This figure reveals that for a small applied ac voltage, the forward current is much smaller than the reverse current. Such a diode is known as a backward diode and can be used as a small signal rectifier. Due to a narrow junction, such diodes

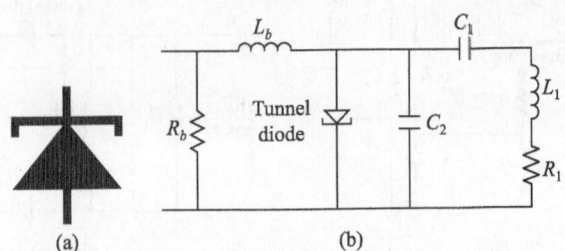

(a)   (b)

Fig. 13.12 Tunnel diode (a) Symbol (b) Equivalent circuit

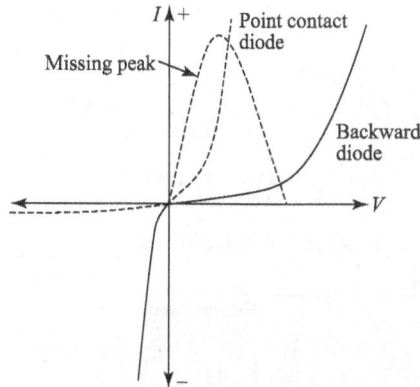

Fig. 13.13 Backward diode I–V characteristic curve

have a high operating speed and a current ratio that is much larger than that of a conventional rectifier.

If GaAs is used as a semiconductor in a backward diode, then a maximum signal of about 0.9 V can be applied to the diode before it starts conducting heavily in the forward direction. For a Ge semiconductor, this limit is even smaller. This means that the backward diode is limited to lower operating levels. Despite this disadvantage, backward diodes are commonly used in video detection and low-level mixing circuits in Doppler radars due to their low noise characteristics.

Characteristics of tunnel diodes The typical characteristics of a tunnel diode are as follows:

1. Frequency range: up to 100 GHz
2. Power: 700 W
3. Tuning range: 1–4.3 GHz

13.3 GaAs Gunn Effect Diodes

Gunn diodes are basically an n-type bulk GaAs semiconductor with an electron concentration of around $10^{14} - 10^{17}/cm^3$ (at room temperature) and a typical dimension of $150 \times 150 \times 30 \, \mu m^3$, as shown in Fig. 13.14. In 1963, J.B. Gunn found that when such a bulk n-type GaAs semiconductor is subjected to a voltage, above a certain threshold value, it produces periodic fluctuations in the current passing through it. Gunn also observed that the drift velocity of the carriers in such a semiconductor initially increases linearly from zero to a certain value when the electric field is increased from zero to a threshold value. However, when the electric field crosses the threshold value the carrier drift velocity decreases and the diode exhibits a negative resistance. These properties of a bulk n-type GaAs semiconductor are referred to as *Gunn effect* and can be explained with the help of the Ridley–Watkins–Hilsum (RWH) theory.

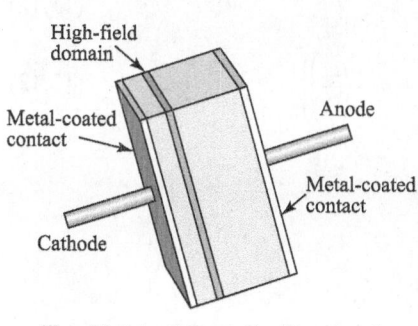

Fig. 13.14 Schematic diagram of Gunn diode

13.3.1 Ridley–Watkins–Hilsum Theory

The RWH theory is based on the transfer of conduction electrons from a low-mass, high-mobility sub-band to a high-mass, low-mobility sub-band of a semiconductor. If the energy level of the former sub-band is lower than that of the latter, then under the influence of an applied electric field of sufficient strength, electrons in the lower valley jump to the upper valley. As a result, their mobility decreases, which in turn decreases the output current. This happens only if the applied electric field is above a threshold value. If the field is increased above this threshold value, more electrons jump to the upper valley, resulting in a further decrease in current. Therefore, the current decreases with the increase in the applied field (or voltage). This gives rise to the concept of negative resistance in Gunn diodes. It may be noted that Gunn effect exists only over

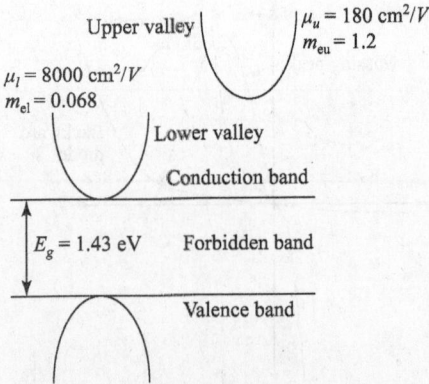

Fig. 13.15 Two-valley model of electron energy versus wave number for n-type GaAs semiconductor

a specified voltage range. If the applied field is very high, then no more electrons are left in the lower valley to jump to the upper valley, and hence a positive resistance region is obtained.

Gunn effect in a bulk n-type GaAs semiconductor can be described as follows. In such a semiconductor, a low-mobility $\left(\mu_u = 180\,\text{cm}^2/\text{V-sec}\right)$, high-mass $\left(M_{eu} = 1.2\right)$ upper valley is separated from a high-mobility $\left(\mu_l = 8000\,\text{cm}^2/\text{V-sec}\right)$, low-mass $\left(M_{eu} = 0.068\right)$ lower valley by an energy gap of $0.36\,\text{eV}$, as shown in Fig. 13.15. Under the equilibrium condition, electron density in both the valleys remains the same. However, if an electric field, which is greater than the electric field of the lower valley but less than that of the upper valley, is applied, then electrons in the lower valley get energy from it and start moving to the upper valley. If the electric field is increased further such that it is greater than the electric field of the upper valley, all electrons will be transferred to the upper valley. On the other hand, if the applied electric field is lower than the electric field of the lower valley, then electrons in the lower valley do not get sufficient energy to jump to the upper valley, and hence no electron transfer takes place. If the electron densities in the lower and upper valleys are n_l and n_u, respectively, then the conductivity of an n-type GaAs semiconductor can be expressed as follows:

$$\sigma = e\left(n_l \mu_l + n_u \mu_u\right) \tag{13.12}$$

or

$$\frac{d\sigma}{dE} = e\left\{\left(\mu_l \frac{dn_l}{dE} + \mu_u \frac{dn_u}{dE}\right) + \left(n_l \frac{d\mu_l}{dE} + n_u \frac{d\mu_u}{dE}\right)\right\} \tag{13.13}$$

If the total electron density is n, then

$$n = n_l + n_u$$

or

$$\frac{dn}{dE} = \frac{d}{dE}\left(n_l + n_u\right) = \frac{dn_l}{dE} + \frac{dn_u}{dE} = 0$$

or

$$\frac{dn_l}{dE} = -\frac{dn_u}{dE} \tag{13.14}$$

Further, if the mobility of electrons in the lower and upper valleys is proportional to E^p, where p is a constant, then

$$\frac{d\mu}{dE} = \frac{d(KE^p)}{dE} = KpE^{p-1} = \frac{p}{E}KE^p = \frac{p\mu}{E} \tag{13.15}$$

Substituting Eqs (13.15) and (13.14) in Eq. (13.13), we get the following relation:

$$\frac{d\sigma}{dE} = e\left\{\left(\mu_l\frac{dn_l}{dE} - \mu_u\frac{dn_l}{dE}\right) + \left(n_l\frac{p\mu_l}{E} + n_u\frac{p\mu_u}{E}\right)\right\}$$

or $$\frac{d\sigma}{dE} = e\left\{(\mu_l - \mu_u)\frac{dn_l}{dE} + \frac{p}{E}(n_l\mu_l + n_u\mu_u)\right\} \tag{13.16}$$

Now, from Ohms law, we know the following:

$$J = \sigma E$$

or $$\frac{dJ}{dE} = \sigma + E\frac{d\sigma}{dE}$$

or $$\frac{1}{\sigma}\frac{dJ}{dE} = 1 + \frac{d\sigma/dE}{\sigma/E} \tag{13.17}$$

To obtain a negative resistance, J must decrease with increases in E. That is,

$$\frac{dJ}{dE} < 0 \tag{13.18}$$

Substitution of Eq. (13.18) in Eq. (13.17) gives the following expression:

$$1 + \frac{d\sigma/dE}{\sigma/E} < 0 \text{ or } -\frac{d\sigma/dE}{\sigma/E} > 1 \tag{13.19}$$

Substituting Eqs (13.12) and (13.16) in Eq. (13.19), we get the following relation:

$$-\frac{e\left\{(\mu_l - \mu_u)\frac{dn_l}{dE} + \frac{p}{E}(n_l\mu_l + n_u\mu_u)\right\}}{e(n_l\mu_l + n_u\mu_u)/E} > 1$$

or $$\left(\frac{\mu_l - \mu_u}{n_l\mu_l + n_u\mu_u}\right)\left(-E\frac{dn_l}{dE}\right) - p > 1$$

or $$\left(\frac{\mu_l - \mu_u}{\mu_l + f\mu_u}\right)\left(-\frac{E}{n_l}\frac{dn_l}{dE}\right) - p > 1 \tag{13.20}$$

where f is a factor and can be expressed as follows:

$$f = n_u/n_l \tag{13.21}$$

Thus, to achieve a negative resistance, Eq. (13.20) must be satisfied by the semiconductor material. Alternatively, we can say that on the basis of the RWH theory, a semiconductor will exhibit a negative resistance if the following conditions hold:

1. Electrons in the lower valley must have a low effective mass and high mobility, whereas those in the upper valley must have a high effective mass and low mobility.

2. The energy gap between the lower and upper valleys must be less than the band-gap energy. Otherwise, before the electrons in the lower valley begin to jump to the upper valley, the valence electrons will jump to the lower valley and the semiconductor will be highly conducting.

3. The energy gap between the lower and upper valleys must be higher than the thermal energy. Otherwise, even without the application of an electric field, electrons in the lower valley will gain thermal energy and begin to jump to the upper valley.

In practice, a negative-resistance device has two modes: voltage-controlled and current-controlled modes. In the voltage-controlled mode, current density can be multivalued, whereas for the current-controlled mode, voltage can be multivalued. This is shown in Fig. 13.16.

In the negative-resistance region of the current–voltage curve, the sample becomes electrically unstable. Therefore, in order to achieve stability, the initially homogeneous sample becomes electrically heterogeneous. In the voltage-controlled mode, this leads the formation of a high-field domain that separates two low-field domains, whereas in the current-controlled mode, a high-density current filament is formed that runs along the field direction. This is shown in Fig. 13.17.

Now, $J = nqv_d$

or $\qquad \dfrac{dJ}{dE} = nq \dfrac{dv_d}{dE}$ $\hfill$ (13.22)

Using inequality (13.18), we can write the following expression:

$$nq \frac{dv_d}{dE} < 0 \text{ or } \frac{dv_d}{dE} < 0 \qquad (13.23)$$

The plot of electron drift velocity as a function of electric field is shown in Fig. 13.18.

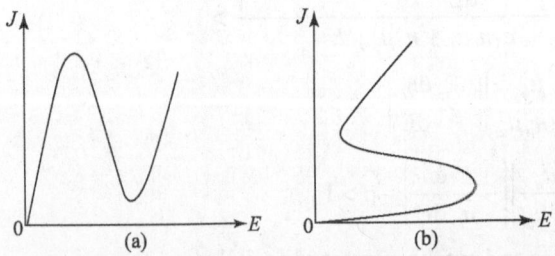

Fig. 13.16 Diagram showing negative resistance
(a) Voltage-controlled mode (b) Current-controlled mode

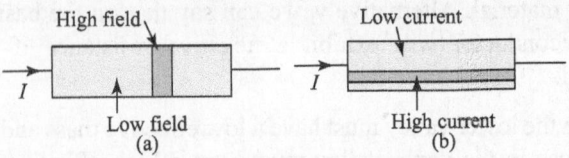

Fig. 13.17 Diagram of high-field and high-current domains

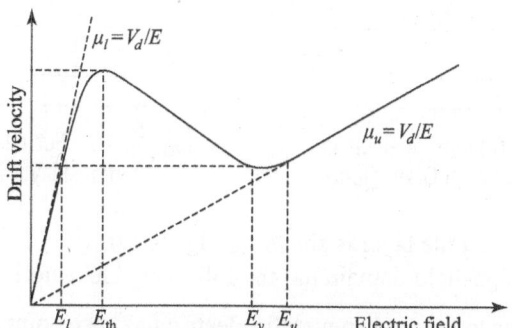

Fig. 13.18 Plot of drift velocity as function of electric field

13.3.2 Formation and Properties of High-field Domain

To understand domain formation, let us assume that at a point 'A' inside the sample, negative charges accumulate due to random noise fluctuation or non-uniformity in doping in a GaAs diode. Accumulated charges create an electric field that is lower at the left of point A than that at the right. Now, if the diode is biased in the negative-resistance region, then the carriers flowing into point A are greater than those flowing out of it (this is because in the negative-resistance region a low field corresponds to a high current density). Thus, there is a further accumulation of space charges at point A, which further lowers the field at the left and increases that at the right of point A. This process is therefore cumulative in nature, and continues until both the low and the high fields attain a value outside the negative-resistance region and the currents in the two regions become equal. The process has been illustrated in Fig. 13.19.

Formation of the aforementioned domain corresponds to the simplest form of space charge instability. When the positive and negative charges inside the domain are separated by a small distance, then a dipole domain is formed, with the field inside the domain being greater than that outside it. Since the diode is operating in the negative-resistance region, the high-field domain will correspond to a lower current than the low-field domain. The two field values will tend towards equilibrium outside the differential negative-resistance region, where the high and low field currents are the same. Under the circumstances, the dipole reaches a stable configuration and moves through the sample towards the terminal. When it is collected at the anode, a new domain starts to form. The width of the dipole domain, so formed, can be estimated using the following relation:

$$V = E_A L = E_2 d + E_1 (L - d)$$

or $$d = L \frac{(E_A - E_1)}{(E_2 - E_1)} \qquad (13.24)$$

In case of current-controlled differential negative resistance, a similar process produces the following relation:

$$a = A \frac{(J_A - J_1)}{(J_2 - J_1)} \qquad (13.25)$$

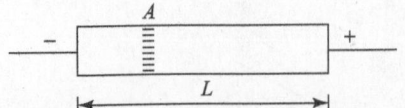

Fig. 13.19 Formation of electron accumulation layer in GaAs diode

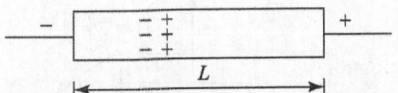

Fig. 13.20 Formation of electron dipole layer in GaAs

The formation of a dipole layer is shown in Fig. 13.20.

In general, a high-field domain has the following characteristics:

1. A domain starts to form whenever the electric field at a point inside the sample crosses the threshold value. The domain will then drift through the device.
2. If the electric field is increased, the domain will also increase in size by absorbing more carriers. Thus, the current will decrease and a negative resistance will occur. An increase in the electric field will also result in the decrease in the carrier drift velocity.
3. The domain drifting through the device will not disappear before reaching the anode, unless the electric field is dropped below the threshold value. Therefore, to avoid the formation of a new domain, the electric field should be decreased below the threshold level. Here, we have assumed that the sample has a uniform cross-sectional area and doping concentration.
4. The length of the domain is generally inversely proportional to the doping concentration. Therefore, the product of doping concentration and domain length is constant. This means that diodes with the same product of doping concentration and domain length will behave similarly.
5. The domain will modulate the current if it passes through regions of different cross-sectional areas or those with different doping concentrations. Alternatively, it may also disappear.

It should be noted that points '3' and '4' are valid only when the domain length is much longer than the thermal diffusion length of the carriers. For a GaAs sample, this is about $1\,\mu m$ for a doping concentration of $10^{16}/cm^3$ and $10\,\mu m$ for a doping concentration of $10^{14}/cm^3$. The presence of a high-field domain at a certain point in the semiconductor can be detected by placing a capacitive contact at that point. When the domain passes the capacitive contact, a certain voltage change occurs in the contact, which will be reflected at the output. The presence of the domain, anywhere in the device, can also be detected by determining the change in the differential increase or decrease in the current.

13.3.3 Modes of Operation

Depending on the material parameters and operating conditions, a Gunn diode can operate in the following four basic modes:

13.3.3.1 Gunn Oscillator Mode

This mode is defined by the region where the product of the frequency and length $(f_0 L)$ is about 10^7 cm/s and the product of the doping concentration and length $(n_0 L)$ is greater than $10^{12}/cm^2$ or more precisely lies in the range $10^{12}/cm^2 \leq n_0 L < 10^{14}/cm^2$.

If in a GaAs sample, $n_0 L$ becomes greater that $10^{12}/cm^2$, space charge perturbation in the sample increases according to the following equation:

$$Q(x,t) = Q(x - vt, 0)e^{t/\tau_d} \qquad (13.26)$$

where

$$\tau_d = \frac{\varepsilon}{e n_0 |\mu_n|} \qquad (13.27)$$

and μ_n is the negative mobility.

It may be noted that the maximum growth is as follows:

$$Q(L, L/v) = Q(0,0)e^{L/(v\tau_d)} \qquad (13.28)$$

Using Eqs (13.27) and (13.28), we get the following relation:

$$\text{Growth factor} = \frac{Q(L, L/v)}{Q(0,0)} = \exp\left(\frac{L}{v\tau_d}\right) = \exp\left(\frac{Ln_0 e |\mu_n|}{\varepsilon v}\right) \qquad (13.29)$$

For growth:

$$\frac{Ln_0 e |\mu_n|}{\varepsilon v} > 1$$

or

$$n_0 L > \frac{\varepsilon v}{e |\mu_n|} \qquad (13.30)$$

For an n-type GaAs sample, the RHS of inequality (13.30) is about $10^{12}/cm^2$, which is the condition for the Gunn oscillation mode.

Note For $n_0 L < 10^{12}/cm^2$, the sample cannot support a travelling dipole domain. However, it can support a growing space charge wave and hence is used as a stable linear microwave amplifier.

For the Gunn oscillation mode, the frequency of oscillation is given by the following relation:

$$f_0 = v_{dom}/L_{eff} \qquad (13.31)$$

where v_{dom} is the domain velocity and L_{eff} is the effective length the domain travels from the time of formation until the generation of a new domain. Since the electron drift velocity is a function of the electric field, Gunn oscillation occurs in three possible domain modes:

Transit-time domain mode   In the transit-time domain mode, $f_0 L \approx 10^7$ cm/s, that is, $v_d \approx v_s \approx 10^7$ cm/s, and the oscillation period is equal to the transit time $(\tau_0 = \tau_t)$. This is shown in Fig. 13.21. Since current is collected only when the domain arrives at the anode, the efficiency of a diode in this mode is very low and is less than 10%.

Delayed domain mode   In the delayed domain mode 10^6 cm/s $< f_0 L < 10^7$ cm/s, and the transit time is chosen $(\tau_0 > \tau_t)$ such that the domain is collected at the anode, whereas the field is below the threshold value $(E < E_{th})$, as shown in Fig. 13.22.

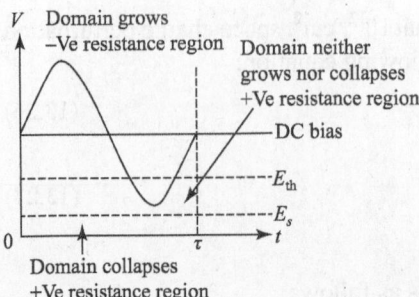

Fig. 13.21 Transit-time domain mode

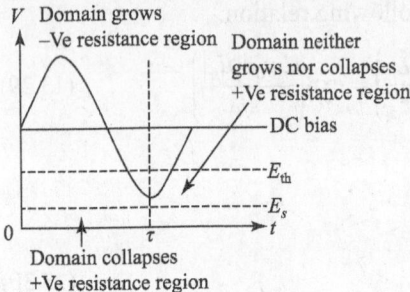

Fig. 13.22 Delayed domain mode

Since the field is below the threshold value, a new domain can form immediately. It forms only after the field value again rises above the threshold value. This results in a time delay between the capture of the present domain at the anode and the formation of the next domain, and that is why it is called the delayed domain mode. The efficiency of a diode in this mode is about 20%. The delayed domain mode is also called an inhibited mode.

Quenched domain mode The quenched domain mode corresponds to $f_0L > 2 \times 10^7$ cm/s. In this mode, the bias field drops below the sustaining field E_s during the negative half-cycle, as shown in Fig. 13.23. This drop leads to a collapse of the domain before collection at the anode. When the field again rises above the threshold value, a new domain starts forming and the process repeats. In the quenched domain mode, the oscillation frequency is determined by the resonant circuit, rather than the transit-time frequency. The efficiency of a diode in this mode is about 13%.

In practice, a Gunn oscillator can be used in several circuit configurations. If the circuit is mainly resistive or the voltage across the diode is constant, then the frequency of oscillation is equal to the inverse of the time taken by the domain to travel from the cathode to the anode. However, it is not very useful in microwave applications. For microwave applications, Gunn diodes are used inside a high-Q microwave cavity in the negative resistive region of the voltage–current graph. The frequency, in such a case, can be tuned to a range of about an octave without any loss of efficiency with the help of a movable short circuit.

The equivalent circuit of a Gunn oscillator in the Gunn oscillator mode is shown in Fig. 13.24.

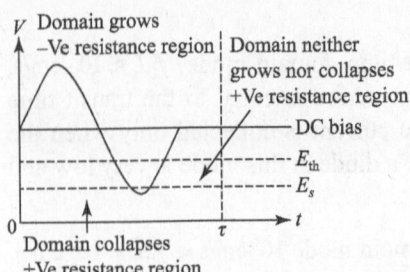

Fig. 13.23 Quenched domain mode

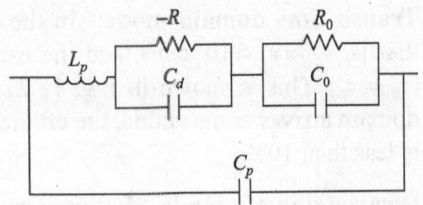

Fig. 13.24 Equivalent circuit of Gunn oscillator in Gunn oscillator mode

13.3.3.2 *Limited Space Charge Accumulation Mode*

The limited space charge accumulation (LSA) mode is defined by $f_0 L > 2 \times 10^7$ cm/s and $2 \times 10^4 < n_0 / f_0 < 2 \times 10^5$, and is the simplest mode of operation. It consists of a uniformly doped semiconductor without any internal space charge. The internal electric field is uniform and proportional to the applied electric field. The device current is also proportional to the drift velocity in this field. In the LSA mode, the frequency of oscillation is very high and the domains do not get sufficient time to form, as depicted in Fig. 13.25. Space charges accumulated near the cathode also have sufficient time to collapse as the field drops below the threshold and sustaining field level. For the LSA mode, the oscillation period should not be more than several times the magnitude of the dielectric relaxation time in the negative conductance region. The efficiency of an LSA diode is about 20%.

In the LSA mode, primary accumulation of space charge occurs near the cathode, whereas the rest of the semiconductor remains fairly homogeneous and behaves as a series negative resistance. This series negative resistance increases the frequency of oscillation in the resonant circuit. It may be noted here that the frequency of oscillation in the LSA mode does not depend upon the transit time and is solely determined by the external resonant circuit.

In the LSA mode, a diode is biased above the threshold field. However, the magnitude of the RF voltage is also large, and it drives the diode below the threshold level over a certain period during its negative half-cycle. When the voltage remains above the threshold level, space charge starts building at the cathode. However, since the oscillation period τ_0 is less than the growth time τ_g, the voltage drops below the threshold level before the domain can form. Further, since the oscillation period is much greater than the dielectric relaxation time, as mentioned, the accumulated space charge is drained in a very small fraction of the RF cycle.

An LSA diode (or Gunn diode in the LSA mode) is very sensitive to load, doping fluctuation, and temperature. The power output of an LSA diode can be expressed as follows:

$$P = \eta M E_{th} n_0 e v_0 LA \qquad (13.32)$$

where η = the DC to RF conversion efficiency

M = multiples of operating voltage above the negative-resistance threshold voltage

$E_{th} \approx 3.4$ kV/cm = the threshold field

$n_0 \approx 10^{15}$ e/cm^3 = the donor concentration

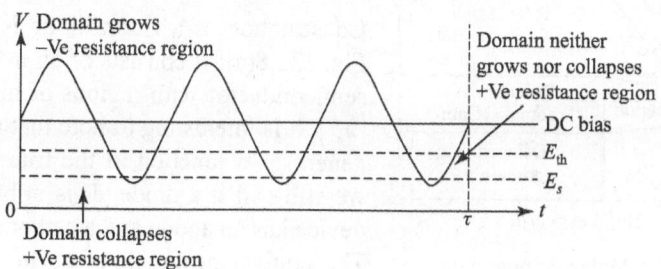

Fig. 13.25 Limited space charge accumulation mode

$e = 1.6 \times 10^{-19}$ C = the charge of electron

$v_0 \approx 10^7$ cm/s = the average carrier drift velocity

$L \approx 10 - 200\ \mu$m = the length of the device

and $A \approx 3 \times 10^{-4} - 20 \times 10^{-4}$ cm^2 = the area of the device

Equation (13.32) reveals that the peak power is proportional to the active device volume. However, this volume cannot be increased infinitely to get very high power because of the skin effect and electrical wavelength. In addition, practical limits exist for thermal dissipation capability, available bias, and technology to maintain uniform doping throughout the length of the device. These facts limit the power capability of LSA diodes.

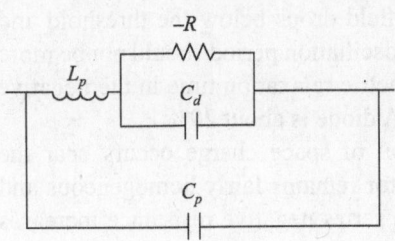

Fig. 13.26 Equivalent Gunn oscillator circuit in LSA mode

The equivalent Gunn oscillator circuit in the LSA mode is shown in Fig. 13.26.

13.3.3.3 Stable Amplification Mode

In the stable amplification mode, $10^{11}/$cm$^2 < n_0 L < 10^{12}/$cm^2 and $f_0 L \approx 10^7$ cm/s. As described in inequality (13.30), under such a situation, the domain cannot grow. In fact, there are too few carriers to form a domain within the transit time. Thus, amplification of signals of near transit time frequency can be accomplished in this mode.

13.3.3.4 Bias Circuit Oscillation Mode

The bias circuit oscillation mode occurs when there is either Gunn oscillation or LSA oscillation. When a Gunn diode is biased at the threshold, Gunn oscillation begins; as a result, the average current suddenly drops, leading to oscillations in the bias circuit in the range from 1 kHz to 100 MHz. Since the frequency of oscillation is small, in the bias circuit oscillation mode $f_0 L$ is too small to appear in the figure.

The region with different modes of operation in an $f_0 L - n_0 L$ plot is shown in Fig. 13.27.

13.3.4 Construction and Equivalent Circuit of Gunn Diode

Construction of a Gunn diode is shown in Fig. 13.28(a). It consists of an n-type GaAs semiconductor with regions of high doping (n^+). It is interesting to note that there is no general p–n junction in the true sense, but we still call it a diode. This is because the device has an anode and a cathode terminal. The equivalent circuit of a Gunn diode is shown in Fig. 13.28(b).

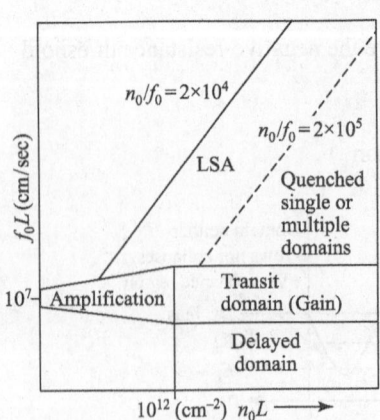

Fig. 13.27 Modes of operation in $f_0 L - n_0 L$ region

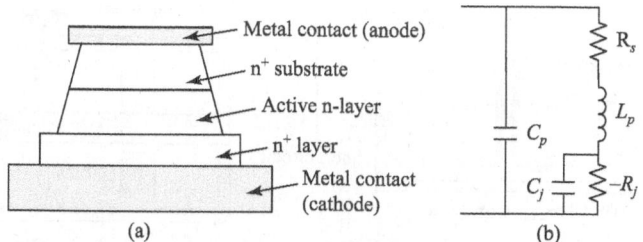

Fig. 13.28 Gunn diode (a) Construction details (b) Equivalent circuit in Gunn oscillation mode

13.3.5 Gunn Oscillator Circuit

A simple Gunn oscillator circuit is shown in Fig. 13.29, where the Gunn device is located inside the rectangular waveguide cavity with the help of a post. The anode of the Gunn diode is connected to the post, whereas the cathode is connected to the cavity. The resonance frequency of the cavity can be adjusted with the help of a movable short circuit located at the end, whereas the degree of the coupling to the waveguide is adjusted by varying the opening of the inductive window, located in between the waveguide and the cavity. In addition, a tuning screw is provided in the system for fine-tuning the resonance frequency of the cavity. Since the supporting post is insulated from the waveguide and a DC biasing of around 12 V is applied between them, the whole combination of the post and the guide behaves as a low-impedance RF bypass capacitor. This bypass capacitance prevents the flow of the RF current through the bias voltage supply.

Since, in the LSA mode, a Gunn diode can be imagined as a parallel combination of a negative resistance $-R_j$ (ranging from $-5\,\Omega$ to $-20\,\Omega$) and a capacitor C_j, as shown in Fig. 13.30, the resistive loading from the cavity and external circuitry should be around 20% more than $-R_j$ to ensure that the parallel combination of the Gunn device resistance and external resistive loading is negative. This, in turn, requires that the cavity must have an impedance transforming circuitry, such as a quarter-wave transformer, to reduce the high impedance of the output waveguide to the appropriate low value. This is shown in Fig. 13.31.

In addition to rectangular cavities, Gunn diodes can be used in coaxial cavities, as shown in Fig. 13.32. In such a case, a Gunn diode is mounted at one end of the cavity in continuation with the central conductor. The output is taken using a coupling loop or a probe. The frequency of oscillation of such an oscillator

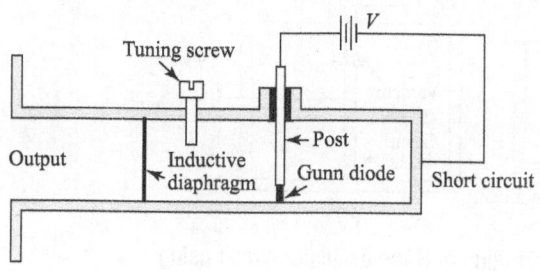

Fig. 13.29 Simple waveguide cavity Gunn oscillator

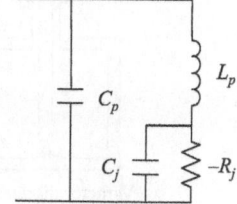

Fig. 13.30 Equivalent circuit of Gunn diode in LSA oscillation mode

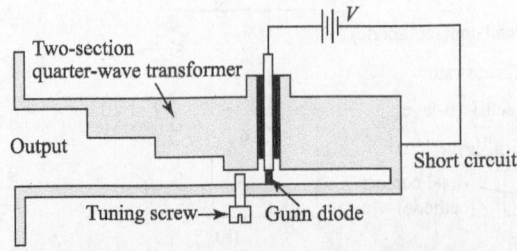

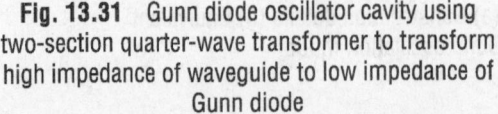

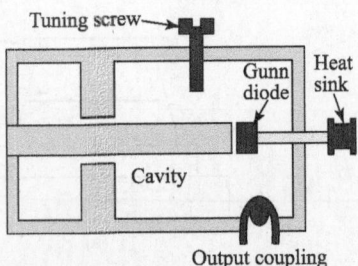

Fig. 13.31 Gunn diode oscillator cavity using two-section quarter-wave transformer to transform high impedance of waveguide to low impedance of Gunn diode

Fig. 13.32 Gunn diode oscillator using coaxial cavity

circuit is determined by the length of the cavity, whereas the location of the loop or probe determines the impedance presented to the Gunn diode. A heat sink, at the end, conducts the power dissipated at the load. Such an oscillator circuit can be fabricated easily. However, it suffers from low Q and the presence of harmonics in the cavity.

In MMIC application, microstrip cavities are also often used for Gunn oscillators. Such cavities have a lower Q than other waveguide cavities, and hence are much noisier and also less stable. Furthermore, mechanical tuning is very difficult in such resonators and we need to use electrical or electronic tuning circuitry. In addition, the diode also has to be electrically isolated from other circuit elements, while maintaining low parasitic and thermal resistance. Conducting out the heat, generated due to power dissipation, is another major challenge in such oscillators. In most cases, the heat is dissipated through the thick GaAs substrate or a thick plated gold layer. In addition, other difficulties exist—for example, fabrication of a diode having a good DC to RF conversion efficiency using an epitaxial layer.

The tuning mechanism, used in Gunn oscillators, may be mechanical or electrical in nature. In mechanical tuning, a screw is used for the tuning purpose. For electrical tuning, we can use a varactor diode, as shown in Fig. 13.33(a). Varactors can also be used in series, as shown in Fig. 13.33(b). Depending on the placement, they are called series tuned or parallel tuned. The tuning range provided by a varactor diode is proportional to the amount of RF power coupled to

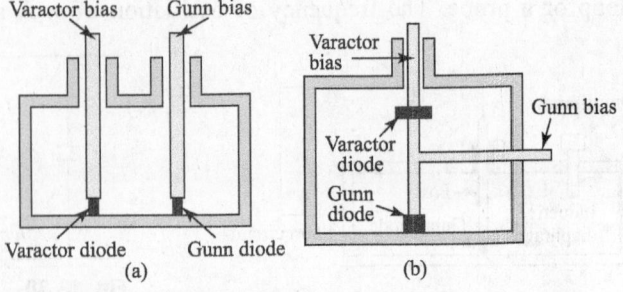

Fig. 13.33 Electrical tuning of Gunn oscillator circuit using varactor diode (a) Parallel tuning (b) Series tuning

it and inversely proportional to the loaded Q of the cavity. A series-tuned voltage–controlled oscillator (VCO) provides more power coupling to the varactor and a wider frequency tuning range. However, such an arrangement is accompanied with higher power loss, as compared with a parallel-tuned Gunn VCO. A varactor diode can also be used for frequency modulation. If a saw-tooth voltage is used as the control voltage, the output from a Gunn oscillator will be frequency modulated. To obtain pulsed output from the Gunn oscillator, we need to apply DC bias in the form of a pulse train of short rectangular pulses. However, this is not common.

In addition to varactor tuning, bias tuning is employed sometimes. In bias tuning, the frequency can be tuned simply by varying the bias voltage. Such tuning is preferred as it results in a wide operating voltage range, low thermal resistance, and good DC to RF conversion efficiency. The characteristics of such tuning depend upon the oscillator circuit, and often results in a discontinuity in power or frequency due to changes in the circuit property. Bias tuning can give a wider tuning range when the oscillator is operated in the second harmonics. However, the tuning range is small as compared to varactor tuning. However, the advantage of bias tuning lies in its simplicity.

Frequency tuning of a Gunn oscillator can be achieved using a magnetically tuned YIG sphere. Such a sphere is loop coupled to a Gunn diode and the load. Such a Gunn oscillator can achieve a wider bandwidth, better linearity, and a super FM noise performance than a varactor-tuned Gunn oscillator. However, it has the drawback of low output power due to weaker coupling. The use of such oscillators is also limited to the high end of the microwave frequency band due to high magnetic field requirements. In spite of these, a tuning frequency of 10 GHz can be achieved in V-band using such a tuning arrangement.

Microwave communication systems often require a highly stabilized, low-noise local oscillator. In such applications, cavity-stabilized Gunn oscillators can be used. Such a stabilization technique offers an efficient and cost-effective way for short-term stability of Gunn oscillators. In cavity stabilization, a Gunn oscillator is coupled to an external high-Q cavity, which in turn determines the oscillation stability. In addition, such stabilization also provides FM noise reduction. Depending on the connection of external high-Q cavity, stabilization may be of reflection, transmission, or reaction type, as shown in Fig. 13.34.

In addition to their use in the first harmonics, Gunn diode oscillators are often used in the second harmonics, especially for frequencies greater than 70 GHz. In practice, the output power should decrease with increase in frequency. However, for a frequency greater than 70 GHz, the output power from the second harmonics is actually greater than that of the fundamental mode. That is why the second

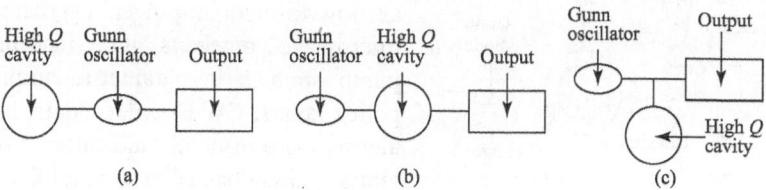

Fig. 13.34 Cavity stabilization (a) Reflection type (b) Transmission type (c) Reaction type

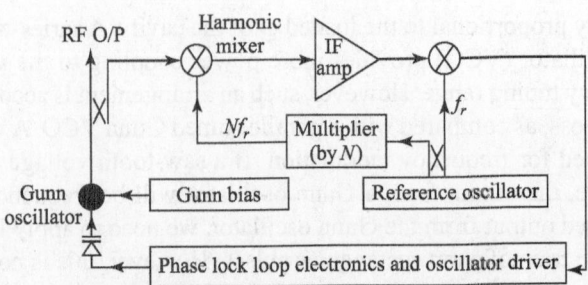

Fig. 13.35 Phase-locked Gunn oscillator

harmonics of a Gunn oscillator are often used to generate oscillations of frequencies greater than 70 GHz. Such use of a Gunn oscillator provides an alternative for manufacturing a W-band (75–110 GHz) GaAs oscillator. However, to use a Gunn oscillator in the second harmonics, waveguide dimensions should be modified so that it can filter out the first harmonics. When varactor tuning is used in such an oscillator, a varactor is connected in series with a Gunn oscillator through a disc resonator.

Gunn diodes can also be injection locked (ILO) or frequency locked to an injected RF source. However, in such a case, the locking frequency should be close to the free running oscillator. The injected signal tends to shift the oscillating frequency towards the injected frequency until they become equal. Such an arrangement considerably improves the FM noise of the Gunn oscillator.

A Gunn oscillator can also be phase locked by means of varactor tuning. Such locking produces a more stable output and reduces output noise as compared to ILO. The circuit diagram of a phase-locked Gunn oscillator is shown in Fig. 13.35.

13.3.6 Applications of Gunn Diode

Gunn oscillators are widely used in microwave circuitry as a source in the frequency range 5–150 GHz. In CW mode, a Gunn diode can provide output power in the range from a few mW to a few hundred mW, whereas for pulsed mode it may be a few kW. In addition, a Gunn diode can be used as an amplifier; however, this is not common because it cannot compete with other semiconductor amplifiers in terms of power output and noise. A simple Gunn amplifier circuit is shown in Fig. 13.36. In designing a Gunn amplifier, care should be taken to stop self-excited oscillation of the diode in the absence of any input signal; otherwise, it will behave as an injection-locked oscillator.

Gunn diodes find wide applications as low-to-medium-power oscillators in microwave receivers and instruments, pump sources in parametric amplifiers, police radars, CW Doppler radars, burglar alarms, rate-of-climb indicators used in aircraft, broadband linear amplifiers, and fast combinational and sequential circuits.

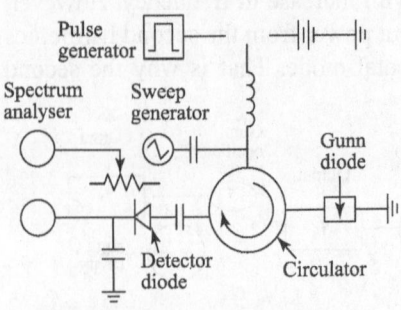

Fig. 13.36 Gunn diode amplifier

Note Due to its operating principle, a Gunn diode is also called a transfer electron device.

13.3.7 Characteristics of Gunn Diode

The typical characteristics of a Gunn diode are as follows:

1. Frequency range: 2–60 GHz
2. Pulsed output power: 1 kW at 2 GHz to 0.5 W at 50 GHz
3. CW output power: 2 W at 80 GHz to 30 mW at 100 GHz
4. Efficiency: 2–12%

13.4 InP Diodes

Indium phospide or InP diodes basically behave similarly to GaAs diodes when a voltage is applied across it. However, the following differences exist between them:

1. An InP diode has a middle valley in addition to the lower and upper valleys, as shown in Fig. 13.37. The lower valley is weakly coupled to the middle valley but strongly coupled to the upper valley.
2. Owing to a larger energy separation between the lower valley and the nearest energy level, the effect of thermal excitation on INP diode characteristics is less. This also results in four times lesser degradation of peak to valley current ratio as compared to an n-type GaAs sample.
3. Since in an InP diode, electron transfer proceeds rapidly as the field increases, it has a larger peak to valley current ratio.
4. Stronger coupling between the lower and upper valleys (as compared to the middle valley) in an InP diode increases electron diffusion coefficient, which, in turn, forces the InP diode to inhibit the formation of domains. As a result, the output waveform of the InP diode becomes transit time dependent.
5. By adjusting the cavity size, InP oscillators can be tuned over a large frequency range. Since the oscillation frequency of an InP diode depends on the active layer thickness, the frequency range is also bound by the layer thickness.
6. Coupling between the lower and upper valleys in an InP diode is weaker than that in a GaAs diode.
7. The middle valley of an InP diode provides an additional energy loss mechanism. This helps the diode to avoid breakdown caused by high energies acquired by the lower valley electrons from weak coupling.

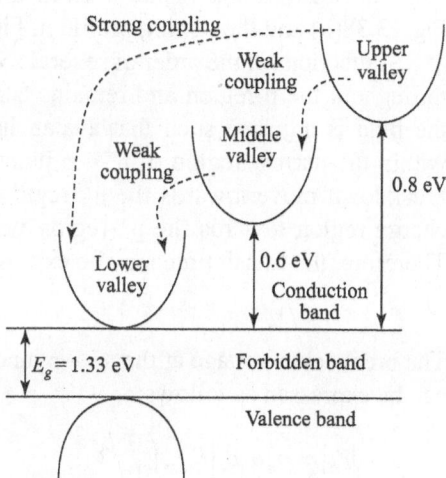

Fig. 13.37 Three-valley model energy level of Gunn diode

Notes 1. In addition to n-type GaAs and n-type InP diodes, alloys of n-type GaAs, n-type GaP, n-type InAs, and n-type CdTe diodes also exhibit Gunn effect. 2. An n-type CdTe diode has a higher threshold field as compared to a GaAs diode. 3. The domain velocity in CdTe and GaAs diodes is the same. Therefore, for the same length, both of them will operate at the same frequency in the transit-time mode. 4. Due to poor thermal conductivity and high threshold voltage, heating is a serious problem in CdTe diodes.

13.5 Read Diodes

A Read diode consists of a highly doped n-region (n^+), a highly doped p-region (p^+), one intrinsic region (i), and one very thin p-region (p) in the $n^+-p-i-p^+$ sequence, as shown in Fig. 13.38(a). However, essentially two of these regions take part in the main operation of the diode. The first region (p-region) is the avalanche region or high-field region. Carriers (holes) moving near the n^+-p junction acquire sufficient energy from the field to knock the valence electrons into the conduction band and thus produce an electron–hole pair. The newly created carrier and the old carriers acquire sufficient energy from the field and create more electron–hole pairs. This process continues and avalanche multiplication occurs in this region. The second region is the intrinsic region or drift region (i-region), through which generated holes drift towards the p^+-region.

A Read diode is operated in the reverse-bias mode. If the reverse biasing is above the breakdown voltage, then the space charge region extends from the n^+-p junction up to the $i-p^+$ junction. Therefore, the whole region between the n^+-p and $i-p^+$ junctions is also called a space charge region. It may be noted that a Read diode can also be $p^+-n-i-n^+$, in which avalanche multiplication occurs in the thin n-region and the carrier drifts through the i-region towards the n^+-region.

13.5.1 Operation of Read Oscillators

The field distribution inside a Read diode, when reverse biased, is shown in Fig. 13.38(b) and the doping profile in Fig. 13.38(c). The field is maximum at the n^+-p junction (of the order of several kV/cm). The field then decreases gradually throughout the p-region and remains constant at the i-region. By proper doping the field is adjusted such that avalanche multiplication occurs and is confined within the narrow region of n^+-p junction. Electrons resulting from avalanche breakdown move towards the n^+-region, whereas holes drift through the space charge region towards the p^+-region with a constant drift velocity of 10^7 cm/s. Therefore, the transit time may be expressed as follows:

$$\tau = L/v_d \tag{13.33}$$

The breakdown voltage at the n^+-p junction and avalanche multiplication factor can be expressed as follows:

$$|V_b| = \rho_n \mu_n \varepsilon_s \left|E_{\max}|_b\right|^2 /2 \tag{13.34}$$

and $\qquad M = \left\{1 - (V/V_b)^n\right\}^{-1} \tag{13.35}$

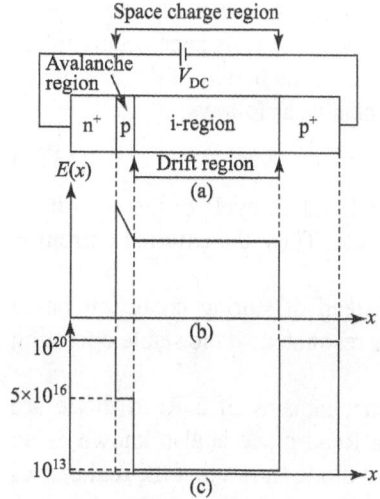

Fig. 13.38 Read diode (a) Structure (b) Field profile (c) Doping profile

where ρ_n is the resistivity, μ_n is the electron mobility, ε_s is the semiconductor permittivity, $E_{max}|_b$ is the maximum breakdown of electric field, V is the applied voltage, and n is 3–6 for Si and depends on the doping of the $n^+ - p$ junction.

If a Read diode is mounted in a microwave resonant circuit and operated above the reverse breakdown voltage, an ac voltage can be maintained in the circuit at a given frequency. The total field across the diode then becomes equal to the sum of the ac and DC fields, as shown in Fig. 13.39. Since the DC voltage alone is above the reverse breakdown voltage, for the complete positive half-cycle of the ac voltage, the diode will be biased above the breakdown voltage and avalanche multiplication will occur at the $n^+ - p$ junction. Since avalanche multiplication grows exponentially, the carrier current $I_o(t)$, developed at the $n^+ - p$ junction also grows exponentially. During the negative half-cycle, polarity of the ac voltage is opposite to that of the DC voltage, and hence the total voltage across the diode drops below the breakdown voltage. Thus, avalanche multiplication stops and the carrier current drops exponentially. It may be noted that the carrier current is the current at the $n^+ - p$ junction only and is in the form of a pulse of a very short duration, as shown in Fig. 13.39. Since the maximum number of carriers is available at the end of the positive half-cycle (due to maximum t), the carrier current reaches its maximum value at the middle of the ac cycle. Alternatively, we can say that the carrier current is delayed by 90° relative to the ac voltage. Now, the generated holes are injected into the space charge region under the influence of the electric field and travel towards the negative terminals. Drifting of these holes through the space charge region induces a current $I_e(t)$ in the external circuit, as shown in Fig. 13.39. The induced current can be expressed as follows:

$$I_e(t) = Q/\tau = Qv_d/L \quad (13.36)$$

The constant current, $I_e(t)$, starts flowing in the external circuit when the carrier current is generated in the $n^+ - p$ junction and continues to flow through the space charge region during the complete transit time τ. Thus, $I_e(t)$ is delayed by 90° relative to the carrier

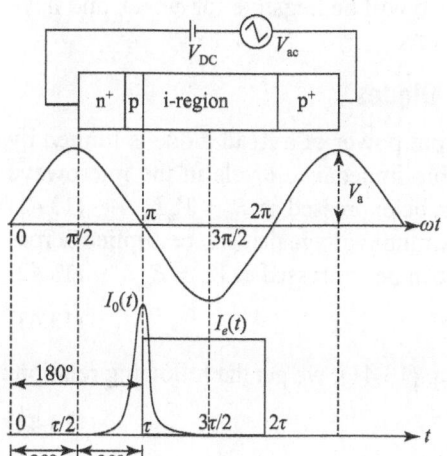

Fig. 13.39 Voltage and current in Read diode

current $I_o(t)$ and hence by 180° relative to the ac voltage. Alternatively, the external current is out of phase with the ac voltage, and a negative resistance occurs in the diode circuit. To achieve the negative resistance, the time period of the ac voltage must be equal to 2τ or the oscillation frequency must be as follows:

$$f = (2\tau)^{-1} \tag{13.37}$$

The external current is negligible during the positive half-cycle of the ac voltage and is fairly constant during its negative half-cycle. Thus, the external current is almost a square wave in shape.

Since a Read diode supplies ac energy, instead of storing energy, it has a negative Q, as compared with the positive Q of a resonator. At the stable operating point, these two Q values must be equal.

As discussed earlier, the basic operating mechanisms of a Read diode are *impact ionization* and *transit time*. Therefore, a Read diode is also known as an IMPATT diode. A small signal analysis of a Read diode shows that the real part of its terminal impedance can be expressed as follows:

$$R = R_s + \frac{2L^2}{v_d \varepsilon_s A} \left[\frac{1}{1-(\omega/\omega_r)^2} \right] \left\{ \frac{1-\cos(\theta)}{\theta} \right\} \tag{13.38}$$

where R_s is the passive resistance of the inactive region, A is the diode cross section,

$$\theta = \omega\tau \tag{13.39}$$

$$\omega_r = \sqrt{\frac{2\alpha' v_d I_0}{\varepsilon_s A}} \tag{13.40}$$

and α' is the derivative of the ionization coefficient with respect to the electric field and is a sharply increasing factor of the electric field. Physically, it denotes the number of ionizations per centimetre produced by a single carrier. Equation (13.38) reveals that the resistance will be negative for $\omega > \omega_r$ and have its peak value at $\theta = \pi$.

13.5.2 Power and Efficiency of Read Diodes

For a given frequency, the maximum output power of a Read diode is limited by the semiconductor materials and attainable impedance levels in the microwave circuitry. The maximum output power can be expressed as $P_m = V_m I_m$ (13.41) where V_m and I_m are, respectively, the maximum voltage that can be applied across the diode and the maximum current, and can be expressed as $V_m = E_m L$ (13.42)

and $\quad I_m = v_d \varepsilon_s E_m A/L \tag{13.43}$

Substituting Eqs (13.42) and (13.43) in Eq. (13.41), we get the following relation:

$$P_m = E_m^2 v_d \varepsilon_s A \tag{13.44}$$

Now, the capacitance across the space charge region can be expressed as follows:

$$C = \varepsilon_s A/L \tag{13.45}$$

Substituting Eq. (13.45) in Eq. (13.44), we get the following expression:

$$P_m = E_m^2 v_d \varepsilon_s A = E_m^2 v_d CL$$

or $\quad 4\pi^2 f^2 P_m = E_m^2 v_d 4\pi^2 f^2 CL = E_m^2 v_d \omega^2 CL = \dfrac{E_m^2 v_d \omega L}{X_c} = \dfrac{E_m^2 v_d^2 \omega \tau}{X_c}$

or $\quad P_m f^2 = \dfrac{E_m^2 v_d^2 \omega \tau}{4\pi^2 X_c} = \dfrac{E_m^2 v_d^2 \pi}{4\pi^2 X_c} = \dfrac{E_m^2 v_d^2}{4\pi X_c}$ $\qquad$ (13.46)

Equation (13.46) reveals that the maximum power decreases with increase in frequency at a rate of $1/f^2$. The efficiency of an IMPATT diode can be expressed as follows:

$$\eta = \frac{P_{ac}}{P_{DC}} = \frac{V_a I_a}{V_d I_d} = \left(\frac{V_a}{V_d}\right)\left(\frac{I_a}{I_d}\right) \qquad (13.47)$$

Now, for a Read diode $V_a/V_d \approx 0.5$ $\qquad$ (13.48)

and $\quad I_a/I_d \approx 2/\pi$ $\qquad$ (13.49)

Substituting Eqs (13.49) and (13.48) in Eq. (13.47), we get the following relation:

$$\eta \approx \pi^{-1} \approx 0.3 \qquad (13.50)$$

Therefore, the theoretical efficiency of a Read diode is about 30%. However, because of the high-frequency skin effect, reverse saturation current effect, ionization saturation effect, and space charge effect, the efficiency of a Read diode is, in practice, less than 30%.

13.5.3 Other IMPATT Diodes

It should be noted that the IMPATT family comprises not only Read diodes, but also several other semiconductor devices that come under IMPATT family. A classification of IMPATT diodes is given in Fig. 13.40.

13.5.3.1 Fabrication and Construction of p⁺–n– n⁺ IMPATT Diodes

IMPATT diodes are generally designed to maximize the output power and efficiency. Since an IMPATT diode is operated in the reverse breakdown region,

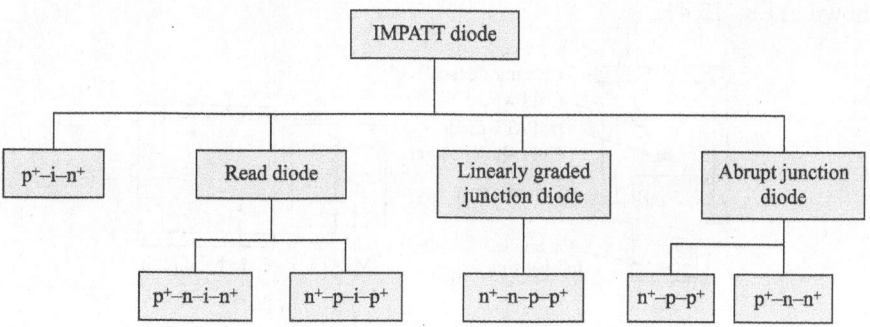

Fig. 13.40 Classification of IMPATT diodes

a huge breakdown current flows through it, which produces enormous amount of heat. This heat must be removed to protect the diode and enhance its efficiency. This can be accomplished by mounting the device on a metal heat sink. Thus, the heat sink should be an integral part of the device. One example of the fabrication process of IMPATT diodes is the Zetler and Couley process, or the integral heat sink process.

To fabricate an IMPATT diode, first an n-type semiconductor is taken and then boron is doped from one end to form a p^+-region and an n-type semiconductor is doped from the other side to form an n^+-region. The ends are then polished and a thin layer of Ti is formed on them by the vapour deposition technique. A layer of gold is electroplated on the Ti surfaces, and thus a metal–semiconductor junction is formed. During this process, the thickness of the epitaxial layer is also controlled so that at breakdown no unswept epitaxial layer is left. For high-frequency operations, the n^+-layer is further thinned down to reduce losses and non-uniformity from the skin effect.

Using this procedure, initially a large semiconductor bar is produced, from which separate IMPATT diodes are formed, using masking and etching techniques. This is known as batch fabrication. The advantage of batch fabrication is that all diodes are formed in an identical condition and hence their characteristics are also almost the same.

IMPATT diodes produced in this manner are also known as Schottky barrier IMPATT diodes. One of the advantages of this Schottky barrier approach is that the device can be fabricated at a relatively low temperature, so that the original high-quality epitaxial layer can be preserved. However, it also has a major disadvantage. The metal electrode can be chemically attacked by the semiconductor under the presence of high-energy carriers. This impairs the long-term reliability of a Schottky barrier.

The fabricated diode is generally available in the market in a package form. During packaging, the metal electrode side is kept in contact with a copper heat sink so that the heat generated at the junction can be readily removed. The copper electrode is first heated to make it soft. Then the device is placed on this copper electrode and pressed with ultrasonic energy to properly attach it to the electrode. Care should be taken so that no air gap is left between the copper and the metal electrodes, and the packaging does not change the characteristic of the device.

The construction of an epitaxed IMPATT diode and its equivalent circuit are shown in Fig. 13.41.

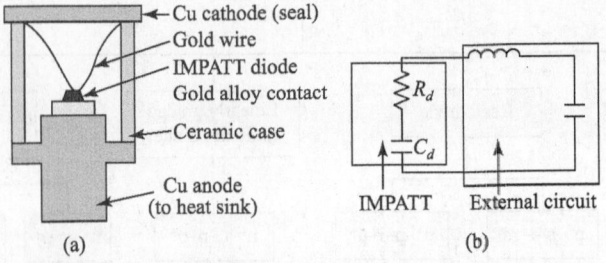

Fig. 13.41 IMPATT diode (a) Construction (b) Equivalent circuit

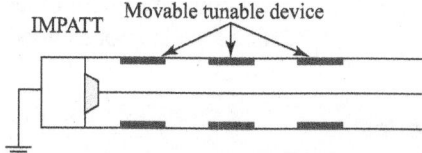

Fig. 13.42 Mounting of IMPATT diode in coaxial line

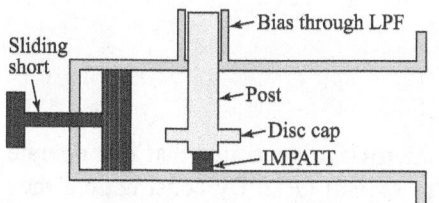

Fig. 13.43 IMPATT diode mounting in resonating waveguide

It may be noted that although an IMPATT diode can be manufactured using Ge, Si, GaAs, or InP, GaAs provides the highest efficiency, highest operating frequency, and least noise figure. However, the fabrication of GaAs IMPATT diode is more difficult and costlier than using Si.

13.5.3.2 Oscillator Arrangement for IMPATT Diodes

To use an IMPATT diode as an oscillator, it is mounted in a cavity and subjected to reverse biasing. The impedance of the diode is capacitive, whereas that of the cavity is mainly inductive. At resonance, the capacitive reactance cancels the inductive reactance, and the whole circuit behaves as a resonator.

IMPATT diodes can be mounted either in a coaxial line (Fig. 13.42) or in a waveguide (Fig. 13.43). By moving the tuning elements, the dimension of the coaxial line and hence the operating frequency can be changed.

In waveguide mounting, an IMPATT diode is mounted centrally on the bottom broad wall of a rectangular waveguide, and a disc–post structure (also known as resonant cap) is placed on the top of the packaged diode. The diode is held in place by making pressure contact with the resonant cap structure. The disc and the bottom broad wall of the rectangular waveguide form a localized open radial cavity, which is coupled to the waveguide through open vertical edges. This coupling between the cavity and the waveguide can be controlled by a sliding short tuner at the end of the cavity. The disc also acts as a quarter-wave transformer between the low-impedance diode and the high-impedance waveguide.

The IMPATT diode is biased through a low-pass filter, which also prevents the generated microwave signal from getting coupled to the bias circuit.

13.5.3.3 Application of IMPATT Diodes

At present, IMPATT diodes are probably the most powerful CW solid-state microwave power sources and are potentially reliable, compact, and inexpensive. They can be fabricated from Ge, Si, GaAs, and other semiconductors. IMPATT diodes are also widely used in modulated output oscillators, receiver local oscillators, and parametric amplifiers. High-Q IMPATT diodes are used in police radars, intrusion alarm networks, and low-power microwave transmitters, whereas low-Q IMPATT diodes are used in FM telecommunication transmitters and CW Doppler radar transmitters.

13.5.3.4 Disadvantages of IMPATT Diodes

The main disadvantage of an IMPATT diode is its noise. Due to avalanche process, it is noisy and the noise figure, which is about 30 dB (and can be as high as 50–60 dB under certain circumstances), is not good compared with other microwave oscillators. The tuning range of an IMPATT diode is also not as good as that of a Gunn diode.

13.5.3.5 Characteristics of IMPATT Diodes

The typical characteristics of an IMPATT diode are as follows:

1. Frequency range: 8 GHz to few hundred GHz
2. Pulsed output power: 100 W at 10 GHz
3. CW output power: 10 W at 10 GHz
4. Efficiency: 15%

13.6 TRAPATT Diodes

A TRAPATT diode is a high-efficiency microwave generator that can operate in the range from several hundred MHz to several GHz. By construction, they are typically $p^+ - n - n^+$ (or $n^+ - p - p^+$) structures, with the width of the n-type depletion region varying from 2.5 to 12.5 µm. Generally, doping in the depletion region is such that the DC electric field in this region, just prior to breakdown, is well above the saturated drift velocity level. That is, the diodes are well *punched through* at breakdown. The p^+-region of this device is kept as thin as possible (typically 2.5–7.5 µm). Diameter of the diode ranges from 50 µm for CW operation to 750 µm at lower frequency for a high peak power operation. The schematic diagram of a TRAPATT diode along with the electric field distribution within it is shown in Fig. 13.44. Although a TRAPATT diode is somewhat similar to an IMPATT diode in structure, the mode of operation of the former is distinctly different from the latter and also exhibits higher efficiency. An IMPATT diode starts from small signal amplitudes and builds up large signals, whereas in a TRAPATT diode, large signal oscillation is produced by triggering pulses.

The basic operation of a TRAPATT diode is based on a semiconductor p–n junction diode, reversed biased to current densities that are well in excess to those encountered in a normal avalanche operation. If the reverse bias across the diode is increased above the breakdown, the diode current will initially increase with the voltage. However, when the current becomes sufficiently high, the n-region (or p-region) is filled with electron–hole plasma produced by the secondary ionization process. This plasma will generate a large potential across the junction, which opposes the applied DC voltage. As a result, the diode voltage will be reduced to a low value. The diode thus exhibits a dynamic negative resistance between a high-voltage–low-current stage and a low-voltage–high-current state. With proper loading, the diode switches back and forth periodically and generates microwave power.

An understanding of the voltage collapse at high current requires the knowledge of the avalanche shock front concept. Let us assume that a sharply rising reverse current pulse is applied to a non-conducting $p^+ - p - n^+$ diode. Until the diode voltage remains below the breakdown voltage, the diode current remains capacitive in nature and the field, everywhere in the p-region, rises at the rate $J(t) = \varepsilon_s \dfrac{dE}{dt}$

$$\text{(13.51)}$$

where $J(t)$ is the time-dependent current density. After the maximum electric field is raised above the breakdown level at $t > t_3$, a huge number of electron–hole pairs will be generated for the current to become conductive even at a very small field. If the rise in current is very fast such that the carrier drift during the current rise time is very small or negligible, then the point at which $E = E_b$ will move rapidly towards the p$^+$-region, as shown in Fig. 13.44 (t_3, t_4). Thus, for sufficiently large $J(t)$, the avalanche shock front moves rapidly through the p-region and fills it with electron–hole plasma. Since no electron–hole plasma is produced at the right side of $E = E_b$, we have

$$\frac{dE}{dx} = \frac{qN_a}{\varepsilon_s} \qquad (13.52)$$

where N_a is the acceptor concentration in the p-region.
From Eqs (13.51) and (13.52), we get the following relation:

$$v_z = \frac{dx}{dt} = \frac{dx}{dE}\frac{dE}{dt} = \frac{\varepsilon_s}{qN_a}\frac{J(t)}{\varepsilon_s} = \frac{J(t)}{qN_a} \qquad (13.53)$$

Since the avalanche zone swipes across the p-region at a speed of $v_z \gg v_s$, Eq. (13.53) implies that

$$J(t) \gg qN_a v_s \qquad (13.54)$$

Due to the collapse of the diode voltage, the electric field also falls below the level required for velocity saturation. Thus, the plasma is trapped. Electrons and holes in the depletion layer then begin to drift in the opposite direction with a low drift velocity due to the low field of the plasma. For a p$^+$–p–n$^+$ structure, hole excitation first starts at the n$^+$ terminal and electron excitation starts at a slightly later time (w/v_z) at the p$^+$ terminal when the shock front has reached the substrate. The low field present in the plasma drives the electrons towards the p$^+$ terminal and holes towards the n$^+$ terminal, increasing the fields. Therefore, at an early stage of plasma extraction, the field profile resembles that shown in Fig. 13.45(a). As the time progresses, the central region remains filled with electrons and holes, whereas at the sides electrons continue to flow towards the p$^+$ terminal and holes towards the n$^+$ terminal. Therefore, the field in the central region remains low due to the presence of the plasma, whereas it linearly rises on either side. The carrier

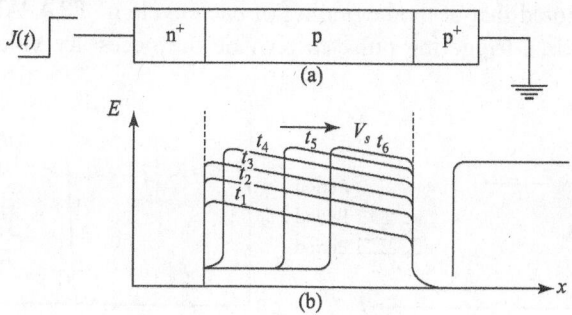

Fig. 13.44 TRAPATT diode (a) Schematic cross section
(b) Field rise before $(t_1 - t_3)$ and after plasma formation $(t_4 - t_6)$

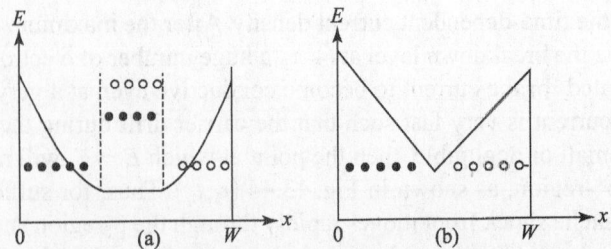

Fig. 13.45   Field and current profiles (a) During slow recovery (b) During fast recovery

velocity is thus low in the central region and rises to the saturated drift velocity in the end regions. Field recovery then becomes fast due to rapid extraction of electrons and holes. The field profile in this second stage is shown in Fig. 13.45(b). Residual carriers are then removed from the active region with a saturated drift velocity, and the original field profile is re-established. Since the total plasma charge is large enough compared to the charge per unit time in the external circuit, a long time is required to complete this process. Hence, a TRAPATT diode operates at a comparatively low frequency. Once the original field profile is re-established, another shock front then passes through the depletion layer and the process is repeated.

The voltage and current waveforms of a TRAPATT diode are shown in Fig. 13.46. During period 1, the diode voltage rises significantly above the breakdown voltage, followed by a sudden drop in the voltage and a rise in the current. This is due to the formation of the electron–hole plasma and filling of the depletion region with the plasma. Recovery of the diode voltage begins in period 2. Initially, the field is low and the carriers move out of the plasma with a velocity lower than the saturation velocity. However, as the plasma is extracted, the diode voltage begins to recover with a subsequent increase in the electric field. This increase in the electric field, in turn, increases the carrier drift velocity. During this period, the current stays close to its maximum value and then suddenly drops to a very low value at the end when almost no carriers are left in the p-region. This sets the beginning of period 3. In period 3, the current remains low and voltage remains high, slightly below the breakdown voltage. At the end of period 3, an over-voltage pulse restores the situation occurring at the start of period 1.

It may be noted that at the beginning of each cycle of TRAPATT oscillation, the device needs a triggering pulse to provide the necessary voltage to initiate

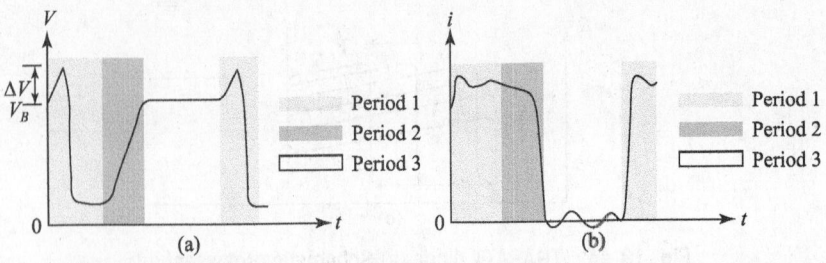

Fig. 13.46   TRAPATT diode (a) Voltage waveform (b) Current waveform

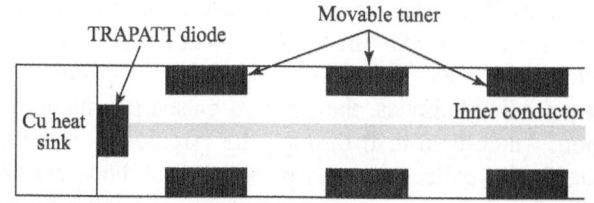

Fig. 13.47 Coaxial circuit TRAPATT oscillator with tuning slugs

the shock front. This can be done using the circuit shown in Fig. 13.47. The slide tuner nearest to the device is kept at a half-wavelength distance at the operating wavelength of the TRAPATT oscillator. Initially, with an increase in the bias current, an IMPATT oscillation is generated by the device, which grows into an RF voltage of appreciable magnitude and provides the first triggering pulse. It produces the avalanche shock front, and the voltage across the device is dropped. The negative voltage pulse then travels towards the first low impedance tuning slug and gets reflected. On reflection, the voltage pulse changes sign to become positive and then starts travelling towards the device. Since the distance between the device and the tuning slug is half-wavelength, the voltage pulse travels a full wavelength when it comes back to the device after reflection. In a time scale, it returns after a complete time period. With proper adjustment, the amplitude of the retuned voltage pulse can be modified such that it can trigger the second avalanche shock front. This again produces a sharp voltage pulse, which similarly travels down the transmission line and comes back after reflection to produce the next trigger pulse. Thus, triggering pulses are produced continuously, resulting in TRAPATT oscillations in the device.

TRAPATT oscillators can produce high output power, with very high efficiency. With a proper design, the efficiency can go up to 80%. The reason for this is that the voltage is very low during the plasma extraction. On the negative side, the TRAPATT mode requires a very high power density $\left(\approx 10^6 \text{ W/cm}^2\right)$ for its operation. Therefore, these oscillators are only suitable for pulse operation. In addition, its maximum operating frequency is low (below 10 GHz). Because of the extreme large fluctuations occurring at the start-up of the current by avalanche multiplication, the noise behaviour of a TRAPATT diode is even worse than that of an IMPATT diode.

13.6.1 Characteristics

The typical characteristics of a TRAPATT diode are as follows:

1. Frequency range: 0.5–10 GHz
2. Pulsed output power: 1.2 kW at 1.1 GHz
3. CW output power: 1–3 W
4. Efficiency: 15–50%
5. Noise figure: >30 dB

13.7 BARITT Diodes

A BARITT diode utilizes the injection and transit-time delay properties of minority carriers to produce a negative resistance at a lower microwave frequency. Since no avalanche is required, such diodes are much less noisy than IMPATT diodes. However, BARITT diodes have relatively narrow bandwidths and the power output is also limited (few mW). By construction, BARITT diodes have different structures such as p–n–p, p–n–metal, p–n–γ–p, and metal–n–metal. For the p–n–γ–p BARITT diodes, the forward-biased p–n junction injects holes into the v-region, which then drift through the v-region with the saturated drift velocity and are finally collected at the p-region. The diode exhibits a negative resistance for the transit angles between π and 2π. The optimum transit angle is approximately 1.6π.

To understand the operation of a BARITT diode, let us consider the metal–n–metal structure shown in Fig. 13.48(a). The n-type semiconductor, with rectifying contacts at the two ends, is uniformly doped. The space charge, electric field distribution, and energy band diagram of the device at zero bias are shown in Figs 13.48(b)–(d). If a DC bias V_B is applied to the device, making the LHS positive with respect to RHS, then junction J_1 is forward biased, whereas junction J_2 is reverse biased. If the bias voltage is now increased gradually, practically all the voltage appears across J_2, and a point is reached when the depletion width of junction 2, W_2, reaches the depletion region of junction J_1. The voltage at which this phenomenon occurs is called the *punched-through voltage* and can be expressed as follows:

$$V_{\text{PT}} \simeq \frac{qN_DW^2}{2\varepsilon_s} - W\sqrt{\frac{2qN_DV_i}{\varepsilon_s}} \qquad (13.55)$$

If the bias field is further increased, W_2 keeps on widening and becomes equal to W, whereas the band at J_1 becomes flat. Such a voltage is called the *flat-band voltage* and can be expressed as follows:

$$V_{\text{FB}} = \frac{qN_DW^2}{2\varepsilon_s} \qquad (13.56)$$

where N_D is the donor concentration and W is the width of the semiconductor.

If the DC bias voltage is greater than V_{PT} but less than V_{FB}, the energy-band diagram will be modified as shown in Fig. 13.49. Now let us assume that an ac voltage is superimposed on the DC bias. In the positive half-cycle of the ac voltage, additional holes will be injected from the metals into the semiconductors. Hole injection will increase exponentially with the applied voltage and will be

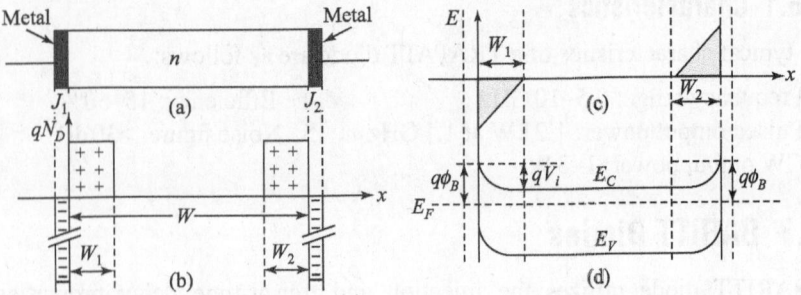

Fig. 13.48 BARITT diode (a) Schematic diagram (b) Doping profile (c) Field profile (d) Energy band diagram

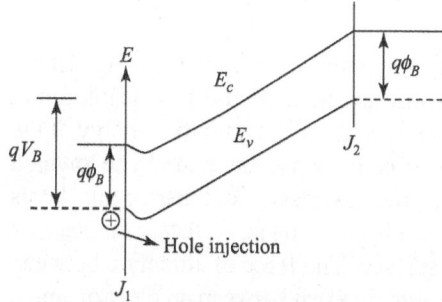

Fig. 13.49 Energy-band diagram of reverse-biased metal–semiconductor–metal structure with DC bias higher than V_{PT} but less than V_{FB}.

maximum at the positive peak of the RF voltage. As the hole moves through the semiconductor, a current will be induced in the external circuit, which will last till the holes are collected at J_2. Figure 13.50 shows the waveform of the ac voltage, injected hole current, and current induced in the external circuit. The figure reveals that holes are injected almost as a delta function when the phase angle of the ac voltage is $\pi/2$. It may further be noted that the ac voltage and induced current are in phase from $\pi/2$ to π (when the holes absorb energy from the ac field), whereas they are out of phase from π to 2π (when the ac field absorbs energy from the holes). Therefore, to maximize ac power, the width of the semiconductor region should be such that the injected charge bunch gets collected at J_2 slightly before the end of the ac cycle. A common practice is to set the transit angle equal to $3\pi/2$. In such cases,

$$2\pi f\tau_d = \frac{3\pi}{2}$$

or

$$f = \frac{3\pi}{2}\frac{1}{2\pi\tau_d} = \frac{3}{4\tau_d} = \frac{3v_s}{4W} \qquad (13.57)$$

13.7.1 Characteristics

The typical characteristics of a BARITT diode are as follows:

1. Frequency range: 4–8 GHz
2. CW output power: 50 mw at 4.9 GHz
3. Efficiency: 18%
4. Noise figure: 9 dB at 6.35 GHz

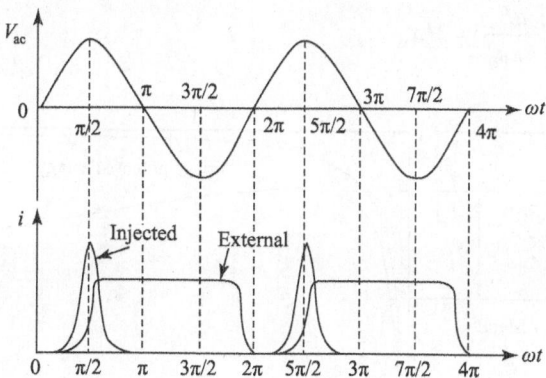

Fig. 13.50 Waveforms of ac voltage, injected charge, and induced current in BARITT diode

13.8 Schottky Barrier Diodes

The energy-band diagram of a metal–vacuum interface is shown in Fig. 13.51, where $q\phi_m$ is the metal work function. The work function is defined as the minimum energy required by an electron to escape into vacuum from an initial energy at the Fermi level and it varies from 2 to 6 eV. The effective work function can be lowered with the application of an electric field at the metal surface. To illustrate this, let us assume that an electron is present at a distance x from the metal surface. This electron will induce a positive charge on the metal surface. The force of attraction between the electron and the positive charge will be equal to that between an electron and a positive charge located at $+x$ and $-x$, respectively, and can be expressed as follows:

$$F_c = -\frac{q^2}{4\pi\varepsilon_0 (2x)^2} = -\frac{q^2}{16\pi\varepsilon_0 x^2} \tag{13.58}$$

The corresponding electric field will be as follows:

$$E_c = \int_0^x F_c \mathrm{d}x = \frac{q^2}{16\pi\varepsilon_0 x} \tag{13.59}$$

If an external electric field E is applied, the total potential energy modifies to the following form:

$$PE(x) = \frac{q^2}{16\pi\varepsilon_0 x} + qEx \text{ eV} \tag{13.60}$$

If the Schottky barrier lowering happens at $x = x_m$, then

$$\frac{\mathrm{d}}{\mathrm{d}x}\{PE(x)\}\Big|_{x=x_m} = 0$$

or $\quad -\dfrac{q^2}{16\pi\varepsilon_0 x_m^2} + qE = 0$

or $\quad x_m = \sqrt{\dfrac{q}{16\pi\varepsilon_0 E}} \tag{13.61}$

The amount of Schottky barrier lowering is given by the following relation:

$$\Delta\phi = \sqrt{\frac{qE}{4\pi\varepsilon_0}} = 2Ex_m \tag{13.62}$$

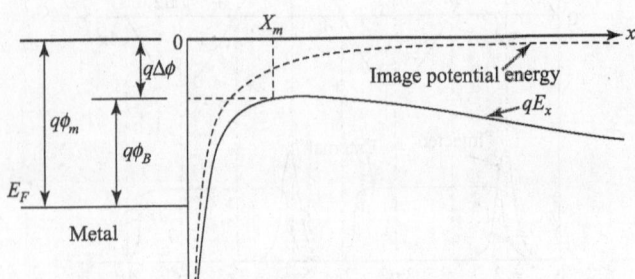

Fig. 13.51 Energy-band diagram between metal surface and vacuum

The lowering of the metal work function as a result of the image force and the electric field is called the Schottky effect.

The result obtained for a metal–vacuum junction can also be applied for a metal–semiconductor junction. However, for such a case, the electric field should be replaced by the maximum field at the interface and the free space permittivity should be replaced by the permittivity of the semiconductor. Since the permittivity of the semiconductor is larger than that of the free space, barrier lowering and the location of the potential maxima in the semiconductor are also much smaller than that due to free space.

If the metal and semiconductor are spaced across a large distance, such that no exchange of carriers can take place then their Fermi level will be different, as shown in Fig. 13.52. However, if they are brought close together, thereby forming a Schottky contact, electrons will be exchanged in an attempt to establish thermal equilibrium. Finally, equilibrium is reached as the Fermi levels become constant throughout the metal and the semiconductor, as shown in Fig. 13.53. This is similar to the situation that arises for a p–n junction. During the process of obtaining equilibrium, the interface becomes polarized due to the transfer of electronic charge across the boundary. The net charge on the metal side resides in a thin surface layer, whereas the opposite charge on the semiconductor side is distributed between surface or interface states. This results in a space charge region, much like the depletion region in a p–n junction, across the boundary.

The I–V characteristics of a Schottky diode are governed by the diode law, but the forward voltage for significant conduction is governed by the work function and electron affinity. Thus, it is possible to produce diodes with a high or low forward voltage. In addition, it is possible to produce a contact where there is no barrier. In such a case, current flow is limited by the resistance of the metal and semiconductor. Such a contact is also called an Ohmic contact. An Ohmic contact is also possible if the depletion width becomes very small due to tunnelling through the energy barrier. The ability to control the forward turn on voltage gives us the opportunity to produce low- and high-power diodes for a variety of applications.

A Schottky device is a majority carrier device, and there are no minority carriers in the reverse-bias condition, that is, no significant current flows from the metal to the semiconductor in the back bias. Due to this the delay present in the junction, because of the hole–electron recombination time, is absent and the diode possesses

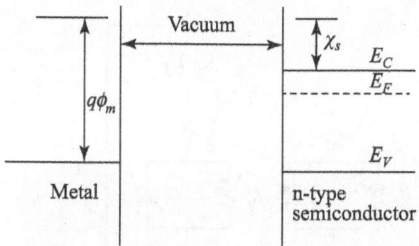

Fig. 13.52 Energy-band diagram of metal and semiconductor when they are spaced across a large distance such that no exchange of carriers can take place

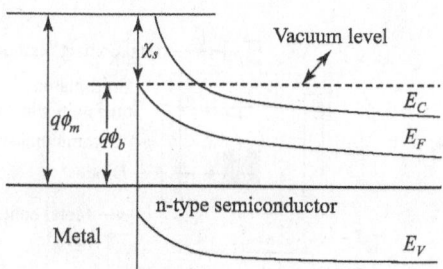

Fig. 13.53 Energy-band diagram of metal and semiconductor when equilibrium is reached

an inherent fast response. Further, being a majority carrier device, it does not exhibit any minority carrier storage effect like diffusion capacitance. In comparison to a p–n junction diode, a Schottky barrier diode has a steeper forward I–V slope, lower series resistance, lower forward turn on voltage, and lower reverse breakdown voltage.

Point-contact diodes were the earliest microwave devices of Schottky barrier type. However, with the recent photo masking technique, evaporated Schottky barrier diodes with small capacitance can be fabricated. Such a device is usually called a Schottky barrier device and is distinctly different from a Schottky barrier diode, which is a point-contact diode. A Schottky barrier device has better I–V characteristics than a point-contact diode, and is also more mechanically rugged and better reproducible. Point-contact diodes are fabricated by pressing a thin pointed tungsten wire on a p-type Si substrate. This is followed by tapping the unit with a small hammer or passing a semi-destructive current pulse through the device. A Schottky barrier device can be obtained by depositing a metal, usually platinum or gold, on a vacuum-cleaned Si surface. Normally, a thin, lightly doped n-type epitaxial layer is grown on an n^+ Si substrate. The thickness of the epitaxial material is maintained at about 1 μm with resistivity of 1 Ohm-cm, whereas the diameter of the contact is maintained at about 25 μm. A guard ring of diffused p^+ material is also used to eliminate unwanted edge breakdown and channelling effect. Such a device does not require any adjustment and hence can be reproduced more easily. Such a device also has a low reverse leakage current.

In principle, the resistance associated with a point-contact diode varies inversely with the contact radius rather than with the area. This relation is valid until the radius is much smaller than the thickness of the semiconductor and other contacts. In contrast, the capacitance is directly proportional to the area or square of the radius. Thus, the capacitance sinks faster than the growth of the resistance and the parasitic RC product drops approximately linearly with the radius, instead of remaining constant. This behaviour differs from that of planar junction devices and produces a superior frequency response.

In addition to Si, GaAs can also be used for fabricating Schottky devices. GaAs has lower noise characteristics and higher frequency limits. In practice, Schottky diodes are available in a package form, as shown in Fig. 13.54(a). Equivalent circuits of an unpackaged and a packaged Schottky diode are shown in Figs 13.54(b) and (c), respectively.

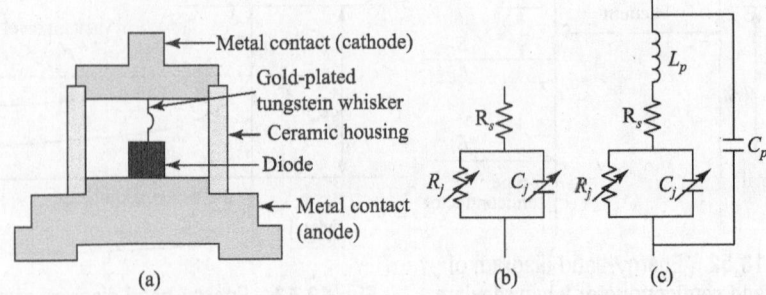

Fig. 13.54 Schottky diode (a) Packaged form (b) Equivalent circuit of unpackaged Schottky diode (c) Equivalent circuit of packaged Schottky diode

Schottky devices are also known as epitaxial Schottky barrier (ESBAR) diodes or hot electron diodes. The latter name is given because electrons that are flowing from a semiconductor into a metal have a higher energy level than the electrons in the metal itself. The situation is just as the metal would have if it were at a higher temperature.

Schottky diodes are available for microwave frequencies up to at least 100 GHz and are widely used as mixers and detectors. The noise figure of mixers, obtained using Schottky barrier diodes, can range from 4 dB at 2 GHz to 15 dB near 100 GHz. The important characteristic parameters of such microwave mixers and detectors are as follows:

$$\text{Cut-off frequency, } f_c = \frac{1}{2\pi R_s C_j} \tag{13.63}$$

$$\text{Mixer conversion loss, } L_c = L_{\text{DC}}\left[1 + \frac{R_s}{R_j} + \omega^2 C_j^2 R_s R_j\right] \tag{13.64}$$

where L_{DC} is the DC conversion loss.

$$\text{Mixer efficiency, } \eta = 1 - \frac{1}{\omega C_j R_j} \tag{13.65}$$

$$\text{Detector microwave current sensitivity, } \beta \simeq \frac{\beta_{\text{DC}}}{1 + \omega^2 C_j^2 R_j} \tag{13.66}$$

where β_{DC} is the detector DC current sensitivity and can be expressed as follows:

$$\beta_{\text{DC}} = \frac{q}{2nkT} \tag{13.67}$$

where n is called the identity factor and depends upon the quality of the metal–semiconductor interface. It is about 1.15 for a beam lead diode and 1.20 for a point-contact diode.

13.9 PIN Diodes

A PIN diode or p–i–n diode is a p–n junction diode with an intrinsic or i-region in between a p-layer and an n-layer, as shown in Fig. 13.55. In practice, the i-region is approximated by a highly resistive p-region, called the π-region, or a highly resistive n-region, called the γ-region; hence, the diode impurity region in the p- or n-layer varies more gradually than shown. Resistivity of this region is about 100–10,000 Ω/cm. The i-region is the active region of the diode. Due to the presence of this region, the breakdown voltage of a reverse-bias PIN diode is very high. Low doping in the i-region also ensures most of the potential drop in this region. The frequency of operation of a PIN diode depends on the thickness of the i-region, cross-sectional area of the diode, and resistivity of the i-region.

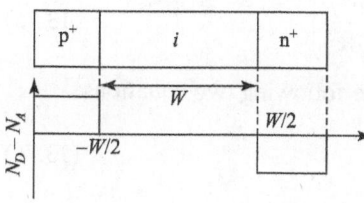

Fig. 13.55 Schematic structure and doping profile of p–i–n diode

In order to find the resistivity of the i-region, let us make the following assumptions:

1. The mobility and lifetime of both electrons and holes are the same, that is, $\mu_n = \mu_p = \mu$ and $\tau_n = \tau_p = \tau$.
2. The number of electrons and holes is the same at any point, that is, $p(x) = n(x)$ and $\frac{\partial}{\partial x} p(x) = \frac{\partial}{\partial x} n(x)$. This corresponds to quasi-charge neutrality.
3. The reverse saturation current is negligible.

These assumptions lead to the following boundary conditions:

$$\frac{\partial n}{\partial t} = (g - R) + \frac{\vec{\nabla} \cdot \vec{J}_n}{e} \tag{13.68}$$

and $$\frac{\partial p}{\partial t} = (g - R) - \frac{\vec{\nabla} \cdot \vec{J}_p}{e} \tag{13.69}$$

where g is the generation rate and R is the recombination rate. Therefore, $(g - R)$ is the net generation – recombination rate, and is the same for electrons and holes. At low injection, that is, if the bias voltage is low or moderate, we can write that

$$(g - R) = -\frac{n - n_i}{\tau} = -\frac{p - p_i}{\tau} \tag{13.70}$$

where n_i and P_i are intrinsic electron and hole concentrations, respectively. The electron and hole current densities are given by the following expression:

$$J_n = ne\mu E + eD_n \frac{dn}{dx} \tag{13.71}$$

and $$J_p = pe\mu E - eD_p \frac{dp}{dx} \tag{13.72}$$

Substituting Eqs (13.70)–(13.72) in Eqs (13.68) and (13.69) and then adding them, we get the following equation:

$$\frac{\partial n}{\partial t} + \frac{\partial p}{\partial t} = -\frac{p - p_i}{\tau} - \frac{n - n_i}{\tau} + D\left(\frac{d^2 n}{dx^2} + \frac{d^2 p}{dx^2}\right) + \mu \frac{\partial}{\partial x}\{(n - p)E\} \tag{13.73}$$

where we have further assumed that $D_n = D_p = D$ \hfill (13.74)

If E is small then $n = p$, and Eq. (13.73) reduces to the following form:

$$\frac{\partial n}{\partial t} + \frac{\partial p}{\partial t} = -\frac{p - p_i}{\tau} - \frac{n - n_i}{\tau} + D\left(\frac{d^2 n}{dx^2} + \frac{d^2 p}{dx^2}\right) \tag{13.75}$$

Equation (13.75) can be written as the sum of the following two equations:

$$\frac{\partial n}{\partial t} = -\frac{n - n_i}{\tau} + D\frac{d^2 n}{dx^2} \tag{13.76}$$

and $$\frac{\partial p}{\partial t} = -\frac{p - p_i}{\tau} + D\frac{d^2 p}{dx^2} \tag{13.77}$$

At the steady state, $\dfrac{\partial n}{\partial t} = \dfrac{\partial p}{\partial t} = 0$ (13.78)

Therefore, $-\dfrac{n-n_i}{\tau} + D\dfrac{d^2 n}{dx^2} = 0$

or $\dfrac{d^2 n}{dx^2} - \dfrac{n - n_i}{D\tau} = 0$ (13.79)

Since $n \gg n_i$, Eq. (13.79) can be written as follows:

$$\dfrac{d^2 n}{dx^2} - \dfrac{n}{D\tau} = 0$$

or $\dfrac{d^2 n}{dx^2} - \dfrac{n}{L^2} = 0$ (13.80)

where $L = \sqrt{D\tau}$ (13.81)

and is called the diffusion constant. In a similar way, we can write that

$$\dfrac{d^2 p}{dx^2} - \dfrac{p}{L^2} = 0$$ (13.82)

The solution of Eq. (13.80) can be written as follows:

$$n(x) = A' \exp(x/L) + B' \exp(-x/L)$$ (13.83)

The variation of electron concentration is shown in Fig. 13.56.
Now at $x = 0$, n will be minimum, that is,

$$\left.\dfrac{dn(x)}{dx}\right|_{x=0} = \left.\dfrac{A'}{L}\exp(x/L) - \dfrac{B'}{L}\exp(-x/L)\right|_{x=0} = \dfrac{A'}{L} - \dfrac{B'}{L} = 0$$

or $A' = B'$ (13.84)

Substituting Eq. (13.84) in Eq. (13.83), we get the following relation:

$$n(x) = A'\{\exp(x/L) + \exp(-x/L)\} = 2A'\cosh(x/L)$$ (13.85)

Since at the $n^+ - i$ junction all the current is due to electrons, at $x = w/2$, n will be maximum and equal to $n(w/2)$. Therefore,

$$n(w/2) = 2A'\cosh\{w/(2L)\}$$

or $A' = \dfrac{n(w/2)}{2\cosh\{w/(2L)\}}$ (13.86)

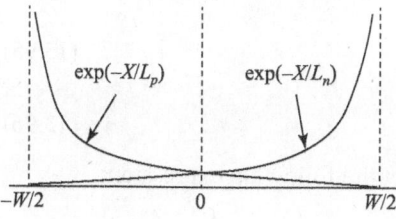

Fig. 13.56 Variation of electron
concentration in intrinsic region

Substituting Eq. (13.86) in Eq. (13.85),
we get the following relation:

$$n(x) = \dfrac{n(w/2)}{\cosh\{w/(2L)\}}\cosh(x/L) \quad (13.87)$$

Further at $x = w/2$, $J_p = 0$ and Eq. (13.72)
gives the following expression:

$$p\mu E = D\dfrac{dp}{dx} \quad (13.88)$$

Again since $n = p$, we can also write, $n\mu E = D\dfrac{dn}{dx}$ (13.89)

Now, $I_0 = J_n A = \left(ne\mu E + eD\dfrac{dn}{dx}\right)A$ (13.90)

Substituting Eq. (13.89) in Eq. (13.90), the following relation is obtained:

$$I_0 = 2eDA\dfrac{dn}{dx}$$ (13.91)

Substituting Eq. (13.87) in Eq. (13.91), we get the following expression:

$$I_0 = 2eDA\dfrac{n(w/2)\sinh(x/L)}{L\cosh\{w/(2L)\}}\bigg|_{x=w/2} = 2eDA\dfrac{n(w/2)\sinh\{w/(2L)\}}{L\cosh\{w/(2L)\}}$$

or $\quad n(w/2) = \dfrac{I_0 L\cosh\{w/(2L)\}}{2eDA\sinh\{w/(2L)\}}$ (13.92)

Substituting Eq. (13.92) back in Eq. (13.87), we get the following relation:

$$n(x) = p(x) = \dfrac{I_0 L\cosh\{w/(2L)\}}{2eDA\sinh\{w/(2L)\}}\dfrac{1}{\cosh\{w/(2L)\}}\cosh(x/L)$$

or $\quad n(x) = \dfrac{I_0 L\cosh(x/L)}{2eDA\sinh\{w/(2L)\}}$ (13.93)

Resistance of the i-region will be as follows:

$$R_i = \dfrac{1}{A}\int_{x=-w/2}^{x=w/2}\dfrac{dx}{\sigma(x)} = \dfrac{1}{A}\int_{x=-w/2}^{x=w/2}\dfrac{dx}{e\mu[n(x)+p(x)]} = \dfrac{1}{2e\mu A}\int_{x=-w/2}^{x=w/2}\dfrac{dx}{n(x)}$$ (13.94)

Substituting Eq. (13.93) in Eq. (13.94), we get the following relation:

$$R_i = \dfrac{1}{2e\mu A}\int_{x=-w/2}^{x=w/2}\dfrac{2eDA\sinh\{w/(2L)\}}{I_0 L}\dfrac{dx}{\cosh(x/L)}$$

or $\quad R_i = \dfrac{D}{\mu I_0 L}\sinh\{w/(2L)\}\int_{x=-w/2}^{x=w/2}\dfrac{dx}{\cosh(x/L)}$

or $\quad R_i = \dfrac{D}{\mu I_0 L}\sinh\left(\dfrac{w}{2L}\right)\tan^{-1}\left\{\sinh\left(\dfrac{x}{L}\right)\right\}\bigg|_{x=-w/2}^{x=w/2}$

or $\quad R_i = \dfrac{2D}{\mu I_0 L}\sinh\left(\dfrac{w}{2L}\right)\tan^{-1}\left\{\sinh\left(\dfrac{w}{2L}\right)\right\}$ (13.95)

Now, $\quad D = \mu kT/e$ (13.96)

Substituting Eq. (13.96) in Eq. (13.95), we get the following expression:

$$R_i = \dfrac{2\mu kT}{\mu e I_0 L}\sinh\left(\dfrac{w}{2L}\right)\tan^{-1}\left\{\sinh\left(\dfrac{w}{2L}\right)\right\} = \dfrac{2kT}{e I_0 L}\sinh\left(\dfrac{w}{2L}\right)\tan^{-1}\left\{\sinh\left(\dfrac{w}{2L}\right)\right\}$$ (13.97)

Equation (13.97) reveals that R_i is a function of I_0. Therefore, by varying I_0 we can change the resistance of a PIN diode.

High resistivity of the i-region in a PIN diode results in low RF loss and the least variation of junction capacitance with the reverse bias voltage. A wide intrinsic region, although producing high breakdown voltage, has low transition frequency and high junction resistance, and requires a large bias voltage. A large area, on the other hand, increases the power-handling capacity and produces large diode junction capacitance. This, in turn, results in a narrow circuit bandwidth and a low diode cut-off frequency. The p^+ and n^+ sides of the diode have a metallic contact for external electrical connections to the circuit. This metal contact, along with the resistance of p^+ and n^+ regions, results in RF loss.

Commercially available p–i–n diodes are found in a package form. Packaged p–i–n diodes are used at low microwave frequencies, whereas beam lead and chip diodes are used at high microwave and millimetre-wave frequencies. This is due to the fact that the latter diodes have high cut-off frequencies because of small junction capacitance and other parasitic elements. The important parameters of p–i–n diodes are as follows:

$$\text{Cut-off frequency: } f_c = \frac{1}{2\pi C_j R_r} \tag{13.98}$$

$$\text{Switching cut-off frequency: } f_{cs} = \frac{1}{2\pi C_j \sqrt{R_r R_f}} \tag{13.99}$$

$$\text{Transition cut-off frequency: } f_t = D_p / w^2 \tag{13.100}$$

where R_r and R_f are the series resistances under reverse- and forward-bias conditions, D_p is the diffusion constant, C_j is the junction capacitance, and w is the width of the intrinsic region.

A typical p–i–n diode and its equivalent circuit are shown in Figs 13.57 and 13.58, respectively. Here L_s is the lead inductance, R_s the series resistance, C_j the junction capacitance, C_f the fringing capacitance, R_i the intrinsic resistance, C_i the intrinsic capacitance, C_p the parasitic capacitance due to packaging, and C_D the diffusion capacitance.

In forward bias, R_j is very small but C_D and C_j are very high. Hence, the forward-bias equivalent circuit modifies as shown in Fig. 13.59(a). In reverse

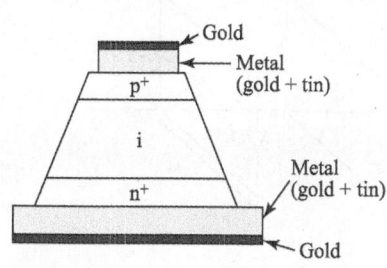

Fig. 13.57 Schematic structure of p–i–n diode

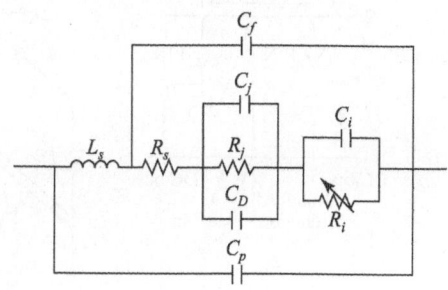

Fig. 13.58 Equivalent circuit of p–i–n diode

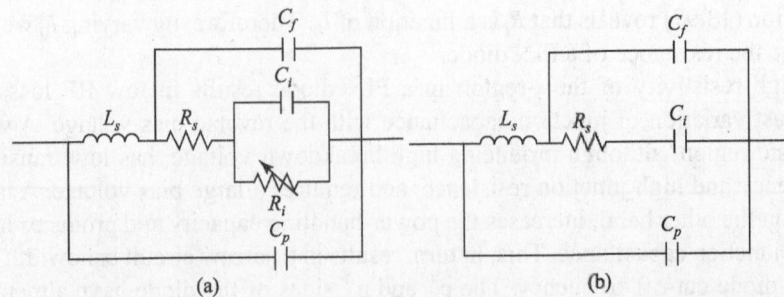

Fig. 13.59 Equivalent circuit of p–i–n diode (a) Forward bias (b) Reverse bias

bias, $R_i \to \infty$ and the reverse-bias equivalent circuit is modified to that shown in Fig. 13.59(b).

The basic operating principle of a p–i–n diode is the variation of diode resistance by varying the bias voltage. In forward bias, the diode resistance is of the order of $1-10\,\Omega$, whereas in reverse bias, it is of the order of $5-10\,\mathrm{k\Omega}$. When the diode is mounted in a transmission line, then the diode impedance can be matched with the impedance of the transmission line by varying the reverse bias voltage.

PIN diodes find wide applications in microwave circuitry as switching elements, as shown in Fig. 13.60. For a shunt configuration, reverse bias to a p–i–n diode results in a high-impedance shunt in the line, thereby making transmission *on*, whereas forward bias to a p–i–n diode results in a low-impedance shunt in the line, thereby making transmission *off*. The reverse is true for series connection. For a shunt configuration, due to non-zero forward bias resistance, isolation between

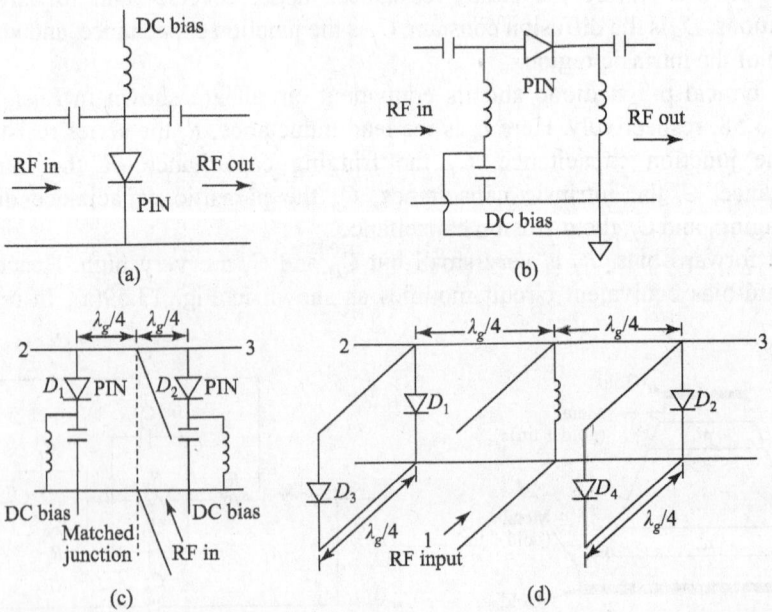

Fig. 13.60 PIN diode as switch (a) Shunt mount (b) Series mount (c) SPDT switch with two diodes (d) SPDT switch with four diodes for better isolation and insertion loss

the input and output is not infinite. Similarly, for reverse bias, shunt impedance is not infinite and a non-zero insertion loss results. The reverse is true for series connection.

The p–i–n diode double-switch circuit shown in Fig. 13.60(c) is also called a single-pole double through switch or an SPDT switch. Diodes are biased through RF chokes to isolate the ac component from the DC source. The quarter-wave line provides impedance matching between the feeder line and the switch. Depending on the bias condition of the diodes D_1 and D_2, the input power can be switched to port 2 or 3. For example, if diode D_1 is reverse biased and D_2 is forward biased then the input power will appear at port 2. Insertion loss and isolation of an SPDT switch can be improved by placing two diodes on both sides at a distance $\lambda/4$ from D_1 and D_2, as shown in Fig. 13.60(d). To get the output at port 3, biasing conditions of the diodes should be as follows: D_1 and D_4, forward bias and D_2 and D_3, reverse bias. For power at port 2, the conditions will be reversed.

A p–i–n diode can also act as a modulator, as shown in Fig. 13.61. In such a case, we apply the modulating signal as a bias current. Since the resistance of the intrinsic layer is a function of the bias current, under such circumstance, the resistance of the intrinsic region of a p–i–n diode is varied according to the modulating signal. As a result, the output signal is amplitude modulated.

A p–i–n diode can also be used in a duplexer circuit, as shown in Fig. 13.62(a). For transmission of RF power, both the diodes are forward biased, which will result in short circuits at A and A'. Under such circumstances, the 3 dB power divider will behave as a hybrid phase shifter, as described in Chapter 10, and all the power will be transmitted through the antenna. Practically, due to non-zero impedance of the diodes at forward bias, some power is transmitted to the matched load, where it gets absorbed. However, power in the receiver port will be zero, which is explained in Fig. 13.62(b). For the reception mode, the RF power enters the receiver from the antenna and the isolator prevents any power from entering the transmitter. Power division in the duplexer, in the receive mode, is shown in Fig. 13.62(c).

PIN diodes are also used in electronic phase shifters for phased arrays. In general, three types of variable phase shifters can be constructed using p–i–n diodes, namely (a) transmission-type switched line phase shifter, (b) reflection-type hybrid coupled phase shifter, and (c) transmission-type loaded line phase shifter, as shown in Fig. 13.63.

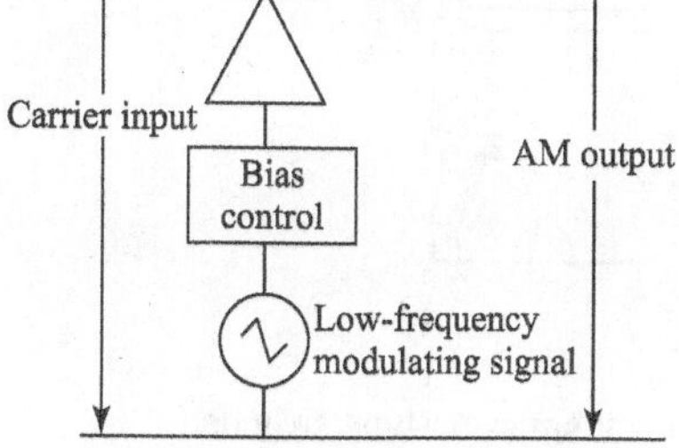

In a transmission-type switched line phase shifter, phase shift is obtained by perturbing the parameters of the transmission line. If D_1 and D_2 are forward biased and D_3 and D_4 are reverse biased, then the signal passes through the l_0 line and produces a phase change of $\phi_0 = \beta l_0$; on the other hand, if D_1 and D_2 are reverse biased and D_3 and D_4 are forward biased, the signal passes through the l_1 line and produces a phase change of $\phi_1 = \beta l_1$. Therefore,

Fig. 13.61 Amplitude modulation using p–i–n diode

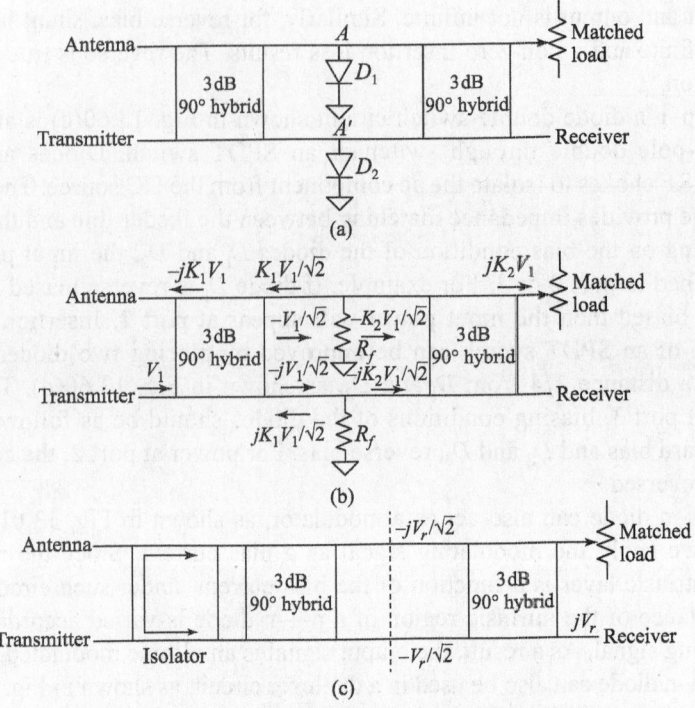

Fig. 13.62 p–i–n duplexer (a) Schematic circuit (b) Transmit mode equivalent circuit (c) Receive mode equivalent circuit

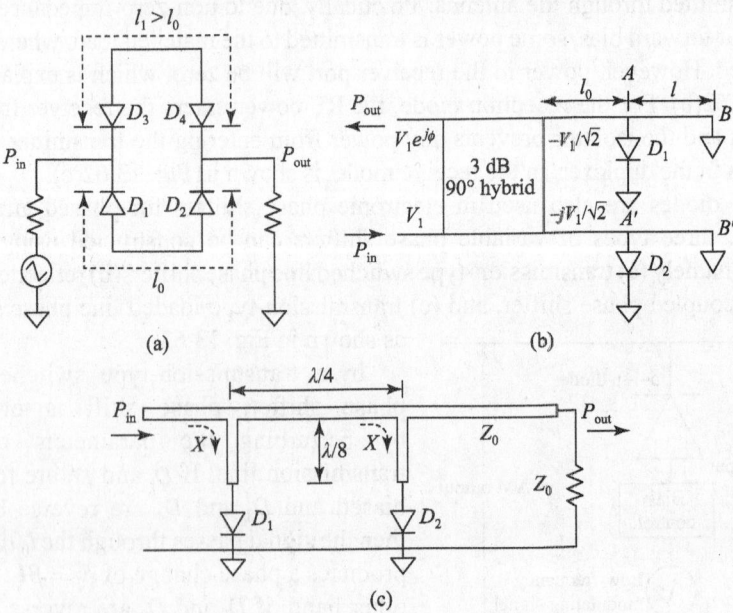

Fig. 13.63 p–i–n diode phase shifters (a) Transmission-type switched line phase shifter (b) Reflection-type hybrid coupled phase shifter (c) Transmission-type loaded line phase shifter

by switching the diodes from one biasing condition to another, a differential phase shift of $\Delta\phi = \beta(l_0 - l_1)$ can be obtained. The phase shift produced by such a phase shifter is fixed, and hence such phase shifters are called 1-bit phase shifters. A 4-bit phase shifter, constructed using the cascaded connections of four 1-bit phase shifters, producing differential phase shifts of 22.5°, 45°, 90°, and 180°, can give the entire 360° phase shift in steps of 22.5°. The disadvantage of such a phase shifter is that due to leakage through the reverse-biased diode and non-zero forward resistance, isolation between the ports decreases, whereas the insertion loss between them increases. In addition, the number of diodes required per bit is also large.

In contrast to a transmission-type switched line phase shifter, the operation of a reflection-type hybrid coupled phase shifter is based on the change in the position of the terminating shorts. Here, the 3 dB coupler behaves as a hybrid phase shifter, as described in Chapter 10. If the diodes are forward biased, then short circuits are implemented at A and A', and a phase change of $\phi_0 = 2\beta l_0$ is obtained. If the diodes are reverse biased, then the physical shorts are implemented at B and B'; thus, a total phase shift of $\phi = \phi_0 + 2\beta l$ is obtained. Therefore, by switching the diodes from one biasing condition to another, a differential phase shift of $\Delta\phi = 2\beta l$ can be obtained. By changing the position of the terminating shorts, variable phase shifts can also be obtained.

In a transmission-type loaded line phase shifter, the phase shift is obtained by loading the transmission line between the input and the output. If the diodes are forward biased, effective short circuits are implemented at respective diode positions and input impedance becomes inductive. This provides a phase shift of ϕ_0. Now, if the diodes are reverse biased, effective open circuits are implemented at the respective diode positions and input impedance becomes capacitive. This provides a phase shift of ϕ_1. Therefore, by switching the diodes from one biasing condition to another, a differential phase shift of $\Delta\phi = \phi_1 - \phi_0$ can be obtained. This phase shift can be maximized by implementing $\omega L = 1/(\omega C)$. The quarter-wave line provides a path difference of $\lambda/2$ between the incident and the reflected signals, so that they get cancelled at the input side.

Since a p–i–n diode behaves as a variable resistor, it can also be used for attenuation purpose. Figure 13.64 shows such a p–i–n attenuator circuit. In a series circuit, since diode resistance decreases with increasing biasing current, attenuation also decreases. The reverse is true for a shunt circuit.

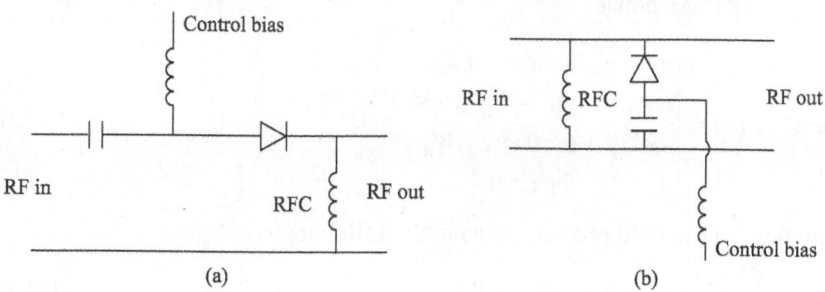

Fig. 13.64 PIN attenuator (a) Series configuration (b) Parallel configuration

A p–i–n diode can also be used as a limiter. If the RF power is sufficiently high, then it will make the p–i–n diode forward biased and the diode will reflect all the power. If the RF power is lower than a specified level, then the p–i–n diode will be off and all the RF power will be allowed to flow.

13.10 Varactor Diodes

It has been described that a p–i–n diode can behave as a voltage variable resistance or a varistor. In contrast, the term *varactor* corresponds to a *variable reactor*, a device whose reactance can be varied in a controlled manner using a bias voltage. Such diodes are widely used in parametric amplification, harmonic generation, mixing, detection, and voltage variable tuning.

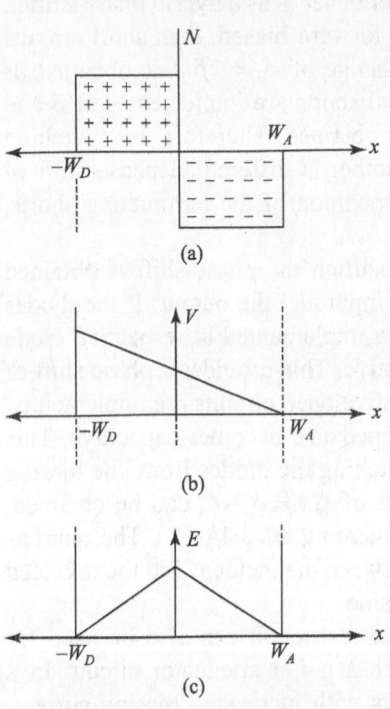

Fig. 13.65   Varactor diode
(a) Doping profile (b) Potential profile
(c) Field profile

In diffused junction diode, a depletion region is formed under the reverse bias, and the diode behaves as a capacitor, with the junction itself acting as a dielectric between two conducting materials. As the applied reverse bias is varied, the width of the depletion region also varies proportionally and hence the diode capacitance. Thus, a reverse-bias p–n junction can be used as a voltage-controlled capacitor. To understand a varactor diode, let us first consider an abrupt p–n junction diode. The charge density profile of such a diode is shown in Fig. 13.65(a), and the potential and electric field profiles in Figs 13.65(b) and (c), respectively. To maintain charge neutrality, the following condition must be satisfied:

$$N_D W_D = N_A W_A \tag{13.101}$$

Charges inside must satisfy the Poisson's equation:

$$\frac{d^2 V(x)}{dx^2} = -\frac{\rho(x)}{\varepsilon_s} \tag{13.102}$$

where $\quad \rho(x) = \begin{cases} 0 & \text{for } x < -W_D \\ eN_D & \text{for } -W_D \le x \le 0 \\ -eN_A & \text{for } 0 < x \le W_A \\ 0 & \text{for } x > 0 \end{cases} \tag{13.103}$

From the electric field profile, we have the following equation:

$$\frac{dV(x)}{dx} = 0 \quad \text{at } x = -W_D \text{ and } W_A \tag{13.104}$$

Further, from the potential profile, we have the following relation:

$$V(x) = 0 \quad \text{at } x = 0 \qquad (13.105)$$

Integrating Eq. (13.102) once, we get the following expression:

$$\frac{dV(x)}{dx} = \begin{cases} -\dfrac{eN_D}{\varepsilon_s} x + C_1 & \text{for } -W_D \leq x \leq 0 \\[2ex] \dfrac{eN_A}{\varepsilon_s} x + C_2 & \text{for } 0 \leq x \leq W_A \end{cases} \qquad (13.106)$$

Substituting the boundary condition of Eq. (13.104) we get,

$$C_1 = -\frac{eN_D W_D}{\varepsilon_s} \qquad (13.107)$$

and $\qquad C_2 = -\dfrac{eN_A W_A}{\varepsilon_s} \qquad (13.108)$

Substituting Eqs (13.107) and (13.108) in Eq. (13.106), we get

$$\frac{dV(x)}{dx} = \begin{cases} -\dfrac{eN_D}{\varepsilon_s} (x + W_D) & \text{for } -W_D \leq x \leq 0 \\[2ex] \dfrac{eN_A}{\varepsilon_s} (x - W_A) & \text{for } 0 \leq x \leq W_A \end{cases} \qquad (13.109)$$

Further integration of Eq. (13.109) gives the following expression:

$$V(x) = \begin{cases} -\dfrac{eN_D}{\varepsilon_s} \left(\dfrac{x^2}{2} + W_D x \right) + C_3 & \text{for } -W_D \leq x \leq 0 \\[2ex] \dfrac{eN_A}{\varepsilon_s} \left(\dfrac{x^2}{2} - W_A x \right) + C_4 & \text{for } 0 \leq x \leq W_A \end{cases} \qquad (13.110)$$

Substituting boundary condition (13.105) in Eq. (13.110), we get the following relation: $C_3 = C_4 = 0$ \hfill (13.111)

Thus, Eq. (13.110) reduces to the following form:

$$V(x) = \begin{cases} -\dfrac{eN_D}{\varepsilon_s} \left(\dfrac{x^2}{2} + W_D x \right) & \text{for } -W_D \leq x \leq 0 \\[2ex] \dfrac{eN_A}{\varepsilon_s} \left(\dfrac{x^2}{2} - W_A x \right) & \text{for } 0 \leq x \leq W_A \end{cases} \qquad (13.112)$$

Thus, the total voltage across the diode is as follows:

$$V_D = V(-W_D) - V(W_A) = -\frac{eN_D}{\varepsilon_s} \left(\frac{W_D^2}{2} - W_D^2 \right) - \frac{eN_A}{\varepsilon_s} \left(\frac{W_A^2}{2} - W_A^2 \right)$$

or $\qquad V_D = \dfrac{e}{2\varepsilon_s} \left(W_D^2 N_D + W_A^2 N_A \right) \qquad (13.113)$

Now substituting Eq. (13.101) in Eq. (13.113), we get the following relation:

$$V_D = \frac{e}{2\varepsilon_s}\left(W_D^2 N_D + W_A^2 N_A\right) = \frac{e}{2\varepsilon_s}\left(\frac{W_D^2 N_D^2}{N_D} + W_A^2 N_A\right)$$

or $\quad V_D = \dfrac{e}{2\varepsilon_s}\left[\dfrac{W_A^2 N_A^2}{N_D} + W_A^2 N_A\right] = \dfrac{eW_A^2 N_A}{2\varepsilon_s}\left(1 + \dfrac{N_A}{N_D}\right)$

or $\quad W_A = \sqrt{\dfrac{2\varepsilon_s V_D}{eN_A\left(1 + \dfrac{N_A}{N_D}\right)}}$ $\qquad\qquad$ (13.114)

Similarly, we can write the following equation:

$$W_D = \sqrt{\frac{2\varepsilon_s V_D}{eN_D\left(1 + \dfrac{N_D}{N_A}\right)}} \qquad\qquad (13.115)$$

The capacitance per unit area can be expressed as follows:

$$C_d = \frac{\partial Q}{\partial V_A} = \frac{\partial}{\partial V_A}\left(-eN_A W_A\right) = -eN_A\frac{\partial}{\partial V_A}\left(\sqrt{\frac{2\varepsilon_s\left(V_D - V_A\right)}{eN_A\left(1 + \dfrac{N_A}{N_D}\right)}}\right)$$

or $\quad C_d = -eN_A\sqrt{\dfrac{2\varepsilon_s}{eN_A\left(1 + \dfrac{N_A}{N_D}\right)}}\dfrac{(-1)}{2\sqrt{V_D - V_A}}$

or $\quad C_d = \sqrt{\dfrac{\varepsilon_s eN_A}{2\left(1 + \dfrac{N_A}{N_D}\right)\left(V_D - V_A\right)}} = \sqrt{\dfrac{\varepsilon_s eN_A}{2\left(1 + \dfrac{N_A}{N_D}\right)V_D\left(1 - \dfrac{V_A}{V_D}\right)}}$

or $\quad C_d = \dfrac{C_0}{\sqrt{1 - \dfrac{V_A}{V_D}}}$ $\qquad\qquad$ (13.116)

where $\quad C_0 = \sqrt{\dfrac{\varepsilon_s eN_A N_D}{2(N_A + N_D)V_D}}$ $\qquad\qquad$ (13.117)

and V_A is the applied voltage. Equation (13.116) reveals that capacitance, and hence reactance, is a non-linear function of the applied voltage.

At high field, an abrupt junction does not work properly, and a graded or hyper-abrupt junction is required. For a graded indexed junction, Poisson's equation modifies to the following form:

$$\frac{d^2 V(x)}{dx^2} = -\frac{eN}{\varepsilon_s} \qquad\qquad (13.118)$$

where N is the generalized doping distribution and is expressed as follows:

$$N = Bx^m \qquad (13.119)$$

and is shown in Fig. 13.66. For $m = 0$, we get a uniform doping profile and for $m = 1$, a linearly graded doping profile. The relation $m < 0$ corresponds to a hyper-abrupt junction. Integrating Poisson's equation with boundary condition, the expression for capacitance per unit area for generalized doping profile can be written as follows:

$$C_d = \left\{ \frac{qB\varepsilon_s^{m+1}}{(m+2)(V_A + V_D)} \right\}^{\frac{1}{m+2}} \qquad (13.120)$$

For a hyper-abrupt junction, when $m = -3/2$, the resonant frequency is linearly dependent upon the bias voltage. This device behaviour is useful in frequency modulation and distortion elimination.

A simplified equivalent circuit of a varactor diode is shown in Fig. 13.67. Here C_j is the junction capacitance, R_s is the series resistance, and R_p is the parallel equivalent resistance. Both C_j and R_s decrease and R_p increases with the reverse-bias voltage. The efficiency of a varactor diode is expressed by the following quality factor:

$$Q \approx \frac{\omega C_j R_p}{1 + \omega^2 C_j^2 R_p R_s} \qquad (13.121)$$

Since C_j and R_p are functions of the applied voltage, Q is also a function of this voltage.

The equivalent circuit for an unpackaged varactor diode is shown in Fig. 13.67. For a packaged varactor diode, the circuit modifies as shown in Fig. 13.68. The introduction of the capacitance C_f and C_p in a packaged diode limits the operation of the diode.

By construction, varactor diodes are of two types: p–n junction and Schottky barrier. Varactor diodes can be fabricated using Si or GaAs. Si varactor diodes generally operate at a low frequency but are easy to fabricate. On the other hand, GaAs varactor diodes operate at a high frequency and possess high Q. The cut-off frequency of a varactor diode can be expressed as follows:

$$f_c = (2\pi C_j R_s)^{-1} \qquad (13.122)$$

In practice, R_s increases with frequency, and the use of a varactor diode gets limited to below $0.2 f_c$.

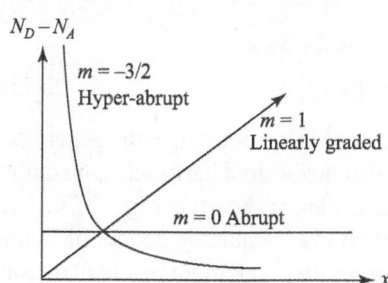

Fig. 13.66 Doping profile of different varactor diodes

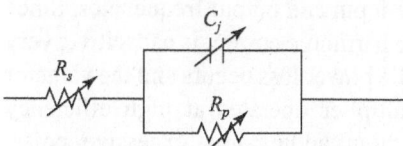

Fig. 13.67 Equivalent circuit of unpackaged varactor diode

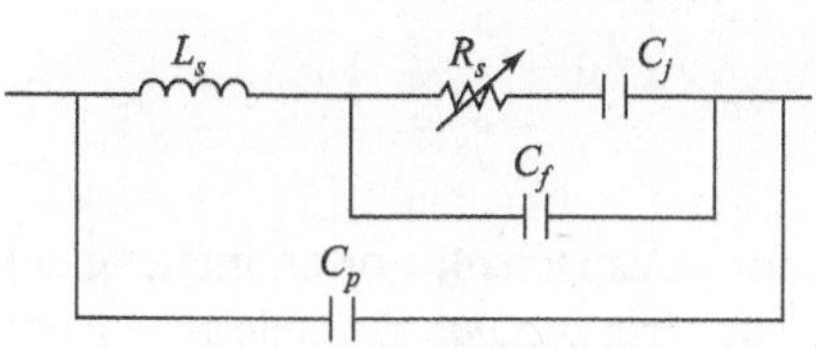

Fig. 13.68 Equivalent circuit of packaged varactor diode

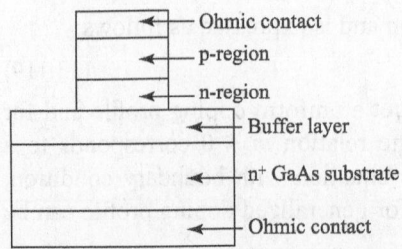

Fig. 13.69 Schematic diagram of varactor diode

The general structure of a p–n junction varactor diode is shown in Fig. 13.69. A buffer layer is first deposited on an n^+ GaAs substrate and then the active n-layer is deposited on this buffer region. A p-region is deposited on this active layer. Finally, Ohmic contacts are made on both sides for electrical contact.

Varactor diodes find wide applications in switching and modulation, harmonic generation, frequency conversion, low-noise amplification, parametric amplification, pulse generation and shaping, tuning stage of radio receiver, etc. A varactor diode can also be used as a frequency multiplier. When an ac pumping voltage $V_{ac} = V_0 \sin(\omega_0 t)$ is applied to a reverse-bias diode, the diode capacitance varies as follows:

$$C_j(t) = C_{j0} \left[1 + \frac{V_R + V_0 \sin(\omega_0 t)}{V_B} \right]^{-n} \qquad (13.123)$$

Expanding Eq. (13.123) in a harmonic series, we get the following relation:

$$C_j(t) = C_0 - C_1 \sin(\omega_0 t) + C_2 \sin(2\omega_0 t) - C_3 \sin(3\omega_0 t) + \dots \qquad (13.124)$$

The corresponding time-varying current will be as follows:

$$i(t) = I_1 \cos(\omega_0 t) + I_2 \cos(2\omega_0 t) + I_3 \cos(3\omega_0 t) + \dots \qquad (13.125)$$

Thus, a varactor diode can be used as a frequency multiplier or a harmonic generator. However, this requires a tuner circuit to be tuned at the desired harmonics. A simple varactor multiplier circuit, also called a varactor tripler, is shown in Fig. 13.70. The input signal is fed from a stable crystal-controlled low-frequency generator to the reverse-biased varactor diode through a buffer amplifier. The input resonant circuit prevents the unwanted frequencies from reaching the diode from the input side. The output resonator is tuned to the third harmonic to produce a frequency of $3f$. The intermediate resonator is called an idler. It eliminates the heterodyning between

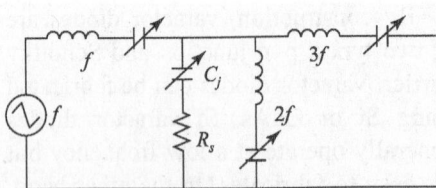

Fig. 13.70 Varactor diode multiplier circuit

the input and output frequencies. Since the harmonic current is capacitive, very little power loss occurs and the varactor multiplier operates at high efficiency without adding any excessive noise. The maximum efficiency of such a frequency multiplier is $1/n^2$, where n is the multiplication factor.

13.11 Parametric Amplifiers

A parametric device uses a non-linear reactance (a capacitance or an inductance) or a time-varying reactance for its operation. The term *parametric* is derived from *parametric excitation* because such reactive parameters can be used to produce capacitive or inductive excitation. Parametric excitations can be classified as parametric amplification and oscillation. Parametric amplifiers utilize ac rather than dc biasing, and hence, in this respect, they are analogous to quantum amplifiers. At present, varactor diodes are most widely used in parametric amplifiers.

The procedure of parametric amplification can be understood with an LC tank circuit, where C is a conventional parallel plate capacitor, except with a variable spacing. For a given plate spacing, the tank will oscillate at its natural resonance frequency. Now, if we want to pull apart the plates of the capacitor up to some distance, just before the capacitor voltage reaches its maximum value or the attractive electrostatic force between the plates is almost maximum, we have to work against the field. With no dissipative mechanism in the tank circuit, this energy is injected into the tank circuit and thus amplifies the amplitude of oscillation. Alternatively, we can say that since $V = Q/C$ or $V \propto 1/C$, a decrease in capacitance due to an increase in plate separation will lead to an increase in V, as Q is constant. We can restore the capacitor by quickly pushing the plates together again $90°$ later, when the tank voltage is zero. Since voltage is zero, no work is done this time. After another $90°$, if we again pull the plates apart, energy is re-injected into the tank circuit, leading to further amplification. Thus, by injecting energy twice in every cycle in this manner, we can continue the amplification of the signal.

In a practical amplifier, energy comes from a separate oscillator, called a pump oscillator, shown in Fig. 13.71. The RF input is fed into the diode through a band-pass filter F_1, whereas the pump input, nominally at twice the input frequency, is fed through another band-pass filter F_2. The pump signal modulates the diode capacitance, causing it to produce negative reactance at the input frequency.

Due to the non-linear effect of the diode, mixing of the pump and signal frequencies takes place and an idler frequency $\left(f_i = f_p - f_s \right)$ is generated. If the pump frequency is more than $2f_s$, then the idler frequency is also greater than the source frequency. Such parametric amplifiers are called parametric up-converters. If the pump frequency is in between f_s and $2f_s$, the idler frequency is less than the source frequency and the parametric amplifier is called a parametric down-converter. Furthermore, if the pump frequency is equal to $2f_s$, then idler frequency is equal to f_s and the parametric amplifier is said to be degenerate.

13.11.1 Manley–Rowe Power Relation

To establish the Manley–Rowe power relation, let us consider the circuit shown in Fig. 13.72. The circuit consists of one signal generator and one pump generator at their respective frequencies f_s and f_p, and resistive loads in series with ideal band-pass filters, connected in shunt with a lossless non-linear capacitance $C(t)$. Each band-pass filter can pass only one harmonic component $mf_p \pm nf_s$. The overall circuit isolates all the harmonics and dissipates their power in separate resistive loads. The Manley–Rowe relations establish two constraints that govern the conversion of input power at frequencies f_s and f_p to frequencies $mf_p \pm nf_s$.

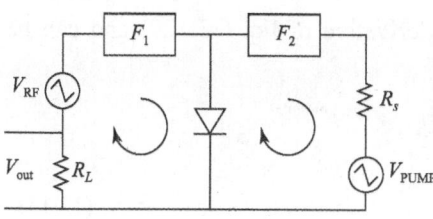

Fig. 13.71 Simplified parametric amplifier

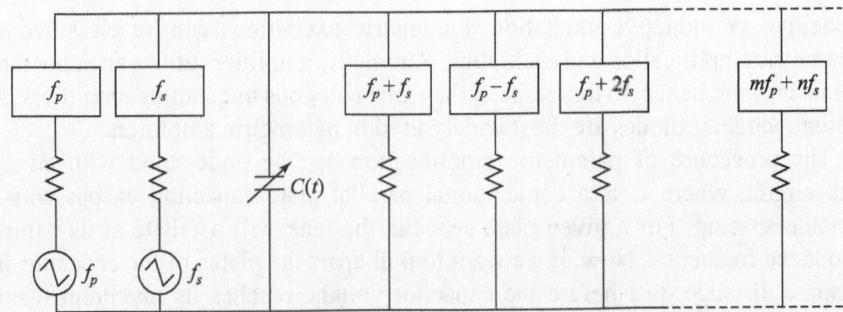

Fig. 13.72 Circuit illustrating Manley–Rowe relations

The total voltage across the non-linear capacitor $C(t)$ is given as follows:

$$v = v_s + v_p = V_s \cos(\omega_s t) + V_p \cos(\omega_p t) \tag{13.126}$$

If the charge on the capacitor is a single-valued function of the applied voltage, then by expanding it in Taylors series around $v = 0$, we can write the following relation:

$$Q(v) = Q(0) + \frac{\partial Q}{\partial v}\bigg|_{v=0} v + \frac{1}{2}\frac{\partial^2 Q}{\partial v^2}\bigg|_{v=0} v^2 + \dots \tag{13.127}$$

Since all powers of v occur, the following equation ensures that the charge Q will have all the harmonics of f_s and f_p:

$$v = v_p + v_s = \frac{V_p}{2}\left(e^{j\omega_p t} + e^{-j\omega_p t}\right) + \frac{V_s}{2}\left(e^{j\omega_s t} + e^{-j\omega_s t}\right) \tag{13.128}$$

If the current of these harmonics passes through the capacitor, the voltage across the capacitor will also contain all possible harmonics. Consequently, the general expression of the charge deposited on the capacitor can be expressed as follows:

$$Q = \sum_{m=-\infty}^{\infty} \sum_{n=-\infty}^{\infty} Q_{m,n} e^{j(m\omega_p t + n\omega_s t)} \tag{13.129}$$

Since the charge Q is real, the following equation must be satisfied

$$Q_{m,n} = Q^*_{-m,-n} \tag{13.130}$$

such that (m, n) and $(-m, -n)$ terms can combine to form a real function of time with frequency $mf_p \pm nf_s$. The total voltage v can be expressed as follows:

$$v = \sum_{m=-\infty}^{\infty} \sum_{n=-\infty}^{\infty} V_{m,n} e^{j(m\omega_p t + n\omega_s t)} \tag{13.131}$$

Further, to make the voltage v real, the following relation must hold:

$$V_{m,n} = V^*_{-m,-n} \tag{13.132}$$

The current flowing through $C(t)$ is the derivative of Eq. (13.129) and can be written as follows:

$$i = \frac{dQ}{dt} = \sum_{m=-\infty}^{\infty} \sum_{n=-\infty}^{\infty} j\left(m\omega_p + n\omega_s\right) Q_{m,n} e^{j(m\omega_p t + n\omega_s t)}$$

or

$$i = \sum_{m=-\infty}^{\infty} \sum_{n=-\infty}^{\infty} I_{m,n} e^{j(m\omega_p t + n\omega_s t)} \tag{13.133}$$

where $I_{m,n} = j(m\omega_p + n\omega_s)Q_{m,n}$ (13.134)

and $I_{m,n} = I^*_{-m,-n}$ (13.135)

Since the capacitance $C(t)$ is pure reactance, there will be no net power into or out of it. Further, if we assume that f_s and f_p are incommensurable, there will be no time-average power due to interacting harmonics. The average power at the frequency $mf_p + nf_s$ is given by the following equation:

$$P_{m,n} = V_{m,n}I^*_{m,n} + V^*_{m,n}I_{m,n} = V_{m,n}I^*_{m,n} + V_{-m,-n}I^*_{-m,-n} = P_{-m,-n}$$ (13.136)

since $\left\{V_{m,n}e^{j(m\omega_p + n\omega_s)} + V_{-m,-n}e^{-j(m\omega_p + n\omega_s)}\right\} \times$

$$\left\{I_{m,n}e^{j(m\omega_p + n\omega_s)} + I_{-m,-n}e^{-j(m\omega_p + n\omega_s)}\right\}$$ (13.137)

$$= V_{m,n}I_{-m,-n} + V_{-m,-n}I_{m,n} = V_{m,n}I^*_{m,n} + V^*_{m,n}I_{m,n}$$

$$= V_{m,n}I^*_{m,n} + V_{-m,-n}I^*_{-m,-n}$$

Therefore, the conservation of power can be written as follows:

$$\sum_{m=-\infty}^{\infty}\sum_{n=-\infty}^{\infty} P_{m,n} = 0$$ (13.138)

Multiplying each term of Eq. (13.138) by $(m\omega_p + n\omega_s)/(m\omega_p + n\omega_s)$ and then separating them, we can write the following equation:

$$\omega_p \sum_{m=-\infty}^{\infty}\sum_{n=-\infty}^{\infty} \frac{mP_{m,n}}{m\omega_p + n\omega_s} + \omega_s \sum_{m=-\infty}^{\infty}\sum_{n=-\infty}^{\infty} \frac{nP_{m,n}}{m\omega_p + n\omega_s} = 0$$ (13.139)

Now from Eq. (13.134):

$$\frac{I_{m,n}}{(m\omega_p + n\omega_s)} = jQ_{m,n}$$ (13.140)

Using Eqs (13.135), (13.136), (13.137), and (13.140), we can write the following relation:

$$\frac{P_{m,n}}{m\omega_p + n\omega_s} = \frac{V_{m,n}I^*_{m,n} + V_{-m,-n}I^*_{-m,-n}}{m\omega_p + n\omega_s}$$

$$= V_{m,n}\frac{I^*_{m,n}}{m\omega_p + n\omega_s} + V_{-m,-n}\frac{I_{m,n}}{m\omega_p + n\omega_s}$$

or $$\frac{P_{m,n}}{m\omega_p + n\omega_s} = V_{m,n}\left(-jQ^*_{m,n}\right) + V_{-m,-n}\left(jQ_{m,n}\right)$$

$$= -jV_{m,n}Q^*_{m,n} + jV_{-m,-n}Q^*_{-m,-n}$$ (13.141)

Equation (13.141) reveals that the factor $P_{m,n}/(m\omega_p + n\omega_s)$ is independent of ω_p and ω_s. For any arbitrary selection of ω_p and ω_s, we can always adjust the network, external to C, so that the resultant current keeps all the voltage amplitudes (V_{mn})

unchanged. Since Q_{mn} is only a function of V_{mn}, it will also remain unchanged. The preceding discussion implies that even with arbitrary changes in ω_p and ω_s, the RHS of Eq. (13.141), and hence the term $P_{m,n}/(m\omega_p + n\omega_s)$, will remain constant. Therefore, to satisfy Eq. (13.139), we must have

$$\sum_{m=-\infty}^{\infty} \sum_{n=-\infty}^{\infty} \frac{mP_{m,n}}{m\omega_p + n\omega_s} = 0 \tag{13.142}$$

and

$$\sum_{m=-\infty}^{\infty} \sum_{n=-\infty}^{\infty} \frac{nP_{m,n}}{m\omega_p + n\omega_s} = 0 \tag{13.143}$$

Equations (13.142) and (13.143) are known as Manley–Rowe relation. These equations can also be written in a different form. For example, the first summation of Eq. (13.142) can be written as follows:

$$\sum_{m=0}^{\infty} \sum_{n=-\infty}^{\infty} \frac{mP_{m,n}}{m\omega_p + n\omega_s} + \sum_{m=0}^{\infty} \sum_{n=-\infty}^{\infty} \frac{-mP_{-m,-n}}{-m\omega_p - n\omega_s} = 0 \tag{13.144}$$

Since from Eq. (13.136), $P_{m,n} = P_{-m,-n}$, the two parts of Eq. (13.144) are same, and we can write the following equation:

$$\sum_{m=0}^{\infty} \sum_{n=-\infty}^{\infty} \frac{mP_{m,n}}{m\omega_p + n\omega_s} = 0 \tag{13.145}$$

Similarly, it can be shown that

$$\sum_{m=-\infty}^{\infty} \sum_{n=0}^{\infty} \frac{nP_{m,n}}{m\omega_p + n\omega_s} = 0 \tag{13.146}$$

Equations (13.145) and (13.146) are the standard forms of the Manley–Rowe relation. It should be noted that the Manley–Rowe relation is a general power conservation relation and does not depend on any specific circuit, as shown in Fig. 13.72.

To illustrate the Manley–Rowe power relation with an example, let us assume that all harmonics, except $f_s + f_p$, are open circuited. Therefore, currents of only three frequencies, f_s, f_p, and $f_s + f_p$, will be present. This corresponds to three combinations of (m, n), namely, $(m = 0, n = \pm 1)$, $(m = \pm 1, n = 0)$, and $(m = n = \pm 1)$, and we can write the following equations:

$$\frac{P_{1,0}}{\omega_p} + \frac{P_{1,1}}{\omega_p + \omega_s} = 0 \tag{13.147}$$

and

$$\frac{P_{0,1}}{\omega_s} + \frac{P_{1,1}}{\omega_p + \omega_s} = 0 \tag{13.148}$$

Since power is supplied by ω_p and ω_s, $P_{1,0}$ and $P_{0,1}$ are positive, $P_{1,1}$ must be negative, and power will be delivered to $\omega = \omega_p + \omega_s$. Therefore we can write the following relation:

$$-\frac{P_{1,1}}{P_{0,1}} = \frac{\omega_p + \omega_s}{\omega_s} = 1 + \frac{\omega_p}{\omega_s} \tag{13.149}$$

A parametric amplifier of this type is called an up-converter. Since losses are always present in a practical amplifier, the gain will always be less than $\left|1+\dfrac{\omega_p}{\omega_s}\right|$.

13.11.2 Linearized Equations for Parametric Amplifiers

For linear capacitance, charge–voltage relationship is given by $Q = Cv$, and hence the current through the capacitor is written as follows:

$$i = \frac{dQ}{dt} = C\frac{dv}{dt} \tag{13.150}$$

Now, if C is a function of time, for example, a parallel-plate capacitor whose plate separation varies with time, Eq. (13.150) is modified as follows:

$$i = \frac{dQ}{dt} = \frac{d}{dt}\{C(t)v\} = v\frac{dC(t)}{dt} + C(t)\frac{dv}{dt} \tag{13.151}$$

Further, instead of time-varying linear capacitance if we have a non-linear capacitance, where the charge is a non-linear function of voltage, then the current can be expressed as follows:

$$i = \frac{dQ}{dt} = \frac{dQ}{dv}\frac{dv}{dt} \tag{13.152}$$

If the voltage is the sum of the pump voltage with frequency f_p and signal voltage with frequency f_s, and the signal voltage v_s is much smaller than the pump voltage v_p, then the total voltage across the non-linear capacitor $C(t)$ is given by the following equation:

$$v = v_s + v_p = V_s\cos(\omega_s t) + V_p\cos(\omega_p t) \tag{13.153}$$

where $V_s \ll V_p$. Expanding the charge on the capacitor by Taylor's series around the point $v_s = 0$, we obtain the following relation:

$$Q(v) = Q(v_s + v_p) = Q(v_p) + \left.\frac{\partial Q}{\partial v}\right|_{v_s=0} v_s + \frac{1}{2}\left.\frac{\partial^2 Q}{\partial v^2}\right|_{v_s=0} v_s^2 + \cdots \tag{13.154}$$

Now since $V_s \ll V_p$, we can neglect the higher-order terms and retain only the first two terms. Therefore,

$$i = \frac{dQ(v)}{dt} = \frac{dQ(v_p)}{dt} + \frac{d}{dt}\left(\left.\frac{\partial Q}{\partial v}\right|_{v_s=0} v_s\right) \tag{13.155}$$

Now, if we denote

$$C(t) = \left.\frac{dQ}{dv}\right|_{v_s=0} \tag{13.156}$$

then Eq. (13.155) is modified as follows: $i = \dfrac{dQ(v_p)}{dt} + \dfrac{d}{dt}\{C(t)v_s\}$ $\qquad$ (13.157)

Comparing Eqs (13.157) and (13.151), it can be concluded that non-linear capacitance behaves like time-varying linear capacitance, provided that the signal amplitude is small compared to the pump signal. The first term on the RHS of Eq. (13.157) gives a current at the pump frequency and is not related to the signal current. If the pump voltage is also small compared to the DC bias voltage, then we can write the following expression:

$$C(t) = C_0 \left\{ 1 + 2M \cos\left(\omega_p t\right) \right\} \tag{13.158}$$

13.11.3 Parametric Up-converter

In a parametric up-converter, a pump voltage of frequency f_p and a signal at frequency f_s is applied to a diode, and the output signal is taken at the output frequency $f_0 = f_p + f_s$. The voltage across the diode at these three frequencies can be expressed as follows:

$$v_s = \left(V_s e^{j\omega_s t} + V_s^* e^{-j\omega_s t} \right) / 2 \tag{13.159}$$

$$v_p = \left(V_p e^{j\omega_p t} + V_p^* e^{-j\omega_p t} \right) / 2 \tag{13.160}$$

$$v_0 = \left(V_0 e^{j\omega_0 t} + V_0^* e^{-j\omega_0 t} \right) / 2 \tag{13.161}$$

Equations (13.157) and (13.158) may be generalized to give the following relations:

$$i = \frac{dQ(v_p)}{dt} + \frac{1}{2}\frac{d}{dt}\Big[C_0 \left\{ 1 + 2M \cos\left(\omega_p t\right) \right\} \big(V_s e^{j\omega_s t} + V_s^* e^{-j\omega_s t} \\ + V_0 e^{j\omega_0 t} + V_0^* e^{-j\omega_0 t} \big) \Big]$$

or

$$i = \frac{C_0}{2}\Big[\left\{ 1 + M\left(e^{j\omega_p t} + e^{-j\omega_p t} \right) \right\} \big(j\omega_s V_s e^{j\omega_s t} - j\omega_s V_s^* e^{-j\omega_s t} \\ + j\omega_0 V_0 e^{j\omega_0 t} - j\omega_0 V_0^* e^{-j\omega_0 t} \big) + jM\omega_p \left(e^{j\omega_p t} - e^{-j\omega_p t} \right) \big(V_s e^{j\omega_s t} \\ + V_s^* e^{-j\omega_s t} + V_0 e^{j\omega_0 t} + V_0^* e^{-j\omega_0 t} \big) \Big] + \frac{dQ(v_p)}{dt}$$

or

$$i = \frac{C_0}{2}\Big[\left(j\omega_s V_s e^{j\omega_s t} - j\omega_s V_s^* e^{-j\omega_s t} + j\omega_0 V_0 e^{j\omega_0 t} - j\omega_0 V_0^* e^{-j\omega_0 t} \right) \\ + Me^{j\omega_p t} \left(j\omega_s V_s e^{j\omega_s t} - j\omega_s V_s^* e^{-j\omega_s t} + j\omega_0 V_0 e^{j\omega_0 t} - j\omega_0 V_0^* e^{-j\omega_0 t} \right) \\ + Me^{-j\omega_p t} \left(j\omega_s V_s e^{j\omega_s t} - j\omega_s V_s^* e^{-j\omega_s t} + j\omega_0 V_0 e^{j\omega_0 t} - j\omega_0 V_0^* e^{-j\omega_0 t} \right) \\ + jM\omega_p e^{j\omega_p t} \left(V_s e^{j\omega_s t} + V_s^* e^{-j\omega_s t} + V_0 e^{j\omega_0 t} + V_0^* e^{-j\omega_0 t} \right) \\ - jM\omega_p e^{-j\omega_p t} \left(V_s e^{j\omega_s t} + V_s^* e^{-j\omega_s t} + V_0 e^{j\omega_0 t} + V_0^* e^{-j\omega_0 t} \right) \Big] + \frac{dQ(v_p)}{dt}$$

or $\quad i = \dfrac{C_0}{2}\Big[\big(j\omega_s V_s e^{j\omega_s t} - j\omega_s V_s^* e^{-j\omega_s t} + j\omega_0 V_0 e^{j\omega_0 t} - j\omega_0 V_0^* e^{-j\omega_0 t}\big)$

$\qquad + M\Big\{ j\big(\omega_s + \omega_p\big)V_s e^{j(\omega_s+\omega_p)t} - j\big(\omega_s - \omega_p\big)V_s^* e^{-j(\omega_s-\omega_p)t}$

$\qquad + j\big(\omega_0 + \omega_p\big)V_0 e^{j(\omega_0+\omega_p)t} - j\big(\omega_0 - \omega_p\big)V_0^* e^{-j(\omega_0-\omega_p)t}$

$\qquad + j\big(\omega_s - \omega_p\big)V_s e^{j(\omega_s-\omega_p)t} - j\big(\omega_s + \omega_p\big)V_s^* e^{-j(\omega_s+\omega_p)t}$

$\qquad + j\big(\omega_0 - \omega_p\big)V_0 e^{j(\omega_0-\omega_p)t} - j\big(\omega_0 + \omega_p\big)V_0^* e^{-j(\omega_0+\omega_p)t}\Big\}\Big] + \dfrac{dQ(v_p)}{dt}$

or $\quad i = \dfrac{C_0}{2}\Big[\big(j\omega_s V_s e^{j\omega_s t} - j\omega_s V_s^* e^{-j\omega_s t} + j\omega_0 V_0 e^{j\omega_0 t} - j\omega_0 V_0^* e^{-j\omega_0 t}\big)$

$\qquad + M\Big\{ j\omega_0 V_s e^{j\omega_0 t} - j\big(\omega_s - \omega_p\big)V_s^* e^{-j(\omega_s-\omega_p)t}$

$\qquad + j\big(\omega_0 + \omega_p\big)V_0 e^{j(\omega_0+\omega_p)t} - j\omega_s V_0^* e^{-j\omega_s t}$

$\qquad + j\big(\omega_s - \omega_p\big)V_s e^{j(\omega_s-\omega_p)t} - j\omega_0 V_s^* e^{-j\omega_0 t} + j\omega_s V_0 e^{j\omega_s t}$

$\qquad - j\big(\omega_0 + \omega_p\big)V_0^* e^{-j(\omega_0+\omega_p)t}\Big\}\Big] + \dfrac{dQ(v_p)}{dt}$

(13.162)

Retaining only the ω_s and $\pm\omega_0$ terms of Eq. (13.162), we get the following equation:

$i_s + i_0 = \dfrac{C_0}{2}\Big[\big(j\omega_s V_s e^{j\omega_s t} - j\omega_s V_s^* e^{-j\omega_s t} + j\omega_0 V_0 e^{j\omega_0 t} - j\omega_0 V_0^* e^{-j\omega_0 t}\big)$

$\qquad + M\big(j\omega_0 V_s e^{j\omega_0 t} - j\omega_s V_0^* e^{-j\omega_s t} - j\omega_0 V_s^* e^{-j\omega_0 t} + j\omega_s V_0 e^{j\omega_s t}\big)\Big]$ (13.163)

where i_s and i_0 are signal current and output current, respectively, and can be written as follows:

$i_s = \big(I_s e^{j\omega_s t} + I_s^* e^{-j\omega_s t}\big)/2$ (13.164)

$i_0 = \big(I_0 e^{j\omega_0 t} + I_0^* e^{-j\omega_0 t}\big)/2$ (13.165)

Comparing Eqs (13.164) and (13.165) with Eq. (13.163) for similar exponential terms, we can write the following relations:

$I_s = j\omega_s C_0 V_s + j\omega_s C_0 M V_0$ (13.166)

$I_0 = j\omega_0 C_0 V_0 + j\omega_0 C_0 M V_s$ (13.167)

Equations (13.166) and (13.167) can be written in the matrix form as follows:

$$\begin{bmatrix} I_s \\ I_0 \end{bmatrix} = \begin{bmatrix} j\omega_s C_0 & j\omega_s C_0 M \\ j\omega_0 C_0 M & j\omega_0 C_0 \end{bmatrix}\begin{bmatrix} V_s \\ V_0 \end{bmatrix}$$ (13.168)

The equivalent circuit of a parametric up-converter is shown in Fig. 13.73. The series model is chosen so that the three circuit loops have resonant frequencies f_s, f_p, and f_0, and current with only these respective frequencies can exist in the loop. Therefore, only I_s is present in the input circuit loop and only I_0 in the output circuit loop. The three circuit loops are coupled together through the time-varying part of capacitance. Therefore, for the two frequencies f_s and f_0, the equivalent circuit can be reduced, as shown in Fig. 13.74. The term $C(t)$ in the figure maintains the relationship given by Eq. (13.168). Once the equivalent circuit has been established in terms of resistance, capacitance, and inductance, the analysis of a parametric amplifier becomes a conventional network analysis problem. In the network, the resonant circuits have been assumed to be lossless and circuit losses have been considered to be included in R_g and R_L. At the end of the analysis, R_g and R_L can be split into two parts so that the circuit losses from the generator and load impedance can be separated. In practice, these circuit losses are small compared with the loss arising from the diode resistance R_s and the external loading represented by R_g and R_L.

Solving Eq. (13.168) for V_s and V_0, we get the following expression:

$$\begin{bmatrix} V_s \\ V_0 \end{bmatrix} = \frac{1}{1-M^2} \begin{bmatrix} 1/j\omega_s C_0 & -M/j\omega_0 C_0 \\ -M/j\omega_s C_0 & 1/j\omega_0 C_0 \end{bmatrix} \begin{bmatrix} I_s \\ I_0 \end{bmatrix} \tag{13.169}$$

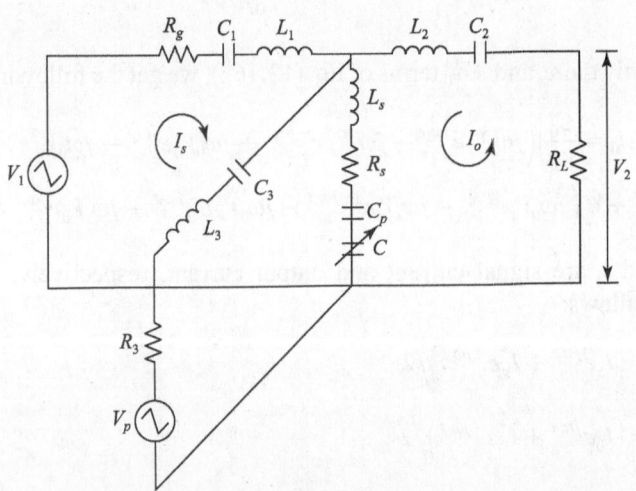

Fig. 13.73 Equivalent circuit model of up-converter

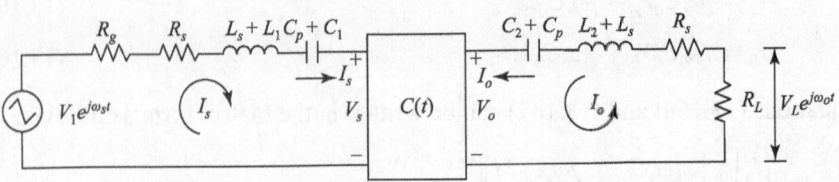

Fig. 13.74 Reduced equivalent circuit model of up-converter

For the input circuit, using Eq. (13.169), we may write the following equation:

$$V_1 = I_s \left[R_g + R_s + j\omega_s (L_s + L_1) + \frac{1}{j\omega_s (C_p + C_1)} \right] + V_s$$

or
$$V_1 = I_s \left[R_g + R_s + j\omega_s (L_s + L_1) + \frac{1}{j\omega_s (C_p + C_1)} \right]$$

$$+ \frac{I_s}{j\omega_s (1 - M^2) C_0} - \frac{MI_0}{j\omega_0 (1 - M^2) C_0}$$

or

$$V_1 = I_s \left[R_g + R_s + \frac{-\omega_s^2 (L_s + L_1)(C_p + C_1)(1 - M^2)C_0 + (1 - M^2)C_0 + (C_p + C_1)}{j\omega_s (C_p + C_1)(1 - M^2)C_0} \right]$$

$$+ \frac{jMI_0}{\omega_0 (1 - M^2) C_0} \tag{13.170}$$

Similarly, for the output circuit, using Eq. (13.169), we can write the following equation:

$$0 = I_0 \left[R_L + R_s + j\omega_0 (L_s + L_2) + \frac{1}{j\omega_0 (C_p + C_2)} \right] + V_0$$

or
$$0 = I_0 \left[R_L + R_s + j\omega_0 (L_s + L_2) + \frac{1}{j\omega_0 (C_p + C_2)} \right]$$

$$+ \frac{I_0}{j\omega_0 (1 - M^2) C_0} - \frac{MI_s}{j\omega_s (1 - M^2) C_0}$$

or

$$0 = I_0 \left[R_L + R_s + \frac{-\omega_0^2 (L_s + L_2)(C_p + C_2)(1 - M^2)C_0 + (1 - M^2)C_0 + (C_p + C_2)}{j\omega_0 (C_p + C_2)(1 - M^2)C_0} \right]$$

$$+ \frac{jMI_s}{\omega_s (1 - M^2) C_0} \tag{13.171}$$

Now, let us assume that the circuit is tuned so that the following conditions are valid:

$$\omega_s^2 (L_s + L_1)(C_p + C_1)(1 - M^2)C_0 = (1 - M^2)C_0 + (C_p + C_1) \tag{13.172}$$

$$\omega_0^2 (L_s + L_2)(C_p + C_2)(1 - M^2)C_0 = (1 - M^2)C_0 + (C_p + C_2) \tag{13.173}$$

Substituting Eq. (13.172) in Eq. (13.170) and Eq. (13.173) in Eq. (13.171), we get the following relation:

$$V_1 = I_s \left(R_g + R_s \right) + \frac{jMI_0}{\omega_0 \left(1 - M^2 \right) C_0} \tag{13.174}$$

$$0 = I_0 \left(R_L + R_s \right) + \frac{jMI_s}{\omega_s \left(1 - M^2 \right) C_0} \tag{13.175}$$

Solving Eqs (13.174) and (13.175) for I_0, we get the following expression:

$$I_0 = -j \frac{\omega_0 C_0 M \left(1 - M^2 \right) V_1}{M^2 + \omega_0 \omega_s \left(R_g + R_s \right) \left(R_L + R_s \right) \left(1 - M^2 \right)^2 C_0^2} \tag{13.176}$$

It may be noted that circuit losses can be included at this point, in the expression of I_0, by replacing R_g and R_L by $\left(R_g + R_{l1} \right)$ and $\left(R_L + R_{l2} \right)$, respectively. However, for simplicity we will consider that $R_{l1} = R_{l2} = 0$.

The maximum power input from the generator can be expressed as follows:

$$P_{in,max} = \frac{1}{2} \frac{V_1^2}{4R_g} \tag{13.177}$$

Similarly, the maximum power developed in the load can be expressed as follows:

$$P_{load,max} = \frac{1}{2} \left| I_0 \right|^2 R_L \tag{13.178}$$

Therefore, the gain is given by the following relation:

$$G_0 = \frac{P_{load,max}}{P_{in,max}} = \frac{4 \left| I_0 \right|^2 R_L R_g}{V_1^2} \tag{13.179}$$

Substituting Eq. (13.176) in Eq. (13.179), we get the following expression:

$$G_0 = \frac{4R_L R_g}{V_1^2} \left[\frac{\omega_0 C_0 M \left(1 - M^2 \right) V_1}{M^2 + \omega_0 \omega_s \left(R_g + R_s \right) \left(R_L + R_s \right) \left(1 - M^2 \right)^2 C_0^2} \right]^2$$

or

$$G_0 = 4R_L R_g \left[\frac{M}{\dfrac{M^2}{\omega_0 C_0 \left(1 - M^2 \right)} + \omega_s C_0 \left(R_g + R_s \right) \left(R_L + R_s \right) \left(1 - M^2 \right)} \right]^2$$

or

$$G_0 = \frac{4R_L R_g M^2}{\omega_s^2 C_0^2 \left(1 - M^2 \right)^2 \left[\left(R_g + R_s \right) \left(R_L + R_s \right) + \dfrac{M^2}{\omega_0 \omega_s C_0^2 \left(1 - M^2 \right)^2} \right]^2} \tag{13.180}$$

To adjust the gain, R_g and R_L need to be adjusted. Since they appear symmetrically in the expression of G_0, the optimized values of R_g and R_L are equal. Therefore, we can write the following equation:

$$G_0 = \frac{4R_L^2 M^2}{\omega_s^2 C_0^2 \left(1-M^2\right)^2 \left[\left(R_L + R_s\right)^2 + \dfrac{M^2}{\omega_0 \omega_s C_0^2 \left(1-M^2\right)^2}\right]^2} \tag{13.181}$$

The maximum gain will be obtained when the following equation is satisfied:

$$\frac{dG_0}{dR_L} = \frac{d}{dR_L}\left[\frac{4R_L^2 M^2}{\omega_s^2 C_0^2 \left(1-M^2\right)^2 \left\{\left(R_L + R_s\right)^2 + \dfrac{M^2}{\omega_0 \omega_s C_0^2 \left(1-M^2\right)^2}\right\}^2}\right] = 0 \tag{13.182}$$

Solving Eq. (13.182) for R_L, we get the following expression:

$$R_L = R_s \sqrt{1 + \frac{M^2}{\omega_0 \omega_s R_s^2 \left(1-M^2\right)^2 C_0^2}} \tag{13.183}$$

The effective Q of the diode can be defined as follows:

$$Q = \frac{1}{R_s \omega_s \left(1-M^2\right) C_0} \tag{13.184}$$

Substituting Eq. (13.184) in Eq. (13.183), we get the following relation:

$$R_L = R_s \sqrt{1 + \frac{\omega_s}{\omega_0}\left(MQ\right)^2} \tag{13.185}$$

Substituting Eq. (13.185) in Eq. (13.181), we get the following expression:

$$G_0 = \frac{4R_S^2 \left[1 + \dfrac{\omega_s}{\omega_0}\left(MQ\right)^2\right] M^2}{\omega_s^2 C_0^2 \left(1-M^2\right)^2 \left[R_s + R_s \sqrt{1 + \dfrac{\omega_s}{\omega_0}\left(MQ\right)^2}\right]^2 + \dfrac{M^2}{\omega_0 \omega_s C_0^2 \left(1-M^2\right)^2}\right]^2} \tag{13.186}$$

Substituting $\delta = \dfrac{\omega_s}{\omega_0}\left(MQ\right)^2$ \hfill (13.187)

in Eq. (13.186), we get the following relation:

$$G_0 = \frac{4R_S^2 \left(1+\delta\right) M^2}{\omega_s^2 C_0^2 \left(1-M^2\right)^2 \left[\left(R_s + R_s \sqrt{1+\delta}\right)^2 + \dfrac{M^2}{\omega_0 \omega_s C_0^2 \left(1-M^2\right)^2}\right]^2}$$

or
$$G_0 = \frac{4R_s^2(1+\delta)M^2}{\omega_s^2 C_0^2 (1-M^2)^2 R_s^4 \left[(1+\sqrt{1+\delta})^2 + \dfrac{\omega_s M^2}{\omega_0 R_s^2 \omega_s^2 C_0^2 (1-M^2)^2}\right]^2}$$

or
$$G_0 = \frac{4(1+\delta)M^2}{R_s^2 \omega_s^2 C_0^2 (1-M^2)^2 \left[(1+\sqrt{1+\delta})^2 + \dfrac{\omega_s}{\omega_0}(MQ)^2\right]^2} = \frac{4(1+\delta)\dfrac{\omega_0}{\omega_s}\dfrac{\omega_s}{\omega_0}(MQ)^2}{\left[(1+\sqrt{1+\delta})^2 + \delta\right]^2}$$

or
$$G_0 = \frac{\omega_0}{\omega_s}\frac{4(1+\delta)\delta}{\left[(1+\sqrt{1+\delta})^2+\delta\right]^2} = \frac{\omega_0}{\omega_s}\frac{4(1+\delta)\delta}{\left[1+2\sqrt{1+\delta}+(1+\delta)+\delta\right]^2}$$

or
$$G_0 = \frac{\omega_0}{\omega_s}\frac{\delta}{\left[\dfrac{2+2\sqrt{1+\delta}+2\delta}{2\sqrt{1+\delta}}\right]^2}$$

or
$$G_0 = \frac{\omega_0}{\omega_s}\frac{\delta}{(1+\sqrt{1+\delta})^2} \tag{13.188}$$

According to the Manley–Rowe relation, the maximum gain can be ω_0/ω_s. Therefore, the factor $\delta/(1+\sqrt{1+\delta})^2$ may be considered as a gain degradation factor. As the diode Q approaches infinity, that is, R_s approaches zero, δ approaches infinity; hence, the gain degradation factor becomes unity, as predicted by the Manley–Rowe relation.

To achieve high gain with an up-converter, the ratio ω_0/ω_s should be large. However, this is not very practical at a microwave frequency. Hence, the use of up-converters is restricted to below 1 GHz. To obtain a higher gain, negative-resistance parametric amplifiers are generally used.

13.11.4 Negative Resistance Parametric Amplifiers

A negative-resistance parametric amplifier, shown in Fig. 13.75, permits current of signal frequency f_s, pump frequency f_p, and idler frequency $f_i = f_p - f_s$. If we replace the voltage v_0 in Eq. (13.161) by

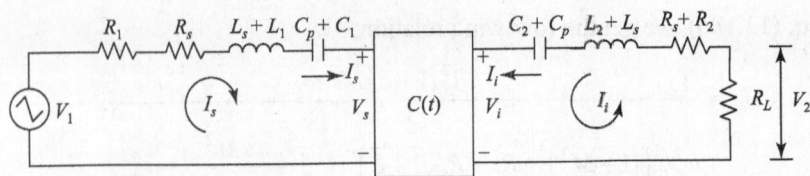

Fig. 13.75 Equivalent circuit of negative-resistance parametric amplifier

$$v_i = \frac{1}{2}\left\{V_i e^{j\left(\omega_p - \omega_s\right)t} - V_i^* e^{-j\left(\omega_p - \omega_s\right)t}\right\} \tag{13.189}$$

and introduce the idler current

$$i_i = \frac{1}{2}\left\{I_i e^{j\left(\omega_p - \omega_s\right)t} - I_i^* e^{-j\left(\omega_p - \omega_s\right)t}\right\} \tag{13.190}$$

then proceeding as before, we can show that

$$\begin{bmatrix} I_i \\ I_i^* \end{bmatrix} = \begin{bmatrix} j\omega_i C_0 & j\omega_i C_0 M \\ -j\omega_s C_0 M & -j\omega_s C_0 \end{bmatrix}\begin{bmatrix} V_i \\ V_i^* \end{bmatrix} \tag{13.191}$$

and $$\begin{bmatrix} V_i \\ V_i^* \end{bmatrix} = \frac{1}{1-M^2}\begin{bmatrix} \dfrac{1}{j\omega_i C_0} & \dfrac{M}{j\omega_s C_0} \\ \dfrac{-M}{j\omega_i C_0} & \dfrac{-1}{j\omega_s C_0} \end{bmatrix}\begin{bmatrix} I_i \\ I_i^* \end{bmatrix} \tag{13.192}$$

For the input and output circuits of Fig. 13.75, we can write the following relations:

$$V_1 = I_s\left[R_1 + R_s + j\omega_s\left(L_s + L_1\right) + \frac{1}{j\omega_s\left(C_p + C_1\right)} + \frac{1}{j\omega_s\left(1 - M^2\right)C_0}\right.$$

$$+ \frac{MI_i^*}{j\omega_i\left(1 - M^2\right)C_0} \tag{13.193}$$

and $$V_2 = I_i\left[R_2 + R_s + j\omega_i\left(L_s + L_2\right) + \frac{1}{j\omega_i\left(C_p + C_2\right)} + \frac{1}{j\omega_i\left(1 - M^2\right)C_0}\right.$$

$$+ \frac{MI_s^*}{j\omega_s\left(1 - M^2\right)C_0} \tag{13.194}$$

As before, using the following tuning conditions

$$\omega_s^2 = \frac{1}{L_s + L_1}\left[\frac{1}{C_p + C_1} + \frac{1}{\left(1 - M^2\right)C_0}\right] \tag{13.195}$$

and $$\omega_i^2 = \frac{1}{L_s + L_2}\left[\frac{1}{C_p + C_2} + \frac{1}{\left(1 - M^2\right)C_0}\right] \tag{13.196}$$

in Eqs (13.193) and (13.194), we get the following expressions:

$$V_1 = \left(R_1 + R_s\right)I_s - \frac{jMI_i^*}{\omega_i\left(1 - M^2\right)C_0} \tag{13.197}$$

and $$V_2 = \left(R_2 + R_s\right)I_i - \frac{jMI_s^*}{\omega_s\left(1 - M^2\right)C_0} \tag{13.198}$$

If we assume that $V_2 = 0$, we can solve Eqs (13.197) and (13.198) for I_i. Substituting this I_i in the expression of gain

$$G_0 = 4R_1 R_2 |I_i|^2 / V_1^2 \tag{13.199}$$

we get the following equation:

$$G_0 = \cfrac{4R_1 R_2 M^2}{\omega_s^2 C_0^2 (R_2 + R_s)^2 (1 - M^2)^2 \left[R_1 + R_s - \cfrac{M^2}{\omega_i \omega_s (R_2 + R_s)(1 - M^2)^2 C_0^2} \right]^2} \tag{13.200}$$

The term $-\cfrac{M^2}{\omega_i \omega_s (R_2 + R_s)(1 - M^2)^2 C_0^2}$ can be interpreted as an equivalent negative

resistance $-R_n$, and thus Eq. (13.200) is modified as follows:

$$G_0 = \cfrac{4R_1 R_2 M^2}{\omega_s^2 C_0^2 (R_2 + R_s)^2 (1 - M^2)^2 (R_1 + R_s - R_n)^2}$$

or $$G_0 = \cfrac{4R_1 R_2 \omega_i^2 C_0^2 (1 - M^2)^2}{M^2 (R_1 + R_s - R_n)^2} \cfrac{M^4}{\omega_i^2 \omega_s^2 C_0^4 (R_2 + R_s)^2 (1 - M^2)^4}$$

or $$G_0 = \cfrac{4R_1 R_2 \left\{ \omega_i C_0 R_n (1 - M^2) \right\}^2}{M^2 (R_1 + R_s - R_n)^2} \tag{13.201}$$

Equation (13.201) reveals that a high gain can be obtained if $R_n \rightarrow R_1 + R_s$. However, if R_n is too close to $R_1 + R_s$ then a small change in the parameters will result in a large change in gain, and oscillation will start.

A parametric amplifier, discussed so far, is also called a negative-resistance converter. It is also possible to take the output at the signal frequency. If we split R_1 into generator internal resistance R_g and load resistance R_L, then the power delivered to load is $R_L |I_s|^2 / 2$ and the transducer gain will be as follows:

$$G_0 = 4R_g R_L |I_s|^2 / V_1^2 \tag{13.202}$$

Solving Eqs (13.197) and (13.198) for I_s and substituting the solution in Eq. (13.202), we get the following expression:

$$G_0 = \cfrac{4R_g R_L}{\left(R_g + R_L + R_s - R_n \right)^2} \tag{13.203}$$

where $$R_n = \cfrac{M^2}{\omega_i \omega_s (R_2 + R_s)(1 - M^2)^2 C_0^2} \tag{13.204}$$

The appearance of negative resistance in the circuit may be described in the following manner: With the application of the pump power to the non-linear capacitance, frequency mixing occurs in the circuit and an idler frequency is generated. Now, if the current with this idler frequency is permitted to exist, further mixing of the pump and idler frequency occurs, which results in further generation of f_s and harmonics of f_p and f_i. When the power generated through frequency mixing exceeds the power supplied at the signal frequency f_s, diodes appear to have a negative resistance.

It may be noted that a negative-resistance parametric amplifier with the same input and output frequencies is not very stable. The reason is that, if the load resistance is not matched then reflection will occur. This reflected wave will return to the amplifier, where they are further amplified and fed to the input and output. This makes the gain to be a sensitive function of the generator and load impedance. The stability of the amplifier, however, may be improved by the use of a circulator, as shown in Fig. 13.76. The circulator makes the load resistance of the amplifier equal to the characteristic impedance of the transmission line, and hence the circuit becomes independent of the generator and load resistance.

Microwave parametric amplifiers can be constructed in various ways. One such typical parametric amplifier, which uses both transmission line and waveguide resonator, is shown in Fig. 13.77. The pump and idler cavities are formed in an X-band rectangular waveguide, whereas the signal cavity is a coaxial transmission line cavity. The varactor diode is mounted on an inductive diaphragm separating the pump and the idler cavity. Coupling with the signal cavity is achieved by terminating the diode in the centre conductor of the coaxial line signal cavity. The pump frequency is coupled to the pump cavity through a coupling aperture. For a pump frequency of 9.2 GHz and an input–output frequency of 1.3 MHz, such amplifiers can provide a gain of 25 dB over a 5 MHz bandwidth.

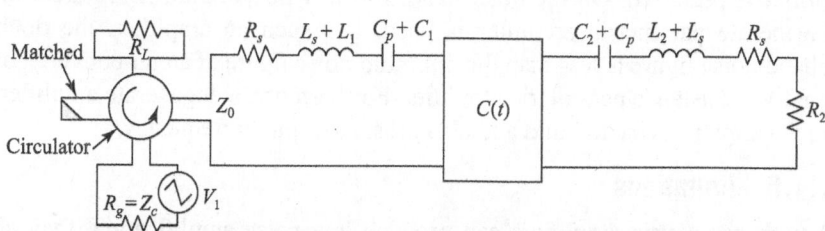

Fig. 13.76 Negative-resistance parametric amplifier using circulator

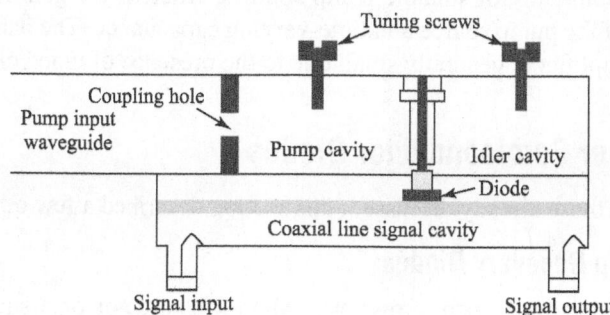

Fig. 13.77 Microwave negative-resistance parametric amplifier

The bandwidth of a negative-resistance parametric amplifier is relatively small. After a detailed analysis, it can be shown that the maximum gain bandwidth product can be expressed as follows:

$$\left(2\sqrt{G_0}\,\frac{\Delta\omega_s}{\omega_s}\right)_{max} = \frac{M}{2}\sqrt{\frac{\omega_i}{\omega_s}} \tag{13.205}$$

The parametric amplifier can be made broadband using broadband circuits at the signal and idler frequency. Alternatively, travelling-wave parametric amplifiers can be used, in which cavities are completely avoided. Instead, a waveguide is periodically loaded with varactor diodes. With the application of signal and pump power, mixing occurs and power is generated at the signal frequency.

13.11.5 Comparison between Different Parametric Amplifiers

The choice of a parametric amplifier depends on the system requirement. An up-converter is a unilateral stable device with a wide bandwidth and low gain. On the other hand, a negative-resistance amplifier is a bilateral and unstable device with low bandwidth and high gain. In general, an up-converter has the following advantages over a negative-resistance parametric amplifier: (a) positive input impedance, (b) power gain is independent of the changes of source impedance, (c) unconditionally stable and unilateral, (d) typical 5% bandwidth, and (e) no requirement of circulator.

At high frequencies, where the use of an up-converter is impractical, a negative-resistance amplifier with a circulator is generally used. Since in a radar system, the frequency of operation is more than the X-band, a negative-resistance amplifier is preferred. On the other hand, if a low noise figure is required, then a degenerate parametric amplifier is used. For such an amplifier, the double sideband noise figure is less than the optimum noise figure of an up-converter and a negative-resistance parametric amplifier. Furthermore, a degenerate amplifier is a much simpler device to build and also uses a low pump frequency.

13.11.6 Limitations

Although parametric amplifiers can produce low noise amplification, they also suffer from some disadvantages in terms of frequency of operation, gain, and bandwidth. The highest operating frequency of a parametric amplifier is restricted by the non-availability of suitable pump sources, whereas the gain is limited by the stability of the pump source and time-varying capacitance. The bandwidth of a parametric amplifier is generally small due to the presence of tuned circuit.

13.12 Other Semiconductor Diodes

In addition to those already discussed, this section described a few other diodes.

13.12.1 Step Recovery Diodes

A step recovery diode, also known as a snap-off varactor or a snap diode, is a p–n junction diode that is designed to exploit the minority carrier storage, inherent in such a diode. It may be recalled that when an initially forward-biased diode is suddenly reverse biased, it does not instantly cause the diode to turn off. To turn off the diode, minority carriers must first be removed. Until then, the diode remains on and conducts a current with a low voltage drop. This restricts the use of such diodes in RF circuits and switching power supplies, where conduction in the reverse direction can result in a significant reduction in the efficiency of the circuit. A step recovery diode converts this drawback into an advantage.

A snap diode is an Si or GaAs p–n junction diode and is similar to varactor diodes in construction. Such a diode stores charge when it is forward biased and briefly discharges this stored charge in reverse bias in the form of a sharp pulse, rich in harmonics. The duration of this discharge pulse depends on the design of the diode and typically ranges from 100 to 1000 picoseconds. Practically, this time period should be smaller than the time period of the output frequency. For a correctly designed circuit, the efficiency of the output device is about $1/n$ where n is the harmonic.

The snap-off speed of a snap diode is a function of the internal electric field and hence is a function of the doping level. Generally, high doping results in a higher built-in field and hence a faster snap-off speed. However, it may be noted that the breakdown voltage of a diode is also a function of the doping level. Since higher doping corresponds to a lower breakdown voltage, it can be concluded that a faster snap-off is associated with a lower breakdown voltage. In designing a snap diode, a doping profile that maximizes the electric field (e.g., exponential doping) can also be used

The reverse recovery of a step recovery diode is a function of the forward current and hence of the bias voltage. Since fast recovery corresponds to a sharp rise time and hence to high spectral content, it can be concluded that the spectral content is also a function of the biasing voltage.

A step recovery diode without a tuned circuit can be used to produce multiple harmonics. Such a circuit is known as a comb generator. In comb generators or frequency multipliers, the bias is often adjusted to optimize some performance parameters like the amount of certain harmonics. Step recovery diodes are also often used in high-speed samplers.

During fabrication, it is possible to stack two or more step recovery diodes in a single package to provide high power capacity.

13.12.2 Noise Diodes

It is well known that Zener diodes are extremely noisy. Noise diodes exploit this noise to make broadband white noise sources with a noise temperature of many thousands of degrees, but without the requirement of actual heating of a resistor to such a temperature. Noise diodes are thus safe and compact. Noise in noise diodes fundamentally arises from the stochastic nature of avalanching. Therefore, as long as the device parasitics are small, the output noise extends over an exceptionally wide frequency range.

The main disadvantage of a noise diode is that noise varies considerably from diode to diode. Therefore, each diode must be calibrated against some primary

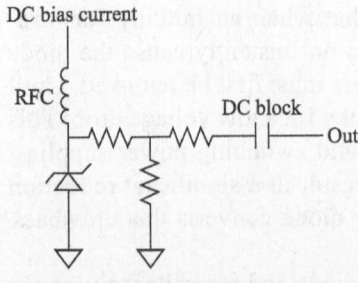

Fig. 13.78 Typical noise diode bias circuit

standard, if the diode is used as a secondary standard in noise measurement. In addition, because of the labour involved in the calibration, these diodes are also very costly. They also require control over the change of characteristics with time and temperature variations. Subsurface or guard-ring-stabilized Zener diodes are generally used to reduce instability over time.

Figure 13.78 describes a simple noise diode circuit. The RF choke prevents shunting of high-frequency components to ground, whereas the DC blocking capacitor blocks the DC current from reaching the load. The T-network provides impedance matching.

Important Formulae

- To achieve a negative resistance, we must have $\left(\dfrac{\mu_l - \mu_u}{\mu_l + f\mu_u}\right)\left(-\dfrac{E}{n_l}\dfrac{dn_l}{dE}\right) - p > 1$.

- For domain growth, $n_0 L > \dfrac{\varepsilon v}{e|\mu_n|}$.

- The Manley–Rowe power relations are $\displaystyle\sum_{m=0}^{\infty}\sum_{n=-\infty}^{\infty}\frac{mP_{m,n}}{m\omega_p + n\omega_s} = 0$ and

$$\sum_{m=-\infty}^{\infty}\sum_{n=0}^{\infty}\frac{nP_{m,n}}{m\omega_p + n\omega_s} = 0.$$

Exercises

Objective-type Questions

13.1   A Gunn diode of length 10 μm is operating at 10 GHz. The electron drift velocity is
(a) 10^5 cm/sec
(b) 10^6 cm/sec
(c) 10^7 cm/sec
(d) 10^8 cm/sec

13.2   Which one of the following is a transferred electron device?
(a) Step recovery diode
(b) BARITT diode
(c) IMPATT diode
(d) Gunn diode

13.3   Indicate which of the three is different from the others?
(a) Schottky barrier diode
(b) Backward diode
(c) ESBAR diode
(d) Hot carrier diode

13.4   Which of the following is used for its negative-resistance characteristics?
(a) Point-contact diode
(b) Step-recovery diode
(c) Schottky diode
(d) Tunnel diode

13.5   The most serious drawback of an IMPATT diode is

(a) high noise
(b) low efficiency
(c) low power-handling capacity
(d) inability to provide pulsed operation

13.6 An LSA diode differs from a Gunn diode in which of the following respects?
(a) It need not be made of GaAs
(b) It prevents formation of domain
(c) It does not require negative resistance
(d) None of these

13.7 Which of the following diodes is not used as a microwave mixer or detector?
(a) PIN diode
(b) Schottky diode
(c) Backward-wave diode
(d) Crystal diode

13.8 Gallium antimonite is preferred to germanium for use in tunnel diode since it permits
(a) low noise
(b) larger voltage swing
(c) better frequency stability
(d) simpler fabrication process

13.9 A Gunn diode is made from
(a) silicon
(b) germanium
(c) gallium arsenide
(d) selenium

13.10 For a Gunn diode, the drift velocity of electrons through the active drift region is 10^7 cm/sec and the active region is 10^{-3} cm. The critical voltage of the diode (critical field is $3\,kV/cm$) is

(a) 0.032 V   (b) 0.32 V   (c) 3.2 V   (d) 32 V

13.11 For Gunn effect, the mobility of the upper valley must be
(a) lower than the lower valley
(b) higher than the lower valley
(c) equal to the lower valley
(d) none of these

13.12 The energy gap between the lower and upper valleys must be
(a) Higher than the band-gap energy and thermal energy
(b) Less than the band-gap energy and higher than the thermal energy
(c) Equal to the band-gap energy or thermal energy
(d) Less than the band-gap energy and thermal energy

13.13 In the voltage-controlled mode of Gunn oscillator
(a) a high-field domain is created
(b) a high-density current filament is formed
(c) either (a) or (b)
(d) both (a) and (b)

13.14 In the current-controlled mode of Gunn oscillator
(a) a high-field domain is created
(b) a high-density current filament is formed
(c) either (a) or (b)
(d) both (a) and (b)

13.15 Theoretically, the efficiency of a read diode is
(a) 10%
(b) 20%
(c) is less than or around 30%
(d) 40%

13.16 By construction, TRAPATT diodes are
(a) $n^+ - p - i - p^+$
(b) $p - i - n$
(c) $p^+ - n - n^+$
(d) p–n

13.17 BARITT diodes produce a negative resistance for the transit angle between
(a) 0 and π
(b) π and 2π
(c) π and $\pi/2$
(d) $\pi/2$ and $3\pi/2$

13.18 In a Schottky barrier device, guard is provided to
(a) improve I–V characteristics
(b) reduce noise
(c) increase operating frequency
(d) eliminate unwanted edge breakdown and channelling effect

13.19 A varactor diode made from GaAs, as compared to that made from Si, has
(a) high frequency and high Q
(b) high frequency and low Q
(c) low frequency and high Q
(d) low frequency and low Q

13.20 For application in parametric amplifiers, a varactor diode is of
(a) p–n junction type
(b) Schottky barrier type
(c) p–i–n junction type
(d) none of these

Review Questions

13.1 With a suitable figure, describe the operating principle of a tunnel diode.
13.2 What is Gunn effect? Describe the RWH theory to explain Gunn effect.
13.3 Classify and describe different operating modes of Gunn diode oscillators.
13.4 With a suitable figure, describe the operation of a Read diode.
13.5 Describe the operating principle of a TRAPATT diode.
13.6 Describe the operating principle of a BARITT diode.
13.7 Write a short note on Schottky barrier diodes.
13.8 Derive an expression for the resistance of the i-region of a p–i–n diode.
13.9 Derive an expression for the junction capacitance of a varactor diode.
13.10 Find the Manley–Rowe equations.

Microwave Solid-state Transistors and MASERs

14.1 Introduction

The first solid-state amplifier that was developed for microwave amplification is a negative-resistance device (e.g., tunnel diode). It was followed by the development of parametric amplifiers that use a varactor diode and a pump frequency for their operation. Parametric amplifiers can produce very low noise when cooled using liquid nitrogen and hence became the most popular solid-state amplifier during 1958–1970. However, by this time material preparation and processing technology improved significantly, leading to the development of n–p–n bipolar junction transistors. Development continued in the next two decades, and other bipolar and field effect transistors were developed.

Like ordinary vacuum devices, ordinary transistors also cannot be used at high frequencies because the inter-electrode capacitance between the base–emitter and base–collector junctions behaves as a short. As a result, signal gets shorted and appears directly at the collector terminal, affecting the gain of the transistor. These inter-electrode capacitances depend on the width of the depletion layer at the junction, which, in turn, depends on biasing of the transistor. However, the advantage of solid-state devices over vacuum devices is that the effect of lead inductance can be minimized by making the transistor leads shorter and packing them in a low-inductance package.

High-frequency microwave transistors can be divided into two categories: (a) bipolar junction transistors (BJT) and (b) Field Effect Transistors (FET). In practice, microwave BJTs are of two types: (a) ordinary BJT and (b) heterojunction bipolar transistor (HBT). Similarly, FETs can also be classified as (a) junction field effect transistors (JFET), (b) metal–semiconductor field effect transistor (MESFET), (c) high-electron-mobility transistor (HEMT), (d) metal-oxide–semiconductor field effect transistor (MOSFET), (e) laterally diffused metal-oxide semiconductor transistor (LDMOS), (f) VMOS, and (g) UMOS.

A high-frequency microwave transistor is a non-linear semiconductor device, and, in principle, its operation is similar to that of other low-frequency transistors. However, due to its operation at high frequencies, requirements of dimensions, process control, heat sinking, and packaging are much more severe. Practically, the key step in designing microwave transistors is miniaturization that is necessary to reduce device and package parasitic capacitance and lead inductance and to overcome the finite transit time of the charge carriers, which is about 5–10% of the speed of light.

Transit time generally depends on the electron velocity and saturation velocity in the semiconductor material. In this regard, GaAs is far better than Si for high-frequency operation. While GaAs transistors can operate effectively at high frequencies, high temperatures, and radiation hardness, silicon BJTs dominate in the frequency range from UHF to about S-band, except for a very-low-noise amplifier design. At these lower microwave frequencies, Si BJTs and Si–Ge HBTs have an edge over GaAs FETs, HEMTs, and HBTs in terms of cost. In addition, Si BJTs are durable and integrative and also offer much higher gain than FETs. However, as the power level of BJT increases, the device draws more current, heats up even more, and then is self-destructed. Thus, BJT requires additional components to limit the current and minimize VSWR. Si BJTs are often used in local oscillators.

While Si BJTs can operate up to S-band, FETs can be used up to X-band for amplifying small signals with low noise figures. Unipolar FETs also have several other advantages over bipolar BJTs: (a) both voltage gain and current gain, (b) low noise figure, (c) higher efficiency, (d) high input impedance, (e) excellent switching characteristics, and (f) higher operating frequency. Further, in many situations, FETs are easier to bias and less sensitive to the load. Due to these features, GaAs FETs are widely used in power amplifiers, and also in hybrid amplifiers, multipliers, mixers, switching circuits, gain control circuits, etc.

Since the heterojunction in an HEMT and HBT device improves the charge transport properties (in case of HEMTs) or p–n junction injection characteristics (in case of HBT), these devices have potential advantages in microwave and millimetre-wave IC applications. Devices such as p-HEMTs, developed using multiple epitaxial III–IV compound layers, exhibit excellent millimetre-wave performance from Ku band to W-band, whereas MMICs, developed using pseudomorphic and lattice-matched HEMTs, have improved noise characteristics and can also work up to 200 GHz. In addition, HEMTs also have a performance edge in ultra-low-noise, low-loss switches. In contrast to HEMTs, HBTs offer better linearity and lower phase noise, and find applications in power devices, when operated using a single power supply.

The requirement of a particular transistor in a circuit depends on the nature of its applications. For example, power amplifiers require transistors capable of operating with a high power density, low-noise amplifiers require transistors having low noise characteristics, and switches require transistors having low *on resistance* and small *off capacitance*. Hence, a transistor is often characterized based on the following features: (a) maximum available gain, (b) cut-off frequency, (c) maximum oscillation frequency, (d) minimum noise figure, (e) output power density, and (f) power-aided efficiency.

It may be noted that to achieve high-frequency performance in transistors, it is necessary to develop a suitable technology that would enable key device dimensions to be less than $1\,\mu m$, such as molecular beam epitaxy (MBE). MBE technology also led to the development of heterostructures, which, in turn, led to the development of HEMT that can operate up to $100\,GHz$. With advances in heterojunction technology, HBTs have also been developed. Such transistors exhibit very low base resistance, high current gain, and an increase in speed by a factor of 2 or 3. An AlGaAs/GaAs HBT can operate with a cut-off frequency of $105\,GHz$ and maximum frequency of oscillation of $175\,GHz$.

14.2 Bipolar Junction Transistors

Bipolar junction transistor (BJT) was developed in 1948 at Bell Laboratory by William Shockley and his co-workers. It is a bipolar device, which means that both majority and minority carriers participate in its operation. By construction, it consists of a thin n-type or p-type semiconductor sandwiched between two p-type or n-type semiconductors respectively. The following subsections describe the properties and features of BJTs.

14.2.1 Physical Structure and Operation

The basic principle of operation of a microwave BJT is that same as that of its low-frequency counterpart. However, their analysis procedure is different due to the fact that the intrinsic device, along with all parasitic elements, requires a more complex equivalent circuit model, as shown in Fig. 14.1. In addition, there is sufficient capacitive feedback from the collector to the base, which makes the device potentially unstable and leads to oscillation in the input and output circuits unless the circuits are properly designed to prevent oscillations. The different circuit elements of Fig. 14.1 are as follows: C_{bp}, base bond pad capacitance; C_{ep}, emitter bond pad capacitance; C_{be}, base to emitter junction capacitance; r_{bc}, base contact resistance; r_{ec}, emitter contact resistance; $R_1, R_2, R_3, \ldots, R_n$, base distributed resistance; $C_1, C_2, C_3, \ldots, C_n$, collector base distributed capacitance; r_e, dynamic emitter base diode resistance; r_c, collector resistance; i_e, emitter current; α, common base current gain; and L_b, L_e, base and emitter bond wire inductance, respectively.

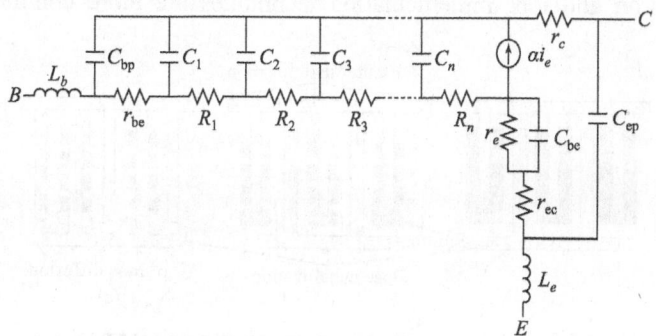

Fig. 14.1 Distributed T-equivalent circuit of BJT

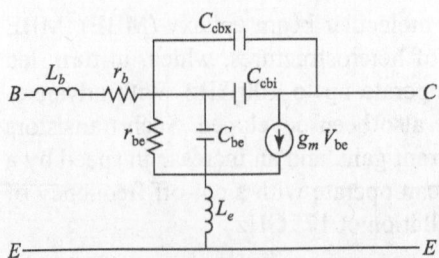

Fig. 14.2 Pi-equivalent circuit of BJT

Figure 14.1 represents the T-equivalent circuit of a BJT, which can also be represented as a pi-equivalent circuit (Fig. 14.2). Here, r_b is the base resistance, r_{be} the base to emitter resistance, C_{cbx} the extrinsic base to collector capacitance, C_{cbi} the intrinsic base to collector capacitance, C_{be} the base to emitter capacitance, and V_{be} the voltage across base to emitter capacitance. Microwave BJTs can be n–p–n or p–n–p types. However, since the electron mobility $\left(1500\ \text{cm}^2/\text{V}\right)$ is much higher than the hole mobility $\left(450\ \text{cm}^2/\text{V}\right)$, n–p–n BJTs are preferred over p–n–p BJTs for microwave applications.

Microwave transistors are planar, in geometry, and its geometry can be classified as (a) interdigitated, (b) overlay, and (c) matrix/mesh or emitter grid types, as shown in Fig. 14.3. In the interdigitated geometry, a large number of emitter strips and base strips are arranged alternately, whereas in the overlay geometry, a large number of segmented emitters are overlaid through a number of wide metal strips. It may be noted that, in the overlay geometry, the emitter metallization runs across the base strips also; however, it makes contact only with the emitter strips as the base strips are separated by an insulating SiO_2 layer. The matrix or mesh geometry is characterized by emitters that form a grid; the base fills the meshes of this grid with a p^+ contact area in the middle of each mess.

The interdigitated geometry is widely used for small signal and power transistors, while the overlay and mesh geometries are used for power transistors only. In addition to these, diamond emitter configuration is also available, which is a modification of the overlay structure and consists of a checker board pattern with alternate squares representing the emitter areas, as shown in Fig. 14.4. In this figure, emitters are marked with E and bases with B. A unit cell has the dimension $4W \times 4W$. The dimension of the emitter is $l \times l$ or $\sqrt{8W} \times \sqrt{8W}$. Metallization strips of width W runs diagonally, and provides emitter and base contact alternatively. For a constant power output, diamond emitter transistors are capable of operating at substantially higher frequencies.

Although microwave transistors can be fabricated using various methodologies, diffusion and ion implementation techniques are more commonly used.

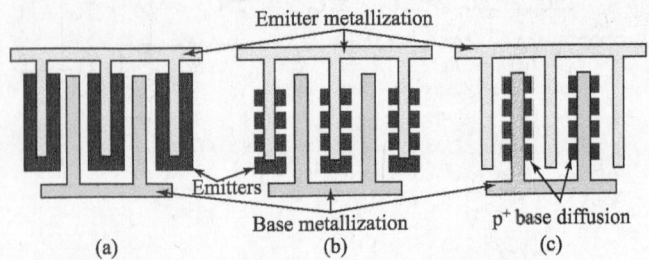

Fig. 14.3 Surface geometries of microwave power transistors
(a) Interdigitated (b) Overlay (c) Matrix

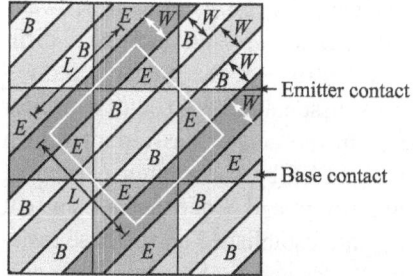

Fig. 14.4 Diamond emitter configuration

Fabrication of such a transistor generally starts with the formation of a lightly doped n-type epitaxial layer that behaves as a collector. The base region is then formed by counter-doping the collector region by p-type diffusion. Finally, the emitter is formed by shallow, heavily doped, n-type diffusion or by ion implementation. Contacts between emitters and base are made on the semiconductor surface in an interdigital planar arrangement. The number of fingers used for emitter/base connection generally depends on the application. For example, if the output power-handling capability of a transistor increases, more fingers will be required. Addition of fingers, however, will increase the device parasitic, thereby degrading the noise and maximum operating frequency limits of the transistor. Schematic diagrams of a discrete n–p–n planar BJT and an integrated-chip-type n–p–n BJTs are shown in Fig. 14.5.

The only device parameters of a transistor that can be measured at microwave frequencies are scattering parameters. They can be measured by embedding the transistor in a 50 Ω transmission line and using a vector network analyser. By measuring the S-parameters over a frequency range, the equivalent circuit parameters can be calculated.

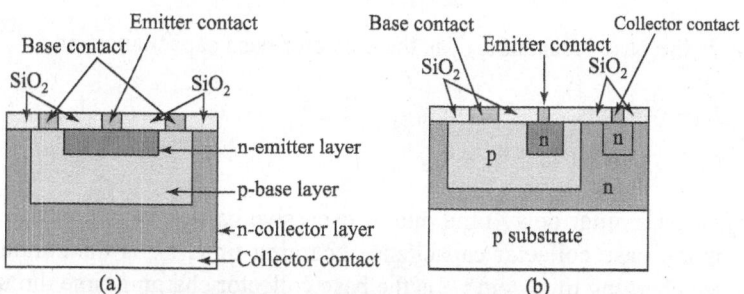

Fig. 14.5 Schematic diagram of BJT (a) Discrete n–p–n planar (b) Integrated chip-type n–p–n

14.2.2 Transistor Biasing

For satisfactory operation, a microwave transistor must be biased properly. Since such a circuit operates at very high frequencies, during designing the biasing circuit, care must be taken that high-frequency signal currents do not flow in the DC biasing circuit. That is, the biasing circuit must be isolated from the high-frequency circuit. Such isolation can be achieved by using low-impedance capacitive bypass circuits to shunt high-frequency currents around the DC circuit elements and by inserting high-frequency, high-impedance circuit elements in series with the DC components. Two bias circuits that can accomplish these are shown in Fig. 14.6. The bias circuit shown in Fig. 14.6(a) is used mainly at low frequencies. However,

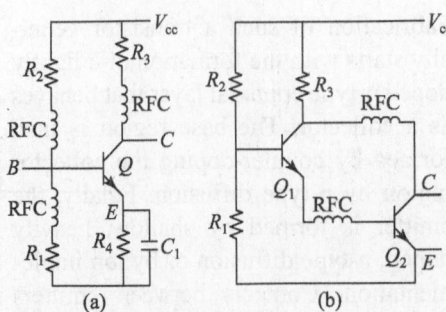

Fig. 14.6 Transistor bias circuit
(a) Passive (b) Active

they can also be used at microwave frequencies but with increased difficulty because of the parasitic inductances associated with the capacitor lead. In the active bias circuit shown in Fig. 14.6(b), the emitter current of transistor Q_1 and hence the base current of Q_2 are established by the base resistors R_1, R_2, and R_3. Since the base current of Q_1 is stable, the base current of Q_2 is maintained at a value that is independent of the transistor parameters. This circuit also consumes less power and requires only RF chokes for isolation.

In designing the bias circuit, care should be taken that oscillation does not occur at any frequency. This can be ensured by properly designing the overall bias circuit and RF matching circuit, so that they provide stable terminations for each active device outside the frequency band of interest.

14.2.3 Important Parameters

The maximum frequency at which the power gain of a transistor has dropped to unity can be expressed as $f_{max} = \dfrac{f_T}{8\pi r_b C_c}$ (14.1)

where r_b is the base resistance, C_c is the collector base capacitance, and

$$f_T = \left(2\pi\tau_{ec}\right)^{-1} \tag{14.2}$$

where $\tau_{ec} = \tau_e + \tau_b + \tau_c + \tau_{eb} + \tau_{bc}$ (14.3)

Here, τ_e is the emitter delay time due to excessive carrier, τ_b is the base transit time, τ_c is the base collector capacitance charging time, τ_{eb} is the emitter–base capacitance charging time, and τ_{bc} is the base collector charging time through the emitter.

14.2.4 High-frequency Noise Characteristics

The high-frequency noise characteristics of BJTs are characterized by three noise sources: (a) base thermal noise (e_b) due to base resistor r_b, (b) shot noise (e_e) due to the forward-biased emitter base junction, and (c) collector partition noise (i_{cp}). These noise sources can be expressed as follows:

$$\overline{e_b^2} = 4kT\Delta f r_b \tag{14.4}$$

$$\overline{e_e^2} = 2kT\Delta f r_e \tag{14.5}$$

$$\overline{i_{cp}^2} = 2kT\Delta f \left(\frac{\alpha_0 - |\alpha|^2}{r_e}\right) \tag{14.6}$$

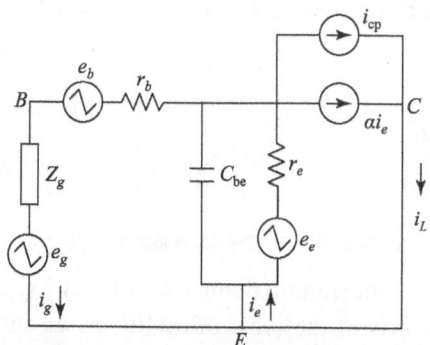

Fig. 14.7 High-frequency noise equivalent circuit of BJT

where $r_e = \dfrac{kT}{qI_e}$ (14.7)

$$\alpha = \frac{\alpha_0}{1+(jf/f_b)}$$ (14.8)

Here, α_0 is the low-frequency common base current gain, f_b is the base cut-off frequency, Δf is the bandwidth, T is the temperature in absolute scale, and K is the Boltzmann constant. A high-frequency noise equivalent circuit is shown in Fig. 14.7, where Z_g is the impedance of the signal source e_g and

$$\overline{e_g^2} = 4kT\Delta f R_g$$ (14.9)

where R_g is the real part of impedance Z_g.
The device achieves the minimum noise figure when it is operated at the optimum input impedance:

$$Z_{\text{opt}} = R_{\text{opt}} + jX_{\text{opt}}$$ (14.10)

where $R_{\text{opt}} = \sqrt{r_b^2 - X_{\text{opt}}^2 + \dfrac{\alpha_0 r_e (2r_b + r_e)}{a|\alpha|^2}}$ (14.11)

$$X_{\text{opt}} = \frac{\alpha_0 \omega C_{\text{be}} r_e^2}{a|\alpha|^2}$$ (14.12)

$$a = \frac{1}{\alpha_0} \left\{ \left[1 + \frac{f^2}{f_b^2}\right]\left[1 + \frac{f^2}{f_e^2}\right] - \alpha_0 \right\}$$ (14.13)

and f_e is the emitter cut-off frequency.
The minimum noise figure can be expressed as follows:

$$F_{\text{min}} = \frac{a(r_b + R_{\text{opt}})}{r_e} + \frac{\alpha_0}{|\alpha|^2}$$ (14.14)

The equivalent noise resistance is given by the following equation:

$$R_n = \frac{r_b}{\alpha} + \frac{r_e}{2}\left\{1 + \left(\frac{fr_b}{f_b r_e}\right)^2\right\}$$ (14.15)

In practice, there are two limiting cases of operation for BJTs: (a) base-limited case when the base time constant τ_b is dominant and (b) emitter-limited case when the emitter time constant τ_e is dominant. For the first case, the expression for the minimum noise figure can be given as follows:

$$F_{\text{min}} = \left(1 + \frac{f^2}{f_b^2} - \alpha_0\right)\left(\frac{r_b + R_{\text{opt}}}{\alpha_0 r_e}\right) + \left(1 + \frac{f^2}{f_b^2}\right)\frac{1}{\alpha_0}$$ (14.16)

The minimum noise figure for the emitter-limited case can be expressed as follows:

$$F_{\min} = \left(1 + \frac{f^2}{f_e^2} - \alpha_0\right)\left(\frac{r_b + R_{opt}}{\alpha_0 r_e}\right) + \frac{1}{\alpha_0} \tag{14.17}$$

14.2.5 Frequency Limitations

To find the power–frequency relation, let us make the following assumptions:

1. Carriers inside a semiconductor have a maximum attainable velocity (v_s), called the saturated drift velocity, which is of the order of 6×10^6 cm/sec for electrons and holes in Si and Ge.
2. A semiconductor can sustain a maximum electric field (E_m) of 10^5 V/cm for Ge and 2×10^5 V/cm for Si without dielectric breakdown.
3. The base width of a transistor sets the maximum value to the current (I_m) that a microwave power transistor can carry.

With these postulates, the basic equations for the power–frequency limitations of microwave transistors can be derived as follows.

Voltage–Frequency limitation

$$V_m f_T = \frac{E_m v_s}{2\pi} = \begin{cases} 2 \times 10^{11} & \text{V/s for Si} \\ 10^{11} & \text{V/s for Ge} \end{cases} \tag{14.18}$$

where
$$f_T = \frac{1}{2\pi\tau} \tag{14.19}$$

$$\tau = L/v \tag{14.20}$$

$$V_m = E_m L_{\min} \tag{14.21}$$

The transit time can be reduced by decreasing the emitter–collector length L until it reaches a value of $L_{\min}$ at which, for a given voltage V, the electric field becomes equal to the dielectric breakdown voltage of the semiconductor. In practice, L is maintained at about 250 µm for interdigitated devices and about 25 µm for overlay and matrix devices. Thus, there is a limitation on the maximum achievable operating frequency of a transistor. Practically, since the saturated drift velocity and electric field intensity are non-uniform, the attainable cut-off frequency is much less than that provided by Eq. (14.18).

Current–Frequency limitation

$$(I_m X_c) f_T = \frac{E_m v_s}{2\pi} \tag{14.22}$$

where
$$X_c = (2\pi f_T C_0)^{-1} \tag{14.23}$$

is the reactive impedance and C_0 is the collector base capacitance.

Power–Frequency limitation

$$f_T \sqrt{P_m X_c} = \frac{E_m v_s}{2\pi} \tag{14.24}$$

Power gain–Frequency limitation

$$f\sqrt{G_m V_{th} V_m} = \frac{E_m v_s}{2\pi} \tag{14.25}$$

where $\quad V_{th} = KT/e \tag{14.26}$

G_m is the maximum gain, K is the Boltzmann's constant, T is the temperature in absolute scale, and E is the charge of electrons.

The maximum available gain of a transistor can be expressed as follows:

$$G_m = (f_T/f)^2 Z_{out}/Z_{in} \tag{14.27}$$

where Z_{out} and Z_{in} are the output and input impedances, respectively. Assuming that the electrode series impedances are zero, we can write the following equation:

$$Z_{out}/Z_{in} = C_{in}/C_{out} \tag{14.28}$$

where C_{in} and C_{out} are the input and output capacitances, respectively, and can be expressed as follows:

$$C_{in} = I_m \tau_b/V_{th} \tag{14.29}$$

and $\quad C_{out} = I_m \tau_0/V_m \tag{14.30}$

where τ_b is the carrier base transit time.

14.2.6 Typical Characteristics

The typical characteristics of a BJT are as follows:

1. Output power: 20 W at 1 GHz and 150 mW at 8 GHz
2. Noise figure: 3.3–14 dB within a frequency range of 4–8 GHz
3. Power gain: 30 ± 2.5 dB within a frequency range of 4–8 GHz

---

EXAMPLE 14.1 A Si microwave transistor has the following parameters: reactance $=$ 0.95 Ω, transit time cut-off frequency $= 3.75$ GHz, maximum electric field $= 1.5 \times 10^5$ V/cm, and saturation drift velocity $= 4.5 \times 10^5$ cm/sec. Calculate the maximum allowable power.

Solution Given: $X_c = 0.95$ Ω, $f_T = 3.75 \times 10^9$ Hz, $E_m = 1.5 \times 10^5$ V/cm, and $v_s = 4.5 \times 10^5$ cm/sec

Therefore,

$$P_m = \frac{1}{X_c}\left(\frac{E_m v_s}{2\pi f_T}\right)^2 = \frac{1}{0.95}\left(\frac{1.5 \times 10^5 \times 4.5 \times 10^5}{2\pi \times 3.75 \times 10^9}\right)^2 = 8.639 \text{ W}$$

Practice Problem

14.1 An Si microwave transistor has the following parameters: reactance $= 0.95 \Omega$, transit time cut-off frequency $= 3.75$ GHz, maximum electric field $= 1.5 \times 10^5$ V/cm, and saturation drift velocity $= 4.5 \times 10^5$ cm/sec. Calculate the maximum allowable current. **[3.0156 A, 44.6751 pF]**

14.3 Heterojunction Bipolar Transistors

In contrast to BJTs, which consist of a junction between two similar materials such as Si or Ge, a heterojunction bipolar transistor (HBT) is characterized by the junction between two dissimilar materials. The following subsections describe the properties and features of HBTs.

14.3.1 Physical Structure

In a transistor, emitter–base or base–collector junction can be created between two similar materials (e.g., Si–Si or Ge–Ge) or two different materials (e.g., Ge–GaAs or GaAs–AlGaAs). The transistor is called a homojunction in the former case and a heterojunction in the latter case. HBTs were first proposed by Shockley in 1948. Later Kroemer developed the theoretical interpretation of the device. In creating HBTs, the lattice constant of the two semiconductor materials must be matched because a lattice mismatch, if any, can introduce a large number of interface states and degrade the junction performance. The lattice constant of Ge (5.646 Å) and GaAs (5.653 Å) can be matched within 1% and hence can be used in the fabrication of HBTs. Such transistors can be developed using MBE or molecular organic chemical vapour deposition (MOCVD) technology. A model diagram of a Ge–GaAs HBT is shown in Fig. 14.8.

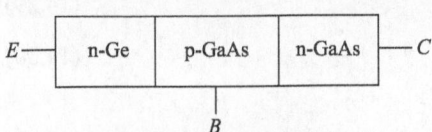

Fig. 14.8 Model diagram of Ge–GaAs HBT

InGaP can also be used in place of AlGaAs to form a heterojunction with GaAs. The use of InGaP avoids the use of Al and enables simple etching of both GaAs and InGaP. Further, it improves the performance of HBTs compared to AlGaAs, while simplifying device fabrication and increasing reproducibility.

14.3.2 Operating Principle

Initially when the n-Ge and p-GaAs samples are isolated, their Fermi energy levels are not aligned, as shown in Fig. 14.9. Considering the vacuum level as a reference, the difference between the conduction and valence band energy of the two samples can, respectively, be expressed as follows:

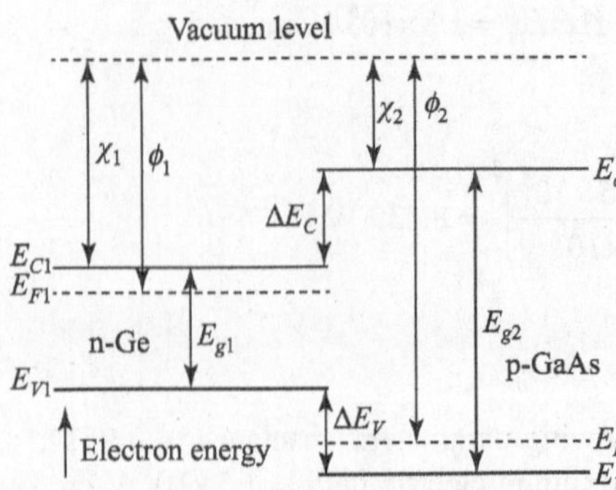

$$\Delta E_c = \chi_1 - \chi_2 \qquad (14.31)$$

$$\Delta E_v = E_{g2} - E_{g1} - \Delta E_c \qquad (14.32)$$

where χ is the electron affinity and E_g is the band gap energy.

Now, if the two semiconductors are brought together, thus forming a junction, electrons in n-Ge will be injected into p-GaAs and holes in p-GaAs will be injected into n-Ge in order to align their Fermi levels. As a

Fig. 14.9 Energy-band diagram for isolated n-Ge and p-GaAs

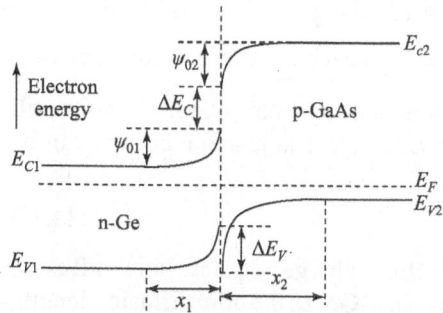

Fig. 14.10 Energy-band diagram for n-Ge and p-GaAs junction

result, their energy bands are depleted or bent at the junction, as shown in Fig. 14.10. The bending energy creates a *built-in voltage* at both sides of the junction. The total built-in voltage can be written as follows:

$$\psi_0 = \psi_{01} + \psi_{02} \tag{14.33}$$

where ψ_{01} and ψ_{02} are barrier potentials or portions of built-in voltage in n-Ge and p-GaAs, respectively. The built-in voltage can be expressed as follows:

$$\psi = \frac{kT}{q} \ln\left(\frac{N_A N_D}{p_0 n_0}\right) \tag{14.34}$$

where $k = 1.38 \times 10^{-23} \text{ m}^2 \text{ Kgs}^{-2} \text{ K}^{-1}$ (Boltzmann constant) $\tag{14.35}$

T is the temperature in Kelvin, N_A is the accepter density, N_D is the donor density, p_0 is the equilibrium hole concentration, and n_0 is the equilibrium electron concentration.
Therefore,

$$\psi_{01} = kT \ln(N_D/n_0)/q \tag{14.36}$$

and $\quad \psi_{02} = kT \ln(N_A/p_0)/q \tag{14.37}$

At the junction, the electric flux is continuous and hence we can write the following equation: $D = \varepsilon_1 E_1 = \varepsilon_2 E_2 \tag{14.38}$

The space charge on both sides of the junction should be equal and hence we can write the following relation: $x_1 N_d = x_2 N_a \tag{14.39}$

where x_1 and x_2 are the depletion widths in n-Ge and p-GaAs, respectively, and N_d and N_a are donor and acceptor concentration in the respective semiconductors.
The electric field on both sides can be written as follows:

$$E_1 = (\psi_{01} - V_1)/x_1 \tag{14.40}$$

and $\quad E_2 = (\psi_{02} - V_2)/x_2 \tag{14.41}$

where V_1 and V_2 are portions of the bias voltages in n-Ge and p-GaAs, respectively. Substituting Eqs (14.40) and (14.41) into Eq. (14.38), we get the following relation:

$$(\psi_{01} - V_1)\varepsilon_1 x_2 = (\psi_{02} - V_2)\varepsilon_2 x_1 \tag{14.42}$$

Further using Eq. (14.39) in Eq. (14.42), we get the following expression:

$$(\psi_{01} - V_1)\varepsilon_1 x_2 = (\psi_{02} - V_2)\varepsilon_2 x_2 N_a/N_d$$

or $\quad (\psi_{01} - V_1)\varepsilon_{r1} N_d = (\psi_{02} - V_2)\varepsilon_{r2} N_a \tag{14.43}$

For the heterojunction shown in Fig. 14.10, the potential barrier across the junction $\left(\psi_{01} + \psi_{02} + \dfrac{\Delta E_c}{q} \right)$ is very large and hence the electron current from n-Ge to p-GaAs is very small. In contrast, the hole current from p-GaAs to n-Ge will be large because of the low potential barrier (ψ_{02}). The junction current can be expressed as follows:

$$I = AqD_p p_{n0} \left(e^{V/V_T} - 1 \right) / L_p \tag{14.44}$$

where A is the cross-section, q the electron charge, D_p the hole diffusion coefficient, p_{n0} the equilibrium hole density in n-Ge, L_p the hole diffusion length, V the bias voltage, and V_T the voltage equivalent of temperature.

HBTs have very low base resistance (due to high doping in the base while maintaining a good β value) and low parasitic capacitance (due to reduced emitter doping). These two parameters reduce the RC time constant and hence increase the speed of the device. In addition, the use of a wider band-gap material in the emitter (e.g., AlGaAs or InGaP) results in an offset in the valence band so that the hole injection from the base into the emitter is retarded, while the electron injection from the emitter into the base is enhanced.

HBTs possess a high current capacity and high transconductance, and hence, a high voltage gain. Since the active region of an HBT is shielded from the output voltage, the output conductance is also low. Compared to Si BJTs, HBTs have a high gain bandwidth product and reduced parasitic. The higher common emitter output impedance, resulting from the high base doping, minimizes the base width modulation. This, in turn, leads to higher linearity and lower harmonic distortion in HBTs. In addition, since the entire emitter area can carry current, due to low emitter resistance, the power-handling capability of HBTs is much higher. The high power-handling capability along with the single power supply operation makes this device most suitable as power amplifiers in hand-held portable communication. HBTs are also widely used as high-speed switching devices.

Note Except Ge–GaAs, GaAs–AlGaAs can also be used for HBTs. For the Si–Ge structure, the lattice mismatch is high (Ge: 5.646 Å and Si: 5.431 Å) and hence cannot be used in HBTs.

14.3.3 Important Parameters

DC current gain of HBTs can be expressed as follows:

$$\beta = \frac{I_n - I_r}{I_p + I_r + I_s} \tag{14.45}$$

where $\quad \dfrac{I_n}{I_p} = \left(\dfrac{N_e}{P_b} \right) \left(\dfrac{v_{nb}}{v_{pe}} \right) e^{\frac{\Delta E_g}{kT}} \tag{14.46}$

I_p is the reverse hole injection current, I_r the bulk recombination current, I_n the injected electron current, I_s the surface depletion region recombination current, N_e/P_b the emitter–base doping concentration ratio, ΔE_g the energy band gap ratio between the emitter and the base material, and v_{nb}/v_{pe} the ratio of electron and hole velocity.

14.3.4 High-frequency Noise Characteristics

At high frequencies, the important noise generators, in HBTs, are shot noise at the base and collector and the thermal noise due to emitter and collector series resistance as shown in Fig. 14.11. In the figure, i_{bs} is the shot noise current at the base, i_{cs} the shot noise current at the collector, e_b the thermal noise generator at the base, e_e the thermal noise generator at the emitter resistance, r_b the base resistance, and r_e the emitter resistance.

The equivalent noise resistance and noise conductance can be expressed as follows:

$$R_n = \frac{1}{2g_m} + (r_e + r_b) + G_n r_b^2 \tag{14.47}$$

$$G_n = \frac{g_m}{2}\left(\frac{1}{\beta^2} + \omega^2 \tau_{ec}^2\right) + \eta\frac{g_\pi}{2} \tag{14.48}$$

where
$$g_\pi = \frac{qI_b}{\eta kT} \tag{14.49}$$

and η is the ideality factor of the gate current (typically between 1 and 2). The minimum noise figure can be expressed as follows:

$$F_{\min} = 1 + 2R_n g_{so} + 2r\sqrt{R_n G_n} \tag{14.50}$$

where g_{so} is the optimum source conductance and r the real part of the complex correlation coefficient.

A further simplified expression of noise current can be given as follows:

$$F_{\min} = 1 + \sqrt{2}\left(\frac{f}{f_T}\right)\sqrt{g_m(r_e + r_b)} \tag{14.51}$$

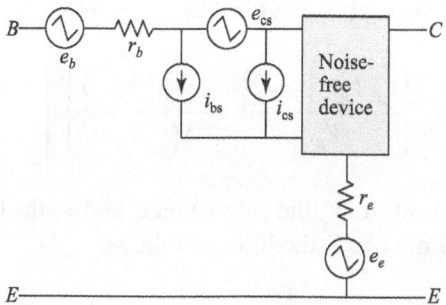

Fig. 14.11 High-frequency noise equivalent circuit of HBT

14.4 Junction Field Effect Transistors

The devices, described in earlier sections are bipolar in nature, that is, both majority and minority carriers are involved in their operation. In contrast, junction field effect transistors (JFETs) are unipolar devices where only the majority carriers take part in the operation. The following subsections describe the properties and features of JFETs.

14.4.1 Physical Structure

The physical structure and operating principle of a JFET are similar to those of its low-frequency version, as shown in Fig. 14.12. However, as before, more control over its dimensions and parasitic is required to make it suitable for operation at microwave frequencies.

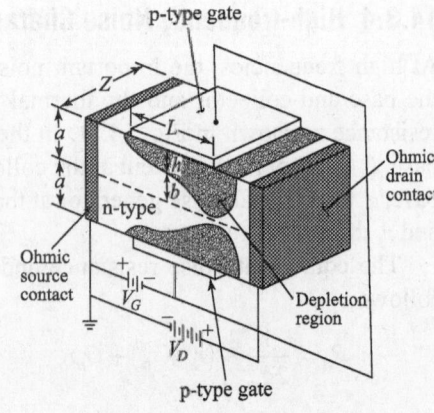

Fig. 14.12 Schematic diagram of n-channel JFET

14.4.2 Important Parameters

The pinch-off voltage of an n-channel JFET can be expressed as follows:

$$V_p = \frac{eN_d a^2}{2\varepsilon_s} \tag{14.52}$$

where e is the charge of electrons, N_d the electron concentration per cubic metres, a the height of the channel in metres, and ε_s the permittivity of the semiconductor in F/m.

The n-channel resistance of a JFET can be expressed as $R = \dfrac{L}{2e\mu_n N_d Z(a-W)}$ (14.53)

where L is the length of the channel, μ_n the mobility of electron, Z the width of the channel, and W the width of the depletion layer.

The pinch-off current can be expressed as $I_p = \dfrac{\mu_n e^2 N_d^2 Z a^3}{L\varepsilon_s}$ (14.54)

The drain current can be expressed as follows:

$$I_d = I_p \left[\frac{V_d}{V_p} - \frac{2}{3} \left(\frac{V_d + |V_g| + \psi_0}{V_p} \right)^{\frac{3}{2}} + \frac{2}{3} \left(\frac{|V_g| + \psi_0}{V_p} \right)^{\frac{3}{2}} \right] \tag{14.55}$$

where V_d is the drain voltage, V_g the gate voltage, and ψ_0 the built-in voltage. In the linear region, Eq. (14.55) modifies as follows:

$$I_d = \frac{I_p V_d}{V_p} \left[1 - \left(\frac{|V_g| + \psi_0}{V_p} \right)^{\frac{1}{2}} \right] \tag{14.56}$$

Therefore,

$$g_d = \left. \frac{\partial I_d}{\partial V_d} \right|_{V_g = \text{constant}} = \frac{I_p}{V_p} \left[1 - \left(\frac{|V_g| + \psi_0}{V_p} \right)^{\frac{1}{2}} \right] \tag{14.57}$$

and
$$g_m = \frac{\partial I_d}{\partial V_g}\bigg|_{V_d=\text{constant}} = \frac{I_p V_d}{2V_p^2}\left[1-\left(\frac{V_p}{|V_g|+\psi_0}\right)^{\frac{1}{2}}\right] \tag{14.58}$$

In the saturation region, the expression for drain current modifies as follows:

$$I_{d,\text{sat}} = I_p\left[\frac{1}{3}-\left(\frac{|V_g|+\psi_0}{V_p}\right)+\frac{2}{3}\left(\frac{|V_g|+\psi_0}{V_p}\right)^{\frac{3}{2}}\right] \tag{14.59}$$

and
$$g_m = \frac{\partial I_d}{\partial V_g}\bigg|_{V_d=\text{constant}} = \frac{I_p}{V_p}\left[1-\left(\frac{|V_g|+\psi_0}{V_p}\right)^{\frac{1}{2}}\right] \tag{14.60}$$

The capacitance between the gate and the source can be expressed as follows:

$$C_g = \frac{LZ\varepsilon_s}{2a} \tag{14.61}$$

The cut-off frequency of a JFET can be expressed as follows:

$$f_c = \frac{g_m}{2\pi C_g} = \frac{(I_p/V_p)}{2\pi C_g} \tag{14.62}$$

Substituting Eqs (14.52), (14.54), and (14.61) into Eq. (14.62), we get

$$f_c = \frac{1}{2\pi}\frac{\mu_n e^2 N_d^2 Za^3}{L\varepsilon_s}\frac{2\varepsilon_s}{eN_d a^2}\frac{2a}{LZ\varepsilon_s} = \frac{2\mu_n e N_d a^2}{\pi\varepsilon_s L^2} \tag{14.63}$$

EXAMPLE 14.2 A Si JFET has the following parameters: donor concentration = 10^{17} cm^{-3}, relative permittivity = 11.8, channel height = 0.18×10^{-4} cm, channel length = 7.5×10^{-4} cm, channel width = 48×10^{-4} cm, electron mobility = 800 cm^2/V.sec, drain voltage = 9 V, and gate voltage = -2 V. Calculate the (a) pinch-off voltage, (b) pinch-off current, (c) drain current, (d) saturation drain current, and (e) cut-off frequency, provided that the built-in voltage is 0.936 V.

Solution Given: $N_d = 10^{17}$ cm^{-3}, $\varepsilon_r = 11.8$, $a = 0.18\times10^{-4}$ cm, $L = 7.5\times10^{-4}$ cm, $Z = 48\times10^{-4}$ cm, $\mu_n = 800$ cm^2/V.sec, $V_d = 9$ V, $V_g = -2$ V, and $\psi = 0.936$ V
Therefore,

$$V_p = \frac{eN_d a^2}{2\varepsilon_s} = \frac{1.6\times10^{-19}\times10^{17}\times(0.18\times10^{-4})^2}{2\times8.8542\times10^{-14}\times11.8} = 2.4809 \text{ V}$$

$$I_p = \frac{\mu_n e^2 N_d^2 Za^3}{L\varepsilon_s} = \frac{800\times(1.6\times10^{-19}\times10^{17})^2\times48\times10^{-4}\times(0.18\times10^{-4})^3}{7.5\times10^{-4}\times8.8542\times10^{-14}\times11.8}$$

or
$$I_p = 7.3163 \text{ mA}$$

$$I_d = I_p \left[\frac{V_d}{V_p} - \frac{2}{3}\left(\frac{V_d + |V_g| + \psi_0}{V_p}\right)^{\frac{3}{2}} + \frac{2}{3}\left(\frac{|V_g| + \psi_0}{V_p}\right)^{\frac{3}{2}} \right]$$

or $\quad I_d = 7.3163\left[\dfrac{9}{2.4809} - \dfrac{2}{3}\left(\dfrac{9+2+0.936}{2.4809}\right)^{\frac{3}{2}} + \dfrac{2}{3}\left(\dfrac{2+0.936}{2.4809}\right)^{\frac{3}{2}} \right]$

or $\quad I_d = 7.3163(3.6277 - 7.0353 + 0.8583) = -18.6514 \text{ mA}$

$$I_{d,\text{sat}} = I_p \left[\frac{1}{3} - \left(\frac{|V_g| + \psi_0}{V_p}\right) + \frac{2}{3}\left(\frac{|V_g| + \psi_0}{V_p}\right)^{\frac{3}{2}} \right]$$

or $\quad I_{d,\text{sat}} = 7.3163\left[\dfrac{1}{3} - \left(\dfrac{2+0.936}{2.4809}\right) + \dfrac{2}{3}\left(\dfrac{2+0.936}{2.4809}\right)^{\frac{3}{2}} \right]$

or $\quad I_{d,\text{sat}} = 7.3163(0.3333 - 1.1834 + 0.8583) = 0.06 \text{ mA}$

$$f_c = \frac{2\mu_n e N_d a^2}{\pi \varepsilon_s L^2} = \frac{2 \times 800 \times 1.6 \times 10^{-19} \times 10^{17} \times (0.18 \times 10^{-4})^2}{\pi \times 8.8542 \times 10^{-14} \times 11.8 \times (7.5 \times 10^{-4})^2} = 4.4924 \text{ GHz}$$

Practice Problem

14.2 A Si JFET has the following parameters: donor concentration $= 1.5 \times 10^{17} \text{ cm}^{-3}$, relative permittivity $= 11.8$, channel height $= 0.21 \times 10^{-4}$ cm, channel length $= 8.5 \times 10^{-4}$ cm, channel width $= 54 \times 10^{-4}$ cm, electron mobility $= 800 \text{ cm}^2/\text{V.sec}$, drain voltage $= 12$ V, and gate voltage $= -1.5$ V. Calculate the (a) pinch-off voltage, (b) pinch-off current, (c) drain current, (d) saturation drain current, and (e) cut-off frequency, provided that the built-in voltage is 0.958 V.

$[5.0651 \text{ V}, \ 25.9486 \text{ mA}, \ -16.1011 \text{ mA}, \ 1.9046 \text{ mA}, \ 7.1409 \text{ GHz}]$

14.5 Metal–Semiconductor Field Effect Transistors

A MESFET may be constructed using either Si or GaAs as a semiconductor material. However, since the mobility of electrons in GaAs is more than that in Si, they are mainly used in fabricating microwave MESFETs. In addition, GaAs has a higher electron drift velocity and can sustain higher electric field than Si. As a result, the output power of a GaAs MESFET is also higher than that of a Si MESFET. Further, GaAs MESFETs also exhibit a lower noise figure than Si MESFETs. Hence, GaAs MESFETs are widely used in microwave integrated circuits for high-power, low-noise, and broadband amplifier designs. GaAs MESFETs work well into the X-band.

14.5.1 Physical Structure

GaAs MESFETs can be developed using either the epitaxial process or the ion implementation method. During fabrication of GaAs MESFETs using the epitaxial process, initially the GaAs substrate is doped with chromium. Since the chromium has an energy level near the centre of the band gap of GaAs, a very-high-resistivity substrate, also called a semi-insulating GaAs substrate, is developed. Next, a thin layer of n-type GaAs is grown epitaxially on this substrate to form the channel region of the n-channel MESFET. Sometimes, a buffer layer, in the form of a high-resistivity GaAs epitaxial layer, is grown in between the n-type GaAs layer and the substrate. Finally, the photolithographic process is used to define the pattern of the source, drain, and gate contacts. Generally, gold is used to make the source and drain contacts, while aluminium is used to make the gate contact.

In ion implementation, a thin n-layer is formed on the surface of the substrate by implementing Si or a donor impurity Se from column VI of the periodic table. Annealing is required to remove the radiation damage.

In either of these processes, an n^+ layer may be implemented in the source and drain regions to improve the source and drain contacts. Generally, such transistors are fabricated in bulk and then separated into individual MESFETs. Such a transistor is called a chip device, which is then alloyed to a header to provide contacts. Al or Au wires are bonded to the metalized regions to act as transistor leads. Such chip devices are called packaged transistors. Packaged and chip GaAs MESFETs are shown in Fig. 14.13. A schematic diagram of a GaAs MESFET is shown in Fig. 14.14.

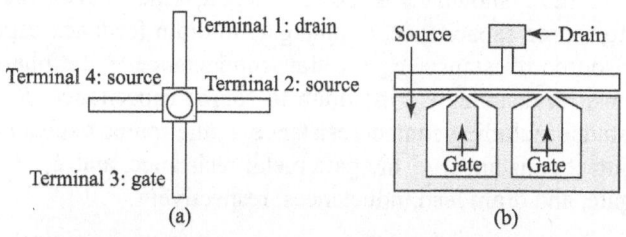

Fig. 14.13 MESFET devices (a) Packaged (b) Chip type

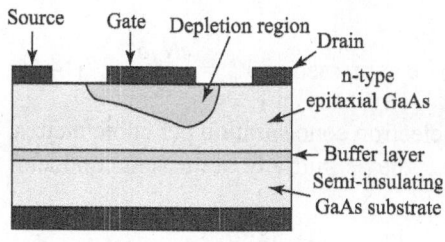

Fig. 14.14 Schematic diagram of GaAs MESFET

The n-type GaAs epitaxial film in a GaAs MESFET is about $0.15 - 0.35\,\mu m$ thick, while the semi-insulating substrate is about $100\,\mu m$ thick. The thickness of the buffer layer is about $3\,\mu m$. The n-channel layer is doped with sulphur or tin in a doping concentration between 8×10^{16} and $2 \times 10^{17}/\text{cm}^3$. This provides an electron mobility of the order of $3000 - 4500\,\text{cm}^2/\text{V}$. The doping concentration of the buffer layer

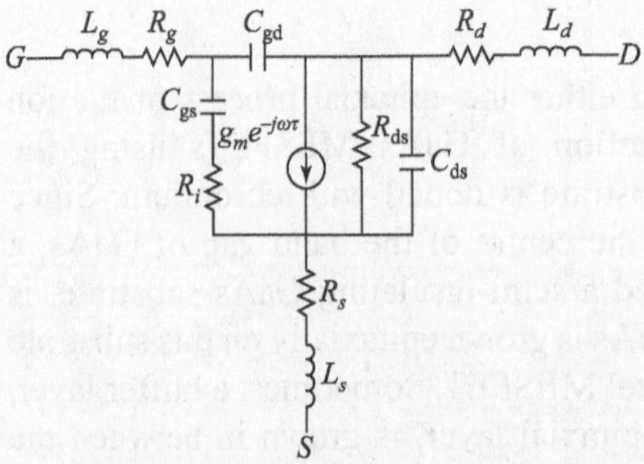

Fig. 14.15 Small signal equivalent circuit of MESFET

is about $10^{15} - 10^{16}/\text{cm}^3$. The source and drain contacts are Au–Ge, Au–Te, or Au–Te–Ge alloys, while evaporated aluminium is used to make the gate contact. A contact metallization pattern of gold is used to bring source, gate, and drain contacts out to bonding pads over semi-insulating surface.

At microwave frequencies, an MESFETs has a very short channel length and its velocity saturation occurs in the channel before reaching the pinched path. Electrical characteristics also depend on both the intrinsic (transconductance, drain conductance, input resistance, gate to source capacitance, and gate to drain capacitance) and extrinsic (load impedance, gate metallization resistance, source gate resistance, drain source capacitance, gate bonding pad parasitic capacitance, and gate bonding pad parasitic resistance) elements. A large value of the extrinsic resistance generally decreases the power gain and efficiency, while increasing the noise figure. Values of these elements depend on the material, structure, dimension, and channel type. In the fabrication of an MESFET, high channel doping is often preferred, because it decreases the effect of gate to drain capacitance and increases transconductance and open-circuit voltage gain. However, it also has a drawback. A high doping concentration is associated with a low breakdown voltage of the gate. A doping concentration of $10^{18}/\text{cm}^3$ generally sets the upper limit. A small signal equivalent circuit of a MESFET is shown in Fig. 14.15, where R_i is the input channel resistance, C_{gs} the gate to source capacitance, C_{gd} the gate to drain feedback capacitance, R_{ds} the drain to source resistance, g_m the transconductance, τ the phase delay due to carrier transit in channel, C_{ds} the drain to source capacitance, R_d the drain to channel resistance including contact resistance, R_s the source to channel resistance including contact resistance, R_g the gate metal resistance, and L_s, L_g, and L_d are the source, gate, and drain lead inductances, respectively.

Note An MESFET is also called a Schottky barrier gate FET.

14.5.2 Important Parameters

The pinch-off voltage of an MESFET can be expressed as $V_p = \dfrac{eN_d a^2}{2\varepsilon_s}$ (14.64)

where e is the charge of electrons, N_d the electron concentration per cubic metres, a the height of the channel in metres, and ε_s the permittivity of the semiconductor in F/m.

The pinch-off current can be expressed as $I_p = \dfrac{eN_d \mu a Z V_p}{3L}$ (14.65) where μ is the low field mobility in m^2/V, Z the channel depth or width, and L the gate length.

The drain current can be expressed as follows:

$$I_d = I_p \left| \frac{3\left(u^2 - \rho^2\right) - 2\left(u^3 - \rho^3\right)}{1 + \eta\left(u^2 - \rho^2\right)} \right| \tag{14.66}$$

where

$$u = \sqrt{\left(V_d + |V_g|\right)/V_p} \tag{14.67}$$

$$\rho = \sqrt{|V_g|/V_p} \tag{14.68}$$

$$\eta = \frac{\mu|V_p|}{v_s L} = \frac{v}{v_s} \tag{14.69}$$

$$v = \mu E_x \Big/ \left(1 + \frac{\mu E_x}{v_s}\right) \tag{14.70}$$

v_s is the saturation drift velocity, and E_x the absolute value of electric field in channel.

In general, the drain current and mutual conductance of a GaAs MESFET can be written as follows:

$$I_{ds} = I_{dss} \left(1 + \frac{|V_g|}{V_p}\right)^2 \tag{14.71}$$

and

$$g_m = \frac{2I_{dss}}{|V_p|} \left(1 + \frac{|V_g|}{V_p}\right) \tag{14.72}$$

The cut-off frequency of an MESFET depends on the way the transistor has been manufactured. For a wideband lumped equivalent circuit, it can be expressed as follows:

$$f_c = \frac{g_m}{2\pi C_{gs}} = \frac{v_s}{4\pi L} \tag{14.73}$$

where C_{gs} is the gate to source capacitance and g_m the transconductance.

The maximum frequency of oscillation can be expressed as follows:

$$f_{max} = \frac{f_c \sqrt{g_m R_d}}{2} = \frac{f_c}{2} \sqrt{\frac{\mu E_p \left(u_m - \rho\right)}{v_s \left(1 - u_m\right)}} \tag{14.74}$$

where R_d is the drain resistance, E_p the electric field at the pinch-off in the channel, and u_m the saturation normalization of u.

The highest frequency of oscillation for the maximum power gain, with both input and output being matched, is given by the following equation:

$$f_{max} = \frac{f_c}{2} \sqrt{\frac{R_d}{R_s + R_g + R_i}} \tag{14.75}$$

where R_s is the source resistance, R_g the gate metallization resistance, and R_i the input resistance.

14.5.3 High-frequency Noise Characteristics

The minimum noise figure of an MESFET can be expressed as follows:

$$F_{min} = 1 + \frac{2\pi C_{gs} k_1 f}{g_m} \sqrt{g_m (R_s + R_g)} \qquad (14.76)$$

The equivalent noise resistance can be expressed as follows:

$$R_n = k_2 / g_m \qquad (14.77)$$

The optimum resistance can be expressed as follows:

$$R_{opt} = k_3 \left(R_s + R_g + \frac{1}{4g_m} \right) \qquad (14.78)$$

The optimum reactance can be expressed as follows:

$$X_{opt} = \frac{k_4}{f C_{gs}} \qquad (14.79)$$

where

$$k_1 = 2 \left(\sqrt{P} - \frac{C_{gs} \sqrt{R}}{C_{11}} \right) \qquad (14.80)$$

$$C_{11} = C_{gs} + C_{gs1} \qquad (14.81)$$

P and R are the numerical factors associated with the drain and gate thermal noise generators, C_{gs} the intrinsic gate capacitance, C_{gs1} the parasitic capacitance from gate to source, and k_2, k_3, and k_4 the numerical fitting factors.

EXAMPLE 14.3 A GaAs MESFET has the following parameters: $R_g = 2.8\,\Omega$, $R_i = 2.45\,\Omega$, $g_m = 54$ m℧, $R_d = 500\,\Omega$, $R_s = 2.45\,\Omega$, and $C_{gs} = 0.5$ pF, Calculate the (a) cut-off frequency and (b) maximum operating frequency.

Solution Given: $R_g = 2.8\,\Omega$, $R_i = 2.45\,\Omega$, $g_m = 54$ m℧, $R_d = 500\,\Omega$, $R_s = 2.45\,\Omega$, and $C_{gs} = 0.5$ pF

(a) The cut-off frequency will be as follows:

$$f_c = \frac{g_m}{2\pi C_{gs}} = \frac{54 \times 10^{-3}}{2\pi \times 0.5 \times 10^{-12}} = 17.1887 \text{ GHz}$$

(b) The maximum operating frequency will be as follows:

$$f_{max} = \frac{f_c}{2} \sqrt{\frac{R_d}{R_s + R_g + R_i}} = \frac{17.1887}{2} \sqrt{\frac{500}{2.45 + 2.8 + 2.45}} = 69.2554 \text{ GHz}$$

Practice Problem
14.3 A GaAs MESFET has the following parameters: $R_g = 3\,\Omega$, $R_i = 2.7\,\Omega$, $g_m = 45$ m℧, $R_d = 500\,\Omega$, $R_s = 2.7\,\Omega$, and $C_{gs} = 0.5$ pF, Calculate the (a) cut-off frequency and (b) maximum operating frequency. **[14.3239 GHz, 55.2558 GHz]**

14.6 High-electron-mobility Transistors

A study on the electron mobility in the GaAs MESFET channel shows that it lies in the range 4000–5000 cm^2/V.s at room temperature for a typical donor concentration of 10^{17} cm^{-3}. Even at 77°K, the mobility does not increase considerably because of ionized impurity scattering. However, in an undoped GaAs sample, at 77°K, the electron mobility increases to $2 \times 10^5 - 3 \times 10^5$ cm^2/V.s. This implies that electrons move faster in a dopant-free region and high-electron-mobility transistors (HEMT) use this fact for their operation.

In practice HEMTs are modulation-doped heterojunction structures that have high electron mobility, shorter gate length, and reduced source and gate contact resistance over FETs. They exhibit lower noise figure and higher gain upto 70 GHz. It is also possible to design HEMT amplifiers that are capable of operating even at a higher microwave frequency. The following subsections describe the properties and features of HEMTs.

14.6.1 Physical Structure

In ordinary MESFET devices, the threshold voltage depends on the doping level, which also determines the carrier mobility. Therefore, the threshold voltage and carrier mobility are coupled to each other. If we want to increase the mobility we must decrease the threshold level. However, we cannot adjust the threshold voltage to arbitrary values without degrading high-frequency performances. Alternatively, we can say that maximization of the high-frequency performance may require an inconvenient or unusable threshold voltage. To decouple these two parameters, one requires a heterojunction between an appropriately doped wide-band-gap material and an undoped low-band-gap material. This attempts lead to the development of an HEMT, as shown in Fig. 14.16(a). HEMTs were developed in 1980 based on the work of the tunnel diode inventor Leo Esaki and the fact that carriers drift fastest in dopant-free regions.

An HEMT or a heterojunction field effect transistor (HFET) is, a selectively doped GaAs–AlGaAs heterojunction structure. In such a structure, an undoped GaAs layer and a Si-doped n-type AlGaAs layer are successively grown on a semi-insulating GaAs substrate using epitaxial growth technologies such as MBE

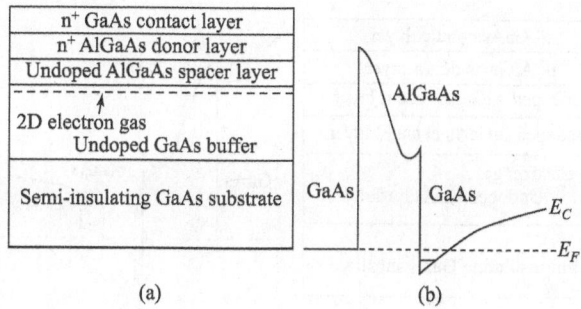

Fig. 14.16 HEMT (a) Schematic structure
(b) Energy-band diagram

or MOCVD. In addition, a buffer layer is also created between the undoped GaAs layer and the semi-insulating surface.

GaAs has a band gap of 1.42 eV, whereas the highly doped n-layer AlGaAs has a band gap of 2.1 eV. The difference in the Fermi energy band between the two materials causes band bending at the heterojunction interface, as shown in Fig. 14.16(b). This bending, in turn, results in a quantum well. Thus, when the junctions are formed, electrons from the n-layer AlGaAs move to the conduction band of the GaAs and are trapped in the potential well. The GaAs layer thus has a high density of electrons and forms a highly conducting channel between the source and the drain. These electrons are confined to the interface of the heterojunction by the attraction of the positive donor ions of AlGaAs, as shown in Fig. 14.17, and form a 2D electron gas. Since the donors and free electrons are in two different media, chances of collision between them are minimized and the drift velocity of the electrons increases. These electrons can easily be modulated by the application of a gate voltage.

In addition to GaAs–AlGaAs, AlGaAs–InGaAs–GaAs double heterojunctions are also used in fabricating HFETs. Such HFETs are called pseudomorphic HEMTs or p-HEMTs. The layer sequence of such a transistor is shown in Fig. 14.18. InGaAs has a lower energy band gap than AlGaAs and GaAs. Thus, when InGaAs is sandwiched between two higher-band-gap semiconductors (AlGaAs and GaAs), the lowest-energy quantum well state resides in the InGaAs. Electrons, provided by the donor atoms, move to this thin layer of InGaAs and get trapped in the quantum well. The p-HEMTs have superior microwave properties

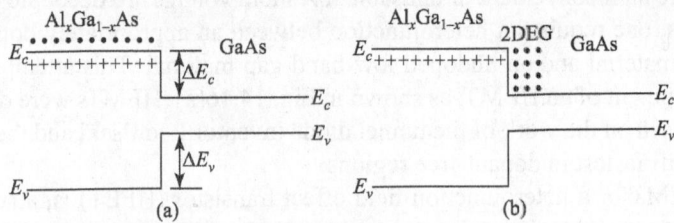

Fig. 14.17 Simplified energy-band diagram of heterojunction between doped $Al_xGa_{1-x}As$ and undoped GaAs (a) Before transfer of free electrons from $Al_xGa_{1-x}As$ to GaAs (b) After transfer of free electrons from $Al_xGa_{1-x}As$ to GaAs

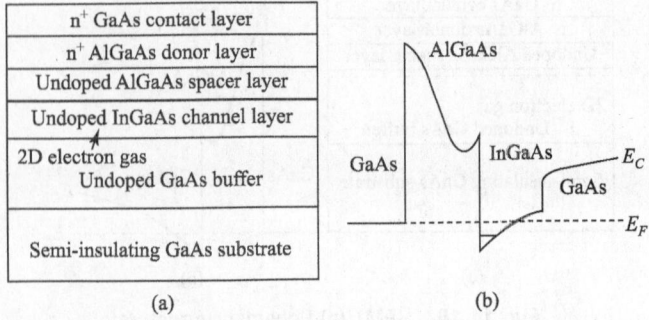

Fig. 14.18 Pseudomorphic HEMT (a) Schematic structure (b) Energy-band diagram

than GaAs HEMTs and hence are rapidly replacing GaAs HEMT devices in many applications.

InP can also be used along with GaInAs and AlInAs to develop HEMTs. Such GaInAs–AlInAs–InP structures also offer many advantages over the conventional GaAs HEMT structures. Here, high-band-gap AlInAs is used as a donor layer. The high conduction band discontinuity of the structure results in a high 2D electron gas concentration, which, in turn, lowers the source resistance and associated thermal noise. High mobility of the electrons in a GaInAs semiconductor, along with the high density of electrons in the channel, results in high channel conductivity. Further, due to a lower inter-valley electron transfer probability, the velocity of the electrons in the channel is also higher and leads to a higher transit frequency (f_T) of the devices. In addition, since the capacitive coupling between the channel and the gate is lower, coupling of noise to the gate is also minimized. InP HEMTs exhibit very high transconductance, a very low noise figure, and a high gain than conventional GaAs HEMTs.

HEMTs can operate in the enhancement mode (E-HEMT) or depletion mode (D-HEMT), and can be fabricated using the integrated circuit technique. The main steps of such fabrication techniques are as follows: (a) Ohmic contact formation, (b) opening gate windows, (c) selective dry etching, (d) gate metallization, and (e) interconnect metallization.

HEMTs can operate as amplifiers up to a frequency of 70 GHz with a low noise figure and at a high speed, and exhibit higher transconductance, lower noise figure properties, and higher frequency of operation compared to GaAs MOSFETs. The high-frequency equivalent circuits of an HEMT and a p-HEMT are similar to that of a GaAs MESFET and are shown in Fig. 14.19.

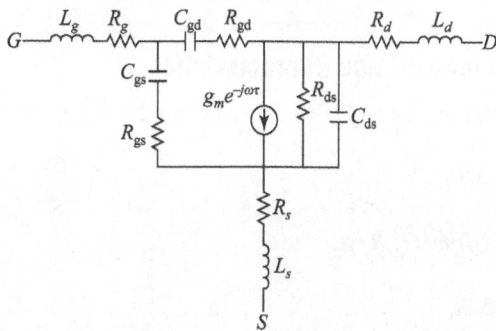

Fig. 14.19 High-frequency equivalent circuit of HEMT and p-HEMT

14.6.2 Important Parameters

The drain current in an HEMT can be expressed as $I_{ds} = qn(z)Wv(z)$ (14.82)

where q is the charge of electrons, $n(z)$ the concentration of 2D electron gas, W the gate width, and $v(z)$ the electron velocity.

The maximum possible frequency of operation is expressed as follows:

$$f_{max} = \frac{v}{2\pi L} \qquad\qquad (14.83)$$

where v is the velocity of the carrier and L is the gate length.

The vertical threshold sensitivity, defined by the differential threshold voltage to the thickness of the $Al_xGa_{1-x}As$ layer, can be expressed as follows:

$$\frac{dV_{th}}{dl} = -\sqrt{\frac{2qN_d\left(\psi_{ms} - \Delta E_c - V_{th}\right)}{\varepsilon}} \tag{14.84}$$

where V_{th} is the threshold voltage in Volt, l the AlGaAs layer thickness in metres, q the charge of electrons, N_d the donor concentration, ψ_{ms} the metal–semiconductor Schottky barrier potential between GaAs and AlGaAs, ΔE_c the conduction band edge difference between GaAs and AlGaAs, and ε the permittivity of AlGaAs.

EXAMPLE 14.4 An HEMT has the following parameters: gate width $= 125\times10^{-6}$ m, electron velocity $= 2\times10^5$ m/sec, and 2D electron gas density $= 5\times10^{15}$ m^{-2}. Calculate the drain current.

Solution Given: $W = 125\times10^{-6}$ m, $v(z) = 2\times10^5$ m/sec, and $n(z) = 5\times10^{15}$ m^{-2}

Therefore,

$$I_{ds} = qn(z)Wv(z) = 1.6\times10^{-19} \times 5\times10^{15} \times 125\times10^{-6} \times 2\times10^5 = 20 \text{ mA}$$

Practice Problem

14.4 An HEMT has the following parameters: gate width $= 175\times10^{-6}$ m, electron velocity $= 2\times10^5$ m/sec, and 2D electron gas density $= 4.5\times10^{15}$ m^{-2}. Calculate the drain current. **[25.20 mA]**

14.6.3 High-frequency Noise Characteristics

The noise generators for an HEMT device can be represented as follows:

$$\overline{i}^2_{nd} = 4kT\Delta fPg_m \tag{14.85}$$

$$\overline{i}^2_{ng} = 4kT\Delta fP\omega^2C^2_{gs}R/g_m \tag{14.86}$$

$$\overline{V}^2_{ns} = 4kT\Delta fR_s \tag{14.87}$$

and $\quad \overline{V}^2_{ng} = 4kT\Delta fR_g \tag{14.88}$

The minimum noise figure can be expressed as follows:

$$F_{min} = 1 + \frac{2f}{f_T}\sqrt{Pg_m\left(R_s + R_g\right)}\left(1 - \frac{C_{gs}}{C_{11}}\sqrt{\frac{R}{P}}\right) \tag{14.89}$$

where $\quad f_T = \frac{g_m}{2\pi C_{gs}} \tag{14.90}$

$$C_{11} = C_{gs} + C_{gs1} \tag{14.91}$$

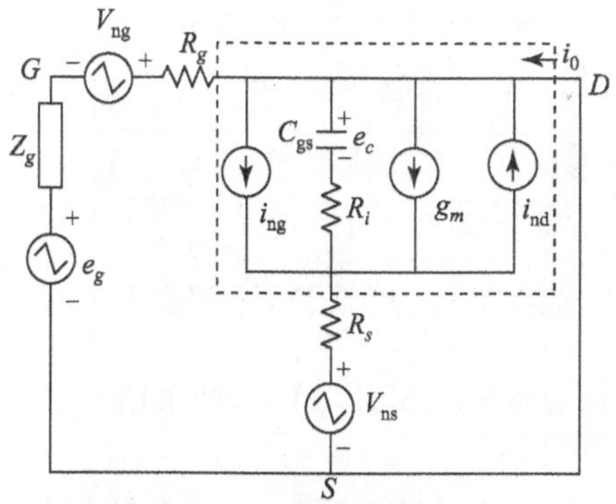

Fig. 14.20 High-frequency noise equivalent circuit of FET

P and R are the numerical factors associated with the drain and gate thermal noise generators, R_g the gate resistance, R_s the source resistance, C_{gs} the intrinsic gate capacitance, and C_{gsl} the parasitic capacitance from gate to source. The high-frequency equivalent circuit is shown in Fig. 14.20.

Electrical characteristics The electrical characteristics of an HEMT are as follows:

1. Frequency range: 1–60 GHz
2. Output power: few watts/mm
3. Gain: 9.1 dB

Note HEMTs are also known as modulation doped field effect transistor (MODFET), two-dimensional electron gas FET (TDEGFET or 2DFET), or selectively doped FET (SDFET).

14.7 Metal-oxide–Semiconductor Field Effect Transistors

Metal-oxide–semiconductor field effect transistors (MOSFETs) are four-terminal devices with the terminals denoted by source, gate, drain, and substrate. Such structures are simple and cost-effective. The following subsections describe the properties and features of MOSFETs.

14.7.1 Physical Structure

The physical structure of an MOSFET is similar to that of its low-frequency counterpart, except that more control over its dimensions and parasitic is required to make it suitable for operation at microwave frequencies. A typical MESFET structure is shown in Fig. 14.21.

Depending on the channel type, MOSFETs can be classified as n-channel (or NMOS) and p-channel (or PMOS) transistors. MOSFET structures can also be classified as enhancement- and depletion-mode structures. In the active region of an enhancement-mode MOSFET, the input capacitance and transconductance are almost independent of the gate voltage, as well as the output capacitance is independent of the drain voltage. Thus, it can be used in the design of a Class A power amplifier. Further, since MOSFETs can be used both in enhancement and depletion modes, the active gate voltage range is very large (from negative to positive). Schematic diagrams of depletion- and enhancement-type MOSFETs are given in Fig. 14.22.

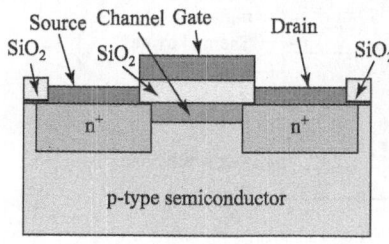

Fig. 14.21 Schematic diagram of metal oxide semiconductor field effect transistor

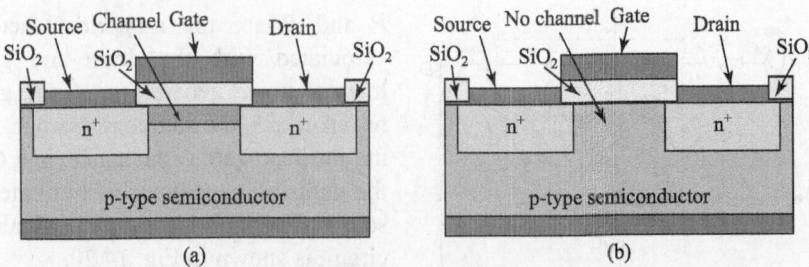

Fig. 14.22 Schematic diagram of various type of MOSFETs (a) Depletion type (b) Enhancement type

In addition to NMOS and PMOS, complementary MOSFETs (or CMOSs) also exist. A CMOS provides n- and p-channel MOSFETs on the same strip. CMOS devices have three structures: (a) n-tub, (b) p-tub, and (c) twin tub. When a tub is formed in a p-type substrate then it is called an n-tub. Similarly, a p-tub can be defined. If an n-tub and a p-tub are combined on the same substrate, then it results in a twin tub. Schematic diagrams of the n-tub, p-tub, and twin-tub CMOS structures are shown in Fig. 14.23.

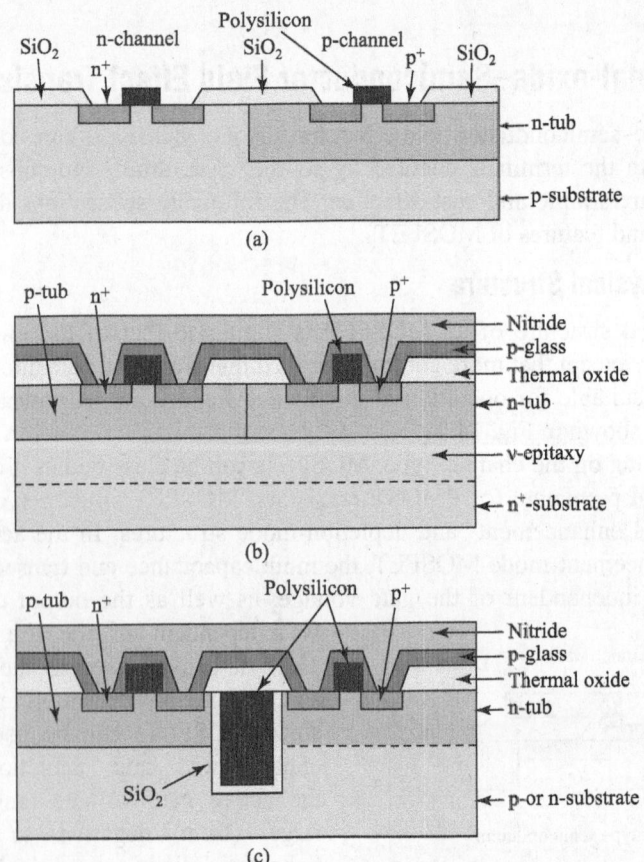

Fig. 14.23 Complementary metal oxide semiconductor structures (a) n-tub (b) p-tub (c) Twin tub

Since the drain, source, and channel of a MOS structure are surrounded by a depletion region, there is no need to isolate individual components separately. This allows the elimination of isolation region and a higher packaging density on a semiconductor chip than is possible with BJTs. The fabrication technology of NMOS is also much simpler than that of BJTs. CMOS circuits have lower power dissipation than both BJT and NMOS circuits.

14.7.2 Important Parameters

The drain current can be expressed as follows:

$$I_d = \frac{Z}{L}\mu_n C_i \left[\left(V_g - 2\psi_b - \frac{V_d}{2} \right) V_d - \frac{2}{3C_i}(2\varepsilon_s q N_a)^{\frac{1}{2}} \left\{ (V_d + 2\psi_b)^{\frac{1}{2}} - (2\psi_b)^{\frac{1}{2}} \right\} \right]$$

(14.92)

where μ_n is the electron mobility, C_i the insulator capacitance per unit area, V_g the gate voltage, ψ_b the potential difference between Fermi level and intrinsic Fermi level, V_d the drain voltage, ε_s the semiconductor permittivity, q the carrier charge, and N_a the acceptor concentration.

In the linear region, Eq. (14.92) modifies as follows:

$$I_d \doteq \frac{Z}{L}\mu_n C_i \left[(V_g - V_{th})V_d - \left(\frac{1}{2} + \frac{\sqrt{\varepsilon_s q N_a / \psi_b}}{4C_i} \right) \right]$$

or
$$I_d \simeq \frac{Z}{L}\mu_n C_i (V_g - V_{th})V_d$$

(14.93)

where
$$V_{th} = 2\psi_b + \frac{2}{C_i}\sqrt{\varepsilon_s q N_a \psi_b}$$

(14.94)

Transconductance in the linear region, therefore, becomes the following:

$$g_m = \frac{\partial I_d}{\partial V_g}\bigg|_{V_d = \text{constant}} = \frac{Z}{L}\mu_n C_i V_d$$

(14.95)

In the saturation region, the drain current can be expressed as follows:

$$I_{d,\text{sat}} \simeq \frac{mZ}{L}\mu_n C_i (V_g - V_{th})^2$$

(14.96)

where $m = 0.5$ is the low doping factor.
Transconductance in the saturation region, therefore, becomes the following:

$$g_m = \frac{\partial I_d}{\partial V_g}\bigg|_{V_d = \text{constant}} = \frac{2mZ}{L}\mu_n C_i (V_g - V_{th})$$

(14.97)

These equations are valid for an idealized n-channel MESFET. For a real n-channel MESFET, the saturation drain current becomes

$$I_{d,\text{sat}} \simeq ZC_i (V_g - V_{th})v_s$$

(14.98)

and the transconductance becomes

$$g_{m,\text{sat}} = \frac{\partial I_d}{\partial V_g}\bigg|_{V_d = \text{constant}} = ZC_i v_s$$

(14.99)

where v_s is the carrier drift velocity.

The threshold voltage can be expressed as follows:

$$V_{th} = \frac{\phi_{ms}}{q} - \frac{Q_f}{C_i} + 2\psi_b + \frac{2}{C_i}\sqrt{\varepsilon_s q N_a \psi_b} \tag{14.100}$$

where ϕ_{ms} is the difference between the metal and the semiconductor work function (in eV) and Q_f is the fixed oxide charge.

The maximum operating frequency of an MOSFET in the linear region can be expressed as follows:

$$f_{max} = \frac{g_m}{2\pi C_m} = \frac{\mu_n V_d}{2\pi L^2} = \frac{v_s}{2\pi L} \tag{14.101}$$

EXAMPLE 14.5 An n-channel MOSFET has the following parameters: channel length $= 3.8\times10^{-6}$ m, channel depth $= 10^{-5}$ m, insulator thickness $= 0.06\times10^{-6}$ m, gate voltage $= 6$ V, doping factor $= 1$, threshold voltage $= 0.12$ V, electron mobility $= 1350\times10^{-4}$ m^2/V.sec, and electron velocity $= 2\times10^5$ m/sec. Calculate the (a) saturation drain current, (b) normal transconductance, (c) saturation transconductance, and (d) maximum operating frequency. The dielectric constant of SiO_2 is 4.

Solution Given $L = 3.8\times10^{-6}$ m, $Z = 10^{-5}$ m, $d = 0.06\times10^{-6}$ m, $V_g = 6$ V, $m = 1$, $V_{th} = 0.12$ V, $\mu_n = 1350\times10^{-4}$ m^2/V.sec, $v_s = 2\times10^5$ m/sec, and $\varepsilon_{r,ins} = 4$

Therefore,

$$C_i = \frac{\varepsilon_i}{d} = \frac{4\times8.8542\times10^{-12}}{0.06\times10^{-6}} = 5.9028\times10^{-4} \text{ F/m}^2$$

$$I_{d,sat} \simeq ZC_i(V_g - V_{th})v_s = 10^{-5}\times5.9028\times10^{-4}\times(6-0.12)\times2\times10^5$$

or $\quad I_{d,sat} = 6.9417$ mA

$$g_m = \frac{2mZ}{L}\mu_n C_i(V_g - V_{th})$$

or $\quad g_m = \dfrac{2\times1\times10^{-5}\times1350\times10^{-4}\times5.9028\times10^{-4}\times(6-0.12)}{3.8\times10^{-6}} = 2.4661$ m℧

$$g_{m,sat} = ZC_i v_s = 10^{-5}\times5.9028\times10^{-4}\times2\times10^5 = 1.1806 \text{ m℧}$$

$$f_{max} = \frac{v_s}{2\pi L} = \frac{2\times10^5}{2\pi\times3.8\times10^{-6}} = 8.3766 \text{ GHz}$$

Practice Problem

14.5 An n-channel MOSFET has the following parameters: channel length $= 4.25\times10^{-6}$ m, channel depth $= 15\times10^{-6}$ m, insulator thickness $= 0.04\times10^{-6}$ m, gate voltage $= 7.5$ V, doping factor $= 1$, threshold voltage $= 0.15$ V, electron mobility $= 1350\times10^{-4}$ m^2/V.sec, and electron velocity $= 2\times10^5$ m/sec. Calculate the (a) saturation drain current, (b) normal transconductance, (c) saturation transconductance, and (d) maximum operating frequency. The dielectric constant of SiO_2 is 4.

$$\left[8.8542\times10^{-4} \text{ F/m}^2, \ 19.5235\text{ mA}, \ 6.2016\text{ m℧}, \ 2.6563\text{ m℧}, \ 7.4896\text{ GHz}\right]$$

14.8 Microwave Amplification by Stimulated Emission of Radiation

The principle of microwave amplification by stimulated emission of radiation (MASER) was developed by Townes in 1954. The basic principle is the excitation of atoms at the ground state (E_0) to a higher energy state (E_2) with the help of a pump source. Under such pumping, the number of atoms at the higher energy state will, at a certain time, be equal to or greater than those in the ground state. This is called population inversion. When these excited electrons simultaneously jump to the ground state, they release the amount of energy they absorbed from the pump source. Now, if the pump frequency is equal to the energy of transition

$$v = (E_2 - E_0)/h \tag{14.102}$$

then the energy will add up and microwave amplification takes place. Such MASERs are called two-level MASERs. Here, h is the Planck's constant.

Another process also takes place, where, instead of jumping to the ground level directly, atoms in the energy level E_2 give up some energy and come to an intermediate energy level (E_1), called the metastable state. Now, from this metastable state they jump back to the ground state, causing microwave amplification. Such MASERs are called three-level MASERs. In practice, most of the atoms follow the second process and are stimulated at the presence of a cavity surrounding the MASER material.

MASERs are widely used in radio telescopes, space probe receivers, and low-noise low-level amplifiers.

14.8.1 Ammonia MASER

Gordon and Jeiger developed ammonia MASER using ammonia gas as a working material. The main advantage of using ammonia is that no separate pumping source is required, because, according to the Boltzmann distribution, ammonia molecules are distributed in two energy levels E_0 and E_1. Ammonia molecules at energy level E_1 are focused into a microwave resonator cavity, having a resonant frequency equal to the natural resonance frequency of ammonia, with the help of a DC magnetic field. These focused ammonia molecules then jump to the lower energy level inside the cavity and release energy. On coupling with this energy, the signal gets amplified. After transition, low-energy ammonia molecules are pumped out of the resonant cavity. It may be noted that the amplification is regenerative because the amplified signal energy stimulates further ammonia molecules to higher levels.

Ammonia MASER is generally operated with beam strength and cavity coupling conditions. If the gain is increased to infinity, the cavity breaks into oscillation. Such oscillation is very stable and of pure spectral density. The frequency stability is of the order of 1 part in 10^6 order in a minute and hence the stability of oscillation or frequency of ammonia MASER can be compared to that of an atomic clock. Thermal noise can be reduced by putting the whole system in a liquid helium bath.

Ammonia MASER has the advantages of low noise, high frequency stability, and spectral purity. However, it suffers from the disadvantages of a low output power, small bandwidth, and lack of tuning facility. Therefore, its use is limited in microwave applications.

14.8.2 Ruby MASER

Instead of ammonia, ruby can also be used as a working element in MASER. Ruby is a crystal of silica with a slight natural doping of chromium. The presence of chromium makes ruby paramagnetic and hence when it is exposed to the input signal under DC magnetic biasing, Zeeman splitting of its energy levels occurs. If the difference between the two levels is 10^{-5} eV, then by applying a suitable microwave signal of the same frequency, signal absorption can be observed.

A ruby MASER is a three-level MASER and its energy-band diagram is shown in Fig. 14.24. Under a proper pumping signal, the chromium in a ruby lattice is moved to the upper most level and then settled in the middle level after losing a certain amount of energy. When sufficient atoms are present in this metastable state, population inversion occurs and the atoms jump simultaneously to the lower level, releasing energy and thereby causing signal amplification. By varying the DC magnetic field, energy levels of the ferrochromium atoms can be adjusted, and therefore it can be used to tune the MASER.

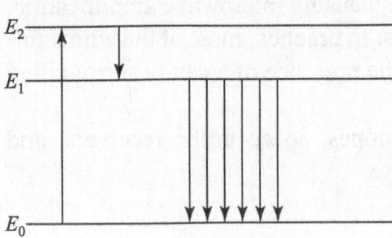

Fig. 14.24 Energy level in ruby MASER

The basic construction of a ruby MASER is shown in Fig. 14.25. Here, a circulator has been used to separate the output signal from the input signal, although it introduces some loss.

Like an ammonia MASER, a ruby MASER is also of narrow bandwidth and governed by the cavity. Although the bandwidth can be increased by compromising with the gain but still it is limited. This problem can be solved by the use of a travelling-wave MASER. In such MASERs, a slow-wave structure of ruby is made and the signal is allowed to pass through it and get amplified at the expense of the pump signal. Traveling-wave MASERs have greater bandwidth and greater stability, and are easily tunable. In addition, they do not require any circulator. Hence, the loss is also small.

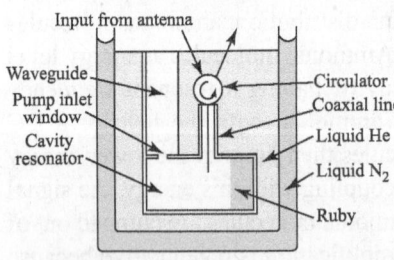

Fig. 14.25 Schematic diagram of ruby MASER

Important Formulae

- The maximum frequency at which the power gain of a BJT has dropped to unity can be expressed as $f_{\max} = f_T/(8\pi R_b C_c)$.

- The cut-off frequency of a JFET can be expressed as $f_c = \dfrac{g_m}{2\pi C_g} = \dfrac{2\mu_n q N_d a^2}{\pi \varepsilon_s L^2}$.

- The pinch-off voltage of an MESFET can be expressed as $V_p = qN_d a^2 / (2\varepsilon_s)$.

- The pinch-off current of an MESFET can be expressed as $I_p = qN_d \mu a Z V_p / (3L)$.

- The cut-off frequency of an MESFET can be expressed as $f_c = \dfrac{g_m}{2\pi C_{gs}} = \dfrac{v_s}{4\pi L}$.

- The maximum frequency of oscillation of an MESFET can be expressed as $f_{max} = f_c \sqrt{g_m R_d} / 2$.

- The highest frequency of oscillation of an MESFET for the maximum power gain, with both input and output being matched, is given by $f_{max} = \dfrac{f_c}{2} \sqrt{\dfrac{R_d}{R_s + R_g + R_i}}$.

- The maximum operating frequency of an MOSFET in the linear region can be expressed as $f_{max} = \dfrac{g_m}{2\pi C_m} = \dfrac{\mu_n V_d}{2\pi L^2} = \dfrac{v_s}{2\pi L}$.

Exercises

Objective-type Questions

14.1 Using a travelling wave structure in a MASER, we can
(a) increase the bandwidth of the operation
(b) increase the power output
(c) stabilize the frequency of operation
(d) minimize the noise level and distortion

14.2 A Schottky diode clamp is used along with a BJT for
(a) reducing the power dissipation
(b) reducing the switching time
(c) increasing the value of β
(d) reducing the base current

14.3 A ruby MASER requires to be cooled
(a) because it generates a lot of heat
(b) because it cannot operate at room temperature
(c) to improve bandwidth
(d) to improve noise figure

14.4 A solid-state MASER is an amplifier of which of the following type?
(a) diamagnetic
(b) paramagnetic
(c) ferromagnetic
(d) electromagnetic

14.5 A ruby MASER is preferred to an ammonia MASER for microwave amplification because it has
(a) a much greater bandwidth
(b) a low noise figure
(c) better frequency stability
(d) no need for circulator

14.6 A MASER RF amplifier is not truly suitable for use in
(a) radars
(b) radio astronomy
(c) satellite communication
(d) troposcatter receivers

14.7 A magnetic field is used in a ruby MASER
(a) to provide sharp focusing of the electron beam
(b) as an energy state separator or focuser
(c) as a resonant cavity and vacuum pump
(d) all of these

14.8 A MASER that uses an ammonia beam consists of
    (a) a beam collector
    (b) an energy state separator or focuser
    (c) a resonant cavity and a vacuum pump
    (d) all of these

14.9 In the X-band, for the best low-level noise performance, an amplifier should use
    (a) a BJT   (c) an IMPATT diode
    (b) an FET   (d) a Gunn diode

14.10 Ordinary transistors cannot be used at high frequencies because, at these frequencies, the inter-electrode capacitance
    (a) behaves as a short circuit   (c) behaves as a complex load
    (b) behaves as an open circuit   (d) is matched

14.11 HBTs exhibit
    (a) very low base resistance
    (b) high current gain
    (c) an increase in speed by a factor of 2 or 3
    (b) all of these

14.12 In creating HBTs, the lattice constant of two semiconductor materials
    (a) must be matched   (c) both (a) and (b)
    (b) may mismatch by a small   (d) must mismatch by a large
        amount   amount

14.13 Compared to Si BJTs, HBTs have
    (a) low gain bandwidth product and reduced parasitic
    (b) high gain bandwidth product and higher parasitic
    (c) low gain bandwidth product and higher parasitic
    (d) high gain bandwidth product and reduced parasitic

14.14 Which of the following junctions is not used to construct a transistor?
    (a) GaAs–AlGaAs   (c) Si–Ge
    (b) AlGaAs–InGaAs–GaAs   (d) GaInAs–AlInAs–InP

14.15 HEMTs are also known as
    (a) MODFETs   (c) SDFETs
    (b) TDEGFETs or 2DFETs   (d) all of these

Review Questions

14.1 Draw and explain the high-frequency equivalent circuit of a BJT.
14.2 With suitable figures, describe different surface geometries of high-frequency BJTs.
14.3 What is meant by heterojunction? With a suitable figure, describe the operating principle of a heterojunction bipolar transistor.
14.4 Describe the physical structure of an MESFET
14.5 Draw and explain the high-frequency equivalent circuit of an MESFET.
14.6 Write a short note on HEMT.
14.7 Write a short note on (a) MOSFET, (b) CMOS, (c) ammonia MASER, and (d) ruby MASER.

Problems

14.1 An n-Ge–p-GaAs–n-GaAs heterojunction transistor has cross-sectional area = 1.5×10^{-2} cm^2, hole diffusion coefficient = 10.40 cm^2/sec, minority hole density = 45 cm^{-3}, and hole diffusion length = 7.9×10^{-3} cm. If the bias voltage at the emitter junction is 0.95 V and threshold voltage is 0.026 V, then calculate the junction current. *[Hint: Use Eq. (14.44)]*

14.2 A Si BJT has a maximum electric field of 2.5×10^5 V/cm, emitter collector length of 4 μm, and saturated drift velocity of $v_s = 5.03 \times 10^6$ cm/sec. Calculate the (a) maximum allowable applied voltage, (b) transit time of a charge to travel the emitter–collector length, and (c) maximum possible transit frequency. *[Hint: Use Eqs (14.18)–(14.21)]*

14.3 A n-channel GaAs MESFET has electron concentration = 7.5×10^{17} cm^{-3}, channel height = 0.12×10^{-4} cm, relative permittivity = 13.1, channel length = 15×10^{-4} cm, channel width = 40×10^{-4} cm, electron mobility = 800 cm^2/V.sec, drain voltage = 6 V, gate voltage = -1.5 V, and saturation drift velocity = 2×10^7 cm/sec. Calculate the (a) pinch-off voltage, (b) velocity ratio, (c) saturation current at zero gate voltage, and (d) drain current. *[Hint: Use Eqs (14.64), (14.69), (14.65), (14.67), (14.68), and (14.66)]*

14.4 An HEMT has the following parameters: threshold voltage = 0.15 V, donor concentration = 2.5×10^{24} m^{-3}, metal–semiconductor Schottky barrier potential = 0.8 V, GaAs band gap = 1.43 V, and AlGaAs band gap = 1.80 V. Calculate the sensitivity of the transistor. The relative permittivity of AlGaAs is 4.43. *[Hint: Use Eq. (14.84)]*

14.5 A p-channel MOSFET has a doping concentration of 2.5×10^{17} cm^{-3} and insulator depth of 10^{-6} cm. If the potential difference between the Fermi level and the intrinsic level of the semiconductor is 0.437 V, then calculate the threshold voltage. The relative permittivity of SiO$_2$ and that of the semiconductor are 4 and 11.8, respectively. *[Hint: Use Eq. (14.94)]*

15

Active Microwave Circuits and Monolithic Microwave Integrated Circuit

15.1 Introduction

In the previous two chapters, we have discussed different types of microwave solid-state devices such as Gunn diode, IMPATT diode, p–i–n diode, varactor diode, and transistors. These devices are widely used in microwave circuits for detection, mixing, switching, frequency multiplication, amplification and as sources. The earliest detector diode was probably the *cat whisker* crystal detector. Later, the advent of tubes as detectors and amplifiers eliminated its use. However, during 1930, Southworth used crystal diodes as detector diodes in his experiment due to the lack of availability of high-frequency tubes. Southworth, King, and Ohl, al Bell Lab, also improved crystal detectors with better materials and packaging. Frequency conversion and heterodyning were first developed during 1920. During World War II, the same technique was followed at MIT Radiation Laboratory for using a crystal diode as a mixer.

In a microwave system, microwave amplification can be achieved in several ways, but overall, they can be classified into three broad categories: (a) reflection amplifiers, (b) parametric amplifiers, and (c) two-port amplifiers. Reflection amplifiers use a device that produces an ac negative resistance such that the ac voltage and current are out of phase. Examples include tunnel diodes, Gunn diodes, and IMPATT diodes with proper terminations. These as well as parametric amplifiers have already been discussed in Chapter 13. Two-port amplifiers include microwave tubes as well as transistor amplifier. Out of these, two-port tubes have been discussed in Chapters 11 and 12. Therefore in this chapter, we will discuss only two-port transistor amplifiers.

With the development of transistor processing technology, since 1970, most of the present-day microwave amplifiers use bipolar junction transistors (BJTs), field effect transistors (FETs), heterojunction bipolar transistors (HBTs), and high electron mobility transistors (HEMTs). Microwave transistor amplifiers are rugged, reliable, and low cost and can also be integrated into both hybrid and monolithic microwave integrated systems. Present-day microwave transistor amplifiers can work up to 100 GHz and have attractive characteristics such as broad bandwidth, low noise figure, and medium power. It may be noted that microwave tubes are still used for high-power and/or high-frequency applications, but continuing development of microwave transistors is steadily reducing the need of tubes.

In contrast to a microwave amplifier, a microwave oscillator converts DC power to RF power. In practice, a microwave solid-state oscillator uses an active non-linear device, such as a diode or a transistor, in conjunction with passive components to produce steady-state sinusoidal RF signals. In practice, oscillation is initially started by transient and noises present in the circuit and after that a particular stable oscillation state is reached, depending on the circuit configuration.

With the advent of latest technologies, the logical trend for microwave circuits is to integrate transmission lines, active devices, and other components on a single semiconductor substrate to form a monolithic microwave integrated circuit (MMIC). The term *monolithic* has been derived from the Greek words *monos* (single) and *lithos* (stone), and hence an MMIC means a microwave integrated circuit built on a single crystal. Such circuits can be fabricated by processes such as epitaxial growth, masked impurity diffusion, oxidation growth, and oxide etching. Though, like conventional ICs, MICs can be made in monolithic or hybrid forms, they are quite different from conventional ICs. One of the main differences is that a conventional IC has very high packaging density, whereas an MMIC has a packaging density that is quite low. An MMIC whose elements are fabricated on an insulating surface, such as glass or ceramic, is called a film integrated circuit, whereas an MMIC that consists of a combination of two or more integrated circuits is called a hybrid integrated circuit.

15.2 Detectors and Mixers

Microwave detectors and mixers use a non-linear device, preferably a diode, to achieve frequency conversion of an input signal. In general, the DC *I–V* characteristic of a diode can be expressed as follows:

$$I(V) = I_s\left(e^{\alpha V} - 1\right) \tag{15.1}$$

where $\quad \alpha = e/(nkT) \tag{15.2}$

Here, e is the charge of an electron, n is the ideality factor, k is the Boltzmann constant, T is the temperature in absolute scale, and I_s is the saturation current.

Now, let us assume that the diode voltage can be expressed as $V = V_0 + v \quad (15.3)$

where V_0 is the DC bias voltage and v is a small ac voltage. Substituting Eq. (15.3) into Eq. (15.1) and expanding it in a Taylor series about V_0, we get the following expression:

$$I(V) = I_0 + v\frac{dI}{dV}\bigg|_{V_0} + \frac{1}{2}v^2\frac{d^2I}{dV^2}\bigg|_{V_0} + \cdots \tag{15.4}$$

where $I_0 = I(V_0)$ is the DC bias current. Further,

$$\left.\frac{dI}{dV}\right|_{V_0} = \alpha I_s e^{\alpha V_0} = \alpha (I_0 + I_s) = G_d = \frac{1}{R_j} \tag{15.5}$$

where R_j and G_d are, respectively, the junction resistance and dynamic conductance of the diode. The second derivative can be written as follows:

$$\left.\frac{d^2 I}{dV^2}\right|_{V_0} = \left.\frac{dG_d}{dV}\right|_{V_0} = \alpha^2 I_s e^{\alpha V_0} = \alpha^2 (I_0 + I_s) = \alpha G_d = G_d' \tag{15.6}$$

Substituting Eqs (15.5) and (15.6) into Eq. (15.4) and neglecting higher-order terms, we get the following relation:

$$I(V) = I_0 + i = I_0 + vG_d + \frac{1}{2}v^2 G_d' \tag{15.7}$$

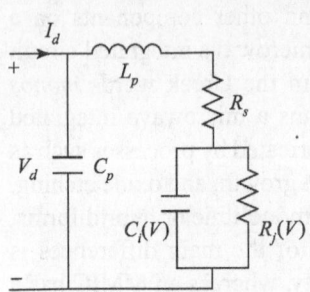

Fig. 15.1 RF equivalent circuit of packaged diode

where i is the ac current. This three-term approximation of diode current, called small signal approximation, reveals that the equivalent diode circuit involves a non-linear resistance. A typical equivalent circuit is shown in Fig. 15.1, where L_p is the package lead inductance, C_p is the package shunt capacitance, R_s is the series resistance, and C_j is the junction capacitance. It may be noted that C_j and R_j are bias dependent.

15.2.1 Diode Rectifier

For rectifier application, we can write $v = v_0 \cos(\omega_0 t)$ $\qquad$ (15.8)

Substituting Eq. (15.8) into Eq. (15.7) we get the following expression:

$$I(V) = I_0 + G_d v_0 \cos(\omega_0 t) + \frac{1}{2}G_d' v_0^2 \cos^2(\omega_0 t)$$

or $\quad I(V) = I_0 + \frac{1}{4}v_0^2 G_d' + v_0 G_d \cos(\omega_0 t) + \frac{1}{4}v_0^2 G_d' \cos(2\omega_0 t) \tag{15.9}$

Equation (15.9) reveals that the output current consists of the DC bias current I_0, DC rectified current $v_0^2 G_d'/4$, and ac frequency components ω_0 and $2\omega_0$. In addition, higher-order harmonics will also be present (due to the terms that we have eliminated because of small signal approximations). These higher-order terms are generally filtered out using a simple low-pass filter. The current sensitivity of the circuit, defined as the ratio of change in DC output current to input power, is given by the following relation:

$$\beta_i = \frac{\Delta I_{DC}}{P_{in}} = \frac{v_0^2 G_d'}{4} \frac{2}{v_0^2 G_d} = \frac{G_d'}{2G_d} \text{ A/W} \tag{15.10}$$

In deriving Eq. (15.10), we have retained only the first term of the input power.

Open-circuit voltage sensitivity, β_v, is defined in terms of voltage drop across the junction resistance when the diode is open circuited and can be expressed as follows: $\beta_v = \beta_i R_f$ $\qquad$ (15.11)

15.2.2 Diode AM Detector

A detector diode provides an output voltage and current that are proportional to the input microwave power (i.e., they follow the square law). A typical microwave detector circuit is shown in Fig. 15.2. The microwave signal, propagating along the transmission line circuit, appears across the diode through an impedance matching network and a blocking capacitor (C_B). The purpose of the impedance matching network is to match the diode impedance with the transmission line impedance so that no reflection can take place, whereas the purpose of the blocking capacitor is to ensure that the rectified current from the diode cannot flow back along the transmission line. This happens because the blocking capacitor behaves as a short circuit at high microwave frequencies and thus allows the microwave signal to appear across the diode whereas the same capacitor acts as an open circuit to rectified DC current and thus prevents the DC current to flow towards the transmission line. Therefore, all the rectified current appears across the load resistance. In practice, the blocking capacitor may also be a part of the impedance matching network. At the output of the diode

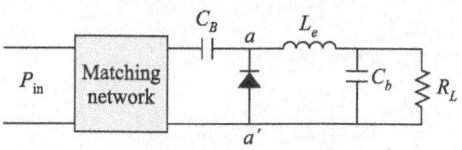

circuit, an LC low-pass filter is used to pass the rectified current to the load resistance and simultaneously to prevent the *leakage* microwave signal from reaching the load.

Fig. 15.2 RF equivalent circuit of detector diode

In detector application, the diode is used to demodulate an amplitude modulated signal, and hence we can write the following equation:

$$v = v_0 \left\{ 1 + m\cos(\omega_m t) \right\} \cos(\omega_0 t) \qquad (15.12)$$

where ω_m is the modulation frequency, ω_0 is the carrier frequency, and m is the modulation index. Substituting Eq. (15.12) into Eq. (15.7), we get the following expression:

$$I(V) = I_0 + G_d v_0 \left\{ 1 + m\cos(\omega_m t) \right\} \cos(\omega_0 t)$$
$$+ \frac{1}{2} G_d' \left[v_0 \left\{ 1 + m\cos(\omega_m t) \right\} \cos(\omega_0 t) \right]^2$$

or $\quad I(V) = I_0 + G_d v_0 \cos(\omega_0 t) + G_d v_0 m\cos(\omega_m t)\cos(\omega_0 t)$
$$+ \frac{1}{2} G_d' v_0^2 \cos^2(\omega_0 t) \left\{ 1 + 2m\cos(\omega_m t) + m^2 \cos^2(\omega_m t) \right\}$$

or $\quad I(V) = I_0 + G_d v_0 \cos(\omega_0 t) + \frac{G_d v_0 m}{2} \left[\cos\left\{ (\omega_0 + \omega_m)t \right\} + \cos\left\{ (\omega_0 - \omega_m)t \right\} \right]$
$$+ \frac{1}{4} G_d' v_0^2 \left\{ 1 + \cos(2\omega_0 t) \right\} + \frac{G_d' v_0^2 m}{2} \cos(\omega_m t) \left\{ 1 + \cos(2\omega_0 t) \right\}$$
$$+ \frac{1}{8} G_d' v_0^2 m^2 \left[\cos\left\{ (\omega_0 + \omega_m)t \right\} + \cos\left\{ (\omega_0 - \omega_m)t \right\} \right]^2 \qquad (15.13)$$

The diode current, therefore, can be expressed as follows:

$$i = G_d v_0 \cos(\omega_0 t) + \frac{G_d v_0 m}{2}\left[\cos\{(\omega_0 + \omega_m)t\} + \cos\{(\omega_0 - \omega_m)t\}\right] + \frac{1}{4}G_d' v_0^2$$

$$+ \frac{1}{4}G_d' v_0^2 \cos(2\omega_0 t) + \frac{G_d' v_0^2 m}{2}\cos(\omega_m t) + \frac{G_d' v_0^2 m}{2}\cos(\omega_m t)\cos(2\omega_0 t)$$

$$+ \frac{1}{8}G_d' v_0^2 m^2 \left[\cos\{(\omega_0 + \omega_m)t\} + \cos\{(\omega_0 - \omega_m)t\}\right]^2$$

or

$$i = G_d v_0 \left[\cos(\omega_0 t) + \frac{m}{2}\cos\{(\omega_0 + \omega_m)t\} + \frac{m}{2}\cos\{(\omega_0 - \omega_m)t\}\right]$$

$$+ \frac{1}{4}G_d' v_0^2 + \frac{1}{4}G_d' v_0^2 \cos(2\omega_0 t) + \frac{G_d' v_0^2 m}{2}\cos(\omega_m t)$$

$$+ \frac{G_d' v_0^2 m}{2}\cos(\omega_m t)\cos(2\omega_0 t) + \frac{1}{8}G_d' v_0^2 m^2 \left[\cos^2\{(\omega_0 + \omega_m)t\}\right.$$

$$\left. + 2\cos\{(\omega_0 + \omega_m)t\}\cos\{(\omega_0 - \omega_m)t\} + \cos^2\{(\omega_0 - \omega_m)t\}\right]$$

or

$$i = G_d v_0 \left[\cos(\omega_0 t) + \frac{m}{2}\cos\{(\omega_0 + \omega_m)t\} + \frac{m}{2}\cos\{(\omega_0 - \omega_m)t\}\right]$$

$$+ \frac{1}{4}G_d' v_0^2 + \frac{1}{4}G_d' v_0^2 \cos(2\omega_0 t) + \frac{G_d' v_0^2 m}{2}\cos(\omega_m t)$$

$$+ \frac{G_d' v_0^2 m}{4}\cos\{(2\omega_0 + \omega_m)t\} + \frac{G_d' v_0^2 m}{4}\cos\{(2\omega_0 - \omega_m)t\}$$

$$+ \frac{1}{16}G_d' v_0^2 m^2 \{1 + \cos\{2(\omega_0 + \omega_m)t\}\} + \frac{1}{8}G_d' v_0^2 m^2 \left[\cos(2\omega_0 t) + \cos(2\omega_m t)\right]$$

$$+ \frac{1}{16}G_d' v_0^2 m^2 \{1 + \cos\{2(\omega_0 - \omega_m)t\}\}$$

or

$$i = G_d v_0 \left[\cos(\omega_0 t) + \frac{m}{2}\cos\{(\omega_0 + \omega_m)t\} + \frac{m}{2}\cos\{(\omega_0 - \omega_m)t\}\right]$$

$$+ \frac{1}{4}G_d' v_0^2 \left[1 + \frac{m^2}{2} + 2m\cos(\omega_m t) + \frac{m^2}{2}\cos(2\omega_m t) + \cos(2\omega_0 t)\right.$$

$$+ m\cos\{(2\omega_0 - \omega_m)t\} + m\cos\{(2\omega_0 + \omega_m)t\} + \frac{m^2}{2}\cos(2\omega_0 t)$$

$$\left. + \frac{m^2}{4}\cos\{2(\omega_0 + \omega_m)t\} + \frac{m^2}{4}\cos\{2(\omega_0 - \omega_m)t\}\right] \tag{15.14}$$

The frequency spectrum of the corresponding output is shown in Fig. 15.3. The desired demodulated output frequency ω_m can be separated using a low-pass filter. It may be noted that the amplitude of this demodulated current is $mG_d' v_0^2/2$, which is proportional to the power of the input signal. This square law behaviour, which is the usual operating condition of a diode detector, is valid only over a restricted range of the input power.[1] For a large input power, the small signal

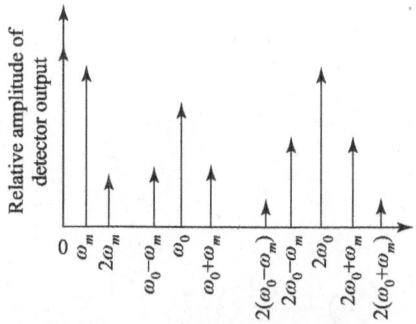

Fig. 15.3 Spectrum of detected amplitude-modulated signal

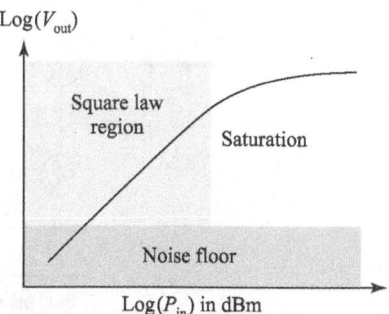

Fig. 15.4 Different regions of typical detector diode

condition will not be valid; the output will become saturated, and approach linear and then constant i–P characteristics. If the input power is too low, the signal will be lost in the noise floor of the device. A typical $v_{out} - P_{in}$ characteristic of a detector diode is shown in Fig. 15.4, where v_{out} is the voltage drop across a resistor in series with diode.

15.2.3 Single-ended Mixer

A mixer uses the non-linearity of a diode for mixing two frequencies, more precisely to produce sum and difference frequencies of two input signals. In receiver applications, an RF local oscillator signal and a low-level RF signal are mixed to produce an intermediate frequency $f_{IF} = f_{RF} - f_{LO}$ (in the range 10–100 MHz) and a much higher frequency $f_{RF} + f_{LO}$. The high-frequency component of the mixer output is then filtered out to get only the intermediate frequency, which is then amplified. Such receivers, called heterodyne receivers, can produce much better noise characteristics and sensitivity than direct detectors. In addition, such systems also provide a means of frequency tuning by simply varying the local oscillator frequency. Typical mixer up- and down-converter circuits are shown in Fig. 15.5.

Apart from being used as a down-converter in a heterodyne receiver system, a mixer can also be used in a transmitter to offset the frequency of an RF signal by an amount f_{IF}. Such a technique allows the use of identical local oscillators in both the transmitter and the receiver.

A number of parameters characterize the performance of a mixer, which are as follows:

Sensitivity Sensitivity of a mixer describes the ability of the mixer circuit to detect low-level signals and hence is related to system noise. Two parameters can affect the sensitivity of a diode: (a) effective noise temperature and (b) conversion Loss. The effective noise temperature is the ratio of the noise power available from the mixer driven by the source at the standard noise temperature to the noise power from a resistance, equal to the output resistance of the mixer, at the same

---

[1] This statement is particularly important for measurement of different microwave parameters using a microwave test bench, as described in Chapter 19.

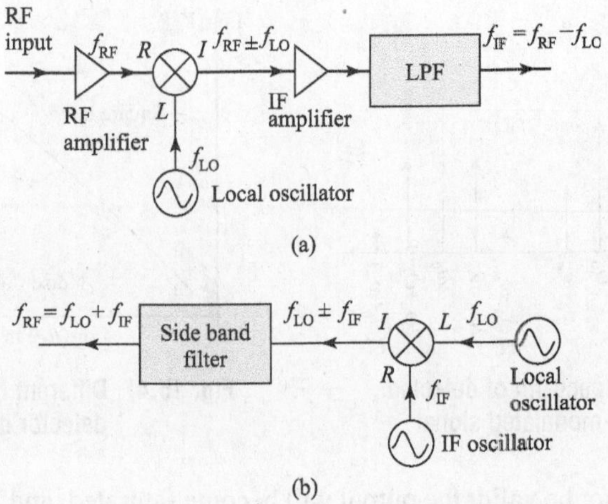

RF
input
f_{RF}
R
I
$f_{RF} \pm f_{LO}$
LPF
$f_{IF} = f_{RF} - f_{LO}$
RF
amplifier
L
IF
amplifier
f_{LO}
Local oscillator
(a)
$f_{RF} = f_{LO} + f_{IF}$
Side band
filter
$f_{LO} \pm f_{IF}$
I
L
f_{LO}
R
f_{IF}
Local
oscillator
IF oscillator
(b)

Fig. 15.5 Typical mixer circuits (a) Frequency down-converter (b) Frequency up-converter

noise temperature. Conversion loss, on the other hand, is the ratio of the input signal power to the output power.

Bandwidth For a given LO frequency, a mixer can convert only a band of RF frequency to a band of IF frequency, and this determines the bandwidth of the mixer circuit. In practice, a mixer is required to convert a wide range of RF frequencies to a fixed IF frequency with the help of a variable LO frequency. The bandwidth of a mixer depends on the intrinsic performance of the diode over the frequency range, parasitics, and matching circuits.

Dynamic range The dynamic range of a mixer is defined by the difference between the lowest and highest power levels that the mixer can handle. The lowest signal level is set by the noise level of the post mixer circuit and different spurious tones generated by the mixer. If only the noise floor is considered, then the dynamic range can be defined by the difference between the IF output power and the noise floor.

Due to the non-linearity of the diodes, theoretically, a mixer circuit contains an infinite number of intermodulation products or spurious tones, each written as $(mf_r \pm nf_0)$, where m and n are positive integers. Out of these infinite spurious tones, only $m = 2$ and $m = 3$ are important, while other higher-order terms can be neglected as their effect is less significant. The effect of this second- and third-order intermodulation distortion or spurious tones can be measured by intercept points, also called second- and third-order intercept points, as shown in Fig. 15.6.

IP_2
P_{1dB}
IP_3
Input power in dBm
Slope 1
Slope 3
DR_2 DR_3
Slope 2
Noise floor
Output power in dBm

Fig. 15.6 Intercept plot of non-linear circuit

For a small input RF power level, the output power of the desired frequency

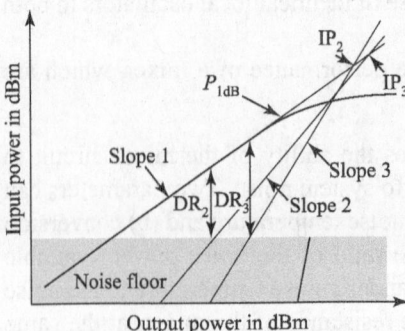

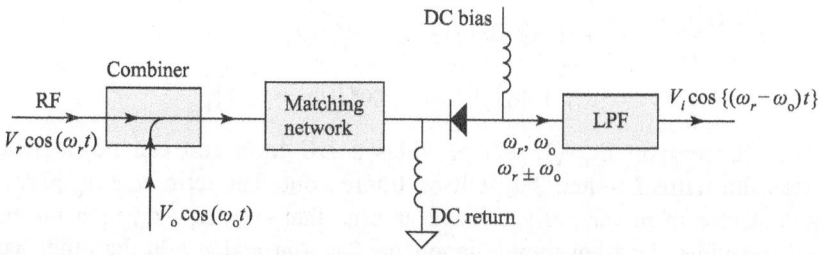

Fig. 15.7 Single-ended mixer circuit

component ($m = 1$ and $n = 1$) is linearly related to the input power by the conversion loss. However, for a high input power, the output power deviates from this linear relationship and becomes saturated. At this stage, conversion loss becomes a function of the RF power. The point at which this conversion loss is changed by 1 dB from its projected low power value is referred to as the 1 dB compression point.

Power-handling capability   For a large RF input power, the dissipated power in the diode is also very large and can heat it up to such an extent that the component can be damaged or the performance of the circuit can get degraded. This sets the maximum power-handling capacity of a mixer circuit.

In practice, there are several types of mixer circuits such as single-ended mixers, balanced mixers, anti-parallel diode mixers, double balanced mixers, and image rejection mixers, of which single-ended mixers are the simplest ones. The circuit of a single-ended mixer is shown in Fig. 15.7. In the circuit, an RF signal

$$v_{RF}(t) = v_r \cos(\omega_r t) \tag{15.15}$$

has been combined with an LO oscillator signal

$$v_{LO}(t) = v_0 \cos(\omega_0 t) \tag{15.16}$$

using a simple T-junction or directional coupler and then fed to a diode. An RF matching circuit precedes the diode. The diode has been biased through a choke that allows only DC current to pass while blocking RF. Equation (15.7) reveals that the diode current will consist of a DC bias term, RF and LO frequencies (due to the linear term in v), and combinations of RF and LO frequencies (due to the non-linear term in v). The current due to the non-linear terms is given by the following expression:

$$i = \frac{G'_d}{2}\{v_r \cos(\omega_r t) + v_0 \cos(\omega_0 t)\}^2$$

or $$i = \frac{G'_d}{2}\{v_r^2 \cos^2(\omega_r t) + 2v_r v_0 \cos(\omega_r t)\cos(\omega_0 t) + v_0^2 \cos^2(\omega_0 t)\}$$

or $$i = \frac{G'_d}{2}\left[\frac{v_r^2}{2}\{1 + \cos(2\omega_r t)\} + v_r v_0 \cos\{(\omega_r + \omega_0)t\}\right.$$

$$\left. + v_r v_0 \cos\{(\omega_r - \omega_0)t\} + \frac{v_0^2}{2}\{1 + \cos(2\omega_0 t)\}\right]$$

or $\quad i = \dfrac{G_d'}{4} \Big[v_r^2 + v_0^2 + v_r^2 \cos(2\omega_r t) + v_0^2 \cos(2\omega_0 t)$

$\qquad\qquad +2v_r v_0 \cos\{(\omega_r + \omega_0)t\} + 2v_r v_0 \cos\{(\omega_r - \omega_0)t\}\Big]$ (15.17)

The DC term in Eq. (15.17) provides a DC level and can be neglected, whereas the terms $2\omega_r$ and $2\omega_0$ will be filtered out. The term $\omega_r \pm \omega_0$ plays an important role in mixer design. If we assume that $\omega_r = \omega_0 + \omega_i$, then the term $\omega_r \pm \omega_0$ provides the frequency components $2\omega_0 + \omega_i$ and ω_i. On the other hand, if we assume that $\omega_r = \omega_0 - \omega_i$, then the term $\omega_r \pm \omega_0$ provides the frequency components $2\omega_0 - \omega_i$ and $-\omega_i$. The latter output is called the image response of a mixer and is indistinguishable from the direct response. Since $\omega_0 \gg \omega_i$, the desired RF frequency $\omega_r = \omega_0 + \omega_i$ is very close to spurious image frequency $\omega_r = \omega_0 - \omega_i$, and hence cannot be eliminated easily using a filter at the input of the mixer. A way to eliminate the image response is to use an image rejection filter.

In a mixer down-converter, $\omega_r - \omega_0$ will be the IF signal. In a mixer up-converter or modulator, the inputs are basically from a local oscillator and an IF oscillator. The IF will be mixed with the LO signal and will produce outputs $\omega_0 \pm \omega_i$. The frequency $\omega_0 + \omega_i$ is called the upper side band and $\omega_0 - \omega_i$ the lower side band. Depending on the number of frequency components present in the final output, the modulation can be double side band (DSB) or single side band (SSB). DSB retains both the side bands, whereas SSB retains only one side band.

Impedance matching of the three ports of a mixer is very difficult due to the presence of different frequency components and their harmonics. Undesired components may be dissipated in resistive terminations or blocked with reactive terminations. However, in practice, resistive termination increases the loss of a mixer, while reactive terminations are very sensitive to frequency.

A mixer is often characterized by a parameter called conversion loss, which is defined as follows:

$$L_C = 10\ \log_{10} \left(\frac{\text{Available RF input power}}{\text{IF output power}} \right) \text{dB}$$ (15.18)

For practical mixer, L_C ranges from 4 to 7 dB. It is strongly affected by local oscillator power level or pump power.

Since the mixer is often the first or second component in a receiver system, its noise properties are of critical importance. In a DSB system, a mixer has the desired signal at both the frequency components as compared to only at one component in an SSB system. Thus, a DSB system has a 3 dB lower noise figure than an SSB system.

In addition to the conversion loss, other critical parameters associated with a mixer are isolation between LO and RF port, cancellation of AM noise from LO, suppression of higher-order harmonics, etc. Though a single-ended mixer performs well in terms of the aforementioned characteristics, a balanced mixer can give better performance, as discussed in the subsequent sections.

15.2.4 Balanced Mixer

A balance mixer generally consists of two or more identical single-ended mixers with a 3 dB hybrid junction (90° or 180°), as shown in Fig. 15.8, to give either better

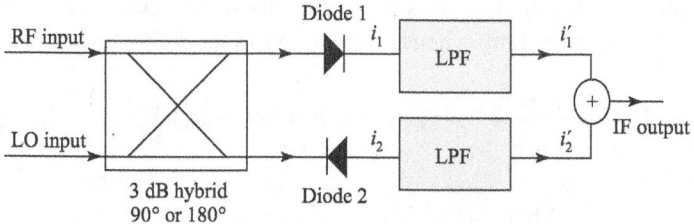

Fig. 15.8 Balanced mixer circuit

input matching or better RF–LO isolation and cancellation of AM noise from the local oscillator. To illustrate this, let us assume that a small random noise $v_n(t)$ is superimposed on a local oscillator signal. Therefore, we can write

$$v_{RF}(t) = v_r \cos(\omega_r t) \tag{15.19}$$

and
$$v_{LO}(t) = [v_0 + v_n(t)] \cos(\omega_0 t) \tag{15.20}$$

where $v_r \ll v_0$ and $v_n(t) \ll v_0$. If 90° hybrid has been used, then the voltage drops across the two diodes are as follows:

$$v_1(t) = v_r \cos(\omega_r t - 90°) + (v_0 + v_n) \cos(\omega_0 t - 180°)$$

or
$$v_1(t) = v_r \sin(\omega_r t) - (v_0 + v_n) \cos(\omega_0 t) \tag{15.21}$$

and
$$v_2(t) = v_r \cos(\omega_r t - 180°) + (v_0 + v_n) \cos(\omega_0 t - 90°)$$

or
$$v_2(t) = -v_r \cos(\omega_r t) + (v_0 + v_n) \sin(\omega_0 t) \tag{15.22}$$

Since only the non-linear terms of diode current gives rise to the desired mixer product, we will consider only this component. Hence, we can write the following relations:

$$i_1 = k v_1^2 = k \{v_r \sin(\omega_r t) - (v_0 + v_n) \cos(\omega_0 t)\}^2$$

or
$$i_1 = k \{v_r^2 \sin^2(\omega_r t) - 2 v_r (v_0 + v_n) \sin(\omega_r t) \cos(\omega_0 t) + (v_0 + v_n)^2 \cos^2(\omega_0 t)\}$$

or
$$i_1 = k \left[\frac{v_r^2}{2} + \frac{(v_0 + v_n)^2}{2} - \frac{v_r^2}{2} \cos(2\omega_r t) + \frac{(v_0 + v_n)^2}{2} \cos(2\omega_0 t) \right.$$
$$\left. - v_r (v_0 + v_n) \sin\{(\omega_r + \omega_0)t\} - v_r (v_0 + v_n) \sin\{(\omega_r - \omega_0)t\} \right] \tag{15.23}$$

and
$$i_2 = -k v_2^2 = -k \{-v_r \cos(\omega_r t) + (v_0 + v_n) \sin(\omega_0 t)\}^2$$

or
$$i_2 = -k \left[\frac{v_r^2}{2} + \frac{(v_0 + v_n)^2}{2} + \frac{v_r^2}{2} \cos(2\omega_r t) - \frac{(v_0 + v_n)^2}{2} \cos(2\omega_0 t) \right.$$
$$\left. - v_r (v_0 + v_n) \sin\{(\omega_r + \omega_0)t\} + v_r (v_0 + v_n) \sin\{(\omega_r - \omega_0)t\} \right] \tag{15.24}$$

After passing through the low-pass filter, the current will consist of only the DC, noise, and IF frequency terms; hence, we can write the following equations:

$$i'_1 = k\left[\frac{v_r^2}{2} + \frac{(v_0 + v_n)^2}{2} - v_r(v_0 + v_n)\sin\{(\omega_r - \omega_0)t\}\right] \tag{15.25}$$

and

$$i'_2 = -k\left[\frac{v_r^2}{2} + \frac{(v_0 + v_n)^2}{2} + v_r(v_0 + v_n)\sin\{(\omega_r - \omega_0)t\}\right] \tag{15.26}$$

Therefore, $i_{\text{IF}} = i'_1 + i'_2 = -2kv_r(v_0 + v_n)\sin\{(\omega_r - \omega_0)t\}$

or

$$i_{\text{IF}} \simeq -2kv_r v_0 \sin\{(\omega_r - \omega_0)t\} \tag{15.27}$$

Equation (15.27) reveals that the IF current is noise free. In practice, a balance mixer can give 15–30 dB noise rejection.

Now, let us consider the reflection of the input RF and LO signals from the diode. If a 90° hybrid has been used, then reflected RF waves from the diode can be expressed as follows:

$$V_{\Gamma 1} = \Gamma V_1 = \Gamma V_r / \sqrt{2} \tag{15.28}$$

and

$$V_{\Gamma 2} = \Gamma V_2 = -j\Gamma V_r / \sqrt{2} \tag{15.29}$$

where Γ is the reflection coefficient and V_r is the phasor RF input voltage. These reflected voltages will then arrive and combine back at the RF and LO input ports with the following amplitudes:

$$V_\Gamma^{\text{RF}} = \frac{V_{\Gamma 1}}{\sqrt{2}} - j\frac{V_{\Gamma 2}}{\sqrt{2}} = \frac{\Gamma V_r}{2} - \frac{\Gamma V_r}{2} = 0 \tag{15.30}$$

and

$$V_\Gamma^{\text{LO}} = \frac{V_{\Gamma 2}}{\sqrt{2}} - j\frac{V_{\Gamma 1}}{\sqrt{2}} = -j\frac{\Gamma V_r}{2} - j\frac{\Gamma V_r}{2} = -j\Gamma V_r \tag{15.31}$$

Equations (15.30) and (15.31) imply that the reflected wave will be absent at the RF input and present at the LO input. Similarly, it can be shown that when LO port is driven, the reflected wave will be present at the RF input and absent at the LO input. Therefore, a balanced mixer, using a 90° hybrid, provides good matching characteristics but poor isolation between RF and LO ports.

Now, if a 180° hybrid is used, with an RF input applied at the sum port and LO input at the difference port, the RF waves reflected from the diodes can be expressed as follows: $V_{\Gamma 1} = V_{\Gamma 2} = \Gamma V_r / \sqrt{2}$ (15.32)

Therefore, the reflected voltage at the RF and LO ports will be as follows:

$$V_\Gamma^\Sigma = \frac{V_{\Gamma 1}}{\sqrt{2}} + \frac{V_{\Gamma 2}}{\sqrt{2}} = \frac{\Gamma V_r}{2} + \frac{\Gamma V_r}{2} = \Gamma V_r \tag{15.33}$$

and

$$V_\Gamma^\Delta = \frac{V_{\Gamma 1}}{\sqrt{2}} - \frac{V_{\Gamma 2}}{\sqrt{2}} = \frac{\Gamma V_r}{2} - \frac{\Gamma V_r}{2} = 0 \tag{15.34}$$

Similarly, when the LO input is applied, $V_{\Gamma 1} = -V_{\Gamma 2} = \Gamma V_r / \sqrt{2}$ (15.35)

Therefore, the reflected voltage at the RF and LO ports will be as follows:

$$V_\Gamma^\Sigma = \frac{V_{\Gamma 1}}{\sqrt{2}} + \frac{V_{\Gamma 2}}{\sqrt{2}} = \frac{\Gamma V_r}{2} - \frac{\Gamma V_r}{2} = 0 \qquad (15.36)$$

and $$V_\Gamma^\Delta = \frac{V_{\Gamma 1}}{\sqrt{2}} - \frac{V_{\Gamma 2}}{\sqrt{2}} = \frac{\Gamma V_r}{2} + \frac{\Gamma V_r}{2} = \Gamma V_r \qquad (15.37)$$

Equations (15.33), (15.34), (15.36), and (15.37) reveal that for both the excitations, the reflected wave appears to the corresponding excitation port. Therefore, a balanced mixer, using a 180° hybrid, provides poor matching characteristics but good isolation between RF and LO ports.

15.2.5 Image Rejection Mixer

It has already been described that two distinct RF signals $\omega_r = \omega_0 \pm \omega_i$ can produce same IF frequency ω_i when mixed with an LO frequency ω_0. Further, if $\omega_0 \gg \omega_i$, the desired RF frequency $\omega_r = \omega_0 + \omega_i$ is very close to the spurious image frequency $\omega_r = \omega_0 - \omega_i$ and hence cannot be eliminated easily using a filter at the input of the mixer. A way to eliminate the image response is to use an image rejection filter, as shown in Fig. 15.9.

To illustrate the operation of an image rejection mixer, let us consider that the input RF voltage can be expressed as follows:

$$v_r = v_U \cos\{(\omega_0 + \omega_i)t\} + v_L \cos\{(\omega_0 - \omega_i)t\} \qquad (15.38)$$

Therefore, the inputs of the two mixers are as follows:

$$v_r^A = \frac{v_U}{\sqrt{2}} \cos\{(\omega_0 + \omega_i)t\} + \frac{v_L}{\sqrt{2}} \cos\{(\omega_0 - \omega_i)t\} \qquad (15.39)$$

and $$v_r^B = \frac{v_U}{\sqrt{2}} \cos\{(\omega_0 + \omega_i)t - 90°\} + \frac{v_L}{\sqrt{2}} \cos\{(\omega_0 - \omega_i)t - 90°\} \qquad (15.40)$$

The output of these two mixers will be as follows:

$$v_i^A = \frac{k}{2\sqrt{2}}(v_U + v_L)\cos(\omega_i t) \qquad (15.41)$$

and $$v_i^B = \frac{k}{2\sqrt{2}}[v_U \cos(\omega_i t - 90°) + v_L \cos(\omega_i t + 90°)] \qquad (15.42)$$

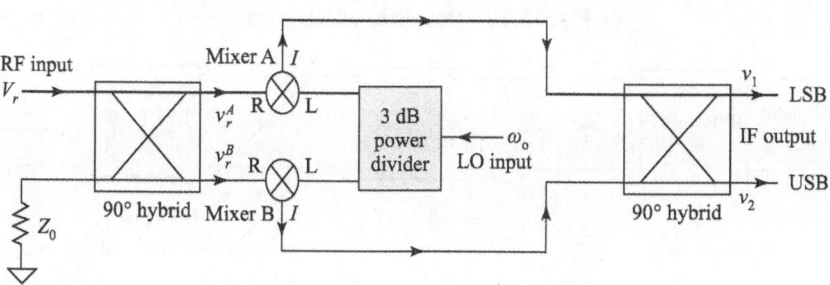

Fig. 15.9 Image rejection mixer circuit

When these two outputs are fed into the input of the second 90° hybrid, then the outputs will be as follows:

$$v_1 = k\left[v_U \cos(\omega_i t) + v_L \cos(\omega_i t) + v_U \cos(\omega_i t - 180°)\right.$$
$$\left. + v_L \cos(\omega_i t)\right]/4 = kv_L \cos(\omega_i t)/2 \tag{15.43}$$

and

$$v_2 = k\left[v_U \cos(\omega_i t - 90°) + v_L \cos(\omega_i t + 90°) + v_U \cos(\omega_i t - 90°)\right.$$
$$\left. + v_L \cos(\omega_i t - 90°)\right]/4 = kV_U \sin(\omega_i t)/2 \tag{15.44}$$

Equations (15.43) and (15.44) reveal that an image rejection mixer can separate the IF frequencies at its two output ports. Typical image rejection or isolation of such a mixer is about 20 dB or more.

15.2.6 Double Balanced Mixer

A balanced mixer, using a 180° hybrid, can eliminate only the even harmonics of LO signals. In contrast, a double balanced mixer, using a 180° hybrid, can suppress all even harmonics of both the RF and the LO signals, thus resulting in very low conversion loss. However, due to the presence of the 180° hybrid, it suffers from poor matching. A typical circuit of a double balanced mixer is shown in Fig. 15.10.

It may be noted that at microwave frequencies, the centre-tapped transformer cannot be formed by conventional means. Hence, balun (balanced–unbalanced) transformers are used for this purpose.

15.2.7 Anti-parallel Diode Mixer

A typical circuit diagram of an anti-parallel diode mixer is shown in Fig. 15.11. In such a mixer, the frequency of the local oscillator is half of the usual LO

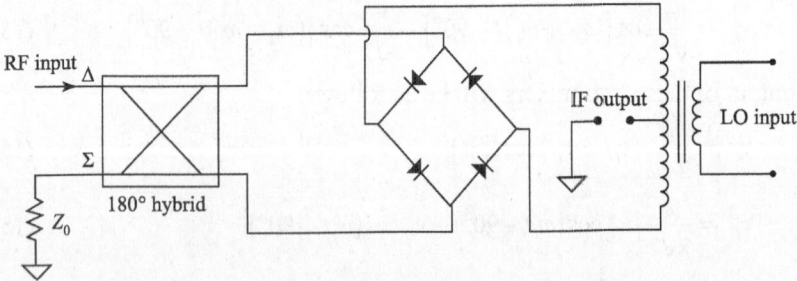

Fig. 15.10 Double balanced mixer

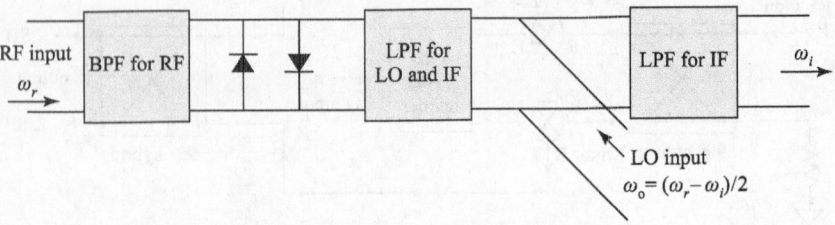

Fig. 15.11 Anti-parallel diode mixer

frequency $(\omega_r - \omega_i)$. The diode nonlinearity generates the second harmonic of the input LO frequency, which is mixed with the RF frequency to produce the desired IF frequency.

In practice, most of the mixers can be used in this manner. In an anti-parallel diode mixer, diode pairs create symmetrical V–I characteristics that suppress the fundamental mixing product of the LO and RF frequencies, which, in turn, leads to better conversion loss. In addition, anti-parallel diode mixers also suppress the AM noise from the local oscillator.

15.2.8 FET Mixers

Diodes are most commonly used non-linear element in mixer circuits because of their certain advantages like low noise. However, a disadvantage of such circuits is that conversion loss cannot be less than 4 dB. In mixer circuits where conversion loss is more critical than the noise properties, transistors can be used as the non-linear elements. In contrast to a diode, which generates conversion loss, a transistor provides conversion gain but is noisier.

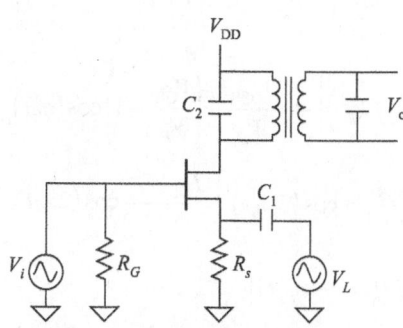

Fig. 15.12 Typical FET mixer circuit

In a transistor mixer, a FET is more preferred than a BJT because it produces less distortion. The input voltage range of a FET mixer is also much wider than a BJT. A typical FET mixer circuit is shown in Fig. 15.12. For the circuit we can write

$$V_{gs} = v_i - v_L + V_{GS} \tag{15.45}$$

where V_{gs} is the total gate to source voltage, v_i is the input voltage, v_L is the local oscillator voltage, and V_{GS} is the gate to source DC bias voltage. Using Eq. (15.45), the drain current can be expressed as follows:

$$i_D = I_{DSS}\left(1 - \frac{V_{gs}}{V_P}\right)^2 = I_{DSS}\left(1 - \frac{v_i - v_L + V_{GS}}{V_P}\right)^2 \tag{15.46}$$

where V_P is the pinch-off voltage and I_{DSS} the saturation drain current.

Now let us assume that

$$v_i = V_i \cos(\omega_i t) \tag{15.47}$$

and $\quad v_L = V_L \cos(\omega_L t) \tag{15.48}$

Substituting Eqs (15.47) and (15.48) into Eq. (15.46), we get the following relation:

$$i_D = I_{DSS}\left(1 - \frac{V_i \cos(\omega_i t) - V_L \cos(\omega_L t) + V_{GS}}{V_P}\right)^2$$

or
$$i_D = I_{DSS}\left[1 - \frac{2}{V_P}\left\{V_i\cos(\omega_i t) - V_L\cos(\omega_L t) + V_{GS}\right\}\right]$$
$$+ \frac{I_{DSS}}{V_P^2}\left[V_{GS}^2 + 2V_{GS}V_i\cos(\omega_i t) + V_i^2\cos^2(\omega_i t) - 2V_{GS}V_L\cos(\omega_L t)\right.$$
$$\left. + V_L^2\cos^2(\omega_L t) - 2V_iV_L\cos(\omega_i t)\cos(\omega_L t)\right]$$

or
$$i_D = I_{DSS}\left[1 - \frac{2}{V_P}\left\{V_i\cos(\omega_i t) - V_L\cos(\omega_L t) + V_{GS}\right\}\right]$$
$$+ \frac{I_{DSS}}{V_P^2}\left[V_{GS}^2 + 2V_{GS}V_i\cos(\omega_i t) + \frac{V_i^2}{2}\left\{1 + \cos(2\omega_i t)\right\} - 2V_{GS}V_L\cos(\omega_L t)\right.$$
$$\left. + \frac{V_L^2}{2}\left\{1 + \cos(2\omega_L t)\right\} - V_iV_L\cos\left\{(\omega_L + \omega_i)t\right\} - V_iV_L\cos\left\{(\omega_L - \omega_i)t\right\}\right]$$

or
$$i_D = \left\{I_{DSS}\left(1 - \frac{2V_{GS}}{V_P} + \frac{V_{GS}^2}{V_P^2}\right) + \frac{I_{DSS}}{2V_P^2}\left(V_i^2 + V_L^2\right)\right\} + \frac{2I_{DSS}V_i}{V_P}\left(\frac{V_{GS}}{V_P} - 1\right)\cos(\omega_i t)$$
$$+ \frac{2I_{DSS}V_L}{V_P}\left(1 - \frac{V_{GS}}{V_P}\right)\cos(\omega_L t) + \frac{I_{DSS}V_L^2}{2V_P^2}\cos(2\omega_L t) + \frac{I_{DSS}V_i^2}{2V_P^2}\cos(2\omega_i t)$$
$$- \frac{I_{DSS}V_iV_L}{V_P^2}\left[\cos\left\{(\omega_L + \omega_i)t\right\} + \cos\left\{(\omega_L - \omega_i)t\right\}\right]$$

or
$$i_D = I_{DC} + a_1\cos(\omega_i t) + a_2\cos(\omega_L t) + b_1\cos(2\omega_i t)$$
$$+ b_2\cos(2\omega_L t) - c\left[\cos\left\{(\omega_L + \omega_i)t\right\} + \cos\left\{(\omega_L - \omega_i)t\right\}\right] \tag{15.49}$$

where
$$I_{DC} = I_{DSS}\left(1 - \frac{2V_{GS}}{V_P} + \frac{V_{GS}^2}{V_P^2}\right) + \frac{I_{DSS}}{2V_P^2}\left(V_i^2 + V_L^2\right) \tag{15.50}$$

$$a_1 = \frac{2I_{DSS}V_i}{V_P}\left(\frac{V_{GS}}{V_P} - 1\right) \tag{15.51}$$

$$a_2 = \frac{2I_{DSS}V_L}{V_P}\left(1 - \frac{V_{GS}}{V_P}\right) \tag{15.52}$$

$$b_1 = \frac{I_{DSS}V_i^2}{2V_P^2} \tag{15.53}$$

$$b_2 = \frac{I_{DSS}V_L^2}{2V_P^2} \tag{15.54}$$

$$c = \frac{I_{DSS}V_iV_L}{V_P^2} \tag{15.55}$$

The ratio of c to the input voltage V_i is called conversion transconductance and can be expressed as follows:

$$g_c = c/V_i = I_{DSS}V_L/V_P^2 \tag{15.56}$$

For a higher conversion gain, g_c must be high. Therefore, it may appear from Eq. (15.56) that a high value of I_{DSS} is preferred, but it is not the case. This is because a high value of I_{DSS} always corresponds to a high value of V_P and, therefore, may result in a lower g_c. Equation (15.56) also reveals that V_L is directly proportional to the gain. Thus, a high local oscillator voltage can increase the conversion gain. Since a FET must be operated in the saturation region, V_L must be less than the pinch-off voltage. As a special case, if we assume that $V_L = |V_P|/2$, then the conversion transconductance can be calculated as follows:

$$g_c = \frac{I_{DSS}V_P}{2V_P^2} = \frac{I_{DSS}}{2V_P} \tag{15.57}$$

Now, the transconductance of a JFET can be written as follows:

$$g_m = \frac{\partial i_D}{\partial V_{GS}} = -\frac{2I_{DSS}}{V_P}\left(1 - \frac{V_{GS}}{V_P}\right) \tag{15.58}$$

Therefore,

$$g_m\big|_{V_{GS}=0} = -\frac{2I_{DSS}}{V_P} \tag{15.59}$$

Comparing Eqs (15.57) and (15.59), we can say that conversion transconductance is one-fourth of the small signal transconductance at $V_{GS} = 0$. Similarly, it can be shown that for a MOSFET, the conversion transconductance cannot exceed half of the small signal transconductance.

A look back at Eq. (15.49) reveals that the output signal consists of the fundamental components of the input signal, their second harmonics, and the desired $\omega_i \pm \omega_L$ frequency components. It further reveals that in a FET mixer, unlike a diode mixer, no higher-order spurious signals are present.

FET mixers also have other configurations, as shown in Figs 15.13–15.15. In Fig. 15.13, a dual-gate FET has been used. The operation of this circuit can be explained by treating the dual-gate FET as two cascaded single-gate FETs. The upper FET works as a source follower, while the lower one works as a mixer. The RF signal is connected at the gate terminal of the FET that is closer to the

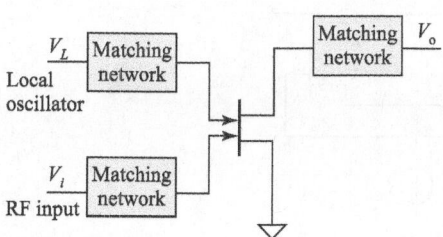

Fig. 15.13   Dual-gate FET mixer circuit

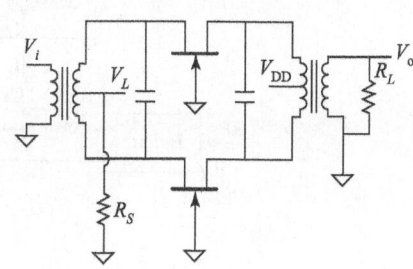

Fig. 15.14   FET balanced mixer

ground, whereas the local oscillator is connected to the gate terminal of the other FET. For the operation of the circuit, the FETs should work in a non-saturation mode. In a practical circuit, bypass capacitors are used at the output to block the local and RF oscillator signals. Further, bypass capacitors may also be required across the local oscillator to block the output frequencies. Such mixers are also called dual-gate FET mixers.

A FET balanced mixer, shown in Fig. 15.14, use two FETs. The phase of the local oscillator signals, fed into the two transistors, are same, while that of the RF signals, fed at the two transistors, are opposite. The output signal is a differential of the two sides; hence, the local oscillator signal is cancelled out, while the RF signals are added. The non-linear properties of the transistors generate the desired upper and lower side bands. The output signal from such a mixer can be expressed as follows:

$$v_o = \left(\frac{4R_L V_i}{V_P} + \frac{4R_L V_i V_S}{V_P^2} \right) \cos(\omega_i t) + \frac{2R_L V_i V_L}{V_P^2} \left[\cos\{(\omega_L + \omega_i)t\} \right.$$
$$\left. + \cos\{(\omega_L - \omega_i)t\} \right]$$

$$(15.60)$$

where V_S is the DC bias at the source.

The Gilbert cell mixer shown in Fig. 15.15 consists of six transistors; transistors Q_1 and Q_2 convert the RF voltage into current, while transistors Q_3 and Q_6 are used for switching. For the circuit, we can write the following expressions:

$$i_{IF1} = i_3 + i_5 \tag{15.61}$$
$$i_{IF2} = i_4 + i_6 \tag{15.62}$$
$$i_1 = i_3 + i_4 \tag{15.63}$$
$$i_2 = i_5 + i_6 \tag{15.64}$$

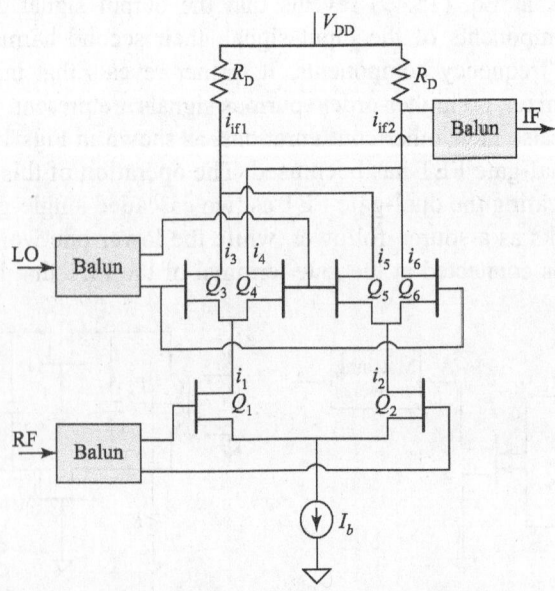

Fig. 15.15 Gilbert cell mixer circuit

Now, if we assume that the local oscillator provides a square wave signal, then transistors Q_4 and Q_5 will be on while Q_3 and Q_6 will be off, and vice versa. Therefore, in the former case,

$$i_{IF1} = i_5 = i_2 \qquad\qquad (15.65)$$

and $\quad i_{IF2} = i_4 = i_1 \qquad\qquad (15.66)$

whereas in the latter case,

$$i_{IF1} = i_3 = i_1 \qquad\qquad (15.67)$$

and $\quad i_{IF2} = i_6 = i_2 \qquad\qquad (15.68)$

EXAMPLE 15.1 A mixer has been designed with a JFET that has $I_{DSS} = 45$ mA and transconductance $g_m = 175$ mS when gate to source voltage $V_{gs} = 0$ V. Find the conversion transconductance. If the load impedance is $R_L = 50\,\Omega$, then calculate the magnitude of conversion gain.

Solution Given: $I_{DSS} = 45$ mA, $g_m = 175$ mS, and $R_L = 50\ \Omega$

Therefore, the pinch-off voltage is calculated to be as follows:

$$|V_P| = \left| \frac{2I_{DSS}}{g_m|_{V_{Gs}=0}} \right| = \frac{2\times45\times10^{-3}}{175\times10^{-3}} = 0.5143 \text{ V}$$

The conversion gain is as follows:

$$g_c = \frac{I_{DSS}}{2V_P} = \frac{45\times10^{-3}}{2\times0.5143} = 43.7488\times10^{-3} = 43.7488 \text{ mS}$$

The magnitude of average gain is expressed as follows:

$$A_v \approx g_c R_L = 43.7488\times10^{-3}\times50 = 2.1874$$

Practice Problem

15.1 A mixer has been designed with a JFET that has $I_{DSS} = 60$ mA and transconductance $g_m = 225$ mS when gate to source voltage $V_{gs} = 0$ V. Find the conversion transconductance. If the load impedance is $R_L = 50\,\Omega$, then calculate the magnitude of conversion gain. **[56.2535 mS, 2.8127]**

15.3 Stability and Gain of Amplifiers

One of the important concerns in transistor amplifier design is its stability and the maximum available gain from it. In the following section we will discuss the different types of transistor amplifier gain and their stability criteria.

15.3.1 Power Gain

Let us consider that an arbitrary two-port network $[S]$ is connected to a source impedance Z_S and load impedance Z_L, as shown in Fig. 15.16. Now three types of power gain can be explained:

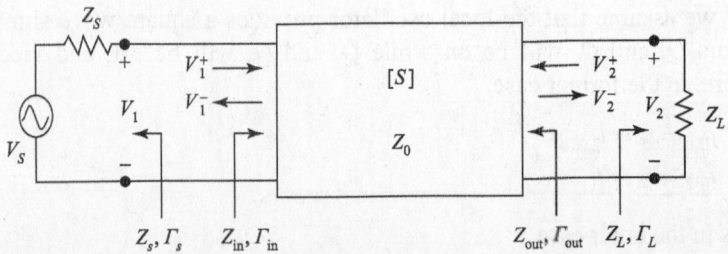

Fig. 15.16 Two-port network with general source and load impedance

Power gain $(G = P_L/P_{in})$ It can be defined as the ratio of power dissipated in the load to the power delivered to the input of the two-port network.

Available gain $(G_A = P_{Avn}/P_A)$ It can be defined as the ratio between the power available from the two-port network and the source, provided both the source and the load are conjugate matched. It depends on the source impedance but not on the load impedance.

Transducer power gain $(G_T = P_L/P_{Avs})$ It is defined as the ratio of power available from the two-port network to the power available from the source. It is a function of both the source and the load impedance.

> **Note** If the source and load are both conjugately matched to the two-port network, then gain is maximum and we can write $G = G_A = G_T$.

To find the expressions for the gains mentioned earlier, let us first consider Fig. 15.16. The reflection coefficient looking towards the load and the source can be expressed as follows:

$$\Gamma_L = \frac{Z_L - Z_0}{Z_L + Z_0} \tag{15.69}$$

and $$\Gamma_S = \frac{Z_S - Z_0}{Z_S + Z_0} \tag{15.70}$$

where Z_0 is the characteristic impedance reference to the S-parameter of the two-port network.

Now from the definition of the S-parameter, we can write the following equations:

$$V_1^- = S_{11}V_1^+ + S_{12}V_2^+ = S_{11}V_1^+ + S_{12}\Gamma_L V_2^- \tag{15.71}$$

and $$V_2^- = S_{21}V_1^+ + S_{22}V_2^+ = S_{21}V_1^+ + S_{22}\Gamma_L V_2^-$$

or $$\frac{V_2^-}{V_1^+} = \frac{S_{21}}{1 - S_{22}\Gamma_L} \tag{15.72}$$

In deriving Eqs (15.71) and (15.72), we have used the following relation:

$$V_2^+ = \Gamma_L V_2^- \tag{15.73}$$

From Eq. (15.71), we can write the following expression:

$$\Gamma_{in} = \frac{Z_{in} - Z_0}{Z_{in} + Z_0} = \frac{V_1^-}{V_1^+} = S_{11} + S_{12}\Gamma_L \frac{V_2^-}{V_1^+} \tag{15.74}$$

Substituting Eq. (15.72) into Eq. (15.74), we get the following expression:

$$\Gamma_{in} = \frac{Z_{in} - Z_0}{Z_{in} + Z_0} = S_{11} + \frac{S_{12}S_{21}\Gamma_L}{1 - S_{22}\Gamma_L} \tag{15.75}$$

where Γ_{in} is the input reflection coefficient and Z_{in} is the input impedance. Similarly, it can be shown that

$$\Gamma_{out} = \frac{V_2^-}{V_2^+} = S_{22} + \frac{S_{12}S_{21}\Gamma_S}{1 - S_{11}\Gamma_S} \tag{15.76}$$

Using the voltage division rule, we can write the following relation:

$$V_1 = V_S \frac{Z_{in}}{Z_S + Z_{in}} = V_1^+ + V_1^- = V_1^+ \left(1 + \Gamma_{in}\right) \tag{15.77}$$

From Eq. (15.75), we can write the following expression:

$$Z_{in} = Z_0 \frac{1 + \Gamma_{in}}{1 - \Gamma_{in}} \tag{15.78}$$

Substituting Eq. (15.78) into Eq. (15.77), the following expression is obtained:

$$V_1^+ \left(1 + \Gamma_{in}\right) = V_S Z_0 \left(\frac{1 + \Gamma_{in}}{1 - \Gamma_{in}}\right) \bigg/ \left[Z_S + Z_0 \left(\frac{1 + \Gamma_{in}}{1 - \Gamma_{in}}\right)\right]$$

or

$$V_1^+ \left(1 + \Gamma_{in}\right) = V_S \frac{Z_0 \left(1 + \Gamma_{in}\right)}{Z_S \left(1 - \Gamma_{in}\right) + Z_0 \left(1 + \Gamma_{in}\right)}$$

or

$$V_1^+ = \frac{V_S Z_0}{Z_S \left(1 - \Gamma_{in}\right) + Z_0 \left(1 + \Gamma_{in}\right)} = \frac{V_S Z_0}{\left(Z_S + Z_0\right) - \Gamma_{in} \left(Z_S - Z_0\right)}$$

or

$$V_1^+ = V_S \frac{\dfrac{Z_0}{Z_S + Z_0}}{1 - \Gamma_{in} \left(\dfrac{Z_S - Z_0}{Z_S + Z_0}\right)} = V_S \frac{\dfrac{1}{2}\left(1 - \dfrac{Z_S - Z_0}{Z_S + Z_0}\right)}{1 - \Gamma_{in} \left(\dfrac{Z_S - Z_0}{Z_S + Z_0}\right)}$$

or

$$V_1^+ = \frac{V_S}{2} \left(\frac{1 - \Gamma_S}{1 - \Gamma_S \Gamma_{in}}\right) \tag{15.79}$$

If peak values are assumed for all voltages, then the average power delivered to the network is given by the following expression:

$$P_{in} = \frac{1}{2Z_0} \left|V_1^+\right|^2 \left(1 - \left|\Gamma_{in}\right|^2\right) \tag{15.80}$$

Substituting Eq. (15.79) into Eq. (15.80), we get the following relation:

$$P_{in} = \frac{1}{2Z_0} \left|V_1^+\right|^2 \left(1 - \left|\Gamma_{in}\right|^2\right) = \frac{\left|V_S\right|^2}{8Z_0} \frac{\left|1 - \Gamma_S\right|^2}{\left|1 - \Gamma_S \Gamma_{in}\right|^2} \left(1 - \left|\Gamma_{in}\right|^2\right) \tag{15.81}$$

The power delivered to the load is expressed as follows:

$$P_L = \frac{1}{2Z_0}\left|V_2^-\right|^2\left(1-\left|\Gamma_L\right|^2\right)$$ (15.82)

Substituting Eq. (15.72) and then Eq. (15.79) into Eq. (15.82), we get the following expression:

$$P_L = \frac{1}{2Z_0}\left|V_1^+\right|^2\frac{\left|S_{21}\right|^2}{\left|1-S_{22}\Gamma_L\right|^2}\left(1-\left|\Gamma_L\right|^2\right) = \frac{\left|V_S\right|^2\left|S_{21}\right|^2\left|1-\Gamma_S\right|^2\left(1-\left|\Gamma_L\right|^2\right)}{8Z_0\left|1-\Gamma_S\Gamma_{in}\right|^2\left|1-S_{22}\Gamma_L\right|^2}$$ (15.83)

Therefore, using Eqs (15.81) and (15.83), the power gain can be expressed as follows:

$$G = \frac{P_L}{P_{in}} = \frac{\left|V_S\right|^2\left|S_{21}\right|^2\left|1-\Gamma_S\right|^2\left(1-\left|\Gamma_L\right|^2\right)}{8Z_0\left|1-\Gamma_S\Gamma_{in}\right|^2\left|1-S_{22}\Gamma_L\right|^2}\frac{8Z_0\left|1-\Gamma_S\Gamma_{in}\right|^2}{\left|V_S\right|^2\left|1-\Gamma_S\right|^2\left(1-\left|\Gamma_{in}\right|^2\right)}$$

or $$G = \frac{\left|S_{21}\right|^2\left(1-\left|\Gamma_L\right|^2\right)}{\left|1-S_{22}\Gamma_L\right|^2\left(1-\left|\Gamma_{in}\right|^2\right)}$$ (15.84)

The power available from the source is the maximum power that can be delivered to the network. Following the maximum power transfer theorem, we can say that this happens when

$$\Gamma_{in} = \Gamma_S^*$$ (15.85)

Therefore,

$$P_{Avs} = P_{in}\big|_{\Gamma_{in}=\Gamma_S^*} = \frac{\left|V_S\right|^2}{8Z_0}\frac{\left|1-\Gamma_S\right|^2}{\left|1-\Gamma_S\Gamma_{in}\right|^2}\left(1-\left|\Gamma_{in}\right|^2\right)\bigg|_{\Gamma_{in}=\Gamma_S^*} = \frac{\left|V_S\right|^2\left|1-\Gamma_S\right|^2}{8Z_0\left(1-\left|\Gamma_S\right|^2\right)}$$ (15.86)

Again the power available from the network is the maximum power that can be delivered to the load. Following the maximum power transfer theorem, we can say that this happens when

$$\Gamma_L = \Gamma_{out}^*$$ (15.87)

Therefore,

$$P_{Avn} = P_L\big|_{\Gamma_L=\Gamma_{out}^*} = \frac{\left|V_S\right|^2\left|S_{21}\right|^2\left|1-\Gamma_S\right|^2\left(1-\left|\Gamma_L\right|^2\right)}{8Z_0\left|1-\Gamma_S\Gamma_{in}\right|^2\left|1-S_{22}\Gamma_L\right|^2}\Bigg|_{\Gamma_L=\Gamma_{out}^*}$$ (15.88)

To find a simplified form of Eq. (15.88), we must evaluate Γ_{in} for $\Gamma_L=\Gamma_{out}^*$. From Eq. (15.75), we get the following

$$\Gamma_{in}\big|_{\Gamma_L=\Gamma_{out}^*} = S_{11} + \frac{S_{12}S_{21}\Gamma_L}{1-S_{22}\Gamma_L}\bigg|_{\Gamma_L=\Gamma_{out}^*} = S_{11} + \frac{S_{12}S_{21}\Gamma_{out}^*}{1-S_{22}\Gamma_{out}^*}$$ (15.89)

Therefore,

$$\left|1-\Gamma_S\Gamma_{\text{in}}\right|^2\Big|_{\Gamma_L=\Gamma_{\text{out}}^*} = \left|1-\Gamma_S S_{11} - \frac{S_{12}S_{21}\Gamma_S\Gamma_{\text{out}}^*}{1-S_{22}\Gamma_{\text{out}}^*}\right|^2$$

or $$\left|1-\Gamma_S\Gamma_{\text{in}}\right|^2\Big|_{\Gamma_L=\Gamma_{\text{out}}^*} = \left|\frac{1-S_{22}\Gamma_{\text{out}}^* - S_{11}\Gamma_S + S_{11}S_{22}\Gamma_S\Gamma_{\text{out}}^* - S_{12}S_{21}\Gamma_S\Gamma_{\text{out}}^*}{1-S_{22}\Gamma_{\text{out}}^*}\right|^2$$

or $$\left|1-\Gamma_S\Gamma_{\text{in}}\right|^2\Big|_{\Gamma_L=\Gamma_{\text{out}}^*} = \left|\frac{(1-S_{11}\Gamma_S)-\Gamma_{\text{out}}^*\left\{S_{22}(1-S_{11}\Gamma_S)+S_{12}S_{21}\Gamma_S\right\}}{1-S_{22}\Gamma_{\text{out}}^*}\right|^2$$

or $$\left|1-\Gamma_S\Gamma_{\text{in}}\right|^2\Big|_{\Gamma_L=\Gamma_{\text{out}}^*} = \left|\frac{(1-S_{11}\Gamma_S)-\Gamma_{\text{out}}^*(1-S_{11}\Gamma_S)\left(S_{22}+\dfrac{S_{12}S_{21}\Gamma_S}{1-S_{11}\Gamma_S}\right)}{1-S_{22}\Gamma_{\text{out}}^*}\right|^2 \qquad (15.90)$$

Substituting Eq. (15.76) into Eq. (15.90) gives the following expression:

$$\left|1-\Gamma_S\Gamma_{\text{in}}\right|^2\Big|_{\Gamma_L=\Gamma_{\text{out}}^*} = \left|\frac{(1-S_{11}\Gamma_S)-\Gamma_{\text{out}}\Gamma_{\text{out}}^*(1-S_{11}\Gamma_S)}{1-S_{22}\Gamma_{\text{out}}^*}\right|^2$$

or $$\left|1-\Gamma_S\Gamma_{\text{in}}\right|^2\Big|_{\Gamma_L=\Gamma_{\text{out}}^*} = \frac{\left|1-S_{11}\Gamma_S\right|^2\left(1-\left|\Gamma_{\text{out}}\right|^2\right)^2}{\left|1-S_{22}\Gamma_{\text{out}}^*\right|^2} \qquad (15.91)$$

Substituting Eq. (15.91) into Eq. (15.88), we get the following relation:

$$P_{\text{Avn}} = \frac{\left|V_S\right|^2\left|S_{21}\right|^2\left|1-\Gamma_S\right|^2\left(1-\left|\Gamma_{\text{out}}\right|^2\right)}{8Z_0\left|1-S_{22}\Gamma_{\text{out}}^*\right|^2} \cdot \frac{\left|1-S_{22}\Gamma_{\text{out}}^*\right|^2}{\left|1-S_{11}\Gamma_S\right|^2\left(1-\left|\Gamma_{\text{out}}\right|^2\right)^2}$$

or $$P_{\text{Avn}} = \frac{\left|V_S\right|^2\left|S_{21}\right|^2\left|1-\Gamma_S\right|^2}{8Z_0\left|1-S_{11}\Gamma_S\right|^2\left(1-\left|\Gamma_{\text{out}}\right|^2\right)} \qquad (15.92)$$

Using Eqs (15.86) and (15.92), the available power gain can be expressed as follows:

$$G_A = \frac{P_{\text{Avn}}}{P_{\text{Avs}}} = \frac{\left|V_S\right|^2\left|S_{21}\right|^2\left|1-\Gamma_S\right|^2}{8Z_0\left|1-S_{11}\Gamma_S\right|^2\left(1-\left|\Gamma_{\text{out}}\right|^2\right)} \cdot \frac{8Z_0\left(1-\left|\Gamma_S\right|^2\right)}{\left|V_S\right|^2\left|1-\Gamma_S\right|^2}$$

or $$G_A = \frac{\left|S_{21}\right|^2\left(1-\left|\Gamma_S\right|^2\right)}{\left|1-S_{11}\Gamma_S\right|^2\left(1-\left|\Gamma_{\text{out}}\right|^2\right)} \qquad (15.93)$$

Similarly, using Eqs (15.83) and (15.86), the transducer power gain can be expressed as follows:

$$G_T = \frac{P_L}{P_{\text{Avs}}} = \frac{|V_S|^2 |S_{21}|^2 |1-\Gamma_S|^2 \left(1-|\Gamma_L|^2\right)}{8Z_0 |1-\Gamma_S \Gamma_{\text{in}}|^2 |1-S_{22}\Gamma_L|^2} \frac{8Z_0 \left(1-|\Gamma_S|^2\right)}{|V_S|^2 |1-\Gamma_S|^2}$$

or
$$G_T = \frac{|S_{21}|^2 \left(1-|\Gamma_L|^2\right)\left(1-|\Gamma_S|^2\right)}{|1-\Gamma_S \Gamma_{\text{in}}|^2 |1-S_{22}\Gamma_L|^2} \tag{15.94}$$

When both input and output are matched, then

$$\Gamma_L = \Gamma_S = 0 \tag{15.95}$$

Substituting Eq. (15.95) into Eq. (15.94), we get the following expression:

$$G_T|_{\Gamma_L=\Gamma_S=0} = |S_{21}|^2 \tag{15.96}$$

Now, let us consider the case for which

$$S_{12} = 0 \tag{15.97}$$

Under such a condition, Eq. (15.74) yields the following relation:

$$\Gamma_{\text{in}} = S_{11} \tag{15.98}$$

Substituting Eq. (15.98) into Eq. (15.94), we get the following expression:

$$G_T|_{S_{12}=0} = \frac{|S_{21}|^2 \left(1-|\Gamma_L|^2\right)\left(1-|\Gamma_S|^2\right)}{|1-S_{11}\Gamma_S|^2 |1-S_{22}\Gamma_L|^2} \tag{15.99}$$

$G_T|_{S_{12}=0}$ is also denoted as G_{TU} and is called the unilateral transducer power gain. Since $S_{12} = 0$ and $S_{21} \neq 0$, the system is non-reciprocal. Such non-reciprocal characteristic is often used in amplifier circuits.

A basic single-stage transistor amplifier circuit is shown in Fig. 15.17. The input and output matching circuit transforms the input and output impedance, Z_0, to source and load impedance Z_S and Z_L, respectively. The most suitable gain definition for this case is that of transducer power gain. Therefore, using Eq. (15.94), we can write the following expression:

$$G_T = G_S G_0 G_L \tag{15.100}$$

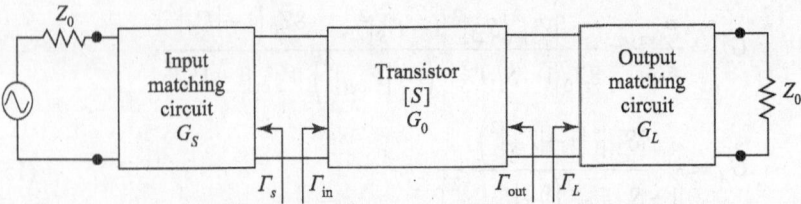

Fig. 15.17 Transistor amplifier circuit block diagram

where
$$G_S = \frac{1-|\Gamma_S|^2}{|1-\Gamma_S\Gamma_{in}|^2} \qquad (15.101)$$

$$G_0 = |S_{21}|^2 \qquad (15.102)$$

and
$$G_L = \frac{1-|\Gamma_L|^2}{|1-S_{22}\Gamma_L|^2} \qquad (15.103)$$

Now, if we assume that the amplifier is unilateral, that is, $S_{12} = 0$ or S_{12} is small enough so that it can be ignored, then $\Gamma_{in} = S_{11}$, and we can write that

$$G_{TU} = G_S G_0 G_L \qquad (15.104)$$

where
$$G_S = \frac{1-|\Gamma_S|^2}{|1-S_{11}\Gamma_S|^2} \qquad (15.105)$$

$$G_0 = |S_{21}|^2 \qquad (15.106)$$

and
$$G_L = \frac{1-|\Gamma_L|^2}{|1-S_{22}\Gamma_L|^2} \qquad (15.107)$$

EXAMPLE 15.2 A microwave transistor has the following S-parameters at $9\,\text{GHz}$, with a $50\,\Omega$ reference impedance: $S_{11} = 0.42\angle120°$, $S_{12} = 0.01\angle-20°$, $S_{21} = 2.15\angle20°$, and $S_{22} = 0.35\angle-120°$. If the source impedance is $25\,\Omega$ and load impedance $40\,\Omega$, calculate the power gain, available gain, and transducer power gain.

Solution Given: $Z_0 = 50\,\Omega$, $Z_S = 25\,\Omega$, $Z_L = 40\,\Omega$,

$$S_{11} = 0.42\angle120° = -0.21 + j0.3637, \; S_{12} = 0.01\angle-20° = 0.0094 - j0.0034$$

$$S_{21} = 2.15\angle20° = 2.0203 + j0.7353, \text{ and } S_{22} = 0.35\angle-120° = -0.175 - j0.3031$$

The source and load reflection coefficient can be calculated as follows:

$$\Gamma_S = \frac{Z_S - Z_0}{Z_S + Z_0} = \frac{25 - 50}{25 + 50} = -0.3333$$

$$\Gamma_L = \frac{Z_L - Z_0}{Z_L + Z_0} = \frac{40 - 50}{40 + 50} = -0.1111$$

The reflection coefficient at the input and output terminated network is

$$\Gamma_{in} = S_{11} + \frac{S_{12}S_{21}\Gamma_L}{1-S_{22}\Gamma_L} = -0.2124 + j0.3636$$

$$\Gamma_{out} = S_{22} + \frac{S_{12}S_{21}\Gamma_S}{1-S_{11}\Gamma_S} = = -0.1826 - j0.3021$$

The power gain is calculated to be the following:

$$G = \frac{|S_{21}|^2\left(1-|\Gamma_L|^2\right)}{|1-S_{22}\Gamma_L|^2\left(1-|\Gamma_{in}|^2\right)}$$

or
$$G = \frac{|2.0203 + j0.7353|^2\left(1-|-0.1111|^2\right)}{|1-(-0.175 - j0.3031)(-0.1111)|^2\left(1-|-0.2124 + j0.3636|^2\right)}$$

or $\quad G = \dfrac{4.5654}{0.7919} = 5.7651$

The available power gain is as follows:

$$G_A = \frac{|S_{21}|^2\left(1-|\Gamma_S|^2\right)}{|1-S_{11}\Gamma_S|^2\left(1-|\Gamma_{out}|^2\right)}$$

or $\quad G_A = \dfrac{|2.0203+j0.7353|^2\left(1-|-0.3333|^2\right)}{|1-(-0.21+j0.3637)(-0.3333)|\left(1-|-0.1826-j0.3021|^2\right)}$

or $\quad G_A = \dfrac{4.1090}{0.77} = 5.3363$

The transducer power gain is as follows:

$$G_T = \frac{|S_{21}|^2\left(1-|\Gamma_L|^2\right)\left(1-|v_S|^2\right)}{|1-\Gamma_S\Gamma_{in}|^2|1-S_{22}\Gamma_L|^2}$$

or $\quad G_T = \dfrac{|2.0203+j0.7353|\left(1-|-0.1111|^2\right)\left(1-|-0.3333|^2\right)}{|1-(-0.3333)(-0.2124+j0.3636)|^2|1-(-0.175-j0.3031)(-0.1111)|^2}$

or $\quad G_T = \dfrac{4.0583}{0.8453} = 4.801$

Practice Problem

15.2 A microwave transistor has the following S-parameters at $10\,\text{GHz}$, with a $50\,\Omega$ reference impedance: $S_{11} = 0.4\angle 130°$, $S_{12} = 0.01\angle -30°$, $S_{21} = 2.25\angle 30°$, and $S_{22} = 0.45\angle -130°$. If the source impedance is $35\,\Omega$ and load impedance $40\,\Omega$, calculate the power gain, available gain, and transducer power gain. **[6.354, 6.7466, 5.653]**

15.3.2 Stability

Stability of an amplifier is a big concern in designing an amplifier circuit. If the real part of the input or output port impedance becomes negative, that is, $|\Gamma_{in}|$ or $|\Gamma_{out}|$ becomes greater than 1, then the circuit results in oscillation. Since Γ_{in} and Γ_{out} depends on source and load matching network, the stability of an amplifier depends on Γ_S and Γ_L, as presented by the matching network. In practice, two types of stability can be defined:

Unconditional stability A network is unconditionally stable if $|\Gamma_{in}| < 1$ and $|\Gamma_{out}| < 1$ for all passive sources and load impedances (i.e., for $|\Gamma_S| < 1$ and $|\Gamma_L| < 1$).

Conditional stability A network is conditionally stable if $|\Gamma_{in}| < 1$ and $|\Gamma_{out}| < 1$ only for a certain range of passive source and load impedance (i.e., for $|\Gamma_S| < 1$ and $|\Gamma_L| < 1$).

Note Stability condition of a network is frequency dependent.

Using these requirements, the condition for unconditional stability can be written as follows:

$$\left| S_{11} + \frac{S_{12}S_{21}\Gamma_L}{1 - S_{22}\Gamma_L} \right| < 1 \tag{15.108}$$

and $$\left| S_{22} + \frac{S_{12}S_{21}\Gamma_S}{1 - S_{11}\Gamma_S} \right| < 1 \tag{15.109}$$

where we have used Eqs (15.75) and (15.76). For a unilateral device, these conditions reduce to the following forms:

$$|S_{11}| < 1 \tag{15.110}$$

and $$|S_{22}| < 1 \tag{15.111}$$

Inequalities (15.108) and (15.109) define a range of Γ_S and Γ_L, where the amplifier will be stable. This range can be found by using a Smith chart and plotting a stability circle. Stability circles are defined as the loci of Γ_L (or Γ_S) plane for which $|\Gamma_{in}| = 1$ (or $|\Gamma_{out}| = 1$). To find the equation of stability circle, let us start with $|\Gamma_{in}| = 1$. Using Eq. (15.75), we can write the following equation:

$$\left| S_{11} + \frac{S_{12}S_{21}\Gamma_L}{1 - S_{22}\Gamma_L} \right| = 1$$

or $$\left| S_{11} - \Gamma_L \left(S_{11}S_{22} - S_{12}S_{21} \right) \right| = \left| 1 - S_{22}\Gamma_L \right|$$

or $$\left| S_{11} - \Delta\Gamma_L \right| = \left| 1 - S_{22}\Gamma_L \right| \tag{15.112}$$

where $$\Delta = S_{11}S_{22} - S_{12}S_{21} \tag{15.113}$$

Squaring both sides of Eq. (15.112), we get the following relation:

$$|S_{11}|^2 + |\Delta|^2 |\Gamma_L|^2 - \left(\Delta\Gamma_L S_{11}^* + \Delta^*\Gamma_L^* S_{11} \right) = 1 + |S_{22}|^2 |\Gamma_L|^2 - \left(S_{22}^*\Gamma_L^* + S_{22}\Gamma_L \right)$$

or $$\left(|S_{22}|^2 - |\Delta|^2 \right)\Gamma_L\Gamma_L^* - \left(S_{22} - \Delta S_{11}^* \right)\Gamma_L - \left(S_{22}^* - \Delta^* S_{11} \right)\Gamma_L^* = |S_{11}|^2 - 1$$

or $$\Gamma_L\Gamma_L^* - \frac{\left(S_{22} - \Delta S_{11}^* \right)\Gamma_L + \left(S_{22}^* - \Delta^* S_{11} \right)\Gamma_L^*}{|S_{22}|^2 - |\Delta|^2} = \frac{|S_{11}|^2 - 1}{|S_{22}|^2 - |\Delta|^2}$$

or $$\Gamma_L\Gamma_L^* - \frac{\left(S_{22} - \Delta S_{11}^* \right)\Gamma_L + \left(S_{22}^* - \Delta^* S_{11} \right)\Gamma_L^*}{|S_{22}|^2 - |\Delta|^2} + \frac{\left| S_{22} - \Delta S_{11}^* \right|^2}{\left(|S_{22}|^2 - |\Delta|^2 \right)^2}$$

$$= \frac{|S_{11}|^2 - 1}{|S_{22}|^2 - |\Delta|^2} + \frac{\left| S_{22} - \Delta S_{11}^* \right|^2}{\left(|S_{22}|^2 - |\Delta|^2 \right)^2}$$

or
$$\left| \Gamma_L - \frac{\left(S_{22} - \Delta S_{11}^* \right)^*}{\left| S_{22} \right|^2 - \left| \Delta \right|^2} \right|^2 = \frac{\left(\left| S_{11} \right|^2 - 1 \right) \left(\left| S_{22} \right|^2 - \left| \Delta \right|^2 \right) + \left| S_{22} - \Delta S_{11}^* \right|^2}{\left(\left| S_{22} \right|^2 - \left| \Delta \right|^2 \right)^2}$$

or
$$\left| \Gamma_L - \frac{\left(S_{22} - \Delta S_{11}^* \right)^*}{\left| S_{22} \right|^2 - \left| \Delta \right|^2} \right|^2 = \frac{\begin{array}{c} \left| S_{11} \right|^2 \left| S_{22} \right|^2 - \left| \Delta \right|^2 \left| S_{11} \right|^2 - \left| S_{22} \right|^2 + \left| \Delta \right|^2 + \left| S_{22} \right|^2 \\ + \left| \Delta \right|^2 \left| S_{11}^* \right|^2 - S_{22} \Delta^* S_{11} - S_{22}^* \Delta S_{11}^* \end{array}}{\left(\left| S_{22} \right|^2 - \left| \Delta \right|^2 \right)^2}$$

or
$$\left| \Gamma_L - \frac{\left(S_{22} - \Delta S_{11}^* \right)^*}{\left| S_{22} \right|^2 - \left| \Delta \right|^2} \right|^2 = \frac{\left| S_{11} \right|^2 \left| S_{22} \right|^2 + \left| \Delta \right|^2 - S_{22} \Delta^* S_{11} - S_{22}^* \Delta S_{11}^*}{\left(\left| S_{22} \right|^2 - \left| \Delta \right|^2 \right)^2}$$

or
$$\left| \Gamma_L - \frac{\left(S_{22} - \Delta S_{11}^* \right)^*}{\left| S_{22} \right|^2 - \left| \Delta \right|^2} \right|^2 = \frac{\left| S_{11} \right|^2 \left| S_{22} \right|^2 + \left| S_{11} S_{22} - S_{12} S_{21} \right|^2}{\left(\left| S_{22} \right|^2 - \left| \Delta \right|^2 \right)^2}$$
$$+ \frac{-S_{22} \left(S_{11} S_{22} - S_{12} S_{21} \right)^* S_{11} - S_{22}^* \left(S_{11} S_{22} - S_{12} S_{21} \right) S_{11}^*}{\left(\left| S_{22} \right|^2 - \left| \Delta \right|^2 \right)^2}$$

or
$$\left| \Gamma_L - \frac{\left(S_{22} - \Delta S_{11}^* \right)^*}{\left| S_{22} \right|^2 - \left| \Delta \right|^2} \right|^2 = \frac{\begin{array}{c} \left| S_{11} \right|^2 \left| S_{22} \right|^2 + \left| S_{11} \right|^2 \left| S_{22} \right|^2 + \left| S_{12} \right|^2 \left| S_{21} \right|^2 \\ -S_{11} S_{22} S_{12}^* S_{21}^* - S_{11}^* S_{22}^* S_{12} S_{21} \end{array}}{\left(\left| S_{22} \right|^2 - \left| \Delta \right|^2 \right)^2}$$
$$+ \frac{-\left| S_{11} \right|^2 \left| S_{22} \right|^2 + S_{11} S_{22} S_{12}^* S_{21}^* - \left| S_{11} \right|^2 \left| S_{22} \right|^2 + S_{11}^* S_{22}^* S_{12} S_{21}}{\left(\left| S_{22} \right|^2 - \left| \Delta \right|^2 \right)^2}$$

or
$$\left| \Gamma_L - \frac{\left(S_{22} - \Delta S_{11}^* \right)^*}{\left| S_{22} \right|^2 - \left| \Delta \right|^2} \right|^2 = \frac{\left| S_{12} \right|^2 \left| S_{21} \right|^2}{\left(\left| S_{22} \right|^2 - \left| \Delta \right|^2 \right)^2}$$

or
$$\left| \Gamma_L - \frac{\left(S_{22} - \Delta S_{11}^* \right)^*}{\left| S_{22} \right|^2 - \left| \Delta \right|^2} \right|^2 = \frac{\left(\left| S_{12} \right| \left| S_{21} \right| \right)^2}{\left(\left| S_{22} \right|^2 - \left| \Delta \right|^2 \right)^2} = \left| \frac{S_{12} S_{21}}{\left| S_{22} \right|^2 - \left| \Delta \right|^2} \right|^2 \tag{15.114}$$

In a complex Γ plane, this equation reveals a circle with centre

$$C_L = \frac{\left(S_{22} - \Delta S_{11}^* \right)^*}{\left| S_{22} \right|^2 - \left| \Delta \right|^2} \tag{15.115}$$

and radius

$$R_L = \left| \frac{S_{12} S_{21}}{\left| S_{22} \right|^2 - \left| \Delta \right|^2} \right| \tag{15.116}$$

The circle is known as the output stability circle. A similar expression can be obtained for the input stability circle. For the input stability circle, we get the following expression:

$$C_S = \frac{\left(S_{11} - \Delta S_{22}^*\right)^*}{|S_{11}|^2 - |\Delta|^2} \tag{15.117}$$

$$R_S = \left|\frac{S_{12}S_{21}}{|S_{11}|^2 - |\Delta|^2}\right| \tag{15.118}$$

Equations (15.115)–(15.118) reveal that if the S-parameter is given, then we can draw the input and output stability circles.

We have $|\Gamma_{out}| < 1$ on one side of the input stability circle, while $|\Gamma_{out}| > 1$ on the other side. Similarly, we have $|\Gamma_{in}| < 1$ on one side of the output stability circle, while $|\Gamma_{in}| > 1$ on the other side. Therefore, it is now important to identify the regions for which $|\Gamma_{out}| < 1$ and $|\Gamma_{in}| < 1$. To do this, let us consider that the output stability circle has been drawn in the Γ_L plane for $S_{11} < 1$ and $S_{11} > 1$, as shown in Figs 15.18(a) and (b), respectively. Now if we set $Z_L = Z_0$, then $\Gamma_L = 0$ and Eq. (15.75) shows that $|\Gamma_{in}| = |S_{11}|$ (15.119)

Equation (15.119) reveals that if $|S_{11}| < 1$ then $|\Gamma_{in}|$ is also less than 1. Therefore, we can conclude that the centre of the Smith chart ($\Gamma_L = 0$) must be in the stable region for $S_{11} < 1$. More precisely, we can say that all the region of the Smith chart that is exterior to the output stability circle, corresponding to $S_{11} < 1$ and defines the stable range of Γ_L. Now, if we set $Z_L = Z_0$ and $S_{11} > 1$, then $|\Gamma_{in}| > 1$ and $\Gamma_L = 0$ must be in the unstable region. Alternatively, we can say that the centre of the Smith chart is in the unstable region. Thus, the region of the Smith chart that is exterior to the output stability circle (corresponding to $S_{11} > 1$) defines the unstable range of Γ_L. A similar argument holds for the input stability circle. The stable and unstable regions of the smith chart for $S_{11} < 1$ and $S_{11} > 1$ are shown in Figs 15.18(a) and (b), respectively.

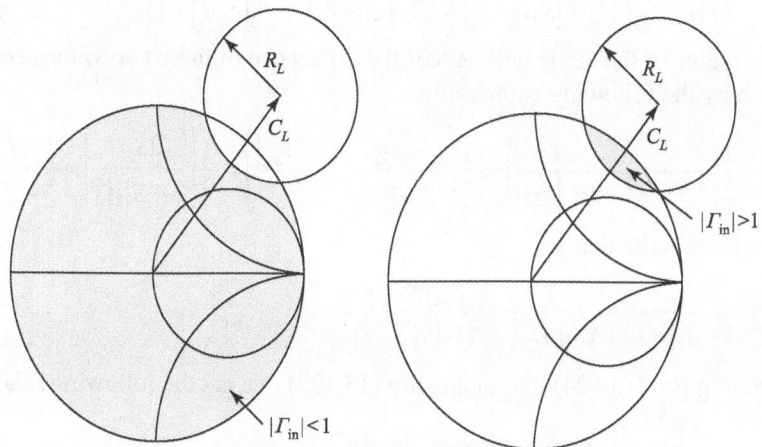

Fig. 15.18   Stable region on Smith chart for conditional stability
(a) $|\Gamma_{in}| < 1$ (b) $|\Gamma_{in}| > 1$

To make a device unconditionally stable, the stability circles must be completely outside of or totally within the Smith chart. These conditions can be expressed as follows:

$$\left\|C_L\right| - R_L\right| > 1 \text{ for } |S_{11}| < 1 \tag{15.120}$$

and

$$\left\|C_S\right| - R_S\right| > 1 \text{ for } |S_{22}| < 1 \tag{15.121}$$

It may be noted that $S_{11} > 1$ or $S_{22} > 1$ can never lead to unconditional stability because we can always have a source and load impedance Z_0 that will result in $\Gamma_L = 0$ or $\Gamma_S = 0$, and hence $|\Gamma_{\text{in}}| > 1$ or $|\Gamma_{\text{out}}| > 1$.

To find the condition of unconditional stability, let us first consider inequality (15.108). For all $|\Gamma_L| < 1$, this inequality can be written as follows:

$$\left|\frac{1}{S_{22}}\left(S_{11}S_{22} + \frac{S_{12}S_{21}S_{22}\Gamma_L}{1 - S_{22}\Gamma_L}\right)\right| < 1 \text{ or } \left|\frac{1}{S_{22}}\left(S_{11}S_{22} - S_{12}S_{21} + \frac{S_{12}S_{21}}{1 - S_{22}\Gamma_L}\right)\right| < 1$$

or

$$\left|\frac{1}{S_{22}}\left(\Delta + \frac{S_{12}S_{21}}{1 - S_{22}\Gamma_L}\right)\right| < 1 \tag{15.122}$$

where we have used Eq. (15.113).

The load reflection coefficient can be written as $\Gamma_L = |\Gamma_L|e^{j\theta}$ (15.123)

For the passive load and matching network, the allowable value of Γ_L must lie within the unit circle defined by $|\Gamma_L| = 1$. For $|\Gamma_L| = 1$, the factor $1/(1 - S_{22}\Gamma_L)$ can be written as follows:

$$\frac{1}{1 - S_{22}\Gamma_L} = \frac{1}{1 - S_{22}|\Gamma_L|e^{j\theta}} = \frac{1}{1 - S_{22}e^{j\theta}} = \frac{1}{1 - |S_{22}|e^{j\phi}}$$

The maximum and minimum values of this factor are $\dfrac{1}{1 - |S_{22}|}$ and $\dfrac{1}{1 + |S_{22}|}$, respectively.

The factor $1/(1 - S_{22}\Gamma_L)$ maps the unity circle intro a new circle. The centre of this circle is the average of these two values and can be given as follows:

$$\frac{1}{2}\left(\frac{1}{1 - |S_{22}|} + \frac{1}{1 + |S_{22}|}\right) = \frac{1}{2}\left(\frac{1 + |S_{22}| + 1 - |S_{22}|}{1 - |S_{22}|^2}\right) = \frac{1}{2}\left(\frac{2}{1 - |S_{22}|^2}\right) = \frac{1}{1 - |S_{22}|^2}$$

The radius of the circle will be half the difference of these two values and can be given by the following expression:

$$\frac{1}{2}\left(\frac{1}{1 - |S_{22}|} - \frac{1}{1 + |S_{22}|}\right) = \frac{1}{2}\left(\frac{1 + |S_{22}| - 1 + |S_{22}|}{1 - |S_{22}|^2}\right) = \frac{1}{2}\left(\frac{2|S_{22}|}{1 - |S_{22}|^2}\right) = \frac{|S_{22}|}{1 - |S_{22}|^2}$$

Thus, we can write that

$$\frac{1}{1 - S_{22}e^{j\theta}} = \frac{1}{1 - |S_{22}|^2} + \frac{|S_{22}|e^{j\psi}}{1 - |S_{22}|^2} \tag{15.124}$$

Substituting Eq. (15.124) into inequality (15.122), we get the following relation:

$$\frac{1}{|S_{22}|}\left\|\left(\Delta + \frac{S_{12}S_{21}}{1 - |S_{22}|^2}\right) + \frac{S_{12}S_{21}|S_{22}|e^{j\psi}}{1 - |S_{22}|^2}\right\| < 1 \tag{15.125}$$

The LHS of inequality (15.125) becomes maximum for the value of ψ that makes phase angles of the two terms identical. For this condition, inequality (15.125) can be simplified to the following form:

$$\frac{1}{|S_{22}|}\left|\Delta+\frac{S_{12}S_{21}}{1-|S_{22}|^2}\right|+\frac{|S_{12}S_{21}|}{1-|S_{22}|^2}<1$$

or

$$0\leq\frac{1}{|S_{22}|}\left|\Delta+\frac{S_{12}S_{21}}{1-|S_{22}|^2}\right|<1-\frac{|S_{12}S_{21}|}{1-|S_{22}|^2} \tag{15.126}$$

Squaring both sides of inequality (15.126), we get the following expression:

$$\frac{1}{|S_{22}|^2}\left[|\Delta|^2+\frac{|S_{12}|^2|S_{21}|^2}{\left(1-|S_{22}|^2\right)^2}+\Delta^*\frac{S_{12}S_{21}}{1-|S_{22}|^2}+\Delta\frac{S_{12}^*S_{21}^*}{1-|S_{22}|^2}\right]<1$$

$$+\frac{|S_{12}S_{21}|^2}{\left(1-|S_{22}|^2\right)^2}-2\frac{|S_{12}S_{21}|}{1-|S_{22}|^2}$$

or

$$|\Delta|^2\left(1-|S_{22}|^2\right)+\frac{|S_{12}|^2|S_{21}|^2}{\left(1-|S_{22}|^2\right)}+\Delta^*S_{12}S_{21}+\Delta S_{12}^*S_{21}^*<|S_{22}|^2\left(1-|S_{22}|^2\right)$$

$$+\frac{|S_{22}|^2|S_{12}S_{21}|^2}{\left(1-|S_{22}|^2\right)}-2|S_{12}S_{21}||S_{22}|^2$$

or

$$|\Delta|^2\left(1-|S_{22}|^2\right)+\frac{|S_{12}|^2|S_{21}|^2\left(1-|S_{22}|^2\right)}{\left(1-|S_{22}|^2\right)}$$

$$+\Delta^*S_{12}S_{21}+\Delta S_{12}^*S_{21}^*<|S_{22}|^2\left(1-2|S_{12}S_{21}|-|S_{22}|^2\right)$$

or

$$|S_{12}|^2|S_{21}|^2+|\Delta|^2\left(1-|S_{22}|^2\right)+\Delta S_{12}^*S_{21}^*$$

$$+\Delta^*S_{12}S_{21}<|S_{22}|^2\left(1-2|S_{12}S_{21}|-|S_{22}|^2\right) \tag{15.127}$$

Using Eq. (15.113) in inequality (15.127), we get the following relation:

$$|S_{12}|^2|S_{21}|^2+|S_{11}S_{22}-S_{12}S_{21}|^2\left(1-|S_{22}|^2\right)+\left(S_{11}S_{22}-S_{12}S_{21}\right)S_{12}^*S_{21}^*$$

$$+\left(S_{11}S_{22}-S_{12}S_{21}\right)^*S_{12}S_{21}<|S_{22}|^2\left(1-2|S_{12}S_{21}|-|S_{22}|^2\right)$$

or $|S_{12}|^2|S_{21}|^2+\left(|S_{11}|^2|S_{22}|^2+|S_{12}|^2|S_{21}|^2-S_{11}S_{22}S_{12}^*S_{21}^*-S_{11}^*S_{22}^*S_{12}S_{21}\right)\left(1-|S_{22}|^2\right)$

$$+\left(S_{11}S_{22}-S_{12}S_{21}\right)S_{12}^*S_{21}^*+\left(S_{11}S_{22}-S_{12}S_{21}\right)^*S_{12}S_{21}<|S_{22}|^2\left(1-2|S_{12}S_{21}|-|S_{22}|^2\right)$$

or

$$|S_{12}|^2 |S_{21}|^2 + |S_{11}|^2 |S_{22}|^2 + |S_{12}|^2 |S_{21}|^2 - S_{11} S_{22} S_{12}^* S_{21}^* - S_{11}^* S_{22}^* S_{12} S_{21}$$

$$- |S_{11}|^2 |S_{22}|^4 - |S_{12}|^2 |S_{21}|^2 |S_{22}|^2 + S_{11} S_{22} S_{12}^* S_{21}^* |S_{22}|^2$$

$$+ S_{11}^* S_{22}^* S_{12} S_{21} |S_{22}|^2 + S_{11} S_{22} S_{12}^* S_{21}^* - S_{12} S_{21} S_{12}^* S_{21}^* + S_{11}^* S_{22}^* S_{12} S_{21}$$

$$- S_{12}^* S_{21}^* S_{12} S_{21} < |S_{22}|^2 \left(1 - 2|S_{12} S_{21}| - |S_{22}|^2\right)$$

or

$$|S_{12}|^2 |S_{21}|^2 + |S_{11}|^2 |S_{22}|^2 + |S_{12}|^2 |S_{21}|^2 - |S_{11}|^2 |S_{22}|^4 - |S_{12}|^2 |S_{21}|^2 |S_{22}|^2$$

$$+ S_{11} S_{22} S_{12}^* S_{21}^* |S_{22}|^2 + S_{11}^* S_{22}^* S_{12} S_{21} |S_{22}|^2 - |S_{12}|^2 |S_{21}|^2$$

$$- |S_{12}|^2 |S_{21}|^2 < |S_{22}|^2 \left(1 - 2|S_{12} S_{21}| - |S_{22}|^2\right)$$

or

$$|S_{11}|^2 |S_{22}|^2 - |S_{11}|^2 |S_{22}|^4 - |S_{12}|^2 |S_{21}|^2 |S_{22}|^2 + S_{11} S_{22} S_{12}^* S_{21}^* |S_{22}|^2$$

$$+ S_{11}^* S_{22}^* S_{12} S_{21} |S_{22}|^2 < |S_{22}|^2 \left(1 - 2|S_{12} S_{21}| - |S_{22}|^2\right)$$

or

$$|S_{22}|^2 \left(|S_{11}|^2 - |S_{11}|^2 |S_{22}|^2 - |S_{12}|^2 |S_{21}|^2 + S_{11} S_{22} S_{12}^* S_{21}^* + S_{11}^* S_{22}^* S_{12} S_{21}\right)$$

$$< |S_{22}|^2 \left(1 - 2|S_{12} S_{21}| - |S_{22}|^2\right)$$

or

$$|S_{11}|^2 - \left(|S_{11}|^2 |S_{22}|^2 + |S_{12}|^2 |S_{21}|^2 - S_{11} S_{22} S_{12}^* S_{21}^* - S_{11}^* S_{22}^* S_{12} S_{21}\right)$$

$$< 1 - 2|S_{12} S_{21}| - |S_{22}|^2$$

or $$|S_{11}|^2 - |\Delta|^2 < 1 - 2|S_{12} S_{21}| - |S_{22}|^2$$

or $$2|S_{12} S_{21}| < 1 - |S_{11}|^2 - |S_{22}|^2 + |\Delta|^2$$

or $$K = \frac{1 - |S_{11}|^2 - |S_{22}|^2 + |\Delta|^2}{2|S_{12} S_{21}|} > 1 \tag{15.128}$$

Since K remains unchanged after the interchange of S_{11} and S_{22}, inequality (15.128) also applies for $|\Gamma_{\text{out}}| < 1$.

Inequality (15.126) implies that $0 < 1 - |S_{22}|^2 - |S_{12} S_{21}|$ \hfill (15.129)

Instead of starting with inequality (15.108), if we start with inequality (15.109) then following the same steps we can get

$$0 < 1 - |S_{11}|^2 - |S_{12} S_{21}| \tag{15.130}$$

Adding inequalities (15.129) and (15.130), we get the following expression:

$$0 < 2 - |S_{11}|^2 - |S_{22}|^2 - 2|S_{12} S_{21}|$$

or $$2|S_{12} S_{21}| < 2 - |S_{11}|^2 - |S_{22}|^2 \tag{15.131}$$

Now, we know that $|\Delta| = |S_{11} S_{22} - S_{12} S_{21}| \le |S_{11} S_{22}| + |S_{12} S_{21}|$

or $$|\Delta| - |S_{11} S_{22}| \le |S_{12} S_{21}| \tag{15.132}$$

Substituting inequality (15.132) into inequality (15.131), we get the following relation:

$$2(|\Delta| - |S_{11}S_{22}|) < 2 - |S_{11}|^2 - |S_{22}|^2 \text{ or } |\Delta| < 1 + |S_{11}S_{22}| - \frac{1}{2}|S_{11}|^2 - \frac{1}{2}|S_{22}|^2$$

or $\quad |\Delta| < 1 - \frac{1}{2}(|S_{11}|^2 + |S_{22}|^2 - 2|S_{11}S_{22}|)$

or $\quad |\Delta| < 1 - \frac{1}{2}(|S_{11}| - |S_{22}|)^2 < 1 \qquad (15.133)$

Therefore, we can summarize that a two-port network will be unconditionally stable if $K > 1$ and $|\Delta| < 1$. Since the $K - \Delta$ test involves constraints on two parameters, it cannot be used to compare the relative stability of two or more devices.

EXAMPLE 15.3 The S-parameters of a certain FET at 3 GHz with a bias voltage $V_{gs} = 0$ and $Z_0 = 50\,\Omega$ are as follows: $S_{11} = 0.87\angle - 57°$, $S_{12} = 0.015\angle60°$, $S_{21} = 3.2\angle125°$, and $S_{22} = 0.75\angle - 30°$. Determine the stability of the transistor and plot the stability circle.

Solution Given: $Z_0 = 50\,\Omega$, $S_{11} = 0.87\angle - 57° = 0.4738 - j0.7296$, $S_{12} = 0.015\angle60° = 0.0075 + j0.013$, $S_{21} = 3.2\angle125° = -1.8354 + j2.6213$, and $S_{22} = 0.75\angle - 30° = 0.6495 - j0.375$.

Now, $\quad \Delta = S_{11}S_{22} - S_{12}S_{21} = 0.082 - j0.6474$ or $|\Delta| = 0.6526 < 1$

and $\quad K = \dfrac{1 - |S_{11}|^2 - |S_{22}|^2 + |\Delta|^2}{2|S_{12}S_{21}|}$

or $\quad K = \dfrac{1 - |0.4738 - j0.7296|^2 - |0.6495 - j0.375| + (0.6526)^2}{2|(0.0075 + j0.013)(-1.8354 + j2.6213)|} = 1.1091 > 1$

Therefore, the transistor is unconditionally stable.

The centre and radius of the stability circles can be found as follows:

$$C_L = \frac{(S_{22} - \Delta S_{11}^*)^*}{|S_{22}|^2 - |\Delta|^2} = 1.0122 + j0.9371 = 1.3794\angle42.7941°$$

$$R_L = \left| \frac{S_{12}S_{21}}{|S_{22}|^2 - |\Delta|^2} \right| = \left| \frac{(0.0075 + j0.013)(-1.8354 + j2.6213)}{|0.6495 - j0.375|^2 - (0.6526)^2} \right| = 0.3513$$

$$C_S = \frac{(S_{11} - \Delta S_{22}^*)^*}{|S_{11}|^2 - |\Delta|^2} = 0.5372 + j1.0267 = 1.1587\angle62.3821°$$

$$R_S = \left| \frac{S_{12}S_{21}}{|S_{11}|^2 - |\Delta|^2} \right| = \left| \frac{(0.0075 + j0.013)(-1.8354 + j2.6213)}{|0.4738 - j0.7296|^2 - (0.6526)^2} \right| = 0.145$$

The stability circles are shown in Fig. 15.19.

Practice Problem

15.3 The S-parameters of a certain FET at 4 GHz with a bias voltage $V_{gs} = 0$ and $Z_0 = 50\,\Omega$ are as follows: $S_{11} = 0.9\angle - 60°$, $S_{12} = 0.01\angle55°$, $S_{21} = 3.25\angle130°$, and $S_{22} = 0.7\angle - 25°$. Determine the stability of the transistor and plot the stability circle.

[**Unconditinally stable, 1.4795∠37.4033°, 0.3531, 1.1153∠62.8375°, 0.0789**]

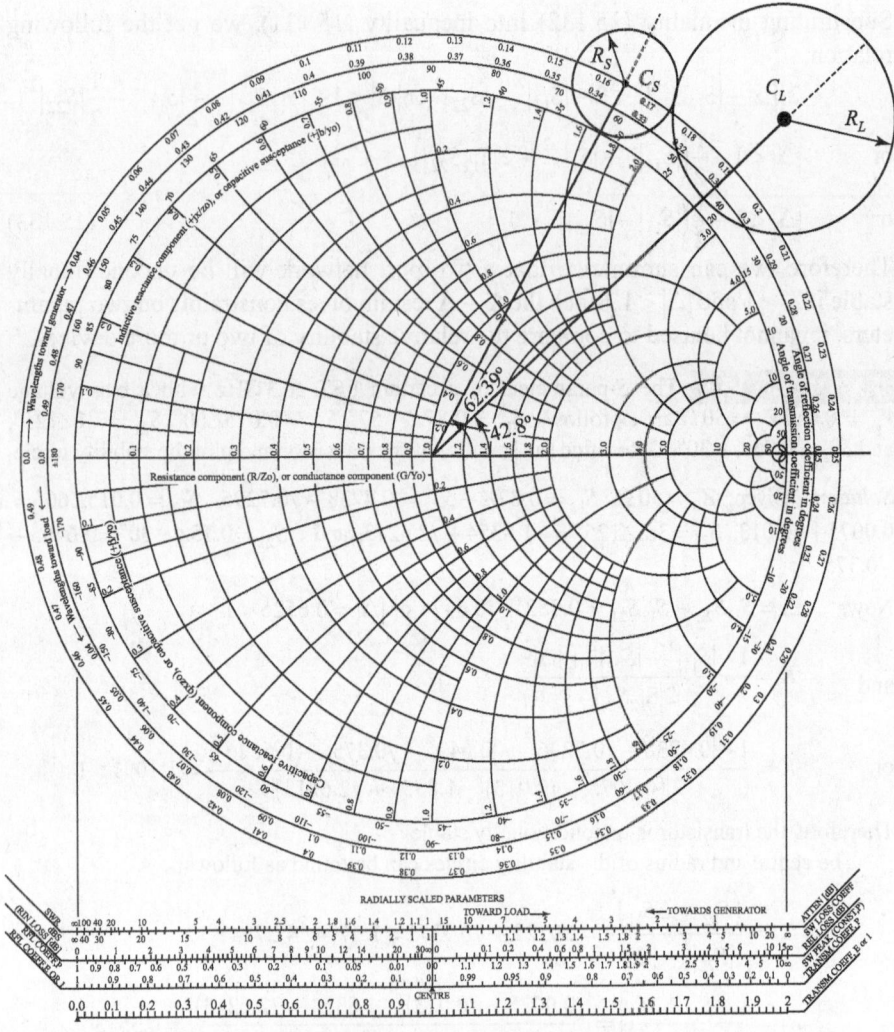

Fig. 15.19 Figure for Example 15.3

15.4 Single-stage Transistor Amplifier Design

In practice, transistor amplifiers can be single-stage or multi-stage. In the following sections we will discuss the different types of single stage amplifier design: (a) design for maximum gain, (b) design for specified gain, and (c) design of low noise amplifier. Multistage/band amplifiers will be discussed in the next section.

15.4.1 Design for Maximum Gain

The design of the input and output matching section of a transistor amplifier plays an important role in the stability of the amplifier as well as in the total gain. Since for a given transistor G_0 is fixed, as evident from Eq. (15.102), the overall gain of the amplifier is controlled by G_S and G_L of the matching section. The maximum gain is realized when these sections provide a conjugate match between the source or load impedance and the transistor.

The maximum power transfer from the input matching section to the transistor occurs when $\Gamma_{in} = \Gamma_S^*$ (15.134)

Similarly, the maximum power transfer from the transistor to the output matching section occurs when $\Gamma_{out} = \Gamma_L^*$ (15.135)

Therefore, from Eq. (15.94) we can write the following expression:

$$G_{T,\max} = \frac{|S_{21}|^2 \left(1-|\Gamma_L|^2\right)\left(1-|\Gamma_S|^2\right)}{\left|1-|\Gamma_S|^2\right|^2 |1-S_{22}\Gamma_L|^2} = \frac{1}{1-|\Gamma_S|^2}|S_{21}|^2 \frac{1-|\Gamma_L|^2}{|1-S_{22}\Gamma_L|^2}$$ (15.136)

Using Eqs (15.134) and (15.135) in Eqs (15.75) and (15.76), respectively, we get the following relations:

$$\Gamma_S^* = S_{11} + \frac{S_{12}S_{21}\Gamma_L}{1-S_{22}\Gamma_L}$$

or $\qquad \Gamma_S = S_{11}^* + \dfrac{S_{12}^*S_{21}^*\Gamma_L^*}{1-S_{22}^*\Gamma_L^*} = S_{11}^* + \dfrac{S_{12}^*S_{21}^*}{\left(1/\Gamma_L^*\right)-S_{22}^*}$ (15.137)

and $\qquad \Gamma_L^* = S_{22} + \dfrac{S_{12}S_{21}\Gamma_S}{1-S_{11}\Gamma_S}$ (15.138)

or $\qquad \Gamma_L^* = \dfrac{S_{22}-S_{11}S_{22}\Gamma_S+S_{12}S_{21}\Gamma_S}{1-S_{11}\Gamma_S} = \dfrac{S_{22}-\Gamma_S\left(S_{11}S_{22}-S_{12}S_{21}\right)}{1-S_{11}\Gamma_S}$

or $\qquad \Gamma_L^* = \dfrac{S_{22}-\Delta\Gamma_S}{1-S_{11}\Gamma_S}$ (15.139)

where we have used Eq. (15.113). Substituting Eq. (15.139) into Eq. (15.137), we get the following relation:

$$\Gamma_S = S_{11}^* + \frac{S_{12}^*S_{21}^*}{\dfrac{1-S_{11}\Gamma_S}{S_{22}-\Delta\Gamma_S}-S_{22}^*} = S_{11}^* + \frac{S_{12}^*S_{21}^*}{\dfrac{1-S_{11}\Gamma_S-S_{22}^*\left(S_{22}-\Delta\Gamma_S\right)}{S_{22}-\Delta\Gamma_S}}$$

or $\qquad \Gamma_S = S_{11}^* + \dfrac{S_{12}^*S_{21}^*\left(S_{22}-\Delta\Gamma_S\right)}{1-S_{11}\Gamma_S-|S_{22}|^2+\Delta S_{22}^*\Gamma_S}$

or $\qquad \Gamma_S - S_{11}\Gamma_S^2 - |S_{22}|^2\Gamma_S + \Delta S_{22}^*\Gamma_S^2 = S_{11}^* - |S_{11}|^2\Gamma_S - S_{11}^*|S_{22}|^2$
$$\qquad\qquad +\Delta S_{11}^*S_{22}^*\Gamma_S + S_{12}^*S_{21}^*S_{22} - S_{12}^*S_{21}^*\Delta\Gamma_S$$

or $\qquad \Gamma_S\left(1-|S_{22}|^2\right) + \Gamma_S^2\left(\Delta S_{22}^*-S_{11}\right) = \Gamma_S\left\{\Delta\left(S_{11}^*S_{22}^*-S_{12}^*S_{21}^*\right)-|S_{11}|^2\right\}$
$$\qquad\qquad +S_{11}^*\left(1-|S_{22}|^2\right)+S_{12}^*S_{21}^*S_{22}$$

or $\qquad \Gamma_S\left(1-|S_{22}|^2\right) + \Gamma_S^2\left(\Delta S_{22}^*-S_{11}\right) = \Gamma_S\left(|\Delta|^2-|S_{11}|^2\right)+S_{11}^*$
$$\qquad\qquad -S_{11}^*S_{22}S_{22}^* + S_{12}^*S_{21}^*S_{22}$$

or $\quad \left(S_{11}-\Delta S_{22}^*\right)\Gamma_S^2 +\left(|\Delta|^2 -|S_{11}|^2 +|S_{22}|^2 -1\right)\Gamma_S +S_{11}^* -S_{22}\left(S_{11}^* S_{22}^* -S_{12}^* S_{21}^*\right)$

or $\quad \left(S_{11}-\Delta S_{22}^*\right)\Gamma_S^2 +\left(|\Delta|^2 -|S_{11}|^2 +|S_{22}|^2 -1\right)\Gamma_S +\left(S_{11}^* -\Delta^* S_{22}\right)=0$ $\quad$ (15.140)

where we have used Eq. (15.113). Solving Eq. (15.140), we can find the expression of Γ_S, as follows:

$$\Gamma_S =\frac{B_1 \pm\sqrt{B_1^2 -4|C_1|^2}}{2C_1} \tag{15.141}$$

where $\quad B_1 =1+|S_{11}|^2 -|S_{22}|^2 -|\Delta|^2$ $\qquad$ (15.142)

and $\quad C_1 = S_{11} -\Delta S_{22}^*$ $\qquad$ (15.143)

Proceeding in a similar way, Γ_L can be expressed as follows:

$$\Gamma_L =\frac{B_2 \pm\sqrt{B_2^2 -4|C_2|^2}}{2C_2} \tag{15.144}$$

where $\quad B_2 =1+|S_{22}|^2 -|S_{11}|^2 -|\Delta|^2$ $\qquad$ (15.145)

and $\quad C_2 = S_{22} -\Delta S_{11}^*$ $\qquad$ (15.146)

For a unilateral case, $S_{12} = 0$ and Eqs (15.137) and (15.138) show that

$$\Gamma_S = S_{11}^* \tag{15.147}$$

and $\quad \Gamma_L^* = S_{22}$ $\qquad$ (15.148)

Using Eqs (15.147) and (15.148) in Eq. (15.136), the maximum unilateral transducer gain can be expressed as follows:

$$G_{T,\max} =\frac{|S_{21}|^2 \left(1-|\Gamma_L|^2\right)\left(1-|\Gamma_S|^2\right)}{\left|1-|\Gamma_S|^2\right|^2 \left|1-S_{22}\Gamma_L\right|^2} =\frac{1}{1-|\Gamma_S|^2}|S_{21}|^2 \frac{1-|\Gamma_L|^2}{\left|1-S_{22}\Gamma_L\right|^2}$$

or $\quad G_{TU,\max} =\frac{1}{1-|S_{11}|^2}|S_{21}|^2 \frac{1-|S_{22}|^2}{\left(1-|S_{22}|^2\right)^2} =\frac{1}{1-|S_{11}|^2}|S_{21}|^2 \frac{1}{1-|S_{22}|^2}$ $\quad$ (15.149)

EXAMPLE 15.4 Prove that $|B_1/(2C_1)|>1$ implies $|K|>1$. Provided $B_1 = 1 + |S_{11}|^2 - |S_{22}|^2 - |\Delta|^2$ and $C_1 = S_{11} -\Delta S_{22}^*$.

Solution $|B_1|>2|C_1|$ or $\left|1+|S_{11}|^2 -|S_{22}|^2 -|\Delta|^2\right|>2\left|S_{11}-\Delta S_{22}^*\right|$

Squaring both sides, we get the following expression:

$$\left|1+|S_{11}|^2 -|S_{22}|^2 -|\Delta|^2\right|^2 >4\left(|S_{11}|^2 +|\Delta|^2|S_{22}|^2 -S_{11}S_{22}\Delta^* -S_{11}^* S_{22}^*\Delta\right)$$

Substituting the expression of Δ in this equation, we get the following expression:

$$\left|1+|S_{11}|^2 -|S_{22}|^2 -|\Delta|^2\right|^2 >4\left\{\begin{array}{l}|S_{11}|^2 +|\Delta|^2|S_{22}|^2 -S_{11}S_{22}\left(S_{11}^* S_{22}^* -S_{12}^* S_{21}^*\right)\\ -S_{11}^* S_{22}^*\left(S_{11}S_{22} -S_{12}S_{21}\right)\end{array}\right\}$$

or $\quad \left|1+\left|S_{11}\right|^{2}-\left|S_{22}\right|^{2}-\left|\Delta\right|^{2}\right|^{2}>4\left(\begin{array}{l}\left|S_{11}\right|^{2}+\left|\Delta\right|^{2}\left|S_{22}\right|^{2}-\left|S_{11}\right|^{2}\left|S_{22}\right|^{2}+S_{11}S_{22}S_{12}^{*}S_{21}^{*}\\-\left|S_{11}\right|^{2}\left|S_{22}\right|^{2}+S_{11}^{*}S_{22}^{*}S_{12}S_{21}\end{array}\right)$

or $\quad \left|1+\left|S_{11}\right|^{2}-\left|S_{22}\right|^{2}-\left|\Delta\right|^{2}\right|^{2}>$

$\quad 4\left\{\left|S_{12}S_{21}\right|^{2}+\left|S_{11}\right|^{2}+\left|\Delta\right|^{2}\left|S_{22}\right|^{2}-\left|S_{11}\right|^{2}\left|S_{22}\right|^{2}-\left(\begin{array}{l}\left|S_{11}\right|^{2}\left|S_{22}\right|^{2}-S_{11}S_{22}S_{12}^{*}S_{21}^{*}\\-S_{11}^{*}S_{22}^{*}S_{12}S_{21}+\left|S_{12}S_{21}\right|^{2}\end{array}\right)\right\}$

or $\quad \left|1+\left|S_{11}\right|^{2}-\left|S_{22}\right|^{2}-\left|\Delta\right|^{2}\right|^{2}>4\left(\left|S_{12}S_{21}\right|^{2}+\left|S_{11}\right|^{2}+\left|\Delta\right|^{2}\left|S_{22}\right|^{2}-\left|S_{11}\right|^{2}\left|S_{22}\right|^{2}-\left|\Delta\right|^{2}\right)$

or $\quad \left|1+\left|S_{11}\right|^{2}-\left|S_{22}\right|^{2}-\left|\Delta\right|^{2}\right|^{2}>4\left\{\left|S_{12}S_{21}\right|^{2}+\left(1-\left|S_{22}\right|^{2}\right)\left(\left|S_{11}\right|^{2}-\left|\Delta\right|^{2}\right)\right\}$

or $\quad \left|\left(1-\left|S_{22}\right|^{2}\right)+\left(\left|S_{11}\right|^{2}-\left|\Delta\right|^{2}\right)\right|^{2}-4\left(1-\left|S_{22}\right|^{2}\right)\left(\left|S_{11}\right|^{2}-\left|\Delta\right|^{2}\right)>4\left|S_{12}S_{21}\right|^{2}$

or $\quad \left|\left(1-\left|S_{22}\right|^{2}\right)-\left(\left|S_{11}\right|^{2}-\left|\Delta\right|^{2}\right)\right|^{2}>4\left|S_{12}S_{21}\right|^{2}$

or $\quad \left|1-\left|S_{22}\right|^{2}-\left|S_{11}\right|^{2}+\left|\Delta\right|^{2}\right|>2\left|S_{12}S_{21}\right|$

or $\quad \dfrac{\left|1-\left|S_{22}\right|^{2}-\left|S_{11}\right|^{2}+\left|\Delta\right|^{2}\right|}{2\left|S_{12}S_{21}\right|}>1$

or $\quad \left|K\right|>1$ (proved)

Practice Problems

15.4[@] Find the maximum gain of an FET amplifier provided that $S_{11}=0.72\angle-104°$, $S_{12}=0.03\angle55°$, $S_{21}=2.75\angle87°$, and $S_{22}=0.72\angle-48°$. The input impedance is 50 Ω.

15.5 Find the maximum gain of an FET amplifier provided that $S_{11}=0.74\angle-94°$, $S_{12}=0.02\angle60°$, $S_{21}=2.9\angle65°$, and $S_{22}=0.8\angle-55°$. The input impedance is 50 Ω.

[17.8853 dB]

15.4.2 Design for Specified Gain

In amplifier design, sometimes it is required to sacrifice the gain for a wider bandwidth or to design an amplifier with a specified gain. This can be done by purposefully introducing mismatches in the input and output matching section to reduce the overall gain.

To understand the design procedure and simplify the analysis, let us assume that $|S_{12}|$ is so small that it can be neglected or assumed to be zero. That is, the device is unilateral. The error introduced in the transducer gain due to such an approximation lies within the following range:

$$\frac{1}{(1+U)^{2}}<\frac{G_{T}}{G_{TU}}<\frac{1}{(1-U)^{2}} \qquad (15.150)$$

[@] Solution in Online Resources

where U is the unilateral figure of merit and can be expressed as follows:

$$U = \frac{|S_{11}||S_{12}||S_{21}||S_{22}|}{\left(1-|S_{11}|^2\right)\left(1-|S_{22}|^2\right)} \tag{15.151}$$

Expressions for G_S and G_L under such an approximation are given in Eqs (15.105) and (15.107), respectively. These gains are maximized when

$$\Gamma_S = S_{11}^* \tag{15.152}$$

and $\quad \Gamma_L = S_{22}^* \tag{15.153}$

Substituting Eqs (15.152) and (15.153) into Eqs (15.105) and (15.107) respectively, we get the following relations:

$$G_{S,\max} = \frac{1-\left|S_{11}^*\right|^2}{\left|1-S_{11}S_{11}^*\right|^2} = \frac{1-|S_{11}|^2}{\left(1-|S_{11}|^2\right)^2} = \frac{1}{1-|S_{11}|^2} \tag{15.154}$$

and $\quad G_{L,\max} = \dfrac{1-\left|S_{22}^*\right|^2}{\left|1-S_{22}S_{22}^*\right|^2} = \dfrac{1-|S_{22}|^2}{\left(1-|S_{22}|^2\right)^2} = \dfrac{1}{1-|S_{22}|^2} \tag{15.155}$

Therefore, the normalized gain factors will be as follows:

$$g_S = \frac{G_S}{G_{S,\max}} = \frac{1-|\Gamma_S|^2}{|1-S_{11}\Gamma_S|^2}\left(1-|S_{11}|^2\right) \tag{15.156}$$

and $\quad g_L = \dfrac{G_L}{G_{L,\max}} = \dfrac{1-|\Gamma_L|^2}{|1-S_{22}\Gamma_L|^2}\left(1-|S_{22}|^2\right) \tag{15.157}$

where $0 \le g_S \le 1$ and $0 \le g_L \le 1$. Equation (15.156) can be written as follows:

$$g_S|1-S_{11}\Gamma_S|^2 = \left(1-|\Gamma_S|^2\right)\left(1-|S_{11}|^2\right)$$

or $\quad g_S\left(1+|S_{11}|^2|\Gamma_S|^2 - S_{11}\Gamma_S - S_{11}^*\Gamma_S^*\right) = 1-|S_{11}|^2 - |\Gamma_S|^2 + |S_{11}|^2|\Gamma_S|^2$

or $\quad \left(g_S|S_{11}|^2 + 1 - |S_{11}|^2\right)|\Gamma_S|^2 - g_S\left(S_{11}\Gamma_S + S_{11}^*\Gamma_S^*\right) = 1-|S_{11}|^2 - g_S$

or $\quad \Gamma_S\Gamma_S^* - \dfrac{g_S\left(S_{11}\Gamma_S + S_{11}^*\Gamma_S^*\right)}{g_S|S_{11}|^2 + 1 - |S_{11}|^2} = \dfrac{1-|S_{11}|^2 - g_S}{g_S|S_{11}|^2 + 1 - |S_{11}|^2}$

or $\quad \Gamma_S\Gamma_S^* - \dfrac{g_S\left(S_{11}\Gamma_S + S_{11}^*\Gamma_S^*\right)}{g_S|S_{11}|^2 + 1 - |S_{11}|^2} + \dfrac{g_S^2|S_{11}|^2}{\left(g_S|S_{11}|^2 + 1 - |S_{11}|^2\right)^2}$

$$= \frac{1-|S_{11}|^2 - g_S}{g_S|S_{11}|^2 + 1 - |S_{11}|^2} + \frac{g_S^2|S_{11}|^2}{\left(g_S|S_{11}|^2 + 1 - |S_{11}|^2\right)^2}$$

or
$$\left|\Gamma_S - \frac{g_S S_{11}^*}{1-(1-g_S)|S_{11}|^2}\right|^2 = \frac{\left(1-|S_{11}|^2 - g_S\right)\left(g_S|S_{11}|^2 + 1 - |S_{11}|^2\right) + g_S^2|S_{11}|^2}{\left(g_S|S_{11}|^2 + 1 - |S_{11}|^2\right)^2}$$

or
$$\left|\Gamma_S - \frac{g_S S_{11}^*}{1-(1-g_S)|S_{11}|^2}\right|^2 = \frac{\begin{array}{c}g_S|S_{11}|^2 + 1 - |S_{11}|^2 - g_S|S_{11}|^4 - |S_{11}|^2 + |S_{11}|^4 \\ -g_S^2|S_{11}|^2 - g_S + g_S|S_{11}|^2 + g_S^2|S_{11}|^2\end{array}}{\left(g_S|S_{11}|^2 + 1 - |S_{11}|^2\right)^2}$$

or
$$\left|\Gamma_S - \frac{g_S S_{11}^*}{1-(1-g_S)|S_{11}|^2}\right|^2 = \frac{2g_S|S_{11}|^2 + 1 - 2|S_{11}|^2 - g_S|S_{11}|^4 + |S_{11}|^4 - g_S}{\left\{1-(1-g_S)|S_{11}|^2\right\}^2}$$

or
$$\left|\Gamma_S - \frac{g_S S_{11}^*}{1-(1-g_S)|S_{11}|^2}\right|^2 = \frac{\left(1-2|S_{11}|^2 + |S_{11}|^4\right) - g_S\left(1-2|S_{11}|^2 + |S_{11}|^4\right)}{\left\{1-(1-g_S)|S_{11}|^2\right\}^2}$$

or
$$\left|\Gamma_S - \frac{g_S S_{11}^*}{1-(1-g_S)|S_{11}|^2}\right|^2 = \frac{\left(1-g_S\right)\left(1-2|S_{11}|^2 + |S_{11}|^4\right)}{\left\{1-(1-g_S)|S_{11}|^2\right\}^2} = \frac{\left(1-g_S\right)\left(1-|S_{11}|^2\right)^2}{\left\{1-(1-g_S)|S_{11}|^2\right\}^2}$$

or
$$\left|\Gamma_S - \frac{g_S S_{11}^*}{1-(1-g_S)|S_{11}|^2}\right| = \frac{\sqrt{1-g_s}\left(1-|S_{11}|^2\right)}{1-(1-g_S)|S_{11}|^2} \qquad (15.158)$$

Equation (15.158) represents a circle with centre

$$C_S = \frac{g_S S_{11}^*}{1-(1-g_S)|S_{11}|^2} \qquad (15.159)$$

and radius

$$R_S = \frac{\sqrt{1-g_s}\left(1-|S_{11}|^2\right)}{1-(1-g_S)|S_{11}|^2} \qquad (15.160)$$

A similar analysis shows that Eq. (15.157) represents a circle with

$$C_L = \frac{g_L S_{22}^*}{1-(1-g_L)|S_{22}|^2} \qquad (15.161)$$

and
$$R_L = \frac{\sqrt{1-g_L}\left(1-|S_{22}|^2\right)}{1-(1-g_L)|S_{22}|^2} \qquad (15.162)$$

The centre of this circle lies on the straight line given by the angle of S_{11}^* and S_{22}^*, respectively. From Eqs (15.159) and (15.160) [or Eqs (15.161) and (15.162)], it may be noted that when $g_S = 1$ (or $g_L = 1$), $R_S = 0$ (or $R_L = 0$) and the centre lies on S_{11}^* (or S_{22}^*). It can also be shown that the 0 dB gain circle ($G_S = 1$ or $G_L = 1$) passes through the centre of the Smith chart.

The results can be used to plot a family of circles of constant gain for input and output sections, respectively. After that, Γ_S and Γ_L can be chosen along these circles to get the desired gain. It may be noted that though the choice of Γ_S and Γ_L is not unique. It is better to choose the points close to the centre of the Smith chart to minimize the mismatch and thus maximize the bandwidth.

If the transistor cannot be represented as a unilateral device, then the design procedure becomes tedious for a gain less than the maximum possible transducer power gain. In such a case, the operating or available power gain approach can be utilized. If we proceed with the operating power gain, it can be shown that

$$\left|\Gamma_L - \frac{g_p C_2^*}{1 + g_p\left(|S_{22}|^2 - |\Delta|^2\right)}\right|^2 = \frac{1 - 2k|S_{12}S_{21}|g_p + |S_{12}S_{21}|^2 g_p^2}{\left[1 + g_p\left(|S_{22}|^2 - |\Delta|^2\right)\right]^2} \tag{15.163}$$

where
$$g_p = \frac{G_P}{|S_{21}|^2} = \frac{1 - |\Gamma_L|^2}{\left(|1 - S_{22}\Gamma_L|^2\right) - \left(|S_{11} - \Gamma_L\Delta|^2\right)} \tag{15.164}$$

and
$$C_2 = S_{22} - S_{11}^*\Delta \tag{15.165}$$

Equation (15.163) represents a circle with centre

$$C_p = \frac{g_p C_2^*}{1 + g_p\left(|S_{22}|^2 - |\Delta|^2\right)} \tag{15.166}$$

and radius

$$R_p = \frac{\sqrt{1 - 2K|S_{12}S_{21}|g_p + |S_{12}S_{21}|^2 g_p^2}}{1 + g_p\left(|S_{22}|^2 - |\Delta|^2\right)} \tag{15.167}$$

For $R_p = 0$, Eq. (15.167) can be solved for g_p, which represents its maximum value:

$$g_{p,max} = \frac{1}{|S_{12}S_{21}|}\left(K - \sqrt{K^2 - 1}\right) \tag{15.168}$$

and
$$G_{p,max} = \frac{|S_{21}|}{|S_{12}|}\left(K - \sqrt{K^2 - 1}\right) \tag{15.169}$$

Similarly, if we proceed with available power gain it can be shown that

$$\left|\Gamma_S - \frac{g_a C_1^*}{1 + g_a\left(|S_{11}|^2 - |\Delta|^2\right)}\right|^2 = \frac{1 - 2k|S_{12}S_{21}|g_a + |S_{12}S_{21}|^2 g_a^2}{\left[1 + g_a\left(|S_{11}|^2 - |\Delta|^2\right)\right]^2} \tag{15.170}$$

Equation (15.170) represents a circle with centre

$$C_a = \frac{g_a C_1^*}{1 + g_a\left(|S_{11}|^2 - |\Delta|^2\right)} \tag{15.171}$$

and radius

$$R_a = \frac{\sqrt{1 - 2K|S_{12}S_{21}|g_a + |S_{12}S_{21}|^2 g_a^2}}{1 + g_a\left(|S_{11}|^2 - |\Delta|^2\right)} \tag{15.172}$$

where

$$C_1 = S_{11} - S_{22}^*\Delta \tag{15.173}$$

and $$g_a = \frac{1-|\Gamma_{\backslash S}|^2}{\left(\left|1-S_{11}\Gamma_S\right|^2\right)-\left(\left|S_{22}-\Gamma_S\Delta\right|^2\right)} \tag{15.174}$$

Practice Problem

15.6[@] Prove Eq. (15.150).

EXAMPLE 15.5 Design a transistor amplifier of 12 dB gain at 3.5 GHz provided that the transistor has the following S-parameters (50 Ω) at 3.5 GHz: $S_{11} = 0.75\angle -110°$, $S_{12} = 0$, $S_{21} = 2.82\angle 85°$, and $S_{22} = 0.63\angle -60°$.

Solution Given: $S_{11} = 0.75\angle -110° = -0.2565 - j0.7048$, $S_{12} = 0$,

$S_{21} = 2.82\angle 85° = 0.2458 + j2.8093$, and $S_{22} = 0.63\angle -60° = 0.315 - j0.5456$

Now since the device is unilateral and $|S_{11}| < 1$ and $|S_{22}| < 1$, the transistor is unilateral and unconditionally stable. The maximum matching section gain of the amplifier is as follows:

$$G_{S,\max} = \frac{1}{1-|S_{11}|^2} = \frac{1}{1-(0.75)^2} = 2.2857 = 3.59 \text{ dB}$$

$$G_{L,\max} = \frac{1}{1-|S_{22}|^2} = \frac{1}{1-(0.63)^2} = 1.6581 = 2.20 \text{ dB}$$

The transistor gain is calculated to be as follows:

$$G_0 = |S_{21}|^2 = 7.9524 = 9 \text{ dB}$$

Therefore, the maximum unilateral gain is expressed as follows:

$$G_{TU,\max} = 3.59 + 9 + 2.2 = 14.79 \text{ dB}$$

Thus, we have 2.79 dB more gain than the specified gain.

Since G_0 is constant, we must reduce the input and output matching section gain to reduce the overall gain. For this, let us choose $G_S = 2$ dB and $G_L = 1$ dB. Now, the total gain will be $2 + 9 + 1 = 12$ dB. It may be noted that we can also chose other combinations of G_S and G_L provided that the condition $G_S + G_0 + G_L = 12$ dB is satisfied.

To proceed further, we need to draw the gain circle corresponding to $G_S = 2$ dB and $G_L = 1$ dB. With this selection,

$$g_S = \frac{G_S}{G_{S,\max}} = \frac{10^{0.2}}{2.2857} = 0.6934$$

$$g_L = \frac{G_L}{G_{L,\max}} = \frac{10^{0.1}}{1.6581} = 0.7592$$

The centre and radius of the corresponding gain circles are as follows:

$$C_S = \frac{g_S S_{11}^*}{1-(1-g_S)|S_{11}|^2} = \frac{0.0.6934(-0.2565+j0.7048)}{1-(1-0.6934)(0.75)^2}$$

or $\quad C_S = -0.2149 + j0.5905 = 0.6284\angle110°$

$$R_S = \frac{\sqrt{1-g_s}\left(1-|S_{11}|^2\right)}{1-(1-g_S)|S_{11}|^2} = \frac{\sqrt{1-0.6934}\left(1-0.75^2\right)}{1-(1-0.6934)(0.75)^2} = 0.2927$$

$$C_L = \frac{g_L S_{22}^*}{1-(1-g_L)|S_{22}|^2} = \frac{0.7592(0.315+j0.5456)}{1-(1-0.7592)(0.63)^2}$$

or $\quad C_L = 0.2644 + j0.458 = 0.5288\angle60°$

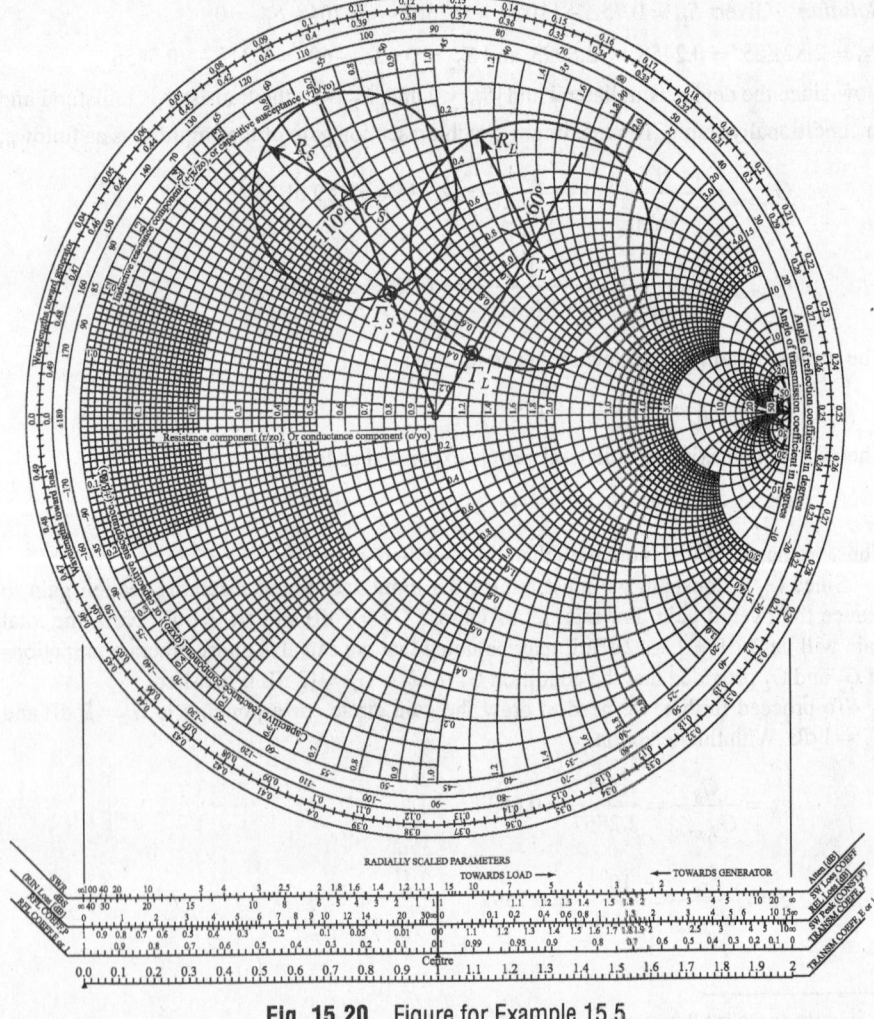

Fig. 15.20 Figure for Example 15.5

$$R_L = \frac{\sqrt{1-g_L}\left(1-|S_{22}|^2\right)}{1-(1-g_L)|S_{22}|^2} = \frac{\sqrt{1-0.7592}\left(1-0.63^2\right)}{1-(1-0.7592)(0.63)^2} = 0.3272$$

Plotting the gain circle we can select Γ_S and Γ_L along the circle (Fig. 15.20). If we choose Γ_S and Γ_L that is closest to the centre of the Smith chart, then we can write that

$$\Gamma_S = 0.35\angle 110°$$

$$\Gamma_L = 0.21\angle 60°$$

From the value of Γ_S and Γ_L, the values of Γ_{in} and Γ_{out} can be easily determined. Once Γ_{in} and Γ_{out} have been found, the corresponding source and load matching network can be designed using the single stub matching technique. This procedure has already been discussed in Chapter 3 and has not been repeated here. Interested readers may proceed further to find the final designed circuit.

Practice Problem

15.7 Find the value of Γ_S and Γ_L of an transistor amplifier of 10 dB gain at 3.5 GHz provided that the transistor has the following S-parameters (50 Ω) at 3.5 GHz: $S_{11} = 0.83\angle -120°$, $S_{12} = 0$, $S_{21} = 2.24\angle 75°$, and $S_{22} = 0.69\angle -65°$. Choose $G_S = 2$ dB and $G_L = 1$ dB. $[0.3\angle 120°, 0.182\angle 65°]$

15.4.3 Design of Low-noise Amplifiers

In addition to the stability and gain, noise figure is another important parameter that is often considered in the design of an amplifier, especially in receiver applications. In practice, it is not possible to achieve both the minimum noise figure and the maximum gain from an amplifier. Therefore, some compromise is required. This can be done by drawing circles of constant gain and constant noise figure.

The noise figure of a two-port amplifier can be expressed as follows:

$$F = F_{min} + \frac{R_N}{G_S}|Y_S - Y_{opt}|^2 \tag{15.175}$$

where Y_S is the source admittance presented to the transistor, Y_{opt} is the optimum source admittance that results in the minimum noise figure F_{min}, R_N is the noise resistance of the transistor, and G_S is the real part of Y_S. Y_S and Y_{opt} can be expressed as follows:

$$Y_S = \frac{1}{Z_0}\left(\frac{1-\Gamma_S}{1+\Gamma_S}\right) \tag{15.176}$$

and

$$Y_{opt} = \frac{1}{Z_0}\left(\frac{1-\Gamma_{opt}}{1+\Gamma_{opt}}\right) \tag{15.177}$$

Note F_{min}, Γ_{opt}, and R_N characterize the noise properties of the transistor being used and are called noise parameters of the device.

Using Eqs (15.176) and (15.177), we can write the following expressions:

$$|Y_S - Y_{opt}|^2 = \left| \frac{1}{Z_0}\left(\frac{1-\Gamma_S}{1+\Gamma_S}\right) - \frac{1}{Z_0}\left(\frac{1-\Gamma_{opt}}{1+\Gamma_{opt}}\right) \right|^2$$

or $$|Y_S - Y_{opt}|^2 = \frac{1}{Z_0^2}\left| \frac{1+\Gamma_{opt}-\Gamma_S-\Gamma_S\Gamma_{opt}-1+\Gamma_{opt}-\Gamma_S+\Gamma_S\Gamma_{opt}}{(1+\Gamma_S)(1+\Gamma_{opt})} \right|^2$$

or $$|Y_S - Y_{opt}|^2 = \frac{1}{Z_0^2}\left| \frac{2(\Gamma_{opt}-\Gamma_S)}{(1+\Gamma_S)(1+\Gamma_{opt})} \right|^2 = \frac{4}{Z_0^2}\frac{|\Gamma_S-\Gamma_{opt}|^2}{|1+\Gamma_S|^2|1+\Gamma_{opt}|^2} \qquad (15.178)$$

and $$G_S = \text{Re}\{Y_S\} = \frac{1}{2}\left\{ \frac{1}{Z_0}\left(\frac{1-\Gamma_S}{1+\Gamma_S}\right) + \frac{1}{Z_0}\left(\frac{1-\Gamma_S^*}{1+\Gamma_S^*}\right) \right\}$$

or $$G_S = \frac{1}{2Z_0}\left\{ \frac{(1-\Gamma_S)(1+\Gamma_S^*)+(1-\Gamma_S^*)(1+\Gamma_S)}{(1+\Gamma_S)(1+\Gamma_S^*)} \right\}$$

or $$G_S = \frac{1}{2Z_0}\left\{ \frac{1+\Gamma_S^*-\Gamma_S-\Gamma_S\Gamma_S^*+1+\Gamma_S-\Gamma_S^*-\Gamma_S\Gamma_S^*}{(1+\Gamma_S)(1+\Gamma_S^*)} \right\}$$

or $$G_S = \frac{1}{Z_0}\left\{ \frac{1-\Gamma_S\Gamma_S^*}{1+\Gamma_S^*+\Gamma_S+\Gamma_S\Gamma_S^*} \right\} = \frac{1}{Z_0}\frac{1-|\Gamma_S|^2}{|1+\Gamma_S|^2} \qquad (15.179)$$

Substituting Eqs (15.178) and (15.179) into Eq. (15.175), we get the following relation:

$$F = F_{min} + R_N \frac{Z_0|1+\Gamma_S|^2}{1-|\Gamma_S|^2} \frac{4}{Z_0^2} \frac{|\Gamma_S-\Gamma_{opt}|^2}{|1+\Gamma_S|^2|1+\Gamma_{opt}|^2}$$

or $$F = F_{min} + \frac{4R_N|\Gamma_S-\Gamma_{opt}|^2}{Z_0(1-|\Gamma_S|^2)|1+\Gamma_{opt}|^2}$$

or $$N = \frac{|\Gamma_S-\Gamma_{opt}|^2}{1-|\Gamma_S|^2} = \frac{F-F_{min}}{(4R_N/Z_0)}|1+\Gamma_{opt}|^2 \qquad (15.180)$$

Equation (15.180) can be rewritten as follows:

$$N(1-|\Gamma_S|^2) = (\Gamma_S-\Gamma_{opt})(\Gamma_S^*-\Gamma_{opt}^*)$$

or $$\Gamma_S\Gamma_S^* - (\Gamma_S\Gamma_{opt}^*+\Gamma_{opt}\Gamma_S^*) + |\Gamma_{opt}|^2 = N - N\Gamma_S\Gamma_S^*$$

or $$(N+1)\Gamma_S\Gamma_S^* - (\Gamma_S\Gamma_{opt}^*+\Gamma_{opt}\Gamma_S^*) = N - |\Gamma_{opt}|^2$$

or $\quad \Gamma_S \Gamma_S^* - \dfrac{\Gamma_S \Gamma_{opt}^* + \Gamma_{opt} \Gamma_S^*}{N+1} = \dfrac{N - \left|\Gamma_{opt}\right|^2}{N+1}$

or $\quad \Gamma_S \Gamma_S^* - \dfrac{\Gamma_S \Gamma_{opt}^* + \Gamma_{opt} \Gamma_S^*}{N+1} + \dfrac{\left|\Gamma_{opt}\right|^2}{(N+1)^2} = \dfrac{N - \left|\Gamma_{opt}\right|^2}{N+1} + \dfrac{\left|\Gamma_{opt}\right|^2}{(N+1)^2}$

or $\quad \left|\Gamma_S - \dfrac{\Gamma_{opt}}{N+1}\right|^2 = \dfrac{N^2 + N - N\left|\Gamma_{opt}\right|^2}{(N+1)^2}$

or $\quad \left|\Gamma_S - \dfrac{\Gamma_{opt}}{N+1}\right|^2 = \dfrac{N\left(N+1 - \left|\Gamma_{opt}\right|^2\right)}{(N+1)^2}$

or $\quad \left|\Gamma_S - \dfrac{\Gamma_{opt}}{N+1}\right| = \dfrac{\sqrt{N\left(N+1 - \left|\Gamma_{opt}\right|^2\right)}}{N+1}$ (15.181)

Equation (15.181) represents a set of circles with centres at

$$C_F = \dfrac{\Gamma_{opt}}{N+1} \tag{15.182}$$

and having radius $R_F = \dfrac{\sqrt{N\left(N+1 - \left|\Gamma_{opt}\right|^2\right)}}{N+1}$ (15.183)

EXAMPLE 15.6 A FET, biased for the minimum noise figure, has the following S-parameters at 3.5 GHz: $S_{11} = 0.62\angle -58°$, $S_{12} = 0.03\angle 30°$, $S_{21} = 2.0\angle 75°$, and $S_{22} = 0.54\angle -62°$. For the circuit $F_{min} = 1.5$ dB, $\Gamma_{opt} = 0.6\angle 105°$, and $R_N = 18$ Ω, calculate the maximum error gain in G_T if we assume the transistor to be unilateral. Now design an amplifier with the maximum gain that is compatible with a 2.1 dB noise figure. Assume that $Z_0 = 50$ Ω.

Solution Given: $S_{11} = 0.62\angle -58° = 0.3285 - j0.5258$, $S_{12} = 0.03\angle 30° = 0.026 + j0.015$, $S_{21} = 2.0\angle 75° = 0.5176 + j1.9319$, $S_{22} = 0.54\angle -62° = 0.2535 - j0.4768$, $F_{min} = 1.5$ dB $= 1.4125$, $\Gamma_{opt} = 0.6\angle 105° = -0.1553 + j0.5796$, $R_N = 18$ Ω, $F = 2.1$ dB $= 1.6218$, and $Z_0 = 50$ Ω.

The unilateral figure of merit is as follows:

$$U = \dfrac{|S_{11}||S_{12}||S_{21}||S_{22}|}{\left(1 - |S_{11}|^2\right)\left(1 - |S_{22}|^2\right)} = \dfrac{0.62 \times 0.03 \times 2.0 \times 0.54}{\left(1 - 0.62^2\right)\left(1 - 0.54^2\right)} = 0.046$$

Therefore,

$$\dfrac{1}{(1+U)^2} < \dfrac{G_T}{G_{TU}} < \dfrac{1}{(1-U)^2} \equiv \dfrac{1}{(1+0.046)^2} < \dfrac{G_T}{G_{TU}} < \dfrac{1}{(1-0.046)^2}$$

or $\quad \dfrac{1}{(1+U)^2} < \dfrac{G_T}{G_{TU}} < \dfrac{1}{(1-U)^2} \equiv 0.914 < \dfrac{G_T}{G_{TU}} < 1.0988$

In dB, $\quad -0.3906 \text{ dB} < G_T - G_{TU} < 0.4092 \text{ dB}$

Now, $\quad N = \dfrac{F - F_{\min}}{(4R_N/Z_0)}|1 + \Gamma_{\text{opt}}|^2 = \dfrac{(1.6218 - 1.4125)|1 - 0.1553 + j0.5796|^2}{(4 \times 18/50)} = 0.1525$

The centre and radius of noise circle will be as follows:

$$C_F = \frac{\Gamma_{\text{opt}}}{N+1} = \frac{-0.1553 + j0.5796}{0.1525 + 1} = -0.1347 + j0.5029 = 0.5206 \angle 105°$$

$$R_F = \frac{\sqrt{N\left(N + 1 - |\Gamma_{\text{opt}}|^2\right)}}{N+1} = \frac{\sqrt{0.1526\left(0.1526 + 1 - 0.6^2\right)}}{0.1526 + 1} = 0.3017$$

The plot of the noise figure circle in Smith chart is shown in Fig 15.21.

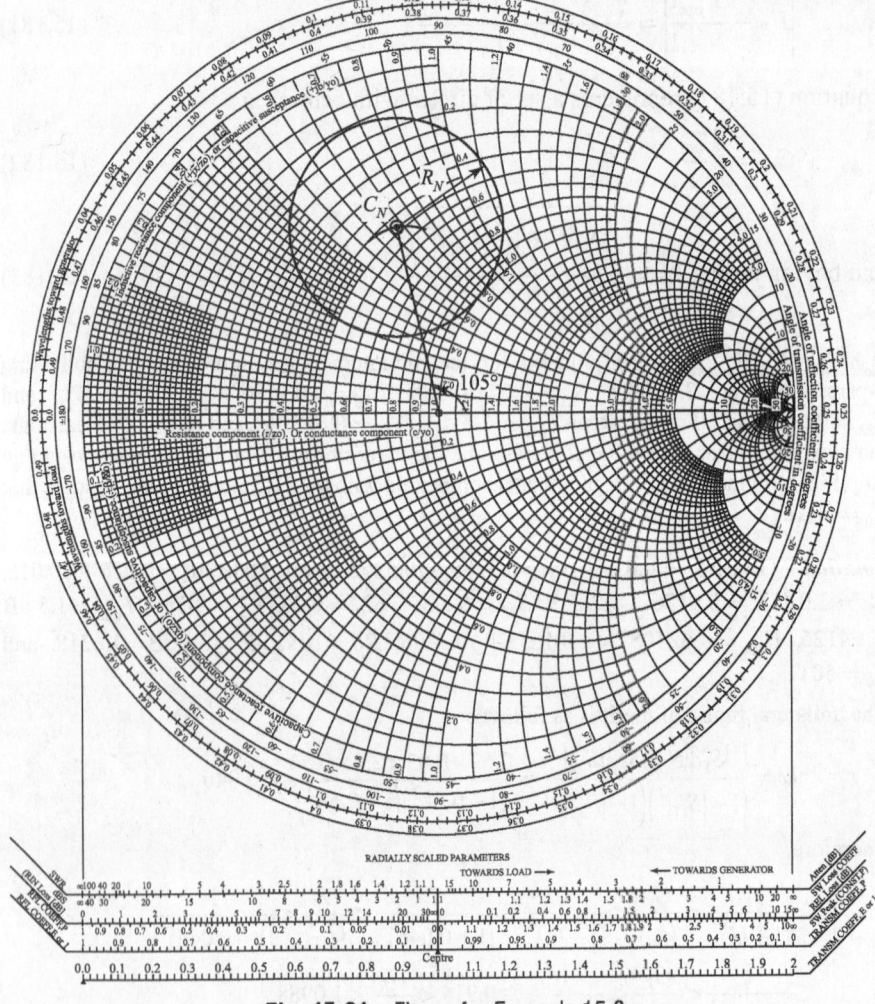

Fig. 15.21 Figure for Example 15.8

To proceed further in this problem, we need to plot a series of source matching network gain circles, starting from a lower gain value towards the higher values, until we find the one that just touches the 2.1 dB noise figure circle. The gain corresponding to this circle is the maximum permissible gain of the source matching network. Any further attempt to increase the gain of this network will deteriorate the noise figure. From the plot of the source matching network gain circle, the value of Γ_S can be easily determined. Further, from the value of Γ_S, the input reflection coefficient Γ_{in} and corresponding source matching network can be designed using the single stub matching technique. For the output section we can choose $\Gamma_L = S_{22}^*$ to obtain a maximum load matching network gain. This part of this problem has not been carried out here and left to the reader as an exercise.

Practice Problem

15.8 A FET, biased for the minimum noise figure, has the following S-parameters at 3.5 GHz: $S_{11} = 0.67\angle-58°$, $S_{12} = 0.02\angle40°$, $S_{21} = 2.3\angle60°$, and $S_{22} = 0.6\angle-75°$. For the circuit $F_{min} = 1.2$ dB, $\Gamma_{opt} = 0.5\angle120°$, and $R_N = 18\,\Omega$. Calculate the maximum error gain in G_T if we assume the transistor to be unilateral. Plot the noise figure circle of an amplifier having a 1.8 dB noise figure. Assume that $Z_0 = 50\,\Omega$.

Limit: $0.9029 < G_T/G_{TU}$, C_F: 0.4538, Angle: 120°, $R_F = 0.2671$

15.5 Broadband Transistor Amplifier Design

An ideal microwave amplifier should have a constant gain and good input matching over the desired bandwidth. In practice, an attempt to increase the gain of the amplifier reduces its bandwidth and vice versa. In addition, $|S_{21}|$ decreases with frequency at the rate of 6 dB/octave. Therefore, special consideration is required for designing a broadband amplifier. The following are some common approaches:

Compensated matching network In such a network, the input and output matching sections are designed to compensate the gain roll-off in $|S_{21}|$, however at the expense of input and output matching.

Resistive matching network In such a network, good input and output matching is obtained with a corresponding loss in gain and an increase in noise figure.

Negative feedback Negative feedback in an amplifier improves the stability, flattens gain response, and improves input and output matching. Bandwidth in excess of a decade can be achieved with such a procedure, however, again at the expense of gain and noise figure.

Balanced amplifier In such a design, two amplifiers have 90° couplers at their input and output. Such a design procedure can provide good matching over an octave bandwidth or more. The gain of the amplifier is equal to that of a single amplifier. The main disadvantage of such a circuit is that it requires two transistors and twice the DC power.

Distributed amplifier In a distributed amplifier, several transistor are cascaded together along a transmission line. Such an amplifier provides good gain, matching, and noise figure over a wide bandwidth. The main disadvantage of this circuit is that it is large and the gain is also less than that of a cascaded amplifier with the same number of stages.

15.5.1 Balanced Amplifiers

A typical circuit diagram of a balanced amplifier is shown in Fig. 15.22. The first 90° hybrid coupler divides the input signal V_1^+ into two equal-amplitude components V_{A1}^+ and V_{B1}^+, with a 90° phase difference between them. These two components drive the two amplifiers A and B of gain G_A and G_B. Finally, outputs of these two amplifiers V_{A2}^+ and V_{B2}^+ are combined by the second 90° hybrid coupler. It should be noted that due to impedance mismatch, reflected waves V_{A1}^- and V_{B1}^- will be present at the input of the two amplifiers. However, due to the phasing property of the 90° hybrid coupler, these reflected waves will be cancelled at the input of the coupler. Thus, it offers a better input impedance match. A similar effect occurs at the output also. It should be noted that these do not improve the gain bandwidth of the amplifier beyond that of single amplifier sections. In addition, due to the use of two 90° hybrid couplers and two amplifier sections, such a circuit is more complex than a single-stage amplifier. A balanced amplifier has also the following interesting advantages:

1. Individual amplifier stages can be optimized for gain flatness or noise figure, without any concern for the input and output matching section.
2. Since the reflections are cancelled by the 90° hybrid coupler, improved input/ output matching as well as improved stability of the amplifier can be achieved.
3. It reduces the gain by 6 dB, if any of the amplifier section fails.
4. It can provide an octave bandwidth or more. The bandwidth is primarily limited by the bandwidth of the 90° hybrid coupler.

The incident voltages on the amplifiers can be expressed as follows:

$$V_{A1}^+ = V_1^+ / \sqrt{2} \tag{15.184}$$

and
$$V_{B1}^+ = -jV_1^+ / \sqrt{2} \tag{15.185}$$

If the gain of the amplifiers are G_A and G_B, respectively, then the output of the amplifiers will be as follows:

$$V_{A2}^+ = G_A V_{A1}^+ \tag{15.186}$$

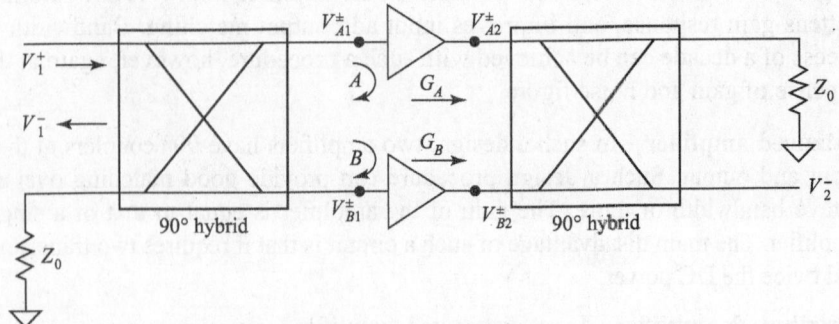

Fig. 15.22   Balanced amplifier circuit

and $\quad V_{B2}^+ = G_B V_{B1}^+$ (15.187)

The output voltage of the circuit will be $V_2^- = -j\frac{1}{\sqrt{2}}V_{A2}^+ + \frac{1}{\sqrt{2}}V_{B2}^+$ (15.188)

Substituting Eqs (15.186) and (15.187) into Eq. (15.188), we get the following relation:

$$V_2^- = -j\frac{1}{\sqrt{2}}G_A V_{A1}^+ + \frac{1}{\sqrt{2}}G_B V_{B1}^+$$ (15.189)

Further substituting Eqs (15.184) and (15.185) into Eq. (15.189), we get the following expression:

$$V_2^- = -j\frac{1}{2}G_A V_1^+ - j\frac{1}{2}G_B V_1^+ = -j\frac{1}{2}(G_A + G_B)V_1^+$$ (15.190)

Equation (15.190) reveals that S_{21} can be expressed as follows:

$$S_{21} = \frac{V_2^-}{V_1^+} = -j\left(\frac{G_A + G_B}{2}\right)$$ (15.191)

Equation (15.191) reveals that the overall voltage gain of the balanced amplifier is the average of individual amplifier voltage gains.

The total reflected voltage at the input can be expressed as follows:

$$V_1^- = \frac{1}{\sqrt{2}}V_{A1}^- - j\frac{1}{\sqrt{2}}V_{B1}^- = \frac{1}{\sqrt{2}}\Gamma_A V_{A1}^+ - j\frac{1}{\sqrt{2}}\Gamma_B V_{B1}^+$$ (15.192)

Substituting Eqs (15.184) and (15.185) into Eq. (15.192), we get the following relation:

$$V_1^- = \frac{1}{2}\Gamma_A V_1^+ + j^2\frac{1}{2}\Gamma_B V_1^+ = \frac{1}{2}V_1^+(\Gamma_A - \Gamma_B)$$ (15.193)

Therefore, S_{11} can be expressed as $S_{11} = \frac{V_1^-}{V_1^+} = \frac{1}{2}(\Gamma_A - \Gamma_B)$ (15.194)

If the transistors are identical, then $\Gamma_A = \Gamma_B$ and we get $S_{11} = 0$. It can be shown that the noise figure of a balanced amplifier will be $F = (F_A + F_B)/2$ (15.195)

where F_A and F_B are the noise figures of individual amplifiers.

15.5.2 Distributed Amplifiers

The basic configuration of an N-stage microwave distributed amplifier consists of N identical FETs connected in cascade, with their gates connected to a transmission line of characteristic impedance Z_g at an interval l_g and drains to a transmission line of characteristic impedance Z_d at an interval l_d, as shown in Fig. 15.23. The input signal propagates along the gate line, and each FET taps off some of the input power and amplifies it. The amplified output signals from the FETs form a travelling wave on the drain line. In practice, the propagation constant and lengths of the drain and gate lines are chosen such that constructive phasing of the output signal can take place. Termination impedances are connected to absorb the waves travelling in the reverse direction. Gate and drain capacitances of the FETs become

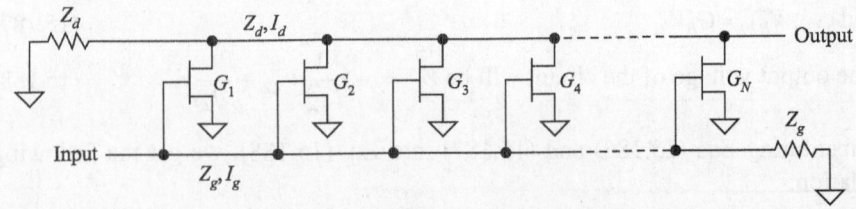

Fig. 15.23 N-stage distributed amplifier

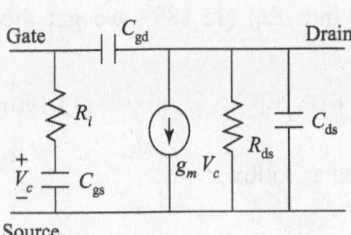

Gate

Drain

Source

Fig. 15.24 Small-signal common-source equivalent circuit of FET

a part of the gate and drain transmission lines, while the gate and drain resistance introduces losses in these lines. Distributed amplifiers are also known as travelling-wave amplifiers.

To analyse a distributed amplifier, we will start with the unilateral version of a FET amplifier $(C_{gd} = 0)$, as shown in Fig. 15.24. Under such a condition, the circuit shown in Fig. 15.23 is transformed into separate loaded transmission lines for gate and drain terminals, as shown in Figs 15.25 and 15.26, respectively. The gate and drain circuits are matched at both ends and are isolated, except for the coupling through the dependent current sources:

$$I_{dn} = g_m V_{cn} \tag{15.196}$$

In the equivalent circuit of the unit cell of gate line, L_g and C_g are, respectively, the per-unit-length inductance and capacitance of the gate transmission line, whereas $R_i l_g$ and C_{gs}/l_g are the equivalent per-unit-length loading in the transmission line due to FET input resistance R_i and gate-to-source capacitance C_{gs}, respectively. Similarly, in the equivalent circuit of the unit cell of drain line, L_d and C_d are, respectively, the per-unit-length inductance and capacitance of

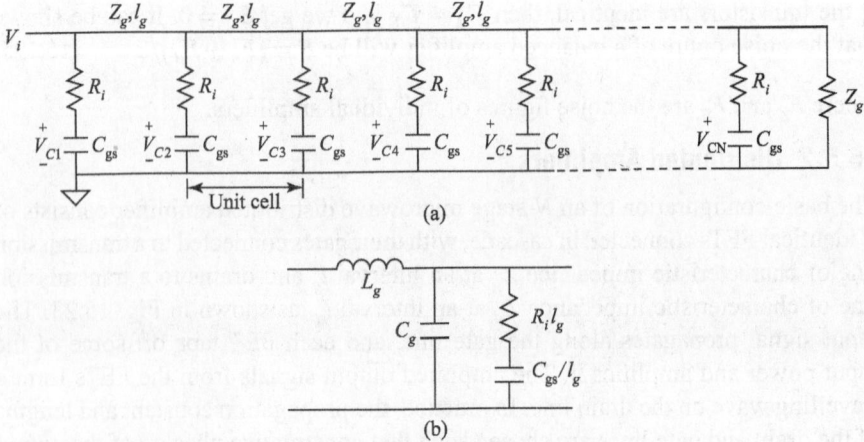

Fig. 15.25 Gate line of distributed amplifier (a) Transmission line circuit (b) Equivalent circuit of unit cell

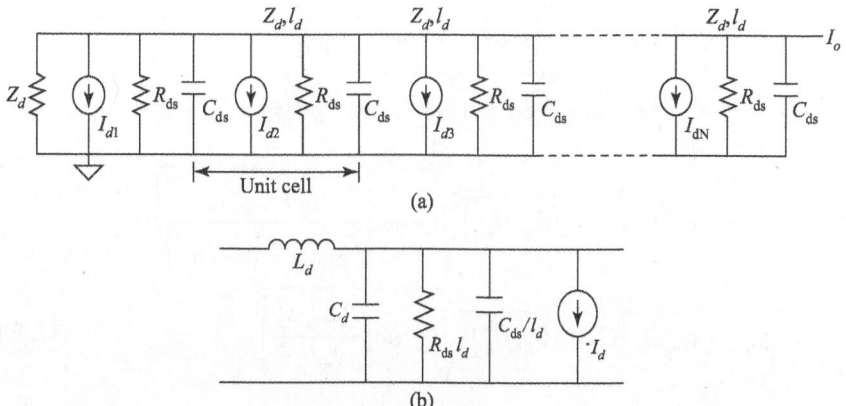

Fig. 15.26 Drain line of distributed amplifier (a) Transmission line circuit
(b) Equivalent circuit of unit cell

the drain transmission line, whereas $R_{ds}l_d$ and C_{ds}/l_d are the equivalent per-unit-length loading in the transmission line due to drain-to-source resistance R_{ds} and drain-to-source capacitance C_{ds}, respectively. Such approximations are generally valid when the electrical lengths of the unit cells are small.

Once the equivalent circuit has been found, the basic transmission line theory can be applied to find the effective characteristic impedance and propagation constant for the gate and drain lines. For the gate line, the per-unit-length series impedance and shunt admittance can be expressed as follows:

$$Z = j\omega L_g \tag{15.197}$$

and
$$Y = j\omega C_g + \frac{j\omega C_{gs}/l_g}{1 + j\omega R_i C_{gs}} \tag{15.198}$$

Therefore, the characteristic impedance of the line can be expressed, using Eqs (15.197) and (15.198), as follows:

$$Z_g = \sqrt{\frac{Z}{Y}} = \sqrt{\frac{j\omega L_g}{j\omega C_g + \dfrac{j\omega C_{gs}/l_g}{1 + j\omega R_i C_{gs}}}} \approx \sqrt{\frac{j\omega L_g}{j\omega C_g + j\omega\left(C_{gs}/l_g\right)}}$$

or
$$Z_g \approx \sqrt{\frac{L_g}{C_g + \left(C_{gs}/l_g\right)}} \tag{15.199}$$

where we have assumed that $1 \gg \omega R_i C_{gs}$.

The propagation constant can be calculated as follows:

$$\gamma_g = \alpha_g + j\beta_g = \sqrt{ZY} = \sqrt{j\omega L_g \left(j\omega C_g + \frac{j\omega C_{gs}/l_g}{1 + j\omega R_i C_{gs}} \right)}$$

or
$$\gamma_g = \sqrt{j^2 \omega^2 L_g \left\{ C_g + \frac{\left(C_{gs}/l_g\right)\left(1 - j\omega R_i C_{gs}\right)}{1 + \omega^2 R_i^2 C_{gs}^2} \right\}}$$

or $\qquad \gamma_g \cong \sqrt{j^2\omega^2 L_g \left[C_g + \left\{ C_{gs}\left(1 - j\omega R_i C_{gs}\right)/l_g \right\} \right]}$

or $\qquad \gamma_g = j\omega \sqrt{L_g \left\{ C_g + \left(C_{gs}/l_g \right) \right\} - j\omega \left(L_g R_i C_{gs}^2/l_g \right)}$

or $\qquad \gamma_g = j\omega \sqrt{L_g \left\{ C_g + \left(C_{gs}/l_g \right) \right\}} \sqrt{1 - j\omega \left[\dfrac{L_g R_i C_{gs}^2/l_g}{L_g \left\{ C_g + \left(C_{gs}/l_g \right) \right\}} \right]}$

or $\qquad \gamma_g = j\omega \sqrt{L_g \left\{ C_g + \left(C_{gs}/l_g \right) \right\}} \sqrt{1 - j\omega \left[\dfrac{R_i C_{gs}^2 Z_g^2}{l_g L_g} \right]} \qquad (15.200)$

where we have used Eq. (15.199). Assuming that $\omega \left(\dfrac{R_i C_{gs}^2 Z_g^2}{l_g L_g} \right) \ll 1$, we can write Eq. (15.200) as follows:

$$\gamma_g = j\omega \sqrt{L_g \left\{ C_g + \left(C_{gs}/l_g \right) \right\}} \left\{ 1 - \frac{j\omega}{2} \left(\frac{R_i C_{gs}^2 Z_g^2}{l_g L_g} \right) \right\}$$

or $\qquad \gamma_g = j\omega \sqrt{L_g \left\{ C_g + \left(C_{gs}/l_g \right) \right\}} - j\omega \sqrt{L_g \left\{ C_g + \left(C_{gs}/l_g \right) \right\}} \dfrac{j\omega}{2} \left(\dfrac{R_i C_{gs}^2 Z_g^2}{l_g L_g} \right)$

or $\qquad \gamma_g = j\omega \sqrt{L_g \left\{ C_g + \left(C_{gs}/l_g \right) \right\}} + \dfrac{\omega^2 R_i C_{gs}^2 Z_g^2}{2 l_g Z_g}$

or $\qquad \gamma_g = \dfrac{\omega^2 R_i C_{gs}^2 Z_g}{2 l_g} + j\omega \sqrt{L_g \left\{ C_g + \left(C_{gs}/l_g \right) \right\}} \qquad (15.201)$

where again we have used Eq. (15.199) and the assumption that $1 \gg \omega R_i C_{gs}$.

Now, for the drain line we can write the following expressions:

$$Z = j\omega L_d \qquad (15.202)$$

and $\qquad Y = \dfrac{1}{R_{ds} l_d} + j\omega \left(C_d + \dfrac{C_{ds}}{l_d} \right) \qquad (15.203)$

Therefore, the characteristic impedance of the drain line can be expressed as follows:

$$Z_d = \sqrt{\frac{Z}{Y}} = \sqrt{\frac{j\omega L_d}{\dfrac{1}{R_{ds} l_d} + j\omega \left(C_d + \dfrac{C_{ds}}{l_d} \right)}} \cong \sqrt{\frac{j\omega L_d}{j\omega \left(C_d + \dfrac{C_{ds}}{l_d} \right)}} \cong \sqrt{\frac{L_d}{\left(C_d + \dfrac{C_{ds}}{l_d} \right)}} \qquad (15.204)$$

where we have assumed that $R_{ds} l_d \gg 1$.

The propagation constant can be expressed as follows:

$$\gamma_d = \sqrt{ZY} = \sqrt{j\omega L_d \left\{ \frac{1}{R_{ds} l_d} + j\omega \left(C_d + \frac{C_{ds}}{l_d} \right) \right\}} = \sqrt{\frac{j\omega L_d}{R_{ds} l_d} + (j\omega)^2 L_d \left(C_d + \frac{C_{ds}}{l_d} \right)}$$

or $\qquad \gamma_d = j\omega \sqrt{L_d\left(C_d + \dfrac{C_{ds}}{l_d}\right)} \sqrt{1 + \left(\dfrac{j\omega L_d}{R_{ds}l_d}\right)\dfrac{1}{(j\omega)^2 L_d\left(C_d + \dfrac{C_{ds}}{l_d}\right)}}$

or $\qquad \gamma_d = j\omega \sqrt{L_d\left(C_d + \dfrac{C_{ds}}{l_d}\right)} \sqrt{1 + \dfrac{1}{j\omega R_{ds}l_d\left(C_d + \dfrac{C_{ds}}{l_d}\right)}}$ $\qquad$ (15.205)

Assuming that $\left[\omega R_{ds}l_d\left(C_d + \dfrac{C_{ds}}{l_d}\right)\right] \gg 1$, we can write Eq. (15.205) as follows:

or $\qquad \gamma_d = j\omega \sqrt{L_d\left(C_d + \dfrac{C_{ds}}{l_d}\right)} \left[1 + \dfrac{1}{2j\omega R_{ds}l_d\left(C_d + \dfrac{C_{ds}}{l_d}\right)}\right]$

or $\qquad \gamma_d = j\omega \sqrt{L_d\left(C_d + \dfrac{C_{ds}}{l_d}\right)} + \dfrac{1}{2R_{ds}l_d}\sqrt{\dfrac{L_d}{\left(C_d + \dfrac{C_{ds}}{l_d}\right)}}$

or $\qquad \gamma_d = \dfrac{Z_g}{2R_{ds}l_d} + j\omega \sqrt{L_d\left(C_d + \dfrac{C_{ds}}{l_d}\right)}$ $\qquad$ (15.206)

where we have used Eq. (15.204).

For an incident input voltage V_i, the voltage on the gate to source capacitance of the nth FET can be written as follows:

$$V_{cn} = V_i e^{-(n-1)\gamma_g l_g}\left(\dfrac{1}{1 + j\omega R_i C_{gs}}\right) \qquad (15.207)$$

where the bracketed factor accounts for the voltage division between R_i and C_{gs}. Since for a typical FET $\omega R_i C_{gs} \ll 1$, the bracketed factor can be approximated as unity over the entire bandwidth. Hence, we can write the following relation:

$$V_{cn} = V_i e^{-(n-1)\gamma_g l_g} \qquad (15.208)$$

Now, since each current generator contributes waves of the form $-I_{dn}e^{\pm\gamma_d z}/2$ in each direction, the output current on the drain line can be expressed as follows:

$$I_0 = -\dfrac{1}{2}\sum_{n=1}^{N} I_{dn} e^{-(N-n)\gamma_d l_d} \qquad (15.209)$$

Substituting Eq. (15.196) into Eq. (15.209), we get the following expression:

$$I_0 = -\dfrac{1}{2}\sum_{n=1}^{N} g_m V_{cn} e^{-(N-n)\gamma_d l_d} \qquad (15.210)$$

Further substituting Eq. (15.208) into Eq. (15.210), we get the following relation:

$$I_0 = -\frac{1}{2}\sum_{n=1}^{N} g_m V_i e^{-(n-1)\gamma_g l_g} e^{-(N-n)\gamma_d l_d} = -\frac{g_m V_i}{2} e^{-N\gamma_d l_d} e^{\gamma_g l_g} \sum_{n=1}^{N} e^{-n(\gamma_g l_g - \gamma_d l_d)} \quad (15.211)$$

The terms within the summation notation will add in phase when

$$\beta_g l_g = \beta_d l_d \tag{15.212}$$

Under such a condition, phase delays on the gate and drain lines are synchronized. In practice, there will be a backward wave on the drain line, but the individual components will not be in phase. The resultant backward wave will be absorbed in the terminating impedance Z_d.

Now using the following formula

$$\sum_{n=1}^{N} x^n = \frac{x^{N+1} - x}{x - 1} \tag{15.213}$$

and assuming that

$$x = e^{-(\gamma_g l_g - \gamma_d l_d)} \tag{15.214}$$

Eq. (15.211) can be written as follows:

$$I_0 = -\frac{g_m V_i}{2} e^{-N\gamma_d l_d} e^{\gamma_g l_g} \frac{e^{-(N+1)(\gamma_g l_g - \gamma_d l_d)} - e^{-(\gamma_g l_g - \gamma_d l_d)}}{e^{-(\gamma_g l_g - \gamma_d l_d)} - 1}$$

$$\text{or} \quad I_0 = -\frac{g_m V_i}{2} e^{\gamma_d l_d} e^{-(N+1)\gamma_d l_d} e^{\gamma_g l_g} \frac{e^{-(N+1)(\gamma_g l_g - \gamma_d l_d)} - e^{-(\gamma_g l_g - \gamma_d l_d)}}{e^{-(\gamma_g l_g - \gamma_d l_d)} - 1}$$

$$\text{or} \quad I_0 = -\frac{g_m V_i}{2} \frac{e^{\gamma_d l_d}\left(e^{-(N+1)\gamma_d l_d} e^{\gamma_g l_g} e^{-(N+1)\gamma_g l_g} e^{(N+1)\gamma_d l_d} - e^{-(N+1)\gamma_d l_d} e^{\gamma_g l_g} e^{-\gamma_g l_g} e^{\gamma_d l_d} \right)}{e^{\gamma_d l_d}\left(e^{-\gamma_g l_g} - e^{-\gamma_d l_d} \right)}$$

$$\text{or} \quad I_0 = -\frac{g_m V_i}{2} \frac{e^{-N\gamma_g l_g} - e^{-N\gamma_d l_d}}{e^{-\gamma_g l_g} - e^{-\gamma_d l_d}} \tag{15.215}$$

If the input and output ports are matched, then the amplifier gain can be calculated as follows:

$$G = \frac{P_{\text{out}}}{P_{\text{in}}} = \left(\frac{1}{2}|I_0|^2 Z_d \right) \Big/ \left(\frac{1}{2}\frac{|V_i|^2}{Z_g} \right) = \frac{|I_0|^2 Z_d Z_g}{|V_i|^2}$$

$$\text{or} \quad G = \frac{g_m^2 |V_i|^2}{4|V_i|^2} \left| \frac{e^{-N\gamma_g l_g} - e^{-N\gamma_d l_d}}{e^{-\gamma_g l_g} - e^{-\gamma_d l_d}} \right|^2 Z_d Z_g = \frac{g_m^2 Z_d Z_g}{4} \left| \frac{e^{-N\gamma_g l_g} - e^{-N\gamma_d l_d}}{e^{-\gamma_g l_g} - e^{-\gamma_d l_d}} \right|^2 \tag{15.216}$$

Now substituting Eq. (15.212) into Eq. (15.216), we get the following relation:

$$G = \frac{g_m^2 Z_d Z_g}{4} \left| \frac{e^{-N\alpha_g l_g - jN\beta_g l_g} - e^{-N\alpha_d l_d - jN\beta_d l_d}}{e^{-\alpha_g l_g - j\beta_g l_g} - e^{-\alpha_d l_d - j\beta_d l_d}} \right|^2$$

or $$G = \frac{g_m^2 Z_d Z_g}{4} \left| \frac{e^{-N\alpha_g l_g - jN\beta_g l_g} - e^{-N\alpha_d l_d - jN\beta_g l_g} e}{e^{-\alpha_g l_g - j\beta_g l_g} - e^{-\alpha_d l_d - j\beta_g l_g}} \right|^2$$

or $$G = \frac{g_m^2 Z_d Z_g}{4} \frac{\left(e^{-N\alpha_g l_g} - e^{-N\alpha_d l_d} \right)^2}{\left(e^{-\alpha_g l_g} - e^{-\alpha_d l_d} \right)^2} \left| e^{-j(N-1)\beta_g l_g} \right|^2$$

or $$G = \frac{g_m^2 Z_d Z_g}{4} \frac{\left(e^{-N\alpha_g l_g} - e^{-N\alpha_d l_d} \right)^2}{\left(e^{-\alpha_g l_g} - e^{-\alpha_d l_d} \right)^2} \qquad (15.217)$$

If the losses are small, then we can approximate that

$$e^{-\alpha_g l_g} - e^{-\alpha_d l_d} \cong \alpha_g l_g - \alpha_d l_d$$

and $$e^{-N\alpha_g l_g} - e^{-N\alpha_d l_d} \cong N\left(\alpha_g l_g - \alpha_d l_d \right)$$

Hence, Eq. (15.217) simplifies to the following form:

$$G \cong \frac{g_m^2 Z_d Z_g}{4} \frac{N^2 \left(\alpha_g l_g - \alpha_d l_d \right)^2}{\left(\alpha_g l_g - \alpha_d l_d \right)^2} \cong \frac{g_m^2 N^2 Z_d Z_g}{4} \qquad (15.218)$$

Equation (15.218) reveals that for a very-low-loss or loss-less amplifier, the gain increases in proportion to N^2. This is in contrast to an N-stage amplifier for which gain increases in proportion to $(G_0)^N$.

If the loss is present, Eq. (15.217) implies that the gain of the amplifier approaches zero when $N \to \infty$. This can be explained as follows. Since the input voltage on the gate line decays exponentially, the FETs at the end of the amplifier receive no signal. Similarly, amplified signals from the FETs near the input of the amplifier also decay along the drain line. The multiplicative increase in gain with N is also not enough to compensate these exponential decays for a large N. Hence, the output voltage and hence the gain become zero. This implies that for a given set of FET parameters, there will be an optimum value of N that maximizes the gain of a distributed amplifier. This value can be achieved by setting

$$\left. \frac{dG}{dN} \right|_{N=N_{\text{opt}}} = 0 \qquad (15.219)$$

Substituting Eq. (15.217) into Eq. (15.219) and simplifying, we get the following relation:

$$N_{\text{opt}} = \ln \left(\frac{\alpha_g l_g}{\alpha_d l_d} \right) \Big/ \left(\alpha_g l_g - \alpha_d l_d \right) \qquad (15.220)$$

Distributed FET amplifiers can provide bandwidth in excess of a decade, with good input and output matching. However, they are not capable of providing a very high gain and very low noise figure, and are also larger in size than amplifiers, having comparable gain over a narrow bandwidth.

EXAMPLE 15.7 Write a MATLAB program to plot gain vs. frequency for a distributed amplifier. Hence, plot gain of a distributed amplifier from 1 to 20 GHz for $N = 2$–18 in steps of 4. Assume that $Z_d = Z_g = Z_0 = 50\,\Omega$, $R_i = 12\,\Omega$, $R_{ds} = 320\,\Omega$, $C_{gs} = 0.23$ pF, and $g_m = 32$ mS. Explain the graph. Now calculate the optimum value of N that will give the maximum gain at 18 GHz.

Solution Given: $Z_d = 50\,\Omega$, $Z_g = 50\,\Omega$, $R_i = 12\,\Omega$, $R_{ds} = 320\,\Omega$, $C_{gs} = 0.23$ pF, and $g_m = 32$ mS

MATLAB Program

```
% Program for Example 15.7
clf;
clear all;
Zd = input('Input Drain Transmission Line Characteristic Impedance
in Ohm (Zd):- ');
Zg = input('Input Gate Transmission Line Characteristic Impedance in
Ohm (Zg):- ');
Ri = input('Input Input Impedance in Ohm (Ri):- ');
Rds = input('Input Drain to Souce resistance in Ohm (Rds):- ');
Cgs = input('Input Gate to Source capacitance in pF (Cgs):- ');
gm = input('Input Transconductance in mS (gm):- ');
fmin = input('Input the minimum frequency in GHz :- ');
fmax = input('Input the maximum frequency in GHz :- ');
Nmin = input('Input the minimum Stages :- ');
Nmax = input('Input the maximum Stages :- ');
Ndiv = input('Input the stage interval :- ');
for N = 2:4:18
    u = 0;
    for f = 0*10^9:0.05*10^9:18*10^9
        u = u + 1;
        Frequency(u) = f;
        omega = 2*pi*f;
        aglg = omega^2*Ri*(Cgs*10^-12)^2*Zg/2;
        adld = Zd/(2*Rds);
        G = (gm*10^-3)^2*Zd*Zg*(exp(-N*aglg)-exp(-N*adld))^2/(4*(exp(-
aglg)-
exp(-adld))^2);
        Gain(u) = 10*log10(G);
    end
    plot(Frequency, Gain);
    hold on;
end
```

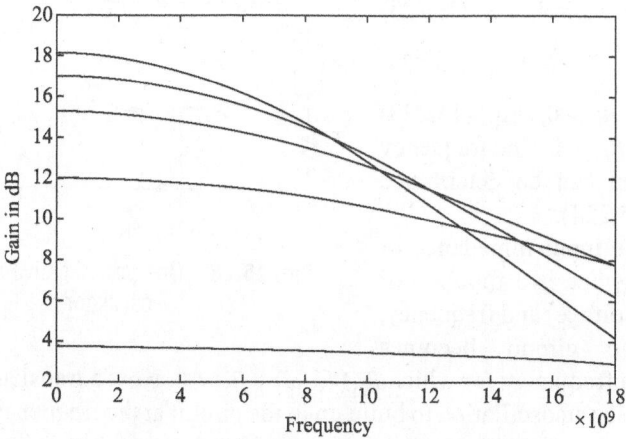

Fig. 15.27 Figure for Example 15.7

Figure 15.27 reveals the following events:

(a) The gain falls with an increase in frequency.
(b) The gain falls more rapidly for larger N values. After a certain frequency, the gain corresponding to a particular N becomes lower than that corresponding to a lower value of N.

Now at 18 GHz, $\alpha_g l_g = \dfrac{\omega^2 R_i C_{gs}^2 Z_g}{2} = \dfrac{\left(2\pi \times 18 \times 10^9\right)^2 \times 12 \times \left(0.23 \times 10^{-12}\right)^2 \times 50}{2} = 0.203$

$\alpha_d l_d = \dfrac{Z_d}{2R_{ds}} = \dfrac{50}{\left(2 \times 320\right)} = 0.0781$

Therefore,

$$N_{opt} \approx \dfrac{\ln\left(\dfrac{\alpha_g l_g}{\alpha_d l_d}\right)}{\alpha_g l_g - \alpha_d l_d} = \dfrac{\ln\left(0.203/0.0781\right)}{0.203 - 0.0781} = 7.6478 \approx 8$$

15.6 Oscillator Design

In the last few sections we have discussed the different types of microwave amplifiers. Now we will briefly discuss the different types of solid state oscillators: (a) one port negative resistance oscillator and (b) dielectric resonator oscillator.

15.6.1 One-port Negative-resistance Oscillators

A typical canonical one-port negative-resistance oscillator is shown in Fig. 15.28. In the circuit, $Z_{in} = R_{in} + jX_{in}$ is the input impedance of the active device and $Z_L = R_L + jX_L$ is the load impedance. Applying KVL in the loop, we get the following relation:

$$\left(Z_L + Z_{in}\right)I = 0 \tag{15.221}$$

If oscillation exists in the circuit then I is non-zero, and we get the following expressions:

$$Z_L + Z_{in} = 0 \tag{15.222}$$

$$R_L + R_{in} = 0 \qquad (15.223)$$

and $\quad X_L + X_{in} = 0 \qquad (15.224)$

Now since $R_L > 0$, Eq. (15.223) implies that $R_{in} < 0$. The frequency of oscillation can be determined from Eq. (15.224).

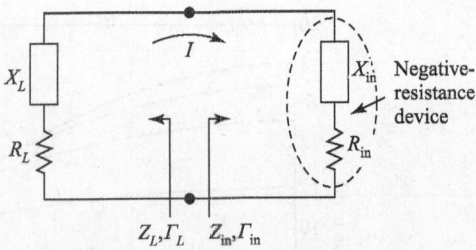

Fig. 15.28 One-port negative-resistance oscillator

Since the input impedance of the active device is a function of current (or voltage) and frequency, initially the circuit becomes unstable at a frequency for which $R_{in}(I, j\omega) + R_L < 0$. Now a transient excitation or noise causes an oscillation to build up in the circuit at the frequency ω, thereby causing I to increase. Now as I increases, $R_{in}(I, j\omega)$ becomes less negative. This is continued until I becomes equal to I_0, at which $R_{in}(I_0, j\omega_0) + R_L = 0$ and $X_{in}(I_0, j\omega_0) + X_L = 0$. Under such conditions, the oscillator runs in a stable state. It must be remembered that the final oscillation frequency ω_0 is different from ω because X_{in} is also current dependent.

EXAMPLE 15.8 A 5 GHz oscillator is to be designed using a negative-resistance diode with a reflection coefficient of $1.25\angle 45°$. Assuming that $Z_0 = 50\ \Omega$, determine the load impedance.

Solution Given: $\Gamma_G = 1.25\angle 45° = 0.8839 + j0.8839$, and $Z_0 = 50\ \Omega$

Therefore,

$$z_G = \frac{1 + \Gamma_G}{1 - \Gamma_G} = \frac{1 + 0.8839 + j0.8839}{1 - 0.8839 - j0.8839} = -0.7078 + j2.2244$$

or $\quad Z_G = 50z_G = -35.389 + j111.22\ \Omega$

Therefore, $Z_L = 35.389 - j111.22\ \Omega$

Practice Problem

15.9 A 3 GHz oscillator is to be designed using a negative-resistance diode with a reflection coefficient of $1.12\angle 30°$. Assuming that $Z_0 = 50\ \Omega$, determine the load impedance.

$$[40.455 - j178.06]$$

15.6.2 Dielectric Resonator Oscillators

The stability of an oscillator can be enhanced by using a high-Q tuning network. Since waveguide cavity resonators have very high Q, they can be used for this purpose; however, the problem is that they are not suitable for integrated circuits. Further, metal cavities produce very significant frequency drifts due to temperature variation. Microstrip resonators are more suitable for integrated circuits, but such resonators have low Q. Under such circumstances, dielectric resonators can be a good alternative. Dielectric resonators have Q values of the order of several thousands and can also be used in integrated circuit technology. Dielectric resonators

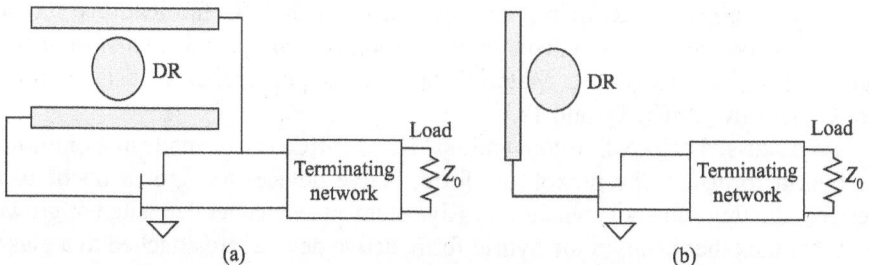

Fig. 15.29 Dielectric resonator oscillator (a) Parallel feedback (b) Series feedback

can be made from ceramic materials that have good temperature stability. A brief discussion on dielectric resonators and their integration in integrated circuits have already been presented in Chapter 6.

A dielectric resonator can be incorporated into a transistor oscillator circuit to provide frequency stability using either a parallel or a series feedback arrangement, as shown in Fig. 15.29. A parallel feedback configuration uses a resonator coupled to two microstrip lines, thus functioning as a high-Q band-pass filter. Amount of coupling is controlled by the spacing between the resonator and the lines, whereas phase is controlled by the length of the line. In contrast to the parallel feedback configuration, a series feedback configuration uses a single microstrip line. The tuning range of a series feedback configuration is less than that of a parallel feedback configuration.

15.7 MMIC—Material, Growth, and Fabrication

The basic materials of an MMIC can be subdivided into four categories: (a) substrate materials, (b) conductor materials, (c) dielectric films, and (d) resistive films. A substrate material is a substance on which electronic devices are built.

Substrate materials   An ideal substrate material should have the following characteristics: (a) high dielectric constant (greater than 9), (b) high purity and constant thickness, (c) high surface smoothness, (d) low dissipation factor, (e) constant dielectric constant over the frequency range and temperature of interest, (f) high thermal conductivity, and (g) high resistivity and dielectric strength. The basic substrate materials that are used in MMIC fabrications are alumina, sapphire, glass, GaAs, ferrite/garner, rutile, and beryllia.

Conductor materials   Conductor materials used in MMIC fabrications should have the following characteristics: (a) high conductivity, (b) low temperature coefficient of resistance, (c) good etchability and solderability, (d) good adhesion to substrate, and (e) potential to be easily deposited and electroplated. Thus, aluminium, copper, gold, and silver are widely used as conductor materials for MMIC applications.

Dielectric materials   The basic characteristics of dielectric materials used in MMIC fabrications are as follows: (a) low RF dielectric loss, (b) capability of withstanding high voltages, (c) reproducibility, and (d) ability to undergo processes without developing pin holes. The commonly used dielectric materials are SiO, SiO_2, Si_3N_4, Al_2O_3, and Ta_2O_5.

Resistive materials Resistive materials used in MMIC fabrications should have (a) good stability, (b) adequate dissipation capability, and (c) low temperature coefficient of resistance. Materials that are used in MMICs as resistive films are Cr, Cr–SiO, NiCr, Ti, and Ta.

As discussed before, like conventional ICs, MMICs can be made in monolithic or hybrid forms. In the monolithic form, active devices are grown on or in a semiconducting substrate, whereas passive elements are either deposited or grown in it. On the other hand, in the hybrid form, active devices are attached to a glass, ceramic, or substrate that contains the passive circuitry. In practice, because of the processing difficulties, low yields, and poor performances, monolithic technology is not suitable for microwave integrated circuits; instead, the hybrid form of technology is exclusively used in the frequency range of 1–15 GHz. Hybrid ICs can be grouped into two categories: hybrid IC and miniature hybrid IC. Hybrid ICs use the distributed circuit elements that are fabricated on the substrate using the single-layer metallization technique. Other circuit elements, such as resistor, capacitor, inductors, and semiconductor devices, are added to the substrate. In contrast, miniature hybrid ICs use multilevel elements. All passive elements such as resistors, capacitors, and inductors are deposited on the substrate, while the semiconductor devices are attached to the substrate.

MMICs can be fabricated using different techniques such as diffusion and ion implementation, oxidation and film deposition, epitaxial growth, lithography, etching and photo resist, and deposition.

15.7.1 Diffusion and Ion Implantation

In a semiconductor device fabrication diffusion and ion implantation are widely used to control the amount of dopant in a semiconductor. Any one of these two processes can be used, as they are complementary to each other. While the diffusion process involves diffusion of impurities into a pure material, ion implantation is used to dope the substrate with high-energy ion impurities. In practice, the ion implantation process has some advantages such as precise control of the total amount of dopants, improvement of reproducibility, and reduced processing temperature.

15.7.2 Epitaxial Growth

The word *epitaxy* is derived from the Greek words *epi* (on) and *taxis* (arrangement). This process provides a means of controlling the doping profile and can be used to grow a single-crystal semiconductor layer on a single-crystal semiconductor substrate. In practice, epitaxy is of three types: (a) vapour-phase epitaxy (VPE), (b) molecular beam epitaxy (MBE), and (c) liquid-phase epitaxy (LPE). Out of these, VPE is perhaps the most important technique for Si and GaAs devices.

MBE involves the reaction of one or more thermal beams of atoms or molecules with a crystalline surface under an ultra-high vacuum condition and can be used for precise control of both chemical composition and doping profile. This epitaxy can also be used to develop single-crystal multi-layer structures with dimensions of the order of atomic layers.

LPE is the growth of epitaxial layers on a crystalline substrate by direct precipitation from the liquid phase. This process is used to grow GaAs and related

III–IV compounds, and is suited for developing thin layers (>2 μm) because of its slow growth rate. Such epitaxy is also useful for developing multi-layered structures in which precise doping and composition control are required.

15.7.3 Lithography

The lithography process is used to transfer the pattern of geometric shapes from a mask to a thin layer of radiation-sensitive material. This pattern (also called resist) is not permanent elements of the final device. In practice, four types of lithography technology are available: (a) electron beam lithography, (b) ion beam lithography, (c) optical lithography, and (d) X-ray lithography.

15.7.4 Etching and Photo Resist

In MIC fabrication, the photoetching method is used to form openings in a SiO_2 layer for impurity diffusion. During this process, the substrate is first coated with a uniform film of Kodak Photoresist, a photosensitive emulsion. Next, a mask of desired opening is placed on it, and the whole thing is exposed to ultraviolet light. This results in a polymerized photo resist over the unmasked region. The unpolymerized portions are then dissolved using trichloroethylene after removing the mask. The SiO_2, which is not covered by the photo resist, can be removed by hydrofluoric acid.

15.7.5 Deposition

In practice, three methods, vacuum evaporation, electron beam evaporation, and DC sputtering, are commonly used for MMIC fabrication. In vacuum evaporation, the impurity material to be evaporated is placed in a metallic boat, through which a high current is passed. The substrate with a mask and the heated boat are located in a glass tube, maintained at a high vacuum level. Heat evaporates the impurities, which are deposited as a polycrystalline layer on the substrate. In contrast, in electron beam evaporation, a narrow beam of electrons is generated to scan the substrate in the boat in order to vaporize the impurity.

In DC sputtering or cathode sputtering, the crucible containing impurity is used as a cathode, whereas the substrate is used as an anode. The whole thing is maintained in a vacuum condition, with a slight trace of argon gas. Application of a high voltage between the anode and the cathode causes a glow discharge of argon gas. As a result, positive argon ions are accelerated towards the cathode and dislodge impurity atoms. Impurity atoms have enough energy to reach the substrate and adhere to it. In MMIC fabrication, the aforementioned processes are used to develop a circuit.

MMICs have several advantages over discrete circuits, such as (a) low cost, (b) light weight, (c) small size, (d) high reliability, (e) improved performance, (f) improved reproducibility, and (g) less parasitic reactance than discrete packaged devices. They also meet the basic requirements for space and military applications, such as the requirements of shock, temperature condition, and severe vibration, and hence are widely used for this purpose. MMICs also suffer from some serious disadvantages: (a) lack of provision of adjustment of circuit parameter, once fabricated; (b) difficulty in integrating tuning screws, variable shorts, etc.; (c) requirement of precise characterization of the semiconductor device; (d) low Q-factor, etc.; (e) waste of large areas of relatively expensive semiconductor substrate for components such as transmission line; (f) critical processing steps and tolerance requirement; (g) small size limiting heat dissipation; and (h) difficulty in implementing high-Q resonators and filters because of the inherent resistive losses.

Important Formulae

- $$L_C = 10 \log_{10} \left(\frac{\text{Available RF input power}}{\text{IF output power}} \right) \text{dB}$$

- $$K = \frac{1 - |S_{11}|^2 - |S_{22}|^2 + |\Delta|^2}{2|S_{12}S_{21}|} > 1$$

- $$\frac{1}{(1+U)^2} < \frac{G_T}{G_{TU}} < \frac{1}{(1-U)^2}$$

Exercises

Objective-type Questions

15.1 A balanced mixer, using a 180° hybrid provides
   (a) Good matching characteristics but poor isolation between RF and LO ports
   (b) Good matching characteristics and isolation between RF and LO ports
   (c) Poor matching characteristics and isolation between RF and LO ports
   (d) Poor matching characteristics but good isolation between RF and LO ports

15.2 To make a device unconditionally stable, stability circles must be
   (a) Completely outside of the Smith Chart
   (b) Totally within the Smith chart
   (c) Completely outside or totally inside the Smith chart
   (d) Partly outside or partly inside the Smith chart

15.3 A two-port network will be unconditionally stable if
   (a) $K < 1$ and $|\Delta| < 1$ (c) $K > 1$ and $|\Delta| < 1$
   (b) $K < 1$ and $|\Delta| > 1$ (d) $K > 1$ and $|\Delta| > 1$

15.4 For a very-low-loss or loss-less N-stage distributed amplifier, the gain increases in proportion to
   (a) N (b) N^2 (c) N^3 (d) $(G_0)^N$

15.5 When $N \to \infty$ and loss is present, the gain of an N-stage distributed amplifier approaches
   (a) 0   (b) 1   (c) any finite value   (d) ∞

Review Questions

15.1 With a suitable figure, describe how a diode can be used as a detector. Derive the necessary equations.
15.2 With a suitable figure, describe the operation of a single-ended mixer.

15.3 Establish that a balanced mixer can give either better input matching or better RF–LO isolation and cancellation of AM noise.

15.4 Show that an image rejection filter is capable of separating image frequencies.

15.5 Write short notes on different types of FET mixers.

15.6 Define different types of amplifier gains. Find out an expression for the power gain of a transistor amplifier.

15.7 Write a short note on the stability of transistor amplifiers.

15.8 Find out the equations of input and output stability circles.

15.9 Describe the design procedure of a single-stage transistor amplifier for maximum gain.

15.10 Briefly describe the properties of different materials used in an MMIC. Give at least one example for each of them.

Problems

15.1 A FET has the following S-parameters at $2\,\text{GHz}$: $S_{11} = 2.1\angle -100°$, $S_{12} = 0$, $S_{21} = 3.9\angle 60°$, and $S_{22} = 0.75\angle -80°$. (a) Determine its input impedance and unstable source impedance region, (b) plot $3\,\text{dB}$ source gain circle, (c) find source impedance that provides $3\,\text{dB}$ source gain with the maximum possible degree of stability, (d) determine the load impedance that gives the maximum load gain, and (e) find the gain of the amplifier circuit. Assume that the characteristic impedance is $50\,\Omega$. *[Hint: Use Smith chart and Eqs (15.154), (15.159), and (15.160)]*

15.2 A FET has the following S-parameters at $9\,\text{GHz}$: $S_{11} = 0.45\angle -170°$, $S_{12} = 0.06\angle 25°$, $S_{21} = 2.3\angle 75°$, and $0.9\angle -110°$. Determine the stability of the transistor and plot the stability circle. Plot the $10\,\text{dB}$ gain circle. *[Hint: Use Smith chart and Eqs (15.113), (15.115)–(15.118), (15.128), (15.164), (15.166), and (15.167)]*

15.3 A BJT has the following S-parameters at $2\,\text{GHz}$: $S_{11} = 0.68\angle -160°$, $S_{12} = 0$, $S_{21} = 4.9\angle 185°$, and $S_{22} = 0.5\angle -25°$. The noise properties of the transistor are $F_{min} = 3\,\text{dB}$, $R_N = 4.2\,\Omega$, and $\Gamma_{opt} = 0.5\angle 185°$. Design an amplifier having a power gain of $15.75\,\text{dB}$ and noise figure of less than $3.5\,\text{dB}$. Assume that $G_S = 1.25\,\text{dB}$, $G_L = 0.75\,\text{dB}$ and $Z_0 = 50\,\Omega$. *[Hint: Use Smith chart and Eqs (15.106), (15.113), (15.149), (15.154)–(15.157), (15.159)–(15.162), (15.180), and (15.182)–(15.183)]*

16

Microwave Propagation and Communication Systems

16.1 Introduction

In the previous few chapters, we have discussed different types of passive and active microwave devices and networks. These microwave devices and networks, when arranged to perform some task, result in a microwave system. In practice, out of a number of available microwave systems, microwave communication systems and radar systems are most important. Other applications of microwave are in the fields of industry, science, and medicine (ISM).

The term *microwave communication* refers to the correspondence between two points, mainly between a transmitting section and a receiving section, with the help of microwave, through atmospheric layers or by some other means. Thus, communication is significantly dependent on propagation of electromagnetic wave. In an ideal case, electromagnetic waves propagate through free space in a straight line without any attenuation. However, free space is an idealized concept, and the propagation of an electromagnetic signal is subjected to propagation effects such as attenuation, reflection, refraction, and diffraction. In this chapter, we will discuss different microwave communication systems, different types of microwave propagation, and some propagation phenomena that can affect the operation of a microwave system. Radar systems and ISM applications will be discussed in the next chapter.

Microwave communication systems can be classified into two categories: terrestrial system and satellite system. A terrestrial system operates on the earth's surface. In practice, terrestrial wave propagations are of five types: (a) ground wave propagation, (b) sky wave propagation, (c) space wave propagation, (d) scattered wave propagation, and (e) duct propagation. Ground and sky wave propagations are over the horizon (OTH) propagation and not much suitable at microwave frequencies. At microwave frequencies, space wave can be used for propagation. However, the main disadvantage of space wave propagation is that it is line of sight (LOS) propagation, that is, the transmitting and receiving antennas must see each other. Due to the radius of curvature of the earth's surface, antennas become invisible after a certain distance, which restricts the space wave propagation coverage area. For communication between two points over a long distance via microwave, satellite communication can be used.

Connectivity between a radio wave transmitter and a receiver is known as a communication link. Microwave communication links are widely used to carry voice, data, or television signals over a distance starting from intercity link to deep space. In general, microwave communication links are of two types: (a) guided wave link where the signal travels over a low-loss transmission line and (b) radio link where the signal propagates through space. In the guided wave mode, power decreases exponentially as $e^{-2\alpha z}$, where α is the attenuation constant of the line. In contrast, in radio link, radiated power decreases at a rate of $1/R^2$ (in the absence of other losses), where R is the distance from antenna. This implies that radio links have less path loss than guided wave modes and also do not require a physical space between the transmitter and the receiver. In addition, guided wave links are not practical for satellite and space applications. A comparison of attenuation for different transmission schemes is shown in Fig. 16.1.

In microwave communication, microwave signals are mainly used as carrier signals, whereas message signals are used as modulating signals. This helps to accommodate a large number of channels within the system bandwidth. Modulation and multiplexing techniques helps the microwave link to carry a large number of individual channels. For example, a 4 GHz microwave carrier with 10% bandwidth can carry 10^5 voice channels or 66 television channels.

Depending on the modulation technique used, microwave communication can be classified as (a) analog communication and (b) digital communication. In analog modulation, signals are frequency division multiplexed (FDM) to form the baseband signal. It is then frequency modulated (FM) onto the microwave carrier for transmission. In digital systems, signals are time division multiplexed (TDM) to form the baseband signal. The baseband signal is then phase modulated (PM) by phase shift keying (PSK) onto the microwave receiver.

Microwaves are widely used in communication to accommodate a large number of channels with a high signal-to-noise ratio. Due to their high frequencies, atmospheric noise and noise due to electric transients and automobile ignition systems cannot easily affect microwaves. Further, since the bandwidth is only a fraction of the carrier frequency, microwaves also correspond to a large bandwidth. In addition, a narrow radiation beam can also be obtained at microwave frequencies with a small antenna. These are a few advantages of microwave communication. However, microwave communication also suffers from some major disadvantages. It can easily penetrate the ionosphere and hence sky wave propagation is not possible. Due to losses, ground wave propagation is also not possible. Therefore, we need to use space wave propagation, which is LOS type. As a result, microwave communication needs a large number of transmit–receive systems or

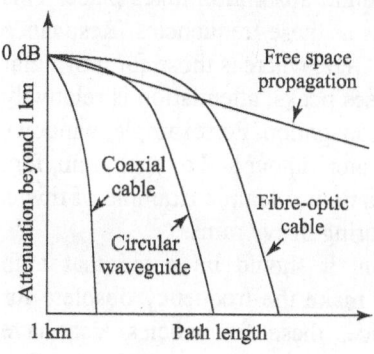

Fig. 16.1 Path length vs attenuation for different transmission schemes

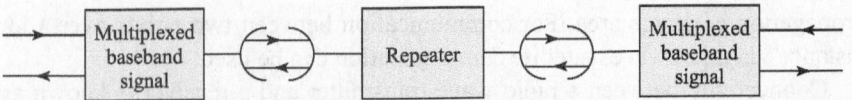

Fig. 16.2 Basic block diagram of terrestrial microwave communication system

repeaters for a long-distance communication, as shown in Fig. 16.2. For proper operation, the distance between two successive repeater stations should be within 32–80 km to make up the transmission loss in space and facilitate communication between two stations that are OTH. This in turn increases the set-up cost of microwave communication.

Microwave communications are widely used in day-to-day life, such as in TV broadcasting and wireless communications. Initially, wireless radio communications were limited between two fixed points. However, later, demand of communication with moving objects such as ships, aircrafts, or automobiles resulted in a whole new communication technology, called the cellular technology. This aided with semiconductor technology resulted in smaller, efficient, and compact communication equipment. Cellular radio systems use better frequency reuse techniques.

In addition to cellular radio systems, another short-range wireless technology, called bluetooth, has occupied the market place since the last decade. The bluetooth system was initiated by a group of four major companies in 1998. It has become so popular that most of the communication devices and other electronic equipment, starting from cellular systems to music systems, are now integrated with this technology.

16.2 Effect of Atmosphere on Propagation

Atmosphere has a large impact on wave propagation, especially at microwave frequencies. Out of different atmospheric effects, atmospheric attenuation is of utmost concern. A plot of atmospheric attenuation versus frequency (Fig. 16.3) shows that the effect of atmospheric attenuation is very little for frequencies below 10 GHz and becomes prominent above it. This is primarily due to absorption of microwave energy by water vapour and molecular oxygen. If the molecular resonance frequency of water vapour or oxygen coincides with the frequency of the signal, then resonance occurs and maximum absorption takes place. This phenomenon results in certain absorption peaks at these frequencies. Resonance peaks for water vapour occur at 22.2 and 183.3 GHz, whereas those for molecular oxygen occur at 60 and 120 GHz. Between theses peaks, attenuation is relatively small and results in a window for microwave propagation. For example, windows exist near 35, 94, and 135 GHz. In addition to water vapour and oxygen, rain, fog, and snow also contribute to atmospheric attenuation. A simple example of this is that DTH televisions sometimes do not work during heavy rains.

In connection to the preceding discussion, it should be noted that high attenuation at a particular frequency does not make the frequency obsolete for communication or other purposes. In practice, these frequencies also have some important applications. For example, atmospheric remote sensing is often performed with radiometers at 20 or 55 GHz. Since atmospheric attenuation is

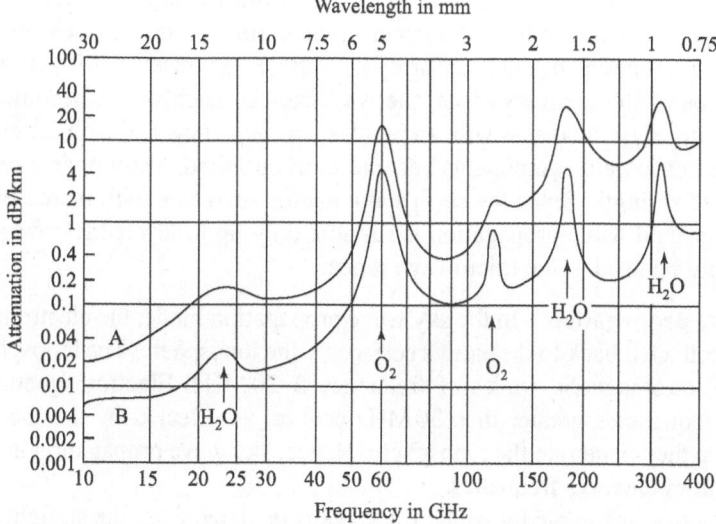

Fig. 16.3 Average atmospheric attenuation at different locations (A) Sea level at $T = 20°C$, $P = 760$ mm, and $H_2O = 7.5$ g/m³ (B) Altitude of 4 km at $T = 0°C$ and $H_2O = 1$ g/m³

high at these frequencies, it maximizes the sensing of atmospheric conditions. Similarly, spacecraft to spacecraft communication is carried out at 60 GHz. A high atmospheric loss at this frequency reduces the possibility of interference, eavesdropping, and jamming from the earth.

In addition to attenuation, several other phenomena, such as ground effect and plasma effect, also affect microwave propagation. The ground reflects microwave signals, which then add with the direct ray, generally with a phase difference. Due to a longer path and non-unity reflection coefficient, the reflected waves are generally of smaller amplitude. In addition to reflection, ground also affects wave propagation through diffraction, especially at the LOS boundary. However, this is relatively small at microwave frequencies. In contrast, diffraction of microwave energy by massive objects like a hill is stronger.

When a spacecraft re-enters the atmosphere, a dense plasma is formed on it as it travels through ionosphere and moves with a high velocity. Plasma can also be formed due to lightning, nuclear explosion, and meteor shower. In such a case, wave propagation is possible only if the frequency of the signal is larger than a particular frequency called the plasma frequency. Lower-frequency waves are completely reflected. The plasma frequency depends on the charge concentration in the plasma. In the presence of the earth's magnetic field, the plasma becomes anisotropic and the situation becomes more complicated. The average plasma frequency is about 8 MHz.

16.3 Ground Wave and Sky Wave Propagation

The following subsection discuss the propagation of ground and sky waves.

Ground wave propagation In the ground wave propagation mode, the electromagnetic wave propagates close to the earth's surface. To avoid short circuiting, polarization of such waves must be vertical. If the wave propagates very close to the ground, then it induces a current in the earth's surface, which results in a loss in wave energy through absorption. Further, because of diffraction by irregularities of the earth's surface, the wavefront of such waves becomes tilted. The amount of tilt increases with an increase in frequency. If the tilt is very large, then its electric field components become short circuited, resulting in a reduction in the field strength. Since the earth's attenuation increases with increases in frequency, ground wave propagation is suitable only up to lower MHz frequencies and cannot be used in the microwave range.

Sky wave propagation In the sky wave propagation mode, the electromagnetic wave is reflected back to the earth's surface by the ionosphere. The ionosphere can reflect electromagnetic waves of frequency 2–30 MHz. Electromagnetic waves having frequencies greater than 30 MHz cannot be reflected by the ionosphere, and hence they penetrate the ionosphere. Hence, sky wave propagation is also not suitable at microwave frequency.

In practice, the angle by which the wave is deviated from the straight path by the ionosphere depends on the (a) frequency of the signal, (b) angle of incidence of the wave at the ionosphere, (c) density of charged particle at the ionosphere at that instant, and (d) thickness of the ionosphere at that point. When the angle of incidence of the electromagnetic wave is large, the sky wave returns to the ground at a long distance from the transmitter. Thus, the region between the transmitter and the point of return remains void of the sky wave (also known as skip distance). To reduce this distance, the angle of incidence must be reduced. However, there exists a limit to which the angle of incidence can be reduced. If the angle of incidence becomes lower than this limit, instead of returning back, the wave penetrates the ionosphere and lost into space, as shown in Fig. 16.4. This sets a maximum usable frequency (MUF), defined as the maximum possible frequency for which reflection towards the earth surface is possible for a given distance of propagation

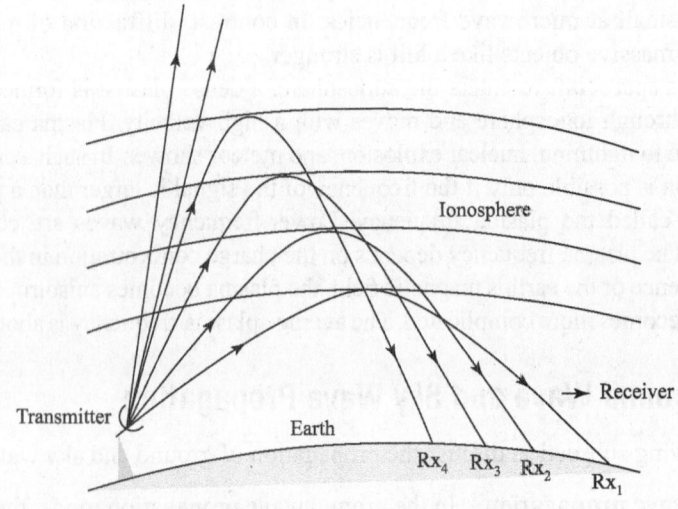

Fig. 16.4 Ionospheric propagation

for the given ionized layer. Electromagnetic waves with frequencies greater than the MUF will not be reflected back to the earth and penetrate the ionosphere.

The MUF can be expressed as $f_{\text{MUF}} = f_c \sec(\theta_0)$ $\qquad$ (16.1)

where θ_0 is the angle of incidence at the ionosphere and f_c is critical frequency given by $f_c = \dfrac{Ne^2}{4\pi\varepsilon}$ $\qquad$ (16.2)

where N is the electron per unit volume, e is the charge of electron, and ε is the permittivity of the medium.

16.4 Space Wave Propagation

It has already been mentioned that propagation of ground and sky waves is not suitable at microwave frequencies, not even at lower microwave frequencies. The mode of propagation mostly used at these frequencies (microwave) is space wave propagation. In contrast to ground and sky wave propagation, where transmitting and receiving antennas are placed at a long distance, space wave propagation is LOS propagation. In space wave propagation, transmitting and receiving antennas can see each other, and the wave travels from the transmitting to the receiving antenna either by a direct path or by reflection from the earth's surface. This mode of propagation is similar to light wave propagation, and hence the laws of geometrical optics can be applied.

Space wave propagation suffers from a serious disadvantage. The presence of a massive obstacle (hill or building) in the path of propagation creates a finite path difference between the direct and reflected waves, producing a *ghost* signal on the receiver. Space wave propagation is generally used for VHF communication employed for television and radar systems. Since space wave propagation occurs through the tropospheric layer of the atmosphere, it is also known as tropospheric propagation.

The property of the medium through which the space wave propagates influences the propagation. For example, as the height from the earth's surface increases, the density of air and distribution of water vapour vary. This changes the dielectric constant and refractive index of the medium. The modified dielectric constant and refractive index can be written as follows:

$$\varepsilon_r = \left\{ 1 + \left(\frac{79P - 11V}{T} + \frac{3.8 \times 10^5 V}{T^2} \right) 10^{-6} \right\}^2 \qquad (16.3)$$

$$n' = n + \frac{h}{R} \qquad (16.4)$$

where P is barometric pressure in millibar, T is temperature in Kelvin, V is the water vapour pressure in millibar, h is the height above the ground, and R is the radius of the earth. The variation of refractive index with height plays an important role in space wave propagation and results in a variety of phenomena such as refraction, reflection, scattering, duct transmission, and fading. Since the decrease of pressure and humidity with height is faster than that of temperature, Eq. (16.3) reveals that permittivity also decreases with increases in altitude.

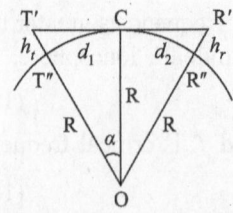

Fig. 16.5 Range of propagation

We can also define an excess *modified index of refractive modulus* as follows:

$$M = (n'-1)\times 10^6 = \left(n-1+\frac{h}{R}\right)\times 10^6 \quad (16.5)$$

To find the expression of the range of propagation, let us consider Fig. 16.5. From this figure, we can write the following expression:

$$\cos(\alpha) = \frac{OC}{OT'} = \frac{R}{R+h_t} = \frac{1}{1+(h_t/R)} \quad (16.6)$$

Since $R \gg h_t$ Eq. (16.6) can be written as $\cos(\alpha) = \left(1+\frac{h_t}{R}\right)^{-1} = 1-\frac{h_t}{R}$ (16.7)

Further, we can write that $\cos(\alpha) \approx 1-\frac{\alpha^2}{2}$ for $\alpha \to 0$ (16.8)

Substituting Eq. (16.8) into Eq. (16.7), we get $1-\frac{\alpha^2}{2} = 1-\frac{h_t}{R}$

or $\qquad \alpha = \sqrt{2h_t/R}$ (16.9)

Further, $\alpha = \text{Arc/Radius} = d_1/R$ (16.10)

Substituting Eq. (16.10) into Eq. (16.9), we get $d_1 = \sqrt{2h_t R}$ (16.11)

Similarly, we can write that $d_2 = \sqrt{2h_r R}$

Therefore, $d = d_1+d_2 = \sqrt{2h_t R} + \sqrt{2h_r R} = \sqrt{2R}\left(\sqrt{h_t} + \sqrt{h_r}\right)$ (16.12)

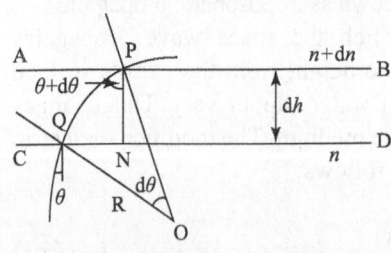

Fig. 16.6 Radius of curvature

In deriving Eq. (16.12), it has been assumed that the ray moves in a straight line. However, due to the variation in the refractive index of medium, the ray does not move in a straight line. To find the radius of curvature, we will assume the following conditions: (a) effect of the curvature of the earth is negligible and (b) tropospheric layers are parallel to the earth's surface. Now, to find the radius of curvature let us consider Fig. 16.6. From this figure, we can write the following expression:

$$\text{Radius}(a) = \text{Arc/Angle} = PQ/d\theta \quad (16.13)$$

From the right angle triangle PQN, we can write the following equality:

$$\cos(\theta+d\theta) = PN/PQ = dh/PQ$$

or $\qquad PQ = dh/\cos(\theta+d\theta) \approx dh/\cos(\theta)$ if $d\theta \to 0$ (16.14)

Substituting Eq. (16.14) into Eq. (16.13), we get the following expression:

$$a = \frac{dh}{\cos(\theta)\,d\theta} \tag{16.15}$$

Now, from Snell's law $\dfrac{\sin(\theta)}{\sin(\theta + d\theta)} = \dfrac{n + dn}{n}$

$$n\sin(\theta) = (n + dn)\sin(\theta + d\theta)$$

or $\qquad n\sin(\theta) = (n + dn)\{\sin(\theta)\cos(d\theta) + \cos(\theta)\sin(d\theta)\} \tag{16.16}$

Now, for $d\theta \to 0$:

$$\sin(d\theta) \cong d\theta \tag{16.17}$$

and $\qquad \cos(d\theta) \cong 1 \tag{16.18}$

Substituting Eqs (16.17) and (16.18) into Eq. (16.16), we get the following expression: $n\sin(\theta) = (n + dn)\{\sin(\theta) + \cos(\theta)\,d\theta\} \tag{16.19}$

For a small variation in n, $dn\,d\theta$ is very small and hence the term $dn\,d\theta \cos(\theta)$ in Eq. (16.19) can be neglected. Hence, we get the following relation:

$$n\sin(\theta) \approx n\sin(\theta) + n\cos(\theta)\,d\theta + dn\sin(\theta)$$

or $\qquad \cos(\theta)\,d\theta = -dn\sin(\theta)/n \tag{16.20}$

Substituting Eq. (16.20) into Eq. (16.15), we get the following expression:

$$a = -\frac{n\,dh}{dn\sin(\theta)} = -\frac{dh}{\sin(\theta)(dn/n)} \tag{16.21}$$

Further, if we assume that θ is small, then $\sin(\theta) \cong 1$ and Eq. (16.21) is reduces to the following form: $a = -dh/(dn/n) = -n/(dn/dh) \tag{16.22}$

Assuming $n \cong 1$, Eq. (16.22) simplifies to the following form: $a = -dh/dn \tag{16.23}$

During sky wave propagation, rays reach the receiving antenna via two path, direct and reflection from the earth's surface, as shown in Fig. 16.7. To find the resultant field strength, let us assume the earth's surface to be flat. Further, we will assume that E_0 is the field strength of the direct ray at the receiving point. If Γ be the reflection coefficient of the earth, then the magnitude of the ground reflected wave will be ΓE_0. The resultant field strength E_r at the receiving antenna will be the vector sum of these two waves and can be written as follows:

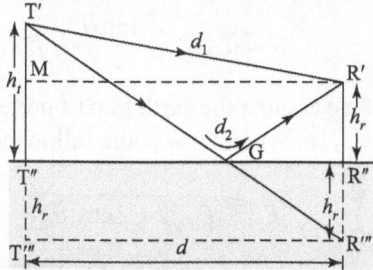

Fig. 16.7 Calculation of field strength

$$E_r = E_0 + \Gamma E_0 e^{j\theta} = E_0\left[1 + \Gamma\{\cos(\theta) + j\sin(\theta)\}\right]$$

or $\qquad E_r = E_0\left[\{1 + \Gamma\cos(\theta)\} + j\Gamma\sin(\theta)\right]$

$$|E_r| = E_0\sqrt{\{1+\Gamma\cos(\theta)\}^2 + \{\Gamma\sin(\theta)\}^2}$$

or $\quad |E_r| = E_0\sqrt{1+\Gamma^2+2\Gamma\cos(\theta)}$ (16.24)

where θ is the phase difference between the direct and reflected waves, and can be derived as follows:

From the right angle triangle T'MR', we can write the following expression:

$$d_1 = \sqrt{(h_t-h_r)^2 + d^2} = d\sqrt{1+\left(\frac{h_t-h_r}{d}\right)^2}$$ (16.25)

Since $d \gg h_t - h_r$, Eq. (16.25) reduces to the following form:

$$d_1 = d\left\{1+\frac{1}{2}\left(\frac{h_t-h_r}{d}\right)^2\right\} = d+\frac{(h_t-h_r)^2}{2d}$$ (16.26)

Similarly, from the right angle triangle T'T'''R''', we can write the following relation:

$$d_2^2 = (h_t+h_r)^2 + d^2$$ (16.27)

where we have used the relations GR' = GR''' and R'R'' = R''R''' = T''T'''. Following the same procedure, as before, and assuming that $d \gg h_t + h_r$, we can write the following expression:

$$d_2 = d+\frac{(h_t+h_r)^2}{2d}$$ (16.28)

Therefore, the path difference between the direct and reflected waves is as follows:

$$\Delta d = d_2 - d_1 = \frac{(h_t+h_r)^2 - (h_t-h_r)^2}{2d} = \frac{4h_t h_r}{2d} = \frac{2h_t h_r}{d}$$ (16.29)

The corresponding phase difference is expressed as follows:

$$\alpha = \frac{2\pi}{\lambda}\Delta d = \frac{4\pi h_t h_r}{\lambda d}$$ (16.30)

Now, if β be the phase difference due to reflection, then

$$\theta = \alpha + \beta = \frac{4\pi h_t h_r}{\lambda d} + \beta$$ (16.31)

If we assume the earth to be a perfect conductor, then $\Gamma = 1$ and $\beta = \pi$. Therefore, Eq. (16.24) reduces to the following form:

$$|E_r| = E_0\sqrt{2+2\cos(\theta)} = E_0\sqrt{2+2\left\{2\cos^2\left(\frac{\theta}{2}\right)-1\right\}}$$

or $\quad |E_r| = E_0\sqrt{2+4\cos^2\left(\frac{\theta}{2}\right)-2} = 2E_0\cos\left(\frac{\theta}{2}\right)$

or $\quad |E_r| = 2E_0\cos\left(\frac{4\pi h_t h_r}{2\lambda d}+\frac{\pi}{2}\right) = 2E_0\sin\left(\frac{2\pi h_t h_r}{\lambda d}\right)$ (16.32)

Equation (16.32) gives the expression of the magnitude of the received field strength. Now, $E_0 = 7\sqrt{P}/d$ V/m (16.33)

where E_0 is the field strength of direct ray and P is the effective radiated power in Watts. Substituting Eq. (16.33) into Eq. (16.32), we get the following expression:

$$|E_r| = 14\sqrt{P} \sin\left(\frac{2\pi h_t h_r}{\lambda d}\right)\bigg/d \qquad (16.34)$$

Now, if $d \gg h_t h_r$,

$$\sin\left(\frac{2\pi h_t h_r}{\lambda d}\right) \simeq \frac{2\pi h_t h_r}{\lambda d} \qquad (16.35)$$

Substituting Eq. (16.35) into Eq. (16.34), the following relation is obtained:

$$|E_r| = \frac{14\sqrt{P}}{d}\frac{2\pi h_t h_r}{\lambda d} = \frac{87.92\sqrt{P}h_t h_r}{\lambda d^2} \qquad (16.36)$$

The LOS propagation concept is based on the assumption that the wavefront of the propagating wave is a plane. However, in practice, this is not true for a finite distance between the transmitter and the receiver. If the transmitter is assumed to be a point source, then the wavefront is spherical. Now from the law of physical optics, each point on this spherical wavefront can be assumed to be a secondary source point, which, in turn, generates a secondary wave. These secondary waves reach the receiver with a phase lag compared to the direct ray. Thus, even an obstacle does not block the direct path; it may be present on the secondary source point and thereby reduce the signal strength at the receiver via interference with the direct path. This is shown in Fig. 16.8.

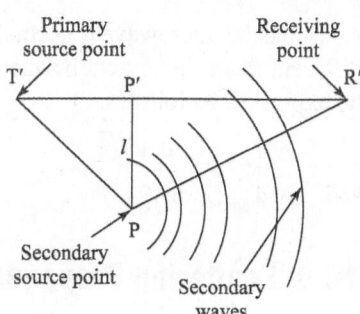

Fig. 16.8 Determination of Fresnel's zone

According to the diffraction theory, the indirect path $T'P + PR' = T'R' + \lambda/2$ must be avoided for obstruction-less transmission. The locus of P that satisfies the equation $T'P + PR' = T'R' + \lambda/2$ (16.37)

is an ellipsoid around the direct path TR. The volume enclosed by this ellipsoidal surface is called the Fresnel's zone, and it must be cleared from obstacles to achieve the free-space propagation condition. The distance PP' in the figure can be expressed as follows: $l = \sqrt{\dfrac{T'P' \times P'R'}{T'R'}\lambda}$ (16.38)

The maximum value of l is obtained when $T'P' = P'R' = T'R'/2$. Therefore from Eq. (16.38), we can write that $l_{max} = \sqrt{T'R' \times \lambda/4}$ (16.39)

The effective height of an obstacle, as seen from a microwave tower, is more than its physical height, due to the earth's curvature and Fresnel's diffraction.

The earth's bulge that adds to the physical height of the obstacle can be expressed as $h_{eb} = 0.078d_1'd_2'$ (16.40)

where d_1' and d_2' are the distances of the obstacle from the two microwave towers, in km.

The LOS communication finds many applications, which are as follows:

1. Fixed point to point, fixed point to multipoint, and movable point to point communications for military applications
2. Point to point link of a large network
3. Point to multipoint communication for television broadcasting and telecommunication systems
4. Transport of audio–video or video signals for community antenna TV head end extensions, broad cast transport, and studio to transmitter link
5. Transport of telemetry, command, and control information by power and pipeline companies
6. Interconnectivity for air traffic control
7. Specialized digital networks
8. Linking of offices and buildings in congested areas and other short-haul applications

Since the microwave is normally bent or refracted, the radio horizon is normally different from the optical horizon. In practice, the optical and radio horizons can be expressed as follows:

$$d_{\text{optical}} = 0.46\sqrt{h} \tag{16.41}$$

and $\qquad d_{\text{radio}} = 0.49h \tag{16.42}$

16.5 Scattering Propagation

Scattering propagation is a kind of space wave propagation and occurs due to continuous state of turbulence between the troposphere and the ionosphere. The turbulence causes a local variation in the refractive index of the atmosphere and hence scattering of high-frequency microwave signal. A typical diagram for scattering propagation is shown in Fig. 16.9.

Scattering occurs in the troposphere up to a frequency of 5 GHz, and the amount of scattering depends on the magnitude of variation of refractive index and the area over which these variations exist. Scattering propagation can be used for wave propagation up to a distance of 750 km, however with a lower signal strength than ordinary space wave propagation.

Troposcatter propagation has several advantages:

1. It reduces the requirement of the number of repeater stations for LOS communication by one-third to one-tenth, in turn reducing the cost.
2. It can be used for radio wave communication across another political territory.
3. It allows multi-channel communication across a large stretch of water, inaccessible terrains, or a territory where repeaters cannot be installed.
4. It requires less maintenance cost.

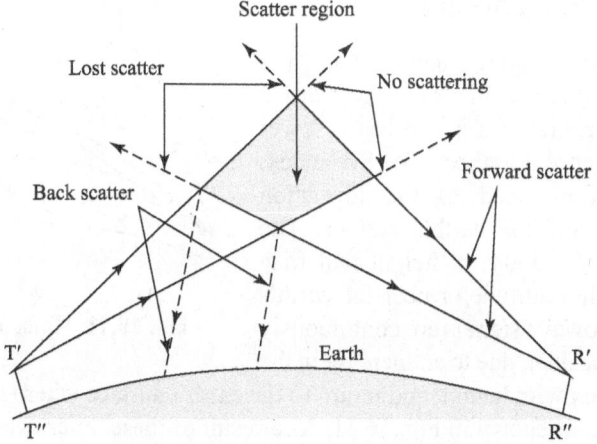

Fig. 16.9 Scattering propagation

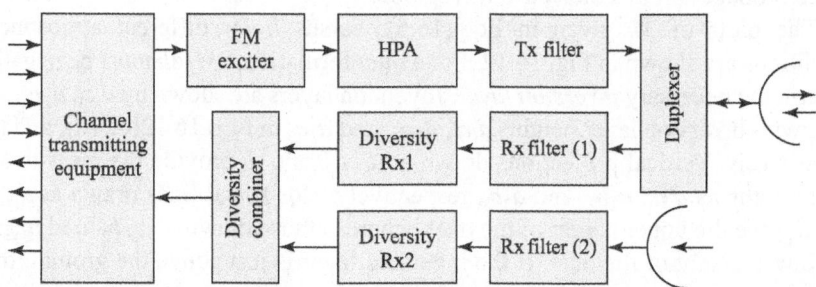

Fig. 16.10 Typical troposcatter communication link, incorporating space diversity system

Scattering propagation also has some major disadvantages:

1. It suffers from larger path losses than LOS propagation.
2. It requires high-power transmitters, large parabolic dish antennas, and sensitive receivers for its operation. This increases the cost.
3. It suffers from undesirable noise and fading.
4. Scatter loss, which depends on fading effect and medium loss due to extended scattered volume and narrow antenna beams, affect the performance of a troposcatter system.

The fading effect in a troposcatter system can be minimized by using diversity reception. A typical troposcatter communication link, incorporating a space diversity system, is shown in Fig. 16.10.

Over-the-horizon microwave propagation can be possible through two modes: (a) tropospheric scattering, as discussed earlier, and (b) diffraction. The distance at which diffraction and tropospheric losses are approximately equal is $65 \times (100/f)^{1/3}$. For a distance less than this, diffraction is the dominant mode, while for a distance greater than this, tropospheric scattering is dominant. For a path having an angular distance of 20 mrad or more, the diffraction mode can be neglected.

16.6 Duct Propagation

Since warm air is lighter than cold air, it is often trapped on the cold air over the earth's surface. In general, warm air has a lower density and hence a higher refractive index. These conditions lead to the formation of a duct around the earth's surface. The duct may be 50–500 ft in height and may form at a higher altitude or near the earth's surface. Microwave signals are continuously refracted in the duct, due to an increase in the

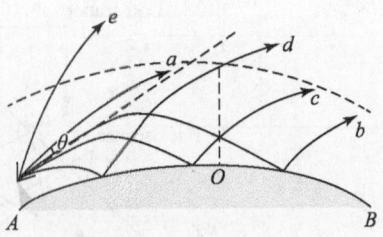

Fig. 16.11 Duct propagation

refractive index with height, and return to the earth's surface where they are again reflected back, as shown in Fig. 16.11. As a result of these, microwave signals are trapped near the earth's surface and conducted over a long distance around the curvature of the earth. Such propagation is called *duct propagation* and is similar to the propagation in a dielectric waveguide.

The plots of M, given in Eq. (16.5), versus h for different atmospheric conditions are shown in Fig. 16.12. For a *duct* formation, dM/dh must be negative to form the necessary *inversion layer*. Inversion layers are shown by d_1c_1, d_2c_2, and d_3c_3, with inversion layer heights d_1a_1, d_2a_2, and d_3a_3 in Figs 16.12(c), (d), and (e), respectively. Vertical projections drawn at d_1, d_2, and d_3 provide the measures of duct widths as d_1a_1, d_2b_2, and d_3b_3, respectively. Horizontal lines drawn at d_1, d_2, and d_3 give the upper height of the duct, whereas those drawn at a_1, b_2, and b_3 give the lower height of the duct. If the inversion layer is just above the ground, then

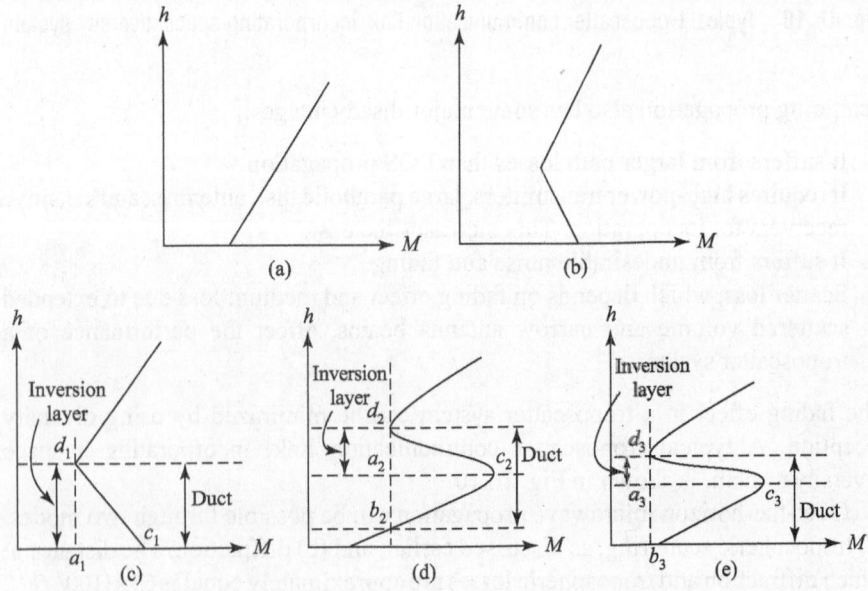

Fig. 16.12 Plot of excess modified index of refractive modulus (M) and atmosphere height (h) (a) Standard atmosphere (b) Refraction at lower height (c) Surface duct (d) Elevated duct (e) Ground-based duct

it is known as a *ground duct*. Figures 16.12(c) and (e) are examples of ground ducts. Alternatively, if the inversion layer is above the ground, then it is called an elevated duct, as shown in Fig. 16.12(d).

In practice, there exists a critical angle below which rays get trapped. Above this critical angle, they penetrate the duct and do not get trapped. In practice, the rays that lie within 1° about the horizon are trapped within the duct. Therefore, for duct propagation the height of the antenna must be lower than that of the duct.

Similar to a waveguide, a duct also behaves like a high-pass filter and allows a signal of wavelength less than a maximum value λ_{max}, given by the following relation: $\lambda_{max} = 0.084 t^{3/2}$ (16.43)

where t is the width of the duct in metres and λ_{max} is in centimetres. Another empirical formula for the cut-off wavelength is $\lambda_c = 2.5 t \sqrt{\Delta \varepsilon} \times 10^{-6}$ (16.44)

where $\Delta \varepsilon$ is the change in the dielectric constant across the duct in esu unit, t is in feet, and λ_c is in centimetres. For a typical case, t is about 400 ft and $\Delta \varepsilon$ about 49 esu. This gives

$$\lambda_c = 2.5 \times 400 \times \sqrt{49 \times 10^{-6}} = 7\text{cm} = 7 \times 10^{-2}\,\text{m}$$

or $\qquad f_c = 4.3 \times 10^9 = 4.3\,\text{GHz}$

Normal duct propagation can extend over 1000 km or more as compared to only 100 km without a duct at this frequency.

16.7 Transmission Interference and Signal Damping

Microwave communication system is interfered by the electromagnetic waves present in the transmission route. In fact, interference may be of several types, such as neighbouring channel or co-channel interference, multipath propagation, fading, and distortion. Some of them are described in the following subsections:

16.7.1 Neighbouring Channel and Co-channel Interference

Neighbouring channel interference occurs due to the lack of a proper guard space or a bad choice of modulation/filter scheme. In this type of interference, harmonics of a particular channel couple with the neighbouring channel, thus causing neighbouring channel interference.

Co-channel interference occurs when the transmitting stations operate at the same frequency in a very short interval. This interference can be prevented by separating the transmitters in space or by reducing transmitter power.

16.7.2 Fading of Space Wave Signals

Certain property of the troposphere (e.g., refractive index) changes with time and causes a variation in the intensity of the received signal. This is known as fading. Fading also depends on meteorological conditions such as season, wind flow, temperature, and humidity. Due to some adverse effects such as ionosphere storms and solar flares, fading may be total and such a condition is called fade-out. The effect of fading on the received signal also depends on the distance travelled by the signal, and is high near the radio horizon and shadow zone. Fading is also pronounced when the received signal is weak and low when a good optical path is present. The following are some fading types encountered in practical applications:

Slow fading   Slow fading occurs due to a change in the atmospheric refractive index and is a long-term variation. Such fading has a log-normal distribution with a standard deviation of 2–10 dBs. The standard deviation depends upon the scattering angle in case of a troposcatter communication system.

Fast fading   Fast fading occurs due to multipath effects and is a short-term variation of the instantaneous values. Fading of a signal is caused by the presence of a large number of components of random and varying phases. Fast fading has a Rayleigh distribution for a sampling time of a few minutes.

Interference fading   Such fading occurs when a signal arrives at the receiver via multiple paths. When the signal penetrates different layers of the ionosphere, it may reach the receiver after several hops between the earth and the ionosphere. Since the strength of the received signal depends on the phases of its various components, variation in the ionosphere condition may lead to alternate cancellation or reinforcement of the signal components and hence variation in the strength of the signal arriving at the receiver. Interference fading is multipath fading and can be of the following types (Fig. 16.13): (a) reflection multipath fading, (b) atmospheric multipath fading, and (c) sub-refraction fading.

Selective fading   This is a multiple-frequency and multiple-path effect, and occurs in case of signals having multiple frequency components. Following interference fading, different components of a certain frequency will be added, while those of the remaining frequencies will be cancelled. Therefore, some frequency components will fade in, while the rest will fade out independently. Selective fading produces serious distortion of a modulator signal. In practice, AM signal with a high percentage of modulation are worst affected by selective fading. Selective fading is predominant at higher frequencies. The effect of selective fading can be minimized using exalted carrier reception and single sideband (SSB) systems.

Absorption fading   This fading results from the absorption of a signal by the medium through which it is propagating. In contrast to interference fading, which is local and occurs temporarily, absorption fading is constant and occurs over the entire length of the link.

Skip fading   This fading occurs near the skip distance. As a result of this fading, the receiving point may move in or out depending on the height and density of the ionosphere layer.

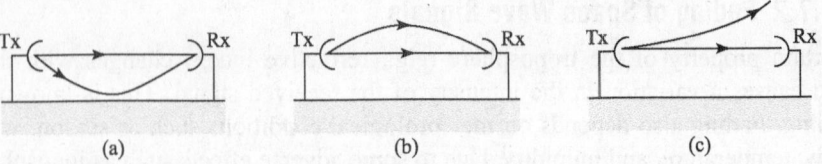

Fig. 16.13   Fading mechanism (a) Reflection multipath (b) Atmospheric multipath (c) Sub-refraction

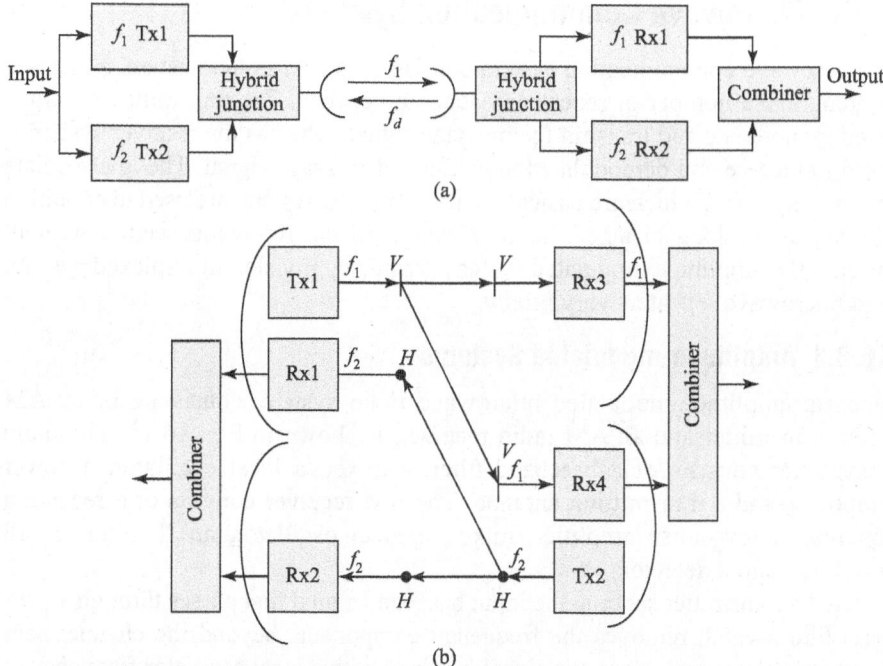

Fig. 16.14 Different systems for reducing fading (a) Frequency diversity system (b) Space diversity system

Polarization fading Such fading occurs due to a change in the polarization of reflected sky wave caused by the superposition of ordinary and extraordinary rays with random amplitudes, phases, and opposite polarization. Since most antennas are polarization sensitive, a change in the polarization of the signal causes reduction in the received field strength, resulting in polarization fading.

Fading of a signal can be minimized using two techniques: (a) frequency diversity and (b) space diversity (Fig. 16.14).

In a frequency diversity system, a signal is sent along the same path at two different frequencies f_1 and f_2, where $f_1 \sim f_2 \geq 100$ MHz. Due to the large frequency difference, it is unlikely that both the signals will fade simultaneously. It should be noted that a frequency diversity system requires two transmitters and two receivers tuned at two different frequencies, and hence is costly to implement. Further, the limitations of spectrum allotment at microwave frequencies also prevent the use of this technique.

In space diversity systems, signals are transmitted from left to right at the same frequency f_1, using a vertically polarized antenna, but via at least two different paths, which requires two receivers. For proper operation, the receivers should be spaced at least 50λ apart. Outputs of these two receivers are then combined to get the desired output. Similarly, right to left transmission can be performed at a different frequency f_2 and horizontal polarization.

In case fading cannot be eliminated, a fading margin of 30 dB can be implemented during system design. This provides 99.99% system reliability even during worst fading.

16.8 Microwave Communication Systems

A microwave communication system consists of a transmitter system, a receiver system, and a number of repeater systems in between. The transmitter system is used to modulate and transmit the message signal, whereas the receiver system is used to receive and demodulate the modulated message signal. The intermediate repeater systems, which are basically a trans-receive system, are used to condition the signal for long distance communication. In the following section we will discuss the amplitude modulated system, frequency division multiplexed system, and microwave repeaters very briefly.

16.8.1 Amplitude-modulated Systems

A basic amplitude-modulated microwave radio system, consisting of an AM radio transmitter and an AM radio receiver, is shown in Fig. 16.15. The radio transmitter consists of a baseband filter, a mixer, a local oscillator, a power amplifier, and a transmitting antenna. The AM receiver consists of a receiving antenna, a low-noise amplifier, mixer, a local oscillator, an IF filter, an IF amplifier, and a detector.

In the transmitter section, the input baseband signal first passes through a low-pass filter, which removes the frequency components beyond the channel pass band from the signal. Next, the signal is mixed with a local oscillator frequency in a mixer. If the signal frequency is f_m and the LO frequency is f_{LO}, then the output of the mixer will contain the mixed frequency components $f_{LO} \pm f_m$. The local oscillator frequency has a much higher power than f_m. The power amplifier then amplifies the signal, which is then radiated into space by the antenna. The signal propagates through space and arrives at the receiving antenna.

In the receiver section, the signal is first received by an antenna and then amplified in a low-noise amplifier (LNA). The LNA is not compulsory for all receivers and may be removed if required. The signal is then mixed with a

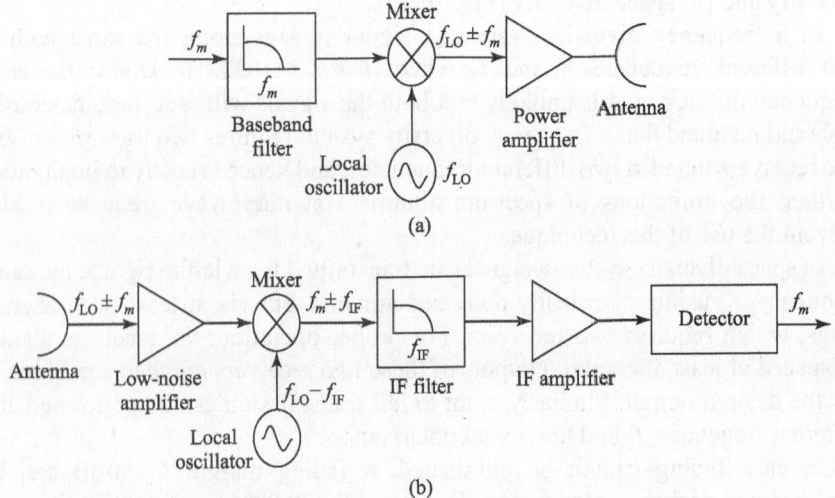

Fig. 16.15 Microwave radio (a) Transmitter (b) Receiver

local oscillator frequency $f_{LO} - f_{IF}$ in a mixer and results in lower frequency components $f_m \pm f_{IF}$. The signal then passes through an IF filter, which eliminates the unwanted frequency components existing at the mixer output. The filtered signal is then amplified by a narrow-band, high-gain amplifier. The IF amplifier also removes the $1/f$ noise. The IF amplifier output then passes through the detector, which recovers the signal f_m. It should be noted that the low-noise amplifier–mixer stage of the receiver is most important and has a prominent effect on the overall noise of the receiver system. The receiver described here is called a super-heterodyne receiver and can be easily tuned by changing the LO frequency.

It should be noted that the aforementioned system is not the only microwave AM communication system available; it has many variations, for example, an SSB system. An SSB modulator generates only one frequency component, either $f_{LO} - f_m$ or $f_{LO} + f_m$, and also uses one-half bandwidth of a double sideband (DSB) system described earlier. An SSB signal can be generated either by filtering out the other sideband using a filter or by using an SSB mixer.

The noise of a system determines the minimum detectable signal at the receiver and hence the range of communication link. In practice, an AM receiver is noisy in character. Better noise characteristics (S/N) can be obtained with an FM system.

16.8.2 Frequency Division Multiplexed System

A simple FDM system is shown in Fig. 16.16. In the transmitter side, a number of narrow-band baseband channels are first multiplexed together to form a group. Next, several such groups are combined into one RF channel, and then several RF channels are accommodated in the same RF carrier. The resultant multiplexed signal is then amplified by a power amplifier and radiated into space by the transmitting antenna.

At the receiver side, the received signal is first amplified by a low-noise amplifier and then fed into channel dropping filters for separating different RF channels. A channel dropping filter generally consists of three- or four-port components; signal is fed at one port, desired channel is filtered at another port, and the remaining signals are passed to the next channel dropping filter through the rest of the port(s). After filtering, each RF channel is down-converted using separate mixers and LO frequencies and is then passed through an IF filter to remove the unwanted frequency components existing at the mixer output. The IF signal is then amplified and demultiplexed into original baseband channels.

16.8.3 Microwave Repeaters

The basic block diagram of a microwave repeater is shown in Fig. 16.17. Horizontally polarized incoming signals of frequency $f_1, f_2, f_3, ..., f_N$ are received by the left-side antenna and then down-converted into IF signals. The IF signals are then amplified and up-converted to different microwave frequencies $f_1', f_2', f_3', ..., f_N'$. These up-converted signals are finally re-transmitted as vertically polarized signals by a transmitting antenna on the next hop after passing through

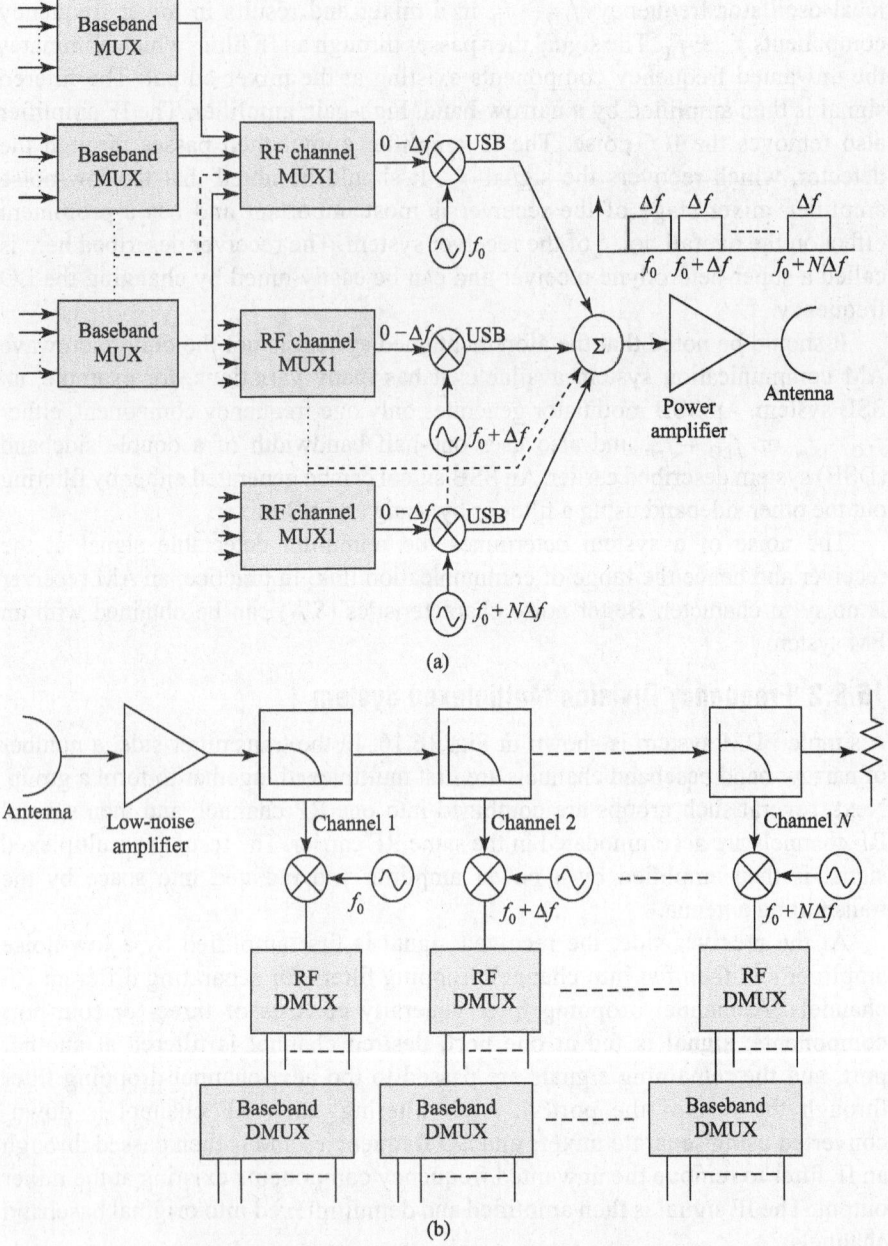

Fig. 16.16 FDM system (a) Transmitter (b) Receiver

branching and polarization filters. Similarly, the repeater receives vertically polarized signals of frequency $f_1, f_2, f_3, \dots, f_N$ from the right-side antenna and retransmits them as horizontally polarized signals with frequencies $f'_1, f'_2, f'_3, \dots, f'_N$ after similar processing steps. These differences in frequencies and polarization of the incident and re-transmitted wave reduce direct coupling between the transmitted and incident signals.

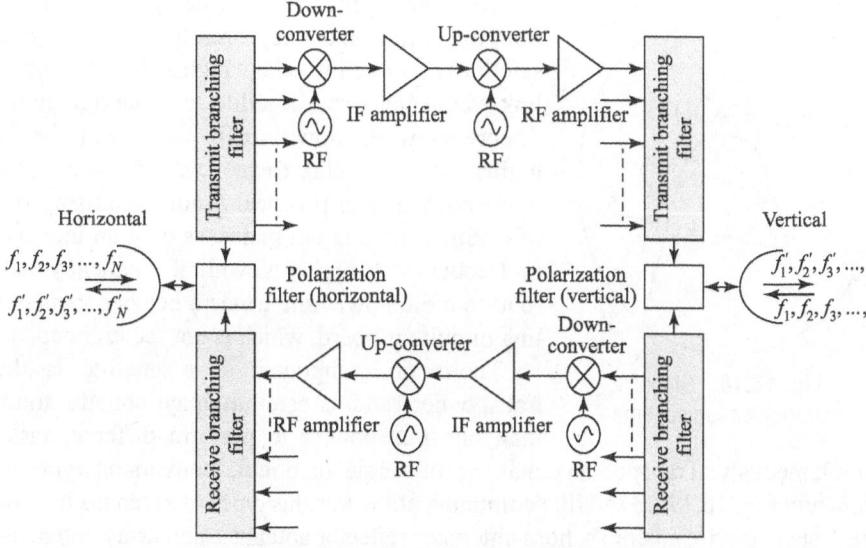

Fig. 16.17   Block diagram of heterodyne repeater

16.9 Satellite Communication

It has already been mentioned that ground and sky waves cannot be used at microwave frequencies. Space wave propagation also has some limitations due to atmospheric effect and the distance over which communication is required. In such cases, satellite communication can be a good alternative. A satellite is a spacecraft rotating around the earth in a fixed orbit. It consists of a transponder (combined transmitter and receiver system) that receives a signal from a ground station, conditions it, and retransmits it to another ground station. Depending on the rotating orbit, artificial satellites can be of two types: (a) orbital and (b) geostationary. Orbital satellites are kept at a lower altitude from the earth's surface. They rotate with a speed higher than that of the earth, and hence revolve around the earth more than once in a day and cover a particular area of the earth at a particular time. Communication via an orbital satellite is possible only during this time period. Orbital satellites may be of lower earth orbit (LEO) or middle earth orbit (MEO) type. LEO satellites are placed within 2000 km from the earth surface, whereas MEO satellites are placed at an altitude of more than 10,000 km (but less than 36,000 km). A geostationary satellite is kept at an altitude of 36,000 km from the earth's surface. It rotates with a speed equal to the speed of the earth and so appears stationary from the earth. The main advantage of a geostationary satellite is that it can be used to communicate with a particular region 24 h/day, but its disadvantage is that it cannot be used to communicate across all regions of the earth and at least three of such satellites, spaced at an angle of 120°, are required to perform this job, as shown in Fig. 16.18. Since the relative motion between the earth and a geostationary satellite is low, the Doppler shift in frequency is negligible. It is also far easier to track a geostationary satellite.

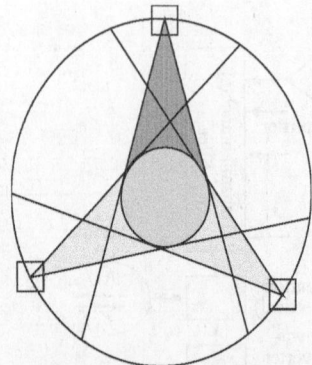

Fig. 16.18 Global communication scheme

To avoid interference in satellite communication, the uplink frequency (from ground station to satellite) of a satellite is kept higher than the downlink frequency (from satellite to ground station). Theoretically, the downlink frequency can also be higher than the uplink frequency, but it is not recommended due to practical problems arising out of it. Since attenuation increases with an increase in frequency, a higher downlink frequency will require more power and hence a heavier transmitting circuit on-board, which is not recommended.

The main component of a satellite is the transponder, and a satellite may contain more than one transponder to perform different tasks simultaneously. Transponders may be of single or double conversion type, as shown in Fig. 16.19. In satellite communication, various types of antennas may be used, such as wire antennas, horn antennas, reflector antennas, and array antennas.

Since the ionosphere is transparent to microwave frequencies, microwave signals from an earth station can easily reach the satellite antenna, or vice versa, without much adverse effect.

Depending on the range of coverage, satellite systems can be classified as domestic, regional, and global. Satellites can also be passive or active. A passive satellite is a metallic balloon or sphere that acts as a passive reflector. In contrast, an active satellite behaves as an active microwave repeater. In addition to the transmitting and receiving antennas and transponders, an active satellite should also contain an accurate positioning system, a control mechanism, and power generating and conditioning systems involving a solar cell array with NiCd battery back-up during solar eclipse.

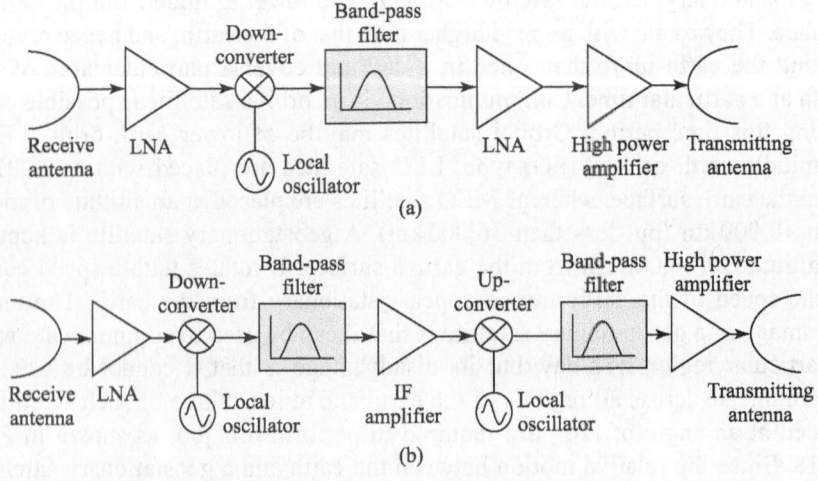

Fig. 16.19 Simplified block diagram of transponder (a) Single conversion (b) Double conversion

16.9.1 Satellite Orbit

Satellite orbits can be divided into three categories: (a) inclined elliptical, (b) polar circular, and (c) geostationary, as shown in Fig. 16.20. The choice of a particular orbit of a satellite is very important in terms of the coverage area, time frame of the availability of the satellite from a particular area of the earth, and path loss. The inclined elliptical orbit is mostly used for coverage of polar regions, whereas the polar circular orbit is mostly used for navigation purpose. The geostationary orbit, also known as the *Clarke orbit*, is mostly used for round-the-clock communication. For the inclined elliptical orbit, the apogee is kept above the polar region so that maximum visibility can be achieved. The geostationary orbit is circular and is fixed at a height of approximately 36,000 km over the earth's surface.

A satellite stays on its orbit due to the balance between inward earth's gravitational force and outward centripetal force. These two forces can be expressed as follows:

$$F_{gravitation} = GMm/r^2 \tag{16.45}$$

and $\quad F_{centripetal} = mv^2/r \tag{16.46}$

where G is gravitational constant, M is the mass of the earth, m is the mass of satellite, and r is the distance of the satellite from the centre of the earth. Equating Eqs (16.45) and (16.46), we get $v = \sqrt{GM/r}$ m/s $\tag{16.47}$

Now, the orbital period of the satellite is as follows: $T = 2\pi r/v \tag{16.48}$

Substituting Eq. (16.47) into Eq. (16.48), we get the following expression:

$$T = \frac{2\pi r\sqrt{r}}{\sqrt{GM}} = \frac{2\pi r^{3/2}}{\sqrt{GM}} s = \frac{2\pi r^{3/2}}{60\sqrt{GM}} \text{ min} \tag{16.49}$$

Substituting $G = 6.67 \times 10^{-11}$ NM^2/kg^{-2} and $M = 5.98 \times 10^{24}$ kg into Eq. (16.49), we get the following relation: $T = 1.66 \times 10^{-4} r^{3/2}$ min $\tag{16.50}$

Equation (16.50) reveals that the choice of r depends on the orbital period chosen. If we choose an orbital period of 24 h, that is, the period of rotation of the earth about its axis, then the satellite will appear stationary with respect to the earth.

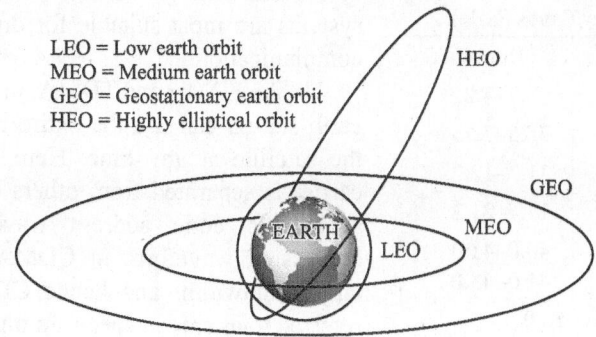

LEO = Low earth orbit
MEO = Medium earth orbit
GEO = Geostationary earth orbit
HEO = Highly elliptical orbit

Fig. 16.20   Satellite orbits

Such a satellite is called a geostationary satellite. For such a case, we get the following expression: $r = 42,180\,km$ (16.51)

If we assume that the radius of the earth is 6370 km, then the distance of the orbit above the earth's surface will be (42,180–6370) or 35,810 km.

16.9.2 Frequencies of Satellite Communication

Commercial satellite communication generally use the frequency range 2–24 GHz, which can be extended to 30 GHz if required. Inter-satellite communication is carried out in the frequency span 54–64 GHz. Table 16.1 shows different frequency bands used in satellite communication.

As the frequency of satellite communication is increased, the corresponding bandwidth also increases. However, higher frequencies are more prone to be affected by atmospheric effects.

In satellite communication, a transmitting station may require to communicate with a large number of receiver stations, simultaneously. Similarly, a receiver station may also require to receive signals from more than one transmitting station at a given instant of time. Hence, a satellite normally handles multiple up- and downlinks simultaneously. This simultaneous accessing and sharing of satellite resources by several earth stations is known as multiple access communication. In multiple access communication, each earth station transmits its individual uplink carriers to the satellite, which relays individual downlink carriers to a number of receiving stations. In practice, different types of multiple access formats are possible, such as (a) frequency division multiple access (FDMA), (b) time division multiple access (TDMA), and (3) code division multiple access (CDMA).

In an FDMA system, each earth station is allotted specific uplink and downlink frequencies, within the allotted satellite bandwidth. FDMA systems are simplest but suffer from some major disadvantages such as inter-carrier interference, crosstalk, and inter-modulation products from nearby carriers.

In contrast to FDMA, in TDMA, each uplink and downlink carrier is assigned a specific time frame, and the earth station can access the satellite during that time frame only. This requires synchronization among all earth stations, which results in a complicated station operation. However, a TDMA system has no inter-carrier interference, crosstalk, and inter-modulation products from nearby carriers. Hence, TDMA systems are most suitable for digital satellite communication.

Unlike FDMA and TDMA, in CDMA each earth station can use the entire bandwidth of the satellite at any time. Here, each uplink carrier is separated from others by assigning a specific code address waveform. This addressing waveform in CDMA occupies a large bandwidth, and hence CDMA is also referred to as spread spectrum multiple access (SSMA). CDMA systems are immune to

Table 16.1 Frequencies of satellite communication

Frequency Bands (GHz)	
Uplink	Downlink
5.9–6.4	3.7–4.2
7.9–8.4	7.25–7.75
14.0–14.5	11.7–12.2
30.0–31.0	17.7–21.2
50.0–51.0	40.0–41.0
	41.0–43.0
54.0–58.0	
59.0–64.0	

interference and are therefore suitable for military applications. Like TDMA, CDMA also requires synchronization and results in complex receiver equipment. However, unlike TDMA, accurate frequency and time interval separation are not required in CDMA.

16.9.3 Satellite Altitude/Station Keeping

The satellite altitude is defined as the orientation of a satellite with respect to the earth. Since satellites normally use linear polarized signals, maintaining a close control on the satellite altitude is required. For proper operation, the highly directional satellite antenna must point towards the earth, with its solar panel oriented towards the sun. This again requires a control of the satellite altitude.

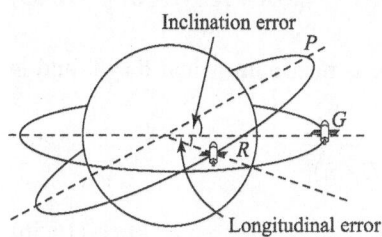

Fig. 16.21 Inclination and longitudinal error

Due to variations in the earth's gravitational forces, solar radiation, magnetic field forces, and gravitational pull of the moon and sun, satellite position and orientation change with respect to the earth, leading to inclination and longitudinal errors, as shown in Fig. 16.21. The angle of inclination (in degrees) of the satellite orbit with respect to a preferred orbit gives the inclination error, which must be kept within $\pm 0.1°$. Longitudinal error, which is a more serious error, also needs to be controlled carefully, as satellites are spaced along the geostationary orbit, which can be maximum $\pm 0.1°$. Control signals that are generated at the ground station to keep the satellite in position are known as station keeping.

16.9.4 Transmission Path

The transmission path corresponds to the distance of a satellite from any point on the earth. To calculate the transmission path, let us consider Fig. 16.22. From this figure, we can write that

$$(R+h)^2 = R^2 + d^2 - 2Rd \cos(90° + \beta) = R^2 + d^2 + 2Rd \sin(\beta)$$

or $\quad R^2 + 2Rh + h^2 = R^2 + d^2 + 2Rd \sin(\beta)$

or $\quad d^2 + 2Rd \sin(\beta) - h(2R + h) = 0$ $\hfill (16.52)$

Equation (16.52) represents a quadratic equation in d whose roots can be expressed as follows:

$$d = \frac{-2R \sin(\beta) \pm \sqrt{4R^2 \sin^2(\beta) + 4h(2R + h)}}{2}$$

or $\quad d = -R \sin(\beta) \pm \sqrt{R^2 \sin^2(\beta) + h(2R + h)}$ $\hfill (16.53)$

Since distance cannot be negative, keeping only the positive roots, we get the following relation:

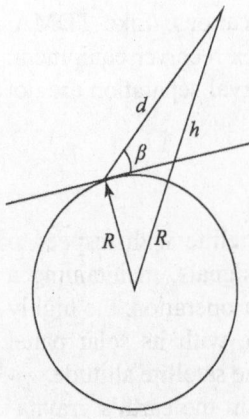

$$d = -R\sin(\beta) + \sqrt{R^2\sin^2(\beta) - R^2 + R^2 + 2Rh + h^2}$$

or $\quad d = -R\sin(\beta) + \sqrt{-R^2\left\{1 - \sin^2(\beta)\right\} + (R+h)^2}$

or $\quad d = -R\sin(\beta) + \sqrt{-R^2\cos^2(\beta) + (R+h)^2}$ $\quad$ (16.54)

Equation (16.54) reveals that the distance is minimum when $\beta = 90°$ and is expressed as follows:

$$d_{\min} = -R\sin(90°) + \sqrt{-R^2\cos^2(90°) + (R+h)^2}$$

or $\quad d_{\min} = -R + \sqrt{(R+h)^2} = -R + R + h = h$ $\quad$ (16.55)

Fig. 16.22 Satellite position with respect to earth

Similarly, the distance is maximum when $\beta = 0°$ and is expressed as follows:

$$d_{\max} = -R\sin(0°) + \sqrt{-R^2\cos^2(0°) + (R+h)^2}$$

or $\quad d_{\max} = \sqrt{-R^2 + (R+h)^2} = \sqrt{h(2R+h)}$ $\quad$ (16.56)

16.9.5 Link Design-Friis Power Transmission Equation

Link design in a satellite system basically corresponds to the calculation of power received by a receiving antenna. Let d be the distance of the satellite transmitter from the earth station and P_T the power transmitted by the antenna uniformly in all directions. We also assume that the beam illuminates an area A_0 in the receiving zone and A_R is the effective area of the receiving antenna. Therefore, the power received by the receiving antenna can be expressed as $P_R = P_T A_R / A_0$ $\quad$ (16.57) Now, the directivity of the transmitting antenna is given by the following relation:

$$G_T = 4\pi d^2 / A_0 \tag{16.58}$$

Using Eqs (16.57) and (16.58), we get $\quad P_R = \dfrac{P_T G_T A_R}{4\pi d^2}$ $\quad$ (16.59)

The product $P_T G_T$ is called the effective isotropic radiated power (EIRP). The gain of the receiving antenna is as follows: $G_R = 4\pi A_R / \lambda^2$ $\quad$ (16.60) Using Eqs (16.60) and (16.59), we get the following relation:

$$P_R = P_T G_T G_R \left(\frac{\lambda}{4\pi d}\right)^2 \tag{16.61}$$

Equation (16.61) is also known as the Friis power transmission equation. Power attenuation is as follows:

$$\alpha = 10\log_{10}\left(\frac{P_T}{P_R}\right) = 10\log_{10}\left\{\frac{1}{G_T G_R}\left(\frac{4\pi d}{\lambda}\right)^2\right\}\ \text{dB}$$

or $\quad \alpha = 20\log_{10}(4\pi) + 20\log_{10}(d/\lambda) - 10\log_{10}(G_T) - 10\log_{10}(G_R)$

or $\qquad \alpha = 21.9842 + 20\log_{10}\left(d/\lambda\right) - \left(G_T\right)_{dB} - \left(G_R\right)_{dB}$ (16.62)

Equation (16.62) reveals that the path loss can be expressed as follows:

$$L_{\text{PS}} = 21.9842 + 20\log_{10}\left(d/\lambda\right)\text{dB}$$ (16.63)

In practice, in addition to the path loss, a variety of other losses are present and hence the total losses can be expressed as follows:

$$L = L_{\text{PS}}L_A$$ (16.64)

where $\qquad L_A = L_{\text{FTX}}A_{\text{AG}}A_{\text{Ra}}L_{\text{PO}}L_{\text{Poin}}L_{\text{FRX}}$ (16.65)

Here, L_{FTX} represents the losses between transmitter output and transmitting antenna, A_{AG} the attenuation in the atmosphere and ionosphere, A_{Ra} the attenuation due to rain, L_{PO} the losses due to polarization mismatch between transmitting and receiving antennas, L_{Poin} the losses due to antenna de-pointing, and L_{FRX} the losses between the receiver antenna and the receiver.

Substituting Eqs (16.64) and (16.65) into Eq. (16.61), we get the following relation:

$$P_R = EIRP + G_R - L_{\text{PS}} - L_{\text{FTX}} - A_{\text{AG}} - A_{\text{Ra}} - L_{\text{PO}} - L_{\text{Poin}} - L_{\text{FRX}}$$ (16.66)

16.9.6 Ground Station

Since there is a limitation in the transmitted power by the satellite and a number of losses are present, the ground station should have a very low noise temperature so that very weak signals can be recognized. A basic block diagram of an earth station receiver is shown in Fig. 16.23. In most of the satellite communication systems, the same antenna is used for both transmitting and receiving signal. Since the up- and downlink frequencies are different, interference can be avoided. The received signals are separated from the transmitted signal using a duplexer switch,

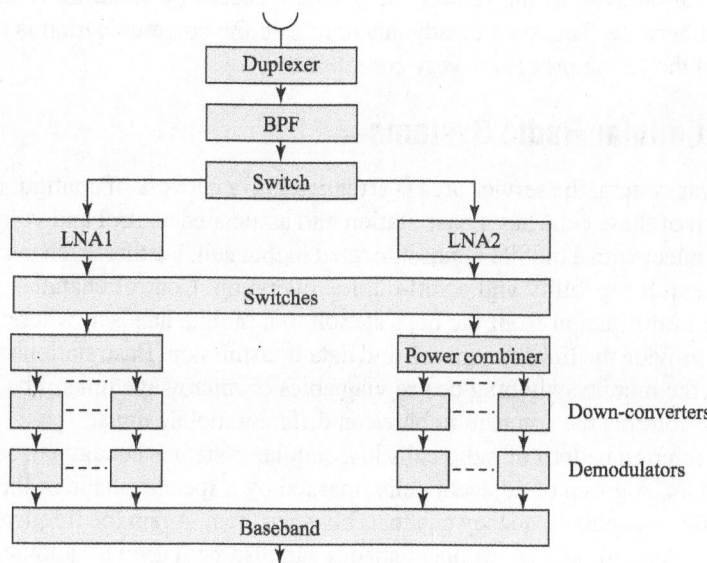

Fig. 16.23 Basic block diagram of ground station receiver

passed through a band-pass filter, and amplified using a low-noise amplifier. The amplified signal is then down-converted and demodulated.

16.9.7 Satellite Antenna

Antennas are an integral part of satellite communication systems and are used to transmit and receive signals. In practice, monopole, dipole, helical, horn, reflector, and microstrip antennas are used as on-board satellite antennas, whereas high-gain large reflector antennas are used as ground station antennas. The choice of an antenna in a satellite system depends on the applications. For example, for UHF/VHF TTC (telemetry, tracking, and command) systems, monopoles and dipoles are primarily used, whereas horn antennas are used when relatively wide beams are required. Reflector antennas are used to produce maximum gain and narrow beam for covering a particular zone on the earth. An array of printed antennas is also used in place of a reflector to reduce the complexity in the mechanical design of the reflector.

A satellite should have one antenna beam for each ground station. Therefore, if the satellite is communicating with multiple earth stations, there should be a multiple number of beams. For proper operation, the beams should also be isolated from each other. This requires one antenna feed per earth station with a single reflector.

Several factors affect satellite communication: (a) noise effect; (b) path loss effect; (c) atmospheric, ionosphere, and rain attenuation; (d) propagation delay due to finite wave velocity; (e) limitations on the number of satellite that can be placed on an orbit; etc.

Satellite communication plays a very important role in present-day communication system and is widely used in mobile communication, TV broadcasting, and telecommunication. In addition, satellites are used for navigation, meteorological services, personal communication system, aeronautic services, maritime services, remote sensing, etc. Due to its high altitude from the earth's surface, a satellite can see a large portion of the earth at a single glance and makes it possible to provide communication even to the remote parts where access by some other means is difficult otherwise. The main disadvantage of satellite communication is that it is costly and the set-up process is very complex.

16.10 Cellular Radio Systems

In a cellular system, the service area is arranged into a network of contiguous radio cells. Each of these cells has a base station and associated control and voice channels to connect with a mobile handset located in that cell. Further, each system has channel search capability and a full-duplex operation. Control channels transfer the control information from the base station to a mobile handset, whereas voice channels provide the links for speech and data transmission. Base stations are connected to the mobile switching centre via cables or microwave links. The switching centre controls the connection between different mobile units.

The preferred pattern of radio cells in a cellular system is hexagonal, as shown in Fig. 16.24. A group of adjacent cells (marked by a specific shade in the figure) share all the available frequency channels between them. Again the frequency used by a particular cell (say 'A' in black shade) can also be reused by another cell in another cell cluster (say 'A' in lighter shade). However, no two such cells should

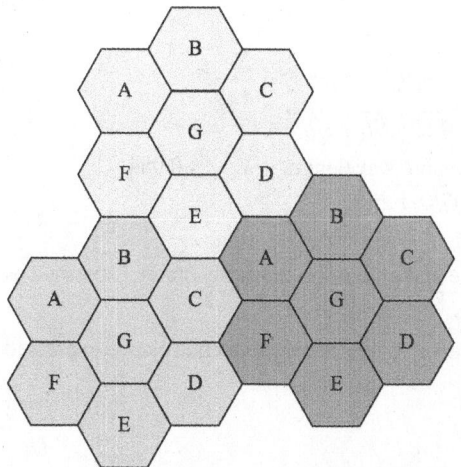

Fig. 16.24 Cell distribution in cellular system

be adjacent to each other. In a cellular system, it is possible to increase the system capability by simply splitting larger cells into smaller ones, which is known as *cell splitting*. In addition, in a cellular system, cells can also be of different sizes depending on the density and traffic distribution.

Another feature of a cellular system is the *hand-off*. This allows the user to roam between different cells without any interruption. During the communication period, the base station always measures the level of the received signal and adjust the transmit power accordingly. If the received signal falls below a certain pre-specified level, then it sends command to its adjacent base stations to monitor the signal strength and communication is handed off to the base station for which the received signal is strongest.

The cell structure shown in Fig. 16.24 can be redrawn as shown by the dotted line in Fig. 16.25(a). Therefore, instead of assuming one base station at the centre of a cell, we can assume three base stations at alternate corners of the hexagon. Each tower in a cellular system consists of three antennas, spaced at an angle of 120°, and transmits or receives signals at three different cells at three different frequencies, as shown in Fig. 16.25(b). This arrangement provides a minimum of three channels per unit cells. Here, the channel frequency is repeated after three alternate cells.

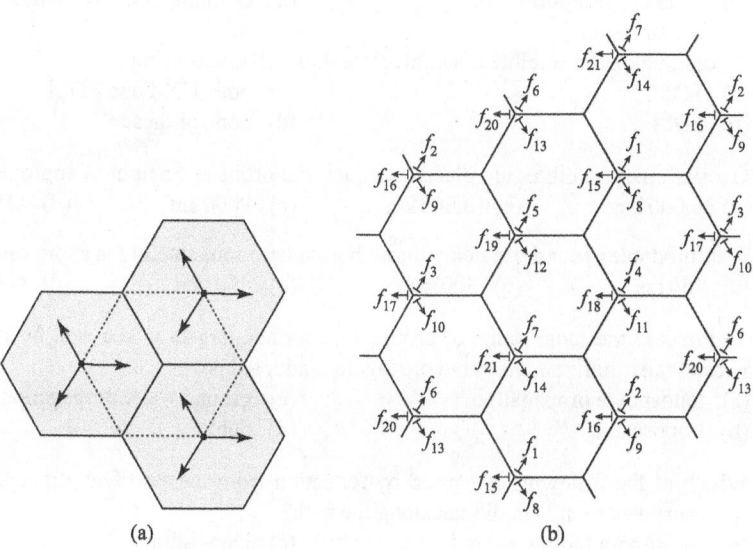

(a)   (b)

Fig. 16.25 Cell structure (a) Alternate cell structure (b) Frequency reuse pattern in alternate cell structure

Important Formulae

- The MUF can be expressed as $f_{MUF} = f_c \sec(\theta_0)$, where $f_c = \dfrac{Ne^2}{4\pi\varepsilon}$.
- Range of space wave propagation is $d = \sqrt{2R}\left(\sqrt{h_t} + \sqrt{h_r}\right)$.
- For duct propagation, the maximum allowable wavelength is $\lambda_{max} = 0.084t^{3/2}$.
- Velocity of a satellite in an orbit is $v = \sqrt{GM/r}$ m/s.
- The time period of a satellite in an orbit is $T = 2\pi r^{3/2}/\sqrt{GM}$ sec.
- Transmission distance between a satellite and a point on the earth can be expressed as
  $$d = -R\sin(\beta) + \sqrt{-R^2\cos^2(\beta) + (R+h)^2}.$$
- In satellite communication the power received by the receiver antenna can be expressed
  as $P_R = P_T G_T G_R \left(\dfrac{\lambda}{4\pi d}\right)^2$.

Exercises

Objective-type Questions

16.1 The front end of an amplifier chain in a communication system is kept immersed in liquid nitrogen to
    (a) reduce distortion by amplifier
    (b) expand its frequency response
    (c) improve its noise figure
    (d) dissipate the heat generated by amplifier

16.2 Which of the following have the noise source in a category different from the other three?
    (a) galactic noise
    (b) solar noise
    (c) atmospheric noise
    (d) cosmic noise

16.3 Short-term fading in microwave communication links can be overcome by
    (a) changing the antenna
    (b) diversity reception and transmission
    (c) increasing transmitted power
    (d) changing the modulation scheme

16.4 A communication satellite assembles telephone channels using
    (a) TDM
    (b) FDM
    (c) both TDM and FDM
    (d) none of these

16.5 Geostationary satellites are placed in equatorial orbits at a height of approximately
    (a) 36,000 km   (b) 18,000 km   (c) 9000 km   (d) 4500 km

16.6 The full duplex round trip delay through a synchronous satellite is approximately
    (a) 150 ms   (b) 300 ms   (c) 450 ms   (d) 600 ms

16.7 To increase the radar range of ground and surface targets to see well beyond the normal radar horizon, the wave propagation adopted is
    (a) ionosphere propagation
    (b) troposcatter
    (c) ground wave propagation
    (d) duct

16.8 Which of the following is caused by reflection from the stratified atmosphere or from surface or land conditions along the path?
    (a) Multi-path fading
    (b) Reflection fading
    (c) Duo-fading
    (d) Selective fading

16.9 A ground wave gradually disappears as one moves away from the transmitter because of
   (a) interference from sky wave
   (b) finite conductivity of the earth's surface
   (c) loss of LOS communication
   (d) maximum single-hop distance limitation

16.10 The skip distance is
   (a) same for each layer
   (b) independent of the state of ionization
   (c) independent of frequency
   (d) independent of transmitted power

16.11 The antenna elevation angle at a ground station for satellite communication is always kept above 5° to
   (a) minimize sky noise temperature
   (b) minimize slant range
   (c) increase the visibility of satellite
   (d) reduce the effect of oxygen and water vapour absorption on the antenna noise temperature

16.12 In microwave communication links, path diversity and frequency diversity are adopted to overcome fading in the path due to
   (a) rain attenuation
   (b) phase lagging
   (c) polarization shifting
   (d) fog accumulation

16.13 A geostationary satellite can cover
   (a) entire earth surface
   (b) one side of the earth
   (c) one hemisphere in one pass
   (d) an area depending on the antenna used

16.14 In cellular communication, the cell geometry is chosen as
   (a) square
   (b) rectangular
   (c) circular
   (d) hexagonal

16.15 In satellite communication,
   (a) Uplink frequency is greater than downlink frequency
   (b) Uplink frequency is equal to downlink frequency
   (c) Uplink frequency is less than downlink frequency
   (d) Any one of these

Review Questions

16.1 Briefly describe ground wave, sky wave, scattering, and duct propagation.
16.2 With suitable approximations, find the expressions for the range and radius of curvature for space wave propagation. In addition, find the expression for resultant field at the receiver.
16.3 What is fading? Describe different types of fading. Describe the methods that can be used to reduce fading.
16.4 Write short notes on (a) AM microwave system and (b) FDM microwave system.
16.5 With a suitable figure, describe the operation of a microwave repeater.
16.6 Derive Friis power transmission equation for satellite communication.

17

RADAR and Other Applications of Microwave

17.1 Introduction

In the previous chapter, we have discussed microwave propagation and different communication systems. In this chapter, we will discuss RADAR, another application of microwave. RADAR (RAdio Detection and Ranging) is a detection system that is used to detect a remote object and find necessary information about it. RADARs have wide applications in fields such as remote sensing, air traffic control (ATC), aircraft safety and navigation, law enforcement in highway traffic, ship safety, space applications, military, and many more, which will be discussed in detail later in this chapter.

The first experiment on RADAR was accidental. In the year 1922, scientists at the Naval Research Laboratory (NRL), Washington DC, were experimenting with a 60 MHz communication system. A transmitter was placed at the Anacostia Naval Air Station and a receiver at Haines Point on the Potomac River, which is half a mile away. During the experiment, they observed a signal fluctuation when a wooden vessel, named USS Dorchester, passed up the river. Later, a further experiment at 300 MHz established that fluctuations were due to the reflection of radio waves by the passing ship. Detection of an object, located at a distance of 3 miles, was achieved in those trials. The first practical RADAR, however, was tested in 1930 by a British team under Sir Robert Watson Watt.

Though the initial works on RADAR were carried out between 1920 and 1930, a major breakthrough was achieved during World War II, with the invention of magnetron in 1940. This high-frequency source allowed a small-size antenna to achieve a high gain. The small antenna size, in turn, allowed the mechanical steering of the antenna. RADARs played an important role in World War II.

The basic operation of a RADAR involves transmitting an electromagnetic signal of a specified pattern [continuous wave (CW), modulated continuous wave, pulsed wave, etc.] and receiving the corresponding echo signal from the target. The ability of a RADAR to send the transmitting signal repeatedly, helps the operator to obtain detailed information about the target, regarding its range, velocity, and direction. For example, as shown in Fig. 17.1, if RADAR pips

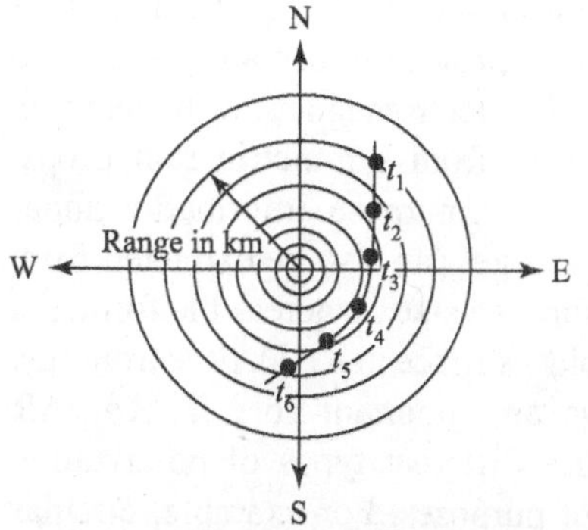

Fig. 17.1 Typical RADAR display

are obtained at different times at different points on the RADAR display, then we can predict the range, velocity, and direction of motion. The size of the pips also provides information on the size of the target.

Generally, a RADAR operates in the range 100 MHz to 36 GHz. However, this limit is not absolute. Over the horizon, RADARs may operate at frequencies as low as a few MHz, whereas in the upper limit, it may be more than 240 GHz. The visible region of electromagnetic spectrum is also used in practice.

Like other systems, a RADAR also has its own advantages as well as limitations. Its advantages are as follows: (a) it can detect targets through darkness, fog, rain, haze, and snow, and (b) it can determine the range and position very accurately. Its limitations are as follows: (a) it cannot resolve the detailed features of the target like human eyes, especially at a short distance, and (b) it cannot recognize the colour of the target.

A RADAR can provide information about a target by comparing the received echo signal with the transmitted signal. This is known as target classification. For example, it can provide the range information by measuring the time delay between the transmission of a signal and the receipt of the echo signal. The range measurement of a target is a unique feature of a RADAR that no other sensor is able to provide. Modern RADAR systems can achieve range resolution better than a fraction of a metre and can measure a range starting from a few metres to astronomical orders.

RADARs generally employ a directive antenna to provide high transmitting gain and hence to detect weak echo signal. The typical beamwidth of a RADAR antenna is 1° or 2°. The beam width of a RADAR antenna also determines its angular resolution. For modern tracking RADAR systems, the rms error in angular accuracy is as low as 0.1 mrad.

When the target moves, it produces a Doppler frequency shift that can be used to measure its relative velocity. The relative velocity of a target can also be measured by noting the rate of change of its range with respect to the RADAR antenna. The latter method is often used in tracking and air surveillance RADARs. In an air surveillance RADAR, Doppler shift is used to separate the moving target from fixed objects. The measurement of relative velocity using Doppler frequency shift is also used for surveillance and tracking of extra-terrestrial objects such as spacecraft and satellites.

The shape of a target can be determined by a RADAR, provided that the target is visible from many directions. Such a RADAR is also known as an *imaging RADAR*. Space object identification (SOI) RADARs and synthetic aperture RADARs (SARs; they map the terrain) are examples of imaging RADARs.

Internal motions of different parts of a target, such as rotation of aircraft engine, spinning of satellites, vibration of vehicles, and rotation of antennas, can also be used by a RADAR to gather information about the target. In addition, polarization-sensitive responses of electromagnetic waves also provide information about the symmetry of the target. In fact this property of the electromagnetic wave permits RADAR systems to separate asymmetric targets from symmetric rain drops. Microwave RADARs use circular polarization for this purpose. Information about dielectric properties and surface roughness of a target can also be extracted from echo signals. The latter method is used to measure sea state, whereas the former is used to probe the nature of the moon's or asteroid's surface in RADAR astronomy.

Polarization of the RADAR signal plays an important role in RADAR operation, and different types of RADARs use different types of polarization (horizontal, vertical, and circular) for different purposes. For example, circular polarization is used to reduce interference by rain, whereas horizontal polarization is used to reduce sea clutter. Polarization of the received signal also provides information about the type of the target. Linear polarization generally indicates a metallic object, whereas random polarization indicates a fractal surface like rock.

Depending on the position of the transmitter and receiver, a RADAR may be categorized as monostatic or bistatic. While the transmitter and receiver are placed at the same place in a monostatic RADAR, they are placed apart in a bistatic RADAR. A bistatic RADAR can also be expanded into a multi-static RADAR using multiple transmitters and a single receiver. Such a RADAR often increases the probability of target detection. A multi-static RADAR can also use multiple receivers and a single transmitter, but it is not preferred because of their high cost. Receivers are more sophisticated instruments that incorporate different subsystems and are often costlier than the transmitter. In addition, a single receiver can also be designed to integrate echoes from all transmitters to provide a single mosaic image of the target.

Apart from wide applications in RADAR and communication, microwave also finds applications in industrial, scientific, and medical fields (ISM applications), for example in microwave cooking, sealing of plastic, curing and breaking of concrete, destruction of fungus and wood worms in timber, diathermy and cancer treatment, thickness measurement, wire diameter measurement, and moisture content measurement.

Like RADARs, microwave heating was also discovered accidentally, during an experiment with an active RADAR set in 1945. Percy Spencer, working with Raytheon at that time, found that a chocolate in his pocket started melting when he was working with an active RADAR set. Later, similar experiments were performed using popcorns and an egg. Popcorns sputtered, crackled, and popped all over the lab, while the egg exploded in the face of one of the experimenters. To verify his finding, Spencer fed high-power microwave from a magnetron to a shielded metallic box containing food and found that the temperature of the food rose rapidly. Raytheon filed a patent for Spencer's microwave cooking process in October 1945, and in 1947 public was able to use this technology for the first time.

The first available microwave oven was named *Radarange*, which was built in 1947 by Raytheon. It was about 1.8 m tall, costed about US$ 5000, and weighed 340 kg. It was used to consume about 3 kW power, about three times that consumed by the present-day microwave ovens. Later, it was modified in several ways to reduce its size, power consumption, and cost, which resulted in present-day microwave ovens.

17.2 Simple RADAR Systems

A simple RADAR system consists of a transmitting and a receiving antenna that are connected to the transmitter and receiver, respectively, as shown in Fig. 17.2. The transmitting antenna transmits a high-power microwave signal, a part of which is intercepted by the *target*. The target then reradiates it in all directions. The receiving antenna then receives a part of this echo signal and processes it to detect the target, its range, velocity, etc. In most of the RADAR systems, the transmitting and receiving antennas are same, and the transmitting and receiving signals are separated with the help of a duplexer, as shown in Fig. 17.3.

In general, RADARs can be classified as CW or Doppler RADARs, and pulsed RADARs. A CW RADAR transmits a CW signal. In contrast, a pulsed RADAR transmits a pulse waveform, generally a train of narrow rectangular pulses modulating a sine wave carrier.

By measuring the time gap between the instants the pulse has been sent and its echo has been received, RADARs determine the distance of the target. If we assume T to be the time gap and c the velocity of the signal in the medium, then the range of the target will be as follows: $R = cT/2$ (17.1)
(Here, cT is the total path travelled by the signal during this period, i.e., the to-and-fro distance between the RADAR antenna and the target).

It should be noted that for unambiguous detection of the RADAR target, the echo must arrive before the transmission of the next pulse. Thus, if T_r is the pulse repetition time, then the maximum unambiguous range can be expressed as

$$R_{un} = \frac{cT_r}{2} = \frac{c}{2f_r}$$ (17.2)

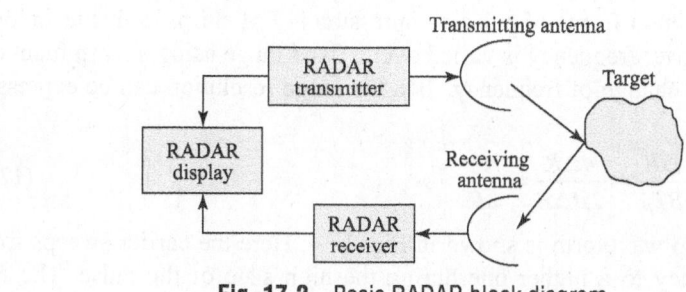

Fig. 17.2 Basic RADAR block diagram

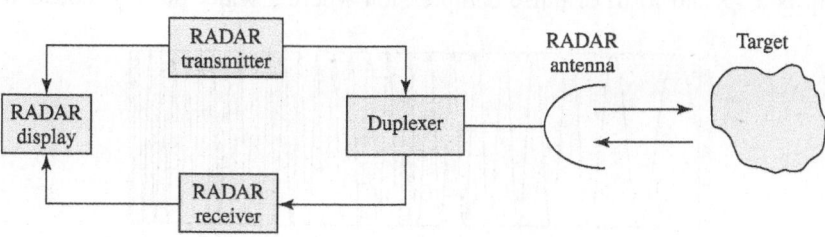

Fig. 17.3 Modified RADAR block diagram using duplexer

where f_r is the pulse repetition frequency. For the given pulse repetition frequency, if the target lies above this range, then the echo signal will arrive after the transmission of the next immediate or next few pulses. In such cases, the RADAR may not be able to correlate the echo time and actual transmission time, and may give a wrong target location. The echoes that arrive after the next immediate pulse are called *second-time around echoes*.

For the proper operation, a RADAR has some basic requirements:

1. The duplexer should be automatic.
2. The RADAR transmitter should be silent during reception.
3. The transmitting signal/pulse should be of sufficient strength so that the received signal is detectable.
4. The receiver should be extremely sensitive and immune to noise so that weak received signals can be processed. The RADAR receiver should have necessary amplification and signal processing circuitry, and should be capable of displaying the information on the RADAR system.
5. The RADAR antenna should have a large gain and highly directive radiation pattern so that most of the transmitted signal is intercepted by the target.
6. The pulse repetition frequency of the transmitted signal should be high compared to the target scanning period.

Another important parameter, associated with a RADAR, is range resolution. It is defined as the smallest distance required between two targets, along the same azimuth line, to detect them separately. If the distance between two targets is less than this value, then the RADAR will not be able to detect them separately and consider them as a single target. Range resolution can be expressed as follows:

$$\Delta R = cT_P/2 \tag{17.3}$$

where T_P is the pulse width.

Range resolution can be decreased using the pulse compression technique. One of the most common forms of pulse compression is FM chirp. In this technique, the RADAR carrier frequency is varied over a fixed range using a ramp function. Now if B is the change of frequency, then the range resolution can be expressed as follows:

$$\Delta R_c = \frac{\Delta R}{BT_P} = \frac{c\Delta R}{2B\Delta R} = \frac{c}{2B} \tag{17.4}$$

A basic chirp waveform is shown in Fig. 17.4. Here the carrier sweeps from a lower frequency to a higher one during the high state of the pulse. The FM chirp is a special form of pulse compression where a wider pulse is coded with

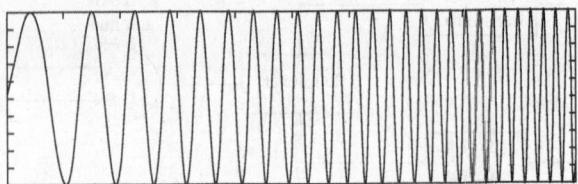

Fig. 17.4 Typical chirp waveform

a distinctive pattern. Since the pulse is wide, it has more power than a narrow pulse; at the same time, due to higher-frequency-component content, the target resolution is also high.

The realization of an optimized receiver network (also called a matched filter) for the chirp waveform is very difficult, both theoretically and in practice. However, in spite of this, many RADARs use this waveform for their operation. In practice, microprocessors and digital signal processing are used to achieve the goal. The received signal is first digitized and then processed by a sequence of calculations to implement the desired matched filter.

17.3 RADAR Range Equation

Let us consider that a RADAR antenna is transmitting some power P_t in space. If the antenna is an isotropic radiator, then the power density at a distance R is expressed as follows:

$$\text{Power density} = \frac{P_t}{4\pi R^2} \ \text{W/m}^2 \tag{17.5}$$

However, RADAR antennas are not isotropic; instead, they concentrate the radiated power P_t in a particular direction. Therefore, if G_t be the directive gain of the antenna, then the power density at a distance R is as follows:

$$\text{Power density} = \frac{P_t G_t}{4\pi R^2} \ \text{W/m}^2 \tag{17.6}$$

The target intercepts a part of the incident power and re-radiates it in all directions. Out of these re-radiated powers, only the part captured by the RADAR antenna is of interest and we will concentrate on that. For this, we need to define the term *RADAR cross-section* (RCS). It determines the power density returned to that RADAR for a particular power density incident on the target. Therefore, if the RCS of the target is σ, then the power density radiated back to the RADAR is given by the following expression:

$$\text{Echo Power density} = \frac{P_t G_t \sigma}{\left(4\pi R^2\right)^2} \ \text{W/m}^2 \tag{17.7}$$

Now, if the effective area of the RADAR antenna is A_e, then the received power is as follows:

$$P_r = \frac{P_t G_t \sigma A_e}{\left(4\pi R^2\right)^2} \ \text{W} \tag{17.8}$$

The value of R for which the received signal P_r just equals the minimum detectable signal $S_{\min}$ is called the maximum RADAR range. The target lying above this range cannot be detected by the RADAR as the received power falls below $S_{\min}$. If $R_{\max}$ is the maximum RADAR range, then $S_{\min} = P_r = \dfrac{P_t G_t \sigma A_e}{\left(4\pi R_{\max}^2\right)^2}$

or
$$R_{max} = \left[\frac{P_t G_t \sigma A_e}{(4\pi)^2 S_{min}} \right]^{\frac{1}{4}} \tag{17.9}$$

Now,
$$G_t = 4\pi A_e / \lambda^2 \tag{17.10}$$

Substituting Eq. (17.10) into Eq. (17.9), we get the following expression:

$$R_{max} = \left[\frac{P_t \sigma A_e}{(4\pi)^2 S_{min}} \frac{4\pi A_e}{\lambda^2} \right]^{\frac{1}{4}} = \left[\frac{P_t A_e^2 \sigma}{4\pi \lambda^2 S_{min}} \right]^{\frac{1}{4}} \tag{17.11}$$

Equation (17.11) can also be written in terms of G_t rather than A_e. For this case, we need to substitute

$$A_e = \frac{G_t \lambda^2}{4\pi} \tag{17.12}$$

into Eq. (17.9), to obtain the following expression:

$$R_{max} = \left[\frac{P_t \sigma G_t}{(4\pi)^2 S_{min}} \frac{G_t \lambda^2}{4\pi} \right]^{\frac{1}{4}} = \left[\frac{P_t G_t^2 \lambda^2 \sigma}{(4\pi)^3 S_{min}} \right]^{\frac{1}{4}} \tag{17.13}$$

Equations (17.9), (17.11), and (17.13) give the expressions for the maximum RADAR range equation. In practice, all the expressions are basically the same; the difference is in their interception. For example, from Eq. (17.11) it may be concluded that $R_{max} \infty 1/\sqrt{\lambda} \infty \sqrt{f}$ and therefore the RADAR range can be increased by increasing frequency, it is conflicting with Eq. (17.13), which shows the $R_{max} \infty \sqrt{\lambda} \infty 1/\sqrt{f}$ and Eq. (17.9) which shows that R_{max} is independent of λ. The correct explanation of these expressions depends on whether the effective area remains constant with frequency, as in Eq. (17.11), or antenna gain, as in Eq. (17.13). To make the RHS of Eq. (17.9) independent of frequency, we must use two antennas, with the gain of the transmitting antenna and effective aperture of the receiving antenna being independent of frequency.

RADAR range also depends on the transmitted power, RCS, and minimum detectable signal. For example, since $R_{max} \infty [P_t]^{1/4}$, to double the maximum RADAR range, the transmitted power must be increased 16 times. The abovementioned transmitted power is usually the peak power. If the transmitter waveform is a train of rectangular pulses of pulse width τ and time period T_p, then the average power can be expressed as $P_{av} = P_t \tau / T_p = P_t \tau f_P$ (17.14) where f_P is the pulse repetition frequency. The factor P_{av}/P_t or τ/T_p or τf_P is called the RADAR duty cycle. For a pulsed RADAR, the typical duty cycle ranges from 0.001 to 0.5.

The RCS, required to calculate the maximum RADAR range, is beyond the control of the RADAR engineer and operator, and depends on the properties of the target, such as its type, size, and angle of arrival. Typical RCSs for different targets is tabulated in Table 17.1. The RCS is also a function of frequency, polarization, and

Table 17.1 Typical RCS of different targets

Target Type	RCS (m^2)
Pickup truck	200
Jumbo jet and automobiles	100
Large bomber or jet	40
Medium bomber or jet	20
Cabin cruiser	10
Large fighter	6
Helicopter	3
Small fighter or four-passenger jet	2
Bicycle	2
Small pleasure boat	2
Small, single-engine aircraft	1
Human being	1
Conventional winged missile	0.1
Small open boat	0.02
Large bird	10^{-2}
Medium bird	10^{-3}
Large insect	10^{-4}
Small insect	10^{-5}

incident and reflected angles relative to the target. If incident and reflected angles are identical, then the RCS is known as a monostatic cross-section; otherwise, it is known as a bistatic cross-section.

It may be noted that a small change in the arrival angle of a target results in a major change in the RCS. For a moving target, this results in RCS fluctuations. In addition, complex RADAR targets are composed of a number of individual scattering centres. For example, an aircraft has several scattering centres such as cockpit, wings, engines, nose, and tail. Echoes from each of these scattering centres have amplitudes and phase that are independent of others. At the RADAR receiver, all these echoes are vectorically combined, resulting in further RCS fluctuation. One of the ways to account for the RCS fluctuation in the RADAR range equation is to select a small value of RCS of the target that has a high probability of being exceeded.

If (S_i/N_i) is the signal-to-noise ratio at the input of the receiver, (S_o/N_o) that at the output of the receiver, and G the overall gain of the receiver, then the noise figure is as follows:

$$F = \frac{S_i/N_i}{S_o/N_o} = \frac{S_i}{S_o}\frac{N_o}{N_i} \tag{17.15}$$

Now, $\quad S_o = GS_i \tag{17.16}$

and $\qquad N_o = \left(GN_i + \Delta N \right)$ (17.17)

where ΔN is the equivalent noise generated in the receiver. Substituting Eqs (17.16) and (17.17) into Eq. (17.15) we get the following relation:

$$F = \frac{S_i}{GS_i} \frac{\left(GN_i + \Delta N \right)}{N_i} = \frac{\left(GN_i + \Delta N \right)}{GN_i} = 1 + \frac{\Delta N}{GN_i}$$

or $\qquad \Delta N = \left(F - 1 \right) GN_i$ (17.18)

Again $\qquad N_i = kT_0 B$ (17.19)

where k is the Boltzmann constant, T is the temperature in Kelvin, and B is the bandwidth of the receiver. Substituting Eq. (17.19) into Eq. (17.18), we get the following relation: $\Delta N = \left(F - 1 \right) GkT_0 B$ (17.20)

If this equivalent noise power at the input of the receiver is larger than the minimum detectable power S_{min}, then the target cannot be detected by the RADAR. Thus, to detect the target, S_{min} must be equal to ΔN and using Eqs (17.11) and (17.20) we can write the following expression:

$$R_{max} = \left[\frac{P_t A_e^2 \sigma}{4\pi\lambda^2 \Delta N} \right]^{\frac{1}{4}} = \left[\frac{P_t A_e^2 \sigma}{4\pi\lambda^2 \left(F - 1 \right) GkT_0 B} \right]^{\frac{1}{4}}$$ (17.21)

If a parabolic dish antenna with diameter D has been used as a RADAR antenna, then $A_e = 0.65\pi D^2/4$ (17.22)

Substituting $k = 1.38 \times 10^{-23}$, $T = 300$, and Eq. (17.22) into Eq. (17.21), we get the following relation:

$$R_{max} = 48 \left[\frac{P_t D^4 \sigma}{BG\lambda^2 \left(F - 1 \right)} \right]^{\frac{1}{4}} \quad \text{Km}$$ (17.23)

It should be noted that the RADAR range cannot be made infinite by controlling the parameters in the RADAR range equation. The RADAR receiver is subjected to interference and jamming signals from enemies. Therefore, the received signal must be high enough to override these interfering and jamming signals. In addition, suitable anti-jamming technique must also be incorporated. In practice, the most suitable method for improving the RADAR range is to increase antenna aperture.

EXAMPLE 17.1 A pulsed RADAR operating at 9 GHz has an antenna gain 30 dB, transmitter power 2.5 kW, and minimum detectable signal −100 dBm. The target is a cabin curser that has an RCS 10 m². Calculate the maximum RADAR range.

Solution Given: $f = 9\,\text{GHz}$, $G = 30\,\text{dB} = 10^{30/10} = 1000$, $S_{min} = -100\,\text{dBm} = 10^{-100/10}\,\text{mW} = 10^{-10}\,\text{mW} = 10^{-13}\,\text{W}$, $P_t = 2.5 \times 10^3\,\text{W}$, and $\sigma = 10\,\text{m}^2$

The operating wavelength is $\lambda = \dfrac{3 \times 10^8}{9 \times 10^9} = 0.0333\,\text{m}$

Therefore,

$$R_{max} = \left[\frac{P_t G_t^2 \lambda^2 \sigma}{\left(4\pi \right)^3 S_{min}} \right]^{\frac{1}{4}} = \left[\frac{2.5 \times 10^3 \times \left(10^3 \right)^2 \times \left(0.0333 \right)^2 \times 10}{\left(4\pi \right)^3 \times 10^{-13}} \right]^{\frac{1}{4}}$$

or $\qquad R_{max} = 19333.021\,\text{m} = 19.333\,\text{km}$

Practice Problem

17.1 A pulsed RADAR operating at 12 GHz has an antenna gain 25 dB, transmitter power 3 kW, and minimum detectable signal −90 dBm. The target is a medium jet that has an RCS 20 m². Calculate the maximum RADAR range.   **[6.5933 km]**

17.4 RADAR System Losses

A RADAR system is subjected to various types of losses, some of which are known and some unknown. RADAR designers make huge efforts to minimize the known losses; however, losses of 10–20 dB are always present in the system. These losses affect the RADAR range equation and hence must be incorporated while deriving this equation. Some of the system losses present in a RADAR system are described in the following subsections:

Plumbing loss   It is related to the transmission line that connects the transmitter/receiver with the antenna and other microwave components present in the RADAR system, such as rotary joint, duplexer, receiver protector, directional couplers, bends, and others. The total microwave plumbing loss is about 3 dB. Antenna mismatch also corresponds to plumbing loss.

Antenna loss   This loss consists of scanning loss, phased array loss, and beam shape loss. Scanning loss occurs due to the movement of the antenna. If the antenna moves rapidly compared to the round trip transit time, then the directivity of the antenna towards the target during reception time is not same as that during transmission time. This is evident because during the round trip transit time, the antenna will change its orientation and hence also the main beam. This introduces a loss known as a *scanning loss*.

A practical RADAR antenna is often a phased array instead of a single element. Such an antenna requires a complex distribution network to connect individual antenna elements, which introduce a loss known as the *phased array loss*.

In addition, a protective shield that is used to protect the RADAR antenna from environment, also introduces some losses, known as *Radom losses*. The overall antenna loss is also about 3 dB.

Signal processing loss   Signal processing plays a very important role in RADAR systems, especially in gathering information in the presence of noise. However, it also introduces losses in the system. The main sources of signal processing losses are different signal processing subsystems such as nonmatched filter and automatic integrator.

Propagation loss   Reflection and refraction of signals by the earth surface and a medium often increase or decrease the free space range. This, in turn, increases or decreases the total atmospheric attenuation loss.

Collapsing loss   This loss occurs due to integration of additional noise samples with the desired RADAR signals.

Loss due to equipment degradation Like other systems, the performance of a RADAR system also degrades with time, which introduces about 1–3 dB losses in that system.

17.5 Detection of Signals in Noise

Detection of a target in the presence of noise signals depends on the establishment of a threshold detection level at the output of the receiver. If the receiver output exceeds this threshold level, then the target is said to be present, whereas in the opposite scenario only noise is said to be present. Such a procedure is called threshold detection. If the threshold level is properly chosen, then the receiver output will not cross it until the presence of the target.

Setting of the threshold level is a tricky job. If it is set too low, noise may exceed it and may be mistaken as an echo from a target. This is called a *false alarm*. On the other hand, if the threshold level is kept too high, then echoes from a weak target may not be able to cross it and hence the target is not detected. This is known as *missed detection*. One of the ways to solve these problems is to set two threshold levels—one low and the other high. If the echo signal is below the lower threshold level it is considered as noise, whereas if it is above the upper threshold level the target is assumed to be present. However, if the echo signal is in between the two levels, more careful observation is required.

To discuss threshold detection in detail, let us consider the time–voltage plot shown in Fig. 17.5. Fluctuations in voltage are due to random receiver and clutter noise. If a large echo signal from a target is present, as shown by point A, then it crosses the threshold level and the target is detected, as discussed. However, an interesting phenomenon occurs when the strength of the echo signal is equal to or near the threshold level (points B and C). The noise present in the system will be added to or subtracted from the echo signal depending on whether its amplitude is positive or negative. If the noise signal is added to the echo signal then the total signal will cross the threshold level and the target will be detected. On the other hand, if the noise signal is subtracted from the weak echo signal, then the total signal will fall below the threshold level and will result in a missed detection.

This discussion reveals that the noise signal may sometimes enhance the detection of the marginal signal or cause loss of detection. Therefore, the threshold level must be chosen very carefully. In military applications, missed detection may cause a severe problem and hence the threshold level is kept low. On the other hand, in traffic, a RADAR false alarm may cause havoc and hence the threshold

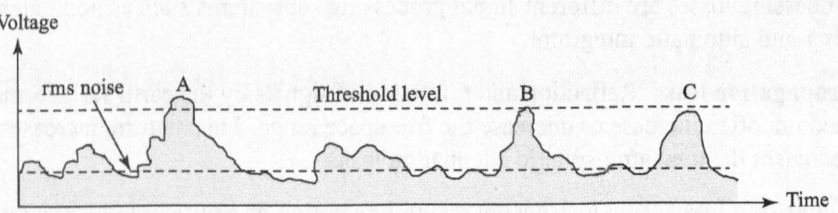

Fig. 17.5 Time–Voltage plot of received signal and noise

level is kept high. The threshold level may be set manually, where the level may be set as required, or electronically, where it is fixed. This is the advantage of a manual threshold level setup. However, this manual set-up is subjected to operational error by the human operator, which is absent in the electronic set-up.

The receiver noise at the input of the IF filter can be described by a Gaussian probability density function. When this Gaussian noise is passed through the IF filter, then the probability density function of the envelope ($\mathbb{R}$) can be expressed as a Rayleigh probability density function, as follows:

$$P(\mathbb{R}) = \frac{\mathbb{R}}{\Psi_0} \exp\left(-\frac{\mathbb{R}^2}{2\Psi_0}\right) \qquad (17.24)$$

The probability that the envelope will exceed the threshold voltage V_T is given as follows:

$$\text{Probability}\,(V_T < \mathbb{R} < \infty) = \int_{V_T}^{\infty} \frac{\mathbb{R}}{\Psi_0} \exp\left(-\frac{\mathbb{R}^2}{2\Psi_0}\right) d\mathbb{R} = \exp\left(-\frac{V_T^2}{2\Psi_0}\right) \qquad (17.25)$$

Equation (17.25) represents the probability of false alarm, since it represents the probability that the noise will cross the threshold level. Therefore,

$$P_{\text{fa}} = \exp\left(-\frac{V_T^2}{2\Psi_0}\right) \qquad (17.26)$$

This equation does not indicate whether or not a RADAR will be subjected to excessive false alarm. In practice, the time between successive false alarms is a better measure of this. To illustrate this, let us consider Fig. 17.6. The average time between crossings of the threshold level in the presence of only noise is known as the false alarm time and can be expressed as follows:

$$T_{\text{fa}} = \lim_{N \to \infty} \frac{1}{N} \sum_{k=1}^{N} T_k \qquad (17.27)$$

where T_k is the time between successive crossings of the threshold voltage by the noise envelope.

The false alarm probability can, therefore, be expressed as follows:

$$P_{\text{fa}} = \sum_{k=1}^{N} t_k \Big/ \sum_{k=1}^{N} T_k = \langle t_k \rangle_{\text{av}} / \langle T_k \rangle_{\text{av}} \qquad (17.28)$$

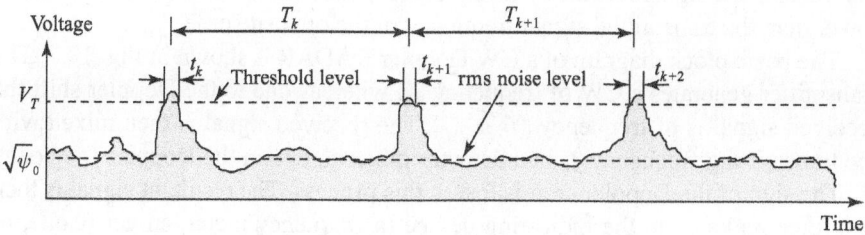

Fig. 17.6 Time–Voltage plot of noise

where t_k is shown in Fig. 17.6. If B is the bandwidth of the IF amplifier of RADAR receiver, then we can write the following equation:

$$\langle t_k \rangle_{av} = 1/B \tag{17.29}$$

Substituting Eq. (17.29) into Eq. (17.28), we get the following expression:

$$P_{fa} = (T_{fa}B)^{-1} \tag{17.30}$$

Further substituting Eq. (17.30) into Eq. (17.26), we get the following relation:

$$\frac{1}{T_{fa}B} = \exp\left(-\frac{V_T^2}{2\Psi_0}\right) \text{ or } T_{fa} = \frac{1}{B}\exp\left(\frac{V_T^2}{2\Psi_0}\right) \tag{17.31}$$

17.6 Doppler Effect and CW RADAR

It is known that a relative motion between a source and a listener results in an apparent shift in the carrier frequency of the received signal. This is known as Doppler effect. In a RADAR system, when the target moves relative to the antenna, this effect comes into play. This Doppler shift in frequency is then measured to find the velocity of the target.

If R is the distance of the target from the RADAR antenna, then the total number of wavelengths contained in the two-way path is $2R/\lambda$, where λ is the wavelength. The corresponding phase shift is as follows:

$$\phi = 2\pi \times 2R/\lambda = 4\pi R/\lambda \text{ rad} \tag{17.32}$$

If the target is moving, then R and hence ϕ are continuously changing with time. The variation of ϕ with time gives the angular frequency:

$$\omega_d = 2\pi f_d = \frac{d\phi}{dt} = \frac{d}{dt}(4\pi R/\lambda) = \frac{4\pi}{\lambda}\frac{dR}{dt} = \frac{4\pi}{\lambda}v_r \tag{17.33}$$

where f_d is the Doppler frequency and v_r the relative velocity of the target. From Eq. (17.33), we can write that $f_d = 2v_r/\lambda$ $\qquad$ (17.34)
If the target is moving at an angle θ, then $v_r = v\cos(\theta)$ $\qquad$ (17.35)
Substituting Eq. (17.35) into Eq. (17.34), we get $f_d = 2v\cos(\theta)/\lambda$ $\qquad$ (17.36)
Equation (17.36) reveals that the Doppler shift is maximum when $\theta = 0°$ and is equal to zero when $\theta = 90°$. The magnitude of Doppler shift is same regardless of whether the target is moving towards the RADAR (positive Doppler shift) or away from it (negative Doppler shift). The received signal frequency will be higher than the transmitted signal frequency if the target is moving towards the RADAR and lower than the transmitted signal frequency in the opposite case.

The basic block diagram of a CW Doppler RADAR is shown in Fig. 17.7. The transmitter generates a CW of frequency f_0, whereas due to the Doppler shift the received signal is of frequency $(f_0 \pm f_d)$. The received signal is then mixed with the transmitting frequency f_0 in a detector mixer to obtain the Doppler frequency f_d. The sign of the Doppler shift is lost in this process. The resultant signal is then amplified and sent to the indicating device (a frequency meter, an ear phone, or a meter calibrated against speed). The beat frequency amplifier in the RADAR

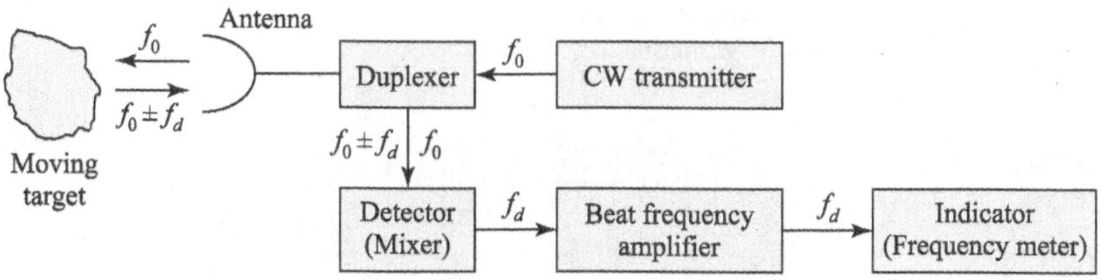

Fig. 17.7 Basic block diagram of CW RADAR

receiver must be properly chosen so that it can detect a slow as well as a fast moving object. Since the maximum velocity achievable by a target is limited, designing a beat frequency amplifier that corresponds to velocity higher than this maximum achievable velocity will be of no use. Thus, the beat frequency amplifier must also perform the job of a band-pass filter. It should be noted that clutters (stationary objects) do not provide any Doppler shift and hence are not detected by the CW Doppler RADAR. This is an inherent advantage of a Doppler RADAR.

A CW RADAR can detect the velocity and presence of the target even in the presence of static objects. In addition, it uses low transmitting power, consumes less power, has a simple circuitry, and is of small size. The following are some of its major disadvantages:

1. If a number of targets are moving with different velocities, then the amount of Doppler shift produced by each of them will also be different, which may lead to confusion.
2. The range discrimination needs the incorporation of complex circuitry.
3. Due to limitation in the maximum transmitting power, the maximum RADAR range is limited.
4. If the transmitter power is large enough, the lack of isolation between the transmitter and the receiver may cause receiver burnout. It can also introduce transmitter noise in the receiver circuitry, thus making detection of the target difficult.
5. Since the receiver is homodyne, flicker noise is introduced.
6. There is a lack of matched filter in the receiver circuitry.
7. It does not provide any information about the direction of target motion.
8. It does not provide any information about the range of the target.

Some of these disadvantages can be overcome by using separate transmitting and receiving antennas and minimizing mutual coupling between them. Absorbing materials or buffers can also be used between them to reduce the coupling. In addition, a small portion of the transmitter signal can be directed to the receiver circuitry to adaptively cancel the coupled signal. The block diagram of such a system is shown in Fig. 17.8. It should be noted that even with perfect isolation, the transmitted signal can enter the receiver circuitry via scattering from nearby clutters and obstructions. Since in most RADAR systems, the same antenna is used for transmission and reception, isolation is required between the transmitter and receiver circuitry. The amount of isolation required depends on the transmitter power, transmitter noise, ruggedness and sensitivity of the receiver, etc. For a

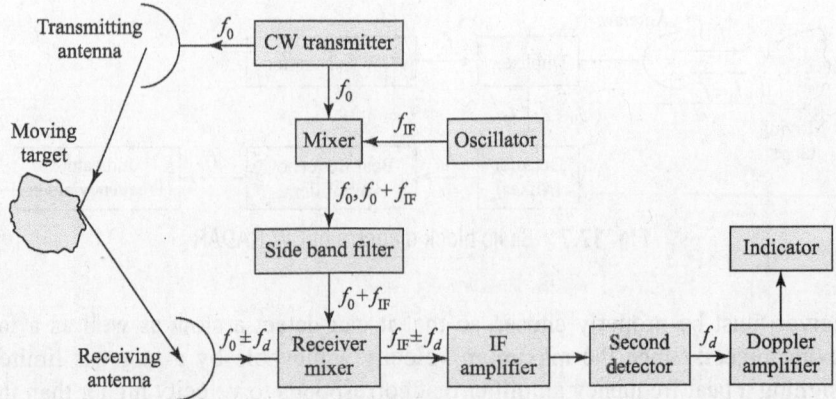

Fig. 17.8 Modified block diagram of CW RADAR using separate transmitting and receiving antennas

long-range CW RADAR, the amount of isolation required is determined by the noise corresponding to the transmitter leakage.

To determine the sign of the Doppler shift, the arrangement shown in Fig. 17.9 can be incorporated. For the two signs of the Doppler shift, the synchronous motor rotates in two opposite directions. To illustrate this, let us further assume that the transmitter signal can be expressed as follows: $E_0 = E_1 \cos(\omega t)$ (17.37)

and echo signal be expressed as $E_r = k_1 E_1 \cos\{(\omega \pm \omega_d)t + \phi\}$ (17.38)

Therefore, the signal at receiver A is $E_A = k_2 E_0 \cos(\pm\omega_d t + \phi)$ (17.39)

and that at receiver B is $E_B = k_2 E_0 \cos\left(\pm\omega_d t + \phi + \dfrac{\pi}{2}\right)$ (17.40)

For the target approaching the RADAR, Eqs (17.39) and (17.40) can be re-written as follows:

$$E_A(t) = k_2 E_0 \cos(\omega_d t + \phi)$$ (17.41)

and $$E_B(t) = k_2 E_0 \cos\left(\omega_d t + \phi + \dfrac{\pi}{2}\right)$$ (17.42)

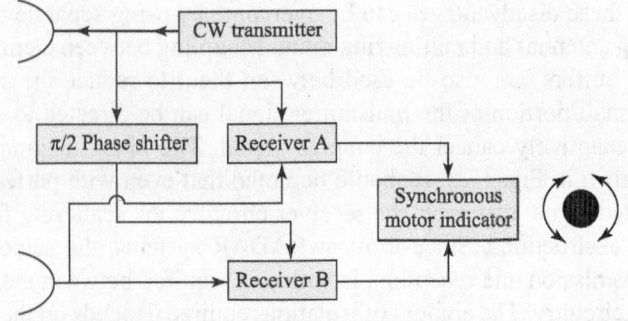

Fig. 17.9 Block diagram of circuit for detection of Doppler shift sign

Thus, the signal from receiver B leads the signal from receiver A by $\pi/2$, and the motor rotates in one particular direction. Similarly, for the target going away from the RADAR, Eqs (17.39) and (17.40) can be re-written as follows:

$$E_A(t) = k_2 E_0 \cos(\omega_d t - \phi) \tag{17.43}$$

and
$$E_B(t) = k_2 E_0 \cos\left(\omega_d t - \phi - \frac{\pi}{2}\right) \tag{17.44}$$

Thus, the signal from receiver B lags the signal from receiver A by $\pi/2$, and the motor rotates in the other direction.

CW RADARs are widely used for the measurement of the speed of a vehicle, in aircraft navigation, and in mobile applications.

17.7 Multiple-frequency CW RADAR

Let us consider that a single-frequency CW RADAR transmits a signal of the form $\sin(2\pi f_0 t)$. If the transit time of the signal is T, then the echo signal will be of the form $\sin\{2\pi f_0(t - T)\}$. Further, if the transmitted and echo signals are compared in a phase detector, then the output of the phase detector will be proportional to the phase difference between these two signals and is given as follows:

$$\Delta\phi = 2\pi f_0 T = 4\pi f_0 R/c$$

or
$$R = \frac{c\Delta\phi}{4\pi f_0} = \frac{\lambda\Delta\phi}{4\pi} \tag{17.45}$$

The phase difference $\Delta\phi$ is unambiguous only if $\Delta\phi \leq 2\pi$. Substituting $\Delta\phi = 2\pi$ into Eq. (17.45), the maximum unambiguous range is obtained as follows:

$$R_{unamb} = \lambda/2 \tag{17.46}$$

At RADAR frequencies, this unambiguous RADAR range is too small to be of practical interest.

The unambiguous RADAR range can be extended by using two CW signals of different frequencies. The unambiguous range in such cases will be half-wavelength at the difference frequency. To elaborate the operation of a multiple-frequency CW RADAR, let us assume that the RADAR transmits two CW signals of frequencies f_1 and f_2, separated by frequency Δf. Let us assume that the amplitudes of the signals are same and equal to unity. Therefore, transmitted voltage waveforms can be written as follows:

$$v_{1t} = \sin(2\pi f_1 t + \phi_1) \tag{17.47}$$

and
$$v_{2t} = \sin(2\pi f_2 t + \phi_2) \tag{17.48}$$

where ϕ_1 and ϕ_2 are arbitrary phase angles. If the target is moving, then the echo signals will suffer Doppler shifts f_{d1} and f_{d2}, respectively. The echo signals can be written as follows:

$$v_{1r} = \sin\left\{2\pi(f_1 \pm f_{d1})t - \frac{4\pi f_1 R}{c} + \phi_1\right\} \tag{17.49}$$

and $\qquad v_{2r} = \sin\left\{2\pi\left(f_2 \pm f_{d2}\right)t - \dfrac{4\pi f_2 R}{c} + \phi_2\right\}$ (17.50)

where R is the range of the target. Since $f_1 \approx f_2$, we can assume that $f_{d1} \approx f_{d2} \approx f_d$.

The RADAR receiver separates the two components of the echo signal and heterodynes each echo signal component with corresponding transmitted frequency to extract the Doppler frequency component. The extracted Doppler frequency component can be written as follows:

$$v_{1D} = \sin\left(\pm 2\pi f_d t - \dfrac{4\pi f_1 R}{c}\right)$$ (17.51)

and $\qquad v_{2D} = \sin\left(\pm 2\pi f_d t - \dfrac{4\pi f_2 R}{c}\right)$ (17.52)

Phase difference between the two Doppler components is as follows:

$$\Delta\phi = 4\pi\left(f_2 - f_1\right)R/c = 4\pi\Delta f R/c$$

or $\qquad R = \dfrac{c\Delta\phi}{4\pi\Delta f}$ (17.53)

Since, as before, the maximum permissible value of $\Delta\phi$ is 2π, we can write that

$$R_{\text{unamb}} = \dfrac{c2\pi}{4\pi\Delta f} = \dfrac{c}{2\Delta f}$$ (17.54)

Equation (17.54) reveals that $\Delta f < \dfrac{c}{2R_{\text{unamb}}}$. A large Δf improves the accuracy of the range measurement, as it implies a large $\Delta\phi$ for the given R. A two-frequency RADAR is a single-target RADAR, since only one phase difference can be measured at a time. In the presence of multiple targets, echo signal becomes complicated.

It can be shown that the rms range error can be expressed as follows:

$$\delta R = \dfrac{c}{4\pi\Delta f \sqrt{2E/N_0}}$$ (17.55)

where E is the energy contained in the signal and N_0 is the noise power/Hz of bandwidth. Equation (17.55) reveals that the greater the Δf, the lower the rms range error. However if Δf becomes too large, unambiguous range will be too small. Therefore, Δf cannot be too large if unambiguous range measurement is required. In practice, both high accuracy and unambiguous range measurement can be achieved by transmitting three or more frequencies. For example, if three frequencies f_1, f_2, and f_3 are used such that

$$\left(f_3 - f_1\right) = k\left(f_2 - f_1\right)$$ (17.56)

where k is a factor of the order 10 or 20, then the pair $\left(f_3, f_1\right)$ provides an ambiguous but accurate range measurement, whereas the pair $\left(f_2, f_1\right)$ resolves ambiguities in the $\left(f_3, f_1\right)$ measurement. For further accuracy, a fourth frequency is to be transmitted.

A multiple-frequency CW RADAR finds applications in surveying, missile guidance, and tellurometers (portable electronic surveying instruments).

17.8 Frequency-modulated CW RADAR

In a frequency-modulated CW RADAR (or FMCW RADAR), frequency of the transmitted signal changes with time in a known manner. Due to this, unlike a Doppler CW RADAR, the frequency of the echo signal and the transmitted signal, at the instant the echo is received, is not same. In practice, this frequency difference will be proportional to the transit time. Since the change in transmitter frequency with time is known, the transit time of the signal and hence the range of the target can be calculated from the frequency difference. In an FMCW RADAR, a greater transmitter frequency deviation in a given time interval corresponds to greater measurement accuracy.

To illustrate the range calculation, let us consider that the transmitter frequency is increasing with time in a linear manner, as shown in Fig. 17.10. If the distance of the target from the RADAR is R, then the transit time of the signal is as follows:

$$T = 2R/c \tag{17.57}$$

If the echo signal is heterodyned with a part of the transmitter signal, then due to the frequency difference between the signals, a beat note f_b is produced, which provides information about the range of the target. If the transmitter frequency changes at a rate of f_0 per unit time, then the total transmitter frequency change during the transit time is expressed as follows: $f_r = f_0 T = 2Rf_0/c$ (17.58)

In any practical RADAR, the frequency cannot be increased or decreased continuously, and hence periodicity in the modulation is required, as shown in Fig. 17.11. The nature of modulation may be saw tooth, triangular, sinusoidal, or some other function. For a stationary target, the beat note is of a constant frequency except at the turn around region, as shown in the figure. If f_m is the frequency of the modulation function and Δf the maximum frequency deviation, then $f_0 = 2f_m \Delta f$ (17.59)

Substituting Eq. (17.59) into Eq. (17.58), we get $f_r = 4Rf_m \Delta f/c$

or $\qquad R = \dfrac{cf_r}{4f_m \Delta f}$ (17.60)

In the previous discussion, we assumed that the target is stationary. If this is not true, then a Doppler shift will be superimposed on the FM beat note. In the frequency–time plot, this Doppler frequency shift will shift the echo signal up or down, as shown in Fig. 17.12. If we further assume that the target is approaching the RADAR, then

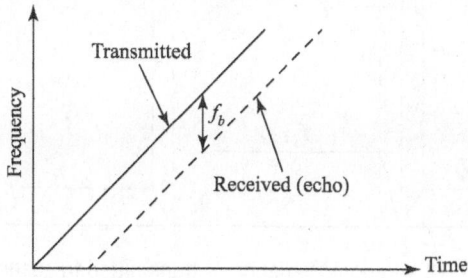

Fig. 17.10 Calculation of range using FMCW RADAR

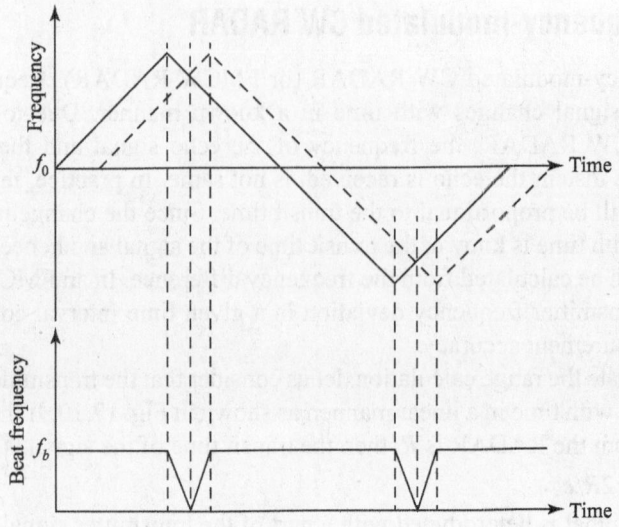

Fig. 17.11 Periodic variation of transmitted frequency and corresponding beat frequency

$$f_{b1} = f_r + f_d \tag{17.61}$$

and $$f_{b2} = f_r - f_d \tag{17.62}$$

where f_d is the Doppler shift due to target motion.

From Eqs (17.61) and (17.62), we get the following expressions:

$$f_r = \frac{1}{2}(f_{b1} + f_{b2}) \tag{17.63}$$

and $$f_d = \frac{1}{2}(f_{b1} - f_{b2}) \tag{17.64}$$

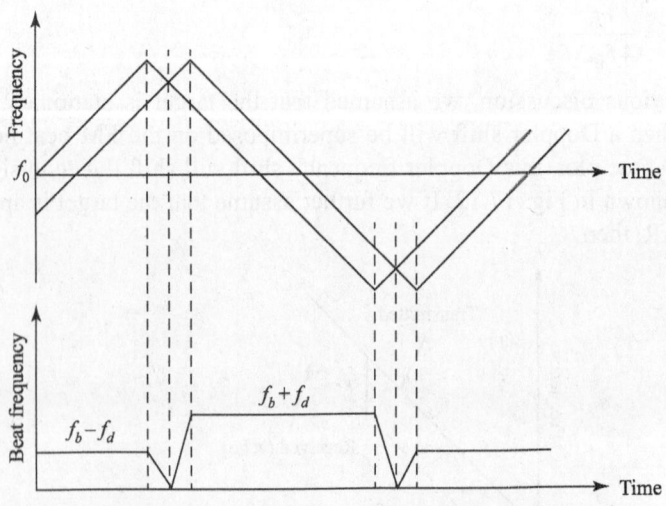

Fig. 17.12 Periodic variation of transmitted frequency and corresponding beat frequency for moving targets

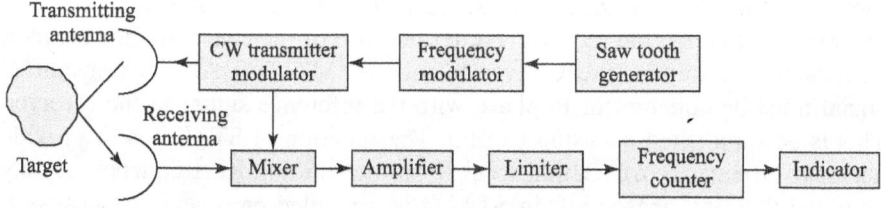

Fig. 17.13 Block diagram of FMCW RADAR

Here, we have assumed that $f_r > f_d$. If $f_r < f_d$, as may be the case for a high-speed target at short range, the role of averaging will be reversed and we can write the following equations:

$$f_r' = \frac{1}{2}(f_{b1} - f_{b2}) \tag{17.65}$$

$$f_d' = \frac{1}{2}(f_{b1} + f_{b2}) \tag{17.66}$$

From Eqs (17.63), (17.64), (17.58), and (17.34), we can write the following expressions: $f_r = \frac{1}{2}(f_{b1} + f_{b2}) = \frac{2Rf_0}{c}$ $\qquad$ (17.67)

and $\qquad f_d = \frac{1}{2}(f_{b1} - f_{b2}) = \frac{2v_r}{\lambda} = \frac{2f_r v_r}{c}$ $\qquad$ (17.68)

In the presence of multiple targets, at different ranges, the RADAR mixer output will contain multiple frequency components. By measuring each frequency component, the range of each target can be determined.

If the targets are moving with different velocities, then the Doppler shift from each target will superimpose on the respective beat note, and the problems of resolving targets and measuring their ranges become more complicated. The complicacy is further increased if the frequency modulation is non-linear or the mixer is operating in its non-linear region. The block diagram of a basic FMCW RADAR is shown in Fig. 17.13.

An FMCW RADAR finds applications in altimeters and airborne Doppler navigations.

17.9 Moving Target Indicator RADAR

A moving target indicator RADAR (or MTI RADAR) utilizes Doppler effect for its operation and is capable of reducing clutter due to stationary objects. Its basic principle is to compare the received echoes with the echoes that were received during the previous sweep, and thereafter to cancel out those echoes whose phase has remained unchanged. For moving targets, the phase of successive echoes, in general, changes and hence they are not cancelled. For stationary targets, however, the phase of successive echoes does not change and hence they are cancelled.

A simple block diagram of an MTI RADAR is shown in Fig. 17.14. In this figure, COHO stands for a coherent oscillator and STALO for a stable local oscillator. The COHO and STALO and the mixer in which they are combined are called the receiver exciter because of the dual role they serve in both the transmitter and the receiver. In coherent MTI RADAR, the transmitted signal must be coherent or in phase with the reference signal in the receiver. This is accomplished by using COHO. The function of STALO is to provide necessary frequency translation from IF to transmitting RF frequency. It may be noted that any phase shift in STALO is cancelled on reception because it simultaneously acts as a transmitting signal generator and local oscillator in the receiver. The amplified IF signal and the COHO signal is fed into a phase detector that provides an output, proportional to the phase difference between its two input signals.

Another block diagram of an MTI RADAR is shown in Fig. 17.15. Here a portion of transmitted signal is mixed with STALO output to produce an IF bit signal, the phase of which is directly related with the phase of the transmitting

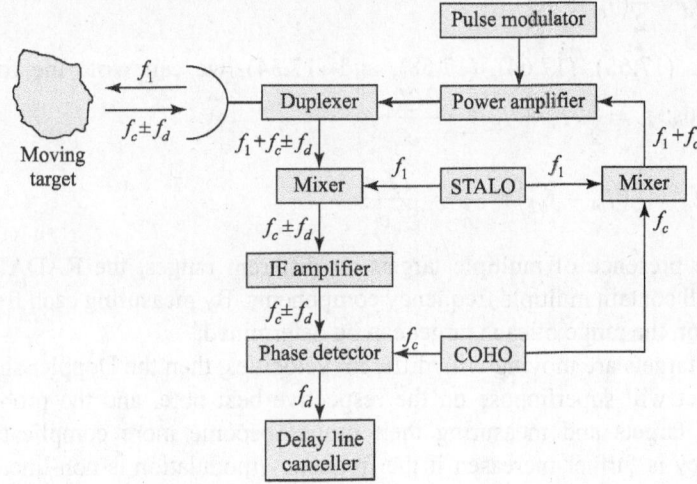

Fig. 17.14 More detailed block diagram of MTI RADAR

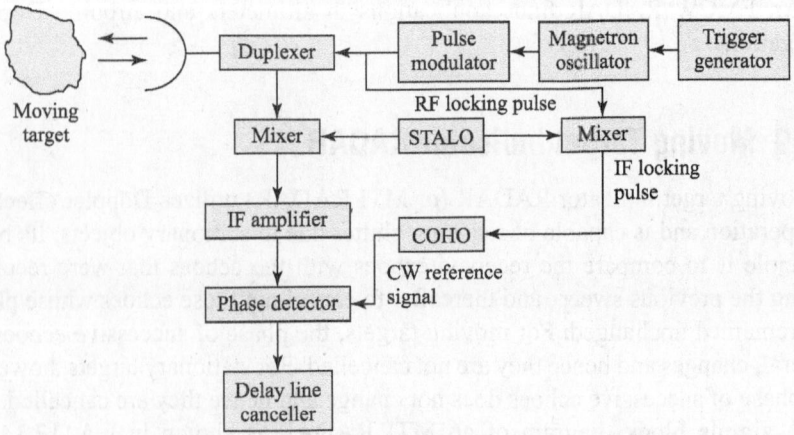

Fig. 17.15 Alternate block diagram of MTI RADAR

signal. The IF bit signal is then used to lock the phase of the COHO oscillation. Since the phase of the IF bit is related to the transmitting signal, the phase of the COHO signal is also related to the transmitted signal. The output of the COHO signal can, therefore, be used as the reference signal for the echoes received from that particular transmitted pulse. The process is repeated for each transmitting pulse.

In an MTI RADAR system, the moving targets can be separated from the stationary targets by observing the output on an A-scope display (amplitude versus range), as shown in Fig. 17.16. It is evident that a single sweep output on the A-scope display, as shown in Fig. 17.16(a), cannot separate the moving object from the stationary one. However, when the successive echoes shown in Figs 17.16(b)–(e) are superimposed, as shown in Fig. 17.16(f), a *butterfly effect* is observed, indicating the presence of a moving target. From the figure, it may also be noted that echoes from the fixed target remain constant throughout, whereas those from the moving target vary in amplitude from sweep to sweep. This amplitude variation corresponds to the Doppler frequency.

Though the butterfly effect is suitable for recognizing moving targets in an A-scope display, it is not suitable for a planned position indicator (PPI) display. For a PPI display, Doppler information must be extracted before displaying it. This job can be accomplished with the help of a delay line canceller, shown in Fig. 17.17. The delay line canceller eliminates the DC component of static objects but passes the ac component of the moving target. To do this, the video output of the receiver is equally divided into two channels; one channel is directly connected to the input of a subtractor, and the other channel is connected to the other input

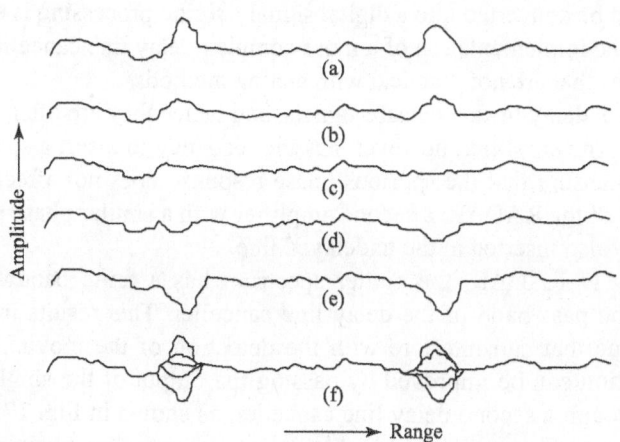

Fig. 17.16 A-scope display (a) At time t_1 (b) At time t_2 (c) At time t_3 (d) At time t_4 (e) At time t_5 (f) Butterfly effect

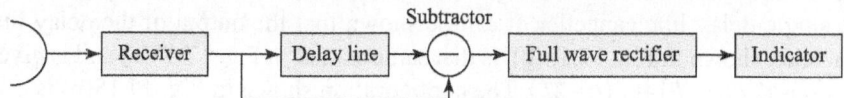

Fig. 17.17 Simple delay line canceller

of the subtractor through a delay line that provides time delay equal to one pulse repetition period. Due to the presence of the delay line, at a given instant, the inputs of the subtractor are the signals from the current and previous sweeps. The subtractor then subtracts these two signals and thus cancels the constant DC component of static objects. Since the amplitudes of the successive moving target echoes are different, the subtractor cannot completely eliminate them and a residue exists at the output of the subtractor. The output of the subtractor is bipolar and hence must be converted to a unipolar output with the help of a full-wave rectifier to avoid intensity modulation of the PPI display.

The delay line canceller, described earlier, is an example of a time domain filter. The capability of such a delay line depends on the quality of the medium used as the delay line. For typical ground-based air surveillance RADARs, the pulse repetition time and hence the delay requirement are of the order of several milliseconds, and the delay magnitude of this order is not achievable with practical electromagnetic transmission lines. For such cases, the electromagnetic signal is first converted into an acoustic signal. Since acoustic waves are 10^{-5} times slower than electromagnetic waves, a delay of this order can be achieved with a delay line of reasonable physical length. After the necessary delay has been obtained with the acoustic line, the acoustic signal may again be converted back to the electromagnetic signal for further processing. In practice, liquid delay lines are large and hence may be inconvenient for use. Hence, they are replaced by fused quartz delay lines. Such delay lines use multiple internal reflections and hence are compact.

In addition to these, digital delay lines can also be used to obtain the necessary time delay. However, such a delay line requires that the output of the phase detector must be converted into a digital signal. Digital processing is compact and also allows the implementation of a more complex delay line canceller with filter characteristics that are not practical with analog methods.

In practice, delay lines attenuate signals, and hence they are often followed by an amplifier. The amplifier, however, has the tendency to insert a spurious phase response. To ensure that the spurious phase response does not affect the normal performance of the RADAR, a second amplifier with a similar phase response but lower gain is also inserted in the undelayed line.

For an MTI RADAR, the clutter spectrum has a finite bandwidth, and it appears in the pass band of the delay line canceller. This results in a non-zero clutter residue that can interfere with the detection of the moving target. The clutter rejection can be improved by passing the output of the single delay line canceller through a second delay line canceller, as shown in Fig. 17.18(a). Such a delay line canceller is called a double delay line canceller or simply a double canceller. A comparison of the relative responses of a double and a single delay line canceller is shown in Fig. 17.18(c), along with the clutter spectrum. The figure shows that the double delay line canceller has better clutter rejection than the single delay line canceller. It can be shown that the output of the delay line canceller, shown in Fig. 17.18(b), is that same as that of Fig. 17.18(a) and is given by $f(t) - 2f(t+T) + f(t+2T)$. The configuration shown in Fig. 17.18(b) is also called a three-pulse canceller.

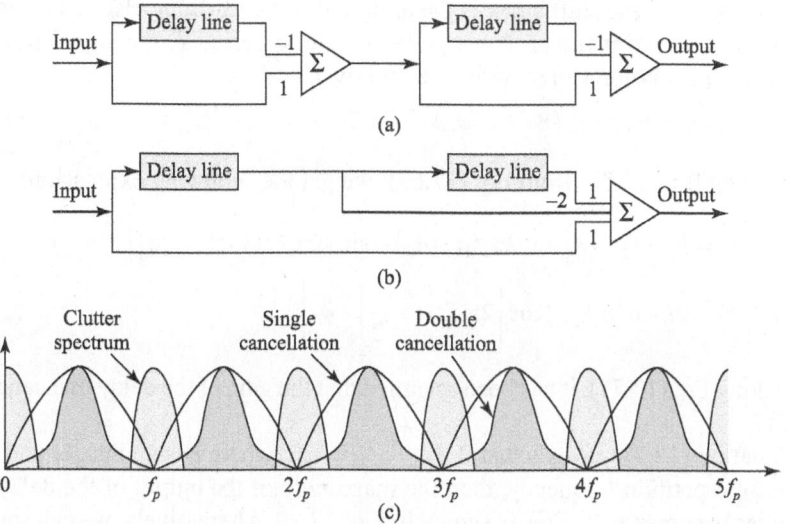

Fig. 17.18 Different delay lines and their responses (a) Single delay line canceller (b) Double delay line canceller (c) Clutter spectrum and relative frequency response of different delay line cancellers

A three-pulse canceller is an example of a *transversal filter*. In its general form, a transversal filter consists of N pulses and $(N-1)$ delay lines, as shown in Fig. 17.19. A transversal filter is also known as a *feed-forward filter*, *non-recursive filter*, *finite memory filter*, or *delay line filter*. For a three-pulse canceller, the weights are chosen as $1, -2, 1$, while for a four-pulse canceller the weights are $1, -3, 3, -1$. The response of a four-pulse canceller is equivalent to that of a triple canceller consisting of a cascaded connection of three single delay line cancellers. It can be shown that the output of a three-pulse canceller is proportional to $\sin^2(\pi f_d T)$, whereas that of a four-pulse canceller is proportional to $\sin^3(\pi f_d T)$.

It has already been mentioned that a delay line canceller acts as a filter that rejects the DC components of the clutter. However, because of its periodic nature, it also cancels energy in the vicinity of pulse repetition frequency and its harmonics. To illustrate this, let us assume that the received signal from a target at time t is given by the following expression: $V_1 = k \sin(2\pi f_d t - \phi_0)$ (17.69)

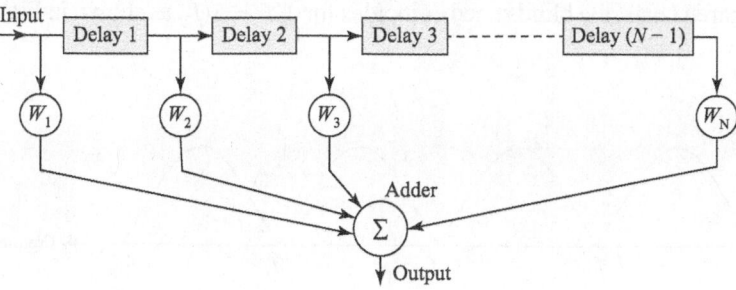

Fig. 17.19 General form of transversal filter

where ϕ_0 is the phase shift and k is the amplitude. If everything else is assumed to remain constant over the previous pulse repetition period T, the signal from the previous transmission can be written as follows:

$$V_2 = k\sin\{2\pi f_d(t-T)-\phi_0\} \tag{17.70}$$

Subtracting Eq. (17.70) from Eq. (17.69), we get the following expression:

$$V = V_1 - V_2 = k\left[\sin(2\pi f_d t - \phi_0) - \sin\{2\pi f_d(t-T)-\phi_0\}\right]$$

or
$$V = 2k\sin(\pi f_d T)\cos\left\{2\pi f_d\left(t-\frac{T}{2}\right)-\phi_0\right\} \tag{17.71}$$

In deriving Eq. (17.71), it has been assumed that the gain of the delay line canceller is unity.

Equation (17.71) reveals that if $f_d = n/T = nf_p$, where n is an integer and f_p is the pulse repetition frequency, then the magnitude of the output of the delay line canceller becomes zero. This is shown in Fig. 17.20. Alternatively, we can say that if the target is moving with a velocity v_r that corresponds to a Doppler shift that is equal to or an integral multiple of the pulse repetition frequency, then the output of the delay line canceller will be zero, and hence the target will not be detected by the RADAR. This particular speed is called the *blind speed*. Using Eq. (17.34), we can write the following expression:

$$f_d = 2v_{r,n}/\lambda = n/T \text{ or } v_{r,n} = \frac{n\lambda}{2T} = \frac{n\lambda f_p}{2} \tag{17.72}$$

where $v_{r,n}$ is the nth blind speed.

The blind speed is one of the limitations of an MTI RADAR and can be avoided by setting its value greater than the maximum probable radial velocity of the target. This requires that λ or f_p or the product λf_p must be very high. Unfortunately, none of these can be made arbitrarily large to meet the requirement. For a given antenna dimension, a high λ corresponds to a wider beam width, which may not be satisfactory where high angular accuracy and angular resolution are required. Similarly, a high pulse repetition frequency (PRF) will decrease the unambiguous RADAR range, which may not be acceptable. One of the ways to avoid blind speed is the use of multiple PRFs. For such a case, zero response is obtained only when the blind speed of each PRF coincides with each other. For example, if an MTI RADAR is operating with two separate PRFs, in a ratio of 5:4, on a time-shared basis, the blind speed coincides for $4/T_1 = 5/T_2$, as shown in Fig. 17.21.

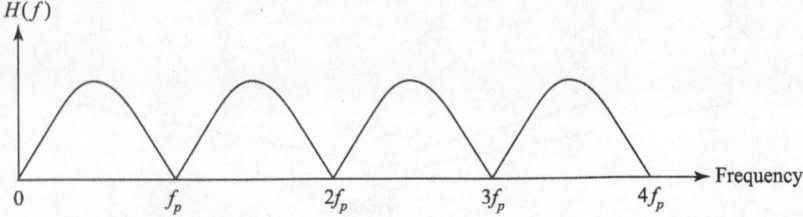

Fig. 17.20 Frequency response function of single delay line canceller

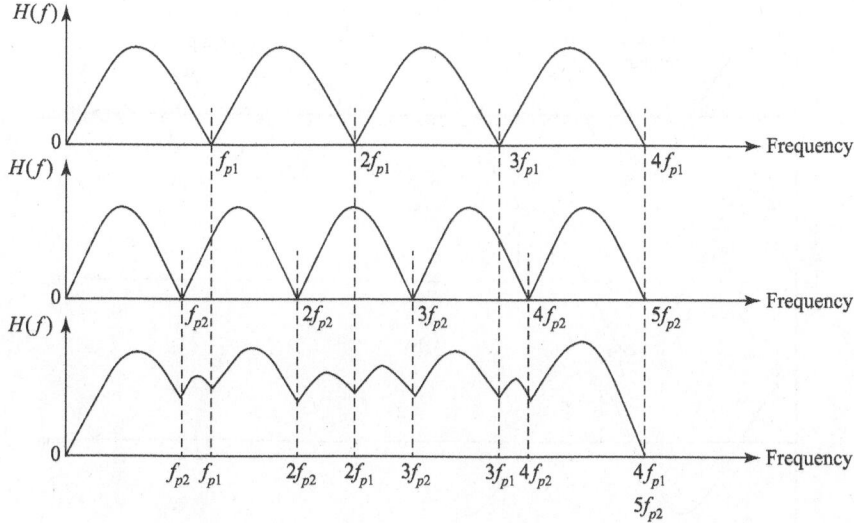

Fig. 17.21 Effect of multiple PRFs on blind speed

In a multiple-PRF MTI RADAR, the ratio $T_1 : T_2$ must be properly chosen because, as this ratio approaches unity, the first blind speed becomes greater, however at the cost of a deeper first null in the vicinity of $f_d = 1/T_1$. It should be noted that multiple or straggled PRFs not only reduce the probability of blind speed, but also allow a sharper low-frequency cut-off in the frequency response than a cascaded single delay line canceller.

The block diagram of a digital delay line canceller is shown in Fig. 17.22. The output of the IF is equally split into an in-phase channel (or I-channel) and a quadrature channel (Q-channel). The signals in these two channels are 90° out of phase. Such an arrangement eliminates the effect of the blind phase.

Before describing how a digital delay line removes the blind phase, let us first define *blind phase* using Fig. 17.23(a). If the pulse sampling occurs at the points P_1, P_2, P_3 in the Doppler cycle, then the phase difference between successive sampled points will be zero, and the target will appear stationary. The introduction of I- and Q-channel removes this shortcoming. Figure 17.23(b) shows the I-channel, whereas Fig. 17.23(c) shows the Q-channel with pulse train. In the I-channel when

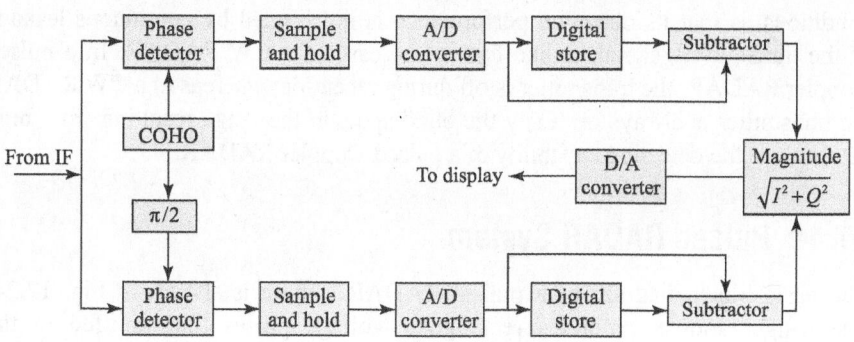

Fig. 17.22 Block diagram of digital delay line canceller

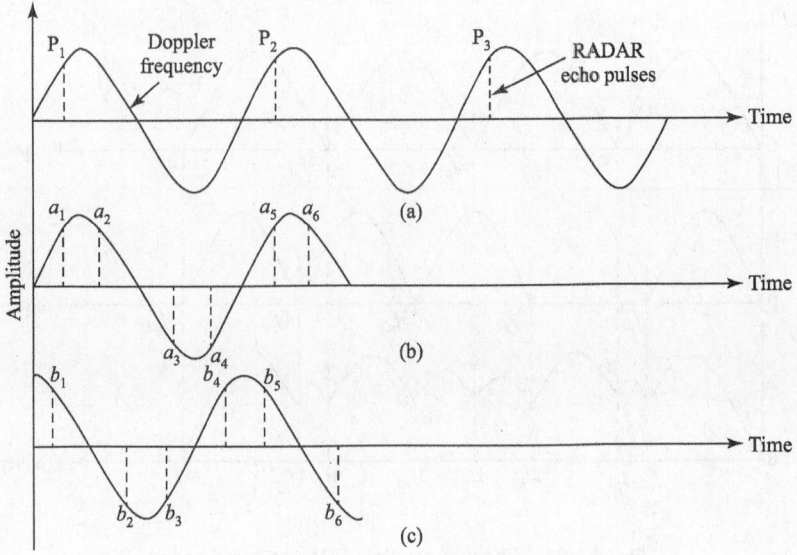

Fig. 17.23 Illustrations of blind phase (a) Normal sampling (b) Sampling in I-channel (c) Sampling in Q-channel

a_1 is subtracted from a_2, the output will be zero, whereas a residue will remain when a_3 is subtracted from a_2. The reverse is true for the Q-channel where, when b_1 is subtracted from b_2, a residue will exist at the output, whereas the output will be zero when b_3 is subtracted from b_2. Thus, the combination of I- and Q-channel will result in a uniform signal with no loss.

It has already been mentioned that the PRF cannot be arbitrarily large to avoid the blind speed. This creates a slight difference between an MTI RADAR and a pulsed Doppler RADAR. In the MTI RADAR, the PRF is chosen low enough to avoid range ambiguities, but it is subjected to the blind speed. In contrast, in the pulsed Doppler RADAR, the PRF is set high enough to avoid the blind speed, but it is subjected to range ambiguities. A pulsed Doppler RADAR operates with a high duty cycle and is also more capable of reducing clutter. If the PRF of a pulsed Doppler RADAR is very high, then the range ambiguities is too large to be resolved easily. The performance of a pulsed Doppler RADAR, in such a case, approaches to that of a CW RADAR. However, the advantage of a pulsed Doppler RADAR over a CW RADAR, in such conditions, is that its detection performance is not limited by continuous leakage of the transmitting signal, unlike that in the case of a CW RADAR. In a pulsed Doppler RADAR, the transmitter is off during reception, whereas in a CW RADAR the transmitter is always on. Only the blind spots in the range resulting from high PRF reduce the detection capability of a pulsed Doppler RADAR.

17.10 Pulsed RADAR System

The basic block diagram of a pulsed RADAR system is shown in Fig. 17.24. The trigger source generates rectangular voltage pulses that are fed to the pulse modulator. The output of the pulse modulator, shown in Fig. 17.25, acts

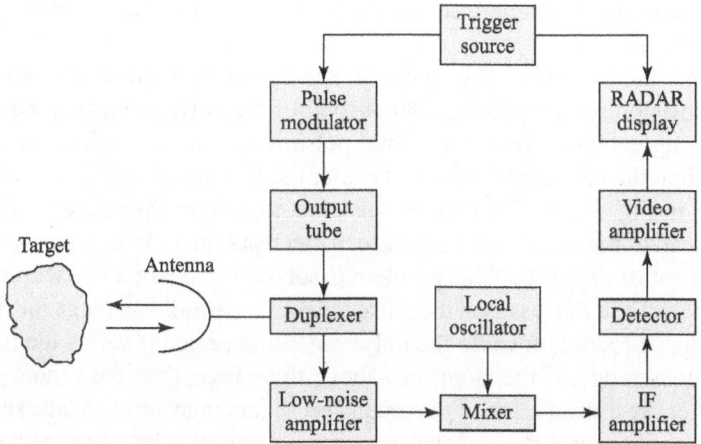

Fig. 17.24 Basic block diagram of pulsed RADAR system

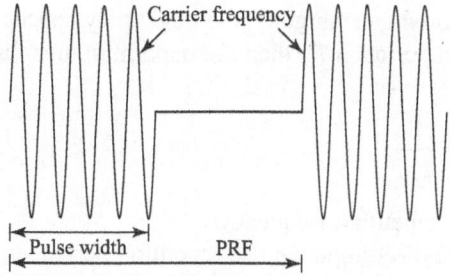

Fig. 17.25 RADAR pulse waveform

as a supply voltage for the output tube and thus switches on or off the tube as required. In practice, the output tube may be an oscillator or an amplifier, like a klystron, TWT, or crossed-field amplifier. If the output tube is an amplifier, then a microwave source is also required. In the case of a low-power RADAR, the source may be an IMPATT oscillator, a Gunn oscillator, or a TRAPATT oscillator. The pulse-modulated sine wave carrier is then radiated into space by the antenna after passing through the duplexer.

The receiver of a pulsed RADAR is of super-heterodyne type, and consists of a low-noise RF amplifier, a mixer, a local oscillator, an IF amplifier, a video detector, a video amplifier, and a RADAR display. For lower RADAR frequencies, a BJT is used in the LNA, whereas at higher RADAR frequencies, GaAs FETs are used. The LNA amplifies the weak received signal, which is then down-converted to an IF signal by a local oscillator and a mixer. The IF signal is then amplified by an IF amplifier to provide the main receiver gain. The IF amplifier consists of a tuned amplifier and is designed to achieve the maximum peak signal to mean noise power ratio. The output of the IF amplifier then passes through a detector circuit, mainly a Schottky barrier diode, which extracts the pulse modulation from the IF signal. The detected output is then amplified by a video amplifier so that it can be properly displayed. In the receiver display, the synchronizing pulses are provided by the trigger source. This sets the range reference value as zero.

In a pulsed RADAR, angle information is obtained from the pointing direction of the antenna.

Due to a finite velocity of the wave in space, a certain time is elapsed between the transmission of the pulse and the arrival of the corresponding echo signal at the receiving antenna. Now if a second pulse is transmitted before the time T is elapsed, then the echo signal will arrive after the transmission of the second pulse, as shown in Fig. 17.26, and may be mistaken to be corresponding to the second pulse instead of the first pulse. Therefore, it can result in an ambiguous or incorrect measurement of the range. This problem is solved if the target lies within a range for which the time T is less than the pulse repetition period. The range for which the travel time T becomes equal to the pulse repetition period is called the maximum unambiguous range. If the target lies above this range, then the echo signal will appear after the transmission of the second pulse (and may be even later) and range ambiguity will occur. Echoes that arrive after the transmission of the next pulse are called second-time around echoes (or multiple-time around echoes if more than one pulse has been transmitted before the arrival of the echo). For second-time around echoes, due to range ambiguity, the target may appear at a closer distance. If the pulse repetition period is T_p, then the maximum unambiguous range can be expressed as follows:

$$R_{\text{un}} = \frac{cT_p}{2} = \frac{c}{2f_p} \tag{17.73}$$

where f_p is the pulse repetition frequency.

It may be noted that setting the pulse repetition period, as per Eq. (17.73), for a given maximum unambiguous range of the target does not ensure that there will be no second- or multiple-time around echoes. Echoes may appear from beyond the maximum unambiguous range, especially from a distant large target, from a large clutter source, or when anomalous propagation occurs to extend the normal RADAR range beyond the horizon. Ambiguous range echoes, though cannot be removed completely, can be separated from the unambiguous range echoes using different pulse repetition frequencies (PRFs). For multiple PRFs, unambiguous echoes will remain at its true range, while ambiguous range echoes will appear at different ranges for each PRF.

Setting the characteristics of a pulsed waveform in a pulsed RADAR is a tricky job. If the PRF is increased, the maximum unambiguous range decreases; on the other hand, if the PRF is decreased, the RADAR will have to scan slowly to ensure that it does not miss the target. This needs a compromise in setting the operating PRF as well as the operating pulse width. A lower pulse width

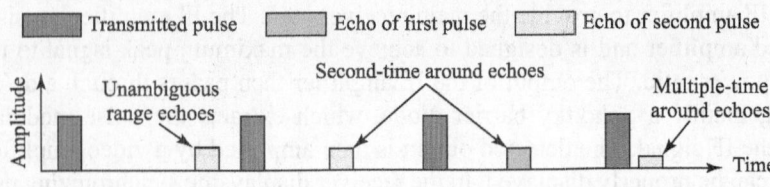

Fig. 17.26 Unambiguous range, second- and multiple-time around echoes

corresponds to a lower average power. On the other hand, if we increase the pulse width, it requires a transmitter that supplies more average power. In addition, an increase in the pulse width also increases the RADAR blind spot. The *blind spot* in the RADAR is the time when no reflection can be detected because of the *on* status of the transmitter. For a wide pulse width, the transmitter remains on for a longer duration and hence increases the blind spot. A wider pulse width also has some advantages like noise reduction. It requires a narrow bandwidth input filter, which, in turn, restricts the amount of external noise received along with the desired signal.

The leading and trailing edges of a transmitted pulse require to be vertical. If the leading edge is not vertical, then ambiguity may arise about the precision instant at which the pulse has been transmitted, which, in turn, may lead to range ambiguity. On the other hand, if the trailing edge is not vertical, then the duplexer will not be able to switch over to the receiver instantly. Further, the trailing edge of the pulse does not contribute much to the total transmitted power but lengthens the pulse width, which, in turn, puts a limit on the minimum range of the RADAR. In addition to having vertical leading and trailing edges, the pulse must also be flat top; otherwise, *frequency pushing* will occur in the magnetron, resulting in a reduction in efficiency.

EXAMPLE 17.2 A RADAR transmits a 2.5 µS pulse at a PRF of 8 kHz. Calculate the range of the target that it can detect unambiguously.

Solution Given: $T_{ON} = 2.5\,\mu S$ and PRF $= 8\,kHz$

$$\text{Therefore, } R_{min} = \frac{cT_{ON}}{2} = \frac{3\times10^8 \times 2.5\times10^{-6}}{2} = 375\,m$$

$$\text{Now, } T_P = \frac{1}{PRF} = \frac{1}{8\times10^3} = 1.25\times10^{-4}\,S$$

$$\text{Hence, } R_{un} = \frac{cT_P}{2} = \frac{3\times10^8 \times 1.25\times10^{-4}}{2} = 18750\,m = 18.75\,km$$

Practice Problem

17.2 A RADAR transmits a 2 µS pulse at a PRF of 12 kHz. Calculate the range of the target that it can detect unambiguously. $[300\,m \leq R \leq 12{,}500\,m]$

17.11 RADAR Displays

The purpose of a RADAR display is to visually represent the information available in an echo signal. If the display is directly connected to the receiver output, then the displayed information is called a raw video; on the other hand, if the receiver video output is first processed by an automatic detection and tracking (ADT) processor before feeding to display, then the output video is called a synthetic video.

Most of the RADAR systems use a cathode ray tube (CRT) as a display. In practice, two types of CRTs are used: deflection modulated and intensity modulated. In a deflection-modulated CRT, a target is indicated by the deflection of an electron beam, whereas in an intensity modulated CRT, the target is indicated by intensifying the electron beam and presenting a luminous spot on the CRT. A-scope display is an example of the former case. PPI display is that of an intensity-modulated CRT. Deflection-modulated displays haves the advantage of simpler circuits, and the targets may be more easily discerned in the presence of noise or interference. Intensity-modulated displays, on the other hand, can present the information in a more convenient form. The spot that indicates the presence of a target on an intensity-modulated display is called a blip.

Focusing and deflection of an electron beam, in a CRT, may be accomplished either electrostatically or electromagnetically. In electrostatic deflection, an electric field is applied between two electrodes/plates to deflect the electron beam. Such tubes are usually longer than magnetic tubes, but their overall size, weight, and power dissipation are less. An electromagnetic-deflection CRT requires magnetic coils/deflection yokes, positioned around the neck of the tube, for its operation. Such tubes are relatively lossy and also require more power than the former. Deflection-modulated CRTs use electrostatic deflection, whereas intensity-modulated CRTs use electromagnetic deflection for their operation.

RADAR displays can be of different types. Few of them are described in the subsequent sections:

17.11.1 A-scope

In an A-scope display, a linear saw-tooth voltage is applied in the horizontal deflection plate in synchronization with the transmitted pulse, and the echo signal is applied in the vertical deflection plate. Thus, A-scope display plots the amplitude of the echo signal as a function of range. In an A-scope display, the first blip appears due to a part of the transmitted pulse that is intentionally applied to the CRT for reference. The distance of the other blips from this reference blip provides information regarding the range of the target, whereas amplitudes of the other blips provide the strength of the echo signal. Since blips at the RHS of the display represent echoes from the far most objects, they are generally of small amplitudes.

A-scope displays have several advantages: (a) the range can be calibrated on a fixed scale or various scales with suitable switching for more accurate determination of close objects, (b) a section of the screen can be zoomed to allow more accurate determination of the targets of a particular area, and (c) sensitivity of measurement can be increased (for weak echoes) or decreased (for strong echoes) by changing the vertical gain. A-scope displays also have some basic disadvantages. For example, it cannot determine the direction of the target. In practice, A-scope displays are suitable for tracking a target rather than searching for it. A basic A-scope display is shown in Fig. 17.27(a).

17.11.2 Planned Position Indicator Display

A PPI display is of intensity modulation type, and indicates both the range and the azimuth angle of the target simultaneously in a polar plot. Here, the demodulated echo signal is applied to the CRT, biased slightly beyond the cut-off. Only when the blip, corresponding to a target, occurs, a saw tooth current flows through a pair of coils. This deflects the beam radially outwards from the centre and also continuously around the tube at the same angular velocity as the antenna. In a PPI display, the presence of a bright spot at any point on the screen indicates the presence of the target there. The distance of the bright spot from the centre of the screen provides the range of the target.

PPI displays are generally circular (diameter 30 cm-40 cm) and use long-persistent phosphors. Resolution of the display depends upon the pulse width, transmitting frequency, bandwidth of the antenna, and diameter of the CRT beam.

PPI displays are used in search RADARs, especially when conical scanning is required. A basic PPI display is shown in Fig. 17.27(b).

17.11.3 B-scope Display

A B-type display is an intensity-modulated rectangular display, where the azimuth angle is indicated by the horizontal coordinate and the range is displayed by the vertical coordinate. A basic B-scope display is shown in Fig. 17.27(c).

17.11.4 C-scope Display

A C-type display is an intensity-modulated rectangular display, where the azimuth angle is indicated by the horizontal coordinate and elevation angle by the vertical coordinate. A basic C-scope display is shown in Fig. 17.27(d).

17.11.5 D-scope Display

A D-type display is an intensity-modulated rectangular display, where the range is indicated by the horizontal coordinate and the elevation angle is displayed by the vertical coordinate. A basic D-scope display is shown in Fig. 17.27(e).

In addition, there are various other types of RADAR displays, such as E-scope, F-scope, G-scope, H-scope, I-scope, J-scope, K-scope, L-scope, M-scope, N-scope, O-scope, R-scope, and range height indicator.

A CRT display has some major disadvantages: (a) the entire display is often big; (b) various associated circuits are needed to display the information in a convenient form, making it costly; (c) limited dynamic range or contrast ratio of the intensity-modulated display may cause blooming of the display by large targets, and therefore blips from nearby smaller targets may be masked; and (d) there is a possibility of collapsing loss.

Availability of colour CRTs has added a new dimension to RADAR displays. In colour CRTs, different RADAR parameters, such as target altitude and cross section, may be coded with separate colours. Alternatively, outputs from different RADARs or those from each beam of a stacked beam RADAR may also be displayed in one CRT, by assigning different colours to different receiver outputs. A four-colour display is often used in ATC, with yellow representing aircraft, green area sector map lines, red navigation aids, and orange showering precipitation.

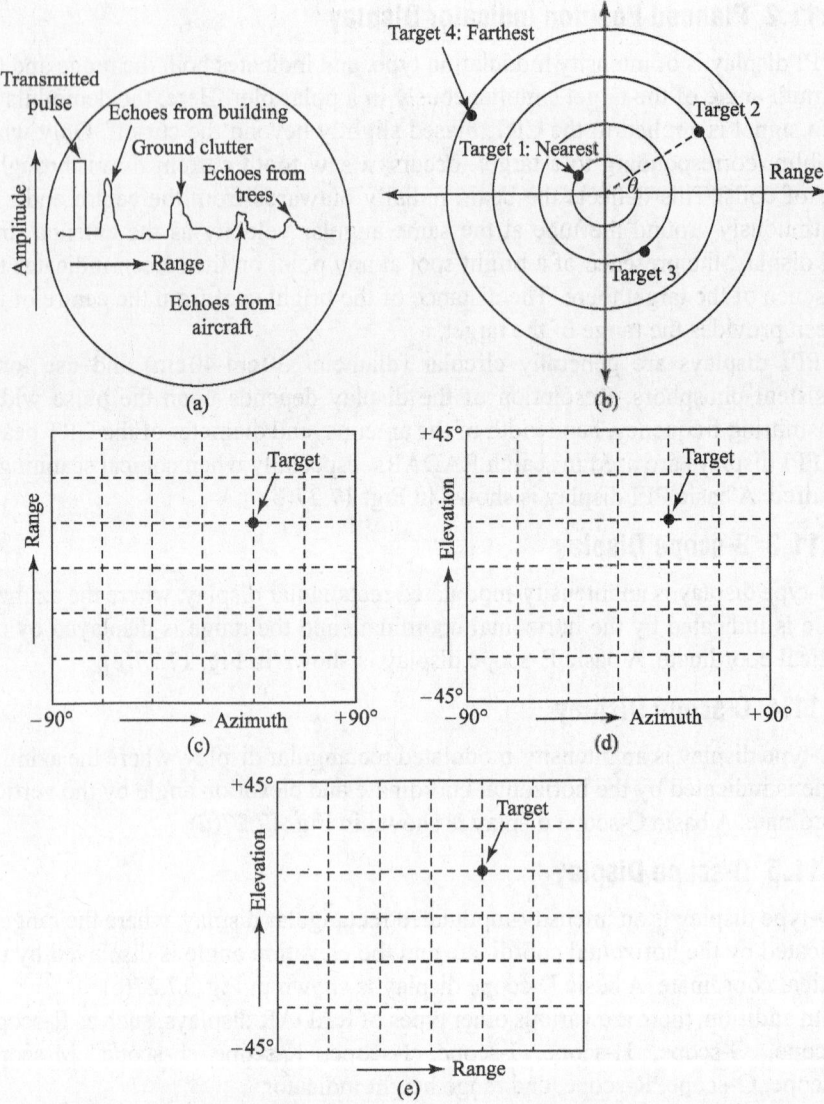

Fig. 17.27 Different RADAR displays (a) A-scope display (b) Plan position indicator (c) B-scope display (d) C-scope display (e) D-scope display

17.12 Tracking with RADAR

Target detection and tracking are among the major applications of a RADAR. A RADAR must find and acquire its target before it can track it. This requires that the RADAR should scan an angular sector of space where there is a probability of the presence of a target. Such a RADAR is known as a search or an acquisition RADAR. In practice, it is possible to use the same RADAR for acquisition and tracking. If so, the RADAR must find the target in the acquisition mode and then switch to the tracking mode. However, this has some major operational limitations.

To accurately track a target, the RADAR antenna should have a narrow pencil beam, with the elevation and azimuth beam width being approximately equal. Now if the search volume is large, a relatively long time will be required to acquire the target. It may be potentially dangerous for some cases, like in defence. Hence, many RADAR tracking systems employ a separate search RADAR. This search RADAR determines the position of the target and then send this information to the tracking RADAR. The tracking RADAR then performs a limited scan around the location provided by the search RADAR and, once located, continuously tracks the target.

A tracking RADAR measures the coordinates (azimuth and elevation angles, and range) and other necessary data, such as Doppler shift, and then determines the target path. It can also predict the future position of the target. Each time the RADAR scans the target, its coordinates are obtained. Now, if the change of coordinates between two successive scans is not too large, then it is possible to track the target by manually joining the pips from each scan on a PPI display. If the traffic is dense, then it may not be possible to track the target manually. In such a case, automatic detection and tracking (ADT) is required. If the outputs from a number of RADARs are automatically combined to provide target tracks, then the process is called automatic detection and integrated tracking (ADIT) or integrated automatic detection and tracking (IADT).

Tracking RADARs may be of two types: track-while-scan (TWS) RADAR and continuous tracking RADAR (CTR). The main difference between them is that a CTR provides continuous tracking data of a particular target, while a TWS RADAR provides sampled data on one or more targets. Landing RADAR, used for ground control approach (GCA) and some missile control, is of TWS type. However, in general, a tracking RADAR is a CTR. In CTR, the antenna beam is positioned by a servomechanism that is actuated by an error signal, as shown in Fig. 17.28. The error signal, in general, corresponds to the angular error that represents the difference between the target and a reference direction. In practice, the reference direction is taken along the axis of the antenna. The tracking RADAR tries to position the antenna so that the angular error becomes zero, that is, the axis of the antenna coincides with the line of sight of the target. A number of methods exist for generating the error signal, such as sequential lobing, conical scan, and simultaneous lobing or monopulse. These are described in the following subsections.

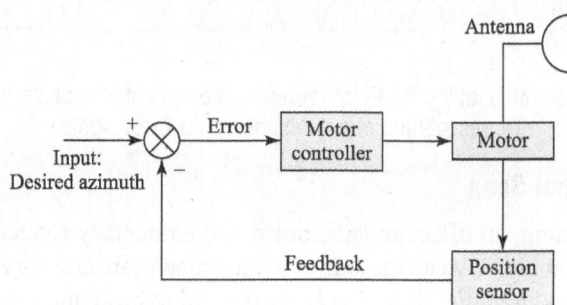

Fig. 17.28 Antenna positioning servo mechanism

17.12.1 Sequential Lobing

In sequential lobbing, angular error in one coordinate is obtained by alternatively switching the antenna beam between two positions. Figure 17.29(a) is the polar plot representation of the antenna beam at two switched positions. The corresponding plot in a rectangular coordinate system is shown in Fig. 17.29(b). Since the target receives more energy from the beam at switching position 1 than at switching position 2, the echo signal strength corresponding to the beam at switching position 1 is also larger than that at switching position 2. The resultant error signal is shown in Fig. 17.29(c). The difference in amplitudes of the voltage signals in Fig. 17.29(c), therefore, represents the angular displacement of the target from the switching axis. If the amplitude of the echo signal at position 2 is subtracted from that at position 1, then, for the given figure, the difference is positive and the antenna axis must be shifted left to reduce the error signal. Similarly, if the sign of the error signal becomes negative, then the antenna axis must be shifted right to reduce the error signal. The amount of shift required is proportional to the strength of the error signal. When the error signal becomes zero, the antenna axis coincides with the line of sight of the target, and the target gets *locked*.

This procedure is suitable for one-dimensional tracking of a target. For two-dimensional tracking, two additional switching positions are required to obtain the angular error in orthogonal coordinate. Thus, two-dimensional sequential lobing corresponds to four switching beam positions from four feed horns illuminating the same antenna. The feed horns should be so arranged that right–left, up–down sectors are covered. Sometimes, a cluster of five feeds is also used, where the fifth feed is used for transmission and the remaining four feeds are used for reception. For a four-feed system, all the feeds should transmit as well as receive.

The fundamental limitation of such a system comes from the system noise caused by mechanical or electrical fluctuations.

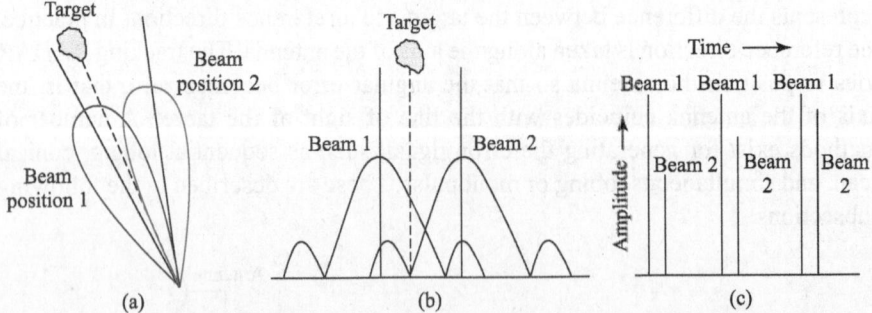

Fig. 17.29 Sequential lobing (a) Polar representation of antenna beam (b) Rectangular representation of antenna beam (c) Error signal

17.12.2 Conical Scan

In conical scanning, an offset antenna beam is continuously rotated, as shown in Fig. 17.30. The angle between the axis of the antenna beam and the axis of rotation is known as a *squint angle*. If the target is assumed to be at point A, then the echo signal will be modulated at a frequency equal to the rotational frequency of the

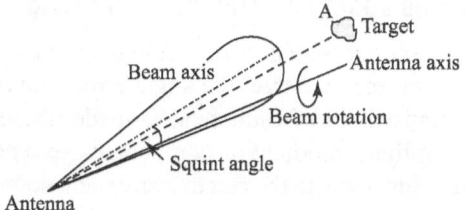

Fig. 17.30 Conical scan and squint angle

beam. The amplitude of the echo signal modulation is a function of the antenna pattern, squint angle, and angle between the target line of sight and rotational axis. The phase of the modulation is a function of the angle between the target line of sight and rotational axis. For operation, the conical scan modulation is first extracted from the echo signal and then applied to a servo control system. The servo control system then continuously positions the antenna on the target so that the target line of sight coincides with the rotational axis and the conical scan modulation becomes zero.

A basic block diagram of a conical scan tracking system is shown in Fig. 17.31. Here, the antenna is mounted in such a way that it can be positioned in both azimuth and elevation directions using two separate motors. The antenna beam can be made offset by tilting the feed. The antenna generates two beams that are 90° out of phase. One of these beams is used as an elevation detector and the other as an azimuth detector.

The receiver of the conical scan system is of super-heterodyne type, and extracts conical scan modulation or error signal at the output of the second detector. The error signal is then compared with the reference elevation and azimuth signal in the respective angle error detector. Angle error detectors are phase-sensitive detectors and produce a DC voltage proportional to the error. The sign of the DC voltage represents the direction of error. This voltage is then amplified by the respective servo amplifier, which, in turn, is used to drive antenna elevation and azimuth servo motors. Elevation and azimuth directions of the antenna axis then determine the target's angular positions.

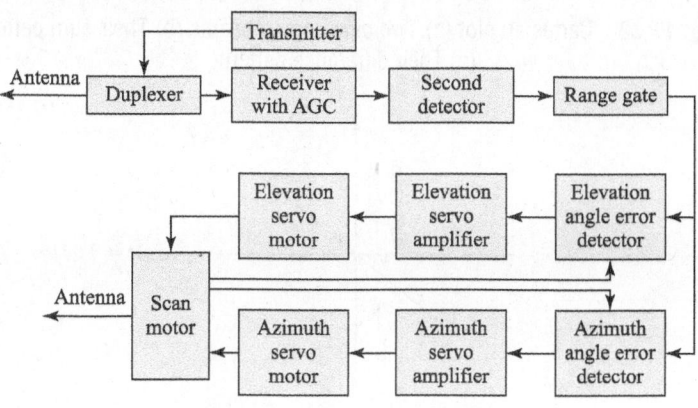

Fig. 17.31 Conical scan system

17.12.3 Simultaneous Lobing or Monopulse Tracking

Tracking techniques described so far have a serious disadvantage—they require a minimum number of pulses to extract the angle error with the condition that in the time interval, during which measurement is made, the train of echo pulses must be void of any amplitude modulation components, except for the modulation provided by scanning. This is due to the fact that any such modulation, for example, by fluctuating RCS, may introduce an error in the tracking accuracy. The situation becomes worse if the frequency components of the fluctuations are near or at the sequential lobing rate or conical scan frequency. This problem can be avoided if the tracking technique uses a single (or mono) pulse instead of a number of pulses. The technique that uses this concept is called monopulse or simultaneous lobing.

Amplitude comparison monopulse Such a system employs two overlapping antenna patterns, as shown in Fig. 17.32(a), to obtain the angular error in one coordinate. The sum and difference patterns are shown in Figs 17.32(b) and (c), respectively. The sum pattern is used for transmission as well as reception, whereas the difference pattern is used only for reception. During reception, the sum pattern provides range information and also acts as a reference for determining the sign of the error signal, while the difference pattern provides the magnitude of the error signal. Signals received from the sum and difference patterns are separately amplified and then fed to a phase-sensitive detector. The detector produces the error signal characteristics shown in Fig. 17.33.

A simplified block diagram of an amplitude comparison monopulse RADAR is shown in Fig. 17.34. The antenna feeds are connected with two collinear arms of a hybrid junction (such as a magic T-junction, rat race junction, or short

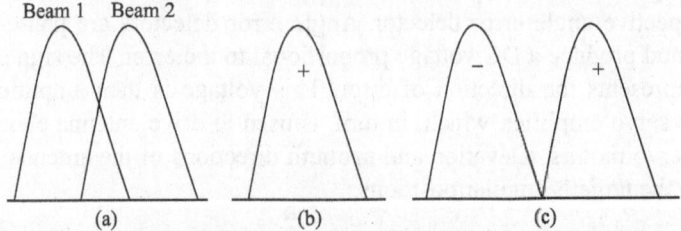

(a)   (b)   (c)

Fig. 17.32 Cartesian plot (a) Two overlapping beams (b) Their sum pattern
(c) Their difference pattern

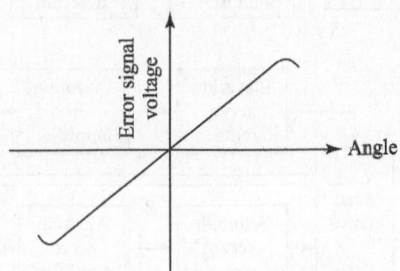

Fig. 17.33 Error signal characteristics

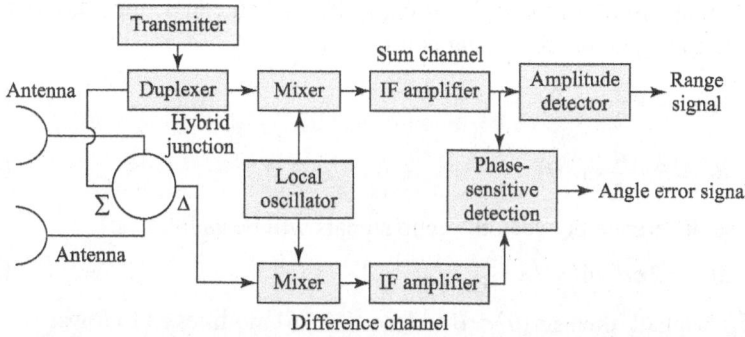

Fig. 17.34 Simplified block diagram of amplitude comparison monopulse RADAR

slot coupler). During reception, the sum and difference outputs from the other two arms of the hybrid junction are mixed with a local oscillator signal to produce the IF signal. The IF signal is then amplified by an IF amplifier. The amplified IF outputs are fed to a phase-sensitive detector to produce the angle error. The angular error signal, obtained from the output of a monopulse tracker, actuates a servo control system that positions the antenna on the target. Sign of the error signal is determined by comparing the phase of the difference signal with the phase of the sum signal. The amplified IF output signal of the sum channel is also fed to an amplitude detector to obtain the range information. The range output signal, from the monopulse system, is fed to an automatic range tracking unit. During transmission, the transmitted signal is fed to the sum channel, so that it can be divided equally between two antennas with same phase. A duplexer, immediately following the sum channel, separates this transmitted signal from the received sum pattern. It should be noted that Fig. 17.34 corresponds to a monopulse tracker that can provide angle error only in one coordinate. To obtain angle error in both coordinates (elevation and azimuth), four antennas are required instead of two, and the hybrid junction should also be replaced with a monopulse comparator, as described in Chapter 8.

Phase comparison monopulse A phase comparison monopulse measures the angle of arrival or angular position of the target by comparing the phase differ-

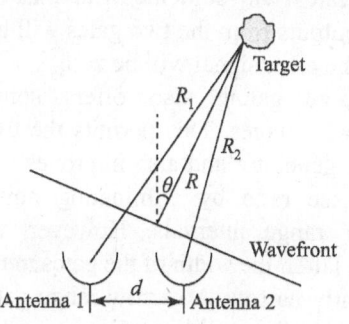

Fig. 17.35 Phase difference between two echo signals at two antenna apertures

ence between the two echo signals at two antenna apertures, as shown in Fig. 17.35. In this system, bore sight axes of the antennas are kept parallel. This results in the illumination of the same volume in space by the two antennas, which, in turn, ensures that the amplitudes of the target echoes, corresponding to the two antenna beams, are essentially the same. However, a phase difference exists between them due to the finite path difference. If the antenna separation is d, the range of the target is R, and the line of sight to the target makes an angle θ with the perpendicular bisector of the line joining the two antennas, then assuming that $R \gg d$, we can write the following expressions:

$$R_1 = R + \frac{d}{2}\sin(\theta) \tag{17.74}$$

and

$$R_2 = R - \frac{d}{2}\sin(\theta) \tag{17.75}$$

The phase difference between the echo signals will be as follows:

$$\Delta\phi = 2\pi d \sin(\theta)/\lambda \tag{17.76}$$

Now, if θ is small, then $\sin(\theta) \approx \theta$ and Eq. (17.76) modifies as follows:

$$\Delta\phi = 2\pi d\theta/\lambda \tag{17.77}$$

Equation (17.77) reveals that the phase difference is directly proportional to the angle of arrival.

Phase comparison monopulse RADARs are widely used in radio astronomy for precise determination of the position of radio stars, and hence they are also called interferometer RADARs.

In practice, most tracking RADARs continuously track the target in angle as well as in range. Range tracking can be accomplished manually using an A-scope or a J-scope display. However, if the target is moving with a high speed, then it becomes very difficult to perform the job manually and automatic tracking becomes necessary.

In a RADAR, automatic tracking in range is based on split range gates. In this procedure, two range gates, known as early and late gates, are generated, as shown in Fig. 17.36(a). Echo pulses are shown in Fig. 17.36(b) and error signals in Fig. 17.36(c). As evident from the figure, signal energy contained in the early gate is less than that in the late gate. Now, if the two outputs from two range gates are subtracted, an error signal will be developed, which can be used to reposition the centre of the gates. The magnitude of the error signal measures the difference between the centre of the gates and that of the pulse, whereas its sign determines the direction in which the centre of the gates must be repositioned. When the centre of the range gates will coincide with that of the pulse, outputs from the two gates will be equal and the error signal will be zero.

The range gating also offers some additional advantages. This permits the use of a boxcar generator and also improves the signal-to-noise ratio by eliminating noise from other range intervals. However, to achieve the latter, the width of the gates must be sufficiently narrow. A narrow range gate can also invite other problems. For example, if the range gate is too narrow, then an appreciable fraction of the echo signal can be

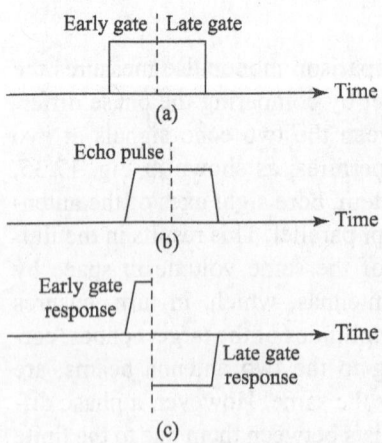

Fig. 17.36 Range tracking (a) Early and late gate (b) Echo signal pulse (c) Early and late gate response

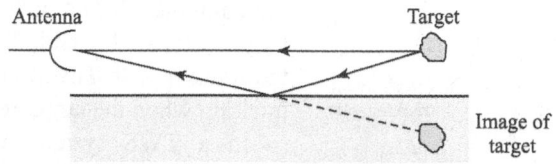

Fig. 17.37 Low angle tracking

excluded. In practice, compromise is necessary and the width of the range gate is often chosen to be of the order of the pulse width.

A RADAR that tracks a target at a low elevation often receives two echo signals, one through direct path and the other through surface reflection. The surface-reflected wave can be thought as an echo signal from a virtual target positioned at the image point of the original target, as shown in Fig. 17.37. These two signals combine in the RADAR receiver along with their phases, resulting in an angle measurement that is different from its real value. The surface-reflected signal is called a multipath signal, and the corresponding error is called the multipath error. The multipath error depends on that part of the antenna pattern that is striking the surface.

17.12.4 Track While Scan

In certain cases, like in ATC, it is undesirable to dedicate the entire RADAR system to track a single target; instead, simultaneous tracking of a number of targets and scanning of the space are more desirable. A TWS RADAR can be used for this purpose. In a TWS RADAR, the antenna scans the space, while the tracking function is accomplished by a digital computer. In addition to scanning and tracking, a TWS RADAR also provides collision warning.

A TWS system may use different types of gates, such as range, Doppler, and angle, to differentiate one target from the others. When a target is first identified, the TWS system assigns an acquisition gate to it. An acquisition gate has fixed boundaries of range and angles, and other possible parameters, depending on the system. When the target is scanned by the RADAR again and if the echo still falls within this acquisition gate, tracking is initiated by a computer.

A TWS system can provide the course and speed of a target by following the history of target position. The combination of course, speed, target position, and angle of a target at any time is known as the target solution. Once the target solution has been determined, the computer issues a tracking gate at its predicted position during the next scan. In the next scan, if the target falls within the tracking gate, the computer refines its solution and tracking is continued. On the other hand, if the target does not fall within the tracking gate, the computer will check to see if the target is within the turning gate. A turning gate surrounds the tracking gate and encompasses all the area where the target may be located since the last observation, as shown in Fig. 17.38. If the target is located within the turning gate,

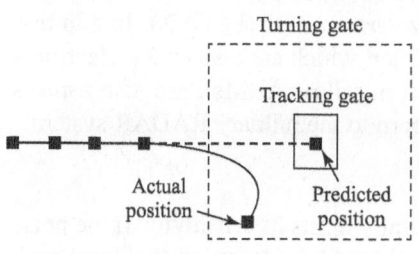

Fig. 17.38 Use of turning gate to continue tracking of target

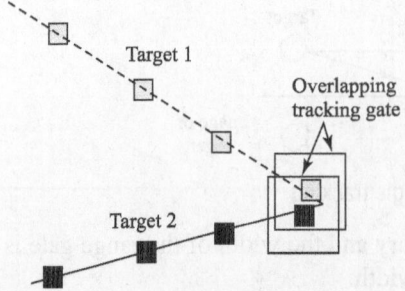

Fig. 17.39 Overlapping of tracking gate when two target paths cross each other

the computer finds a new solution. If the target is located outside the turning gate, the track is lost. The system again starts tracking when the target re-appears.

In a TWS system, the process of assigning observations with an already established track is known as correlation. During each scan, the system attempts to correlate all echoes with the existing tracks. If the echo cannot be correlated with any of the existing tracks, the computer issues a new acquisition gate to it.

In some cases, the echo from a new target may fall within an already existing tracking gate. In such a case, the TWS system attempts to separately identify the echoes of the existing and new targets, but may also fail to do so. A TWS system also finds it difficult to separately identify the targets when the number of targets is large or the tracks of existing targets cross each other, as shown in Fig. 17.39. In all such cases, the observer may need to intervene to correct the problem.

A TWS system uses a track file that contains all the observations that are correlated with the target for each establishment. Depending on the system, the track file may also contain other information such as current solution and target type.

17.13 Electronic Countermeasure and Electronic Counter Countermeasure

In real life applications, all RADARs must be able to operate in the presence of other electromagnetic signals. In addition, the transmitted RADAR signal should also not cause trouble to other RF systems. This requirement is known as electromagnetic compatibility (EMC). However, in practice, various methods result in electromagnetic interference with RADARs, known as *electronic countermeasures* (ECMs). ECMs are very big hurdles for a RADAR system, especially for military RADARs, where the ECMs are introduced intentionally and purposefully. In general, ECMs can be classified into two categories: active and passive ECMs. An active ECM is also known as *jamming*. In a passive ECM, chaffs, decoys, etc. are used to scatter the incident RADAR signal to create a clutter and simulate a false target. The methods used by RADARs to combat ECMs are known as *electronic counter countermeasures* (ECCMs). In addition to ECMs, intercept receiver and direction finder, which are also called electronic support measures (ESMs) and anti-radiation missiles (ARMs), are also aspects of electronic warfare (EW), which is of concern to the military RADAR system.

17.13.1 Noise Jamming

The receiver noise of a RADAR system generally limits its sensitivity. If the noise level is very high, then the sensitivity of the RADAR is degraded to a level such that the RADAR may not be able to detect the target. This drawback of a RADAR is often used in ECMs to degrade the RADAR performance. In noise jamming, the receiver noise is raised by some external means, such as with a noise source, to limit the sensitivity of the RADAR. The noise signal can enter the RADAR receiver through the main- or side-lobe direction of the RADAR antenna. If it enters through the main-lobe direction, then the narrow sector, in the direction of jammer, appears as a narrow strobe on the PPI display. In such a case, though the direction of the jammer can be determined, the noise strobe may mask the range of the jammer and the targets. In contrast, if the noise signal enters through the side lobes, then the entire display may be obliterated and no target information can be obtained. Thus, in ECCM design, it is essential to prevent the noise signal from entering through side lobes. The noise power, entering through the side lobes, can be reduced by using coherent side lobe cancellers. Side lobe cancellers consist of one or more omnidirectional antennas and cancellation circuitries. The noise signal received by the omnidirectional antenna is used to cancel the jamming noise entering the RADAR receiver through the main RADAR antenna side lobe. A RADAR antenna with the minimum possible side lobe level can also be used, so that the effect of side lobe jamming gets reduced. The effect of main lobe jamming can be reduced using a narrow beam lobe. If the main lobe beam width cannot be reduced, due to the associated increase in antenna size or for some other reasons, an auxiliary antenna can be used to create a notch in the main lobe of the main RADAR antenna in the direction of the jammer. With adaptive circuitry, similar to side lobe cancellers, the direction of the main beam notch can be adjusted towards the jammer.

A jammer that concentrates its noise energy within the RADAR receiver bandwidth is called a *spot jammer*. Since a spot jammer can insert high noise power in the RADAR receiver, it is a potential threat to RADAR systems. To get rid of spot jammers, the RADAR system designer must force the jammer to spread its power over a wider band by some means, such as by using frequency agility, that is, by changing the RADAR frequency from pulse to pulse in an unpredictable manner. A frequency agile RADAR often employs a pre-look receiver to examine the jammer spectrum and select a particular frequency for the next pulse for which the noise power is minimum. Pre-look sampling occurs during the RADAR inter-pulse period. RADAR engineers can also force the jammer to widen its bandwidth by placing many RADARs, each operating with a different frequency, at different locations. A jammer that radiates over a wide bandwidth is called a *barrage jammer*. Though the pulse compression technique can also be used to force the jammer to spread its energy over a wider band than a normal spot jammer, it is seldom used for this purpose. Instead, such a technique is mainly used to achieve good range resolution with a long pulse. In practice, in ECCMs, it is generally required to force the jammer to spread its energy over a much wider band than the spectral width of most pulse compression RADAR systems. It may be noted that spreading the jammer energy over the entire available band of a RADAR may also not be sufficient in some cases, and the jammer may be required to force to spread the noise energy over more than one RADAR band. This can be accomplished using the frequency diversity technique.

One of the ECCM techniques that can be used against noise jamming is *radiation homing*. In radiation homing, the seeker is programmed to detect and take advantage of the attempt of ECM. Present-day ARMs are competent enough to home in directly on the source of RADAR jamming if the jamming is too powerful to allow them to find and track the target. This is known as *home on jam* (HOJ). HOJ makes the missiles' job easier because the jammer now behaves as a beacon, and indicates the presence of the radiation source and hence the target. The presence of an ARM makes noise jamming more difficult to use, because it may put the jamming vehicle or target at risk of being targeted and hit by ARMs.

In practice, high-frequency RADARs are less vulnerable to jamming than low-frequency RADARs due to their wide bandwidth, high gain, and lower side lobes. Therefore, high-frequency RADARs can be used to reduce the effect of jamming. Further, since it is unlikely that jamming power can be distributed uniformly in space, multiple RADARs viewing the same coverage area in a coordinated manner can also provide some benefit over ECMs. Another way to reduce the effect of jamming is to increase the RADAR energy so that the RADAR echo power becomes larger than the jamming noise power. This is known as *burn through* and can be accomplished with reserved transmitter power or by dwelling longer in the direction of the jammer. Dwelling longer, however, reduces data rate, thereby degrading the overall performance.

It may be noted that several other ECCM techniques can be used effectively in RADAR systems against noise jamming. However, due to lack of scope they have not been discussed here.

17.13.2 Repeater Jamming

In repeater jamming, the target generates a false echo by delaying the received RADAR signal and retransmitting it at a slightly later time. This delay basically increases the to-and-fro travel time of the RADAR signal and simulates the target at a different location. In principle, this virtual target cannot be separated from the true target by the RADAR. A true repeater is one that retransmits the same signal that the target would radiate in normal cases.

In contrast to a true repeater, a transponder repeater transmits a stored replica of the RADAR signal after it is triggered by the RADAR. The transponder repeater can be programmed to remain silent when illuminated by the main lobe and transmit when illuminated by the side lobe. This creates spurious targets on RADAR display at some other direction. The transmitted signal of the transponder repeater is made to resemble the original RADAR signal as closely as possible.

Another way of repeater jamming is to use range gate stealers. This causes a tracking RADAR to beak the lock on the target. The jammer operates initially by sending pulses in synchronism with each pulse received from the RADAR, thereby strengthening the echoes. The repeater jammer then slowly shifts the timing of its own pulse transmission to simulate an apparent change in the target range. Now, if the jamming signal is of sufficient strength compared to the echo signal, then the RADAR tracking circuit will start to follow the jammer signal and ignore the true echo signal. This is continued till the range servo limits in the RADAR are exceeded, and then the repeater is turned off, leaving the tracker without a target. Thus, the repeater *steals* the RADAR tracking circuit from the *locked* target.

In addition to these, velocity gate stealers are also available to falsify the target speed and represent it as a stationary object. Repeater jammers can also be used to break the conical scan angle track by transmitting a signal at the conical scan frequency.

It should be noted that a repeater jammer may be very effective against a non-prepared RADAR but not much against a prepared one. Such jammers are easier to counter than noise jammers. For example, a monopulse tracker can be used in place of a conical scan tracker to avoid the latter's vulnerability, discussed earlier. RADAR echo signals from a target can generally be considered as linearly related to the RADAR signal. On the other hand, a repeater jamming signal is not generally linearly related with the RADAR signal. Thus, a RADAR signal, with a form of identification that is difficult to *mimic* by the *non-linear* repeater, can be used to recognize the echo signal. By changing the PRFs, pulse width, internal pulse modulation, and polarization, it is also possible to unmask false echoes. Side-lobe blankers, as discussed before, can easily be implemented to prevent repeater signals from entering through side lobes. It is also difficult to design a repeater jammer that can produce an exact replica of the RADAR signal.

17.13.3 Passive ECM

Noise and repeater jammers, discussed previously, are examples of active ECMs. They either generate or amplify electromagnetic energy for ECM purpose. Passive ECM techniques, in contrast, do not generate or amplify electromagnetic signals. Instead, they act in a passive manner to modify the reflected energy. Examples of passive ECMs are chaffs, decoys, RCS reduction, etc.

A chaff consists of a large number of dipole reflectors in the form of metallic foil strips. On being released by an aircraft in bundle, they are scattered by the wind over a large area and form a reflecting cloud with a cross-section that is comparable with a large aircraft. They are used either to deceive (known as *spot chaff*) or to confuse (known as *corridor chaff*) the RADAR. Spots chaffs are dropped as individual bundles and appear as additional targets on the RADAR, whereas corridor chaff is produced by continuously releasing chaffs from aircraft to form a corridor-like cloud through which following aircraft can fly undetected. Thus, corridor chaffs basically mask the target. Chaffs can also be used to break the lock of a target on the tracking RADAR. In the presence of a chaff, the tracking RADAR may confuse it as a target and can start following it instead of the original target, resulting in break lock. Chaffs are slow moving targets, with their vertical decent being a function of gravity and drag characteristics of individual foil strip. Its horizontal velocity is a function of wind speed. The slow velocity of chaffs makes them distinguishable from high-speed targets, and hence they can easily be detected with a properly designed MTI RADAR.

In contrast to chaffs, *decoys* are small aircraft-like vehicles that appear as real targets to RADARs. Decoys can be fitted with signal enhancement devices such as corner reflectors, Luneburg reflectors, or active repeaters, so that their RCS becomes comparable to that of a large aircraft. In addition, they can also be loaded with a jammer to mimic the jammers on a real target. This makes the decoys apparently identical with the real target. They can easily be carried on-board an attack aircraft and launched outside the normal RADAR detection range. They can also carry bombs on-board.

Another passive ECM technique that is practised is to reduce the RCS of the target. This can be achieved by properly shaping the target or using an absorbent material on the body of the target. One of the absorber designs is based on destructive interference. The destructive absorber is quarter-wavelength thick, and hence the energy reflected from the front and back surfaces of the absorbers are out of phase and cancel each other. This type of absorbers is relatively narrow band due to the requirement of quarter-wavelength thickness. There is also another type of absorber that operates on a relatively wide band. Such absorbers are relatively thicker than destructive absorbers and dissipate the energy incident upon them. Relatively thin absorbers that use magnetic materials with appropriate dielectric properties are also used.

If absorbers are not used, then the body of the target should be designed such that there is no flat, cylindrical, or conical surfaces that may be illuminated by the RADAR signal from a direction along the normal of the surface. A target with a double-curved (curvature in two dimensions) surface has a relatively small RCS. Cone–sphere is also a good example of target shape that results in a low RCS.

17.14 Modern RADAR Systems

The sensitivity of a RADAR system can be expressed as follows:

$$\text{Sensitivity} = K_1 \frac{(\text{Antenna area} \times \text{Frequency})^2 \times \text{Transmitted tower}}{(\text{Distance to target})^4 \times \text{Receiver noise temperature}} \quad (17.78)$$

whereas the maximum tracking range can be expressed as follows:

$$\text{Maximum tracking range} = K_2 \sqrt{\text{Antenna area} \times \text{Frequency}}$$

$$\left(\frac{\text{Transmitted power}}{\text{Receiver noise temperature}} \right)^{\frac{1}{4}} \quad (17.79)$$

These equations reveal that to increase both the sensitivity of the system and the maximum tracking range, a larger antenna area and a higher transmitted power are required. These requirements can be achieved by using an array of antennas, instead of a single antenna, as shown in Fig. 17.40. If the RADAR pulses reach all the elements at the same instant of time, then a parallel wavefront is launched in a particular direction. The direction of the wavefront can be controlled by introducing a linearly increasing phase shift/time delay in the arrival of the pulses at each element. Therefore, electronic scanning and beam steering, which are used during the target searching period, become possible. Once the target has been detected, a high-speed digital computer takes over and controls the orientation of the beam to track the target continuously. Since a number of antenna elements are being used, the transmitted power is also large and may be of the order of a few MW.

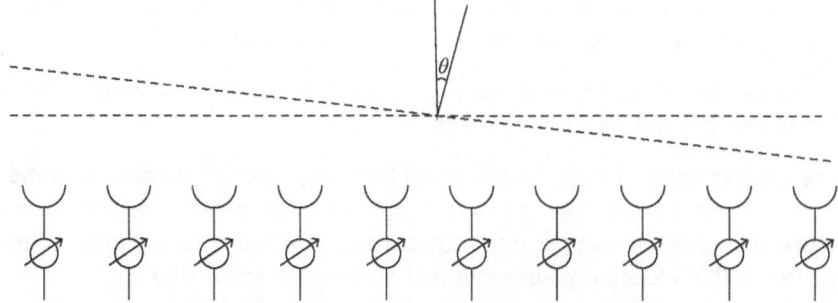

Fig. 17.40 Modern RADAR antenna

The total noise temperature of a RADAR system is a resultant combination of different contributions, such as input noise temperature at the receiver, component effect, and areal noise including the effect of any side-lobe pickup. Nowadays, helium-cooled MASERs are used that have noise temperatures as low as 20K. If the antenna system is looking into the cold sky instead of 290K temperature then the effective noise temperature comes down below 100K. This provides 30-fold improvement and hence 30 times less power.

In a modern RADAR system, the maximum advancement has been made in its signal processing and control function. The system has now gone completely automatic, and the human operator just monitors the operation to ensure that the correct action is being taken.

17.15 Applications of RADAR

A RADAR has a number of applications, some of which are described in the following subsections:

Remote sensing RADARs are widely used to remotely sense the weather conditions. Depending on the condition of the weather, the amplitude of the echo signals change, and hence the received signal gives an idea about the weather condition. RADAR remote sensors are also used in planetary observation, short-range below-the-ground probing, mapping the sea ice, etc.

Air traffic control One of the important applications of a RADAR is safe controlling of the air traffic in the vicinity of the airport, aircraft *en route* from one airport to another, taxing aircraft on the ground, and also ground vehicle traffic. In addition, an air surveillance RADAR maps the weather condition so that aircraft can be diverted. RADARs are also used to aid aircraft in landing at night, during poor visibility, and under adverse environmental conditions.

Aircraft safety and navigation Air-borne weather RADARs are employed in aircraft to determine weather conditions in the surrounding area. Such RADARs outline the region of precipitation and dangerous wind shear, and warn the pilot about the hazardous environmental conditions. Low-flying military aircrafts often employs RADARs to detect high terrains so that collision can be avoided. It also employs a radio altimeter, which is a kind of a RADAR, to indicate the height of the aircraft. In addition, military aircraft also employ a ground imaging RADAR to image the ground so that the required target area can be identified.

Law enforcement in highway traffic RADARs are widely used by traffic police to measure the speed of a vehicle.

Space applications Large ground-based RADARs are used to detect and track satellites. They are also used to detect and study other space objects such as meteors, moon, and planets, and to measure astronomical units accurately. Space vehicles use RADARs for rendezvous and docking the space ship.

Ship safety RADARs are used in ship for collision avoidance, to observe navigation buoys, and to detect icebergs. Shore-based RADARs are used to control river traffic and the surveillance of harbours.

Military applications RADARs form an integral and very important part of the present-day military systems. They are widely used to detect and track enemy planes, missiles, and other fixed or moving objects in the battle field. They are also used to control their own military aircraft, missiles, and fuze weapons. A missile RADAR guides the missile towards the target. High-resolution imaging RADARs are used for reconnaissance purposes. Many of the civilian applications of a RADAR are also used by military.

Other applications In addition to these, RADARs are also used in several other applications such as non-contact measurement of speed and distance in industry, and oil and gas exploration. Entomologists and ornithologists use RADARs to study the movement of insects and birds.

17.16 Radiometry

Radiometry is the technique that is used to gather information about a target using solely the microwave portion of the blackbody radiation (noise) that reaches the antenna either directly or by reflection from surrounding bodies. A radiometer is an equipment that is used to measure this noise power.

According to Plank's radiation law, a body, in thermal equilibrium radiates some energy. The amount of power P radiated by the body can be expressed as follows: $P = kTB$ (17.80)
where k is the Boltzmann constant, T the equilibrium temperature, and B the system bandwidth. The expression assumes that the target is a perfect blackbody. A blackbody is an idealized material that absorbs the entire energy incident on it and also emits at the same rate to maintain thermal equilibrium. A target, however, is not a perfect blackbody and hence reflects some power incident on it. The ratio of power emitted by the body to that by a perfect blackbody at the same equilibrium temperature is known as emissivity, and can be expressed as follows:

$$e = \frac{P}{kTB}$$ (17.81)

The value of e ranges from 0 to 1 (for a perfect blackbody).

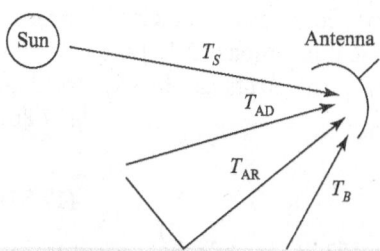

Fig. 17.41 Different noise power sources in radiometric applications

Emissivity, being unit less, when multiplied by the physical temperature of a body results in another temperature known as the brightness temperature. Since the value of e is less than 1, the brightness temperature is also less than the physical temperature. The brightness temperature is widely used in radiometry to characterize the target and is expressed as follows:

$$T_B = eT \qquad (17.82)$$

It may be noted that since noise power can be represented in terms of temperature, brightness temperature also corresponds to noise.

Since a radiometer antenna is subjected to different noise powers from different sources, as shown in Fig. 17.41, the total brightness temperature, as measured by the radiometer antenna, is a function of the scene under observation, frequency, polarization, observation angle, atmospheric attenuation, and antenna patterns. The main objective of radiometry is to gather information about the scene from this measured brightness temperature.

The main challenge in radiometry is to design the receiver itself. In the receiver circuit, the measured noise is added to the inherent receiver noise. In some cases, the receiver noise is much higher than the measured noise. The radiometer receiver should be able to distinguish the measured noise from the receiver noise and process it efficiently.

17.16.1 Total Power Radiometer

The basic block diagram of a total power radiometer is shown in Fig. 17.42. If the antenna measures a brightness temperature T_B, then the following will be the antenna power: $P_A = kT_B B$ (17.83)

Further, if T_R is the receiver noise temperature, then the receiver noise power is as follows: $P_R = kT_R B$ (17.84)

Therefore, the output voltage of the radiometer will be as follows:

$$V_o = G(T_B + T_R)kB \qquad (17.85)$$

where G is the overall gain constant of the receiver. The system is calibrated by replacing the antenna with two calibrated noise sources, from which system gain constants GkB and $GT_R kB$ can be determined. Once these factors have been determined, the brightness temperature can also be measured easily.

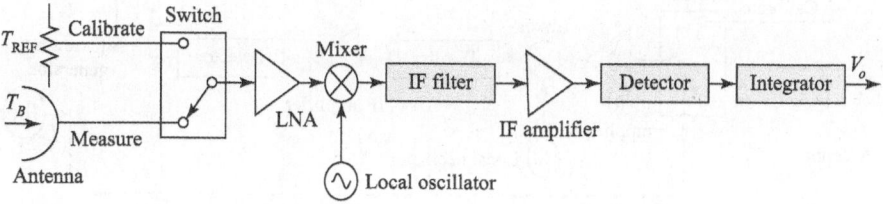

Fig. 17.42 Block diagram of total power radiometer

A total power radiometer is subjected to two types of error, namely, (a) error in the measured brightness temperature due to noise fluctuations (ΔT_N) and (b) error due to random variation in system gain G (ΔT_G). The errors can be expressed as follows: $\Delta T_N = (T_B + T_R)/\sqrt{B\tau}$ (17.86)
where τ is the measurement time and

$$\Delta T_G = (T_B + T_R)(\Delta G)/G \tag{17.87}$$

where ΔG is the rms change in gain. Equation (17.86) reveals that the error due to noise fluctuation can be minimized if a larger measurement time is permissible.

17.16.2 Dicke Radiometer

The basic block diagram of a Dicke radiometer is shown in Fig. 17.43. The super-heterodyne receiver is, to some extent, similar to a total power radiometer, but the difference is that here the input is periodically switched between the antenna and a variable noise power source. The switch is known as a *Dicke switch*. The output of the square law detector drives a synchronous demodulator that consists of a switch and a subtractor circuit, and that operates in synchronism with the Dicke switch. This ensures that the output of the subtractor is proportional to the difference between the noise power available from the antenna (T_B) and the noise power of the reference source (T_{REF}). The output of the subtractor (V_0) is used as an error signal to a feedback control circuit, which, in turn, generates a control signal (V_C). The control signal changes the level of the variable noise power in such a way that $V_0 \rightarrow 0$. Once $V_0 = 0$, a balanced state is achieved that corresponds to $T_B = T_{REF}$. Thus, T_B can easily be determined from the known T_{REF}. In a Dicke radiometer, the sampling frequency is chosen to be much faster than the drift time of the system gain, so that gain error is virtually removed.

For typical cases, the brightness temperature, measured by the antenna, varies between 50K and 300K. Therefore, the reference noise source should also cover this range. However, it is very difficult to achieve this requirement in practice, and several modifications in this radiometer receiver have been proposed. One possible modification is to use a fixed reference noise temperature, greater than the brightness temperature, and then to control the amount of reference

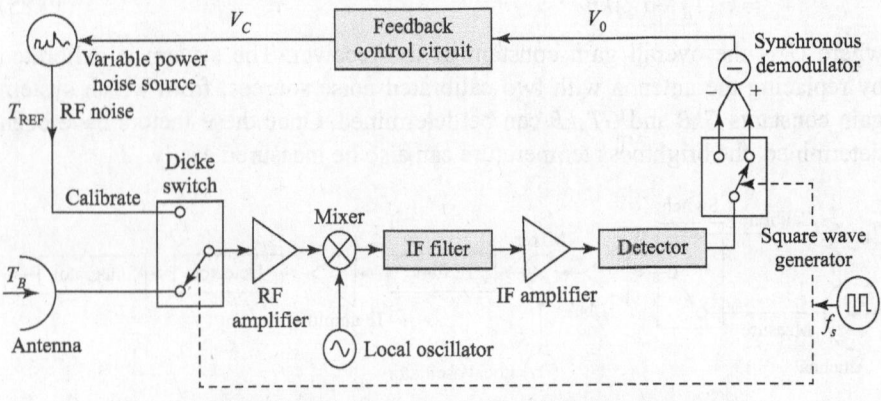

Fig. 17.43 Block diagram of balanced Dicke radiometer

noise power delivered to the system by varying the pulse width of the sampling waveform. An alternative method is to use fixed reference noise power and then to vary the gain of the IF amplifier during the reference sample time so that a null is achieved.

17.17 Other Applications of Microwave

In addition to those already stated, various other applications are discussed in the following subsections.

17.17.1 Microwave Heating

Though theoretically an ideal dielectric should have infinite resistance and zero conductivity, practically no dielectric is ideal and each has some finite non-zero conductivity. Microwave can penetrate into such materials and then dissipate as heat due to Ohmic losses.

In conventional heating, the outside of a material gets heated by convection and the inside by conduction. In contrast, in microwave heating, the inside of a material is heated first. The loss tangent of many materials decreases with an increase in temperature. In such cases, microwave heating is self-regulatory. The efficiency of microwave heating, defined as the ratio of power converted into heat to the power supplied, is generally less than 50%. However, it is better than the efficiency of a conventional oven.

Microwave oven A microwave oven consists of a magnetron tube operating at 2.45 GHz/915 MHz, a waveguide feed operating in the TE_{10} mode, an oven cavity, and a mode stirrer (basically a metallic fan blade), as shown in Fig. 17.44. The rotating mode stirrer reflects the microwave energy in all directions, and statistically a uniform field is produced. The uniform field, in turn, reduces uneven heating caused by standing waves in the oven. The food table is also rotated with a motorized platter to ensure further uniform heating. A plastic ceiling shield is placed between the stirrer and the oven cavity for environmental protection. Since microwave cooks the food from inside to outside, an electric heater is often used to brown the food so that it appears as conventionally cooked food from outside. The typical input power to the microwave cavity is about 1–1.5 kW.

The heating of food in a microwave oven is not due to Ohmic loss, as described previously. Instead, under the influence of microwave frequency, molecules of the moisture contained in the food start vibrating. This vibration results in friction, which, in turn, produces necessary heat for cooking.

In a typical microwave cavity, both the cooking time and the temperature can be controlled using a microprocessor-based control circuit. While the temperature is sensed from the exhaust air outside the oven cavity, the time of cooking is predetermined for certain ranges of food. During preparation of food, the user initially sets the temperature and time of cooking. Once the time is elapsed, the microwave is switched off automatically.

Safety is a big concern in microwave ovens, especially since high power is used. Therefore, the magnetron, feed waveguide, and oven cavity must be shielded properly

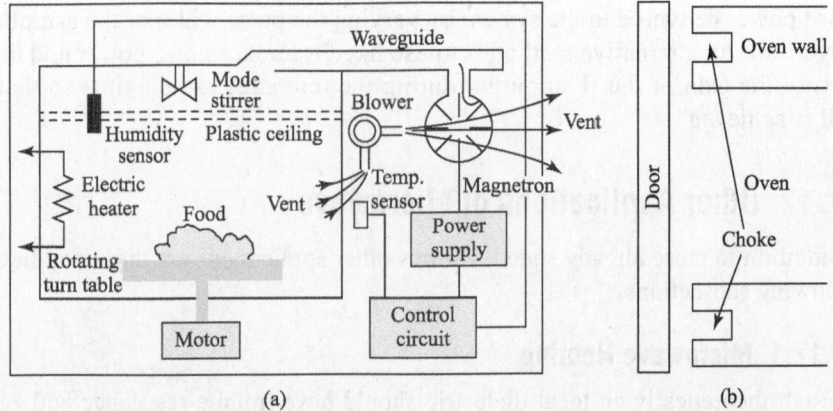

Fig. 17.44 Microwave oven (a) Typical structure (b) Choke in oven door

so that microwave cannot leak out from the equipment. Special attention is also required in the design and setting of the door, as it is the most probable area through which leakage can occur if the normal mechanical door setting is used. In practice, a microwave quarter-wave choke is designed in series with the microwave door gap, as shown in Fig. 17.44(b). Due to it, the door gap sees an open circuit towards the oven cavity and, in turn, prevents leakage. The choke is also filled with ferrite materials that readily absorbs microwave and also prevents the entry of any foreign material. In addition, an additional safety lock is also provided that automatically switches off the microwave if the door is opened when the microwave is still on.

Medical application Microwave heating is also used in many medical applications as it is less harmful than radioactive or X-ray exposure. Microwave can easily penetrate into the human body to reach bones and deep muscles, and hence localized heating can be achieved. In such applications, microwave is launched into the required part of the patient's body using a focusing device, such as a dielectric lens. One of the most common applications of microwave in the medical field is diathermy, where microwave heating is used to relieve pain. Microwave heating, in combination with radioactive exposure, is also used to destroy cancerous cells.

17.17.2 Thickness Measurement

Microwave is often used in industries to measure the sheet thickness of rolling metal sheets. The measurement setup is shown in Fig. 17.45. The metal sheet, during the rolling process, is passed between two horns placed face to face. The horns are connected to two waveguides that are again connected to two ports of a three-port circulator. The third port of the circulator is connected to a reflection-coefficient-measuring circuit. A phase shifter is also inserted in one of the waveguide arms, connecting the horn with the circulator. The waveguide along with the circulator, phase shifter, and horns forms a cavity whose resonance frequency can be varied by adjusting the phase shifter. In practice, the phase shifter is adjusted so that the resonant cavity has a resonance frequency equal to the signal frequency for a given metal thickness. This ensures the minimum reflection coefficient at

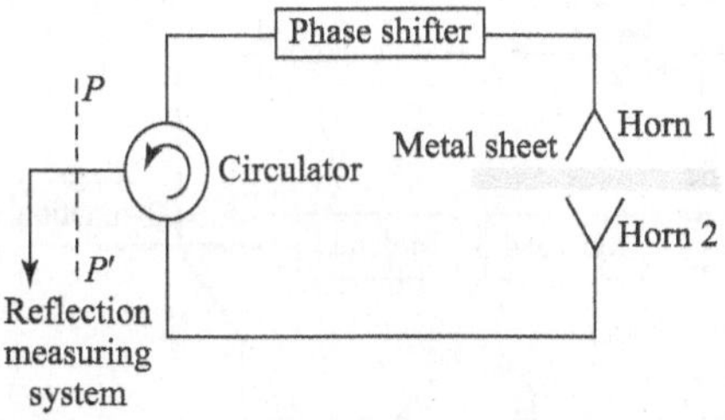

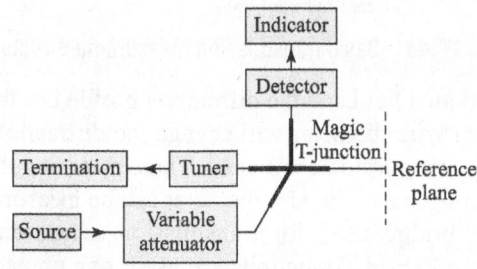

Fig. 17.45 Basic thickness measurement system

Fig. 17.46 Basic reflection measurement system

the measurement plane. Now, if there is any change in the metal thickness, then the measured reflection coefficient deviates from its minimum value. By noting the variation in the measured reflection coefficient, the variation in sheet thickness can be estimated. If the metal thickness is uniform, then any change in the position of the metal plate will not change the overall cavity length and hence have no effect on the measurement. If the signal frequency is changed, the resonant frequency of the cavity can also be changed by adjusting the phase shifter.

A microwave bridge, used for reflection measurement, is shown in Fig. 17.46. Here, the source is connected to the H-plane arm and the detector is placed in the E-plane arm. The plane for which the reflection coefficient is to be measured is connected to one of the collinear arms, while a tuner and a matched termination are connected to the other collinear arm. For a given reflection, the tuner is initially adjusted such that the reflected fields in the two collinear arms are equal in magnitude and phase, and a null is detected. Now, if the reflection coefficient of the test arm is changed from the initial value for any reason, then the null condition becomes void and a deflection in the indication meter is observed. The meter can be calibrated in terms of the measurement parameter so that the value of the parameter can be measured directly.

17.17.3 Measurement of Wire Diameter

The experimental setup to measure the diameter of a wire is shown in Fig. 17.47. A wire, the diameter of which is to be measured, is passed through the space between two closely spaced horns, placed face to face, and the corresponding transmission coefficient is measured. Since the

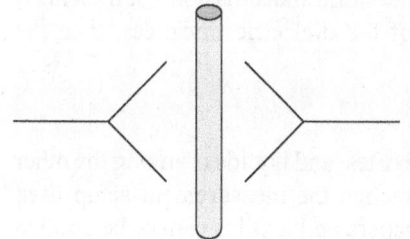

Fig. 17.47 Basic transmission measurement system

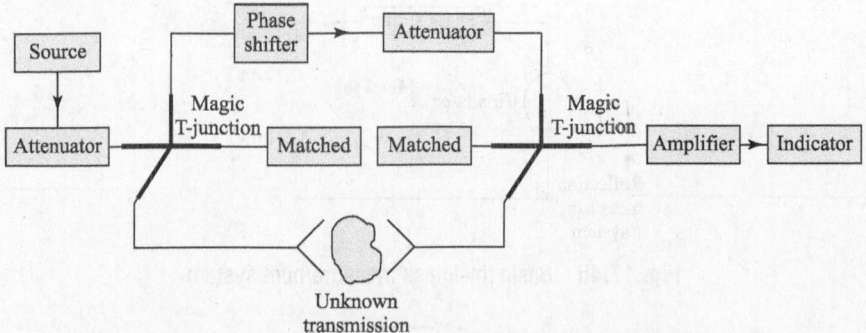

Fig. 17.48 Basic transmission measurement system

wire diffracts the radiated field and the diffraction profile is a function of the wire diameter, a change in wire diameter will change the diffraction profile, which in turn will change the transmission coefficient. Thus, by noting the change in transmission coefficient, the change in wire diameter can be measured.

The microwave bridge used for transmission measurement is shown in Fig. 17.48. Here, two hybrid T-junctions are used, one on each side. The input power, fed at the E-plane arm of the first T-junction, is divided equally between its two collinear arms. These powers reach the two collinear arms of the second T-junction via two paths, one through the transmitting and receiving horns and the other through the phase shifter and attenuator. The power reaching the second T-junction through the phase shifter and attenuator behaves as a reference signal. For a given transmission coefficient, the attenuator and phase shifter are so adjusted initially that a null is obtained at the detector connected to the H-plane arm of the second T-junction. Now, if the transmission coefficient of the test arm is changed from the initial value for any reason, then the null condition becomes void and a deflection in the indication meter is observed. The meter can be calibrated in terms of the measurement parameter so that the value of the parameter can be measured directly. The H-plane arm of the first T-junction and the E-plane arm of the second T-junction are terminated with matched loads.

17.17.4 Measurement of Thickness of Dielectric Sheet

The thickness of a dielectric sheet can also be measured using the aforementioned setup. The only difference is that here the dielectric sheet, instead of a wire, is passed through the space between the two closely spaced horns. Since the measured transmission coefficient is a function of the dielectric thickness, a change in the dielectric thickness will be reflected in the measured transmission coefficient. If the indicating meter is calibrated in terms of the dielectric thickness, then the dielectric thickness can be measured directly.

17.17.5 Moisture Content Measurement

Measurement of moisture content in papers, textiles, and liquids is among the other applications of microwave in industry. In practice the measurement setup used for the measurement of moisture content in papers and textiles cannot be applied for that in liquids as the latter has no specific shape. The following subsections describe the measurement of moisture content in various materials.

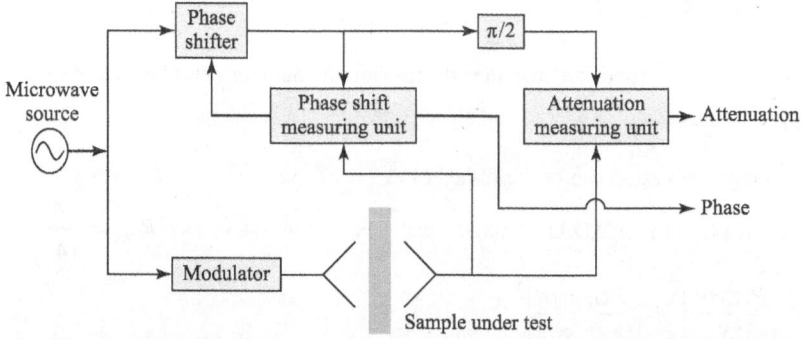

Fig. 17.49   Measurement of moisture content in solid material

Moisture content of papers and textiles   Moisture content of papers or textiles can be measured using a microwave bridge. The necessary measurement setup is shown in Fig. 17.49. The required information can be obtained by measuring attenuation and phase shift of the wave passing through the material or by measuring the complex reflection coefficient of the wave reflected from the material.

Moisture content of liquids   The aforementioned measurement set-up is not suitable for the measurement of moisture content of liquids. One of the possible measurement set-ups to perform this task is shown in Fig. 17.50. Here, the microwave source is connected to the E-plane arm of a magic T-junction, whereas the detector is placed in the H-plane arm. The collinear arms of the magic T-junction is connected with two identical cavities (with the tube through which liquid is allowed to pass as an integral part of the cavity). A moisture-free reference liquid is allowed to pass through one of the cavities and the liquid under test to pass through the other cavity. The cavity is designed in such a way that the electric field is maximum along the axis of the tube. If the liquid under test is moisture free, then the resonance frequencies of the cavities will be identical and the detector will show a null. On the other hand, if moisture is present in the test liquid, then the resonance frequencies and hence the reflection coefficient of the cavities will be different, and the detector output will deviate from its null condition. If the output meter is calibrated in terms of moisture, then the moisture content can be measured directly.

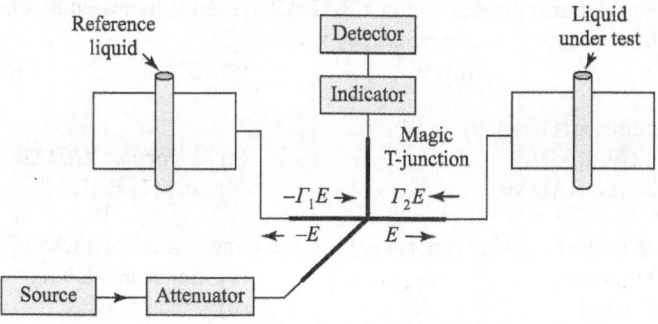

Fig. 17.50   Measurement of moisture content in liquid material

Important Formulae

- If T_r is the pulse repetition time, then the maximum unambiguous range can be expressed as $R_{un} = \dfrac{cT_r}{2} = \dfrac{c}{2f_r}$.

- The range resolution can be expressed as $\Delta R = cT_P/2$.

- The maximum RADAR range can be expressed as $R_{max} = \left[\dfrac{P_t G_t \sigma A_e}{(4\pi)^2 S_{min}} \right]^{\frac{1}{4}}$

  $= \left[\dfrac{P_t A_e^2 \sigma}{4\pi \lambda^2 S_{min}} \right]^{\frac{1}{4}} = \left[\dfrac{P_t G_t^2 \lambda^2 \sigma}{(4\pi)^3 S_{min}} \right]^{\frac{1}{4}}$.

- The average power can be expressed as $P_{av} = P_t \tau / T_p = P_t \tau f_P$.

- The Doppler frequency can be expressed as $f_d = 2v\cos(\theta)/\lambda$.

- In a CW RADAR, the range can be predicted using the expression $R = \dfrac{\lambda \Delta \phi}{4\pi}$.

- In a multiple-frequency CW RADAR, the range can be calculated as $R_0 = \dfrac{c\Delta\phi}{4\pi \Delta f}$.

- The maximum unambiguous range that can be measured by a multiple-frequency CW RADAR is $R_{unamb} = \dfrac{c}{2\Delta f}$.

- In an FMCW RADAR, the range can be calculated using the equation $R = \dfrac{cf_r}{4f_m \Delta f}$.

- The beat frequency and Doppler shift in an FMCW RADAR can be expressed as $f_b = \dfrac{1}{2}(f_{b1} + f_{b2}) = \dfrac{2Rf_0}{c}$ and $f_d = \dfrac{1}{2}(f_{b1} - f_{b2}) = \dfrac{2v_r}{\lambda} = \dfrac{2f_r v_r}{c}$, respectively.

- The blind speed in an MTI RADAR can be expressed as $v_{r,n} = n\lambda f_p/2$.

Exercises

Objective-type Questions

17.1 In a microwave RADAR, PRF is used to resolve range and Doppler ambiguities using a
   (a) pulsed Doppler RADAR
   (b) pulsed RADAR
   (c) MTI RADAR
   (d) CW RADAR

17.2 If the peak transmitted power in a RADAR system is increased 81 times, then the maximum range will be increased by a factor of
   (a) 3   (b) 9   (c) 27   (d) 81

17.3 An altimeter is basically
   (a) a CW RADAR
   (b) an FM RADAR
   (c) a Doppler RADAR
   (d) an MTI RADAR

17.4 Which of the following is not suitable for automatic satellite tracking?
   (a) Monopulse
   (b) Conical
   (c) Sequential lobing
   (d) Step-back

17.5 A Radom is a
    (a) dome-shaped RADAR antenna
    (b) RADAR housed in a dome
    (c) protective cover for the RADAR antenna
    (d) dome-shaped RADAR antenna

17.6 For tracking the RADAR antenna beam pattern is
    (a) omni directional   (c) highly directive
    (b) isotropic   (d) none of these

17.7 A non-zero Doppler shift represents
    (a) a static target
    (b) a target moving towards the RADAR
    (c) a target moving away from the RADAR
    (d) either (b) or (c)

17.8 The Doppler frequency provides the
    (a) range of the target   (c) direction of the target
    (b) velocity of the target   (d) number of targets

17.9 If both the transmitter and the receiver are placed at the same place, then the RADAR system is called
    (a) monostatic   (c) multi-static
    (b) bistatic   (d) none of these

17.10 Second-time around echoes corresponds to an
    (a) echo that is received two times
    (b) echo that arrives after two unit times
    (c) echo that arrives after the transmission of the next pulse
    (d) echo that arrives after the transmission of the next two pulses

17.11 The most suitable way to reduce the blind speed is
    (a) increasing the wavelength of the carrier
    (b) increasing PRF
    (c) using multiple PRF
    (d) both (a) and (b)

17.12 A scope display plots
    (a) range versus amplitude   (c) elevation versus range
    (b) azimuth versus range   (d) elevation versus azimuth

17.13 The jammer that concentrates its noise energy within the RADAR receiver bandwidth is called
    (a) a spot jammer   (c) a repeater jammer
    (b) a barrage jammer   (d) none of these

17.14 PPI display is
    (a) intensity modulation type   (c) either (a) or (b)
    (b) deflection type   (d) none of these

17.15 Chaffs and decoys are used
    (a) as active jammers   (c) as ECCMs
    (b) as passive jammers   (d) for HOJ

Review Questions

17.1 Derive the basic RADAR range equation as a function of the effective aperture of an antenna and as a function of gain. Explain the result.

17.2 Describe different RADAR system losses.

17.3 How can the sign of a Doppler shift be detected? Explain with a suitable figure.

17.4 Draw and explain different block diagrams of a CW RADAR.

17.5 Draw the block diagram of an FMCW RADAR. How can it be used to obtain information about a target?

17.6 Draw and explain the block diagram of an MTI RADAR. In addition, draw other available bock diagrams of an MTI RADAR.

17.7 With a suitable block diagram, describe the operation of a pulsed Doppler RADAR.

17.8 Write a short note on RADAR displays.

17.9 Describe sequential lobing and conical scan. What are the main disadvantages associated with them. How can those be removed?

17.10 With a suitable figure, describe monopulse tracking.

17.11 Describe the TWS RADAR system.

17.12 Write a short note on noise jamming and repeater jamming.

17.13 What is radiometry? With suitable figures, describe different radiometers.

17.14 How can microwave be used for heating? Describe different applications of microwave heating.

17.15 How can microwave be used to measure the (a) thickness of a metal and dielectric sheet, (b) diameter of a wire, and (c) moisture content?

Problems

17.1 A RADAR is operating with a 3 cm wavelength, 500 kW peak power, and 10^{-12} W and minimum detectable signal. If the RADAR antenna has an effective area of $6 m^2$ and the cross-section of the target is $10 m^2$, then calculate the maximum RADAR range. *[Hint: Use Eq. (17.11)]*

17.2 A military RADAR operates at 6 cm wavelength, 3 MW power output, 1.5 MHz receiver bandwidth, and 10 dB noise figure. If the diameter of the RADAR antenna is 6 m and RCS of the target is $\sigma = 0.8 m^2$, then calculate the maximum RADAR range. Assume the gain of the antenna to be 3500. *[Hint: Use Eq. (17.23)]*

17.3 A RADAR, operating at 10 GHz, has a maximum range of 45 km for an output power of 300 kW and an antenna gain of 3800. If the minimum detectable signal strength by the RADAR is 10^{-12} W, then determine the minimum RCS of the target that can be detected by the RADAR. *[Hint: Use Eq. (17.13)]*

17.4 Calculate the power received by the receiver antenna when a target of RCS $2 m^2$ is illuminated by a 3 GHz, 5 kW signal at a distance of 5 km. Assume the antenna gain to be 35 dB. *[Hint: Use Eq. (17.13)]*

17.5 A RADAR gun operates at 10 GHz. If an automobile directly approaches the RADAR at a speed of 120 km/h, then calculate the Doppler shift. *[Hint: Use Eq. (17.34)]*

17.6 A CW RADAR has been designed to cover target velocities of ± 750 m/s. The RADAR operates at a wavelength of 0.1 m. A bank of narrow-band filters is used after the detector. If the bandwidth of each filter is 50 Hz, then calculate the number of filters needed. *[Hint: Use Eq. (17.34) and $N = f_0/\Delta f$]*

17.7 Prove that the product of maximum unambiguous range and maximum unambiguous radial velocity is independent of the RADAR PRF and is equal to $K\lambda$, where K is a constant.

Microwave Antennas

18

18.1 Introduction

In the previous two chapters, we have discussed different applications of microwave. In most of these applications, especially in communication and RADAR, microwave systems require one or more antennas for transmission and reception of signals. In this chapter, we will discuss different types of antennas that are commonly used in microwave systems.

An antenna is basically a metallic structure that is capable of radiating or receiving electromagnetic signals. In addition, they also transform a signal from the circuit domain (electrical signal) to that in the field domain (electromagnetic signal) so that the signal can propagate through space or vice versa. Thus, an antenna behaves as a transducer between a guided wave and a propagating wave. It also behaves as an impedance matcher between the transmission line and the free-space impedance. Antennas generally follow the law of reciprocity, and hence a transmitting antenna can also be used to receive signal. However, due to some practical and mechanical requirements, some antennas are preferred for transmission, while others are preferred for receiving signals.

In general, antennas can be classified into three categories: (a) conduction current or wire antennas, (b) displacement current or aperture antennas, and (c) array antennas. Array antennas consist of a number of radiating elements that may be of either conduction or displacement current type. Examples of conduction-type antennas are dipole antenna, loop antenna, helical antenna, etc., and those of displacement-current-type antennas are horn antennas, slot antennas, etc. In practice, microwave antennas are of displacement current or aperture type, though conduction-type antennas also find applications at microwave frequencies. Displacement or aperture antennas can be classified into two categories: primary and secondary. Primary antennas are excited directly by a source and can be used independently. Examples of such type of antennas are horn antennas. A secondary antenna, on the other hand, requires a primary antenna as its feed and hence cannot be used independently. Reflector antennas are examples of secondary antennas.

When an antenna is excited by a source, two types of fields are generated in its surroundings: non-propagating and propagating fields. The non-propagating field remains close to the antenna and does not contribute to energy propagation, whereas the propagating field remains far from the antenna and contributes to energy

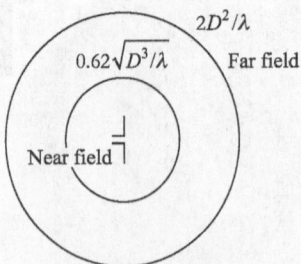

Fig. 18.1 Different field domains surrounding an antenna

propagation. Propagation fields are also known as radiation fields, far fields, or inverse square fields. Non-propagating fields are reactive in nature and are also called near fields. The approximate distance of the boundary between the near and far fields from the antenna can be approximated as $2D^2/\lambda$, where D is the maximum dimension of the antenna and λ is the operating wavelength. In practice, total space surrounding the antenna can be divided into three regions, as shown in Fig. 18.1. In the first region, a reactive field dominates, which extends up to a distance of $0.62\sqrt{D^3/\lambda}$. This reactive field then begins to diminish but continues up to a distance of $2D^2/\lambda$. After this distance, the reactive field vanishes and a far field starts.

Since microwaves are generally used in communication and RADAR systems, microwave antennas are not omni-directional. Instead, they are highly directive. In addition, microwave antennas are very small in size, since the dimension of an antenna is proportional to its wavelength. This is one of the advantages of microwave antennas. These antennas also suffer from some disadvantages, like high power requirement. Since power attenuation increases with frequency, microwave antennas require high power and also high gain. Large power is also required to overcome the received and system noises.

18.2 Radiation Mechanism

One of the interesting questions related to antenna theory is how an antenna radiates. To answer it, let us start with a two-wire transmission line connected to a time-varying source at one end and left open circuit at the other end, as shown in Fig. 18.2(a). This will result in a standing wave in the line, with current in each wire being same in magnitude but opposite in phase. As the wires are in close proximity, the field of one wire will cancel that of the other, and no radiation will take place. Now if the open terminals of the transmission line are bent, as shown in Fig. 18.2(b), spacing between the wires becomes significant in comparison to wavelength, and complete field cancellation does not take place. This results in a net radiation by the system. The separation between the wires can be increased by further bending the wires, and in the limiting case, it will result in a dipole.

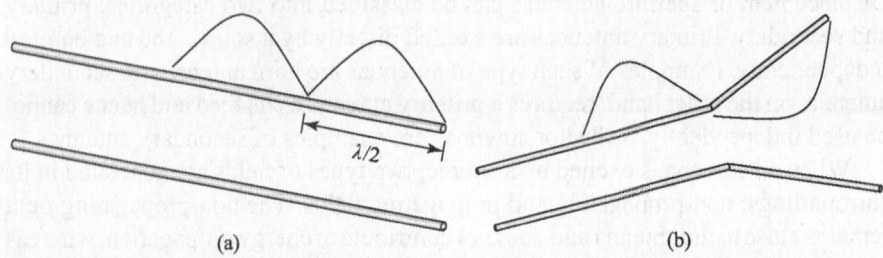

Fig. 18.2 Standing wave (a) Normal transmission line (b) Flared transmission line

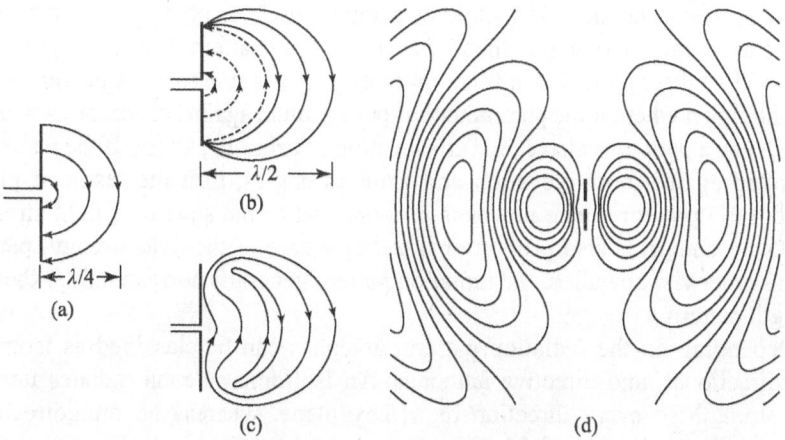

Fig. 18.3 Radiation mechanism of dipole (a) Field distribution at the instant $T/4$
(b) Field distribution just before $T/2$ (c) Field distribution just after $T/2$
(d) Field distribution at an arbitrary time

To understand field detachment from an antenna, let us consider Fig. 18.3. The lines of force created between the arms of a small centre fed dipole, in the first quarter of the period, are shown in Fig. 18.3(a). If sinusoidal excitation is assumed, then at the end of this quarter, the charge will be maximum and field lines will travel a radial distance of $\lambda/4$ outwards. During the next quarter, charge density on the conductor begins to diminish. At the end of the quarter, field lines of the first quarter travel an additional $\lambda/4$ and hence a total of $\lambda/2$ distance from the antenna. The decline in charges is thought to be accomplished by introducing opposite charges in the line, which at the end of the first half of the period, have neutralized all the charges on the conductors. The lines of force created by the opposite charges will be of same strength as the original field lines in the first quarter and will travel a distance of $\lambda/4$ at the end of the second quarter. This has been shown in Fig. 18.3(b). Therefore, at the end of the first half, the net result is that some lines of force point upwards within 0–$\lambda/4$ distance and the same number of lines direct downwards within $\lambda/4$–$\lambda/2$ distance. Since at the end of the first half the net charge on the antenna is zero, the lines of force will detach from the conductors and unite together to form closed loops, as shown in Fig. 18.3(c). The process will be repeated in the remaining second half, but now in the opposite direction. Field distribution surrounding the dipole at any arbitrary time is shown in Fig. 18.3(d).

18.3 Antenna Parameters

An antenna is characterized by its different parameters such as radiation pattern, direction of major lobe, 3 dB beam width, direction of nulls, half-power beam width, first null beam width, gain, directivity, polarization, efficiency, impedance, bandwidth, etc. In this section we will define these terms.

18.3.1 Radiation Pattern

The radiation pattern of an antenna is the graphical representation of the filed strength or power per unit solid angle, radiated by the antenna at a given distance, as a function of the elevation angle or azimuth angle or both. If field strength is plotted as a function of angle, then it is called a field pattern, whereas if power per unit solid angle is plotted as a function of angle, then it is called a power pattern. Depending on whether the parameter has been plotted against elevation or azimuth angle, such a pattern is also called an elevation or azimuth pattern. If the parameter is plotted against both elevation and azimuth angles, then the resultant plot is called a 3D pattern. Since power is proportional to the square of field strength, the power pattern is also proportional to the square of the field strength pattern. Unless otherwise specified, the radiation pattern generally corresponds to the field strength pattern.

Depending on the radiation pattern, antennas can be classified as isotropic, omnidirectional, and directive antennas. An isotropic antenna radiates uniform field strength in every direction or in any plane, whereas an omnidirectional antenna radiates the same field strength in a particular plane only. For a directional antenna, the radiated field varies with direction, and such an antenna generally concentrates its radiated field in one or a few particular directions only. It may be noted that an omnidirectional antenna is also a kind of a directional antenna. In practice, in electromagnetics, there is no such antenna that truly radiates uniform power in all directions, and hence no true isotropic antenna exists. However, the concept of an isotropic antenna is widely used in the antenna theory to define other antenna parameters.

The typical radiation pattern of a directive antenna is shown in Fig. 18.4. It is generally characterized by different parameters such as major lobe, minor lobe, side lobe, back lobe, half-power beam width (HPBW), and first null beam width (FNBW).

Radiation lobe   A radiation lobe is a part of the radiation pattern that is surrounded by different regions of relatively weak radiation intensities. Depending on the maximum radiated field strength, lobes can be classified as the major lobe and minor lobes. The major lobe contains the direction of maximum radiation, whereas a minor lobe is any lobe except a major lobe. Minor lobes can be classified as side and back lobes. A back lobe has a direction just opposite to that of a main lobe, whereas a side lobe has a direction different from that of a main lobe and a back lobe. Minor lobes generally correspond to undesired radiation, and hence for a properly designed antenna, they should be minimized.

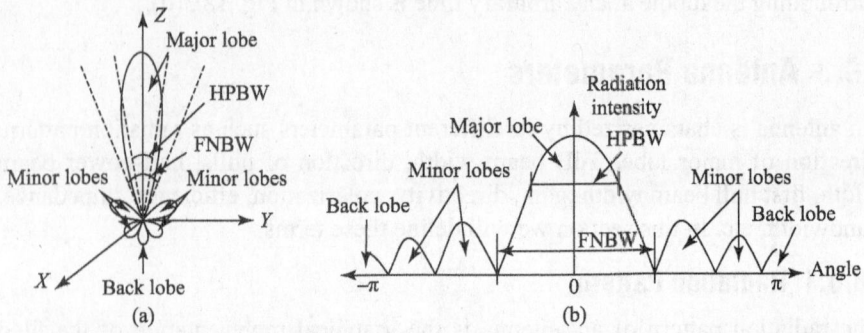

Fig. 18.4   Radiation lobes (a) Polar representation (b) Cartesian representation

Half-power beam width The HPBW corresponds to the angle between the two directions in the main lobe at which the field intensity becomes half of the maximum field intensity. This is also known as the 3 dB beam width.

A radiation lobe is often characterized by its 3 dB beam width. If the antenna has a relatively narrow beam width, in both elevation and azimuth planes, then it is called a pencil beam antenna. On the other hand, if the beam width of an antenna is relatively narrow in one plane but much broader in the orthogonal plane, then it is called a fan beam antenna.

First null beam width The FNBW corresponds to the angle between the directions of the nulls immediately following the major lobe.

18.3.2 Radiation Intensity

The power radiated per unit area by an antenna in a given direction can be calculated from the Poynting vector P and can be expressed as follows:

$$P = E^2/\eta_0 \text{ W/m}^2 \tag{18.1}$$

where we have assumed that $\vec{E} = \eta_0 \vec{H}$ $\tag{18.2}$

and $\vec{E}$ and $\vec{H}$ are orthogonal in a plane normal to radius vector. Since the surface area per unit solid angle (or sterradian) at a distance r is r^2 m^2, the radiation intensity $\Phi(\theta, \phi)$ in the given direction can be expressed as follows:

$$\Phi(\theta, \phi) = r^2 P = r^2 E^2/\eta_0 \text{ W/unit solid angle} \tag{18.3}$$

The total power radiated can therefore be expressed as follows:

$$P_r = \int \Phi \, d\Omega \text{ W} \tag{18.4}$$

Since there are a total of 4π sterradians in the total solid angle, the average power radiated per unit solid angle is given by the following expression:

$$P_{av} = \frac{P_r}{4\pi} \text{ W/sterradian} \tag{18.5}$$

In general, P_{av} represents the radiation intensity of an isotropic radiator radiating the same total power.

18.3.3 Directive Gain and Directivity

The directive gain of an antenna, in a given direction, is defined as the ratio of radiation intensity of the antenna in that direction to the average radiated power or radiation intensity of an isotropic radiator. Therefore,

$$g_d(\theta, \phi) = \Phi(\theta, \phi)/P_{av} = 4\pi\Phi(\theta, \phi)/P_r = 4\pi\Phi(\theta, \phi)/\int \Phi \, d\Omega \tag{18.6}$$

The directive gain in a dB scale is denoted by G_d, where

$$G_d = 10 \log_{10}(g_d) \tag{18.7}$$

The maximum directive gain of an antenna is known as directivity (D_0). The directivity of an isotropic radiator is unity.

18.3.4 Power Gain

The power gain of an antenna, in a given direction, can be expressed as follows:

$$g_p = 4\pi\Phi(\theta, \phi)/P_t \tag{18.8}$$

where P_t is the total input power and can be expressed as follows:

$$P_t = P_r + P_l \tag{18.9}$$

In Eq. (18.9), P_l is the Ohmic loss in the antenna. Using Eqs (18.6), (18.8), and (18.9), we get the following expression:

$$\frac{g_p}{g_d} = \frac{P_r}{P_r + P_l} \tag{18.10}$$

Equation (18.10) is a measure of the antenna efficiency. For suitably designed antennas, the Ohmic loss must be very small and hence the efficiency should be nearly 100%. For 100% efficient antennas, power and directive gains are equal.

18.3.5 Effective Area and Aperture Efficiency

The effective area or effective aperture of an antenna can be defined as follows:

$$A_e = \frac{\lambda^2}{4\pi} g_d \tag{18.11}$$

The effective area is basically the ratio of power available at the antenna terminal to the power per unit area of the appropriately polarized incident wave.

If A be the physical area of the antenna aperture, then the ratio of the effective aperture and physical aperture of the antenna is called the aperture efficiency. It can be expressed as follows: $\eta_a = A_e / A$ (18.12)

18.3.6 Antenna Efficiency

If an antenna is not matched with the source impedance, then reflection will take place at the antenna input and reflection loss will occur. As a result, the power radiated from the antenna will not be equal to the power available from the source. The ratio of these two powers is called antenna efficiency. The efficiency of an antenna, in the presence of reflection, is called reflection or mismatch efficiency and can be expressed as follows: $\eta_r = 1 - |\Gamma|^2$ (18.13)
where Γ is the reflection coefficient at the antenna input.

In addition to reflection, power is also lost at the imperfect antenna conductor and antenna dielectric. These losses are very difficult to calculate but can be found by measuring the input and radiated power. The combined conductor dielectric efficiency can be expressed as follows: $\eta_{cd} = P_r / P_{in}$ (18.14)
Sometimes, the combined conductor dielectric efficiency is also called the radiation efficiency. The overall efficiency of an antenna can therefore be expressed as follows: $\eta_o = \eta_r \eta_{cd}$ (18.15)
It may be noted that η_r depends on the external circuitry and not solely on the properties of the antenna.

18.3.7 Antenna Gain

It is defined as the product of radiation efficiency and directivity of an antenna, and can be expressed as follows: $G = \eta_{cd} D$ (18.16)
where D is the directivity of the antenna. It may be noted that directivity depends only on the shape of the radiation pattern and not on the efficiency, whereas gain includes the loss.

18.3.8 Effective Noise Temperature

The effective noise temperature of an antenna can be expressed as follows:

$$T_e = \eta_{cd}T_b + (1-\eta_{cd})T_p \qquad (18.17)$$

where T_b is the brightness temperature and T_p the physical temperature of the antenna.

18.3.9 Antenna Polarization

The polarization of an antenna is the same as the polarization of the field it radiates and can be classified into the following categories: (a) linear (horizontal or vertical), (b) circular (LHCP or RHCP), and (c) elliptical. For the maximum response, the polarization of a receiving antenna must be same as the incident wave; otherwise, polarization loss will take place. The polarization loss can be measured as the square of cosine of the angle between the polarization of the antenna and that of the incident wave.

18.3.10 Antenna Impedance

The impedance of an antenna can be represented as $Z = R + jX$ (18.18) where the reactive part corresponds to the electric and magnetic fields close to the antenna, which returns energy to the antenna during every cycle. The resistive part of the impedance consists of two parts: (a) Ohmic resistance (R_0) and (b) radiation resistance (R_r). The Ohmic resistance results due to heating of the antenna conductor by the current flowing through it, whereas radiation resistance results due to radiation of energy. Since Ohmic resistance causes losses, it must be very low. In terms of loss resistance and radiation resistance, radiation efficiency can be expressed as follows:

$$\eta_{cd} = \frac{P_{rad}}{P_{in}} = \frac{R_r}{R_r + R_0} \qquad (18.19)$$

18.3.11 Bandwidth

The bandwidth of an antenna is defined as the frequency range over which certain antenna characteristics satisfies a given standard. For example, if antenna impedance is of concern, then impedance bandwidth is defined as the frequency range over which antenna impedance lies within a specified value. It may be noted that since different antenna characteristics vary differently with frequency, there is no unique definition of antenna bandwidth.

18.3.12 Effective Isotropic Radiated Power

The effective isotropic radiated power (EIRP) is the power needed by an isotropic antenna to provide the same radiation intensity at a given direction as that of the directional antenna. It can be expressed as follows: $EIRP = P_{ant}G_t$ (18.20) where P_{ant} is the power fed to the antenna and G_t is the antenna gain.

18.3.13 Space Loss

A transmitting antenna radiates power in all directions as per its power pattern. However, since the receiving antenna is positioned at a particular direction with respect to the transmitting antenna, only a part of the transmitted power is accepted by it, while the rest of the radiated power is lost. This loss can be represented by space loss, which can be expressed as follows:

$$\text{Space loss} = 20\log_{10}\left(\frac{\lambda}{4\pi R}\right)\text{dB} \qquad (18.21)$$

where R is the distance and λ the operating wavelength.

EXAMPLE 18.1 Assume that the radial component of the radiated power density of an antenna can be expressed as $A_0\sin(\theta)/r^2$, where A_0 is the peak power density. Determine the (a) total radiated power, (b) directivity, and (c) directive gain.

Solution Given: $P = A_0\sin(\theta)/r^2$ W/m²

(a) The radiated intensity of the antenna can be expressed as follows:

$$\Phi(\theta,\phi) = r^2 P = A_0\sin(\theta)\,\text{W/unit solid angle}$$

Therefore, the total radiated power is given by the following expression:

$$P_r = \int \Phi d\Omega = \int_0^{2\pi}\int_0^{\pi} A_0\sin^2(\theta)d\theta d\phi = \pi^2 A_0\ \text{W}$$

(b) The maximum radiation intensity is expressed as follows:

$$\Phi_{\max} = A_0$$

Hence, the directivity is as follows:

$$D = \frac{4\pi\Phi_{\max}}{P_r} = \frac{4\pi A_0}{\pi^2 A_0} = \frac{4}{\pi} = 1.2732$$

(c) The directive gain is expressed by the following equation:

$$g_d(\theta,\phi) = \frac{4\pi\Phi(\theta,\phi)}{P_r} = \frac{4\pi A_0\sin(\theta)}{\pi^2 A_0} = 1.2732\sin(\theta)$$

Practice Problem

18.1 Assume that the radial component of the radiated power density of an antenna can be expressed as $A_0\sin^2(\theta)/r^2$, where A_0 is the peak power density. Determine the (a) total radiated power, (b) directivity, and (c) directive gain.

$$\left[8\pi A_0/3\ \text{W}, 1.5, 1.5\sin^2(\theta)\right]$$

18.4 Half-wave Dipole and Quarter-wave Monopole Antennas

A half-wave dipole is perhaps the most basic and simple antenna that is still used at high frequencies. It consists of two wires, bent at right angle and fed by a voltage source. The total length of the antenna is approximately half-wavelength and so it is called a half-wave dipole.

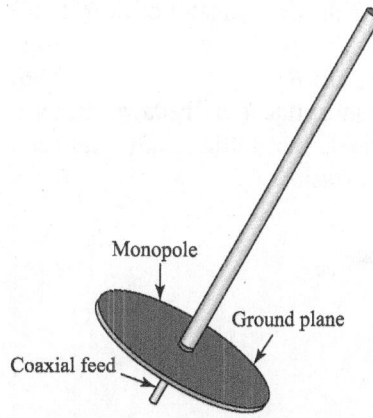

Fig. 18.5 Monopole antenna

In addition to a dipole, a monopole is also often used as an antenna element. A monopole is basically a quarter-wavelength wire mounted on a large ground plane. If the ground plane is large, then the radiation pattern of such an antenna is similar to that of a half-wave dipole, but its input impedance is half of that of a half-wave dipole. A basic monopole antenna is shown in Fig. 18.5.

Current distribution along a half-wave dipole or monopole can be assumed to be sinusoidal, as shown in Fig. 18.6. The corresponding expression can be written as follows:

$$I = I_m \sin\{\beta(H-z)\} \qquad z>0$$
$$I = I_m \sin\{\beta(H+z)\} \qquad z<0 \qquad (18.22)$$

where I_m is the current maximum. The expression for the vector potential at a point P due to the current element Idz can be written as follows:

$$dA_z = \frac{\mu I}{4\pi R} e^{-j\beta R} dz \qquad (18.23)$$

where R is the distance between the current element and the point P. The total vector potential at P due to all the current element can be obtained by integrating Eq. (18.23) over the antenna length and can be expressed as follows:

$$A_z = \frac{\mu}{4\pi}\left[\int_{-H}^{0} \frac{I_m \sin\{\beta(H+z)\}e^{-j\beta R}}{R} dz \right.$$
$$\left. + \int_{0}^{H} \frac{I_m \sin\{\beta(H-z)\}e^{-j\beta R}}{R} dz \right] \qquad (18.24)$$

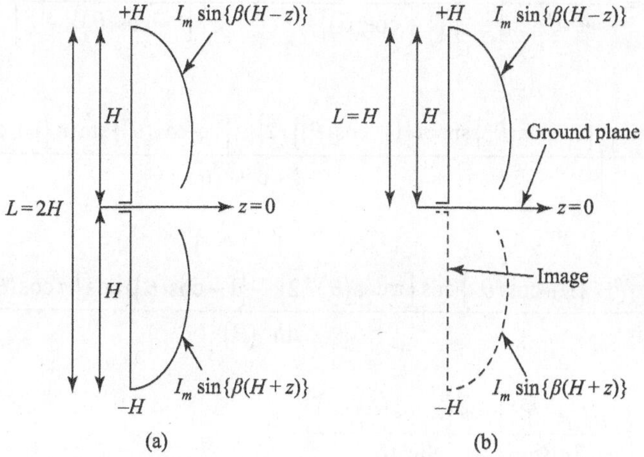

Fig. 18.6 Current distribution (a) Along half-wave dipole (b) Along monopole

If the observation point is sufficiently large, then in the denominator of Eq. (18.24) we can substitute $R \approx r$ (18.25)

whereas for numerator we can substitute $R = r - z\cos(\theta)$ (18.26)

It may be noted that, at the numerator, we cannot substitute $R \approx r$ because the term $z\cos(\theta)$ corresponds to the phase of the radiated field. Substituting Eqs (18.25) and (18.26) into Eq. (18.24), we get the following expression:

$$A_z = \frac{\mu I_m e^{-j\beta r}}{4\pi r} \left[\int_{-H}^{0} \sin\{\beta(H+z)\} e^{j\beta z \cos(\theta)} dz \right.$$

$$\left. + \int_{0}^{H} \sin\{\beta(H-z)\} e^{j\beta z \cos(\theta)} dz \right] \qquad (18.27)$$

For a half-wave dipole $H = \lambda/4$, and we can write the following relation:

$$\sin\{\beta(H \pm z)\} = \sin(\beta H \pm \beta z) = \sin\left(\frac{\pi}{2} \pm \beta z\right) = \cos(\beta z) \qquad (18.28)$$

Substituting Eq. (18.28) into Eq. (18.27), the following expression is obtained:

$$A_z = \frac{\mu I_m e^{-j\beta r}}{4\pi r} \left[\int_{0}^{\lambda/4} \cos(\beta z)\left\{ e^{j\beta z \cos(\theta)} + e^{-j\beta z \cos(\theta)} \right\} dz \right]$$

or $$A_z = \frac{\mu I_m e^{-j\beta r}}{4\pi r} \left[\int_{0}^{\lambda/4} 2\cos(\beta z)\cos\{\beta z \cos(\theta)\} dz \right]$$

or $$A_z = \frac{\mu I_m e^{-j\beta r}}{4\pi r} \int_{0}^{\lambda/4} \left[\cos[\beta z\{1+\cos(\theta)\}] + \cos[\beta z\{1-\cos(\theta)\}] \right] dz$$

or $$A_z = \frac{\mu I_m e^{-j\beta r}}{4\pi\beta r} \left\{ \frac{\sin[\pi\{1+\cos(\theta)\}/2]}{\{1+\cos(\theta)\}} + \frac{\sin[\pi\{1+\cos(\theta)\}/2]}{\{1-\cos(\theta)\}} \right\}$$

or

$$A_z = \frac{\mu I_m e^{-j\beta r}}{4\pi\beta r} \left[\frac{\{1-\cos(\theta)\}\sin[\pi\{1+\cos(\theta)\}/2] + \{1+\cos(\theta)\}\sin[\pi\{1+\cos(\theta)\}/2]}{1-\cos^2(\theta)} \right]$$

or

$$A_z = \frac{\mu I_m e^{-j\beta r}}{4\pi\beta r} \left[\frac{\{1+\cos(\theta)\}\cos\{\pi\cos(\theta)/2\} + \{1-\cos(\theta)\}\cos\{\pi\cos(\theta)/2\}}{\sin^2(\theta)} \right]$$

or $$A_z = \frac{\mu I_m e^{-j\beta r}}{2\pi\beta r} \left[\frac{\cos\{\pi\cos(\theta)/2\}}{\sin^2(\theta)} \right] \qquad (18.29)$$

If the current is entirely in the z-direction, then $H_\phi = -\dfrac{1}{\mu}\dfrac{\partial A_z}{\partial r}\sin(\theta)$ (18.30)

Substituting Eq. (18.29) into Eq. (18.30) and keeping only the inverse distance term, we get the following relation:

$$H_\phi = j\frac{I_m e^{-j\beta r}}{2\pi r}\left[\frac{\cos\{\pi\cos(\theta)/2\}}{\sin(\theta)}\right]$$ (18.31)

Electric field strength for the radiation field can therefore be expressed as follows:

$$E_\theta = \eta H_\phi = j\frac{60\,I_m e^{-j\beta r}}{r}\left[\frac{\cos\{\pi\cos(\theta)/2\}}{\sin(\theta)}\right]$$ (18.32)

The time average value power will therefore be the following:

$$P_{av} = \frac{1}{2}E_\theta H_\phi^* = \frac{\eta I_m^2}{8\pi^2 r^2}\left[\frac{\cos^2\{\pi\cos(\theta)/2\}}{\sin^2(\theta)}\right]$$ (18.33)

If now a monopole is considered, then using Eq. (18.33), the total power radiated through a hemispherical surface of radius r can be derived as follows:

$$\oint P_{av}\,da = \int_{\theta=0}^{\pi/2}\int_{\phi=0}^{2\pi} r^2\sin(\theta)\,d\theta d\phi = \frac{\eta I_m^2}{4\pi}\int_{\phi=0}^{\pi/2}\frac{\cos^2\{\pi\cos(\theta)/2\}}{\sin(\theta)}\,d\theta$$ (18.34)

Applying numerical methods, it can be shown that

$$\int_0^{\pi/2}\frac{\cos^2\{\pi\cos(\theta)/2\}}{\sin(\theta)}\,d\theta = 0.609$$ (18.35)

Substituting Eq. (18.35) into Eq. (18.34), we get the following expression:

$$\text{Power} = \frac{0.609\eta I_m^2}{4\pi}$$ (18.36)

In Eq. (18.36), I_m is the maximum or peak current. In terms of effective current, the radiated power is expressed as follows:

$$\text{Power} = \frac{0.609\eta I_{m(\text{eff})}^2}{2\pi} = 36.5 I_{m(\text{eff})}^2$$ (18.37)

Therefore, the radiation resistance of a quarter-wave monopole antenna is 36.5 Ω.

For a half-wave dipole, power would be radiated through a complete spherical surface and hence the upper limit of θ in Eq. (18.34) will be π instead of $\pi/2$. Therefore, for the same current, the power radiated by a half-wave dipole will be twice that of a quarter-wave monopole, and the radiation resistance for the half-wave dipole becomes 73 Ω.

For a half-wave dipole,

$$g_d = \frac{4\pi\Phi(\theta,\phi)}{\int\Phi d\Omega} = 4\pi\left[\frac{\eta I_m^2 r^2}{4\pi r^2}\left[\frac{\cos^2\{\pi\cos(\theta)/2\}}{\sin^2(\theta)}\right]\right] \Bigg/$$

$$\int_{\theta=0}^{\pi}\int_{\phi=0}^{2\pi}\left[r^2\frac{\eta I_m^2}{4\pi r^2}\left[\frac{\cos^2\{\pi\cos(\theta)/2\}}{\sin^2(\theta)}\right]\right]\sin(\theta)d\theta d\phi$$

or $\qquad g_d = 4\pi\dfrac{\cos^2\{\pi\cos(\theta)/2\}}{\sin^2(\theta)}\dfrac{1}{2\pi\times1.2188} = 1.64\dfrac{\cos^2\{\pi\cos(\theta)/2\}}{\sin^2(\theta)}$ $\qquad(18.38)$

Dipoles and monopoles also have several other configurations, such as, folded dipole, folded monopole, inverted L-antenna, and inverted F-antenna, which are used at microwave frequencies. Basic structures of these antennas are shown in Fig. 18.7. A folded half-wave dipole is formed by joining the ends of a half-wave dipole, whereas a folded monopole antenna is formed by bending a monopole antenna. Inverted L-and inverted F-antennas are low-profile antennas that are formed by bending quarter-wavelength monopole elements into L and F shapes, respectively.

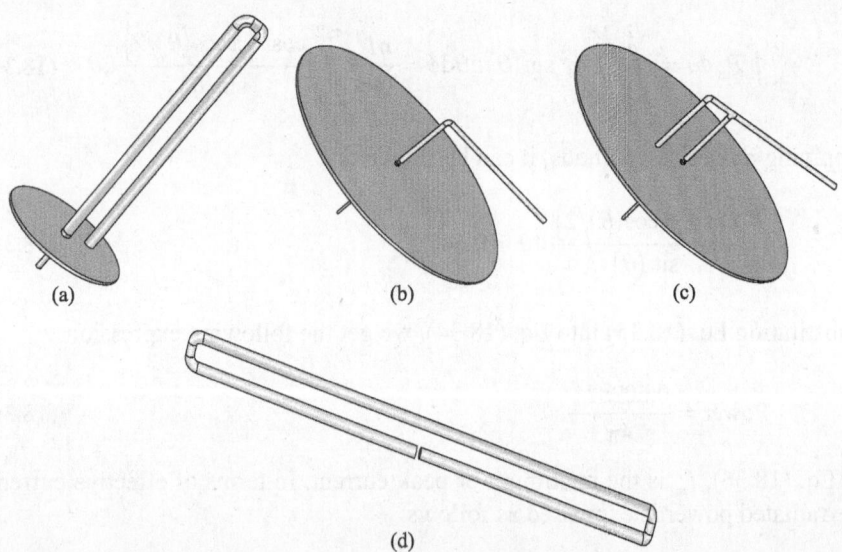

Fig. 18.7 Different dipole and monopole antenna structures (a) Folded monopole antenna (b) Inverted L-antenna (c) Inverted F-antenna (d) Folded dipole antenna

EXAMPLE 18.2 A magnetic field of strength 6 μA/m is required at a point $\theta = \pi/2$; $R = 2.5$ km from a half-wave dipole antenna in air. If the antenna is assumed to be loss less, then calculate the transmitted power.

Solution Given: $|H_\phi| = 6\mu\text{A/m}$, $\theta = \pi/2$, and $R = 2.5$ km

Now,

$$|H_\phi| = \frac{I_m}{2\pi r}\left[\frac{\cos(\pi\cos(\theta)/2)}{\sin(\theta)}\right] = \frac{I_m}{2\pi \times 2.5 \times 10^3}\left[\frac{\cos(\pi\cos(\pi/2)/2)}{\sin(\pi/2)}\right]$$

or $\quad |H_\phi| = \dfrac{I_m}{2\pi \times 2.5 \times 10^3} = 6 \times 10^{-6}$

or $\quad I_m = 6 \times 10^{-6} \times 2\pi \times 2.5 \times 10^3 = 0.0942$ A

Therefore, $P_{\text{rad}} = 73 I_m^2 = 73 \times (0.0942)^2 = 0.6478$ W

Practice Problem

18.2 A magnetic field of strength 5 µA/m is required at a point $\theta = \pi/2$; $R = 1.5$ km from a quarter-wave monopole antenna in air. If the antenna is assumed to be loss less, then calculate the transmitted power. **[0.081 W]**

18.5 Loop Antennas

A loop antenna is basically a loop of wire that can take several forms such as circular, ellipse, rectangular, square, and triangle. Out of them, circular loops are most common and are shown in Fig. 18.8.

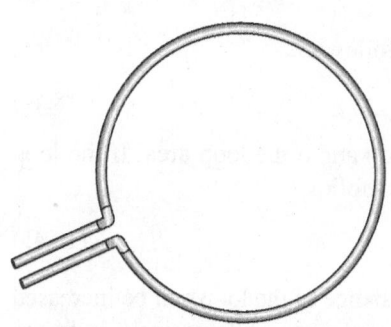

Depending on the electrical length of the loop, such antennas are classified into two categories: (a) electrically small and (b) electrically large. For electrically small antennas, the overall antenna length is usually less than about 1/10th of a wavelength, whereas for electrically large loops the loop circumference is about a free-space wavelength. It can be shown that an electrically small loop is equivalent to that of an infinitesimal magnetic dipole, with its axis perpendicular to the plane of the loop.

Fig. 18.8 Loop antenna

For such loops, the radiation resistance is usually smaller than the loss resistance, and hence they are poor radiators. Such antennas are generally used as receiving antennas in portable radios and pagers, as probes for field measurements, and in other applications where antenna efficiency in not as important as the signal-to-noise ratio. Since an electrically small loop is equivalent to an infinitesimal magnetic dipole, the field pattern of an electrically small loop antenna is also similar to that of an infinitesimal dipole. If the overall length of the dipole is increased, the direction of major lobe gradually shifts from the plane to the axis of the loop.

The radiated power of an electrically small loop antenna, carrying a current I_0, at a distance r, can be expressed as follows:

$$P_r = \eta\left(\frac{\pi}{12}\right)(ka)^4 |I_0|^2 \left\{1 + j\frac{1}{(kr)^3}\right\} \tag{18.39}$$

For the near field, $kr \ll 1$, and the second term within bracket of Eq. (18.39) is dominant. This implies that the power is mainly reactive. In the far field, $(kr \gg 1)$; the second term within the bracket is negligible and the power becomes real.

The radiation intensity can be expressed as follows:

$$\Phi(\theta,\phi) = r^2 P_r = \frac{\eta}{2}\left(\frac{k^2 a^2}{4}\right)^2 |I_0|^2 \sin^2(\theta) \tag{18.40}$$

Equation (18.40) reveals that the maximum radiation intensity occurs at $\theta = \pi/2$, and is given as follows:

$$\Phi_{max} = \Phi|_{\theta=\pi/2} = \frac{\eta}{2}\left(\frac{k^2 a^2}{4}\right)^2 |I_0|^2 \tag{18.41}$$

Using Eqs (18.39) and (18.41), the directivity of the loop can be written as follows:

$$D_0 = 4\pi\Phi_{max}/P_r = 1.5 \tag{18.42}$$

The maximum effective aperture of an electrically small loop antenna can be expressed as follows:

$$A_{em} = \left(\frac{\lambda^2}{4\pi}\right)D_0 = \frac{3\lambda^2}{8\pi} \tag{18.43}$$

The radiation resistance can be determined as follows:

$$R_r = 20\pi^2 (C/\lambda)^4 \approx 31,171(S^2/\lambda^4) \tag{18.44}$$

where $C = 2\pi a$ is the circumference of the loop and S the loop area. If the loop consists of N turns, then Eq. (18.44) modifies as follows:

$$R_r = 31,171 N^2 (S^2/\lambda^4) \tag{18.45}$$

Equation (18.45) reveals that the radiation resistance of the loop can be increased by increasing its perimeter and/or the number of turns. Alternatively, a ferrite core of very high permeability can be inserted within the loop circumference. The ferrite core will increase the magnetic field intensity and hence the radiation resistance. Such loops are also called ferrite loops. The radiation resistance of a ferrite loop can be calculated using the following relation:

$$R_f/R_r = (\mu_{ce}/\mu_0)^2 = \mu_{cer}^2 \tag{18.46}$$

where R_f is the radiation resistance of the ferrite loop, R_r is the radiation resistance of the air core loop, μ_{ce} is the effective permeability of the ferrite core, μ_0 is the permeability of free space, and μ_{cer} is the relative effective permeability of the ferrite core. Radiation resistance of a single-turn ferrite loop can be written as follows:

$$R_f = 20\pi^2 (C/\lambda)^4 (\mu_{ce}/\mu_0)^2 = 20\pi^2 (C/\lambda)^4 \mu_{cer}^2 \tag{18.47}$$

If there are N turns in the loop, then Eq. (18.47) modifies as follows:

$$R_f = 20\pi^2 N^2 (C/\lambda)^4 (\mu_{ce}/\mu_0)^2 = 20\pi^2 N^2 (C/\lambda)^4 \mu_{cer}^2 \tag{18.48}$$

where $\quad \mu_{cer} = \dfrac{\mu_{ce}}{\mu_0} = \dfrac{\mu_{fr}}{1+D(\mu_{fr}-1)}$ $\qquad$ (18.49)

and D is the demagnetization factor. The relative intrinsic permeability μ_{fr} is very large and hence the relative effective permeability of the ferrite core μ_{cer} is approximately inversely proportional to the demagnetization factor, which is again a function of the geometry of the ferrite core.

It can be shown that for constant current and large loop approximation $(a \geq \lambda/2)$, radiated power, maximum radiation intensity, directivity, effective aperture, and radiation resistance can be expressed as follows:

$$P_{rad} \approx \frac{\pi(a\omega\mu)^2 |I_0|^2}{4\eta ka} \qquad (18.50)$$

$$\Phi_{max} = \frac{(a\omega\mu)^2 |I_0|^2}{8\eta}(0.584)^2 \qquad (18.51)$$

$$D_0 = 0.682(C/\lambda) \qquad (18.52)$$

$$A_{em} = 5.43 \times 10^{-2} \lambda C \qquad (18.53)$$

$$R_r = 60\pi^2 (C/\lambda) \qquad (18.54)$$

Most of the applications of loop antennas are in the HF, VHF, and UHF bands.

EXAMPLE 18.3 A small loop antenna of circumference 0.3λ is operating at $1\,\text{GHz}$. Calculate its radiation resistance and maximum effective aperture.

Solution   Given: $C = 0.1\lambda$ and $f = 1\text{GHz}$

Therefore, the operating wavelength is $\lambda = c/f = 3 \times 10^8/10^9 = 0.3\,\text{m}$
Therefore, its radiation resistance is as follows:

$$R_r = 20\pi^2 (C/\lambda)^4 = 20\pi^2 (0.3)^4 = 1.5989\ \Omega$$

The effective aperture is expressed in the following way:

$$A_{em} = \frac{3\lambda^2}{8\pi} = \frac{3 \times 0.3^2}{8\pi} = 0.011\,\text{m}^2$$

Practice Problem

18.3 A loop antenna of circumference $1.5\pi\lambda$ is operating at $1\,\text{GHz}$. Calculate its radiation resistance, directivity, and maximum effective aperture.

$$\left[2790.5649\ \Omega,\ 3.2138,\ 0.023\,\text{m}^2\right]$$

18.6 Helical Antennas

A helical antenna is one of the simplest antennas that are widely used at microwave frequencies. It consists of a wire wound in a helical shape in the presence of a ground plane. Such antennas are generally fed with a coaxial probe, with its centre conductor connected to the helical structure and the outer conductor to the ground plane.

In practice, the ground plane may be of various shapes, but the most popular of them is circular.

A helix consists of N turns of diameter D and inter-turn spacing S, as shown in Fig. 18.9. The total length of the helix is, therefore,

$$L = NS \qquad (18.55)$$

The total length of wire required to construct a helix is calculated as follows:

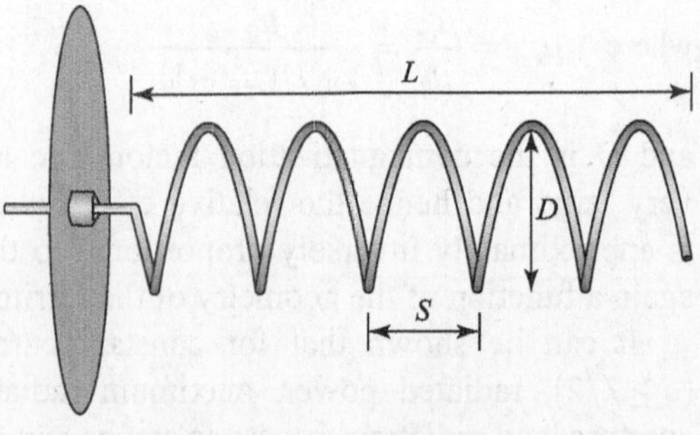

Fig. 18.9 Helical antenna

$$L_w = N\sqrt{S^2 + C^2} \qquad (18.56)$$

where $\quad C = \pi D$ (18.57)

is the helix circumference. Another important parameter, associated with a helix structure, is the pitch angle α that defines the angle formed by a line tangent to the helix wire and a plane perpendicular to the helix axis. The pitch angle is defined by the following formula:

$$\alpha = \tan^{-1}\left(\frac{S}{\pi D}\right) = \tan^{-1}\left(\frac{S}{C}\right) \qquad (18.58)$$

For $\alpha = 0°$, a helical antenna reduces to a loop antenna.

By controlling the electrical dimensions of different helix parameters, the radiation characteristic of a helix can be varied. In general, polarization of the radiated field of a helical antenna is elliptical. However, circular and linear polarizations can also be achieved at different frequency ranges. The input impedance of a helical antenna is a function of the pitch angle and wire size, especially near the feed point.

Although a helical antenna can operate in different modes, normal (broadside) and axial (end fire) are the principal modes. Out of these two principal modes, axial mode is most commonly used because it can provide circular polarization over a wide bandwidth. It is also more efficient.

18.6.1 Normal Mode

A normal mode in a helix is characterized by a radiation pattern in which the maximum field is radiated in a plane perpendicular to the helix axis and the minimum field radiated in a direction along the axis. Thus, the radiation pattern is of figure *eight* shape rotated about its axis. The normal mode of helix takes place when the helix dimension is small compared to the operating wavelength.

In the normal mode, a helical antenna can be assumed as a combination of N small loops and N short dipoles connected together, with the planes of the loops parallel to each other and perpendicular to the axis of vertical dipole, and the axis of both the loop and dipole coinciding with the axis of the helix. This is shown in Fig. 18.10. Further, in the normal mode, the current throughout the helix length can be assumed to be constant and its relative far-field pattern can be assumed to be independent of the numbers of loops and short dipoles. Thus, the total radiated field of a helix can be described by the sum of the fields radiated by a small loop

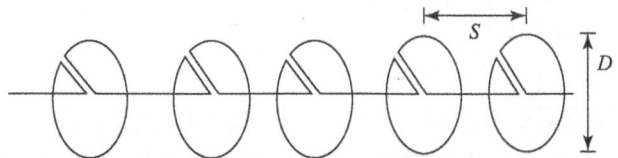

Fig. 18.10 Normal mode equivalent of helical antenna

of diameter D and a small dipole of length S. In practice, the dipole radiates a field in θ direction, and the loop radiates a field in ϕ direction. The ratio of $|E_\theta|$ and $|E_\phi|$ describes the axial ratio and is given by the following equation:

$$AR = 2S\lambda/(\pi D)^2 \tag{18.59}$$

The axial ratio ranges between '0' and '∞'. If AR $= 0$, that is, $|E_\theta| = 0$, then we get horizontal polarization and the helix behaves as a loop antenna, whereas if AR $= \infty$, that is, $|E_\phi| = 0$, then we get vertical polarization and the helix behaves as a vertical dipole. A special case happen when

$$AR = 1 \text{ or } 2S\lambda/(\pi D)^2 = 2S\lambda/C^2 = 1$$

or $\qquad C = \sqrt{2S\lambda} \tag{18.60}$

Therefore, $\tan(\alpha) = \dfrac{S}{C} = \dfrac{C^2}{2\lambda C} = \dfrac{C}{2\lambda} = \dfrac{\pi D}{2\lambda} \tag{18.61}$

When this condition is satisfied, the radiated field is circularly polarized in all directions except $\theta = 0°$. If AR has other values except 0, 1, and ∞, the radiated field is elliptically polarized, with its major axis either horizontally polarized (for AR lying between 0 and 1) or vertically polarized (for AR lying between 1 and ∞). For the normal mode operation, the bandwidth is small and efficiency is also poor.

18.6.2 Axial Mode

The axial mode is more practical for application and also can be generated very easily. Such a mode is characterized by one major lobe along the axis of the helix. The diameter D and spacing S of the helical structure, in this mode, is a large fraction of the wavelength, and the circumference lies in the range $3\lambda/4 < C < 4\lambda/3$. The range of pitch angle lies within $12° < \alpha < 14°$, and the diameter of the ground plane is kept at least $\lambda/2$.

The input impedance of a helical antenna operating in the axial mode is nearly resistive and lies within $100–200\,\Omega$. However, with proper design, an input impedance of nearly $50\,\Omega$ can also be achieved. The terminal impedance can be expressed as follows:

$$R = 140C/\lambda\,\Omega \tag{18.62}$$

For a helical antenna, operating in the axial mode,

$$\text{HPBW} = \frac{52\lambda^{3/2}}{C\sqrt{NS}} \text{ degree} \tag{18.63}$$

$$\text{FNBW} \approx \frac{115\lambda^{3/2}}{C\sqrt{NS}} \text{ degree} \tag{18.64}$$

$$\text{Directivity} \cong \frac{15NC^2S}{\lambda^3} \tag{18.65}$$

$$\text{AR} = \frac{2N+1}{2N} \tag{18.66}$$

This relation assumes that $3\lambda/4 < C < 4\lambda/3$, $12° < \alpha < 14°$, $N > 3$, turns are identical, and the spacing between turns is uniform.

Since an elliptical polarization can be represented as a sum of two orthogonal linear components in time-phase quadrature, a helical antenna operating with an elliptical polarization can be used as a receiving antenna for a rotating linear polarized wave. This enables the helical antenna to be used as space probes in ballistic missiles and as a ground base antenna for receiving signals that have undergone Faraday rotation during its travel through the ionosphere in space telemetry applications of satellites.

EXAMPLE 18.4   A helical antenna has diameter 0.3λ, 12 turns, and turn spacing $S = 0.25\lambda$. If the antenna is operating at 3 GHz, calculate the (a) directivity, (b) axial ratio, (c) HPBW, and (d) FNBW.

Solution   Given: $D = 0.3\lambda$, $N = 12$, $S = 0.25\lambda$, and $f = 3$ GHz

The operating wavelength is expressed as follows:

$$\lambda = \frac{c}{f} = \frac{3 \times 10^8}{3 \times 10^9} = 0.1\text{m} = 10\text{cm}$$

The diameter is written as follows: $D = 0.3 \times 10 = 3$ cm
The spacing is given by the following equations:

$$S = 0.25 \times 10 = 2.5 \text{cm}$$

and

$$C = \sqrt{2S\lambda} = \sqrt{2 \times 2.5 \times 10} = 7.0711$$

(a)   The directivity is calculated as follows:

$$D = \frac{15NC^2S}{\lambda^3} = \frac{15 \times 12 \times 7.0711^2 \times 2.5}{10^3} = 22.5 \text{dB}$$

(b)   The axial ratio is estimated to be the following:

$$\text{AR} = \frac{2N+1}{2N} = \frac{2 \times 12 + 1}{2 \times 12} = 1.0417$$

(c)   The HPBW is as follows:

$$\text{HPBW} = \frac{52\lambda^{3/2}}{C\sqrt{NS}} = \frac{52 \times 10^{3/2}}{7.0711\sqrt{12 \times 2.5}} = 42.4576°$$

(d)   The FNBW is calculated as follows:

$$\text{FNBW} = \frac{115\lambda^{3/2}}{C\sqrt{NS}} = \frac{115 \times 10^{3/2}}{7.0711\sqrt{12 \times 2.5}} = 93.8967°$$

Practice Problem
18.4 A helical antenna has diameter 0.5λ, 12 turns, and turn spacing $S = 0.4\lambda$. If the antenna is operating at 5 GHz, calculate the (a) directivity, (b) axial ratio, (c) HPBW, and (d) FNBW.   [72 dB, 1.0333, 23.7345°, 52.4897°]

18.7 Yagi–Uda Antennas

A Yagi–Uda antenna consists of a half-wave dipole or folded dipole as a driven element and several parasitic elements parallel to it. The parasitic element in front of the dipole is called a director and those at the back of the dipole are called reflectors. The length of the dipole is usually kept in the range $0.45\lambda - 0.49\lambda$ (resonant value), and the length of the directors is kept in the range $0.40\lambda - 0.45\lambda$. The separation between the directors is kept in the range $0.3\lambda - 0.4\lambda$. The length of the reflector is maintained at a value greater than that of the dipole, and the separation between the dipole and the reflector is kept at a value smaller than the spacing between the dipole and the nearest director. The separation is found to be optimum at 0.25λ. Approximate design formulas of the Yagi–Uda antenna are as follows:

$$\text{Reflector length} = 500/f\,(\text{MHz})\ \text{ft} \tag{18.67}$$

$$\text{Driven element length} = 475/f\,(\text{MHz})\ \text{ft} \tag{18.68}$$

$$\text{Director length} = 455/f\,(\text{MHz})\ \text{ft} \tag{18.69}$$

The effects of different parameters associated with a Yagi–Uda antenna are summarized in Table 18.1.

Equations (18.67)–(18.69) show that the lengths of the reflector and the director are, respectively, larger and smaller than that of the driven element. Since the length of each director is smaller than its corresponding resonant length, they have capacitive impedance and hence its current leads the induced emf. The progressive phase shifts reinforce the field of the energized element towards the directors. The impedance of the reflectors is inductive, and the phase of the currents lags those of the induced emfs.

In a multi-reflector Yagi–Uda antenna, the major role is played by the first element next to the driven elements. Successive reflectors have very little role in the performance of the antenna. That is why most of the Yagi–Uda antennas contain a single reflector. In contrast to reflectors, performance of the Yagi–Uda antenna can be improved considerably if more directors are used. However,

Table 18.1 Effects of different parameters associated with a Yagi–Uda antenna

Parameters	Effects
Reflector spacing and size	1. Negligible effects on the forward gain 2. Large effects on the backward gain 3. Can be used to optimize antenna parameters without affecting the gain significantly
Feeder length and radius	1. Small effect on the forward gain 2. Large effect on the backward gain 3. Large effect on the input impedance
Size and spacing of the directors	1. Large effect on the forward gain 2. Large effect on the backward gain 3. Large effect on the input impedance

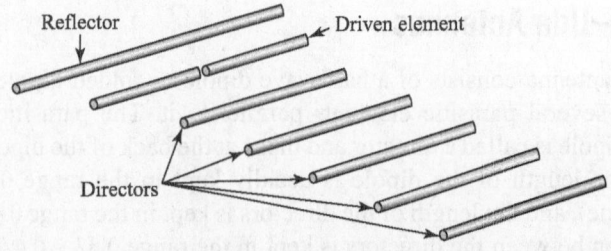

Fig. 18.11 Yagi–Uda antenna

there is also a limit to the number of directors that can be used. If the number of directors is more than this value, very little improvement is achieved because of the progressive reduction in magnitude of the induced currents on the more extreme director elements. In general, most Yagi–Uda antennas have about 6–12 directors. A schematic diagram of a Yagi–Uda antenna is shown in Fig. 18.11.

A Yagi–Uda antenna is a very-high-gain antenna (of the order of 14.8–17.3 dB), and has low input impedance and relatively narrow bandwidth (about 2%). Improvement in both can be achieved at the expense of other parameters such as gain and magnitude of minor lobes.

EXAMPLE 18.5 Design a Yagi–Uda antenna at 1 GHz.

Solution Given: $f = 1$ GHz

The reflector length will be as follows:

Reflector length $= 500/f \, (\text{MHz}) = 500/1000 = 0.5 \, \text{ft} = 15.24 \, \text{cm}$

The driven element length will be as follows:

Driven element length $= 475/f \, (\text{MHz}) = 475/1000 = 0.475 \, \text{ft} = 14.478 \, \text{cm}$

The director length will be as follows:

Director length $= 455/f \, (\text{MHz}) = 455/1000 = 0.455 \, \text{ft} = 13.8684 \, \text{cm}$

Practice Problem

18.5 Design a Yagi–Uda antenna at 3 GHz.   [5.08 cm, 4.826 cm, 4.6228 cm]

18.8 Log-periodic Antennas

A log-periodic antenna consists of a number of dipoles of different electrical lengths and spacing, as shown in Fig. 18.12, and is fed by a balanced two-wire transmission line. Different geometrical parameters associated with a log-periodic antenna are related by the following formulas:

$$1/\tau = f_2/f_1 = l_{n+1}/l_n = R_{n+1}/R_n = d_{n+1}/d_n = s_{n+1}/s_n \qquad (18.70)$$

where d is the diameter of the dipole. Rests are shown in the figure. Since it is very difficult to obtain wires of different diameters and simultaneously to maintain tolerance of very small gap spacing, constant-diameter wires are also used in a

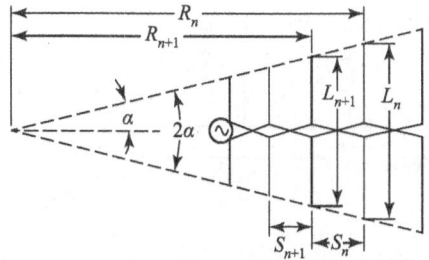

Fig. 18.12 Log-periodic antenna

log-periodic dipole array. This approximation does not degrade the overall performance by a considerable amount. The spacing factor, σ, of a log-periodic antenna can be expressed as follows:

$$\sigma = \frac{R_{n+1} - R_n}{2l_{n+1}} \qquad (18.71)$$

The relation between the apex angle (α) and τ can be expressed as follows:

$$\alpha = \tan^{-1}\left[\frac{1-\tau}{4\sigma}\right] \qquad (18.72)$$

For typical log-periodic dipole arrays, the half apex angle lies in the range $10° \le \alpha \le 45°$ and τ in the range $0.95 \ge \tau \ge 0.7$. Larger values of α or smaller values of τ require a larger number of elements that are close together.

In practice, the total length of a log-periodic array can be divided into three different regions, namely, (a) loaded transmission line region, (b) active region, and (c) reflection region. In the loaded transmission line region, dipole elements are short compared with the resonant length and present relatively high capacitive impedance. Therefore, the small dipole current leads the voltage by approximately 90°. In addition, due to the phase reversal introduced by the transposition of the transmission line, adjacent elements are nearly 180° out of phase. In practice, each dipole current leads the preceding dipole currents by approximately $\phi = \pi - \beta d$, where d is the element separation and β the phase shift constant along the line.

In contrast to the transmission line region, element lengths in the active region approach the resonant length and present a resistive component. The dipole current in this region is large and nearly in phase with the base voltage. More precisely, the dipole current is slightly leading just below resonance and slightly lagging just above resonance. As before, the phase of current in a given element leads that in the preceding element by an angle $\phi = \pi - \beta d$. In practice, this phase difference may be approximated as $\pi/2$ rad.

In the reflection region, dipole lengths are greater than the resonant length and hence present inductive impedance. The dipole currents, therefore, lag the voltage and is quite small because all the energy transmitted down the line has been radiated by the active region. It can be shown that the transmission line characteristic impedance becomes reactive in this region, and hence the incident energy is reflected back towards the source. The equivalent circuit of a log-periodic dipole array is shown in Fig. 18.13.

The input impedance of a log-periodic antenna repeats periodically with logarithm of the frequency, and hence it is called log-periodic. In practice, to achieve a truly log-periodic configuration, infinite dipole elements will be required. However, this is not a practical condition, and the structure is truncated at both ends. This limits the infinite bandwidth of a log-periodic antenna to a finite, but still large, bandwidth. The cut-off frequencies or wavelength of the bandwidth can be determined by the electrical lengths of the longest and shortest dipole elements

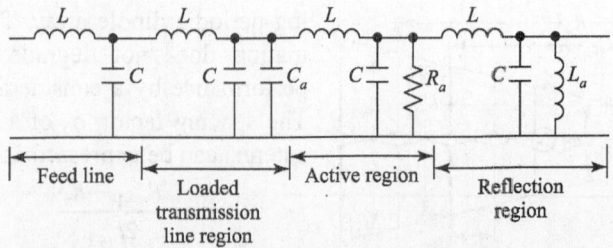

Fig. 18.13 Equivalent circuit of log-periodic dipole array

of the structure. In general, the lower cut-off wavelength is approximately double the length of the longest dipole element, whereas the high cut-off wavelength is approximately double the length of the shortest dipole element. The latter is valid when the active region of the antenna is very narrow. The active region of a log-periodic dipole array is determined by the elements whose lengths are nearly equal to or slightly smaller than a half of the operating wavelength. Therefore, the role of active elements is passed from the shorter to the longer elements as the frequency decreases. In practice, the energy travelling towards the longer inactive elements from the shorter active elements decreases very rapidly. Thus, a negligible amount of energy is reflected from the truncated end.

It has already been mentioned that the input impedance of a log-periodic dipole array repeats periodically with the logarithm of the frequency. In practice, periodicity of the structure does not ensure broadband operation, but if the impedance variation within one cycle can be kept within an acceptable range for the corresponding bandwidth, then a broadband response can be achieved. The total bandwidth of an antenna is determined by the number of repetitive cycles for the given truncated structure. The relative frequency span Δ of each cycle can be expressed, based on Eq. (18.70), as follows:

$$\Delta = \ln(f_2) - \ln(f_1) = \ln(1/\tau) \tag{18.73}$$

In a log-periodic dipole array, the phase velocity of the wave travelling along the structure is about $0.6c$ where c is the free-space velocity. This corresponds to about $150°$ phase change for every $\lambda/4$ free-space length of the transmission line. The smaller phase velocity, mentioned previously, results from the shunt capacitive loading of the line by the smaller elements in the active region. It may be noted that since the spacing between longer elements is larger, loading per unit length is almost constant.

Apart from the log-periodic dipole array, different other log-periodic antenna structures also exist, as shown in Fig. 18.14. For such log-periodic antennas, we can write the following equations:

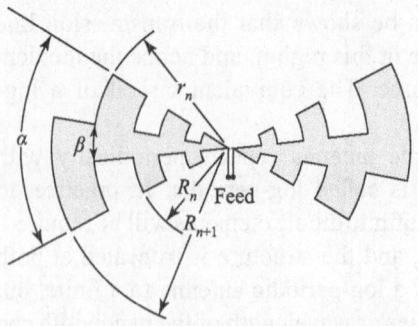

Fig. 18.14 Microstrip log-periodic antenna

$$\tau = f_1/f_2 = R_n/R_{n+1} \tag{18.74}$$

$$\text{and } \chi = r_n/R_{n+1} \tag{18.75}$$

Log-periodic antennas are widely used for TV reception and all-round monitoring.

18.9 Horn Antennas

Unlike an open-ended transmission line that reflects all the incident power at the open termination, an open-ended waveguide behaves as an antenna and radiates some power into space. However, due to a significant amount of reflection, only a small amount of incident energy can be radiated into space in this way. In addition, diffraction around the open edges of the waveguide makes its radiation pattern non-directional. These problems can be overcome if the open end of the waveguide is flared in the form of a horn. Such an antenna is called a horn antenna and was developed by Sir Jagadish Chandra Bose in the year 1897.

In practice, several types of horn antennas are available, such as E-plane horn, H-plane horn, pyramidal horn, conical horn, corrugated horn, and hog horn. The basic structures of some of these horn antennas are shown in Fig. 18.15. In E- and H-plane horns, also known as sectoral horns, the waveguide flares in the respective plane, whereas for a pyramidal horn, the waveguide flares in both planes. A conical horn can be obtained by flaring the open face of a circular waveguide.

For the horn antenna shown in Fig. 18.16, we can write the following equations:

$$\cos(\phi/2) = \frac{L'}{L' + \delta} \qquad (18.76)$$

$$\sin(\phi/2) = \frac{D'}{L' + \delta} \qquad (18.77)$$

where ϕ is called the flare angle.

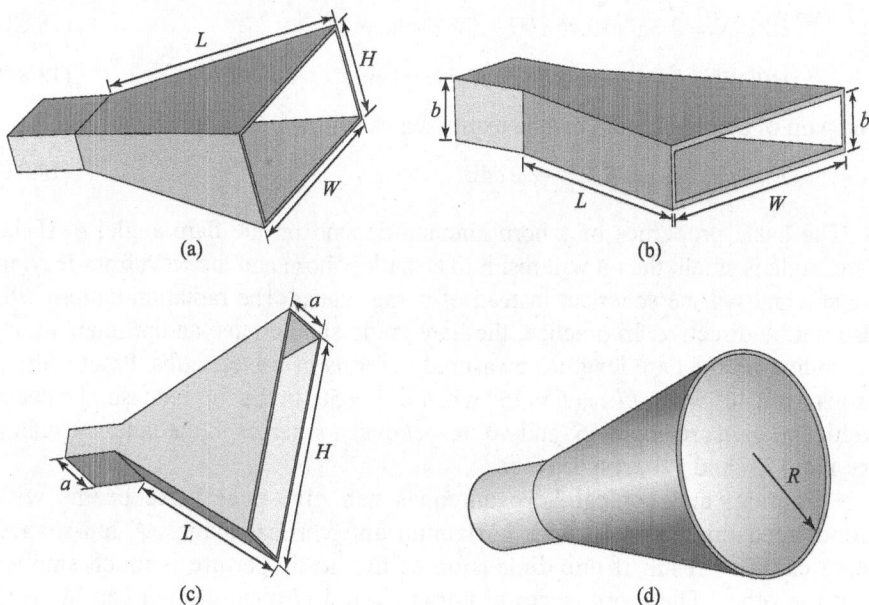

(a)

(b)

(c)

(d)

Fig. 18.15 Horn antennas (a) Pyramidal (b) H-plane (c) E-plane (d) Conical

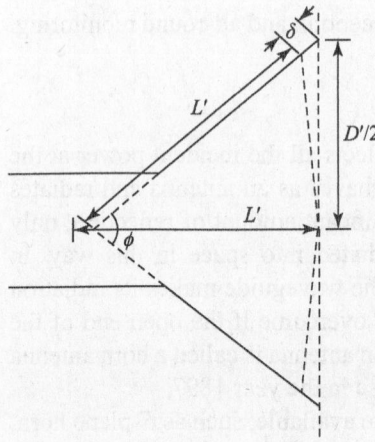

Further, $\quad (L'+\delta)^2 = \dfrac{(D')^2}{4}+(L')^2$

or $\quad (L')^2+\delta^2+2L'\delta = \dfrac{(D')^2}{4}+(L')^2$

or $\quad 2L'\delta = \dfrac{(D')^2}{4}-\delta^2 \approx \dfrac{(D')^2}{4}$ as δ is very small

or $\quad \delta = \dfrac{(D')^2}{8L'}$ (18.78)

In practice, the optimum value of δ and L' are as follows:

Fig. 18.16 Different horn dimensions required for designing optimum horn

$$\delta_0 = \dfrac{L'}{\cos(\phi/2)}-L' \quad (18.79)$$

and $\quad L' = \dfrac{\delta_0 \cos(\phi/2)}{1-\cos(\phi/2)}$ (18.80)

For an optimum designed horn, δ_0 must lie in the range 0.1λ–0.4λ. The directivity and power gain of pyramidal horn can be expressed as follows:

$$D_0 = 8.1+10\log_{10}\left(HW/\lambda^2\right) \text{ dB} \approx 7.5A/\lambda^2 = 7.5HW/\lambda^2 \quad (18.81)$$

and $\quad g_P = 4.5A/\lambda^2 = 4.5HW/\lambda^2$ (18.82)

The E-and H-plane HPBW can be expressed as follows:

$$\text{HPBW} = 2\sin^{-1}\left(0.45\lambda/H\right) \text{ for } E\text{-plane} \quad (18.83)$$

$$\text{HPBW} = 2\sin^{-1}\left(0.63\lambda/W\right) \text{ for } H\text{-plane} \quad (18.84)$$

The gain of a circular horn can be expressed as follows:

$$D_0 = 20\log_{10}\left(\pi R/\lambda\right)-2.82\,\text{dB} \quad (18.85)$$

The basic properties of a horn antenna depend on the flare angle, ϕ. If the flare angle is small, then it will result in a shallow horn and the wavefront leaving the antenna will be spherical instead of being planar. The radiation pattern will also not be directive. In practice, the flare angle should have an optimum value, depending on the flare length L measured in terms of wavelengths. Practically, ϕ varies from $40°$ when $L/\lambda = 6$ to $15°$ when $L/\lambda = 50$. In the former case, the beam width and gain are about $66°$ and 40, respectively, whereas in the latter case they are about $23°$ and 120, respectively.

Pyramidal and conical horn antennas can give pencil-like beams with pronounced directivity in both horizontal and vertical planes. A fan-shaped beam can also result if one dimension of the horn aperture is much smaller than the other. Therefore, sectoral horns (E-and H-plane horns) can be used for this purpose.

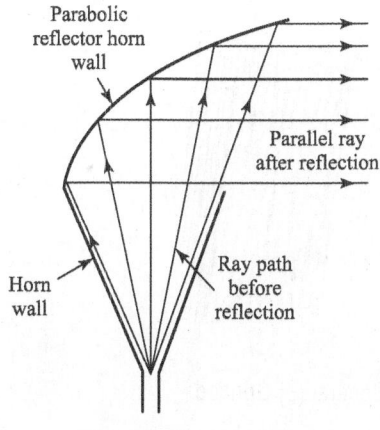

Fig. 18.17 Hog horn

In general, wavefronts leaving the horn are not plane wavefronts. A plane wavefront, however, can be obtained using a hog horn. A hog horn, shown in Fig. 18.17, is basically a combination of a parabolic reflector and a horn antenna. In a hog horn antenna, the receiving point does not move when the antenna is rotated about its axis.

Horn antennas are often used as standard antennas for laboratory measurement and reflector antenna feed.

EXAMPLE 18.6 For an *E*-plane sectoral horn with horn length 6λ and $\delta = 0.22\lambda$, find the flare angle.

Solution Given: $L' = 6\lambda$ and $\delta = 0.22\lambda$

Therefore, $\cos(\phi/2) = \dfrac{L'}{L'+\delta} = \dfrac{6\lambda}{6\lambda + 0.22\lambda} = 0.9646$

or $\quad \phi = 2\cos^{-1}(0.9646) = 30.5815°$

Practice Problem

18.6 For an *H*-plane sectoral horn with horn length 12λ and $\delta = 0.1\lambda$, find the flare angle. [**14.7743°**]

18.10 Reflector Antennas

Reflector antennas basically consist of a feed antenna, such as a horn and dipole, and a passive reflector that collimates or directs the beam in a specified direction. Depending on their geometrical shape, reflectors can be classified as (a) corner reflectors and (b) parabolic reflectors.

18.10.1 Corner Reflector Antenna

A corner reflector generally consists of two flat metallic plates placed at an arbitrary angle, as shown in Fig. 18.18(a). A special case occurs when the angle between the plates becomes equal to 90°. Under such a circumstance, the reflector reflects the signal exactly in the same direction as that of the received signal. Such reflectors are often used as passive targets for RADAR applications. Again due to this reason, it prompts military ship and vehicle designers to minimize the number of sharp corners to reduce their detection by enemy RADARs.

Generally, half-wavelength dipoles or an array of collinear dipoles, placed parallel to the vertex and at a distance *s* away, as shown in Fig. 18.18(a), is used as a feed. If the wavelength is large compared to tolerable physical dimensions, the surface of the corner reflector can be replaced by grid of wires, as shown in Fig. 18.18(b). Such modification reduces wind resistance and also overall system

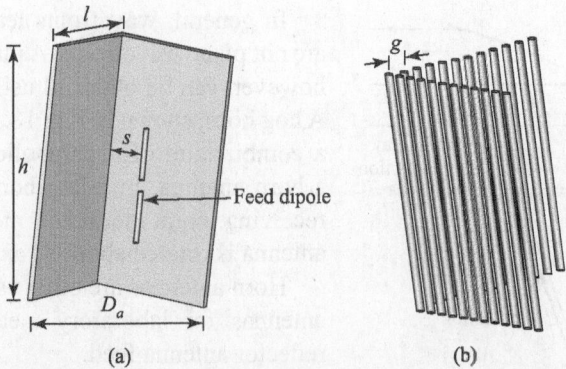

Fig. 18.18 Corner reflector (a) General (b) Gridded

weight. However, to simulate the solid, grid spacing g must be less than 1/10th of the operating wavelength.

For a corner reflector, the aperture dimension (D_a) is kept between one and two wavelengths, whereas the length of the sides (l) is kept twice the distance between the vertex and the feed. For reflectors with smaller included angles, the sides may be much longer. The feed to vertex distance (s) is generally kept at an optimum value between $\lambda/3$ and $2\lambda/3$, because if this spacing becomes too small then the radiation resistance will decrease and become comparable to the loss resistance of the system. This will make the antenna inefficient. On the other hand, if this spacing becomes very large, then the system will produce undesirable multiple lobes, and hence its directional characteristics will be lost. Increasing the dimension of the sides of the reflector, in general, increases the bandwidth and radiation resistance, and does not affect the beam width and directivity greatly. In order to reduce radiation towards the back region from the ends, the height of the reflector (h) is usually taken to be about 1.2–1.5 times greater than the total length of the feed element.

18.10.2 Parabolic Reflector Antenna

From the general property of a parabola, it is known that if a point source is placed at the focal point of a paraboloid reflector, then all the waves coming out of the reflector will be in phase and parallel to each other along the axis of the parabola, after reflection from the paraboloid reflector. Alternatively, if a parallel beam or a plane wave is incident on the parabolic reflector along its axis, then it will be concentrated at the focal point of the paraboloid after reflection from its surface. Waves reflected towards or coming from other directions will cancel each other because of path differences. These properties of a parabola is used to design a highly directive transmitting or receiving antenna, known as a parabolic reflector antenna or a dish antenna. The basic operation of a parabolic dish antenna is shown in Fig. 18.19. It may be noted that in a dish antenna, the paraboloid structure is not the main radiator, but it

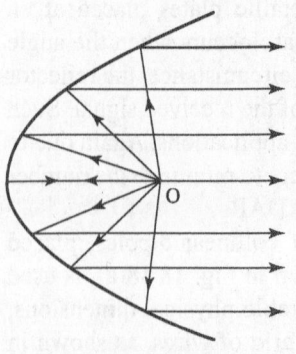

Fig. 18.19 Principle of parabolic reflector

behaves as a reflector. Therefore, all dish antennas need at least an antenna, such as a horn or dipole, to serve as a feed to the reflector.

For a non-directional feed or primary antenna, a parabolic reflector produces the following half-power beam width: $\phi = 70\lambda/D_m$ (in degrees)   (18.86)
where λ is the operating wavelength and D_m is the mouth diameter of the reflector. The beam width between the nulls will be as follows:

$$\phi_0 = 2\phi \tag{18.87}$$

Note If the focus and vertex of a parabola are $n\lambda/4$ apart, where n is an even number, then the direct radiation in the axial direction will be in opposite phase of the reflected wave and will be cancelled.

The maximum power gain of an antenna using a parabolic reflector and resonant half-wave dipole feed, is given by the following equation:

$$g_d = 6(D_m/\lambda)^2 \tag{18.88}$$

A more exact formula is given later. These expressions assume an omni-directional feed antenna, so that the whole reflector is illuminated uniformly. If the intensity of illumination at the reflector falls off at the edges, then Eq. (18.88) is modified as follows: $g_d' = 4.2(D_m/\lambda)^2$ (18.89)

Note For effective operation, the mouth diameter of a parabolic reflector should be at least 10λ. That is why parabolic reflectors are not so common in lower microwave and VHF regions.

It has already been mentioned that a dish antenna requires a feed or a primary antenna as a source of radiation, and the most preferred location of the feed is at the focal point of the reflector. This should result in a narrow beam and hence a high overall directivity of the system. However, in many cases, radiation from the feed tends to spread out in all directions. This radiation is added to the preferred reflected fields from the reflector and thus spoils the overall directivity. In practice, several techniques can be used to avoid this problem, for example, to use a spherical/parabolic sub-reflector that redirects all such radiations back to the main parabolic reflector. This is shown in Fig. 18.20.

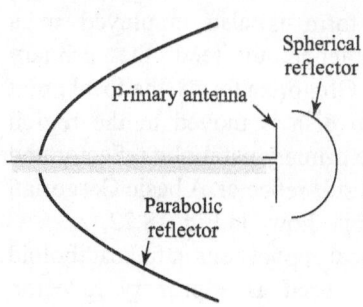

Fig. 18.20 Parabolic reflector with spherical sub-reflector

In addition to these feeds, a Cassegrain feed is also sometimes used in parabolic reflector antennas. In such a feed system, shown in Fig. 18.21, the feed antenna generally lies behind the main reflector, and a hyperboloid surface is used as a sub-reflector. One of the foci of the sub-reflector coincides with the focus of the primary paraboloid reflector. Radiations from the feed antenna fall on the hyperboloid sub-reflector and are reflected from it. Since the focus of the hyperboloid overlaps with that of the parabola, this reflected beam appears

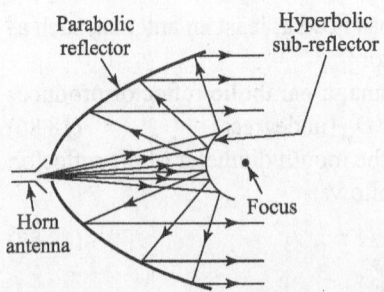

Fig. 18.21 Parabolic reflector with Cassegrain feed

to the primary reflector as coming from a virtual source placed at its focus point. The primary reflector then produces a plane wave, as desired. A Cassegrain feed system is often used for low-noise receivers, where losses in the transmission line are not tolerable. A Cassegrain feed system also has some other advantages: (a) ability to reduce spill over and minor lobe radiation, (b) ability to obtain an equivalent focal length that is much greater than physical length, and (c) ability to place the feed in proper location so that the rays do not suffer any interference. In addition, beam broadening can also be achieved with a Cassegrain feed by moving one of the reflecting surfaces.

It may be noted that a Cassegrain feed system suffers from a serious disadvantage—aperture blockage by the sub-reflector. This can cause a major problem in the antenna operation, especially when small parabolic reflectors are used. In practice, the dimension of the hyperboloid is determined by the distance of the feed horn from it and the mouth aperture dimensions of the feed horn. Since the mouth aperture dimensions of the feed horn depends on the operating wavelength, the size of the hyperboloid is also a function of the operating wavelength. To avoid this problem, a large paraboloid reflector can be used, with the horn placed as close as possible to the sub-reflector. However, this may not be applicable in certain cases. In such cases, a vertically polarized wave source (feed) can be used along with a vertical bar hyperboloid sub-reflector. If the field reflected by the sub-reflector can be changed to horizontal polarization by some mechanism at the primary reflector, then the wave reflected by the primary reflector will pass freely through the vertical bars of the sub-reflector and the problem of aperture blockage will be removed. Like a Cassegrain feed system, the feed horn in an ordinary feed system also causes some aperture blocking, which is negligible if the horn is placed at the focus of the parabolic reflector.

Note In addition to its classical forms, a Cassegrain feed system also employs a variety of main reflector and sub-reflector surfaces such as concave, convex, and flat shapes.

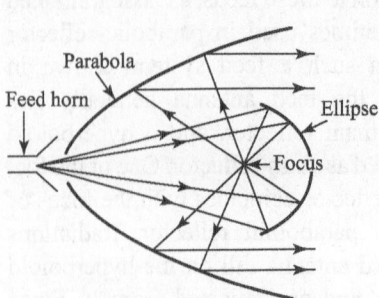

Fig. 18.22 Parabolic sub-reflector with Gregorian feed

In addition to the Cassegrain form, Gregorian form is also employed in a reflector antenna to feed the primary reflector. In Gregorian forms, the focal point of the main dish is moved in the region between the primary parabolic reflector and an elliptical sub-reflector. A basic Gregorian feed system is shown in Fig. 18.22.

In practical applications, a full paraboloid is often not used as a primary reflector. Instead, several other types of paraboloids, shown in Fig. 18.23, can be used. Each of

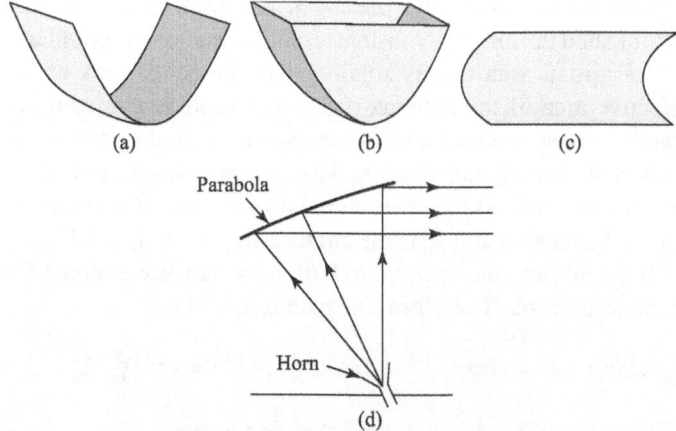

Fig. 18.23 Different types of paraboloids (a) Cut parabola (b) Pill box
(c) Cylindrical parabola (d) Offset paraboloid

them has their own advantage, but at the cost of a non-directional beam in one of the plane. For example, for a pill box reflector the beam is very narrow in the *H*-plane but not in the *E*-plane. This property may be a disadvantage in some applications but may be advantageous in applications like ship to ship communication/RADAR, where the azimuth directivity must be very good but the elevation directivity is immaterial. In contrast to a pill box reflector, a cylindrical paraboloid reflector produces a narrow beam in the *E*-plane but a wide beam in the *H*-plane.

Note The field reflected by the main antenna may be coupled to the feed antenna and may thus cause antenna mismatch. To get rid of it, offset feed techniques are often used. An offset feed also solves the problem of aperture blockage.

In addition to its several advantages, dish antenna systems also have the following limitations:

1. Theoretically, the beam of the reflector system should be a pencil beam, but in practice, the main lobe is surrounded by several side lobes.
2. In practice, the fabricated reflector is not a true paraboloid; rather, its surface deviates from the required paraboloid shape. For a satisfactory operation, this deviation should not exceed $\lambda/16$.
3. In parabolic reflector antennas, diffractions also occur around the edges of the reflector. This adds to the diffraction caused at the horn supports.
4. The primary size of the feed antenna influences the beam width.
5. Since the feed antenna is not a true point source, it cannot be placed precisely at the focus of the parabolic reflector. This causes aberration, that is, broadening of main beam and reinforcement of side lobes.
6. Since the primary antenna does not radiate evenly at the reflector, distortions occur in the system. This is prominent if a dipole is used as a primary antenna, since it radiates more in one plane than in the other. Such a problem can be avoided if a circular horn is used as a feed, but even then, because of gradual

tapering of illumination towards the edges, the entire surface of the paraboloid is not illuminated uniformly. Non-uniform illumination of the antenna aperture results in a virtual area that is smaller than the actual area of the antenna. This effective area of the antenna is called a capture area, and is calculated based on the power received and its comparison with the power density of the received signal. The capture area simulates an antenna surface that is uniformly illuminated and produces the same signal power at the primary antenna. If A is the actual antenna area and A_e is the capture area, then $A_e = kA$ (18.90) where k is a constant, and is equal to 0.65 for a half-wave dipole feed and 0.9 for a mattress antenna. Therefore, the gain is as follows:

$$g_d = 4\pi A_e/\lambda^2 = 4\pi kA/\lambda^2 = 4\pi k\pi D_m^2/(4\lambda^2) = k\pi^2 D_m^2/\lambda^2 \qquad (18.91)$$

For $k = 0.65$, Eq. (18.91) reduces to the following form:

$$g_d = 0.65\pi^2 D_m^2/\lambda^2 = 6.4152(D_m/\lambda)^2 \qquad (18.92)$$

For a uniformly illuminated circular aperture $k = 1$, and we get the following equation:

$$g_d = \pi^2 D_m^2/\lambda^2 = 9.8696(D_m/\lambda)^2 \qquad (18.93)$$

For uniform illumination and a large circular aperture, we can write the following expressions:

$$FNBW = 140\lambda/D_m \text{ (degree)} \qquad (18.94)$$

$$HPBW = 58\lambda/D_m \text{ (degree)} \qquad (18.95)$$

Note The distance of a parabola from its focus is not uniform; this leads to amplitude tapering at the edges. To make power distribution at the reflector aperture uniform, a special feed arrangement is required.

EXAMPLE 18.7 A parabolic dish antenna has mouth diameter 12λ at 3 GHz. Assuming uniform illuminations, calculate the directive gain, FNBW, and HPBW.

Solution Given: $D_m = 12\lambda$ and $f = 3$ GHz

Therefore, the directive gain is as follows:

$$g_d = 9.8696(D_m/\lambda)^2 = 9.8696(12)^2 = 1421.2224 = 31.5266 \text{ dB}$$

The following is the FNBW: $FNBW = 140\lambda/D_m = 140/12 = 11.6667°$
The HPBW is calculated as follows: $HPBW = 58\lambda/D_m = 58/12 = 4.8333°$

Practice Problem

18.7 A parabolic dish antenna has mouth diameter 23λ at 6 GHz. Assuming uniform illuminations, calculate the directive gain, FNBW, and HPBW.

[**37.1775 dB, 6.087°, 2.5217°**]

18.11 Microstrip Patch Antennas

A microstrip antenna is basically a very thin metallic patch placed above a ground plane. Such antennas are low weight, highly compact, and efficient, and are widely used at microwave frequencies for low-power applications. In practice, out of different types of patch antennas available, rectangular and circular patches are simpler and also common. A rectangular patch antenna, with a rectangular patch of length L and width W, is shown in Fig. 18.24. The thickness of the dielectric is kept below $0.02\lambda_g$, where λ_g is the guided wavelength in the dielectric.

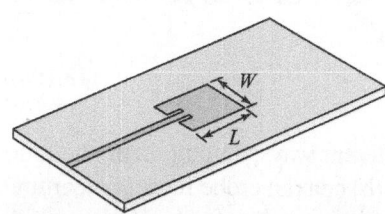

Fig. 18.24 Microstrip patch antenna

Radiation edges of a microstrip patch antenna can be assumed as radiating slots connected to each other by a microstrip transmission line. Radiation conductance and susceptance of such a slot can be expressed as follows:

$$G_r = \begin{cases} \dfrac{W^2}{90\lambda^2} & \text{for } W<\lambda \\[2mm] \dfrac{W^2}{120\lambda^2} & \text{for } W>\lambda \end{cases} \tag{18.96}$$

and

$$B_r = \frac{k(\Delta l)\sqrt{\varepsilon_{\text{eff}}}}{Z_0} \tag{18.97}$$

where

$$Z_0 = \frac{120\pi h}{W\sqrt{\varepsilon_{\text{eff}}}} \tag{18.98}$$

$$\varepsilon_{\text{eff}} = \frac{\varepsilon_r+1}{2} + \frac{\varepsilon_r-1}{2\sqrt{1+(12h/W)}} \tag{18.99}$$

$$\Delta l = 0.412h\left(\frac{\varepsilon_{\text{eff}}+0.3}{\varepsilon_{\text{eff}}-0.258}\right)\frac{(W/h)+0.264}{(W/h)+0.8} \tag{18.100}$$

If Y_{slot} is the slot admittance, then the input admittance of the antenna can be calculated as follows:

$$Y_{\text{in}} = Y_{\text{slot}} + Y_0\frac{Y_{\text{slot}}+jY_0\tan\{\beta(L+2\Delta l)\}}{Y_0+jY_{\text{slot}}\tan\{\beta(L+2\Delta l)\}} \tag{18.101}$$

where

$$\beta = 2\pi\sqrt{\varepsilon_{\text{eff}}}/\lambda \tag{18.102}$$

At resonance, $L+2\Delta l = \dfrac{\lambda_g}{2} = \dfrac{\lambda}{2\sqrt{\varepsilon_{\text{eff}}}}$ (18.103)

Substituting Eq. (18.103) into Eq. (18.101), it can be shown that,

$$Y_{\text{in}} = 2G_r \tag{18.104}$$

From Eq. (18.103), the resonance frequency can be calculated as follows:

$$f_r = \frac{c}{2(L+2\Delta l)\sqrt{\varepsilon_{\text{eff}}}} \tag{18.105}$$

In a patch antenna, W is not a very critical factor; however, its value can be obtained using the following equation:

$$W = \frac{c}{2f_r\sqrt{(\varepsilon_r+1)/2}} \tag{18.106}$$

Microstrip patch antennas can be fed in different ways; among them, the four most popular feeds are (a) microstrip line feed, (b) coaxial probe feed, (c) aperture coupling, and (4) proximity coupling. Both the microstrip line feed and the coaxial probe feed have inherent asymmetries, and hence generate higher-order modes, which, in turn, produce cross-polarized radiation. To overcome some of these problems, an aperture coupling feed is used.

In addition to rectangular patches, circular patches are also widely used as radiating elements. To design a circular patch antenna, the following equations can be used:

$$F = \frac{8.791\times10^9}{f_r\sqrt{\varepsilon_r}} \tag{18.107}$$

and

$$a = \frac{F}{\left[1+\dfrac{2h}{\pi\varepsilon_r F}\left\{\ln\left(\dfrac{\pi F}{2h}\right)+1.7726\right\}\right]^{1/2}} \tag{18.108}$$

where h is in cm and f_r is in Hz.

Rectangular and circular patch antennas generally produce linearly polarized waves with a conventional feed arrangement. However, using modified feed arrangements, circular and elliptical polarizations can also be obtained.

EXAMPLE 18.8 Design a circular patch antenna at 8 GHz. Assume that the substrate has a dielectric constant of 2.2 and thickness of 0.381 mm.

Solution Given: $f = 8$ GHz, $\varepsilon_r = 2.2$, and $h = 0.381$ mm $= 0.0381$ cm

Therefore,

$$F = \frac{8.791\times10^9}{f_r\sqrt{\varepsilon_r}} = \frac{8.791\times10^9}{8\times10^9\sqrt{2.2}} = 0.7409$$

Hence, the radius of the circular patch will be as follows:

$$a = \frac{F}{\left[1+\dfrac{2h}{\pi\varepsilon_r F}\left\{\ln\left(\dfrac{\pi F}{2h}\right)+1.7726\right\}\right]^{1/2}}$$

or

$$a = \frac{0.7409}{\left[1+\dfrac{2\times0.0381}{2.2\pi\times0.7409}\left\{\ln\left(\dfrac{0.7409\pi}{2\times0.0381}\right)+1.7726\right\}\right]^{1/2}} = 0.7138 \text{ cm}$$

Practice Problem

18.8 Design a circular patch antenna at 10 GHz. Assume that the substrate has a dielectric constant of 2.2 and thickness of 1.59 mm. **[0.5249 cm]**

18.12 Lens Antennas

The geometrical theory of optics describes that diverging rays, coming out of an optical source, placed at the focus of a convex lens, become parallel after passing through it, and alternatively parallel rays, coming from an optical source placed at infinity, get collimated at the focal point of a convex lens after passing through it. This property of light wave is used in microwave to design lens antennas. A lens antenna thus can be used as a plane wave source in the transmitting mode and as a collimator in the receiving mode, as shown in Fig. 18.25.

The function of a lens antenna is based on the property that electromagnetic waves travel relatively slowly through a material of a higher dielectric constant. This implies that for a specified time, the ray passing through the centre region of a dielectric lens will travel less distance than those travelling through the edges of a convex lens. Therefore, if a spherical wave approaches the lens as shown in Fig. 18.26, it will be straightened by the lens. Lens antennas are generally made of polystyrene materials, though other materials can also be used for their fabrication.

One of the major drawbacks of a lens antenna is its excessive thickness, especially below 10 GHz. To avoid this problem, *zoning* or *stepping* is often done, as shown in Fig. 18.27. This also reduces lens weight and total absorption of the field passing through it. However, it is associated with a reduced antenna bandwidth.

Note Unstepped dielectric lens antennas can provide 12% bandwidth, whereas zoned lens antennas provide only 5% bandwidth.

It may be noted that a lens antenna cannot directly radiate electromagnetic energy, but it can correct a curved waveform or collimate electromagnetic energy at a point. Therefore, lens antennas are always associated with a primary antenna like a horn. They are often directly mounted at the face of the horn to correct the curved wavefront there. A lens antenna can also be used in place of a parabolic reflector at higher frequencies. Such a system does not face any aperture blockage due to the feed.

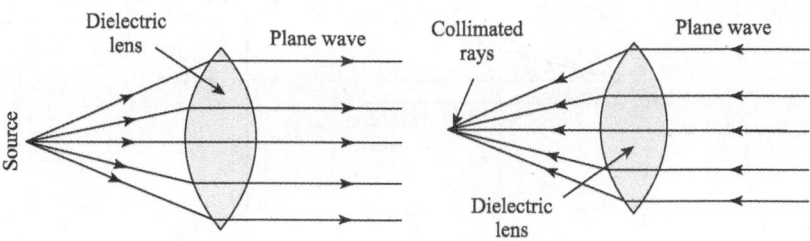

Fig. 18.25 Properties of dielectric lens antenna

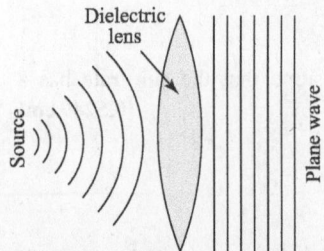

Fig. 18.26 Generation of plane wave using lens antenna

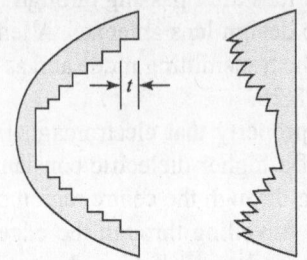

Fig. 18.27 Zoned lens antenna

In general, lens antennas can be divided into two categories: delay lenses and fast lenses. The electrical path length is increased by the lens medium in delay lenses, whereas the reverse is true for fast lenses. Dielectric lenses and H-plane metal plate lenses are examples of the delay-type lenses, whereas E-plane metal plate lenses is an example of fast lenses. Dielectric lenses also may be divided into two groups: (a) lenses constructed of non-metallic dielectrics and (b) lenses constructed of metallic or artificial dielectrics.

18.12.1 Non-metallic Dielectric Lens Antenna

To illustrate the function of such a lens, let us consider Fig. 18.28. To achieve a plane wave at the plane AB, the electrical path lengths of the waves reaching this plane from the source O must be equal. Therefore,

$$OP + PP' = OQ + QQ' + Q'Q'' \tag{18.109}$$

Now $\quad PP' = Q'Q'' \tag{18.110}$

Therefore, $OP = OQ + QQ' \tag{18.111}$

If c is the velocity of wave in air and v the velocity of wave in the lens medium, then from the figure and Eq. (18.111), we can write the following expressions:

$$\frac{R}{c} = \frac{L}{c} + \frac{x}{v}$$

or $\quad R = L + x\left(\dfrac{c}{v}\right) = L + nx \tag{18.112}$

where n is the refractive index of the lens medium and $OQ = L$. Now, from the triangle OPQ',

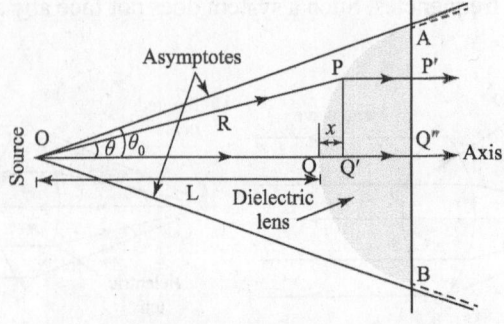

Fig. 18.28 Demonstration of equal path length

$$\cos(\theta) = (L+x)/R$$

or $\qquad x = R\cos(\theta) - L$ $\qquad\qquad$ (18.113)

Substituting Eq. (18.113) into Eq. (18.112), we get the following relation:

$$R = L + n\{R\cos(\theta) - L\}$$

or $\qquad R\{1 - n\cos(\theta)\} = L(1-n)$

or $\qquad R = \dfrac{L(1-n)}{1 - n\cos(\theta)} = \dfrac{L(n-1)}{n\cos(\theta) - 1}$ $\qquad$ (18.114)

Equation (18.114) defines the curvature of a plano-convex lens. In general, this equation represents a hyperbola, with the asymptotes of the hyperbola at an angle θ_0 with respect to the axis. The value of θ_0 can be determined by letting $R \to \infty$. For this assumption,

$$n\cos(\theta_0) - 1 \approx 0$$

or $\qquad \theta_0 = \cos^{-1}(1/n)$ $\qquad\qquad$ (18.115)

For a point source at O, the 3D lens surface should be a spherical hyperbola, whereas for an in-phase line source, the lens surface should be a cylindrical hyperbola.

For zoned lenses, shown in Fig. 18.27, the thickness of dielectric can be expressed as follows: $t = \lambda/(n-1)$ $\qquad\qquad$ (18.116)

It may be noted that since there is an impedance mismatch at the lens surfaces, reflection results from them. These reflections may directly approach the feed and may thus cause an interference problem. In practice, reflections from the convex surface does not return to the feed, except from points at or near the axis, as shown in Fig. 18.29. Thus, it does not cause a major problem. The main problem is associated with reflections at the flat ends, which directly collimate at the feed, as shown in the same figure. In general, the reflection coefficient can be expressed as follows:

$$\Gamma = \frac{n-1}{n+1}$$ $\qquad\qquad$ (18.117)

Thus, to reduce the effect of reflection, n must be close to unity. Reflections can also be minimized by adding a quarter-wavelength plate of refractive index $\sqrt{n}$ at the flat lens surface. This plate behaves as a quarter-wave transformer between the lens and air, and thus reduces reflection from the flat surface, as shown in Fig. 18.30(a).

Another alternative way is to tilt the lens slightly, as shown in Fig. 18.30(b), so that the reflected waves collimate at some offset point.

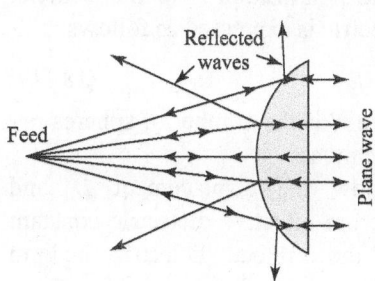

Fig. 18.29 Reflections at lens surfaces

18.12.2 Artificial Dielectric Lens Antennas

In addition to the use of ordinary dielectrics as lens materials, metallic or artificial dielectrics can also be used. An artificial dielectric consists of discrete metal particles of macroscopic size. Such particles are

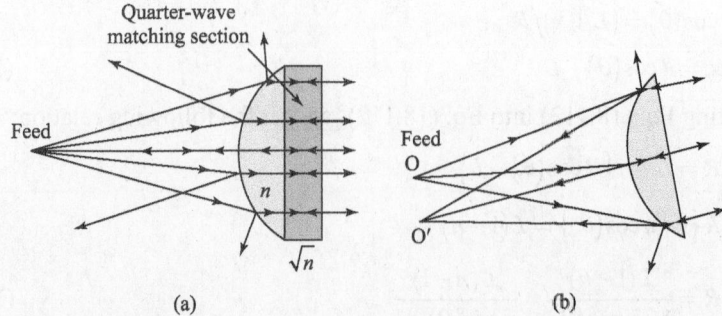

Fig. 18.30 Methodologies used to avoid reflection interference
(a) Quarter-wavelength matching section (b) Tilting lens

generally small compared to the operating wavelength to avoid resonance effect and are placed less than a wavelength apart to reduce diffraction effect.

The basic structure and operation of an artificial dielectric lens antenna, made using metal spheres, are shown in Fig. 18.31(a). The EM wave from the feed antenna induces surface current on the metal spheres while passing through the lens. The metal sphere thus resembles the oscillating molecular dipoles of ordinary dielectrics. It may be noted that disks, strips, or rods can also be used in place of metal spheres for the fabrication of an artificial dielectric. In practice, disks are often preferred over spheres due to their less weight. If strips are used, they may be continuous in a direction perpendicular to the electric field, as shown in Fig. 18.31(b).

Although designing of an artificial dielectric lens is similar to that of an ordinary dielectric lens, determination of the refractive index of the artificial dielectric material is difficult. However, this can be determined experimentally using a slab of this artificial dielectric or can be calculated approximately. To do this, let us consider that an initially uncharged conducting sphere has been placed in an electric field $\vec{E}$. As a result, positive and negative charges will be induced in the spheres. These induced charges may be represented by point charges $+q$ and $-q$ separated by a distance l ($l \ll r$). This simulates an electric dipole of dipole moment ql. Since $r \gg l$, the potential due to the dipole is given by the following equation:

$$V = \frac{ql\cos(\theta)}{4\pi\varepsilon_0 r^2} \qquad (18.118)$$

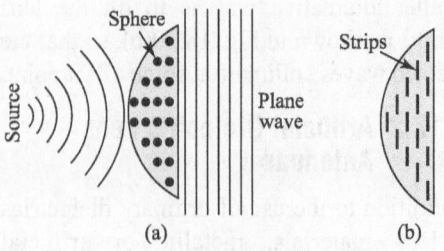

Fig. 18.31 Artificial dielectric lens antennas
(a) Made of metallic spheres
(b) Made of metallic strips

where polarization $\vec{P}$ of the artificial dielectric is expressed as follows:

$$\vec{P} = Nq\vec{l} \qquad (18.119)$$

where N is the number of spheres per cubic metre.

Now the displacement $\vec{D}$ and therefore effective dielectric constant ε of the artificial dielectric medium can be expressed as follows:

$$\vec{D} = \varepsilon\vec{E} = \varepsilon_0\vec{E} + \vec{P} \qquad (18.120)$$

or $\qquad \varepsilon = \varepsilon_0 + \dfrac{\vec{P}}{\vec{E}}$ (18.121)

Substituting Eq. (18.119) into Eq. (18.121), we get the following relation:

$$\varepsilon = \varepsilon_0 + N\dfrac{\vec{ql}}{\vec{E}} \qquad (18.122)$$

Equation (18.122) reveals that if the number of spheres per unit volume and the dipole moment of one sphere per unit applied field are known, then the effective dielectric constant can be determined. To determine the dipole moment per unit applied electric field, we will start with the following equation:

$$\vec{E} = -\vec{\nabla}V \qquad (18.123)$$

Therefore, in a uniform field, the potential is given by the following equation:

$$V = -\int_0^r E\cos(\theta)\,dr = -Er\cos(\theta) \qquad (18.124)$$

where θ is the angle between the radius vector and the field. Therefore, the potential outside the sphere can be expressed as follows:

$$V_0 = -Er\cos(\theta) + \dfrac{ql\cos(\theta)}{4\pi\varepsilon_0 r^2} \qquad (18.125)$$

At the surface of the sphere, the potential must be zero, and hence we can write that

$$0 = -Ea\cos(\theta) + \dfrac{ql\cos(\theta)}{4\pi\varepsilon_0 a^2} \qquad (18.126)$$

or $\qquad \dfrac{ql}{E} = 4\pi\varepsilon_0 a^3$ (18.127)

where a is the radius of the sphere. Substituting Eq. (18.127) into Eq. (18.122), we get the following expression:

$$\varepsilon = \varepsilon_0 + 4\pi\varepsilon_0 Na^3 \qquad (18.128)$$

or $\qquad \varepsilon_r = 1 + 4\pi Na^3$ (18.129)

where ε_r is the effective relative permittivity of the artificial dielectric.

It can be shown that the effective relative permeability of an artificial dielectric of a conducting sphere can be as follows:

$$\mu_r = 1 - 2\pi Na^3 \qquad (18.130)$$

Therefore, the effective index of refraction of the artificial dielectric of a conducting sphere can be expressed as follows:

$$n = \sqrt{\varepsilon_r \mu_r} = \sqrt{\left(1 + 4\pi Na^3\right)\left(1 - 2\pi Na^3\right)} \qquad (18.131)$$

For disk or strip-type artificial dielectrics, the relative permeability can be approximated as unity, and hence the refractive index of artificial dielectrics made from such particles can be expressed as follows:

$$n = \sqrt{1 + 4\pi N a^3} \qquad (18.132)$$

Artificial dielectric lens antennas are of lesser weight than ordinary dielectric lens antennas.

18.12.3 E-plane Metal Plate Lens Antennas

The dielectric lens antennas discussed earlier rely on the fact that wave propagates slowly when passing through it. The opposite is true for E-plane metal plate lens antennas. The velocity of the TE_{10} wave propagating through the space between two parallel metal plates separated by a distance b and oriented parallel to the E-plane can be expressed as follows:

$$v = c \bigg/ \sqrt{1 - \left(\frac{\lambda}{2b}\right)^2} \qquad (18.133)$$

where we have assumed that $b \geq \lambda/2$. Therefore, the refractive index of the medium between the parallel plates can be expressed as follows:

$$n = \frac{c}{v} = \sqrt{1 - \left(\frac{\lambda}{2b}\right)^2} \qquad (18.134)$$

Equation (18.134) reveals that $n < 1$ or $c < v$. This property of parallel metal plates can be used to design a lens antenna; however, this time, since the wave is propagating at a velocity higher than its free-space velocity, geometry of the lens should be plano-concave rather than being plano-convex, as shown in Fig. 18.32. The figure shows that to obtain a plane wave at the output, we must have

$$OPP' = OQQ'' \text{ or } \frac{R}{c} = \frac{L}{c} - \frac{x}{v} = \frac{L}{c} - \frac{L - R\cos(\theta)}{v}$$

or $\qquad R = L - \frac{c}{v}\{L - R\cos(\theta)\} = L - n\{L - R\cos(\theta)\}$

or $\qquad R\{1 - n\cos(\theta)\} = L(1 - n)$

or $\qquad R = \frac{L(1 - n)}{1 - n\cos(\theta)} \qquad (18.135)$

Since the velocity of the wave inside the metal plate depends on the plate separation b, two types of E-plane metal lens can be designed: (a) lens designed

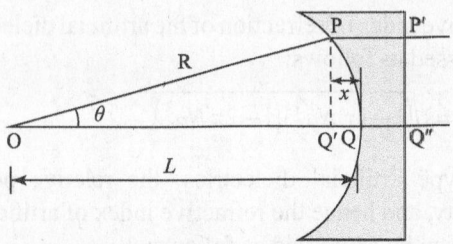

Fig. 18.32 Demonstration of equal path length for plano-concave lens

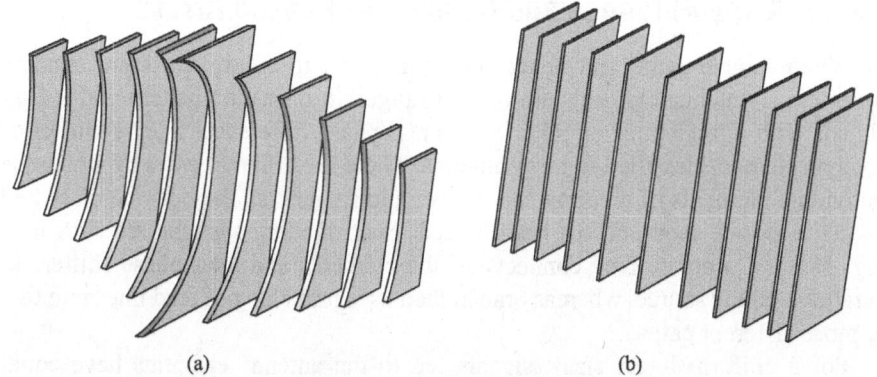

Fig. 18.33 *E*-plane metal lens antenna (a) Varying strip thickness
(b) Varying plate separations

with metal plates of varying thickness but same spacing and (b) lens designed with metal plates of same thickness but varying spacing. These are shown in Fig. 18.33.

It should be noted that since the refractive index n depends on the operating wavelength λ, E-plane metal plate lenses are of narrow bandwidth. The bandwidth of such lenses can be expressed as follows:

$$B = \frac{2n}{\left(1-n^2\right)} \frac{\delta_\lambda}{t_\lambda} \qquad (18.136)$$

where δ_λ is the maximum tolerable path difference in free-space wavelengths and t_λ is the thickness of lens plate at the edge of lens in free-space wavelengths. In contrast to dielectric lenses, zoning in E-plane metal plate lenses decreases its frequency sensitivity. Therefore, zoning is desirable for E-plane metal plate lens, both to save weight and to increase the bandwidth of such lenses. The bandwidth of a zoned E-plane metal plate lens can approximately be expressed as follows:

$$B = \frac{50n}{1+Kn} \qquad (18.137)$$

where K is the number of zones, the zone on the axis of the lens being counted as the first zone.

18.13 Slot Antennas

A slot milled on the broad or narrow wall of a waveguide behaves as an antenna at microwave frequencies. Such an antenna is called a slot antenna. A single slot, however, is seldom used as an antenna. Instead, an array of slots is used to achieve the desired radiation pattern. Such an antenna is called a slot array antenna. A properly designed slot array antenna can radiate 90% of the feed power and has a main lobe surrounded by very narrow side lobes. A typical slot array antenna is shown in Fig. 18.34.

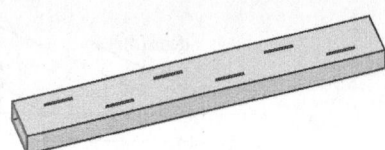

Fig. 18.34 Waveguide slot array antenna

18.14 Array Antennas and Concept of Phased Arrays

To obtain higher gain, and hence directivity, and to control different antenna parameters, antenna elements are often arranged in a specific pattern and fed by signals with a predefined amplitude ratio and phase difference. Such a cluster of antenna elements is called an array antenna. There are different ways by which the individual elements of an array can be fed, and in general, they can be classified as (a) corporate feed and (b) branch feed. Such feeding systems are shown in Fig. 18.35. Corporate feed connects all the elements and their phase shifters in parallel with the source, whereas branch feed uses branching of feed lines into two or more different paths.

For a uniform linear array, signals fed to the antenna elements have equal magnitude and uniform progressive phase shift. Therefore, for an N-element array, the total field can be expressed as follows:

$$E_T = E_0 \left| 1 + e^{j\psi} + e^{j2\psi} + e^{j3\psi} + \ldots + e^{j(N-1)\psi} \right| \tag{18.138}$$

where $\quad \psi = \beta d \cos(\phi) + \alpha \tag{18.139}$

where α is the progressive phase shift between the elements and $\beta d \cos(\phi)$ is the phase shift due to the path difference, as shown in Fig. 18.36. Equation (18.138) may be viewed as a geometric progression and can be re-written in the following form:

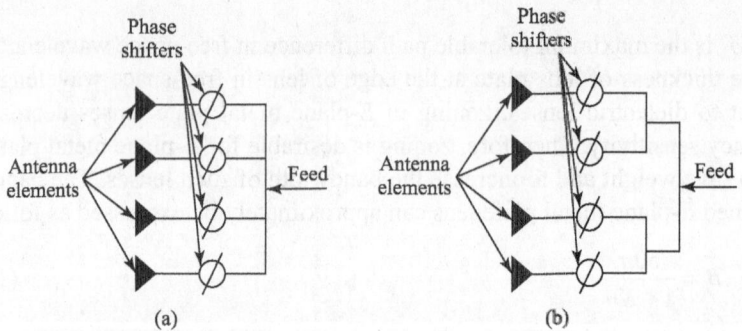

Fig. 18.35 Array feeding (a) Corporate feed (b) Branch feed

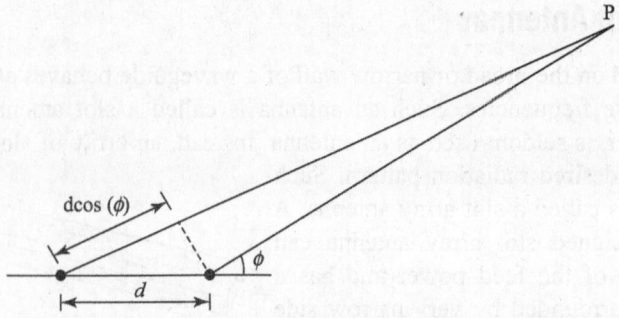

Fig. 18.36 Demonstration of path difference between observation point and array elements

$$\frac{E_T}{E_0} = \left|\frac{1-e^{jN\psi}}{1-e^{j\psi}}\right| = \left|\frac{e^{jN\psi/2}\left(e^{-jN\psi/2}-e^{jN\psi/2}\right)}{e^{j\psi/2}\left(e^{-j\psi/2}-e^{j\psi/2}\right)}\right|$$

or
$$\frac{E_T}{E_0} = \left|\frac{e^{j(N-1)\psi/2}\left(-2j\right)\sin\left(N\psi/2\right)}{\left(-2j\right)\sin\left(\psi/2\right)}\right| = \left|\frac{\sin\left(N\psi/2\right)}{\sin\left(\psi/2\right)}\right| \tag{18.140}$$

Equation (18.140) reveals that the maximum value of this expression is equal to N and occurs when ψ becomes zero. This corresponds to the principal maximum of the array. Recalling the condition of principal maxima, we can write the following equation: $\psi = \beta d \cos(\phi) + \alpha = 0$

or
$$\cos(\phi) = -\frac{\alpha}{\beta d} \tag{18.141}$$

Equation (18.141) reveals that by changing the progressive phase shift between the antenna elements, direction of the major lobe can be changed. This is the principle of phased array antennas, where beam steering is accomplished by changing the progressive phase shift between antenna elements, instead of mechanically rotating the whole antenna structure. The movement of direction of the main lobe by changing the progressive phase shift is known as electronic scanning.

Note If the inter-element spacing, d, is equal to or greater than λ, there may be more than one principal maximum.

The direction of maximum radiation is perpendicular to the line of the array at $\phi = 90°$ for a broadside array, whereas it is along the line of the array at $\phi = 0°$ for an end-fire array. Substituting these values of ϕ in Eq. (18.141), we get $\alpha = 0°$ for a broadside array and $\alpha = -\beta d$ for an end-fire array.

Equation (18.140) becomes zero when

$$N\psi/2 = \pm k\pi \qquad k = 1, 2, 3, \ldots \tag{18.142}$$

and these values correspond to the nulls of the pattern. The first null occurs when $k = 1$, and we get the following expression: $\psi_{\text{first null}} = \pm 2\pi/N$ (18.143)
The following is the condition for the secondary maxima:

$$N\psi/2 = \pm(2m+1)\pi/2 \qquad m = 1, 2, 3, \ldots \tag{18.144}$$

The first secondary maximum corresponds to $m = 1$, and we get the following relation: $\psi_{\text{first maxima}} = 3\pi/N$ (18.145)
The amplitude of the first secondary lobe can be obtained from Eq. (18.140) after substituting Eq. (18.145), as follows:

$$\left|\frac{1}{\sin(\psi/2)}\right| = \left|\frac{1}{\sin(3\pi/2N)}\right| \approx \frac{2N}{3\pi} \qquad \text{for large } N \tag{18.146}$$

The amplitude ratio of the first secondary maximum to the principal maximum is, therefore, $\dfrac{2}{3\pi} = 0.212$. Therefore, the first side lobe is about 13.5 dB below the principal maximum.

Beam width of the principal lobe can be calculated as twice the angle between the principal maximum and the first null. If for a broadside array, the angle between the principal maximum and the first null is assumed to be $\Delta\phi$, then Eq. (18.139) can be written as follows:

$$\psi_{\text{first null}} = \beta d \cos(\phi_{\text{null}}) + \alpha = \beta d \cos\left(\frac{\pi}{2} + \Delta\phi\right) \text{ as } \alpha = 0°$$

and principal maxima occur at $\phi = 90°$

or $\qquad \cos\left(\dfrac{\pi}{2} + \Delta\phi\right) = \dfrac{\psi_{\text{first null}}}{\beta d} = \dfrac{2\pi}{N}\dfrac{\lambda}{2\pi d} = \dfrac{\lambda}{Nd}$ \hfill (18.147)

where we have used Eq. (18.143). If $\Delta\phi$ is small then Eq. (18.147) can approximately be written as follows:

$$\Delta\phi = \frac{\lambda}{Nd} \hspace{4cm} (18.148)$$

and the width of the principal lobe can be expressed as follows:

$$2\Delta\phi = \frac{2\lambda}{Nd} \hspace{4cm} (18.149)$$

Similarly, for the end-fire array we can write the following relation:

$$\psi_{\text{first null}} = \beta d \cos(\phi_{\text{null}}) + \alpha = \beta d \cos(\Delta\phi) - \beta d \text{ as } \alpha = -\beta d$$

and principal maxima occur at $\phi = 0°$

or $\qquad \psi_{\text{first null}} = \beta d \{\cos(\Delta\phi) - 1\}$

or $\qquad \cos(\Delta\phi) - 1 = \dfrac{\psi_{\text{first null}}}{\beta d} = -\dfrac{2\pi}{N}\dfrac{\lambda}{2\pi d} = -\dfrac{\lambda}{Nd}$ \hfill (18.150)

If $\Delta\phi$ small, then Eq. (18.150) can be approximated as follows:

$$\cos(\Delta\phi) - 1 = 1 - 2\sin^2\left(\frac{\Delta\phi}{2}\right) - 1 = -2\sin^2\left(\frac{\Delta\phi}{2}\right) = -\frac{\lambda}{Nd}$$

or $\qquad 2\dfrac{(\Delta\phi)^2}{4} = \dfrac{(\Delta\phi)^2}{2} = \dfrac{\lambda}{Nd}$

or $\qquad \Delta\phi = \sqrt{\dfrac{2\lambda}{Nd}}$

or $\qquad 2\Delta\phi = 2\sqrt{\dfrac{2\lambda}{Nd}}$ \hfill (18.151)

EXAMPLE 18.9 An array consists of 10 isotropic radiators with an inter-element spacing of 0.5λ. Calculate the progressive phase shift so that the beam points at $\phi = 60°$. In addition, calculate the FNBW of such an array.

Solution Given: $N = 10$, $d = 0.5\lambda$, and $\phi = 60°$

For principal maxima, we must have

$$\psi = \beta d \cos(\phi) + \alpha = 0$$

or $\quad \alpha = -2\pi \times 0.5\lambda \times \cos(60°)/\lambda = -1.5708 \text{ rad} = -90°$

Now, let us assume that the first null is at an angle $\Delta\phi$ with respect to principal maxima. Therefore,

$$\psi_{\text{first null}} = \beta d \cos(\phi_{\text{null}}) + \alpha$$

or $\quad \psi_{\text{first null}} = \dfrac{2\pi}{\lambda} \times 0.5\lambda \times \cos(60° + \Delta\phi) - 1.5708 = \pi \cos(60° + \Delta\phi) - 1.5708$

Further, we know that $\psi_{\text{first null}} = \pm 2\pi/N$

Considering the negative value and equating both equations, we get the following relation:

or $\quad \pi \cos(60° + \Delta\phi) - 1.5708 = -\dfrac{2\pi}{10}$

or $\quad \cos(60° + \Delta\phi) = \dfrac{1}{\pi}\left(-\dfrac{2\pi}{10} + 1.5708\right) = 0.3$

or $\quad 60° + \Delta\phi = \cos^{-1}(0.3) = 72.5423°$ or $\Delta\phi = 12.5423°$

Therefore, the FNBW is $25.0846°$

Practice Problem

18.9 An array consists of 20 isotropic radiators with an inter-element spacing of 0.5λ. Calculate the progressive phase shift so that the beam points at $\phi = 30°$. In addition, calculate the FNBW of such an array. $\qquad$ [$-155.88°$, $20°$]

Important Formulae

- The directive gain of an antenna is given by $g_d(\theta, \phi) = \Phi(\theta, \phi)/P_{\text{av}}$.
- The power gain of an antenna, in a given direction, can be expressed as $g_p = 4\pi\Phi(\theta, \phi)/P_t$.
- The effective area or effective aperture of an antenna can be defined as $A_e = \dfrac{\lambda^2}{4\pi} g_d$.
- The antenna gain can be expressed as $G = \eta_{\text{cd}} D$.
- In terms of loss resistance and radiation resistance, radiation efficiency can be expressed as $\eta_{\text{cd}} = \dfrac{R_r}{R_r + R_0}$.
- For an antenna, EIRP $= P_{\text{ant}} G_t$.
- For an electrically small loop antenna, $A_{\text{em}} = \left(\dfrac{\lambda^2}{4\pi}\right) D_0 = \dfrac{3\lambda^2}{8\pi}$.

- For an electrically small loop antenna, $R_r = 20\pi^2(C/\lambda)^4 \simeq 31,171(S^2/\lambda^4)$.
- For an electrically large loop antenna, $D_0 = 0.682(C/\lambda)$.
- For an electrically large loop antenna, $A_{em} = 5.43 \times 10^{-2}\lambda C$.
- For an electrically large loop antenna, $R_r = 60\pi^2(C/\lambda)$.
- The normal mode axial ratio of a helical antenna is $AR = 2S\lambda/(\pi D)^2$.
- For an axial-mode helical antenna, $R = 140C/\lambda\,\Omega$.
- For an axial-mode helical antenna, $HPBW = \dfrac{52\lambda^{3/2}}{C\sqrt{NS}}$ degree.
- For an axial-mode helical antenna, $FNBW \approx \dfrac{115\lambda^{3/2}}{C\sqrt{NS}}$ degree.
- For an axial-mode helical antenna, Directivity $\cong 15NC^2S/\lambda^3$.
- For an axial-mode helical antenna, $AR = \dfrac{2N+1}{2N}$.
- For a Yagi–Uda antenna, Reflector length $= 500/f\,(\text{MHz})$ ft.
- For a Yagi–Uda antenna, Driven element length $= 475/f\,(\text{MHz})$ ft.
- For a Yagi–Uda antenna, Director length $= 455/f\,(\text{MHz})$ ft.
- For a pyramidal horn, $D_0 = 8.1 + 10\log_{10}(HW/\lambda^2)$ dB $\approx 7.5A/\lambda^2 = 7.5HW/\lambda^2$.
- For a pyramidal horn, $g_P = 4.5A/\lambda^2 = 4.5HW/\lambda^2$.
- For a pyramidal horn, $HPBW = 2\sin^{-1}(0.45\lambda/H)$ for E-plane and $HPBW = 2\sin^{-1}(0.63\lambda/W)$ for H-plane.
- The gain of a circular horn can be expressed as $D_0 = 20\log_{10}(\pi R/\lambda) - 2.82\,\text{dB}$.
- For a parabolic antenna excited by a dipole, the maximum gain is $g_P = 6(D_m/\lambda)^2$.
- For a uniformly illuminated parabolic antenna, $g_d = 9.8696(D_m/\lambda)^2$.
- For a uniformly illuminated parabolic antenna, $FNBW = 140\lambda/D_m$ (degree).
- For a uniformly illuminated parabolic antenna, $HPBW = 58\lambda/D_m$ (degree).
- For a uniform array antenna, $\psi = \beta d\cos(\phi) + \alpha$.
- The principal maxima of a uniform array is located at $\cos(\phi) = -\dfrac{\alpha}{\beta d}$.
- For a broadside uniform array, $\Delta\phi = \dfrac{\lambda}{dn}$.
- For an end-fire uniform array, $\Delta\phi = \sqrt{\dfrac{\lambda}{nd}}$.

Exercises

Objective-type Questions

18.1 A standard reference antenna for the measurement of directive gain is
    (a) an elementary dipole
    (b) a horn antenna
    (c) a half-wave dipole
    (d) an isotropic radiator

18.2 Side lobes of an antenna pattern causes
    (a) ambiguity in direction finding
    (b) increased antenna gain
    (c) reduced bandwidth
    (d) reduced antenna gain

18.3 Radiation from a helical antenna is
    (a) horizontally polarized
    (b) vertically polarized
    (c) circularly polarized
    (d) randomly polarized

18.4 The antenna most commonly used for TV broadcasting in UHF band is a
    (a) parabolic dish antenna
    (b) Yagi–Uda antenna
    (c) dipole antenna
    (d) patch antenna

18.5 If the operating frequency of a helical antenna is doubled, then the gain will
    (a) be half of the original
    (b) be double the original
    (c) be eight times the original
    (d) remain unchanged

18.6 A dipole antenna is a
    (a) conduction current antenna
    (b) displacement current antenna
    (c) includes both displacement and conduction current
    (d) none of these

18.7 The FNBW of an antenna
    (a) is less than its HPBW
    (b) may be equal to its HPBW
    (c) is greater than its HPBW
    (d) any one of these

18.8 The radiation pattern of a normal mode helix is
    (a) isotropic
    (b) of figure *eight* shape rotated about its axis
    (c) directional along its axis
    (d) none of these

18.9 In a Yagi–Uda antenna, the directors show
    (a) capacitive impedance
    (b) inductive impedance
    (c) resistive impedance
    (d) characteristics of an LC tank circuit

18.10 In the active region of a log-periodic antenna, the dipole presents
    (a) inductive loading
    (b) capacitive loading
    (c) resistive loading
    (d) tank circuit

18.11 Theoretically, the bandwidth of a log-periodic antenna is
    (a) zero
    (b) finite but narrow
    (c) finite but wide
    (d) infinite

18.12 Which of the following horns provides a plane wave at the antenna aperture?
    (a) pyramidal horn
    (b) *E*-plane horn
    (c) *H*-plane horn
    (d) hog horn

18.13 For ship to ship communication using a parabolic reflector antenna, the preferred shape of the reflector is
    (a) cylindrical paraboloid
    (b) cut paraboloid
    (c) pill box
    (d) normal paraboloid

18.14 Zoning in a lens antenna is done mainly to
    (a) reduce its weight
    (b) increase its bandwidth
    (c) make the antenna more directive
    (d) increase the gain

18.15 In a broad side array, the antenna radiates
    (a) parallel to the line of the array
    (b) perpendicular to the line of the array
    (c) at an angle of $45°$ with respect to the array axis
    (d) isotropically

Review Questions

18.1 Define an antenna. How does radiation take place from a dipole? Explain with a suitable figure.

18.2 Define (a) radiation pattern, (b) major lobe, (c) minor lobe, (d) back lobe, (e) HPBW and (f) FNBW with suitable figures.

18.3 Define (a) radiation intensity, (b) directive gain, (c) directivity, (d) power gain, (e) effective area, (f) aperture efficiency, (g) antenna efficiency, (h) effective noise temperature, (i) antenna polarization, (j) effective isolated radiated power, and (k) space loss.

18.4 Assuming sinusoidal current distribution, find the radiation resistance of a quarter-wave monopole.

18.5 What is a helical antenna? Describe its different modes of operation.

18.6 Describe the construction of a Yagi–Uda antenna and its principle of operation.

18.7 Describe the geometry of a log-periodic antenna and its limitations as an infinite bandwidth antenna.

18.8 What is a horn antenna? Describe different types of horn antennas.

18.9 State the principle of a parabolic reflector. Describe different types of feed systems used with parabolic reflectors. In addition, mention their advantages and disadvantages.

18.10 Describe the designing of a rectangular microstrip patch antenna for a given substrate and operating frequency.

18.11 Describe the principle of operation of a lens antenna. What is zoning of lens antenna?

18.12 Describe the construction of a non-metallic lens antenna. What is the major disadvantage of such an antenna? How can this disadvantage be overcome?

18.13 What is meant by artificial dielectric? Describe the principle of an artificial dielectric lens antenna.

18.14 What is a metal lens antenna? Describe its principle of operation.

18.15 Describe the theory of array antennas. Find the conditions of the maxima location, first null, and first secondary maxima for an N element uniform linear array. In addition, show that for such an array the first side lobe is 13.5 dB below the main lobe.

18.16 What are broadside and end-fire arrays? Find out the positions of the first null and expressions of beam width for a uniform linear broadside and end-fire array.

Problems

18.1 A parabolic reflector has an aperture efficiency of 0.6 and directivity of 32 dB at 250 MHz. Calculate the diameter and half-power beam width. *[Hint: Use Eqs (18.11), (18.12), and (18.86)]*

18.2 Calculate the beam width between the first null and maximum power gain of a 3 m paraboloid reflector used at 5 GHz. *[Hint: Use Eqs (18.88) and (18.86)]*

18.3 An electric field of 12 μV/m has been measured at a distance of 500 km and observation point $\theta = \pi/2$ from a half-wave dipole at 6 GHz. Calculate the (a) length of the dipole, (b) current fed at the antenna, and (c) average radiated power. *[Hint: Use Eq. (18.32) and $P_{rad} = 73\,I_m^2$]*

18.4 Calculate the radiated electric field intensity at a distance of 15 km from an antenna of gain 10 dB. Assume that the total radiated power is 25 kW. *[Hint: Use the relation $|E|^2 / (2\eta) = g_d P_r / (4\pi r^2)$]*

18.5 Determine the directivity of an antenna for which the radiation intensity can be expressed as $\Phi(\theta,\phi) = \begin{cases} 2\sin(\theta)\sin^3(\phi) & \text{for } 0 \le \theta \le \pi, 0 \le \phi \le \pi. \\ 0 & \text{elsewhere} \end{cases}$ *[Hint: Use Eqs (18.4)–(18.6)]*

18.6 An antenna radiates into only the upper hemisphere, and its normalized radiation intensity is given by $\Phi(\theta,\phi) = \begin{cases} \cos^2(\theta) & \text{for } 0 \leq \theta \leq \pi/2,\, 0 \leq \phi \leq 2\pi \\ 0 & \text{elsewhere} \end{cases}$. Calculate the (a) pattern solid angle, (b) directivity, and (c) HPBW. *[Hint: Use Eqs (18.4) and (18.6)]*

18.7 Calculate the effective area of a half-wave dipole at 1 GHz. The directivity of the antenna is 1.64. *[Hint: Use Eq. (18.11)]*

18.8 For a pyramidal horn with $L' = 12\lambda$ and E- and H-plane dimensions 4λ and 6λ, respectively, examine whether it is an optimum horn or not. *[Hint: Use Eq. (18.78)]*

18.9 Design a microstrip antenna at 2 GHz using a substrate of dielectric constant 2.2 and thickness 1.6 mm. The antenna should be fed with a source of 50 Ω. *[Hint: Use Eqs (18.106), (18.99), (18.100), (18.103), (18.96), and (18.104), and the relation $Z_T = \sqrt{Z_0 R_{in}}$]*

18.10 If the mouth diameter of a parabolic reflector is doubled, then calculate the improvement in gain. *[Hint: Use Eq. (18.88)]*

18.11 Find the diameter and effective aperture of a paraboloid reflector that produces an FNBW of $12°$ at 3 GHz. *[Hint: Use Eq. (18.94)]*

18.12 A pyramidal horn antenna has a dimension of $9 \text{cm} \times 8 \text{cm}$ and is operating at 6 GHz. Calculate the (a) directivity, (b) gain, and (c) E-and H-plane HPBW. *[Hint: Use Eqs (18.81)–(18.84)]*

18.13 An antenna has a radiation efficiency of 95% and directivity 7.5 dB. Calculate its gain. *[Hint: Use Eq. (18.16)]*

18.14 Calculate the effective area of a half-wave dipole at 1 GHz. *[Hint: Use Eq. (18.11)]*

18.15 Determine the magnitudes of the electric and magnetic fields of a half-wave dipole operating at 500 MHz at a distance of 100 m in the broadside direction. Assume input current to the terminals to be $12\angle 0°$ A. In addition, calculate the time average radiated power. *[Hint: Use Eqs (18.32) and (18.33)]*

18.16 A certain antenna has a physical temperature of 280 K. When aimed at the overhead sky, the effective noise temperature measured is 30 K. If we assume that the antenna has sufficient gain so that it sees an essential uniform brightness temperature of 5 K, calculate the efficiency of the antenna. *[Hint: Use Eq. (18.17)]*

18.17 At a distance of 3 km from an antenna, the radiated power density in the main beam is $1.4 \ \mu\text{W/m}^2$. If the input power is 100 W, then calculate the gain of the antenna. *[Hint: Use Eqs (18.3) and (18.8)]*

18.18 The mouth diameter and the HPBW of a parabolic reflector are 5 m and $2.5°$, respectively. Calculate the operating frequency and power gain of the antenna. *[Hint: Use Eqs (18.95) and (18.93)]*

18.19 Calculate the dimension of each side of a square horn antenna having a power gain of 75 dB. *[Hint: Use Eq. (18.81)]*

18.20 Calculate the gain of a parabolic dish operating at a wavelength of 2.5 cm and having a reflection efficiency of 0.6 and diameter 12 m. *[Hint: Use Eq. (18.91)]*

18.21 Calculate the effective aperture of a perfectly reflective parabolic dish antenna having a diameter of 12 m and operating at 10 GHz. *[Hint: Use Eq. (18.90)]*

18.22 A geostationary satellite operating at 4 GHz is 35,860 km away from the earth's surface. Calculate the space loss ratio. *[Hint: Use Eq. (18.21)]*

18.23 Calculate the number of elements in a linear array that provides a broadside FNBW of $12°$. Assume the spacing between the elements as $\lambda/2$. Repeat the same for an end-fire array. *[Hint: Use Eqs (18.149) and (18.151)]*

18.24 Calculate the first side lobe level of a 20-element uniform array. *[Hint: Use Eq. (18.146)]*

Microwave Measurements

19.1 Introduction

Till now, we have discussed different types of microwave transmission lines, components, circuits, antennas, and some of the major applications of microwaves. In this chapter, we will concentrate on microwave measurements. At low frequencies, we generally measure terminal voltages, terminal currents, frequency, power, etc., from which impedance, phase angle, power factor, and other related parameters can be calculated. However, at microwave frequencies, amplitudes and phase of voltages and currents are functions of distance and cannot be measured easily. In some single-conductor transmission lines, like a waveguide, there is also no option for measurement of voltage and currents, as such measurements require two-conductor systems. Therefore, microwave measurements are different from their low-frequency counterparts. At microwave frequencies, generally power is measured instead of voltages and currents, provided the line is lossless. In addition, measurements of the S-parameters, phase shift, VSWR, noise figure, etc. are also useful for providing further information.

One of the major barriers for measurements at microwave frequencies is the cost involved. Due to a small wavelength and the requirement of a high-precision and complicated network structure, direct measuring instruments at microwave frequencies, such as vector and scalar network analysers, spectrum analysers, and power meters, are very costly. Therefore, in laboratories, microwave measurement is carried out using a low-frequency-tuned receiver and a VSWR meter, after modulating the signal with a 1 kHz square wave.

In the following sections, we will discuss some of these high-end and low-frequency equipment. Measurement of few microwave parameters will also be discussed.

19.2 Tuned Detectors

Tuned detectors are specially designed point contact or metal–semiconductor Schottky barrier diodes that are mounted in a microwave transmission line to detect low-frequency square-wave-modulated microwave signals. The tuned detector rectifies the received signal and produces a DC current proportional to the received power. Hence, tuned detectors are also called square-law detectors.

The basic details of such diodes have already been provided in Chapters 13 and 15 and will not be repeated here. The use of such detectors, in a transmission line, requires a tunable stub for matching. Some such matching techniques are shown in Fig. 19.1. By moving the tuning plunger, the distance between the short end and the diode can be varied, thus optimizing the detector performance at the desired frequency.

The performance of a detector diode is defined in terms of its current sensitivity. The current sensitivity of a detector diode is defined as $\Delta\beta = \Delta i/P$ (19.1) where Δi is the increase in short-circuit current due to available input power. It may be noted that the current sensitivity of a detector diode is a function of bias level and usually lies in the range $1-15\,\mu A/\mu W$. The maximum current sensitivity can be obtained when the bias level is in the range $10-100\,\mu A$. For simplicity, however, most of the detector diodes are operated without any bias level.

A detector diode, when mounted in a transmission line, provides a current that is proportional to the incident power. This output current is of the order of few micro-amperes and can be measured using a DC micro-ammeter.

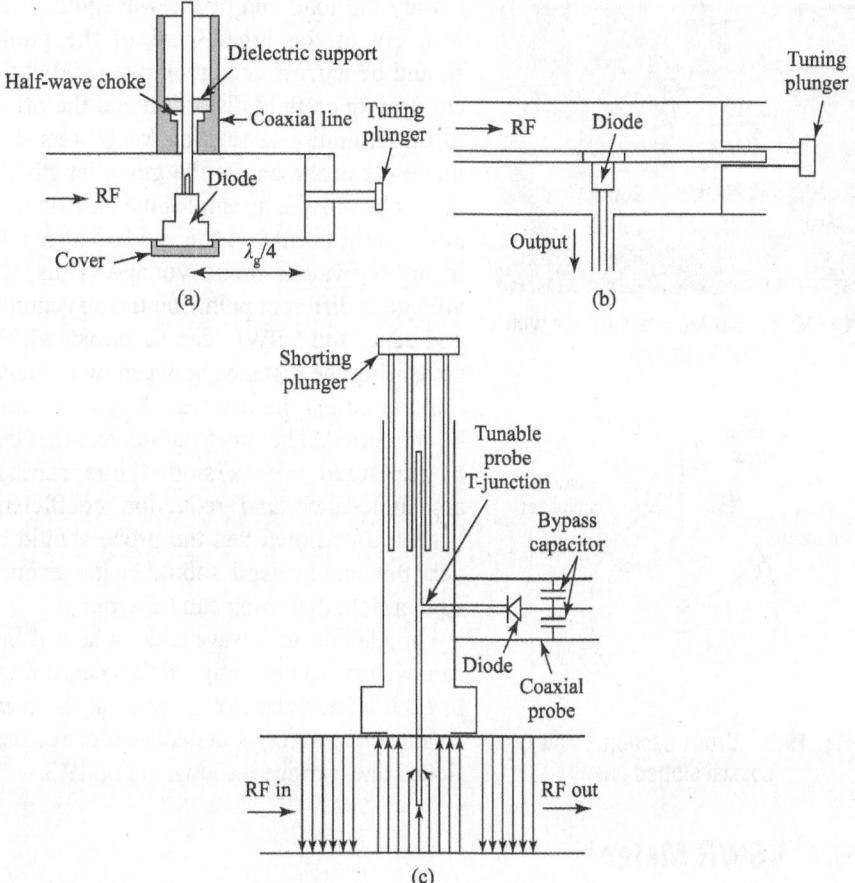

Fig. 19.1   Schematic diagrams (a) Tunable waveguide detector (b) Coaxial detector (c) Tunable probe detector

To increase the detector sensitivity, the microwave power is sometimes modulated with a 1 kHz square-wave signal, and the output of the detector diode is amplified using a tuned amplifier. The instrument that is used to amplify and display the output of the detector diode is called a detector amplifier or a standing wave detector, and will be discussed later in this chapter. It may be noted that, in principle, a DC signal, out of the detector diode in absence of modulation, can also be amplified. However, because of the drift problem, the instrument becomes more complicated in such cases.

19.3 Slotted Line Carriage

A slotted line carriage, shown in Fig. 19.2, consists of a coaxial E-field probe that is inserted within a waveguide through a centred narrow longitudinal slot cut in the broad wall of the waveguide. The probe is able to move longitudinally along the length of the guide through the slot. The centred slot ensures that radiation will not occur. A Vernier scale is placed at the side of the slot to measure the position of the probe from a reference point, mainly the load end of the waveguide. The slot, cut at the broad wall of the guide, should be narrow enough and tapered at the end to reduce field distortion and the effect of discontinuity. A rotating knob is used to move the probe across the guide length. As the probe moves, it samples the electric field at different points on the guide and results in an equivalent probe voltage. Thus, the voltage at different points on the waveguide, and hence the VSWR, can be measured. By measuring the distance between two voltage minima points, guided wavelength can also be measured. The other parameters that can be measured with a slotted line carriage are impedance and reflection coefficient. It should be noted that the probe should be very thin and its depth should be low enough so that field distortion can be avoided.

Fig. 19.2 Slotted line carriage with tuned detector

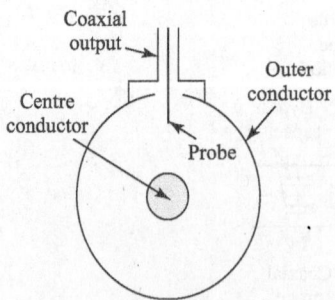

Fig. 19.3 Cross-sectional view of coaxial slotted line

In addition to a waveguide, a slotted line carriage can also be designed for coaxial lines. In such a case, the slot is made at the outer conductor. The cross-sectional view of a coaxial slotted line carriage is shown in Fig. 19.3.

19.4 VSWR Meter

A VSWR meter is basically a high-gain, high-Q, and low-noise voltage amplifier that is tuned at the modulating frequency of the microwave signal (generally

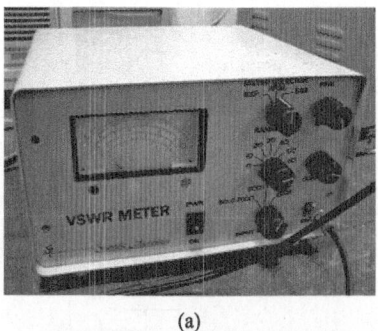
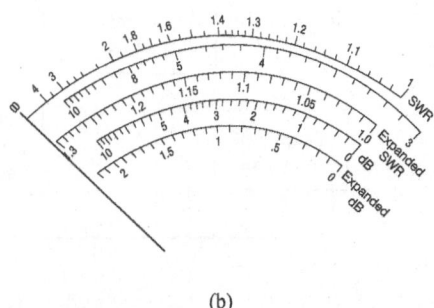

(a)   (b)

Fig. 19.4   VSWR meter (a) Laboratory model (b) Expanded view of its scale

1 kHz). The overall gain of the amplifier is about 125 dB, which can be altered at a step of 10 dB by a gain control panel provided with the meter. The display panel of the meter consists of three scales: normal, expanded, and dB. If the VSWR is between 1 and 4, then the top *normal SWR* scale can be used; on the other hand, if the VSWR is between 3 and 10, then the lower scale of the *normal SWR* can be used. If the VSWR is smaller than 1.3, then the *expanded SWR* scale is used. The rest scale is used for measurements in dB scale.

To measure the VSWR, the meter needle is initially adjusted to 1 after placing the probe in V_{max} position within the guide. The gain control panel is used for making this adjustment. Now, for any input voltage, the output of the amplifier is measured with a square-law calibrated voltmeter directly in terms of VSWR. The input of the VSWR meter is basically the detected output voltage of the tuned detector that is fed by a coaxial cable. A basic VSWR meter and its scale are shown in Figs 19.4(a) and (b), respectively.

19.5 Spectrum Analysers

A spectrum analyser is an instrument that is able to measure received signal in the frequency domain. It consists of a local oscillator that can be electronically swept back and forth at a linear rate between two frequency limits (start and stop frequencies), with a saw-tooth-type swept voltage. The fly back time of the saw tooth swept voltage is kept zero. The saw tooth voltage is also applied at the horizontal plates of a CRT. The input RF signal is first mixed with the local oscillator signal to produce an intermediate frequency (IF) signal. The IF signal is then passed through a narrow-band IF amplifier. The amplified IF is then detected by a detector, and the output of the detector is then fed into the vertical plates of the CRT. To achieve better resolution, the bandwidth of the IF amplifier should be as small as possible. In addition, the swept speed should also be low so that voltage can build up in the receiver circuit. In order to avoid image response, the IF should be chosen as high as possible. The simplified block diagram of a spectrum analyser is shown in Fig. 19.5. Figure 19.6 shows a spectrum analyser that is used in the laboratory.

In addition to displaying frequency versus amplitude, a spectrum analyser is able to display the exact frequency of a signal, its stability against time, presence

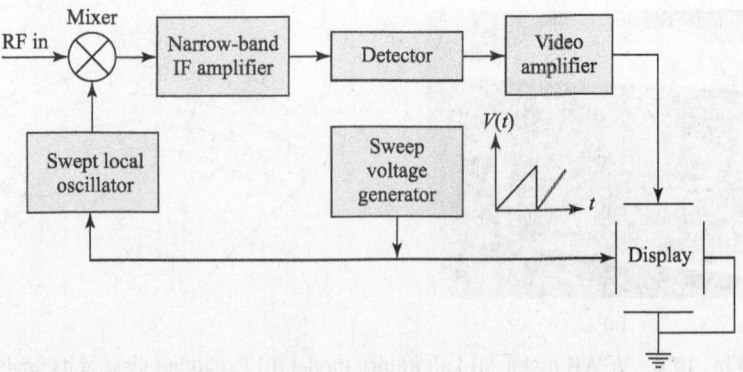

Fig. 19.5 Simplified block diagram of spectrum analyser

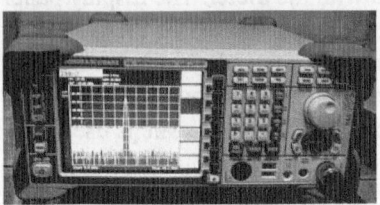

Fig. 19.6 Spectrum analyser

of spurious oscillations and interfering signals, noise and distortion, and effect of modulation on carrier.

The performance of a spectrum analyser is generally defined by the following terms:

Frequency range This is defined as the overall frequency band that the instrument can receive and analyse. The frequency range of a spectrum analyser generally depends on the mixer and the local oscillator system.

Frequency span This is defined as the range of frequencies that can be displayed on the screen at a particular time. Depending on the instrument and setting, it can be full frequency range or some part of it. Modern spectrum analysers have the advantage that the user can set the start and stop frequencies, both within the frequency range, to define a suitable frequency span for a specific use. Alternatively, the user can also define the centre frequency and frequency bandwidth required to set the frequency span. If the signal frequency is unknown, then it is a usual practice to start with a wide frequency span and then zoom in on the signals of interest by reducing the frequency span.

Frequency resolution Frequency resolution of a spectrum analyser is defined as the minimum frequency difference between two signals required to identify them separately using a spectrum analyser. The frequency resolution of a spectrum analyser is set by the bandwidth of the swept bandpass filter used in the spectrum analyser. The frequency resolution of a spectrum analyser has a direct relation with the sweep speed. If the resolution is high, then it will correspond to a longer sweep time.

Sensitivity Due to the electronic components present, spectrum analysers are always associated with a noise floor. This noise floor sets a minimum signal level that can be measured by the instrument. This defines the sensitivity of the instrument. The noise floor of a spectrum analyser is directly related with the bandwidth of the swept bandpass filter—the smaller the bandwidth, the lower the noise floor and the larger the sensitivity. Therefore, to achieve the highest sensitivity, we must configure the instrument at the lowest possible bandwidth.

Dynamic range The dynamic range of a spectrum analyser is defined as the difference between the maximum usable signal power and the minimum detectable signal power. The maximum usable signal power is limited by the non-linear distortion of the components and damage issues, whereas the minimum detectable signal power is set by the system sensitivity, defined previously. The dynamic range of a spectrum analyser requires to be high to study the sidebands or spurious signals that are generally 60–80 dB below the carrier frequency.

19.6 Network Analysers

The slotted line measurement technique suffers from the disadvantage that the measurement of amplitude and phase of signals is restricted to a single frequency. Therefore, if we want to measure the amplitude and phase over a large bandwidth, which is more often the case, such technique will be tedious and time consuming. Hence, some other measurement technologies must be adopted so that the job can be accomplished in an easier way. Network analysers are most preferred in such cases. A network analyser can, in general, measure network characteristics (such as scattering parameters) of a microwave network. Network analysers used nowadays are much advanced and can measure several other parameters such as time domain and noise figure. Depending on the measurement capability, network analysers can be classified as vector network analysers (VNAs) and scalar network analysers (SNAs). VNAs can measure both the amplitude and the phase, whereas SNAs can measure only amplitudes. Needless to say, VNAs are much more expensive than SNAs.

The block diagram of a network analyser is shown in Fig. 19.7. Its basic operation involves the generation of an accurate reference signal and its comparison with a test signal. The signal generated from the sweep generator is first split into two signals, namely, test and reference signals, which are then fed into the device under test (DUT) and a length equalizer, respectively. The length equalizer takes care of the phase changes caused by the length of DUT. The outputs of the DUT and length equalizer are then converted into an IF signal, as the processing of microwave signal is much more difficult. A harmonic frequency converter is used for this purpose, which uses a phased lock loop that helps the local oscillator to track the reference channel frequency. The block diagram of a harmonic frequency converter is shown in Fig. 19.8. As shown in the figure, frequency conversion takes place in two steps.

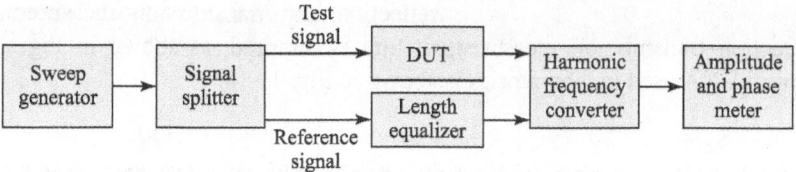

Fig. 19.7 Schematic block diagram of VNA

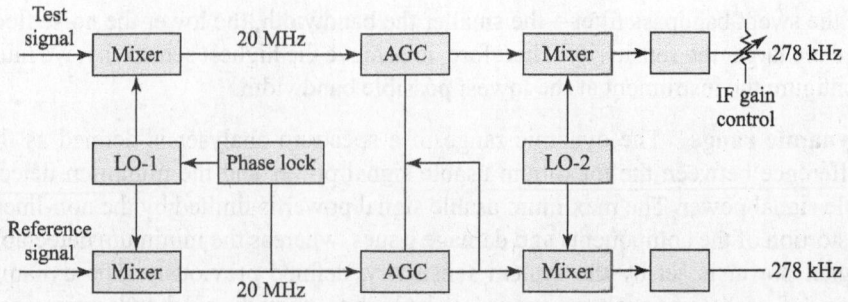

Fig. 19.8 Schematic block diagram of harmonic frequency converter

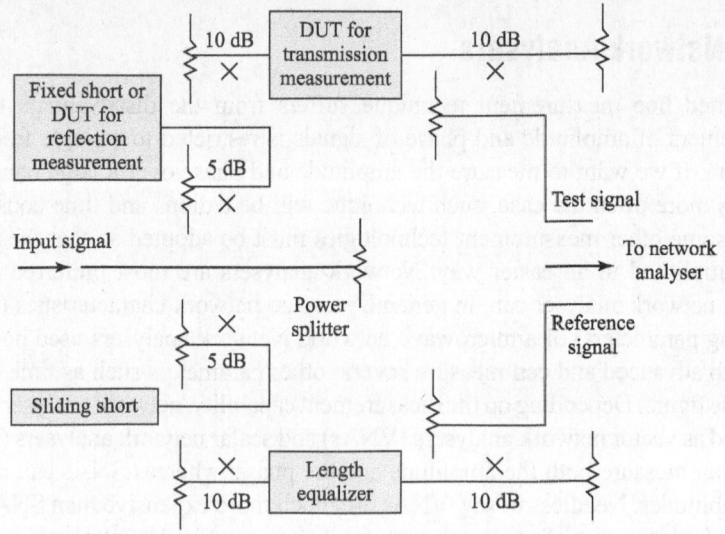

Fig. 19.9 Schematic diagram of reflection transmission test unit

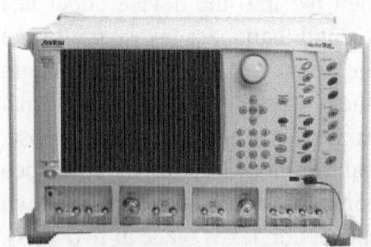

Fig. 19.10 Typical VNA

Initially, the test and reference signals are fed to the first mixer. The first mixer converts these signals into MHz signals, which, after amplification, are fed to the second mixer. The second mixer converts the MHz signals into IF signals having frequency in the kHz range. Outputs of the harmonic frequency converter are finally compared to determine the amplitude and phase of the test signal. For reflection and transmission measurements, a reflection–transmission measurement unit is required, as shown in Fig. 19.9. A typical VNA used in laboratories is shown in Fig. 19.10.

Note Directional couplers used in the bridge should be accurately matched so that a good balance can be obtained between the two channels.

19.7 Power Meters

Microwave power meters are widely used in microwave instrumentation to measure the power carried by a wave. A microwave power meter generally consists of a power sensor that converts the microwave power into equivalent heat energy. This conversion results in a temperature rise, which, in turn, changes electrical parameters. As a result, an electrical current flows in a low-frequency circuit. This flow of current is measured and calibrated to represent power in a display meter. In practice, microwave powers are divided into three categories: (a) low power (<10 mW), (b) medium power (10 mW–10 W), and (c) high power (>10 W). High microwave power is often measured using a calorimeter, in which the rise of temperature of the load directly indicates the power carried by the wave. In addition, different sensors are used in microwave power measurements. Some of them are described briefly in the following subsections.

19.7.1 Schottky Barrier Diode Sensor

A zero-biased Schottky barrier diode sensor is used to measure microwave power as low as −70 dBm. Such a diode behaves as a square-wave detector and provides

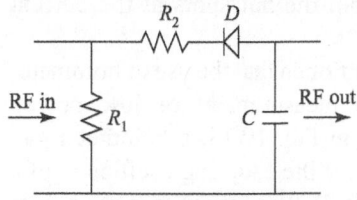

Fig. 19.11 Schottky barrier diode sensor circuit

an output current proportional to the input power. Since diode resistance is a function of the diode temperature, and hence the input power, an input power has a tendency to cause a mismatch at the input circuit. Therefore, care must be taken to minimize this effect. A typical Schottky barrier diode sensor circuit is shown in Fig. 19.11.

19.7.2 Bolometer Sensor

The operation of a bolometer is based on the change in resistance due to rise in temperature caused by absorption of power. In practice, there are two common types of bolometers: barretters and thermistors. A barretter uses a short, thin platinum wire and has a positive temperature coefficient. Such a device is used for very low power measurement (less than a few mW). A thermistors, on the other hand, uses a semiconductor bead for its operation and has a negative temperature coefficient. The bead is composed of a mixture of nickel oxides and manganese with finely divided copper particles, and has a diameter of about 0.05 cm. In addition to beads, thermistors are also available in the form of washers, disks, rods, or flakes. Thermistors can be used to measure medium and high power after suitable attenuation, can easily be mounted in a transmission line, can provide good isolation from physical and thermal shock, good shielding against energy leakage, and also good match. They are also low-loss devices.

Barretters can be prepared by etching the silver from the platinum core of a Wollaston wire. This results in a very narrow platinum wire. The desired resistance of the platinum wire is then achieved by adjusting its length. For microwave applications, resistance of the barretter should be such that it can be matched with the system for efficient absorption of energy. The overload characteristic of a barretter is not as good as that of a thermistor that can withstand severe overloads

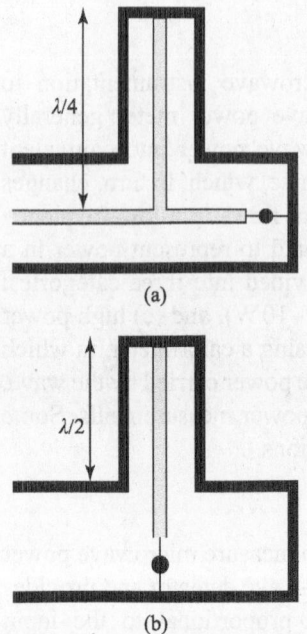

Fig. 19.12 Bolometer mounts
(a) Coaxial line (b) Waveguide

without any change of its characteristics. However, it has a fairly low thermal time constant (about 0.3 msec). The detector sensitivity of a thermistor is about $60\,\Omega/mW$, whereas that for a barretter is about $5.2\,\Omega/mW$.

The impedance matching of a bolometer with a waveguide/coaxial line is an important concern. Sometimes tuners are used to obtain such matching, but at the cost of a reduced bandwidth. The schematic diagrams of bolometer mounts in a coaxial line and a waveguide are shown in Fig. 19.12. For a waveguide, the bolometer should be placed at the point of maximum electric field. If a standing wave exists along the length of the barretter wire, then it should be placed such that the centre of the barretter wire coincides with the midpoint between current maxima and current minima; this arrangement will reduce the error resulting from the hot spots at the current maxima.

It has been mentioned that the use of bolometer is limited to the measurement of low power. However, with suitable arrangements, as shown in Fig. 19.13, a bolometer can also be used to measure high power. For example, if the coupling coefficient of a coupler is 20 dB and the attenuator is providing 20 dB attenuation, then the power measured by the bolometer will be 40 dB below the actual power. This method can further be extended to measure a higher power than is possible with the use of a single attenuator and a directional coupler. However, such attempts suffer from a practical limitation induced by the power-handling capability of the directional coupler and the attenuator.

Note Like a detector diode, a barretter is also a square-law device and produces a current that is proportional to the applied power.

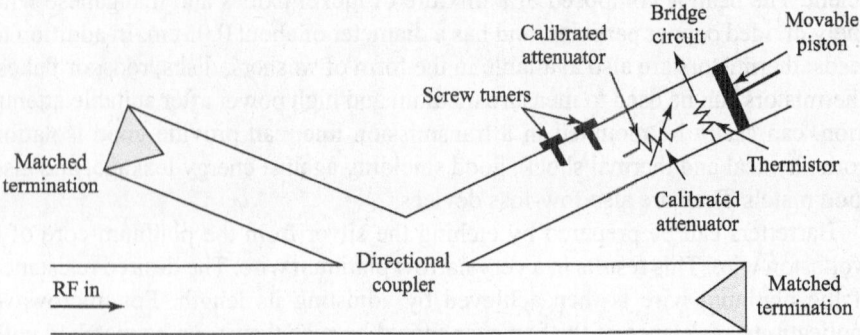

Fig. 19.13 Extension of power measurement capability of bolometer using directional coupler

19.7.3 Bolometer Bridge

Microwave power meters generally use a balanced bridge network, with a bolometer as one of its arm, as shown in Fig. 19.14. The bridge is energized with an adjustable regulated DC power supply. Since a bolometer is a temperature-sensitive device, its resistance may be controlled by the heat produced by the current passing through it and hence by the variable DC power supply. Thus, by changing the DC voltage, the bridge can be brought into balance. Now, if a microwave power is applied to the bolometer, its resistance is changed as a result of heating by the unknown power, and the balance of the bridge is destroyed. This, in turn, results in a non-zero output at the meter connected with the bridge. The meter reading is calibrated to measure the incident power directly.

The unbalanced bridge technique, described earlier, is rarely used in commercial microwave power meters, where the bridge circuit is modified to include self-balancing. In a self-balancing bridge, initially some DC and some audio frequency (AF) powers are applied to the bolometer to balance the bridge. Now, when microwave power is incident on the bolometer, balance of the bridge is disturbed. The amount of unbalance is now sensed by an electronic circuit. Next, an equivalent amount of AF power is removed from the bolometer so that the balance is restored. The amount of AF power withdrawn to restore the balance of the bridge is equal to the unknown RF power, which is measured and displayed.

A single bridge network, which is used to measure RF power, generally comes with a temperature compensation network, as shown in Fig. 19.14. Thermistors R_4 and R_6 are identical and placed close to each other so that they are subjected to the same ambient temperature. A change in the ambient temperature will affect both thermistors identically. If, due to a change in temperature, the resistance of R_4 and R_6 is decreased, then the current through R_7 will increase (as voltage across the series of R_6 and R_7 is constant). This, in turn, will reduce the current through the bridge network and hence through R_4. Thus, balance of the bridge will be restored. A typical power meter used in laboratories is shown in Fig. 19.15.

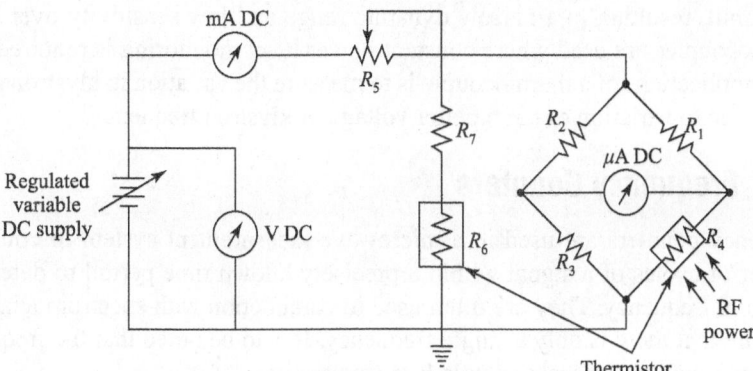

Fig. 19.14 Balanced bolometer bridge

Fig. 19.15 Typical power meter with power sensor

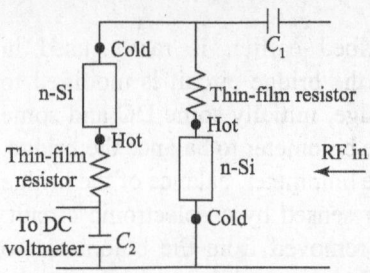

Fig. 19.16 Thermocouple power sensor circuit

19.7.4 Thermocouple

A thermocouple consists of two different metals or semiconductors that contact each other at two or more different spots. When the microwave power is incident on one of the junctions, heat is developed in it. Now, if the other end is maintained at a lower temperature, then due to the temperature difference of the two ends, an emf is developed in the circuit. This generated emf is proportional to the temperature difference of the two ends. Therefore, by measuring the generated emf, temperature of the hot end, and hence the incident power, can be measured, provided that the temperature of the cold end is known. For microwave applications, a thin-film tantalum nitride resistive load deposited over n-Si is used to form the thermocouple junction.

A basic thermocouple power sensor circuit is shown in Fig. 19.16. The emf's generated by the two parallel thermocouples add and appear across the RF bypass capacitor C_2. The output, in the form of a DC voltage, is measured using a DC voltmeter. It may be noted that the output leads that connect the DC voltmeter should be at the RF ground, so that the measured DC voltage is proportional to the incident microwave power. If the incident microwave signal is square-wave modulated, then the peak power can be calculated using the relation $P_{peak} = P_{av}T/\tau$ (19.2)
where P_{av} is the average power, T is the time period, and τ is the pulse width.

Thermocouples are very sensitive to temperature and also easy to use, since only a DC voltmeter is required. However, they are rarely used because their resistance changes with temperature, making the matching difficult. In addition, a thermocouple also needs a small-diameter resistance heater to remove skin effect and frequency dependence. This forces the thermocouple to work close to its limit, resulting in a narrow dynamic range and low sensitivity over loads. Thermocouples are used where constant power level monitoring is required. One of the applications of a thermocouple is to measure the variation in klystron power output due to variation of the repeller voltage or klystron frequency.

19.8 Frequency Counters

Frequency counters are used in a microwave measurement system to count the number of cycles of a signal within a precisely known time period to determine the signal frequency. They are often used in conjunction with spectrum analysers to ensure that there is only a single frequency. It is to be noted that the frequency counters are able to measure single frequency only.

19.9 Microwave Sources

A basic microwave source produces CW carrier oscillation of a desired frequency at a calibrated power level. In modern microwave sources, the achievable signal frequency ranges from a few hundred kHz to a few hundred GHz, with a high-frequency resolution. The achievable power level ranges from about $-100\,dBm$ to $15\,dBm$. An advanced microwave signal generator can also produce AM, FM, or PM signal for testing purpose in addition to pulses.

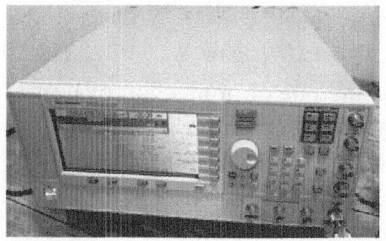

Fig. 19.17   Typical microwave signal generator

A sweep oscillator can produce a frequency sweep during measurement; that is, they start at one frequency and smoothly automatically sweep it through an entire range of frequencies. Such oscillators are widely used to test the frequency response of amplifiers, attenuators, filters, and other microwave circuitries. A typical microwave signal generator is shown in Fig. 19.17.

19.10 Microwave Amplifiers

A microwave amplifier is often used in a microwave measurement system, especially for antenna measurement, to increase the signal strength. Depending on the construction, they can be classified as variable gain amplifiers and fixed gain amplifiers. Output powers of such amplifiers range from a few mW to a few watts. Amplifiers of much larger output powers are also available, but they are generally not required for normal measurement purpose.

A basic setup for testing the performance of a microwave amplifier is shown in Fig. 19.18. A CW signal is applied at the input of the amplifier, whereas the output of the amplifier is fed to a signal splitter. The splitter splits the signal into two channels; one is fed to a power meter and the other to a spectrum analyser. The power meter measures the overall gain of the amplifier, while the spectrum analyser checks for distortion, if any. If the gain of the amplifier is high, then the output power may be outside the dynamic range of the spectrum analyser. In such a case, an attenuator is placed between the signal splitter and the spectrum analyser. The CW source may be replaced by a sweep oscillator to test the linearity of the amplifier and by a signal generator that can produce the modulated waveform to test the modulated waveform performance. A medium-power, fixed-gain, solid-state microwave amplifier, used in laboratories, is shown in Fig. 19.19.

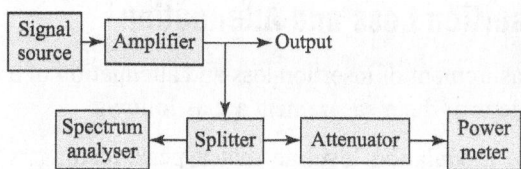

Fig. 19.18   Basic set-up for testing performance of microwave amplifier

Fig. 19.19   Typical medium-power, fixed-gain, solid-state amplifier

19.11 High-power Measurement by Calorimetric Method

The calorimetric method for measuring high power is based on the measurement of rise in temperature of a fluid, usually water, due to absorption of microwave power. In the measurement setup, shown in Fig. 19.20, the calorimetric fluid is passed through a waveguide, where it is subjected to the high-power microwave. The fluid absorbs a fraction of the power and gets heated. If T is the temperature rise in °C, v the rate of flow of calorimeter fluid in cc/sec, C_p the specific heat in cal/g, and d the specific gravity of the fluid in g/cc, then the average power in the microwave signal can be calculated as follows: $P = 4.187C_pTvd$ W $\qquad$ (19.3)

This method is called the direct method, as the rate of production of heat is measured by observing the rise in temperature of the calorimetric fluid directly. In contrast, in the indirect method, heat of the dissipating medium is transferred to another medium for measurement.

The aforementioned calorimeter is called a circulating calorimeter, as the dissipating medium circulates through the setup. In a static calorimeter, a $50\,\Omega$ coaxial cable is filled with a dielectric that has high hysteresis loss. Dissipation of microwave power in the dielectric load results in a rise in temperature of the load, which is measured and used to calculate the average power carried by the microwave. If T is the temperature rise in °C in t sec, m the mass of the thermometric medium in g, and C_p the specific heat in cal/g, then the average power can be calculated using the relation $P = 4.187C_pmT/t$ W $\qquad$ (19.4)

The calorimeter method suffers from a disadvantage that there is a time lag between the application of microwave power and measurement of the parameter.

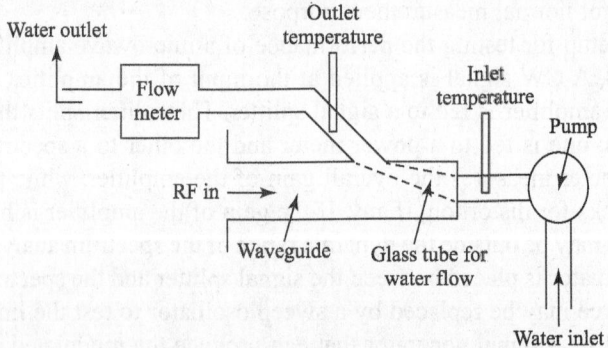

Fig. 19.20 Calorimetric method to determine high microwave power

19.12 Measurement of Insertion Loss and Attenuation

The experimental setup for the measurement of insertion loss and attenuation of a DUT is shown in Fig. 19.21. The steps of the measurement are as follows:

1. Adjust the 1 kHz square-wave modulation level so that a peak reading is observed in the VSWR meter, connected at position A, with the minimum input attenuation level.

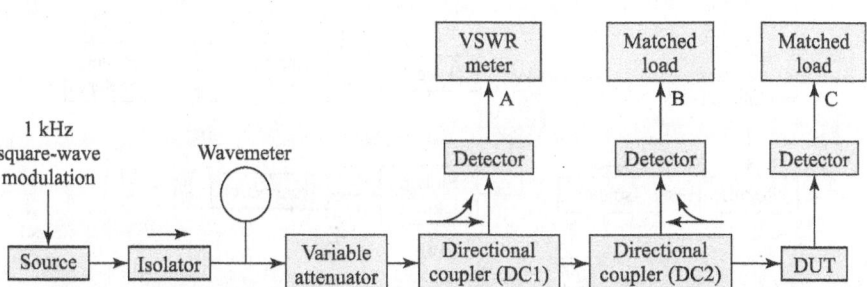

Fig. 19.21 Experimental set-up to measure insertion loss and attenuation

2. Adjust the power level so that a reading in the 30 dB range is obtained in the VSWR meter. This will ensure that the crystal detector is operating in the square-law region.

3. Adjust the gain control of the VSWR meter so that the input power at A is 0 dB or 1.

4. Measure the frequency using a cavity wavemeter. At the signal frequency, a dip will be observed in the VSWR meter.

5. Connect the VSWR meter at point B and a matched load at point A, without disturbing other setups. Since directional couplers are identical and the reference input power has been adjusted to 1, the measured reading in the VSWR meter directly gives the ratio of reflected and incident power (P_r/P_i). This ratio is called the reflection coefficient. If measured in a dB scale, this will give the return loss.

6. From the measured value of P_r/P_i, calculate $\{1-(P_r/P_i)\}$. In the dB scale, this is called reflection loss.

7. Adjust the variable attenuator to obtain an input attenuation that is equal to the dB coupling of the directional coupler.

8. Connect the VSWR meter at point C and a matched load at point B, without disturbing other setups. Since the attenuation level has been increased to the dB coupling value of DC-1 and the reference input power was adjusted to 1, the measured reading in the VSWR meter gives the ratio of output power to incident power (P_0/P_i) directly. If measured in a dB scale, this will give the insertion loss.

9. Subtract reflection loss from insertion loss. This will give attenuation loss, as insertion loss is the sum of reflection loss and attenuation loss.

During these measurement steps, the following care should be taken:

1. All the measured powers are in the square-law region of the crystal detector.
2. Characteristics of both the directional couplers are identical.

As an alternative, we can also measure attenuation loss using the substitution method, as follows:

1. Measure the output power from the DUT.
2. Remove the DUT and insert a calibrated variable attenuator in its place.
3. Increase attenuation of the attenuator until the output power becomes equal to that provided by the DUT.

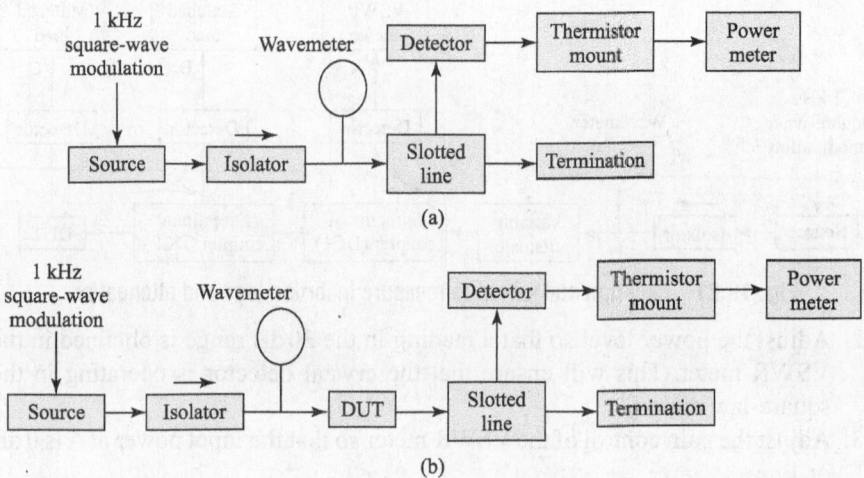

Fig. 19.22 Experimental set-up to measure insertion loss using power ratio method (a) Without DUT (b) With DUT

4. Under such a circumstance, attenuation provided by the attenuator will be equal to that of the DUT.

In addition to the RF substitution method, power ratio method is also sometimes used to measure the attenuation of a DUT. In this method, output powers with (P_2) and without (P_1) the DUT are measured. The ratio of these two powers (P_1/P_2) in dB gives the attenuation of the network. Experimental setups for the power ratio method are shown in Fig. 19.22. This method suffers from the drawback that the powers are measured at two different levels. Therefore, if attenuation is large, that is, the two power levels are wide apart, then it may be possible that any of them is not in the square-law region of the detector. In such a case, error will be introduced.

19.13 Measurement of VSWR

The experimental set-up for the measurement of VSWR is shown in Fig. 19.23. If the load impedance (i.e., input impedance of the DUT) is different from the characteristic impedance of the line, a reflected wave will result. This reflected wave will superimpose on the forward travelling wave and result in a standing wave pattern in the slotted line. Depending on the value, VSWR of the line can be classified as either low $(\rho < 20)$ or high $(\rho > 20)$ VSWR.

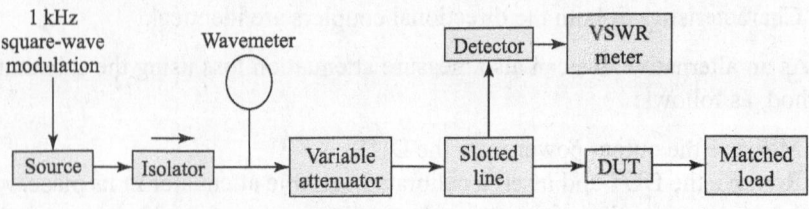

Fig. 19.23 Experimental set-up to measure VSWR

19.13.1 Measurement of Low VSWR

If the VSWR in the line is low, then it can be directly measured from the VSWR meter using the experimental setup shown in Fig. 19.23. The steps are as follows:

1. Adjust the variable attenuator to 10 dB.
2. Adjust the microwave source at the desired frequency.
3. Tune the probe carriage stub for the maximum reading in the VSWR scale.
4. Adjust the modulation (1 kHz AM) for the maximum reading on the VSWR meter in a 30 dB scale.
5. Move the probe carriage along the slotted line to obtain a peak at the VSWR meter.
6. Adjust the meter gain control to get the maximum meter reading as 1.0 or 0 dB.
7. Move the probe carriage along the slotted line to obtain the adjacent minima at the VSWR meter. The corresponding VSWR meter reading at the normal SWR scale directly indicates the VSWR, provided that $1 < \rho < 4$. For $1 < \rho < 1.33$, use the expanded scale.
8. For $3.2 < \rho < 10$, select a 10 dB lower range and take the reading corresponding to the minima position on the VSWR normal scale, second from the top.
9. For $10 < \rho < 40$, select a 20 dB lower range. Now take the reading corresponding to the minima position on the top VSWR normal scale (i.e., 1–4) and multiply it by 10 to get the actual VSWR.
10. For $32 < \rho < 100$, select a 30 dB lower range. Now take the reading corresponding to the minima position on the second VSWR normal scale, second from the top (i.e., 3.2–10), and multiply it by 10 to get the actual VSWR.

The following precautions should be taken while carrying out the aforementioned measurement:

1. Probe thickness and penetration may cause distortion in the field and reflection in the slotted line waveguide. To avoid these, probe penetration should be kept small.
2. The VSWR of the connector contributes a significant error to the measurement when the load VSWR is very low ($\rho < 1.05$). In such cases, very low VSWR connectors should be used to reduce the measurement error.
3. The 1 kHz modulating signal should be a perfect square wave; otherwise, frequency modulation will take place. For a frequency modulated wave, standing waves will be generated for each frequency component, which, in turn, will reduce the sharpness of voltage minima. Therefore, an error will be introduced in the reading of the position of actual voltage minima. A flat standing wave pattern is shown in Fig. 19.24.
4. Any harmonics or spurious signals from the source must be eliminated; otherwise, an error will be introduced in the measured result.
5. The mechanical slope between the slot geometry and probe movement should be removed; otherwise, the measured VSWR values will be different at different probe positions along the slot width. The standing wave patterns that are subjected to such an error are shown in Fig. 19.25.

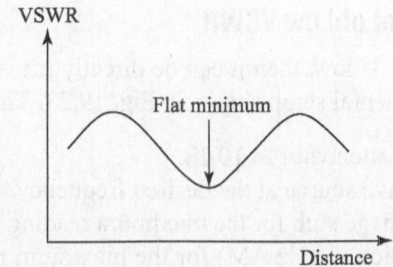

Fig. 19.24 Flat minimum due to imperfect square-wave modulation

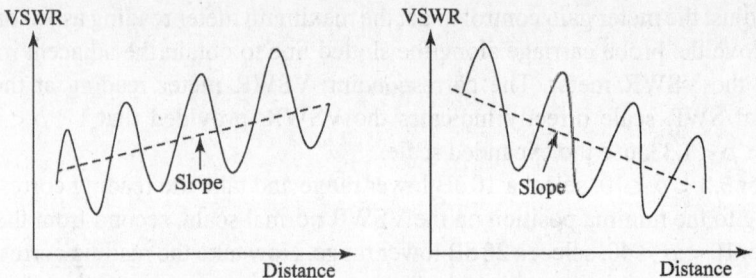

Fig. 19.25 Slanted VSWR pattern due to mechanical slope error

6. The voltage maxima and voltage minima should be measured in the square-law region of the crystal detector.
7. The slot structure itself launches a standing wave pattern in the line, which introduces an error in the measurement.

19.13.2 Measurement of High VSWR

If the VSWR of the line is high enough, the difference between the measured readings of voltage maxima and voltage minima is also large. This high value of voltage maxima may put the detector diode outside of its square-law region, as the diode current may exceed 20 μA. Therefore, an error may be introduced in the measured result. Thus, to measure high VSWR, the double-minima method, shown in Fig. 19.26, is used.

In the double-minima method, measurement is carried out at two points around the voltage minimum. If $|V(x)|$ is the magnitude of the voltage at a point x and $|V_{\min}|$ that of the voltage at the minima, then there ratio can be expressed as follows:

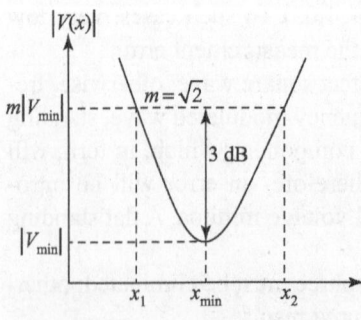

Fig. 19.26 Determination of VSWR using double-minima method

$$r = |V(x)|/|V_{\min}|$$

(19.5)

Now if V_{inc} is the incident voltage and $\Gamma = |\Gamma| e^{j\phi}$ the reflection coefficient, then we can write the following equation:

$$|V(x)| = |V_{inc}| \left|1 + |\Gamma| e^{j(\phi - 2\beta x)}\right| = |V_{inc}| \sqrt{1 + 2|\Gamma| \cos(\phi - 2\beta x) + |\Gamma|^2} \quad (19.6)$$

Therefore,

$$|V_{min}| = |V_{inc}| \sqrt{1 - 2|\Gamma| + |\Gamma|^2} = |V_{inc}| (1 - |\Gamma|) \quad (19.7)$$

Now, let us assume that x_1 and x_2 are two measurement points around the minima position such that $|V(x_1)| = |V(x_2)| = m|V_{min}|$ $\qquad (19.8)$

where $\quad m = |V(x_1)| / |V_{min}|$ $\qquad (19.9)$

Substituting Eqs (19.6) and (19.7) into Eq. (19.9), we get the following relation:

$$m = \frac{|V(x_1)|}{|V_{min}|} = \frac{|V_{inc}| \sqrt{1 + 2|\Gamma| \cos(\phi - 2\beta x_1) + |\Gamma|^2}}{|V_{inc}| (1 - |\Gamma|)}$$

or $\qquad m = \dfrac{\sqrt{1 + 2|\Gamma| \cos(\phi - 2\beta x_1) + |\Gamma|^2}}{(1 - |\Gamma|)}$ $\qquad (19.10)$

Now substituting $|\Gamma| = (\rho - 1)/(\rho + 1)$ in this equation and carrying out necessary manipulation, we get the following relation:

$$\rho = \frac{\sqrt{m^2 - \cos^2\{\beta(x_1 - x_{min})\}}}{\sin\{\beta(x_1 - x_{min})\}} = \frac{\sqrt{m^2 - \cos^2\{2\pi(x_1 - x_{min})/\lambda_g\}}}{\sin\{2\pi(x_1 - x_{min})/\lambda_g\}} \quad (19.11)$$

where we have substituted $\beta = 2\pi/\lambda_g$. Further substituting

$$\Delta x = x_2 - x_1 \approx 2(x_1 - x_{min}) \quad (19.12)$$

into Eq. (19.11), we get the following expression:

$$\rho = \frac{\sqrt{m^2 - \cos^2(\pi\Delta x/\lambda_g)}}{\sin(\pi\Delta x/\lambda_g)} = \sqrt{1 + \frac{m^2 - 1}{\sin^2(\pi\Delta x/\lambda_g)}} \quad (19.13)$$

Further, if we assume that measurements have been taken at 3 dB point, that is, $m = \sqrt{2}$, then Eq. (19.13) reduces to the following form:

$$\rho = \sqrt{1 + \frac{1}{\sin^2(\pi\Delta x/\lambda_g)}} = \sqrt{1 + \operatorname{cosec}^2(\pi\Delta x/\lambda_g)} \quad (19.14)$$

Again if $\lambda_g \gg \pi\Delta x$, Eq. (19.14) reduces to the following form:

$$\rho \approx \operatorname{cosec}(\pi\Delta x/\lambda_g) \approx \frac{\lambda_g}{\pi\Delta x} \quad (19.15)$$

Equation (19.15) reveals that the VSWR can be measured by measuring the distance between the 3 dB point around a voltage minima and the guided wavelength. The guided wavelength can be measured by measuring the distance between two successive voltage minima and multiplying it by 2. The measurement procedure is as follows:

1. Adjust source frequency, probe carriage, modulation, etc. for the maximum reading.
2. Position the probe at a voltage minima point. Note the position of the voltage minima.
3. Move the probe at any side of the voltage minima to find the position of successive minima. Again note the position.
4. Find the distance between the positions of two successive minima and multiply it by 2. This will give the guided wavelength.
5. Position the probe at any suitable voltage minima point.
6. Adjust the gain control of the meter so that a 3 dB reading is obtained.
7. Now move the probe on one side of the minima until a 0 dB reading is obtained on the meter. Note the first 0 dB position.
8. Now move the probe on the other side of the minima until again a 0 dB reading is indicated on the meter. Note the second 0 dB position.
9. The distance between these two positions will give the value of Δx.
10. Use Eq. (19.14) or (19.15) to calculate the VSWR.

19.14 Measurement of Return Loss by Reflectometer

A reflectometer, shown in Fig. 19.27(a), consists of two identical directional couplers, connected opposite to each other. The first coupler is coupled to the forward wave, whereas the second coupler is coupled to the reflected wave.

The experimental setup for the measurement of return loss using a reflectometer is shown in Fig. 19.27(b). Here we will assume the following conditions:

1. Directivity of the directional couplers is infinity.
2. VSWR in the mainline is 1.
3. Detectors have perfect impedance matching with the coupled arm.
4. Coupling is extremely small.

The measurement procedure is based on the following facts:

1. A certain fraction of the incident wave at port 1 will appear at port 4.
2. A certain fraction of the reflected wave at port 2 will appear at port 3.
3. If the coupling coefficient is very small, then the ratio of the amplitude measured at port 3 to that at port 4 directly gives the reflection coefficient.

To illustrate this, let us assume that the amplitude of the incident wave at port 1 is unity and the coupling coefficient of the couplers is C. If b_2 and b_4 are the wave amplitudes at ports 2 and 4, respectively, then

$$b_2 = \sqrt{1-C^2} \tag{19.16}$$

and $\quad b_4 = C \tag{19.17}$

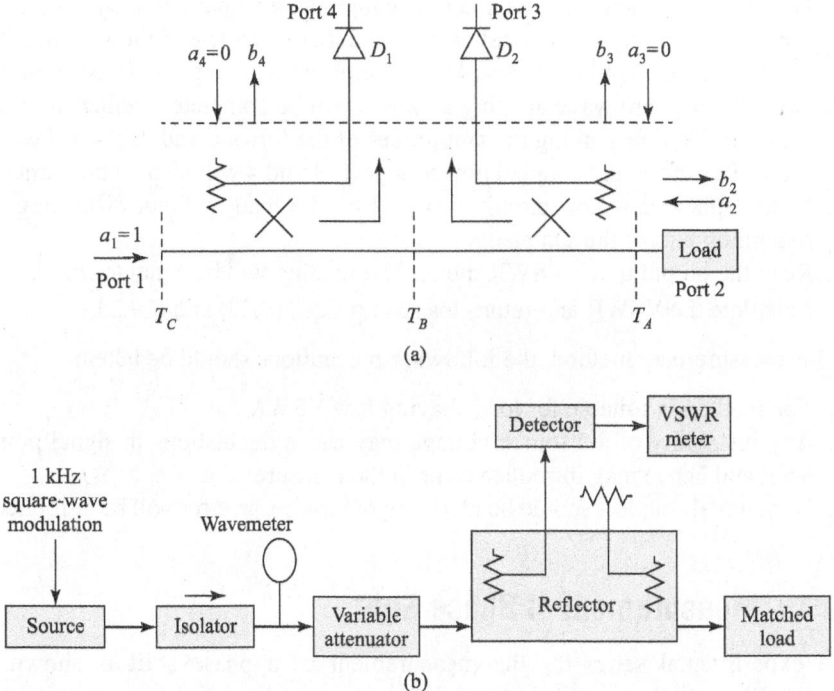

Fig. 19.27 Schematic diagrams (a) Reflectometer (b) Experimental set-up to measure return loss using reflectometer

The incident voltage at port 2 will suffer reflection by the unmatched load. If Γ_L is the reflection coefficient, then the amplitude of the reflected wave at port 2 is as follows: $a_2 = |\Gamma_L|\sqrt{1-C^2}$ (19.18)

Therefore, the amplitude of the wave at port 3 will be as follows:

$$b_3 = |\Gamma_L|C\sqrt{1-C^2}$$ (19.19)

Therefore, $|b_3/b_4| = |\Gamma_L|C\sqrt{1-C^2}\big/C = |\Gamma_L|\sqrt{1-C^2}$ (19.20)

Now, since $C \ll 1$, Eq. (19.20) reduces to the form $|b_3/b_4| \approx |\Gamma_L|$ (19.21)

Once the reflection coefficient is known from Eq. (19.21), VSWR and return loss may be calculated using the following relations:

$$\rho = \frac{1+|\Gamma_L|}{1-|\Gamma_L|}$$ (19.22)

and $\quad RL = -20\log_{10}\left(|\Gamma_L|\right)$ (19.23)

The detailed experimental procedure is as follows:

1. Terminate port 2 by a short circuit and port 3 by a matched load.
2. Adjust the detector at port 4 so that a unit reading is displayed at the VSWR meter.
3. Interchange the detector at port 4 and matched load at port 3.

4. Note the output at port 3. It should be equal to the output obtained at port 4 in Step 3 (i.e., unity) in the ideal case. The explanation is as follows. Since the coupling is small, all the forward wave amplitudes at port 1 will pass to port 2. Now, the forward wave arriving at port 2 will be completely reflected by the short circuit, thus making the amplitudes of the forward and backward waves same. Therefore, the coupled power at ports 3 and 4 will also be the same.
5. Now, replace the short circuit at port 2 by the load, without disturbing the remaining part of the setup.
6. Read the output in the VSWR meter. The reading will be equal to $1/|\Gamma_L|$.
7. Calculate the VSWR and return loss using Eqs (19.22) and (19.23).

In this measurement method, the following precautions should be taken:

1. The method is suitable for loads having low VSWR.
2. Any instability of the source voltage may cause fluctuations in signal power level and hence may introduce error in the measurement.
3. Directional couplers should be identical; otherwise, an error will be introduced.

19.15 Measurement of Phase Shift

The experimental setup for the measurement of a phase shift is shown in Fig. 19.28. Initially, the 1 kHz modulated wave is divided in two channels of equal amplitude and phase using an H-plane T-junction. One of these two signals is then passed through the DUT, while the other is passed through a precision phase shifter. After that they are combined using another H-plane T-junction and is displayed in a CRO. Since the phase shift produced by the phase shifter will be different from that produced by the DUT, signals in the two channels will not be added in phase in the second T-junction. Now, the phase shift produced by the precision phase shifter is varied until it becomes equal to the phase shift produced by the DUT, and the two waves add in phase in the second T-junction. Reading of the precision phase shifter now directly provides the phase shift produced by the DUT.

In this method, we need an approximate idea about the electrical length of the DUT, since it is not possible to distinguish one wavelength and from any integral multiple of the wavelength.

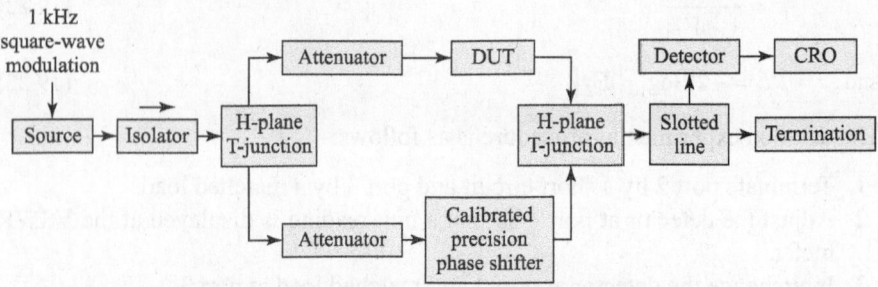

Fig. 19.28   Experimental set-up for measuring phase shift

19.16 Measurement of Impedance

Various methods of impedance measurement are discussed in the following subsections.

19.16.1 Slotted Line Method

The input impedance of a load can be calculated by measuring the complex reflection coefficient, using the following formula:

$$Z_L = Z_0 \frac{1+\Gamma_L}{1-\Gamma_L} = Z_0 \frac{1+|\Gamma_L|e^{j\phi_L}}{1-|\Gamma_L|e^{j\phi_L}} \qquad (19.24)$$

Magnitude of the reflection coefficient can be measured by measuring the VSWR in the line and the phase by measuring the position of the first voltage minima from the load. The formulas to be used for this purpose are as follows:

$$|\Gamma_L| = \frac{\rho-1}{\rho+1} \qquad (19.25)$$

$$\phi_L = 2\beta d_{min} - \pi \qquad (19.26)$$

$$\beta = 2\pi/\lambda_g \qquad (19.27)$$

λ_g can be calculated by measuring the distance between the positions of two successive minima and multiplying it by 2.

The experimental setup for the measurement of unknown impedance by the slotted line method is shown in Fig. 19.29. The measurement steps are as follows:

1. Measure the VSWR using the technique described in Section 19.13. Hence, calculate $|\Gamma_L|$ using Eq. (19.25).
2. Find the distance between the positions of two successive minima and multiply it by 2. This will give the guided wavelength. Hence, calculate β using Eq. (19.27).
3. Locate the position of the voltage minima nearest to the load (i.e., the first voltage minimum) by moving the probe along the slotted line. This will give the value of d_{min}.

   At this point, it may be noted that it is not always possible to position the probe exactly at the first minima due to the short guided wavelength. In such a case, to measure d_{min}, a short circuit load may be used. In this method, identify the position of a voltage minimum when the line is terminated by the given load and then find the shift of that voltage minimum, towards termination end, after replacing the given load by a short. This has been demonstrated in Fig. 19.30.

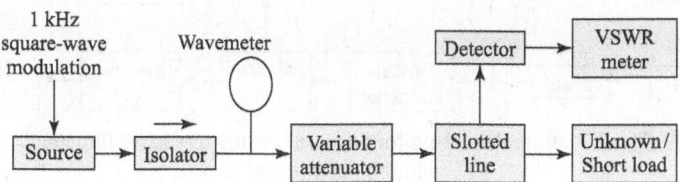

Fig. 19.29 Experimental set-up to measure unknown impedance using slotted line method

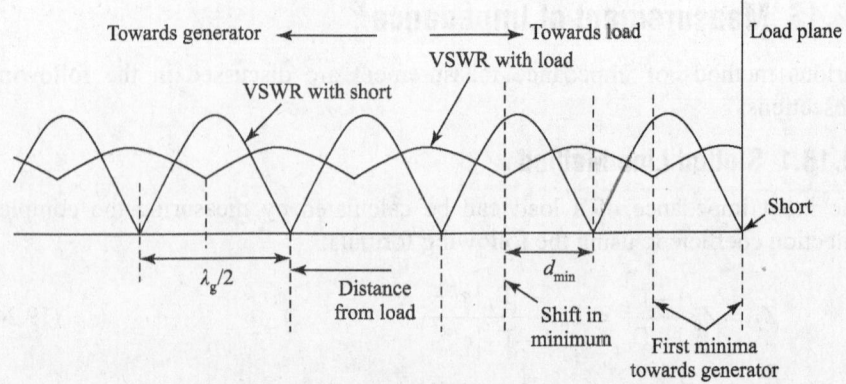

Fig. 19.30   Determination of shift in minima

Alternatively, since the successive voltage minima are $\lambda_g/2$ apart,

$$d_{\min} = d - \frac{N\lambda_g}{2} \qquad (19.28)$$

where d is the position of the minima nearest to the load that can be located with the setup and N is an integer such that $\left| d - \dfrac{N\lambda_g}{2} \right| < \dfrac{\lambda_g}{2}$. For example, if the position of the minima nearest to the load that can be located by the setup is 7 cm from the load and the guided wavelength is 3 cm, then $N = 4$ and $d_{\min} = 1$ cm.

4. Once $d_{\min}$ and β are known, use Eq. (19.26) to calculate ϕ_L.

5. Now use Eq. (19.24) to calculate Z_L.

It may be noted that instead of manual calculation, Smith charts can also be used. The details of how to use a Smith chart to find the unknown impedance has been discussed in Practice Problem 2.13.[1]

19.16.2 Impedance Measurement of Reactive Discontinuity

To measure the impedance of a shunt reactive discontinuity in a transmission line, such as due to a waveguide step/diaphragm/post, the measurement setup is to be modified as shown in Fig. 19.31. If jX is the reactance of the discontinuity and Z_0 the characteristic impedance of the line, then the equivalent load impedance can be expressed as follows:

$$Z_L = \frac{jZ_0 X}{Z_0 + jX}$$

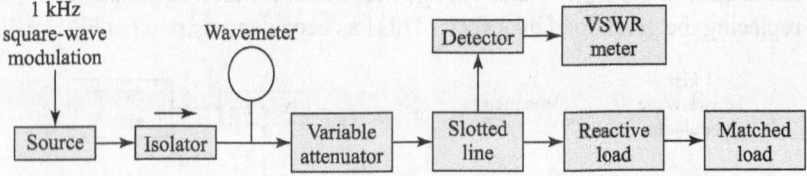

Fig. 19.31   Experimental set-up to measure unknown reactive impedance using slotted line method

---

[1] Solution in Online Resources

or $\quad z_L = \dfrac{Z_L}{Z_0} = \dfrac{jX}{Z_0 + jX} = \dfrac{jX(Z_0 - jX)}{(Z_0 + jX)(Z_0 - jX)} = \dfrac{X^2 + jXZ_0}{Z_0^2 + X^2} = x + jy$ $\quad$ (19.29)

where $\quad x = \dfrac{X^2}{Z_0^2 + X^2}$ $\hspace{6cm}$ (19.30)

$\quad y = \dfrac{XZ_0}{Z_0^2 + X^2}$ $\hspace{6.3cm}$ (19.31)

Taking the ratio of Eq. (19.30) to Eq.(19.31), we get the following relation:

$$\frac{x}{y} = \frac{X^2}{Z_0^2 + X^2}\frac{Z_0^2 + X^2}{XZ_0} = \frac{X}{Z_0}$$

or $\quad X = Z_0 x/y$ $\hspace{6.5cm}$ (19.32)

Now using the aforementioned experimental procedure and Smith chart, z_L (and hence x and y) can be found. Now taking the ratio of x to y and multiplying it by Z_0, the unknown reactance can be calculated.

19.17 Measurement of Frequency

There are different techniques to measure the frequency of a microwave signal. Few of them are described in the following subsections:

19.17.1 Slotted Line Method

In slotted line method for measuring the unknown frequency, the distance between two successive minima is first obtained by changing the probe location along the line. Next, this distance is multiplied by 2 to find the guided wavelength. Once the guided wavelength is obtained, the wavelength of the microwave signal can be calculated using the following relation:

$$\lambda = \lambda_g \Big/ \sqrt{1 + \left(\lambda_g / 2a\right)^2}$$ $\hspace{3cm}$ (19.33)

To measure the guided wavelength, the experimental setup of Fig. 19.29 can be used after replacing the unknown load by a short circuit.

19.17.2 Wavemeter Method

A cavity wavemeter is basically a cylindrical cavity resonator, with one of its ends being terminated in a sliding short circuit. As the position of the short circuit is mechanically moved in or out, the effective length of the cavity changes, which, in turn, changes its resonance frequency. The most suitable mode of operation, for a cavity wavemeter, is TE_{011} because of its higher Q and the absence of axial current. However, TE_{011} is a higher-order mode, and hence it will be associated with lower-order modes. Due to this reason, a cavity wavemeter generally operates in the dominant TM_{010} mode. To measure the unknown frequency, the wavemeter is arranged as shown in Fig. 19.32(a). Due to magnetic field coupling, such an arrangement excites the dominant TM_{010} mode inside the cavity. Once the required

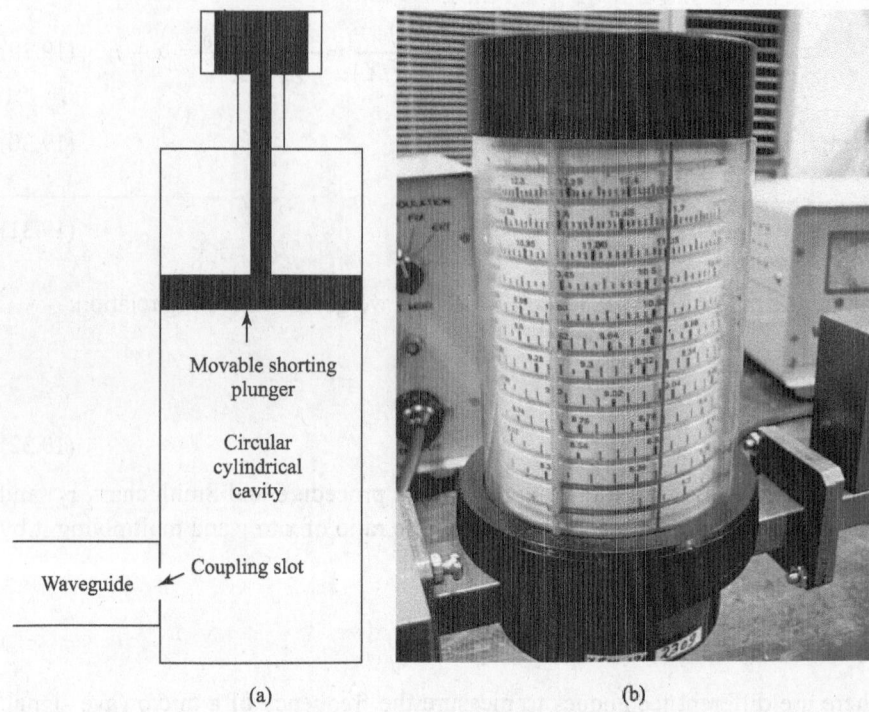

(a)   (b)

Fig. 19.32 Cavity wavemeter used in laboratory (a) Schematic diagram (b) Photograph

mode has been excited, the position of the short circuit is varied slowly so that resonance takes place inside the cavity. The resonance point can be identified by noting a dip in the power meter that is attached to the waveguide at the output side. In practice, the depth of the short circuit, for which resonance occurs, is calibrated in terms of the corresponding resonance frequency, as shown in Fig. 19.32(b). Once the resonance point has been identified, the corresponding resonance frequency can be read directly from the calibrated scale. The accuracy obtained using a cavity wavemeter lies the range from 1% to 0.005% for available Q of 1000–50,000. It may be noted that a block of absorbing material, such as polytron, is also placed at the back of the tuning short circuit to prevent oscillation at the top.

Note Since, at resonance, power is absorbed by such a wavemeter, it is also called an absorption-type wavemeter.

19.17.3 Transfer Oscillator Method

The basic block diagram of a transfer oscillator is shown in Fig. 19.33. Initially, the signal from a stable frequency source is amplified and fed to a harmonic signal generator, to generate comb frequencies in the desired microwave range. The harmonic output is then mixed with the unknown frequency signal in a mixer. The mixer output is connected to an indicator through a detector so that a beat frequency can be obtained. Now, the frequency of the stable frequency source is varied until a null-beat condition is achieved. Under such a circumstance, the unknown frequency

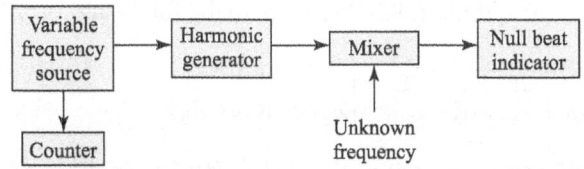

Fig. 19.33 Schematic diagram of transfer oscillator circuit

will be an integral multiple of the frequency of the stable frequency source. To find the integer multiple, the frequency of the stable source is further varied until the next null-beat condition is achieved. If f_1 and f_2 are the two frequencies corresponding to these null conditions and f is the unknown frequency, then we can write the following equations:

$$f = nf_1 \tag{19.34}$$

and $\quad f = (n-1)f_2 \tag{19.35}$

Eliminating n from these two equations, we get the following relation:

$$f = \frac{f_1 f_2}{f_2 - f_1} \tag{19.36}$$

Since f_1 and f_2 are known, f can be calculated using Eq. (19.36).

19.18 Measurement of Cavity Q

Similar to the measurement of frequency, there are also several methods to measure the Q of a cavity. Some of them are discussed in the following subsections:

19.18.1 Slotted Line Measurement Technique

This measurement technique is used to measure the Q of a reflection cavity by measuring the VSWR in the line at resonance and half-power frequencies. The measurement setup is shown in Fig. 19.34. The half-power points, where the equivalent resonator reactance is assumed to be equal to the magnitude of the equivalent resonator resistance, are directly found from the measured VSWR. If $Z_{in} = R + jX$ is the input impedance of the cavity in the vicinity of resonance frequency and Z_0 the characteristic impedance of the line, then the VSWR in the line can be expressed as follows:

$$\rho = \frac{|Z_{in} + Z_0| + |Z_{in} - Z_0|}{|Z_{in} + Z_0| - |Z_{in} - Z_0|} \tag{19.37}$$

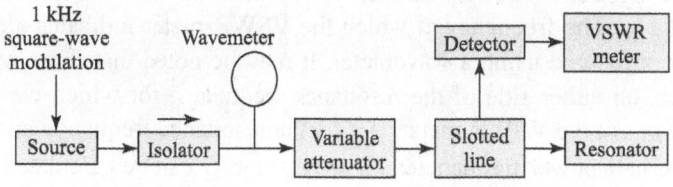

Fig. 19.34 Experimental set-up to measure cavity Q using slotted line method

At resonance, $X = 0$ and Eq. (19.37) reduces to the following form:

$$\rho_R = \frac{|R+Z_0| + |R-Z_0|}{|R+Z_0| - |R-Z_0|} = \begin{cases} R/Z_0 & \text{if } R > Z_0 \\ Z_0/R & \text{if } R < Z_0 \end{cases} \tag{19.38}$$

Now, at the half-power point $R = X$, and Eq. (19.37) modifies to the following form:

$$\rho_{HP} = \frac{|Z_{in} + Z_0| + |Z_{in} - Z_0|}{|Z_{in} + Z_0| - |Z_{in} - Z_0|} = \frac{\sqrt{(R+Z_0)^2 + R^2} + \sqrt{(R-Z_0)^2 + R^2}}{\sqrt{(R+Z_0)^2 + R^2} - \sqrt{(R-Z_0)^2 + R^2}}$$

or

$$\rho_{HP} = \frac{\left[\sqrt{(R+Z_0)^2 + R^2} + \sqrt{(R-Z_0)^2 + R^2}\right]^2}{\left[\sqrt{(R+Z_0)^2 + R^2} - \sqrt{(R-Z_0)^2 + R^2}\right]\left[\sqrt{(R+Z_0)^2 + R^2} + \sqrt{(R-Z_0)^2 + R^2}\right]}$$

or

$$\rho_{HP} = \frac{(R+Z_0)^2 + R^2 + (R-Z_0)^2 + R^2 + 2\sqrt{\{(R+Z_0)^2 + R^2\}\{(R-Z_0)^2 + R^2\}}}{(R+Z_0)^2 + R^2 - (R-Z_0)^2 - R^2}$$

or

$$\rho_{HP} = \frac{4R^2 + 2Z_0^2 + 2\sqrt{4R^4 + Z_0^4}}{4RZ_0} = \frac{R}{Z_0} + \frac{Z_0}{2R} + \sqrt{\frac{4R^4 + Z_0^4}{4R^2 Z_0^2}}$$

or

$$\rho_{HP} = \frac{R}{Z_0} + \frac{1}{2}\frac{Z_0}{R} + \sqrt{\frac{R^2}{Z_0^2} + \frac{1}{4}\frac{Z_0^2}{R^2}}$$

or

$$\rho_{HP} = \begin{cases} \rho_R + \dfrac{1}{2\rho_R} + \sqrt{\rho_R^2 + \dfrac{1}{4\rho_R^2}} & \text{if } R > Z_0 \\[3mm] \dfrac{1}{\rho_R} + \dfrac{\rho_R}{2} + \sqrt{\dfrac{1}{\rho_R^2} + \dfrac{\rho_R^2}{4}} & \text{if } R < Z_0 \end{cases} \tag{19.39}$$

Therefore, if the VSWR in the line at resonance is found, the VSWR at the half-power points can be calculated using Eq. (19.39).

Once the VSWR at the half-power points has been calculated using Eq. (19.39), change the klystron frequency in such a way that the VSWR meter indicates a reading of ρ_{HP}. The frequency at which the VSWR meter indicates a reading of ρ_{HP} can be measured using a wavemeter. It may be noted that there will be two frequencies, on either side of the resonance frequency, for which we will get a reading of ρ_{HP} at the VSWR meter. If f_0 is the resonance frequency and f_1 and f_2 are the two half-power frequencies $(f_1 > f_2)$, then Q can be calculated using the following equation: $Q = f_0/(f_1 - f_2)$ (19.40)

In this measurement, the following precautions should be taken:

1. The detector operates in the square-law region.
2. Frequency of the source is stable.
3. The generator is matched with the line.

19.18.2 Measurement of Q from Transmitted Power

The measurement technique, discussed previously, is suitable for reflection-type cavities. In contrast, the present method is suitable for both reflection- and transmission-type cavities and uses the power transmitted through the cavity, as a function of frequency, to measure the loaded $Q(Q_L)$. This method has the advantage of measuring Q_L directly, regardless of the existing coupling losses. However, to measure the unloaded $Q(Q_0)$, some additional measurements are required. If $T(f_0)$ is the transmission loss of the cavity at the resonance frequency f_0, and β_i and β_o are the input and output coupling factors, respectively, then the unloaded Q can be expressed as follows:

$$Q_0 = Q_L (1 + \beta_i + \beta_o) \tag{19.41}$$

where

$$\beta_o = \beta_i \rho_0 - 1 \tag{19.42}$$

$$\beta_i = \frac{4}{4\rho_0 - (\rho_0 + 1)^2 T(f_0)} \tag{19.43}$$

$$T(f_0) = P_{\text{out}} / P_{\text{in}} \tag{19.44}$$

Therefore, the values of ρ_0, $T(f_0)$, and Q_L are required for the measurement of unloaded Q. The present method has three sources of error:

1. Mismatch between the source and the load between which the cavity is inserted
2. Measurement of the bandwidth of the cavity frequency response curve
3. Inaccuracies introduced in the relative power measurement due to deviation of the detector from the square-law response, imperfect calibration of attenuators and power meters, reading errors, etc.

19.18.2.1 CW Measurement

The setup for the measurement of cavity Q using CW is shown in Fig. 19.35. The power transmitted through the cavity can be directly measured with a power meter. If the signal is square-wave modulated, a tuned amplifier or VSWR meter can also be used. A directional coupler is used to couple a part of the incident energy in the

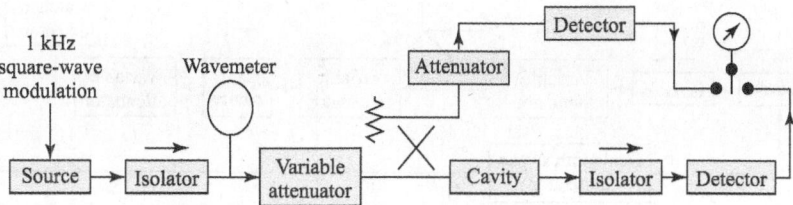

Fig. 19.35 Experimental set-up to determine Q from CW transmitted power measurement

auxiliary arm, which is measured using a power meter or VSWR meter (if square-wave modulated), to calculate the incident power. The ratio of these two powers gives the transmission loss of the cavity. During the measurement, the source should be tuned at the resonance frequency of the cavity. Once the transmitted power at the resonance frequency has been obtained, the source frequency is varied on either side of the resonance frequency to obtain the half-power points. At the half-power points, the power measured by the power meter or VSWR meter will be 3 dB lower than that at resonance. The half-power frequencies can be measured with a wavemeter. During the measurement, an error may be introduced due to the deviation of detectors from the square-law response. Hence, an attenuator is inserted before the directional coupler to ensure that the detectors operate in the square-law region. The VSWR at resonance can be measured using the procedure described before.

19.18.2.2 Swept-frequency Measurement

One of the major difficulties in measuring Q using the CW measurement technique comes from the unstable microwave source. In that respect, swept-frequency Q measurement has the advantage of the requirement of less frequency stability of the RF source than the aforementioned cases. The basic measurement setup for swept-frequency Q measurement is shown in Fig. 19.36. Crystal detectors, used in the measurement setup, should be matched and have identical response over the power and frequency range of the measurement. Two traces, one proportional to the incident power and the other to the power transmitted through the cavity, are displayed simultaneously in a dual-beam CRO. These power displays are used to measure the cavity Q. The detailed procedure is as follows:

1. Apply a linear saw tooth voltage to the repeller of the reflex klystron to frequency modulate the output signal. The frequency swing of the FM signal should be much larger than the cavity bandwidth, so that the input voltage to the cavity at different frequencies around the resonance frequency of the cavity can be assumed to be constant. The same saw tooth voltage is also applied

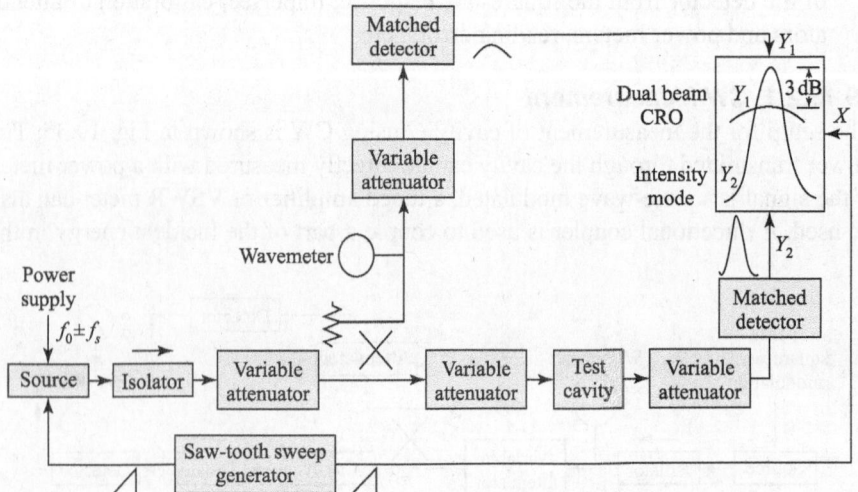

Fig. 19.36 Experimental set-up to determine Q using swept-frequency measurement

to the horizontal deflection plates of the CRO. If power transmitted through the test cavity is applied at the Y_2 terminal of a dual-beam CRO, the frequency response of the cavity will be displayed on the CRO screen.

2. Feed the output of the matched detector, connected to the auxiliary arm of the directional coupler, to the Y_1 terminal of a dual-beam CRO so that the klystron power output mode characteristics are also displayed on the CRO screen.

3. Tune the klystron frequency so that the cavity response is maximized, and the peak of the klystron mode characteristics and the cavity peak response coincide.

4. Adjust the wavemeter to obtain a dip at the peak of the cavity response. Reading on the wavemeter, at this point, will be the resonant frequency of the test cavity (f_0).

5. Adjust the flattop of the klystron mode curve at the 3 dB point of cavity response by varying the attenuator connected to the auxiliary arm of the directional coupler.

6. Measure the frequency span between the intersection points of the klystron mode curve and cavity response curve (i.e., between the 3 dB points). This will give Δf. The frequency span can be measured by obtaining dips at the two 3 dB points by adjusting the wavemeter and noting the corresponding frequency values.

7. The ratio of f_0 to Δf gives the loaded Q of the cavity (Q_L).

8. Terminate the cavity with a matched load and measure the VSWR in the line at resonance.

9. Measure the incident power (P_{in}) and output power (P_{out}) of the cavity.

10. The ratio of P_{out} to P_{in} gives the transmission loss $T(f_0)$.

11. Use Eqs (19.41)–(19.44) to calculate the unloaded Q.

In this measurement technique, the following precautions are to be taken:

1. Q_L of the wavemeter should be much higher than the test cavity Q_L, so that marker positions can be determined accurately.

2. The electronic tuning range of a reflex klystron is narrow (10–15 MHz at 10 GHz centre frequency). The cavity resonance curve should be completely accommodated within this frequency range.

3. Input power to the cavity should be maintained constant within the cavity bandwidth. This restricts the measurement of low Q using this method.

4. Resolution of the frequency meter should be very high.

19.18.2.3 Swept-frequency Measurement Using Electronic Frequency Marker

It should be noted that in the preceding cavity Q measurement process, the loaded Q of the wavemeter should be very high compared to the test cavity, and the resolution of the meter should also be very high. If the test cavity is operating at 9 GHz and has a Q of 9000, then its 3 dB bandwidth is 1 MHz. It is very difficult to tune the reflex klystron for such a small frequency difference. Further, measurement of the frequency span of 1 MHz is also very difficult because standard wavemeters have the smallest scale division of 1–5 MHz. Therefore, in such cases, the aforementioned measurement technique will not work and we need an electronic frequency marker. The experimental setup for swept-frequency measurement using an electronic frequency marker is shown in Fig. 19.37. The

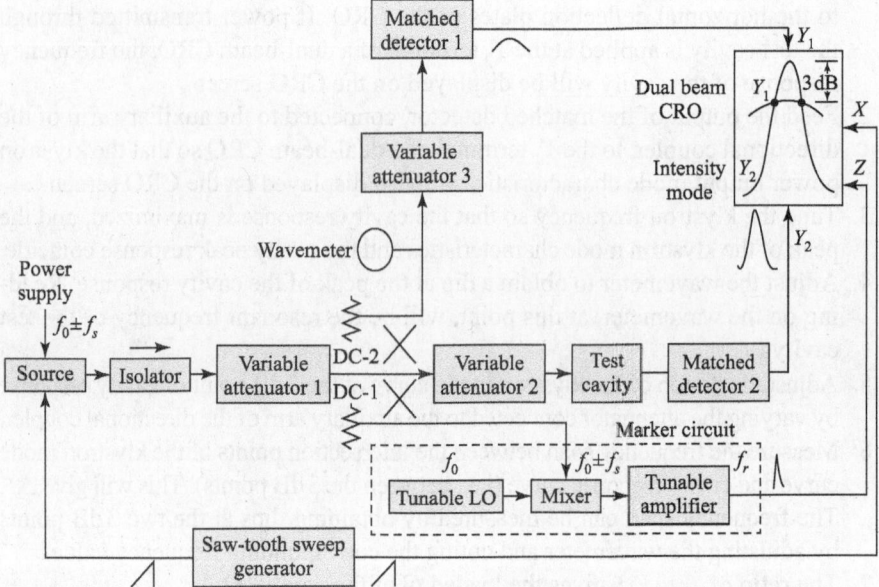

Fig. 19.37 Experimental set-up to determine Q using swept-frequency measurement with frequency marker

setup is almost the same as that of the previous one, except for the introduction of a marker circuit that is connected to the main line through a directional coupler at the input side and to the intensity grid of the dual-beam CRO at the output side.

In the marker circuit, markers are generated with the help of an auxiliary frequency stable low-frequency oscillator and klystron signal that is frequency swept from $f_0 - f_s$ to $f_0 + f_s$. A portion of the frequency-swept klystron signal, coupled into the auxiliary arm of DC-1, is mixed with the output of the fixed frequency oscillator (~ 800 MHz) in a harmonic crystal mixer, which produces sum and difference frequencies of the two signals. The output of the harmonic mixer circuit is then fed to a super heterodyne receiver, which produces bright spots in the display whenever the input signal equals its tuned frequency (f_r).

The spots occur twice during a saw tooth sweep, once above the resonance frequency when $f_0 + f_s(t) - f_0 = f_r$ and once below it when $f_0 - f_0 + f_s(t) = f_r$. Frequency separation between the markers is therefore $2f_r$. If the receiver is tuned such that $f_1 - f_0 = f_0 - f_2 = f_r$, where f_1 and f_2 are two 3 dB points, and f_0 is the resonance frequency of the cavity, as shown in Fig. 19.38, then the markers (or bright spots) will appear on the 3 dB points. Therefore, the 3 dB frequency span can easily be found from the tuned frequency of the receiver.

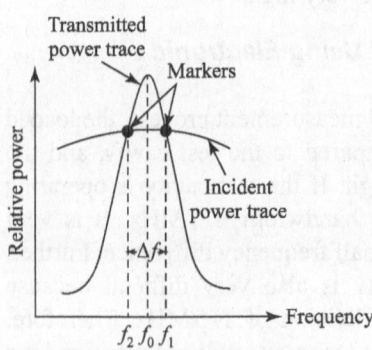

Fig. 19.38 Input and output power traces at display

In the setup, the crystal detectors (D_1 and D_2) must be matched to the line and must have identical response over the required frequency and power range. The RF power level at both crystals should also be the same. The latter requirement implies that the sum of the cavity insertion loss and attenuation of attenuator 2 should be equal to the coupling factor of directional coupler 2 and attenuation of precision attenuator 3. Additional requirements are as follows:

1. The directional coupler should have high directivity in the required frequency range.
2. The isolators should have isolation better that 20 dB and VSWR less than 1.2.
3. VSWR of the attenuators should be less than 1.1.

To obtain good matching, additional isolators may be inserted before the cavity and detectors. Broadband matching of the crystal detectors can also be achieved using 10 dB fixed pads in tandem with the crystal mounts. To check the crystals for identical response over the required frequency and power range, the following procedures are followed:

1. Tune the klystron at the cavity resonance frequency f_0 by noting the maximum transmitted signal through the cavity.
2. The cavity transmission trace Y_2 should be centred with respect to the incident power trace Y_1.
3. Remove the RF power and adjust the vertical centring controls of the dual-trace oscilloscope so that the base lines of the two traces (Y_1 and Y_2) coincide with each other.
4. Apply the RF power and set attenuator 3 to zero attenuation. Now adjust attenuators 1 and 2 so that the peak of Y_2 touches Y_1.
5. Now remove the cavity and adjust attenuator 2 so that traces Y_1 and Y_2 coincide. This will ensure that the RF power levels at both crystals are same.
6. Increase the attenuation of attenuator 1 by approximately 6 dB. The two traces should still coincide over this range. This is to check that the crystal detectors have identical response over the required power range. If there is any discrepancy, it should be noted and later applied as a correction. To determine the discrepancy, set attenuator 3 at some initial attenuation and observe the change in setting required to align both the traces.
7. Repeat these steps for a few frequencies near f_0. This is to check that the crystal detectors have identical response over the required frequency range.

The measurement procedure is as follows:

1. After removing the cavity from its position, adjust the frequency sweep so that a sufficient portion of the cavity resonance curve can be covered.
2. Adjust attenuators 2 and 3 so that both the traces coincide. This should be done after the base lines of the two traces with zero RF power have coincided.
3. Insert the test cavity in its position and tune the klystron at the cavity resonance frequency f_0 by adjusting klystron repeller voltage, so that both the incident and transmitted power traces are centred on the oscilloscope display.

4. Adjust attenuator 2 so that the incident power trace (Y_1) touches the peak of the transmitted power trace of the cavity (Y_2). The change in attenuator 2 setting is equal to the transmission loss at resonance $(T(f_0))$.
5. Measure the resonance frequency using the wavemeter.
6. Lower the incident power trace by 3 dB by adding extra 3 dB attenuation to attenuator 3.
7. Tune the receiver of the marker circuit such that the markers are placed at the point of intersection of traces Y_1 and Y_2. The tuning frequency will be half of the 3 dB bandwidth of the cavity.
8. Calculate the loaded Q using Eq. (19.40).
9. Measure the VSWR at resonance using the procedure described before.
10. Use Eqs (19.41)–(19.44) to calculate the unloaded Q.

19.18.2.4 Q Measurement by Measuring Phase Shift of Modulation Envelope of Transmitted Signal

When a microwave signal of frequency f_0 is amplitude modulated by a sine wave of frequency f_1 and is applied to a resonant cavity with resonance frequency f_0, its side bands are symmetrically shifted in phase on transmission through the cavity. One side band is advanced while other is delayed with respect to the input wave. This, in turn, results in a phase shift in the modulation envelope, which can be expressed as follows: $\phi = \tan^{-1}(2Q_L f_1/f_0)$ (19.45)
Now, if f_1 is set such that $\phi = 45°$ (19.46)
then Eq. (19.45) gives the following relation:

$$2Q_L f_1/f_0 = 1 \text{ or } Q_L = \frac{f_0}{\Delta f} = \frac{f_0}{2f_1}$$

or $f_1 = \Delta f/2$ (19.47)

Since f_1 can be known accurately from the modulating signal generator, by satisfying Eq. (19.46) we can easily calculate Δf using Eq. (19.47). The corresponding measurement setup is shown in Fig. 19.39. Phase shift of the modulation envelope is maximum when the carrier frequency and resonance frequency are same.

The present method is less affected by the frequency variation of the generator but is susceptible to the errors in phase measurement instruments.

19.18.3 Decrement Method for Q Measurement

For a high-Q resonator, the measurement of VSWR, at the resonance frequency, is very difficult. To overcome this problem, a method based on the measurement of the time rate of decay of microwave energy stored in the high-Q cavity is adopted. The rate of decay of microwave energy stored in the high-Q cavity can be expressed as follows: $U = U_0 e^{-\omega_0 t/Q}$ (19.48)
If the cavity is excited by a pulse, and U_1 and U_2 are the stored energy at time t_1 and t_2, respectively, then using Eq. (19.48), we can write the following expression:

$$10 \log_{10}(U_1/U_2) = 4.343\omega_0 (t_2 - t_1)/Q$$

or $$Q = \frac{8.686\pi f_0 (t_2 - t_1)}{10 \log_{10}(U_1/U_2)}$$ (19.49)

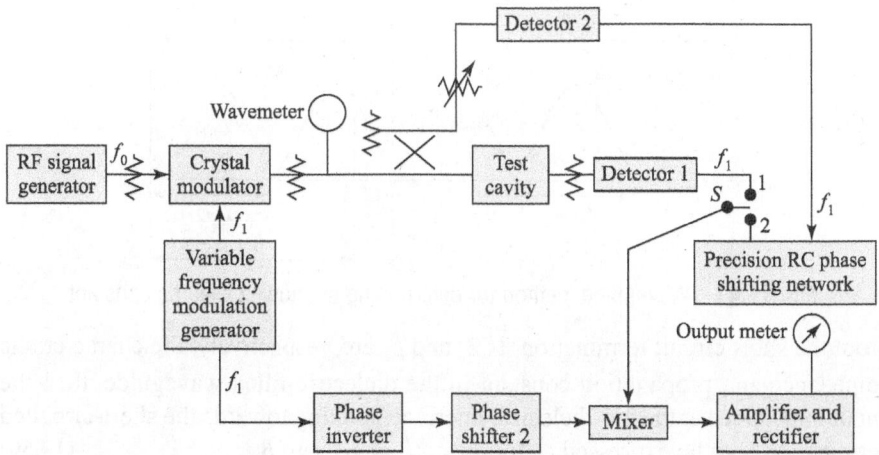

Fig. 19.39 Experimental set-up to measure Q by measuring phase shift of modulation envelope of transmitted signal

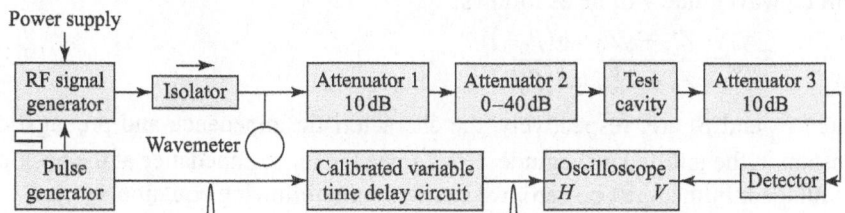

Fig. 19.40 Experimental set-up for measuring Q by decrement method

The measurement setup for the present method is shown in Fig. 19.40. Though the fluctuation of the source frequency does not affect the rate of energy decay in the resonator, it affects the level of energy build-up in the cavity. This results in fluctuations in the decay curve trace on the oscilloscope screen. To avoid this problem, the trace due to a single pulse is photographed and then the photograph is used to measure the decay time.

19.19 Measurement of Dielectric Constant

The following subsections describe various methods used to measure the dielectric constant of a material.

19.19.1 Waveguide Method

The waveguide method is used to measure the dielectric constant of a lossless or very-low-loss solid dielectric material that completely fills a length l_e of a waveguide. The end of the sample is terminated in a short circuit, as shown in Fig. 19.41. Due to the termination, a standing wave is formed inside the waveguide. Let us assume that the first voltage minimum in the unfilled waveguide occurs at a distance l_0 from the air–dielectric interface, as shown in the same figure. Therefore, the position of the first voltage minimum in the unfilled waveguide is at a distance $(l_e + l_0)$

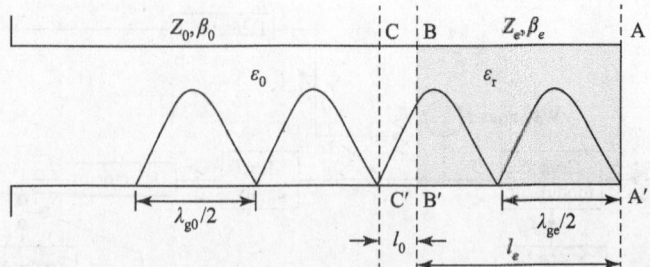

Fig. 19.41 Waveguide method for determining unknown dielectric constant

from the short circuit termination. If Z_e and β_e are, respectively, the characteristic impedance and propagation constant in the dielectric-filled waveguide, then the input impedance at the air–dielectric interface, looking towards the short-circuited termination, can be expressed as follows: $Z'_{in} = jZ_e \tan(\beta_e l_e)$ (19.50)

This input impedance behaves as an effective load impedance of the unfilled waveguide. Therefore, the input impedance at the first voltage minima in the unfilled waveguide will be as follows:

$$Z_{in,0} = \frac{Z'_{in} + jZ_0 \tan(\beta_0 l_0)}{Z_0 + Z'_{in} j \tan(\beta_0 l_0)}$$ (19.51)

where Z_0 and β_0 are, respectively, the characteristic impedance and propagation constant in the unfilled waveguide. Now, since the input impedance at the position of voltage minima must be zero, we can write the following equation:

$$Z'_{in} + jZ_0 \tan(\beta_0 l_0) = 0$$ (19.52)

Substituting Eq. (19.50) into Eq. (19.52), we get the following relation:

$$jZ_e \tan(\beta_e l_e) + jZ_0 \tan(\beta_0 l_0) = 0$$

or $\quad Z_e \tan(\beta_e l_e) = -Z_0 \tan(\beta_0 l_0)$ (19.53)

Now, if the dielectric is non-magnetic, then we can write the following equation:

$$Z_0 = Z_e \beta_e / \beta_0$$ (19.54)

Substituting Eq. (19.54) into Eq. (19.53), we get the following expression:

$$\frac{Z_e \beta_e}{\beta_0} \tan(\beta_0 l_0) = -Z_e \tan(\beta_e l_e) \text{ or } \frac{l_0}{\beta_0 l_0} \tan(\beta_0 l_0) = -\frac{l_e}{\beta_e l_e} \tan(\beta_e l_e)$$

or $\quad \dfrac{l_0}{l_e} \dfrac{\tan(\beta_0 l_0)}{\beta_0 l_0} = -\dfrac{\tan(\beta_e l_e)}{\beta_e l_e}$ (19.55)

Now for the dominant-mode excitation, $\beta_0 = 2\pi/\lambda_{g0}$, where λ_{g0} and hence β_0 can be measured by measuring the distance between two successive minima in the unfilled waveguide. l_0 can be measured using a slotted line waveguide. Further, since l_e is known from the length of the dielectric sample, the LHS of Eq. (19.55) is known, and we can write the following equation:

$$\frac{\tan(\beta_e l_e)}{\beta_e l_e} = -\alpha$$ (19.56)

where α is a known value. Solving this transcendental equation, $\beta_e l_e$ and hence β_e can be found. Now,

$$\beta_e = \frac{2\pi}{\lambda_{ge}} = \frac{2\pi}{\lambda_0}\sqrt{\varepsilon_r - \left(\frac{\lambda_0}{\lambda_c}\right)^2} = \frac{2\pi}{\lambda_0}\sqrt{\varepsilon_r - \left(\frac{\lambda_0}{2a}\right)^2} \qquad (19.57)$$

where a is the dimension of the waveguide broad wall. Since all the parameters, except ε_r, are known in Eq. (19.57), it can be solved to find ε_r.

It may be noted as the preceding equation is transcendental, an infinite number of solutions exist for ε_r. To find the actual solution, two samples of different lengths are taken and the corresponding solutions are found. If ε_{r1}, ε_{r2}, ε_{r3}, .. be the solutions for sample length l_e and ε'_{r1}, ε'_{r2}, ε'_{r3}, .. for sample length l'_e, then the common term between them will be the actual solution.

19.19.2 Cavity Perturbation Method

The cavity perturbation method is used to accurately determine the dielectric constant of a material having a small loss tangent. In this method, a small volume of the test material is inserted into a cavity resonator at the position of the maximum E-field. Due to its small volume, the field within the cavity may be assumed to be undisturbed, and hence changes in the field can be neglected. However, changes occurring in the resonance frequency and dielectric loss cannot be neglected, as they result in a change in the cavity Q. In this method, these changes are measured to determine the dielectric constant of the material.

If the unperturbed cavity has resonance frequency f_0, permittivity ε_0, and permeability μ_0, and the dielectric material has permittivity $\varepsilon = \varepsilon'_r + j\varepsilon''_r$ and permeability μ_0, then for sufficiently small perturbation we can write the following equations:

$$\omega - \omega_0 = \omega\left(\varepsilon'_r - 1\right) \int_{V_s} |E_0|^2 \, dv \Big/ \int_{V_c} |E_0|^2 \, dv \qquad (19.58)$$

and $\qquad \dfrac{1}{Q} = \varepsilon''_r |E_0|^2 V_s \Big/ \int_{V_c} |E_0|^2 \, dv \qquad (19.59)$

where V_s and V_c are the volumes of the sample and the cavity, respectively, and E_0 is the field of the original cavity. For small perturbation, ω can be approximated as ω_0 and the preceding equations can be written as follows:

$$\frac{\Delta\omega}{\omega_0} = \frac{(\varepsilon'_r - 1)|E_0|^2\Big|_{V_s} V_s}{2\int_{Vc} |E_0|^2 \, dv} \qquad (19.60)$$

and $\qquad \dfrac{1}{Q} = \varepsilon''_r |E_0|^2 \Big|_{V_s} V_s \Big/ \int_{V_c} |E_0|^2 \, dv \qquad (19.61)$

where we have assumed that $|E_0|\big|_{V_s}$ is constant throughout the sample and

$$\frac{1}{Q} = \frac{1}{Q_c} \sim \frac{1}{Q_s} \tag{19.62}$$

Here, Q_c is the Q of the unperturbed cavity and Q_s is that of the perturbed cavity.

In practice, any type of cavity, either rectangular or circular, can be used with a suitable mode. However, the following criteria should be satisfied:

1. The cavity Q should be as high as possible, so that a high accuracy can be achieved, especially for low-loss materials.
2. The operating mode should be such that the test sample can be placed at a uniform field E_{max}, where $H = 0$. The condition $H = 0$ is particularly important for magnetically lossy materials; otherwise, complex permeability will shift the resonance frequency. RF losses due to complex permittivity will also be added to the losses due to complex permeability, introducing an error in the measurement. Due to this, the TE_{103} mode of rectangular cavity and TM_{010} mode of circular cavity are preferred. The placement of the dielectric inside a cavity is shown in Fig. 19.42.

Now, if we consider the TE_{103} mode in a rectangular waveguide, then

$$E_y = \frac{\lambda}{2a} \sin\left(\frac{\pi x}{a}\right) \sin\left(\frac{3\pi z}{d}\right)$$

and the sample should be placed as $x = a/2, z = d/2$. Therefore, within the sample,

$$\left|E_0\right|^2\Big|_{V_s} = \left|E_y\right|^2\Big|_{V_s} = \left(\frac{\lambda}{2a}\right)^2 \sin^2\left(\frac{\pi}{2}\right)\sin^2\left(\frac{3\pi}{2}\right) = \left(\frac{\lambda}{2a}\right)^2 \tag{19.63}$$

$$\int_{V_c}\left|E_0\right|^2 dv = \int_{V_c}\left|E_y\right|^2 dv = \left(\frac{\lambda}{2a}\right)^2 \int_0^a \int_0^b \int_0^d \sin^2\left(\frac{\pi x}{a}\right)\sin^2\left(\frac{3\pi z}{d}\right) dx\,dy\,dz$$

or $\qquad\displaystyle\int_{V_c}\left|E_0\right|^2 dv = \left(\frac{\lambda}{2a}\right)^2 \frac{abd}{4} \tag{19.64}$

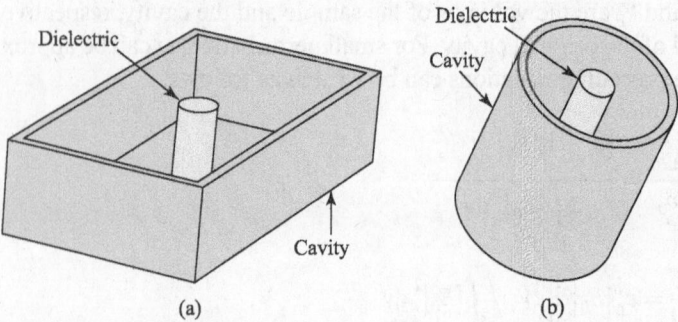

Fig. 19.42 Cut plane view of placement of dielectric material (a) Inside rectangular cavity operating at TE_{103} mode (b) Inside circular cavity operating at TM_{010} mode

Substituting Eqs (19.63) and (19.64) into Eqs (19.60) and (19.61), we get the following relations:

$$\frac{\Delta\omega}{\omega_0} = \frac{f - f_0}{f_0} = \frac{(\varepsilon_r' - 1)|E_0|^2\big|_{V_s} V_s}{2\int_{V_c} |E_0|^2\, dv} = \frac{(\varepsilon_r' - 1)\left(\dfrac{\lambda}{2a}\right)^2 V_s}{2\left(\dfrac{\lambda}{2a}\right)^2 \dfrac{abd}{4}} = \frac{2(\varepsilon_r' - 1)V_s}{abd}$$

or

$$\frac{\Delta\omega}{\omega_0} = 2(\varepsilon_r' - 1)\frac{V_s}{V_c}$$

or

$$\varepsilon' = 1 + 0.5\left(\frac{V_c}{V_s}\right)\left(\frac{f - f_0}{f_0}\right) \tag{19.65}$$

and

$$\frac{1}{Q} = \frac{\varepsilon_r''|E_0|^2\big|_{V_s} V_s}{\int_{V_c} |E_0|^2\, dv} = \frac{\varepsilon_r''\left(\dfrac{\lambda}{2a}\right)^2 V_s}{\left(\dfrac{\lambda}{2a}\right)^2 \dfrac{abd}{4}} = \frac{4\varepsilon_r''V_s}{abd} = 4\varepsilon_r''\frac{V_s}{V_c}$$

or

$$\varepsilon_r'' = 0.25\left(\frac{V_c}{V_s}\right)\frac{1}{Q} \tag{19.66}$$

Substituting Eq. (19.62) into Eq. (19.66), we get the following relation:

$$\varepsilon_r'' = 0.25\left(\frac{V_c}{V_s}\right)\left(\frac{1}{Q_c} \sim \frac{1}{Q_s}\right) \tag{19.67}$$

If we consider the TM_{010} mode in a circular cavity, then it can be shown that

$$\varepsilon' = 1 + 0.539\left(\frac{V_c}{V_s}\right)\left(\frac{f - f_0}{f_0}\right) \tag{19.68}$$

$$\varepsilon_r'' = 0.269\left(\frac{V_c}{V_s}\right)\left(\frac{1}{Q_c} \sim \frac{1}{Q_s}\right) \tag{19.69}$$

Accuracy of the perturbation method depends on the sample smoothness and its fitness inside the cavity. Therefore, the sample should be machined very carefully to attain appropriate smoothness and size.

19.20 Measurement of Scattering Parameters Using Network Analysers

The basic setup for the measurement of scattering parameters of an N-port network using a network analyser is shown in Fig. 19.43. Here, the source is connected to the nth port and the receiver to the mth port. The rest of the ports are kept matched terminated. Directional couplers at the nth and mth ports sample a part of the incident and transmitted power, respectively. These two signals are fed to the analyser that displays the complex ratio of these two signals, that is, S_{mn}. For the

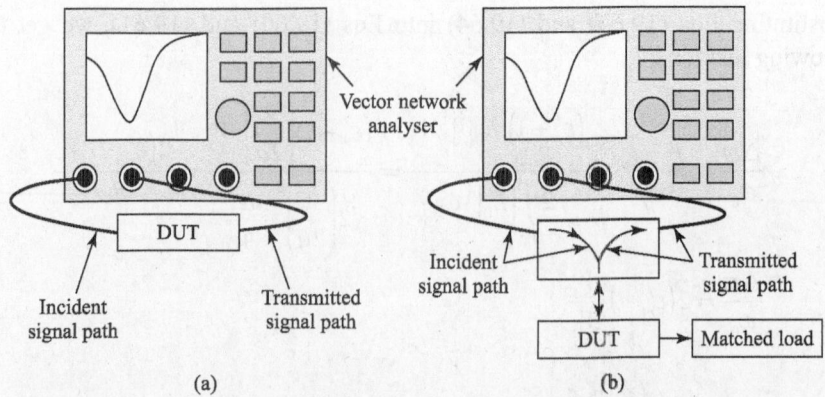

Fig. 19.43 Measurement of scattering parameters using VNA (a) S_{mn} (b) S_{nn}

measurement of S_{nn}, all the ports, except n, are matched terminated and signal is fed to port n. As before, both the incident and the reflected powers are separated and compared to display S_{nn}.

Network analysers are used to measure the scattering parameters over a frequency range. For this, a sweep-frequency oscillator is used. The output frequency of such an oscillator is varied by a voltage and current drive. The output of the sweep oscillator is then passed through an automatic level controller circuit (ALC), which maintains a constant level of its output signal. An ALC circuit consists of a variable attenuator, a directional coupler that samples the output power, and a feedback circuit consisting of a detector and a DC amplifier. If the level of the output signal of the ALC goes down, then it results in a reduced detector current in the feedback loop. The reduced feedback current, in turn, reduces the attenuation to such a value that the output of the ALC comes back to the desired level. The scattering parameter, at this condition, becomes proportional to the output of the mth port.

19.21 Measurement of Noise Factor

The steps to measure the noise factor are as follows:

1. Measure the noise power output in the absence of a source (P_N).
2. Measure the power output in the presence of a source (P_T).
3. Find the available power from the source (P_A).
4. Calculate the power amplification of the network using the following equation:

$$K = (P_T - P_N)/P_A \tag{19.70}$$

5. Plot K as a function of frequency.
6. Calculate the noise bandwidth, which is equal to the area under the curve plotted in step 5, divided by the height of the curve at the frequency of interest, multiplied by the frequency scale in cycles per second per unit length.
7. Calculate the average noise factor using the following formula:

$$F = \frac{7.3P_N \times 10^{22}}{BT_0 K} \tag{19.71}$$

where $T_0 = 290K$ is the standard noise temperature.

19.22 Antenna Measurements

The basic antenna parameters that are measured to determine the antenna characteristics are radiation amplitude pattern, radiation phase pattern, absolute gain, directivity, beam width, radiation efficiency, input impedance, bandwidth, and polarization. These measurements are generally carried out in antenna test ranges. There are two types of antenna test ranges: (a) indoor test ranges and (b) outdoor test ranges.

The indoor test range is generally an anechoic chamber, which is an electromagnetically shielded room. It is covered by a thick, good conductor so that external radiations cannot penetrate the room wall and contaminate the internal fields caused by the test antenna. The room is well grounded so that the current induced by the external RF field is grounded properly. Internally, the room is covered with absorbers to prevent reflection from the walls. This ensures the measurement of the direct field only. Microwave absorbers are generally made of carbon-impregnated polyurethane. The design and size of such pyramids depend on the frequency of interest. The size and design of the room depend on the measurement required, size of the test antenna, etc. It may be noted that these antenna measurements have to be carried out in a far-field region, that is, the transmitting and receiving antennas should be kept in the far-field regions of each other. This requires a large anechoic chamber, which is costly. The space may also be constrained. In such a case, compact test ranges can be used. In compact test ranges, a plane wave is produced at a smaller distance using an offset fed reflector antenna having a special edge geometry. An alternative method of compact test range is to perform near-field measurements. The far-field data are obtained from these measured near-field data using mathematical computations.

These measurements are costly and, at the same time, have several limitations. If such measurements are not possible, outdoor measurements are used. A basic outdoor measurement setup is shown in Fig. 19.44. In outdoor test ranges the antennas are mounted on tall towers with proper care such that the line of sight between the antennas is obstacle free. Care is also taken so that reflected fields can be avoided.

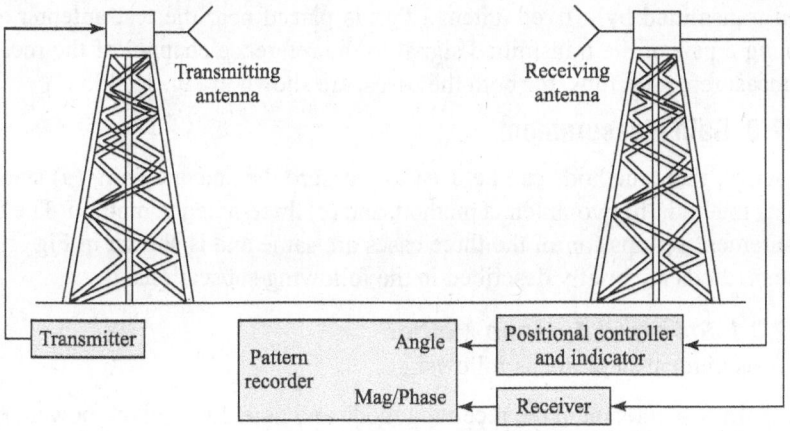

Fig. 19.44 Outdoor test range set-up

19.22.1 Measurement of Radiation Pattern

Since antennas are reciprocal in nature, their transmitting and receiving patterns are the same. Therefore, without any loss we can measure its receiving pattern to replicate its transmitting or radiation pattern. During such measurement, the transmitting antenna is fed by a stable source. The received signal is generally measured using a spectrum analyser or power meter. The measurement method is as follows:

1. Initially both the transmitting and receiving antennas are aligned for the maximum signal by adjusting their height and angle.
2. Next, the antenna is rotated by a certain angle in a particular plane (θ, ϕ), keeping the rest of the setup unchanged, and the corresponding received signal strength is noted.
3. Step 2 is repeated until the antenna makes a complete 360° rotation.
4. The measured signal strength is plotted as a function of antenna angle, to obtain the radiation pattern of the antenna in that particular plane.

Radiation pattern can also be measured automatically. To do this, the output of the receiving antenna is fed to the Y-axis and the angle information is fed to the X-axis of an XY recorder. The recorder automatically plots the radiation pattern. In this measurement process, the following precautions are to be taken:

1. Mutual coupling between the two antennas due to reflection, scattering, etc. should be removed.
2. Antennas should be in the far field of each other. Otherwise, the curvature of the incident phase front will cause phase variation over the aperture of the receiving antenna, thus introducing an error in the measurement.
3. If the illuminating field over the receiving antenna aperture is not uniform, then amplitude tapering will occur. Amplitude tapering can cause deviation of the measured pattern from the actual pattern.
4. Interference from the nearby radiating elements should be eliminated.

19.22.2 Phase Measurement

The phase of the radiated field is a relative parameter and hence should be measured with respect to a reference. The reference may be obtained either by receiving the signal transmitted by a fixed antenna that is placed near the test antenna or by coupling a part of the transmitted signal to the reference channel of the receiver. The measurement setups, for both the cases, are shown in Fig. 19.45.

19.22.3 Gain Measurement

In practice, three methods can be used to measure the antenna gain: (a) standard antenna method, (b) two-antenna method, and (c) three-antenna method. The basic measurement set-ups for all the three cases are same and is shown in Fig. 19.46. These methods are briefly described in the following subsections.

19.22.3.1 Standard Antenna Method
The measurement steps are as follows:

1. Set the test antenna in the receiving mode and note the received power (P_r) in the receiver.

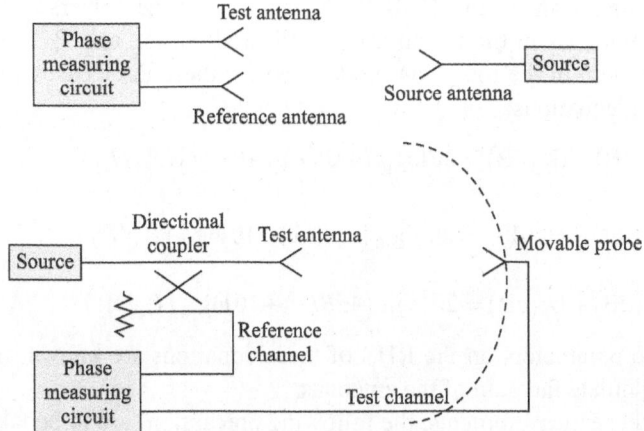

Fig. 19.45 Different phase pattern measurement set-ups

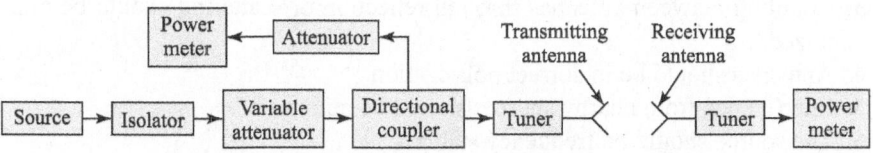

Fig. 19.46 Experimental set-up for measuring antenna gain

2. Replace the test antenna with a standard gain antenna, keeping rest of the measurement setup unaltered. Now again note the received power (P_s).

3. If G_r and G_s are the gain of the test and standard gain antennas, respectively, then G_r can be calculated using the following relation:

$$G_r\,(\text{dB}) = G_s\,(\text{dB}) + 10\log_{10}\left(P_r/P_s\right) \qquad (19.72)$$

Two-antenna method This method requires two similar antennas, one for transmitting the signal and the other for receiving it. The measurement steps are as follows:

1. Set one antenna for transmitting the signal and the rest for receiving the signal.
2. Measure the frequency and wavelength of the signal.
3. Note the transmitted (P_t) and received (P_r) powers.
4. Since gains of both the antennas are same, they can be calculated using the following relation:

$$G_r\,(\text{dB}) = 10\log_{10}\left(4\pi R/\lambda\right) + 5\log_{10}\left(P_r/P_t\right) \qquad (19.73)$$

where R is the distance between the antennas and λ is the wavelength of the signal.

Three-antenna method In this method, three antennas with gains G_1, G_2 and G_3 are used. During measurement, any two of the three antennas are used for transmitting and receiving the signal, and the corresponding transmitted and received powers are noted. Measurements are performed for all the three combinations

(1 and 2, 2 and 3, and 3 and 1). If P_{t1}, P_{t2}, and P_{t3} are the powers transmitted by these three antennas in the transmitting mode and P_{r1}, P_{r2}, and P_{r3} are the power received by them in the receiving mode, then for these three cases we can write the following equations:

$$G_1 (\text{dB}) + G_2 (\text{dB}) = 20 \log_{10} (4\pi R/\lambda) + 10 \log_{10} (P_{r2}/P_{t1}) \tag{19.74}$$

$$G_2 (\text{dB}) + G_3 (\text{dB}) = 20 \log_{10} (4\pi R/\lambda) + 10 \log_{10} (P_{r3}/P_{t2}) \tag{19.75}$$

$$G_3 (\text{dB}) + G_1 (\text{dB}) = 20 \log_{10} (4\pi R/\lambda) + 10 \log_{10} (P_{r1}/P_{t3}) \tag{19.76}$$

Since all the parameters on the RHS of these equations are known, they can be solved to calculate the gain of the antennas.

During these measurements, the following precautions are to be taken:

1. Antennas should be in the far field.
2. Antennas should be aligned face to face.
3. Coupling between antennas through reflection or scattering should be minimized.
4. Antennas should be in correct polarization.
5. Interference from nearby sources should be minimum.
6. The source should be frequency stable.

19.23 Measurement of Radar Cross Section

The radar cross-section (RCS) of a target is defined as follows:

$$\sigma = \frac{4\pi \times \text{Power re-radiated per unit solid angle}}{\text{Incident Power Density}} \tag{19.77}$$

If P_t is the transmitted power, G_t is the gain of the transmitting antenna, $\sigma(\theta)$ is the RCS of the target where θ is the aspect angle of the target, A_e is the effective area of the receiving antenna, and R is the distance of the target from the antenna, then the received power $(P_r(\theta))$ can be expressed as follows:

$$P_r (\theta) = \frac{P_t G_t A_e \sigma(\theta)}{\left(4\pi R^2\right)^2} \tag{19.78}$$

Equation (19.78) can be re-written as $\sigma(\theta) = K P_r(\theta)$ (19.79)
where K is a constant. If $\sigma(0)$ is the RCS when $\theta = 0$ and $P_r(0)$ the corresponding received power, then we can write the following equation:

$$\frac{\sigma(\theta)}{\sigma(0)} = \frac{P_r(\theta)}{P_r(0)} \tag{19.80}$$

In practice, there are two basic RCS terms: (a) monostatic or back scattering and (b) bi-static. In monostatic RCS measurement, the reflected signal is measured by the common transmitting–receiving antenna, whereas in the bi-static case, the scattered field is measured in the forward direction instead of measuring towards the source end. In this section, we will describe the former one.

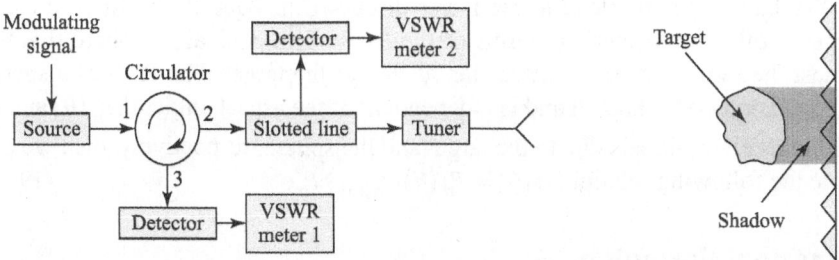

Fig. 19.47 Basic set-up for measuring RCS

The experimental setup for RCS measurement is shown in Fig. 19.47. The measurement procedure is as follows:

1. Adjust the tuner in the absence of the target to obtain the minimum reading in VSWR meter 1. This is to reduce the error due to background reflections.
2. Without disturbing the tuner, position the target in its place.
3. Adjust the received signal to read 1 in VSWR meter 2 for $\theta = 0$.
4. Rotate the target in the azimuth plane and note corresponding readings in VSWR meter 2.
5. Plot these data as a function of angle to obtain the angle versus normalized RCS plot.

This measurement procedure may be subjected to errors due to the following factors:

1. Isolation between the ports of the circulator is finite.
2. In Step 1, in which background cancellation was performed, the target was not present, and so reflections from the shadow region of the target were also taken into account. In the presence of the target, these reflections do not exist and therefore background cancellation is not perfect.

To reduce the error introduced due to the factor referred in Point 1, the circulator can be replaced by arranging a separate receiver system very close to the transmitting antenna. Such an RCS is called a quasi-monostatic RCS. The corresponding measurement setup is shown in Fig. 19.48.

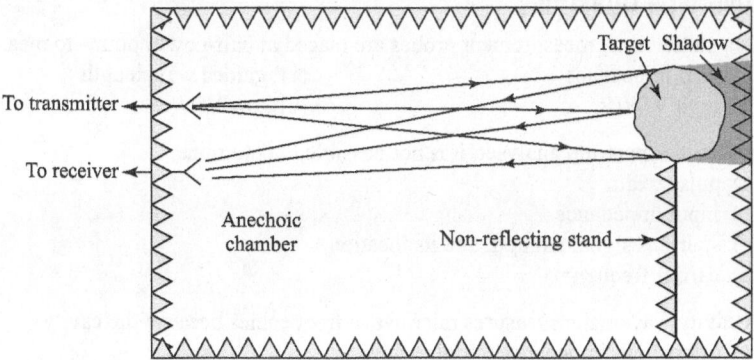

Fig. 19.48 Measurement of quasi-mono-static RCS in anechoic chamber

To reduce the error due to the factor discussed in Point 2, the RCS of a test target is often represented in terms of the RCS of an equivalent sphere that has almost the same vertical cross-section as that of the target. The RCS of a sphere (σ_{sphere}) is precisely known and is independent of the aspect angle. If $P_{\text{rt}}(\theta)$ and P_{rs} are the received signals due to the target and the sphere, respectively, then we can write the following relation: $\sigma(\theta) = P_{\text{rt}}(\theta)\sigma_{\text{sphere}}/P_{\text{rs}}$ (19.81)

Important Formulae

- The current sensitivity of a detector diode is defined as $\Delta\beta = \Delta i/P$.
- In the circulating calorimetric method, RF power can be calculated using the relation $P = 4.187 C_p Tvd$ W.
- In the static calorimetric method, RF power can be calculated using the relation $P = 4.187 C_p mT/t$ W.
- In the double-minima method, VSWR can be calculated using the relation $\rho \approx \operatorname{cosec}\left(\pi\Delta x/\lambda_g\right) \approx \dfrac{\lambda_g}{\pi\Delta x}$.
- Insertion loss is defined as $10\log_{10}(P_0/P_i)$.
- Reflection loss is defined as $10\log_{10}\left[1 - \dfrac{P_r}{P_i}\right]$.
- Attenuation loss is defined as $10\log_{10}\left(\dfrac{P_0}{P_i - P_r}\right)$.
- Return loss is defined as $10\log_{10}(P_r/P_i)$.
- In the dielectric medium, the phase constant can be expressed as $\beta_e = \dfrac{2\pi}{\lambda_0}\sqrt{\varepsilon_r - \left(\dfrac{\lambda_0}{2a}\right)^2}$.
- $G_r(\text{dB}) = G_s(\text{dB}) + 10\log_{10}(P_r/P_s) = 10\log_{10}(4\pi R/\lambda) + 5\log_{10}(P_r/P_t)$
- $\sigma(\theta)/\sigma(0) = P_r(\theta)/P_r(0)$
- $\sigma(\theta) = P_{\text{rt}}(\theta)\sigma_{\text{sphere}}/P_{\text{rs}}$

Exercises

Objective-type Questions

19.1   In a slotted line, measurement probes are placed at half-power points to measure
    (a) load impedance   (c) guided wavelength
    (b) high VSWR   (d) low VSWR

19.2   Without a spectrum analyser, it is not possible to determine
    (a) pulse width
    (b) input impedance
    (c) spurious signal strength and its location
    (d) carrier frequency

19.3   A cavity wavemeter measures microwave frequencies because the cavity
    (a) offers high impedance to microwave
    (b) offers resistive load

(c) has more volume for microwave to interact

(d) has resonance with one frequency of microwave signal

19.4 A power–resistance plot of a device shows a linear increase of resistance with power. The device may be
(a) a thermistor   (b) a barrater   (c) a photodiode   (d) an LDR

19.5 Large microwave power may be measured with a
(a) calorimeter   (c) thermistor
(b) barrater   (d) thermocouple

19.6 The impedance or admittance of a microwave circuit can be determined by measuring
(a) the maximum and minimum values of crystal current
(b) short circuit current
(c) none of these
(d) all of these

19.7 Barraters are used for the measurement of
(a) VSWR   (b) impedance   (c) power   (d) frequency

19.8 For measurement of high VSWR, the proper method is
(a) single-minima method   (c) either (a) or (b)
(b) double-minima method   (d) none of these

19.9 In general, most of the microwave power-measuring devices actually measure
(a) instantaneous power   (c) peak power
(b) average power   (d) none of these

19.10 A microwave device used in the measurement of VSWR is
(a) a crystal   (c) a travelling detector
(b) a thermocouple   (d) none of these

Review Questions

19.1 What is a bolometer? How is it used to measure an unknown power?

19.2 Describe the calorimetric method for the measurement of high power.

19.3 How can you measure attenuation loss and insertion loss of a network?

19.4 Describe a suitable method to measure a low VSWR produced by a DUT.

19.5 Describe a suitable method that can be used to measure a high VSWR.

19.6 What is a reflectometer? How can it be used to measure return loss?

19.7 Describe the slotted line method for the measurement of unknown impedance.

19.8 How can you measure the impedance produced by a reactive load?

19.9 Describe the different methods used for the measurement of the frequency of a signal.

19.10 Describe the slotted line method for the measurement of cavity Q.

19.11 Describe the swept-frequency measurement techniques for the measurement of cavity Q.

19.12 Describe the waveguide method for the measurement of unknown dielectric constant.

19.13 How can you measure the radiation pattern of an antenna?

19.14 Describe the different methods used for the measurement of antenna gain.

19.15 Describe the measurement of RCS of a target.

20.1 Introduction

In present days, microwaves are widely used in different applications—from personal equipment, like mobile, to military applications, such as RADAR and missiles. Consequently, almost the entire environment is now filled with microwave signals resulting from radiations from mobile antennas, satellite antennas, RADAR antennas, and other RF systems and antennas. With the rapid growth in wireless communication technology and systems, the intensity of microwave field, existing in the environment, is also increasing rapidly. This has raised serious concerns among microwave and electromagnetic engineers about the adverse effects of these increasing electromagnetic fields on electronic systems, fuels, and especially living beings. This is because if microwave can cook food then it can also cook brains.

In the following sections, we will briefly discuss about the hazards of electromagnetic radiation to ordnance (HERO), hazards of electromagnetic radiation to fuel (HERF), and hazards of electromagnetic radiation to personnel (HERP).

20.2 Hazards of Electromagnetic Radiation to Ordnance and Fuel

Microwave signals are dangerous to different electronic equipment used in households, industries, as well as battle fields. Out of them, the ordnance used in battlefields is probably most affected because they are subjected to the high-power microwave radiations from RADAR antennas. Such high-power radiation easily affects weapon system, electro-explosive devices (EEDs), missile control system, and emergency devices, in addition to the attending personnel. These radiations can trigger EEDs and also missiles. One common example of such an event is the destruction of the US aircraft carrier Forrestal, which was deployed to the coast of North Vietnam on 29 July 1967. The carrier had on its deck numerous attack aircraft, which were fuelled and loaded with 1000-pound bombs, and air-to-air and air-to-ground missiles. On that day, certainly one of the aircraft missiles was deployed and stroked another aircraft, causing explosion of its fuel tank and subsequent death of 134 servicemen. It was later found out that the accident occurred due to the generation of an RF voltage, across the contacts of one of the shielded connectors of the missile, by the ship's high-power search RADAR. Another example includes the crash of several US attack helicopters

(UH-60 Black Hawk) due to interference with citizen band transmitters, RADAR antennas, and radio antennas. These accidents caused the death of 22 servicemen.

It should be noted that ordnance is more sensitive to electromagnetic signals than living beings like humans, as the former does not have a circulatory system to dissipate internal heat. In addition, ordnance responds to peak power, whereas human beings respond to average power over some time.

Microwave and electromagnetic signals also have adverse effects on fuels. They have a tendency to ignite fuel vapours through RF-induced sparks, mostly during fuel handling in the presence of high-level RF fields.

20.3 Hazards of Electromagnetic Radiation to Personnel

Microwave radiation of proper strength may cause different types of health hazards, from sterility to leukaemia. Research conducted in this field has shown that a mobile phone user's brain is exposed to an electromagnetic field that is almost 10^{10} times stronger than neuronal waves. Further studies have revealed that microwave signal can open the blood–brain barrier (BBB)[1] within just 2 min of mobile phone usage and can also break chromosome within 10 min of usage.

In general, electromagnetic radiations, causing adverse effects on personnel, can be classified into two categories: (a) ionizing radiation and (b) non-ionizing radiation. Ionizing radiations are capable of causing characteristic changes in atoms or molecules in the human body and, hence, can result in radical tissue changes. On the other hand, non-ionizing radiations can prompt molecules to vibrate. This does not cause immediate changes, but can lead to non-thermal oscillatory and long term changes, causing various slow toxic types of effects in the body. Examples of ionizing radiation include X-ray and gamma rays, whereas those of non-ionizing radiation include ultra-violet ray in sunlight, visible light, light bulb, infrared radiation, and mobile phones.

In general, HERP is caused by the thermal effect of electromagnetic energies. Biological substances such as blood, brain, fat, muscle, and bone behave as conductive dielectrics, and scatter, reflect, and absorb the microwave energy depending on the frequency and field strength of the signal, dimensions of the body, and electrical properties of the tissues. If the biological substances cannot dissipate this heat, then the internal temperature of the body rises, which may damage these biological substances. A body part that is easily affected by microwave signals is the eye. The circulatory system of eye cannot provide sufficient blood flow to dissipate the heat developed in it when it is exposed to a microwave field for a longer time. This may result in the development of cataract. Other body parts that are sensitive to high-power microwave fields are stomach, intestine, and bladder.

In addition to these, a microwave signal with a wavelength of the order of the dimension of a body part couples with it very efficiently and generates a large amount of heat, which may be sufficient to damage the body part. In general, the coupling between a body part and a microwave signal is efficient when the dimension of the body part becomes equal to or more than one-tenth of the wavelength of the microwave signal.

---

[1]The BBB protects the brain from different viruses and toxins like prions (basically a protein that causes rare and incurable disease called Creutzfeldt-Jakob or CJD).

The most common parameter that is used to measure the radiation level is the average power level of a plane wave in free space. However, in practice, most of the radiations that are hazardous to health are not simple plane waves, and have complicated amplitude, phase, and polarization distribution due to their modulation characteristics, near-field characteristics, and tendency to form standing waves. Therefore, the specific absorption rate (SAR) is used to specify the hazardous limit. The SAR is defined as the amount of energy absorbed per unit mass of substance, and it depends on the magnitude of the field, and density and conductivity of the biological substance. Mathematically, it can be expressed as follows: $\text{SAR} = \sigma E^2/m_d$ W/kg   (20.1)
where the electric field E is in V/m, conductivity σ is in S/m, and mass density m_d is in kg/m^3. The SAR of a human being is about 0.03 W/kg at 700 MHz for an incident power density of 1 mW/cm^2.

When microwave energy penetrates the body, it gets absorbed by body substances. The extent of penetration of the microwave into the body depends on the characteristics of the body substances and the frequency and magnitude of the field. Since the conductivity and dielectric constant of fat are less than the rest of the body substances (such as blood and muscle), penetration of microwave into fat is also much larger than in other parts over most of the frequency band. However, since the conductivity of the other body substances (such as blood and muscle) is more, they absorb more energy, resulting in a higher rate of heating. In practice, inside the body, where the dielectric constant is higher, microwave penetration generates hot spots due to the focusing action and the cavity nature of the shape, formed by bones. Theoretical investigations have revealed that hot spots occur within the human skull in the frequency range 918–2450 MHz (used for mobile communication) and in the eyeball at around a frequency of 1500 MHz. It has also been revealed that exposure to a 830 MHz EM field for 72 h in the temperature range 34.5C–37.5°C may cause losses and gains of chromosome (aneuploidy), a major *somatic mutation*, leading to genetic instability and thereby cancer in the human blood lymphocytes. Another study revealed that 24 h of continuous exposure of leukaemia cells to 1 mW, 900 MHz radio waves can turn on the suicide genes.

The rate of temperature rise in body substances due to microwave heating can be expressed as follows: $\dfrac{dT}{dt} = Q/C_p$ (°C/S)   (20.2)
where C_p is the specific heat of the biological substance, and Q is the sum of SAR and metabolic rate of heat production per unit mass. In practice, due to absorption of microwave energy, the temperature of the body initially increases rapidly for a few minutes, after which the thermo regulatory system of the body takes control of it and tends to stabilize the temperature. However, if the system fails and cannot remove the excess heat at the same rate, then the body temperature starts rising again, causing a probable health hazard.[2]

20.4 Radiation Hazard Limits and Regulations

The International Radio Protection Association (IRPA) has set the permissible exposure level (PEL) for general public. The guideline is divided into two

---

[2]For more information, visit www.who.int

categories. The first one is for the personnel who work in the vicinity of RF fields for 8 h per day (occupational PEL) and the second one is for the residents who are exposed to an RF environment throughout the years (general public PEL). These limits have been prescribed considering skin depth and SAR of body substances.

The Federal Communications Commission (FCC) has also established RF radiation human exposure limits to protect personnel. Their guidelines and rules regarding RF exposure are based upon standards developed by the Institute of Electrical and Electronics Engineers (IEEE), the National Council on Radiation Protection and Measurements (NCRP), and inputs from other federal agencies, such as the Food and Drug Administration (FDA), Environmental Protection Agency (EPA), Occupational Safety and Health Administration (OSHA), and National Institute for Occupational Safety and Health (NIOSH). As per the FCC, public exposure to cellular telephones has been restricted to an SAR level of 1.6 W/kg. They have also mandated cell phone manufacturers to ensure that their phones comply with these limits for safe exposure.[3] Therefore all the wireless devices, marketed within the USA, must satisfy the FCC limits when operating at its highest possible power level.

20.5 Radiation Protection

For protection from hazardous radiations, RF exposure should be avoided as much as possible. In areas of high-power radiation, servicemen and personnel must use microwave-absorbing suits. As an alternative, dress materials made of RF blocking fabrics can also be used. Nowadays, paints are available in the market that can reflect or absorb microwave radiation. Painting of buildings with such paints can also provide protection to some extent. In addition, RF-blocking curtains, window shields, mats, etc. can be used.

Mobile phones are one of the major sources of hazardous RF fields. As per the World Health Organization (WHO), by the end of 2009, there were about 4.6 billion mobile phone subscribers globally. At the end of 2013, the number of mobile phone users in India was around 0.9 billion. The figures are increasing rapidly, day by day. Such phones operate at frequencies between 450 and 2700 MHz, with peak powers in the range of 0.1–2 W. Therefore, excessive use of mobile phones should be avoided. If this is not possible, then the following measures may be taken: (a) use of speakerphone, earpiece, or headset; (b) increasing the distance between the mobile phone and ear; and (c) use of text messages rather than talking.

Review Questions
20.1 Describe the effect of radiation on health.
20.2 Describe some means to stay protected from radiation.

---

[3]For more details, visit www.fcc.gov

Wheeler's Incremental Inductance rule is widely used in microwave engineering for the evaluation of attenuation due to conductor loss in TEM or quasi-TEM lines. The inductance and resistance per unit length of a transmission line can be expressed as follows:

$$L = \frac{\mu}{|I_0|^2} \int_S \vec{H}.\vec{H}^* ds \text{ H/m} \tag{A.1}$$

and

$$R = \frac{R_s}{|I_0|^2} \int_{C_1+C_2} \vec{H}.\vec{H}^* dl \text{ }\Omega/\text{m} \tag{A.2}$$

where

$$R_s = \frac{1}{\sigma\delta_s} \tag{A.3}$$

is the surface resistance of the conductor and $C_1 + C_2$ is the integration path over the conductor boundaries. The power loss per unit length of a transmission line is given by the following equation:

$$P_l = \frac{R_s}{2} \int_C |\vec{H}_t|^2 dl \text{ W/m} \tag{A.4}$$

For a lossy conductor, magnetic field intensity in the conductor is non-zero. This contributes an additional incremental inductance ΔL. Since the field decays exponentially and we have $\int_0^\infty e^{-2z/\delta_s} dz = \delta_s/2$, this incremental inductance can be obtained from Eq. (A.1) as follows:

$$\Delta L = \frac{\mu_0\delta_s}{2|I_0|^2} \int_C |H_t|^2 dl$$

or

$$\int_C |H_t|^2 dl = \frac{2\Delta L|I_0|^2}{\mu_0\delta_s} \tag{A.5}$$

where $\delta_s = \sqrt{1/(\pi f \mu_0 \sigma)}$ is the skin depth. Substituting Eq. (A.5) into Eq. (A.4) and using Eq. (A.3), we get the following relation:

$$P_l = \frac{R_s}{2}\frac{2\Delta L|I_0|^2}{\mu_0\delta_s} = \frac{R_s\Delta L|I_0|^2}{\mu_0\delta_s} = \frac{1}{\sigma\delta_s}\frac{\Delta L|I_0|^2}{\mu_0\delta_s} = \frac{|I_0|^2\Delta L}{\mu_0\sigma\delta_s^2}$$

or

$$P_l = \frac{|I_0|^2\Delta L}{\mu_0\sigma}\frac{\omega\mu_0\sigma}{2} = \frac{|I_0|^2\omega\Delta L}{2} \tag{A.6}$$

Now

$$P_0 = |I_0|^2 Z_0 / 2 \tag{A.7}$$

Therefore, using Eqs (A.6) and (A.7), the attenuation constant due to conductor loss is expressed as follows:

$$\alpha_c = \frac{P_l}{2P_0} = \frac{1}{2} \frac{|I_0|^2 \omega \Delta L}{2} \frac{2}{|I_0|^2 Z_0} = \frac{\omega \Delta L}{2Z_0} \tag{A.8}$$

Now

$$Z_0 = \sqrt{\frac{L}{C}} = \frac{L}{\sqrt{LC}} = L v_p \text{ or } \omega L = \frac{\omega}{v_p} Z_0 = \beta Z_0$$

or $\quad \omega \Delta L = \beta \Delta Z_0 \tag{A.9}$

Substituting Eq. (A.9) in Eq. (A.8), we get the following expression:

$$\alpha_c = \frac{\beta \Delta Z_0}{2Z_0} \tag{A.10}$$

where ΔZ_0 is the change in characteristic impedance due to incremental inductance. Alternatively, it is the change in characteristic impedance when all conductor walls are reduced by $\delta_s / 2$. Now using Taylor series expansion of Z_0 around $\delta_s / 2$ and keeping the first two terms (since δ_s is small), we can write that

$$Z_0 \left(\frac{\delta_s}{2} \right) \approx Z_0 + \frac{\delta_s}{2} \frac{dZ_0}{dl}$$

or $\quad \Delta Z_0 = Z_0 \left(\frac{\delta_s}{2} \right) - Z_0 = \frac{\delta_s}{2} \frac{dZ_0}{dl} \tag{A.11}$

Substituting Eq. (A.11) into Eq. (A.10), we get the following expression:

$$\alpha_c = \frac{\beta \delta_s}{4Z_0} \frac{dZ_0}{dl} \tag{A.12}$$

ABCD Parameters of Some Two-port Networks

S. No.	Circuit	ABCD Parameters
1.		$\begin{bmatrix} 1 & Z \\ 0 & 1 \end{bmatrix}$
2.		$\begin{bmatrix} 1 & 0 \\ Y & 1 \end{bmatrix}$
3.		$\begin{bmatrix} \cos(\beta L) & jZ_0\sin(\beta L) \\ jY_0\sin(\beta L) & \cos(\beta L) \end{bmatrix}$
4.		$\begin{bmatrix} N & 0 \\ 0 & 1/N \end{bmatrix}$
5.		$\begin{bmatrix} 1+\dfrac{Y_C}{Y_B} & \dfrac{1}{Y_B} \\ Y_A+Y_C+\dfrac{Y_A Y_C}{Y_B} & 1+\dfrac{Y_A}{Y_B} \end{bmatrix}$
6.		$\begin{bmatrix} 1+\dfrac{Z_A}{Z_C} & Z_A+Z_B+\dfrac{Z_A Z_B}{Z_C} \\ \dfrac{1}{Z_C} & 1+\dfrac{Z_B}{Z_C} \end{bmatrix}$

Interrelation between Different Two-port Network Parameters

	[S]	[Z]	[Y]	[ABCD]
S_{11}	S_{11}	$\dfrac{(Z_{11}-Z_0)(Z_{22}+Z_0)-Z_{12}Z_{21}}{(Z_{11}+Z_0)(Z_{22}+Z_0)-Z_{12}Z_{21}}$	$\dfrac{(Y_0-Y_{11})(Y_0+Y_{22})+Y_{12}Y_{21}}{(Y_0+Y_{11})(Y_0+Y_{22})-Y_{12}Y_{21}}$	$\dfrac{A+(B/Z_0)-CZ_0-D}{A+(B/Z_0)+CZ_0+D}$
S_{12}	S_{12}	$\dfrac{2Z_{12}Z_0}{(Z_{11}+Z_0)(Z_{22}+Z_0)-Z_{12}Z_{21}}$	$\dfrac{-2Y_{12}Y_0}{(Y_0+Y_{11})(Y_0+Y_{22})-Y_{12}Y_{21}}$	$\dfrac{2(AD-BC)}{A+(B/Z_0)+CZ_0+D}$
S_{21}	S_{21}	$\dfrac{2Z_{21}Z_0}{(Z_{11}+Z_0)(Z_{22}+Z_0)-Z_{12}Z_{21}}$	$\dfrac{-2Y_{21}Y_0}{(Y_0+Y_{11})(Y_0+Y_{22})-Y_{12}Y_{21}}$	$\dfrac{2}{A+(B/Z_0)+CZ_0+D}$
S_{22}	S_{22}	$\dfrac{(Z_{11}+Z_0)(Z_{22}-Z_0)-Z_{12}Z_{21}}{(Z_{11}+Z_0)(Z_{22}+Z_0)-Z_{12}Z_{21}}$	$\dfrac{(Y_0+Y_{11})(Y_0-Y_{22})+Y_{12}Y_{21}}{(Y_0+Y_{11})(Y_0+Y_{22})-Y_{12}Y_{21}}$	$\dfrac{-A+(B/Z_0)-CZ_0+D}{A+(B/Z_0)+CZ_0+D}$
Z_{11}	$Z_0\dfrac{(1+S_{11})(1-S_{22})+S_{12}S_{21}}{(1-S_{11})(1-S_{22})-S_{12}S_{21}}$	Z_{11}	$\dfrac{Y_{22}}{Y_{11}Y_{22}-Y_{12}Y_{21}}$	$\dfrac{A}{C}$
Z_{12}	$Z_0\dfrac{2S_{12}}{(1-S_{11})(1-S_{22})-S_{12}S_{21}}$	Z_{12}	$\dfrac{-Y_{12}}{Y_{11}Y_{22}-Y_{12}Y_{21}}$	$\dfrac{AD-BC}{C}$
Z_{21}	$Z_0\dfrac{2S_{21}}{(1-S_{11})(1-S_{22})-S_{12}S_{21}}$	Z_{21}	$\dfrac{-Y_{21}}{Y_{11}Y_{22}-Y_{12}Y_{21}}$	$\dfrac{1}{C}$
Z_{22}	$Z_0\dfrac{(1-S_{11})(1+S_{22})+S_{12}S_{21}}{(1-S_{11})(1-S_{22})-S_{12}S_{21}}$	Z_{22}	$\dfrac{Y_{11}}{Y_{11}Y_{22}-Y_{12}Y_{21}}$	$\dfrac{D}{C}$

	[S]	[Z]	[Y]	[ABCD]
Y_{11}	$Y_0\dfrac{(1-S_{11})(1+S_{22})+S_{12}S_{21}}{(1+S_{11})(1+S_{22})-S_{12}S_{21}}$	$\dfrac{Z_{22}}{Z_{11}Z_{22}-Z_{12}Z_{21}}$	Y_{11}	$\dfrac{D}{B}$
Y_{12}	$Y_0\dfrac{-2S_{12}}{(1+S_{11})(1+S_{22})-S_{12}S_{21}}$	$\dfrac{-Z_{12}}{Z_{11}Z_{22}-Z_{12}Z_{21}}$	Y_{12}	$\dfrac{AD-BC}{B}$
Y_{21}	$Y_0\dfrac{-2S_{21}}{(1+S_{11})(1+S_{22})-S_{12}S_{21}}$	$\dfrac{-Z_{21}}{Z_{11}Z_{22}-Z_{12}Z_{21}}$	Y_{21}	$-\dfrac{1}{B}$
Y_{22}	$Y_0\dfrac{(1+S_{11})(1-S_{22})+S_{12}S_{21}}{(1+S_{11})(1+S_{22})-S_{12}S_{21}}$	$\dfrac{Z_{11}}{Z_{11}Z_{22}-Z_{12}Z_{21}}$	Y_{22}	$\dfrac{A}{B}$
A	$\dfrac{(1+S_{11})(1-S_{22})+S_{12}S_{21}}{2S_{21}}$	$\dfrac{Z_{11}}{Z_{21}}$	$-\dfrac{Y_{22}}{Y_{21}}$	A
B	$\dfrac{(1+S_{11})(1+S_{22})-S_{12}S_{21}}{2S_{21}}$	$\dfrac{Z_{11}Z_{22}-Z_{12}Z_{21}}{Z_{21}}$	$-\dfrac{1}{Y_{21}}$	B
C	$\dfrac{(1-S_{11})(1-S_{22})-S_{12}S_{21}}{2S_{21}}$	$\dfrac{1}{Z_{21}}$	$-\dfrac{Y_{11}Y_{22}-Y_{12}Y_{21}}{Y_{21}}$	C
D	$\dfrac{(1-S_{11})(1+S_{22})+S_{12}S_{21}}{2S_{21}}$	$\dfrac{Z_{22}}{Z_{21}}$	$-\dfrac{Y_{11}}{Y_{21}}$	D

Answers to Objective-type Questions

Chapter 1

1.1 [d]	1.2 [d]	1.3 [d]	1.4 [d]	1.5 [c]

Chapter 2

2.1 [c]	2.2 [b]	2.3 [d]	2.4 [a]	2.5 [c]	2.6 [a]	2.7 [b]
2.8 [d]	2.9 [a]	2.10 [a]	2.11 [a]	2.12 [b]	2.13 [c]	2.14 [a]
2.15 [c]	2.16 [b]	2.17 [d]	2.18 [b]	2.19 [a]	2.20 [b]	2.21 [a]
2.22 [a]	2.23 [a]	2.24 [c]	2.25 [d]	2.26 [b]	2.27 [b]	2.28 [a]
2.29 [c]	2.30 [a]					

Chapter 3

3.1 [b]	3.2 [c]	3.3 [d]	3.4 [a]	3.5 [c]	3.6 [c]	3.7 [b]
3.8 [b]	3.9 [c]	3.10 [b]				

Chapter 4

4.1 [a]	4.2 [c]	4.3 [d]	4.4 [b]	4.5 [d]	4.6 [c]	4.7 [c]
4.8 [a]						

Chapter 5

5.1 [b]	5.2 [a]	5.3 [b]	5.4 [c]	5.5 [c]	5.6 [d]	5.7 [b]
5.8 [b]	5.9 [c]	5.10 [a]	5.11 [a]	5.12 [d]	5.13 [d]	5.14 [d]
5.15 [b]	5.16 [d]	5.17 [c]	5.18 [a]	5.19 [a]	5.20 [c]	

Chapter 6

6.1 [a]	6.2 [b]	6.3 [c]	6.4 [c]	6.5 [c]	6.6 [c]	6.7 [b]
6.8 [d]	6.9 [b]	6.10 [b]				

Chapter 7

7.1 [a]	7.2 [b]	7.3 [c]	7.4 [a]	7.5 [c]	7.6 [d]	7.7 [b]
7.8 [d]	7.9 [c]	7.10 [a]	7.11 [a]	7.12 [d]	7.13 [a]	7.14 [c]
7.15 [b]						

Chapter 8

8.1 [b]	8.2 [c]	8.3 [c]	8.4 [c]	8.5 [d]	8.6 [a]	8.7 [a]
8.8 [b]	8.9 [d]	8.10 [a]	8.11 [a]	8.12 [d]	8.13 [a]	8.14 [a]
8.15 [c]						

Chapter 9

9.1 [c]	9.2 [a]	9.3 [a]	9.4 [a]	9.5 [c]	9.6 [d]	9.7 [d]
9.8 [b]	9.9 [a]	9.10 [d]				

Chapter 10

10.1 [a]	10.2 [d]	10.3 [a]	10.4 [a]	10.5 [a]	10.6 [c]

Chapter 11

11.1 [c]	11.2 [a]	11.3 [a]	11.4 [b]	11.5 [c]	11.6 [b]	11.7 [a]
11.8 [c]	11.9 [d]	11.10 [d]	11.11 [b]	11.12 [b]	11.13 [c]	11.14 [d]
11.15 [d]	11.16 [b]	11.17 [d]	11.18 [a]	11.19 [c]	11.20 [b]	

Chapter 12

12.1 [b]	12.2 [d]	12.3 [d]	12.4 [c]	12.5 [b]	12.6 [c]	12.7 [d]
12.8 [b]	12.9 [a]	12.10 [d]	12.11 [d]	12.12 [c]	12.13 [a]	12.14 [b]
12.15 [c]						

Chapter 13

13.1 [c]	13.2 [d]	13.3 [b]	13.4 [d]	13.5 [a]	13.6 [b]	13.7 [a]
13.8 [b]	13.9 [c]	13.10 [c]	13.11 [a]	13.12 [b]	13.13 [a]	13.14 [b]
13.15 [c]	13.16 [c]	13.17 [b]	13.18 [d]	13.19 [a]	13.20 [b]	

Chapter 14

14.1 [a]	14.2 [b]	14.3 [b]	14.4 [b]	14.5 [a]	14.6 [b]	14.7 [b]
14.8 [d]	14.9 [b]	14.10 [a]	14.11 [d]	14.12 [c]	14.13 [d]	14.14 [c]
14.15 [d]						

Chapter 15

15.1 [d]	15.2 [c]	15.3 [c]	15.4 [b]	15.5 [a]

Chapter 16

16.1 [d]	16.2 [c]	16.3 [b]	16.4 [b]	16.5 [a]	16.6 [b]	16.7 [c]
16.8 [a]	16.9 [c]	16.10 [d]	16.11 [c]	16.12 [b]	16.13 [d]	16.14 [d]
16.15 [a]						

Chapter 17

17.1 [c]	17.2 [a]	17.3 [b]	17.4 [d]	17.5 [c]	17.6 [c]	17.7 [d]
17.8 [b]	17.9 [a]	17.10 [c]	17.11 [c]	17.12 [a]	17.13 [a]	17.14 [a]
17.15 [b]						

Chapter 18

18.1 [d]	18.2 [d]	18.3 [c]	18.4 [b]	18.5 [c]	18.6 [a]	18.7 [c]
18.8 [b]	18.9 [a]	18.10 [c]	18.11 [d]	18.12 [d]	18.13 [c]	18.14 [a]
18.15 [b]						

Chapter 19

19.1 [b]	19.2 [d]	19.3 [d]	19.4 [b]	19.5 [a]	19.6 [a]	19.7 [c]
19.8 [b]	19.9 [b]	19.10 [c]				

Answers to Problems

Chapter 2

2.1 [1.151 cm]; 2.2 [58.887 pF/m, 0.331 μH/m, 0.467 m℧/m, 2.625 Ω/m];

2.3 [33.115 pF/m, 82.787 nH/m]; 2.4 [50$\angle$1.1°];

2.5 [570.144 + $j(6.989 \times 10^{-4})\,\Omega$, 2.371×10^{-4} Np/m, 3.762×10^{7} m/s];

2.6 [156.248 rad/m, 1.086×10^{8} m/s, 1.086×10^{8} m/s];

2.7 [0.25 μH/m, 0.482 m℧/m, 1.205 Ω/m];

2.8 [$3.684\cos(3\pi \times 10^{8}t - 37.699z)$V, $0.074\cos(3\pi \times 10^{8}t - 37.699z)$ A];

2.9 [1.638]; 2.10 [10.61 MHz]; 2.11 $[703.132\,\Omega]$; 2.12 [33.333 Ω, 30$\angle$ − 90°, 12 W];

2.13 $[117 - j67.5\,\Omega]$; 2.14 $[44.876 - j72.895\,\Omega]$; 2.15 [8.54 mm]; 2.16 $[0.029\lambda]$;

2.17 $[35.89 - j6.176\,\Omega]$; 2.18 $[3.802 - j10.857\,\Omega]$; 2.19 $[j13.457\,\Omega]$;

2.20 $[28.284\,\Omega]$; 2.21 $[53.548 - j27.687\,\Omega]$; 2.22 $[0.143\lambda]$; 2.23 $[39.308 + j70.755\,\Omega]$;

2.24 $[61.562 - j59.133\,\Omega]$; 2.25 $[28.548 - j10.846\,\Omega]$; 2.26 $[0.708, \; 0.007\,\text{Np/m}]$;

2.27 [6.581 mW]; 2.28 $[0.245\angle 48.739°, \; 0.021\,\text{m}, \; 0.1\,\text{m}, \; 97.896 - 49.995\,\Omega]$;

2.29 $[1.334, \; 14.859\,\text{V}, \; 11.141\,\text{V}, \; 0.198\,\text{A}, \; 0.148\,\text{A}, \; 2.208\,\text{W}]$;

2.30 $[8\,\text{cm}, \; 0\,\text{cm}, \; 2.216\,\text{V}, \; 2.956\,\text{V}, \; 56.25\,\Omega, \; 100\,\Omega]$; 2.31 $[0.016 + j0.013\,℧, \; 3.69\,\text{mm}]$;

2.32 $[66.5 - j26\,\Omega, \; 1.74]$; 2.33 $[2\,\text{GHz}, \; 5.615, \; 50 + j97.5\,\Omega]$; 2.34 [1.48 cm, 8 cm];

2.35 $[j130\,\Omega, \; 0.395\lambda]$; 2.36 $[21 + j23.5\,\Omega]$; 2.37 $[69.5 - j14.5\,\Omega, \; 0.2, \; 33.5 + j4\,\Omega]$;

2.38 $[968\,\text{m}, \; 30.008\,\Omega, \; 12\,\text{V}]$; 2.39 $[484\,\text{m}, \; 30\,\text{V}, \; 75\,\Omega]$;

2.40 [75 Ω, 22.727 Ω, 500 km from source]

Chapter 3

3.1 $[(-j0.17, -j1.3, -j0.44)/(-j0.17, -j0.55, j0.45)/(j0.83, j56, -j0.44)/ \\ (j0.83, j1.31, j0.45)]$;

3.2 $(-j0.68, j0.15, -j0.48)/(-j1.54, -j0.15, j0.48)/(-j0.68, j0.95, -j0.48)/ \\ (-j1.54, -j0.95, j0.48)]$;

3.3 [0.375λ, 0.125λ, Yes]; 3.4 [0.02$\angle$ − 108°]; 3.5 $[64e^{9.708x}]$; 3.6 [9.718];

3.7 $[0.001\angle 107°]$; 3.8 $\left[\begin{cases} 45e^{0.6729\{x/(1.23\lambda)\}^2} & \text{for } 0 \le x \le 0.0615 \\ 45e^{0.3365\{4x/(1.23\lambda) - 2x^2/(1.23\lambda)^2 - 1\}} & \text{for } 0.0615 \le x \le 0.123 \end{cases} \right]$;

3.9 $\left[1.23\,\text{m}, \; 87e^{0.346\left(3.25x - 1.322x^2 - 1\right)} \quad \text{for } L/2 \le x \le L \right]$; 3.10 $[0.069, \; 0.046\angle 33.414°]$;

3.11 [0.2812, 1.7667]

Chapter 4

4.1 [0 GHz, 43.03 GHz, 86.06 GHz, 129.09 GHz and 172.12 GHz]; 4.2 [1.035 mm];

4.3 [3.944 mm, 2.206, 0.066 Np/m]; 4.4 $[0.269\,\mu\text{H}]$

Chapter 5

5.1 [1.807 dB]; 5.2 $\left[173.684 \text{ rad/m}, \ 36.229 \text{ mm}, \ 4.347 \times 10^8 \text{ m/s}, \ 2.07 \times 10^8 \text{ m/s}, \ 546.315 \, \Omega\right]$;

5.3 [13.607 GHz]; 5.4 [88.909Np/m]; 5.5 [12.5 mm × 6.25 mm];

5.6 $\left[0.598 R_{\text{air}}\right]$; 5.7 [2.288 cm × 1.144 cm]; 5.8 [9.965 GHz]; 5.9 [7.5 GHz];

5.10 $\left[224.055 \text{ rad/m}, \ 28.043 \text{ mm}, \ 1.402 \times 10^8 \text{ m/s}, \ 1.019 \times 10^8 \text{ m/s}\right]$;

5.11 [10.982 mm, 10.456 GHz]; 5.12 [168.672 V/m]; 5.13 $\left[5.796 \, \mu\text{S}\right]$; 5.14 [22.894 mm];

5.15 $\left[\text{TE}_{12}, \ 13.952 \text{ GHz}, \ 250.074 \text{ rad/m}, \ j250.074 \text{ rad/m}, \ 1026.603 \, \Omega, \right.$

$\qquad \left. \text{TE}_{10}, \ \text{TE}_{20}, \ \text{TE}_{01}, \ \text{TE}_{11}/\text{TM}_{11}\right]$;

5.16 [7.218 GHz to 11.811 GHz, 4.177 GHz]; 5.17 [7.577 GHz];

5.18 [13.198 mm]; 5.19 [0.262];

5.20 $\left[\dfrac{1}{\mu\omega}\left[20\pi\cos\left(30\pi y\right)\exp\left\{-j\left(20\pi z-\omega t\right)\right\}\hat{a}_y - \dfrac{30\pi}{j}\sin\left(30\pi y\right)\exp\left\{-j\left(20\pi z-\omega t\right)\right\}\hat{a}_z\right] \text{H/m}\right]$;

5.21 [9.714 GHz]; 5.22 [22.499 mm]; 5.23 [8.434 GHz]; 5.24 [10.764]; 5.25 [2.156];

5.26 $\left[\text{(i)}\dfrac{0.5c}{a}, \dfrac{c}{a}, \dfrac{2c}{a}, \dfrac{2.062c}{a}\text{(ii)}\dfrac{0.5c}{a}, \dfrac{c}{a}, \dfrac{c}{a}, \dfrac{1.118c}{a}\text{(iii)}\dfrac{0.5c}{a}, \dfrac{c}{a}, \dfrac{0.667c}{a}, \dfrac{0.833c}{a}\right.$

$\qquad \left.\text{(iv)}\dfrac{0.5c}{a}, \dfrac{c}{a}, \dfrac{0.5c}{a}, \dfrac{0.707c}{a}\right]$;

5.27 [32.03 mm]; 5.28 $\left[\varepsilon_r < 4\right]$; 5.29 $\left[f_c < f < \dfrac{c\sqrt{\varepsilon_1}}{2b\varepsilon_2}\right]$; 5.30 $\left[\dfrac{c}{2b}\dfrac{\sqrt{\varepsilon_1^2+\varepsilon_2^2}}{\varepsilon_2\sqrt{\varepsilon_1}}\right]$;

5.31 [19.495 mm, 12.292 mm]

Chapter 6

6.1 [5.066 nH, 5 pF, 309.938 kΩ]; 6.2 [0.296 GHz]; 6.3 [4000]; 6.4 [6.218 GHz];

6.5 [15.908 mm]; 6.6 [4.055 mm, 509.92]; 6.7 [19.703 mm]; 6.8 [805.536, 773.975];

6.9 [0.017]; 6.10 $\left[R' = \dfrac{Z_0^2}{R}, \ C' = \dfrac{L}{Z_0^2}, \ L' = C Z_0^2\right]$; 6.11 [0.1225 pF];

6.12 [8.25 GHz, 11.96 GHz, 15.587 GHz, 16.156 GHz, 16.372 GHz]; 6.13 [21.213 mm];

6.14 [Decrease by 4.951 mm]; 6.15 [7.682 GHz to 8.912 GHz]

Chapter 7

7.2 $\left[Y\right] = \begin{bmatrix} (Z_A+Z_B)/(Z_A Z_B) & -1/Z_B \\ -1/Z_B & (Z_A+Z_B)/(Z_A Z_B) \end{bmatrix}$,

$\quad [Z] = \begin{bmatrix} Z_A(Z_A+Z_B)/(2Z_A+Z_B) & Z_A^2/(2Z_A+Z_B) \\ Z_A^2/(2Z_A+Z_B) & Z_A(Z_A+Z_B)/(2Z_A+Z_B) \end{bmatrix}$,

$\quad [Z] = \begin{bmatrix} (Y_A+Y_B)/(Y_A Y_B) & 1/Y_B \\ 1/Y_B & (Y_A+Y_B)/(Y_A Y_B) \end{bmatrix}$,

$\quad [Y] = \begin{bmatrix} Y_A(Y_A+Y_B)/(2Y_A+Y_B) & -Y_A^2/(2Y_A+Y_B) \\ -Y_A^2/(2Y_A+Y_B) & Y_A(Y_A+Y_B)/(2Y_A+Y_B) \end{bmatrix}$;

7.4 $\begin{bmatrix} 0.3\angle195° & 0.7\angle95° & 0.6\angle225° & 0.25\angle120° \\ 0.7\angle95° & 0.4\angle30° & 0.5\angle145° & 0.32\angle60° \\ 0.6\angle225° & 0.5\angle145° & 0.25\angle185° & 0.57\angle100° \\ 0.25\angle120° & 0.32\angle60° & 0.57\angle100° & 0.72\angle60° \end{bmatrix}$;

7.5 $\begin{bmatrix} 0.6180\angle81.6168° & 1.3981\angle125.6715° \\ 1.3981\angle125.6715° & 0.6180\angle81.6168° \end{bmatrix}$;

7.6 $\begin{bmatrix} \begin{bmatrix} Z/(Z+2Z_0) & 2Z_0/(Z+2Z_0) \\ 2Z_0/(Z+2Z_0) & Z/(Z+2Z_0) \end{bmatrix}, & \begin{bmatrix} -YZ_0/(2+YZ_0) & 2/(2+YZ_0) \\ 2/(2+YZ_0) & -YZ_0/(2+YZ_0) \end{bmatrix} \end{bmatrix}$;

7.7 $[60°]$; 7.8 $\begin{bmatrix} 0.586 & 0.801 \\ 0.801 & 0.586 \end{bmatrix}$;

7.9 $\begin{bmatrix} 10 & 10 \\ 10 & 10 \end{bmatrix}, \begin{bmatrix} 10 & 0 \\ 0 & 5 \end{bmatrix}, \begin{bmatrix} 25 & 20 \\ 20 & 30 \end{bmatrix}, \begin{bmatrix} Z_0\coth(\gamma L) & Z_0/\sinh(\gamma L) \\ Z_0/\sinh(\gamma L) & Z_0\coth(\gamma L) \end{bmatrix}$;

7.10 $\begin{bmatrix} 0.07 & -0.07 \\ -0.07 & 0.07 \end{bmatrix}, \begin{bmatrix} 0.144 & -0.141 \\ -0.141 & 0.148 \end{bmatrix}, \begin{bmatrix} 0.164 & -0.161 \\ -0.161 & 0.168 \end{bmatrix},$

$\begin{bmatrix} \{Z_0\tanh(\gamma L)\}^{-1} & \{Z_0\sinh(\gamma L)\}^{-1} \\ \{Z_0\sinh(\gamma L)\}^{-1} & \{Z_0\tanh(\gamma L)\}^{-1} \end{bmatrix}$;

7.11 $\begin{bmatrix} 1 & 10 \\ 0 & 1 \end{bmatrix}, \begin{bmatrix} 1 & 0 \\ j\omega & 1 \end{bmatrix}, \begin{bmatrix} 1+j10\omega & 20+j100\omega \\ j\omega & 1+j10\omega \end{bmatrix}, \begin{bmatrix} \cosh(\gamma L) & Z_0\sinh(\gamma L) \\ \sinh(\gamma L)/Z_0 & \cosh(\gamma L) \end{bmatrix}$;

7.12 $\begin{bmatrix} -Y/(Y+2Y_0) & 2Y_0/(Y+2Y_0) \\ 2Y_0/(Y+2Y_0) & -Y/(Y+2Y_0) \end{bmatrix}, \begin{bmatrix} (N^2-1)/(N^2+1) & 2N/(1+N^2) \\ 2N/(1+N^2) & (1-N^2)/(1+N^2) \end{bmatrix},$

$\begin{bmatrix} 0.4472\angle116.5651° & 0.8944\angle-63.4349° \\ 0.8944\angle-63.4349° & 0.4472\angle116.5651° \end{bmatrix}, \begin{bmatrix} 0 & e^{-\gamma L} \\ e^{-\gamma L} & 0 \end{bmatrix}$

Chapter 8

8.1 $[19.8\,\text{mW}, 2\,\mu\text{W}, 0.2\,\text{mW}]$; 8.2 $[30\,\text{dB}, 5\,\text{dB}]$; 8.3 $[1\,\text{W}, 0.3162\,\text{W}, 3.1622\,\mu\text{W}]$;
8.4 $[0.5\,\text{W}, 0.5\,\text{W}, 0\,\text{W}]$; 8.5 $[972.9233\,\text{mW}, 1.624]$; 8.6 $[40\,\Omega]$;
8.7 $[87.4874\,\Omega, 58.3292\,\Omega]$; 8.8 $[3.1616\,\text{mm}, 4.5598\,\text{mm}, 4.5598\,\text{mm}, 3.1616\,\text{mm}]$

Chapter 9

9.1 $[100\,\Omega, 3, 37.5\,\Omega, 3]$; 9.2 $[20.5237\,\Omega, -0.2954]$;
9.3 $[0.6366\,\text{pF}, 0.7958\,\text{nH}, 0.6366\,\text{pF}]$;
9.4 $[0.2074\,\text{nH}, 4.8843\,\text{pF}, 13.8842\,\text{nH}, 0.073\,\text{pF}, 0.2074\,\text{nH}, 4.8843\,\text{pF}]$

Chapter 11

11.1 $[1.4095\times10^8\,\text{rad/s}, 0.5638\times10^8\,\text{rad/s}, 0.4875\,\text{A}, 14.625\,\text{kV}, 7.13\,\text{kW},$
   $\text{n } 59.3237\,\text{dB}, 22.63\%]$;

11.2 $\left[0.8914 \times 10^9 \text{ rad/s}, 0.624 \times 10^9 \text{ rad/s}, 2.4227 \text{ A}, 36.3405 \text{ kV}, 88.0421 \text{ kW}\right]$;

11.3 [11.3505 MHz]; 11.4 [16.2019 V, 128.6959 V, 0.3681 W];

11.5 [12.5 W, 2.4361 W, 19.49%]; 11.6 [3.9566 ns]; 11.7 [1.1182 mm];

11.8 [242.0839 MHz]; 11.9 [49.075 cm];

11.10 [19.49%, 9.745 mW, 8.2832 mW]; 11.11 [478.777 m/s]

Chapter 12

12.1 [502.6548, 448.799, 4188.7902, 10.71%, 43.33%];

12.2 $\left[8.7950 \times 10^7 \text{ m/s}, 8.7950 \times 10^7 \text{ m/s}, 83.5525 \text{ kV}\right]$;

12.3 [100 MHz, 6.6667 MHz, 6.6667 kHz, 2.25 ms, 222.2233 Hz]

Chapter 14

14.1 [1.05 A]; 14.2 [100 V, 0.0795 nS, 2 GHz];

14.3 [7.4489 V, 0.1986, 7.6277 mA, 3.7922 mA]; 14.4 [−75.5697 mV/nm];

14.5 [0.8748 V]

Chapter 15

15.1 $\left[-27.7716 - j33.6855 \,\Omega, \text{ Within } r = 0.56 \text{ circle}, 0.4276 \angle 100°,\right.$
$\left. 0.2851, 58.5 + j35 \,\Omega, 16 + j56 \,\Omega, 18.4115 \text{ dB}\right]$;

15.2 $\left[\|\Delta\| = 0.2793, K = 0.2373, \text{ Potentially Unstable}, 1.0607 \angle 108.4324°, 0.1885,\right.$
$1.6602 \angle 158.1523°, 1.1087, \text{ Stable Region: Outside the stability circles,}$
$\left. 0.6157 \angle 108.4326°, 0.4076\right]$;

15.3 $\left[0.5609 \angle 159.9982°, 0.3291, 0.4582 \angle 24.9968°, 0.2541, 0.4224 \angle -175°, 0.35\right]$

Chapter 17

17.1 [355.185 km]; 17.2 [99.2588 km]; 17.3 $\left[2.0872 \text{ m}^2\right]$; 17.4 $\left[8.0629 \times 10^{-10} \text{ W}\right]$;

17.5 [2.2222 kHz]; 17.6 [300]

Chapter 18

18.1 [19.6316 m, 4.2788°]; 18.2 [2.8°, 41.7609 dB]; 18.3 [2.5 cm, 0.1 A, 0.73 W];

18.4 [0.2582 V/m]; 18.5 [6]; 18.6 [2π/3, 6, 90°]; 18.7 $\left[0.012 \text{ m}^2\right]$;

18.8 [0.1667λ, 0.375λ, Optimum]; 18.9 [4.9804 cm, 5.9293 cm, 120 Ω];

18.10 [6.0206 dB]; 18.11 [1.1667 m]; 18.12 [21.6 dB, 12.96 dB, 32.6696°, 40.9746°];

18.13 [7.2774 dB]; 18.14 $\left[107.4296 \text{ cm}^2\right]$; 18.15 [7.2 V/m, 0.0191 A/m, 0.0688 W];

18.16 [90.91%]; 18.17 [1.5834]; 18.18 [1.5464 GHz, 37.2512 dB]; 18.19 $\left[3.1623\lambda\right]$;

18.20 [61.3493 dB]; 18.21 $\left[452.3893 \text{ m}^2\right]$; 18.22 [−195.5752 dB]; 18.23 [20, 365];

18.24 [−13.3842 dB]

Bibliography

Balanis, Constantine A., *Antenna Theory*, Second Edition, John Wiley and Sons, 2005.

Bhusan, Sunil, *Fundamentals of Engineering Electromagnetics*, Oxford University Press, 2012.

Carr, Joseph J., *Microwave & Wireless Communication Technology*, Elsevier, 2012.

Chang, Kai, Inder Bahl, and Vijay Nair, *RF and Microwave Circuits and Components Design for Wireless Systems*, John Wiley and Sons, 2002.

Collin, Robert E., *Foundations for Microwave Engineering*, Second Edition, Tata McGraw Hill, 1992.

Collin, Robert E., *Foundations for Microwave Engineering*, Second Edition, John Wiley and Sons, 2012.

Das, Annapurna and Sisir Das, *Microwave Engineering*, Tata McGraw-Hill Publishing Company Limited, 2001.

Ganesh Rao, D., *Electromagnetics and Transmission Line*, Pearson, 2011.

Gupta, K.C., Ramesh Garg, Inder Bahl, and Prakash Bhartia, *Microstrip Lines and Slot Lines*, Second Edition, Artech House, 1996.

Gupta, K.C., *Microwaves*, New Age International Pvt. Ltd, 1997.

Harish, A.R. and M. Sachidananda, *Antennas and Wave Propagation*, Oxford University Press, 2009.

Harrington, Roger F., *Time Harmonic Electromagnetic Fields*, IEEE Press, 2001.

Hayt, W.H. and J.A. Buck, *Engineering Electromagnetics*, Seventh Edition, Tata McGraw-Hill, 2006.

Jordan, Edward C. and Keith G. Balmin, *Electromagnetic Waves and Radiating Systems*, Second Edition, Prentice-Hall of India Private Limited, 2002.

Kraus, John D., Ronald J. Marhefka, and Ahmad S. Khan, *Antennas and Wave Propagation*, Fourth Edition, Tata McGraw-Hill, 2010.

Liao, Samuel Y., *Microwave Devices and Circuits*, Third Edition, Prentice-Hall of India Private Limited, 2000.

Maral, G. and M. Bousquet, *Satellite Communications Systems*, Second Edition, John Wiley and Sons, 1998.

Misra, Devendra K., *Radio Frequency and Microwave Communication Circuits—Analysis and Design*, Second Edition, John Wiley and Sons, 2004.

Paul, Claytron R., Keith W. Whites, and Syed A. Nasar, *Introduction to Electromagnetic Fields*, Third Edition, Tata McGraw-Hill, 2007.

Pennock, S.R. and P.R. Shepherd, *Microwave Engineering with Wireless Applications*, Macmillan Press Ltd, 1998.

Pozar, David M., *Microwave Engineering*, Fourth Edition, John Wiley and Sons, 2013.

Pratap, Rudra, *Getting Started with MATLAB*, Oxford University Press, 2010.

Raju, G.S.N., *Electromagnetic Field Theory and Transmission Line*, Pearson Education, 2011.

Sadiku, Matthew N.O., *Principles of Electromagnetics*, Fourth Edition, Oxford University Press, 2010.

Skolnik, Merrill I., *Introduction to RADAR Systems*, Third Edition, Tata McGraw-Hill, 2001.

Ulaby, Fawwaz T., *Electromagnetics for Engineers*, Pearson Education, 2011.

Van De Roer, Theo G., *Microwave Electronic Devices*, Chapman & Hall, 1996.

Yadava, R.L., *Antenna and Wave Propagation*, Prentice-Hall of India Private Limited, 2011.

Index

RELATED TITLES

Microprocessors and Microcontrollers
(9780198066477)

N. Senthil Kumar, Mepco Schlenk Engineering College, Sivakasi; **M. Saravanan**, Thiagarajar College of Engineering, Madurai; **S. Jeevananthan**, Pondicherry Engineering College, Puducherry

Microprocessors and Microcontrollers is designed as a comprehensive textbook for undergraduate engineering students to lay the foundation of the basic principles, functioning, and applications of microprocessors and microcontrollers.

Key Features
- Includes case studies on traffic light control, washing machine control, and elevator control to enable students appreciate the applications of processors
- Has numerous programming examples (in assembly language) which can also be used for conducting lab sessions and a section on high-level programming in 8051 processors using C
- Provides discussions on advanced processors, such as 80186, 80286, 80386, 80486, Pentium, PowerPC, and PIC 16F877
- Comes with a companion CD containing simulator and ALP codes for 8085, 8051, and 8086

Microprocessors and Interfacing
(9780198079064)

N. Senthil Kumar, Mepco Schlenk Engineering College, Sivakasi; **M. Saravanan**, Thiagarajar College of Engineering, Madurai; **S. Jeevananthan**, Pondicherry Engineering College; and **S.K. Shah**, The Maharaja Sayajirao University of Baroda

Microprocessors and Interfacing is a textbook designed for engineering courses covering a study of various microprocessors, microcontrollers, their interfacing, programming, and applications.

Key Features
- Provides assembly language program codes with corresponding comments
- Discusses case studies on electronic weighing machine, temperature monitoring and control, and DC motor speed monitoring and control as an appendix.
- Provides numerous programming examples, think-and-answer-type questions, design-based exercises, programming exercises, and 200+multiple-choice questions on 8085, 8086, and 8051 processors

Wireless Communication
(9780198060666)

Upena Dalal, faculty member, Electronics Engineering Department of SVNIT, Surat

Wireless Communication provides an exhaustive coverage of the basic topics and the latest developments in the field of wireless communication.

Key Features
- Contains the latest coding techniques, such as wavelet and DCT, as well as the most recent standards and systems, such as DAB, DVB, WiMAX, and UWB
- Contains solved examples and case studies to reinforce the textual concepts discussed in the chapters
- Includes chapter-end review questions, numerical problems, and multiple-choice questions for practice, and model question papers for self-practice

Digital Image Processing
(9780198070788)

S. Sridhar, Associate Professor, Department of Information Science and Technology, College of Engineering, Guindy Campus, Anna University, Chennai

Digital Image Processing is a basic textbook designed to cater to the needs of undergraduate engineering students of computer science, information technology, electronics and communication, and electrical engineering. The book adopts algorithmic approach to illustrate image processing and aims to provide an understanding of the principles and processing techniques of digital images

Key Features
- Includes solved numerical examples interspersed throughout the book
- Contains appendices that discuss the basics of MATLAB programming and ImageJ, and provides information on other public domain image processing software
- Includes a laboratory manual with simple examples illustrated through MATLAB and augmented through ImageJ, a GUI-based public domain software
- Comes with a companion CD containing MATLAB programs, some images of the book reproduced in colour, and additional test images for performing laboratory exercises

Other Related Titles

9780195696134	**Bell**: *Operational Amplifiers and Linear ICs*, 3e
9780195670929	**Moorthi**: *Power Electronics*
9780195684513	**Nagsarkar, Sukhija**: *Power System Analysis*
9780198075509	**Bhalja**: *Protection and Switchgear*
9780195669305	**Khare**: *Fiber Optics and Optoelectronics*